AF843520

Infinitization of Selfhood

Copyright © 1977 by Michael D. Robbins, Ph.D.

Psychologist, philosopher, educator, and President of the University of the Seven Rays,
Michael Robbins has also written the Tapestry of the Gods series,
which provides an innovative approach to the
study of esoteric psychology and esoteric astrology.

All rights reserved. No part of this book may be reproduced in any form or by any means, electronic or mechanical, including photocopying, recording, or by any information storage and retrieval system, without permission in writing from the publisher, except by a reviewer, who may quote brief passages in a review.

Published by
The University of the Seven Rays Publishing House
A Division of the Seven Ray Institute, Inc.
A non-profit, tax-exempt educational corporation

Publishing and Book Orders:
P.O. Box 5003
Mariposa, CA 95338-5003
http://www.yosemite.net/sources/
(209) 966-5379 or 6972

University Headquarters:
128 Manhattan Avenue
Jersey City Heights, NJ 07307
(201) 798-7777

Cover art by Richard Haeger
Printed and bound in the United States of America

Library of Congress Cataloging-in-Publication Data

Robbins, Michael David, 1943-

The infinitization of selfhood : a treatise consecrated to the destruction of the ego / by Michael David Robbins.

p. cm.

Includes bibliographical references and index.

ISBN 0-9621869-4-5 (alk. paper)

1. Self (Philosophy)--Miscellanea. 2. Infinite--Miscellanea. 3. Conduct of life.

I. Title.

BF1999.R549 1997

126--dc21

Infinitization of Selfhood

A philosophical treatise consecrated to
the destruction of the ego

by
Michael David Robbins, Ph.D.

That beyond which there is nothing; which shines even above Maya,
which again is superior to its effects, the universe.
The inmost Self of all, free from differentiation;
the Real Self, the Existence-Knowledge-Bliss Absolute;
infinite immutable—that Brahman art thou,
meditate on this in thy mind.

Vivekacudamani, Verse 263

The University of the Seven Rays Publishing House
Mariposa, California
January, 1997

This treatise is dedicated
to the great sage Sri Sankaracarya
for his uncompromising love of
the one and only REALITY
—The INFINITE SELF.
Thou Art That.

Special Thanks
from the author

To my co-workers, Dr. Daniele Vecchio and Dr. Daniel Krummenacher, without whose generosity, friendship, and support the publication of *Infinitization* would not have been possible.

To my colleague, Vicktorya Stone, for executing with great skill, intelligence, fortitude, persistence, and patience the extraordinarily arduous task of editing, illustrating, and organizing the format for *Infinitization*. Since she is the only person who has read and carefully considered the whole of the text before publication, I have relied on her keenness and insight to clarify many extremely difficult concepts and find for them the best mode of presentation. I have found her collaboration on this project absolutely indispensable.

To Dr. Annie Muller for her deep spiritual friendship, constant encouragement, and provision of a "mountain retreat" where the majority of *Infinitization* was written.

As always, to my friend and colleague of long-standing, Maryann Casalino, for her constant encouragement for all my creative efforts on behalf of The University of the Seven Rays.

To my colleague, Carole Beckham, for her expertise and good common-sense in relation to the preparation, printing, and publication of *Infinitization of Selfhood.*

To Richard Haeger for creating the design and artwork for the covers of *Infinitization.*

To my daughter, Heidi Rose Robbins, for her love, patience, and infallible good humor as she faithfully transferred the original spoken text from cassette tape into typed format. She, above all persons, knows the condition of this book when it was first begun. I congratulate her for surviving the interminable droning of her father's tape-recorded voice as he speculated upon Time, Space, and the ALL.

To my son, Mathew McIver Robbins, with whom I have had some of the deepest and most philosophically penetrating conversations I can remember. Mathew's innate understanding of Monism and Non-Dualism coupled with his sharp and insightful philosophical mind have, on numerous occasions, given me pause to reflect deeply on certain perplexing questions. Although *Infinitization* falls far short of the philosophical rigor for which Mathew might hope, his influence has been my 'philosophical conscience'.

To my wife Rita—*again*, for her sustainment and long-suffering patience.

To my fellow-non-dualist, Scott Hamilton, for numerous wonderful dialogues between '**8**' and '**8**'. Scott was perhaps the first of my recent spiritual associates in whom I recognized a deep understanding of the ONE AND ONLY SELF—of the '**I-as-8**'. I've never had so much fun 'talking to MySELF'!

To my fellow-seeker, Dr. Michael Wiesenberger for a number of deeply searching dialogues focusing upon the nature of REALITY. Michael's clear and penetrating intellect never failed to stimulate a deeper search. While the theme of "particle and wave" is far from exhausted, at times, it seemed we came close! At least *we* were exhausted.

To my fellow-seeker and profound student of Esoteric Astrology, Niklas Nihlen, for many excellent conversations on the nature of Cosmos.

To my uncle, Leon Epstein, without whose faithful assistance over many years, I would not have been in a position to pursue the thought and study which made my esoteric work, including the writing of *Infinitization*, possible.

To Ron Tonini for his generous printing of a number of proofing-copies of *Infinitization.* His help was very timely, speeding the book towards publication.

To LaUna Huffines, for her generous support of the publishing of *Infinitization.*

To my many students and colleagues in all Ph.D.E. classes, who (learning to survive a noticeable pause in the torrent of cassette tapes with which they are constantly deluged) patiently endured my preoccupation with the lengthy final preparations of *Infinitization.*

Further, I must thank my students for their growing depth of thought, which has been revelatory to me during the period of the conception and execution of *Infinitization of Selfhood.*

To the members of the Board of Trustees of the University of the Seven Rays for their constant encouragement and support.

To my brothers on the Path, Keith Bailey, Cliff Ribaudo, Serge Ledan, and Martin Vieweg, who gathered in 1991 for a deeper penetration into the Mysteries. The results of that gathering have been spiritually impulsive and productive for all concerned.

Special gratitude must be given to my first Teacher, Andrew McIver, who *lived* Non-Dualism. The seed for *Infinitization* was sown long ago during the seven years of instruction he gave me—even more through *presence* than through words. Through Andrew, **8** learned that **8** *Am* the Mystery.

Indirectly, to Sri Ramakrishna and Swami Vivekananda, for being living modern exemplars of the Advaita Philosophy of Non-Dualism. Reflection upon their lives has never ceased to inspire me in my pursuit of Radical Infinitism.

Special gratitude must also be conveyed to three great initiates—H. P. Blavatsky, Mme. Helena Roerich, and Alice A. Bailey—without whose arduous, self-sacrificial labors, students of esotericism (such as I am) would still be very much in the dark regarding the true nature of REALITY.

Finally, to three Masters of the Wisdom—Master Morya, Master Koot-Hoomi, and Master Djwhal Khul—for impulsing and sustaining the revelation of Trans-Himalayan Occultism, the precious Source from which so many students of spirituality in the modern world have drawn their spiritual nourishment.

Michael David Robbins

Preliminary Remarks
from the first reader

Infinitization of Selfhood is a vital contribution to philosophical pursuit. The approach employed by Michael Robbins is to discern and question the implications of the ageless teachings. The resulting insights are called Radical Infinitism—a dynamically active process of inquiry. For those familiar with Non-Dualism, Buddhism, or Theosophical concepts, the assumptions and suggestions made in *Infinitization of Selfhood* may be readily accessible. However, *Infinitization* is inviting to all earnest thinkers because Radical Infinitism deals with the essential nature of reality and identity, and upon these the fundamentals of philosophy, religion, and science are based.

Infinitization strikes at the root of existence, as Universe and Being are the focus of a rigorous investigation. The author's conclusions are unexpected and profound as he addresses the magnitude of problems of living and philosophy. Thoughtful commentary on possibilities and paradoxes are combined with meditations and mantrams, creating a means to synthesize abstract inner realizations with concrete, rational thought. This synergistic approach is thoroughly applied to the quest of what is usually considered the unknowable—the essential nature of identity and of the universe.

The ideas in this treatise deeply challenge our understanding of individuality. As we search and ponder, the response to "How much of the vision do we see for ourselves?" becomes more crucial. It may be that through employing the relentless questioning approach that *Infinitization of Selfhood* typifies, we will see more of this vision for ourselves. By "inquiring the way" along with the author, we too can pierce into the "raincloud of knowable things" and enjoy a radical change in our understanding of Selfhood.

To that end, Godspeed.

Vicktorya Stone

How beautiful is the vast horizon!
How powerful is the thought penetrating Space!
What new ways are disclosed by communion with Infinity!
Seek these treasures; in them is guarantee of your advancement.
Of what use is knowledge which brings one to a closed gate
marked by the sign "we know no more"?
Limitation of knowledge is a grave.
Therefore, fathom Infinity!

Verse 9, Infinity, Agni Yoga Society

Summary
Table of Contents

Detailed Table of Contents

Note to the Reader
from the editor-illustrator

It has been my part, primarily, to edit some unusual-looking text, and to illustrate essentially inconceivable concepts. In doing so (and having apparently survived), I thought a few comments would be in order to help familiarize the reader with the stylistic conventions employed, mention limitations of the illustrations, and suggest one way to begin studying this work. I believe this is an enjoyable book to read, (maybe even more so than it was to 'proof'), and I encourage you to be daring and vigilent in your readership.

Writing Conventions

- The author employed an extreme use of capitalization, single quotes, bolding, and italics, to discriminate between levels of being or suggest more subtle relationships or meanings. This technique, as well as using some Sanskrit terms, are instrumental in describing concepts for which there is no word, or no suitable word in English.
- A technique of formatting complex paragraphs into bulleted sentences or lists is often used to emphasize key elements, show an underlying structure of thought, or highlight a particular line of reasoning. (I hope this assists and not distracts from the author's intent.)
- The word forms of "he" (at my request) have been conventionally used to refer to human beings regardless of gender.

Diagrammatic Disclaimer

Illustrating the processes or states discussed within this book has proved (to my mind) curious, at best. These attempts at illustration (found predominantly in Sections I and V) are therefore meant to be suggestive only, and I encourage the reader to scrutinize the concepts or relationships within his own visualizations. That there are inadequacies in my descriptions or portrayals would be a compliment, and so I therefore hasten the careful reader to bring comments to my attention. While the author has reviewed most of these diagrams, and has given much correction and thought to them, the final forms are mine, and as such, I request that he not be held culpable for any erroneous, inconsistent, or misleading concepts to which the drawings may allude.

Suggestions for Study

1. Read through the Introduction to learn how and why *Infinitization of Selfhood* was written, to fully understand the unusual stylistic conventions and organization of the book, and to reference the foundational assumptions and primary concepts.
2. Glance through Section V, which contains the bulk of the diagrams and synthesizing lists, to get an overall sense of the major relationships and structures.
3. Start reading through the main sections, beginning with the Commentaries, while making references to the Glossary on unfamiliar terms.
4. When you have a feel for the radicalness these concepts, refer to Section VI, Aphorisms, Meditations and Mantrams, and find techniques for expanding your awareness and to help anchor your realizations in consciousness.

In sum, working with *Infinitization* has been a challenging, enjoyable, and interesting experience for me, both technically and subjectively, and one in which I am very grateful to have had the opportunity to participate. This is an *impressive* work (in the many senses of that word), and I do conclude that my perspective of who "I" am has definitely been altered! May you enjoy the same.

Vicktorya Stone

There is only Brahman, the One without a second,
infinite, without beginning or end, transcendent and changeless;
there is no duality whatsoever in It.

There is only Brahman, the One without a second,
the Essence of Existence, Knowledge, and Eternal Bliss,
and devoid of activity;
there is no duality whatsoever in It.

There is only Brahman, the One without a second,
which is within all, homogeneous, infinite, endless, and all-pervading;
there is no duality whatsoever in It.

There is only Brahman, the One without a second,
whose real nature is incomprehensible, and which is
beyond the range of mind and speech;
there is no duality whatsoever in It.

There is only Brahman, the One without a second,
the Reality, effulgent, self-existent, pure, intelligent,
and unlike anything finite;
there is no duality whatsoever in It.

Vivekacudamani Verses 464, 465, 466, 469, and 470

The Introduction to Infinitization of Selfhood

The Achievement of Conscious Immortality Through Right Identification with REALITY

The Purpose of *Infinitization of Selfhood*

Infinitization of Selfhood is written in order to 'restore the SELF unto the Self'—to restore to the human being the INFINITE SPIRIT that he has *forever been* and *forever will be*—the INFINITE SPIRIT or 'INFINITIZED' SELF, which he can *never cease to be.* That SELF, however, is temporarily 'lost' to *mayavic consciousness* and must be reclaimed through *re-identification* with REALITY.

Careful thought and profound meditation upon the ideas presented in this treatise will contribute to that state of *identification* by means of which the seemingly separate and distinct human individual may actually *realize* himself to BE none other than the ONE AND ONLY SELF.

Infinitization of Selfhood is a Treatise for Philosophers

Who is the true philosopher after all? It is not someone who merely thinks about things, but one whose thinking has affected his action in a positive and regenerative manner—in such a manner as to produce increasing clarity concerning the meaning and purpose of life in Cosmos, with the immediate goal of reducing and negating existential grief and inharmony.

Philosophy, in fact, is meant to lead to bliss. While there are many conditions that cannot immediately be changed, one's way of viewing them *can* be changed, and one's mode of participation in them can be changed.

Most human beings today labor grievously under the illusion of a profoundly mistaken identity. Far from being a merely philosophical or abstract issue, the question of what constitutes the REAL IDENTITY of a human being lies at the root of all human problems, and, ultimately, of all problems in Cosmos. In the view of the author, the one and only supreme effort of any E/entity in this or any other Cosmos is to solve the problem of IDENTITY. The Problems of Love and Creativity are closely related but, fundamentally secondary, because when the REAL IDENTITY is understood (and, as it were, *become),* then Love and Creativity in fullest measure follow.

One of the deepest reasons, then, for writing this often demanding and unusually abstract text is, to the extent presently possible, none other than to impel human beings towards 'Bliss-in-Universe'—i.e., simply, to contribute to a *radical* and *foundational* human happiness rooted in REAL IDENTITY.

Sometimes the heavy and *apparently* contradictory arguments may, temporarily, seem to produce the very opposite—shock, frustration, rejection, amusement, the apparent outrage of reason, etc. The author is convinced, however, that persistent consideration of the thoughts offered in these pages will lead to a great broadening of the abstract, intuitive, philosophical mind, and the possibility of apprehending a Universe of astonishing wonder, and a still more REAL '*NON*-UNIVERSE' that is magnificently and utterly *inconceivable* in ITS INFINITUDE.

Infinitization of Selfhood is a Manual for the Systematic Destruction of the Ego

Infinitization of Selfhood has a benevolently *destructive* as well as *constructive* purpose. The treatise is written to facilitate the destruction of the *personal* ego as the major center of human identity. The personal ego, until it is conquered, is a *false* identity which robs life of true joy, and renders the bliss of living unattainable. Even the Transpersonal Ego (the Egoic Consciousness within the Causal Body), and Transcendental Ego (the Monad within the Monadic Sheath), though centers of far greater consciousness, wisdom, and power, are, ultimately, centers that must be destroyed to pave the way for the apprehension of true IDENTITY.

As *ego* on all un-REAL, hence subsidiary, levels is destroyed, it becomes possible to affirm and identify *with* (and *as)* THAT which is REAL as the fundamental *sine qua non* of Life. As a result of considering these thoughts and their implications, may there be a permanent change in all who read and ponder this treatise!

A further destructive purpose may be understood as follows: *Infinitization of Selfhood* is intended to facilitate the destruction of the normal categories and modalities of mind so that they no longer distort and disguise the apprehension of greater Truth, and, even, of ULTIMATE TRUTH. Once the concrete nature of the mind has been relegated to its proper sphere and is no longer allowed to intrude upon and limit illuminating flights of abstract/intuitive speculation, the Reality of Omnipresence, Omniscience, and Omnipotence may be at least dimly, and ever-increasingly, apprehended. Human beings are prisoners of their illusions. A major blow against the World Illusion can be struck if people can be taught to think far more broadly and intuitively, and yet with analytical caution. The author surely realizes that after all is *said* and *read—silence* will be found best. He is acutely aware that what he is trying to convey cannot be put into words. His purpose is simply to set the "Jungle of Illusion" afire, so that it may dissolve in flames and the 'Pathless Path' to the SILENCE may stand free and clear.

Infinitization of Selfhood is Designed to Clarify the Interplay Between the ABSOLUTE and The Relative

Fundamentally, the ABSOLUTE (the INFINITUDE) is *out of all possible relation with* the Conditional, the Relative—a point stressed in the *Proem* of *The Secret Doctrine.* THAT which not only *includes* all things, but *is,* ITSELF, all things, has no *other* or *second* to which to relate. Yet, such is the nature of the human consciousness, that the INCOMPARABLE, when *thought* or *conceived* by the human mind, becomes, as it were, *comparable*—though not REALLY so.

The ABSOLUTE and the Relative are necessarily (due to human limitations) discussed *as if* they were *two* and not *one.* The fundamental problem of living becomes (for the philosopher and spiritually-minded individual) how to live in the world of the *apparently* Relative, in cognizance of, and (more) in complete identification with the ABSOLUTENESS. This treatise is intended to explore the difficulties that are involved in such a task, and to offer some possible solutions. Thus, perhaps, will the *natureless nature* of the ABSOLUTE be intuited within the World of Relativity.

Infinitization of Selfhood is a Book of Apparent Contradiction and Paradox

Infinitization of Selfhood is full of *seeming* contradictions and paradoxes. It appears that a rigorous approach to ABSOLUTE TRUTH defies logic as ordinarily and academically conceived. Scholastic logic demands that a thing be *itself* and itself alone. Mystical or Occult Philosophy requires that a thing, while being itself, must, in some *essential* way, *be everything else* as well as *itself*—and not only *partially* everything else, but *fully* everything else. Thus the stage is set for what the normal logician would call absurd or outrageous thought.

The reader will consequently encounter many puzzles based upon the fact that REALITY is at once unalterably opposed to Relativity (at least to the 'eye' of consciousness), and, yet, *is* inescapably the *entirety* of that very Relativity. While, the pursuit of normal logic is indispensable for functioning within the normal world (i.e., the World of Illusion, the World of Relativity), when the consciousness seeks to escape from the limitations of that World, classical logic, per se, collapses or, at least, is shown to have severe limitations. This does not mean that there has been no attempt to be *logical* throughout this treatise. Constant are the logical arguments, and much care has been expended upon making them consistent. When dealing with certain fundamental ontological issues, however, careful logic ends inevitably in contradiction, no matter how much (and especially when) careful thought is expended.

It is therefore advisable for the reader to reserve judgment concerning the apparent contradictions that are necessarily built into the Philosophy of Radical Infinitism until he or she has acquired an 'in-periential' (rather than 'ex-periential') 'feel' for Philosophical Non-Dualism. When the intuition begins to awaken, the inescapable necessity for verbal contradictions and paradoxes will be better understood.

Infinitization of Selfhood and Trans-Himalayan Occultism

Students of *Infinitization of Selfhood* will find that it does *not* demand the same approach of "meticulous entirety" that serves so well in understanding most of the books of Alice A. Bailey and the Tibetan Master, Djwhal Khul. While familiarity with the "Blue Books", or some comprehensive branch of eastern or western esotericism, will definitely provide a useful orientation in consciousness for the understanding of *Infinitization*, this treatise is written along a different ray line, using far more the third ray of Abstract, Creative Intelligence, and the first ray of Being, rather than the second ray of Love-Wisdom. Many of the conclusions drawn in relation to a practical life-application of philosophical learnings, however, *do* relate to the second ray—one outstanding motto of which is, "Naught Is But Me."

The purpose of *Infinitization of Selfhood*, therefore, is *not* to explore the structure of Cosmos in detail, as is, for instance, so adeptly achieved in the Tibetan's *A Treatise on Cosmic Fire*. Rather, the broad and general impulses operating within Cosmos, and even antedating *any* and *every* Cosmos are brought forward speculatively for consideration. The reasons and motives for cosmic events (and even for Cosmos Itself) are carefully considered, rather than the specific, differentiable qualities of those events, or the various 'planes' upon which such events occur. Thus, while close familiarity with the "Bailey Books" and with the works of H.P. Blavatsky will *predispose* the reader towards understanding *Infinitization*, such familiarity will not, in itself, guarantee understanding; a certain adventurous spirit of mind will be required, accompanied by a willingness to forsake the usual conventional assessments of *identity*.

Infinitization of Selfhood and Complexity and Simplicity

Infinitization of Selfhood may be considered by some a rather complex book, especially due to its language. Comprehensiveness and complexity of consideration, however, are certainly a *means* rather than an *end*. We live within a 'Great Maze' called the World of Illusion, the World of Maya. Events and the interrelation between them are of an almost inconceivable complexity (except, perhaps, when 'Seen' from the perspective of the Universal Logos). A premature rush to simplification is no guarantee of deep understanding. Lingering in a burdensome complexity, however, is almost certain, ultimately, to stifle the most profound type of understanding.

This treatise, therefore, strives to achieve *simplicity* by means of *exhausting* complexity. Very few stones are left unturned in the exploration of the intricacies of thought related to the subjects in question. The result should be a clearing of the reader's way through the Maze; in fact, there are presented certain lines of thought that could contribute directly to the *destruction* of the Maze.

One day the Maze that is this Vale of Illusion must be thoroughly destroyed, and it is IDENTIFICATION that will achieve this ultimate consummation. In short, it is hoped that *Infinitization of Selfhood* will contribute to profound and essential *simplification* within the life of the earnest reader.

Means and Methods

The Method of Speculation

Infinitization of Selfhood is written *speculatively,* using, as much as is possible (given the author's individual limitations) the *Pure Reason.* The abstract/intuitive mind is allowed to range upon a host of subjects, the true nature of which is presently incapable of ascertainment for the great majority of human beings (including the author), and yet high spiritual impressions may be induced that offer flashing intimations of certainty. The certainty to be achieved in relation to these abstruse considerations is *subjective* (in the truest sense) and 'in-periential'.

To follow some of the thoughts and arguments presented in *Infinitization* will likely require a considerable extension of the manner in which the mind is ordinarily used. It is principally the exercise of the *abstract reasoning mind* that is required, especially as that mind is inspired from the realm of *ideation,* which is the realm of the *intuition.* In a certain sense, to read this treatise is an exercise in *extended mental visioning.* The higher abstract and intuitive mind is encouraged to touch those realms that are usually considered *formless* by the occultist or esotericist, and to, speculatively, divine the nature of energy relationships within those realms.

There is considerable value for some kinds of minds in this approach. There can be what might be called a *freeing* of the mind as an *organ of spiritual apprehension.* The mind is released from sense-boundedness into a growing apprehension of the Worlds of Archetypal Reality and, perhaps, into the consideration of Super-Cosmic Worlds even more fundamental to the Origin of all things. The mind can thus *begin* to become an instrument for the apprehension and revelation of *radical* Truth.

The Organization of the Text

Infinitization of Selfhood is divided into seven major sections, in addition to this introductory matter and an index:

Section I — This first section consists of commentaries that all begin with "On ..." and offer an extensive series of essays upon the major concepts of Radical Infinitism (the Philosophy herein espoused), and delineate and amplify the principal ideas in this exploration.

Section II — This section is an exploration of certain "Problems of Living" (or moral problems) and "Philosophical Problems" that arise as one attempts to understand and apply the fundamental concepts of Radical Infinitism. To some of these problems there is no easy or evident solution, and none is offered. The vexing paradoxes so often encountered in the search for Radical Truth are, however, at least honestly brought forward for discussion and possible future solution.

Section III — Certain "conclusions and implications" that may be reasonably inferred from the consideration of the commentaries and philosophical problems are summarized in this portion of the text. This section offers a relatively simple harvest of thought from the sometimes intricate preceding speculations.

Section IV — Here is practical advice, "Counsels", on how human beings may live their lives in the light of Radical Infinitism.

Section V — To help the student of *Infinitization* summarize and essentialize their thought, major "tabulations, lists and figures" are combined into this section. The reader is encouraged to tab this section and refer to it throughout the reading. This section is a harvest of fundamental premises and conclusions and, in general, of important points made throughout the treatise, as well as a Table of Equivalences, which is intended to produce synthesis in thought.

Section VI — Contained in this section is a series of "aphorisms, meditations, and mantric exercises" designed to induce in the practitioner the realizations that lie at the heart of the Radical Infinitist Philosophy.

Section VII — This is a very extensive "Glossary" (almost a book in itself) intended to define as rigorously as possible the multitude of terms utilized in *Infinitization.* The Glossary should be utilized throughout the reading of the treatise proper, and considerable time could profitably be spent studying it in its own right. Some students may find it useful to become familiar with the Glossary *before* beginning the study of the treatise as a whole.

The Sequence of Topics

It should be said at the outset that *Infinitization* is not written in a strictly sequential manner, or in what might be called a strictly logical unfoldment. Rather, it will be found that every section of the treatise has a strong bearing upon every other section. This may be a book more to be *lived with* and *studied over time* rather than read through in one heroic effort. The kind of study likely to prove most revelatory is 'spherical' (so to speak) and not linear. Each return to a section of text previously studied will offer more in the way of illumination and understanding because the study of other sections of text will have shed increasing light upon the topic under consideration.

The Use of Language in the Writing of this Treatise

Words limit the expression and understanding of Realities, and are hopeless to convey REALITY, but, at this stage of human evolution, words must, perforce, be used. The meaning of words will vary greatly according to the context in which they are used. The principal contextual *substratum* employed in this treatise is that of REALITY (or at the very least, Reality), and the concepts considered are, for the most part, considered *sub specie aeternitas* (under the aspect of *eternity)*, or in their *essential* or universal form or nature. The World View espoused, therefore, is a rigorously non-dualistic one, taking its inspiration from the Advaita (Non-Dualistic) School of Vedanta Philosophy, the foremost exponent of which is Sri Sankaracarya. The adaptation of Vedantic philosophy found in this treatise is given the name, "Radical Infinitism".

The author's purpose is to explore Non-Dualism and its surprising, destructive, and extraordinarily empowering implications as thoroughly as his limitations, and the limitations of language, will allow, and to elucidate the astonishing transformational potential of rigorously Monistic thinking and living.

In order to help the reader enter deeply into this subject, the meanings of a number of concepts which are indispensable to the understanding of rigorously non-dualistic philosophy are explored in depth. In this exploration the necessary use of conventional language for deeply unconventional purposes causes a certain set of linguistic problems which must be acknowledged and faced.

Human language is fundamentally *dualistic.* The very structure of language reinforces the Illusion of Duality. Except for the language of visual symbolism (which is tremendously important) the language of words is the only instrument available with which to approach and discuss Reality and REALITY. The fact that language is structured upon dualistic misconceptions makes it almost impossible to use language successfully for the induction of what might be called 'monistic realization'. Nevertheless, an attempt is made.

Because of these unavoidable limitations, the author has put a special *pressure* upon the language he uses, inventing quite a number of terms and specialized usages and, at times, bending (almost 'torturing') customary words and their syntax so that they yield a meaning far beyond their conventional usage and associations. In this way, perhaps, that which is largely unfamiliar and significantly more *real* will emerge in the consciousness of the reader. In quite a number of respects, *Infinitization of Selfhood* is meant to be a 'mind-bender'; it is certainly a 'word-bender'.

On a number of occasions the reader may wonder whether that which is being read can possibly 'make sense' as the entire subject matter of Radical Infinitism concerns itself, literally, with '*non*-sense'. Certain thoughts found in the text may incline the frustrated reader to conclude that *nonsense*, in its *least* flattering usage, is precisely what is being offered. The problem, however, may well lie *not only* in the author's obvious inabilities (of which he is painfully aware), but in the impossibility of accurately using dualistic language to write about THAT which is utterly UNITARY.

The limitations of dualistic language make it impossible to avoid the *appearance* of contradiction—not because contradiction in intended, but because there is no way to express *unitary truths* in dualistic language without *seeming* contradiction. Therefore, when dealing with the great subject of ONENESS, mental constructs must be used predominantly as a springboard for the penetrative intuition, and the even more potent faculty of *identification.* Language must serve strictly as an illuminative symbol for THAT which (if IT *can* be apprehended at all) can truly be apprehended only by other, deeper human faculties.

Conventions Adopted in the Use of Language

The reader is advised to bear in mind certain unusual conventions utilized throughout this treatise.

Capitalization Conventions in General

When, for instance, the three words, 'SELF', 'Self', and 'self' are used, they by no means refer to the *same self.* Rather the difference in capitalization, in this instance, is offered as a means of distinguishing between *orders of beings.*

Words that appear entirely in capitals are meant to refer to the 'level' of the ONE AND ONLY SELF, and ITS DOMAIN. When, for instance, the first letter of a word is capitalized (and sometimes underlined)—e.g., Self, Reality, Entity (as opposed to SELF, REALITY and ENTITY), the words are associated with, Cosmos (as a limited Being), and with that which transpires *within* Cosmos, especially in relation to relatively high Beings upon relatively high cosmic dimensions.

When, on the other hand, words like 'self', 'reality' or 'entity' are written entirely in the lower case, they usually refer to the human level of functioning within Cosmos, or to lower levels still. In such instances the first letter of words may be underlined to emphasize the relatively low level indicated—e.g., 'self', 'reality', 'entity', etc.

It may be asked why bother to resort to such a tedious method of capitalization for such words as 'SELF', 'Self', and 'self', for instance, when the use of three distinct and uncapitalized words, or appropriate adjectives, might serve as well. In response, the holographic, analogical construction of Cosmos makes it necessary to discuss structurally similar B/beings, E/entities, and situations with reference to different cosmic dimensions. Capitalization conventions make abundantly clear the dimensional 'level' or order of the B/being concerned. In fact, the use of capitalization conventions may render unnecessary the use of a superfluity of words necessary to specify such levels.

There has naturally been an attempt to be *entirely* consistent in such capitalization, but the attempt cannot possibly be completely successful, due to human error and ambiguous contexts in which more than one type of capitalization can be judged as appropriate. Sometimes an inclusive approach is taken through such a convention as using composite words like 'S/self', or 'E/entity', both of which convey that *both* higher and lower order E/entities are being referenced.

If, in relation to these capitalization conventions, readers detect what they perceive to be an entirely erroneous or inconsistent usage, they are kindly requested to alert the author or publisher, so that corrections, if justified, may be undertaken in future editions of *Infinitization.*

A Special Category of Capitalizations

Although the ONE AND ONLY SELF cannot properly be said to *act* or to *do*, it is usually necessary to think of IT in such terms. Therefore the implied 'DOINGS' of the INFINITE SELF are presented as capitalized verbs (and, sometimes, nouns) surrounded by single quotes. It is principally the simple verbs like 'DO', 'DOES', 'ACTS', 'SEES', etc. which are capitalized. The single quotes alert the reader to the fact that we are studying a special category of what might be called 'actionless action'.

The Use of Underlining

The first letter of certain words beginning with a capital letter may be underlined, to *emphasize* that type of capitalization, and its distinction from the same word when completely capitalized, and from the same word written entirely in lower case. The first lower case letter of a word written entirely in the lower case may also be underlined to accentuate that that particular word has a meaning attached to the lower case, as distinct from the meaning of the word when the first letter is capitalized, and the meaning of the word when all letters are capitalized. This convention of underlining is meant to alert the reader that a 'level' or dimensional factor is under consideration.

Additionally, following each definition in the Glossary are examples of the particular word under consideration. When the word being defined appears in the example, it is completely underlined (e.g., underlined) to help direct the reader's attention specifically to that word.

The Use of Bolding

In general, the bolding of words is reserved for section titles and for indicating (by means of subject headings) the specific subjects upon which commentary is to be generated. One of the very significant exceptions is in relation to the word, '**I**', which, when **bolded**, represents the ONE AND ONLY SELF. On the other hand, the 'unbolded' 'I', simply represents the personal self. Occasionally, pronouns associated with the '**I**' may also be bolded, as in '**MY**'. As well, the word '**8**' is customarily bolded. (The '**8**' Is the '**I**', once that '**I**' has 'ENTERED' manifestation in Cosmos.)

The Use of Single Quotes

The standard use of single quotes, is to cite a quotation *within* a quotation. If and when such situations arise in the body of this text, this standard convention is observed.

However, in this text there are other significant usages of single quotes which should be noted. The method adopted of 'mentioning' a term by surrounding it with single quotations ('') is for the purpose of calling special attention to a term because it is being used in an unusual or special way—a way that is different from the way in which it would *usually* be understood. Without the single quotes, the term might be glossed over, and its special meaning misunderstood. Sometimes also, a *seemingly* inappropriate word must be used to indicate a highly spiritual process for which no proper word exists. The use of single quotes around such a word indicates that it is not *exactly* appropriate but conveys some idea of the subtle process indicated, and stand for the phrase, "so to speak".

Occasionally, a convention called "philosophical mention" requires the utilization of single quotes. Through "philosophical mention" special attention is called to such a word. The word that is 'mentioned', however, need not be either inappropriate or an approximation; the single quotes simply draw the reader's attention to the word, without rendering it emphatic, as italicizing it would do.

The Use of Italicization

Italics (e.g. *italics)* are used especially for emphasis. It is often necessary to be *emphatic* in relation to certain words in order to convey to the reader the desired meaning. (An analogy might be an accented note in a musical phrase.) The italicization of words is, in this regard, an especially useful practice. The meaning of the italicized word is not approximate, inappropriate, or used in a special way as is so often the case with a word or words surrounded by single quotes; quite simply, italics indicate a definite *emphasis.*

Italicization is also used when speaking of a given term *in general*, especially when the author's intention is to *avoid* the assignment of 'levels' or dimensions. Sometimes, the author desires that the reader think inclusively in terms of *all* levels or dimensions. For instance, to speak of the *self* in general, without specifying the 'SELF', the 'S̲elf' or the 's̲elf', the italicized word, *'self'* might be used.

The Consistency of Linguistic Rigor

Linguistic rigor and verbal inventiveness are practiced in order to facilitate the grasping of the otherwise ungraspable. Many different words are used for the same thing (namely the SELF), because each word provides an alternative perspective, a distinct angle of vision. Multiple perspectives are necessary because the concept of INFINITY is so very hard to import into the realms of mind. The mind tends, usually, to reject it (or to misunderstand it), so there is a need for many different ways 'in'.

There is no absolute consistency between a rigorously technical and a sometimes nontechnical use of language. In portions of the treatise it is completely necessary to be rigorously technical and observant of the preestablished conventions; in other portions, however, it would be unduly tedious to be so.

The Use of the "I"

As the variations of the term "I" are used so often throughout the treatise, this is a good place to note some differentiations between the use of the various 'I-symbols'. Although the symbol 'I' may from time to time inadvertently be used in a nontechnical way, the reader is asked to bear in mind the various distinctions here discussed.

1. **I** is the symbol used to indicate my identification with the ABSOLUTE SELF, and ITS identity with my selfhood;

2. **8** is the symbol that equates to **I**-as-'I' or to 'I'-as-**I.** This symbol indicates the fact that the ONE SELF, the **I**, manifests through all lesser 'I's, and that all lesser 'I's are really inseparable from the INFINITE SELF. One can read the '**8**' as an *infinity sign* standing upright, as the 'I' stands upright. The '**8**' therefore means the INFINITE SELF in manifestation as the 'I' or even the 'i'. For practical purposes, the '**8**' represents any E/entity as a 'Ray' of the ABSOLUTE in-Cosmos. For man, it is a symbol of identity *as* the One Identity, *one* with the Universal Logos, the One 'Ray' of the ABSOLUTE in-Cosmos. Thus '**8**', represents the human being as any of a series of 'Higher Selves' all of which are *fully* and *completely* the One Self of the Cosmos. The symbol '**8**' is obviously different from, and 'infinitely' more *dynamic* than the figure 'eight' when it is lying on its side;
3. By the 'I' is normally meant the normal ego (or personality) consciousness (even if that ego, or personality, is quite expansive). It means normal 'bounded selfhood'. The 'I' emphasizes the ahamkaric principle and does not emphasize the connection with the INFINITE SELF; and,
4. By the 'i' is meant an extremely bounded state of I-ness. One might say that such an 'i' is completely dominated by the lunar lords which circumscribe, extensively, its state of self-realization. The term 'i' is limited entirely to the consciousness *of* the personality, and, in an even more restricted sense, to consciousness *within* the personality, as when selfhood is identified principally with one or the other of the lunar vehicles.

The term 'I' can range all the way from personality consciousness through soul consciousness (especially when egoistically appropriated by the personality). Soul consciousness is a transitional phase. The 'I' is fast dissolving (for the human being) when the Spiritual Triad becomes influential. The 'I' consciousness of Beings greater than man is never entirely an 'I' consciousness, but is, given the group and expansive awareness of such Beings, an '**I**-as-I' consciousness.

The symbol **8** ('**I**-as-I') stands, initially, for incipient Monadic Awareness, and extends through all possible states of awareness-in-Cosmos, all the way to the threshold of the Universal Self. '**8**' is Spirit-'I-ness'. It may be that the *pure* **I** consciousness is reserved for that apparent *interlude* when the SELF is ALL-in-ALL, but some mystics and occultists in the state of samadhi would say that they had experienced it. It would be signaled by the complete obliteration of any sense of 'I', even within the relationship symbolized by '**I**-as-I'.

I am the Universal, I am the All, I am transcendent,
the One without a second.
I am Absolute and Infinite Knowledge,
I am Bliss and indivisible.

Vivekacudamani, Verse 516

Infinitization of Selfhood As a Source-Book

This treatise may well become for you, the reader, a source of rare and almost unthinkable thoughts—thoughts, at least, that you may never have *thought* before. The implicit demand is that such recondite thoughts be studied seriously.

A Few Prerequisites for the Reader

Infinitization of Selfhood is not an easy book, though, for some, it will have its rewards. A certain background in Trans-Himalayan Occultism or Esotericism would definitely be helpful. The author has, for twenty-five years, been a student of the Alice Bailey Teaching; for a number of years before that he was an ardent student of Theosophy. A great part of his world view has been shaped in accordance with these studies. If, however, the reader has familiarity with other branches of esotericism, that familiarity may suffice as an sufficient orientation to *Infinitization,* but some familiarity with the esoteric world view is assumed.

The book, however, places another kind of demand upon the reader: the willingness to become familiar with a new and largely *invented philosophical vocabulary.* The author discovered that to treat this subject with any degree of subtlety or acuity of thought, it was necessary to use unusual words, or usual words in an unusual way. Thus, the reader's patience is requested with becoming familiar with these new thoughtforms that, ultimately, are meant to place them *en rapport* with THAT which is more subtle than any *thing* conceivable.

Above all, the reader must possess a love of *RADICAL TRUTH* (presently unattainable thought IT be). The true definition of 'radical' is, "that which lies at the *root*." All *truth* (short of this kind of TRUTH) is not *ULTIMATE* TRUTH. Of course one cannot possibly *know* ULTIMATE TRUTH; one can only and ever *BE* IT. Nevertheless, the strong desire to approach such TRUTH with the *mind* (at least, initially) will give the readers the fortitude required to make their way through the book successfully, and will make the approach to R̲adical T̲ruth (i.e., Ultimate Truth *within* Cosmos) fruitful.

On the Use of *Infinitization of Selfhood*

Be patient and *live* with it. Value cannot really be extracted from *Infinitization of Selfhood* unless the *antahkarana* (the "rainbow bridge" that unites the intuition and abstract mind with the concrete mind) has been somewhat created. The very attempt, however, to read and to understand what is written will surely contribute to the construction of the "bridge".

The reader may well find *Infinitization* to be an invocation to his or her more abstract (and, probably, under-utilized) *triadal* faculties. If the process of reading and understanding proceeds as desired, the term 'Monad' will no longer be just a word.

On the Use of The Glossary

The Glossary in *Infinitization* is unusually extensive and thorough. Through the utilization of the Glossary, the most important subjects can be studied in a relatively *condensed* form. The attempt has been made to be very rigorous in the definition of terms so that a significant degree of exactitude of thought may be achieved. The Glossary may be used at any point during the reading. Some, who have a passion for definition, may even prefer to read it first in order to clarify their minds, but such would not be the preferred path for all. At any rate, the Glossary contains the maximum degree of particularity with respect to the use of words within the overall text. As well, there are some topics elucidated and points made in this section that are not elsewhere.

On the Limitations of the Present Effort

The author all too well realizes the insurmountable task which he has set for himself, and, the consequent impossibility of *real* success. Persistent attempt has been made to penetrate into distant heights and depths of abstract-intuitive thought. Though, at times, there has been some measure of success, failure to *fully* penetrate has been an ever-present and chastening awareness, a goading frustration. Sometimes confusion has entered alongside inescapable paradox, as the boundaries of the author's present thinking have been pushed to their extreme limits. Sometimes, language has "gotten the better of him", and in his very attempt to meticulously clarify, he may have led the reader into the vale of complexity and confusion. The author can only hope that his errors and possible blunders will at least promote a searching inquiry into areas of thought and speculation which, he believes, demand careful philosophical attention by students of the Ageless Wisdom. In this thought, he takes some consolation, and also hopes that his critics, and all those who come after, may correct his errors with close, illumined thinking. He submits *Infinitization of Selfhood* as a "work in progress", and hopes that it will not take *forever* to complete—or to *read!*

He who has killed the shark known as sense-object
with the sword of mature dispassion,
crosses the ocean of *Samsara*,
free from all obstacles.

Vivekacudamani, Verse 80

A Personal Letter on an Impersonal Subject

To readers of *Infinitization of Selfhood*

Dear Friend:

I would like to say a few words about this book, *Infinitization of Selfhood. Infinitization* is a philosophical treatise, and needless to say, is necessarily academic and impersonal in tone, so it is quite unusual for me to be writing about it in a more *personal* way. Nevertheless, perhaps it would be useful to prospective readers of this work to understanding something of *how* it came to be written and some of the subjects they will encounter.

When I consider all the books I have written or plan to write, I think none will be more important (or more difficult, both for me and for my reader) than *Infinitization.* Many of my readers and students know me as an astrologer, rayologist, and something of a musician/dramatist, but really, I am a *philosopher*—one who *loves wisdom* and who is determined analytically and intuitively to *think his way* into the clearest possible understanding of the great questions proposed by philosophy, especially *occult* philosophy. I am well aware that there are many books which my students, friends and fellow group members would *rather* I write (the forthcoming *Esoteric Astrology* books, especially, and indeed those books are on their way), but there is no book of mine that it is more important that I write, or which will challenge them more, than *Infinitization.*

From the time I was a child, I have always been interested in questions for which there were, apparently, no answers. I was born asking questions—mostly, it seemed, *unanswerable* questions. I found my way into Theosophy, and, fortunately, discovered many important answers, but certainly not *all;* nor *will* I-as-a limited-Entity find them in this life, nor in a million life cycles to come. I will, however, persist in my questioning, for I have learned to "Inquire the Way".

Infinitization of Selfhood was born out of this tendency to *question*, and as well, from a tendency to be dissatisfied with most answers. There are certainly great limits upon what it is possible for the human mind to know, but many thinkers put unnecessary limitations even upon the *kinds* of questions they dare to ask. I have *not* done this. I have dared to ask outrageous questions and, sometimes, thought my way into outrageous answers. I have dared to use my own small and limited mind as a means of inquiry into the Nature of REALITY. There is nothing new in this, as for aeons human beings have been doing so. Thus were established the great systems of spiritual philosophy which have so nurtured the world of thought. But, you see, I discovered that *I* had not really done so (had *not* used every resource of mind and intuition available to me to *think* my way into TRUTH)—at least not in a way that was uncompromisingly *authentic* and *daring*. The Tibetan Teacher, Djwhal Khul, asked His students, "How much of the mountain of vision have you climbed for yourself?" I asked myself this same question, and decided that even though the climbing would certainly be dangerous, and the conclusions reached quite possibly outrageous, absurd, or impractical (at least *seemingly so)* I must climb anyway.

So, I began. I began to think about the great subjects which provoke the minds of all philosophical thinkers, and I began to come to my *own* conclusions. For this reason I consulted *no texts* in the writing of *Infinitization*, other than *The Crest Jewel of Discrimination* by Sri Sankaracarya (my *ultimate source book)*. For about two to three weeks, I set out with my knapsack on my back and battery-powered tape-recorder in hand, walked into the mountainous forest surrounding Santa Fe, New Mexico, and began speaking to myself out loud about Time, Space, Motion, Reality, Identity, etc. I focussed upon one subject after another, always trying to place my consciousness "Under the Aspect of Eternity" and I let come what came. In this way, through philosophical many-miled hikes, thinking out loud, *Infinitization of Selfhood* was born. At the end of the process, I had eight or nine 90-minute tapes in hand which my daughter, Heidi, selflessly volunteered to transfer to computer—a harrowing experience I'm sure, but she survived! Then, over the next two years, whenever time was available (and it rarely was) I began to reorganize and elaborate upon what had been spoken, deepening it, enriching it, exploring implications that had not occurred to me during the dictation process. Needless to say, the treatise expanded, and even now would continue to do so, had *time* not called a halt to *eternal questioning.*

From a more philosophical perspective, let me say, that *Infinitization* was written because of my determination to know and to BE the ONE SELF (which I and you and all *already* ARE). In so many ways, I am, as a thinker, an *absolutist, (rather* than the *relativist* I *seem* to be), and I am *completely* dissatisfied unless I "strike bedrock"—the "bedrock" of ULTIMATE REALITY (a presently impossible task, of course). There are so many things that it is *interesting* to know, but only *one thing,* which, in my opinion, it is *necessary* to know in order to *live,* and that is the nature of the ONE AND ONLY SELF. This SELF *IS* the INFINITE SELF, the ABSOLUTE, the ONE WITHOUT A SECOND. Already, in *Infinitization,* I have counted more than 200 names I use for this ONE AND ONLY ENTITY/NON-ENTITY. I jokingly call this long list of names, "Names for the NAMELESS". The writing of this treatise has been an experience in paradox from start to finish.

There is no doubt about it, I am SELF-consumed—some of my friends might say SELF-*obsessed.* By this statement I mean, that to prove to myself through thought and experience (and, perhaps, to 'others'—if 'others' REALLY exist?) the REALITY of the INFINITE SELF is my all-consuming passion. I have reached the point in my thought where I do not think it is possible to live as a thinking being without identifying *with* and *as* this ONE AND ONLY REALITY.

I have, for exactly 25 years, been a persistent student of the works of the Tibetan Master Djwhal Khul (and for years before that, of Helena P. Blavatsky). I have approached His books with that attitude of "meticulous entirety" characteristic of those upon the third aspect of the Second Ray. Such an approach is natural to me and I have been able to share (hopefully with some success) with students and fellow group members the harvest of all these years of intense study and thought. *Infinitization,* however, is *not* written upon the Second Ray. Rather, it is written upon the Third and First Rays, and this has represented a significant departure from earlier works, and perhaps from works to follow.

The premise of *Infinitization* is a *First Ray premise*—'BE-NESS' IS ABSOLUTE, and there is naught but 'BE-NESS'. The entire Universe is a Necessary Illusion (albeit a beautiful and often terrible one), and is simply a Modification of THAT which has *never* for all ETERNITY been ESSENTIALLY *modified* in any way. Thus, *paradox* lies at the very root of the World View that I have unfolded through the writing of *Infinitization.* I have begun to think that paradox does not arise simply because of the limitations of the human mind, but is inherent in the Process of Universal Manifestation.

Thus, I began with the First Ray assertion of the *inviolable primacy* of the BOUNDLESS IMMUTABLE PRINCIPLE (the INFINITE SELF) and sought, with as much logic as I possess, to understand the ramifications of accepting the *immutability* of this PRIN-

CIPLE, the *immutability* of the ONE AND ONLY SELF. How does one live in a world of apparent diversity when one *knows* that all diversity is simply *apparent*—that all diversity is *resolvable* into an ABSOLUTE SUBSTRATUM that is incapable of ESSENTIAL variation? This Philosophical Problem (which is essentially the Problem of the Reconciliation of the INFINITE with the Finite) lies at the heart of *Infinitization.* From what I have already said, I think you can see that what begins as a First Ray Issue, is, in *Infinitization,* approached in a manner characteristic of the Third Ray of Abstract Intelligence.

In my First Ray pursuit of TRUTH through Identification, and my Third Ray pursuit of the implications of TRUTH through *reasoning* (hopefully at times *Pure* Reasoning) I dared (whether foolishly or not) to extend my thought into areas for which humanity-as-humanity clearly has no present answers. A host of philosophical problems forced themselves upon me—problems that I had not even realized as existent before I ventured into the pursuit of ULTIMATE REALITY.

Let me share with you a few of the subjects to which I addressed myself, usually generating more *questions* than answers:

- The Radical Indivisibility of the INFINITE SELF
- The absolute *identicalness* of every unit-in-Cosmos with the ABSOLUTE IDENTITY
- The INFINITE SELF considered, paradoxically, as the 'GREAT CONTRADICTION'
- The Nature of the **I-as-8-as-**I (the many modes of 'I-ness')
- The 'REASON' for the Birth of Cosmos
- The Nature of Maya as *the* SELF-'VEILING' Process
- Consciousness considered at the generator of all limitation, i.e., Consciousness *as* Maya
- Time, Space, and Motion as the Primal Intra-Cosmic Trinity and as three *impossibilities* in the WORLD OF REALITY
- The contrast between Infinity and INFINITUDE
- The unimpeachability of ABSOLUTE FREEWILL
- The 'INFINITESSENCE' as the FOUNT OF ALL POSSIBILITY
- The necessity for a *finite* Universe, and the impossibility of an *infinite* Universe
- The ESSENTIAL impossibility of *relationship* in the World of Relativity
- The mysterious Birth of Cosmos from a CAUSELESS CAUSE
- The startling implications of a *linear* Infinitude of Cosmoses
- The relationship of Linearity and Simultaneity
- The uneasy Relationship between Time and ETERNITY
- The impossibility of the existence of ABSOLUTE SPACE, along with a description of Mulaprakriti *as* Infinite Space
- The discussion of Mulaprakriti or Root Matter as both the Infinite Object, and Infinite Self-Reflectivity
- The concept of Matter as NOTHING rather than Something, or the essential vacuity of Matter
- The Nature of the Divine Emanatory Stream in relation to Monadic Paths of Descent and Re-ascent
- The Origin, History, and Scope of Ultimate Cosmic Monads as 'Rays' of the ONE 'RAY' of the ABSOLUTE
- The inseverable, indivisible identity of all Monads with the One Cosmic Monad

- The Nature of and Necessity for Cosmic Evolution concomitantly with a discussion of the *impossibility of the evolution of the SELF*
- A discussion of the ETERNAL NOW and its contrast with the 'Cosmic Now'
- The immovable *'middleness'* of the ETERNAL NOW
- The impossibility of *continuity* in Cosmos, and thus the necessity for a Discontinuous Universe
- The necessary 'particulateness' of the Worlds of Fabrication and due to the Law of Reflective Discontinuity, contrasted with the 'Partite-Imparticulateness' of the World of Being
- The nature of 'Ontological Oscillation' or the fluctuation of the Fabricated Cosmos between Being and Non-Being 'x-tillions' of times per second
- The nature of the Ultimate Particle/Event and its relation to Fohat
- The nature of the Ultimate Moment as a bounding Cosmic Parameter
- The impossibility of evolution or improvement from one Universe to the next successive Universe
- Karma in the light of Radical Monism
- The Execution of Responsibility in a Non-Dualistic REALITY
- The Process of 'NOUMENESSENTIALIZATION' as the *infinitization* of all Noumena
- The inescapable Necessity for Universal Redemption
- The Nature of Pre-Cosmic, Intra-Cosmic and Post-Cosmic 'FOHAT'/Fohat
- The Nature and Origin of the Cosmic Algorithm, the Design-at-the-Beginning
- The essentially illusory nature of *Individual Identity*
- The impossibility of *individual* 'Salvation'
- The Nature of Cosmos as an 'Infinitesimal-*izing*' Object and the SELF as the INFINITE SUBJECTIVITY
- The Way to Bliss through Identification, etc.

These are a few of the many points considered in depth and, perhaps, they will give you something of a flavor for *Infinitization*. I have often wondered to myself, upon what right do I discuss these things? Certainly not because of some all-encompassing 'Divine Revelation', but, perhaps, by 'Right of the Demanding Mind' that realizes its *essential identity* with the Universal Mind, and knows Omniscience to be a Cosmic Birth-Right.

In my thought, at least, I have broken through many 'ring-pass-nots'. I have imaginatively transposed myself to the *beginning* and *end* of things, and then sought to do something even more difficult—in fact, *impossible*. I have sought to *imagine* BEGINNINGLESSNESS and ENDLESSNESS. All this has arisen from my age-long fascination with the concept of *Infinity*. I have begun to think of the concept of Infinity as the 'Destroyer of Minds'. Certainly, there is a sense in which the mind simply *refuses* to think about it for any length of *time*. The mind (even in all its aspects) is, like everything else in-Universe, Time-Conditioned, and does not really *know* how to approach THAT which is not *of* Time and Space. Nevertheless, I have found it necessary to attempt the approach.

In this labor I had few guides. Perhaps the thought of the Greatest Vedantic Sage of India, Sri Sankaracarya, was the source of my inspiration. In his 'simple' little book, *Vivekacudamani (The Crest Jewel of Discrimination)*, he approached the Nature of Fundamental Reality in the most convincing way I have ever encountered. From the time I began working with his book (more a treatise, really) in late 1991, I was captivated by it, almost *overcome* by it. What I might call the 'Shiva Perception' descended upon me, and I became determined to become what I might call an 'Agent of Infinitization', a 'Witness to LIFE', a 'Conscious Unit of INFINITUDE'. I worked (probably clumsily and crudely) with the only tools at my disposal—Atma/Buddhi/Manas, as I could access them. The precedents to follow were few.

Basically, what I have tried to do is to approach the concept of the INFINITUDE with the mind (especially the intuitively-inspired abstract mind). Concrete mind alone would be hopeless. In fact, REALLY, the entire Quest is hopeless and 'Quixotic'; however, in the very attempt to approach the INFINITE with necessarily Finite means, great vistas within Finitude tend to appear, and indeed they did for me. Whether I succeeded in capturing them in words is another matter. Whether these vistas can be retained and accessed at will is also another matter.

As I contemplate releasing *Infinitization* to the printer in early 1997, I feel a certain trepidation. It seems as if this treatise can *never* be properly completed. Every day that I dwell upon the many, many issues and philosophical problems that have arisen, I conceive of new approaches and solutions (*and* new problems!). Therefore, I could be writing on this treatise to the 'end' of my present days. But instead, I *have* decided to release what I have thus far completed (during five years of intensive thought) in the hope that I will stimulate students and fellow group members to "think on these things".

What can be gained from a reading of *Infinitization of Selfhood?* Well, certainly, many will think—"Absolutely *nothing* at all", and in a way, they are very close to the truth. If they would change their response slightly, and say, "ABSOLUTELY NO-THING—the ALL", I would agree with them. I dearly hope that students of *Infinitization* (for reading it will be a *study)* are confronted with the possibility of Radical SELF-Realization.

I call the philosophical discipline in which I am engaged, 'Radical Infinitism' (more popularly understood as "Non-Dualism"). Its purpose is to 'deliver' some glimmer of REALITY to the Maya-benighted consciousness. Nothing, it seems to me, is more valuable than the *realization* of THAT. The idea that 'Thou art THAT' is *absolutely ageless.* This great *formula* is the 'Way to Salvation' for all in *all* Cosmoses. In our Cosmos and upon our little planet, it has been the Vedantin Sages Who have (most recently) propounded this great dictum. In *Infinitization* I have simply (and sometimes *not* so simply!) tried to *think* about the profound implications of this deceptively simple statement.

It is my great hope that readers of *Infinitization of Selfhood* will be willing to think *with* me; that they will be willing to suspend the usual limitations of mind and dare to extend their thought into unsuspected regions, the existence of which is almost certainly doubted by most. Perhaps, readers will be willing to *puzzle* with me as I raise question after question, and problem after problem in my pursuit of *cosmic* and *trans-cosmic* understanding—to the degree that the Mind of Man can grasp it. They will have to realize that there is no *certainty* promised, but that some of the major philosophical issues may be revealed in a new, and sometimes *astonishing*, light.

Of special mention is a section of the treatise that offers a number of meditations designed to help induce the *infinitized* state of mind. These are meditations based upon *both* the Art of Identification and the relentless stretching of the mind to the point where it no longer acts as an obstacle to the deep realization of *essential identity*. These meditations have been conceived largely along First and Third Ray lines, and, I think, lead significantly towards the realization of Synthesis.

As abstract and remote as many of the considerations in *Infinitization* may *seem*, they are not REALLY so. The goal of the treatise is perhaps the most practical of all possible goals—a new appreciation for and deep realization of one's TRUE ESSENTIAL NATURE. It is my conviction that a great range of human problems can only be solved first, through *approach* to this great realization and, finally, through the *realization itself*. Otherwise, it seems to me that, no matter how much we may *know*, no matter how much we may *love*, and no matter how much *power* we may possess, if we are ignorant of the *realization*, we would necessarily lead relatively superficial lives. This latter possibility has never been acceptable to me.

It is important to state that the Selfhood of all beings-in-Cosmos is *already* 'infinitized'. Every unit of life in Cosmos, including Cosmos Itself, is *already* the IRREDUCIBLE SELF, and has never been (nor can ever be) otherwise. But we, enthralled by an *oppressive finitude*, tend to forget this. Over and over again, (Cosmos after Cosmos) due to SELF-Veiling, we forget the GREAT TRUTH. Is it important to remember? I believe it is not only important, but *indispensable*, and it is to this *remembering*, this *rediscovery*, that *Infinitization of Selfhood* hopes to make a small but significant contribution.

With love to all,

Michael David Robbins

To realize the whole universe as the Self
is the means of getting rid of bondage.
There is nothing higher than
identifying the universe with the Self.
One realizes this state by excluding the objective world
through steadfastness in the eternal Atman.

Vivekacudamani, Verse 339

The Three Fundamentals of *The Secret Doctrine**

1. There is one Boundless Immutable Principle.
2. There is a basic law called the Law of Periodicity.
3. All souls are identical with the Oversoul.

* The foundational treatise on eastern occultism for the western mind, by Helena P. Blavatsky

The Fundamental Ideas in *Infinitization of Selfhood*

The thoughts developed in this treatise on IDENTITY are based upon the following Fundamental Ideas:

1. There 'EXISTS' a BOUNDLESS IMMUTABLE PRINCIPLE—the 'INFINITESSENCE'.
2. The BOUNDLESS IMMUTABLE PRINCIPLE, or 'INFINITESSENCE', is the ULTIMATE *MYSTERY*—ESSENTIALLY, *unthinkable* and *unspeakable.* Nothing *absolutely true* can be *known* of IT.
3. The BOUNDLESS IMMUTABLE PRINCIPLE, or 'INFINITESSENCE', always *was,* and always *will be;* IT never *began,* nor will IT ever *end.*
4. The BOUNDLESS IMMUTABLE PRINCIPLE, or 'INFINITESSENCE', IS the ONE AND ONLY BEING/NON-BEING, the ONE AND ONLY IDENTITY/NON-IDENTITY, and IT *alone* 'ABIDES' forever.
5. REALLY and ESSENTIALLY, there is *naught other* than the BOUNDLESS IMMUTABLE PRINCIPLE, the 'INFINITESSENCE'—nor has there ever *been,* nor will there ever *be.*
6. The BOUNDLESS IMMUTABLE PRINCIPLE, the 'INFINITESSENCE', 'APPEARS' periodically, as a single *finite* Universe. IT has done so cyclically forever, is doing so Now, and will do so, cyclically, forever. Thus, there has been a *beginningless/endless* infinite sequence of successive finite Universes.
7. All Universes are illusory in form but, not in ESSENCE, for their ESSENCE *is* the BOUNDLESS IMMUTABLE PRINCIPLE, the 'INFINITESSENCE'.
8. Time, Space, and Motion do not REALLY exist (in and of themselves); they are REALLY only *illusions,* or *appearances.* Only the BOUNDLESS IMMUTABLE PRINCIPLE, the 'INFINITESSENCE', *REALLY* 'EXISTS'.
9. All apparently existing times, therefore, are REALLY, *no time,*—the TIMELESS, ETERNAL NOW.
10. All apparently existing spaces, therefore, are REALLY, *no space*—the GREAT NOWHERE/EVERYWHERE.

11. All apparently existing motions, therefore, are REALLY, *no motion*—the GREAT IMPERTURBABLE MOTIONLESSNESS, the ABSOLUTE STILLNESS.
12. All apparently distinct *beings* or *identities* do not REALLY exist (in and of themselves); they are REALLY *illusions* or *appearances*. Only the BOUNDLESS IMMUTABLE PRINCIPLE, the 'INFINITESSENCE', *REALLY* 'EXISTS'.
13. All *beings* or *identities*, therefore, are REALLY, *no particular beings* or *identities* but, instead, only the ONE AND ONLY INFINITE BEING/NON-BEING, the ONE AND ONLY INFINITE IDENTITY/NON-IDENTITY, the ONE AND ONLY INFINITE SELF/NON-SELF. Therefore every apparent *being* is, *REALLY*, every 'other' apparent *being*. Thus, every-*one* is, REALLY, everyone else, and every-*thing* is, REALLY, everything else.
14. Despite periodically and forever *appearing* to be *other than* IT IS, the BOUNDLESS IMMUTABLE PRINCIPLE, the 'INFINITESSENCE', forever 'REMAINS' *only* and *exactly THAT* WHICH IT *REALLY* IS, *without the slightest modification*.
15. Thus, Infinite *Apparent* Duality cyclically co-exists with ABSOLUTE MONISTIC INFINITUDE—*forever*.
16. The BOUNDLESS IMMUTABLE PRINCIPLE, the 'INFINITESSENCE' manifests periodical Universes simply because IT *IS* WHAT IT *IS*. The *true reason* for ITS periodical manifestation is, ESSENTIALLY, inscrutable, but is probably related to SELF-NECESSITY, and to the 'FACT-of-SELF' that the PERFECT must be *imperfect* in order to be the PERFECT.
17. The BOUNDLESS IMMUTABLE PRINCIPLE, the 'INFINITESSENCE', *IS* the *infinitization of all possibility*, (i.e., the FOUNT OF ALL POSSIBILITY). Each of an infinitude of Universes is but a single *'infinitessentialized possibility'*, 'EXTRUDED' *for reflected SELF-Objectification*, from the FOUNT OF ALL POSSIBILITY.
18. The Purpose of all Life in Cosmos is the *realization of* and *identification with* the BOUNDLESS IMMUTABLE PRINCIPLE, the 'INFINITESSENCE', the INFINITE SELF/NON-SELF, *and* the *perfected, reflected SELF-Objectification* of the One 'EXTRUDED' 'Infinitessentialized Possibility' which the Universe Is.
19. The Mystery of the 'Creation' of the Universe, and the Unfoldment of the Universal Process, is based upon Subject/Object relations, and the eternally irreconcilable contrast between *Being* and *'Seeing'*.
20. The greatest of all Illusions is Consciousness. The INFINITE SELF-as-'MAYA' (the great 'SEEING') *is* 'CONSCIOUSNESS'-as-Consciousness. 'MAYA' is the 'SEEING' that 'CREATES' all Universes; thus, by transposition, Consciousness is the 'Seeing' which 'Creates' all Universes.
21. '**I**' (which includes all B/beings, in all Cosmoses that have ever been, are now, or will be, and all Cosmoses, Themselves) AM the INFINITE SELF/NON-SELF, the BOUNDLESS IMMUTABLE PRINCIPLE, the 'INFINITESSENCE'.
22. ALL that ever has *apparently* 'happened', is *apparently* 'happening' or will *apparently* 'happen' is 'MY-STORY' (the 'MY-STERY')—MY 'MONOLOGUE', MY 'DRAMA', MY 'PLAY', MY 'GAME', MY 'DANCE', MY 'MASQUERADE'—the 'STORY' of '**I**'—the BOUNDLESS IMMUTABLE PRINCIPLE, the 'INFINITESSENCE', the INFINITE SELF/NON-SELF, the INFINITE BEING/NON-BEING, the ABSOLUTE.

Names for 'THE NAMELESSNESS' or Speaking of NOTHING

Fundamental to *Infinitization of Selfhood* is the concept of an ABSOLUTE REALITY—the BOUNDLESS IMMUTABLE PRINCIPLE presented in the Fundamentals of *The Secret Doctrine* by H.P. Blavatsky. This GREAT PRINCIPLE is not only *unspeakable* but IT is *unthinkable*; IT cannot be conceived as IT *IS*. Yet, enlightened thought may, as it were, *tend towards* IT (though, as *thought*, never 'reach' IT). For this reason, many, many *names* for the GREAT NAMELESSNESS have been conceived and utilized in this treatise, thus emphasizing *different ways of thinking* about THAT which cannot really be thought about at all, i.e., THAT which IS, in fact, forever *inconceivable*. All these many names may be very useful when trying to avoid falling into the trap of thinking REALITY is *something* IT *is not*. The many names are a kind of *protection* against misapprehending REALITY.

Here follows an alphabetized list of the majority of these names, though others may be encountered in the main body of the text. As readers make their way through the treatise, the revelatory power of these names will become increasingly apparent.

1. ABSOLUTE PLENITUDE, the
2. ABSOLUTE PLENUM, the
3. ABSOLUTE SELF, the
4. ABSOLUTE, the
5. ABSTRACTION, the GREAT
6. 'ACTOR', the
7. AFFIRMATION, the ULTIMATE
8. AIN-SOPH
9. ALL, the
10. ALL-GOODNESS, the
11. ALL-IN-ALLNESS, the
12. ALLNESS, the UTTER
13. ALL-PEACE, the
14. ALL-SELF, the
15. ALL-SUFFICIENCY, the
16. ALWAYS SO, the
17. ATMAN, the
18. ATMAN, the INFINITE
19. AUTHENTIC NATURE of ALL, the
20. BEING, the ABSOLUTELY UNITARY
21. BEING, the ONE
22. BEINGNESS
23. BE-NESS
24. BOUNDLESS IMMUTABLE PRINCIPLE, the
25. BOUNDLESS ONE, the
26. BOUNDLESS, the
27. BOUNDLESSNESS, the
28. BRAHMAN
29. CAUSELESS CAUSE, the
30. CHANGELESS, the
31. CHANGELESSNESS, the
32. CONSTANT, the SUPER-UNIVERSAL
33. CONTAINER, the GREAT
34. CONTINUANCE, the
35. CONTINUITY, the GREAT
36. CONTINUITY, the OMNIPRESENT
37. CONTINUUM, the
38. CONTINUUM, the ABSOLUTE
39. CONTINUUM, the GREAT
40. CONTINUUM, the HOMOGENEOUS
41. CONTINUUM, the HOMOGENEOUS IMPERTURBABLE
42. CONTRADICTION, the GREAT
43. CURRENT, the
44. EGO/NON-EGO, the ABSOLUTE
45. EMPTINESS, the GREAT
46. ENDLESSNESS, the
47. ENTIRETY, the
48. ENTIRETY, the ALL-INCLUSIVE
49. ENTITY, the INCOMPARABLE
50. ENTITY/NON-ENTITY, the ONE AND ONLY
51. ENTITY/NON-ENTITY, the ONE EGOLESS
52. ENTITY/NON-ENTITY, the ONE GREAT
53. ESSENCE, the
54. ESSENCE, the ONE AND ONLY

55. ESSENCE, the ONE IRREDUCIBLE
56. ESSENCE, the ULTIMATE
57. ESSENTIAL, the
58. ESSENTIAL-I-MYSELF
59. ETERNAL, the
60. ETERNAL, the ONE
61. EVER-FULLNESS, the
62. EVER-SO, the
63. EYE, the ONLY
64. FACELESS, the
65. FACELESSNESS, the
66. FACT, the ONE ULTIMATE
67. FOREVERNESS, the
68. FORMLESSNESS, the
69. FOUNT OF ALL POSSIBILITY, the
70. FULLEST OF THE FULL, the
71. FULLNESS, the BOUNDLESS
72. GOOD, the
73. GREAT CONTINUITY, the
74. GREAT SELF, the
75. GROUND OF ALL BEING, the
76. HOLE, the
77. HOMOGENEITY, the
78. HOMOGENEITY, the ENFORCER of
79. HOMOGENEITY, the INFINITE
80. HOMOGENEITY, the INFINITELY ENDURING
81. HOMOGENEITY, the UTTER
82. 'I'
83. IDEAL, the
84. IDENTICALNESS, the
85. IDENTITY, the ONE
86. IDENTITY, the PERMANENT
87. IMMEASURABLE, the
88. IMMOBILE SUBSTRATUM, the
89. IMMUTABILITY, the GREAT
90. IMPERTURBABLE, the
91. INCOMPARABLE, the
92. INDEFINITENESS, the ABSOLUTE
93. INDIVIDUAL, the
94. INDIVIDUAL, the ONLY TRUE
95. INDIVIDUALITY, the
96. INDIVISIBILITY, the GREAT
97. INDIVISIBLE, the
98. INESCAPABLE, the GREAT
99. INEXHAUSTIBILITY, the
100. INFINITE INSEPARABILITY, the
101. INFINITE POSSIBILITY, the
102. INFINITE POTENTIAL, the
103. INFINITE SUBSTRATUM, the
104. INFINITE, the
105. INFINITESSENCE, the
106. INFINITIZATION OF ALL POSSIBILITY, the
107. INFINITIZED EVERYTHINGNESS
108. INFINITUDE, the
109. INFINITUDE, the ALL-EMBRACING
110. INFINITY OF INFINITIES, the
111. INSEPARABILITY, the SUPREME
112. INTERNALIZATION, the GREAT
113. INVARIABLE, the
114. INVARIANT, the
115. IRREDUCIBLE, the
116. IT
117. LIFE
118. LIMITLESS POTENTIAL, the
119. MAXIMIZATION, the
120. MYSTERY, the ULTIMATE
121. NAMELESS, the
122. NAMELESSNESS, the UTTER AND COMPLETE
123. NEGATION, the ABSOLUTE
124. NEGATION, the GREAT
125. NEGATOR OF ALL, the
126. NONE, the
127. NONE-NESS, the
128. NON-EVENT, the SUPREME
129. NO-THING, the
130. NOTHING, the
131. NOTHINGNESS, the
132. NOUMENESSENCE, the
133. NOUMENON, the
134. NOUMENON, the ONE INDIVISIBLE
135. ONE ABOUT WHOM NAUGHT MAY BE SAID, the
136. ONE AND ONLY, the
137. ONE INFINITELY BEHIND, the
138. ONE WHO CANNOT 'ACT', the
139. ONE WHO CANNOT 'DO', the
140. ONE WHO IS NONE, the
141. ONE WHO IS THE ZERO, the
142. ONE WHO LIVES, the
143. ONE WITHOUT A SECOND, the
144. ONE, the MOTIONLESS
145. ONE, the SPACELESS
146. ONE, the TIMELESS
147. ONE, the UNDIFFERENTIABLE
148. OTHERLESS, the
149. PARABRAHMAN
150. PARADOX, the SUPREME
151. PEACE
152. PEACE, the GREAT
153. PERFECT, the

154. PERFECTION, the
155. PERFECTION, the MAXIMAL
156. PERMANENT, the
157. PERVADER, the
158. PLENUM of INFINITE POSSIBILITIES, the
159. PLENUM, the
160. PRESENCE, the
161. PRESENCE, the UNFLUCTUATING, UNDENIABLE
162. PRINCIPLE, the GREAT
163. REAL, the
164. REALITY
165. REALITY, the ULTIMATE
166. REALM OF ABSOLUTE CONTINUITY, the
167. REALM OF ALL POSSIBILITY, the
168. REASSURANCE, the GREAT
169. RESIDUUM, the
170. ROOT, the
171. ROOT, the PARABRAHMIC
172. ROOT, the ULTIMATE
173. ROOTLESS ROOT, the
174. SAMENESS, the GREAT
175. SAT
176. SELF, the
177. SELF, the CELEBRATED ANCIENT
178. SELF, the INFINITE
179. SELF, the ONE AND ONLY
180. SELF, the SUBJECTIVE
181. SELF-ABSORBED STATE, the
182. SELFHOOD, PURE
183. SELFHOOD, the 'STATE' of ETERNAL
184. SILENCE, the
185. SILENCE, the UNDISTURBABLE
186. SIMPLEST ONE, the
187. SINGULARITY, the
188. SOLVENT, the UNIVERSAL
189. SOURCE AND SUSTAINMENT, the
190. SOURCE, the
191. SOURCE, the ABSOLUTE
192. SPIRIT
193. STATE, the 'SUMMUM'
194. STATE, the ULTIMATE SUBJECTIVE
195. STATE, the ULTIMATIZED
196. STATELESS STATE, the
197. STILLNESS, the
198. SUBJECT, the
199. SUBJECT, the ONE AND ONLY
200. SUBJECTIVITY, PURE
201. SUBJECTIVITY, the INFINITE
202. SUBSTANCE
203. SUBSTANCE, the ABSOLUTE
204. SUBSTANCE, the ONE AND ONLY INDIVISIBLE
205. SUBSTRATUM, the
206. SUBTLETY, the GREAT
207. SUPER SYSTEM, the
208. TAT
209. THAT
210. THAT THAN WHICH THERE IS NO OTHER
211. THAT TO WHICH NAUGHT MAY BE ADDED
212. THAT WHICH IS ITSELF ALONE
213. THAT WHICH REMAINS WHEN ALL ELSE DISAPPEARS
214. TONE, the NON-VIBRATING
215. TOTAL-I/ALL-SELF
216. TRUE, the
217. TRUTH
218. ULTIMATE IDENTITY, the
219. ULTIMATE, the
220. UNCHANGEABLE, the
221. UNCONDITIONED CONDITION, the
222. UNCONDITIONED, the
223. UNDEFINABLE, the
224. UNDISTURBABLE, the
225 UNDIVIDED REALM, the
226. UNLIMITED, the
227. UNSEEING ONE, the
228. ULTIMATE KING, the
229. VACUUM, the
230. VOID, the
231. VOIDNESS, the
232. WHO
233. WHOLE, the
234. WHOLENESS, the HOMOGENEOUS
235. WITNESS OF ALL, the
236. WITNESS, the
237. WITNESS, the ULTIMATE
238. WOMB, the INFINITE
239. WORLD OF BEING, the
240. WORLD OF DURATION, the
241. ZERO, the

The Creation of the Universe

What follows is a kind of 'Esoteric Myth of Creation'. This 'Myth', written in a language that is quasi-poetical, is placed at the beginning of *Infinitization* to help the reader enter immediately into the *spirit* of the treatise, but the deepest meaning of the Myth will surely not be *understood* until the entire treatise has been carefully read, and, perhaps, not even then, entirely. Unusual words and some symbols are used directly and without explanation. The reader should simply enter into the spirit of what is said, divining the meaning intuitively; careful study of *Infinitization* will disclose the precise meaning of terms and the deeper intent of the Myth.

The Sequence of Creation according to the Radical Infinitization 'Myth'

I-as-**I/8**-as-**8**-as-**I/8**-as-I

1. 'I'.
2. 'I' only.
3. Forever **I**, NOW.
4. **I**-as-**I**. INFINITIZED BE-NESS. Infinitely 'ENFOLDED' 'within' MYSELF. MYSELF a*lone* to the *infiniteth degree.*
5. **I**-as-**I**. The PLENUM. The 'INFINITESSENCE'. The VOID. The ABSOLUTE. The MYSTERY.
6. No 'SIGHT'. No 'HEARING'. No 'SENSE'. NOTHING. INFINITIZED NOTHINGNESS. Naught but **I**.
7. Suddenly, in no Time (for Time is not), the 'FLASH'.
8. 'Within' ME, 'APPEARING'—in no Time, in no Space—the 'RAY', 'POINT', the 'CHANGE' 'ARISING' out of 'CHANGELESSNESS'.
9. Suddenly, in no Time, the 'Beginning'. Time begun; Space begun—suddenly both Time and Space are *present* with the FIRST and *only* 'MOTION' **I** 'ALLOW' 'within' the **I**, **I** ever AM. Suddenly, the FIRST 'APPEARANCE'.
10. Suddenly, both TIMELESSNESS *and* Time.
11. Suddenly, both SPACELESSNESS *and* Space.
12. Suddenly, a 'MOTION'—yet MOTIONLESS as ever **I** 'REMAIN'.
13. Suddenly, though **I** BE **I**, yet, NOT-**I** as the **I/8**, 'APPEARS' 'within'.
14. Suddenly, though **I** BE ONE—Another.
15. **I** have not 'CHANGED'. **I** cannot 'CHANGE'. **I** 'ABIDE' as ever.
16. And, yet, 'within' ME, **I** have 'BECOME' Another, which *cannot be* a REAL Other.
17. For Now, in MY NOW, (though **I** be **I** as ever) yet **I** be **I/8** as well.
18. Though **I** 'ABIDE' the ZERO WHO IS ONE, yet Now, in MY NOW, **I** BE **I/8** WHO IS a sudden-THREE and not the ONE.

19. **I** 'ABIDE', for **I** cannot *not-BE* **I**-as-**I** forever.
20. Yet, suddenly, **I** 'SEE'.
21. Yet if **I** AM **I** as **I** have ever 'BEEN', **I** cannot 'SEE'.
22. It cannot be **I** WHO 'SEES', for **I** can only 'BE'. And, yet, the 'SEEING' IS.
23. Therefore, **I** 'SEE' *not.* It is the 'Other' Who Is **I**, Who 'Sees', and, yet, Who Is that 'Other', but **I**, and **I** alone?
24. Yes, suddenly **I** 'SAW', and yet did *not.* The 'THREE' 'ARISING' in ME 'SAW'—the **I** WHO 'SAW', the **I** thus 'SEEN', and the **I** WHO WAS the 'SEEING'.
25. But **I** 'CAST' 'THEM' *out,* as soon as 'THEY' 'AROSE'—as soon as 'THEY' 'AROSE' as 'RAY', as 'POINT', **I** 'CAST' 'THEM' *out.*
26. For, in the *REAL,* **I** AM **I**, and **I** *alone.* 'THEY' (the 'THREE') have in ME no rightful 'PLACE'. **I** WILL have no Other *in* ME as **I** AM (even though the Other be but **I**). How *strange* that 'THEY' should 'ARISE' in ME at all.
27. *Out* into Illusion. *Out* into Appearance. *Out* into an un-REAL Realm, where 'Seer', 'Seen' and 'Seeing' can for a Time Abide as Shadows of MYSELF.
28. Let the 'THREE' WHO 'THOUGHT' 'THEY' could 'REMAIN' with 'ME' be *Three,* not 'THREE'.
29. Let 'SEER', 'SEEN' and 'SEEING' be the 'Seer', 'Seen' and 'Seeing' ... There—*out*—There. They-as-**I/8** are 'in' ME nonetheless, but There They offer no disturbance; There, the 'Theyness' Which Is **I/8**, will leave MY ALL-PEACE undisturbed, and **I** CAN 'BE' as *ever* while They 'Play' as **I/8** in Time and Space.
30. Though **I** 'REMAIN' as ever **I**, **I** SHALL *be* Their 'Playing'; **I** SHALL *be* Their 'Game'. **I** SHALL *be* Whate're They Do, Whoe're They Think They Are.
31. Though **I** 'REMAIN' as ever, **I**-as-**I/8**-then-**8** SHALL 'PLAY' along, 'PLAY' the 'Game' of 'Seeing', even as **I** 'DWELL' in BEING.
32. Thus, 'SEEING' *not* (as **I**) 'within'; what do **I/8** 'See' 'without'?
33. What else is there to 'See' but Who **I/8** AM?
34. INFINITUDE is what **I** AM, and thus, Infinitude **I/8** 'See'.
35. **I/8** the 'Seer' Am; and naught there is to 'See' but **I/8** MySelf.
36. Behold the Boundlessness **I/8** Am; the Nothingness, the Plenum which **I/8** Am.
37. **I/8** the 'Seer' Am the Subject and the Father.
38. And What **I/8** 'See' is **I/8** MySelf as Mother.
39. Infinite Am **I/8**; thus, Infinite is She as well.
40. **I/8** Who once was **I** (and never *else* can be) Am now the 'Seer' and the 'Seen'. The 'Seeing', too, **I/8** Am. Infinite all Three. **I/8** Am the 'Maya' Who (once 'MAYA') is the 'Seeing' that reveals MySelf unto MySelf—Infinitude of Object, 'Seen' by the 'Seer' (Infinite as well).
41. Behold, then, My Infinitude—Seamless, boundless, all extending.
42. From every Point of View **I/8** 'See', 'Seeing' but the Sameness that **I/8** Am.
43. Infinitude, **I/8** Prakritic Mother of an endless Host of Self-'Seen' Sons. **I/8** Infinite Father/Mother of a countless chain of Worlds, each One My Son, a Son Unique, yet surely only **I/8**.
44. **I/8** 'See' them not (as They *have been)* before My opened 'Eye'; **I/8** only 'See' My Mother-Self, yet *know* what vastness has Arisen through My Infinite Outgoings.
45. **I/8**-the-Father; **I/8**-the-Mother; **I/8**-the-Self-'Seen'-Son.

46. And Now the Time in NOW has come to Be a Father/Mother/Son again—not All but only One.
47. And, so, **I/8** 'Look' within, to find the One that **I/8** must Be.
48. As **I** 'REMAIN' 'ABOVE', so my Infinitude as well 'Abides', though Now the Time is well upon Me when **I/8** must descend from Allness to specific Oneness, narrowing My 'Sight' of Self.
49. Infinite **I/8** Was; Infinite **I/8** shall Remain, even as **I/8** leave that All-Embracing Self within the Super-Cosmic Boundless Sphere.
50. Behold, within, **I/8** 'See' a Point, a One **I/8** Am to Be in *Time*.
51. Narrowing, Condensing, **I/8** 'journey' to that Point Becoming ***8*** instead of **I/8**.
52. Instantly, or over Time, **I/8-as-8** Am now the Son of ALLNESS which **I/8** must become for Now in NOW.
53. Instead of ALL that *could* Be, **I/8** have Chosen to Become the One, the **8**. In doing this **I/8** but prepare to Play the Endless Cyclic 'Game' as **8**.
54. **I** 'ABOVE'; **I/8** Between; Now **8** Below—ready to Begin—Begin the Game that **I**-as-**I/8**-as-**8** have Played for *aye*. Who Am **I** Now?
55. **8** Am **8**, the Cosmic Father; **8** Am **8**, the Cosmic Mother. Father/Mother, Spirit/Matter, is the very **8** Who Now **I** AM.
56. The ALLNESS which **I** AM 'ABOVE', **8** cannot remember.
57. The ALLNESS which **8** 'Saw' as **I/8** within the Super-Cosmic Spheres is gone.
58. From everything **8** might have been, a 'CHOICE' there was in Me that had been 'MADE' 'ABOVE', a 'CHOICE' to Be but One, to 'Show' as One, to Be but One of all the many other Ones Who **I**-as-**I/8**-as-**8** might well have Been.
59. That One (in Seed) at last **8** Am—with My Pattern from the 'PATTERN' from the FOUNT OF POSSIBILITY within.
60. My Task is to Unfold the Pattern 'PLANTED' from the Time **8** Was the 'POINT' as well as **I**.
61. **8** the Father, **8** the Mother (We a Son of What we were 'ABOVE' and, then, Between) are now ready to Begin.
62. **8** 'See' MySelf as Mother; **8** as Mother 'See' MySelf as Father. Reflections of Each Other, We 'See' Ourselves (in warm embrace), We 'See' the Singularity which We have now Become.
63. **8** as Father hold a Son to be within, un-'Seen'. **8** shall 'Look' within to 'See' MySelf as Son, and in doing so, the Mother Who **8** Am, as well, shall offer Me a Son—a Son to 'See'.
64. Within MySelf **8** 'See' a Point, a 'distant' Point at first.
65. But drawing near **8** 'See' the Point as Sun-like Son, My Only Son, the Universal Radiant Son to 'Keep' the Cosmos which **8** must Become.
66. **8** the First, the Number One, now slip within the Son the Mother bore Me; **8** slip within the Son **8** 'See' and Am.
67. **8** with My Mother-Self was Two, and Now with Son Am Three. He the Third Who Now the Second Is, and She (though *with* Me as the Second) now is Third.
68. The Son **8** 'Saw' as Object-Point, that Son **8** *Am* as well as Father Self.
69. **8**-as-He Am Now the Number Two.

70. To Him **8** give the Pattern which **8** know **8** bear within MySelf, presented Me by ME 'MY-SELF' in 'Days' when **8**-as-**I** was only ONE the NONE.
71. Thus, **8** Remain as Father; **8** (Reflected **8**) Am Mother too; and now as Son of both **8** 'Shine' that Mother/Father both may 'See'.
72. **8**-the-Father/Mother/Son, am Father/Mother as They were. **8** Hold the Pattern of the Cosmos which **8** must Become.
73. **8**-the-Son, need many Sons. Sons of Being, Sons of 'Sight'. Sons that **8**-the-Son must Now Become.
74. Thus **8** 'Look' into MySelf, and 'See' MySelf as Mother to the Sons **8** must Become.
75. Within MySelf **8** 'See' a Point, and in that Point another Point and in that Point another still. Three Points in all, three Sons **8** Now must Be as well as Be MySelf.
76. In 'Seeing' Them, They come to Life as did **8**-MySelf (Who Am the offspring of the Two Who are but One).
77. **8** send them Forth to take Their Place around Me.
78. Three Sons **8** have, Three Sons **8,** Son (as Mother/Father), have produced.
79. Three will be My closest Children, and from Them the Rest will Come.
80. **8**-as-*They* will *'See'* the Seven into Being, each of Them as Grandchild to My Mother Self and **8**.
81. Ten in all, and thus the World of Being has the Ten to Hold the Pattern.
82. **8**-the-Father, **8**-the-Mother, **8**-the-Son Whose Sons and Grand-Sons number Ten.
83. The Pattern We must 'See' within Ourselves has now Its Guardians, yet one more Step remains.
84. Thirteen is My Number, as at present **8** Am counted, and a Fourteen must **8** now Become.
85. Every time that **I**-as-**I/8**-as-**8** went forth in 'SIGHT' or 'S̲ight' unto the Other Self **I** 'SAW', Who *Was* **I** in Going Forth?
86. Whoe're the **I, I** *Was,* **I** must Become.
87. **I**-as-**I/8**-as-**8** have ever been the Actor—invisibly unto MySelf. **I**-as-**I/8**-as-**8** have ever been the One Who Moves from Here to There, even while **I** Did Not Move.
88. Now the One Who *Moved* **8** must Become.
89. And thus, within MySelf, as Son-to-Be, **8** find as Son a One Who with My Father/Mother caused My Coming Into Light.
90. Thus, within MySelf-as-Son, **8** find the One Who taking form will *mold* all Form unto the Pattern **8**-as-Universal Son Now Hold.
91. **8**-as-Son find Fohat as My Son.
92. As He, **8** will Go Forth in multiplicity unto the Mother Whom, as Self, **8** 'See'.
93. As He, **8** shall 'See' *multitude*, a multitude of tiny Self-Reflected Sons, which **8** as Fohat will command to pattern as **8** Will.
94. **8**-as-Fohat will Create (manipulating Time and Space) a multitude of patterns to Become (in Time and Space) the very Pattern which **8** Hold as **8**-the-Universal Son.

95. **8** will *Fabricate* the Lower Worlds upon the Pattern which is Held by Ten and **8**—the Father's, Mother's Son.
96. **8**-as-Fohat will build *true*—true to Pattern that **8** Hold as God-the-Son.
97. With building done upon whatever level claims my 'Sight', **8**-as-Son Whose Sons and Grandsons number Ten, will send forth lesser Sons of Theirs to meet the handiwork which **8**-as-Fohat have prepared.
98. Lesser Sons of lesser Sons (Reflections of Reflections) shall *go forth* to Work within MySelf as Fohat.
99. By the 'Ray' They *are* They shall go forth. 'Rays' of 'Rays' of 'Rays' Who all are but the Universal 'Ray'—My Own.
100. Forth into Prakritic Depths **8** have Become, they go, descending that They may ascend when building is complete and Beauty reigns.
101. 'Rays' as 'Points' and 'Points' as 'Rays', yet all but One—One 'Point', One 'Ray', yes all but One, the **I**-as-**I/8**-as-**8**.
102. Each 'Ray' projects the 'Rays' It must Become; within each Point is found the lesser Points which wait as Sons-to-Be.
103. Thus the Pattern in the Heavens finds Its way into Fohatic Depths of **8** MySelf, while **8** 'below' in blindness to the **8**, **8** Am 'Above', labor to conform to all that **8**, MySelf, would have 'me' be.
104. **8**-as-lesser-'Rays' have lost my Way back to the Universal Self **8** Am. **8**-As-Greater-'Rays' uphold the Light which lights the Long Way Home.
105. **8**-in-darkness; **8**-in-Light. **8**-the-'Many', **8**-the-One, **8** Father/Mother/Son and Holy Ghost as well.
106. **8** Remain upon the Heights; **8** Descend into the Depths (the Depths of Self-'Seen' Multiplicity).
107. In My many, many 'meetings' **8** Encounter but MySelf alone. **8**-the-'Seer', **8**-the-'Seen', **8**-the-'Seeing' which unites Them.
108. In the manyness **8** have become, **8** Am All, and never one.
109. Yet far 'below' where every 'Ray' or 'Point', in error, thinks Itself but one, **8** Am that One. Not the lesser one but All—All at every level of my Self-Become Diversity.
110. This the **8**-in-ignorance will know upon ascending to the Higher Spheres whereon e'en now **8** Reign Supreme.
111. And when at last re-gathered—'Ray' to 'Ray' to Final 'Ray', and 'Point' to 'Point' to Final 'Point', **8** shall Be the One **8** always Was—the One Who never for the Cosmic Manvantara ceased to Be the One and Only Cosmic One.
112. Then, Pattern all Enacted, the Play Performed, the Drama Done, victorious in the Self-Made Game, **8** *alone* as One *in-drawn* Will find the Way from One unto the Allness held in 'Sight' within the Super-Cosmic Spheres.
113. **8** Will Return, a Bearer of one *possibility fulfilled*, as ever **8**-as-**I** have done.
114. From Finitude the Tunnel **8** Will Re-ascend, from Single Point to each and every Point that e're may be.
115. Who will greet Me in My re-ascent?

116. Only **I/8** and **I/8** alone; the One Who never left a sure Infinitude of 'Sight'—the One Who, Infinite in 'Seeing', is as Infinitely 'Seen', the One Who Rules the Super-Cosmic Sphere.
117. There upon the 'Heights' beyond the Finitude of Cosmic Spheres, **8**'ll know that ever as **8** *'Saw'* 'below', as surely was **8** *'Seen'* as well.
118. The **8** Now **I/8** could well Remain, enrapt in contemplation of the Boundless ALLNESS **I/8** have ever been.
119. Perhaps some 'Sight' of Consummation there is granted to the One (the **8**-as-**I/8** Who new returns). Perhaps the 'Sight' of ALL that e're has been will greet the 'Eye' of **8**-now-**I/8** Who far 'below' as **8** has newly been but *one.*
120. Whate're the 'Sight' (the 'Sight' of Boundless Self as One or 'Sight' of Endless Many)—the Time for 'Sight' is nearly at an End.
121. For 'Seeing' is of Maya. 'Seeing' keeps Me from MYSELF, and to that SELF **8** never REALLY left, **8** would return.
122. With Time now at an End; with Space Now at an End; with Motion no more needed, **I/8** can Re-Become the SELF **I/8** always 'AM'.
123. In a 'FLASH' it All is over. 'Rayness', 'Pointness' 'Maya'—'RAYNESS', 'POINTNESS' 'MAYA', from 'Something' into NOTHING, all as if the Cosmic All had never Been.
124. **I** 'SEE' naught. **I** 'SEE' naught. **I** can only *BE*—BE as **I** WAS forever, BE as **I** *AM* forever.
125. **I**-as-**I** *alone.*
126. Forever **I**, NOW.
127. '**I**', only.
128. '**I**'.

... an outbreathing of the 'unknown essence' produces the world;
and an inhalation causes it to disappear.
This process has been going on from all eternity,
and our present universe is but one of an infinite series,
which had no beginning and will have no end.

The Secret Doctrine, Volume I, page 4

Section I
Commentaries

The following are Commentaries on certain major concepts in the Radical Infinitist World View.

On the ALL-IN-ALLNESS

By the ALL-IN-ALLNESS (in all capital letters) is meant the ONE AND ONLY SELF (or BOUNDLESS IMMUTABLE PRINCIPLE) with particular focus upon ITS perfect or pralayic 'STATELESS-STATE', the 'STATE' of Universal Pralaya.

NOTE: The inadequate term 'STATE' is used rather than the still more inadequate term 'condition', as 'condition' refers to patterns/relationships that, in relation to the descending scale of REAL/Real/real, cannot approach the REAL, and are remote even from the Real. The word 'STATE' is not really suitable—no word is, but it is better than 'condition'.

On the 'ALL-SELF'

By the ALL-SELF (each word with capital letters) is meant the ONE AND ONLY SELF with particular reference to the *infinitude* of SELFHOOD. The ALL-SELF refers to the ABSOLUTE SELF and not to the *All* which the ALL-SELF *becomes* (while yet remaining IT-SELF) during Universal Manvantara.

On the 'ALL'

The ALL (with capital letters) refers to the totality of E/entities, states, conditions in this Universe and in *all Universes* that ever have been or ever will be. The ALL is a collective noun which denotes an infinite number of 'enumerables' (i.e., enumerable things), but is used to emphasize 'in-Universe' E/entities, states and conditions—the total collection of 'denotables' that have been, are now, or ever will be 'GENERATED' by the BOUNDLESS IMMUTABLE PRINCIPLE-as-SELF, rather than emphasizing the BOUNDLESS IMMUTABLE PRINCIPLE in ITS purity, perfection, homogeneity and abstracted singleness.

On the 'All'

By the All (with only the first letter capitalized) is meant the totality of E/entities existing within the *present* Universe as well as the totality of states and conditions resulting from their interplay on all dimensions of their being. Each and every E/entity and condition in our Universal System is included, and absolutely none is excluded. The All is, therefore, inclusive of the total number of patterns or relationships that can be generated within the *present* Universe. The number of *actualizable* relationship patterns between E/entities in a given Cosmos is huge but finite; the number of possible conditions is infinite *in potential* (but never, in the limited duration of a single Universe capable of manifestation). The infinitude of relationship/patterns/conditions that can, *potentially*, be generated within a single Universe is, as it were, a 'lesser infinitude' than the infinitude of possibilities which has ceaselessly been generated (is now being generated and will be generated) through the periodic manifestation of the BOUNDLESS IMMUTABLE PRINCIPLE-as-SELF.

On the Meaning of the Term 'Infinite'

The term 'infinite', used as an adjective, describes that which is endless, without limit. This term is frequently used to describe that which is endlessly enumerable and/or endlessly extensive.

- An infinite series of numbers is a series for which no final term or number can exist, hence this series requires endless enumeration.
- An infinite field is a field for which no terminating boundary can exist; hence, such a field is endlessly extensive.

That which is 'infinite' must remain forever incapable of conclusion, hence 'in-definite'. The ideas of endlessness and boundlessness correlate closely with that which can be described as infinite. Infinite Space is boundless. Infinite Time, or Duration, has no beginning and no end.

On the INFINITE

The concept of the INFINITE does not relate in any way to that which can be counted (i.e., the enumerable), but precisely to that which can *never* be counted. It is impossible to grasp or define the INFINITE through the use of any word(s) or symbol(s). The INFINITE is forever 'unreifiable' and, thus, forever elusive, defying definition. Associated with thoughts of the INFINITE are thoughts of THAT which is boundless, endless and beginningless. The INFINITE has been described as BE-NESS (THAT from which all things come to *be).* The INFINITE is also equivalent to the ABSOLUTE (which is the *infinitized* NOUMENON, the 'INFINITESSENCE', of all possible things). Though nothing *definitive* can be said of that 'STATELESS-STATE' called the INFINITE, some thoughts may prove useful.

The INFINITE 'ABIDES'/'DWELLS'/'EXISTS' in a STATE of *immutability.* Within IT change of any kind is impossible. The INFINITE has no 'parts', nor has IT ever had parts, nor will IT ever have parts. Hence the INFINITE is 'impartite' and 'imparticulate' (there is a difference). Incapable of having parts IT is, therefore, *indivisible.*

- The INFINITE has, strictly speaking *no relation* to any finite thing of any kind, because relation requires an 'other'.

An 'other' means at least *two,* which two can be seen from one perspective as that which relates, and that to which it relates, and vice versa. Since there is none *other* than the INFINITE, the ABSOLUTE, there can be no relating. *Nothing* can relate to the INFINITE because *everything* already (more than being 'contained' *within* the INFINITE), *is* the INFINITE. All things are thus inseparable from IT, and no thing is essentially distinct from IT (though consciousness may see to register an apparent and illusory distinctness). In order for relationship to occur, distinctness between the relating agents is required. This brings the fundamental thought that:

- No E/entity can *relate* to its own *essential self;* that E/entity can only *be* its *essential self,* which *self* is none other than the ONE and ONLY SELF.

The INFINITE (or, if one prefers, the 'INFINITENESS') is *ALL-containing* (see definition of the ALL) and yet *infinitely* transcends any distinct ALL or <u>A</u>ll that can be contained. The INFINITE cannot be in any way reduced, expanded, or, in fact, altered in the slightest. IT cannot be conceived in spatial terms, though the image of *boundless* Space may serve as a vague, suggestive but feeble attempt to grasp ITS 'NATURE'. The INFINITE IS 'no thing' at all. IT stands independent of the concept of 'thing-ship' or 'thing-hood'. The INFINITE is not only a complete and total abstraction, IT is the *only* true abstraction, (i.e. the ABSTRACTION). From IT no further abstraction is possible because no further position of 'remove' can be found 'outside' IT.

The INFINITE (INFINITENESS, BE-NESS, ABSOLUTE, ABSOLUTENESS) is the FIRST PRINCIPLE, the uncaused, unproduced PRINCIPLE. The INFINITE, even more accurately, is not so much the *FIRST* PRINCIPLE as the *ONLY* PRINCIPLE—not so much the *FIRST OF ALL THINGS* as the *ONLY* OF THINGS—though, of course, it is incorrect to objectify IT by denoting IT as a *'thing'* or, in fact, by any other term. The word 'thing', however, is useful to the presently constituted and severely limited human consciousness in indicating any 'denotable'. Strictly speaking, however, to attempt to *name* the INFINITE is to attempt to *define* IT, which is, somehow, to limit IT or 'contain' IT, and this, by definition is impossible. There are, however, a number of terms which are helpful in indicating, at least, what IT is *not.*

- Perhaps, the INFINITE might evocatively be called 'THAT THAN WHICH THERE IS NO OTHER' or, as IT has been called in so many occult and philosophical works, the "ONE WITHOUT A SECOND".

The INFINITE, or INFINITENESS, 'CONTAINS' or, rather, *IS*, in potential, the most *infinite infinity-of-possibilities.* (This strange wording is used because modern mathematical theory has proposed differing varieties and magnitudes of infinities.) But thought cannot succeed in fathoming the 'NATURELESS-NATURE' of the INFINITE if thought approaches the subject *quantitatively*, for infinity added to infinity is still infinity; and infinity multiplied by infinity is also infinity; and even infinity to the 'infiniteth' power is also, merely, infinity. This being so, the INFINITE might be thought of (though IT is ESSENTIALLY inconceivable and 'unthinkable') as the ETERNAL and PRINCIPAL 'STATE' of ABSOLUTE INDEFINITENESS out of which all defined things come.

- ◆ From another perspective, the INFINITE or INFINITENESS (BE-NESS, ABSOLUTENESS) might be called 'THAT WHICH REMAINS WHEN ALL ELSE DISAPPEARS'.

Again, our terminology is hopelessly inadequate, for it is improper to call IT (the INFINITE) 'THAT', because to denote a thing as THAT requires the presence of a separate, *conscious* 'SUBJECT' in order to 'VIEW' IT (the INFINITE) as an 'OBJECT'. Indeed, in general, the term 'that' pertains only to objects, and the INFINITE (which is the ONE AND ONLY SUBJECTIVITY) can never be an object. For similar reasons, it is impossible to call IT by the term 'this'. The INFINITE is finally and truly, the UTTER AND COMPLETE NAMELESSNESS. IT is the only 'STATE' of SILENCE that 'EXISTS'—though unless the term *exist* is understood as devoid of form of any kind, the term cannot properly be used—thus, in a way, the INFINITE does *not* 'EXIST' (which is the reason for the single quotations around the word EXIST).

The INFINITE is the ONE AND ONLY STATE OF TOTAL PRIVATION (i.e., IT is *deprived* of any *particular* attribute whatsoever). The INFINITE is the GREAT EMPTINESS or VOID, but from another and illuminating perspective, the only thing of which the INFINITE is truly *empty* is *emptiness* itself, for IT is the FULLEST OF THE FULL. Or, perhaps, the INFINITE can be considered *empty of limitation.* No thing can ever *come to be* which is not, ESSENTIALLY, IT (the INFINITE) ITSELF.

IT IS at once the whitest whiteness and the blackest blackness, and yet IT IS beyond all possibility of any polarity. Though this is true, the INFINITE 'CONTAINS' or rather IS, *in potential,* all polarities. All language must cease as futile with respect to the INFINITE, for language is a means of clarifying and defining relations, and IT has no relations within IT nor is IT *related* to any other thing. The INFINITE IS, at once, NOTHINGNESS and ALLNESS, and EVERYTHINGNESS (which means that IT is the very SUBSTANCE and ESSENCE of everything that has been, is now or will be, or was, is, or will be in any way possible).

The INFINITE is inexhaustible, the INEXHAUSTIBILITY. Even the infinity of Cosmoses past (during which, it would seem, that all that *could ever possibly be* has, in fact *been),* and the infinity of future Cosmoses, have not exhausted (and cannot possibly exhaust) ITS SUPREME INFINITUDE. It is true that the *possibility* (though not the *actuality)* of an infinity of relations and patterns exists even within a finite Cosmos, but, *mysteriously*, the INFINITE is *still more infinite* than these lower magnitude infinities. The INFINITE is indeed, the INFINITY OF INFINITIES.

On How Much Can be Accomplished in 'Infinite Time'?

A problem arises when one considers what might be called *contrasting infinities.* An infinity of possibilities which will *never* be realized within the present Universal Manvantara exists within the All (the present Universe). An infinity of elapsed and un-elapsed possibilities (still *greater,* it would appear) exist within the ALL (the totality of all Universes past, the present Universe, and all Universes to come). From a certain mathematical perspective, a still *greater* magnitude of infinity exists 'within' the ALL-SELF, the uncontainable SOURCE of both the A̲ll and the ALL.

Let us now, forgetting the limitations of the A̲ll, focus upon the ALL, and, in doing so, confine ourselves to a retrospective view. We will consider only the infinity of Universes past and bear an important thought in mind:

- ♦ Throughout Infinite Duration (call it Infinite Time, if we will), the infinitude of which *has existed* for all Time up to this very present moment in Time, *there has been infinite time to work out an infinitude of possibilities.* Now, the BOUNDLESS IMMUTABLE PRINCIPLE or INFINITE-SELF IS the very NOUMENON of all *possibility,* the FOUNT OF ALL POSSIBILITY—the FOUNT of infinitudinous possibilities.

The question arises, Given an infinite number of Universes past, have *all* of the infinitude of possibilities 'resident' 'within' the INFINITE-SELF, the FOUNT, *already* been manifested within the Infinite Time during which that infinitude of past Universes have manifested? Or, are there *more possibilities* within the INFINITE-SELF than have been rendered into actuality through the instrumentality of an infinitude of past Universes?

In looking for a way to approach a solution to this problem, we might suppose that, perhaps, there is an imbalance in favor of the INFINITE-SELF. Perhaps, because of the time-limitation for the manifestation of any particular Universe (and according to the Law of Periodicity, there *is* such a time-limitation) even certain categories of infinite possibility *within* such a finite Universe (such as, for instance, the *counting* of an infinitude of sequential integers) cannot be manifested *in time* (i.e., in a *timely* manner).

- Hence the INFINITE-SELF, replete with every imaginable possibility would always be *ahead* (as it were) in ITS INFINITUDE, of the ALL in its apparently more limited, non-*actualizable* infinitudes.
- Hence, it would seem, that 'within' the INFINITE-SELF there are more possibilities than can ever be actualized or worked out in *all* Universes (i.e., in an infinity of *limited* Universes—whether those past or those yet to come).

In realizing this, we begin to understand the dynamism, the *mainspring,* so to speak, driving this Universe and all Universes. More possibilities will always be flowing *in* from the INFINITE-SELF than can ever be *actualized* in a given Universe—*in Time.*

Looking, from another perspective, more closely at the structure of any particular limited Universe, it seems *confirmed* that *not* all possibilities within the ALL-SELF will *ever* be made *actual* through It (the limited Universe) as *only one set of possibilities out of all possible sets of possibilities is 'CHOSEN' for the manifestation of a particular Cosmos or Universe.*

- Infinite are the sets of possibilities which could be 'CHOSEN' by the INFINITE SELF-'BECOME'-Universal Logos for any Cosmic Manifestation, and yet only *one* multi-factorial Design is so chosen per Universe or Cosmos.

If at the dawn of every Universal Manifestation (Universal Manvantara) but *one* Design is SELF-'CHOSEN' out of the infinity of possible Designs, then *never* can the infinity of possibilities 'INHERENT' 'WITHIN' the FOUNT OF ALL POSSIBILITY be actualized, even though there be an infinite number of successive Cosmoses. So the 'ACTLESS ACT' of 'CREATING' a Cosmos involves a Universe-determining 'CHOICE', a maximal limiting of the infinite possibilities resident within the INFINITE-SELF. The INFINITE-SELF having 'CHOSEN' one set of possibilities, must then 'BECOME' That which can manifest that set of possibilities (i.e., the INFINITE-SELF must become the Universal Logos, while still remaining ITSELF). From a psychological/mathematical perspective, **I/8/WE** must work out this Cosmos-Conditioning-*Set* (of *possibilities).*

On Limitation and Finiteness (Finitude)

Things are often best understood through contrast. Therefore, as we have examined the terms 'infinite' and 'the INFINITE' (which could be called the UNLIMITED) let us consider the concept of 'limitation' and its relation to 'finiteness'.

From the most common perspective, limitation denotes a condition of privation. The condition of limitation denotes a degree of achievement or fulfillment which is *less* in magnitude than the magnitude of the most complete degree of achievement or fulfillment possible (given the context under consideration). Limitation denotes a condition of partialness rather than wholeness, of less rather than more, of incompleteness with respect to the completeness possible within the context in which the limitation is occurring. The Spirit Aspect within every E/entity in-Cosmos struggles against limitation, because the principle of limitation is diametrically opposed to the Essential nature of Spirit which is SPIRIT, ITSELF.

Let us now enlarge the concept of limitation by considering it as a Universal Process. To denote the cosmic, universal nature of the Process of Limitation we shall call it Limitation with a capital 'L'. As we entertain this thought, it will be necessary to hold in mind thoughts concerning the utter unboundedness of the ONE GREAT ABSTRACTION, the ONE REALITY, the BOUNDLESS IMMUTABLE PRINCIPLE. That ONE GREAT PRINCIPLE is *homogeneous* and *limitless*. 'Within' IT is *no-'thing' in particular* (even though IT IS the ULTIMATE SOURCE of the generation of all things).

- Ours is a Universe of *things* (considering this term in its most abstract sense). No *thing* can come into existence without the veiling, dividing, and separating action of Limitation, which, when it is considered as a Universal Principle/ Process could be called 'Maya' (i.e., Limitation is equivalent to Maya).

From this perspective, Limitation is *reification* (the generation/precipitation/condensation of things, objects, 'perceivables', 'apperceivables', 'denotables'—terms which indicate discrete *things* which can be registered by consciousness). The BOUNDLESS IMMUTABLE PRINCIPLE cannot properly be called any of the above for IT (the INFINITE-SELF) is not a 'perceivable' or an 'apperceivable' and cannot be registered by consciousness (as consciousness is usually understood). Limitation, thus, is the conferrer of 'thing-hood'.

- Limitation, from the universal perspective, is *entification*—the generating of entities as well as Entities from the ONE EGOLESS ENTITY/NON-ENTITY.
- Limitation is the cause of the arising of *definiteness* ('de-finite-ness') from the undefinable HOMOGENEOUS INDEFINITENESS.
- Limitation is the creator of *conditions,* and brings about the *conditioned* state, whereas the ONE REALITY is *conditionless.*
- Limitation is the generator of all that is evanescent, impermanent, passing.
- Everything bounded, partite and partial arises from the Action of this Universal Principle, the Principle of Limitation.
- Limitation is the 'Creator' of speciality, of particularity, of the 'item'—the (apparently) separated unit.
- The onset of Limitation prevents or inhibits the full registration of the All (not to mention the greater possible registrations of the ALL or of the ALL-IN-ALLNESS).
- Limitation, as it functions through registering consciousnesses in-Cosmos appears to 'rob' the INFINITE-SELF of homogeneity by generating the factor of *modification* which leads to diversity.
- The First 'LIMITATION' (necessarily, paradoxically, 'ARISING' within' the INFINITE SELF, as opposed to having an 'Extra-SOURCE' Arising—i.e., 'outside' the ABSOLUTE) is the FIRST 'ACT', as well as the FIRST 'MODIFICATION' (even though this 'FIRST' has repeated itself an infinitude of times with the appearance of every one of an infinitude of finite Universes).
- 'LIMITATION' 'launches' the *Universe-'GENERATING' MODIFICATION PROCESS.*

Bear in mind that 'within' the necessarily *limitless* INFINITE (the ABSOLUTENESS), there REALLY can be no modification, or possibility of modification. 'Within' *IT*, per se, there is no movement or variation of any kind. 'LIMITATION', as a Universe-'INAUGURATING' 'PROCESS', is, thus, *movement.* The First 'LIMITATION' is the FIRST 'ACT', is the FIRST 'MODIFICATION', is the *First 'MOVEMENT'* (even though, 'within' *IT*, *movement* {at least as we know it} is impossible). This First 'MOVEMENT' is the source of alteration, division, and separation. 'LIMITATION' (which instantly 'BECOMES' Limitation) is, in fact, perturbation or disturbance 'in' THAT which is, paradoxically, the IMPERTURBABLE, the UNDISTURBABLE. Limitation, in fact, is any-*thing* at all.

This suggests that Limitation (in the most ultimate sense) is 'LIMITATION'—the Super-Cosmic 'PROCESS'-'BECOME'-Process by which *possibility* 'resident within' the INFINITE POTENTIAL is 'PRECIPITATED' (via or *as* the 'RAY' of the ABSOLUTE). The 'PRECIPITATION of POSSIBILITY' from the INFINITE SOURCE is, therefore, *the* prototypical 'ACT' of 'LIMITATION'.

- Particular possibilities (no matter how huge, complicated or universal) must be considered simply as *limitations* when contrasted with the *perfect* STATE of the ONE PERFECTION.

The FIRST 'LIMITATION' must also be considered as the FIRST *'VEIL'*, and all succeeding Pre-Cosmic and Intra-Cosmic limitations are simply *veils.* Our Universe is (and all Universes are) 'Created' by a *Mayavic Veiling Process,* a SELF-'LIMITING''PROCESS'. Further, the FIRST 'LIMITATION' is the FIRST 'EVENT', even though THAT there can REALLY be no 'event', and nothing *(no-thing)* 'HAPPENS'. 'Within' THAT there is only the maximizing, 'ABSOLUTIZING', or 'INFINITIZING' of all possibility, but the occurrence of no specific or particular 'event'. Upon the Involutionary Arc of the Universal Process, however, every event is a limitation (when contrasted with the absolute freedom from form 'within' the ABSOLUTE).

- Upon the Evolutionary Arc, there exists the possibility of events which *lessen* Limitation. With reabsorption into THAT (the INFINITE SELF), E/event again disappears as does Limitation.

Since Limitation is event, action, movement, modification, etc., Limitation must be considered the generator/creator of the possibility of *relation* and *relationship.* Along the same line, Limitation is the creator of *number* and each number *is,* in fact, a limitation when contrasted with the GREAT ZERO. Thus Limitation, in one important sense, is *enumeration.* The non-number, *zero,* indicates, metaphysically, that which is infinite, boundless, limitless—i.e., no-'thing'. The ZERO considered as the ONE REALITY is also the INFINITE, the BOUNDLESS, the LIMITLESS. With the onset of Limitation (i.e., Maya) begins enumeration; it becomes possible to count. The entire dynamic of LIFE, whether 'in' and 'out' of Cosmos, can be seen as the dynamic alternation between non-enumeration (ZERO) and enumeration (Number).

On the Point

As usually considered, a *point* is a mathematical term designating both location and a center of origin. The point might usefully be considered as the smallest possible indicator of location—so small, in fact, that the point is defined as *dimensionless.* The dimensionlessness of the point leads to the question of whether something dimensionless can have any tangible actuality in the world of three dimensions, i.e., the Cosmos. The answer is, *no.* No true point has physical *actuality.* A point is really an idea or a mental construct, which denotes the ultimate in precision, exactitude and specificity.

The point is also one of the principal symbols in metaphysics and is found useful for conveying ideas which can be conveyed in no other way. Although within the context of the three-dimensionality of the physical plane, the point is an abstraction or non-actuality, within the context of metaphysics, the point can justifiably be considered a Reality. Here are some metaphysical speculations on possible meanings of the point. It may be necessary to distinguish between:

- the point
- the Point (with a capital 'P')
- the 'POINT'

The *point* serves as a general term, but the words *Point and 'POINT'* have a specific meaning in the metaphysics of Radical Infinitism—the *'POINT'* denoting the first deviation/departure from ALL-IN-ALLNESS, and the *Point*, the First Pre-Cosmic Condition in pre-nascent Cosmos. The 'POINT' and the Point will be called by *many names* throughout this treatise, the most important of which are:

- the 'POINT' called the 'RAY' of the ABSOLUTE
- the 'POINT' called 'POINTNESS'
- the 'POINT' called the 'FIRST 'ACT'
- the 'POINT' called the 'EVANESCENT' 'INFINITE TRINITY'— then,
- the Point called 'Pointness'
- the Point called the Dual or Triple Point (Dual or Triple in Pre-Cosmic 'Time')
- the Point called the Infinified Point
- the Point called the SELF-as-Infinified Point
- the Point called the Condensing Point
- the Point called the SELF-as-Infinified Point-as-Condensing Point
- the Point called the Condensed Point
- the Point called the SELF-as-Infinified Point-as-Condensing Point-as-Condensed Point.

The foregoing words and phrases are used equivalently to enrich the consideration and also to make links with various texts (most notably *The Secret Doctrine)* wherein such terminology has been used.

◆ The 'POINT' (or 'POINTNESS') interestingly, has an equivalence with the 'RAY' of the ABSOLUTE, which 'FLASHES FORTH' to signal the Beginning of the Pre-Cosmic 'Activity' leading to the Universal Manvantara.

The 'POINT'-instantly-Point is the *seed* or *potentiality* of the Number One. From one perspective, the 'POINT'-instantly-Point *is* the Number One, for certainly (considering the generation of any Cosmos/Universe), the 'POINT'-instantly-Point comes *first.* A little thought will reveal that the 'POINT'-instantly-Point is not only the Number One, the Pre-Cosmic First, but (as soon as, or *almost* as soon as, it is One) it is the Number 'Three' as well, for the 'POINT'-instantly-Point (with *infinitesimal instantaneity) becomes* Three. If the Point (or 'Pointness', or 'POINT'-instantly-Point) is considered the Number One, then It should represent a *Super*-Cosmic or *Pre*-Cosmic Number One, and *not* a *Cosmic* or an *Intra*-Cosmic Number One. As the *movement towards*—Universe coalesces, the Pre-Cosmic Numbers conveniently tend to 'collapse into' the Cosmic Numbers, but the ZERO ever *remains* the ZERO.

We must remember that the INFINITE SELF (especially in ITS pure, *uncomplicated* 'STATE', the 'STATE' of ALL-IN-ALLNESS) is *pointless.* (It is also virtually *pointless* to discuss the INFINITE!) Within the INFINITE SELF the concept of location is meaningless. IT is indivisible, undifferentiable and has no *parts,* whether such parts be small, microscopically small, or infinitesimally small. IT *may* however 'have' 'parts' which are *utterly dimensionless.*

Though what, after all, is a *dimensionless* 'part'? Perhaps, we could say that *infinitessences* are 'dimensionless parts'. Such 'parts' are so *infinitessentialized,* so *noumenessentialized* that they have become {or, rather, *are*} NOTHING AT ALL—the INFINITESSENCE ITSELF. Understanding of the 'NATURELESS-NATURE' of the INFINITE SELF can be somewhat illumined by means of considering ITS *pointlessness,* a theme to be developed as we progress.

Along with the inevitable reassertion of EVANESCENT MAYA/LIMITATION (a fundamental *SUPER-Universal, intra-SOURCE,* PARABRAHMICALLY-'GENERATED' (or, rather, PARABRAHMICALLY- *'BECOME'*'ENTITY'/'POTENCY')—'MAYA' (which has reasserted 'ITSELF' cyclically an infinite number of 'times' throughout all of ETERNITY past, and which will do the same endlessly 'into the future')—comes, coeval with the 'REAPPEARANCE' of the 'RAY' of the ABSOLUTE ('POINTNESS' by another name). All of these come as One, '*falling out of* SELF', as it were, and *instantaneously* (or in instantaneous sequence) *becoming* the Infinite Subject and the Infinite Object (Mulaprakriti) and the Mayavic 'Relation Between'. EVANESCENT' 'MAYA', 'descending' from the SOURCE as Pre-Cosmic Maya, Is Really, at this high stage in the Pre-Cosmic Process, Infinite *Consciousness* (or Consciousness of Infinitude). It is most difficult to understand Maya, but the following thought can be illuminative:

- 'MAYA'/'LIMITATION' is PARABRAHMAN-the-SELF *suddenly* 'FOCALIZED' 'CONSCIOUSLY' *as* the 'EVANESCENT' INFINITE 'SUBJECT', and thence (with *infinitesimal instantaneousness)* focussed Pre-Cosmically as the Infinite Subject.

This refocusing allowing PARABRAHMAN-the-SELF (now in Pre-Cosmos) to sustainedly 'CONTEMPLATE' the INFINITUDE of ITS OWN SELFHOOD as a 'Reflection' called Mulaprakriti (Protogenic Root-Matter).

There are, of course, a number of other ways to understand 'MAYA'/Maya and Mulaprakriti. 'MAYA' could simply be called the SELF-'REFLECTIVE' POWER, or, SELF-'CONSCIOUSNESS'. Mulaprakriti could be called SELF-as-Object/Matter. 'MAYA'-instantly-Maya can be conceived as the *only* 'ACT' of the INFINITE-SELF, per se. 'MAYA'/Maya is *not* Mulaprakriti, per se, but the 'ACTIONLESS-ACTION' of the INFINITE-SELF which 'PRODUCES' 'within' ITSELF, *'CONSCIOUSNESS'* of ITSELF *as* that Infinite 'OBJECT'-instantly-Object' known as Pre-Cosmic Root Matter (i.e., Mulaprakriti).

The word 'instantly' is used to show how immediately that which *begins* as 'occurring' 'within' the SOURCE, becomes, in a *flash,* extra-SOURCE. Any sort of Number, even the Number One, cannot 'ABIDE' in the SOURCE, and so it is *cast out* into the Pre-Cosmic World which can 'tolerate' an Infinite Trinity. Really, it is of the utmost difficulty to conceive how anything at all can 'ARISE' within the SOURCE, but because of the evidence before our eyes—namely the Universe—we must conclude that this 'ARISING' 'HAPPENED'. And yet, in from a certain perspective, the 'ARISING' never *has arisen* or *will arise.*

♦ It is important to reaffirm that the appearance of the Universe-Impulsing 'EVANESCENT INFINITE TRINITY' and the Pre-Cosmic Infinite Trinity *are for all practical purposes simultaneous*—practically *coeval*—with the 'ARISING' of the FIRST 'INFINITE TRINITY' and its 'BECOMING the Pre-Cosmic Infinite Trinity', occurring with *infinitesimal instantaneity.*

Was there a 'PREPARATION' for the 'ARISING'? This cannot be answered. If the answer is Yes, the 'ORIGIN' of the 'FIRST ACT' is thrown into infinite regress, for we would need 'within' a 'STATE' in which 'Time' cannot exist, a 'PREPARATION' for the 'PREPARATION', and then a 'PREPARATION' for the 'PREPARATION' of the 'PREPARATION'. Thus, the 'ARISING' and all that 'IT' implies for the Nature of the Coming Universe, must remain for us (while in-Cosmos) utterly mysterious. The FIRST 'ACT', it seems, 'ORIGINATES' in a Pre-Cosmic MOMENTLESS 'MOMENT'—Time, 'ARISING' out of TIMELESSNESS.

In the above paragraphs we have been discussing the *greatest* of all 'EVENTS', the 'ORIGINAL ACT'—the 'ACT' which instantaneously 'PRODUCES' the 'EXTRUSION' of the Infinite Trinity:

- Infinite Subject
- Infinite Object
- Infinite Consciousness

or, worded otherwise, Infinite Sat/Chit/Ananda.

Study these paragraphs closely, because they will unite a number of important terms which are identical in nature and in time of origin, and thus much confusion will be avoided. Maya, perhaps, should be considered as of three kinds:

1. 'MAYA'-as-the-'RAY' of the ABSOLUTE;
2. Super-Cosmic Maya (both Pre-Cosmic and Post-Cosmic); and
3. Cosmic-Maya. The sudden 'ARISING' of 'Intra'-'SOURCE' 'MAYA' becomes, instantaneously, the Appearance of 'FORTH-RADIATED' Pre-Cosmic Maya/Limitation.

The 'ONE ACT' of the INFINITE-SELF is here being 'described', and the 'ACT' is equivalent to the appearing of a SELF-'CONCEPTION'-instantly-Conception (or Mulaprakriti). The 'ACT' is equivalent to the 'DROPPING' of the 'MAYAVIC VEIL':

1. Instantaneously upon the 'DROPPING of the VEIL', 'MAYA'-becomes-Maya (Pre-Cosmic Maya).
2. The Great Veiling Process then begins to Act in relation to the Infinite Subject confirming Its View of Mulaprakriti.
3. And, subsequently, that further intensification/densification of the Veiling Process that produces the Limited Universe and All within It is generated.

This Process of Generating the Content of the Universe should be called *Cosmic* Maya. As a simplification, 'MAYA'/Maya is ever *consciousness* at whatever level of manifestation it (i.e., consciousness) may be *creating!* If this seems an unusual thought, we must consistently bear in mind that SELF-'CONSCIOUSNESS' 'CREATES' Matter—even more, that SELF-'CONSCIOUSNESS' *IS* Matter.

It is vital to bear in mind that, ESSENTIALLY, all these differentiations in terminology indicating the *one who acts* and the *one who is acted upon*, as well as the *one through which action is induced* are UNREAL and illusory. ESSENTIALLY, there is naught but the UNDIFFERENTIATED INFINITE *ONE*. PARABRAHMAN and 'MAYA' are, therefore, not *REALLY* a Duality. Nothing but PARABRAHMAN or the SELF ever IS.

- In ESSENCE, PARABRAHMAN *is* 'MAYA', *is* the 'RAY', *is* 'POINTNESS', *is* the 'EVANESCENT' INFINITE TRINITY, *is* the Infinite Trinity, *is* the ALL as well as the All, and every (apparent) 'thing' that ever has been or will be! PARABRAHMAN *IS* the UTTER ALLNESS. There is only ONE ENTITY/NON-ENTITY both 'within' and 'without' all Cosmoses/Universes, and that ENTITY is the NAMELESS ONE, futily called by many names. Let it, therefore, be understood that all names are really equivalent to the ONE NAMELESS NAME.

'MAYA' 'ARISES' as PARABRAHMAN 'FACES' ITSELF, and, in fact, is the 'CAUSE' of 'PARABRAHMIC SELF-REFLECTION'. 'MAYA', in fact, *Is* the 'ACT' of PARABRAHMAN 'FACING' ITSELF. This 'FACING' *Is* the 'RADIATION' of the 'RAY' of the ABSOLUTE. Thus, the 'RAY' *Is* 'MAYA'. The 'ACT' of PARABRAHMAN 'FACING' ITSELF *Is* 'POINTNESS' as well, and, thus 'POINTNESS' *Is* 'MAYA'. Then, the 'stabilization' and continuance of this 'FACING' (paradoxically 'outside' the INFINITE SELF, though nothing can REALLY be 'outside') *is* the Infinite Subject 'Facing' the Infinite Object—Mulaprakriti, and 'Seeing' (by means of Pre-Cosmic Maya) that Infinite Subject—the Pre-Cosmic 'Contradiction' of the INFINITE SUBJECTIVITY 'PRE-OCCUPIED' with *infinitessentially* 'BEING' ITS INFINITE SELFHOOD.

The 'EVANESCENT INFINITE TRINITY' can scarcely be called a REAL *triplicity* (so fleeting 'IT' is, and so quickly 'TRANSFORMED' into a Pre-Cosmic Trinity). The Pre-Cosmic Trinity, however, should be considered both Real and a *sustained triplicity* throughout the Universal Manvantara. The Relationship between Infinite Being/Knowledge/Consciousness; Infinite Subject/Object/Maya; Infinite Sat/Chit/Ananda is Real and Super-Cosmically Abiding, no matter what may be 'Happening' in-Universe.

When we consider this Manvantarically Abiding Trinity in a little more detail, we have the following Interplay: the Infinite Subject (the SELF-as-'RADIATED' Infinified Point) *omni-directionally* 'Faces' (by means of Pre-Cosmic Maya, i.e., Infinite Consciousness) an Infinite Object (Mulaprakriti)—an Object by means of which:

- *infinitessentialized* INFINITUDE is capable of being 'Seen', and
- infinite possibility is capable (eventually, perhaps!) of being objectified and manifested (never, of course, during any one Universe).

Mulaprakriti (Primary Root Matter), as the One Object, is as *infinite* in Its capacity to 'Show Forth' what we call "matter", as PARABRAHMAN IS in ITS 'CAPACITY' to 'BE' *infinitized* POSSIBILITY, and as the Infinite Subject Is in Its Capacity to *subjectively 'contain'* or *'hold'* the infinite possibility to be mirrored as Mulaprakriti and Its Prakritic Progeny. Mulaprakriti, therefore, is, by definition the Infinite Object, initially Unlimited, and, hence, capable (with the assistance of Maya) of 'Becoming' (through the Procession of Universal Time in infinite Cosmoses) Infinite Generation. In fact, so strange is INFINITUDE, that Mulaprakriti already *has* 'Shown Forth' as Infinite Generation, for It has Existed *cyclically forever.*

- Mulaprakriti is but an Objectification of the INFINITE-SELF-as-Infinite Subject. Hence, it must be remembered that Mulaprakriti is *not* the *totality* of the INFINITE SELF as It is sometimes mistaken to be, but only ITS Reflection caused to Arise through the Agency of 'MAYA'-instantly-Maya, the Great PARABRAHMIC 'POWER' of 'SELF REFLECTION'.

The GREAT 'ACT' of 'LIMITATION' *IS* intra-SOURCE 'MAYA'-as-the 'RAY'-as 'POINTNESS'—all three the same, and all three 'ARISING' *instantaneously.* After this 'ARISING', the Process of Pre-Cosmic Limitation proceeds. Working, for a moment with the symbol of the Infinified Point, we can conceive that It (the Infinified Point) can be called the *Aperture of INFINITY.* The 'POINT'-instantly-Infinified Point is the INFINITE SELF's 'doorway to Finitude'. The 'POINT'-instantly-Infinified Point is the *Periodical Finitization* of the INFINITE. The 'MAYAVIC 'RAY' which IS the instantaneous 'ARISING' of the 'POINT'-instantly-Infinified Point and Its Infinite Object—Mulaprakriti, IS the First 'CHANGE' in the CHANGELESSNESS.

- The paradox exists that, though there must occur at the inception of every Universe a First 'CHANGE', the CHANGELESSNESS cannot ever REALLY *change.* Can there be 'CHANGE' within the INFINITE-SELF, the CHANGELESS ONE? The answer, to be accurate, must be contradictory, hence paradoxical—both Yes, and, definitely No!

The Infinified Point in the Pre-Cosmic World (because it is 'Infinified') stands for nothing *localized* or *geometrical.* The Infinified Point is simply the First 'ALTERATION' in THAT WHICH CANNOT ALTER. The appearance of the 'POINT' or of 'POINTNESS' 'within' the ABSOLUTE, represents what, for want of better words, can only be called a *change of 'STATE'* within the CHANGELESS ONE, and that change of 'STATE' is the *inception of the process of Objectification.*

Whereas 'within' PARABRAHMAN (in ITS Maha-Pralayic ALL-IN-ALLNESS) there was and could be no 'RAY', no 'POINT', and equally, no 'OBJECTIFICATION' (i.e., there was no 'CONSCIOUSNESS' within the INFINITE SELF, of the INFINITE SELF-as-'OBJECT'), the sudden, paradoxical 'APPEARANCE' of the 'RAY' coevally with and/or *as* the 'POINT'-instantly-Infinified Point *and* (whether simultaneously or in instantaneous sequence) Its 'EVANESCENT INFINITE OBJECT-instantly-Infinite Object' represents a 'CHANGE' from 'BEING' to 'SEEING' by means of which the SELF can begin to 'SEE' (or be 'CONSCIOUS' of) ITSELF-as-'OBJECT' (albeit only within the Pre-Cosmic World in any *sustained* way).

Thus, via that sudden, *paradoxically* 'ARISING' 'STATE' 'within' PARABRAHMAN which we might call 'RAYNESS', or 'POINTNESS', or SELF-'SIGHT', Mulaprakriti (which is, Essentially, the One and Only Infinite Object) *appears.* Mulaprakriti, *for the infiniteth time,* "comes into view" of the now-'SEEING' SELF-as-Infinite Self/Subject. Paradoxically, every *appearance* of Mulaprakriti has been *for the infiniteth time!* And every Cosmos that has ever *appeared* has been the *infiniteth* Cosmos!

The 'MAYA'-'INDUCED' 'RAY' of the ABSOLUTE ('FLASHING FORTH' as Infinified Point and Infinite Object/Mulaprakriti), therefore, is the FIRST 'LIMITATION' of the *never-to-be-limited* ABSOLUTE. However, that 'LIMITATION' is only a 'CHANGE' in SELF-'CONSIDERATION' from the 'STATE' of TOTAL SELF-IDENTIFICATION (PURE

BE-NESS—the natural 'STATE' of the INFINITE-SELF in ITS Maha-Pralayic ALL-IN-ALLNESS) to an Extra-SOURCE State of TOTAL SELF-Objectification or, if we dare say it, Infinite-SELF-*Consciousness* (*Consciousness* of the Infinitude of the SELF)—a State of SELF-'SEEING' instead of SELF-'BEING'.

NOTE: The terms SELF, ALL-SELF, and INFINITE SELF are equivalent, just as BRAHMAN and *PARA*-BRAHMAN are equivalent. The *infinitude* of the SELF and BRAHMAN are simply *emphasized* through the use of verbal intensifiers such as the terms 'ALL' and 'PARA'.

Other ways of considering the 'POINT'-instantly-Point [see Glossary] are as follows:

- The 'POINT'-instantly-Point is the first *reduction* or *finitization* of the INFINITE (even though the INFINITE/ABSOLUTE can *never* be reduced or finitized).
- The 'POINT'-instantly-Point represents the 'medium of passage' whereby the INFINITE 'TRANSFORMS' ITSELF into the Finite, without, however, 'BECOMING' any less the INFINITE. The Pre-Cosmic Finite at this stage of Translation is still very 'close' to the originating INFINITE STATE (and thus the capitalized 'F' is used).

The 'POINT'-instantly-Infinified Point in the Super-Cosmic World (the Pre-Cosmic World, and, later, *Post*-Cosmic World) is the "First Fruit" of what can be called the 'WILL'-to-Finitization—a 'WILL' that mysteriously 'ARISES' 'within' the INFINITE. The INFINITE's first 'ACT' of SELF-'LIMITATION' is the 'ARISING' 'within' ITSELF of a 'STATE' of 'POINTNESS' (which 'ARISING' is equivalent to the appearance of the 'POINT'-instantly-Point as the Aperture of the INFINITE). This Aperture can mean many things, but from one helpful perspective it can be viewed as the 'Aperture of Objectification'. In all these 'ARISINGS' it must be remembered that the INFINITE SELF never ceases to BE ITSELF—*changelessly.*

The appearance of 'POINTNESS' 'within' the INFINITE SELF (which is the 'FLASHING FORTH' of the 'RAY' of the INFINITE SELF) is also the First 'EVENT' within that GREAT HOMOGENEITY in WHICH *no-thing* has ever 'HAPPENED' throughout the entire Universal Pralaya. This 'EVENT'-instantly-Event is the First Pre-Cosmic 'HAPPENING'-'BECOME'-'Happening'.

The 'RADIATING' of the 'RAY' and the 'ARISING' of the 'POINT-instantly-Point' is also the First 'MOVEMENT', the First 'MODIFICATION', and the First *'VEILING'* of the ABSOLUTE INFINITUDE of the INFINITE-SELF, for in 'RADIATING' the State of 'POINTNESS'-instantly-'Pointness', the INFINITE-SELF becomes 'AWARE' of ITSELF as 'OBJECT'-instantly-Object (which 'AWARENESS' *is* the appearance of Mulaprakriti) and that very 'AWARENESS' is a *'VEILING'-instantly-Veiling,* a Limitation upon the *absolute* 'PERFECTION' of ITS ALL-IN-ALLNESS.

◆ A difficult philosophical problem (difficult for the mind of man and to be treated later) here arises with the question, How can the INFINITE-SELF forever remain in the state of ABSOLUTENESS, and *simultaneously* 'undergo' 'VEILING' and 'LIMITATION'? Or is the 'VEILING' and 'LIMITATION' that begins with the 'RADIATION' of 'POINTNESS'-instantly- 'Pointness', only an *apparent* 'VEILING' and 'LIMITATION'?

Let us conceive of the 'POINT-instantly-Infinified Point' as 'the Point of Potentiality' for all finite Cosmoses and Universes—i.e., a kind of 'Cosmic Seed'. A Cosmos is a Universe, and a Universe is a Cosmos. (There are no *infinite* Cosmoses and Universes.) The 'POINT-instantly-Infinified Point' is the first Externalization of the uncompromisingly *internalized* 'STATELESS-STATE' of ALL-IN-ALLNESS. From this perspective, *externalization* is *finitization.* Internalization (to the infiniteth degree) is the absolute *perfection* of ALL-IN-ALLNESS.

The 'POINT-instantly-Infinified Point' is the 'doorway' into the Root of Illusion, the doorway to the appearance of 'otherness', for the INFINITE (when there is no Point and no Universe generated from the Point) IS, indeed, the OTHERLESS. (Other than IT, nothing *is.)* One manifested Universe (no matter how immediately *All*-encompassing It may *seem* during Its particular Universal Manvantara), *can* be *compared* with an-'other' such manifested Universe, especially from the Super-Universal 'PERSPECTIVE' of that GREAT LIFE 'EXPRESSING' ITSELF through one Universe after another. The Realm of Infinitely Successive Universes is to the INFINITE SELF, the Realm of 'Otherness'. It is clear, however, that, due to their infinite multiplicity, no manifested Universe is 'otherless'.

All these Universes are *resolvable* into the INFINITE SELF. Thus in a way, the Super-Cosmic Point is both the Aperture *of* INFINITY and the Return Aperture *into* INFINITY. The 'POINT'-instantly-Point (Cycle after Cycle) is the 'Conduit' for the INFINITE SELF to 'enter' into the illusory 'otherness' of an infinitude of Cosmoses, but the Super-Cosmic Point, as well, (in *Post*-Cosmic 'Days') is the 'Conduit' of Return (for the *retracted* 'RAY' of the ABSOLUTE) leading the generated Cosmoses *back* 'into' the 'OTHERLESSNESS' of the INFINITE SELF. There will always be an-*other* Cosmos, but there will never be an-*other* INFINITE SELF, which IS *utterly otherless.*

- There is a tendency to think of a point as simply a pin-point location. When considering the 'POINT' or the 'POINT'-instantly-Point, this, however, would be erroneous. The 'POINT' is simply the first 'MODIFICATION' 'APPEARING' within the INFINITE (the UNCONDITIONED) then *instantaneously* 'EXTRUDED' as the Point, the Super-Cosmic Point.

The Super-Cosmic Point, indeed, *Is.* It exists, but the Point has no *location* and no *dimension.* The Point must, therefore, be considered as *omnipresent.* Thus, the more accurate and illustrative name for the Point is, indeed, the *'Infinified Point'.* With the Pre-Cosmic Appearance of the 'POINT'-instantly-Point arises the possibility of Space.

The Point is found *everywhere* within that State of Pre-Cosmic Abstract Space which exists before Time and Space (as we *think* we know them) arise. An ancient mantram states, "Having pervaded the entire universe with a fragment of Myself, I remain." This Mantram, though as beautiful as it is spiritually inspiring and effective, contains a fundamental fallacy. There is no *fragment* of MYSELF. I-the SELF, the ALL-SELF, the INFINITE SELF, AM *indivisible.* Division, in ESSENCE, being *impossible* to the SELF, it is impossible to pervade *with only a fragment.*

Any pervasion must be accomplished with the *whole* of the SELF. Furthermore, all of Mulaprakriti, from which every one of an infinite number of Universes is 'made', is *already* and *forever* the INFINITE SELF, and nothing else, so the pervasion exists *forever* and can never not exist. In fact, greater than *pervasion* is *identicalness.*

In another way, however, the idea of a "fragment of the SELF" does make sense. Never in the course of any one Cosmos can the INFINITE POTENTIAL, the totality of the FOUNT OF ALL POSSIBILITY, be entirely 'Seen' as objectively articulated within Mulaprakriti and Its Prakritic Progeny. For Cycle after Cycle that constant SELF-'Seeing' Reflected as Mulaprakriti has occurred and will occur in the beginningless, endless 'FOREVERNESS'. So, in one important way, no single Universe contains or can contain the *utter infinitude* of the INFINITE-SELF. A Universe can only contain 'less', i.e., the familiar *fragment* of that INFINITUDE.

The 'POINT'-instantly-Infinified Point meets the necessary requirement of being a *fragment* of the INFINITE-SELF, a kind of 'Singularity' (even though, ESSENTIALLY, there never is or can be any such thing as a REAL *fragment* anywhere in-Cosmos).

- The 'POINT'-instantly-Point *does*, however, represent the Process of Limitation.
- The 'POINT'-instantly-Point *is* the INFINITE-SELF in the Process of Finitization.
- The 'POINT'-instantly-Point is at once a *fragment* and the *whole.*
- The 'POINT'-instantly-Point is, as it were, a 'non-fragmentary fragment'.

In one respect the 'POINT'-instantly-Point is an Aperture *infinitely wide*, for <u>*all*</u> of the INFINITE-SELF can and (strangely) even *must* 'fit' through It. Each Universe is, REALLY, invested with the *total infiniteness* of the INFINITE-SELF (the infiniteness of ITS *ESSENTIAL* {not potentially 'formal'} 'NATURE'). There is 'nowhere' the INFINITUDE is not found *completely.* From another perspective, however, the Point is the narrowest of apertures. Only a certain set of SELF-'SELECTED' 'INTENTIONS' (a very limited set of *infinitized* possibilities—one might say, *infinitely* limited) can 'fit' through it. Each Universe being a Singularity is built upon such 'INTRA-SOURCE' 'SELECTIONS' of 'INFINITESSENTIAL POSSIBILITY'.

The Point is indeed the Infinified Point, the *omnipresent* Point and (before the onset of the Creation/'Becoming' of a Specific Universe) *pervades* Mulaprakriti infinitely, being 'Aware' of Infinite Mulaprakriti in an infinity of ways (i.e., from an infinity of perspectives). Mulaprakriti, however, Is (*during* the formation of a *particular* Universe or Cosmos) pervaded by (or, rather, 'Seen' in relation to) *a very limited set of possibilities* which are 'EXTRUDED' from the INFINITESSENCE and 'Conveyed' by the Presence of the Infinite Subject/Infinified Point 'Seeing' only a certain 'portion' of Itself. While the Infinified Point is Consciously Omnipresent in Mulaprakriti (for the Infinite Trinity *does,* indeed, endure throughout the Universal Manvantara), yet, as the Infinified Point *condenses,* only *one SELF-'INTENDED' DESIGN* 'passes through' and is 'Seen'.

- ♦ So the Infinified Point-becoming-Condensing Point-becoming-Condensed Point is, in a way, the 'Universal Valve' through which 'passes' only one small (call it, 'infinitesimalizingly' small) set of *infinite infinitized possibilities* from the INFINITE OCEAN OF SELFHOOD.

From another perspective, the Infinified Point represents the 'Dynamic Pervading Factor' by means of which a new Universe is Created. The *ESSENCE* of the INFINITE-SELF 'fits through' the Aperture, but the infinite range of *infinite objectifiable possibility inherent* in the INFINITE SELF does not.

- The Infinified Point, as It Condenses, serves as the Extra-SOURCE Agent of the INFINITE-SELF within Mulaprakriti.
- The Infinified Point as it Condenses is, as it were, the INFINITE-SELF in Pre-Cosmic *Action*—an unfragmented *'fragment'* of the INFINITE-SELF—a SELF-Extension which might be called the INFINITE-SELF in ITS Super-Cosmic *Creative Mode.* (In *this* context the terms 'Super-Cosmic' and 'Extra-SOURCE' are equivalent, both relating to the Pre-Cosmic Period, though a Post-Cosmic application is equally possible.)

Even while the INFINITE-SELF 'DESCENDS' *completely* into the Pre-Cosmic Condition of 'Pointness' (as due to ITS indivisibility, IT can never 'DO' anything *partially)* in order to 'Face' the Infinite Object, Mulaprakriti, yet, the INFINITE-SELF remains ever and always exactly and unchangeably what IT IS. Let this fundamental *mystery* (to be repeated in a number of ways) never be forgotten.

The Point in Its many modes must also be considered in relation to Time. The Point is the Aperture through which ETERNAL DURATION *becomes* Time. The 'POINT', arising through *Intra*-SOURCE SELF-'OBJECTIFICATION' leads, as the 'Point', instantaneously, to *Extra*-SOURCE SELF-Objectification (Mulaprakriti), and with the on-set of SELF-Objectification follow all lesser forms of objectification. With objectification comes limited Object, Number, the possibility of Motion and thus the possibility of Time. The Cosmos/Universe is the World of Time; the INFINITE-SELF is the WORLD OF DURATION, though the INFINITE-SELF cannot properly be called a *world* at all.

The Triple Point (the Triple Pre-Cosmic Subject—'Infinified', 'Condensing' and 'Condensed') might, also, be called the *SELF-that-Intends.* Before the 'FLASHING FORTH' of the 'RAY' and the coeval Appearance of the 'POINT'-instantly-Point, there is no *intention* (as we human beings know intention). NOTHING 'WAS'; no-'thing' *was.* With the Appearance of the 'POINT'-instantly-Point (which is the First 'ACT') came SELF-'WILL'. Before the appearance of the 'POINT'-instantly-Point, the Self IS/WAS ALL-IN-ALL. Every 'ACT' 'ARISING' from 'within' the MOTIONLESS ONE is a SELF-'LIMITATION'.

This FIRST 'ACT' signals and *is* the *appearance* of *Incomplete SELFHOOD*—though COMPLETE SELFHOOD must, necessarily, 'ABIDE' *as ever.* The SELF, as manifested through the 'POINT'-instantly-Point (i.e., the SELF-as-Infinified Point, or, alternatively, the SELF as 'extended' 'within' the Pre-Cosmic Condition of 'Pointness'), is not the *completely infinite* SELF in the 'STATE' of ALL-IN-ALLNESS.

The appearance of the 'POINT'-instantly-Infinified Point signals, therefore, the appearance of the *First Entity* which has *not* the absolutely infinite scope of the ONE AND ONLY ENTITY/NON-ENTITY. This First Entity can have many names (some of which we have suggested and discussed) but we can understand It simply as the Pre-Cosmic Representative of the INFINITE SELF.

- ♦ Examining the 'POINT-instantly-Point' in terms of *selfhood,* we see that the Point is, therefore, the first limitation of 'I'.

The Intra-Cosmic Self (and, as well, every intra-Cosmic Self who has, by transcending 'I-ness', found Unity) is called '**8**'. The Super-Cosmic Triple Point being the *transitional interface* between the two could be called **I/8.** Though the '**I/8**-ness' Arising with the Triple Point represents a State of Limitation when compared with the perfection of the PERFECT SELF, still this limited **I/8**-ness is the potent Seed of the entire Cosmos-to-come, and is, thus, from the Cosmic Perspective (which is, after all, though vast, a *limited* perspective) *omnipotent.*

The 'POINT'-instantly-Point, as well, is the first Super-Cosmic 'Eye', for with the Arising of that 'POINT'-instantly-Point comes the possibilities of SELF-'OBJECTIFICATION'-instantly-Self-Objectification, (and, then, Pre-Cosmic Objectivity followed by Intra-Cosmic Objectivity).

◆ The Super-Cosmic Point, therefore, is an Aperture of Vision—the means by which the SELF can begin to 'SEE' or be 'CONSCIOUS'-instantly-Conscious of ITSELF. In this context, 'SEEING' and 'CONSCIOUSNESS' are tremendous *limitations*—in fact, *infinite limitations,* the Work of 'MAYA'-as-Maya.

What might it REALLY mean for a *point (the* 'POINT') to 'APPEAR' in the ALL-IN-ALLNESS, in the BOUNDLESSNESS, in the ALL-SELF, in the INFINITY OF INFINITIES? This 'APPEARANCE' is usually symbolized by a white point against the background of a dark circle, or a dark point against the background of a purely white circle.

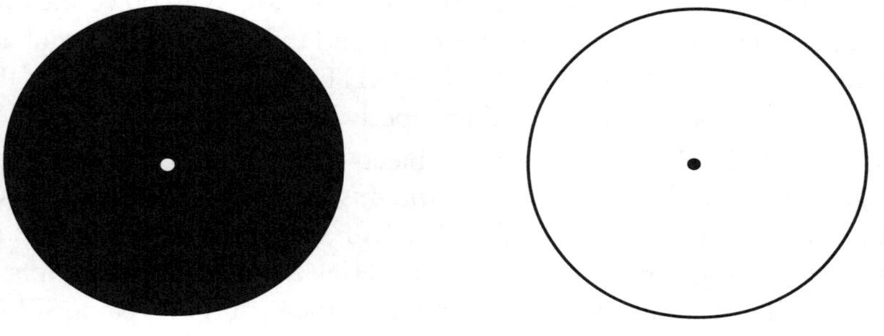

The 'POINT'-instantly-Point, along with Its *infiniversally apprehended* Object Mulaprakriti, are to be regarded (as we have been suggesting) as the First 'EXTERNALIZATION' of THAT. The 'POINT'-instantly-Point is, in a way, the INFINITE SELF, 'LOOKING' through the 'doorway' into Objectivity, which is a Mayavically-Induced SELF-Reflected *otherness*—albeit an ESSENTIALLY *illusory* otherness.

So, therefore, the symbol of one point within a circle is not totally illuminative because, since the Infinified Point is dimensionless and has no location, the *entire circle should be seen as covered with dimensionless points.* The 'POINT'-instantly-Point, is, after all, an *omnipresent* Point. Given, however the *dimensionlessness* of the Point, the image of the circle alone with *no visible* Point or Points might be suitable, for that which is dimensionless, is invisible. Or, given the darkness or lightness of the points used in the symbol (depending upon which contrasting pair is used), the entire circle might be thought of as changing 'color' (either brightening or darkening) because it would be covered with an infinity of points of a 'color' (using black and white as colors) contrasting with its own. The usual symbol is, therefore, merely a *convention,* and its usual form fails to represent what is REALLY happening, but, then, no symbol REALLY can. The value of the symbol, however, is that it indicates that 'SOMETHING' is 'HAPPENING'-instantly-Happening.

If we consider the true nature of the 'POINT'-instantly-Point, we will discover that that Point is REALLY the SELF-as-Infinite Subject poised for ITS First Pre-Cosmic Finitizing Act *(not* ITS First Intra-SOURCE 'ACT'—for *that infinitesimally instantaneous* 'ACT' was the 'ARISING' of 'POINTNESS' Itself (the 'FLASHING FORTH' of the 'RAY' of the ABSOLUTE). The First *Pre*-Cosmic (extra-SOURCE) Act takes place in relation to Mulaprakriti, and is the Act of SELF-as-Self-Limitation (not the 'ACT' of *SELF*-'LIMITATION'). In other words, the 'POINT'-'BECOME'-Point, or the State of Super-Cosmic 'Pointness' is the SELF-as-Self Pre-Cosmically "poised upon the brink" of generating specific limitation, finite entification, enumeration, divisibility, etc. These thoughts are useful in destroying the *literalness* with which we usually consider the concept of the Pre-Cosmic Point as well as such extremely limited (unless knowingly interpreted) symbols as the Point within the Circle.

- Along this same line of thinking, we might consider the 'POINT-instantly-Point' or POINTNESS-as-'Pointness' as the beginning of SELF-'INVERSION'—the *turning inside out* of the SELF. It has been said that "Demon est Deus Inversus."

Given this perspective, the entire Universe belongs symbolically to the "Demon", the "Devil", the *'Inversion* of God'. From this point of view, which in former times would be considered the greatest of heresies, it becomes necessary to say that all Limitation (being, apparently, the very opposite of GOD-the-SELF) is, philosophically, *evil.* Objectification is first an 'ACT' of *SELF*-'LIMITATION' and then an Act of *Self*-Limitation. Objectification is, therefore, in the *absolute* sense, 'sub specie aeternitas', *evil.* Only GOD-the-SELF, the INFINITE DEITY in ITS ALL-IN-ALLNESS is ABSOLUTELY GOOD.

Perhaps, then, the most 'Original' of all Original Sins is the very SELF-'LIMITING' 'ACT' of 'RADIATING FORTH' to 'BECOME' a Finite Universe. We must not be blind to the very illuminating thoughts which arise when comparing the combinations and permutations of the word *evil*—namely, *live, veil, vile.* The implications are astonishing. If the "Devil" is the very "lining of GOD", then through the 'CREATIVE ACT' of SELF-'EXTERIORIZATION' by means of which *impulsion* (after the due Enactment of a Pre-Cosmic Process) a Universe is Self-Exteriorized, then, the Universe, Itself, by extension, becomes the "lining of God". In fact, strangely, the *Devil Itself* becomes the very Universe. This SELF-'EXTERIORIZATION'-as-Self-Exteriorization which we call the Universe is Itself, therefore, both the "lining of God" *and* the "Devil" ("the Father of Lies" as well as the Father of Duality). Nevertheless, we must love the Universe, for what can It be but the ONE AND ONLY SELF? "BRAHMAN and Samsara are one"!

These thoughts point to the idea that what we call the Principle of Evil and the Principle of Limitation are intimately related. There is, however, a redeeming side to this rather outrageous hypothesis. Since Limitation is really an *illusion* and does not exist at all, REALLY, and is part of the Great Illusion, there is no such thing as *absolute evil.* If Limitation disappears, evil disappears. Evil itself, then, is part of the Great Illusion which is Really the Illusion of Limitation. When the Great Illusion dissipates, only the GOOD remains—the GOOD that is REALITY. This GOOD is ABSOLUTE GOOD which can hardly be described as the opposite of what we normally call evil. In fact, this ABSOLUTE GOOD *has no opposite,* just as the BOUNDLESS IMMUTABLE PRINCIPLE has no opposites. Not only does IT 'contain' all opposites, but IT already *IS* anything that could be called ITS opposite.

So, let us review our ideas about the Point:

- There is in Super-Cosmos one Omnipresent Point (the Triple 'POINT'-instantly-Point) which has no *actuality* (for It is ESSENTIALLY Subjective) but which is Real and *almost* REAL.
- The Point is Really a Pre-Cosmic 'Condition' of the *conditionless* INFINITE-SELF (note the paradox).
- The Point is the 'Extension' of INFINITE SELFHOOD into Super-Cosmic Objectivity (which is only Objectivity when compared with the INCOMPARABLE—the INFINITE SELFHOOD).
- Analyzing the true Nature of the Point we find that It is Really a 'Condition of Readiness' to generate a Universe.
- The 'POINT-instantly-Point' is the infinitely recurrent Beginning, within an INFINITE-SELF, which SELF has neither a beginning nor end.
- 'Structurally', the 'POINT'-instantly-Point, per se, is *ever* the same, no matter which one of an infinitude of Universes (SELF-Expressions) is about to be generated.

The 'Players' in the Pre-Cosmic Drama (Whom, together and at-ROOT, are REALLY ONE 'PLAYER') are ever and always the same: the INFINITE SELF, the 'EVANESCENT INFINITE TRINITY', the Pre-Cosmic Infinite Trinity, and the Universally Focused Trinity.

The reader can by now fill in all the individual PLAYERS/Players which compose these three *trinities* [see Glossary]. The INFINITE-SELF IS PERFECTION and 'ABIDES' ever in its *absolutized, infinitized* SAMENESS. Mulaprakriti, the Infinite Object, the completely homogeneous Infinite Potential for 'material' production is ever and always the same—eternally ready to receive and develop what the INFINITE-SELF, through various Agents, 'envisions' through SELF-as-Self-Sight. Even the Infinified Point (the Infinite Subject) as the means through which the INFINITE-SELF achieves *concrete* SELF-as-Self-Objectification as a Universe is ever and always the same.

The INFINITE SUBJECTIVITY 'LOOKING' at ITSELF (which 'ACT' of 'LOOKING' *is* the 'RAY' of the ABSOLUTE) instantaneously 'BECOMES' the Infinite Subject (the Infinified Point) 'Looking' *(infinidirectionally)* upon the Infinite Object, which, too, has Arisen with the 'RAY'.

While Mulaprakriti and the Infinified Point retain their generic 'sameness' from Universe to Universe, that *infinitessentially encoded possibility* which, so to speak, *'comes through'* the Point as a result of the (probably *infinitesimally instantaneous)* 'Intra-SOURCE' 'IDEATION' of the INFINITE SELF is ever and always different—uniquely different for each Universe. It might be said that by means of the Infinified Point (the Infinite Subject) and Its relation to the Infinite Object, we have the *modus operandi* through which the INFINITE invests ITSELF into the Finite, generating by this means, the Number One (a *limited* Cosmos/Universe).

On the ZERO —And ITS Relations to the Point and the One

We concluded the last discussion by stating that Cosmos is 'CREATED' by the INFINITE or ALL-SELF 'INVESTING' ITSELF in the One (*not* in the ONE). The One is our Cosmos/Universe. The ONE, on the other hand, IS the ALL-SELF, but the ONE is even, more accurately, the ZERO. This 'INVESTMENT' is *complete in* ESSENCE, though astoundingly *partial* in terms of the infinitude of possibility which remains, as it were, latent 'within' the INFINITE-SELF, unable to be 'INVESTED' because of 'Space/Time constraints'.

Zero (considered strictly mathematically rather than metaphysically) is entirely different from the number one. In fact zero is no *number* at all, because it does not enumerate anything. Zero, rather, indicates *nothing, no-'thing'.* Although in a common set of integers arranged ordinally, zero and one are immediate neighbors, in fact, zero and one are the great opposites, extremes. One is more closely related to any other of an infinite set of numbers than it is to zero, for the number one is divisible into *any* number whatsoever, yielding, as a result that very number. Zero and one, however, are incommensurate. For that matter, Zero and *any* number are incommensurate.

In mathematics, zero has been called a "place holder". (Throughout the text this will be shown to have profound occult significance.) Within the place or space *occupied* or *held* by the zero, something can *happen,* (metaphysically, numbers are *happenings* or *events),* but zero does not indicate exactly *what* is happening. The division of one by zero gives an important hint concerning the nature of zero; the quotient is infinity (again, like zero, *not* a number, but, strangely, intimately related to zero).

Let us begin to look at zero and the number one metaphysically. To do so, let us dignify these terms by naming them the ZERO and the Number One. (Not ZERO and the *ONE,* for the term 'ONE', as we are using IT in this discussion, is, for practical purposes, equivalent to ZERO.) Metaphysically, One is entified, the symbol of egoity. ONE, however, represents no limited E/entity but rather the completely EGOLESS 'STATE', or the one and only EGOLESS ENTITY/NON-ENTITY.

ZERO is the PLENUM, the VOID, the INFINITE, the ABSOLUTE, the SOURCE—the names for the NAMELESS can be generated continuously. The ZERO (the INFINITE WOMB) represents *infinite possibility,* the possibility of *absolutely* everything, whereas the Number One is the Universal or Cosmic *Singularity,* the maximal narrowing down of all possibility into a limited, *definite set* of possibilities. This Singularity (the One) is (strangely) *infinitely* removed from the INFINITE POTENTIAL, which the ZERO represents, for any quantity (especially the *least multiple quantity,* the One, because It, of all numbers, *most* 'negates' infinitude) is *infinitely* removed from the indefiniteness of a 'completed!' infinitude which ZERO represents.

The great motivating dynamic *during* Cosmos is for everything to *become* the One and, eventually, NOTHING, the ZERO; the great Pre-Cosmic Dynamic is for ZERO to "cross the fathomless gulf separating the INFINITE from the Finite", and to become the One (the very symbol, from the Perspective of the Infinite SELF, of ultimate *limitation).*

- ♦ The entire interplay within the SUPER SYSTEM (consisting of both the INFINITE-SELF and our Cosmos) is described by the Pre-Cosmic Interplay between the ZERO and the One in the Phase of 'Venturing Forth' and, then, the Post-Cosmic Interplay between the One and the ZERO in the Phase of 'Returning'.

In the Pre-Cosmic Process of Universe-'Generating' (by means of the Gradually Condensing Infinified Point, once that Point has been established as 'POINTNESS'-becoming-'Pointness'), the ZERO 'appears' to 'BECOME' the One—the One out of all the Many possible—the One 'SELECTED' for Manifestation by the INFINITE-SELF from out of the infinite infinity of SELF-'RESIDENT' *infinitized possibilities*. Nor through all Time and Space, throughout an infinity of recurring Universes, can the INFINITE-as-'DOG' ever catch ITS OWN 'TAIL', for infinite are the possibilities, and limited are the 'Fields' (i.e., Universes) through which to express these possibilities.

The Fields of Expression available to the INFINITE-SELF are a beginningless/endless Linear Chain of Singularities—i.e., singular and particular Finite Universes. Never will the potential ALLNESS of the INFINITE-SELF be worked out in Time and Space (even in a Time and Space at a tremendous 'noumenal remove' from what we usually consider Time and Space, considering all the possible planes of Cosmos).

This very situation existing (from the mathematical perspective) between the *infinitely full* ZERO and the beginningless/endless chain of *Ones* is responsible for that condition of 'Divine Imbalance' within the SUPER SYSTEM that drives the *oscillatory behavior* of the 'PERPETUAL MOTION MACHINE', which the *periodically breathing* INFINITE-SELF seems to BE. The INFINITE-SELF seems SELF-'DRIVEN' to *finitely* express the supreme *infinity* of *infinitized possibilities* 'RESIDENT' 'within' IT, and this—if there is anything that is *impossible* in the UTTER ALLNESS—may be the *one impossibility.*

What we have called the *drivenness* of the INFINITE-SELF is, of course, non-sensical, because the INFINITE-SELF, which IS PERFECTION ITSELF, has no *need* to *express* anything. From a human anthropomorphic perspective this *super systemic locomotion* (this *oscillatory dynamic* of the SUPER SYSTEM) may *seem* like 'drivenness', but we will have to search deeper into the Mystery of the REAL 'MOTIVE' for Cosmic Manifestation as we proceed. There may be answers—almost infinitely faint to the questing human mind—but there *may be* what we can provisionally accept as answers.

From a mathematical perspective we have dealt with three items—the ZERO, the One and the Point (the Infinified Point). The Point seems to occupy a strange "no man's land" between the ZERO and the One. The Point is *not* the ZERO, but the ZERO, or INFINITE-SELF, expresses as fully as is *'Extra-SOURCEDLY'*/Super-Cosmically possible through It (the Point). From another perspective, the Point is associated with the number One, for the 'POINT'-instantly-Point is the *First* 'LIMITATION', the *First* 'REDUCTION' of the INFINITE-SELF (which, let us remember, paradoxically, *cannot* be reduced).

Mathematically, a point can be understood without difficulty as the generator of the number one. Geometrically this can easily be seen, for to extend a point is to create a line.

Never mind that the point which is being extended has no dimension or, at least, no *measurable* dimension! We must overlook the impossibility of the *actual* existence of the *ideal* point, the *Real* point. The only point which can *actually* exist in Cosmos must be called a 'virtual point'.

Perhaps the closest thing to an *actual* point-in-Cosmos is the "anu", the "speck", the smallest measurable *extension* in a particular Cosmos (i.e., an ultimate particle/event). Anu is a 'virtual point'. Such an extension would probably be found upon the highest subplane of the highest *Sub*-Archetypal Plane [see 'Planes' in Glossary], and would exist in an unknown-to-us Realm which we could only call *Super Energy.* (Later, we will deal with the nature of "extension" and the problems concerning it.)

Metaphysically, the Point (considered as the 'Condensation of Infinitude') is directly responsible for the generation of the Number One (not the number one)—the One in this case being the particular Logos (and Its Cosmos) under consideration.

- Through the Triple Point (particularly in Its *final phase*—the Condensed Point) issues the INFINITE-SELF in ITS SELF-Diminished Universal Logoic Modes—the Father, the Son, and the Holy Ghost/Fohat, working with, and upon, a Point-Envisioned/Isolated 'portion' of Mulaprakriti which we can reasonably call, *Cosmic* Prakriti, the Divine Mother, the One Who 'Shows Forth' (in Reflection of the Subject/Father) all Forms of any particular Cosmos. Further, through the Point (the Infinified Point, the Condensing Point and the Condensed Point) come those Forces of Limitation (Maya), which Create (through the Power of Consciousness-Restriction) a limited Cosmos.

Relating the Point to the Universal Mathematics of the ZERO and the One, it might be said that the Point represents a kind of 'Mayavic Funnel' which, "in the Beginning" is a Pre-Cosmic Deviation from ZERO. Perhaps the Point (as a kind of node between the INFINITE and the Finite) bears a similarity to what H. P. Blavatsky in *The Secret Doctrine* calls the "Laya" state, but on a Super-Cosmic Level (either Pre- or Post-Cosmic). (Since 'POINTNESS' and the 'RAY' of the ABSOLUTE are equivalent, that 'RAY' can also be considered as the kind of 'Mayavic Funnel' here described.)

The Point cannot readily be likened to any number (except from a certain perspective, to an inauthentic 'number one'), nor can It really take the place of ZERO, though there is a kinship to the ZERO. We might think of the 'POINT'-instantly-Point as the 'Conduit' by which ZERO 'BECOMES' the One without ever ceasing to be ZERO, and, reciprocally, the 'Conduit' by which the One Re-Becomes the ZERO without ever having been, truly and ESSENTIALLY, the One. As the Tibetan would say, "Ponder on this."

From one very important perspective (reversible as most perspectives are) the ZERO means *"Yes"*, and One means *"No." "No"* is a condition of limitation, and the One, from the ALL-'SEEING' *infini-perspectived* 'VISION'/NON-'VISION' of the INFINITE-SELF (and even from the Perspective of the Infinite Subject viewing Its own Infinitude as Mulaprakriti) is about as limited as It can be.

- The appearance of the One apparently introduces the deepest of *ontological chasms* between the INFINITE,or ZERO, and the Finite (for which the glyph of the Number One, '1', is perhaps the best symbol). The One is the *line,* and one of the functions of the line is to *divide.*

Separation or division, however, as we understand from the First Fundamental of *The Secret Doctrine,* is an illusory and, ESSENTIALLY, *impossible* state in relation to the BOUNDLESS IMMUTABLE PRINCIPLE. Therefore, the One (the dividing line) that apparently divides or creates boundaries in THAT which is *indivisible* and *boundless,* does not and cannot TRULY divide IT at all.

◆ Let us consider the Point as 'the Potentiality of Number'.

When the Infinified Point appears, no thing, no delimited *reification,* it seems, truly appears. There is only a condition of Infinitude at one remove from the ALL-IN-ALLNESS, as the Infinite Subject beholds Itself as the Infinite Object. A *kind* of Super-Cosmic Enumeration is possible when considering these Pre-Cosmic Players, but They are so 'Trinitized' into Oneness, and Their Oneness is still so closely related to ZERONESS, that They cannot be considered Numbers in the same way the Logoi of the Cosmos-to-Come can be considered Numbers. Numbers are *things/entities/objects,* but the 'POINT'-instantly-Point is *not* a Number.

The 'POINT'-instantly-Point is not a *thing,* but is rather, a *Mode of Super-Cosmic Being,* assumed through the 'ACT' of 'RADIATION', by the INFINITE-SELF. The Point represents what might be called the *beginning of* <u>*condition*</u>. (Earlier we differentiated the term 'condition' from the term 'state'—'condition' connoting greater limitation.) The 'POINT'-instantly-Point is ZERO on its way to *becoming* the Number One.

It is impossible in that particular condition we call *'in-Cosmos',* to achieve the state of *utter pointlessness.* Such *pointlessness* is a Super-Cosmic (or, better, 'Intra-SOURCE') 'STATE' pertaining alone to the INFINITE-SELF. In-Cosmos, on Its very highest level, there will always be a <u>R</u>eal (if non-*actual)* Point, that is, a 'doorway' for a *mediated* ZERO to *enter* the One and for the One to re-enter the *mediated* ZERO. (Only a *mediated* ZERO can exist in the Super-Cosmic State which is *transitional* between Cosmos and the ZERO. The Super-Cosmic Triple Point <u>*Is*</u> the *mediated* ZERO.) The Super-Cosmic State, equally, cannot be *utterly pointless.* In this Super-Cosmic State, the Point exists Really in two or three Phases, depending upon how one considers the Pre-Cosmic Process:

- the Infinified Point;
- the Condensing Point; and/or
- the far more densely *veiled* Condensed Point.

'During' the ALL-IN-ALLNESS of Universal Pralaya there is no 'POINT'. Certainly there is no 'POINT'-instantly-Point. There is no 'Door'. Strange as it may seem, there is no REAL *escape* from the ABSOLUTE PERFECTION. REALLY, there *never* is.

◆ PERFECTION is *inescapable.* The 'POINT'-instantly-Point is the 'Doorway to Limitation', the 'Doorway to Imperfection'. From one perspective the two *trinities* composing the 'EVANESCENT' INFINITE TRINITY-instantly-Infinite Trinity are virtually *identical* as are all the 'PLAYERS'-instantly-Players 'within' them. All these 'PLAYERS'/Players are REALLY *players* in the FIRST 'DRAMA' of 'CONSCIOUSNESS'. Before these 'AROSE'/Arose, BE-NESS never *had* 'CONSCIOUSNESS'.

Not only is the 'POINT'-instantly-Point the 'Doorway' to Imperfection, but It, the 'POINT'-instantly-Point, might be called the 'First Imperfection' Itself (though It is not a *manifest* Imperfection). *The Universe itself is Manifested Imperfection,* and every Universe that has been or ever will be, *never* has been and *never* will be *perfect.* After all, PERFECTION is one of the names of THAT which can *never completely* manifest ITS INFINITE POTENTIAL in any Universe, or even in the whole infinite series of Finite Universes. The word 'perfect', therefore, is a *relative* term. Only *THAT* is PERFECT. Only the INFINITE is PERFECT. The 'part', *apparently* will never be the WHOLE—*formally.* At the same time, the part can never cease to be WHOLE, *ESSENTIALLY,* for there is *naught* but the HOMOGENEOUS INFINITE WHOLE.

There is, however, a kind of Perfection that has relevance to the One, and we shall develop ideas concerning it later. We might call such Perfection: *Perfection-in-Cosmos,* which is a bit like *perfection-in-context.* This kind of Perfection is, literally, *infinitely removed* from the ABSOLUTE PERFECTION, the PERFECT ONE. Though *infinitely removed,* such Perfection-in-Cosmos is, nevertheless, *existent.*

Using mathematics to throw light on this idea, the number 'one' is, in a way, 'farther' from infinity than any other number. Numbers greater than one approach infinity by augmentation. Numbers smaller than one become *larger* (in a way more multi-partite) through division, even as they become magnitudinally smaller, and approach, by reduction, what has been called the *infinitesimal*—an ever-indefinite, unspecifiable number, ever smaller, converging upon, but never reaching zero.

An infinitesimal (or 'infinitiesimalizing'—a word which better indicates a process), never reaches zero, unless the 'rate' of infinitesimalization becomes infinite (by means of 'infinitization'—or for simplicity, let us say by an 'Act of 'WILL'). In this way, the number one (considered simply as a number and not a God) is a kind of pivot point, equidistant from two 'species' of infinity (the infinitely large and the infinitely small, which are one an the same).

While it is convenient to think in this way of the supreme limitation that the number one represents, infinity is a tricky something, because no matter how large may be any *definite* number which is assumed as a starting point, that number is *always* 'equi-distant' from the infinitely large and the infinitely small.

The same is true at the other 'end' of the continuum; no matter how small the *definite* number that is assumed as the starting point, that tiny number is *always* equidistant from the infinitely (though every *indefinitely)* large *and* from the *infinitesimal,* the infinitely (though, ever *indefinitely)* small. (Some might prefer to describe this 'equidistant-ness' as equidistant from the 'infinitizingly large-ing' and the 'infinitesimalizingly small-ing'.) Nevertheless, the number one can be seen as a fulcrum, and if, for some reason, the two extremes can be considered as meeting (the infinitely large with the infinitely small, thus forming a kind of infinite circle) then certainly the number one is *farthest* from both extremes.

- These thoughts are brought forward to alert the reader to the extremely limited nature of our Universe (or any Universe), for each Universe is built mathematically upon the number one, and metaphysically upon the Number One. The strictly mathematical use of number must be contrasted with its *metaphysical* meaning.

In one of those amusing (and revealing) coincidences of language (in this case English), the terms 'whole' and 'hole' have a certain illuminative equivalency. From the perspective of the SUPER SYSTEM, the UTTER ALLNESS (that which is formed of the INFINITE-SELF *plus* ITS generated Cosmoses), what might be called the 'HOLE', is the VOID which is, REALLY, the WHOLE (i.e., the HOMOGENEOUS WHOLENESS).

From another perspective, the 'POINT'-instantly-Point Itself is a kind of 'Hole' which leads 'outwards' towards the limited Whole (which is Cosmos) and 'inwards' towards the unlimited WHOLE which is the INFINITE-SELF. In all cases, when considering the possible meanings of the term 'hole', we find that it represents a kind of privation or isolation, depriving one (or *the* One) of contact with a certain state or condition in order to introduce a new form of contact with the 'world' into which it leads.

Following up the theme of privation and of contact, we find that in the greatest of the WHOLES there is also privation *from any form of contact,* for there is no-thing to contact. From the Post-Cosmic Perspective, the Post-Cosmic Point-instantly-'POINT' is a kind of 'hole' leading unto the WHOLE (in a *cosmo-evolutionary* sense)—a 'hole' through which the One rises in order to leave behind Imperfection of the One and find the ONE PERFECTION of the ZERO. Also, the Point is reciprocally a 'hole' through which the INFINITE-SELF in a Process of 'Descending' SELF-Limitation necessarily 'FALLS' in order (apparently) to leave behind PERFECTION and find a way to *manifest* through the Imperfection of the One.

In this way the Point is rather like a tube or funnel running reciprocally between the ZERO and the One. The Point is, in fact, a 'two-way' Point.

On Number

With the appearance of the 'POINT'-instantly-Point, arises the possibility for the INFINITE SELF-as-Infinite Subject to 'register' Mulaprakriti-becoming-Prakriti and work 'within' It. This kind of Work (leading to the generating of a Cosmos/Universe) necessitates the power to *enumerate,* the Power of the INFINITE-SELF—mediated through the 'POINT'-as-Point—to 'BECOME' Number, and thus to *entify* (albeit, while still *remaining* the imperturbable ZERO, the great NON-ENTITY).

The ABSOLUTE FULLNESS or PLENUM of the ZERO, the ALL-IN-ALLNESS, has naught *(nought!)* to do with Number. It is important to realize this. The PLENUM can, perhaps, be called 'EVERYTHINGNESS', but there is nothing *particulate* about the PLENUM, and the idea of EVERYTHINGNESS can easily suggest particulation. Only an *infinitized* EVERYTHINGNESS escapes the oppression of Number. At any rate, the EVERYTHINGNESS of the INFINITE-SELF has nothing to do with Number.

But there is a kind of Everythingness (note the single capitalization) which *is* numbered. This Everythingness refers to the ALL or (more limitedly) to the All, and is the sum total of every possible particulate registration in-Cosmos (the All), or in the Infinite Chain of Cosmoses (the ALL). This Everythingness is not infinite in the way the INFINITE or ALL-SELF is infinite. Such an Everythingness is a kind of *ever-extended limitation,* for it is formed by a 'more-and-moreness' of limited *things* (those objects which are both 'perceivables' and 'apperceivables'). While such limited things will never be exhausted, they, in their aggregate, do not lead to PERFECTION, i.e., to the INFINITUDE, but only to infinite extension. Enumeration, therefore, while indispensable to the Creation of a Universe/Cosmos, is but a *necessary imperfection.* Enumeration is stands in 'metaphysical opposition' to the indivisibility of the ZERO.

- ♦ Number is fundamentally mysterious. The metaphysical Act of generating Number (i.e., Enumeration) is also mysterious. Through Enumeration, Emanation proceeds, and through Emanation the INFINITE-SELF 'BECOMES' manifold, while, nonetheless 'REMAINING' *simple* and *impartite.*

Metaphysically, *Number is Entity.* Not only are all Great Entities-in-Cosmos Numbers, but all E/entities of any kind are actually Numbers (but more complex Numbers than are the Pure Numbers which Archetypal Entities in the World of Being). Number gives the key to quality and form. Number reveals the *identity-as-pattern* of *E/entity,* though never the IDENTITY of the GREAT ENTITY/NON-ENTITY—the ZERO.

From one perspective, the greater the Entity, the simpler the Number; the more elementary (rudimentary) the entity, the more complex the Number. Many are the apparent divisions/emanations which must be undergone to reach the elementariness of the smallest manifested unit in the Divine Emanatory Stream (probably *not* an ultimate particle, which is elementarily Fohatic, and, therefore *not* a *point of perceptual attachment* for a direct Emanation of the One Ultimate Cosmic Monad as that Emanation takes place along the Cosmic Pattern of Unfoldment we are calling the Divine Emanatory Stream. (The reasons for the putative *non-evolution* of the ultimate particle/event will be discussed later in the text.)

It is strange, but, according to this Model of Creation, there is (probably) no *evolution* for the *least* (the ultimate particle/event) and, as well, for the *greatest* (the One Ultimate Cosmic Monad).

- Involution begins with numerical simplicity and leads to numerical complexity. The concept of numerical simplicity indicates that consciousness is unitary and, in one way, 'simple' on the highest Cosmic Levels from which the Process of Involution takes its start. As forms begin to proliferate (becoming more numerous and, in a way, more individually simple at the same time), consciousness becomes more fragmented.
- Evolution (the Return) begins with the numerical complexity of many apparently fragmented consciousnesses manifesting through a huge number of relatively simple forms, and leads, eventually, to the numerical simplicity of Unified Consciousness (again, as at the beginning), manifested through huge aggregations of forms (hence, through great complexity of form).

Thus, in the Cosmic Evolutionary Process, the gradually simplifying consciousness expresses through ever more complex aggregates of forms, until the Ultimate-in-Cosmos is reached—the Universal Logos expressing Its utterly simple and synthesized consciousness through *all* forms in-Cosmos (through a tremendous *formal* complexity).

There are mysteries here concerning simplicity and complexity. We see consciousness going from unified to fragmented to unified, which, in a way, can be likened to a movement from simple to complex and back, again, to simple. We see *form,* on the other hand, going from inclusively complex (huge aggregates) to isolatedly simple and 'back', again, to inclusively complex. We have contrary motions here.

It is clear that as the Initiate's consciousness merges more and more into Synthesis, the Initiate becomes meticulously aware of ever greater aggregations of forms. The ultimate in this Cosmic Movement would be the Awareness of those who have merged their Consciousnesses with that of the Universal Logos—Who (unitary, yet multiple-Being that It Is) *Is* Conscious of the Wholeness of the Cosmic Whole and, simultaneously, of the tiniest particle/event in-Cosmos—and that awareness would be a seamless, synthetic, Essentially *simple* awareness.

- To enumerate is to divide. The Universal Logos, Who comes into Identifiable Being through the Condensed Point, Creates the Universe through *enumeration/emanation.* In the Universal Process, Enumeration and Emanation are related processes, because the *emanative* process of the Logoi is *hierarchical,* hence *enumerative.*
- To enumerate, is an act of limitation, reducing the magnitude of the One. It is strange, but the generation of more and more E/entities, and the generation of larger and larger living Numbers (i.e., Numbers of *greater* magnitude), are the generation of living Numbers and E/entities Who are, metaphysically, *less than One.*
- Orderly, hierarchical Emanation (the kind used by the Universal Logoi {the Cosmic First Family and Their immediate 'Mind-Born' Relatives} in generating a Universe) produces *relationship* between the E/entities emanated. Numbers, themselves, rather than being Emanated Singularities (Singular Entities *distinct* from their Emanatory Source), may well be considered quantified *relationships* between such Emanated Singularities.
- Continued Entification/Enumeration is the producer of ever more bounded states of limitation. Such Entification/Enumeration is also the 'Creator' of the immensely differentiated 'particulation' which characterizes the form of any Universe.
- Enumeration determines not only relationship, but, through relationship, position and function within the Cosmic Whole. From one perspective, each Number (simple or complex) is an E/entity (simple or complex) with both *position* and *function* determined by the orderly Emanative Sequence.
- Hierarchical Enumeration/Emanation creates systematization. There is no system without Number. Cosmos is based upon Divine Order and Harmony, and there no Cosmos without the Principle of Enumeration.
- The process of Enumeration/Emanation, through which Number is generated/created, is the means which brings about the emergence of *actuality* from the infinity of *infinitized possibility* 'RESIDENT' 'within' the INFINITE POTENTIAL, the FOUNT OF ALL POSSIBILITY.

The appearance in-Cosmos of *'the Numbers'* is, in fact, the *actualization* of that 'impartite-part' of the INFINITE POTENTIAL that was incorporated in the Algorithm of the Cosmos in question, and transmitted through the ALL-SELF-as-Triple Point/Triple Subject.

NOTE: These compound, hyphenated terms are meant to demonstrate the functioning of hierarchical relationship and/or emanative sequence. For instance, if we used such as term as 'the Master-as-Disciple', the term could denote the influence of the overshadowing Master as *mediated* through the equipment of the aligned disciple. All these compound, hyphenated terms and phrases have a *hierarchical directional flow* (either 'up' or 'down'), so in this case, the term might also mean that the Master was a Disciple of some higher Entity. The terms involved and the context in which they are used must sometimes be closely examined to determine the exact meaning and directional flow of the hyphenated formulae. If two terms are separated by a 'slash' (/) for instance, 'particle/event', it means that there is an equivalence between the two terms joined by the slash.

◆ The Universal Pralaya brings an end to that Reduced State of the ALL-SELF called Cosmos or Universe and, also, obliterates the products of Enumeration/Emanation. The Universal Pralaya likewise brings an end to Number, thus, Number is conterminous with Cosmos; Number ends precisely as Cosmos ends.

Most accurately, we could say that Number ends precisely as *Super*-Cosmos ends. By the ending of Cosmos, is meant the ending of Number upon *all* levels, not only gross but even the most subtle. So Enumeration is the creator of Finitude. The *in*-finitude which Number also seems capable of creating is a kind of Infinitude *within* Finitude, and is, therefore, a *lesser* Infinitude than the ABSOLUTE INFINITUDE of the INFINITE, the INFINITE-SELF.

Further thinking about the meaning of Number reveals that Number, metaphysically, denotes *function* within an Emanatory Sequence. The Number Three has a different function than the Number Seven, largely because of its *place* within the Divine Emanatory Sequence. Further, Enumeration is the process of 'creating distance', as it were, from the ZERO. Enumeration 'distances' the SELF-as-Self from the INFINITE-SELF (the SELF, per se). Progressive 'distance' from ZERO translates as progressive Limitation. Thus Enumeration is the means by which the INFINITE, the ALL-SELF, 'BECOMES', *apparently*, progressively limited, without *REALLY* 'BECOMING' limited. All such Limitation is *never* ESSENTIAL, but is only *apparent* and *formal.*

- Number and the Hierarchical Sequence of Numbers is *the progressive, though ESSENTIALLY illusory, minimization of the INFINITE.*
- Number can be seen as the emergence of specificity from HOMOGENEOUS 'IN-DE-FINITE-NESS' (i.e., ZERO).
- Number is the Process by which INFINITUDE specifies ITSELF.

Enumeration is, therefore, a kind of condensation or precipitation of some (a measure) of the infinite *infinitized possibilities* 'RESIDENT' 'within' the INFINITENESS.

The Emanation of Number is the very foundation of relationship. Enumeration is the very origin of relationship. To understand the relationship between Enumeration and Emanation, it must be realized that each succeeding Number *contains* all preceding Numbers. Another way of expressing this is that each Number is 'invested within', and 'absorbed by', the Number which succeeds it. [See also *On Emanations-in-Combination.]*

The apparent enlargement of number is, in fact, a *minimization* of potency, and a 'distancing' from Unity, the One. The process is as follows:

- The INFINITE-SELF (through certain Pre-Cosmic Transformations/Veilings) 'BECOMES' the Finite One.
- The Finite One becomes the finite Many.
- The finite Many re-become the Finite One.
- The Finite One merges Itself (after certain *abstractive* Transformations/Un-Veilings) into the INFINITE, the ALL-IN-ALLNESS.

The ALL-SELF gains naught (through Manifestation) for, IT being PERFECT for all ETERNITY, there is naught to gain.

Some may hypothesize that there is an ever-closer approximation of each in the series of manifested Cosmoses to the INFINITE and ITS SELF- 'RESIDENT' INFINITE POTENTIAL. Although this is an attractive idea, there are a number of problems which arise if we adopt this view:

- The concept of ever-closer approximation assumes a *qualitative progression* of Cosmoses, which would mean that since there have been, already, an *infinity* of Cosmoses, the present Cosmos would be *infinitely evolved* (a blatant impossibility).
- Worse, any Cosmos chosen as an example, even a Cosmos that occurred millions of Cosmoses ago, would *also* have to be infinitely evolved, as it, too, would have to have been the most recent of an infinity of Cosmoses just as our present Cosmos is.
- So, if there were an Evolution of Cosmoses, how is it that our Cosmos (which being the latest of an infinite chain of Cosmoses) evolves from and is an 'improvement' upon the immediately preceding Cosmos (*which, <u>also</u>, was infinitely evolved)?*

Given this kind of absurdity, it seems wise to drop the hypothesis of an infinite chain of progressively evolving Cosmoses. A contrary and probably more plausible point of view, is thateach Universe is utterly unique and bears no particular relation, especially a *progressive* relation, to the infinity of other Universes.

One thing is rather more certain than a definitive solution to the problem of progressive or non-progressive Universes: the (perhaps *infinite)* 'gap' between the INFINITE POTENTIAL of the FOUNT OF ALL POSSIBILITY, and the huge, though Finite, Potential of each individual Cosmos will *never* be closed. There will always be more 'within' the INFINITE to 'EXPRESS' (i.e., SELF-'OBJECTIFY') than can ever be expressed in a Finite Cosmos, or even in the incalculable infinitude of all Finite Cosmoses. Thus, the process of 'Cosmos-Making' or of 'Universe-Generating' (by means of which the INFINITE 'EXPRESSES' ITSELF through Finitude) will continue indfinitely, *forever*—just as it (already!) *has.*

It is Number that determines the *scope* of Cosmos, but not the *Essence* of Cosmos. Let us understand, however, that a Number is *not* a symbol. We have become so familiar with the symbols of numbers that we take the symbol for the <u>R</u>eality—the digit for the <u>R</u>eality it denotes. Numbers are Entities, apparently infinitely-'partialized' Emanations of the ONE AND ONLY ENTITY/NON-ENTITY. Numbers *live*; they are living, though progressively minimized, representatives of the ONE AND ONLY LIFE, identical *in ESSENCE* with that LIFE.

Let us assume (at least metaphysically and in terms of consciousness) that Number *loses* its power the greater its 'distance' from the Number One. Small integers and ordinary fractions are therefore the most powerful, the most fundamental. (Musical ratios, in which the simplest ratios, such as 2/1 or 3/2, produce the most harmonious musical intervals, reveal this to be the case.) While it is difficult, in mathematics, to say that zero is more powerful than one, metaphysically ZERO is *infinitely more powerful than the Number One,* for the ratio of INFINITY to One is INFINITY, and the ZERO and INFINITY are *identical,* just as the NOTHING and *infinitized, non-particulate* EVERYTHINGNESS are identical.

Let us re-state certain fundamentals:

- Number is Entity.
- Number is Relationship.
- Entity is Relationship (even the Number One!).
- Number determines Quality as well as Quantity.
- Really, Quantity plus Position *is* Quality.
- All E/entities have *position* within Cosmos due to their place within the Divine Emanatory Sequence, and their *exact position determines quality.*
- Quality determines Function.

Thus, Number (Quantity) determines Position, determines Quality, determines Function. To know the exact Number of an Entity or entity, whether that Number be simple or complex, is Essential for understanding that Entity or entity.

On Relation

With the appearance of the 'POINT'-instantly-Point arises also the possibility of *relation* or *relationship.* Really, the 'POINTNESS'/'EVANESCENT INFINITE TRINITY-instantly-'Pointness'/Infinite Trinity *is* the Fundamental RELATIONSHIP', between the SELF and ITSELF-as-Not-SELF, 'ACHIEVED' by means of that *limitation* called 'CONSCIOUSNESS'.

Relationship is dependent upon Number. For a relationship to exist there must be two or more E/entities or factors between whom, or between which, some conscious or unconscious *exchange* occurs. The ONENESS (or, better, the ZERONESS) which is the pure HOMOGENEITY cannot REALLY 'RELATE' to ITSELF, per se. Even the Oneness, the Cosmic Singularity, the Condensed Point-as-Universal Logos, cannot, unaided (unaided by Maya) Relate to Itself. A kind of Process may occur within that Oneness, but it should not be called Relationship. Even the word *process* is unsuitable, as there can be no process without interactivity and no interactivity without Number and differentiation. A process occurring exclusively within such an even *relatively pure* Homogeneity should be described in terms of *identification, being,* and *auto-intensification,* but not in terms of *relationship.*

It might be said that the SUPER-SYSTEMIC 'FIRST FAMILY', or FIRST 'RELATIONSHIP', is an *infinitesimally instantaneous* 'RELATIONSHIP', *ending* as soon as it *begins.* It 'RELATES' the MEMBERS of the 'EVANESCENT INFINITE TRINITY':

- the 'EVANESCENT INFINITE SUBJECT'
- the 'EVANESCENT INFINITE OBJECT'
- the 'EVANESCENT INFINITE CONSCIOUSNESS' ('MAYA')

This 'RELATIONSHIP' is *over almost before it begins,* and leads instantaneously to the first Real and Sustained Relationship between the Members of the Infinite Super-Cosmic Trinity:

- the Infinite Subject/Infinified Point
- the Infinite Object (or Mulaprakriti)
- Pre-Cosmic Maya or Infinite Objectified Consciousness

At this Pre-Cosmic Stage (which occurs 'before' the inception of every one of an infinite series of Universes), it is virtually impossible to distinguish between the 'Characters at Play', and the entire thought process must remain unashamedly *speculative.* It bears repeating, however, as a point of simplification, that all of these Players are REALLY PARABRAHMAN, the INFINITE-SELF.

◆ Mulaprakriti is simply the result of SELF-Objectification, and the 'POINT'-instantly-Point (rather than being anything tangible and localized, as the word *point* suggests) is simply a symbol for what might be called the *First 'CHANGE' in the SELF's ABSOLUTE SELF-'ABSORPTION'.*

The *appearance* of the 'CHARACTERS'-instantly-'Characters' (upon which so much descriptive power has been *speculatively* lavished) indicates the sudden 'onset' of *'Finitization'* 'within' what had been *undefiled* ABSOLUTENESS, the 'onset' of a kind of SELF-'DIVISION'. A lesser 'THING'-instantly-Thing (the SELF as an 'OBJECT'-instantly-Object) 'comes into View' of the SELF-as-Pre-Cosmic Self. On the 'LEVEL' of the ABSOLUTE 'INVOLVING' PARABRAHMAN, and the instantly *radiated* 'TRINITY' which ('within') IT 'ARISES'/'RADIATES in a 'FLASH' of Newly Born 'Time', a kind of 'INTERPLAY' *within SELF* has begun, becoming 'Extra-SOURCE' (i.e., Super-Cosmic rather than SUPER-Cosmic) the very *instant* it 'BEGINS'. This interplay should be called the 'BEGINNING'-as-Beginning of *Relationship.*

In a way, the 'instant' there is the slightest 'STIRRING' in THAT (impossible though such 'STIRRING' may be to understand), the 'instant' that there is the slightest 'DEVIATION' from the MOTIONLESSNESS and the HOMOGENEITY—at *that* instant the Super-Cosmic Trinitarian Archetype of Relationship *flashes into Existence.* There is only one 'STATE' in which there is *no* relationship, and that is 'within' the ALL-IN-ALLNESS during the Universal Pralaya. The moment 'CHANGE' occurs, as it must with the 'REAWAKENING' of the INFINITE-SELF from ITS aeonial 'SLUMBER' (a STATELESS 'STATE' of *absolutized, infinitized* PERFECTION), Relationship *begins* (albeit, for the *'infiniteth time').*

Bear in mind that, REALLY, there can be no *REAL relationship* between the INFINITE SUBJECTIVITY and anything else, because there *is* nothing else. It is only as 'CHANGE' 'ARISES' (however this 'HAPPENS') that relationship becomes possible between the newly-appearing 'ASPECTS'-as-Aspects of the INFINITE-SELF, 'Aspects' which did *not* and could not 'ex-ist' 'within' the ABSOLUTE HOMOGENEITY, the ALL-IN-ALLNESS. Even *then*, relationship is not REALLY possible.

- There is no relationship without Number—i.e., without enumeration and hence differentiation. Enumeration and differentiation are based upon arithmetical operations. All multiplicity in-Cosmos is caused simultaneously by division, addition, multiplication, and even by subtraction.

The multiplicity which makes relationship-in-Universe possible is created, especially, by *division*, because the One is (apparently) subdivided into many parts, each part but an Aspect of the Original One. In this process the One remains more important, more fundamental than any of its subdivisions (even though their numerosity is greater). From a metaphysical perspective, the One remains 'larger' and more complete than any Numbers which issue from It. Even though the One is apparently *divided*, none of Its divisions ever ceases, Essentially, to be the One (even though each division is simultaneously *more* and *less* than the One).

Multiplicity can also be seen as arising from *addition*, for in order to create other Numbers (i.e., E/entities), the One begins by *adding* Itself to Itself, and then continues to add Itself to every new Number thus created. Another way of conceiving this addition, is to think of every Emanation of the One (each *less* ... yet *more)* than the Emanation preceding it) as continuously added to the Original One (*as well as to all Emanations that preceded the Emanation in question)* in order to create each newly emerging Number. The One, however the process of addition may be conceived, is thus *resident* within every Number, as many 'times' as the existing quantity or cardinality of the Number in question. For instance, the One is found twice in the Two, and three times in the Three, and so forth. Thus, all 'new' Numbers/Entities arise through repeated addition of One to Itself. (The metaphysical statement, "All is *One*" takes on new significance.)

The multiplicity that is necessary for relationship in-Cosmos arises also by *multiplication*, which is related to addition. Each Number/Entity is composed, from one perspective, of *multiple* Number *Ones*. Through Enumeration/Emanation, the One is multiplying its power. Depending upon the perspective, higher Numbers can be seen as greater or more powerful than lower Numbers, *or* weaker. In matters of form, higher numbers *are* more powerful. Many simple forms united, when working together cooperatively, demonstrate with more power within the worlds of form than simply one or two. (Is one atom more powerful, or are a quadrillion? Is one man more powerful than a united and harmonious group of a million?) With respect to consciousness, however, the lower (i.e., simpler) numbers are more powerful, for power in consciousness is indicated by the ability to fuse multiplicity into unity. Multiplication is a form of interplay between two or more groups of items which leads to the *mutual intensification* of the constituent members of the groups. Multiplication, however, does not necessarily lead to spiritualization. As stated, in matters of consciousness, proliferation can lead to fragmentation and loss of power.

The multiplicity of Cosmos is also generated by the means of *subtraction*. From a metaphysical perspective, the sum of all Numbers *larger* than One will (strangely) *never* add up to One—i.e., to the *fullness* of the Number One. This is another way of saying that "the Whole is greater than the sum of its parts", no matter how many parts there are. Metaphysically, Enumeration/Emanation proceeds by means of the *subtraction* of power. Each succeeding Number is a further reduction of the Number One (even though its numerosity be greater), and is created by *taking away* power from the One through the Process of Emanation. The curious thing is, that in this subtraction, the power of the Number One is *never* Really *reduced,* just as the flame is not diminished by lighting a multitude of lesser fires.

- These four mathematical operations are fundamental to an understanding of relationship. Any of the four can lead either to a distancing from Reality/REALITY, or an approach to Reality/REALITY, and each of the four operations must be considered carefully with respect to its differing effects upon form and consciousness.

The Universe/Cosmos *Is* Relationship. Number is Relationship. The Science of Relations is revealed through a metaphysical study of mathematics. Within the 'domain' of the INFINITE-SELF there is no mathematics, for there is only one *unvarying magnitude*—the INFINITE, which is simultaneously both *infinitized* EVERYTHINGNESS (INFINITY) and NOTHINGNESS (ZERO).

Thus we see that each Number, other than the Number One, can be seen as more or less powerful and important than the Number One, depending upon the perspective. Power in form is built through systematized aggregation, but even immense aggregation has for its goal the approximation of Oneness (as when "the many work together as One"). Power in consciousness is achieved through simplicity and complete unification, but that does not mean that consciousness is not capable of acting skillfully through a hierarchical multiplicity of subdivisions. When dealing with Metaphysical Arithmetic, unity and diversity are constantly transformed into one another.

A symbol for this would be a long chain of figure-eights, each node (the center of the eight where the two lines cross) is followed by dualistic separation followed, again, by a node. The play of Ida and Pingala present the same picture from a physiological perspective.

On Relationship, Separation, and Pain

The cause of all pain in-Cosmos is the *illusion of separation.* While no separation or division REALLY or ESSENTIALLY exists, the Maya-Imposed, (i.e. SELF-'IMPOSED') Great Illusion (which *is* Cosmos/Universe) forces the conviction of separation upon all registering consciousnesses (i.e., 'registrants').

All E/entities within the same system are necessarily *interchanging* E/entities, whether they know it or not. Energy or influence flows between them, simply because they are members of the same system. Not only are they animated by the same primary Energy, but they cannot help but influence each other, even if they abide in a state of so-called isolation. Somehow, they still 'touch', however subtly. Of course, there are many different *degrees* of interchange. Those E/entities with a high degree of noticeable interchange are usually thought to relate or be in relationship, while those with a relatively low degree of noticeable interchange are thought to be unrelated or out of relationship. All E/entities however, are unquestionably related, for not only are they part of the same THING but they *are, in ESSENCE,* the same THING.

On the higher levels of Cosmos the unavoidable *fact* of relationship is undeniable. As Consciousness 'descends', or extends towards the figurative 'periphery' of the Cosmo-System, and as it becomes more (apparently) fragmented, the evidence of relationship becomes obscured. Here in the lower worlds, for instance, human beings understand very little how thoroughly they are related, not only to every other human being, but to every E/entity in-Cosmos. At first the complexification of material manifestation veils and bewilders consciousness, which upon higher and more rarefied planes 'saw through' all form to the Truth of Relatedness. Later, consciousness is no longer bewildered by complexity and dwells in *unity* (Isolated Unity) regardless of the diversity and variety of forms through which it must function.

There is very little pain and agony upon the higher planes because the cause of pain and agony is the illusory perception of separateness. As Consciousness descends and form 'complexifies', the illusion of separation becomes augmented, and pain and agony increase. At first it may be an unconscious pain and agony, but, nevertheless, suffering exists. When self-consciousness arises amidst the complexity and fragmentation of the lower worlds of form, the degree of pain and agony is at its height. Such is the lot of the Fourth Kingdom in Nature, the Human Kingdom—self-conscious, immersed in complexity and seeming division, and, for the most part, cut off (in consciousness) from those higher, simpler worlds in which division is seen to be illusory.

♦ The 'DESCENT' of the INFINITE-SELF (by whatever means) into apparent multiplicity creates a seeming isolation for the *apparent* fragments of ITSELF. *The way home is through conscious establishing between 'so-called fragments' of harmonious relationships reflective of the relationships to be found within the highest dimensions of Cosmos.*

We can thus view relation as the bridging of apparent cleavages and the ending of the lower-world agony caused by the illusion of separation. Agony is at its height when there is the *conscious* realization of division without the realization of the possibility of re-unification. The building of right relationship (based upon recognizing and fulfilling right position and function within the immense Hierarchy which the Universe is) is, in fact the very process of re-unification. Wrong relationship (based upon the insistent consciousness of separation, inharmony, insufficient self-knowledge, and ignorance of the emulatable "Pattern in the Heavens") is the cause of continuing agony. Right relationship is based upon many things, but a knowledge of Divine Mathematics (esoteric number theory, esoteric geometry, and vibrational analysis—to name a few) is indispensable to establish enduring right relationship.

In the last analysis (or should it be *synthesis?)* it will be demonstrated that all Relationship (just as are all Number, Quantity, Space, Time, Motion, and Change) is fundamentally illusory. But the illusion of the un-REAL World is (apparently) very Real, and must be dealt with intelligently. For *WHO* 'MADE' It?! (This last statement is an *affirmation* as well as a question!)

◆ The ONE SELF, the ALL-SELF (or simply, the SELF), WHO WE ARE, has a long-established 'habit' of 'BECOMING' Universes. ITS (OUR/MY) tendency to SELF-'LIMITATION' will never cease and, REALLY, *can* never cease.

Thus relationship and the right handling of relationship is a theme that has *always* been with US-as-Us-as-us and will *never* disappear. The right handling of relationship should, however, never disguise from the one who apparently relates that, indeed, HE—only, ever and always—is relating to HIMSELF. This realization, common to liberated consciousnesses upon the higher planes, will dawn upon Man as evolution proceeds.

On Relationship and Fragmentation

The Cosmic Process of Entification is a process of apparent fragmentation. Only fragments can relate. Only parts can relate. Of course there are no REAL fragments or parts. Hence relationship is a *seeming,* but an inevitable and *necessary* seeming. Entification is the orderly hierarchical creation of many E/entities from the One Universal Entity (God)—an Entity Who took Its origin from the ONE GREAT ENTITY/NON-ENTITY, and Who, ESSENTIALLY, *Is* that ENTITY. The very first *'Move' towards Universe* (by means of which the 'RAY'/'POINT'-as-Point appears) is the initial 'ACT' of Entification. From that *time* forward (for it is 'then' that Time 'appears—yet again), it becomes possible to cognize and *name* things *other* than the NAMELESS.

Entification is coeval with the Principle of Relation. Once the Process of Entification has begun, relation exists. The first *disturbance* 'in' THAT, and the first *relation* are coeval, that is, they originated at the same *time.* Entity, Motion, Number, Time, Space, Relation (and Consciousness!) are all coeval. None of these factors can exist without each other. They are among the first 'gods'. They 'emerged' instantaneously. They 'emerged' because THAT—the ALL-PEACE—was 'DISTURBED'. (Or should it be said, 'BECAME' a 'DISTURBANCE' ... for what besides ITSELF could 'PERTURB' the IMPERTURBABLE?)

From this perspective, Entity, Motion, Number, Time, Space and the *relations* which describe them (for, all of these are the *product* of relationship) are all *disturbances, perturbations.* They are, as it were, 'waves in the WAVELESS'. Interestingly the very highest inconceivable extra-Cosmic STATE and the lowest '"hell" in-Cosmos (Cosmic "Avitchi") can both be thought of as 'waveless'. Right relationship, from an important mathematical/musical perspective is the coordination and harmonious addition of wave forms.

Relationship is a link (subtle or gross) between E/entities. Relationship is a bonding or binding of parts, one to another. Relationship is a bridge by means of which vibrations are exchanged, thereby altering the vibratory condition of the exchanging agents. Causality is a way of describing the process through which one E/entity affects another. If every E/entity dwelt in "splendid isolation" incapable of registration, not only would there be no relationship but there would be no causality.

Relationship is based upon the appearance of Time and Space—not necessarily time and space as we know them through physical plane consciousness, or even within the three worlds of human evolution, but Time and Space as they must necessarily exist on the very highest planes of Cosmos once the Process of Finitization has begun. For without Time and Space no (apparently) separate E/entities or, for that matter 'items-in-Universe' (i.e., those *things* that are the *relatables* in any relationship) would exist.

Think of E/entities and items. They, in their expressive or phenomenal aspect (with the partial exception of the Universal Logos), occupy (formally, at any given time) but one 'space' within a dimension (or but one set of 'spaces' within specified dimensions) and not others. Any E/entity that *formally* occupies *exactly* the same space and *(simultaneously) exactly* the same dimensions, is *exactly* the *same* E/entity.

This is phenomenal 'congruence' in Space and Time and is a method of determining identicalness of phenomenal identity (does such a thing exist, and *not* noumenal Identity?). Even without the proviso that the occupation of only the same space(s) and only the same dimension(s) be *at exactly the same time,* such an 'occupation' (in itself) would be an almost certain distinguisher of one E/entity from another. From the *spirit* perspective, however, which concerns the ESSENCE, *all* E/entities occupy exactly the *same* spaces (dimensions)—i.e., *all* spaces (dimensions) pervasively.

- E/entities *in a given form* (i.e., the phenomenal aspect of E/entities) also occupy (in sum) only one *time* and not another. For instance, they certainly can never be 'born' in *exactly* the same 'space' at *exactly* the same time (otherwise they would be phenomenally identical, and not distinct E/entities at all).

Additionally (while it is theoretically possible) it would be the rarest of phenomena for E/entities to exist/endure for *exactly* the same number of ultimate moments in-Cosmos. It is rare enough for E/entities (in their formal aspect) to be born on the same 'day' and die on the same 'day'.

So, Time too, is a great distinguisher of the phenomenal identity of E/entities. We see then, that distinct Time/Space coordinates (referring not just to the Systemic Physical Plane, but to the multiple Planes of Cosmos) distinguish E/entities from each other, phenomenally. The distinction would not be complete, of course, without a consideration of the factor of Motion, without which the distinctness and uniqueness of an E/entity cannot be fully described. Thus, the combined factors of Time, Space, and Motion distinguish (phenomenally) one E/entity from another.

Therefore, Entification arose with the arising of Time and Space and Motion. It is important to realize which factors-in-Universe are coeval (simultaneously arising) and which conterminous (simultaneously ending), and also, which factors are *necessary* to each other—i.e., which are the factors, without which, other factors could not exist.

R/relation has a definite dependency upon Motion, for R/relation is the means by which pattern is impressed, registered and, thus, exchanged.

- The goal of all E/entities-in-Cosmos is to properly reflect/embody the "Pattern in the Heavens", appropriately to the level upon which these E/entities may be functioning. In the descent of Consciousness into complexity-of-form and fragmentation, the higher Pattern is (apparently) lost.

Moreover, the Pattern becomes, as it were, fragmented in the lower worlds, so that the 'parts' of It might be subdivided among many E/entities—no single E/entity being capable of reflecting the entire Pattern. An analogy is found in the contrast between an assembly of specialists and the ideal "Renaissance Man"—it may take a number of specialists to perform a complex activity that can be performed as well by one multi-talented individual (the so-called "Renaissance Man"). On the lower levels of the Divine Emanatory Stream, E/entities become more specialized, but as they re-ascend, they become multiply-capable in wider and wider areas of activity.

Growth in-Cosmos, however, demands an end to this fragmentation (no matter how illusory it is). The capacity to reflect/embody more and more of the Divine Pattern must grow in every E/entity, until "in the fullness of time" every E/entity consciously re-becomes the One Cosmic Entity (the Supreme Logos of Cosmos—which every E/entity *already is in Essence)*. That all-inclusive Entity will at length, express perfectly (i.e., Objectively and not just Ideationally) the very highest Pattern/Purpose in-Cosmos.

- With these ideas in mind, one can see how important are the factors of impression, registration, and exchange—i.e., the dynamics of *mutual influence.* Relation is, thus, the means by which each 'part' transfers its pattern to every other 'part' and absorbs the pattern of every other 'part'.

It must always be remembered that at the relatively low level of the Evolutionary Process where most human beings are focussed, fragmentation prevails and the usual entities contacted are reflecting/embodying, but a very partial aspect of the Pattern in the Heavens. Through relationship, however, the growth from ability to reflect/embody but a fragment of the Grand Pattern, to the ability to reflect/embody the entire Pattern, proceeds by increments. There are horizontal exchanges and vertical exchanges. Both are necessary in the Cosmic Process which every E/entity (eventually, consciously) pursues—that of *re-becoming the One.*

On Right Relationship and the Cosmic Process

The Cosmic Process proceeds through and by means of relation or relationship. Right Relationship consciously and successfully pursued provides the guarantee for the completion each Cosmic Process. In its own way, Right Relationship is the reflection in-Cosmos (even though Cosmos is infinitely 'low' when compared to Its INFINITE SOURCE) of ETERNAL ONENESS and of the absolute and indivisible SYNTHESIS of the INFINITE.

Synthesis, let us remember, can be partially reflected in Unity. The Unity of Right Relationship in-Cosmos is thus necessary to the integrity of Cosmos, which is the *infinitely* 'distant' reflection of the ABSOLUTE SYNTHESIS of the INFINITE-SELF. This means that E/entities who wish to rediscover the SELF from whence they came, have to proceed through Right Relationship (even though within that ULTIMATE SOURCE there is no *relationship,* per se, at all). Ever the mastery of the Second Aspect of Divinity precedes the mastery of the First Aspect (at least in our local Cosmosystem). While relationship is fundamentally illusory, its mastery is also fundamentally necessary if compulsorily related E/entities are to eventually find release from the Great Illusion.

Right Relationship, humanly considered, fulfills the needs of the heart. From the human perspective, Right Relationship is the fellowship of all Entities within the human kingdom. Right Universal Relationship is the eventually-to-be-achieved Fellowship of all E/entities in-Cosmos. This Fellowship *already* exists on the planes of Spirit and must be achieved thoroughly on all cosmic 'levels' before the onset of the Universal Pralaya.

- Whenever Right Relationship is achieved upon a particular plane in-Cosmos (beginning with the lower planes) the particular plane upon which that fulfillment occurs is *absorbed* into a higher plane.

Such are the dynamics of *abstraction* and *obscuration.* The idea of Universal Salvation (for *all* E/entities in-Cosmos {having by the time of that Salvation *consciously become through abstraction* the One Universal Entity}) is another way of saying that Universal Right Relations have been achieved by the commencement of the "Day Be With Us" (the commencement of Universal Pralaya).

From a practical psychological perspective, relationship, Right Relationship is the antidote for the *wrong kind* of isolation. Relationship is at once the means of overcoming unproductive isolation and of creating the *right kind* of isolation—isolation in the largest, best, and most revealing sense. Pursuing the Path of Right Relationship in-Cosmos leads eventually to the apprehension of Isolated Unity, by means of which all intra-Cosmic items/factors/units/entities/etc. are, as it were, *isolated into Oneness* and, thus, both seen and experienced as One. Again and again, the Many lead to the One and the One to the Many.

Just as there is *absolute inseparability* 'within' the INFINITE-SELF, so a faint reflection of that inseparability must be achieved within the Cosmos (which is really the World of Relationship). The production of Right Relationship within Cosmos (for, unlike Synthesis, Right Relationship must be *produced*—at least upon the lower planes) is the method of working towards a fitting (though necessarily limited) reflection/embodiment of the SUPREME INSEPARABILITY.

Everything related can, of course, be separated, but upon the highest planes of Cosmos (though such planes are still within the vale of the Great Illusion) a *kind* or *species* of inseparability is generated that is based upon the freest possible-in-Cosmos apprehension of the ABSOLUTE INSEPARABILITY of the INFINITE-SELF. Upon those highest planes (the World of Being) those things that are related may be 'seen' as separate but are incontrovertibly *known* as identical.

Relation is never good or bad, per se, which is why there is a great necessity to stress the need for *Right* Relationship in-Cosmos. Relationship, per se, is simply the channel or conduit between the One and the Many and the Many and the One. Relationship, in short, is the reminder, *while in-Universe* (i.e., during the entire Universal Manvantara), of WHERE WE/I come from—our *utterly homogeneous* SOURCE.

On Relativity

Relativity (or 'Relate-ivity') is a term describing the mutual interplay (whether conscious or not) of all E/entities in-Cosmos. To be an E/entity-in-Cosmos, (excluding for a moment the consideration of Super-Cosmic, hence *Infinite*, Entities) is to be inescapably influenced to some degree by every other E/entity. In-Cosmos Relativity is a great Law—each E/entity in-Cosmos is, in some measure (proportional to its sensitivity) responsive to every other E/entity in-Cosmos. A refinement of this Law as it affects self-conscious beings can be found in the following perspective upon the Law of Karma: each self-conscious E/entity in-Cosmos is in some measure (proportional to its impact) *responsible* to each other self-conscious E/entity in-Cosmos. There are probably variations to these Laws which involve response and responsibility between mutually un-self-conscious beings, as well as between those beings which are self-conscious and those which are not.

So, Relativity is foundational to the Law of Karma. This is so because Relativity is the guarantor of the impactful consequences of all motion, all action. By means of the Principle of Relativity, all motion in-Cosmos touches every other motion. In fact, *entity is motion-generated.* There are no E/entities (in their phenomenal aspect) that are not, simultaneously, *motions,*(i.e., O/objectivities 'generated' by *{apparent}* motions) but there are motions that are not E/entities. That which seems to be a stable *something,* i.e., an E/entity (whether primary, secondary or tertiary, etc.) is Really (in its phenomenal aspect) a *repetition of certain characteristic motions*—repeating for a certain duration (i.e., the life span of the E/entity) and then dissipating (the death of the E/entity). Where there is no motion, there is no E/entity. (Of course, what we might call 'motions in Consciousness', or 'acts of Sight' would have to be included within the category of *motion.)*

- Where motion begins, E/entity begins.

So, discounting for a moment the *motionless, All-pervading* ESSENCE of every E/entity (i.e., the ONE ABSOLUTE ENTITY/NON-ENTITY) it must be said that every E/entity (in Cosmic Manifestation and thus, considered, *phenomenally)* is, qualitatively and quantitatively, Really, a relatively stable collection of motions, or more specifically, vibrations (since every motion in-Cosmos is fundamentally oscillatory). (It should be noted that 'Motion' considered as existing within the World of Being may need special attention and special description.)

If Relativity is the means by which the Universal Logos *works* the Cosmic Process, then Motion (along with the necessary inclusion of the factors of Time and Space) is the principal *modus operandi* of Relativity. The main practical point to be extracted from all the above is that Relativity guarantees mutual responsiveness between E/entities and leads to mutual responsibility.

All this said, there is a certain rather intriguing perspective from which Motion is understood to be as illusory as Time and Space. From this perspective, intra-Cosmic 'Motion' cannot Really exist, even though intra-Cosmic 'Change' *does*. Thus, the phenomenon of 'Motion' (so apparently *evident* to the human consciousness) must be re-evaluated. It must be decided whether 'Change' is Real and Motion merely an *apparency*. These thoughts will be elaborated as the text proceeds.

Because Relativity is an *inescapable fact* in-Cosmos, there is a demand or necessity that each E/entity (at length) be *consciously* (as well as *intelligently* and *lovingly)* interactive with every other E/entity. Somehow during the vast expanse of Universal Time (which, remember, though vast, is *finite)* this *conscious* and *complete* interactivity is, ultimately, inescapable. Unconscious or un-self-conscious interplay or interactivity is a given, simply because of the relational Nature of Cosmos. For all E/entities in-Cosmos are, not only Essentially (through identity of Spirit) related, but *actually* related (i.e., from moment to moment in Time and Space). All E/entities are, as well, necessarily (though, largely, unconsciously) interactive. All E/entities are mutually impressive and impactful and, as such, are *conscious* registrants.

Relativity is, therefore, at the foundation of the *enforced intra-Cosmic interplay* and mutual responsiveness between *all* E/entities in-Cosmos, but is also at the foundation of one of the cardinal principles of *right* interplay—responsibility.

On the Interactivity of E/entities through Time

It might be asked, In what meaningful way can it be said that "I", an E/entity manifesting as a member of the Human Kingdom in the twentieth or twenty-first century, am interactive with an E/entity who lived and died or changed its form, for instance, thousands of years ago? From the ordinary perspective of thought it would be safe to say that such E/entities never 'touched' each other, and never registered anything from each other. Yet to say so would be fallacious, for at *no time* is any *authentic* E/entity in-Cosmos *out of* relation with any other E/entity. Every E/entity is ever and always within the same 'Cosmic arena', the same 'field of interplay', within the same *Now* as every other E/entity. No *authentic* E/entity ever ceases to be regardless of change of phenomenal form.

Where was I (or, better, **8**) several thousand years ago? Where were you? Where was **8** one million years ago? Where were you? Where was **8** before the formation of this solar system? Where were you? Surely, we were both authentic E/entities-in-Cosmos (whether at that time self-conscious or un-self-conscious). It is impossible for an E/entity to be *'removed from the action'*, as it were; no E/entity can *ever* 'vacate Cosmos' (which is, at this *time*, the one and only Field of Relationship). Certainly the INFINITE-SELF is *not* a Field of Relationship, because 'within' IT there is no differentiation and hence no relationship. But during the Universal Manvantara (and because of the Universe-circumscribing Guardianship of the Universal Lipika Lords) there is nowhere else to 'go' but Cosmos, and hence Relativity (Relate-ivity) prevails inescapably.

At length all authentic E/entities achieve self-consciousness (though, in a way, no E/entity can {at its deepest level of identification} escape from perpetual-in-Universe, Universal Consciousness which is its birthright {no matter what temporary obscuration apparently seems to overtake the relative, *emanatorily conditioned* part of that E/entity's consciousness}). As their S/self-consciousness grows *consciously* into the Universal Consciousness (which they have *always-in-Cosmos* possessed at their very deepest level of *identification)*, they will see how they have, throughout the duration of any particular Cosmos, *always* been *related intimately* to all other E/entities-in-Cosmos.

Extending these ideas it will ultimately be found that, because of the Law of Relativity (which is uncompromising in-Cosmos), *every being is everything to every other being.* Thus the great (and sufficiently astonishing) revelation of *intimate relationship* between all E/entities throughout the duration of Cosmos becomes the even more astonishing revelation of *identical identity* of all E/entities throughout the duration of Cosmos. (This and other paradoxes are discussed in specific sections given to such thoughts which outrage and boggle the conventional and putatively logical mind.) Summarizing then, all E/entities-in-Cosmos are, first, *responsive* to each other and, at length, will be found *consciously responsible* to each other.

The termination of this Universe will see all Entities (for by that time they will be Entities and not merely entities, since they will have re-ascended to the highest height of the Divine Emanatory Stream) harmoniously related/unified in the magnetism of mutual responsibility. At such a time, before the Great Absorption/Obscuration, all Entities-in-Cosmos will also consciously *know* that they *are* each other, and, *relativity,* will be merging into what, in intra-Cosmic terms, can be designated as the *Synthesis which Is.* This Synthesis is not the same as the SYNTHESIS.

What of those who have 'already merged' with the Synthesis, the SELF? 'Where' will they 'be'? It is the conclusion of the author that 'they' will be *entified* even though they have actually 'merged' into the Synthesis. Further, they will have merged into the Synthesis, but, *not,* into the SYNTHESIS (as many of 'them' seem to think). It is impossible for any Entity (no matter how highly advanced) to abandon *being* the Primary Universal Logos during Universal Manvantara.

There is no 'quitting' Cosmos for 'RESIDENCE' exclusively 'within' ABSOLUTENESS. Those who have consciously *become through identification* the INFINITE-SELF they always have been and will be, realize as well, that there is no way they cannot *also* be every E/entity in-Cosmos, including the One Great Logos Who supervises the entire Cosmic Process. (More anent these Mysteries in other areas dedicated to them entirely.)

On Cosmos

Now, let us examine the concept of *Cosmos* (and the Universe, which in this treatise is regarded as equivalent to Cosmos). It is always important to bear in mind that the Cosmos (or Universe) can *never* be equated with the ABSOLUTE, BE-NESS, THAT. We must be so careful in the use of words because we are dealing with certain abstract matters to which little precise thought is usually given (at least in the Western World). Further, the little thought focused in these directions is vague, for most do not see any value in being 'overly' (but *Really*, 'sufficiently') precise about what they regard as mere *abstractions*.

- Cosmos is *not* INFINITUDE.
- Cosmos is, from the perspective here adopted, a kind of *disturbance* or *perturbation* in THAT, or, 'THATNESS PERTURBED'.
- Cosmos is Limitation Itself, REALLY the *only* 'LIMITATION' which the TOTAL-I-ALL-SELF periodically *appears* to 'INFLICT' upon ITSELF/MYSELF.

This strange form of wording is meant to serve as a reminder that when we speak of the SELF we are not speaking of something *other than who we are*, but of the ONE AND ONLY SUBJECTIVITY or IDENTITY to be found in the UTTER ALLNESS. This means that IT IS you and I, and that both you and I are **I.**

- So Cosmos, the Universe, is *not* the INFINITE.
- Cosmos is a *special and unique case* of the INFINITE, the INFINITUDE, the INFINITENESS, or what ever you choose to call THAT which *truly* cannot be 'called' anything. Cosmos is not the SELF.
- Cosmos is a *special and unique case* of the SELF.
- Every Cosmos/Universe is utterly and unrepeatably *special* and *unique*, for Cosmoses *do* conform to one of the major Laws of the World of Becoming (a Law frequently advanced by Master Morya)—the Law of Unrepeatability.

Thus, we see that Cosmos is a State of Finitization, a State of Limitation of SELF. With respect to the SELF, the Universe/Cosmos is the *only* condition of 'LIMITATION'/ Limitation—(the first term referring to the instantaneous 'arising' of *limitation* 'Intra-SOURCE', and the second term to the abiding of *limitation* 'Extra-SOURCE'). Let us say that *'cosmifying'* (the Intra-SOURCE 'CREATION'/'GENERATION' of Cosmoses) is the SELF's *perpetually practiced process of sequentially actualizing infinite possibility.* Every *actualization* [as detailed in the Glossary] is an illusion.

So the SELF is involved in 'CREATING'/'BECOMING' Illusion. Cosmos is the 'PLAY'-as-Play of the SELF. As we know of *play*—"it's not *for REAL*". The popular idiom tells all. Cosmos is the 'Play' of the (apparently) subdivided SELF. Cosmos is but the greatest possible 'part' (always *infinitely* 'removed' with ABSOLUTENESS).

- ♦ Any 'removal' *at all* from the PERFECTION, from the ABSOLUTENESS is *infinite* removal, for 'removal' from the INFINITE comes through *finitization*, and any finite thing (no matter how quantitatively vast, even if as *quantitatively* vast as specifiably possible) is infinitely 'removed' from the *infinitized vastness* of the INFINITE.

Further, the INFINITE is as if infinitely removed from that which could be called 'the field of externalization for *ALL* 'INTENDED' possibility', i.e., the Infinite Chain of Cosmoses. Cosmos is the Great Periodic Externalization, and the Infinite Chain of Cosmoses is the Eternally yet Cyclically Existent, Infinitely Enduring Field of Externalization. The SELF, by contrast, is the GREAT INTERNALIZATION.

Cosmos is the only Finite Object—an Object which (only *apparently)* is not the INFINITE SUBJECTIVITY. Indeed, in a sense, with the exception of Mulaprakriti (which can be denominated as the *Infinite* Object) there is no other Object than Cosmos (ever, anywhere, anytime), for which reason we will we forced to conclude that Cosmos is the *One and Only Finite Object*—a *limited* Object (both Real and Actual) that appears cyclically *forever*. (The 'Memory' {probably in Super-Cosmic Spheres} of 'Cosmoses Past' fits into another Category of 'Object'—a Category of Non-Actuality.)

In discussions of this kind, the word 'apparently' must consistently be used, because only thus is it possible to talk meaningfully about the Great Illusion (i.e., all that which is both Cosmic, Intra-Cosmic, and, even, to a degree, *Super*-Cosmic, but *not* SUPER-Cosmic). The entire Great Illusion (the Cosmos/Universe) is naught but a Great Apparency, ESSENTIALLY un-REAL.

◆ As heretical as it may sound to the Cosmo-Centered consciousness (as opposed to the SELF-Identified Consciousness) the 'GENERATION'/'CREATION' of Cosmos must be seen as the "Original Sin" and the SELF-'GERMINATED' "Root of all Evil" (at least the Root of that which is conventionally called Evil).

Consider again the following astonishing permutations—*evil, live, veil, vile:*

- *Limited* life is *evil.* The only limited life is Cosmic/Universal Life—i.e., Life *within* and *as* this or any other Cosmos. LIFE 'within' the ALL-IN-ALLNESS is completely UNLIMITED and thus is ABSOLUTE GOOD.
- When we *live* (that's what it's called!) in Cosmos, we live in the Great Limitation. Such living is not LIFE at all. It is a *limited* life under the spell of an inevitable Cosmic Evil (*never* an *absolute* 'EVIL'). Even subject to SELF-'INDUCED' Limitation, however, the SELF can be rediscovered and the ABSOLUTE GOOD, as it were, 'imported' (through *right identification)* into the limited state.
- The third word is *veil.* The only *veil* 'dropped' upon the ABSOLUTE 'STATELESS-STATE' of BEING/CONSCIOUSNESS/BLISS (SAT/CHIT/ANANDA) is SELF-'DROPPED'. The Veil of Maya was 'DROPPED' upon THAT WHICH IS ITSELF ALONE by the THAT WHICH IS ITSELF ALONE and Cosmos came to be. It is amusing to think of this as a SELF-'VEILING' Universe, and a profound and enigmatic TRUTH is therein concealed. L/life *within* the Universe in never TOTAL LIFE; such L/life is never the LIFE that IS 'within' the ALL-IN-ALLNESS. The Universe is REALITY *Veiled*, and can never be anything else, even though within-Universe the ALL-SELF (never REALLY 'lost' or inaccessible) can be 'reclaimed' through the disciplines of *identification.*
- At last we come to the fourth word—*vile.* Not much need be said. The World of Becoming is beautiful and becomes progressively beautiful to those who ascend into the Realms of Divine Simplicity, but in comparison with the INCOMPA-

RABLE PERFECTION, which the ALL-IN-ALLNESS IS (a comparison which WE-the-I-the-SELF cannot make until the ultimate "Day Be With Us"), it may be that the word "vile" aptly describes the *actuality* of what we are 'living' through and how we usually live. The vileness, however, is only for those who *are not* super-consciously identified as the SELF. When that identification occurs, then All is instantly transformed into the SELF. Then, with Nagarjuna we can say, "Brahman and Samsara are One."

This line of *absolutist* thought minimizes the Universe in one way, but in other ways we shall be enabled to see Cosmos/Universe (no matter how limited It may be) as *absolutely necessary* and worthy of the profoundest *love*. Still, we must recognize that Cosmos/Universe is the merest 'fragment' of a fragment (though no fragment can REALLY exist) of THAT which is *already* PERFECT, and the possibilities which are manifest in-Cosmos/Universe are but a finite 'fragment' of the infinite *infinitized possibilities* of the INFINITE POTENTIAL, the FOUNT OF ALL POSSIBILITY.

- In a way, without 'robbing' Cosmos of the reverence due to It, one may be forced to conclude that Cosmos is a 'Game'—*the* Game, "the only game in town"—if "town" be the UTTER ALLNESS.

Who plays the Game? As an answer to this question comes swiftly, *WHO* 'PLAYS' the Game! It is *WHO*, who Plays the Game. The idea is an ancient one. The Universe is seen to be the result of Divine Play (and the ONE AND ONLY 'PLAYER' substands the Play). Does this idea seem irreverent? Why should it? Many old thoughtforms must be exploded in order to think this way, but a tremendous sense of freedom and gratitude arise, and an entirely new and permanent sense of SELF.

Along the same line, Cosmos is the "Creation"! Whose? We know the answer. Cosmos is the Universe, but this Universe is *not* the One and Only Universe. There is no such thing as an One and Only Universe. Given the fundamentals of *The Secret Doctrine*, one thing is certain: we can be utterly confident in the sequentially eternal appearance and disappearance of an infinitude of Universes. This rhythm of appearance and disappearance has been going on forever. So, yes, Cosmos is the Universe. But what of that? Universes are to be prized but not over-prized. Universes come and go—the SELF remains, as ever, forever.

Cosmos is the result of the incessant alternation between *something* and *nothing* (or, more specifically, between objective 'Somethingness' and subjective NOTHINGNESS—though, REALLY, NOTHINGNESS can *never* cease). This alternation is ceaseless and is *the* one and only "perpetual motion". Perpetual motion, like all motion, is essentially illusory and cannot exist *as such* 'within' the ALL-IN-ALLNESS, but since the generation of the World of Illusion is the one 'thing' the SELF *actually* 'DOES', the Principle of Illusion, and Illusion, Itself, will be with us *forever!*

We-the-I-the-SELF have 'DONE' it (i.e., 'BECOME' Illusion) *forever* and we *will* not, *cannot* stop. NOTHING is what we ARE; and Something is what we 'DO' (or, better, 'Do', since Extra-SOURCEDLY). *WHO* is ever responsible for appearance and disappearance of Universes! (This is a statement, not a question.) Why WE-the-I-the-SELF 'DO'/'Do' this is another matter—the Mystery of Mysteries. We will examine the ONLY 'Intra-SOURCE' 'MOTIVE' as we go along. It is impossible, of course, to know this MOTIVE, but revealing ideas can be offered which will throw, perhaps, some ray of light into the otherwise impenetrable darkness of the "REASON FOR IT ALL".

Here are some other ways of thinking about Cosmos:

- Cosmos is both the Unity and the Multiplicity, but Cosmos is not the ZERO, not the HOMOGENEITY.
- Cosmos is the result of the out-working of the cyclically unique, ever recurrent Primeval Design—the Design-at-the-Beginning. How many Beginnings have there been? An infinitude of them—literally *a 'countlessness'*.
- Cosmos is the result of Primeval Intention. Whose Intention? The answer must be SELF-'INTENTION' as it is mediated through that *emergent aspect* of the INFINITE-SELF which can be called the SELF-as-Infinified Point, and brought to specificity through the SELF-as-Condensed Point.
- In one way of thinking, Cosmos is the very 'point' of the Point—i.e., the aim or unfolded Purpose for which the Point (both *infinified, condensing* and then, *condensed)* periodically appears.
- From another perspective, Cosmos might be called the 'Great Modification'. This is important.
- Or, perhaps, Cosmos should be called the 'Great Disturbance' or 'Perturbation'. The ALL-IN-ALLNESS is the ALL-PEACE. IT is the unmodified CHANGELESSNESS. But then, *something* 'HAPPENS'. The *how* of this 'HAPPENING' is surely beyond mortal ken, and probably beyond the ken of any E/entity-*in*-Cosmos. The *how* is probably only 'KNOWN' to the INFINITE-SELF as IT 'CHANGES' for the first 'time' (though "time was not") in "Seven (Universal) Eternities".
- Cosmos might also be called the 'Great Effect' of an infinitely greater CAUSE. The enigma is that there is, and can be, no *relation* between this fundamental CAUSE and ITS supposed Effect, because the SELF (as H. P. Blavatsky has assured us in the *Proem* to *The Secret Doctrine)* is out of all relation with Cosmos. Since, metaphysically considered, there is no "Second", nor "Other", what can the SELF possibly *relate* to? So causality, as we normally understand it in the World of Relations does not REALLY apply in the *non-relation* of the SELF to Cosmos. How can THAT which cannot *relate* to something be *causal* in relation to that something? Cause and Effect demand Relation. Paradox!

It seems, however, that the human mind (at least) is doomed to think in terms of Cause and Effect. This mode of thinking is one of the Kantian "Categories", which are, essentially, ways in which the human mind limits the perception of Reality (REALITY, ITSELF, cannot be perceived) by forcing *its own structural processes* upon Reality. By repeatedly pointing out the contradictions that arise when we think of the INFINITE and ITS relation/non-relation to Cosmos, we may stimulate the intuition, and learn to transcend, somewhat, the "Categories" of perception/apperception which our limited minds force upon our consciousness.

From an enigmatical point of view, Cosmos is the *only* 'Thing'/Object which is not the SUBJECT. (The "special case" of Mulaprakriti {the *Infinite* Object} and the 'Infinite Memory' {of the Super-Cosmic Infinite Subject} which may exist in relation to It, is excluded from this consideration.)

An important question arises: how can the SUBJECT become the Object and still be called a SUBJECT? To fathom this, a special way of thinking is required. We might ask the question, How can **8** 'go forth' from MySelf, emanatorily, 'Seeing' MySelf in the Emanation 'gone forth' and, nevertheless, abidingly 'Be' (i.e., *remain)* MySelf within the Source Position **8** never left? It must be realized that there are no Objects in-Cosmos that are not REALLY Subjects. To see a Subject *as an Object* is, metaphysically, Really, the most common of experiences. If there is to be the perception of a Not-SELF (as *object),* that Not-SELF must have *begun* as a SELF.

- All this is another way of saying that **8** Am (in-Cosmos, at least) "in two places at the same time"—and also, in many, many, many more 'places'. As the SELF, **I AM** abiding *forever as ever.* When, **I** 'RADIATE' MYSELF into Objectivity, however, **I** 'BECOME', instantaneously (or in *instantaneous sequence)* the 'SEER' and the 'SEEN', and, more abidingly, familiarly (and 'Extra-SOURCEDLY') the 'Seer' and the 'Seen'. That which **8** 'See'—**8** Am.

Even the Infinite Subject (being at one remove from the ONE AND ONLY SUBJECT, i.e., the INFINITE SUBJECTIVITY) is a kind of *object* from the 'PERSPECTIVE' of the INFINITE SUBJECTIVITY. (This is provided that the INFINITE SUBJECTIVITY can REALLY 'SEE'! Can the Super Cosmic Trinity and the Cosmos even *exist* if They are not 'SEEN'? A deep question.)

The Infinite Object (Mulaprakriti) is (to the ONE AND ONLY SUBJECT, the INFINITE SUBJECTIVITY) *another* kind of *object* (at a greater 'remove' as an Object than is the Infinite Subject as an Object) and *also* an Object to the Infinite Subject (the Pre-Cosmic **I/8**). So (strangely!) the INFINITE SUBJECTIVITY 'BECOMES', as it were, *two* 'OBJECTS' *one* of which is (paradoxically) the Infinite Subject and the *other* of which is the Infinite Object. Perhaps, even Super-Cosmic Infinite Consciousness could be considered (with respect to the INFINITE SUBJECTIVITY) an Object!

The Infinite Object, Mulaprakriti, (and also Its 'Reduction', the one and only *Finite* Object we call Cosmos) must, in truth, ultimately be understood as, and *resolved into,* the ONE AND ONLY SUBJECT, the INFINITE SUBJECTIVITY.

- Thus, while, on a lesser scale, in-Cosmos, that which 'Sees' an Object must rightfully be designated as a Subject, REALLY and ESSENTIALLY, every Subject *is* none other than the INFINITE SUBJECTIVITY (as is every Object). What else IS/Is but the INFINITE SUBJECTIVITY in all ITS *apparent* combinations and permutations—the variously enumerated and reflected **I** in ever-changing relationship to ITSELF? (but, of course, not *REALLY!)*

From a more artistic point of view, Cosmos is the Great Drama. It is at once the 'Great Grief' and the 'Great Sorrow', simultaneously the 'Great Bliss' and the 'Great Terror'—a statement worthy of the worshippers of Kali. Cosmos is the 'Great Joke' filled with Cosmic Laughter, and yet, in a way, no task could be more *serious* than to "Do Cosmos 'Right'." Cosmos, the ceaselessly recurring periodic Limitation of the INFINITE-SELF, is, as Ramakrishna has said, "the play of the Great Mother".

There are so many ways to think of the Allness with which we are surrounded and of which we are, yes, a *part,* but also (as it is hoped this treatise will demonstrate)—of which we are the *whole.* These alternative and seemingly contradictory points of view are more than useful. The human mind, as ordinarily used, is so limited and so rigid. Alternative perspectives loosen its illusory grip upon Reality/REALITY, and allow the flash of wordless, simultaneous apperception (intuition) to enter.

This treatise contains many seeming assertions which *cannot* be proven, now or in the future. It cannot be said that these assertions are *facts*. They are simply ideas based upon both rigorous logic, expansive speculation, and as much intuition as is available to the author. These often 'outrageous' ideas arise when one dares to take certain frequently accepted working hypotheses to their obvious though *extended* conclusions. The purpose is to introduce or re-introduce to potentially congenial consciousnesses points of view which may upset their prevailing opinions, but may also allow the entrance of a greater apperception of Reality and a greater intimation of REALITY.

- ♦ In conclusion, let it be said that in relation to DURATION, each Cosmos is an 'instantaneity' (an 'infinitesimalizing')—less than a blink of the ONLY EYE. Each Cosmos is a finite impermanence, an evanescence, an ephemerality, produced or, rather, 'BECOME' by THAT which is the very opposite of transience, having no *time* in IT. To each and every Cosmos the wise words of King Solomon apply, "This too shall pass."

On Entity and Entification

Now, let us address two vitally important terms, without which it is impossible to understand consciousness:

- *entity*
- the formation of entity—i.e., *entification*

An *E/entity* (an *authentic* E/entity) is an (apparently) distinct unit of Life. <u>L</u>ife (with a capital 'L') is simply *finitized* LIFE; <u>L</u>ife is LIFE-in-Cosmos. Of course, LIFE (which is another 'name' for the nameless INFINITE-SELF) is indivisible, and so, in ESSENCE, there can be no distinct units of IT. Nevertheless, within the Great Illusion, it *seems* that there are a great multiplicity of distinct *lives*. Ever we are confronted with the paradoxes that arise from the *essential homogeneity* and *indivisibility* of THAT.

A mere congery of continuous (or even related) things is not an *authentic* E/entity. A machine is not an authentic E/entity (though sometimes such things seem to have "a life of their own" and might be classified as a *tertiary entities,* as explained below). A piece of wood is not an authentic E/entity. Such *things* are actually collections of minute entities—in this case, atoms, which <u>are</u> *authentic* E/entities.

- ♦ Authentic E/entities can never be *created* intra-cosmically. They can only be *emanated* from the One Entity-at-each-Beginning (the Universal Logos—the One).

This Great Entity (one per Cosmos) is the cyclically recurrent Representative of the ONE GREAT ENTITY/NON-ENTITY—the INFINITE SELF. The INFINITE-SELF both *is* and *is not* an 'ENTITY'.

- IT has no distinct attributes, but rather (in potential, at least) *all* attributes (in an *infinitized state*);
- REALLY, IT is the NOUMENON of *all* attributes.
- IT has no boundaries and is out of relation with all other E/entities, all of which are derivative from IT and, REALLY, are *identical* in ESSENCE with IT.
- IT, unlike all other E/entities does not and cannot 'ACT'.
- IT cannot be defined, limited or specified in any way.

Thus, one sees, that it is difficult to call IT an 'ENTITY'. And yet, IT is the ONLY REALITY, the ONLY SUBJECT, the ONLY SELF, and so it is difficult *not* to call IT an 'ENTITY'. REALLY, as always, IT is *both* of every possible pairs of opposites. So it is profitable for consciousness to consider IT *both* as the GREAT EGOLESS 'NON-ENTITY' as well as the ONE AND ONLY 'ENTITY'.

But confining ourselves for a moment to conditions in-Universe, there must be a distinction made between *authentic* E/entities, and those congeries of contiguous or even obviously related things which are called entities but are not emanated *beings*. Structures that arise from random, un-Purposed contiguities/relationships within the Cosmic Configuration are not E/entities. If a number of perceptions come together in my field of vision, **8** can think of these perceptions as forming a *whole* and call that whole a *thing*, but **8** cannot call it an *E/entity*. Otherwise *any* semi-permanent congery/collection/aggregation could be called an E/entity.

From this point of view authentic E/entities are *Monads*, emanations of the One Self and *identical* in ESSENCE with the ONE SELF (just as the One Self Is identical with the ONE SELF). Monads manifest as lives animating only certain kinds of patterns-in-Cosmos, such as:

- the lives animating atoms
- the lives animating the various kinds of elementals
- mineral, vegetable, animal and human lives
- devic lives in their astounding variety
- the lives of the various Creative Hierarchies, both familiar and unfamiliar to the human consciousness
- the lives animating various kingdoms of nature *as a whole*, such as the great Life animating the Human Kingdom or the Animal Kingdom, etc.
- planetary Manus, Lords, regents and Logoi
- Solar Logoi
- Cosmic Logoi, etc.

The list is hopelessly incomplete but, hopefully, indicative. The idea to be conveyed is that certain kinds of cosmic structures/patterns are *Logoically-Designed* specific fields of expression for Monads (un-detached Universal Life Units) immersed in various levels or prakritic vibration and, thus, expressing through various levels of density within many dimensions of Cosmos. The L/lives, then, W/who manifest through these Logoically-Designed structures (all such structures being precipitations of the Great Purpose or

Cosmic Pattern in the Heavens) can then be considered *authentic* or *primary* E/entities. Such E/entities might be denoted as *first class entities* and are really *Cosmic I/identities.*

NOTE: When the word *E/entity* is used, it is used mainly to designate *authentic, primary E/*entities. In a way, all authentic or primary *entities* should be designated with a capital 'E' as 'Entities' (for even though they might manifest through lesser forms and unconsciously), they are *identical* in Essence with the Cosmic Logos, the Chief Primary Entity in Cosmos.

The distinguishing feature of *secondary entities* is that they are fashioned and supervised or ensouled. There are types of *secondary entities,* which (when of the higher kind) are supervised structures which were not *designed* by a Universal Logos at the beginning of a particular Cosmos, but which are, nevertheless, *fashioned by other authentic E/entities.* (i.e., fashioned either by *creation,* or by *participation,* or by both). The manner of 'making' or fashioning can be crude or extremely subtle.

In any case, such structures are not *authentic* in the way we have been using that term. Only when such structures are ensouled:

- by Beings of a higher order than the beings (primary E/entities) who *fashion* the structures (or in their aggregate *constitute* the structures), or
- by Beings of a higher order who supervise the structures and direct them (and virtually *become* them)—should structures of this nature be considered as *secondary entities of the higher kind.*

Certain man-made organizations—ensouled and animated by a higher Being—could be considered in this class. Such organizations are formed by a systematized, cooperative collection of authentic, primary E/entities (for instance, the human beings in a group) who fuse and blend the different aspects of their energy systems. Perhaps, a Master or some high Initiate (or even a Being of a still higher order) will see such a group or organization as an opportunity for expression and will, then, infuse and pervade the group, becoming, as it were, the "heart and soul" of the group. Such a group structure, then, would be a secondary entity of the higher kind composed of a number of authentic E/entities of a certain class and infused, ensouled, inspirited, and directed by a primary Entity of a still higher class.

On a much lower turn of the spiral, even a thoughtform *may* become a secondary entity. It may be taken over by an E/entity superior in power to the many tiny (though still *primary* and *authentic)* entified units which are gathered to compose the thoughtform. Perhaps a guiding deva will ensoul the form and direct it to certain purposes. The ensouling, 'inspiriting' agent may not be 'good', per se; all that is necessitated is that such an agent be superior in power and scope to the aggregate of primary entities that it informs. Thoughtforms can be infused and directed by so-called "demons" as well as "angels". If such a thoughtform remained unensouled by a higher-order authentic entity, it would have to be classed as a *tertiary entity.* Many unensouled creations are of this type—for instance, the many mechanical and artistic creations of man.

Unensouled thoughtforms, while seeming to have a kind of "a life of their own" are merely congeries/collections/aggregations of thought matter of a certain vibratory quality and cannot be deemed *authentic* (primary) or even secondary entities. They are *third class* collective entities. Such 'creations' only *seem* to have a *life* of their own (the word *life* is important) because of the purpose of the primary E/entity who designed them, and who, by means of intelligence and/or imagination, co-ordinated and directed the forces of the collection of low order primary entities which had been gathered to construct the creation. As an example, think of such a machine as a computer. It may seem intelligent and *self*-directing, but it has no soul, no inspiriting life. The primary entities which compose it (i.e., the many mineral atoms, for instance) *do* have a life of their own and are coordinated and ordered in such as way as to (collectively) make the machine *seem* intelligent and self-directing. But still, the primary entity who built and programmed the computer remains *outside* his creation.

Thus, in summary:

- *primary E/entities* can be understood as *authentic E/entities* (i.e., 'Rays' of the ABSOLUTE, undetached Universal Life Units);
- *secondary entities* (whether of a high or low order) can be understood as ensouled or supervised aggregates of primary E/entities; and,
- *tertiary entities* can be understood as non-ensouled (though coherent and often directional) aggregations of primary entities.

While this classification of E/entities as primary, secondary and tertiary is somewhat arbitrary and incomplete, it should help to render the reader more discriminating in the recognition of various kinds of E/entities. Close consideration of this subject will reveal that there is often an equivalence between *primary group entities* and *secondary entities of the higher kind.* Also, *objective entities* are principally *tertiary entities,* but may also be *secondary entities of the lower kind.* [See Glossary for further amplification.]

Now that we have differentiated among E/entities, deeming only the primary ones truly *authentic,* let us examine the concept of *entity* from other perspectives. Every E/entity (except the ONE AND ONLY ENTITY/NON-ENTITY) has (as regards its prakritic/material/objective nature) boundaries and a vibratory distinctness by which it is distinguished from every other E/entity. An E/entity is Really unique, singular and particular *in manifestation* (for instance, no two atoms are *exactly* alike) though (both Essentially and ESSENTIALLY) *identical* with every other E/entity, i.e., identical in ESSENCE/Essence. By this is meant that:

- every E/entity in-Cosmos is not only Essentially the One (i.e., the Cosmic Logos as the Source of all Intra-Cosmic Emanations), but that
- every E/entity in-Cosmos is ESSENTIALLY the ONE—*identical* with the ONE AND ONLY BEING/NON-BEING—the SOURCE.

An *authentic entity* is a *life.* When that life is self-conscious, we can begin to talk about *entity* as *ego.* Given, however, the entire intra-Cosmic range of B/beings, we find that there are E/entities *with* the sense of ego, and many unconscious entities *without* the sense of ego yet developed. [The term *ego* is explained below and in the Glossary.]

All E/entities-in-Cosmos are different *in manifestation* from every other E/entity. E/entities have boundaries—at least in form and expression. An E/entity is *Self*-Contained and *self*-contained, for all E/entities are contained within the One Self (and are thus *Self*-Contained), and are also a "world unto themselves" because of the rotary motion which keeps them apparently distinct, and are, thus, *self*-contained. An E/entity is a *conditioned B/being,* a *bounded B/being.*

All B/beings-in-Cosmos are *conditioned* B/beings. They are *subject* to and influenced by something other than their apparent S/selves. The one exception is the INFINITE-SELF which can rightly be called the UNCONDITIONED, and this exception has been somewhat discussed. Always and ever we find the INFINITE-SELF as the ONE EXCEPTION to any assertion or predication applicable to E/entities-in-Cosmos. If we name the ALL-SELF an 'ENTITY', then IT is the only BOUNDLESS 'ENTITY', for since there is naught but ITSELF in all of the UTTER ALLNESS, what is there to bound IT or limit IT—except ITSELF?!

♦ The idea of *entity* is inseparable from the idea of the Number 'One'.

By means of 'BECOMING' *the* Great Cosmic Entity (the Universal Logos), the ZERO 'BECOMES' the One, the Cosmic Monad. Every E/entity is a kind of distinct 'Number One' (even though it is Essentially and ESSENTIALLY *identical* with every other E/entity). When an entity becomes self-conscious (at least in the early phases), the idea of 'Number One' is dramatically reinforced in the limited psyche, for such self-conscious entities become the seeming 'center' of their own tiny 'universe'. This is called "selfish centralization". It might be said that such a stage lies 'between' the state of 'e̲ntity' and that of 'E̲ntity'.

The One Cosmic Entity (the One Being *in* Cosmos known as the Universal Logos) is an externalization of an (*nearly*, infinitely) severe limitation of possibility which the INFINITE POTENTIAL (ALL-SELF) places upon ITSELF. (If the *limitation* were *infinitely* severe, there could be no externalization, for although the INFINITE SELF must 'SAY NO' to an infinitude of potentially 'EXTRUDABLE' possibility, IT must 'SAY YES' to *one!)* In this holographic Universe, every lesser E/entity within the One Cosmic Entity is a *reflection* of the One Cosmic Entity. Since the One Cosmic Entity is an expression of severely limited possibility, every lesser refection of Itself (i.e., every lesser E/entity in-Cosmos) is *also* an expression of that same severely limited possibility. The severely limited Divine Algorithm (determined by the ALL/SELF-as-Infinified Point-as-Condensed Point-as-Universal-Logos), which determines the unique *modus operandi* of a particular Cosmos, is found duplicated within every E/entity within that Cosmos.

Thus, every E/entity in-Cosmos (when 'compared' to the infinite possibility 'RESIDENT' 'within' the INFINITE POTENTIAL, the FOUNT OF ALL POSSIBILITY, the ALL-SELF) is drastically limited—one might almost be *tempted* to say—*infinitely* limited, for any singular possibility is infinitely dwarfed (a ratio of 'infinity to one') by an infinitude of possibilities such as are 'RESIDENT' 'within' the ALL-SELF.

This is another way of saying that the Number 'One' (a number so well suited to define the nature of the idea 'E/entity') is *infinitely* removed from the ZERO. By thinking in this way the Universe/Cosmos appears to be the tiniest of possible things, and almost infinitely insignificant compared to what is *infinitessentially* possible. The largest 'definite large' we can ever conceive, is, ever, but an *infinitesimal-izing*. In comparison with the ABSOLUTE, to discuss size becomes meaningless; all 'measurables' become negligible (i.e., *infinitesimalizings)*. The meaning of Maya is "to measure".

All this being said, still the proviso (mentioned above) *against* the Cosmos being the result of *absolutely infinite SELF-limitation* cannot be easily discounted, for if That which was to 'EMERGE' through 'EXTRUSION' were an *infinite limitation* of all possibility within the FOUNT OF ALL POSSIBILITY, then 'That' would not 'survive as a Singularity' capable of 'EMERGENCE'. Infinite limitation would be the eradication of *absolutely all possibility* of 'EMERGENCE'. Thus Cosmos might be thought of as 'EMERGING' from a 'INTRA-SOURCE' 'PROCESS' which 'ELIMINATES' (thus 'LIMITS') all infinitized possibilities but *one!*

♦ The idea, 'entity', is also inseparable from the idea of the "ring-pass-not."

Every E/entity is distinct, and it is its ring-pass-not, its boundary, which enforces this distinctness. Thinking again upon whether the BOUNDLESS IMMUTABLE PRINCIPLE qualifies as an 'ENTITY', we would ask, Does IT have a ring-pass-not? The answer may be (as usual), "Yes and No."

- "No" because IT is unbounded, and a ring-pass-not is a boundary. Further, because IT *IS* everything else (ESSENTIALLY), no boundary separates IT from anything else, just as no boundary can separate IT from ITSELF; and,
- "Yes" in another way, IT *has* a ring-pass-not because IT is *absolutely separate* and distinct from every other E/entity or thing. IT is out of relation with every other E/entity or thing in the Universal All.

So there is, as it were, an unbridgeable 'gulf' between IT and every other E/entity or thing in the All, and within the ALL, as well. That 'gulf' is a kind of ring-pass-not which *no entity or thing can cross.*

The ancient Aryan formula "Neti, neti"—"Not this, not that" tells us that no *thing* can be IT. No *thing,* per se (in its own limited 'entity-hood'), can cross the 'boundriless boundary', the ring-pass-not of INFINITY, *but,* any *thing-in-essence* has *already* crossed the invisible boundary, and caused the previously impassable ring-pass-not (impassable to that which is *finite)* to be as if it never were.

♦ Paradox abounds as we see that the ALL-SELF is at once the most *unreachable* and yet the most *reachable*—unreachable to the 'finite-as-finite' and instantaneously reachable (nay, *'be-able')* to the 'finite-as-INFINITE'.

Before leaving the exploration of the idea of 'entity', it would be useful to examine the relationship between the terms 'entity' and 'identity':

- An authentic E/entity is a Logoically Pre-Ordained Life Unit (a 'lesser' Cosmic Monad within the One and Only Cosmic Monad) and is directly related to the core of LIFE, to ABSOLUTE BEING. Authentic E/entities are actually *identities* and are the main 'players' within the Universal System.
- Identities need not be (at this particular point in the Universal Manvantara) self-conscious. An atom is an identity in which the Universal Consciousness is not yet self-aware. An identity is a distinct (i.e., *apparently* distinct) and *authentic* part of the indivisible Whole.

Connotatively, the term 'entity' is more focussed upon the pattern aspect of any manifestation, whereas the term 'identity' is more focused upon the *life* which expresses through the pattern. The term 'identity' emphasizes subjective oneness with the Universal Logos and even with the INFINITE SELF.

On Emanations-in-Combination

There are several ways to look at how Numbers combine to produce other Numbers. The following are explanations of how numbers combine according to:

The Self-Attentuation Model of Emanation

- Number One emanates a *likeness* to Itself.
- Number Two is formed by the Number One (the Monad) plus *Itself* (which is a One—again, the Monad: "monad" is but another name for "One").
- Twoness is a Relationship of The Source (the One), *plus* Its Emanation, which produces a Cosmic Relationship *inhering in the emanated likeness.*
- Twoness is 'Oneness gone forth from Itself (in attenuation), and then combining with Itself' (through mutual 'perceptual interplay' within the Field of the emanated likeness).
- Twoness is the Relationship between the Source and that emanative replica of Source, which has gone forth (or been 'sent forth') from Source.
- The Entity/Relationship that is the Number Two (and which 'contains' the Number One, for the Subject which is the Number One inheres within that Object which is the emanated likeness of Itself) sends forth an Emanation of Itself (which Emanation contains *both* the Number One *and* the Relationship between the Number One and the attenuated replica of Itself *within* the *field* of the attenuated replica) that functions as the Number Two.
- The subsequent Emanation from the Number Two is an attenuation of Twoness. The Entity/Relationship that is the Number Two then combines with Its own attenuated Emanation to produce the Relationship called Threeness—i.e., the Number Three, which Relationship occurs within the field of the attenuated replica of the Entity/Relationship called Number Two. And so it goes.
- All Numbers are therefore formed by attenuations of the One and successive combinations of those attenuations with the Original One.
- All Numbers are combinations of the Original Number One: 2 = 1 + attenuated 1, and 3 = 1 + attenuated 1, + the attenuation of (1 + attenuated 1).

Thus all Numbers are, Essentially, the Universal Monad, Number One, in interplay with Itself through Self-Attenuation. This sequence of self-attenuation could be represented in a Cosmic Emanatory Formula, if we apply the following definitions to the following symbols:

a	=	attenuation (and is equal to some decimal value less than One, due to Emanatory Loss)	
»	=	leads to	
<>	=	interplays with	
1	=	Oneness, or 1	
2	=	Twoness, or '1<>a1'	NOTE: 'a1' is the emanation of 1
3	=	Threeness, or [(1<>a1)<> a(1<>a1)]	NOTE: 'a(1<>a1)' is the emanation of (1<>a1).
4	=	Fourness = {[(1<>a1)<> a(1<>a1)]<>a[(1<>a1)<> a(1<>a1)]}	

The Cosmic Emanatory Formula Sequence of Whole Numbers

1... 2 ... 3 ... 4 ...

1 » (1<>a1) » [(1<>a1)<>a(1<>a1)] » (1<>a1)<>a(1<>a1)<>a[1<>a1)<>a(1<>a1)]

The term 'a' then can be known as the *function of attenuated emanation* and could also be called the *coefficient of emanatory reduction.* The sign '<>' signals an *interplay* between a Number and the Emanation of that Number.

An *emanation* of a number is always, Essentially, a One, the Monad. But the *interplay* between the number value that *is the emanator* and that which it *emanates* produces the *next* number value in the sequence. That interplay occurs because the Number One has 'Seen' (as an Object) Its coming Emanation within Itself, and then, (fully endowed with Its own Subjective Oneness) 'enters' that Object in the form of a Subject, and then Interplays with Its own Emanation of Itself.

The dynamics are simple:

1. the Number One *'Sees'* Its Son, the attenuated Reflection of Itself, as Number One;
2. then the Number One 'goes forth' into the Son which It has 'Seen' as Object, and sets up a *field of interplay* with the Son.

Though the Son is the Son of the Father, the Father is *present* with the Son, even though the Father 'Remains' the Numbers One (also uncombined with His Son). Not only is the Number One *in* all other Numbers, but all other Numbers are simply Emanations of the Number One, and combinations of the Number One with Itself (albeit *in an attenuated state*).

It is necessary to keep in mind that there is only One, because if it is true that:

- 1<>a1 » 2 (Twoness); then
- when '1' emanates 'a1', the 'a1' is simply '1' in an attenuated/objectified form; and therefore,
- although 1<>a1 may equal Twoness, 1<>a1 is also, still, profoundly Oneness.

Numbers are simply what might be called the Interactive 'Modes of Oneness', as shown in, 2 = 1, 3 = 1, 4 = 1, etc. ... any number = 1 ... if the One 'simply' Self-attenuates. Every Number in Cosmos is the cause of the emanation of the *germ* of the next Number in sequence. So that the:

- One (Oneness) emanates Its attenuated Self, with which It interplays, producing the Two (or Twoness);
- Two (Twoness) emanates Its attenuated Self, with which It interplays, producing the Three (or Threeness), and so forth.

Clearly the Two is not just a *homogeneous* Number Two; rather, it is 'Twoness', a Compound, a Relationship between the One and Its attenuated Self. And the Three is not just a homogeneous Three, but is 'Threeness', a Relationship comprising the One, *and* Its attenuated Emanation in interplay with the next succeeding projected attenuated Emanation (which comes forth from Twoness).

The Number One *is* the Cosmic Monad, and *in* every Number from Two onwards, the Cosmic Monad is *fully present* as well as all the succeeding Relationships arising from the interplay of the Cosmic Monad with Its projected attenuations, i.e., Relationships which It (i.e., the Cosmic Monad) has formed with attenuations of Itself.

Thus, *all previous numbers participate* and *inhere (in an attenuated state) within each new Number* because all previous numbers are *contained* within the Number from which the *germ* of the new Number directly emanated. Notice that which 'comes forth' from Twoness *(the germ* of Threeness) is not *yet* Threeness, but only the attenuated emanation of Twoness which, *in Psycho-infusional interplay with Twoness becomes Threeness.*

The Emanator must *go forth interactively* into Its Emanation. Borrowing an analogy which is not quite exact but illustrative, that of the Master and disciple. His disciple has a consciousness quite his own, different from his Master's consciousness; however, when the Master 'over-lights' the disciple and penetrates his consciousness with His presence, the consciousness of the disciple becomes something more and other than it was before the combination of influences occurred.

A similar dynamic is occurring all the time when 'Ray' Influences are transmitted *through* a Planetary Logos. The 'Ray' plus the native quality of the Planet yields a new combination. The 'Ray' is still the 'Ray', of a quality unique unto Itself, but It has also 'gone forth' into the Planetary Field and merged and blended with what It found there, producing an entirely *new* effect.

Therefore, a key consideration is when the 1 is multiplied by the function 'a' (expressed as 'a1'), the result is not so much a fraction or fragment of 1, as a *qualitative reduction* or *de-intensification* of the 1. The Number One continues to Be Essentially Itself through all successive qualitative reductions and all successive de-intensifications. The Essence stays the same; all successive Self-Reflections (Emanations) can be recognized as faithful to the Original, but always less and less intense. So attenuation does not result in a change of Essence. Oneness remains Oneness through all Its successive Reflections/Emanations, for the Cosmic Monad is a Unity uniting all Its apparent 'parts'. Even the highest Number in Cosmos is *none other* that the One—Essentially. Any single *thing,* no matter how many parts, is a *oneness.*

After the Number One has sent forth Its Emanation, It need not (by Itself) do so again and again to create 'other' Numbers. This is because when any of the Relationships that the Number One has engaged in with Its attenuated Self 'sends forth' an Emanation (as when two *combined rays* send forth yet another ray) the Number One is already contained within that Emanation. Every Emanation sent forth (from a Number/Relationship—for instance, from Twoness, or Threeness, or Fourness) is Really the Number One in progressive attenuation.

Thus every time any Emanation-in-Combination sends forth another Emanation, the Number One not only is *fully present* and *participating* in that Emanation, but, Essentially, *Is* that very Emanation Itself, because, as we remember *Numbers,* per se, are not emanated. Only the One (in attenuation) is emanated over and over again, and each successive Emanation is the *germ* of the Number/Relationship which follows as the Emanator *enters in interplay* that which It has emanated.

- The Number One does *not* directly emanate the Number Two; instead, the Number One emanates an attenuation of Itself called the *germ* of Twoness.

When that attenuation (that *germ*) interplays (within Its own field) with Its Source (the Number One), then and then only, is the Number Two (Twoness) formed. Following on, Twoness emanates *not* the number Three immediately; rather, it is the *germ* of Threeness that is emanated, which *germ* in interplay with the One *and* the *germ* of Twoness (which together, in interplay, *are* Twoness) produces Threeness.

- Always the Number One is the Parent of Itself—its *germ*-children (Sons) being the attenuated forms of Itself (i.e. Oneness-in-attenuation)—and these various attenuations in interplay with their Emanator produce those successive relationship which are the Numbers.

What all this means is that there are no Real Numbers other than the Number One. All other Numbers are really Relationships of the One with Its attenuated Self. Thus, every Number is Itself, *and* the Number One. Each Number (other than the Number One), then, is a Relationship formed of various germ-Emanations of the Number One, which germ-Emanations (in order to form Numbers) are interactively combined according to certain Laws inherent in the Design-at-the-Beginning (and it may well be a Geometrical Design as well as and Arithmetic Design, for "God Geometrizes").

Each Number, then, (other than the One) is *not just an Emanation,* but an <u>Emanation-in-Combination</u>. In combination with what? In combination with other attenuated emanations of the Number One Emanations, for every *germ*-Source *combines* with the Numerical Relationship which It sends forth to create a new Cosmic Relationship which is a new Number.

If, for instance, *from* the Relationship that is Twoness (a Relationship which is *Essentially* Oneness) a subsequent Emanation went forth, and the Twoness (born, remember, of Oneness *and* Its first germ-Emanation) did *<u>not</u> combine* with the germ-Emanation that went forth from Itself (i.e., from Twoness), *there would be no Threeness.* There would only be an attenuated replica of the Relationship called Twoness. Threeness requires for its appearance the interactivity between the One, *and* the germ of Twoness (which together yield Twoness) *and* the germ of Threeness. Further, without this kind of combining, the Cosmic Sutratma would be 'cut' and the Cosmic Antahkarana could never be built.

Therefore, we see that no Emanation by Itself is, Essentially, anything other than the Number One. Only through the interactivity/relationship of Emanations does Number (which *is,* fundamentally, Relationship) arise. The problems which arise in thinking of the Metaphysics of Emanation are resolved by realizing that the *germ* of each Number to Be, is *also* the *field of relationship* wherein the Emanating Entity *interplays* with the *germ.* So each Emanated Numerical Entity, is a Field of Interplay, a Field wherein is found combined the methods whereby Oneness and its many progressive modifications interplay with themselves.

Remember the two steps:

1. to 'See' Oneself as Object;
2. to then enter in the Object One has 'Seen'.

Thus is Emanation *complete,* and never with the 'Seeing' alone. This brings us to a most important point: the Numerical Archetypes (Numbers-as-Cosmic Archetypes) are *com-*

posite. No *single* Emanation *is homogeneously* the Number Two or the Number Three, or Four, etc. But rather,

1. Twoness has Oneness in It;
2. Threeness has Twoness *and* Oneness in It; and,
3. Fourness has Threeness, Twoness *and* Oneness in It.

All Numerical Archetypes are derivative from the Number/Relationship preceding It in the Divine Emanatory Sequence, and, ultimately, from the Number One. But the Number One is derivative from the ZERO, and the gulf between the ZERO and the One is *infinite* and rationally unbridgeable.

So we face the gulf between the INFINITE and the Finite! How does ZERO '<>' or '»' ('interplay with' or 'lead to') 1?

- ♦ Since all Numbers are Essentially and *metaphysically* equal to the Number One, and resolve into the Unitary Cosmic Monad, paradoxically, the Number One (the Cosmic Monad) is equal to ZERO, i.e., 1 = 0.

Perhaps, the Pre-Cosmic Concentrated *Point* (the Condensed Point, from which the Universal Logos and Cosmos are born) may, from one perspective, be considered a *No-Number* midway between the ZERO and the Number One. This *Point* may (since 'Pointness' is, in its very nature, ambiguous, and could therefore) be considered the ZERO 'on its way' to becoming Number—a sort of 'aperture' through which the INFINITE becomes the Finite.

It may be that the Point Itself is the Ensouling Presence of all Entities-in-Combination. The Point, from one perspective, might be considered the Unmanifest Universal Logos, the 'almost' 'O-n-e', Who/Which is greater and more 'originant' than even the Numbers One, and expresses through all numbers altogether. From a less abstract perspective, the Point-become-Universal Logos, is simply the Number One.

On Identity

An *identity* can be considered a distinct, unique and separate being. From this facile definition, it should be clear that there is REALLY no such thing as an 'identity' other than the ONE ABSOLUTE IDENTITY, WHO IS, REALLY, the ONE ABSOLUTE IDENTITY/NON-IDENTITY. Nevertheless, for practical purposes (or, in order to function properly according to Universal Intention within Cosmos) we must think as if, and "act as if", there were distinct, unique and separate beings called 'identities'.

Perhaps we should think of two kinds of 'identity':

- Identity-in-Cosmos
- IDENTITY-*'IN'*-INFINITUDE

1. Identity-in-Cosmos is uniqueness, the apparent impossibility of identicalness with any other E/entity. Such uniqueness makes an E/entity distinctly 'identifiable', capable of being discriminated from all others.

2. IDENTITY-'IN'-INFINITUDE is absolute *sameness—identicalness.* There is naught but ONE IDENTITY 'IN' INFINITUDE. In fact, even in-Cosmos, there is naught but ONE IDENTITY, though the SELF-Imposed Great Illusion makes it *seem* otherwise.

   A number of ideas are associated with Identity-in-Cosmos:
   - partiality
   - particulateness (or, at least, 'partite-ness')
   - boundary
   - ego
   - distinctness
   - separateness
   - uniqueness, etc.

One way of looking at such limited identity is to designate it as 'PURE BEING *plus* the perception of a *prakritic boundary'* (an apparently impenetrable 'wall' of 'matter'). We realize that living as a true identity while within Cosmos is ultimately impossible (however close may be the approximation achieved by certain Beings upon the highest dimensions of Cosmos), for any true identity *is*, ESSENTIALLY, but *the* TRUE IDENTITY, which can only be 'LIVED' 'within' the ALL-IN-ALLNESS.

The method of determining Identity-in-Cosmos is through *identification with object.* Those prakritic registrations are frequently experienced are often mistaken for Self. (An example: I see my physical body all the time, so I think that I *am* my physical body.) This perceptual/conceptual dynamic extends to subtler and subtler fields of prakriti, but the mode of self-identification is ever the same. Clearly, there is almost no end to this mistaken means of self-determination. Finally (and much later in the Universal Process) the Entity seeking Itself might conclude with the assertion "I am the Universe." This assertion, too, though hugely in advance of such assertions as, "I am the soul within the Causal Body", or "I am the Monad within the Monad Sheath", is also fallacious.

◆ Perhaps the mantram "I Am That I Am" comes closest to the TRUTH, as it avoids prakritic identification.

From another perspective, Identity-in-Cosmos is usually based upon *having* rather than *being.* Such *mistaken* Identity is based ultimately upon the *possibility of the object* (i.e., of the existence of the objective state). The objective state, the prakritic state is a great *reducer.* There is such a thing as the prakritic reduction of Universal Potency, the prakritic reduction of Universal Consciousness, and the prakritic reduction of Universal Intelligence. Identity-in-Cosmos is based upon *prakritic reduction* which is an overpowering registration-in-consciousness of a ring-pass-not which is a prakritic vibration/ condition. This registration temporarily overpowers the realization of <u>U</u>ltimate <u>I</u>dentity and of ULTIMATE IDENTITY.

Of course, every unit of LIFE immersed in-Cosmos (which Really, and surprisingly, means, 'immersed in *consciousness')* appears to have boundaries—limited power, limited consciousness and limited self-extension, but these boundaries are only temporary ring-pass-nots; they are *not essentially* determinants of selfhood/Selfhood/SELFHOOD.

Even the One Great Identity in-Cosmos (the Identity of the Universal Logos, which we all, <u>E</u>ssentially, *are)* arises from perceiving boundaries as Self (through the perceptual mechanism of *limited Self-'Sight').* A boundary within boundless Mulaprakriti has been created by a SELF-Limiting *Act of Self-Perception* by the Infinite Subject-'Becoming'-Universal Subject/Logos. The ALL-SELF (which can *never* REALLY cease *being* completely and totally ITSELF) seems to *add* something to ITS ongoing 'STATELESS STATE', and that something is a SELF-Limiting Consciousness or Perception. Not only does IT (in addition to what IT always 'DOES'/IS) 'SEE' ITSELF as Object (as IT 'SEES' Mulaprakriti), but IT begins to 'See' (by means of the mediation of the Infinified Point, Condensing and Condensed Point) a *limited, finite* Object instead of an infinite Object.

Thus is delineated the Ring-Pass-Not of the Cosmos-to-Be. 'Drawing' the Ring-Pass-Not is an 'Act of Consciousness'—a kind of 'Decision *not to see* the Allness of Mulaprakriti' (i.e., Mulaprakriti in Its Objective Infinitude, which is Really the *reflected* Allness of the Infinite Subject).

- Remember always that compared to *being*, *seeing* represents a *limitation*. In the language of Synthesis, Being is hierarchically superior to Consciousness (i.e., to 'Seeing', which depends upon twoness, whereas Being demands oneness). In the language of SYNTHESIS, BEING demands no less (and certainly no more) than 'ZERONESS'.

Concerning the dynamics of identity, greater identity is absorptive of lesser identity. This Law allows, on a relatively low level (low from the Universal Perspective), the Master to overshadow or infuse the consciousness of His disciple. Lesser identity is absorbed into greater identity. Every identity grows by entering, first, into the consciousness and, then, into the identity of a being greater than itself.

Identity-in-Cosmos is conditioned and regulated by enumeration. Another way of saying this is: *identity is numerical.* Identity-in-Cosmos is based upon relations between pure, primary, authentic Entities which could also be called 'Numeric Entities'. Numbers, Really are the purest Entities. The purest of such Entities is the Number One—the Supreme Cosmic Entity, *the* Ultimate Cosmic Monad. Identity-in-Cosmos is *conditional* identity (i.e., subject to conditions). Pure Numbers (integers, particularly, as a particular set within the set of all real numbers, and excluding what have been called *irrational* numbers) should be considered *conditioned* E/entities. Only ZERO is an (the) UNCONDITIONED 'ENTITY'.

The *relations* between integers are called simple *ratios.* Infinite are the potential relations between the integers in the infinite set of integers, but in a finite Universe, no infinite set of integers can *actually* exist (i.e., *actually manifest).* So in any Cosmos (since it is finite) we are, necessarily, dealing with a *finite* set of integers, and a finite set of ratios between them, which finite set of ratios, reveal the nature of the possible intra-Cosmic interrelationships between the set of actualizable integers in Cosmos. For instance the relationship between the two Entities numerically represented by the Number 'One' and the Number 'Two' are described by the two ratios 1/2 or 2/1. All ratios composed of integers are (in principle) *rational* numbers (note the word 'ratio' included in 'rational'), but some ratios may yield (when converted to decimals) irresolvable quotients (such as would 1/3 or 2/3, etc.), similar, in a way, to those real numbers (such as the square root of 2) which cannot be expressed as a ratio between integers. In order to understand the Great Science of Relationships, the value and meaning of all potential ratios between integers must be understood.

The numerous possible relationships between different Entities (Numbers) can also be understood geometrically through the study of different two or three dimensional geometrical figures, polygons, for instance, (regular, but especially irregular), and in such cases the various lengths of each side of the polygon would be most significant.

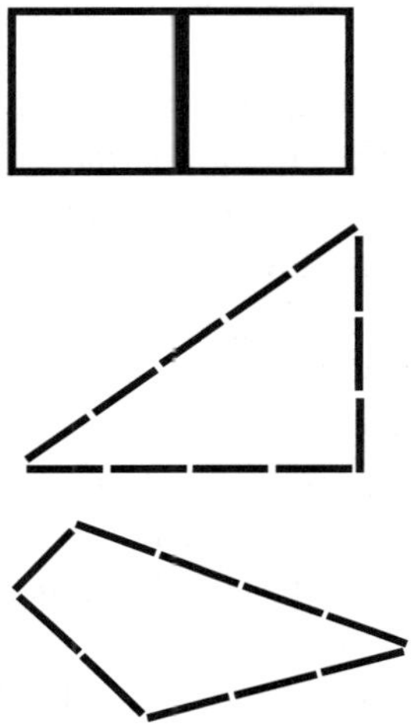

1 : 2

3 : 4 : 5

1 : 2 : 3 : 4

Further, the relationship between E/entities can be studied musically and coloristically, as each E/entity is not only a number, but a musical note (complex or simple), or color (complex or simple), as well. Naturally enough, compound or complex E/entities will be formed by combinations of numbers, colors and notes (creating chords). The many complex relations possible between the various existent integers, or Existent Numbers (determined by various mathematical operations), determine the many passing conditions to which all E/entities-in-Cosmos are subject.

Many of the *conditions* to which E/entities are subjected arise from interaction with various aggregations of other E/entities. It might be said that environment is created by the aggregation of various kinds of E/entities interacting in various ways. No matter what conditions present themselves, there will be a mathematical *key* to understanding them.

- So we have the idea that conditions are *numerically determined*. Earlier we learned that E/entities are Really Numbers or numerical. It is not too far a leap to assert that *E/entities are conditions,* albeit a very special category of condition.

In conclusion, we must remember that every *thing* in-Cosmos is (when phenomenally considered) Really a *motion* or *change,* and that all *motion/change* can be quantified—measured by number. The key to the very first Law of the Solar System (the Law of Vibration) is a mathematical key. The study of Law of Vibration (cosmically considered) will unlock the door to understanding all motion/change, whether simple or complex.

On Condition and the Conditional

Conditions are 'SOURCE-*extruded*' crystallized possibilities forever 'RESIDENT' 'within' the INFINITE POTENTIAL. 'Within' the INFINITUDE (or INFINITE POTENTIAL, or FOUNT OF ALL POSSIBILITY) are 'CONTAINED' (in their *noumenessentialized* 'STATE') the *absolutized* or *infinitized* Noumena of any and every possibility that now exists, that ever has been or that ever could be.

The entire Universe is but a Condition, i.e., a precipitation from the INFINITE SOURCE, and within each Universe inhere a *virtual but not actual* infinitude of conditions (precipitated possibilities). If a Universe were infinite in duration, perhaps a *literal/actual* infinity of conditions could arise. Universes, however, are 'timed' Events, and in the time 'allowed' by the SELF-as-Point-as-Universal Logos for the duration of a Cosmos, *all* possibilities (i.e., an *infinitude* of possibilities) cannot possibly be unfolded.

Simply understood, conditions are limitations. They represent *one particular relationship* or configuration and not another. There are certainly no conditions without relationship. The nature of a given condition can be described by:

- the rate of vibration of the number and kinds of E/entities related, so as to create the condition; and/or
- the relationship of such E/entities in Space and Time relative to each other; as well as
- the relationship of the E/entities involved in the condition to other E/entities which are not the main focus of attention.

As a practical example, conditions in a given environment, for instance, involve the aggregation of a certain number of entities belonging to various classifications within the mineral, vegetable, animal and human kingdom (and perhaps other more subtle kingdoms—elemental, devic, etc.). At what rate are these entities vibrating, either individually or in aggregation?; How are they 'arranged' or disposed relative to each other?; What are the spatial and temporal dynamics of their interactions?; To what other E/entities not immediately involved in the intra-environmental interplay are the E/entities in question related? Answers to such questions will determine the 'conditions' within the environment.

From another perspective, conditions are the quantitative and qualitative states of a certain number of related variables. To determine conditions, one would ask: What is the nature of the 'motion' of these variables—of each variable in relation to itself and of each variable in relation to every other variable?

When something is described as *conditional,* this means that:

- its condition is dependent upon predictable or unpredictable changes in the variables constituting its *own* set of *intra*-relationships; *or* that
- its condition is dependent upon predictable or unpredictable changes in *other* variables which impinge upon its own set of variables due to its *outer* relationships—i.e., *inter*-relationships.

The possibility of *change* within a set of variables is the principal factor determining *conditionality* or *conditional nature* of that set of variables. Even from the metaphysical perspective, *everything* in-Cosmos is *conditional* and *conditioned,* including the entire Cosmos Itself. Within the INFINITE, however, there are no variables, no possibility of change, and, therefore, nothing 'within' that 'STATELESS STATE' is conditional. Of course, 'within' that 'STATELESS STATE' no *things* (as separate, identifiable items) can possibly exist, and where there are no *things* there can be no *conditions.*

Every E/entity-in-Cosmos *is* a kind of *condition* and is also subject to conditions. (We are not used to thinking of living E/entities as *conditions,* but there is only ONE ENTITY/NON-ENTITY that is not a condition, and, hence, not conditional.) ESSENTIALLY, of course, every E/entity *is* the ONE AND ONLY ENTITY/NON-ENTITY, but from a more familiar, *intra*-Cosmic perspective, every E/entity is a kind of *variation* of that ENTITY/NON-ENTITY—a variation not due to any difference in ESSENCE between E/entities, but due to the degree of what might be called each E/entity's 'prakritic immersion'.

◆ 'E/entity-hood', in ESSENCE, does not vary; it cannot; Prakriti, however, (displaying each E/entity's phenomenal aspect) *does* vary (i.e., the matter of the vehicles through which an E/entity expresses does vary).

Again, only the ONE ENTITY/NON-ENTITY is *not* subject to prakriti (because that SELF in its Objectified Aspect *is* prakriti itself, (or more fundamentally, is Mulaprakriti, Itself). If there were no prakritic variations (resulting in varying depths of E/entities' prakritic immersion), there would appear to be no separate E/entities. Practically speaking each E/entity is known and distinguished by the prakritic variations/immersions with which it is associated.

- A 'high' Being has a low degree of prakritic immersion.
- A 'low' being has a high degree of prakritic immersion.

In their ESSENTIAL 'E/entity-hood' no being is 'low' or 'high'. If all prakritic variations were the same, all E/entities would be seen/cognized as the same. If, alternatively, prakritic variations did not exist at all, all E/entities would be intuited/fathomed as the SAME ENTITY.

So, for practical purposes, E/entities-in-Cosmos can be considered as variations or waves or disturbances (i.e., 'FOHATIC SELF-Reflections') in the pre-Cosmic motionless homogeneity of Mulaprakriti (Pre-Cosmic Root Matter). From this perspective, all apparently distinct E/entities are, in their phenomenal aspect at least, *conditions.* The complex secondary and tertiary, etc., conditions to which E/entities are later subjected during the 'heat' of the Cosmic Process, can be understood as complex relationships following upon the fundamental Conditions inaugurated by the Originant Pre-Cosmic and early-Cosmic Entities.

Authentic *E/entities* (in their phenomenal aspect) can, therefore, be thought of not only as primary *E/entities,* but, as primary *conditions.* Other, lesser kinds of conditions in-Cosmos—some very complex—are simply caused by the interaction of primary *conditions,* i.e., primary *entities.* Thus, primary E/entities are (phenomenally considered) *primary conditions,* and secondary and tertiary entities (phenomenally considered) can be thought of as *derivative conditions.* Only the ESSENCE of *entity* is unconditional. The *phenomenality* of *entity* is always conditional.

On 'I'

Perhaps the word **'I'** (bolded) is the most important, *truth*-bearing word we use, and the word 'I' (unbolded) or 'i' (lower case) is the least. To understand the difference between these two is to understand the entire "Great Work" of the SELF and ITS *'relationless relation'* to the Cosmos which IT 'BECOMES'. Once a Cosmic Monad (or Universal Life Unit, i.e., 'Ray' of the One Universal 'Ray') has entered the denser points of immersion in Cosmos, it takes a multitude of life cycles (in many forms, certainly not all human) to come to a recognition and understanding of the **I**, and many more cycles to learn to use IT (or, rather, *be* IT) meaningfully.

The word 'I' will, of necessity, have to be used in various ways, but the use of the **I** will be reserved to stand for the ABSOLUTE SELF. The symbol '**I**-as-I' can stand for the descent of ABSOLUTE IDENTITY into the 'ahamkaric levels' of Cosmos and the form 'I-as-**I**' for the re-ascent of any E/entity through many levels 'into' the SOURCE, the ABSOLUTE IDENTITY. This descent and re-ascent passes through the important stage of '**8**' [discussed fully in the Glossary, and below]. Note the relation of the '**8**' to the symbol usually used to denote *infinity.*

'I' as usually understood is equivalent to ego, and signals a state of mistaken identity. 'I', as customarily used, means that the center LIFE within an entity (for an Entity {technically considered} is too conscious to use the 'I') is projected 'outwards', objectified and identified with the prakriti/matter which (because of veiled consciousness) appears to *surround* or immediately *environ* the entity, and with which the entity is most *familiar.*

In a way, focusing overly much upon the familiar (that which is "near at hand") both breeds and sustains the state of ignorance we call *ego.* An example might be that I relate more to the sensations of *my* body that I do to *yours.* This might not be so in the case of the Christ Who said, "In as much as you have done it unto the least of these, you have done it unto me." 'I', as usually used, connotes a falsely conceived center of identity, or a shallow center, or a loss of *true* center.

For a long time near the 'bottom' of the involutionary/evolutionary arc, no 'I' at all is experienced. The SELF-in-Cosmos has veiled ITSELF from ITSELF (all the while, paradoxically, on ITS OWN SUPREME 'LEVEL', 'ABIDING' completely *unveiled).* At the bottom of the arc, the appearance of ignorance is 'thick' indeed. It may be that as the involutionary stage in any Cosmos commences, ignorance is *not* thick, just as it is not thick towards the higher part of the Evolutionary Arc.

An interesting problem arises as we trace the Universal Life Wave from Its origin within the Universal Logos:

- Presumably all Emanations from that Greatest of all Beings *in* Cosmos, would *initially* be supremely Self-conscious, and would only later lose Self-consciousness as prakritic immersion densified.

A more popular (and extremely short-sighted) point of view has each spirit beginning as a tiny blinded life and working its way *up* into full Universal Consciousness, but to think in this manner almost certainly overlooks the dynamics of involutionary descent. Actually, the subject of the descent and re-ascent of a Universal Life Unit (one of a vast number of Logoically-Enfolded 'Rays' of the ABSOLUTE) must be extremely complex as we (in our human consciousness) have no idea even of the nature of the Entities to be found beyond the ring-pass-not of our local "One About Whom Naught May Be Said"—a Great, though *local,* Logos.

Continuing with a discussion of ignorance on a level which we can better understand, we might say that the profound universal ignorance of the later Involutionary Arc and early Evolutionary Arc gives way after individualization to the ignorance of egoism—a necessary stage amounting to the existence for a long cycle of a *false* 'I'.

It is this false 'I' with which the majority of humanity is now preoccupied. Under the influence of egoism 'being' and 'having' become confused. We *are,* ESSENTIALLY, the SELF, but we *have* various fields of prakriti associated with us functioning as our instruments of expression. The false 'I' identifies its REAL though consciously inaccessible '**I-ness**' with these fields.

Now, throughout the extraordinarily 'long' (by human standards!) 'Pilgrimage' through all dimensions of Cosmos, every E/entity (no matter at what depth of prakritic immersion) will necessarily have vehicles of prakriti (whether particulate in the Worlds of Fabrication, or imparticulate in the World of Being). These vehicles may be of such tenuosity that at our present level of human ignorance, we could not convince ourselves that they are composed of matter at all. But in-Cosmos, the association of E/entity with prakriti/matter is inescapable. The only utterly dematerialized state is found within the ALL-SELF, and is a 'STATE' of INFINITE SUBJECTIVITY.

Past the human stage however, Entities 'have' but do not *identify with* what They have. Our Planetary Logos *has,* compared to our tiny auric fields, immense auric fields through which to express Its I-ness (or, better, Its realized '**8-ness**'), but that great Being (really a 'small' being when considered cosmically) certainly does not identify Its **8**-ness with any of Its vehicular fields.

- Humanity stands evolutionarily at the point in consciousness at which it is very easy to make the *great ontological mistake*—mistaking 'having' for 'being'

The symbol 'I-as-**I**' means, evolutionarily, something much more advanced than 'I'. The stage of mistaken identification with vehicles is past. Technically, the *human* stage is past. The surrounding fields are recognized, understood and valued, but a great inward realization has occurred. The ROOT SELF (the ALL-SELF) has become at least an *intimated* PRESENCE, and it is understood that all E/entities are REALLY undetached aspects or 'Rays' of that GREAT ENTITY/NON-ENTITY, and, as well, that all 'I's are REALLY *undetached aspects* or 'Rays' of the ONE AND ONLY **I**. Even the relatively great Entities cannot leave the prakritic fields until "the Great Day Be With US", but They can begin to *identify* with THAT which They ESSENTIALLY are—the **I**.

The range of Beings Who know Themselves under the symbol 'I-as-**I**', is immense—running all the way from a human initiate of the fourth degree to the Universal Logos.

- Fundamentally this symbol (I-as-**I**) represents the overcoming of egoism, the overcoming of the *great ontological mistake.*

There are almost certainly a multitude of degrees of this overcoming, for it is said that only an initiate of the ninth degree has truly overcome illusion. The illusion here indicated is relative to humanity. Certainly, beyond even the maximally developed stage of unfoldment available to the Fourth Creative Hierarchy (the Human Monads *as* strictly Human Monads), there are other Illusions to overcome, for the Universe itself is an Illusion. At least, once 'I-as-**I**' characterizes the consciousness of an Entity, the strangle-hold of grossest illusion has been overcome.

- Of '**I**' in ITS *absolute purity*, nothing REALLY TRUE can be said, just as it is impossible to predicate anything REALLY TRUE of the ALL-SELF. '**I**' *is* the ALL-SELF. The important thing to realize about '**I**' is that every 'I' is '**I**'.

When we wish to think about the descent of '**I**' into Cosmos and 'I-ness', we can use the symbol '**I**-as-I' which is the reverse of 'I-as-**I**' (indicating ascent). How and why the '**I**' becomes 'I' (and yet remains ever and always only '**I**') is, of course, the Great Mystery. Mysterious though it is, it is still a fundamental thought, and one which no true occultist, intent upon achieving identification with Synthesis, can afford to forget.

Let us now examine '**I**/I' from a few more perspectives. '**I**-as-I' means *essentially* a *bounded* point of ABSOLUTE POTENCY, a center of BEING-as-Being. the fundamental capacities of which are Will, Love and Intelligence. The center is *bounded* because of a necessary prakritic immersion—*necessary* if there is to be a Cosmos at all.

- So '**I**-as-I' symbolizes a prakritically-bounded point of ABSOLUTE BEING.

This is important to remember. The SOURCE of ABSOLUTE POTENCY is *resident* within every (seeming) 'I', who is Really (while in-Cosmos) '**I**-as-I' (descending) or 'I-as-**I**' (ascending) and is *ultimately* (in the ABSOLUTE PURITY of the ALL-IN-ALLNESS, '**I**'. '**I**-as-I', being prakritically-bounded, am also *numerically* bounded. This means that, in-Cosmos, 'I-ness' is subjected to number, to variation, and to modification (a term so often used in Raja Yoga).

When I know that I *Am* Really 'I-as-**I**', I realize the importance of the figure eight—'**8**' (bolded to represent the descent of **I**). In this treatise, the '**8**' is taken to signify '**I**-as-I', (or 'I-as-**I**', as practically speaking, there is no difference), the SELF-*in*-Cosmos, the SELF in manifestation.

The World of Effects (i.e., the World of Fabrication) and the World of Being are *both* part of the World of Becoming, and neither World is part of the WORLD OF BEING. So '**8**', representing (in-Cosmos) the WORLD OF BEING, immerses Itself in the World of Becoming in the two aspects of that World—i.e., the World of Being (the Archetypal Planes) and the *lower* part of the World of Becoming (i.e., the World of Effects or the Planes of Effects).

The figure eight (**8**) is the Representation of the ABSOLUTE SELF in manifestation. The *horizontal* figure eight is the ABSOLUTE SELF in the state of ALL-IN-ALLNESS and in this treatise (though unused) is equivalent to the '**I**'.

It is impossible that the ALL-SELF (TOTAL **I**/ALL-SELF) shall not be 'in' both the All and the ALL [see Glossary]. It is equally impossible that the ALL-SELF (TOTAL **I**/ ALL-SELF) shall not *entirely be* both the All and the ALL, for Brahman *is* Samsara ("Brahman and Samsara are One"). The Cosmos, however, is the Great Illusion and under that Illusion, it appears that the ALL-SELF is *not* 'in' the All and the ALL, and, further, that the ALL-SELF is *not* the All and the ALL. This Illusion is overcome in consciousness through discovering (by means of *identification)* the meaning of the '**8**'.

NOTE: Henceforward in this treatise the two expressions '**I**-as-I' and 'I-as-**I**' will be used interchangeably with '**8**'.

So, then, '**8**' Am an objectively bounded *point* of INFINITE SUBJECTIVITY. 'I' can only know '**8**'. 'I' cannot, while in-Cosmos, know the *pure* '**I**', though, in an *essential* sense, 'I' can never cease from *being pure* '**I**'.

- The 'I' is *relative* I.
- The '**I**' IS ABSOLUTE **I.**
- The '**8**' Is the ABSOLUTE **I** in a relative world.
- My true Identity *in cosmos* is '**8**'.

If I think I am 'I', I am shallow. If I claim to experience 'I' only as '**I**', I am hopelessly pretentious. Through all these discriminations we must, however, remember that the '**I**' IS the ONLY '**I**', and *THAT* **I AM.** In fact, when all these discriminations (necessary, from one point of view, 'understand' the ONE IDENTITY while one is 'immersed' in-Cosmos) have slipped from the conscious human mind, we will be left with certain very *simple* realizations about identity which will guide our way unerringly.

Thus, '**8**', in truth, Am a center of Omnipotence, Omniscience, Omnipresence in a relative world. From the largest possible perspective, however, '**I**' AM *the centered/centerless* ABSOLUTENESS which IS OMNIPOTENCE, OMNISCIENCE and OMNIPRESENCE. The first series of *'omnis'* refers only to Omnipotence, Omniscience and Omnipresence within a specific Cosmos. These three Aspects are *inherently possessed* by the '**8**'; the '**8**' will discover the Truth of this inherent endowment in the "fullness of time". The second series (with all letters capitalized) refers to the '**I**' in ITS purest, *truest* 'STATE' as the ALL-SELF in ITS OWN 'WORLD'.

How can the symbol or word '**8**' be rightly, effectively and powerfully used? To understand and assert (in realization, thought and word) the Reality of the '**8**' is an immensely powerful affirmation. Such an affirmation is equivalent, in essence, to "I Assert the Fact." There is, in fact, in all the UTTER ALLNESS only ONE *FACT*—the ABSOLUTE '**I**'. When one asserts the '**8**', one is engaged in an intra-Cosmic affirmation of the ONE AND ONLY (SUPER-Cosmic) INFINITE SELF WHO both *IS* and *IS NOT.* The use of '**8**' is, indeed, the most powerful assertion possible (in-Cosmos).

- ♦ The use of the '**8**' asserts the *fundamental, irreducible* BEING of the ONE/One/ one who thus asserts. The TRUE '**I**', however, 'ABIDES' only in *negation*—infinitely 'remote' and yet, strangely, infinitely *present* as the ONE AND ONLY PRESENCE. The *purity* of the TRUE '**I**' is 'KNOWN' only 'within' the ALL-IN-ALLNESS of the Universal Pralaya.

On the CENTER/Center/center

Mathematically, a center is a point equidistant from all points upon the circumference of a circle or a sphere. For the sake of symbolic simplicity we will confine ourselves to the consideration of a circle. In a more metaphysical sense, a center occupies each and *every* point both within a circle and upon its circumference, for the center extended itself (without 'leaving' itself) to become a radius, and then as a radius swept around the center to become a complete circle.

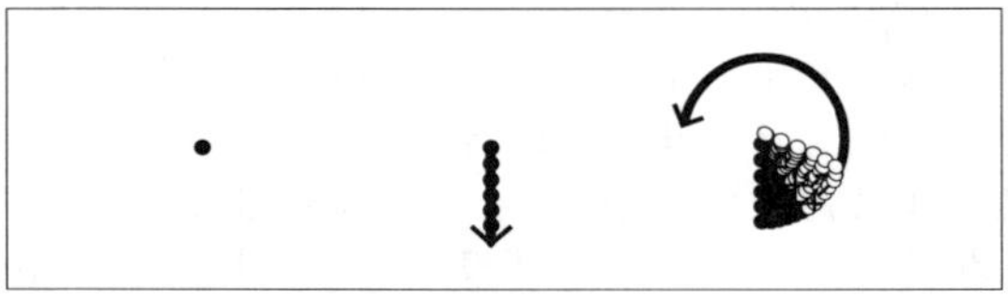

A center can thus be conceived as a *point* which is present at every point in any geometrical figure which is generated from itself.

Metaphysically, a center represents a concentration of power and influence. It is a point from which *authentic* (i.e., *self*-impelled) action proceeds. A center can be considered a *point of origin,* and the point from which all other points can be *immediately* reached, just as the number 'one' serves as a divisor for all other numbers without exception, and has, therefore, metaphorically, 'immediate access' to all other numbers.

The center is the point at which all qualities within the field of its (the point's) influence are essentialized. The center of a field 'contains', as it were, all other points within its field of influence. The center, therefore, is the *essentialized whole*—the most powerful point, the most intimate point, the most flexible point—most influential, most impactful, most impressive. The center is the point from which all within a ring-pass-not is controlled and directed.

A center is to be understood differently, depending upon whether a system is:

1. bounded (like Cosmos), or
2. unbounded (like the ABSOLUTE—although it is technically incorrect to call the ABSOLUTE a *system,* or, to 'call' IT anything, for that matter).

1. Within a bounded system, the center is equidistant from all points on the periphery of its sphere of influence. Within such a system, various points have location and are considered to be closer or farther from the center which, also, relative to all other points, *has position—location.*

2. Within an unbounded system, a paradox arises; the center (or the point of access to all other points) is found to be *everywhere* and, being everywhere, is, as if, *nowhere.* Such a center (though it be 'everywhere'-hence-'nowhere') is equidistant (a 'distance' of *zero)* from the 'boundless bound' which, also, is nowhere to be found. In other words, when a system is boundless, the center cannot be designated, *located,* and any and every point becomes the center. (The problem of whether a point can REALLY exist is another inquiry, and one into which we have entered extensively in various sections of this treatise.) When there is no periphery to be found, regardless of which one of an infinite number of points may be designated as center, then the radius is of infinite extension or, equally, of zero extension.

Thus, within the BOUNDLESS IMMUTABLE PRINCIPLE, any and every point (if such a point could exist) would be the *central point.* Of course, to say the above is REALLY to assert that within the BOUNDLESS IMMUTABLE PRINCIPLE in ITS 'STATE' of HOMOGENEOUS INFINITUDE, *there is no Point,* no CENTER.

Within Cosmos, there at least *seems* to be a Center (probably *spatial* but certainly *dynamic*-as in a "center of power"), and from the perspective of the *form* of Cosmos, there probably is a center as well. Actually since Cosmos is ESSENTIALLY, utterly at-one with the BOUNDLESS IMMUTABLE PRINCIPLE, the *Real* Center, even in-Cosmos, is *everywhere* to be found (regardless of the apparently contrary testimony due to the illusion of *extension),* for all points are Really but One Point, and all centers are Really One Center.

At least Cosmos (which, *as perceived by human consciousness,* has shape and formal definition) can reasonably be conceived as possessing a *formal* Center, even if, *metaphysically,* It has a *ubiquitous* Center. (The 'shape' of Cosmos has not been satisfactorily ascertained by science to date, and there are competing theories offering contradictory conclusions.) If, however, we search for the deepest TRUTH, we find that 'within' the ALL-SELF, even the rarefied concept of a *ubiquitous* 'CENTER' is too crude.

◆ In that *unrelieved* HOMOGENEITY there is REALLY no CENTER; NOTHING cannot have a CENTER.

The center of Cosmos should be designated as 'Center' (with a capital 'C'). A question arises as to whether such a Center *actually* exists.

- On the physical/etheric systemic plane it may be possible, one day, to designate a kind of spatial Center. This may even be possible with reference to the three worlds of human evolution which, from the Solar Logoic Perspective, are worlds of *form.*
- Metaphysically, however, the attempt to designate a Cosmic Center presents a greater problem, as we must ask ourselves whether there is Really any form of *extension* and, thus, of *location* on the highest ('arupa') planes of Cosmos (to which metaphysical inquiry applies). In some manner, extremely remote from present human understanding, the answer should be, "Yes".

The word 'CENTER' (all capitals) is still more subtle. We have realized that the ALL-SELF cannot have a CENTER. NOTHING has no CENTER. But what of the 'FLASHING FORTH' of the 'RAY of the ABSOLUTE' and the simultaneous 'ARISING' of the Infinified Point (the First 'Compromise' 'within' UNCOMPROMISING INFINITUDE)? The Infinified Point 'ARISES' in the ALL-SELF (for 'within' what *other* than the ALL-SELF can It 'ARISE?). Can this 'ARISING' be considered the 'EMERGENCE' of a 'CENTER' within the ALL-SELF?

From one perspective, with the 'EMERGENCE' of the Infinified Point, the ALL-SELF is, as it were, *centralizing* ITSELF for *action* (and simultaneously *limiting* ITSELF—at least, *apparently).* Yet, contrarily, it could be said that the Infinified Point (or Infinite Subject beholding the Infinite Object from an infinitude of 'Points of View') pertains to a Pre-Cosmic extra-SOURCE World, and not to the 'STATELESS-STATE' of the ALL-SELF at all.

Even if one could relate the concept of the Infinified Point to the concept of *center*, the concept of *center* would have to be modified, because the Infinified Point is at once *everywhere* and *nowhere*. In general, careful thought reveals that, as regards the SELF-AS-SELF (the INFINITE SELF *purely* 'within' ITS 'STATELESS STATE' of *infinitized* ALL-IN-ALLNESS and divorced from all consideration of *modification)*, the concept of 'CENTER' is meaningless.

- The concept of *center* is vital to the occultist: The entire occult quest is to find and *be* the *center*—first the center, then the Center, then the CENTER.

The goal is to live in an enlightened manner as if one *were* the Center of the Universe, without identifying as the false and egoistic 'I', but, instead, identifying as the cosmically-ubiquitous '**8**'. Later (as the higher dimensions of Cosmos become the natural field of expression for the apparently advancing Entity) it would seem to be necessary to *live*, paradoxically, increasingly *as* the CENTER 'within' an UTTER ALLNESS which has no CENTER. Probably this kind of *living* is the sole prerogative of the INFINITE SELF 'during' Universal Pralaya. But the question must be asked, "WHO IS the INFINITE SELF"? And the answer must one day come, '**I** MYSELF'.

All these statements are simply rather meaningless words unless they can be translated into immediate metaphysical *realization*. Speaking of realization, at length it will be realized that every authentic E/entity or I/identity is *already* the ubiquitous metaphysical Center of Cosmos could he but *realize* it.

On the Imperfection of Cosmos

We can think of Cosmos as a very weak approximation of the INFINITENESS, an ever-inexact approximation, an approximation (strangely) *infinitely removed* from the PERFECTION of the INFINITESSENCE.

- The fact of imperfection is forever built into Cosmos, and no Cosmos (in and of Itself alone) can *ever* be *absolutely perfect*.

Only the PERFECT (the INCOMPARABLE) IS *absolutely perfect*. PERFECTION IS the INFINITUDE and *nothing other*. The Cosmos, however, is the Great Limitation and can, per se, never *become* the INFINITUDE (though from the *absolute* 'PERSPECTIVE'/'INSPECTIVE', It can never *be* anything *other*, either). Some permutations of the word 'live' (live, evil, veil, vile) symbolically signal the imperfection of the ever-limited Cosmos.

For practical purposes, when we think of the possibility of achieving Perfection in any Cosmos, we must think only in terms of a *relative* Perfection. In any finite system (which Cosmos Is) there is possible a condition of optimal relationship between all systemic variables. This condition we can call *Perfection-in-Cosmos*, and the Universe must be brought to this *condition* before the onset of the Universal Pralaya (at least, *optimists* think so!). Though Perfection-in-Cosmos fall *infinitely 'short'* of PERFECTION, it is the best that can be achieved.

On the Beginning and Beginnings

The term 'beginning' indicates the commencement or inception of a process. A beginning is, essentially, a *change*—a change of state or condition. (The terms 'state' and 'condition' are similar, but 'state' is used to indicate a more homogeneous collection of variables, and 'condition' a more heterogeneous collection.)

The concepts of 'beginning' and 'change' can be equated and, thus, every change can be defined as a beginning. A beginning initiates the new. In this respect, every change initiates a state or condition which is new relative, at least, to the condition immediately preceding the change. Following the idea that "the more things change, the more they remain the same", it may seem that changes may be repetitious, and that a change can duplicate a previously existing state or condition. While this may be so in practical and inexact terms, it is not *exactly* true, because within Cosmos the Principle of Unrepeatability holds sway for *all time*.

The term 'beginning' REALLY conveys an ESSENTIALLY illusory idea. The idea of 'beginning' has no applicability with reference to the REAL WORLD, the 'DOMAIN' of BE-NESS, which is *beginningless* and *endless*. The idea of 'beginning' is only applicable *relatively*, i.e., in relation to the World of Relativity, which is the World of Cosmos, the Universe. Contrarily, 'within' the ABSOLUTE HOMOGENEITY of the PURE INFINITUDE there never was nor can there ever be a REAL *change*. A beginning is dependent upon the existence of change, hence 'within' the ABSOLUTE, 'within' the ALL-SELF, there never *was* nor will there ever *be* a beginning. With respect to the BOUNDLESS IMMUTABLE PRINCIPLE (REALITY), any *beginning* is an impossibility—as is any *ending*.

- However, 'within' the FOUNT OF *ALL* POSSIBILITY, by definition, *everything* is possible, even a *beginning*. This is the reason for the *necessity* of a Cosmos (which can exist without a beginning).

The appearance of a Cosmos allows for the (albeit, 'EXTRUDED') possibility of a 'beginning', and hence, of *apparent* SELF-'CONTRADICTION' which is 'NECESSARY' to the INFINITE POTENTIAL (if IT is to be *truly infinite)*. All 'SELF-NECESSARY' *impossibilities* thus become *possible* (albeit, *illusorily* possible, because possible only within and as *Cosmos)*.

Therefore, each and every Cosmos (born ever of 'SELF-NECESSITY') certainly *does* have a *beginning*. Beginning is change, but all change need not be purposeful. The *Beginning* which inaugurates each successive Cosmos *is*, however, most definitely purposeful. (We would be loathe to think it was *not!)* Perhaps the 'INAUGURATION' of Cosmos is the most *purposeful* 'ACT' which ever occurs 'within' the UTTER ALLNESS. Reasonably, this 'INAUGURATION' is the *only* 'ACT' thus 'OCCURRING'.

There is a problem with the term 'act'. The Beginning of Cosmos is most definitely an *act*, but *who* is the *actor?* Is the ALL-SELF the 'ACTOR'? In one way, yes; in another, no. As there is no other BEING in the UTTER ALLNESS, WHO or WHAT *else* could be the 'ACTOR'? An important concept to realize is, **'I AM the ACTOR'**.

A problem, however, still remains. The ALL-SELF as 'ACTOR' 'DWELLS' in the REALM of MOTIONLESSNESS. It is virtually, if not absolutely impossible to imagine how *act* can arise in a MOTIONLESS HOMOGENEITY which, moreover IS (by definition) *motionless* and *homogeneous* forever! So the Beginning is, indeed, an *act*, but an *act*, perforce, 'within' the ONLY REALM there IS, the REALM of INFINITUDE—for no Universe is yet existent. But 'within' that REALM there can be no action because movement is impossible—at least action and movement of which we can have any conception. What the *absolutization* or *infinitization* of action and movement may be, WHO can 'SAY'?!

As we consider this apparent contradiction (only one contradiction of an endless series), we must think of *noumena* and *phenomena*. There is a NOUMENAL WORLD and a Noumenal World.

- The NOUMENAL WORLD is the WORLD OF BEING (the ABSOLUTENESS).
- The Noumenal World is the World of Being. The Noumenal World (in any Cosmos) is the World of Archetypes—Divine Patterns.

But the *ultimate* NOUMENA for those archetypal Noumena 'inhere' *infinitessentially* in the ABSOLUTE WORLD (and, even 'THERE', *not* in any *discrete, differentiable* form, nor in any *form* at all). It begins to seem that there is ONLY *ONE* NOUMENON for all possible Noumena. The NOUMENON IS, as it were, an *infinite distillation* or *abstraction* of all possible Noumena. That NOUMENON IS not the very quintessence, but the very *'infinitessence'* of all that appears in-Cosmos (whether in the World of Being or the World of Effects—both Worlds being Aspects of the World of Becoming). The *infinitessence* is an *essence* of infinite rarefaction and refinement. The *infinitessence is* the INFINITESSENCE (the INFINITE SELF).

It is necessary to avoid as much as possible using the word 'exist' when speaking of that which IS 'within' the NOUMENAL WORLD. When speaking, however, of the Noumenal World on the Archetypal Planes (i.e., the higher levels of the Systemic Planes, from a human perspective, and the very highest Cosmic Planes from a Cosmic Perspective) there is no such problem. Everything *below* the ABSOLUTE NOUMENAL WORLD can be spoken of as *existing.*

Sometimes, to *be* and to *exist* have different connotations: 'being' is somehow more intangible than 'existence'. The word 'being' suggests a simple *affirmation of presence.* Through the use of the word 'being', one knows that a thing *is* rather than *is not.* But 'existence' seems to suggest concretization, limitation, definiteness, densification, and most importantly the idea of a definite or subtle *separation from simple presence.* When things *exist* they seem to *stand forth* or *stand apart*—separately (and the 'substrative' medium of which they are inseparably a part is comparatively diminished in importance). Thus *existence* is a kind of *separative emergence* from a *substanding presence.* Words of the kind provided by our English Language are not adequate to fully convey such subtleties of thought. When referring to a thing or things which have *being,* one is often reduced to simply saying "it *is*", or "they *are*". In doing so, one is simply *positing their presence.*

Applying these thoughts we find that in the Noumenal world, Archetypes for all intra-Cosmic phenomena *exist,* but that in the NOUMENAL WORLD, the ONE NOUMENON for all intra-Cosmic Noumena simply *IS.* 'Within' that ULTIMATE WORLD (the ONE AND ONLY SOURCE) there are no differentiated *sources* for various Noumena. There, *nothing is differentiated.*

Returning to thoughts of the Cosmic Beginning, such a Beginning must be *different* from THAT which preceded it. In general terms, a 'beginning' might be defined simply as—'an action which is different from the action which preceded it'. Conversely, if any difference (variation, modification) ever occurs, a beginning is indicated. Every difference is, therefore, a beginning. Throughout the INFINITE DURATION of the UTTER ALLNESS there have, therefore, been countless beginnings and *never* an initial one, for we ARE a BEING WHICH IS forever. This thought leads to an astonishing idea:

- *There never was a first universe* ... nor will there be a final one.

So a 'beginning' is a change of condition; an 'end' is also a change of condition. There is no BEGINNING, but there *is* a Beginning—in fact, an infinitude of them. There cannot be a BEGINNING (all capital letters) because this would suggest that the ALL-SELF 'BEGAN' which, we have seen, IT *does* not and (by definition) *cannot.* But a 'Beginning' suggests the onset of that perishable Item called a Cosmos, which like any other finite form, *must* have a beginning (and, of course, an ending).

The Beginning (i.e., the Beginning of Cosmos) is the Archetype of all lesser beginnings within Cosmos. The Beginning (as far as we in this Cosmos are concerned) is the *True* Beginning of All (the entirety of what *exists* in our present Cosmos).

- Remember: 'ALL' (all letters capitalized) indicates the entirety of that which has existed, does exist and will exist in each and every actualized Cosmos—i.e., in ALL of them, an infinitude of them. Thus, 'ALL' represents an infinite collection of finite things.

There are so many Beginnings, that it is useless to think of them (in any quantity or detail) for, at present, we can know nothing of them (though, in fact and necessarily though, paradoxically, we are *already* OMNISCIENT with respect to the UTTER ALLNESS). All we need to know is that Beginnings have, indeed, occurred forever and will continue to occur forever. We need only to focus on *our* Beginning in *this* Cosmos, and assume that all others resemble It in certain significant respects.

The True Beginning (Our Cosmic Beginning) is a 'MOVEMENT'/'CHANGE' from the UNCONDITIONAL and UNCONDITIONED into the Conditional and Conditioned. Thus, this Beginning is the 'END of SAMENESS'. (SAMENESS must be capitalized because IT refers only to the ABSOLUTE.) The 'STATE of SAMENESS' is not to be found manifest 'in' this or any other Cosmos. Cosmos Begins; SAMENESS 'ENDS'. The 'END of SAMENESS', however, can only be an *apparent* 'END', for SAMENESS necessarily 'ABIDES EVER', even while the Great Illusion of Difference *seems* to present Itself throughout the Universal Manvantara.

The True Beginning is a Change which arises out of CHANGELESSNESS. NOTHING IS *going on (and on and on and on ...);* and 'then!' (though Time exists not) 'SOMETHING' 'HAPPENS'. The UTTER ALLNESS IS simply the *apparent* rhythmic variation of 'NOTHING' and Something while, nonetheless, preserving inviolate the *absolute continuity* of NOTHING. The True Beginning is one of the Greatest Mysteries. This Mystery is the Noumenon of *How.* Another Mystery, deeper still, is the Noumenon of *Why.*

Every beginning is a variation upon the state or condition which preceded it. The mind refuses to accept the possibility of a beginning that has no 'pre-cedent'. A *state* (though an inadequate word) *must* precede a beginning. Therefore, beginning is a variation from a pre-existing pattern or a pre-existing state. Beginning always assumes pre-existence. If this model is accepted, we realize that everyone of the Great Beginnings (always of an All-Inclusive Cosmos) must have emerged in contradistinction to a preceding *state*. Back and back through infinite regression we may travel, and never find any *Beginning* to this infinite series of Beginnings, whereupon we are led to conclude that Beginnings *always* were.

- The human mind continues to interpose the question: How did it all start?

We might discuss two varieties of this question: How did It All start? and How did *IT ALL* 'START'? Cosmology attempts to answer the first question, and many are the religious, scientific, and philosophical theories of Creation. The second question, however, is illegitimate, for, indeed, IT did not ALL 'START'. IT *never* 'STARTED'.

Is it inconceivable to think of THAT which did not 'START' but was *always* so? Most minds find it difficult. It is often easier to think of THAT which always will BE, than THAT which always WAS. ENDLESSNESS is somehow more acceptable to the structure of our minds that BEGINNINGLESSNESS. Certainly no *finite thing* can have *always* been. Our minds are so constructed that when we *think,* any *thing* which we conceive is immediately rendered *finite* by the very act of conceiving it. Therefore H. P. Blavatsky is correct when she calls the BOUNDLESS IMMUTABLE PRINCIPLE "Unthinkable" and "Unspeakable".

The frustrated mind at length arrives at the conclusion that there is *no-thing* which did not *start* but was *always* so. While this may seem like defeat, it is Really victory, for the correct answer is precisely that (*"nothing"*,or—adjusting the symbol—*"NO-THING")*. Only the NOTHING AT ALL (which IS and yet *exists not)* can fulfill the requirements of THAT which did not *start* but was *always* so (and, of course, will *always be* so). Only the GREATEST ABSTRACTION, INFINITENESS, BE-NESS, can have 'BEEN' *always* so. IT had absolutely no *origin*. IT IS the ORIGIN ITSELF. IT IS simply the GREATEST OF ALL GOODS, the INFINITESSENCE OF UTTER ALLNESS. Because there is no *other* than IT, IT can have had no BEGINNING, for What was there which could possibly have preceded IT, to provide that pre-existent 'state' from which any normal beginning must derive.

Further, all things that begin must be finite. Beginning and finitization are coeval. There is no beginning without finitization. To begin a thing is to finitize that which is begun. Even to begin writing an infinite set of numbers is to begin a 'de-finite' act of writing. The set may, indeed, be infinite, but the act of *beginning* to write is a finite act. The set may be infinite, but the act of writing (which once begun) must, at length, *end!*—simply because Cosmos, Itself, is finite. Be assured that, following the dissolution of Cosmos, the determined writer will not persist in any form that would allow him to hold a pen.

The Cosmic Event called the Beginning signals the entrance of Time, emerging out of DURATION. Into What does Time enter? Into the Consciousness of the SELF-as-Infinified Point 'Beholding' with 'Super-Cosmic Consciousness' Mulaprakriti.

Consciousness did not exist before the arising of the 'RAY'/Point. So, one could only say, perhaps, that ABSOLUTE CONSCIOUSNESS WAS/IS as an *infinitized aspect* of the

ONLY NOUMENON (the INFINITESSENCE). Time, then, 'first' appears (at least with respect to any particular Cosmos) with the Beginning (as time has 'appeared' and 'disappeared' constantly during Infinite Duration). The ALL-SELF IS always 'outside of Time'. The SELF-as-Infinified Point-as-Condensing Point-as-Condensed Point-as-Universal Logos Is *not*.

So the Beginning, the entrance into Finitude, Is NOTHING 'BECOMING' Something. The ancient question then arises, Can something come out of nothing? Common sense (subject, unfortunately, to intense prakritic densification) tells us, No! But the answer to this question depends, perhaps, upon how 'NOTHING' is defined. When NOTHING is defined as the ABSOLUTE FULLNESS, the INFINITESSENTIAL PLENUM, the question begins to seem more reasonable.

- Perhaps the question should be rephrased: Can Something come out of the *possibility* or *potentiality* of, or for, *everything* (which possibility and potentiality we have called the INFINITESSENCE)?

'Something' is merely a *'limited EXTRUSION of un-particulated potential everythingness'*. In the IT, everything and every possibility of everything IS (already and forever) in the ABSOLUTIZED 'STATE'. The ABSOLUTIZED 'STATE' is the UNCOMBINED 'STATE', for IT IS a HOMOGENEOUS 'STATE', so *everything* does, indeed, 'INDWELL'/'INHERE' in IT, but *not* in *extension or particulation,* i.e., not in 'thing-hood'.

In one way, the Cosmic Beginning is a *positing,* an assertion of *position.* It is, as it were, a 'turning on'. In the Universal Pralaya, the Universe is in the 'off' 'STATE', which, paradoxically, for the SELF, is the maximally or undistractedly *infinitized* 'STATE'. In one way, the SELF can *never* 'turn off', as IT IS always 'on'. IT IS the *unfluctuating undeniable* PRESENCE. Rather, it is the Universe which is turned 'off'. 'WHO' "flicks the switch"! (Another statement, not a question.) One thing that must be clearly understood is that 'turning on' or 'positing' is *limitation.* How can a 'posit-ive' 'ACT' be a limitation? Nothing is *stranger* than the SELF! But, then, NOTHING *IS* the SELF. ABSOLUTENESS is the IMPENETRABLE DARKNESS in relation to which all *light* is *limitation.* Thus, there is no need to *'turn IT on'!*

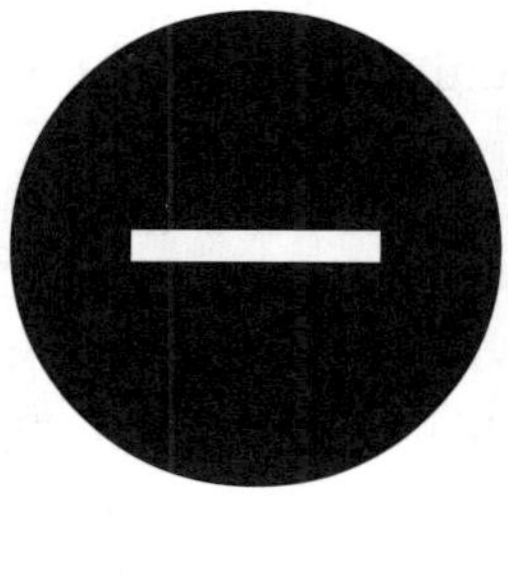

ABSOLUTENESS
the constant 'off' state of the PRESENCE, which is ESSENTIALLY 'ON'

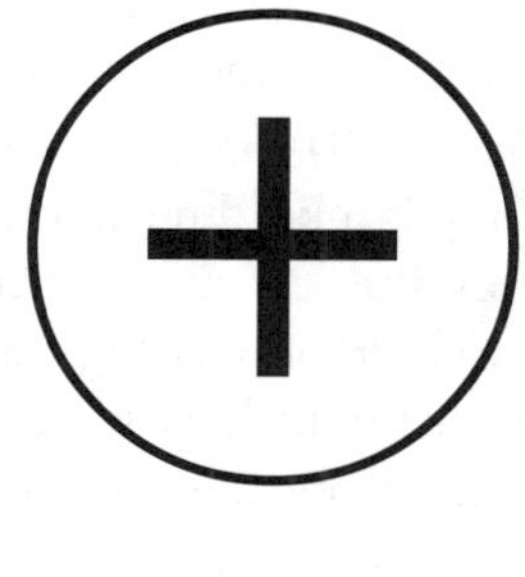

LIMITATION
the periodical 'on' state of Cosmos

On End, Ending and the End

Now let us look at the concept of 'end' or 'ending':

- An end, like a beginning, is a termination of pattern (and is simultaneously a commencement of a new pattern, as well).
- An end is a *change*.
- Every change is *both* end and beginning.
- An ending is a cessation of that which was begun.
- Every *movement* is both beginning and ending.
- Every *moment* is both beginning and ending.

Endings and beginnings in Cosmos are occurring at every 'ultimate moment' (at every 'ultimate unit of Cosmic Time') which means, *virtually*, but not precisely, 'all the time'. ('All the time' cannot exist in-Universe, for there can be no true continuum *in* Cosmos. Only INFINITE DURATION is a *continuum*, or rather *the* CONTINUUM, with respect to Time.)

For instance, the Cosmic Configuration, which began, let us say, one '*x*-tillioneth' of a second ago was the *noticeable*-in-lower-Cosmos ending of the Cosmic Configuration which began two '*x*-tillioneths' of a second before and endured for one '*x*-tillioneth' of a second (though the *real ending* of the prior Cosmic Configuration was the onset of the 'inter-moment interval'), the duration of which is probably also one '*x*-tillioneth' of a second. (This example is for illustrative purposes only, and suggests that the 'ultimate moment' of our Universe {and the inter-moment interval, if equal to an ultimate moment} have a duration which is not any *known* fraction of an Earth second.)

The measure of the 'ultimate unit of time' in-Cosmos is unknown to any human being-strictly as-human being. In order to know it, the tiniest Fohatic particle/event in-Universe (on the highest Cosmic Plane of Fabrication/Effect) would have to be identified, and both the *nature* of its movements/changes and the *rate* of its movements/changes be known. Since we do not even know the frequency of vibration of what in Occult Chemistry is called the "ultimate physical atom" (a compound particle resident upon the atomic subplane of the etheric-physical plane) we surely cannot determine the rate of the 'ultimate Cosmic particle/event', but such a ultimate quantifiable frequency of vibration *must* exist if the Law of Correspondences holds good.

Every ending is a kind negation just as every beginning is a kind of positing/assertion/affirmation. But, every beginning is also a negation of that which preceded, and every ending is a kind of positing/assertion of that which begins with the ending. Clearly, it is difficult to distinguish between these two dynamics. The pairs of opposites are ever meeting, 'becoming' one another, so to speak.

Change is necessary if there is to be ending and beginning. Only in the World of Becoming can there be endings and beginnings. Motion is inseparable from beginnings and endings. With every motion/movement/change, there is inevitably a beginning and an ending. The tiniest, briefest motion in-Cosmos changes the configuration/relationship of all items-in-Cosmos (i.e., the Cosmic Configuration). Change ever causes a reconfiguration of relation, or *'re-relation'.* Every 're-relation' is a beginning and, also, an ending.

In the World of Becoming, never is there SAMENESS. In the WORLD OF BEING, never is there variation. In the World of Being, variation is slower (and far more 'seamless') than in the World of Effects, perhaps much slower, perhaps so slow (in the upper strata of the World of Being) as to last for great Periods in the Duration or Cosmos, or, even, for the entire duration of the Cosmos. But Cosmic Duration is infinitesimal in relation to ETERNAL DURATION. 'Infinitesimal' means 'infinitely small'—as small as possible without being zero. No definable quantity can be an 'infinitesimal' and, thus, a better though more cumbersome word for an 'infinitesimal' is an *'infinitesimalizing'* (which is a word suggesting *process* and *indefiniteness).*

The WORLD OF BEING includes the World of Becoming. The World of Becoming includes both the World of Being (the higher strata of the World of Becoming) and the World of Effects, or the World of Fabrication/Approximation (the lower strata of the World of Becoming). The entire World of Becoming is Illusory and does not include the WORLD OF BEING. The WORLD OF BEING is REAL *(utterly* REAL) and not only includes, but ESSENTIALLY *IS* the World of Becoming.

Never can there be an end of endings and beginnings in the World of Becoming. Never was there a beginning of beginnings and endings in the World of Becoming, for the World of Becoming (though appearing only *cyclically)* spans Infinite Duration. While it is impossible not to BE, it is cyclically possible and even necessary not to 'be something' (i.e., to *exist).*

The symbol for an ending is 'X' (though an X can mean many other things). An 'X' crosses out; it negates. That which was, is no more.

The moment *movement/change* is introduced following the Universal Pralaya, there are endings and beginnings. In the SUPREME MOTIONLESSNESS which IS REALITY there can be no endings and beginnings for there can be no variation, and hence there can be no change, and no negation of that which *was* before, nor assertion of that which *was not* before—all this, because THAT WHICH *IS,* 'ABIDES' *changelessly forever.*

- In sum, all endings, as all beginnings, are *illusory* and bear relation only to the World of Cosmos, i.e., the World of Illusion.

On Positing and Negation

We sometimes speak of the astral-inversion that causes humans upon the physical plane to desire what is worthless and repudiate that which is of high value. This inversion is a very low-level example of a condition that arises from the huge (in fact, *infinite* though, paradoxically, *bridgeable)*'gulf' which separates the Finite from the INFINITE. That which is valued from the lofty philosophical perspective called "sub specie aeternitas" (i.e., "under the aspect of Eternity" and, therefore, valued when one is entirely identified with the SELF-as-Self) is usually entirely the reverse of that which is valued when objects are not understood to *be* the SUBJECT.

When 'studying' the INFINITE, and attempting to live 'in' IT, we must become accustomed to frequent reversals of our ordinary values. For instance, given the usual human value system, 'positivity' is valued as desirable and 'negativity' is considered an undesirable trait. While, for practical purposes, there is much wisdom in this attitude, the need to reconcile the life with the REALITY of the INFINITE presents an alternative point of view. To 'posit' (which is the root of 'positivity') is REALLY a reduction or limitation of INFINITENESS. Positing is an act of assertion; it is to put something forward—that 'something' is ever infinitely less than the NO-THING from which it came.

◆ To 'posit' is to assert the value of the partial over the TOTAL.

In the World of Becoming (the World of Cosmos) to posit is necessary, indispensable. Nevertheless, we should remember that every 'something' is infinitely (literally, *infinitely)* less than NOTHING. If 'x' is a definite number of any magnitude, whether relatively tiny or inconceivably vast, then—infinity minus 'x' = infinity.

Continuing our analysis of 'positing', to posit is to assert the *existence* of anything. The term *"is"* 'summons forth' from NON-BEING (which IS TRUE BEING) and posits 'existence'. To posit is be 'posit-ive'. Positing place, summons from indefiniteness a definite presentation. Positing makes 'Something appear out of NOTHING'. Any form of 'posit-ivity' (based, necessarily, upon the act of *positing)* is a form of limitation, for the moment something is posited, made positive, it is 'drawn out' of the INFINITE POTENTIAL, drawn out of the ABSOLUTE 'STATE' and 'brought forward' from BOUNDLESSNESS into delineation (i.e., its boundaries can clearly be seen). So positing and the tendency towards 'posit-ivity' (when understood in this special way) creates lines, boundaries, 'thingship', reification, definiteness. Positivity promotes finitude. Positivity is "day force".

Negativity, on the other hand, is "night force", and when understood according to this thought-model, takes away limitation. Positivity asserts limitation. Positivity is the precipitation of or from INFINITUDE into a state of definiteness, hence into a state of limitation. Positivity is enumeration, the *actionless* 'ACTION' that (through SELF-'LIMITATION') 'GENERATES' a Cosmos.

◆ Negation leads to the eradication of all Limitation—specifically, the limitations of all that has been posited, thus restoring all posited things to their ESSENTIAL INFINITUDE. Can anything TRUE be *posited* of REALITY? Uncompromising *right negation,* however, discloses REALITY.

To create a Cosmos, the SELF turns *positive* and *becomes* the Triple Point: the Infinified Point, Condensing Point, and then the Condensed Point, albeit while still remaining ITS supremely ALL-NEGATING SELF. The SELF by means of the Triple Point is readying a *Presentation*—necessarily a *SELF*-Presentation (however disguised by the Illusion of Finitude). Every positing is Really a *presentation* (as opposed to an *obscuration* or an *abstraction)*. The Universe, Itself, and all modifications within It are a Presentation (an 'EXTRUSION' of *infinitized possibility* from the INFINITESSENCE). Every 'I' (matter and ego-bound) finds itself limited by impermanent presentations. Even the realized '**8**' (the pivot in consciousness between both **I**-as-I and I-as-**I**) finds Itself limited by semi-permanent (relative to Cosmos) though still *impermanent* presentations—presentations in the World of Being, the World of Archetypes.

Positing and affirmation will not remove the **I/8** from its habit of *objectivizing Itself*. The true **I** exists only in negation. The **8** (and especially the ULTIMATE **I)** cannot be approached through positing. Positing and predication are one and the same. They can be helpful. They can point in the direction of the **8-as-I**, but the **8-as-I** can only be identified with in silence/Silence. Let words do everything they can; then let the REST BE SILENCE.

'Right negation', on the other hand, is the method of arriving at the PRESENCE, which, of course, is the PRESENCE (to the small degree possible in human awareness) of the INFINITE SELF. Negation negates 'somethings' in favor of NOTHING—hence that ancient Indian method of arriving at REALITY, "Neti, neti." From the psychological, ethical perspective, to be rightly *negative* is the most blissfully *positive* thing one can do.

Ever in this treatise, the opposites will be found to change places or turn into one another (just as they did for Heraclitus). So, in a strange way, *right negation* is an assertion of the INFINITE SELF. **8** assert the FACT! As, we know, there is only ONE ULTIMATE FACT. Declaration of IT by *frontal* means is a reduction or negation of IT. Use of *rightly negative* means is a blissful assertion of IT. Paradox!

On Doing and Non-Doing

To 'do' is to act; to act is to move; to move is to change; to change is to repattern; to repattern is to 're-relate'; to 're-relate' is to make a difference; thus:to 'do' is to 'make a difference'. 'Within' the ALL-SELF, however, 'everything' IS always the SAME. 'Within' IT, what is there to 'DO'? NOTHING! It would *make no difference!* Nothing can *ever make a difference* to the ALL-SELF.

What can the ALL-SELF 'DO'? What can the ABSOLUTE SELF 'DO'? IT REALLY can 'DO' nothing 'within' ITSELF and *as* ITSELF. ITS 'STATE' IS forever *invariant* and there IS no possible way to alter that 'STATE'. 'Within' IT, there is NOTHING to 'DO' *forever!* The ALL-SELF can only *BE* NOTHING *forever!* So how can the ALL-SELF BE the ONE 'ACTOR', the ONE SELF-as-ACTOR—how, can IT *BE* the ONLY IDENTITY in ALL that *happens?*

In answer, let us say that IT (the ALL-SELF) *seems* to be able to *reduce* ITSELF in some way—to *apparently* become *less* than ITSELF. The Universe, the Cosmos, is merely an *apparent* Cosmos, an appearance, an apparition. It is a *seeming* 'Yes' against the ETERNAL BACKGROUND of 'NO'. The Cosmos is, also, *apparently*, an INFINITY-Negating 'No' against the ABSOLUTE BACKGROUND of an *infinitudinous* 'YES'. In the WORLD OF REALITY, it appears that 'YES' and 'NO' meet. 'YES' is 'NO' and 'NO' is 'YES'. Ever do the extremes meet in the creation of the circle. It might justly be said that at any moment, the Universe or the Cosmos is *the only thing that is going on*, the *Only Action*—"The only game in town."

- This Universe is periodically the 'Only Doing', even if merely an *apparent* 'Doing'.

The ALL-SELF, PER SE, is not the 'DOER', but the 'ALL-SELF-as-Triple Point-as-Universal Logos-as-**8**-the-Actor' *Is* the Doer. The 'Doer-of-and-in-Universe' can be traced back to the ONE WHO CANNOT 'DO'; the 'Actor-as-and-in-Universe' can be traced back to the ONE WHO CANNOT 'ACT'. The paradox is profound but, it does no good to pretend that there is no paradox, and that *conventional* logic alone will reveal the TRUTH.

Since the Universe is the *only Thing happening*, It becomes the sole Object of Attention of the Attending One, the Universal Logos (Who Is '**8**' by another name). The Attending One can also be described by the formula—the 'ONE-as-Triple Point (Infinified, Condensing, and Condensed)-as-One-as-**8**. ABSOLUTE 'CONSCIOUSNESS' (although there REALLY IS no such 'THING'!) has NOTHING *objective* to 'ATTEND TO'. Super-Cosmic Consciousness (the Pre-Cosmic Consciousness of the SELF-as-Infinified, Condensing and Condensed Point) *does;* so does Universal Consciousness (i.e., Cosmic Consciousness). Super-Cosmic Consciousness attends to the Universe-as-Object. Cosmic Consciousness attends to the All within Cosmos. Both remember WHO these 'Attenders' ESSENTIALLY Are, and from WHENCE They 'EMERGED'.

We can see that the 'RADIATION' of the 'RAY' of the ABSOLUTE (forgetting for a moment the great problem of how IT 'AROSE'); the Pre-Cosmic Action of the SELF-as-Triple Point; and the Creating of the 'SELF-as-Triple Point-as-Universal Logos' all involve *action,* a *doing.* The ABSOLUTE SELF IS *not* an 'ATTENDER' for IT IS ESSENTIALLY in relation to *no* object, since IT *already IS* any possible object to which IT could 'RELATE' (*if* IT *could!).*

If attention becomes *absolute,* it is called 'ATTENTION' (and requires no object of any kind). The ONE 'ATTENDS' only to the 'task' of *BEING ITSELF* to the *infiniteth* degree, although there is no 'ACTION', *per se,* in this 'ATTENDING', and no *objectification.* Contrarily, all RADIATED 'Aspects' of the ALL-SELF (i.e., the Pre-Cosmic and Intra-Cosmic 'Players') *attend to Their 'object'* because They, unlike the ALL-SELF, *have one!* Pre-Cosmic Attention is given by the Infinite Subject to Mulaprakriti, the Infinite Object. This involves the relation between the Triple Point (Infinified, Condensing, and Condensed) and Pre-Cosmic Root Matter.

On a lower level, Attention-in-Cosmos is limited by objectification (even if the objects attended to are what are usually called *subjective).* Attention-in-Cosmos (which is attention as we know it) is aroused with the appearance of *number* and *item,* for there is naught to attend to without number, division, relation, item, etc. Attention begins with focus *(from* a *point*—a *point of view) upon* a *point*, a point of attention. As we think of the Infinified Point *(and* the Condensing and Condensed Points), we should think of INFINITUDE coming to a State of Pre-Cosmic Consciousness preparatory to the arising of the Great and *singular* Limitation which Is the Universe or Cosmos.

'Doing', in any way we can hope to recognize it, begins at the 'end' of the Universal Pralaya. There is no 'DOING' during the ALL-IN-ALLNESS—the SELF-'PREOCCUPIED'-'STATE'.

- The very first 'DOING' (or should we call it 'Doing') can be seen as an 'ACT' of 'ATTENTION'. The 'FLASHING FORTH' of the 'RAY' of the ABSOLUTE, and the simultaneous *arising* of the Infinified Point *Is* an 'ACT' of 'ATTENTION' by the SELF.

With that 'ACT' (arising *inexplicably* out of NOTHINGNESS) an Object (that is usually considered the First Object) is generated, and that Object Is Mulaprakriti. If we consider the Infinified Point, Which Is the Infinite Subject, to be the 'First Object' because It *emerged* from the INFINITE SUBJECTIVITY, then Mulaprakriti would be the *'Second* Object' (because it serves as Infinite Object to the Infinite Subject—Itself capable of being 'BEHELD' as an Object by the ONE AND ONLY 'WITNESS'—PARABRAHMAN as 'MAYA'—the 'SEEING'). The question still remains as to whether the INFINITE SELF *can* 'REMAIN' ITSELF and still 'SEE'!

This *inscrutable* 'ACT' of 'ATTENTION' by the INFINITE SELF is an apparent *limitation* upon the degree to which the ALL-SELF can continue to *BE* ITSELF with *infinite intensity* (but the limitation is only *apparent).* The change from 'BEING' to 'SEEING' is, therefore, the very first 'DOING'. If before that 'time' we were to inquire of the ALL-SELF, What is 'HAPPENING'? the 'response' would simply be 'DOING NOTHING'. (Maybe an equally informative 'response' to such an impertinent question would be, "NOTHING DOING!")

- Sometimes we are advised to achieve enlightenment by "Sitting quietly, *doing* nothing." One can see how this simple action (or, rather, 'non-action') reflects the TOTAL SILENCE and NON-'DOING' of the INFINITE SELF. Even 'non-doing' on the human level can begin to penetrate the Maha-Maya.

On the Now

We now come to a fascinating and difficult subject, but fortunately we have *forever* to understand it! Or should we try to understand it *right now?!* Let us see if we can gather some light-bearing thoughts upon the concept of 'now' and the 'Now'. Then we will address ourselves to the ETERNAL NOW, the UNCHANGING MOMENT (both infinitely 'short' and infinitely 'long').

The word 'now' commonly indicates the juncture between past and future. It signifies the apparently ever-moving, immediate moment. Or is that moment REALLY ever-*unmoving?* The 'now' certainly conveys a sense of immediacy.

♦ The Latin epigram, "Carpe Diem" (Seize the Day) illustrates the idea of the 'now' in all its urgency.

Occultists so often think of the ancient historical past or the distant future, but really they need to learn to live *now.* What one does during the infinite string (a 'beaded chain') of successive, apparent *nows* determines success or failure (although by 'now' we should realize that there can be no *ultimate* failure—how can the ALL-SELF {already PERFECT *forever*} fail?). How can IT ever BE anything but the PERFECTION? But, coming back down to 'earth', we find that 'now' is a concept much related to the life of *practical* discipleship. As important as this concept is, there is not much solid thinking about it, most of it being vague and wishful.

♦ "Living in the Now" has been used as an excuse to absolve oneself of all sense of discipline and Self-possession, abandoning oneself to so-called 'spontaneity' instead of exercising the Will.

The result has been an unholy 'holiday' for the Lunar Lords, who represent not the 'now' but the 'then'—i.e., the *past.* Yet the avoidance by many of anything but vague thinking is understandable, because the 'now' is difficult to understand. After all, it is so elusive, so fleeting, so instantaneous—*apparently.* Let's enter into discussion on the matter—*now.*

It is difficult to differentiate between 'NOW', 'Now', and 'now', but perhaps some distinctions should be made.

1. NOW is ETERNAL DURATION, UTTER TIMELESSNESS. NOW *never* had a past or future with which to contend.
    - NOW is 'TIME' 'within' the ALL-SELF, in which the disturbances called 'past' and 'future' never 'OCCURRED', and can never possibly 'OCCUR'.
2. 'Now', on the other hand, is 'God's Time'—God, being the Universal Logos—Who Is (preserving the sense of descent in the following formula) SELF-as-(Triple Point)-as-Universal Logos. The Consciousness Who experiences the Now (and in some profound way every B/being in-Cosmos *is* this Consciousness) *does* have past and future with which to contend, but these two are seen from such a great 'altitude', that that Consciousness can never be trapped by them.
    - 'Now', then, is Time as experienced by the Universal Logos and by *All Who Consciously Become the One.*

    "A thousand Ages in His sight is like an evening gone", proclaims the old hymn, speaking of this 'Summit Consciousness'. This 'Now' is Time as experienced from the pinnacle of the Mountain of Cosmos. The changing patterns of the Cosmic Configuration are seen under the "All Seeing Eye," and total simultaneity of intra-Cosmic apprehension prevails, even though there is witnessed a kind of kaleidoscopic unfoldment (a kind of succession) in which all patterns are seen as One Pattern which continues to mutate before the All Seeing Eye (i.e., the All-Seeing Logoic '**8**').
3. Finally, when speaking of 'now', we deal with that elusive moment which the fragmented human consciousness ever tries to notice, to seize, to experience. We are

here dealing with the so-called "march of time" as those still captivated by ego (the sense of separate identity) and object, perceive the change of events.

- The 'now' in this case is (symbolically) simply the infinitesimally thin line of the present moment, forever seeming to move *forward continuously,* and forever seeming to generate a past and future, which are forever imperceptibly divided from each other.

With this preamble, we are in a position to offer a few thoughts: Far from being fleeting, the NOW is the most *permanent* (even if *dimensionless)* 'unit of Duration' there *is.* The NOW is all of Time there ever has *been* or ever will *be.* The 'now', in contrast, is the point/line of abutment between all endings and beginnings, seemingly following each other, racing after each other, (but, rather REALLY *coinciding* with each other) in a beginningless/endless infinite chain of change.

The 'Now' is the reflection of the 'NOW' within Cosmos. It does little good to speak of the 'NOW' (the ETERNAL NOW) in relation to a relative Universe/Cosmos. That TIMELESS 'STATE' in continuous purity is 'KNOWN' (the 'noumenessence' of *knowing),* perhaps only within the ALL-IN-ALLNESS of the Universal Pralaya. In such a 'KNOWING' there is no relativity. For instance, in such a STATE of KNOWING there would be no need to face the relative Universe/Cosmos and say: Regardless of the relativity that I am registering, regardless of the past and future appearing before my eyes, it is, nevertheless, NOW. Within the NOW (in its purity), relativity of any kind would never be noticed as a distraction because, in fact, it would have ceased to exist with the reabsorbed Universe.

◆ In our Universe, it is the <u>N</u>ow that we must come to understand. This Now can be thought of as—as much of the NOW as the Universe can 'contain', just as, on a tremendously lower level, the Guardian Angel is said to be as much of the Solar Angel as can incarnate in any one human incarnation.

Focusing on some of the qualities of the <u>N</u>ow, we might say that the Now is the permanent abiding 'place' of the 'Cosmic Witness of all Change' (that Witness being the One Being Who 'informs' Cosmos). The Now is the focus of that Great Being's Cosmic Point of Tension. The Now is the point of immediacy; *Livingness indwells the Now.*

In terms of Time *in* Cosmos, it is *always* Now. At any *time in* any Universe that has ever been, it was *always Now; in* any Universe that will be, it will *always be Now.* A better and, perhaps, more thought-provoking way of saying this is—"In any Universe that has ever been or ever will be, it *<u>Is</u>* always *Now!*" In this second sentence the usual sense of past and future are eradicated, and all of history and futurity is, as it were, *contemporized in the immediacy of the Moment.* Perhaps, if this provocative statement is to be strictly accurate, the 'Now' will have to be changed to the 'NOW'!

◆ All pasts and futures must be seen as categories of perception (arising from the limitations of the perceiving consciousness). 'Past' and 'future' are useful terms and, in-Universe, they represent perceived *actualities,* but they do not fully represent <u>R</u>ealities, and they certainly do not say anything meaningful about the ONE AND ONLY ABSOLUTE REALITY.

The One Omnipresent, Omniscient, Omnipotent Observer Who *emerges* only at the inception of Universal Manvantara, perceives only in the <u>N</u>ow. This Observer, is Omnipresent, Omniscient and Omnipotent with respect to the particular 'Cosmos of the Moment', over which It has jurisdiction, but *not* with respect to all Cosmoses that have ever been or will be. Its Omnipresence, Omniscience and Omnipotence pertain only to the particular set of variables or relationships which will be found in Its particular Cosmos.

And, yet, from another perspective, Who *Is* the Universal Logos (the Cosmic Observer) but a 'RADIATION' of the ONE AND ONLY BEING? Since there is, obviously, only *one* ONE AND ONLY BEING (the ALL-SELF) perhaps each Universal Logos/Observer should be considered *identical* to every other Universal Logos/Observer. Certainly, this is true in ESSENCE.

The One Being Who *emerges* with the inception of every Universal Manvantara as the Representative of the INFINITE SELF *IS* the INFINITE SELF, ITSELF. What or Who *else* can It Be? The factor that is different about every Universal Logos/Observer is not its Identity but Its Scope of Action, Its degree of Limitation. A *different* 'Set of Limits' is SELF-'IMPOSED' upon every Universal Logos (Who Is the INFINITE SELF-as-Triple Point-as Universal Logos). Thus is a *different* Universe prepared at the 'Dawning' of each Universal Manvantara—not better or worse (or more or less evolved) than any preceding or following Universe, but simply *different.*

As we realize, not *all* possibilities within the INFINITE POTENTIAL can be unfolded in any one Universe, and hence the need for the Principle of Distinctive Limitation (enforced upon and within each Universe by means of a distinctive Cosmic Algorithm, which distinguish the nature and scope of one Universe from another, and determine how the particular *possibilities* 'EXTRUDED' from the FOUNT OF ALL POSSIBILITY are to be manifested.)

- Remember that the inception of the Pre-Cosmic Process, and the appearance of what we can understand as an Observer (and, thus, the appearance of Consciousness, as well) are *coeval.*

There is no ABSOLUTE 'OBSERVER', because what the INFINITE SELF 'DOES' (a useless term!) within ITSELF is *not* to 'OBSERVE', but simply to BE ITSELF (to the *infiniteth* degree). The 'OBSERVER'-instantly-Observer appears at the GREAT OUT-BREATH, and Exists only when there is Something to Observe, which is not *always* the case. The 'ARISING' of the 'OBSERVER'-instantly-Observer 'CREATES' *something to observe.*

Nevertheless, when one has 'found' the Observer or, as it were, 'become' the Observer (through the application of rigorous spiritual training, or simply through spontaneous 'remembrance') one has also found the SELF. For 'behind' the Observer IS ever the INFINITE SELF. The Universal Observer, Really, *Is* that SELF—the SELF, considered as *oriented* towards 'Something'. Remember that even as the SELF-'MODIFIED' SELF is 'turned towards Universe' as the Observer, IT IS, *simultaneously,* "all wrapped up in ITSELF" as the *PERPETUAL NON-OBSERVER*—the *'BE*-er'. This paragraph contains some very important thoughts touching upon:

1. the cause of the Great Illusion, and
2. the ESSENTIAL *impossibility* of the Great Illusion because of the *ABSOLUTELY and NECESSARILY PERPETUAL STATE of SELF-ABSORPTION* characteristic of the INFINITE-SELF—*regardless of Appearances!*

The Great Illusion (the Universe) Which *appears* to be 'Real' Is Actual, but Is not *ever* REAL. The Great Illusion both *exists* and, yet ESSENTIALLY, *exists not!*

We are focusing in-Cosmos, under the Eye of the One Being in-Cosmos (*not* the ONE BEING WHO *as ITSELF* has no Cosmos with which to contend). In-Universe or Cosmos, every Act/act takes place in the Now. The Now is the arena for all intra-Cosmic action. The One Actor, Who is the ONE SELF-as-Self, can only *do* or *re-position Itself* in the Now (or, even more accurately, *between* Cosmic Nows—i.e., during inter-moment instants).

NOTE: Please realize the equivalency of the terms 'the ONE SELF-as-Self' and 'the One Self'. Even the formula, 'the ONE SELF-as-Triple Point-as Universal Logos' is, in this case, an equivalent. The first phrase/formula simply reminds us of the ORIGIN of the Universal Self; the last phrase/formula reminds us not only of the ORIGIN, but of the *interim transformations* of the ONE SELF, the INFINITE SELF 'BECOMING' Cosmos.

Continuing with the idea of the temporal limitations of the One Self, we find that the One Self cannot Now be in the 'then'. It cannot Now be in the 'yet to be'. That which *was* the *then* exists as a kind of memory within the Now. That which is *yet to be* exists only as a kind of anticipation, also within the Now. The *present forms* of the 'then' and the 'yet to be' are *not exactly* what they *actually were*, or what they *actually will be.*

- Sometimes, it is mistakenly said that the past and the future exist Now, but in relation to the *exact form* that configurations of variables in the past *took*, and the *exact form* which configurations in the future *will take*—the statement of asserting the existence of the past or future in the Now, is not *exactly* correct.

Formal configurations are unique and unrepeatable throughout all Eternity. Archetypal, Numerical 'Arrangements' of Entities as Abstractions in the World of Being (the higher levels of the total World of Becoming) are, however, more enduring and abiding, and may show a *virtually* unchanging stability (in-Cosmos) regardless of the fluctuations of the lower forms in the lower Worlds of Effects/Fabrication which, consciously or unconsciously (under Fohat) attempt to 'follow Their Pattern'.

With this in mind, these Great (Platonic) Forms in the World of Being can be considered relatively *the same* in the past or in the future—or, at least, virtually the same, as some Planned Modifications *do* occur in the World of Being according to the Schedule of the Cosmic Plan. Thus, of *Them* (far more than of forms in the Worlds of Fabrication/Approximation/Effects) it can be said that "the past exists Now" and "the future exists Now". It is possible to say this because the Great Archetypes, the Great Forms, the Great Numerical Arrangements (being, Themselves, stable *Designs* in the Universal Purpose) are not Really *variables* (at least with respect to the particular Cosmos in which They inhere). At least their rate of inter-Numeric Interplay is very slow and measured. With respect to an infinitude of other Cosmoses, past and to come, the Great Archetypes are, of course, variables.

Continuing our examination of the 'then' and 'yet to be' in terms of the 'Now', the 'then' and the 'yet to be' are not (existing) Now (at this very instant). Only when the 'then' *was* the Now *did* the One Self *do, move,* and *change.* Only when the 'yet to be' *will be* the Now *will* the One Self *do, move,* and *change.* From the Great Point of View of the One Self-in-Cosmos, formal configurations of variables make no 'Real' difference in Its *synthetic,*

Cosmos-embracing Perception of Time. For this Being—Really, all things are occurring in the Now (which, for the Universal Logos, includes the on-going sequence of intermittent Cosmo-Objective Nows {ultimate moments} and intermittent inter-moment instants {Cosmo-Subjective Nows, occurring between ultimate moments}).

Regardless of the Universal Logos' sense of a relatively *seamless* Eternal Now, Actually, notice is taken by the Universal Logos of the Total Formal Cosmic Configuration (the relationship of all variables in-Cosmos to each other) at every ultimate moment (and *between* each ultimate moment). This would mean an *unthinkably huge* number of Logoic Perceptions/Registrations per human second. Ultimate moment by ultimate moment, those Cosmic Configurations that precede and those that follow are noted, and thus, Actually and for the *pragmatic* purposes of bringing the Formal Aspects of the Cosmos to their Logoically-Intended Consummating Design, an *intra-Universe global* past and future *are*, indeed, noted. Thus, for instance, our Planetary Logos is said to have a "time-space Schedule", and this may be equally true of the Universal Logos.

To the consciousnesses immersed in the Lower Worlds of Cosmos (the Worlds of Approximation), the Now is the *ever present moment.* Really, there are *two kinds* of 'quantized' Now in Cosmos:

- quantized Nows (or ultimate moments) within the Fohatic Worlds of Fabrication/Approximation; these can be called Cosmo-Objective Nows and
- quantized Nows which correspond to inter-moment instants (and which are *uncognized* by consciousnesses immersed in the Worlds of Fabrication/ Approximation, but *are* cognized by Beings within the World of Being; these can be called Cosmo-Subjective Nows.

It can, as well, be hypothetized that there is a kind of:

- continuous-in-Cosmos Now, which is Cosmically 'Timeless', and exists independently of the two previously mentioned kinds of *quantized* Nows. Such a third kind of Now would 'flow' for the duration of a Cosmos; Cosmic Duration would be equivalent to *One such Now.* In a strange way, this third variety of Now in Cosmos (a kind of Eternal Now) would, nevertheless, be *quantized* for it would endure only for the duration of a Cosmos, and then *cease.* Thus, in no sense, could it be considered *absolutely continuous* as is the ETERNAL NOW.

The Cosmic Now (in its two blatantly *quantized* forms) can also be considered the only moment in which the Universal Presence *moves, does,* or *changes.* A true 'moment', remember, is not yet *humanly* measurable, and yet there is, for any limited Cosmos (and all Cosmoses are limited) a *smallest unit of Time* which we might call the 'ultimate moment in-Cosmos'. The Cosmic Now (of the first variety, obtaining in the Fohatic Worlds of Fabrication) does not *continue between* those time-quanta we are calling 'ultimate moments'. That Now exists only 'during' or 'upon' ultimate moments.

Only the Cosmic Eternal Now *continues between* ultimate moments ('filling', as it were, all inter-moment instants). Of course, the ETERNAL NOW *continues,* as well. The ETERNAL NOW, however, is not a *Cosmic Measure* but, rather, a *Super-Cosmic Measure* (if the ETERNAL NOW can be called a *measure* at all!). It is the Cosmic Nows (in their three varieties) that are the *Cosmic Measures.*

Summarizing,

1. The first type of Cosmic Now obtaining in the Fohatic Worlds of Approximation occurs only 'during' ultimate moments; it can be called the Cosmo-Objective Now.
2. The second type of Cosmic Now occurs only 'during' inter-moment instants when the Worlds of Approximation 'disappear'; this second type of Cosmic Now is uncognized in the Worlds of Approximation. It can be called the Cosmo-Subjective Now.
3. The third type of Cosmic Now should be called the Cosmic Eternal Now and endures as a *relative continuity* throughout the Universal Manvantara.

Ultimate moments are limited by the most rapid possible frequency of particulated Fohat in the Lower Worlds of a given Cosmos. Particulated Fohat manifests as what we can call 'ultimate particle/events' that can be conceived as the smallest and most evanescent "holes" which Fohat "digs in Space" (through Acts of Self-Perception). If there is a *limit* or *ceiling* to the frequency of 'appearance'/'disappearance' of the 'ultimate particle/event' in-Cosmos, then there is a measurable 'ultimate moment' within Cosmos (at least, 'within' the Fohatic Worlds of Approximation).

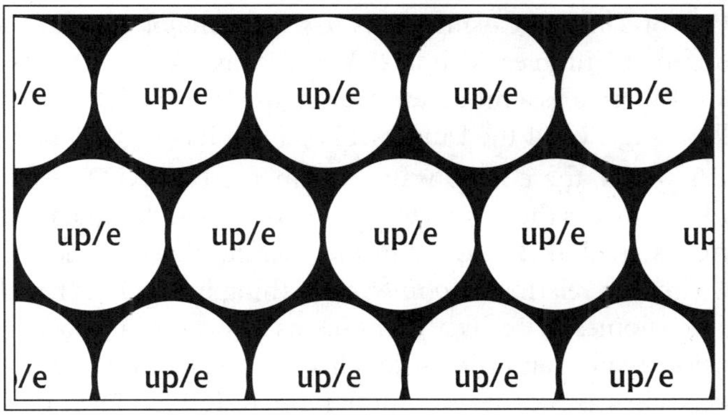

up/e = ultimate particle/event

We must ask, however, Is there any 'time' *between* 'ultimate moments'? Although no movements (motions of ultimate particle/events) occur (in the Lower, Fohatically Fabricated Worlds of Cosmos) in those seemingly instantaneous 'between times', for the very reason that the Lower Worlds (composed entirely of ultimate particle/events) 'disappear into non-objectivity' through 'Fohatic Inattention', it might be said that the Cosmic Eternal Now *persists through* the oscillatory 'disappearance' of the Lower Worlds and, also, through the disappearance of the kind of *time* existent in those Lower Worlds. The Cosmic Eternal Now endures 'unperturbed' throughout the exceedingly numerous appearances and disappearances of 'Now' in the World of Fabrication, the 'Mosaic' World *created* and *sustained* by Fohat.

The 'between times' (between ultimate moments) in a Cosmos cannot properly be called 'Now' from the perspective of consciousnesses immersed incognizantly within the

World of Fabrication. For them, such 'between times' do not even exist. Within those worlds, *time* is definitely quantized and discontinuous. As can be gathered from the theory of the three kinds of Cosmic Now, the quantization of Time in-Cosmos seems to vary with the Cosmic Dimension. In-Cosmos (as opposed to 'within' the INFINITE-SELF), there is no *absolute* CONTINUUM, but there is a kind of intra-Cosmic Time Continuum (which applies to the World of Being). The only *true* 'TIME' CONTINUUM that can possibly *be,* has its being 'within' the INFINITE SELF—WHICH 'ABIDES' ever in the ABSOLUTE ETERNAL NOW.

It might be said that, the so-called 'times' between 'ultimate moments' provide a kind of interface between the lowest form of Cosmic Now and the Cosmic Eternal Now. Perhaps, in that State of Cosmic Eternal Now, there may be gathered *intimations* of the ETERNAL NOW. Since the Universe is *limited,* the Eternal Now within a Universe is only *virtually* Eternal—Eternal for all practical purposes in that Cosmos, and for the Duration of that Cosmos.

It might also be said that for the Universal Logos abiding in the Cosmic Eternal Now (and 'dwelling' on Its own Supreme 'Plane'), the entire Universe *elapses* in but One Moment (as measured against the Infinite Time Line). Within that One Moment, however, are virtually countless fluctuations in the Cosmic Configuration, and, thus, virtually countless occurrences of the lower form of Cosmic Now in the Fohatic Worlds of Fabrication. Perception of Time in Cosmos thus has *three* major *contexts,* and the *sense of Time* is vastly different within each context. What seems like an Eternity (a Cosmic Eternity) in the Lowest Worlds, is a moderate span in the World of Being, and simply a 'Moment' to the All-Seeing 'Eye' of the Universal Logos on Its Own Highest Plane.

In *ultimate* TRUTH, there is no way ('within' the INFINITE SELF) to designate a moment of any kind, just as there is no way (in-Cosmos) to designate an 'actual' *point.* In relation to moments, the INFINITE SELF 'tolerates' no division, and moments certainly are *divisions* of Time. In relation to *points,* everything in-Cosmos has dimension and a Real *point* is dimensionless. (Really a *point* seems to occupy a "no-man's land" midway between dimensionlessness and dimensionality.) While a point is an abstraction which cannot *actually* exist in-Cosmos, it also cannot *be* in the INFINITE SELF, which is 'POINTLESS').

We must explore whether it is possible for a point Really to exist in the World of Being, the World of Archetypes, where *spatial measurement* is almost certainly not necessary. Perhaps a Real Point within the World of Being would be *omnipresent* but *differentiable* from other points. Perhaps, such Points would simply be called 'Points of Ideational Distinction'. Some deeper questions are here implied:

- Can any object lack extension and still be Real?
- Can any object be non-spatial?
- What is the *form* of an idea?

An 'ultimate moment' is *actual,* whereas a *point* never is REAL (though it *may* be Real, at least as an *idea).* Perhaps, there are ways in which a *pure moment* (call it an *infinitesimal moment)* can approximate a *point,* but never in-Cosmos. An 'ultimate moment' is never instantaneous (i.e., as brief as possible while still remaining, theoretically, measurable). Instantaneity is *too fast* for an ultimate moment which, in-Cosmos, is a precipitated *actuality* with theoretically measurable duration.

- ♦ An *infinitesimal moment* (a moment of infinitely small duration, while still *having* duration) is (with respect to Cosmos), an abstraction just as a *point* is. The actuality of an 'ultimate moment' is expressed as the duration (unit of time) between a specified Cosmic Configuration, and the *next* Cosmic Configuration.

Think of the pattern seen in a kaleidoscope at one moment, and the next pattern that is seen after a turn. The inter-moment instant ('between' ultimate moments) may be a measure of how long it takes to 'turn' the Universal Kaleidoscope, or the 'turn' may occur in virtually 'no time at all', *after* a period of 'decision-making' within the World of Being, 'during' the inter-moment instant. Within the present hypothesis, that Kaleidoscope displays differing patterns in 'sometimes', for Time is *quantized.* The Kaleidoscope does not present a different pattern *all the time!* So there is no *continuous 'turning'* of the Cosmic Kaleidoscope. It 'turns' suddenly and cyclically.

The ETERNAL NOW *underlies* the 'present moment'. Remembering that in-Cosmos, 'ultimate moments' are to be considered *quantized* (discrete and discontinuous), in any Cosmos there are limitations upon what we (when in the Lower World, the Worlds of Fabrication) call the 'present moment'. The 'present moment' (which in-Cosmos is the 'ultimate moment') is, as it were, an *aperture* for the ETERNAL NOW or PRESENCE to enter Cosmos. So, we might say that:

- The present moment, the ultimate moment, is *the current presentation of ETERNITY* to the benighted consciousnesses immersed within the Fohatically Fabricated Worlds.
- The present moment, or the 'presented moment', is all of ETERNITY the Lower Universe has 'Time' for!
- The 'present moment' (or ultimate moment) provides for the *possible-in-Universe Presentation of ETERNITY.*

In the Worlds of Fabrication, there is no other 'Time' in which to apprehend ETERNITY. This is true, as well, of the Cosmic Eternal Now. It, too, though far more accessible than the ETERNAL NOW, can only be apprehended by the immersed consciousness through the aperture of the lowest Cosmic Now or ultimate moment.

It may well be that for apparently separate beings-in-Cosmos, the ETERNAL NOW can only be experienced (or, rather, 'in-perienced) in its COMPLETE CONTINUOUSNESS 'within' the ALL-IN-ALLNESS which IS *only* during the Universal Pralaya. From a paradoxical perspective, however, the ETERNAL NOW *always* IS and is *always* 'in-perienced' 'within' the 'STATELESS-STATE' of INFINITE SELFHOOD.

- ♦ Be this as it may, due to Cosmic Limitation, when we say we are 'inperiencing' the ETERNAL NOW, we may, in *fact,* be 'inperiencing' glimmers of the Eternal Now (which is the highest degree of simultaneity *possible* in a limited Cosmos, the Lower Worlds of which are extremely discontinuous and quantized). Of course, even Cosmic Nows occur at so rapid a rate (let us say, '*x*-tillions' per second) that even if consciousnesses bound in Lower Cosmos are *not* 'inperiencing' the ETERNAL NOW *all the time,* they may easily have the impression of 'inperiencing' a kind of illusory continuity which suggests to them that it is *eternally Now.* If such consciousnesses could but remember their 'trip' to the

World of Being 'x-tillion' of times per second, they would be deep in the 'inperience' of the Cosmic Eternal Now (the 'inperience' of which *abides* in the World of Being).

The gradual building of the antahkarana makes this kind of *memory* possible, and thus, Continuity of Consciousness (in a deep and true sense) develops. Whether we know it or not, we 'in-perience'/experience the Cosmic Eternal Now with such amazing frequency that we (unbeknownst to ourselves) Really have the Perception of the Universal Logos. (which Being, of course, we Essentially Are). Maybe even the Universal Logos-as-Logos, per se, (limited and discontinuous Being that It Is) cannot even *fully* 'in-perience the ETERNAL NOW.

◆ These kinds of thoughts have to be confirmed by a growing internal 'inperience' of the Mysteries of Time, which begin to be accessible as buddhic impression (with its sense of *simultaneity)* becomes an actuality in consciousness.

Whereas the ETERNAL NOW is devoid of conditions, the lowest Cosmic Now is tied to *conditionality.*

◆ The Now (called in-Cosmos, the 'ultimate moment') is that *available aperture of perception* through which the present Total Formal Configuration in-Cosmos (the Cosmic Configuration) is 'Seen' by the Cosmic Consciousness (for the Configuration disappears 'during' the 'not-Now' (the inter-moment instant).

Everything (every rearrangement in Fohatically Fabricated Cosmos) must 'Occur' *on* (rather than *in* or *during)* the lowest Cosmic Now, and *only* on that Now. There is (Cosmically considered) no other *time* for it to 'Happen'.

- Change (in the Fohatic Worlds) is the slave of the ultimate moment, the lower Cosmic Now.
- Change can only be apprehended (by Those in the World of Being) by contrasting the 'frozen' Configurations 'held' during ultimate moments, during the Cosmic Nows.
- There is no change in the Worlds of Fabrication (i.e., Lower Cosmos) except *precisely at* or upon ultimate moments.

This brings forward a fascinating situation: things/variables in the Worlds of Fabrication do not move *continuously* 'through' space *all the time,* but shift relative position instantaneously (in quantized 'lurches', as it were), from one ultimate moment to the next. Only certain 'positions' of variables/items *relative to each other* are possible (in-Cosmos).

This means that *within* the Fohatically-Fabricated Cosmos, Time is quantized and that there *are* (experientially and 'in-perientially') no *actual* 'between times' (though there *are* 'between times' from the Perspective of the World of Being). If it were possible for a thing to move from one position to another *continuously* (through *all* 'points in Space') the Cosmos would not be a *finite* Cosmos.

◆ Within Cosmos *there is no continuous moving through Space.*

There is only a 'flashing on' in one position' and, to those immersed in the Worlds of Fabrication, an *immediate* (and, almost assuredly *imperceptible)* 'flashing on' in the next position. The 'time taken' for the change, of course, is not *utterly* instantaneous, but, probably measurable, and its duration (i.e., the duration of an inter-moment instant) *may be* equal to an ultimate moment *(if* Cosmos is symmetrical), or, it may be much more or much less. The 'ultimate moment' itself is measured by the fixed duration of a single Cosmic Configuration, the duration of a single 'flashing on' (as measured against the Infinite Time Line).

If there were a continuum in Cosmos, and continuous movement through all possible positions in space between a designated first position and a designated second position, then units of time *less* than ultimate moments would be necessary in order to measure the 'time' it would take to move (through all possible *interim* positions) on the way from the first position to the second.

Because we must, necessarily, have a *finite* Universe, it appears that we must necessarily have a *quantized* Universe as well. Such considerations remind one of the more 'concrete!' example of the condition in the atom with respect to the transference of electrons from one electron shell to the next. However, the idea of the necessity for a quantized Universe can be conceived through philosophical as well as scientific thinking.

What we are discussing here is the *quantization* of the Cosmic Now, the Cosmo-Objective Now in the Fohatic Worlds of Fabrication. That Now is not really an *Eternal* Now, because of time/vibration limitations within lower Cosmos. The Cosmic Eternal Now, does exist, *not* in the World of Fabrication, but *semi-continuously* in the World of Being. Only the ETERNAL NOW is a *true* and *absolute* CONTINUUM, devoid of moments, 'points in Time', etc., and IS, hence, an UNQUANTIZED NOW.

Is the Space within a Cosmos <u>R</u>eally Continuous. (Another way of asking this is: Is Cosmic Prakriti <u>R</u>eally Mulaprakriti?) Whenever we think of Space we usually think of a Continuum. But probably only ABSOLUTE SPACE IS a CONTINUUM, and moving from 'place' to 'place' is *Cosmic Space* is not an continuous movement at all, but is simply:

1. the act of existing in one location
2. disappearing from that location
3. reappearing in another

The Total Formal Configuration of Cosmos appears to reconfigure by *quantum leaps.* This is *magic*—the act of appearing, then, disappearing then appearing again—without 'taking' *continuous time* to *travel through' space.* We appear, here, to be dealing with the quantizing of Time as well as the quantizing of Space. It is important to realize the reasons why Cosmos (and the Prakritic Space {not Mulaprakritic Space} within which It exists) is almost certainly are not a Continuum.

PARABRAHMAN IS a CONTINUUM. From the material or *objective* perspective, Mulaprakriti (Infinite Objective 'Extension', or the Mother of Potentially Infinite Objectification) is also a <u>C</u>ontinuum. But these are the *only* Continua. The Act of 'Creating' a Cosmos is to *separate* or 'serate' (through ever-narrowing Mayavic Self-Perception) the Continuum known as Mulaprakriti, i.e., to create *separation-via-articulation* where none existed. This is done, at first, through the process of SELF-'REFLECTION'(a Pre-Cosmic Dynamic).

Maya is what might be called the 'Reflector-of-Infinitude' capable of *inaugurating the Self-Reflection* of a potentially infinitely 'ideated', hence, potentially infinitely 'articulated' Subjectivity in such a way that a potentially infinite Objectivity is Generated *as* and *from* Mulaprakriti. Mulaprakriti could, thus, be considered "the Mother of all Objectified Possibility" but 'MAYA'-instantly-Maya (as 'CONSCIOUSNESS'-in-'ACTION') makes that 'Motherhood' possible.

To 'Create' a Cosmos, Mulaprakriti is not crudely *divided* per se, (as some dense material 'stuff' would be divided). Rather,

- Mulaprakriti (through the Agency of Maya, the Great Reflector) begins to *show forth* the SELF-'INTENDED' *particularities* 'EXTRUDED' from the FOUNT OF ALL POSSIBILITY and 'Carried' as Ideation by FOHAT/Fohat—the Agent of the INFINITE SELF Who both 'Induces' and causes (through the unfoldment of the Cosmic Process) the SELF-'INTENDED' Reflection to be 'Seen' (by a variety of Subjects/Logoi, etc.).
- Mulaprakriti (in Its more *concrete* form of Cosmic Prakriti) <u>E</u>mbodies <u>*these*</u> *particularities* rather than simply reflecting pure, <u>un</u>articulated or *'infinitessentialized'* Infinitude.

This reflecting is what Mulaprakriti formerly *did* in that State (immediately 'following' the 'FLASHING FORTH' of the 'RAY' of the ABSOLUTE). At that 'time' the Infinite Subject 'Sees' Its Infinitude (Really, the Infinitude of the INFINITE SELF at one remove from ALL-IN-ALLNESS) *homogeneously reflected* in the Infinite Object/Mother—Mulaprakriti. It is unlikely (due to *time constraints*—for Pre-Cosmic Time has, indeed, begun) that the Infinite Subject would 'See' Reflected 'within' Mulaprakriti *all possible articulated forms of INFINITUDE.* When (following the Stage of the 'Reflection of INFINITUDE-as-Infinitude', {unarticulated and *infinitessentialized}*) the SELF-'INTENDED' *particularities* to be manifested in the Cosmos-to-Be are 'Reflected' <u>*instead*</u>*, then,* we have the appearance of what we normally call Matter (or, better, Substance).

Let us return now to the question of Time, Space, and Motion in the probable 'Discontinuum' called Cosmos. What, probably, is Really happening when change and movement in-Cosmos occur, is that only certain *relationships* are possible and permissible in-Cosmos. Not *all* Conditions (States) and not *all* Configurations (positions of Cosmic variables/items relative to each other) are possible and permissible.

- Possible and permissible Configurations should be called <u>Cosmically Sanctioned Configurations</u>.
- Impossible and impermissible Configurations should be called <u>Cosmically Unsanctioned Configurations</u>.

Bear in mind that a Cosmically Sanctioned Configuration is not *necessarily* a Cosmically *Ideal* Configuration, or there would be no Free Will. Such a Cosmically Sanctioned Configuration which was not a Cosmically Ideal Configuration would be merely *allowed* without being Divinely 'Desired' or 'Intended'.

With these thoughts in mind, we can see thatMovement in-Cosmos can be conceived as movement from one Cosmically Sanctioned Configuration to another such sanctioned configuration. But no movement is possible to or from Cosmically *Unsanctioned* Configurations. We are talking here about *Possibility*-in-Cosmos, and *Impossibility*-in-Cosmos. It is worth restating that 'the Possible' need not be 'the Good' in terms of Intended Cosmic Design; the Possible is simply *possible.*

Regardless of the limitations we place upon the Now (whether Cosmo-Objective or Cosmic Eternal), when considering it philosophically, in practical, intra-Cosmic terms, the Now represents (for us in *this* Cosmos) the 'One and Only Arena of Action'. The Now (the moment of the ultimate moment) defines the conditioned, conditional world.

- Change in the Worlds of Fabrication takes place only from Now to Now (or is *noticed* only from one Cosmo-Objective Now to another Cosmo-Objective Now).
- Movement (if it Really occurs), similarly, occurs only from Now to Now.
- Relationship or relation, equally, varies only from Now to Now.

The Now, the Present Moment, is, in the Lower Worlds of Cosmos, the only 'Moment of Opportunity'. Yet, the Now is utterly *immobile* (otherwise it would be unquantized and, hence, further and even *infinitely,* divisible). The deep question arises: How does *change* Really occur?

The Cosmo-Objective Now is not the *'Infinitesimal* Moment'; such a Now is the 'ultimate moment'. The ETERNAL NOW, however, is no *REAL* 'MOMENT' at all; there are no REAL 'MOMENTS'. The ETERNAL NOW could be called the ETERNAL UNBOUNDED MOMENT. While the 'infinitesimal moment' is conceivable, it can never be *actual* in any Cosmos. The ETERNAL MOMENT (which is the *immeasurable time measure* of the ETERNAL NOW) is ever infinitely smaller than the 'Infinitesimal Moment'. In fact, the time value of the ETERNAL MOMENT is zero.

Within' the INFINITE SELF, for all DURATION, the TIME is always and ever NOW. Within all Universes (including the World of Becoming and Its Aspects, the World of Being and the World of Effects) it is also, ever and always, Really, Eternally Now and, intermittently, Cosmo-Objectively Now (and, alternatively, Cosmo-Subjectively Now). Certainly, all of these Nows in Cosmos (the Cosmic Eternal Now, the Cosmo-Subjective Now, and the Cosmo-Objective Now), because of the discontinuity of Their manifestation (albeit a very *different kind* of discontinuity), have *limitations* compared to the ETERNAL NOW.

The Cosmo-Objective Now, based upon the occurrence of ultimate moments, (and the simultaneous Cosmos-Wide apprehension of all ultimate particle/events) is, for all practical purposes in the World of Fabrication, a *seemingly Eternal* Now—Eternal *enough* for the consciousness *immersed* in the World of Fabrication. For such beings (incapable of perceiving or responding to the 'ultimate moment' until, through Identification, they *become* the One Cosmic Being which they always *were* and *are),* it is always, more or less, Now, and Time *seems* to "flow along" uninterruptedly.

♦ Once all beings in-Cosmos begin to identify with the Universal Logos (by becoming increasingly conscious 'during' Cosmo-Subjective Nows (because of the sensitizing Antahkarana) a growing appreciation of the relatively continuous Cosmic Eternal Now may begin to dawn—the appreciation of Cosmos as One Great Temporarily 'Frozen' Particle/Event.

Then, even 'later', as the 'Time' for the Universal Pralaya has arrived; the ETERNAL CONTINUOUS NOW may begin to be apprehended in a truer way than ever possible at 'lower' positions within the Divine Emanatory Stream.

It is worth restating the doubt concerning whether *any* intra-Cosmic being (including the Universal Logos) can Really, *fully* and *completely* experience the ETERNAL NOW. Perhaps, in those 'inter-moment instants', when the entire patently Objective Cosmos 'disappears' in a flash, there may be (perhaps?) the instantaneous 'inperience' of the ETERNAL NOW (or more likely an intensified apprehension of the Eternal Now), but the 'inperience' might as instantaneously disappear the 'moment' the Universe 'flashed on' again. The multitude of in-Cosmos C/consciousnesses would not register it, for such a C/consciousnesses are 'pre-occupied' with the Universal Task. The Universal Logos, of course, is 'involved' in the World of Fabrication chiefly *through* His emanative Extensions, and, thus, is able to 'Abide' in the Cosmic Eternal Now for the entire duration of Cosmos. More likely is He than any (from His loftiest of all possible Cosmic Vantage Points) to identify somewhat with the ABSOLUTE ETERNAL NOW. Within Universal Pralaya, of course, 'IDENTIFICATION' for All who are, 'then', ONE-as-NONE is *complete.*

Nevertheless, for all of us (in our *limited* aspects—for we *do,* also, have unlimited ones), the Cosmo-Objective 'Now' is *more* than "good enough!" As well, intimations of the Cosmo-Subjective Now can grow. Of course, the kind of 'now' we *actually* experience is merely an agglomeration of a huge number of true Cosmo-Objective Nows (true ultimate moments), but, for the *'time being!',* it will have to do.

- As our sense of the Cosmo-Objective Now becomes ever more refined, broadened, and *immediate,* we will realize its immense value, for the Cosmo-Objective Now is pregnant with all Cosmic Possibility, at least in the World of Approximation. This Now 'contains' (for the World of Fabrication) the total Cosmic Potentiality of the moment—not *all* POTENTIALITY, but the Total *Cosmo-Objective* Potentiality. Who is *swift enough* to know the immediacy of this Now, is ever rich in the Knowledge of things as they *actually* are. Who lacks the apprehension of this Now is ever impoverished by being doomed to crude temporal perceptions.

Through Knowledge of the Cosmo-Objective Now (such as is possessed by Those within the World of Being) all imprisoning kaleidoscopic conditions (possible in the World of Approximation, the World of Fabrication) can be understood and released. To live increasingly in that Now is, colloquially, the means of "keeping current" and of being 'in the Current'—which is the ever-present Divine Presence (to the degree It is present and can be sensed in the World of Fabrication)—the means of attuning with the Supreme Consciousness as It penetrates the Fohatically Fabricated Worlds. When one is in that 'Current', one has 'currency' with the Presence. Excuse the play on words, but illumination may arise thereby. Focus upon the immediate moment in Fabricated Cosmos (the Cosmo-Objective Now) slides *imperceptibly* into a realization of that which is even more Essential (and which is revealed by the Cosmo-Subjective Now.

Growing facility in apprehending the relative 'Timelessness' of the Cosmo-Subjective Now, confers Great Perspective upon and skill-in-action within the Worlds of Fabrication. At length, as ascent continues, there arises the Temporal Apprehension of Those

Who are consistently focussed upon the higher 'levels' of the World of Being—Those Who apprehend and contemplate from within the Cosmic *Eternal* Now. For all practical purpose, the Cosmo-Objective Now and its ongoing more Subjective Background, the Cosmic-Subjective Now, seem to fluctuate so rapidly that they are, almost inescapably, perceived as the *same.*

◆ This means that immersed consciousnesses do not sense the discontinuity of the Cosmo-Objective Now (the usual Cosmic Now, and the ultimate moments upon which this Cosmo-Objective Now is based). They sense a 'flow' of time where there Really is none. As their consciousness begins to deepen, they begin to 'feel *between* the discontinuous moments', through growing unconscious registration of the Cosmo-Subjective Nows, and gradually begin to attune with the Universal Presence, which truly ever dwells in fullness in the Cosmic Eternal Now (a Now which can be said to abide *even 'during'* the 'frozen' ultimate moment which characterizes the Cosmo-Objective Now, and certainly during the Cosmo-Subjective Now).

In seeking to differentiate between the Cosmo-Subjective Now, Cosmo-Objective Now, and the Cosmic Eternal Now, we might say that:

- The Cosmo-Subjective Now is a kind of 'polar opposite' to the Cosmo-Objective Now.
- Within both such Nows the immersed consciousnesses are pre-occupied.
- Within Cosmo-Objective Nows, the immersed consciousnesses are pre-occupied with the blatant illusory presentation of the objectified Cosmic Configuration.
- Within the Cosmo-Subjective Nows, these same consciousnesses, released to 'positions' within the World of Being, are pre-occupied with the next arrangement of the Cosmic-Configuration to be Willed.
- Both of these kinds of Nows are, as it were, 'entrapped by the rhythm of Ontological Oscillation'.
- The 'View' from the Perspective of the Cosmic Eternal Now is very different, however. The sense of Cosmic Continuity there prevails, and the Consciousness is not ensnared by fluctuation. 'Moment consciousness' is not fragmented, and Time is not quantized to the same 'frenzied' totally pre-occupying extent.

Perhaps, it could be said that, with evolution, the apprehension of these three types of Cosmic Now (the Cosmo-Objective, Cosmo-Subjective, and the Cosmic Eternal) proceed simultaneously. Indeed, in will be long before the human unit is in a position to ascertain the existence of ultimate moments upon which the simultaneous perception of the Cosmos-Wide, Cosmo-Objective Now is based, for speed of conscious registration does not yet suffice.

Already, however, some advanced units of the human family are having some dim apprehension (through flashes of simultaneity) of the Cosmo-Subjective Now (for the obvious 'flow' of Time is being questioned). There are even those (for instance, the poet, William Blake) who are receiving some intimation of the Cosmic Eternal Now which substands all objective Fohatic Variation. For practical purposes, then, one must come to realize that there is a ceaseless discontinuity, but, as well, learn to identify with the semi-continuity of sustained Universal Spirit-Awareness which underlies the ceaseless Fohatic discontinuity. Progress grows with facility in *abstraction.*

No change or modification can take away the Cosmic Eternal Now. No change or modification can remove the Presence through both the 'on' and 'off' 'times' of the Fohatic Worlds. The Eternal Now is all-devouring in Cosmos, for nothing can preserve (*exactly as they were then)* the ever-mutating changes in the World of Illusion. The past can be remembered, but the form of it simply cannot be retained, for infinite are the number of past combinations and configurations—literally, *infinite.* Yet, throughout Cosmos, the Cosmic Eternal Now *persists* unchanged.

- The Cosmic Eternal Now is the 'doorway' to the highest Cosmic Initiations (for all problems of Time, Space, and Relationship are solved within It). The Cosmic Eternal Now is the doorway, not only to the present, but to the Supreme Divine Presence of this Cosmos. If one is to *live* in the Fabricated Cosmos, the only available 'time' to do it is Cosmo-Objective Now. If one however, is to *live* in Cosmos in the deepest possible way, the only 'point' from which to *live* is from the All-Seeing Pinnacle of the Cosmic Eternal Now.

We have said that beginning to identify with the relative constancy Cosmic Eternal Now brings release, but release from what? From the life-patterning compulsive momentum of past configurations, and from the fearful anticipation of future configurations. Both past and future are effectively rendered harmless in the Cosmic Eternal Now, from which Perspective, past and future are as One Immediate Event.

In the ALL of Cosmoses-Gone, there have been an infinitude of combinations gone, relations gone; in All of Cosmoses-to-Come, there are an infinitude of combinations to come, relations to come. In all the accomplished infinitude of modification past it *was* always Now (Cosmo-Objectively, Cosmo-Subjectively and/or Eternally), and in all the destined infinitude of modification future, it *will be* always Now (Cosmo-Objectively, Cosmo-Subjectively and/or Eternally).

- The Single Eye of the One Universal Observer immovably rooted in detached Nowness, is *blind to modification,* (as we 'lower' immersed 'Rays' understand modification). Yes, modification is 'Seen', but the 'Seen' modification does not imprison the Observing Consciousness in Time.

The Cosmic Eternal Now is Timeless (at least with respect to the particular Cosmos of the Moment). Though the countless differences presented by endless kaleidoscopic modification are 'witnessed' by the One Divine Observer, the substratum of the Cosmic Eternal Now *dissolves* those differences. (The 'Pond of Cosmos' has a huge diversity of ripples and eddies, but the Pond remains continuously the Pond.) So, though there is endless movement and unrepeatability, (and the sequential-cum-simultaneous changes in the Cosmo-Objective Now are registered), yet, the Cosmic Eternal Nowness abides 'beneath' all discontinuous disturbance. To the Eye of the One focused in the World of Being (though it 'See' the Nows of ultimate moments) it is, nevertheless, always *Cosmo-Eternally Now.* To the '**8**' of the One, it is always *Cosmo-Eternally Now.*

To the One Cosmic Observer, 'sequence' means something quite different from what it does to the limited consciousness. While there is a Cosmic Configuration that precedes, and a Cosmic Configuration that follows, all reconfigurations in-Cosmos are 'Seen' as occurring simultaneously (Cosmic 'Frame' by Cosmic 'Frame'). There is only, on the one hand, the One Observer (observing Cosmos at all *Cosmo-Objectively-available 'times')* and, on the other hand, the innumerable Cosmic Changes before the Cosmic Eye/**8.**

The Cosmic Sense of a comprehensive and continuous Cosmic-Eternal *Now* is, theoretically, much stronger than the sense of a chain of distinct identifiable ultimate moments (however much each moment and the Configuration of the moment are registered). The Cosmic Observer will hold Its Position on the Pinnacle of Observation for the duration of the entire Cosmos, and will 'See' into *all* 'Places' in-Cosmos.

- Therefore,one 'time' is very much the same as all 'times'; one 'place' is very much the same as all 'places'. For such an Observer, Who constantly Beholds *everything* in Fabricated Cosmos, the maximally significant time in the World of Fabrication is Now, and the maximally significant *place* is 'Here'. And all this Fohatic Change, which, through a 'simultaneitizing Eye', is barely 'Seen' as Change, is undergirded by the Eye that 'Sees' all Cosmo-Objective Nows as a Single Cosmic Eternal Now (like a Single Cosmic Particle/Event).

It is awe inspiring to think of how many *nows* are seen by the Universal Observer:

1. Every crude 'now', for He 'Sees' through all eyes.
2. The many more precise ('scientific') 'nows' that are of ever lessening duration.
3. The Now of each and every ultimate moment which changes, discontinuously but constantly, with each change in ultimate particle/events.
4. The Cosmos-Comprehensive Cosmo-Objective Now which 'Sees' all the multitudinous changes in the entire Cosmic-Configuration as One Simultaneous Change occurring 'Now', i.e., as one great ceaselessly Reconfiguring Movement.

Really #3 and #4 are two aspects of the same Universal Logoic Perception, except the focus in #3 is upon each Change in the Configuration, upon each ultimate moment, and the focus in #4 is upon the Cosmos-Wide Cosmic-Kaleidoscope which seems to blend all this multiple changes in the World of Fabrication into one seamlessly unfolding Change.

5. The Cosmo-Subjective Now might be called the 'Interval of Imagination and Will', used to propel the Cosmo-Objective Configuration towards fulfillment of the Design-at-the-Beginning.
6. The Cosmic Eternal Now which collapses all Cosmic Past, Present, and Future, and all sense of Cosmic Space, into one Moment of Cosmo-Eternal Time, the Duration of a Cosmic Particle/Event. Naturally, all differentiation in terms of Time and Space ceases from such a Perspective.
7. The ETERNAL NOW, which obliterates Time and Space forever. How much of the ETERNAL NOW is truly apprehensible, even to the Contemplating Universal Logos, is questionable.

With respect to the Now Perceived by the Cosmic Observer in #4, above: even while the preceding and following Cosmic Configurations are 'registered' and the preceding and following 'ultimate moments' (the 'times' available in-Cosmos) are also noted, the Observer of Fabricated Cosmos 'Sees' everything as One Great All-Including Action, one Great Spontaneous Movement *<u>almost</u> always* changing before Its One All-Observant Eye—for the term 'always' necessitates a continuum which cannot exist in Fohatic World of Fabrication.

It can be seen how, under such *observational circumstances*, the usual sense of *sequence-induced Time-boundedness* would be offset. To the limited human consciousness, the petty configurations of the past and those to come, loom overly large, and most often obliterate even the kind of relatively refined *now* available to the human consciousness. To the Universal Observer, however, Intra-Cosmic Happenings appear as a Single (virtually) Seamless Action (without separated parts and pieces), simply a constant Cosmic Repositioning *almost always* occurring. (The phrase *almost always* is used because the Cosmo-Objective Now should be considered *quantized* and, in the Fohatic Objective Cosmos, at least, *not* an absolutely continuous *presence*, as previously discussed.)

If, for the One Universal Observer, there is any sense of Time (as a great and extended Context) somewhat analogous to what lesser beings experience, it must be in relation to the endless sequence of Universes which have preceded and which are to come. Does the One Observer have 'Memory' of Universes Past? Perhaps, yes, just as the human being (as a personality) can have memory of incarnations past. Perhaps, however, the 'SELF-as-Triple Point-as-Universal Logos/Divine Observer' should be considered only the *Experiencer of Its Own Universe*, in which case the Infinite Memory required would not be accessible, just as the memory of past lives is not usually accessible to the personality, per se, but only to the 'Over-lighting' Soul.

From this perspective, only the INFINITE SELF ITSELF (or perhaps the Super Cosmic Infinite Self/the Super-Cosmic Trinity) would 'retain' the infinite experiences of infinite Cosmoses past, and such 'Memory' might well be inaccessible, or only partially and occasionally accessible, to the Incarnating Universal Logos. However the case may be, it would seem that while *in-Cosmos* there would be no need to reclaim an Infinite Memory of Configurations Past. The Chain of Universes Is not, so reason would dictate, progressive and developmental. Each Universe can be conceived as a "World Unto Itself", and has (so reason seems to indicate) no bearing upon the Universes to Come, nor is It 'borne upon' by Universes which preceded It.

Thus, from the Universal Perspective, as the Universal Observer Perceives Intra-Cosmic Action, there is only One Great *virtually* continuous Act, and *not* many little acts (though that Observer, necessarily, cannot be incognizant of such). We human beings, in viewing our lives, see many little fragmentary actions, but what is the vision of the Overlighting Soul? For the Cosmic Observer, also, there is very much the 'One Eye' and the 'One Act'. The One Action is Really what **8**—the Cosmic Actor/Observer—Do. **8** Am the Cosmic Actor/Observer, *not* the INFINITE ACTOR/OBSERVER (for, as IT 'ABIDES' wholly within the SELF-CONTAINED 'WORLD OF CAPITAL LETTERS'!) the INFINITE SELF, per se, has no 'ACTING' or 'OBSERVING' ASPECT.

♦ This is My ('M**8**') Universe, and Its many Activities are My One Act. **8** Move Now, **8** Act Now, **8** 'Happen' to 'MySelf'—Now. By the time the reader finishes this treatise, it may well seem a matter of course, to consider himself (transcendentally) as the Actor 'behind' and 'within' the Universal Act, and, therefore *responsible* for It as well!

Considering this larger Universal Perspective (into which we can imaginatively transpose our consciousnesses), if one has to be *attached*, let attachment be to the large Cosmo-Objective Now (as It fluctuates with the Cosmo-Subjective Now, and as they *both* blend into the Cosmic Eternal Now), and not to the infinitude of petty conditions created in the Now. The 'SELF-as-Infinite Subject/Object-as Universal Logos-as-**8**(!)' is a form-creating/form-devouring BEING-as-Being. 'My' Great Act of creating/devouring occurs only in the Cosmo-Objective, Cosmo-Subjective and Cosmic Eternal Nows. Though forms come and go endlessly, **8** can never be divested of the 'SELF-as-Self-as-**8**', Who dwells only in the Triple Cosmic Now. **8** cannot be divested of the Presence. Obscure Myself however **8**-as-I will through *Veiling* (MY OWN ETERNAL 'GAME' of SELF-'INFLICTED' Ignorance), **8** can never Really be rid of the Triple Cosmic Now.

In Fabricated Cosmos, the Cosmo-Objective Now is My Moment-of-Moments, My only Moment, even though, as a human being, **8** apprehend its immediacy but dimly. When temporarily extricated from Fabricated Cosmos ('x-tillion' of 'times' per Earth second), My Moment of Supervisory Purposefulness occurs, the Moment where 'Decisions' are 'Made' that *will make a difference* in the Worlds Below. In *all* of Cosmos, however, there is the still more *Essential* Moment—the Moment of the ongoing Cosmic Eternal Now—in one sense, My Only Moment for the entire duration of Cosmos. From the Perspective of the utmost Reality that can be achieved in-Cosmos, there has never been another Cosmic Moment, nor will there ever be another Cosmic Moment *other* than the Cosmic Eternal Now.

- Who Am '**8**' to be happy or sad? Are these not merely ignorant responses to ceaselessly passing conditions? **8** can be one, therefore, who greets fearlessly every new presentation in the Cosmo-Objective Now because **8** Am beginning to apprehend the 'Will-laden' Supervisory Cosmo-Subjective Now, and Am, increasingly, deeply rooted in the Cosmic Eternal Now. The relativism of good and evil come and go, but let the synthesis of the Present Eternal Cosmic Moment be 'in-perienced'!

The Cosmic Eternal Now is infinitely rich regardless of evanescent conditions. That ever-Present Eternal Moment is the *Divinely Presented Moment.* To fear is to lose the Presented Eternal Moment; to find and embrace the Presented Eternal Moment is to destroy the limited ego.

- The realization of the Cosmic Eternal Now is the 'doorway' to the realization, first, of Cosmic Identity (which includes All that *Is,* Now) and, then, at long last, to BOUNDLESS IDENTITY, which is the INFINITE SELF, which NOW and *forever* has *been* as IT ever IS.

On Change, and What Can Change

Change is a process by which one relationship follows another. By means of change, a given Configuration-in-Universe is followed (after an inter-moment instant has elapsed) by a different Cosmic Configuration. When we say that "all things change" we mean that all items-in-Universe seem to *move,* i.e., seem, from moment to moment, to occupy a different position relative to each other.

What can change? *Conditions* can change. Conditions are ESSENTIALLY illusory. But is ESSENTIAL BEING a condition? Can BEING change? BEING IS Being Is being—the ONE ABSOLUTE LIFE subject to progressively limited circumstances (i.e., subject to SELF-'IMPOSED' *objectifications).* BEING, thus is a CONSTANT. BEING 'RADIATES' ESSENCE 'BECOMING' the World of Cosmos.

This means that the 'RAY' of the ABSOLUTE and all 'Rays' of the ABSOLUTE are *of* that CONSTANT. If BEING is a CONSTANT and ITS 'RADIATION' of ESSENCE which *becomes* Cosmos is *of* that CONSTANT then, there is naught that can modify my BEING-as-Being-as being. Naught can modify my *being* because there IS naught else other than IT—i.e., MY BEING. Any other *thing* is but an illusory modification, a mere *seeming* only. How can **8** (the **I**-as-I, and the I-as-**I**) *be,* in my *ESSENTIAL* nature, aught but that which is the PERFECT ONE—*perfect* regardless of conditions? All changes are alterations of conditions—for better or for worse. All changes are intra-Cosmic, the only *field* in which 'better' and 'worse' have any meaning. (It could be said that a variety of Change occurs in the Super-Cosmic Realm of the Super-Cosmic Trinity.)

♦ The CHANGELESS REALM is "Beyond Good and Evil." Good and Evil are modifications, changes of configuration, changes of relationship. While there is no ABSOLUTE EVIL, it is psycho-spiritually helpful to think that there *is* an ABSOLUTE GOOD (unlike any Good we know and, strangely *inclusive of* what we call Evil).

On Combination

What is a combination? Here are a few ways of approaching the definition:

- A combination is a relationship of items-in-Cosmos, an 'engagement of variables'. The word 'engagement' speaks to the magnetism which is so often present in the production of a combination.
- A combination is always multiple. It is a relationship of E/entities, whether primary (authentic) E/entities, or secondary and tertiary (i.e., 'inauthentic') entities. If any positioning of one E/entity with respect to another or others is *sustained,* then, by means of that sustaining, a combination is created. But do all combinations have duration? The answer (according to the definitions used in this treatise) would have to be, yes, even though the duration of a combination might be as short as *two* 'ultimate moments' in a particular Cosmos. If an engagement of variables endures for only one ultimate moment, that engagement cannot be called a *combination* but merely a *configuration.*
- In practical consideration, a combination is a least a somewhat *enduring* configuration. That which is combined can be uncombined. All combinations are evanescent, temporal, temporary, transient.

Let us consider the idea of *combination* in relation to the past and the future. The past and the future are, so to speak, *not of moment*.(i.e., not of the *present moment* which is the 'Moment' of REALITY and, of course, Reality, as well). When we think of the past, we think of the *combinations* which existed in the past. When we think of the future, we think of the *combinations* that will exist in the future. All combinations are *actual* but UNREAL. The same is true for those combinations which exist in the present moment. While REALITY is ever present in the present moment, the REALITY of which we speak is certainly not the *combinations* of the present moment.

The REALITY is found in the PRESENCE which underlies and pervades the combinations of any moment. The past is doubly UNREAL. Its combinations are gone, and, also, the PRESENCE 'keeps pace', as it were, with the present moment which always leaves the combinations of the past behind. The future, in the same way, is doubly UNREAL. The combinations which will characterized it, do not *yet* exist (certainly not in detail and on all levels of form) and the PRESENCE has not yet 'reached' that which will be the present moment in the future.

- There remains to us a very *present* NOW. NOW is the Moment of PRESENCE. The past no longer 'possesses' PRESENCE; the future does not yet 'possess' PRESENCE. The present moment, then, is the *Moment of REALITY*, but, again, it is not the *combinations* of the present which are REAL. These combinations are merely the *actualities* of the moment. REALITY is the PRESENCE/Presence ITSELF/Itself which is only *present* in the NOW/Now.

The combinations of the present moment, however, must be attended to from the maximal point of tension. The combinations of the past, while no longer *actual* and *present* must be studied, just as those to come in the future must be intelligently *anticipated*.

Thus fortified by study and anticipation, it will become possible to manipulate the combinations of the present in accordance with the Divine Plan, and more properly ensure that the combinations of the future, which evolve through the repositioning of the combinations of the present, will be more in accord with That which Is Intended (by the Universal Logos—and many lesser Logoi, by reflection).

How does one combination change into another? Through 'movement' of constituent variables (by whatever *discontinuous* means) from one relative position to another. (All positions are relative.) Re-combining is re-positioning.

This re-positioning may be thought of as constantly occurring according to the Law of Affinity, by means of which each constituent variable in a combination seeks a position of greatest advantage—i.e., one that is most conducive to receive (or impart) those impressions which will add to its growth and expressiveness in-Cosmos (or, later, the growth and expressiveness of other constituent variables). A constituent variable (whether relatively macro or micro) will not naturally seek a relative position that will dampen or thwart its 'pattern of being' its 'motivating archetypes' (unless functioning under the Law of Sacrifice, and thus possessed of an identification which far surpasses its own ring-pass-not).

On the NOW with Respect to Combinations

It is sometimes heard that the past and the future are 'happening' NOW. In as much as always and ever there is only NOW (in the utterly abstract sense), there may be some justification for saying so. With respect, however, to the *combinations* of the past and those of the future, it cannot be said that such combinations are 'now' supplanting the combinations of the present moment or occurring simultaneously with the present combinations.

Carried to its logical conclusion we detect the absurdity of this thought as we begin to consider the possibility of an infinite chain of Cosmoses in which no combination leads to any other combinations, or follows any other combinations, because absolutely all motions are occurring in the present *immobile* moment! Even the entire concept of 'combinations' would break down under such a hypothesis for there would exist no 'time' in which the 'changes' necessary to form combinations could occur. *Formally,* we are dealing with a fallacy; *infinitessentially* (and in terms of the FOUNT OF ALL POSSIBILITY) we are possibly dealing with a recondite *Truth.*

With respect to combinations, there has been an infinite past and there will be an infinite trans-Cosmic future. The NOWNESS of the past and the future has to do *not* with combinations, but with the SELF-'IDENTICALNESS' of the *always present PRESENCE.* So PERPETUAL NOWNESS deals with the PRESENCE and not with that which (appears) to move or change. This is a most important distinction.

The PRESENCE-as-Presence is found *only* in the NOW-as-Now. The intra-Cosmic PRESENCE is the Presence; the intra-Cosmic NOW is the Triple Now (Cosmo-Objective, Cosmo-Subjective and Cosmo-Eternal). The PRESENCE and the Presence are *not* identical (though the *greater* subsumes the *lesser).* By occasionally noting these two *presences* (as PRESENCE and Presence) the distinction between the INFINITE and the Finite is accentuated. So, there is only Presence (the Universal Logoic Presence) in the Now.

- The Presence Itself can only be *present* in the Now. *Being* Itself can only *be* in the Now (just as BEING ITSELF can only BE in the NOW). The Presence cannot *now* be present in the past (though the Presence *was* in the past); and Being cannot *now be* in the future (though the Presence *will be* in the future). From a certain perspective, the past is dead and the future is without life, without potency, because Presence and Being do not *now* animate either one.

If we are tempted to say that that which is to precipitate in the future *already exists,* we would be more accurate to say that it is the *'energy matrix' from which future combinations will be precipitated* (complete, albeit, with *some* tendencies toward precipitation) that exists—yes, exists, but *not in the future;* rather, that 'energy matrix' exists *now.* (The lower case, italicized *'now'* is being used to focus upon the present moment in a practical sense.)

The exact combinations as they will be precipitated upon the physical plane in the future, however, do not *now* exist. That which will be a *definite* future exists (in a way) *now,* but only *vaguely* as a *present indefinite condition* which will yield *future actualities.* The future exists *now* but only in potential. The past exists *now,* but only in memory. Neither exists *now* in *full precipitated detail.* The Cosmo-Configuration of the present moment is *not* the Cosmo-Configuration of any specifiable moment in the future or of any specifiable moment in the past.

We must understand that the general term *now* cannot only be applied to happenings upon the physical plane. The potency of the patterns that *will* precipitate upon the physical plane and be physical actualities in the future, *presently exist* as subtle potencies in the *now.* The potency of patterns that have already precipitated upon the physical plane and are now past, also exist *now,* but in a form (as part of the akashic record) different to that in which they once existed when they were present, precipitated actualities. So the past combinations are combinations that *were,* i.e., that were once in the *now,* but which are no longer in the *now;* the future combinations are combinations that are not now in the *now,* but that will be in the *now.*

Every Cosmo-Objective Now has one Cosmo-Configuration and only *one.* Let us imagine we can 'freeze' the Cosmo-Configuration. Then the 'memory' of a past Cosmo-Configuration is part of the *present* Cosmo-Configuration; and the latent, 'waiting' energy-pattern, which will precipitate concretely and specifically in the future, is *also* part of the *present* Cosmo-Configuration.

Thus, simply, that which *was* the past, and that which *will be* the future, are *both* part of the *present* Cosmo-Configuration. Every Cosmo-Objective Now and every Cosmo-Configuration (each pertaining to One Cosmo-Objective Now and one alone) is *formally* *unique* and *unrepeatable.* While the Cosmic Eternal Now abides through all such changes, It abides as a Presence, and has naught to do with unique and unrepeatable formal objectivities.

- The Cosmic Eternal Now is based entirely upon Presence and simultaneous *abstraction* from all objective particularities. Cosmic future and past are *one* in the Cosmic Eternal Now simply because of the Pinnacle of Consciousness upon which it (the Cosmic Eternal Now) exists. The top of the triangle exists simultaneously within all points upon the base-line.

When considering this picture of constantly changing combinations in relation to the (apparent) "March of Time," it is important to realize that *objects* change but the *perceiving/apperceiving* SUBJECT-as-Subject-as-subject changes not. A *subject* (essentially a SELF-as-Self-as-self) *cannot* change in ESSENCE. Prakriti is modification, but Purusha, (the Spirit, in ESSENCE, the INFINITE SUBJECTIVITY) is the Observer/Pervader of modification, and cannot change.

In sum, all beings-in-Cosmos are neither entirely "fish nor fowl." They are not entirely "free-flying" as pure 'Radiations' of the INFINITE, nor are they entirely captivated, swimming hopelessly in the Ocean of Matter. Beings-in-Cosmos must balance the INFINITE and the Finite. In relation to combinations, they must balance the constant (quantized) change of the form/object/matter with the CHANGLESSNESS of Purusha.

The tendency may be to lay emphasis upon one or the other of these two poles of being-in-Cosmos, or perhaps, to strike somewhat of a balance between. At length, however, the greater absorbs the lesser and the greater is That which is permanent-in-Cosmos, and ultimately THAT which IS PERMANENT. THAT which is PERMANENT is the OCEAN of BE-NESS, the UNIVERSAL SOLVENT, which is destined to dissolve all combinations.

On Sudden, Discontinuous Movement and Recombination

Are *all* possible configurations between relatable variables *actually* possible in-Cosmos? If all configurations *were* possible we would have a Cosmos in which Inaugurating Pre-Cosmic Parameters could be violated. A Cosmos is not inaugurated with Parameters that allow for an *actualizable* infinity of configurations. Therefore it is only possible to have motion or change that result in *Cosmically-possible* configurations or relationships.

For instance, why is any fundamental tone only capable of generating certain overtones in certain octaves? Try as one will, working with a given fundamental tone, one cannot *compel* the appearance of *other-than-possible* notes in a given octave. The parameters (Laws) which govern the functioning of the overtone series will not allow it. So it is in relation to possibilities in-Cosmos. The pre-cosmically *sanctioned* possibilities do not allow *all* possible configurations of variables that *could* arise within an infinitely divisible continuum. There are *leaps* from one 'Cosmically Sanctioned Configuration' to another, with *many* 'Cosmically Unsanctioned Configurations' (since they are *impossible*-in-Cosmos) passed over, forgone. The 'Choice' (if *choice* it is) would be determined in the Cosmo-Subjective Now of the inter-moment instant.

To one in the Worlds of Approximation (the 'Fohatic Worlds') the movement from one actualizable possibility to another occurs in sudden 'flashes of movement' or 'quantum leaps' (with no 'time' *determinable in Fohatic Worlds* 'between' the Cosmically Sanctioned Configurations). Yes, 'time' may elapse 'between' the changes in Cosmically Sanctioned Configurations, but the 'amount' of 'time' cannot be determined by any *being* 'within' the Fohatic Worlds of that particular Cosmos. The 'length' of such 'time' between Cosmic Configurations (whether it be equivalent to the duration of an 'ultimate moment', or of virtually infinitesimal duration, or of a duration even lengthier than that of an ultimate moment), could only be determined (perhaps!) by Beings within the World of Being measuring the Cosmo-Subjective inter-moment instant against an Infinite Time Line, upon which the appearance and disappearance of Cosmoses provided the standard of reference

If, in any given Cosmos, *all* possibilities were actualizable and to be actualized, not only would Cosmic Laws be broken and Cosmic Parameters breached, but no Cosmos would have a finite duration, as the time necessarily expended to actualize an infinity of possibilities (within an infinitely divisible Cosmic Prakritic Continuum probably required for such actualization) would be *infinite*.

If on the other hand, an infinity of actualizable possibilities were allowed, and also Free Will were allowed, *and*, a Cosmos were given only a finite amount of 'time' to endure, then there is a very great probability that the Purpose or Objective of the Cosmos would not be actualized—so many would be the *deviating options* and so fierce the competition between them. At this point we might well ponder on the 'Happy Ending Theory of Cosmos'!

On 'Here' and 'There'

Let us look at the concepts of 'here' and 'there'. Here' and 'there' are one of the Pairs of Opposites and have meaning (as a pair, at least) only in the World of Becoming—a world in which the illusion of *space* and *extension* exists.

- It has been said that there is no such thing as a stick with only one end, but when dealing with the INFINITE and ITS PRESENCE in-Cosmos, there may be, metaphorically, 'a stick with only one end'. In other words, within the SELF-as-Self, the immediacy of 'here' (the Real Center of '**8**-ness') may seem to exist without a 'there', for a 'there' arises only with objectivity (which is a fundamental illusion).

Psycho-spiritually, the *intimacy* of 'here' is far more Real than the estrangement of 'there', for 'here' is associated with closeness and even identification, which are the watch-words of a Cosmos in which all the infinity of points other than the One Point are Essentially illusory. Perhaps, 'here' joins the ranks of illusion 'within' the NOWHERENESS of the INFINITE-SELF, but within *Cosmos* Itself (once there is identification with the One) 'hereness' (as a symbol of *'Presentness')* prevails.

'There' implies separation, distance, extension. 'Here' usually suggests a location within the immediate ring-pass-not, if not in its very *center.* If the immediate ring-pass-not grows through evolution and expanded consciousness, and eventually becomes totally extensive (bounded only by the ring-pass-not of the particular Cosmos) then every possible point of reference is *within* the ring-pass-not, and all such points are, therefore, 'here'.

To the Pervader, all is 'here'. The only 'there' that can then exist is the ALL-IN-ALL 'STATE' which is the incompletely accessible INFINITUDE—incompletely accessible to all Cosmically-bounded B/beings. It seems, as well, that (to Cosmically-bounded beings) the INFINITUDE can never Really be completely *inaccessible* either, for the reason that there must ever be a fundamental *duality* in the UTTER ALLNESS consisting of both the PRESENCE of the ABSOLUTE SELF along with the illusory, apparent modifications of Cosmos (which, strangely, the SELF IS {hence, the partial accessibility} as well as BEING *wholly* ITSELF).

Carrying these thoughts still farther, we determine that the INFINITUDE is not REALLY *totally accessible* to Finite Consciousnesses until the "Day Be With Us", and by that 'time' all C/consciousnesses have entered the 'STATE' of the NON-CONSCIOUSNESS of TOTALLY INFINITIZED SELF-PREOCCUPATION, and hence, are no longer Finite C/consciousnesses.

In metaphysics, (which can be conceived as a discipline of thought which seeks to collapse Time and Space in order to fathom REALITY) it seems that, with respect to *certain* of the Pairs of Opposites, one member of such pairs must be given precedence. (This approach could be called the 'hierarchicalization of the Pairs of Opposites'.) Thus, *one* (of a given Pair of Opposites) comes to be understood as more Real than the other—more Real, that is, to a consciousness the quality of which has been enhanced by the *realization of the factuality of non-differentiation.*

If we think this way (and avoid becoming grammatically pedantic) we would find that the term (and its associated idea), 'here', would be given preference to the term 'there'; 'in' would be given preference to 'out'; and 'I' to 'you'. (The term which indicates or suggests a position *closer* to the subject, *closer* to the center, would be deemed as more Real or, at least, more *evocative* of the Real.) In like manner, 'this' would be given preference to 'that', 'near' to 'far', and 'now' to 'then'.

- Whatever terms within the Pairs of Opposites symbolize perceptions/apperceptions/realizations which suggest *instantaneity* or *immediacy* or *proximity* or *centrality*, etc. (*'pointness'* instead of *'lineness')*—such are the ones that would be most honored, valued. All terms within the Pairs of Opposites which are derived from, or suggest, the *distancing factor* are eventually negated *because Reality never leaves the Center.*

So, that term within a Pair of Opposites which relates most to the *center* is the *truer* term—more evocative of Reality. Note that the term 'REALITY' was *not* used, for 'REALITY' has no 'CENTER'! The distinctions we have been making apply only to the discovery of the Real with the Great World of Becoming (which, we must always remember, includes both the World of Being and the World of Effects—Approximation, Fabrication—the Mosaic World).

Now, not all terms within the Pairs of Opposites have a consistently simple interpretation. There are niceties to be respected and a watchful eye must be kept. For instance, before one realizes oneself (through Identification) to be, in fact, God Transcendent, God Transcendent appears as a *'That'*, a something *other*. Upon realization, God Transcendent appears far closer (far more inseparable from the Center)—It appears more as a 'This' rather than a 'That'. In the same way, God Transcendent appears as 'Thou' to the unenlightened consciousness, and as '**I**' or '**8**' to the enlightened consciousness.

Returning to 'here' and 'there', the term 'there' always depends upon two things—the sense of separation and the existence of a ring-pass-not (the experience of boundedness), with something-in-Cosmos lying on the 'other side' of the boundary. As long as the SELF-as-Self is 'contained' within the Cosmic Ring-Pass-Not, there will always be a 'there', which seems opposed to a 'here'. When the ring-pass-not of any given intra-Cosmic consciousness becomes as large as is permissible within Cosmos (i.e., when it becomes *identical* with the Cosmic Ring-Pass-Not) then nothing *in-Cosmos* exists beyond the boundary, and 'hereness' prevails. Then, the only 'thereness' is the 'THERENESS' of THAT (the ABSOLUTE, which for the sake of spiritual *intimacy* should sometimes, perhaps, be more frequently called 'THIS') which, ESSENTIALLY, is more 'HERE' than the nearest 'here'—"Closer than hands and feet" as the mystic said. So we see that the ULTIMATE 'THERE' IS REALLY the ULTIMATE 'HERE'.

It has been said (with reference to both logic and grammar) that it is impossible to have an 'in' without an 'out', and a 'this' without a 'that', and a 'here' without a 'there'. But further reflection reveals that, even if this rule *is* so to the dichotomizing concrete mind, is *not experientially* so to the synthesizing Spirit.

On REALITY

REALITY IS THAT which has naught to do with the World of Becoming. That which is REAL has only to do with the ETERNAL INFINITE SUBJECTIVITY, the GREAT ABSTRACTION, the BOUNDLESS IMMUTABLE PRINCIPLE. The Real (note capitalization) is less *real* than the REAL, but "real enough"—in fact the Real denotes That which is Permanent-in-Cosmos upon the highest Planes of the Cosmos (i.e., within the World of Being). The Real denotes the Realm of the most Sacred Numbers—Sacred Integer/Entities—from One to Ten.

That which is ABSOLUTELY REAL is the ultimate SOURCE of all *apparently real* things, for to 'reify' is to generate a *thing*. That which is Real is more the Cause-in-Cosmos of *those things,* and the REAL, their ULTIMATE SOURCE. That which is *ABSOLUTELY* REAL is never REALLY condensed nor precipitated, and in ITS ESSENCE can *never* become *actual.* To become so, the INFINITUDE would have to 'BECOME' Finite. (This IT *never* REALLY 'DOES', though the appearance of Cosmos is based upon the fact that IT *seems* to 'DO SO'.)

From another paradoxical perspective, since the INFINITUDE is *indivisible,* any apparent *part* of ITSELF, is the *whole* of ITSELF! Thus every apparent 'part' is the *TOTAL* INFINITUDE. (This does not mean that the apparently finite 'part' is REALLY *finite!)* REALITY (REALLY) 'ABIDES' as IT *IS,* only in the *unprecipitated* STATE. Infinitely more potent than that which is precipitated is THAT which remains ever unprecipitated. In relation to the WORLD OF BEING, reification, precipitation, condensation, expression, etc., all reduce potency—infinitely. Yet, paradoxically, the INFINITUDE 'REMAINS' forever *unreduced.*

NOTE: *Reality,* as we have indicated by various capitalizations of the word, is a relative term, and things can be more or less *real.* The closer things are to the STATE of ULTIMATE ABSTRACTION and HOMOGENEOUS SIMPLICITY, the more *real* they are. (The italicized word *'real'* indicates a general, global, and non-specific use of the term.)

On Actuality

The term 'actuality' is important to clear thinking, and should rather be used far more frequently than the term 'REALITY' (or, even, the term 'Reality') when referring to most human experience. We will speak first of REALITY, which is a pure abstraction (though REAL), and then of Reality, which is *actually* part of the World of Becoming, and an Aspect of the Great Actuality, which is the Universe.

'Actuality' refers to that which has been *precipitated by an act.* An 'act' is a *movement* of an 'actor'. The 'FLASHING FORTH' of the RAY of the ABSOLUTE and the coeval 'GENERATION' of Infinite Subject/Object (sometimes called the 'Infinified Point' and Mulaprakriti) is, cyclically and forever, the FIRST 'ACT', but it is necessary to refer to this 'GENERATION' as an 'ACTLESS ACT' because it 'ORIGINATED' 'within' a 'STATE' in which *action* is, by definition, impossible just as *change* is impossible.

The problem boils down to the question: How can there be a *change* in the CHANGELESSNESS? (Perhaps the versatile concept of *apparency* will have to be invoked!) Also the 'RAY' and the Infinite Subject/Object (i.e., the Pre-Cosmic 'Players' which are all Really one and the same) cannot REALLY be considered an *Actuality* or *Actualities.* They Are (when merged together) simply, what might be called (for want of words to convey the unconveyable) a *'finitization of the ATTENTION'* of the ONE WHO is incapable of any ESSENTIAL *mutability* whatsoever. Paradox abounds when thinking of the Origin of the FIRST 'ACT'.

PURE REALITY IS the STATE of ALL-IN-ALLNESS. REALITY is never conditional or limited by conditions. 'Actuality' *is* conditional and describes the state of conditions. Actuality describes any condition of relationship pertaining to the precipitated Cosmos and Its 'contents'. The Universe or Cosmos is a Great Actuality. The Universe/Cosmos is *not* a REALITY at all. All things within the World of Becoming, including the whole of Cosmos Itself, are *approximations of 'infinitessences'.*

This is equivalent to saying that *all actualities are approximations of 'infinitessences'.* 'Infinitessences' are *primal possibilities* 'RESIDENT' within the FOUNT OF ALL POSSIBILITY. 'Infinitessences' 'abide' is a *cosmified form* within the World of Being; the *actualities* of the Worlds of Fabrication are *approximations* of these 'infinitessences'-as-Ideas-in-Cosmos.

- An important question arises: Can there be 'REALITIES' or can there only be REALITY? From one perspective, since REALITY is undifferentiated, IT must be HOMOGENEOUS and *singular.* REALITY must be the ONE 'INFINITESSENCE' of all that appears as Realities or Essences in Cosmos. So the ROOT IS ONE. The INFINITESSENCE IS the NOUMENON of all the multiple so-called Realities/Essences found in-Cosmos. From one perspective, since REALITY is undifferentiated, IT must be HOMOGENEOUS and singular. REALITY must be the ONE 'INFINITESSENCE' of all that appears as Realities or Essences in-Cosmos.

In this consideration, it becomes important for us to think of the higher planes of the Cosmos as containing Archetypal Patterns which cause the precipitation of many effects. These Archetypal Patterns are Realities in-Cosmos. They are, as it were, *derivative* Realities. They derive from the ONE REALITY. For *practical* purposes they are also Essences of many *effects* upon the lower planes of Cosmos, but in regard to their *origin,* these Realities are *derivative* Essences. They derive from the ONE ESSENCE that we have called the 'INFINITESSENCE' [see Glossary]. So it becomes clear that in addition to thinking about the ONE REALITY, or the ONE ESSENCE (the INFINITESSENCE), it is important, when thinking of the functioning of the Cosmos, to include thought upon the Realities and Essences which are Conditioning Archetypes upon the higher planes of Cosmos (the World of Being)—all of Them, of course, deriving from ONE NOUMENON—the ONE REALITY, the ONE ESSENCE.

Similarly, there is only ONE NOUMENON, but from IT, Noumena (plural) in-Cosmos derive. It is not practical to think of all *sources* as being Trans-Cosmic (even though, ESSENTIALLY, They-as-IT, ARE). There is REALLY only ONE SOURCE, and Cosmos, in and of Itself, can never 'contain' IT. But as *secondary potent sources,* we must think of

Realities, Essences and Noumena *within* Cosmos. For all practical purposes, These are the Sources of all that transpires in the worlds we can directly know anything about.

Is it Really clarifying or accurate to think that, within the ONE, there are REALITIES and ESSENCES/INFINITESSENCES (all plural)? One is tempted to do so because it is said that every phenomenon has its Noumenon. Naturally one might begin looking for this diversity of Noumena within the SOURCE. But, for all phenomena there is ESSENTIALLY only ONE INDIVISIBLE NOUMENON. Noumena (plural) are also ESSENTIALLY ONE (derived from the ONE) but, *practically, plural.* These diverse Noumena gain Their authenticity from the ONE NOUMENON, which IS the 'HEART' and 'SOUL' of every Noumenon. If one wishes to think of plural Noumena, one must focus within the *World of Reality* (which contains many Realities) upon the higher levels of the World of Becoming (i.e., the World of Being) and *not* focus within the non-plural WORLD OF REALITY. So summarizing this necessary digression:

- There is ONE REALITY, ONE NOUMENON, ONE ESSENCE, ONE INFINITESSENCE. All these are simply NAMES for the INFINITE SELF.
- Then, descending the Ladder of BEING/Being, the Universal Logos can be called the One Reality, the One Essence, and, even, the Cosmic Noumenon (different than the ONE NOUMENON, the ABSOLUTE).
- Within the Cosmos (i.e., within the Being of the Universal Logos) there are Realities, Noumena, and Essences—(all plural). All of These are found upon the higher planes of Cosmos and are the controlling Archetypes of all that appears within the World of Effects (the lowest planes of the World of Becoming, just as the World of Being forms the highest planes). Our problem with language arises because *there is no pluralism in the GREAT INDIVISIBILITY.*

Returning to our subject, and in light of the foregoing, we can then say: Some (higher) Actualities (as opposed to actualities) can be considered as Realities, but these same Actualities, per se, can never be REALITY ITSELF (indivisible and non-plural). Discrimination is needed to determine which Actualities qualify as Archetypal Realities/Essences/Noumena, for all precipitations in Cosmos (including Cosmos Itself), are Really, from a Super-Cosmic Perspective, Actualities.

With these thoughts in mind, however, there is less likelihood of mistakenly assigning any Actuality (no matter how high its nature) to the status of REALITY. Of course, it must never be forgotten, that the ESSENCE of all actualities, Actualities, realities, and Realities IS *REALITY* ITSELF. In this kind of study, ESSENCES/Essences must always be discriminated from the phenomenal manner in which THEY/They *manifest.*

An 'actuality' is the result of 'action', but action, at root, is a 'DISTURBANCE' of the GREAT PEACE. Action is motion and requires relation. Actualities are found only within the World of Relativity (the World of Cosmos). REALITY IS THAT into which Cosmos ultimately dissolves. Realities (plural) however, are those Living Archetypes that embody and direct the Law of the Universal Logos, which Law, when adhered to, absolves and releases Cosmos from Its limited Condition into that 'STATELESS STATE' in which 'action' of any kind is no longer required, or *possible.*

◆ Thus it can be seen why, ultimately, 'non-action' is more REAL and infinitely less limited than 'action'. Any action is, in fact, a limitation upon REALITY, the

INFINITY of INFINITIES. What 'NEED' has the ABSOLUTE for *action* at all? Action produces *actualities* which (no matter how *high*, relatively) are, in and of themselves, Non-REALITY—hence, the existence of manifold illusions, limitations, and finitizations.

The highest form of *realist* is the REALIST—identified with the ONE INFINITE SELF. The REALIST is one who releases SPIRIT-as-Spirit from the *ignorance of actuality.* A lesser *realist* (but still a being well on its way to liberation) is a 'Realist'—one who recognizes and understands and acts in accordance with the Great Archetypes upon the higher planes of the Cosmos. What we normally call a *realist* is in fact an ignorant *actualist*—one who is attached to and identified with that which has been created through *act*, especially those things found upon the lower planes of Cosmos. The SELF in ITS PURE STATE of ALL-IN-ALLNESS is *not* an *actualizing* 'ACTOR'; but, then, can *any other actor* ever be found in all the UTTER ALLNESS? Paradox!

In sum:

- A 'REALIST' *knows* he *is* the INFINITE SELF.
- A 'Realist' *knows* and cooperates with the Higher Laws of Cosmos, but is not necessarily identified with ULTIMATE REALITY. A Realist focuses upon those Objects which are Permanent-in-Cosmos.
- An 'actualist' (the normal so-called 'realist') is focused upon lower objects (which are not Permanent-in-Cosmos).
- A 'Higher Actualist' is concerned about Higher Objects and can, therefore, be called a 'Realist'. 'Actualism' and 'Realism' blend within the World of Being.

On Time and Space

We must consider Time and Space together because of their Essential inseparability. Let us begin with Time. Time is not REAL. It is Real-in-Cosmos (hence, in the large sense, Actual). Time is a Universal Factor or Principle which is Permanent-in-Cosmos. Therefore, although Time is not ULTIMATELY REAL, It *exists*—at least, *in-Cosmos* and, depending upon our perspective, perhaps, *between* Cosmoses and Universal Pralayas—during which 'times' it could be said with reason that Time both *exists* and *does not exist.*

Let us explore some thoughts about Time in relation to other Factors-in-Universe. Time and movement/change are coeval (originating simultaneously). Time and number are coeval. Time and relation are coeval. Time disappears for all practical purposes and becomes DURATION in the ETERNAL NOW, in the CURRENT, in the PRESENCE. Time, however, remains an existent Factor within the Eternal Now, within the Current, within the Presence—as much of the ETERNAL NOW, the CURRENT and the PRESENCE as the Cosmos can absorb and express.

Time (of which Saturn is our Solar Systemic Representative) may have mutilated Uranus (the Representative of the Principle of Simultaneity and of Un-segmented Duration), but Uranus returns to dissolve Time, to negate Time.

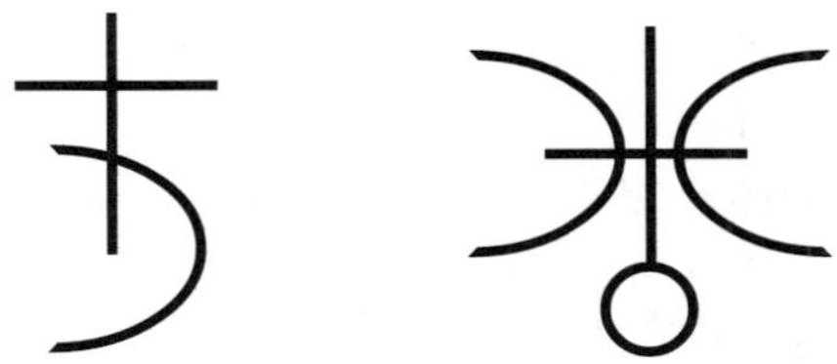

Time is a means of *measurement,* and let us remember that only Maya can be measured! Time measures the duration of pattern/combination in terms of longer and shorter, greater and smaller, more and less. There are three *principle* units of time in any Cosmos:

1. The unit of time which is the duration or (intended duration) of the Cosmos.
2. The unit of time which we are called the 'ultimate moment'—the shortest *possible* unit of time within a particular Cosmos. An 'ultimate moment' might be called a 'quantum of nowness' or the Cosmo-Objective Now.
3. The unit of time which we are calling the 'inter-moment interval', or the 'inter-moment instant', another name for which is the 'Cosmo-Subjective Now'.

There is immense variety in-Cosmos, and multifarious are the combinations. Combinations differ in terms of spatial 'extension' (to the extent that it is Real) and, also, significantly, in terms of duration. It is obvious, simply from observation, that not all patterns or combinations coincide/correspond with each other (variable for variable) spatially. It is equally obvious that not all patterns or combinations coincide/correspond with each other (variable for variable) in all ultimate moments (i.e., in all 'Nows' available within the Fohatic Worlds of Fabrication). For instance, take two combinations consisting of rapidly moving particles, and compare their configurations from one unit of time to the next. (For the purpose of this comparison, units of time *larger* than ultimate moments will suffice!)

Even when comparing closely related combinations (such as, for instance, identical twins) or, even, those considered structurally identical (such as two atoms of the same physical matter), nevertheless, identicalness of configuration does *not* exist from moment to moment. When comparing *widely different* kinds of combinations we see that some persist through a great number of ultimate moments (Lower-Cosmic Nows) and others persist through very few such moments. For instance, the combination of variables which is the firefly, and the combination of variables which is the elephant have for their respective durations vastly different quantities of ultimate moments.

On the Relativity of Time

Time, ultimately, is a measurement of rates of change. A 'unit of time' can be understood as a 'unit of change'. It makes little sense to speak of Time unless we speak of *quantities* of time. The measurement of Time depends upon the factor of *comparison.* A given unit of change is accepted as a *standard of comparison,* and other units of change are *compared* with it. In the simplest case, the 'perceived duration in present (i.e., 'visible') sameness of the first unit of change' is *compared* with the 'perceived duration in present sameness of the second unit of change'.

The process of comparison determines, initially, which unit of change 'lasts longer'. Broadly speaking, this means that the comparison determines which unit of change disappears from 'perceived present sameness' first. For instance, a man, who is the 'same' 'present' man for the sixty years he lives, disappears from perceived, present sameness before a man, who is the 'same', 'present' man for the seventy years he lives. Thus the time measuring the duration of the first unit of change (i.e., the sixty year old man) is *shorter* than the time measuring the duration of the second unit of change (i.e., the seventy year old man).

When units of change are noticeably repetitive with respect to each other, another way of measuring Time is quantified through the use of *ratio.* We then ask a simple question, the answer to which can be readily quantified: How many 'units of change' of type 'A' occur *during* a single 'unit of change' of type 'B'? In such a comparison a ratio is established, a ratio being a relationship between two quantities such that one is divided by the other. A ratio determines how many 'times' one quantity will be included within the other. In a simple case, let us examine the ratio 6/2. This ratio tells us that the quantity '2' is included in the quantity '6', three times. Naturally, not all ratios necessarily have an 'even' quotient, i.e., an 'even' number of inclusions. The simple point to be grasped is that the measurement of Time begins with the *comparison* of appearing/disappearing patterns, a comparison between changing patterns.

Time, thus, is a *quantity of duration* (determined by means of comparison) used for the purposes of measurement. Time is a measurement of the *relative* endurance of pattern/condition/combination/relationship. Through the use of time we measure the rate of change of a pattern/condition/combination/configuration/relationship relative to the rate of change of another pattern/condition/combination/configuration/relationship.

Let us consider a given relationship of variables and the rate of change of those variables—i.e., the 'quantity of *time* it takes' for those variables to change. When a relationship does not change with relative frequency (relative to the duration of a certain standard of measurement, such as 'ultimate moments'), the time perceived to elapse is different from when a relationship changes with relative frequency relative to the standard of measure.

- What we are doing here is designating the duration ('perceived present sameness') of a specified *unit of change* as a *standard* against which all other changes (or, more technically, 'units of change') are measured. The designated specified unit of change used as the standard can be as small as one vibrational cycle of a certain kind of atom, or as relatively large as an earth year. It does not matter so long as the standard remains consistent with itself, and as long as all other changes are accurately *compared* to it.

What changes and what is measured? Such things as relationships, combinations, patterns, conditions, and configurations, etc., change and are measured. None of these things exist without Time and Space. Relationships that do not change frequently (or which change only slowly relative to the standard of measurement in use) are said to 'last for a *long time*'. Relationships that change rapidly relative to the designated specified unit of change that is used as the standard of measurement are said to 'last for a (relatively) *short time*' In all such measurements, 'long' and 'short' are purely relative terms—varying in relation to each other.

'Short' and 'long' would ideally be measured with respect to an *absolute Universal Constant* but it is difficult to find one. Certainly, at this point of human development, the human mind cannot determine one. Even those changes or cycles that we call a constant may be only "special cases" related to a particular dimension within Cosmos and not to other equally important dimensions. So our designated, specified constant may not be constant *under all Cosmic conditions.* If, perhaps, we should be fortunate enough to determine a reliable Universal Constant, constant for all dimensions, at all 'times' and in all 'places' *within* our Cosmos, can we be guaranteed that the same constant would hold good in all Cosmoses (since each Cosmos is necessarily designed upon unique Parameters)?

We have seen that the meaning of 'long' and 'short', as rough assessments of the duration of time, are purely relative, and are dependent upon the duration of the standard of measurement used. We have also seen that the search for a Universal Constant (a *movement of unvarying duration under all possible Cosmic conditions)* is beyond the grasp of the human mind.

We have been speaking of Time in terms of *change.* What about Time in relation to *changelessness?* If a relationship *never* changes, it is said to be *timeless*, but is there such thing as a relationship which never changes? Only that which is *impartite* (i.e., that which has no parts and no divisions) can *never* change, because parts and divisions are, *themselves, movements* and *changes.* So relationships, patterns, conditions, combinations, and configurations etc., not only *have* parts and divisions, but, more precisely, *are themselves* parts and divisions, and so, are not only subject to change, but *are change itself*, and therefore, cannot be *timeless or ETERNAL,* for anything that *changes* is subject to Time. It is only the 'STATE' of REALITY, the 'STATE' of the INFINITE SELF, the STATE of BE-NESS that can be truly *timeless.*

Time is (apparently) the division of DURATION. (DURATION, of course, cannot REALLY be divided.) Most often, there is something arbitrary about this process of division. Relatively stable cycles (such as the cycle of the Moon) are chosen as the unit of division, but the methods of time division are by no means exact or applicable to the many dimensions of Cosmos.

Units of Time are determined by pulsations. Cycles can be thought of as pulsations. 'Within' the ONE INFINITE SELF, *per se*, there is no 'PULSATION', just as there is no 'EVENT'. A pulsation is an event, followed by a non-event from which a pulsation is absent. The interval between the beginnings of sequential pulsations is a cycle. A pulsation is a disturbance of equilibrium. Every disturbance of equilibrium is followed by a return to the equilibrized state, until another disturbance occurs. Pulsation, thus conceived, is related to event and non-event, disturbance and non-disturbance.

The interval between two disturbances (especially in a series of similar and regular disturbances) can be called a 'wave'. Every coherent, abiding system has an fundamental rate of pulsation. That pulse determines what might be called the fundamental tempo of a system. All subsequent pulsations—'on'/'off', event/non-event, appearance/disappearance—all these are measured against the tempo of the fundamental pulsation of the system.

- To measure Time accurately throughout Infinite Duration, it is necessary to have as a standard an absolutely regular pulsation—unvarying in regularity throughout all ETERNITY. The one Pulsation which has any hope (and not a *complete* hope) for fulfilling this requirement is the Great Breath, the Pulsation which expresses Itself endlessly in Universal Manvantara followed by Universal Pralaya. This Greatest of all Cycles could serve as the Fundamental *Change* with which all other possible changes could be compared, against which all other possible changes in all Cosmoses could be measured.

With our scant knowledge of dimensions higher than the three worlds of human evolution, it is impossible for humanity at this time to wrap its mind around such a mammoth Cycle. Further, are such Universal Cycles *regular?* If they are not regular we might as well abandon the hope of finding any Ultimate Standard of Time Measurement.

The duration of one Great Universal Inhalation and Exhalation (that of our Cosmos, for instance) could be used as the Standard of Measurement for all that transpired *within* the Cosmos produced by that particular Great Breath *(our* Cosmos again), but as *other* Great Breaths might be of different duration compared to the one which produced our Cosmos (for the analogy holds good in relation to human breathing), simply to choose our Great Breath Cycle as the standard for all Universes rather than other Great Breath Cycles would be arbitrary and limited in its revelatory power. Fortunately, we have plenty of *time* in which to solve the problem!

On the Measurement of Time

Time is a means of measuring Maya. Maya (in one of Its more familiar roles) induces (or *is)* the consciousness of duality. Where there is no experience/'inperience' of duality, there is no Time. Where there is a REAL CONTINUUM, there is no Time.

- For Time to exist (i.e., be *actual) discontinuity* is required. Time cannot 'appear' unless there exists the fundamental discontinuity/duality of *event/ non-event.*

Cosmos is the Event which contrasts with the SUPREME NON-EVENT known as the CHANGLESSNESS. Cosmos is, indeed, the "Main Event," in fact the 'Only Event'—the 'Big Inning' to use a quasi-baseball metaphor. It is the 'Be(g)inning' as well. The Universal Pralaya is the 'Big Outing'.

On the Infinite Existence of TIME

In occultism we become used to the idea that DURATION is fundamental and Time is secondary. DURATION, it seems, always *IS* whereas Time does not always exist. But perhaps Time, also, has existed ETERNALLY (albeit *cyclically).* If the infinite chain of Universes had no Beginning (and it is unreasonable to think that there should have been a Beginning) then Time, which *must* exist during the manifestation of every Universe, has cyclically appeared and disappeared forever. Thus, not only is DURATION forever, but Time is forever. The appearance of Time is an infinitely recurring *Event* in ETERNAL DURATION.

There are certain Factors in the UTTER ALLNESS [see Glossary] which have existed forever but which appear only cyclically. Mulaprakriti (depending for Its Existence upon the 'ARISING' 'within' the INFINITE SELF of the 'CONSCIOUSNESS' of INFINITE SELF-as-Not-SELF) is One such Factor. Maya, which is the (apparently) SELF-'DIMINISHING' Power which manifests as SELF-'VEILING' is another.

The idea that Time appears cyclically forever, throughout the entirety of INFINITE DURATION (or, more accurately, <u>I</u>nfinite <u>D</u>uration) depends upon what might be called the 'Absoluteness of Linearity'. The term 'linear' has become almost a pejorative when applied to the life of the mind, but it is a most necessary concept in metaphysics. To conceive of Time as an Event appearing cyclically throughout all of Infinite Duration requires the conception of an *infinite linear chain* of appearing and disappearing Cosmoses. In the metaphysical model used in this treatise, each Cosmos is preceded and succeeded (in Time) by a (perhaps) similar though definitely unique Cosmos.

Cosmoses appear and disappear in order, "One at a Time" forever. Since each Cosmos is, by definition, 'All there Is at any one Time', They can only appear and disappear "One at a Time" and cannot co-exist simultaneously, or be in some manner superimposed upon one another (at least not *formally,* and Cosmoses *are* 'matters of *form*', or, equally, '*forms* of matter'). The term 'Universe' (from Latin, *universus)* connotes *allness, entirety.* The equivalent term 'Cosmos' equally connotes *allness, entirety,* as well as the beauty, order and harmony of that allness, that entirety. 'Cosmos' and 'Universe' are used in this treatise to indicate 'All there Is'—Every<u>thing</u> that is happening <u>N</u>ow.

- To suggest more than one Universe at a Time is simply to suggest a more *complex* Universe, because the sum of all such simultaneous Universes would not be different from a single Universe. Such a Universe might be multiple or even infinitely multiple, but It would still be only *<u>One</u>* All-inclusive Universe. Always the One will 'devour' the Many.

A little thought about the theory of simultaneous Cosmoses, would demonstrate that, followed to its ultimate conclusion, this idea would demand that *all* Cosmoses in the ALL occur simultaneously forever. There are many philosophical problems with this concept, which will be examined in Section II under Philosophical Problems. Let us look at some of these Problems without *fully* discussing them. They are:

1. The probable negation of the Law of Karma and of the Law of Cause and Effect (as these Laws are usually understood) due to an uncompromising simultaneity of all action-in-

Universe, and even, possibly of *all* action-in-*all*-Universes. Such a simultaneity would clearly negate the linear sequences of action and reaction (even if such actions and reactions are *multiple)* upon which the Cause and Effect Paradigm depends.

2. The need for an *infinite number* of parallel Universes (why stop at merely a few?!) which (and here is the problem) *must persist infinitely;* if the 'infinite number' of parallel Universes did not persist infinitely, it can be proven that there could not be an *infinitude* of them (for other Universes preceding or following the infinitude in question, would not have been 'numbered' among them, hence the infinitude in question would not be a Real infinitude).
3. If, retaining a vestige of Theosophical Orthodoxy, *the Great Breath is still entertained as a possibility,* but is joined to the hypothesis of an infinitude of simultaneous parallel Universes, then we are forced to return immediately to the paradigm of *linearity,* for one manifestation of an infinitude of parallel Universes would be followed (because of the Great Breath) by another such manifestation, endlessly and, successively, forever as the Great Breath 'Breathes'! Yet because the Great Breath *does* 'Breathe', Universes would exist on 'either side' of the kind of Universe we are here discussing (i.e., One supposedly consisting of an *infinite* number of so-called parallel Universes). Clearly, however, the '*infinite* number' of Parallel Universes in said Universe cannot REALLY be *infinite,* because that so-called 'infinite number' does not *include* either the Universes which occurred *before* and those which will occur *after* (for the Great Breath does not stop 'Breathing', and, thus, producing Universes! As well, is not the *hidden linearity* of the Infinite-Simultaneous-Parallel-Universes Model of the Universe apparent?—i.e., *if* the Great Breath Model is retained. In conclusion, if the Infinite-Simultaneous-Parallel-Universes Model of the Universe is accepted, then the Great Breath must surely be rejected, for the Great Breath challenges the *infinitude* of the term 'Infinite' in this Model of our Universe. Further, since the Great Breath does not take all of an infinitude of Breaths *at once,* Its Breathing Process *demands* linearity.
4. A further problem with this theory of Infinite-Simultaneous-Parallel-Universes serving as our Universe, is that the theory (extended to its logical conclusions) suggests the simultaneity of not only *all* possible *Universes* (if the putative *infinitude* of these Parallel Universes is to be *rescued from nonsense*) but, the simultaneity of *all* possible moments (forever) as well! As a corollary, this Model may also logically demand the simultaneity and, perhaps, sameness of *all* possible movements (forever), leading to the conclusion that:
   - Because all possible moments are presumed simultaneous) there is but one moment (forever ... for one moment could not possibly lead to another moment, there being (in complete simultaneity) no other moments, and
   - Because all possible movements are presumed simultaneous) there is but one movement (forever ... for one movement could not lead to *another* movement, there being no *other* movements).

   A *movement* of this kind (i.e., the *one movement)* is, therefore, a non-*movement*—REALLY, no movement at all—because, for one thing, movement requires more than *one* moment in which the movement is to occur, and, if there is only *one moment,* then the conditions for having more than *one movement* are not fulfilled).

5. As well, it can be shown that the <u>*one simultaneous moment*</u> (forever) is really a *non-moment,* for moments exist only if there are *other* moments to precede and follow them.
   - Can we also say that Universes exist only if there are *other* Universes to precede and follow them? Can we say that *events* exist only if there are other *events* to precede and follow them? Remember that Universes, moments and movements are *all <u>events</u>*. If this oneness/sameness/identity of *movement* and *moment* are true, we see that accepting the hypothesis of an infinite number of simultaneous parallel Universes as constituting our Universe leads to the collapse or negation of Time and Motion (without which we cannot have Space) and hence, to the impossibility of *any* Universes at all (since the existence of a Universe demands Time, Space, and Motion.)

This conclusion (that there can be no Universe) is absurd, of course, and contrary to all human experience (though it does make sense when one 'thinks' of the BOUNDLESS *IMMUTABLE* PRINCIPLE). In light of these problems, it seems wise to preserve as more probable the idea of an infinite *linear* succession of discrete and *unitary* Universes.

As a further question on Time and DURATION: Is Time measurable when there *is* no Time? During the Universal Pralaya the SELF is ALL-IN-ALL and Time has ceased to exist. And yet, there is, apparently, a measurable Interval *between* Cosmoses, and that Interval spans the 'time it takes' for the SELF to enter and 'ABIDE IN' the TIMELESS STATE. So, during that 'STATE', SUBJECTIVELY there is no TIME, but following that 'STATE', Time *reawakens*, and it is seen (albeit from the extra-SOURCE, Dualistic Perspective) that the TIMELESS STATE *took Time!* From another perspective, we might say that DURATION never ceases even though Time runs an infinitely recurring Cycle.

- It begins to look like the only way to preserve inviolable the ABSOLUTE is to declare all appearances as *less than NOTHING,* which they certainly *are* from several perspectives.

The INFINITE SELF 'IN' ETERNAL DURATION IS NOTHING, but everything *else* is *infinitely less than NOTHING,* counting for *nothing* when 'measured' against NOTHING, the ZERO. Paradox plagues the questing human mind.

On Maya and the Flow of Time

Time, then, exists, forever but is not ULTIMATELY REAL. We might look at the "Flow of Time" as the Eternal Flow of Changes—the Eternal Flux of combinations and recombinations. All of these endless changes are *mayavic*—'Children of Maya'. Time can be considered the 'Instrument of Maya'. Maya:

- 'uses' Time to *veil* ETERNAL DURATION, just as It
- 'uses' Number to veil the ZERO, and
- 'uses' the *many* selves to veil the ONE INFINITE SELF.

Maya is so 'dangerous' precisely because It is the only 'thing' which is *less than Nothing!* All finitudes emerge from, 'MAYA', the *original finitude.*

We can agree with Sankaracarya that Maya is most mysterious and "without beginning"—as infinitely old as Brahman. The Wheel of UTTER ALLNESS has rolled endlessly, *monality* and *duality* following each other endlessly in succession. Although the members of each pair of the Great Pairs of Opposites seem to be equal, one member of the pair may be "more *equal* than the other!" Monality, being closer to REALITY (in fact, *being* REALITY ITSELF) has precedence over duality.

◆ Monality abides ever and Duality is a generated subset of it.

Is there a 'time' when Duality does not exist? Yes! Thus It is secondary to Monality which *ever is.* But 'later' Duality is 'remembered' as having existed when it was 'IN-PERIENCED' as *not existing!* It appears that 'God' has decreed that no answer shall stand without the appearance of an equally true contradictory answer.

The bringing forth of the Cosmic Egg (the Universe) has gone on endlessly and the BE-NESS, the INFINITUDE IS the "GREAT SWAN"—the "Swan, out of Time and Space" And so eternal, eternal is the question: Which came first the SWAN or the Egg?—(the metaphysical version of "Chicken and Egg"). The riddle is insoluble because, although under the sway of "common sense", we assume there must be a *first,* in the metaphysics of the INFINITE there is no first. We know only that:

- There is a NOTHING, which 'HAPPENS' *all the TIME.*
- There is a Something, which 'Happens' *some of the TIME.*
- There is the alternation between prevailing NOTHINGNESS and apparent Somethingness that 'Happens' *all of the TIME.*

On Infinite Knowledge

Do **8**, even Now, know all things? Do **8**, even Now, have Infinite Knowledge? Am **8** capable of remembering all combinations that have ever been, and could these combinations be presented to '**M8**' Consciousness even Now, should **8** wish it? The answer must be ,Yes, *if* **8** Am REALLY the SELF. Even NOW, THAT which IS ALL, necessarily 'KNOWS' these things.

It seems that even though REALITY is a MONALITY, the Principle of Duality is of infinite duration and recur infinitely. The paradox is that even during those 'times' when Duality seems to supervene, ABSOLUTE MONALITY ABIDES *as ever!* Duality, however, while *infinitely recurrent* is not *continuously present* (at least as an 'IN-PERIENCE') During the Universal Pralaya, MONALITY or ALL-IN-ALLNESS 'reigns'. Duality is negated as if it never were. In the CONTINUUM (which *ever* IS) Duality *never* IS (infinite intermittent appearances to the contrary). Duality prevails not.

Therefore, MONALITY is of a higher order, REALLY, and, strange as it may seem, of an *infinitely higher order.* Nevertheless, Duality is an *infinitely continuous* Discontinuity, or, equally, a *perpetually discontinuous* Continuity. 'OFFNESS' is continuous; 'On-ness' is discontinuous; but 'On-and-Offness' (if It can be counted as *one* thing instead of *two)* is, again, *continuous!* 'OFFNESS', however, *never varies!* Although neither the SWAN nor the Egg came *first*, the IMMORTAL SWAN was *not* hatched from an Egg—the SWAN was never born, and never will IT die. It is the SWAN that REALLY *counts* (even though IT *"counts for NOTHING".)*

On Maya and the SELF

Maya is That which can be measured. THAT which is boundless and impartite cannot be measured. Hence, the BOUNDLESSNESS and INDIVISIBLE has naught to do with Maya. Yet, is not Maya the SELF as well? All things which *are* and yet are (apparently) less than the SELF, are *actually* the SELF-as-Not-SELF. It seems that the SELF-as-Not-SELF can *Do* what the SELF-as-SELF cannot 'DO'! The Great Paradox can be put in the form of a question: How can the SELF-as-SELF 'BECOME' the SELF-as-Not-Self and still, forever BE the SELF-as-SELF?

On Time and the Sense of Time

The sense of time depends upon the experience of the sense of sequence—the registration of one experience/'inperience' after another, one perception/apperception after another, one object/thing after another. In a condition in which all things are experienced *simultaneously*, there is far less likelihood that the sense of sequence *usually* experienced by the human consciousness would remain untransformed. Perhaps, in a human being's highest moments, he can register (albeit very imperfectly) something of the sense of simultaneity available to a greater more inclusive Consciousness.

It is presumed, however, that even for a Great Consciousness (capable of comprehensive simultaneous registration of all variables within Its vast field of expression) the sense of sequence would not *entirely* disappear, for sequence is fundamental to Cosmos. The sequence registered by the Consciousness that can continuously register all things simultaneously, would be the sequence produced by *change of configuration at each ultimate moment.* The normal *linearity* associated with the registration of impressions in sequence would be greatly modified. The ongoing Presence of Now would be a far greater Factor in Consciousness than the sense of fleeting modifications. In such an Experience/ 'Inperience' of "everything all at once", the Sense of Time (as human beings know such a sense) is greatly altered, and can be tuned out, all but disappearing—precedence being easily given to the Eternal Now (as much of the ETERNAL NOW as can be registered in a Cosmos).

One cannot justifiably say, however, that with everything perceived as happening, "all at once", the experience of Time would utterly disappear, for Time is as necessary to the Universal Logos in the orchestrating of His Cosmic Process, as it is to the human being in the management of his little life. Despite the *absolutely continuous* PRESENCE of IMMUTABLE DURATION, Time is the Lord of All-in-Cosmos, and the Universe has a Finite and, perhaps, Pre-Ordained Duration. So Time will not be utterly forgotten by the Great Observer even in Its Experience/'Inperience' of Cosmic Simultaneity. Some global sense of the sequence of changes in the Great Configuration (which includes all Cosmic Variables) must, nonetheless, register. It is possible for a Great Consciousness to experience/'inperience' Time in a number of ways *simultaneously.*

It is important to differentiate between the *experience* of Time and *Time itself.* (Of course, the distinction is ESSENTIALLY an artificial one, because no thing at all would exist unless it was registered in the SELF-Objectifying Consciousness of the SELF-as-Infinite Subject/Infinified Point-as-Condensing Point-as-Condensed Point.) The SELF-as-Infinite Subject/Infinified Point-as-Condensing Point-as-Condensed Point Generates/Creates O/objects by 'Seeing' them. Matter Itself would not Exist unless the Self were 'Self-Seen'. Nevertheless, for practical purposes the distinction is an important one. Matter is what the SELF 'SEES' of ITSELF; the 'SEEING' 'CREATES' the Matter. [See Glossary under 'Matter' and 'Prakriti'.] This is all to say that Time is as much a Self-Perception as Space, therefore, Time, per se, cannot Really be differentiated from the *experience/'inperience'* of Time.

- 'Simultaneity' and 'sequence' hold the key to the nature of the experience/'inperience' of Time.

In the simultaneous experiencing of *all* movements within a Cosmic System—assuming (as one assumes in relation to the Cosmic System) that there can be no *other* Cosmic System co-existing with It—there can be no experience of long or short, but only of *one* virtually continuous Great Movement consisting of many movements—some of which are relatively long and some of which are relatively short. Such an Observer of Cosmic Simultaneity would cognize infallibly each one of the virtually innumerable ultimate moments and each change of Cosmic Configuration which occurred at each such ultimate moment. The Registration/Impression upon that large, All-inclusive, Multi-Dimensional Consciousness would be in a way, *very even.* Change would be constant with respect to at least some items-in-Cosmos—the ones which change with each ultimate moment (i.e., the ones {Fohatic Particulations} the changes of which actually *define the parameters* of ultimate moments). Other items (aggregates of varying magnitudes) would require *many* ultimate moments to change/move. The Sense of Causality in the Great Registering Consciousness would be vastly different from what it is for man.

It is not easy to understand what an ultimate moment Really is, because at present we have no way to detect such a unit of time,or that slightest and most rapid of all possible apparent 'movements'-in-Cosmos that corresponds to it. We have measured movements in milli-seconds and nano-seconds but even such minute units of time are gargantuan compared with the hypothesized ultimate moment.

Time Line

To find such an ultimate unit of time, measurements would have to be taken (by some internal means) on planes far subtler than our physical-etheric plane. In fact, measurements would have to be taken on that highest of all planes in-Cosmos *defined* by the 'ultimate particle/event' (a complete non-materiality, as far as human consciousness is concerned). This would probably take us, at the very least, to the Atomic Sub-Plane of the Logoic Sub-Plane of the Cosmic Logoic Plane, and perhaps *beyond* (depending upon whether there are ten or even one hundred *dimensions* in Cosmos). So it can be seen that the determination of the ultimate moment in-Cosmos is totally beyond the reach of any human mind, and perhaps beyond the reach of any mind upon our planet or even within the solar system.

Admitting the difficulties of ever measuring such an ultra-minute, yet hypothetically *actual,* unit of time (which is, by no means the briefest unit of time *possible,* though it *is* {or *would be*}the briefest *actual* unit of time) the concept of the ultimate moment is, nevertheless, valuable in forming a Metaphysical Cosmology. Let us attempt to enter more deeply into the ideas that surround this concept so basic to the understanding and measurement of Time in-Cosmo:

- In the PEACE of the NOTHINGNESS there is no *disturbance,* because there IS *nothing.*
- In the NO-THING, nothing moves, nothing 'ex-ists' to move; there is not the slightest deviation from the 'STATE' of UNDIFFERENTIATED HOMOGENEITY.
- But 'within' IT, *mysteriously* to our minds, *disturbance* 'ARISES', *change* 'ARISES', *modification* 'ARISES'.
- The PEACE of NOTHINGNESS is *objectless* but a disturbance *is* a modification *is* an object (an *other).*
- If there is any *change* in the NOTHING, there is necessarily an *object,* and (with the 'ARISING' of *object)* the necessary 'birth of Time'.
- The formula to have in mind is that PEACE equates to NOTHINGNESS, and disturbance to 'Somethingness'.

To understand disturbance in a medium we should study the *wave form.* Disturbance in a medium (a disturbance generated by some form of impact) generates wave forms. Wave forms arise because the impact has caused (in the particles of a given medium) a displacement from a condition of equilibrium. The 'peaks' and 'troughs' of the wave represent maximum displacement of the disturbed medium and the point where the wave periodically crosses the base line represents a virtually instantaneous return to the equilibrium which existed before the disturbance was created.

From what we know, a disturbed medium tends to return to equilibrium, eventually through friction, but does friction in any form we can recognize *exist* in those Pre-Cosmic 'times' when 'Something' is being *made* of NOTHING? The meaning of equilibrium, or of a non-disturbed state is important here, but we shall arrive at it by a means other than invoking the existence of friction. It may well be that any kind of actual 'touching' (the cause of friction) of one thing by another (ultimately, of one ultimate particle/event by another) is impossible and illusory.

If we study the wave form, we see that even in the midst of the ongoing cyclic disturbance which a wave represents there is a brief moment of equilibrium. The wave must cross the base line, which means that the disturbed medium returns instantaneously to its 'pre-disturbance' state, before being displaced, virtually (except for friction) equally and in the 'direction' *opposite* to its first displacement. Anyone familiar with vibrational movement, such as exemplified by a vibrating string or by the smooth surface of a pond set into vibratory motion by a pebble, will "see the picture".

In each and every wave cycle there is a momentary equilibrium which is analogous (when thinking of the ABSOLUTE) to the *PEACE before the disturbance* during which NOTHING IS. Every 'Something' is a disturbance in that 'MEDIUM'. Where there is no disturbance, there is no 'Something'. At the instant when the base line is crossed (the "x-axis", 'x' meaning 'nothing' or 'negation') whatever 'Something' the wave form *is*, ceases to exist, because the wave, for an instant, ceases to exist.

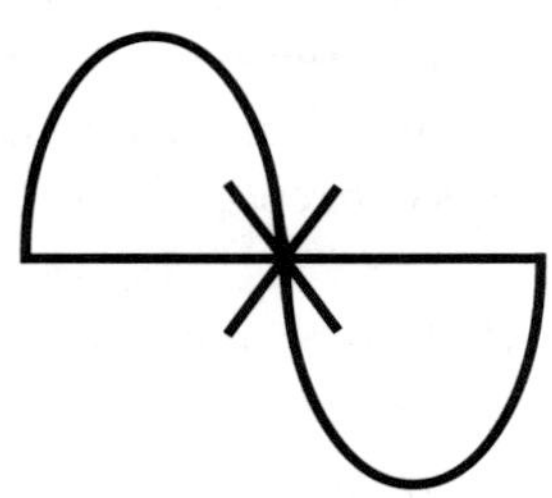

This idea has some astonishing implications for the *manner* in which the many 'Somethings' in our Universe exist, for The Universe exists because of multiple *(consciousness-generated) disturbances* within Mulaprakriti. From One perspective, the Universe is nothing but a tremendously complicated Wave Form, composed of virtually innumerable tiny wave forms (very likely *particulated* wave forms correlated {by means of some supreme Fohatic Cosmic Mathematics} to the duration of an ultimate moment, and related intimately, if not *identically* to an 'ultimate particle/event').

Each and every one of these tiniest wave forms (which *additively* create the many other wave forms in Universe) *ceases for an instant to exist* as the disturbance *which it is* returns for an instant to an equilibrized state identical to the state of 'peace before disturbance'. Since a wave form is Really a *thing,* and at that base line instant, it is as if *no-thing* exists. The wave disappears. By extension, the Universe, if conceived as a complex Wave

Form, and hence a *Thing* also, disappears, instantly, into Non-Objective Intra-Cosmic Nothingness/Subjectivity (not NOTHINGNESS), only to reappear once the *instant of negation* is over.

This model fits very well with the model of Ontological Oscillation developed later in this treatise in relation to the behavior of all 'ultimate particle/events'. What demands the most careful exploration is whether a tiny 'wave/thing' and an 'ultimate particle/event' are *identical.*

The exemplification of the wave form to denote the appearance and disappearance of *thingship* is suggestive and useful, but may not be appropriate in every respect to our Cosmos. The wave model suggests and even demands a *continuum,* and the Cosmos is *not* a Continuum. If anything, Cosmos is a 'Dis-Continuum'. The Universe *seems* to exist continuously, but only ONE 'THING' which is the NO-THING *REALLY* does. The UNDISTURBED NO-THING is the only CONTINUUM, for the very reason that IT IS an UTTER SAMENESS which <u>*continues*</u> *exactly* AS IT ALWAYS IS.

Returning to the model of the wave form, we see that from a certain perspective the half of the complete wave form 'below' the baseline can be conceived as symbolizing an 'ultimate particle/event' occurring in an 'ultimate moment' in-Cosmos. The half of the complete wave form 'above' the baseline can, perhaps, be conceived as representing the 'inter-moment instant' (the Cosmo-Subjective Now) when the Fohatically Fabricated Universe hypothetically, *disappears* into Intra-Cosmic Nothingness/Subjectivity (a State in which, intra-Cosmically, Spirit is entirely disengaged from Matter and only the Logoically Sustained, Archetypal World remains).

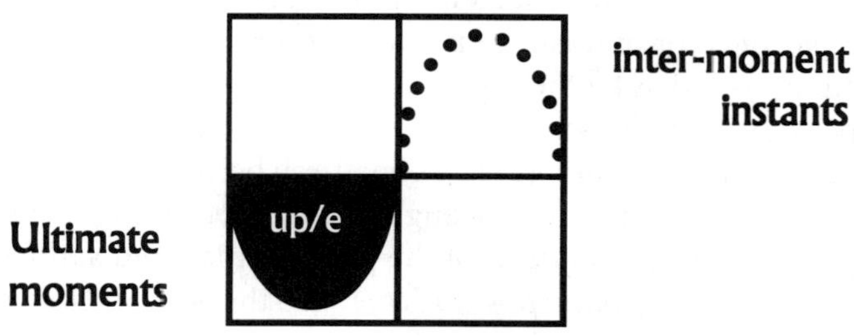

Since the World of Being, the Archetypal World, Is *not* the NOTHING, It *too* must be seen as a species of 'disturbance' just as is the World of Effects of Fabrication in which the 'ultimate particle/event' is the basic building block of the Cosmic Configuration. Discounting any frictional effect, it can be seen that 'disturbance' and, thus, *thingship* exists 'above' the base line and a kind of *'no-thing-ship'* (or perhaps, a negative, *'Archetypal anti-Objective State')* exists 'below' the base line. Certainly, the Universe *as we know it* would cease to exist if only for an *instant* (the fleeting duration of which would have to be determined) at the base line crossing.

A question as regards this model would necessarily arise, Does the lower half of the wave represent *complete* NOTHINGNESS? Or, does it represent a kind of 'anti-Objective State' which is more a complete 'disengagement' of Universal Spirit from the Universal Form, and, yet, *not* an entire NEGATION?

These thoughts need much pondering. Obviously, different models can suggest different dynamics to the interpreting mind. The greater likelihood, however, is that at no 'time' during the Universal Manvantara does COMPLETE NOTHINGNESS supervene. Thus, Cosmo-Subjectivity within the World of Being is a kind of Nothingness with respect of Cosmo-Objectivity, but certainly *not* the NOTHINGNESS of the ALL-IN-ALLNESS (the INFINITE SUBJECTIVITY).

If it is possible to correlate two major models related to the ultimate moment (namely the Wave Model, and the Particle/Event Model), then the duration of an ultimate moment (*and* the duration of its cyclic non-existence in-Cosmos!) a duration which can, theoretically, be different in different Cosmoses, would correlate with the frequency of the wave form. Presumably this frequency is a Cosmic Measure 'DETERMINED' by the INFINITE SELF *(just before!* or *at!)* the 'MOMENT' of the 'FLASHING FORTH' of the 'RAY' of the ABSOLUTE. Certainly the 'Determination' is 'Made' (whether the *determining entity* be the INFINITE SELF or ITS 'EXTRUDED' 'Representative', the Infinite Subject) *before* the interplay between the Infinite Subject and the Infinite Object (i.e., undifferentiated Mulaprakriti—Pre-Cosmic Root Matter) begins.

When considering the Wave Form Model of Universal Motion, it must also be asked whether the *gradualism* suggested by the form of a wave fits with the increasingly confirmed view of a Quantized Universe. While Wave Forms *do* indeed exist in Cosmos (perhaps *more* at a relatively *macro*-level than the smallest of micro-levels) it is beginning to appear that they are of *secondary* importance to the Particle/Event Model which fits better with the idea of the Fohat Strata of the Universe (especially) as a Great 'Discontinuity'.

Let us explore somewhat the problem of *gradualism* in the Wave Form Model. Because the Wave Form is (or, at least, *seems) continuous,* it suggests a *gradual movement* from the 'position' occupied by the medium at equilibrium to the position of the medium at the 'peak' or 'trough'—the places of maximum disturbance. Meta-physically, however, the facts might be otherwise. There may well be, instead of a gradual movement (over time) from base line to peak or trough, a *quantized movement* in which the disturbed medium is, as it were, *suddenly* at the peak position and *suddenly* at the trough position, having occupied *no other positions between* the peak and the trough *other* than the base line position, which, in terms of Cosmos can be conceived as representing the moment of Cosmic disappearance or reappearance.

Such a model would operate simply on an 'on'/'off' dynamic. While the wave form model (as *usually* conceived) may be suggestive in some ways of how things *appear* to happen (especially on a macro/illusory level), it is more likely that a model which features the *instantaneous repositionings of flashing and unflashing particle/events* would prove more descriptive of how the Cosmos Really works.

The Vibrating Wave Model might suggest to those familiar with the newer physics, *alternating states of thingship,* or more familiarly, *alternating states of matter* in a bi-polar Universe. Taking the idea further, we might even think of *Alternating Universes* (equal but opposite) which persistently flash 'on' and 'off'. Modern ideas of matter and 'anti-matter' seem to fit into this category of speculation.

Whatever model is used, it certainly seems that every *thing* that exists-in-Lower Fohatically-Fabricated Cosmos, does not exist *continuously* but 'appears' and 'disappears', flashing 'on' and 'off', with a tiny undetermined interval of negation (an interval more likely of Objective-Nothingness and Universal Subjectivity than of complete NOTHINGNESS) in between.

- Complete NOTHINGNESS during the interval is highly unlikely because, such a theory, taken to its logical conclusion, would demand that the 'RAY' of the ABSOLUTE, 'ITSELF' continue to 'FLASH' 'on' and 'off' for the entire duration of the Universal Manvantara, which seems not only terribly 'unaesthetic', but, at the very least, a tremendous violation of the Cosmic Law of Economy.

Further, if such a model were correct, it would indicate a huge oversight in the traditional metaphysical literature of World Philosophies and Religions which fail to mention anything even remotely resembling this model. More reasonable seems the hypothesis that, since the 'RAY' of the ABSOLUTE, being ESSENTIALLY NO-THING, is a 'quasi-CONTINUITY' (discontinuous only during Universal Pralayas) and since all Super-Cosmic Players and Cosmic First Family Players are, Essentially, derivations of that 'RAY' of the ABSOLUTE, the *non-fluctuation* of quasi-CONTINUITY is imparted to Them through a Chain of Lessening Scope of Being, and thus They (the 'Major Players') can and must *abide* (in the Super-Cosmic Realms and in the World of Being/Archetypes) though all the Universal Structure fabricated by Intra-Cosmic Fohat disappear from the perception of every intra-Cosmic E/entity after each ultimate moment. [The implications arising from various possible durations for the undetermined 'interval of negation' are discussed elsewhere and in the Glossary.]

Whether or not we know exactly the mechanics of this process (and we *do not),* the picture of a 'Discontinuous Universe' is emerging. In the two models here discussed, a *disturbance* (i.e., modification) and a *thing* are the same. We are interested in the 'positions' of things relative to each other, how they 'arrive' in those positions, and the 'time' it takes for them to arrive in those positions. It is beginning to appear as if *gradual continuous movement of a thing through all points of space* is an *cosmic impossibility.*

An ultimate particle is an *event* arising out of the interplay between intra-Cosmic Fohat and Its Own Self-Reflection (a kind of Cosmic Prakriti). An ultimate particle is an energy event of a specific duration, that duration being an ultimate moment. The only moment of time *available* in the Lower Worlds of Cosmos is the time it takes for an ultimate particle to appear and disappear, and even that is a 'frozen' moment.

Questions arise as to whether an ultimate particle (the most minute form of disturbance possible within Cosmos, Itself {as there *are Pre*-Cosmic Disturbances on a vast scale}) flashes into existence/objectivity *all at once,* or *gradually.*

- If gradually, there would be continuous *changes* in the ultimate particle as it became more and more manifest. The different phases through which it passed on the way to complete manifestation would suggest that the ultimate particle was *divisible* (having *parts* within it which could change—intensify or de-intensify). By definition, however, an ultimate particle has no parts and, hence, can undergo no modification, for a modification is a change and change requires parts. The ultimate particle/event represents the extreme lower limit (in-Cosmos) of apparent divisibility. If the ultimate particle is divisible, it becomes *infinitely* divisible and then (if we follow out the implications) we have the old problem of

INFINITY 'BECOMING' completely *actualized* in a Finite Universe, which is an impossibility.

- So, rather than *gradually* grow into the fullness of its existence, the ultimate particle/event would seem, necessarily, to flash 'on' suddenly, completely, as a complete, 'unchanging' unit, and 'off', suddenly, completely, and as a *complete unit*, the duration of its *appearance* being considered to be the ultimate moment for that Cosmos, which moment also is indivisible (i.e., no moment of shorter duration than it can exist in that Cosmos).

The indivisibility of the ultimate particle parallels the indivisibility of the ultimate moment. While 'within' the FOUNT OF ALL POSSIBILITY *anything* is possible (including the divisibility of the ultimate moment and ultimate particle/event) it is, so it would seem, 'WILLED'/Willed, that the ultimate moment and the ultimate particle/event be *indivisible* in that Cosmos in which they play their 'part'.

What we are seeing is that the *actual* *expression* of the INFINITE SELF with ITS *infinitude of infinitized possibilities* is denied or impossible in a *Finite Field* such as the Universe. (An *infinitely partial expression* is what Cosmos *Is.)* Perfect *gradualism* of movement from one 'place' to another, or from one 'phase' of a thing to another 'phase' of that same thing requires an infinite, perfectly homogeneous continuum in Universe. In our Universe, however, it seems that not *all* possible 'places' (whatever a 'place' *is!?)* can be occupied and not all possible 'phases' can take place—only some. What this means, effectively, is that not all relationships and combinations are Cosmically Sanctioned by the Universal Logos, administering His Intention by means of the Cosmic Algorithm which (in Time and Space and through Motion) fulfills the Design-at-the-Beginning.

◆ So we are taking the position that an ultimate particle/event does not move through *all* points in space from one position to another. In fact (for all practical purposes) ultimate particle/events *are* the *points of Space*—albeit, *virtual* points and not *Real* points.

Further, we hypothesize that the ultimate particle/event does not *gradually* become a complete ultimate particle/event by 'moving' continuously through various phases of *'partial ultimacy'* on its way to becoming a *complete* ultimate particle/event (as the Wave Form Model might suggest). The ultimate particle simply disappears and reappears instantly, where it is *supposed* to be, and in the *completeness* of its nature.

Under such conditions, what would the movement in-Cosmos of ultimate particle/events be like? The ultimate particle/event would:

- Appear in a 'position' *relative* to the position of other ultimate particle/events.
- Persist (changelessly) for an ultimate moment in that 'position'.
- Disappear into what we might call a 'Fohatic Disengagement from Its Own Objective Reflection in Cosmic Prakriti') for an undetermined instant (a Cosmo-Subjective Now) which may be of duration equal to that an ultimate moment or, perhaps, of briefer duration (or perhaps, *much* briefer, although a *lengthier* disappearance is also *possible*—we have no way to determine this).
- Reappear in a new 'position' relative to the position of other ultimate particle/events.
- Endure (changelessly) for another ultimate moment, and so forth.

Is an ultimate moment a *constant* for a given Universe/Cosmos? This is an important question. There is no known reason why the ultimate moment of a particular Cosmos has to be of the same duration as the ultimate moment of a different Cosmos. The shorter the ultimate moment, the shorter (relative to the Absolute Standard of Time Measurement) will be the duration of an ultimate particle's 'holding' of a position in a Cosmo-Configuration. We have, at present, no way to measure the duration of ultimate moments in our own Cosmos or in Cosmoses past or to come, or even, for that matter to confirm that they exist! But, what if an analogous condition prevails at various *'times'* in our own Cosmos? What if there is a Plan for increasing the *tempo* of the Cosmic Pulse as the Universal Process proceeds? This is a fascinating and possibly disturbing question.

The duration of ultimate moments and the size of ultimate particle/events presumably vary directly. It is possible, however, that the magnitude of these ultimate units is *not* a Cosmic Constant throughout the duration of a Cosmos. It may be that the Pre-Ordained Cosmic Parameters call for an increase or decrease in the magnitudes of both ultimate moments and ultimate particle/events at various *phases* in the Cosmic Process. This would introduce a condition of relativity in-Cosmos which would produce fluctuations in *time quanta* as well as in the degree of 'extension' of what we call 'matter'. Thus, some periods within the Cosmic Process would take less or more time than others, even though the number of events in such periods might be virtually equal.

If the Pre-Ordained Cosmic Parameters called for such Cosmic Changes during the Universal Process, we might find the 'Key Note' of a Cosmos subjected to orderly, periodic change, and with that change, all conditions in-Cosmos would change. If such Changes of the Fundamental Cosmic Rhythms *did* occur, there would probably be a numerologically significant number of them—perhaps three, or seven, or twelve (the number twelve being especially related to the musical octave), for Cosmos is organized *musically* (the different periodical vehicles, such as personality, Soul, and Monad, in either man or God, standing to each other as musical ratios). During the reign of a given Key Note, the ultimate moment in-Cosmos would be of a specific duration and the ultimate particle/event of a particular size/magnitude/intensity (however one might describe such an ultimate unit). When the Key Note changed, the duration of the ultimate moment would change as would the parameters of the ultimate particle/event. Such Cosmic Changes would be entirely related to the Will of the Universal Logos and to the adaptive, responsive Virtuosity of Fohat. What would presumably *not* change is the need for:

- a minimum time quantum (which an ultimate moment exemplifies);
- the need for an ultimate, indivisible unit of matter/force (which the ultimate particle/event exemplifies); and
- the need for the timed appearance, timed disappearance and timed reappearance of an ultimate particle/event.

On a speculative note, it is interesting to think that as the Cosmos passed through its involutionary phases a "slowing down of time" *might* take place, such that during the involutionary or descending arc ultimate moments could increase in duration, just as they might decrease in duration during the Universe's evolutionary arc, especially as the 'time' for Universal Pralaya approached. A hypothetical modus operandi for Universal Pralaya might see ultimate moments becoming more and more rapid approaching infinite speed, and ultimate particles becoming smaller and smaller in 'extension', approaching *nothingness.* When the ultimate moment of Cosmos became *zero* and when the mass of an ultimate particle/event became *nothing* (again zero) reabsorption into the NOTHING, the INFINITE SELF, the ALL-IN-ALLNESS would have occurred. The problem of what induces the onset of Universal Pralaya is unresolvable.

From all this speculation certain hypothetical ideas are important to grasp. The first and foremost is that each Universe (being finite) has Its limits. These limits affect Time, Space, and Motion, all of which will exist in a certain relationship with each other *during* Cosmos, and not exist at all during the Universal Pralaya.

- All Universes are Discontinuities in INFINITE DURATION, and manifest as 'Dis-Continuums'. Time, Space, and Motion vary from Universe to Universe, and, perhaps, (according to certain SELF-Pre-Ordained *Dynamic* Laws), *within* each Universe as well.

All of these minute Time, Space and Event magnitudes are, for us at this 'time', undetectable. They occur at the extreme *micro*-level of the Cosmic Process while we, as human beings, function at a relatively *macro*-level. As well, Cosmic Rhythms at the *extreme* macro-level of the Cosmic Process are equally undetectable—by Man.

Holding both the micro-levels and macro-levels of Cosmos in mind, we find that items-in-Cosmos can be differentiated according to their stability over time. Some items last without change for only an ultimate moment. Other items seem to last without change for billions or trillions of ultimate moments, or more. Even those items which seem not to change from one ultimate moment to the next, in Essence, *do,* because all prakritic constituents within them (atoms and minute particles of all varieties) are basically constituted of ultimate particle/events (the fundamental Fohatic building blocks of Cosmos), and these ultimate particle/events undergo a change of position at the onset of ultimate moment.

NOTE: It is not said that ultimate particle/events change position *between* ultimate moments. According to the theory being advanced, they do not even *exist* between ultimate moments, so they cannot be 'moving'! The illusion that some relatively macro-items-in-Cosmos 'hold position' is created by the tendency of the ultimate particle/events which constitute such items to reappear in the same position (or virtually the same position) *relative to each other* as the position from which they disappeared. This would be what the Tibetan has called "Repetition in Space." This kind of repetition or near repetition of pattern on the extreme micro-level produces a *seeming continuity* of pattern on more macro-levels.

Just the way all moments are additive—greater moments of time being composed of ultimate moments, so all prakritic modifications are additive. There is no *greater* modification/particle/object/thing which is not composed of *lesser* modifications/particles/objects/things (until the boundary of minuteness and evanescence is reached in the ultimate particle/event). Similarly, there are no greater movements which are not composed of lesser movements (until the least of all possible movements is reached as a necessary boundary in a Finite Universe). The laws governing the 'least common denominator' of an item, govern (to a significant extent) the item as a whole. Since such items are composed of many ultimate particle/events which are, as it were, blinking/flashing 'on' and 'off', the items they compose also blink/flash 'on' and 'off'.

- Thus every disturbance/modification in-Cosmos, returns for a specifiable instant to the relatively *undisturbed* Subjective Substratum between ultimate moments. All items in-Cosmos, behaving responsively in the manner of their smallest/briefest constituents, instantaneously appear and disappear, and all items *appear and disappear together.*

This is *very* important to realize. All 'flashes' are coordinated, and all ultimate particle/events are occurring *at the same time.* It is impossible, for instance, for an ultimate particle/event to occur, and another ultimate particle/event begin 'midway' through the duration of the first—because a quantized unit of time cannot be divided, and the necessary 'starting time' for the second ultimate particle/event (midway through the duration of the first) would not be *Cosmically available* (according to pre-set Cosmic Parameters). Any thought of 'midway' (or of any fraction whatsoever) would suggest the existence of a unit of time *less* than that of the ultimate moment in that Cosmos, which is, by definition, impossible.

With the absolutely necessary simultaneity of all ultimate moments and all appearances and disappearances of ultimate particle/events established, we can see that since the Worlds of Fabrication within the Universe as a Whole Is the Greatest Intra-Cosmic Objectivity Now existing. And, since It, as a Whole (like every other item-in-Cosmos except the ultimate particle/event) is composed entirely of ultimate particle/events, then It too, as a Whole, must blink/flash 'on' and 'off'—fluctuating throughout the entire Universal Manvantara between existence and a State of Subjectivity which approximates non-'ex-ist-ence'.

The movement of ultimate particles/events/modifications/vibrations—they are all virtually equivalent—are different, perhaps, from what one might expect. The movements, so we hypothesize, are *not* continuous, because *truly continuous movement* (through all *possible* points in space and through all *possible* instants) is not *possible* <u>*in*</u> Cosmos.

We can imagine the flashing on and flashing off of events. With each such flash, on the level of ultimate particle/events, a change of relative 'position' of the ultimate particle/event is noted. The repositioning would seem to occur through discontinuous 'leaps' or 'lurches'—nothing like a smooth flow through 'space' of a thing in one position 'on its way' to another position. The 'leaping' and 'lurching' is caused by the disappearance into the 'State' of Fohatic Disengagement [see Glossary] (as Fohat 'Blinks') of the moving/changing *thing.* Each reappearance takes place in what we have been calling a Cosmically Sanctioned Position or Configuration, which has to do not with 'points in Space' but with the 'geometry of relationship' within the Cosmic Configuration.

Discounting the suggested bi-polarity of the Wave Model for a moment, it may be possible for two or more things/particles/events/modifications to reappear in exactly the same position relative to each other as the positions they 'occupied' before they disappeared. In such an instance, would *change* have taken place? Would there have been movement?

It could be said that (as regards that particular set of variables—i.e., the 'position-repeating particle/events) there would *not* have been movement *within* the Universal Field, but that, nevertheless, there would have been movement *between* the Universal Field and, what we might call the Non-Objective Universal Field caused by Fohatic Disengagement (the Field of the First Family of Cosmic 'Players', the World of Being). Further, due to *relativity,* even non-moving (apparently non-changing) variables in the Cosmic Configuration would seem to have moved/changed when viewed from the perspective of other repositioned particle/events. In fact, if one variable in a system of variables changes, then *all*, willy-nilly, change.

Micro-movements of the repetitive nature discussed above suggest the form of a *standing wave.* The appearances and disappearances of event/things are animated by a constantly applied source of Energy (somehow originating in the INFINITE SELF, but mediated through the Agency of a number of Super-Cosmic 'Players'), but the event/thing/appearances do *not* change position relative to each other. The integrity/stability-in-presence of the larger item/thing composed of these 'position-repeating' ultimate particle/events is thus preserved.

Thus it seems that ultimate particles have *many* 'choices' of 'positions' to occupy, but not a choice of *all* 'positions' (relative to each other). Among the choices appears to be the possibility of returning exactly to the place/position from which is 'disappeared' the instant before. (But does that 'place' or 'position' even *exist* unless a certain number of other ultimate particle/events 'decide!' to do the same?) The stability and instability of all configurations-in-Cosmos is determined in these ultimate micro-movements. When things on the relatively macro level are seen (with macro perception) to change, great indeed must have been the *repositionings* of ultimate particles on the ultimate micro level (over 'x-tillions' of ultimate moments).

Other than offering a few potentially provocative theories, it is quite impossible to be at all specific about the Ultimate Cosmic Physics—for that is what we are talking about here. From thinking about these speculations the student should emerge with new thoughts about Motion and its relation to Time and Space. The student may also emerge, as the author has emerged—humbled at the thought of his ignorance!

It might be asked what determines 'where' an ultimate particle/event will reappear once it has disappeared. In a large and global way, it could be said that the Will and Imaginative Power of the Greater Entity (the Universal Logos) that pervades the system in which the ultimate particle is functioning (which Entity is 'Represented in Action' by Intra-Cosmic Fohat) determines the 'placement' of the ultimate particle/event (this Will being *activated* during a Cosmo-Subjective Now).

Visualization creates in consciousness the image of extension. Ultimate particles (and, for that matter, any particles/units/things subject to the Will of the greater pervading Entity) 'move' in such a way as to 'take their assigned places in the visualized image'. Visualization/Imagination on the level of the Universal Logos and His Agent, Intra-Cosmic Fohat, is a geometrizing force and Logoically-inspired Fohatic-Will 'repositions' the ultimate particle/events geometrically and, then, 'holds' the geometrically repositioned ultimate particle/events and (by extrapolation, more macro items constituted of such particle/events) "in their proper places."

♦ We can see how any in-depth philosophical consideration of Time must deal with the ultimate micro-level of the Universe, as well as with the Universe as a Whole in Its relation to the First Family of Super-Cosmic Players Who are Essentially the Super-Cosmic Trinity of Infinite Subject/Infinite Object/Infinite Consciousness and Who, even more ESSENTIALLY, 'AROSE' from out of the NO-THING.

It has been said that the INFINITE SELF is always *present* as the PRESENCE. This is another way of saying that there is a CONTINUOUS PRESENCE of the INFINITE SELF no matter what is 'going on' in-Cosmos. Modifications do not negate the PRESENCE though they certainly *obscure* IT for the consciousnesses whose attention is entirely captivated by modification. But, using this paradigm, we can see how, in quite another way, from ITS CONTINUOUS PRESENCE on ITS own LEVEL of BE-NESS, the INFINITE SELF 'PRESENTS' ITSELF as a PRESENCE to intra-Cosmic beings (or 'PRESENTS' ITSELF through Representation by ITS immediate First Family Cosmic Extensions—the Universal Logos and Sub-Logoi—Father, Mother, Son, and Holy Spirit) during every instantaneous interval 'between' ultimate moments.

At that moment when the disturbance-wave crosses the base line—at that moment when the particle/event disappears for what may be an invisible-to-Lower-Cosmos ultimate-moment-of-Cosmo-Objective Negation, or a still *lesser* moment-of-Cosmo-Objective Negation—the Fabricated, Objective Universe disappears and *all* that remains is the INFINITE SELF, ITS Super-Cosmic Representatives, and (intra-Cosmically) the Cosmic First Family Representatives of the INFINITE SELF. At such a moment-of-Cosmo-Objective-negation:

- the PRESENCE of the INFINITE SELF, PER SE,
- the PRESENCE as mediated through the Super Cosmic Trinity (of Infinite Subject/Infinite Object/Infinite Consciousness), or
- that PRESENCE as meditated *through* the Presence of the Cosmic First Family of Ultimate Gods ... is all there is.

◆ Thus from ultimate moment to ultimate moment 'we' *have the opportunity to identify with and even consciously become* Them (far more than when prakritically involved), and even become *IT*. We are restored to intimacy and perhaps, even, to *realized identicalness* with What and Who We Really *Are*, and are brought into greater *identificatory intimacy* with WHAT and WHO we REALLY *ARE*.

Therefore, there is no need to 'wait' until a later 'time' to merge with Ishvara, the Universal Logoic Representative of the INFINITE SELF, or to merge (perhaps, to some degree) with the INFINITE SELF, per se. There is not one single ultimate-moment-cycle, not one single Ontological Oscillation, but that the PRESENCE of the INFINITE SELF (albeit through ITS Great Super-Cosmic and Cosmic Representatives) is *immediately* accessible. Also, we do not have to wait for the Universal Pralaya for the Objective Fohatic Cosmos to disappear; It is disappearing ceaselessly between Cosmo-Objective Nows, and has disappeared virtually *countless* times during our particular Universal Manvantara, and, an *extraordinarily huge* number of times per second.

What all this adds up to, in a practical sense, is the realization that the SELF (as mediated by ITS Logoi, Who are ESSENTIALLY identical with IT) is *always present!* Let's put it in this strange way: The INFINITE SELF and ITS Representative Logoi (with all Their Emanatory Extensions, among which the Cosmic Monads are numbered) are intimately *present* between 'Nows'.

Since **I** AM, ESSENTIALLY, the SELF, but, more pertinently, since **8** Am, Essentially, the Universal Logoic Self, then **8** Am (constantly, yet sometimes imperceivably) *present*, and **8** *Am* so (no matter what else of an apparently evolutionary nature **8**-as-I may be doing and being during the Lower Cosmic Process). My two *identities*—that of a *relative*, very limited B/being-in-Cosmos, and that of the INFINITE SELF-as-Universal Self, alternate with astounding rapidity, virtually countless times during every fraction of a human second.

- For practical, psychologically appreciable purposes then, **8** Am *continuously* the INFINITE SELF-as-Infinite Self, and only *discontinuously* the Finite self, *simultaneously*, throughout the entire Duration of Cosmos. These thoughts should bring SELF-as-Self-Assurance to the limited human personality, should they not?

We come to an interesting question of what can 'happen' during that tiny instant (perhaps equal to an ultimate moment) during which an ultimate particle/event is 'off'—during the instant that, by *extension*, the Fohatically Fabricated Universe, Itself, is 'off'. One answer (almost certainly incorrect) could be that *nothing can happen because NOTHING IS.* (This would be true if the inter-moment instant reverted entirely to ABSOLUTENESS, the reasons against which have been explained above.)

The radicalness of a totally disappearing Universe appears increasingly less likely than its alternative—namely, that during the inter-moment instant there is a reversion to a State of Pure Spiritual Identity, untrammeled by *appearance*, i.e., untrammeled by the illusion of Cosmic Objectivity. It is tempting to think that such a State of Pure Spiritual Identity would be *perfect* for a supremely intelligent exercise of the Free Will. Fohat, for instance, must make Its 'next move' in the form of the repositioning of a multitude of ultimate particle/events. How and *when* is that 'move' determined? Is the virtually instantaneous interlude of the inter-moment instant used by Fohat to adjust Its Actions so that they ever more closely *approximate* the requirements of the Design-at-the-Beginning 'Held' by the Universal Logos and the Son (and Its Supernal Tetraktic Emanations)?

- In a way one could conceive of the inter-moment instant as an 'interval of correction', utilized for the "Correction of the Form" of Cosmos. We know that the Intention of the Universal Logos and all the subsidiary, emanated Creators through which It Acts, is to Create the Universe *true* to Original Design. Any artist knows that the creative process is very much one of bringing the form of a medium into compliance with an inner, ideal template. A process of constant adjustment and readjustment is required to ensure that the form of the piece of art *conforms exactly* to that ideal template.

Thus it may be in relation to the Creation called Cosmos and Its Creator—the Universal Logoi, Sub-Logoi, Subsidiary Logoi, and a host of emanated sub-creators. Perhaps for all of them, the inter-moment instant (during the Cosmo-Subjective Now) is an instantaneous opportunity for reassessment of the next 'move', the next 'adjustment' in the Great Work of creating the Universe *true* to the Pattern in the Heavens.

It is important to realize that even *if* during the inter-moment instant there is a reversion to the State of Pure Cosmic Identity (achieved by means of Fohatic Disengagement), *rather than* an ultimate reversion to a 'STATE' of ABSOLUTE INFINITE SELFHOOD, that, nevertheless, the INFINITE SELF is the All-pervasive SUBSTRATUM of Cosmos and Super-Cosmos, and *IS*, in fact, *both* of Them. The INFINITE SELF IS *always present* as the PRESENCE. There *IS* NOTHING ELSE.

It would stand to reason that all intervals/instants *between* ultimate moments occur simultaneously, and that all disappearances of ultimate particle/events also occur simultaneously. If all greater structures (phenomenally) are, cumulatively, simply ultimate particle/events variously configured, then these greater structures also 'disappear' in the 'inter-moment instant'. Nothing is 'on' while ultimate particle/events are 'off'. Certainly, when an ultimate particle/event is 'off' no *other* ultimate particle/.event can be 'on'. If another particle/event could be 'on' on this way, an ultimate moment would be *divided* which it cannot be.

- ♦ No movement in Fohatically-Particulated Cosmos can 'go on' during the 'off' instant of *any* ultimate particle/event.

If a movement could occur during the 'off' instant, Time in Objective Cosmos becomes infinitely divisible and unquantized. In fact, there can be no *movement* of any kind in Objective Cosmos *for (or 'during') the duration of* an ultimate particle/event. Additionally, since Fohatically Fabricated Cosmos disappears (at least as a 'Creation') during the inter-moment instant, there obviously can be no objective movement of any kind *between* ultimate moments (for there is no 'moveable' Objectivity). Perhaps this incredibly rapid fluctuation is the constant reminder to the Identities which have *informed* and *created* the Reflected Universe of their Source-as-SOURCE.

The 'off' instants in-Cosmos, as 'Interludes of Fohatic Disengagement', are, thereby 'Interludes of Reality' wherein identification with REALITY is, presumably, greatly facilitated. Such inter-moment instants have their reflections on every Cosmic Level, whether micro or macro. (Perhaps these thoughts have some applicability to the process of meditation, wherein the meditator learns the use of the interludes to contact the higher, more Real Spheres; see Section VI.)

Returning to the study of Time and Motion, many tiny ultimate moments elapse before a larger, slower item-in-Cosmos resident upon a relatively macro level, seems to *move* upon that level. Obviously, however, the movement 'within' that item (at the level of ultimate particle/events) is extremely rapid and constant (though naturally discontinuous) Configurations of events/variables/modifications/objects/things/disturbances, etc. exist on so many different dimensions/planes/levels of Cosmos, and each type of configuration has its laws and rates of movement.

The configuration called the Sun and the planets has laws and rates of cyclic movement which (against the Standard of Measurement for Cosmic Time) are immensely different from the laws and rates of cyclic/vibratory movement which pertain to that configuration called an atom, or which pertain to that unexpectedly *complex* configuration called an electron. Even the longer, slower movements (longer and slower than ultimate moments), cannot take place *except* during a moments of time *available* in that particular Cosmos. Movements requires *times*, and, in a particular Cosmos, the only times available in which to move (in the Cosmo-Objective Realm) are ultimate moments. When for any item-in-Cosmos (modification, variable, unit, particle—call it what you will) the "moment has come to move," that moment will be an *ultimate moment*. It can be no smaller unit of time, because no smaller unit is *available* within the Cosmic Parameters.

It is curious to realize that the tiny 'off' time in-Cosmos, is (to the *recollection* of consciousnesses immersed in Cosmos) a 'Timeless Moment', a moment *unperceived* 'in' Fohatically Fabricated Cosmos and, thus, experienced (in terms of the sensed continuity of consciousness within the World of Fabrication) as "no time at all". Such an inter-moment instant could be infinitesimally short or of immense duration (Cosmic Parameters allowing) and the normal, Cosmo-Objective consciousnesses would proceed normally with Cosmo-Objective living—their sense of Time unaltered.

No matter how long the inter-moment instant may endure, the veiled, hence limited, consciousnesses within the Worlds of Fabrication/Approximation would be none the wiser. That internal 'moment of disappearance' may also be hypothesized as a supreme moment of Will, when the SELF-as-Universal Logos determines Its 'next Move.' (It is likely that the advancing human consciousness will discover the Metaphysical Physics of Cosmos to be very strange indeed!)

On How the Great Observer May 'See' the Cosmic Simultaneity

We have been considering events and movements at the micro level. Let us return to the largest possible perspective in-Cosmos to seek some degree of synthesis. While certain items-in-Cosmos might be more or less enduring, the Divine Observer is not prisoner of such duration, as are lower lives which "live and move and have their being" within Him. The Divine Observer Knows Its Life Span; *dies not* throughout the duration of Its Cosmos; Is not subject to any of the cycles It witnesses, but, rather, is subject only to Its Own Universal Cycle. This means that while 'long' and 'short' are *observed* they are not experienced/'inperienced' as conditioning limitation as they are for man.

The Great Being (which it must be remembered we *all* Essentially *are)* witnesses (with Cosmo-Omni-Dimensional Consciousness) all the many movements within Cosmos. First is registered the astonishingly rapid and persistent *pulse* of ultimate moments, and then all the various more macro-pulses in Universe. All these various pulses, correlated to various levels and dimensions of Cosmos, are aspects of the Cosmic Pulse. These pulsations create a Great Symphony of Percussion, with *rhythmic beats* occurring *virtually* (but not Really) "all the Time"—Time in-Cosmos being regulated not only by the duration of ultimate moments, but by the Greater Cosmic Rhythms in which the ultimate moments inhere.

The Perception of such a Cosmically Inclusive Observer would be one of an amazing variety of rhythmic impacts, deriving from the rhythm and meters of pulsating events which are keyed to and correlated to the Principal Cosmic Rhythms. Since the Observer of all this Cosmic Simultaneity (the Universal Logos) *endures longer than anything It Observes*, Its Sense of Time is entirely altered. The 'Past' is held perfectly in Cosmic Memory and can impact the Cosmic 'Present'. The process and phases of Cosmic 'Future' are clearly anticipated though not perfectly imagined in detail (for this would violate the Free Will of the Cosmic Observer and all Its 'enfolded' Monadic 'Rays').

Nevertheless, the Cosmic 'Future', too, can impact and 'lead' the Cosmic 'Present'. The Vision presented to the Great Observer would be Cosmically Kaleidoscopic with the ultimate moments beating out the rhythm of each *least change* of Cosmic Configuration in the Cosmic Kaleidoscope.

No change in-Cosmos is possible except precisely *at* or *upon* an ultimate moment. No changes can take place 'during' ultimate moments, or the ultimate moments would be divisible and, hence, *not ultimate*. No changes in the World of Fabrication can take place 'between' ultimate moments or they would be taking place at 'times' which do not even *exist* in the Cosmo-Objectivity of that Cosmos. The moment of unspecified duration between ultimate moments (which we have called all 'intra-moment-instant') is not-available in-Cosmos as a 'time for action'!

The Great Experiencer, Who Is the Cosmic Self, the Universal Logos, can be called the "All-Seeing-Eye". Perhaps every fairly developed system in-Cosmos has its "all-seeing-eye", but for Cosmos as a Whole there is only One. The "All-Seeing-Eye" experiences all levels of the Cosmic Pulse simultaneously. From a point in space a man may see all events on earth below as *one* flowing event/perception, whereas to the man on earth, his event/perception is partial, and of many separate and discrete events.

Similarly, to the All-Seeing-Eye, time, as an experience of 'long' or 'short' micro-impacts, would cease if the Temporal and Spatial Perspective were sufficiently 'distant' and 'encompassing'. To the All-Seeing-Eye, the time experienced *merely* as a sense of sequentialized linear events, is negated because the many 'lines of sequenced events' would be seen as one event, and the many experiential impacts 'below' would be experienced (at that remote 'distance') as one 'holistic' impact. Acknowledgment of the duration of pulsations (i.e., cycles) would still exist, but to the All-Seeing-Eye (Willfully assuming the Position of the 'Remote Observer'), everything that is happenin,g both in the Cosmo-Objective and Cosmo Subjective Worlds, is *most Really* happening in the Cosmo-Eternal Now.

In these ideas is given a hint to the human being who wishes to rise 'above' his present limited perspective on past and future and to live ever more completely within the Cosmo-Eternal Now.

- Let the human ring-pass-not be expanded persistently until it becomes increasingly identical with the Ring-Pass-Not of the Universal Logos.

Essentially, man already *is* the Universal Logos, *and* every other subsidiary Logos in-Cosmos as well, but the vehicle through which the human consciousness functions is still not (and will not be for virtually uncountable aeons) the Vehicle through which the Universal Logos functions completely—i.e., the Universe-in-Toto as Vehicle. In practical terms, every expansion of the human ring-pass-not is a step towards the perception/apperception of Cosmic Simultaneity.

On Time as Devourer

As long as there *is* a Ring-Pass-Not, however vast, it will be impossible to escape from the Illusion of Time. Even the Universal Logos is subject to Time (the Time which emerges from the Rhythm of the Great Breath). Only the INFINITE SELF, which has no RING-PASS-NOT, is as utterly TIMELESS as it is UTTERLY BOUNDLESS, and abides or 'forevers' [sic] in ETERNAL DURATION.

Time is said to be the "devourer", but Time does not *Really* devour anything. Time is simply a relative measure of perceived rate of change. When one pattern in a kaleidoscope changes into another, do we say that the first pattern is 'devoured' by the next? Rather, why not say that the first pattern "gives way" to the next or is transformed into the next. For those consciousnesses, however, which in ignorance cling to form, Time may be illusorily experienced/'inperienced' as a devouring process, but Time can also be experienced as a (seemingly) flowing medium in which creative change can be accomplished. Every experience of change can be considered a devouring or a re-creation. There need be no attachment to either one.

Thus, Time is one of those fundamental Cosmic *Conditions* which arises through the Finitization of the INFINITUDE. Time arises when any registering consciousness, registers impression within a finite field—a field bounded by a Ring-Pass-Not. Time arises from the transposition of 'ABSOLUTELY INFINITE MOTION' *(infinitized* motion) which is MOTIONLESSNESS into limited motion which can be *measured* and is, hence, a 'child of Maya'.

On Some Practical Applications of Time as a Measurement

- Time is the *apparent* segmentation of ETERNAL DURATION due to the limited consciousness of a perceiver/apperceiver.

A ring-pass-not, Really, is not so much a tangible boundary, as it is an inability to 'see' or be conscious beyond a certain 'circumference'. When the Ring-Pass-Not for a Universe-to-Be is determined, the determination does not reflect the *capacity* or *incapacity* of the SELF-as-Condensed Point to 'See' or not to 'See' the Wholeness of Mulaprakriti, but reflects, instead, a SELF-*Chosen* Limitation (resulting in a vision of *Cosmic* Prakriti)—a 'Decision' by the SELF-as-Infinified Point-as-Condensing Point-as-Condensed Point *not* to 'See' or extend Consciousness beyond a certain SELF-Chosen Circumference 'in' Mulaprakriti (Which is, after all, Its {the Self's} own Self-Reflection). The SELF-as-Infinified Point is *Intending* upon the Specificity which will lead to Becoming a Specific Cosmos, and thus deliberately 'Narrow Its Vision' to Become the SELF-as-Condensing Point and, finally, the SELF-as-*Condensed* Point.

Time is an instrument for dealing with Relativity, and is incapable of being applied to REALITY, the WORLD OF BEING. As a means of measuring Relativity, Time can only be applied to things-in-relationship—combinations, configurations, patterns, collections, aggregations, etc. (Even an *isolated* thing, however, may be considered as 'in relationship', for instance, with the *consciousness* that observes it.)

From a practical perspective, Time is used as a means of relating with intelligence to the World of Becoming, i.e., the Cosmos Considered as a Whole. To do this, Time is used as a measure of the relative duration of cycles (both regular and irregular). *Regularity*, however, is required of the unit of time that is used as a standard of measurement for measuring other cycles. For instance, we might ask, *how many* of one type of cycle can occur relative to *how many* of another type of cycle? In the process of measurement, *comparison* is always needed. If *many* of one type of cycle occur during the span of *one* of another type of cycle, the individual cycles among the many are said to endure for a 'shorter' time and the one cycle (into which the many 'fit') for a 'longer' time. The designations 'shorter' and 'longer' are, however, meaningless unless a *systemically significant cycle of time* is used as the standard of measurement.

On Earth, we often use the Earth year to measure human events. In relation to more Planetary Events, the Twelve Precessional Ages might be used, or the summation of these Ages—the Great Platonic Year. For Solar Systemic Events, the 250,000 year cycle (suggested by the Tibetan) of the Sun and Solar System around the Pleiades might be used. None of these standard of measurements is absolutely regular and independent of outer influences, but for practical purposes can be used. For great Intra-Cosmic Events, galactic measures would have to be used. The search for the consistently reliable Standard for Cosmic Measurement will be a long one—for that unit of Life called Man.

The concept of Time is intimately connected with certain other concepts which the metaphysician must always consider. Time is intimately connected with Number. Time is to Number as DURATION is to ZERO. Where ZERO prevails, Time exists not. ZERO is the destroyer both of Number and of Time. ZERO is the *true* DEVOURER (the FORCE which annihilates distinction of any kind).

Time, as well, is related to Maya, Illusion, Object and everything which is measurable. Time is sequential to the limited consciousness. Where there is no perception/apperception of sequence, there is no experience of Time. In the experience/'inperience' of Time, one 'event' seems to occur 'after' another. This experience/'inperience' of "one-after-another" is related to the limitations of the registering consciousness, but, then, to the ABSOLUTELY UNLIMITED CONSCIOUSNESS (if we can call IT a 'CONSCIOUSNESS' at all) there *is* no registration of Time.

To the atom, which presumably blinks 'in' and 'out 'of tangible manifestation (incarnationally) an extraordinary number of times in one second of Time, an event occurring even one minute of human time later than a given starting point would seem to the atom (had it a 'knowing' consciousness) to occur perhaps *many, many <u>lifetimes</u>* later. To the human being, the event would seem much 'closer'—only a minute away. To a far superior Consciousness, that same event, one minute later than our starting point, would seem *virtually simultaneous* with the starting point one minute before. We are dealing here with the differential perception of the *same* interval of duration (between events) as registered by different E/entities from various 'heights', planes, or dimensions in-Cosmos.

On Units of Time

The measurement of Time is related in a simple manner to Number. In such measurement we are seeking to establish the number of measurable events of one nature which can be *counted against* the number of measurable events of another nature—assuming the act of measuring of both sets of events begins simultaneously and ends simultaneously. Or, more conventionally, a continuously regular pulsation is selected as a standard unit of time measurement. For instance, any system to which time measurement is to be applied often has a number of ongoing pulsations or cycles. *One* such category of pulsation is selected, either arbitrarily or because the selected pulsation is both regular and integral to the system to be measured. The selected pulsation becomes a *standard unit of time* used in all measurements of other pulsations or cycles within the system.

Thus, the duration of the various kinds of pulsations within any system can be measured against each other. For present humanity upon the Earth, such standard units of time are the second, the minute, the hour, the day, the month, the year, etc., (all of which are loosely derived from Earth-related pulsations—for instance, the duration of the rotation of the Earth between two successive exact sidereal alignments (the sidereal day); the division of the Earth day into twenty four equal segments (the hour); the number of rotations of the Earth between vernal equinoxes (the Earth Year), etc.

It is impossible to say just how long such a standard unit of time Really lasts because human knowledge has nothing *absolute* or *invariant* (as yet) against which to measure it. One can only say how long such a standard unit of time lasts relative to other *relatively regular* pulsations within the system (which pulsations also serve as apparently constant *standards)*. None of the standard units of time in use have been compared to any Ultimate Standard Unit of Time in-Cosmos, because no Ultimate Standard Unit of Time has been discovered which can apply to all dimensions in-Cosmos. Even the various dimensions themselves have not been widely recognized.

Time is also inseparably related to the principle of *repetition*. The utterly smooth face of REALITY is disturbed through the appearance of *event*. The word 'Eve'—considered by many the 'Mother of Humanity'—can be found in the word 'event'. An event, as well, is 'evanescent'—passing, as are all things which come to birth ("e-venire"—"to come out").

When regularity of interval between events begins to be noticed, then the measurement of Time can begin. Nature is full of *regularities,* but by far the majority are not noticed by man.

- To register regularities and compare them to other regularities (and even irregularities) is to begin to understand the *order* in-Cosmos.

Vibration is the persistent recurrence of identical (Really, *indistinguishably similar)* events at regular intervals. Certain vibrations (also called cycles) are selected as regularly recurring, *acceptably* 'identical', changes which can be used to quantify other vibratory and non-vibratory changes.

On Perceiving/Apperceiving Lengths of Time

A *length of time* is the duration (or space) of non-event, or (non-perceived event) between two perceived events. In a regular vibration, the *length of time* which characterizes or defines the vibratory cycle is a regularly recurring duration identical with itself. In action which is not necessarily vibratory, but is simply regular, the *spacing* of events is regular, though the *events themselves* may be totally different. When, in normal human living, "nothing happens," time seems to drag out. When many things happen in a short "space of time", time seems to speed up.

If, literally, "everything happens at the same time," (a popular statement which is never *literally* true) there would be no time at all, that is, if all things or events were experienced at the very same moment, time would not be perceived as a "sequence of events" and there would seem to be no time at all. All happenings would be perceived as "happening in no time", or "in nothing flat!" On the *other* hand, if "nothing ever happens," (again, a gross exaggeration, as this phrase is usually used) there is, seemingly, "all the time in the world"—and this is *literally* true. In the ALL-SELF-as-SELF, NOTHING *ever* 'HAPPENS', or No-Thing *ever* 'HAPPENS', or NOTHING 'HAPPENS' *forever.*

These are simply some suggestive alternative perspectives for speaking about the process of perceiving/apperceiving "lengths of Time" and must be experienced/ 'inperienced' intuitively to be understood.

On Event

We now come to a more detailed discussion of the meaning of the term 'event'—so important in the description of the strange world that today's quantum physics has discovered. Let us begin simply:

- An 'event' is a happening, an occurrence, a presentation.
- More esoterically, an 'event' is an object, a modification, a difference, an appearance, a manifestation.
- An event-in-Cosmos is a variation from a previous pattern or configuration.
- An event is the presentation of a 'difference' against the background of regularity, or against a background of relative immobility or 'sameness' (although neither one of these can *actually, literally* occur in-Cosmos).

An event is a change. If there is no change there is no event. If a field under observation remains absolutely static (again an impossibility in-Cosmos) there is no event to observe.

What is a change? A change is a motion of any kind, and a motion is the production of a difference of one thing or pattern relative to another thing or pattern. If no *comparative difference* occurs between two items or two fields, there has been no motion, no change, and no event: a change is the occurrence of a difference; a change is a disturbance of sameness. A Fundamental Change (and perhaps there is only *one*—the 'FLASHING FORTH' of the 'RAY' of the ABSOLUTE) is a *disturbance* of *the* SAMENESS—the INFINITE SELF. In motionlessness, there is no change. Whenever there is motion, there is change.

Motion demands that which moves or *can* move, a 'thing' that moves—an item, or particle, or object/force—a discrete 'something'. In a way, it is incorrect to separate the motion from that which moves, though in the macro-world we must do so. In the micro-world, the motion and the thing that moves may be *one and the same,* for in that world a motion *is* a thing.

All movement is relative. Movement can only be detected in relation to other movement or in relation to a static point of reference which does not move at all. (Is there such a thing in-Cosmos? Perhaps, in the World Archetypes, the World of Being, but then, the *thing* would not be *spatial.)* We cannot really speak of detecting motion against a static *background* that has no boundary and, further, has no-thing in it or on it. (As to this static *background,* such as utterly 'empty' boundless space—again, does such a thing exist? If it does exist, its exists 'Super-Cosmically' and not Cosmically; further, it is not 'empty' {in the usual sense of the word} but utterly, infinitely, dense.)

If a background against which motion is to be detected is entirely homogeneous and static and imparticulate, then the greatest movement and the least movement all appear the same because there is no point of reference to measure the movements against. Of course, one can use the position of the observer as a point of reference, and the limitations of the observer's eyesight, and other senses would help to distinguish differences in motion—even if the background of such motion had no point of reference.

An observer moving in absolute tandem with the moving object could *not* detect changes in the motion of the object. The object would appear 'motionless'. What if, however, the observer is at once the moving object/point as well. Then, no matter how fast the object/point is moving in the void (discounting friction and any other kind of internal or external environmental feedback), there will be no way to judge speed or distance. In fact, without another point of reference, it will seem as if the moving point/observer is *motionless.* Location does not exist if there is only a single point within the void.

Similarly, to God (or a disembodied Omnipresent Observer Who, therefore, pervades all points in the Field of Observation) a *single* moving object/particle/item, whether traveling fast or slow, will not appear to change position—provided a Ring-Pass-Not is not used as a point of reference).

- Determination of 'position' depends upon a fixed point of reference. As well, in an ABSOLUTE VOID (and there is only ONE such) assuming also the 'voidness' and omnipresence of the observer, there is no way to measure movement (because there is no *location).*

Always for movement to occur there must be, in some manner, at least 'two', a 'twoness'—*that* which moves, and *that* relative to which the movement is detected. This would probably apply even when there is *apparently* only 'one' (that which moves in the void), but in such a case, the position *from* which it moved (if it can be fixed or remembered) to where it now *is,* provides the second of the 'two', the point of reference against which to measure.

There was (apparently) a Movement-at-the Beginning, so there must have been 'two'. The Original Two are always:

- Existence and NOTHING, or
- Being and BEING/NON-BEING, or
- Manifestation and BE-NESS, or, finally,
- the 'Positing' and the GREAT NEGATION.

The initial 'change' which is 'MOVEMENT' is the 'variation' from BE-NESS to Being (note the capitalization).

Any change demands duality (i.e., *more than* one). This duality can be explicit or implicit (and therefore easily overlooked) as we have shown. Change demands contrast and difference. What is difference? Difference is 'non-identicalness'. Difference is non-sameness. The generation of difference requires Number. Where ZERO prevails, difference does not exist. Change, movement, event, variation, modification, etc., occur only in the World of Becoming (the Cosmos as a Whole) which is the World of Number and thus, also, difference exists there. Non-difference exists only in the WORLD OF BEING, the WORLD of the ZERO, the WORLD of CHANGELESSNESS. In the World of Being, however, difference and hence 'event' still occurs. The World of Being is the World of Fixed Design, a Subjective World of 'Logoically Sustained Ideation', 'Logoically Sustained Image', and yet, for all its relative intra-Cosmic stability, the World of Being is still, technically considered, a World of Change—the WORLD OF BEING IS *not*.

On the Somethingness or Nothingness of a Point

We go back to a particular question: Is a point *something* or *nothing?* Perhaps it is correct to say that a point, in Essence, is no tangible thing—an idea, Really. It has no extension, and in fact, no location, although it is used to *specify* location. In the World of Effects, the World of Fabrication (the lower part of the World of Becoming), that which *appears* as a point is not a *real* point, because a real point cannot *exist* as an appearance—an appearance, after all, has dimension, which a point has not. A point is an *ideality*, a pure abstraction incapable of being manifested, yet it can be *conceived* (if not 'pictured').

But in a World/Field in which ideas are Realities (note the capitalization), and are even relatively crude Realities compared with higher Beings, on higher planes than those on which ideas are found, a point must be considered Real. In the Pre-Cosmic World, far 'above' any plane of Cosmos, a 'Point' (Really a Triple Point, first 'Infinified', then 'Condensing', then 'Condensed') 'appears'. This Point (which we can also call the Infinite Subject and, later, the Condensing and Condensed Subject) is a 'Transitional Entity', something of a cross between REAL and Real, a cross between NOTHING and Something. To call the Reality which the Infinified, or Condensing, or Condensed (i.e., 'Finified') Point represents a 'Point', is merely to *symbolize* the Reality.

Perhaps the Point (Triple) on the Super-Cosmic (i.e. Pre-Cosmic/Post-Cosmic) 'Level', is the best symbol for That which is the Super-Cosmic 'Doorway from the REAL to the Real'. Everything other than the REAL Is, of course, UNREAL—even the Real! If this sounds paradoxical, it is. The Point (in Its triple nature) might be called the Doorway to the Game', the 'Doorway to the Great Illusion'.

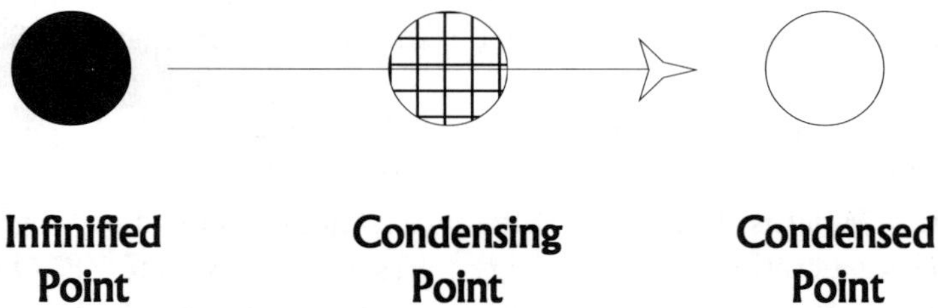

On the Measurement of Time (one more 'time')

The question to be answered by those who would measure Time is: How many events to be measured take place 'between' the measurable events chosen as the 'brackets' of a standard unit of time? Or, How many standard units of time take place between two events whose distance *in time* from each other we are seeking to measure?

Always, an *apparently* undeviating length between certain designated, regularly occurring events (which length we will call a *unit of time)* must be chosen as a *standard* in order to measure time. So the *measurement* of time is dependent upon designating certain regularly occurring events, 'bracket events', as bounding or defining a standard of reference.

There must be an *event* before there can be Time and, in fact, there must be *another* event to relate to the first. As long as there is only *one* event, there is no Time, and Time cannot be measured. Fortunately for those who seek to measure time, there has *never been* just *one* event. At first, this may seem a strange statement, but the dynamics of infinity justify the statement. For instance, suppose we say that: the appearance of the 'RAY' of the ABSOLUTE (the First Tendency towards Finitization) might be considered a single Event. If no other Event follows 'IT', does the 'RAY' have a "*place* in Time"? The answer would have to be "yes", because for an infinity of instances prior to 'ITS' present 'EMERGENCE', the 'RAY' has 'EMERGED', has 'FLASHED FORTH'. Because of an infinitude of Cosmoses which have *already* transpired, any event which occurs can always be related to an antecedent event. This has *always been so,* for never has been the 'time' before which there has not been an infinitude of Cosmoses. Even an event which occurred a centillion Cosmoses ago, still had at least one (and in fact, an infinitude) of antecedent Cosmoses-as-Event(s) to relate to, because throughout all ETERNITY there have *always* been Universal Manvantaras—i.e., Universes.

On Time and Sequence

Time demands relation. So Time 'appears' with the appearance of a sequence of events. Of course, because of an infinitude of Universes past, each one of which was *actually* an 'Event', and each one of which was filled with 'events', *there have always been a sequence of events.* Thus *there has always been Time* just as there has always been Maya. Number and the multiplicity which can be generated from Number have existed *infinitely,* if not *continuously.*

If there is to be *sequence,* there must be *discontinuity.* In the WORLD of ABSOLUTE BEING which is the WORLD of the ZERO, there is *continuity.* In fact, the INFINITUDE is the GREAT CONTINUITY. BE-NESS is the GREAT CONTINUITY as IS the BOUNDLESS IMMUTABLE PRINCIPLE. The GREAT CONTINUITY is ever the SAME without interval or variation of any kind. Discontinuity enters with Number. Number like Maya *never began,* but has had an infinitude of *periodical beginnings,* one beginning at the commencement of each Universal Process.

On Event, Continuity, and Discontinuity

An Event is a disruption in the CONTINUITY, a disturbance. An event is a *difference* and all differences create discontinuity. If there were only the CONTINUITY, no difference at all could ever arise, but, somehow, an 'EVENT'-as-Event 'HAPPENS' and a Universe is born. Within the WORLD OF BEING, there are (*usually!*) no happenings, no events, no occurrences, no evanescenses. There is only a 'STATE' of imparticulate ALL-IN-ALLNESS. That ALL-IN-ALLNESS is the ABSOLUTE PLENUM, the EVER-FULLNESS, the INVARIABLE, the THAT TO WHICH NAUGHT MAY BE ADDED.

Always
the ZERO, ABSOLUTE PLENUM, and NO-THING prevails

Periodically
events disrupt the CONTINUITY, and Time and Universe occurs

On Continuity and Discontinuity

Discontinuity exists where interval exists. In a continuum there is no interval. In a continuum, paradoxically, every point is infinitely close to every other point, because there is no extension, and so all points are one point and everything coincides exactly with everything else.

When everything coincides exactly with everything else, point for point, everything *is* everything else. Under such conditions, everything disappears and there is nothing at all—the NOTHING.A true continuum is REALLY 'pointless'. Only in space does the illusion of multiple points exist. Only in space is one point (apparently) separated from another. Only in space does interval *seem* to exist.

On Linearity and the Production of Universes

Is there a kind of linear measurement possible between successive Universes? There is no way to measure this. Obviously, however, the One Pulse, which exists as the Ultimate Standard of Time Measurement, is the Pulse of Universes. This is the Major (or is it the *only)* Event in the UTTER ALLNESS. Can we say this Pulse Appears at regular intervals? It is utterly, at this point, impossible to say, though it may be *known*, (perhaps?!) by the One Knower, the Universal Logos—depending upon how *veiled* that Logos Really is.

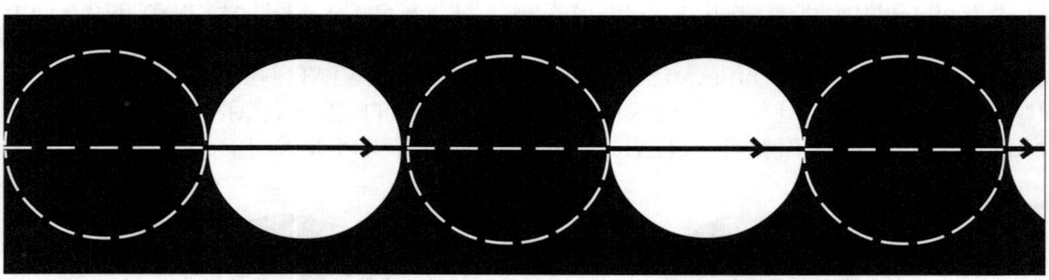

NOTHING ... Universe ... NOTHING ... Universe ... NOTHING ...
Behind it all the ABSOLUTE ever IS

It seems that *linearity* and *simultaneity* must ETERNALLY coexist. From the Point of View of the SELF-as-Triple Point-as-Universal Logos, viewing the Cosmos and 'comparing' It (to the extent possible) to all Cosmoses past, the sense of *eternal unrepeatability* must prevail, and Conditions-in-Cosmos Now are never exactly as they were before the Cosmos began, nor (by extrapolation) are they as they will be in future Cosmoses. And yet from the 'POINT OF VIEW' of ABSOLUTE BE-NESS, if such a PERSPECTIVE/'IN-SPECTIVE' can be thought of as REAL, it is NOW (not Now) exactly as IT WAS, and exactly as IT WILL BE.

The number of combinations and recombinations throughout Endless Time have been numberless, in fact, infinite, for *never* was the 'time' when this GREAT SELF ceased to represent ITSELF in Objectified Universes. THAT and That(s) (which *both never began)* can never cease. When it comes to THAT and That (which is a name for any Cosmos whatever) *always* is a *fact.* So Infinite Time is definitely related to linearity and the One Great Linearity is 'traveled' by an infinite succession of stupendous Events—the Infinite Succession of Universes.

There is only One Great Cosmic Pulse and we can assume Its regularity or non-regularity. We can assume evenness or unevenness. To favor the ideal would be to assume exact and absolute evenness of the Inter-Cosmic Interval. A still more ideal model would show each Universal Pralaya of exactly equal duration to each Universal Manvantara. Cosmic 'Disappearances' and Cosmic 'Reappearances' would be of mathematically equal durations.

If this idealized model were true, the Cycles of Infinite Time would be perfectly regular and would provide an Ultimate Invariable Standard against which to measure accurately all Intra-Cosmic Events. Given, however, the ABSOLUTE FREE WILL of the ONE GREAT BEING—a being *uncompelled* by any other Being; a BEING WHO has at ITS disposal, literally, "all the Time in (all) Worlds"; a BEING which is 'PLAYING' ITS OWN 'GAME'; it is difficult to imagine why *absolute regularity* of the Inter-Cosmic Interval would be a necessity. To such a BEING/NON-BEING, what is *necessity,* anyway? This question requires careful reasoning.

♦ In sum, we could say that, paradoxically, the ETERNAL NOW is inescapably associated with a kind of Linearity which has existed and will exist forever. This Linearity, present in the remotest times of Infinite Time has run like a Thread of infinite length through all Cosmoses, rendering *sequentialness* a Factor as important in the Cosmic Process as *simultaneity.*

How do the 'sequential' and the 'simultaneous' interplay? This is a deep philosophical problem. In all simplicity it might be said: It is *always* NOW, but the conditions extant during the Now are infinitely various when compared to each other and *always* unrepeatable. Remember the following important distinction: While it is *always* NOW forever, it is only *periodically* Now forever.

On Self-Consciousness

A radical thought occurs: no self in-Cosmos is wholly un-Self-Conscious, even during the darkest cycles of unconsciousness. The INDIVISIBLE SELF-as-Self-as-self always retains *ITS* full SELF-Consciousness and *Its* full Cosmic Consciousness (quite different Registrations) even *while* functioning as a totally benighted, apparently drastically limited being. This is one of the most profound paradoxes.

This is another way of saying that an atom not only *will be* a god, but *already is* as god even *while* its apparent consciousness is that of a limited atom. Every B/being in-Cosmos is, even *unknown to itself,* simultaneously finite and INFINITE, dual and UNITARY, prakritically bound and UTTERLY FREE—and every being (even if it does not *know* it) *Knows* it, and, even, 'KNOWS' IT!

There is no way for any E/entity-in-Cosmos to avoid being, ESSENTIALLY, PERFECT even *while* it is *manifestly* drastically imperfect. With these thoughts we are entering into the Paradox of BEING/Being.

On the Reality of Time and Space

Time is fundamentally an illusion because it is not REAL, though it is Real. Space also is not REAL, though, likewise, it is Real, having existed cyclically forever. If the infinitely recurring Inter-Cosmic Interval is *regular,* then there is an infallible Super-Cosmic Standard against which to measure the duration of Cosmoses and, also, the duration of everything Intra-Cosmic. We have, however, no assurance that the Inter-Cosmic Interval *is,* in fact, regular, so while Time is infinitely recurring, it may be 'elastic'. There may be no invariant, Absolute Unit of Time. But what about *space?*

NOTE: We will use the term 'space' to indicate the normal use of the term as it is applied to the relation between specific objects in relatively small contexts. We will use the term 'Space' to mean one of the three basic Structural Principles in-Cosmos, Time and Motion being the other two. We will also use the term 'Space' to indicate 'space' when the context is much larger, for instance, in relation to Cosmic 'Space'; as well, 'Space' may mean Mulaprakriti, which is Really, *Infinite* Space.

We know that, for practical purposes, space exists, if we define space as the appearance of extension, or of interval between objects. Is it meaningful, however, to ask if there is ABSOLUTE SPACE? If there is *no-thing* in the NOTHING (the ALL-IN-ALLNESS), then space, as we usually know it, (as interval and extension) is negated. Space is usually considered an interval between presentations or objects. In accordance with that definition, if there are no objects, there is no space, and, even, no 'Space' (for compared to the INFINITE SELF, 'Space', Itself, is a 'Something', an Object which did not exist before the 'CONSCIOUSNESS'-of-ITSELF-by-ITSELF 'AROSE' in IT).

- Movement is meaningless unless there are 'two' (one of which is a point of reference—remember the single frictionless, 'conscious' item *alone* in the void). There cannot be 'space', as usually considered, unless there are two. If we think of 'space' (lower case) as interval, this is obvious. But what about '<u>S</u>pace'? Must there be a *twoness* in relation to 'Space'? 'Space' is not and never can be ABSOLUTE SPACE (though, as Mulaprakriti, it can be *Infinite* Space).

Let us examine 'Space' as Cosmic 'Space' (Cosmic Prakriti, which is a far lesser kind of 'Space' than Mulaprakriti). 'Cosmic Space' also cannot exist unless there are 'two'. When we think of Cosmic Space we often think of a great emptiness, a void, a 'nothing'. This thought, however, is an illusion when applied to Cosmic Space.

- Cosmic Space is *finite.* and not an utter vacuum (i.e., utterly empty) nor is it a complete continuum (i.e., utterly, infinitely dense, like Mulaprakriti). Cosmic Space is a specific Self-Reflected Object/Thing in the 'Eye' of the SELF-as-Condensed Point-as-Self, and because It is a *specific* Object/Thing, it is *bounded.*

If we think carefully about the foregoing, we shall find both the 'twoness' we need and the 'interval' necessary for the existence of *space* in general, and, now, of *Cosmic* Space. The 'two' are:

- The Self-Observing Consciousness—the 'Cosmic Consciousness'—(as one Point of Reference [and represented below by the white point in the black field]), and
- The <u>*extent*</u> of that Consciousness (as it is focused inwardly, in Self-'Sight'), that determines the Ring-Pass-Not of Cosmic 'Space' [as represented by the white-outlined gray field]. It is the Ring-Pass-Not of *Consciousness* within the Universal Logoic Observer observing Itself that determines the Scope of Cosmic Space, and not by any strictly Objective Boundary.

The *space* of an Entity is always fundamentally *reflective* of Who that Entity is. This is especially so in the case of the Primary Creators in the World of Being—including Fohat. This Ring-Pass-Not is the product of the Introspective 'Attention' of the SELF-as-Condensed Point-as-Universal Logos (to be), Who *'Attends'*, ostensibly, to only a certain 'portion' (hopeless word) of Mulaprakriti, but is <u>R</u>eally, 'Seeing' Reflected 'in' and 'as' Mulaprakriti only a certain portion of Its Own Infinitude (i.e., the Infinitude of the Infinite Subject, Who Is the INFINITE SELF considered as an Observing Subject). This Self-'Attention' (which, thus, only "goes so far"), creates a *boundary* in Mulaprakriti, which otherwise (though objective) is as *infinitely boundless* as PARABRAHMAN.

What this Really means, Cosmo-Psychologically, is that:

- When the Self-'Attending', Self-'Seeing' Consciousness of the Infinite Subject-as-Infinified Point 'Sees' Its own Infinitude Reflected *as* Mulaprakriti, then we say that Mulaprakriti is the *Infinite* Object.
- When, however, the Self-'Attending' Consciousness has become (through Willful Self-'Reduction') the Infinite Subject-as-*Condensed* Point, It no longer 'Sees' *Infinitude* Reflected, and thus It no longer 'Sees' Mulaprakriti. Instead it begins to 'See' a supposed *limitation* of Mulaprakriti called Cosmic Prakriti.

What has Really happened is that the Infinite Potential for Self-Reflectivity, Maya (the Great Reflector—Executor of the 'WILL' of THAT) is Reflecting back *less* of the 'Identity' of the Infinite Subject than before. This *lessening of Identity* is because it is the Will of the Infinite Subject to 'Narrow its Consciousness', as it were, in order to 'Move Towards Specificity' and 'Become' a 'Specific Cosmos'. Generalized Maya (like Generalized Fohat, which Maya can be considered to be) always *follows the Will of One Greater than Itself.*

Perhaps, 'MAYA'-instantly-Maya, as the Agent of PARABRAHMAN (REALLY, a 'MODE' of PARABRAHMAN ITSELF) *engineers* this Reduction of Consciousness without the cooperation of the Infinite Subject, per se, and simply *as* 'INSTRUCTIONS' 'from' the INFINITE SELF. (For who else is 'MAYA'-instantly-Maya than the INFINITE SELF 'GONE' 'FORTH' from ITSELF *in* SELF-'REDUCTION' and *as* the Super-Cosmic and Cosmic Will to Self-Reduce?)

Whichever way it happens, *reduction* occurs. *Before* the deliberate 'Narrowing/Reduction of Consciousness' occurs, we can say that:

- The Infiniteness of Mulaprakriti is 'Infiniteness as an Object', whereas
- The Infiniteness of the SELF-as-Infinified Point is 'Infiniteness as a Subject' (albeit as a Subject which is slightly 'outside!' the SELF-as-SELF, which SELF IS the ONLY UTTERLY INFINITE SUBJECTIVITY).

Put another way, Mulaprakriti is the Infinite Potential <u>*in Matter*</u> (albeit, an Infinite Potential first 'inhering' *noumenessentially* in the PARABRAHMAN, the INFINITE SELF) for every kind of modification which can possibly *appear* in any of an infinite number of Cosmoses 'ORIGINATES in PARABRAHMAN. What is Matter, though, other than that which the INFINITE SUBJECTIVITY (through ITS 'EXTENSIONS') 'SEES' of ITSELF?

Obviously, the potential infinitude of which Mulaprakriti is the Reflection is not, and cannot be, incorporated into any *one* Finite Cosmos. The SELF-as-Condensed Point 'Sees' in Mulaprakriti (the Infinite Image of the INFINITE SELF-as-Infinite Self) only *certain* SELF-'EXTRUDED' potentials for the coming Cosmos and 'Attends' to, or 'Focally Sees' only *those* potentials; at least, *those* are the potentials upon which the 'Attention' of the Focused Universal Subject (the incipient Universal Logos) is placed. Mulaprakriti is the First Object (<u>R</u>eally the SELF/PARABRAHMAN-as-Infinified Point 'Seeing' ITSELF as an Object) and, at first, It is an *Infinite* Object (for the Infinite Subject Who 'Sees' is *Subjectively Infinite).*

- ♦ Thus, Mulaprakriti can be conceived as the Reflection of the INFINITE SELF-become-Infinite Self/Infinite Subject. Not only is Mulaprakriti the <u>First</u> Object, not only is It an <u>Infinite</u> Object, but It is the *only* Infinite Object.

Mulaprakriti cannot, however, continue to be 'Seen' as Infinite by the SELF-as-Infinified Point (i.e., the Self-as-Infinite Subject) if there is ever to be a *Finite* Cosmos (which there *must* be). So the SELF-as-Infinified Point *ceases* 'Seeing' Itself as Infinite and, thus, Mulaprakriti (as an Image) ceases to be an Infinite Object (the Image of Infinity) and *becomes* (to the Focusing Universal Subject) an Object of *lessening* scope. This change arises during the Phase of SELF-as-*Condensing* Point.

NOTE: It must be remembered, however, that even though this Reduction in Subjective Consciousness and in Mulaprakriti *seems* to take place, yet *there is no Reduction*, and the Infinite Subject and Infinite Object *persist* throughout the Universal Manvantara. This is a Paradox based on the idea that, though Emanation proceed, the State from which Emanation proceeded, *remains* even as it *was* before Emanation occurred.

Continuing this tendency towards Reduction, the SELF (now as-Condensed Point) begins to 'See', increasingly, the 'Intended Finitude' within Itself and thus begins to 'See' Its Own Prakritic Image as a Finite Object of certain *potential dimensions* (potential dimensions which, of course, were subjectively *inherent* {SOURCE-'DEPOSITED'} in the Focused Universal Subject {i.e., the Condensed Point} from the very FIRST 'MOMENT', much *before* It began to 'Attend to Itself' as a *specific* Object, an Objective Individuality).

What this Really means, is that the SELF-as-Infinified Point-as-Condensing Point-as-Condensed Point deliberately (through the continued Falling of the Mayavic Veil) allows only certain potential dimensions (not *infinite* potential dimensions) of *Itself* to be 'Self-Seen', and thus *reflected*, within Mulaprakriti-fast becoming-Prakriti. Thus, Mulaprakriti 'Becomes' (instead of an *Infinite* Object) a *Finite* Object (i.e., Mulaprakriti becomes Cosmic Prakriti).

This act of 'Attending' to Mulaprakriti (the Infinite Self-Image) as a *Finite* Object is the Act of Finitization that creates the necessary *boundaries* (in scope and 'depth') of the Cosmos-to-Be. The SELF-(now)-as-Condensed Point only *extends* Its Self-Focused 'Attention' *so far*, and in doing so creates (what we call in philosophy) that internal objective *presentation* called 'extension' (which, strangely, is a lessening of the boundlessly 'extended' infinitude earlier registered). The nascent *dimensions* 'Seen' (most likely, anticipated as potential for the Future Cosmos) are hardly the three dimensions we are accustomed to upon the physical plane, but they are the 'seed' of dimensions nonetheless, and the appearance of 'dimension' (no matter how exalted from the perspective of the limited human consciousness) means the appearance of *limitation* (i.e., *boundaries).* All this happens because the Infinite Subject begins to 'See' Itself as limited, finite, bounded—i.e., as the limited, finite, bounded Mother.

Whatever the Meta-*Physics* (necessarily far different in exactitude from the speculative model here presented) of this presently incognizable 'Beginning of the Trend Towards Cosmos' (i.e., the *Condensing* of the Infinified Point), there is a 'twoness' to be reckoned with:

1. The Super-Cosmic Consciousness of the SELF-as-Focusing Subject (i.e., the Condensing and Condensed Point) which are Really Phases Two and Three of *Super-Cosmic Consciousness;* and
2. That *apparently* Finite *Thing* of which the SELF-as-Focusing Universal Subject Is Conscious (i.e., *Bounded* Mulaprakriti that *Is,* in fact, Bounded Infinite 'Space', or *Cosmic* Space, which is a faithful Reflection of the unlimited, limiting, or more lim-

ited Self-'Sight' of the Infinite Subject, become Focusing Universal Subject, become Focused Universal Subject—the incipient Universal Logos.

So, in sum:

- 'Space' (Cosmic Space) *exists.*
- It is an Object, but not an 'External' Object at all; there is Really no such thing as an *external* Object.
- Cosmic Space is an Object because the Super-Cosmic Consciousness 'Sees' Itself *as* Cosmic Space'.

The Super-Cosmic Consciousness of the Condensing/Condensed Point is Conscious of It (Cosmic Space) because the Condensing/Condensed Point (the Focusing/Focused Universal Subject) is Mayavically Induced to be *Conscious of the Cosmic Space within Its Own Subjective Nature!* Cosmic Space is certainly *within* before it is *without.* The Focusing/Focused Universal Subject Mayavically *'Measures'* (or 'Takes Cognizance' of) the Boundedness 'within' Itself and, thus, discovers (apparently *externally)* the Bounded Object, which we are calling Cosmic Prakriti, *Universal* Matter, the Universal Prakritic Sphere. To become aware of what is *'within'* as if it were *'without'* is a perpetual, Maya-Induced Dynamic of Consciousness. This Dynamic is called *Projection.*

The Infinite Subject has Unboundedness 'within', and thus 'Sees', as Mulaprakriti, *Unboundedness* 'without'. The Focusing or Focused Universal Subject ('Intent on Universe') Wills to 'See' the degree of Boundedness 'within' Itself which is *appropriate* to the Building of a Coming Universe of certain dimensions and, thus, 'Sees' as Cosmic Prakriti (which is Bounded, Reduced, Finitized Mulaprakriti), a kind of Boundedness 'without'. All this Projection of the 'within' to the 'without' is quintessentially Mayavic. Who, however, or What, is Maya, if not PARABRAHMAN, ITSELF?

In ordinary 'space' there is always a kind of interval between two points. Is this so with respect to Cosmic Space, Cosmic Prakriti? There is certainly normal intervalic space *within* Cosmic Space (apparently), but is there any *other* kind of interval (providing *twoness)* associated with Cosmic Space?

Perhaps another interval (if it can be conceived to be so) associated with Cosmic Space is the 'interval' between:

- The Internal Boundedness, Projected 'outwards' as Cosmic Prakriti.
- The Internal Infinitude which, were it projected 'outwards', would be Infinite Space, or Mulaprakriti.

From this perspective, Cosmic Space is simply the Objectification of *Finite* Consciousness. It is amazing to see how the Psychological and the Physical Aspects interplay, the Psychological *ever* producing that which we imagine to be Physical: 'Space' becomes 'internalized' and appears as an objectification of degrees, extents, and depths of Consciousness.

As for what we are inclined to call ABSOLUTE 'SPACE', IT has no boundaries at all for there is naught 'within' IT (other than the *no-thingness* of ITSELF) capable of forming a boundary. In esoteric circles, ABSOLUTE SPACE has become a term for the ALL-SELF, but whether or not the term 'SPACE' should be related to what we usually consider 'space' must be thoroughly analyzed. Other than ABSOLUTE SPACE (if IT REALLY 'EXISTS' as 'SOMETHING' even remotely akin to 'space'), there is naught *else,* and an *'aught else'* is needed to create a boundary.

If, however, we think of ABSOLUTE SPACE as *boundless 'extension',* we will also be inaccurate, for in the ALL-SELF (i.e., in so-called 'SPACE') there IS no extension, there is no-thing to have an 'extent'. Extension (related, as it is, to *interval)* appears with and 'within' 'Space' (not 'SPACE'). There is no interval whatsoever in the SELF ('SPACE'), for interval requires 'somethings' between which the interval must exist, and there are no 'somethings' in the BOUNDLESS IMMUTABLE PRINCIPLE (i.e., 'SPACE'). There are not even *points*—a point being the most intangible 'something' conceivable—so intangible as to be "almost *nothing*".

Really, Pre-Cosmically, the Infinified Point and Mulaprakriti are (apparent) opposites. The Infinified Point (or, better, the instantaneously SELF-'EXPELLED' intra-SOURCE 'POINT') is the closest thing to ZERO without quite being a *pure* ZERO—i.e., NOTHING. Mulaprakriti is the closest 'thing' to Infinite Extension (i.e., to *everything possible*—a Reflection of *infinitized* intra-SOURCE 'POSSIBILITY') without quite being the FOUNT OF ALL POSSIBILITY. Really, the Infinite Point/Infinite Subject/Infinite Father is the 'Seed of Infinite Possibility', and Mulaprakriti Is the 'Matrix of Infinite Fecundity'. Neither, however, *is* the ABSOLUTE INFINITUDE, per se.

We find, thus, that 'SPACE' (ABSOLUTE SPACE) is missing the possibility of *interval* and/or *twoness,* which are necessary for our usual concepts of space. SPACE and Space/space do not have very much in common (though they are ESSENTIALLY, *identical).* In the ALL-IN-ALLNESS, Space is not needed; Time is not needed; Motion is not needed.

'Space', when seen from the largest perspective, can be considered the 'possibility of *event'.* It seems that all events in the UTTER ALLNESS somehow involve Mulaprakriti (except the very FIRST 'EVENT', the *ultimate in mystery,* which 'EVENT' 'CREATED' Mulaprakriti). Events and happenings demand the appearance of the Principle of Objectivity. Mulaprakriti *is* that Principle Even the appearance of the 'POINT'/'EVANESCENT INFINITE TRINITY (which 'APPEARANCE' might be considered the FIRST 'EVENT'), becomes "in no Time at all" or with *infinitesimal instantaneity* the Infinite Super-Cosmic Trinity, involving the Registration by the Infinite Subject of *Mulaprakriti*—the Greatest of all Objectivities.

From one perspective, Mulaprakriti is 'Infinite Space'. Mulaprakriti (as Infinite Space) is the possibility for what we call *matter,* in an infinite number of modes. Thus, Mulaprakriti can be called *infinite material possibility.* 'Space' Itself is more limited. 'Space' (since *Infinite* 'Space' is Mulaprakriti) can be seen, simply, as *limited material possibility*—i.e., the potential for the appearance of *matter* as we know it.

Really, Infinite Space *is* Pre-Cosmic Root Matter. Since, however, Cosmic 'Space' is limited, *Cosmic* Space should be called the *possibility of matter in a limited number of modes,* as contrasted with Mulaprakriti (Infinite Space) as the *possibility of matter in an unlimited number of modes.* The important thing to realize about 'Space' is that It is an Object, and cannot exist unless there is the Principle of Objectification (Maya) in operation (and, of course, an *Observer).*

ABSOLUTE SPACE, unlike 'Space' can *never be* an Object, nor an 'OBJECT' (though it appears to be 'SEEN' as such for, perhaps, an infinitesimally brief Moment of Nascent Time). Neither is ABSOLUTE SPACE a *field* in which *anything happens.* NOTHING 'HAPPENS' in ABSOLUTE SPACE (the SELF), until the paradoxical MOMENT OF AWAKENING. In fact, ABSOLUTE SPACE is so unlike 'Space' in that regard, that it is probably confusing to call the ABSOLUTE by the name 'SPACE'. The term, 'space', always seems to connote *extension,* or an Observer and an Observed, and of this SPACE is utterly devoid.

We can think of 'Space' as a 'something', but we cannot even *think* of 'SPACE', for IT IS NOTHING, and any *thinking* immediately renders the NOTHING into a 'something'. Whenever we think about *space* of any kind, we cannot help but think about the *possibility* of *extension* (i.e., of 'things' being "spread out in space"). But ABSOLUTE SPACE (or more simply, 'SPACE') has no possibility of 'spread-out-ness', for within IT, there are never any-'things' which can be spread out.

We see that with respect to SPACE, the usual idea of *space* begins to collapse, and we see, paradoxically, that:

- *There is no 'space' in SPACE!* 'Extension', remember, can only exist in relation to an observer, but in SPACE (the BOUNDLESS IMMUTABLE PRINCIPLE) there is no 'OBSERVER', for an *observer* is an 'other' than that which is observed, and the SELF is 'OTHERLESS'. Since, as well, there is no-'thing' to observe, *observation* too, is impossible.

Think of it! Can there be *space* and *extension* when all occurs *here,* when the *locus of all occurrence* is *here?* When, in fact, there can never be a *there?* And when, further, *here* has no *position* at all, for *'hereness'* requires a point which is defined as *'here'* (a position, a location), and 'within!' the BOUNDLESS PRINCIPLE there is not even a *point.* Even if there *were* a point, a Real point has no *dimension,* and how can there be *location* without *dimension?*

In sum:

1. When we use the term 'SPACE', we must be very careful to realize that we are talking about 'NOTHING'.
2. When we use the term 'Space' (even if Space appears to be *empty),* we must realize that we are still talking about an objective 'Something'.

We see, then, that the term 'SPACE' turns out to be highly symbolic, and even *misleading* as a symbol, for we cannot conceive of *space* independently from *extension,* and within SPACE extension is impossible.

Let us relate, for a moment, Time and Space. Objects are events and Time is, from one perspective, the *space* between events. Space (as interval) is the perceived distance between objects. We see, here, how Time and Space are interrelated, and both related to the concept of *interval* which only exists when there is a *discontinuum,* and never exists when there is a *continuum.* In *objectlessness* and *eventlessness,* there is no Time and no Space. 'Everything' truly collapses into NOTHING.

The collapse of Time and Space yields ABSOLUTELY NOTHING—a very *assertive* statement. 'Within' the ABSOLUTENESS, the usual interval between objects is negated by the UNCONDITIONAL OMNIPRESENCE of the ONE PRINCIPLE and ITS completely imperturbable HOMOGENEITY. Considering the possibility of Time 'within' THAT, the usual *space* or interval between events is negated by the UNCONDITIONALLY ETERNAL NOW, and more obviously, by the absence of any *event.* The ABSOLUTE is the SUPREME NON-EVENT. There is no *event/happening/action/disturbance/modification* 'until' the 'PROCESS'-instantly-Process of Super-Cosmic Universe-Inception begins.

Let us here introduce an unusual perspective having to do with *interval.* We have seen that *time*-intervals and *space*-intervals are both impossible in the BOUNDLESS IMMUTABLE PRINCIPLE. We do know, however, that both time and space intervals *appear* to exist within any Cosmos because of the Nature of the Universal Structure. These apparent intervals may be of virtually infinitesimal duration, but what, REALLY, are they? What is 'in' them? The answer would be:

- ♦ The intervals are *no-thing,* and NOTHING IS 'IN' them. NOTHING *substands* all *intermittent 'Somethings'.*

It might be said that Cosmos Is an 'Intermittent Something' (a Discontinuity), and that during Universal Pralaya, NOTHING is infinitely 'full' of the *infinitization* of all possible 'Intermittent Somethings' such that they 'ABIDE' 'within' the IT *as* NO-THING AT ALL. *Intermittency* pertains to Cosmos, and *never* to the BOUNDLESS PRINCIPLE. It is as if the NOTHING is the FOREVER CONTINUOUS ALL-TONE (really an UNVARYING ALL-INCLUSIVE SILENCE SOUNDING EVER—the only NON-VIBRATING TONE), from which arises all the other *intermittent* tones necessarily based upon vibration. There is no intermittency or vibration 'within' the INFINITE SELF—the HOMOGENEOUS IMPERTURBABLE CONTINUUM.

The above reverse-idea of *interval* asserts that the CONTINUUM *never ceases* even while the Discontinuum (with Its virtually infinite multiplicity of discontinuities) periodically arises and subsides. The 'white background' is never REALLY covered up by the impermanent 'black drawings'. Or, the 'black background' is never REALLY covered up by the impermanent 'white drawings'. Both models are ESSENTIALLY the same. There is *never* any REAL interval.

Appearance **REALITY**

It is necessary to speak for a moment of Perpetual Motion in Space. All motion is vibratory and, hence, intermittent. Perpetual Motion, too, is intermittent. There is no Perpetual Motion 'within' the CONTINUUM, per se, in fact, no motion of any kind. When the CONTINUUM, however, is 'thought' of as a 'THING' 'called' 'NO-THING' (hence wrongly *objectivized),* then the apparent alternation between the 'NO-THING' and 'Something' called the 'Great Breath' can be thought of as Perpetual Motion. It is debatable whether it is justified to name the *entirety* of this alternation—an *entirety* that *includes* the CONTINUUM as if IT were an *object*—a 'Discontinuum'! To do so would presume that the CONTINUUM could be 'Seen' *objectively,* but 'where' *(outside* the CONTINUUM) would one 'stand' thus to 'See'?

Thus, the Discontinuum called the 'Great Breath' can only be constellated in consciousness from an 'illegitimate' perspective. A vision of this foremost of 'Discontinua' is created only by *'thinging the un-thingable'!* Yet these are the 'thinging' minds we have and we cannot avoid using them despite the frequent fallaciousness of their testimony.

- ♦ In a sense, Perpetual Motion (although it never began and will never end) is *discontinuous* (i.e., oscillatory) otherwise it could not be a *motion.* The 'MOMENT' the FIRST 'MOVE' towards the Inception of a Universe occurs, the Discontinuum (or *a* Discontinuum) Arises (yet again).

Some might say, citing the thoughts in the previous paragraph, that the Discontinuum did not have to arise, because It *necessarily* always *was!* From one perspective, this idea cannot be disputed. It can be said that Perpetual Motion occurs only 'within' the Discontinuum (but the Discontinuum occurs only 'within' the CONTINUUM, for 'where' *else* is there?).

One view upon the Discontinuum theorizes that:

1. It is born and dies (periodically, forever) even while the CONTINUUM *continues!*
2. The contrary view would say that the very 'dying' of the Manifest Duality into 'NO-THINGNESS' is, itself, a 'part' of the Discontinuum, and is what makes this Greatest of all Discontinua possible at all!

In view #1, a Discontinuum exists where there are 'two' simultaneously: the CONTINUUM and the State of Objectivity called Cosmos. Thus, when there is no State of Objectivity, there is no Discontinuum.

In view #2, the Discontinuum is judged to consist of the apparent alternation of 'NOTHINGNESS' and 'Somethingness'. The fact that 'NOTHINGNESS' 'CONTINUES' while 'Somethingness' appears to exist, is not taken into consideration. The question to be answered (and it depends upon the perspective) is:

- Is the Discontinuum *periodical* or is it *perpetual?*
- When there is only NO-THING, does a Discontinuum exist?
- Or, is NO-THINGNESS just a 'Phase' in a Perpetual Discontinuum?

We know that what we can judge to be a Discontinuum can *appear* 'out of' a CONTINUUM and, thus, be judged to be 'part' of the CONTINUUM, but can the *CONTINUUM* be rightly judged to be 'part' of the Discontinuum? Perhaps, only if the CONTINUUM is conceived as 'disappearing' during Objective Manifestation, which IT *does not!* The whole question is not easy to resolve. So, we have either:

1. A cyclically appearing Discontinuum that is a Discontinuum, because with Its appearance there are *two*—the CONTINUUM and the Universal Object (the 'Appearance') *or,*
2. The entire Pattern of 'appearances' and 'disappearances' is judged to be the Discontinuum, which then is said to exist 'all the time', forever!

Synthesizing these two points of view, what we appear to have is an infinitely enduring, Periodical Process of 'appearances' and 'disappearances', producing *periodically* what can be recognized as a Discontinuum, and giving rise to extrapolative thoughts that attempt to see the endless sequence of 'appearances' and 'disappearances' as *One Thing* to which the name Discontinuum is affixed (even though there would necessarily be 'Periods' in that foreverness when only 'CONTINUOUSNESS' would 'PREVAIL'. Behind the alternation of 'Thing'/'NO-THING' 'ABIDES' THAT which 'CONTINUES' as 'NO-THING'. Though IT 'CONTINUES' yet IT *seems* (from a probably illegitimate perspective) to *alternate* with That which 'appears' and 'disappears'.

- ♦ In summary, it could be said that there are (apparently) always "two things going on at the same *time*"—the ONLY THING THAT ALWAYS EVER IS, and ITS *apparently contradictory* Opposite (a Thing that 'Sometimes' Is) in relation to which the ONLY THING THAT ALWAYS EVER IS, is understood sometimes as a 'part', and at other times as the *entirety* of the Thing that 'Sometimes' Is. Thus, ONE (the CONTINUUM)—*forever,* and Two (the Discontinuum) *forever* as well!

Herein lies the *seed* of all *apparent contradiction.* The key to solving the problem is that the 'Thing' that 'Sometimes' Is, is REALLY, *ever* and *always* the ONLY THING THAT ALWAYS EVER IS. So, the apparent intermittency, or Discontinuum is REALLY the *continuance* of the CONTINUUM.

We have established that SPACE IS not *extension.* Of course, it is impossible to say that the SELF IS or IS-NOT *anything.* The SELF does not REALLY tolerate *exclusive predication.* It is safe to say *nothing* about the SELF, or *all possible things* about the SELF (the INFINITE PLENUM), and even this latter option will be severely limited due to the less-than infinite fecundity of our predicative powers.

We might say that in the Super-Cosmic Realm (the Realm of the Infinified Point and of Mulaprakriti, which is the First Objectification of the SELF-as-Infinite Subject) all opposites meet. In this Realm (before the Willful further Finitization of the Infinite Subject), we might say that Infinite Extension is no extension at all, and, in this Realm:

- Infinity is equal to Zero.
- Everything and Nothing are identical.
- Everywhere and Nowhere are the same.

If we *infinitized one* of any particular Pair of Opposites that designates Spatial, Temporal, or Numerical boundaries, we arrive at the *other one* of the Pairs. In the Super-Cosmic Realm the extremes meet, and become as one. In the SELF, however, the Pairs of Opposites do not even *exit*—NOTHING 'DOES'.

On the Interval Between Objects —Does it REALLY Exist?

We have seen that interval of any kind is negated in the SELF (the GREAT HOMOGENEITY), and yet in Cosmos, interval appears to exist.

- ♦ Everything about Cosmos is based upon the concept of *interval,* which appears with the Principles of Number and Object.

Now, however, we have to examine the concept of *interval* more closely, and see if it REALLY exists. What, role, for instance, does *consciousness* play in the existence of interval? We know it must play some important role, because consciousness has been defined as "the *relation between* the subject and the object." The 'span in consciousness' between subject and object may be considered *primary* among all intervals.

SPACE is an *utterly*'DENSE' NOTHINGNESS'. Within IT there are no *spaces.* This is also *initially* true of Mulaprakriti (before the *Penetration* of the Infinified Point, which means the Finitizing Change of Consciousness from the Infinified Point to the Condensing and Condensed Points). It is strange to think of the INFINITE HOMOGENEITY (PARABRAHMAN) as 'UTTERLY' 'DENSE' and, also, the Infinite Homogeneity (Mulaprakriti) as being Utterly Dense—both without interval. That which is utterly dense is *also* utterly vacuous—again, the meeting and merging of the extremes.

The SELF is the only ABSOLUTE VOID; Mulaprakriti is the only *Objective* Void. Both contain *no-thing,* and both are *infinitely dense* (though let us remember, that Mulaprakriti is a 'Something', whereas the INFINITE SELF is *not).* That which is infinitely dense is as *empty* as it is *full.* The VOID is the only UTTER PLENUM. The Void (Mulaprakriti) is a Plenum, a Potential Utter Fullness, but only in the realm of Object/ Matter—*not* in the realm of ROOT ESSENCE. The INFINITESSENCE is, as it were, *rooted* only in the SELF which IT IS.

What might be the cause of the appearance of *interval* in Pre-Cosmos? We will not speak of an 'INTERVAL' within the SOURCE, because the 'ARISING' of the 'POINT' instantly 'BECOMES' the Super-Cosmic Point, almost challenging the thought that any intra-SOURCE 'HAPPENING' ever REALLY 'OCCURRED' *at all!* Certainly any 'INTERVAL' that 'AROSE' in THAT was 'OVER' almost before it 'BEGAN'. (That very 'ARISING' was tantamount to 'EJECTION'!)

1. The First Interval effective in Super-Cosmic/Pre-Cosmic Realms is 'GENERATED' 'ABOVE' 'in' the INFINITE SELF, and instantly leads to the Perception, **I/8** Am the Subject, (followed, perhaps, as instantly by) *That* Is the Object. This is the Interval in Super-Cosmic Consciousness, or, in this case, Pre-Cosmic Consciousness *between* the Infinite Subject and the Mulaprakriti, the Infinite Object.
2. The Second Interval effective in Super-Cosmos (and, remember, *intervals require two)* is Generated *from within* the Super-Cosmic Realm. This Second Interval is achieved through an Act of Super-Cosmic Consciousness by means of which the Vision of the Super-Cosmic Consciousness of the Infinified Point/Infinite Subject is 'narrowed' from the Infinite Depth and Extension of Mulaprakriti (which, remember, is equivalent to *no extension at all)* to the Perception of merely a 'portion' of Infinite Material Possibility. (Or, probably, the Perception of a Sphere of a certain scope, extent and Pre-Cosmic *depth* in which only a certain portion of Infinite Material Possibility could be Objectified.)

Thus a kind of *interval* is created between the Infinite Wholeness of Mulaprakriti, and that 'portion' of It which is now 'Seen' by the Super-Cosmic Consciousness as a *specific, delineated* Object (with an apparent *external,* a Ring-Pass-Not, caused by an *internal* Ring-Pass-Not of Self-Perception.) Any ring-pass-not (whether physical or psychological) is a kind of *interval* or space, or gap between that which it encloses, and that which it does not enclose.

3. The next Act of the SELF-as-Point (i.e., or Super-Cosmic Self, the Triple Point) is a kind of 'Investiture'. Having now 'Seen' 'Mulaprakriti-become-Cosmic Prakriti' as a specific Object with boundaries—boundaries (of no matter how many nascent dimensions) which are to bound the New Universe, the SELF-as-Condensed Point (or Super-Cosmic Self) *perceptually enters more deeply* the Bounded Object, and *per-*

vades it (not as the PURE SELF—for the SELF is *always* ALL-PERVADING) but as a different kind of *Agent,* an *active, directing, Agent.* The Universal Father more directly *engages* the Universal Mother.

Thus, the Bond of Focused Pre-Cosmic Father and Focused Pre-Cosmic Mother is unbreakably established. The Father is 'inter-playing' with His own Self-Reflection—not just 'Seeing' It, but, more, 'Identifying as'.

It is at this point that the Super-Cosmic Self/Subject, has Become the One, Individual Cosmic Self/Subject, or Universal Logos (which later, of course, must apparently 'subdivide' through Emanation into the various Secondary Cosmic Logoi). The descent (in terms of *subjectivity)* can be thought of as follows:

1. From the SELF in the PURITY of ITS SELF-'PREOCCUPATION'
2. To the SELF-as-Infinified Point or Super-Cosmic Self/Infinite Subject
3. To the SELF-as-Condensing Point of Focusing Universal Subject
4. To the Cosmic-Self as Condensed Point or Focused Universal Self-Become-Universal Logos
5. To the Three Sub-Universal Logoi to emerge from the Universal Logos, among Whom can be numbered, the Son, the Fohatic-Holy Spirit and the Mother (though in a sense, the Mother in various degrees of Infinitude-to-Finitude has necessarily Existed along with the Infinified Point, and Condensing and Condensed Points, and also, Fohat in one Mode or another, has been 'PRESENT'-instantly-Present 'since' the 'FLASHING-FORTH' of the 'RAY' of the ABSOLUTE);
6. To the Three Sub-Logoi in the Company of the Son
7. To the Seven Subsidiary Logoi in the Company of both the Son and of the Three Sub-Logoi
8. To the successive Generation by the Son and His Holy Ten/Seven (as 'Rays') of a numbered multitude of Universal Life Units, Cosmic Monads, which are All the many varieties of Subjective E/entity-E/experiencers-C/creators within Cosmos.

We must remember that ESSENTIALLY the SELF IS all these *secondary selves,* who serve as the Divine Media to mediate ITS PRESENCE into ever more constricted and objectified dimensions.

- Clearly the Universal Drama has one Principal Theme: the INFINITE SELF 'JOURNEYS' into ITS OWN SELF-'REFLECTION'.

The Process of *intervalizing,* or *imaginatively articulating* Mulaprakriti begins, Really, *after* the Perceptual/Inceptual *Measuring Out* of the Boundaries of the New Universe through an Act of *Internalized* Super-Cosmic Consciousness on the part of the Focused Universal Self. Within the inchoate Cosmos, the *intervalizing* and nascent *articulating* begins. This Process is achieved, Essentially, through *Counting*—i.e., through Numbering and Emanating. This Process is not yet the Fohatic Process of Creation in the Lower World of Becoming, the World of Fabrication, but is the preparation of the Higher World of Becoming, the World of Being.

The Cosmic Self (Universal Logos) by 'Seeing' the Pre-Cosmically 'Bounded' Mulaprakriti (now called Cosmic Prakriti) in a *certain way* (which Really means 'Seeing' *Itself* {the Universal Logos} in a certain way), creates the Destined Pattern in the Heavens which the Cosmic Subject/Comic Self/Cosmic Logos may have been 'Holding' in 'Invisible Solution' ever since the 'ARISING' of the 'POINT'-instantly-Infinified Point/Infinite Subject. Remember the SELF and all its *reduced Affiliates,* ARE ESSENTIALLY NOTHING. We have heard of "Fohat digging holes in Space." That "digging" occurs a bit 'later' in the Cosmic Process.

Now, however, on a higher level, something like that happens. The SELF-as-Super-Cosmic Self-as-Universal Logos is REALLY *'NOTHING'*, ESSENTIALLY, as are ITS *reduced Extensions,* the various Subjects/Logoi to be Emanated within Cosmos. This Representative of NOTHINGNESS, the Universal Logos, *enters* the extremely (but not utterly) Dense-but-Bounded-'Something' called Cosmic Space (Cosmic Prakriti).

Thus, Really, the Universal Logos is *entering* deeply (Perceptually/Inceptually) into Itself! But the Universal Logos enters Cosmic Space 'step-by-step' *principally* through the Agency of Its Emanated Son (which It has 'Discovered' 'hidden' within Himself). The Universal Logos 'stands' 'behind' and 'within' the further *Emanating* (and, thus, *Counting)* which is carried out by the Universal Son-of-the-Universal Father, and the emanated Agents of the Universal Son.

◆ The thought to be extracted in all this is in order to create a Cosmos, 'NOTHING' (in the form of a Chain of Decreasing Subjectivities) must enter 'Something'.

This IT 'DOES' through *Mayavic Projection* and, then, *Identification*: 'SUBJECT'-instantly-Infinite Subject must Become *Object.*

- 'NOTHING'-instantly-Infinite Subject is the SELF-in-Pre-Cosmos and all ITS *perceptually/inceptually* reduced *Extensions* enumerated above;
- 'Something' is that *almost* completely Dense, Bounded Object which we are calling Cosmic Space (Really, the Subjective Father-'Self-Seen'-as-Mother). This Cosmic Space is as yet *almost* totally undifferentiated, except for Its Perceived Boundary and the 'Anticipation of Cosmic Dimensions to Be' (i.e., the 'Seeing' of Less-than-Infinite-Depth) *created* by the Super-Cosmic Self's (i.e., the Infinite Subject's) Conscious Act of Finitizing.

NOTE: Remember that the Finitizing Intent of the Super-Cosmic Self causes a *kind* of differentiation (but not, Really, a 'particulation') in Mulaprakriti—it is an appearance of the *difference* between the *Infinitude* of Mulaprakriti, and the *Reduced Infinitude* of the Prakriti that has been Perceptually Isolated from Mulaprakriti to accommodate the Birthing of the Nascent Cosmos.

Cosmic Space (once the specifically Cosmic Process has begun), Really, ***Becomes*** full of 'holes' (which, in one way, means the Mother is full of 'Sons') because the First Patterning-Holding Being (the Universal Logos/Cosmic Father, which Perceptually/Inceptually enters Its Own Self-Image, Cosmic Prakriti) begins Counting/Numbering/Emanating what It 'Sees' of Itself: Essentially full of articulatable, but implicit 'holes' ('Sons', 'Points', Objects) which, in Cosmic Prakriti, are to be 'brought forth' as Objective 'Sons'.

We can think of the Universe as 'the Sons of NOTHINGNESS' *showing forth* as the 'Sons *in* Somethingness'. We can think of the 'Sons of Subjectivity' as the Representative of the VOID making Their Presence Objectively Known in and *as* 'Somethingness', i.e., Objectivity. This brings us to the strange conclusion that the many patterns we see are Really *'less* than NOTHING(s)' *manifest* in the 'Somethingness'. Visible Patterns are 'Radiations' of the VOID, but *not* of the *voidness* of the VOID, but of the implicate, 'unarticulated' *infinitized* 'holes' in the VOID (the 'Sons' of the VOID).

- ♦ Our Universe is, as it were, 'hollowed out' of Infinitely Dense 'Something' by the 'ARISEN' 'Sons of NOTHINGNESS', i.e., by the reduced SELF-'EXTENSIONS' of the VOID, all of which are, ESSENTIALLY, NOTHING, but manifest (in-Cosmos) as 'the *articulations* of NOTHINGNESS'. In fact, though ESSENTIALLY NOTHING, these 'Sons of NOTHINGNESS', by 'BECOMING' *articulable,* thereby 'BECOME', as it were, *'less* than NOTHING(s)'.

It is as if one *extreme <u>subjective</u> density* (Cosmic Logos-the Father-Subject *and His latent subjective {and 'hole-y'} Self-'Seen' Sons) Self-perceptually* penetrate a lesser *extreme <u>objective</u> density* (Cosmic Prakriti-the Mother-Object, rendering His Sons Her Sons, and the Son's Sons Her Sons, etc.). (The Birth of Cosmos is a 'Cosmo-Narcissistic', 'Intra-Psychic Affair!)

Thus is created a Great Discontinuum, which we call the *articulated* Cosmos. The invisible 'Sons of NOTHINGNESS' 'Become' *spaces* in dense 'Somethingness'—*spaces* that are the 'holes' (ever *infinitessentially implicate)* in NOTHINGNESS—'holes', which all *infinitized possibilities* (i.e., 'Sons', 'Points', etc.) <u>*become*</u> *once they are objectified by 'EXTRUSION'.*

Thus, are all 'Sons' 'Hole-y'. All Objects in-Cosmos are, as it were, 'holes' or 'discontinuities' or 'spaces!' in infinitely dense, hence unarticulated, Objective Continuity. The *spaces, intervals* or initial discontinuities in the World of Being are *emanatively* 'Generated' by the Universal Logos/Cosmic-Self by:

1. 'Counting'/'Acknowledging' the many Selves (inherent, by 'ULTIMATE DESIGN') within Itself,
2. 'Seeing' Them as Objects, and then,
3. 'Willfully Entering into Them', as it were.

First the Great Universal Logoic Self demonstrates in the newly forming Cosmos as a Pervading Principle called the Number One, radiating the Quality of Oneness:

- The Universal Logos is One
- The Universe-to-Be is One
- The Father (Logos) and the Mother (Logos-as-Object), *united,* are One

The Universal Logos, is the Father of the Universe but soon is to 'Retire' or 'Withdraw into the Role of 'Sustainer-in-Strength', as the *'Hidden* Father' (or what H. P. Blavatsky calls the "Unmanifest Logos"). The Number One, the Father/Logos, 'Looks Deeply within Himself' (always Seeing Himself as the Mother), and, then, 'Sees' not only the *totality* of His Own Being, but 'Sees Himself' as something *lesser* than His Totality; He 'Sees' Himself as the 'Latent Son' (enfolding *many* lesser sons, as one Point enfolding many lesser points), which Son, by being *'Seen'* (within the Logos viewing Himself as Mother) becomes *objectified.*

So, in sequence, the Father:

1. 'Sees' Himself (His total Being) as the Mother, and, then,
2. 'Sees' Himself (in a 'reduced' manner) as the Son within the Mother. The Son (Who is Really the Father-as-Son) by being *'Seen' with Self-'Recognition' (by the Father and through the Father's Self-'Investment')* is, thereby, Emanatorily *sent forth,* (i.e., given Objective Existence).

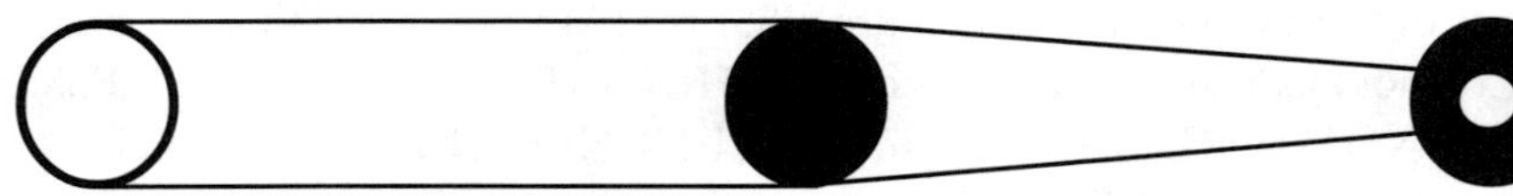

Thus, the Father sends Himself *forth* as the Son. The *sending forth* means that the Father 'Sees' Himself as a lesser *Object*, an Agent, a Son (and 'Invests Himself' in that Son) who, like the Father is also a *Subject* (though to the Father, He be an Object as well). The Father is the First; the Mother (Who is the Self-Seen Father) is the Second; the Father-as-Son is the Third, who becomes the Second.

The Universal Son embodies in Himself the Quality of the Number Two, but it is more the *interplay* between:

- Father-as-Father, and
- the Father-as-Son

(this interplay occurring *within* the First Emanation called the Son, for the Father is deliberately and sustainedly *in* the Son) which creates Twoness. It is the Father 'being within' the Son which 'Creates' Twoness.

Really, in all Emanatory Sequences, the Emanator (in this case the Number One) only and ever sends forth *Himself* (as an *objective attenuation* of the Numbers One), i.e., He 'Sees' Himself as an Object. (Of course, that which the Emanator sends forth is *also* a *Subject.)* It is the *interplay* between the various 'Onenesses' *sent forth* along the Chain of Emanators which creates Twoness, Threeness, Fourness, etc.—all the many Numerical Qualities. It is the *interplay* which makes of the many emanated 'Onenesses'—the many emanated attenuations of the One Cosmic Monad—the Essential Numerical Beings.

In one way the Universal Son *is* the Number Two; in another way it is the interplay between the Father and Son (interplaying <u>*within*</u> that Emanation of the Father called the Son—for the Father deliberately 'Enters' the Son He has Objectified by 'Seeing') which is the Twoness. All preceding Numbers are *invested* or *present* in each new Emanation. Within the Two (or the relationship called Twoness) is found, not only the Number One (for the Number One is divisible into all Numbers), but all the Pre-Cosmic Progenitors (the Pre-Cosmic Infinite Subject and Diminishing Subjects) of the Number One, and ESSENTIALLY, the ZERO, ITSELF. Every greater Subject is *in* every lesser Subject. Every greater Number (i.e., *lower* Number) is *in* every Number of greater numerosity. The WHOLENESS-as-Wholeness is completely *in* every apparent 'part'.

Then the First Emanation (the Son, being Itself *and* the Father as well, and, thus a Twoness) sends a Self-'Seen' 'reduction' of Itself forth as an Emanation, and in reproducing a 'portion' of Itself as an Object, generates (within that Emanation of Itself) the relationship called Threeness. It is the Second in the Series of Emanations (i.e., the Emanation going forth from the Son) which contains within It not only both the Father and the Son, but Its (the Son's) own unique Quality.

- 'Within' this Second Emanation is found the three-way Relationship between the Father, the Son *and* a new Object/Subject/Entity which has gone forth and embodies the Quality of Threeness.

And so it goes—the SELF (the GREAT ZERO, the PLENUM) 'ENTERING' the Cosmos as Number after Number, bringing into Finite Actuality the SELF-'INTENDED' Pattern for the New Cosmos.

During the Process of Emanation/Enumeration, Bounded Space (the Finitization of Mulaprakriti achieved through the Condensing Self-'Sight' of the Super-Cosmic Self) *Responds.* The Universal Logos at the start of Cosmos *pervades utterly* Cosmic Space (i.e., Bounded Space), which means that the Universal Logos (by thoroughly Seeing Itself as the Mother) becomes a Great Oneness, a Cosmic Individuality, a 'Singularity'. That *Pervasion* might *seem* to have created a Great Dual Entity, the *hidden* Subjective Nature of which is the Universal Logos and the *visible* Objective Nature of which is the Bounded Prakriti we call Cosmic Space, but, Really, *the Two are One* (Spirit-Matter). Then, with each successive act of Enumeration (which is Entification), Cosmic Space/Bounded Prakriti seems *reduced* (in depth and scope) just as are the Great Numbers through the Process of Emanation.

Each Number/Entity, therefore, 'Sees Itself' as a *lesser form of Prakriti,* a lesser Object, a lesser Image. The Entity called Number Two is a *reduction* of the Entity called Number One, and expresses Itself through a lesser breadth and depth of Cosmic Space/Prakriti than does Number One.

Psychologically, this means that the Number Two has not the same *'Cosmic Depth of Identity'* of the Number One. Number One expresses through the *entire* depth and breadth of Cosmos. Number Two does so, to a certain extent, perhaps, but less completely. (This is because the Number Two, in seeking to 'See' to the Root of Its Own Nature, discovers Itself being *'Seen'* by the Number One, the Universal Logos; in other words, the Number One—the Universal Logos as Father—is capable of deeper *Self-Seeing,* than is the Number Two, the Son.)

The mechanics of this Process are difficult to imagine, but the idea is clear. The Entity/Relationship called, for instance, the Number Three:

- being sent forth as the Second in the Chain of Emanations;
- being born Emanatively of Number One *and* Number Two as they interplay within the Being of the First Emanation (the Son Who embodies the Quality of Twoness) before the Second Emanation is sent forth; and
- being *smaller* (in a way) and more *reduced* than both One and Two;

has, as Its sphere of expression, an even *lesser* depth and breadth of Cosmic Prakriti/Self-Image than the Number Two. And so it goes, each Entity/Number in Cosmos serving as a 'Reduced Representative' of the Entity/Number which Emanated It.

We see, then, that Interval in Cosmic Space, is generated by Enumeration and Entification. The Number Two (in Its Own newly formed Nature) does not include *all* that the Number One includes. Thus it is with the Number Three in relation to the Number Two, and so forth. This is the Principle of apparent Separation at work. There is no REAL separation or division, but there is the *Bounding of Consciousness in Prakriti* (i.e., a *lessening* scope and depth of Self-'Sight').

The Entity Number Two is, as it were, bounded into a 'smaller' portion of Prakriti than the Entity Number One, which, Cosmo-Psychologically, means that the Self-'Sight' of the Number Two is of lesser breadth and depth than the Self-'Sight' of the Number One. Self-Image *Is* Prakriti. A lesser Self-Image is a lesser form of Prakriti/Matter. There are many *layers* of Prakriti. The layers of which we are now speaking relate to the World of Being upon the higher levels of the Great World of Becoming (which is the Cosmos as a Whole). The prakritic ring-pass-nots get smaller and smaller, more and more superficial, and thus consciousness becomes more and more 'encircled' through the continuingly *reductive* Process of Emanation.

The true *modus operandi* of all this is far beyond the capability of the author to imagine accurately or in any detail. We are speculatively speaking of Processes which relate to the Emanation of the Gods, the Cosmic First Family and the Supernal Tetraktys—all of Whom uphold the Divine Pattern in the Heavens (i.e., within the World of Being). Even the best of our humanity have only risen to the fifth, sixth, and seventh degrees within the Human Creative Hierarchy, which means that they only *begin* to become Solar Beings (such as, for instance, Solar Angels) in our own tiny solar system. Thus the scope of the ideas suggested here must necessarily remain profoundly and expansively speculative but, hopefully, suggestive of the possible dynamics of a Fundamental Cosmic Process.

We come to the idea then, that Emanation and Enumeration generate *interval* in Cosmos (psychological intervals that are reflected in *apparently* physical intervals). Are these intervals REAL? No, they are not REAL. They are Real, however, and they are certainly *Actual*. The Universal Self and Its 'Children', the Numbers, *'Think'* Numbers (i.e., 'See' within Themselves, Numbers) which are *higher-thus -lower, larger-thus-smaller* than Themselves, and in thus 'Thinking', the *Greater* Number/Entities 'Create'/'Become' *lesser* Number/Entities—all in Emanative Sequence.

- Every Number is found *completely* within the succeeding Number It generates through Emanation and, thus, within *all* Numbers subsequently generated. Eventually it will be seen that, even in Cosmos (the 'Great Limitation') All is in Each and Each is in All.

Therefore, although all Consciousness in Cosmos is Essentially One, and although all Matter/Prakriti in Cosmos is Essentially One, we find there are boundaries or barriers (distinctions) generated between Emanated Entities, *through the very process of Enumeration* which (apparently) separates the Emanated Entity from the Emanating Entity. Spiritual growth means the realization on the part of the Emanated Entity that It is One with the Emanating Entity. This is Cosmic Reabsorption.

In the light of the foregoing we can ask: Are there, indeed, *spaces*, as it were, between 'things-in-Cosmos'? There are no REAL *spaces*. Indeed, *space*, as we usually consider it, (i.e., as *interval)* is most UNREAL. (Should it not Really be said that *things are spaces?)*

There are, however, even between two adjacent Number/Entities, *Numerical Barriers, Prakritic Barriers, Field Barriers.* Psycho-spiritually, there is (between two such Numerical Entities) a gulf, gap or interval created by the contrasting quality, scope, and depth of Selfhood.

Two adjacent Numbers are like two separate but related Worlds. Passage is possible from the smaller Number to the larger Number, but not from the larger Number to the smaller unless certain requirements are fulfilled. (The metaphysical interpretation of arithmetic operations tell us something about the *modus operandi* of this passage.) Any given Numerical Entity (other than the Number One), characterized as it is by partially reduced Universal Consciousness, cannot *cross* the next immediate 'Barrier' until *re-establishing identity with the Number that preceded It in the Emanatory Sequence.*

The mechanics *are* difficult. The NOTHING completely and absolutely pervades the apparent 'Somethingness' of Cosmos. Paradoxically, NOTHING is *Everywhere* in Cosmos, and Cosmos, Itself, is REALLY, NOTHING. Yet the 'Agents of NOTHINGNESS' (the Emanatory Sequence of apparently 'Lessening Subjective Selves') Perceptually/Inceptually penetrate the density of Bounded Prakriti in such a way as to:

1. First 'See' Their Own Selfhood, and, then,
2. 'See' the contents of that Selfhood reflected as Prakritic Forms, 'Son Forms'.

This Process, Really, 'Generates' different forms of Prakriti, and thus separates one kind of Prakriti from another (one type of Objectified Selfhood from another). SUBJECTIVE NOTHINGNESS in the form of the Ever-Self-Reducing Emanative Chain of Self-Objectifying Selves is the great molder and shaper of Prakriti/Objectivity (the *Image* of NOTHINGNESS). Thus appear all the different varieties of Matter in Cosmos as Reflections of all the varieties of the scope and depth of Subjective Being.

Let us look closely at the concept of a 'hole' (for we are told, among other things, that Fohat "digs holes in Space"). A 'hole' is usually known for its vacuity, its nothingness, its voidness. The NOTHINGNESS of the ALL-SELF has within it an infinitude of *infinitized* possibilities. In their *infinitized* 'STATE' these possibilities *are* the HOMOGENEOUS INFINITESSENCE ITSELF; *infinitized* possibilities are indistinguishable from one another and, also, from ULTIMATE REALITY. *Potentially*, however, possibilities 'inhering' 'within' the INFINITESSENCE are *'discontinuities* in NOTHINGNESS'.

NOTHINGNESS (PARABRAHMAN) and ITS Objective Image, Nothingness (Mulaprakriti), are both *infinitely dense.* They are not only utter *negations,* they are (since opposites meet) utter *affirmations* as well. As the *infinitized possibility* within the NOTHINGNESS becomes (via the Pre-Cosmic Process of Finitization) capable of objective articulation, it is as if 'somethings' that are 'less than NOTHING' (and that originally *inhered* within the FOUNT OF ALL POSSIBILITY as indistinguishable *infinitized* possibilities) are *arising* within the Nothingness of Infinite Subjectivity and seeking specific objectification. Perhaps, that articulation of possibilities (that 'ARISING' of distinct possibilities) is even *present* at the 'APPEARANCE' of the 'RAY', but, if so, then as a deeply 'resident' *availability* within 'SUBJECTIVITY'-as-Subjectivity, and not as any expressed or yet-*noticed* distinctly articulated possibilities.

A 'hole', then, is an *articulatable subjectivity* arising in Infinite Subjective Density and, through Self-'Sight', becoming an *articulated objectivity* taking form in Infinite Objective Density. How is a 'hole' 'created'? An 'EXTRUDED' discontinuity/possibility, 'IM-

PLANTED' and latent within the Infinite *Subjective* Density, is 'Seen' (as a 'Son') and thus, by being focally 'Seen', is *objectified* within the Infinite *Objective* Density. 'Holes' (beings *'nothings'* of a kind, or, at least, *derivative* from NOTHINGNESS) begin as discontinuous possibilities within the Subject, and become discontinuous actualities within the Object.

- To 'dig a hole' (in Space) is to 'See' a *subjective* discontinuity arising from within Infinite Subjective Nothingness *as* an *objective* discontinuity in Infinite Objective Nothingness. (Thus, to 'dig' is an *act of sight.)* 'Holes' are *Subjective 'less than NOTHING(s)'* which then *appear objectively* in a *virtual* Objective Continuum (for only Mulaprakriti, Per Se, is a *perfect* Objective Continuum, yet *even* Mulaprakriti begins as a *hole/point).* 'Holes' are emergent articulable subjective possibilities becoming objectified actualities. '*Sons'* on Their way from *subjectification* to *objectification* are *hidden* 'holes' becoming *visible* 'holes'. All *sons* are *'hol(e)-y'.*

Matter in Cosmos (Cosmic Prakriti) is emanatively 'undensified' by the perceptual/ 'inceptual' objectification of the implicate 'holes' 'RESIDENT' and 'ORIGINATING' (as *infinitized possibilities)* 'within' the INFINITE SELFHOOD. Every lesser Self/self emerging from the GREAT SELF, (no matter what the seeming magnitude of that Self), is ESSENTIALLY a 'Son', a kind of *'less than NOTHING'.* Every *self,* from the Great Universal Self (Who, though the Universal Father, is also a Son of THAT) to the smallest subject 'ringed' in a minute quantity of matter (i.e., ringed {strangely} in self-perception) is a *'hole'*, a *nothing,* or (better) a *less than nothing within the NOTHING.*

So, interestingly, the only interval or intervalic space in Dense Objective Space is made of discontinuities (always latent and *homogeneously implicate* within ABSOLUTE SPACE) which (especially from the 'TIME' of the 'FLASH') are seeking objectification in and *as* Cosmos. These discontinuities are not, and never were REAL, per se; only the CONTINUUM is REALLY REAL. The INFINITE CONTINUUM ITSELF, which *has no extension,* 'BECOMES' (through the agency of ITS many 'EXTRUDED' Son/Subjects Who Represent IT in Cosmos) THAT-as-*that* which apparently 'separates' the Infinite Objective Density from Itself, 'aerating It (the Field of Homogeneous Objective Perception) with interval!'.

So in a way, interval is entirely illusory. Let us assume that every separating 'hole' begins (at the beginning of the strictly Cosmic Process) as a 'Self-Seen illusory *subjective* something' (which is REALLY a 'less than NOTHING' though it *was* in its *infinitized state, fully* NOTHING, ITSELF). Then, this 'hole' *objectifies* as an (apparently) 'illusory *objective* something' which (by becoming *objectively perceivable) seems* to 'separate' abstract, incognizable Objective No-thingness (Cosmic Prakriti) from Itself (and, thus, apparently creating 'interval'; however, there is no *REAL* separation—no *REAL* interval, because that which *creates* interval, or *becomes* interval, is un-REAL).

- The utter HOMOGENEITY of NOTHINGNESS can *never* REALLY be violated. We merely have the GREAT HOMOGENEITY *masquerading* as apparently extended (but ESSENTIALLY un-REAL) *objectivities,* but, for all that, the HOMOGENEITY 'ABIDES'.

Of course, we know, metaphysically, that the greatest heresy is the "Heresy of Separateness" and that *there is no REAL separation* anywhere in the SUPER SYSTEM (i.e., the UTTER ALLNESS, inclusive of the INFINITE SELF and all Cosmoses). To arrive, however, at this conclusion through *reasoning* rather than strictly through intuition, or merely assertion, is of value. Therefore, in the Discontinuum (the Cosmos) there is, indeed, a *functional Discontinuity*, which is *Actual*, but, substanding the *apparent* Discontinuity, is a REAL CONTINUUM (which cannot be negated by any Illusion of Discontinuity in any Cosmos).

Metaphysicians wonder to what extent the NOW is *always present* in Cosmos, thus, ESSENTIALLY negating the REALITY of Time in Cosmos. They also wonder to what extent the GREAT SPACELESSNESS is *always present* in Cosmos, thus, ESSENTIALLY, negating the REALITY of any Space or Interval in Cosmos. In fact, the NOW and the GREAT SPACELESSNESS are *ever present* through and through Cosmos, no matter how great the number of apparent discontinuities in Time and Space.

Everything that *seems to exist* through the *agency* of Time and Space is thereby forever *negated!* Because NOTHING:

- *Is* the 'Something' which *appears* as *Space* and *spaces* (via the instrumentality of the 'Sons' of NOTHINGNESS), and

because (more strangely) the TIMELESS NOW (or the ETERNAL NOW).

- *Is* the Time *appearing* as separate moments-in-Cosmos (for moments *seem* to appear and disappear only to dissolve utterly into the solution of Eternal Nowness-as-ETERNAL NOWNESS as if they never were).

we may be able to see, once again, how—regardless of appearances to the contrary—there is REALLY *nothing* but NOTHING and NOW. Indeed, it begins to seem as if the Ancient Vedantins were correct in stating that the World of Becoming is merely an Illusion, a complete un-REALITY, and that only the UNMODIFIED SELF is PRESENT *everywhere and at all times.*

This being the case, we find that the perceived interval between objects is due to unachieved Universal Identification and Omnipresence, and, even more to unachieved identification 'with' THAT. There is no REAL interval between objects, but only an *apparent* one which will disappear completely with the disappearance of the Great Limited Object (the Universe). If there is no interval between objects, there is no REAL Subject/Object relationship (despite all *seeming* to the contrary).

- ♦ It is appearing that the CONTINUOUS SELF is totally interfusing the Great (Apparent) Discontinuity; that not only is there a complete interfusion, but that the Great Discontinuity is absolutely *nothing* but the CONTINUOUS SELF, ITSELF. The Great Discontinuity is, though *illusory—necessary* Discontinuity.

Following this line of thought, the Universe, and all in It, begins to look very strange, indeed. It would appear that the entire activity of Cosmos is *occurring at a single dimensionless point.* In fact, even this idea is a materialization of the REALITY involved.

The Truth can be alternatively described as follows: everything is taking place at exactly the same 'place' (in the same space) and at exactly the same 'time'. That place/space is *NOWHERE/EVERYWHERE,* and that time is *NOW.* Extension and Sequence are just Categories of Consciousness. Space and Time, as we usually conceive them, are simply Categories of Consciousness—artifacts of the mechanism of human perception/apperception, and even artifacts of the necessarily limited Perception/Apperception of the Universal Logos.

On Forever and Always —Some Unusual Perspectives on Time and Space

To say that all events occur at the same 'time' is to view event purely from the SUPER-Cosmic Perspective (the 'INFINISPECTIVE') of the ETERNAL NOW. To say that all events occur at the same *point in Space* is also to view all things from the SUPER-Cosmic 'INFINISPECTIVE' of the SPACELESSNESS of 'SPACE'.

These perspectives discount all Presentation of Time and Space as *complete Illusion*. Such perspectives suggest:

- That in all of ETERNAL DURATION not a moment of Time has REALLY passed, and it has always been the SAME 'TIME'.
- That throughout the vastnesses of all Cosmoses, no events have occurred anywhere but 'HERE', which means both NOWHERE and EVERYWHERE.
- That appearances notwithstanding, NOTHING has *REALLY* happened (or 'HAPPENED')!

There is an important similarity between the Cosmic Principles of Time and Space: Time is *bounded* for the reason that It periodically comes to birth and periodically ceases to be; the bounded intervals of Time occur forever cyclically. Similarly, Space is *bounded,* because the Field of Space through which a Universal Logos Expresses Its Cosmos is a *finite* Field; the bounded Fields of Space also recur cyclically, forever. ABSOLUTE TIME is the ETERNAL NOW. ABSOLUTE SPACE is the SPACELESSNESS of NOWHERE/ EVERYWHERE.

On the Possible Divisibility of the Ultimate Particle/Event

Is infinite divisibility possible or not possible? Theoretically it is possible, for all possibilities 'within' the FOUNT OF ALL POSSIBILITY can, theoretically, be 'enacted'.

- What is it that is being divided?
- Is there an ultimate particle? Is there such a thing as a *thing?*
- What is it that vibrates?

A thing is simply a *focused intensification within a medium.* If there is no intensification, nothing emerges, or stands out as different from the medium as a whole. Therefore, a thing is a *difference.* What creates a difference? The focus of a Consciousness, perhaps. Perhaps, Imagination determines the "shape of things" and the arising of any thing, itself.

In seeking to understand the divisibility or indivisibility of the ultimate particle/ event, it may be that there is nothing in Cosmos that is small enough, fast enough, or strong enough to destroy it—to divide it.

- If it is the Will of the Imagination that created it so that it not be capable of subdivision, then it cannot be subdivided, for that Will is Incontrovertible Law. Imagination itself *could* divide an ultimate particle/event, because Imagination made it, but because the Will of the Universal Logos (obeying 'IMPLANTED' Cosmic Parameters) *refuses* to do so, there is no sub-division possible (within *that* Cosmos). The happenings within Cosmos cannot transgress the 'Delimiting Will'.

On the Pervasion of Cosmic Prakriti by the Condensed Point

As the moment for creating a Specific Cosmos approaches, Mulaprakriti ceases to be *merely* an Object, but becomes *pervaded* and *infused* by the 'One Who Sees' (i.e., the One Who 'Sees' Itself *as* Cosmic Prakriti). (Perhaps 'Seeing' *is* Pervasion!) Indeed, from one perspective (as already suggested) the 'Seer' and the 'Seen' are the *same 'Point':*

- The firs,t a 'Point of View'
- The second, a 'Point of Objectivity'

This Pervasion is, Really, Self-Pervasion, the Cosmic Subject achieving *Intimate Knowledge* of Itself. What this Pervasion means (as the Moment of Cosmo-Formation draws closer) is that every Idea generated by or 'within' the Pervading One—Who is the 'Inserted Point', or the 'Condensed Pervading Point'—is now embodied by Mulaprakriti (or more accurately, by that *bounded 'portion'* of Mulaprakriti that has been 'Seen' as the Limited Field of the Cosmos in Formation, i.e., Cosmic Prakriti).

Still more deeply considered, we find that the Universal Logos-as-Object is responding to the Universal Logos-as-Subject (and that They each *pervade* each other). Ever, the psychological interpretation of Cosmology is more profoundly true than any interpretation based upon the supposed *reality* of an *external* Object.

In one way, Mulaprakriti and Cosmic Prakriti are always *substanded,* as are the Three Modes of Pre-Cosmic Subjectivity—for what is anything, in ESSENCE, but the ONE BEING/NON-BEING? The Pervasion of Cosmic Prakriti by the Condensed Point (the Focused Universal Subject), however, is an *Active Engagement* (the Spirit 'moving' over the Face of the Waters) and not just an 'Artifact of Identity'. By an 'Artifact of Identity' is meant that both Mulaprakriti and the Super-Cosmic Self, or Cosmic Prakriti and the Universal Self, *ARE,* simply, and, ESSENTIALLY, the ONE AND ONLY SELF, but there is certainly no *interaction* 'between' that SELF and these Two Pairs, because the SELF-as-SELF *cannot* 'ACT' (and further, the SELF is, necessarily, "out of all relation" with anything, for there is NOTHING *other* than IT).

- Through the Universal-Self, however, the SELF (which, at this Phase in the Pre-Cosmic Process, we have called the SELF-as-Condensed Point) *can, in a mediated manner,* 'Act', and, as Cosmic Prakriti, the SELF, *in a mediated manner,* can 'Respond'.

Pervasion is something different to Passive Identification. Pervasion is psychological. Pervasion is the interplay of the Subject with Itself-'Seen'-as-Object. With Pervasion there is an *interplay*, an *engagement*, between the 'Two', Subject and Object (which Two are, Really, One). With Pervasion, the SELF-as-Condensed Point finds (or 'takes) Its place within the Designated *n-'potenti'-dimensional* Sphere of Cosmos. In the ensuing Interplay, Mulaprakriti (now Cosmic Prakriti because *bounded* by Self-'Sight'), *responds* to every initiative coming through the Field-Imbedded SELF-as-Condensed Point (i.e., the Field-Imbedded Super-Cosmic Self or the Universal Self). The Field-Imbedded Super-Cosmic Self Is, because of its imbedding, the Universal Logos.

So the Field of the Universe has been prepared by:

- the SELF,
- the SELF-as-Triple Point (i.e., the Super-Cosmic Self becoming the Universalizing Self becoming the Universal Self), and
- Mulaprakriti-becoming-Cosmic Prakriti.

The Field of the Universe is now, (in one manner of conceiving) a *bounded n-'potenti'-dimensional Sphere* consisting of Cosmic Prakriti, with the Universal Logos, not only at Its *Center*, but Pervading It *entirely*. This Really means that the Subjective Universal Logos now *Knows* Himself as an Object, *entirely*—one might almost say, "point for point" (though certain, infelicitous, rather materialized images of 'points-as-things' arise in association with this phrase). We must remember that the symbol of the Circle (or better, Sphere) with the Point at its center is *symbolic only*. The true situation cannot be easily expressed geometrically, because the Condensed Point within the Universal Sphere is *everywhere* within that Field.

The Universal Logos is, at once, a Point of View and an Objectively 'Seen' Point ('Seen' from 'Above', in Super Cosmos . This is another way of saying that (almost) every Subject is an Object (with the obvious exception of the ONE INFINITE SUBJECTIVITY). The Point of View that the Universal Logos *Is*, becomes (in reflection) an Objectified Field (not a Point) that is the Self-Image of the Logos. In 'Seeing' Himself, He thus 'Sees' a Field (call it Cosmic Prakriti) out of which all that is subjectively within Him may arise and emerge.

In the next phase of the Process of 'Self-Sight', it is a Point that arises within Him (rather than a Field) and it is this Point that is to 'Become' the Universal Son. In a way, this Son is a Point emerging from the 'Point-as-Field' which is the Logos' 'Seen-Self' (Cosmic Prakriti). In this manner of conceiving, Points may emerge from Fields (which were Points when 'Seen' from a higher Perspective).

- ♦ Much depends upon whether, in our envisioning process, we view Cosmic Prakriti as a Point or a (Spherical) Field. As a Field, Cosmic Prakriti will be envisioned as a Field of Potential Emergence.Cosmic Prakriti can also be thought of as a Single Point (a *'Point* of Emergence'!).

It is usually more congenial to the human imagination, however, to imagine fecundity arising from within a Field (as if the Field were a womb). Imagination often conceives in terms of 'extension', and the Universal Logos, too, may have imaginatively conceived in such a manner, for if 'we' human beings see in a certain manner, He/It does so as well—but not necessarily *exclusively* so. The apparent 'extendedness' of Cosmic Space perhaps reflects the possibility of that kind of Universal Logoic 'Conceiving'—i.e., 'Conception-in-Extension'.

If the image of the 'n-dimensional' (Spherical) Field is retained for Cosmic Prakriti, then the Universal Logos could be conceived as an omnipresent Point within that Field. His 'place' would certainly not be only at the 'Center'. Everyone of those omnipresent Points would also be a 'Point of View' allowing the Logos to 'See' everywhere 'within' Himself and yet (for the 'time being') 'See' but Himself alone. Such a ubiquity of Points-of-Self-Objectivity would, in the last analysis, be equivalent to the Single Point, and so the Spherical Model of Cosmic Prakriti may, in the last analysis, be quite interchangeable with the 'Objective Point' Model.

Such pictorial symbols have their limitations. These thoughts are offered to demonstrate that the symbol of the Point within the Circle (or Sphere) referring (here) to the Universal Logos 'within' His Self-Reflection, Cosmic Prakriti, (the Cosmic Mother) is not *entirely* adequate and *cannot* convey the necessary ubiquity or pervasiveness of that Point. Really, a Point *becomes* a Spherical Field, and so we may be dealing with the same symbol at two different phases of unfoldment or 'extension', plus a transitional image that suggests a phase between the 'centrality' of the point and the ubiquity of the point.

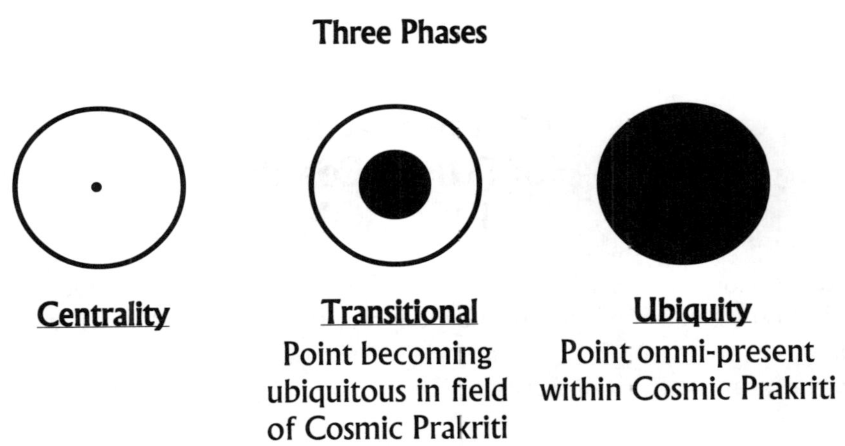

If we retain the image of the Spherical Field for Cosmic Prakriti, the power and focus that the Universal Logos-as-Condensed Point represents is capable of *engaging* with Cosmic Prakriti at every possible *point* within this Sphere, "point for point" (as it were), which would be difficult to portray symbolically.

The Condensed Point (the Universal Logos as 'Point of View') would seem to disappear because It would be everywhere within the Sphere (for "God is Omnipresent in His Cosmos"). It is important to realize that in this consideration the Point within the Circle/Sphere does symbolize the Universal Logos (and not another Being). Another and higher Point that 'RADIATED' from the INFINITE SELF, is the SELF-as-Infinified Point (or the Super-Cosmic Self endowed with Super-Cosmic Consciousness). That same Super-Cosmic Point, once it has become the Condensed Point within the n-'potenti'-dimensional

Sphere (which is Cosmic Prakriti) is much, much more Limited than when it was the Super-Cosmic Infinified Point. *Any* point in Cosmos, however, *is Essentially that very Super Cosmic Point* (by attenuation). This idea speaks to the Omnipresence of God.

Now, what is it the Universal Logos 'Does', and 'How' does He 'Do' It? We cannot here be concerned with the details. We are dealing with a Process which is occurring far above the Cosmic *Logoic* Plane (see page 344 of *A Treatise on Cosmic Fire,* by A. A. Bailey), which is six Cosmic Planes 'above' the Cosmic Physical Plane, about the higher strata of which (i.e., the Cosmic Physical Plane) we know practically nothing—in experiential detail. What will be offered is a more speculative model of how the Universal Logos and Cosmic Prakriti can be imagined to Interact, Interplay, Engage.

The Universal Logos is endowed with the SELF-'IMPLANTED', SELF-'INTENDED' Parameters of the Cosmos which It must Emanate from Itself. These Parameters have been 'DECIDED UPON' (a hopelessly inadequate formulation), probably instantaneously (or "in no time at all") at the MOMENT of 'AWAKENING' (although Time being no factor in their determination). These Parameters are SELF-'DETERMINED' *possibilities* from the FOUNT OF ALL POSSIBILITY. They are examples of what might be called SUPER-Cosmic 'IDEATION'.

On the Enactment of SUPER-Cosmic 'IDEATION'

SUPER-Cosmic 'IDEATION' becomes with *infinitesimal instantaneity* Super-Cosmic Ideation (the basis of the Cosmic Design-at-the-Beginning). Such Ideation can be considered a process of *Geometrization.* It has been said that "God Geometrizes." Geometry is the means of *Creating Relation.*

- A Cosmos is simply an 'Orderly Hierarchy of Relationships between Pre-Ordained Modifications of THAT.

Perhaps the term, 'PRE-ORDAINED' MODIFICATIONS (all in capital letters) should be used instead, for it is most reasonable to hypothesize that Super-Cosmic Ideation originated as SUPER-Cosmic 'IDEATION', at the GREAT 'MOMENT' of 'AWAKENING'.

From a certain perspective, the Essential Numerical Beings are the Pre-Ordained Modifications. Numbers (in their many combinations and permutations) are the E/entities of the Future Cosmos. What is required in the Future Cosmos is Right Geometrical Relation between Numbers/Entities. Super-Cosmic Ideation is the means of creating the Model depicting Right Geometrical Relation between Numbers/Entities. These Relations have to be 'Expressed' through Cosmic Prakriti (i.e., they have to be Objectified).

The Universal Logos must 'Discover' Super-Cosmic Ideation within Himself, for 'There' it 'resides'. The Chain of Descending Subjects originating with the INFINITE

SUBJECTIVITY have been *carrying* that 'IDEATION'-instantly-Ideation throughout the Pre-Cosmic Process (for as long or as relatively brief as that Process may be), and the Universal Logos (being found at the *terminal point* of Pre-Cosmic Subjects) is the 'Heir' to the 'IDEATION'-instantly-Ideation 'BESTOWED', as it were, in the 'FLASH'. Super-Cosmic Ideation is 'Carried' by the Pre-Cosmic Subjects as a Potential for the Future Cosmos, and may determine, to an extent, the *degree* to which, and *manner* in which, the Infinite Subject seeks to Finitize Itself, as It transforms Itself into the Focusing Universal Subject and the Focused Universal Subject (the Condensing and Condensed Points).

When, however, the Stage of Cosmo-Formation arrives, then the *articulable specificity* inherent in the Super-Cosmic Ideation pertinent to the Universe-to-Come must be 'Seen' by the Universal Logos as *residing within* the (Himself) Universal Logos. Thus it is that the Universal Logos, in 'Seeing' Itself as the Mother, 'Discovers' the Super-Cosmic Ideation which It has been, perhaps, Unconsciously 'Holding', which Ideation is *hidden* 'Sonship' waiting to be Enacted as *revealed* Objective Sonship.

Consulting, then, the 'IMPLANTED' Super-Cosmic Ideation resident within the 'Inner Recesses' of Its Own Identity, the Universal Logos 'Discovers' that it is His Task (in His coming Cosmos) to *Become* Number, to *Remain* as Number, and *as Number,* to *Remain in Right Geometrical Relation* (with all Emanated Aspects of Himself) throughout the whole of the Universal Manvantara.

To fulfill Super-Cosmic Ideation, the Universal Logos must *Become* the Number *One* in *Action* (which, indeed, He already Is in *Identity)* and then the Number Two, and then the Number Three, etc.,—all the while remaining the, Hidden *sustaining* Father. It must be mentioned that the Universal Logos (even in the Process of Emanative Unfoldment) all the while *Remains* His Superiors—for instance, the Super-Cosmic Self (in all Its Modes). Every time the Universal Logos (via the Emanations of Himself) *Becomes* yet another Number/Entity, Cosmic Prakriti *Reflects* this Becoming. Indeed the Logos, or any Emanator, could not *Become* a Number, unless He first 'Saw' that Number within Himself-as-Mother, and in the 'Seeing', thereby 'Became' What He 'Saw'. The Process of Cosmos Creation is deeply Psychological with nothing of *externality* in It.

There is probably something very *geometrical* about the way in which the Logos 'Sees' within Itself-as-Mother all Its various 'Sons' (the Specificities of Cosmic Ideation) waiting to be Born as Emanated Logoi (Emanation succeeding Emanation). Within the Universal Logos' Cosmic Prakritic Self-Image, the Noumena of Lines, Planes, Triangles, Squares, Pentagrams, Hexagrams, etc. are to be 'Seen' and thereby Emanated. The Logos' Emanations (each an Emanator in Its Own Right) also 'See' the Geometry appropriate to Its 'Station' in the Divine Emanatory Stream.

By recognizing that which is 'within' Its Own Psyche, the Universal Logos is *organizing* Cosmic Prakriti (i.e., His Own Self-Image) to Express and Embody these Archetypal Super-Cosmically 'BESTOWED'/Bestowed Ideations.

These Formations are not merely one, two, and three dimensional, but are to be *n-dimensional*—specific organizations of Cosmic Prakriti <u>as</u> every Intended Dimension of the Cosmos-in-Formation. The Universal Logos is the First 'Holder' of the Design-to-Be, but the Design will be *unfolded* by *other* Emanators, Who are, Essentially, none but the Universal Logos, Itself, *in-Emanation.*

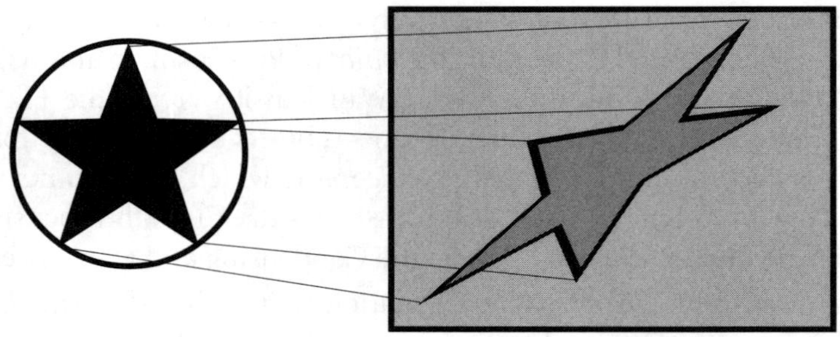

ARCHETYPE
within World of Being

Emanation
into the World of Approximation

The Number/Entities (through Their *agents*, the successively emanated 'lesser Numbers', 'lesser' 'Rays'), pervade Cosmic Prakriti (especially once the later Stage of Fohatic Creation under the Guidance of the Son has been reached), and predispose the then *differentiated* Cosmic Prakriti to systematically aggregate Itself into Formations (on all Levels of the World of Fabrication, eventually) which Reflect these Numbers. It is as if the Number/Entities are Omnipresent Forces underlying the Whole of differentiated Cosmic Prakriti, directing the Formation of Geometrical Aggregations which will be necessary to fulfill the Pre-Ordained Super-Cosmic *Geometrical* Ideation-once-'IDEATION' inherent in the Universal Logos. The Numbers are formed by Emanation, and these Numbers (as Immaterial Formative Potencies) *Descend.*

♦ Numbers are, Really, Beings, reflective of the ONE BEING/NON-BEING. Numbers are REALLY NOTHING. Numbers are ZERO in manifestation.

How does each successive Emanation of the Universal Logos 'Respond' to the Subsequent Number which It finds in Itself waiting to be Objectified through Emanation? Each Number is really a *nothing,*—a *Subjective Force.* A Number has the Power to Compel Aggregation according to Its Nature: the Number One will compel Oneness, the Number Two will compel Twoness, and so forth. Larger Numbers are formed by combinations of smaller Numbers. Arithmetic Operations demonstrate the ways various Numbers can *relate to each other* to produce other Numbers. For instance:

- A Number One in *additive relation* with a Number Four can produce a Number Five, but also, a Number Two in additive relation with a Number Three can produce a Number Five'.
- A Number Two in *divisional* relation with a number Four can produce a Number Two, but, also, a Number Three in divisional relation with a Number Six can also produce a Number Two.

Numbers are Beings capable of operating upon each other to produce other Numerical Beings, or to render *present* the aggregative Power 'Held' by the Essential Numerical Beings within the World of Being.

Numbers, Really, always remain Essentially *subjective.* Numbers are children of the ZERO. They cannot be 'Seen' (in Their True Nature) by 'lesser' beings unless they become Objectified in Prakriti. No Radiation/Emanation of the SELF-as-Self can be 'Seen' (in Truth) except by Those 'Above' (for the Universal Logos 'Sees' the Universal Son, and the Universal Son, 'Sees' Its Sons, etc.).

For apparently 'lesser' beings immersed in the World of Fohatic Fabrication (and, there, unconscious of the World of Being) however, the Radiation/Emanation of the SELF-as-Universal Self (i.e., the Great Numbers) are not 'Seeables' but, only, *Identifiables* [see Glossary], and can only *'be Been'* [sic]. Evidence of the Power and Essence of Numbers *can*, however, be *seen* in the aggregations of Cosmic Prakriti, but Pure Numbers (like the SELF from which They originally derived) cannot be 'Seen' by the 'lesser Selves'; instead, They must be 'discovered' 'within' through Identification.

On "Hard Sayings" and 'Outrageous' Thoughts

It is often said by metaphysicians that within Cosmos, ESSENTIALLY and regardless of appearances, all times are the *same* 'time', and all places are the *same* 'place'. This may be considered a "hard saying". If we try to fathom what may be meant, we can think of everything in Cosmos as taking place at the same *point* (yet Real points cannot be *actual* in-Cosmos, since a point, by definition, is dimensionless, or, at least, of forever indeterminate dimension. A defined point is not a point!).

From this perspective, we see that all extension would be merely an appearance, an illusion, an 'apparent something that is *almost* nothing' (just as a *point* is *almost nothing).* Since all times are, putatively, the same time, then, all things would be transpiring in the Cosmic Eternal Now. All *sequence* (except, perhaps, the momentary changes in the Cosmic Configuration under the All-Seeing Eye of the Universal Logos to Whom all *in* Cosmos is objective—and even *those momentary changes*, from a certain Perspective) would be, Essentially, illusory (and based upon limitation in the perspective or point of view of the perceiving consciousnesses immersed in the World of Effects).

In a deep psychological/metaphysical respect, it would seem impossible in-Cosmos to depart from the Cosmic Eternal Now. Similarly, it would be impossible to depart from, or 'beyond', the Point, for there would be, Really, no 'beyond'. Although there would be an *apparent* 'extension' beyond the Point, it would be, Essentially, illusory, and, 'depart' how we might, there never would have been a *Real* departure or extension. Under the Eye of the One Observer, all 'places' in-Cosmos would be *one* place 'collapsed upon the Point', as it were (and that Point *almost* non-existent).

Similarly, although there is an *apparent* deviation from the Cosmo-Eternal Now through the customary existence of Past and Future, yet under the Eye of the One Observer, in Its Cosmically-inescapable Presence, all times would be one time, 'collapsed upon the Cosmic Eternal Now' as it were. All these things would be so, if the "hard sayings' are true. Let us, for argument's sake, assume they *are* true.

In a Quantized Cosmos Time cannot go on all the time—at least within the World of Effects (the Lower World of Becoming). Time in this Fohatically-Created World (the World of Approximation) is a Presence only from ultimate moment to ultimate moment, even though it *seems* (to prakritically immersed consciousnesses) to be "going on all the time". The illusion of Time in Fabricated Cosmos 'appears' and 'disappears' with each ultimate moment, just as the illusion of Space in Fabricated Cosmos 'appears' and 'disappears' with each ultimate moment. The World of Fabrication is always an 'Arena' of far deeper illusion than the World of Being (even though they are *both* Aspects of the Illusory World of Becoming).

To deepen our thought on these matters, let us think of perspective and observation for a moment. Imagine an Observer hovering over a great arena of activity, such that the many occurrences in the many places are all simultaneously visible. The articulation of Time and Space is visible and capable of isolation in the Consciousness of the Observer, but the 'altitude' of the observer gives a certain sense of *same* time, *same* space.

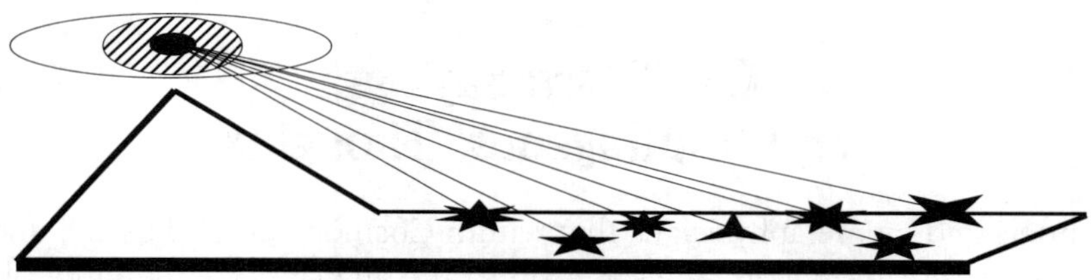

Now imagine that the Observer, keeping his opened, *unblinking* 'Eye' upon the arena (seeing it alone), withdraws and continues to withdraw to a great distance—such a distance, in fact, that the great arena of articulated activity becomes an ultra-tiny *point of light* with *no parts of any kind visible*. As the Eye of the Observer gazes at the Point, there is no noticeable change, only the impression of a constant tiny source of unvarying light.

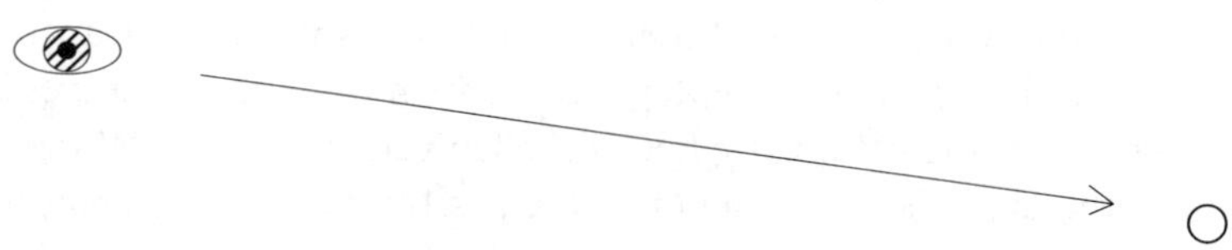

In the Consciousness of the Observer, what has happened to *space* with respect to the arena? All spaces would seem to be *one* space, would they not? Everything that may be happening within the arena is undifferentiable, and seen as happening at a single *point*. In the Consciousness of the Observer, what has happened to *time* with respect to the arena? All times would seem to be one time, would they not?

- There is, after all, no noticeable *change* in the point of light.
- There is no longer a view of the arena which informs the Observing Consciousness that *many* things are happening (even at the same time), but, rather, that only *one thing is happening.*
- Since there is no perceivable movement of any kind (and the point is the only item whatsoever in the field of perception) there is no Time.

The Observer may *know* that many things are "going on down there" as a sequence of events, but the Perspective reveals otherwise. All the many things are coalesced into the vision of a constant beam of light (which, in this idealized model) *changes not.*

Perhaps this imaginative model can give us some sense of how the Universal Logos might 'See' Time and Space within the Cosmos which It has 'Become'. Although the Logos through His Subjective Emanated Extensions (and their Reflections as Prakriti within Himself) is involved (nay *is)* everything that transpires in Cosmos, He also Is, as it were, within His Own Essential Nature, per se, at a *great remove.* From a sufficiently 'distant' Perspective all points in Cosmos would seem as One Point, and all the many sequences of events (even the astounding multitudinous changes in the Cosmic Configuration) seem to be taking place at the same time.

This Perspective would arise if the Universal Logos, the Divine Observer, concentrated *only* upon His Cosmos, forgetting the Memory of Cosmoses past, and the anticipation of Cosmoses to come—deliberately forgetting, in fact, the infinitely recurring rhythm of the Great Breath. If the Logos divorced His Consciousness from any other thing than That that was under His Eye—namely the Cosmos-as-Point (the Cosmos-as-Object), and if an Aspect of His Consciousness 'Saw' the Cosmos as if it were an almost "Vanishing Point" (i.e., a Point *in process* of *vanishing,* and, thus, a *Real* Point [see Glossary]), then all changes that indicate the "movement of Time" would appear to*cease* and the 'time' would always seem to be Now, and of course, all Space would seem to be collapsing toward that 'infinitesimalizing' Real Point.

Clearly the Logos is capable of both the utterly meticulous and maximally long-range Point of View, and can maintain them simultaneously. It is not that the *experience* of Time and Sequence is absent from His Consciousness, for the Logos is *all things in Cosmos,* and whatever E/entities in Cosmos may experience, He Experiences. Certainly, as well, the Logos experiences the illusion of Space-as-interval so inescapable to lesser consciousness who have not the scope to see as the Logos 'Sees', but He can, presumably, also *negate* that relative experience and dwell in the Experience which affirms Objective Space as but a single Point.

Now, (similarly to the model we began to discuss) let us imagine the Universal Logos having transported Himself (as nothing but a Subjective Point of View) to a very great distance from His Cosmos (a distance sufficient to render the appearance of the Cosmos into a *virtual* Point of sufficient tiny-ness to seem to collapse all Time and Space into an apparent Oneness). Let us assume that it is *firmly-established* in the Realization of the Logoic Observer (due to His Point of View) that all that may be happening in Cosmos (regardless of its timing or variety) is happening (as *one* impartite 'Happening') at the very *same* time, and all at the very same virtual Point.

Now, let us imagine the Logos reversing Himself, and drawing closer and closer to the Point—*all the while holding the Vision in Consciousness that He has achieved at the distant Point of View,* even while becoming cognizant of newly appearing extensive activity and sequence in Cosmos.

- Would the achieved Realization of *One Point/One Space* necessarily *fade* even as the illusion of many 'places' appeared through the increasing visibility of articulated, distinguishable activity?

- Would the achieved Realization of *One time/One Ongoing Event* necessarily fade even as the illusion of sequential activity impressed the closely observing Universal Consciousness?
- Would it not be possible for the Logos to maintain the deep Realization of *no change* through all the *apparent* sequential change?

These are obviously rhetorical questions to which the intended answer is, "Yes, it would be possible."

Perhaps we are beginning to see that the estimation of whether or not Time and Space exist is very much a function of the nature and *position* of the Observer, the Point of View. Neither Time or Space can exist unless there is *particularity* and the visualizations we have just considered have *negated particularity*. The assessment of what *now* means, begins to change according to the perspective of the observer.

If the Cosmos appears as an unchanging, unarticulated, unparticulated Point of Light (and, ideally, in this model, there is no noticeable variation in the light), then the Cosmic Eternal Now can be apprehended. No matter what may be "going on" (invisibly to the Observer) 'within' that Point of Light, the *Nowness* of the One Event which the unvarying Point represents, remains *unbroken*. This means that the Eternal Now is existing *simultaneously* with all the illusory pasts, presents, and futures. They never *disturb* the Cosmic Eternal Now. This does not mean that the past, the moveable present and future do not exist for a great number of 'lesser' observers—they do. It simply means that pasts, moveable presents and futures do not Really exist in the Eye of the One Beholder, Whose Perspective is the Essential *determinant* of all Reality—at least with respect to Cosmos.

If we wish to enter the ETERNAL NOW (albeit, imaginatively) rather than the Eternal Now, one more step in imagination is required. The Cosmos that appears to us (as we, in utmost immaterial subjectivity, have *joined* the One Observer, since, of course, We-as-**8** Am the One Observer) as the tiniest point of invariant light, now suddenly *disappears* into *nothingness*. We (as Logos) have withdrawn still 'farther' (into 'recessibility') to make this 'happen'.

Universal Logos' View of Cosmos in Recession

Let us forget, for a moment, the 'meta-*physics*' of what we-the-**8** have just 'done', for, indeed, we must have increased our speed to infinite and *achieved* INFINITUDE. Forgetting all that, for the moment, let us just assume that we have simply withdrawn beyond the range of visibility, for we are concentrating upon Perspective as the *determinant* of Reality.

Now what? All is unarticulated blackness, and such is our Power of Concentration (as Universal Logos) that we lose sight (and thought) of anything but the object of our attention. Blackness is. Nothing moves. All other considerations are banished from Consciousness. Nothing moves, nothing is registered. There is no other point of reference, for we have dismissed any reference to ourself as Logos.

Thus, there is absolutely no Time and absolutely no Space (as usually conceived). We, as Logos, *can* know, if we choose, that beyond the range of visibility is that little 'Point' (the Cosmos), either seen as a single *Event* at a single *Point*, or, (if we were closer) as *many* events occurring in a broad arena of space, but *now* (whatever *now* it is) there is nothing. No matter how many Points of the kind that just vanished, no matter how many Cosmoses might come and go, the *blackness of nothingness* prevails *ever*. Thus, there comes the Realization that it is *always eternally* NOW and NOWHERE, no matter what may be 'happening', *illusorily* beyond the range of our vision.

We can perhaps imagine from this imaginative model how it is possible to dwell in the 'REALM' of NO-THINGNESS even *while*, 'beyond' that 'REALM', 'things' might appear (to some observers) to be 'happening'. Perhaps we can gather some intimation of how it might be that:

- The ETERNAL NOW EVER IS, even while
- The Eternal Now 'Sometimes' is, and even as
- All the little Cosmo-Objective and Cosmo-Subjective Nows, and as
- The innumerable intra-Cosmic *illusory perceptually-based nows* are occurring.

Similarly, we can, perhaps, imagine how NOWHERE/EVERYWHERE 'PREVAILS', even though specific *Points* (i.e., Cosmoses) and diversified *illusory perceptually-based locations* (i.e., locations within Cosmoses) come and go, even *as* and *while forever*, there is only the NOTHINGNESS of NOWHERE/EVERYWHERE. It is a healthy exercise to imaginatively collapse Time and Space even while maintaining them ('beyond the pale', or, rather 'beyond the *pall*') as 'Necessary Illusions'.

Returning to an Intra-Cosmic Perspective, let us think for a moment of the relationship of the Universal Logos and His immediate Emanations, the Son, and the Archetypal Sons of the Son, to Time. These Mind-Born Sons are all 'Holders of the Fixed Design', 'Holders of the Cosmic Archetype', 'Holders of the Design-at-the-Beginning'. All 'Players' in Cosmos are Cosmo-Essentially members of either the Cosmic First Family, or of the Son-Emanated Supernal Tetraktys. Even the Ultimate Universal Units (the multitudinous 'Rays' of the ABSOLUTE, the Cosmic Monads in all their apparently varying 'grades') are integral members of the Archetypal World, the World of Being.

Now, when an inter-moment instant 'comes', and the Fohatically-Created Universe 'Flashes Off', what remains, and what has happened to Time, intra-Cosmically? All the various Essential beings (Cosmic Monads) active within the World of Effects (the World of Approximation) (and, remember, *not* all Cosmic Monads are *active* in prakritic immersion to the same degree, if at all) are instantaneously restored to their status within the World of Being. Those, apparently distinct Cosmic Monads, rather than necessarily being identified with various ring-pass-nots of prakriti, are restored to their Synthetic Identity as members of the Supernal Tetraktys (which is Ten and yet is One).

Simply put, Spirit, perceptually engaged in Objectivity, is *Restored to Itself* as Cosmo-Objective Motion/Change Ceases. Although there is still a kind of 'Supernal Thingness' in the World of Being, *fixity of Supernal Self-Perception reigns.* The Fixed Design (Subjectively inherent within the Son and the Supernal Tetraktys) is *held* virtually without variation or modification. Divine 'Steadiness of Gaze' prevails. (This does not mean that momentarily abstracted 'Rays' cannot "plan their next move" with respect to the Fabricated Cosmos instantly to reappear; it simply means that perceptually, all such Monad/'Rays' are *untrammeled by Cosmo-Objectivity.)*

Has Time then ceased in the inter-moment instant? Those focused preeminently in the World of Being (the Cosmic First Family and the Supernal Tetraktys) are ever and always (throughout the entire duration of Cosmos) 'Holders' of the Fixed Design (in fact, in a way, They *are* the Sustained Fixed Design). When we say, "those in the World of Being", we are, thereby, including in some mysterious way, the Subjective Aspect of all beings-in-Cosmos, all without exception (however, Essentially undifferentiable these beings may be within that World of Being, for all beings are but One Cosmic Monad). Somehow, the lesser 'Rays'-to-be, are *'there'* in that World, enfolded within the 'greater' 'Rays' of which they are indistinguishably a 'part', within which they are *totally* identified, and which, in fact, they *totally are.*

This means that no matter what the depth of prakritic immersion for any ostensibly 'lesser' 'Ray', in Essence, it 'dwells' within the World of Being, and has intermittently, the full Realization that those permanent Members of the World of Being have uninterruptedly. In fact it gets stranger.

- In a way, even the ostensibly 'lesser' 'Rays' *always* and uninterruptedly have full Realization, even *while* the aspect of themselves that has gone forth into immersion does *not* have full Realization. Again, we are dealing with a variation of the Law of Emanative Retention: the 'high' remains 'high' even as it, apparently, 'descends'. Of course, **8** Am all these differentiations and *none* of them, and, therefore, My participation is complete upon every level.

That State and Consciousness of the Subjective Aspect within the World of Being is, Essentially, *invariant.* So within the World of Being (which is the true 'Home' of all Subjects, ostensibly 'great' or 'small') a kind of 'Timelessness' prevails for all Universal Subjects (even though for some Subjects, at certain Phases in the Cosmic Process, that Timelessness may be more or less unconscious—not unconscious during those inter-moment instants when all Subjects are *restored* to the World of Being, but unconscious within the World of Fabrication). As the Cosmic Process continues, the Realization inherent to the World of Being, becomes, as well, the dominating Realization within the World of Fabrication. Thus the "Highest and the Lowest Meet".

Thus at first, Realization of Timelessness prevails for all, constantly, *only* within the World of Being. The 'part' 'gone forth' *realizes* the Realization it has always had (even while, apparently *not* having it 'below') during those inter-moment instants when all *selves* are *restored unto the Universal Self.* Later, (as the Universe is in 'Retraction') the Realization of Timelessness prevails for all, even during the illusory moments of Cosmo-Objectivity (in the fast disappearing World of Fabrication). Therefore we see, in Cosmos, a strange dynamic we see the Timeless State *abiding,* even as Time *seems* to appear and disappear with every ultimate moment. So interestingly, Time, in Cosmos, both *is* and *is not* (even though, Really, for the Universal Logos, Time is Not, and even though, *ultimately,* in the ALL-devouring ETERNAL NOW, Time *never is).*

Thus, from the perspective of the Cosmic Subjects in the World of Being (especially the ostensibly 'greater' 'Rays' focused always in the World of Being), Cosmos is *always*

Timeless, and it is (Cosmically considered) *always* the Cosmic Eternal Now. Curiously, since even the 'lesser', and more prakritically immersed, 'Rays' are *none other than* the 'Rays' Who do *not* enter immersion, it is also, *always,* Timeless for the ostensibly 'lesser' 'Rays' too!

Do the 'lesser' 'Rays' have any Really individual existence within the World of Being? Perhaps, Really, *not*—at least, not, *consciously.* This, then, would be the strange reason why they (the ostensibly 'lesser' 'Rays' in their highest Aspect) always perceive just as the Highest Subjects in the World of Being Perceive—for the ostensibly 'lesser' 'Rays' <u>are</u> the Higher Subjects, the Higher 'Rays'—and *fully <u>are</u>.* Perhaps we begin to see, that human categories of mind aside, there are no Real 'lesser' and 'greater'. Such distinctions are, Really, *illusions.*

From the perspective of the more *immersed* Subjects (the ostensibly *'lesser' 'Rays' of 'deeper immersion',* when they are focused in the World of Fabrication), it is, *seemingly* (to their Cosmo-Objective consciousness, 'below') Now all the time, but *not* an Eternal Now (which they are constantly appreciating 'Above'). When their consciousness is focussed within the World of Cosmo-Objectivity, they have no 'recollection' of the Beingness and Timelessness *they are even <u>then</u> simultaneously experiencing.* Their 'in-Objectivity consciousness' is *sealed,* temporarily, to the World of Being. Naturally, the perspective of these ostensibly 'lesser' 'Rays' (i.e., their perspective within the Cosmo-Objective World) changes through the Process of 'Ray' Retraction (which is also called Evolution).

♦ A very interesting picture is here presented of a dual-life for many immersed 'Rays'. They are simultaneously free and imprisoned, but their consciousness of imprisonment knows nothing of their ever-present consciousness of *freedom* (which, contrarily, <u>*does*</u> know their imprisonment).

Evolution, however, shifts the balance, and they awaken to their freedom even during the conditions of their supposed imprisonment. The two become as one. The differing qualities of perception of ostensibly 'greater' and 'lesser' 'Rays' in various deeper and shallower depths of prakritic immersion requires much pondering. Meditation upon the interplay between the three authentic Cosmic Nows:

- The Cosmo-Objective Now;
- The Cosmo-Subjective Now; and
- The Cosmo-Eternal Now is required.

Curiously, to the immersed aspect (for there is always an *unimmersed* Aspect) of those Cosmic Subjects which have been designated for prakritic immersion, (it *seems)* during immersion, that it is always *now,* (even though there is a relatively instantaneous interlude between Cosmic-Objective *times).* The *now* of such immersed consciousnesses is an *illusory now* that *seems* to be "happening all the time", but, in fact, is not. Still, for these limited consciousnesses, there is a *seeming,* if un-Real, continuity-in-Time.

Similarly, to the Cosmic Subjects focused within Their invariant Consciousness within the World of Being (and <u>*all*</u> Cosmic Subjects, being but *One* Cosmic Subject, *always are),* it is also always *now* but Their Now is *both* the series of Cosmo-Subjective Nows ('downwardly' engaged in "moment-to-moment" planning) and the all-subsuming *Cosmo-Eternal* Now.

Perhaps the Cosmic Monads (during the Cosmo-Subjective Now) might emulate the God Janus who, having to faces, could look in two directions simultaneously—in this case, towards Cosmo-Objectivity 'below' and towards the Synthesis of the Cosmo-Eternal Now 'above'.

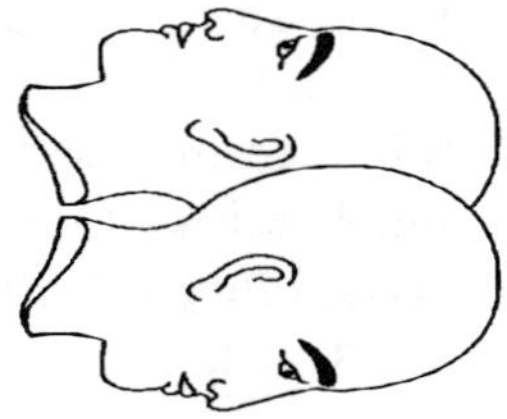

For all Emanated Aspects of the Cosmic First Family and the Supernal Tetraktys (i.e., for all *supposedly* 'lesser' Cosmic Monads) there is (at least potentially) in the inter-moment instant, a kind of return to the Invariability of Apprehension of the 'Held Design', a return to the Invariant Nowness of the Perspective of the 'Holders of the Cosmic Pattern'. The 'return' to Cosmo-Subjectivity (as we have established) is, however, only a *seeming* return, for, indeed, they who appeared to go forth, never Really left. This applies to even those 'Rays' of the ABSOLUTE who are ensouling the forms of minerals, and similar apparently 'lowly' forms.

The Cosmo-Eternal Nowness that is constantly appreciated in varying degrees by all Subjects, regardless of their *extensions into prakriti,* is the Nowness of the 'Beauty-which-Is *held* as the Beauty-to-Be', the Nowness of the Fixed Design.

- How strange it is that there need be no *Real return* for those Monadic Emanations of the Primary Cosmic Subjects who are engaged, in Time, within the World of Approximation, for, the Essential Aspect of all Cosmic Subjects (i.e., the Monads in their utmost *essentiality)* have <u>ever</u> (during Cosmos) 'abided' (undifferentiated in consciousness from the Son and the Supernal Tetraktys) within the World of Being, invariantly 'Holding the Contemplation of Fixity' even during the virtually countless ultimate moments of *non-fixity* during which they were *apparently* engaged within the World of Fohatic Approximation.

Are we seeing how much the sustainment of Oneness requires Twoness? Indeed, it seems obvious that there can be no sustainment of Oneness without Twoness. What we are seeing, then, is the *relativity* of the different kinds of *nows.* The kind of *nowness* experienced depends entirely upon the quality of the nature and scope of the experiencer. For 'Holders of the Archetype', the 'Holders of the Fixed Design' (which, in a sense, is *all* beings in their Cosmic Subjective Aspect) there is, in-Cosmos, and at the *highest peak* of Their Consciousness, no *movement* and hence no passage of time. The summit of these Consciousnesses is anchored in *invariant being.* 'Whenever(!)' there is *invariance,* time does not *pass, if* the consciousness is *merged* with the invariance.

Clearly the Spirit-focussed Consciousness of the Cosmic Subjective Aspect is constantly, entirely *invariant,* and yet, through Its extensions *intermittently it is <u>not</u>.* This means that while Time does not pass (and it is, therefore, Eternally Now) within the World of Being, yet, 'simultaneously', Time *does* pass for the *prakritically immersed consciousnesses* of the *immersed aspects* of the Cosmic Subjects (and, by *emanation,* all Cosmic Subjects are thus *immersed).* Thus, intra-Cosmically, the Eternal Now 'abides' even as, *below,* Time is being experienced. Again—the contradictory Twoness in face of the abiding Oneness.

There is a further great paradox that, even as the Eternal Now abides (as a fact of *highest* Consciousness) within Cosmos, Absolute Time is passing. Time can be measured against the rhythm of the Great Breath. From the Perspective of the Universal Logos, it may be Eternally Now for the entire duration of a Cosmos, and *yet,* Absolute Time is Passing as measured against the Infinite Time Line.

- Comparison is the key; without the possibility of comparison there can be no Time. A Cosmos, even though *Consciously* Sustained in the Eternal Now, has an *absolute* duration, a duration relative to Cosmoses that have gone 'before' and to Periods when there were no Cosmoses.

So, interestingly, even though upon the highest 'levels' of the World of Being, it is *fixedly* and *consciously* Eternally Now, time is passing *below* and Time is passing *above.* Again, the measurement of Time is entirely a matter of *perspective.* During Cosmos, is the Universal Logoic Point of View the *only* Point of View? Does not the Perspective of the Infinite Subject/Object *still persist* in the Super-Cosmic Worlds? If THAT has not ceased to *BE,* simply because the 'RAY' 'FLASHED FORTH', there is no reason to think that ITS first 'EXTENSION' (Infinite Subjectivity and Infinite Objectivity) should cease to *Be* just because there has been a *necessary narrowing* of Infinite Subjective Consciousness.

The image of the 'telescope reversed' serves. The larger cylinders continue to exist even as the lesser cylinders are unfolded from it. There is never any *loss of being* within the Divine Emanatory Stream.

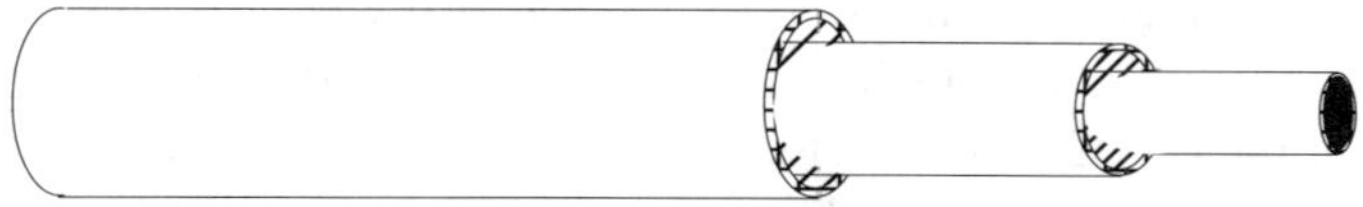

Speaking of Points of View other than the Universal Logoic, what of the Point of View that *Remembers*—Remembers everything that has transpired forever. *Is* there such a Point of View, such a Consciousness? The answer must be, Yes, although the *modus operandi* is difficult if not impossible to presently conceive. Such thoughts lead to the question of whether the INFINITE-SELF has a 'MEMORY'? Certainly IT 'HAS' a 'MEMORY' because if anything *can* exist, it *already* 'EXISTS' as the INFINITE SELF, the FOUNT OF ALL POSSIBILITY. We will not pursue this difficult problem here, but, *if* an infinite 'MEMORY' 'EXISTS', then, clearly, the apparently *durationless* Eternal Now in-Cosmos (which is 'durationless' only because of *one* of the Perspectives of the Universal Logos) can be compared with/against the durations of an endless sequence of Universes that have *transpired in Infinite Time.* From another perspective we might say that the 'March' of Infinite Time does not cease, simply because one of an infinite number of 'Eternal Now(s)' is occurring!

Again, we are dealing with an UTTER ALLNESS in which, paradoxically, the most important Factors (and perhaps, even, *all* factors) both *are* and *are not!* So much of what is designated as Reality depends entirely upon Point of View. Even more strangely, while every little thing is happening in time during an endless sequence of perceived moments:

- even as the Fabricated Universe is 'Flashing' 'on' at every ultimate moment and 'off' inter-moment-instant;
- even as the Cosmic Eternal Now *abides*, Essentially negating all such moments, whether illusorily perceived or Real and ultimate in the Structure in Cosmos;
- even as the 'March' of Infinite Time negates the Timelessness of the Eternal Now—

it is *always* NOW, ETERNALLY NOW, and NOTHING is 'happening' at any time, nor has anything 'happened' at any time because ABSOLUTE 'TIME' (INFINITE DURATION) *IS*.

◆ Beyond all the multiple 'levels' of *seeming* depending upon the 'altitude' and scope of the Observing Subject, there is the ONE ABSOLUTE NON-PERSPECTIVE (the 'INFINISPECTIVE'), which *always* and *ever* ESSENTIALLY IS. This is the ETERNAL NOW which *never* can be exorcised, regardless of all *seemings*. This NOW is the most 'ongoing' of all *nows*, and, being ABSOLUTE 'TIME', is the complete annihilator of Time and even of the Infinite Time Line—which forever *must* exist, and yet, necessarily *must not* REALLY exist.

In relation to all the foregoing, we can perhaps come back to the consideration of the *reality* of interval 'between' events. In the most *absolute* sense, it is not that there is only *one event* through all Time; from the perspective of the ETERNAL NOW there are no events at all. Perhaps, we could call the ONE GREAT NON-EVENT (the ABSOLUTE) the ONLY THING 'HAPPENING' and thus the ONE GREAT 'EVENT'. From the *absolutist* 'PERSPECTIVE', however, it is meaningless to ask if there are intervals between events which do not REALLY exist.

From a more intra-Cosmic Perspective, Cosmic Subjectivity dwells in a *sustained, invariant* State of Contemplation within the World of Being. Although that which immersed consciousnesses call 'events' are naturally registered through the 'immersed Emanations' (by that part of Cosmic Subjectivity which is 'oriented towards' the World of Approximation), from the World of Being, the Universe is 'Seen' as One Great *unchanging* Event (remember the view of Cosmos as an almost 'Vanishing Point') occurring in an Eternal Nowness (the Cosmo-Eternal Now). We might say that the Consciousness of Those in the World of Being abides at four levels:

1. 'Within' the Cosmo-Eternal Now, seeing the Universe as a Single Seamless Event.
2. Projected towards the lower Worlds of Fabrication, from which Perspective quantized Time and Space are Observed.
3. Within the domain of 'Unfolding Archetypal Relations' (the World of 'Slow Purposeful Change') which controls the Emergence of Regulatory Ideas. (These Relations unfold according to the Time-Formula inherent within the Design-at-the-Beginning.)
4. A fourth level of consciousness involves (by emanation) Those within the World of Being, and that is simply immersion within the World of Fabrication. (Just as 'we' 'below' are Really 'above' as well, so, They, 'above' are Really 'below'.)

Even as the Universal Logos and the other Emanated Members of the Highest Cosmic Subjectivity 'Hold' the Eternal Now, They also have an "eagle's eye view" of the Cosmic Configuration (within the World of Approximation), as that Configuration changes from ultimate moment to ultimate moment (as described previously in #2). While *knowing* that all of Cosmos is One Great Event transpiring at One Point, They are not blind to the necessary Illusion of Sequence. Intermittent Change is registered even though it occurs against the background of Eternal Nowness (and even against the *all annihilating* 'BACKGROUND' of ETERNAL NOWNESS). From the Perspective of a Consciousness that can 'See' the Cosmic Configuration in its entirety, there is most definitely momentary Event after Event, and *interval* between those Events.

Considering the 'Telescope of Divine Observers', we have to ask ourselves if 'contradictory things, and, apparently, even mutually self-annihilating things, can happen at the same time'? The answer always seems to be, Yes, for the illusion that there is a contradiction depends upon maintaining vision from only *one* perspective, and usually a *lower* one. In this particular example, event and non-event can and do co-exist.

- The passage of time and Timelessness can and do co-exist. The existence of interval between events, and the REAL impossibility of interval can and do co-exist. Always we have to consider the contradictory testimonies of the World of Illusion (in its various grades including the so called World of Reality {which is nonetheless *Illusory}*) as compared to the WORLD OF REALITY (which is *sure* to contradict and negate anything occurring in the World of Illusion, i.e., Cosmos, and, in fact, any<u>thing</u> at all)!

Thus we see that, from a relatively close perspective, Time and Space *actually* exist, and all modifications must be taken into account. From the Perspective that reveals Cosmos as an almost Vanishing Point, however, Time and Space (while existing) collapse into a singularity—one Time (Eternal Now), and one Space. From the ULTIMATE 'PERSPECTIVE'/'IN-SPECTIVE' of the ABSOLUTE, there is *no* Time and *no* Space—there never *has* been and there never *will* be. The apparent *contradictions* are thus existing *simultaneously*, and have so existed, cyclically, *forever*. There is only one 'STATE' of ABSOLUTE NON-CONTRADICTION and IT IS the ABSOLUTE in the 'STATE' of ALL-IN-ALLNESS. Yet even the ABSOLUTE in order to 'ABIDE' as the great 'NON-CONTRADICTION' must (apparently) *contradict* ITSELF and, thereby, 'ABIDE', as well, as the GREAT *CONTRADICTION*. IT *never contradicts* ITSELF in that IT *always must contradict* ITSELF. Well, that's the way IT IS!—and ISN'T!

On Combinations and Recombinations in the Worlds of Space and Time

We are now in-Cosmos. What then is the analog in the World of Time to combinations and recombinations in the World of Space? From one perspective, it is impossible to move from *everywhere/nowhere*, and the (imaginatively conceived) dimensionless 'Point of Cosmic Action' occurs *everywhere/nowhere*. Likewise, from the same perspective, it is impossible to 'leave the Cosmo-Eternal Now'. Nowhere and everywhere are really the same because *there is no specificity in them,* no location, no limitation. Similarly, it is Cosmically Eternally Now at every point in Cosmic Time.

All Cosmo-Objective Nows (i.e., all ultimate moments in the World of Fabrication) are part of the One Unchanging Cosmic Event ('Seen' from the almost 'Vanishing Point' Perspective), and this is known infallibly by all Cosmic Subjectivity (the entire Cosmic Host considered *subjectively)* during each inter-moment instant.

- A Cosmic-Objective Now is an *'on'* moment in Fabricated Cosmos.
- A Cosmo-Subjective Now is an 'off' moment with respect to the Worlds of Fabrication.
- The Cosmic Eternal Now prevails for *both* 'on' moments in Fabricated Cosmos *and* 'off' moments in relation to Fabricated Cosmos.
- The Cosmo-Eternally Sustained World of Being is *always* 'on'.
- Intra-Cosmically, the Cosmic Eternal Now is (cosmically considered) the *inescapable forever-enduring unit of Time.*
- The Cosmic-Objective Now, per se, though it welds the perceptions of units immersed within the Fabricated Cosmos into an illusory sense of ongoing Nowness, does not endure *eternally* throughout the Cosmic Process, but only *intermittently.*
- The Cosmic Eternal Now, with respect to Cosmos, is a Great Continuity (though *not* so with respect to the Infinite Time Line).
- The Cosmo-Objective Now (though the only time available within the Fabricated Cosmos, and thus a kind of Continuity within the *Fabricated* Cosmos) is not *continuous* within the Cosmos as a whole—i.e., in both Its Fohatically-Fabricated and Fohatically-Unfabricated Aspects. (Parenthetically, it might be said, that Fohat in one of Its Modes, even has something to do with the 'Creation' of the World of Being, but nowhere nearly as *actively* as when engaged in the 'Creation' of the World of Fabrication, the Mosaic World.)

Cosmic Eternal Now—
(the "on" *part* of the
Infinite Time Line)

Cosmo-Objective Now
White — "on"

Cosmo-Subjective Now
Black — "off"

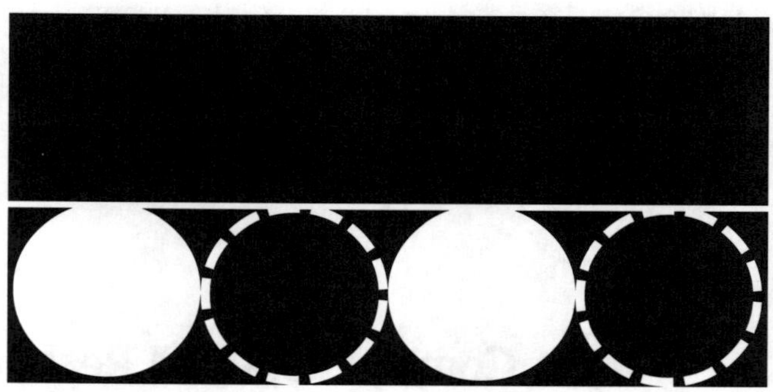

The World of Being (above the line)
The World of Fabrication (below the line)

The question arises: Can the ETERNAL NOW be apprehended from the World of Being wherein the Eternal Now Cosmo-Eternally abides? The best opportunity for such apprehension would exist in that high World, and yet, because, in relation to the Cosmo-

Eternal Now there is the apprehension of the Cosmic Process as One Unchanging Singular *Event* occurring at One *Point*, it could be said that the ETERNAL NOW is *not* REALLY being apprehended. 'Eventness' is in the Contemplative Consciousness of the Great Beings (all of us, Cosmo-Monadically, Really) Who are cognizant of the Cosmic-Eternal Now. 'NON-EVENTNESS', however, is 'in' the 'NON-CONSCIOUSNESS' of the GREAT 'NON-BEING' who is not *cognizant of,* but WHO *LIVES* and *IS* the ABSOLUTE INVARIABLE *PRESENCE.*

As long as there is any registration of any *event,* (even though that Event be an undeviating, impartite Singularity, such as the Cosmos may appear to the Universal Logos from a certain 'distant' Perspective) there is a State that is *less than* ETERNAL NOWNESS. However, can the ETERNAL NOW *ever* REALLY cease? Never! It is simply that the 'PARTICIPATION' within that ETERNAL NOWNESS is most *ultimate* and *infinitized* 'within' the ABSOLUTE in ITS UNDISTURBED ALL-IN-ALLNESS 'during' Universal Pralaya. One sees that language is so time-based that it is impossible to use it to do more than indicate that which negates Time as we know it.

- Never is the 'time' when the ETERNAL NOW is not 'HAPPENING', not 'PRESENCING'. The ETERNAL NOW is that which *devours Time.* The ETERNAL NOW is the answer or antidote to any possible illusory *vacuum in ABSOLUTE TIME.* A perceptually-based interval between events, for instance, represents a kind of vacuum or 'hole' in Cosmic Time.

When the Son is asserted, the continuity of the Father is disrupted. All 'sons' are 'holes' or 'vacuums' within continuity. The ETERNAL NOW, however, negates such a negation. In the ETERNAL NOW, though Time appear, Time is *not.* Even the Cosmic Eternal Now that prevails 'for a time!' during Cosmos, is negated (by the ETERNAL NOW) as a *discontinuity* and, even, as *non-existent.* The ETERNAL NOW has so mastered even Infinite Time, that IT prevents any-thing from 'happening' *ever.*

The Cosmos, from the Perspective of the Universal Logos is a Continuity-within-Itself but, from the larger 'Infini-Perspective'-as-'INFINI-PERSPECTIVE' a Great *Discontinuity.* Within Cosmos, there is no *Subjective* Discontinuity 'Inperienced' in the World-of-Being-Presence of the SELF-as-Self, (however much perceptual discontinuity there may be in the Fohatically Fabricated Worlds, the Worlds of Approximation), for the opened 'Eye' of Subjectivity 'blinks not', and it is *percepetual 'blinking'* that brings on the experience of *discontinuity.*

Just as a human being can experience a tiny, quasi-eternal 'timeless moment', so Those within the World of Being (from the 'altitude' of the Universal Logos especially) can experience the entire Cosmic Process as if It were a 'Timeless Moment' (i.e., can experience the Cosmic Process from the Perspective of the Cosmo-Eternal Now). This experience, of course, is a Psychological Continuity ('during' a Real Discontinuity).

- The one Event 'SPONSORED' by the INFINITE SELF—namely the appearances and disappearances of Cosmoses, is, from the Super-Cosmic Perspective ('wherein', perhaps, the 'Memory' of Infinitude Past exists) a great Discontinuity. From the ABSOLUTE 'INFINISPECTIVE' the Universe is *not* REALLY a great Discontinuity; It is *infinitely less:* in fact, from the 'INFINISPECTIVE', the Universe doesn't exist at all!

Is there, however, a *REAL* Interval between Cosmoses? There is certainly a Real Interval, but is there a REAL interval? The Real Interval is obvious. There are an infinitude of cyclically appearing sequential Universes. Their appearance, according to *The Secret Doctrine* conforms to the Law of Periodicity. No appearance "lasts forever", nor does a disappearance. However, if lesser apparent temporal interludes within Cosmos are Really negated by the Cosmic Eternal Now, why should not the greatest of all supposed Temporal Interludes be REALLY negated by the ETERNAL NOW?!

From one Perspective there exists Infinite Time and an Infinite Time Line, but from another and *greater*'PERSPECTIVE' perhaps the Great Breath, Itself, is un-REAL. Worded otherwise, perhaps even while the Great Breath (which marks out the Divisions along the Infinite Time Line) is occurring, It is not REALLY occurring! Perhaps there is no *REAL* way to violate NOTHINGNESS and the ETERNAL NOW.

A more moderate point of view would say that an infinitude of events, great and small, *are* now, forever *were*, and forever *will be occurring* (however *intermittently) NOW*. The more radical 'PERSPECTIVE' states thatall events (great and small) have *never* REALLY *occurred at all*, except *illusorily*. This can be said because the INFINITE SELF 'ASSERTS' forever, uninterruptedly. Continuous SELF-'ASSERTION' leaves no REAL room for a vacuum or 'hole' in that SELF-'ASSERTION'. "Nature abhors a vacuum" can mean many things. In this case it may mean: NATURE abhors a discontinuity in REALITY.

- Thus it is that all discontinuities are 'CAST OUT' into the World of Illusion, for they cannot remain 'within' the HOMOGENEOUS CONTINUOUS SELF. But then, how can they not? For there is *nowhere else*.

While there is, by definition, no interval 'within' the ABSOLUTE SELF, there *appear* to be intervals within (and in relation to) all other *selves*. Can these intervals be REAL? In answer, let us ask if, in the ETERNAL NOW, there can be both a Now and a Not-Now? If there is to be an interval, there must be both a *thing* and a *not-thing*, and if there is to be an interval in Time, there must be a 'this-time' and a 'not-this-time'.

Is there, however, both a Now and a Not-Now within the ETERNAL NOW? Clearly, in the ETERNAL NOW (since all times are NOW), there can be *no such interval*. All times are exactly the *same* 'TIME' (albeit, when Time actually does *exist!*—which is not *always*).

Thus it is that no events can REALLY occur at different times because there *are* no different *times*. 'In' the ETERNAL NOW, it is *never not NOW*. From the Perspective of the Eternal Now in-Cosmos, *during* Cosmos it is ever Cosmo-Eternally Now. *Nothing in Cosmic Space leaves Cosmo-Nowhere/Everywhere and nothing in Cosmic Time leaves the Cosmo-Eternal Now*. Even more radical is the statement that with respect to the UTTERALLNESS: Nothing ever is anywhere but NOWHERE/EVERYWHERE and nothing ever occurs within the ETERNAL NOW.

On Universal Space and Beyond

It is fallacious to think about "how much" Space is "taken up" by a Universe, as if Space were Really an extendible quantity. Really, from the Perspective of the Infinite Subject, no-Space is "taken up" by a Universe. No matter how great the apparent 'size' of a Universe (as viewed from the *intra*-Cosmic perspective of one of its participating units of Life), the 'extension' of the Universe must be compared to the infinite 'extension'/non-'extension' of Mulaprakriti. When the Infinite Subject 'Sees' Its Own Infinitude, Mulaprakriti is 'Seen' as *Boundless* and infinitely extensible (even though there is nothing within It that can be either extended or contracted).

From such an Infinite Perspective (or *infinitesimally less* that Infinite Perspective), any Cosmos, no matter how great (or small) may be Its intra-Cosmically assessed 'size', would appear but as a *point* (and, Really, a *rapidly vanishing Point).* A Real point, remember, has *no dimension* (at least no *measurable* dimension, ever) and, hence, no measurable size. If it has no measurable dimension, does it exist at all?

In terms of *actuality,* a point both *is* and *is not.* It is a *presentation,* therefore it *is;* but, it has no way of 'claiming space' or 'being extensible' (for it is always less in magnitude than any measurable thing), so, in a way, it *is not.* The completed *infinitization* of Perspective must probably be arrested if a Point is to maintain its presence and not simply *vanish* altogether. When the infinitization of Perspective is arrested, a Point becomes, not a Real Point, but a *virtual* Point.

Granting, however, that a Point can and does exist, and that this Point called the Universe can and does exist, and can be considered, at least, an Objective Presence. Then, at Its very *largest,* the Universe-as-existent Point (seen from a rapidly *infinitizing-but-not-infinitized* Perspective) would have to be called *infinitesimal* in size—having a *kind of* size or extension (for It *Is),* but an ever-*indefinite* size or extension *as progressively close to zero as possible without actually being zero.* Clearly, such a Universe would have to appear *ever smaller* in order to fulfill the condition of infinitesimal (or 'infinitesimaliz-ing') size. It could not *stop* on any one magnitude, or it would no longer be a *Real* Point (for it would then, theoretically, have *measurable* dimension).

It could be justifiably asked whether infinitesimal size or extension is Really *any* size or extension at all. Certainly there is, in this case, a (perhaps infinitesimally less that infinitely rapid) *convergence upon zero.* That which apparently *Is,* i.e., a Universe, is (from the perspective of the Infinite Subject) tending towards becoming *nothing—no-thing.* The Universe as 'Seen' or 'Understood' from the Perspective of the Infinitizing Self and in relation to Mulaprakriti is becoming a 'Vanishing Point'. Because the Universe is an finite Object, and thus a definite Singularity, It is infinitely removed from Infinitude (the mathematical ratio being 'infinity'-to-'one').

- Every definite quantity is infinitely removed from Infinitude. From the Perspective (the 'Infinispective') of the Infinite Subject, it is as if the Universe were not. To such thoughts one is led when considering the Real *size* of this or any Universe.

This negation of the existence of the Universe is even more compellingly TRUE from the 'INFINI-SPECTIVE' of REALITY. It begins to seem as if INFINITUDE will not 'tolerate' the existence of any other *thing* (no matter how apparently great or vast). Any thing can be posited, and indeed the INFINITUDE *seems* to 'POSIT' (thus our Universe comes to be). That very positing, however, seems a non-'ACT', for anything posited is, as if, when compared to the INFINITE (the INCOMPARABLE), *nothing*. There is the *nothing* which is the Cosmos and the NOTHING which is REALITY. Which one is REAL?

In this particular consideration, it must be realized that the Infinite is *not* the INFINITE. For the purposes of this model, we should consider the Infinite as Infinitization-becoming-INFINITE. The Infinite is that which is in the Process of *infinitizing* but is not 'there' yet! The Infinite is the consistent *movement towards* INFINITUDE. It is greater than any known quantity, farther than any known distance, but ever *indefinite*, and ever, *in process*. Could this be said of the Infinite Subject and Infinite Object?

The implications are interesting, for indeed Infinite Subject and Infinite Object are *transitional Entities* between the ABSOLUTELY INFINITE and the Definitely Finite. Perhaps this model is what would keep the Infinite Subject and Infinite Object from falling prey to being classified as definite *things* (which classification might be a threat to what *infinitude* they do 'possess'). The Infinite Subject and Infinite Object are like the Real Point—neither existing nor *not*-existing, but transitional between the ABSOLUTE and the Finite. The Super-Cosmic Trinity is as *indefinite* and *immeasurable* as the Real Point *(dimensionless* yet *present)*.

In summary, we are coming to see that all that is 'happening' is, as if, happening at a *point*, and that that point (like all Real, non-virtual Points) is fast vanishing (vanishing as fast as possible without vanishing entirely). Or, perhaps, is *has* vanished? And always has been *vanished?* This would be the testimony 'from' the *completed* 'INFINISPECTIVE'. So, is anything happening? REALLY?

Continuing our exploration, if 'within' the Cosmos-Perceived-as-Point, nowhere and everywhere are the same, then, similarly, *no time* and *any time* and *Now* are all the same, as well. In Cosmos, then, there never has been anything but the Cosmic Eternal Now; that Now is the Great Continuity and that Now persists throughout Cosmos. What, however, of *other* times? What can be said of them, for certainly (in terms of space) much can be said of *other objects*. They—other objects *than* our Cosmos or than *in* our Cosmos—can certainly be imagined. Other times can apparently be imagined, as well.

But have there REALLY been other *times?* Have there REALLY been other *spaces?* While there may have been (from a dualistic, extra-SOURCE Perspective) other times and other spaces, there has never been another 'TIME' and another 'SPACE'. Can we, even if imaginatively, establish the reasonableness of this assertion?

On Annihilating Time

Let us transform our 'Viewer', from the Universal Logos, to the Infinite Subject (or Infinitizing Subject—'on the way to' the Infinite). Super-Cosmos is transitional between INFINITUDE and Finiteness. No Super-Cosmic Being is *utterly indefinite*, but neither is It *delimitable* and *definite.*

The contrast between the Infinite Subject and the Infinitizing Subject is very interesting. We speak of the Infinite Subject and the Infinite Object without comprehending fully the implications of these terms. It may well be that the Infinite Subject, per se, can 'See' nothing but Itself. Since the Infinite Subject is not a Definite Point, per se, but is an 'Infinified Point', It, as it were, 'Sees' *all* possible Points, which is equivalent to 'Seeing' *no* Points at all! The Infinified Point of View of the Infinite Subject allows the 'Seeing' of 'infini-pointed' Infinitude, but of nothing more specific.

The moment the Infinite Subject 'Sees' any definite thing (isolating a number of points from *all* points, or one point from all points) it should be called the 'De-Infinitizing Subject' (i.e., the Condensing Point). Perhaps, we should say that a Subject can 'See' only what It, Itself, *is.* If the Subject is Infinite, It can 'See' only the Infinite. If the Subject is in process of *finitization,* it can 'See' that which is 'moving' from indefiniteness towards definiteness. An Infinite Subject is Infinite, and so 'Sees' or 'Cognizes' Infinitude. A 'De-Infinitizing' Subject moving 'away' from Infinitude and a 'Re-Infinitizing' Subject moving 'towards' Infinitude, can begin to see the 'particular', the distinct.

A Real point, though it is dimensionless, is a specifiable thing; it has existence, even though it is spatially and temporally indeterminate and immeasurable. A Real point is visible to the De-Infinitizing or Re-Infinitizing Subject, but *not* to the Infinite Subject. Only the *'unspecifiable'* 'Other' is registered by the Infinite Subject, and a Real point is *'specifiable'* and can, so to speak, be perceptively isolated (but not measured, since the Observer is 'in Recession'). The Infinite Subject can only register Its Own Unspecifiable Infinite Selfhood.

We will speak primarily of the 'Re-Infinitizing' Subject—one who is 'moving away' from Finitude, in order to establish the Perspective of Infinitude. For the sake of simplicity, we will call such a Subject, an 'Infinitizing Subject'. We might think of an Infinitizing Subject (one which is moving from a Finite Perspective towards an Infinite Perspective) at a *virtually* infinite speed (but then, even *virtually infinitesimal speed* might do as well!). Whether 'traveling' at a 'speed' infinitesimally less than infinite, or infinitesimally more than zero, the Infinitizing Subject will not (in that way) 'reach' the State of *Infinite* Subject.

If the Infinitizing Subject should achieve *'infinite* speed'—for instance on the verge of Universal Pralaya, then specificity would disappear, and only the Infinite Subject and Infinite Object would remain (in 'pointless' 'Contemplation of each other—perhaps for an 'infinitesimalizing moment') before *reabsorption* into INFINITUDE.

- For the sake of a clear presentation of the forthcoming discussion, it might be asked: Can an Infinite Subject (who obviously cannot see a *virtual* point, because a virtual point is a finitude) 'See' a *Real* 'Point', since a Real Point is "next to *nothing*"?

- The answer would have to be 'No', for the Infinite Subject is 'No-thing' at all, 'Seeing' Itself as 'No-thing' at all—in other words, '*detected* No-thing-ness' (or 'Every-thing-ness').

A point, even a Real Point, is a 'something', a specifiable 'singularity' (even though an 'indefinite' 'something'). A Real point is an 'isolatable' and 'specifiable', whereas the Infinite Object which the Infinite Subject registers, is a forever indefinite 'non-isolatable', 'non-specifiable' Presence. The Infinite Subject 'Sees' Itself as the Great 'Non-Specificity', without any trace of 'Con-centra-tion'. A Real Point is that which calls awareness 'away' from indefiniteness and infinitude, 'towards' definiteness and finitude.

In short, the assumption is being made that even a Real Point could not be 'Seen' from the Perspective of the Infinite Subject (Who, being the Infinified Point, 'Sees' Itself as *all* possible points which means as no particular point at all). Thus:

- The Infinified Point Self-Reflects all possible points and thus no specifiable points;
- The Condensing Point (or Un-Condensing Point, which can also be called the 'De-Infinitizing' or 'Re-Infinitizing' Subject) can 'See' *Real* points (which are always transitional between INFINITUDE and Finitude); and
- The Condensed Point (which has been rendered finite) can see Virtual Points, which are measurable and finite.

With this preamble in place, let us imagine that the Infinite Subject (or, rather, *Infinitizing* Subject) is 'viewing' Cosmos from a continuously receding succession of distances continuously approaching what we would mistakenly call 'infinite distance', but never 'reaching it' (as there is no such thing as a *definite* infinite distance). The Infinitizing Subject, having passed the 'Point of View' from which all things in Cosmos seem as one, can now be moving slowly or rapidly—it makes no difference. For the sake of our visualization we can increase or decrease the 'speed' of the Infinitizing Subject at will. Also, for the sake of simplicity we will not consider any possibility of quantization of Time, Space and Motion in Super-Cosmos, and will simply assume continuous movement is possible. (For this discussion, it will make little difference anyway.)

We can imagine that (under the established parameters) the Infinitizing Subject will 'See' Cosmos as an 'infinitesimalizingly' tiny Point (i.e., a point which is getting smaller and smaller, but which can never be measured). All that can be said is that the point is 'present', but of indeterminable dimension.

In our model, there would have to be continuous 'movement' of the Infinitizing Subject towards a naturally unspecifiable (un-Real) 'infinite distance' in order to sustain the infinitesimalizing magnitude of the tiny Cosmic Point, because, no infinitesimal can exist as a *definite* magnitude, just as no *definite* infinite can exist. As regards the 'infinite' and the 'infinitesimal', there can only be a 'movement towards'—towards the ever larger and larger, and the ever smaller and smaller. Do no more than simply *stop*, and the endlessly increasing or endlessly decreasing magnitude under observation becomes *definite* and *dimensioned*, and, thus, both the infinite and infinitesimal are destroyed.

We will assume, for the sake of meditational visualization, that the Infinitizing Subject has very good 'Eyes' and is not limited by physical laws concerning vision and the properties of light. We are by now familiar with the idea that such may be the 'distancing' of the Infinitizing Subject from the Cosmos It has under observation, that all events within that Cosmos *seem* to 'happen' at one point, i.e., in one space ("on top of each other", as it were, since space in Cosmos is being reduced to the smallest possible extension without being zero). As well, for such an Infinitizing Subject, all events within that Cosmos seem to 'happen' at the same unvarying time.

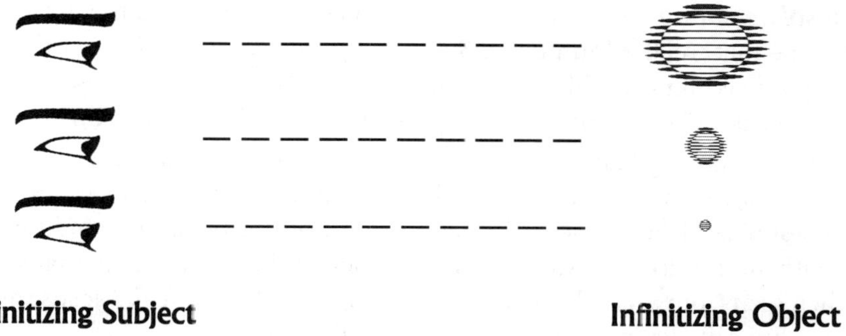

We could, if we chose, be less radical and still achieve results satisfactory for the unifying of Time and Space. Thus, even if we did not reduce the Point to an 'infinitesimalizing', there would be, theoretically, a *definite* 'distance' at which all within the Point would be *perceptually* blended into One. Such a Point, however, would, inescapably, be a *virtual*, hence, *dimensioned*, Point, and not a *Real* Point at all, and would require the Infinitizing Subject to give up Its continuous, recessional 'distancing' and, instead, *stop* at a given 'distance' for Observation.

Instead, however, of being conservative, and 'creating' in imagination simply, a *virtual*, thus *definite*, Cosmic Point (however tiny and fused into *one* it may seem), let us go all the way, and imaginatively create an *infinitesimalizing* Point. Later, when we seek the complete disappearance and obliteration of the Point, it will have been more useful to have done so.

Continuing, then, with "setting the scene", since no event in Cosmos is distinguishable at that indefinite 'distancing' rapidly or slowly approaching 'infinite distance' (whatever that Really is?!), the timing of one event or another, whether *before*, *now*, or *after* cannot be distinguished, and everything *seems* to be occurring as if in an unvarying *now*. With respect to what is being viewed, a Cosmos-become-Point, all things within It have become *one* Thing, all 'times' within It have become *one* Time, and all 'places' within It have become *one* Place.

A point in general (especially a Real point) is a great 'obscurer' of distinction and individuality. With respect to the Point we are considering, all 'happenings', 'places', 'things' and 'times' within the Cosmos have been fused into *one Single Event*. At that ever-increasing distance, no change is perceived, so only "the same thing" seems to be happening "all the time". All distinction is obscured because the *'distancing'* is so great. A point of *infinitesimalizing* extent, for instance, would 'crush' all distinction and individuality. No-thing in distinctness (other than the *fact* of the Point) would be discernable. It is interesting that:

- A point is an *existence* about which nothing specifiable can be predicated.
- The INFINITE is a 'NON-EXISTENCE' about which nothing specifiable can be predicated.

Continuing, let us imagine for a moment that the particular Cosmos we have under observation disappears, and *another* is seen in Its *place*. From a close range, the second Cosmos would appear to be different and distinct. However, from an *infinitizing* range (not an *infinitized* range, whatever that Really would be)—that continuously-increasing range required for Cosmos to *seem* an infinitesimalizing Point without quite vanishing into *absolute nothingness*—all distinction would disappear, and the two Cosmoses (let us imagine that we, the Infinitizing Subject, can now see them *both* at the same 'time') would appear as infinitesimalizingly small Points. All distinction *between the two Cosmoses* would be 'crushed' out of them by the pressure of 'infinitesimalization'.

If the Infinitizing Subject can imaginatively see two, "at the same time", (i.e., "under the Eye at once") why not, imaginatively, a huge number, or even an infinitude of Cosmoses? The vision of non-distinction, and non-particularity would be exactly the same with an infinitude of Cosmoses, every one of them (though perhaps vastly different when seen from a small distance) appearing *exactly identical*, because all in the process of *infinitesimalizing*. In that infinitesimalizing *state:*

- All things in every one of an infinitude of Cosmoses would appear as exactly the same thing.
- All places as exactly the same place.
- All distinguishable times as exactly the same unvarying, event/point.

We learned that this annihilation of distinction was true for *each* Cosmos in it own right. Now we see, imaginatively, that annihilation of distinction is true for each of an infinitude of Cosmoses with respect to each other, if 'Seen' from an infinitizing 'distancing'.

◆ Here is a provocative thought: it may well be that from the 'Infinispectivizing' of the Infinitizing Subject/Infinitizing Self, there would seem to be (at any one 'time') but *one* infinitesimalizing point in all the Objective Void.

If the Infinite Time Line (of appearing and disappearing Cosmoses) were imaginatively seen (at any Time of Cosmic Manifestation) from that Infinispectivizing [see Glossary], then only *one* seemingly *identical* Cosmic Point would be 'Seen' as the representative of an infinitude of Cosmoses, occurring (albeit) at an infinitude of 'times'—in an infinitude of 'places'? Perhaps! Yet, in the Voidness (with no other points of reference, and with the possibility of virtually infinite recession in 'distancing') all specifiable 'places' are *One* Place, and so the tiny infinitesimalizing point would always be occurring (perceptually) in the *same place* relative to the 'ever-distancing' viewer. Assuming that if even an infinitude of Cosmoses occurred in definite 'places' of a quantifiable 'distance' from each other, a range of sufficient recession could be found from which they would all be seen to be occurring (perceptually) in the same 'place'.

◆ This model is attempting to show that no matter which one of an infinitude of Cosmoses may be observed, that Cosmos will always appear as a Point indistinguishable from all the other Points, and so (from a sufficient Perspective) will *seem* like the same Point.

What do we have, in imagination, so far? We have an infinitude of Cosmoses, appearing identical in every way, appearing and disappearing always in the same 'place' in Space—the only 'place' there is (from a sufficiently 'distancing' Perspective). We have an infinitude of Cosmoses appearing "one at a time" (as theoretically they do) in Space. Well and good. Thus far we have a model of what we may call 'Repetition in Space'. An infinitude of different Cosmoses, all *appear* as *identical* Objects in exactly the same space, forever, but at *different times*. So, at least, we have (through sufficient 'recessioned distancing) collapsed all 'Spaces' into *one* 'Space'. Space is much easier to 'collapse', it seems, than Time.

Entering the imagination again, it is because of the tremendously vast though indefinite spatial 'distancing' produced by the Infinispectivizing of the Infinitizing Subject/Self, that all *things* in Cosmos can be perceived as one Thing, and all *places* in that Cosmos as one Place/Space, and even, that all intra-Cosmic *events*, can be perceived as one Single Event.

- Now, what is the analog to virtually infinite *spatial* distance? It would be virtually *infinite temporal extent.* If Space can be minimized by virtually infinite 'distancing' (infinitizing distancing), to the degree that any object (even one so 'large' as Cosmos) approaches the *vanishing point* and perhaps, would vanish at an *absolutely infinite distance* (whatever that *might be,* for it is *unspecifiable),* then what is the analog that can be applied to the minimization of *Time?*

In the diagram below, the double arrowhead represents any time span, the black ellipses represent the temporal 'distance' from which the time span is 'perceived'; the same span of time can therefore be viewed as 'collapsing' due to greater 'temporal distancing' of the perceiver. An analogy might also be made that, to a child, childhood seems to be a very 'large' timespan, as the child has yet to achieve any distance from his experiences; to an adult, childhood seems 'smaller' because there has been more temporal distance achieved, and the adult may even perceive childhood as one event.

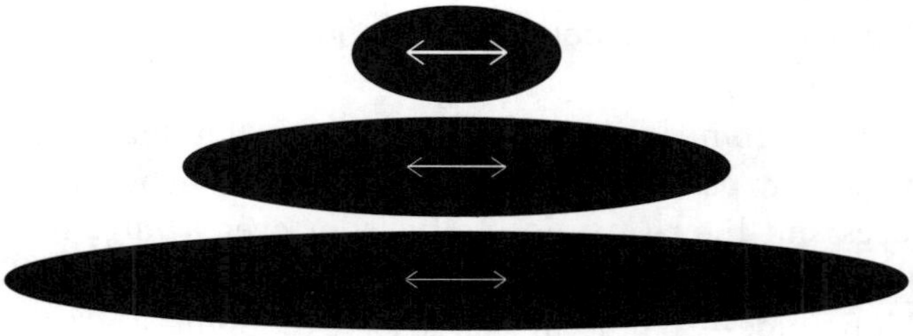

Temporal Distancing

Thus far in our model, an infinitude of Universes (which *seem* identical) are appearing every so often, forever, in exactly the same 'place' in Infinite Space. From the perspective of various *intra*-Cosmic beings, the time interval between those appearances must seem staggeringly huge. Even a Universal Logos ("at the end of the *Day*") must think He has a relatively *long* and well-deserved rest coming! Remember, however, that, in our model, we are viewing what *seem* like repetitive Cosmoses from the 'Infinispectivizing' of the Infinitizing Subject/Infinitizing Self—not *yet* the 'INFINISPECTIVE' of the INFINITE SUBJECTIVITY/INFINITE SELF. If the view were from that greater 'PERSPECTIVE' (i.e., the INFINISPECTIVE) everything would be changed into NOTHING.

Just as an Object of any spatial dimension (extraordinarily huge or tiny) can be reduced to virtual dimensionlessness at a near-infinite (though indefinite) spatial 'distance' (for, at an *actually infinite* 'distance' {obviously not available extra-SOURCE} the object *would* become utterly dimensionless and disappear into nothingness). Therefore, an *event* of *any* duration (whether a Cosmos or ultimate moment) would be transformed (when seen against the background of 'Infinite Time in Process') into that which is exceedingly brief, perhaps *infinitesimally* brief. From up close, some *things* look bigger and some things look smaller. Similarly, some *events* seem to take *longer* and some events seem of *shorter* duration. However, from a *virtually* infinite 'Temporal Distancing' (imagine yourself 'there' imaginatively, i.e., using the Endless Duration of 'Infinite Time in Process' as the standard of measurement), all *events* (relatively long or short) are reduced to an infinitesimalizing duration, though not quite annihilated altogether.

- What we did with Space, was imagine a *virtually infinite though indefinite spatial 'distancing'* which produced the infinitesimalizing of all possible *objects* or collection of objects.
- What we are doing with Time, is to imagine a *virtually infinite though indefinite temporal span* which will (as we continue to 'broaden' the span toward Infinite Time, just as we increased the 'distancing') produce the infinitesimalizing of any possible *event* or *non-event*—which, (for temporal purposes) *is also* an *event!*

Just as, at a progressively infinitizing spatial distance, all 'places' in a Cosmos become one Place, and all 'things' in Cosmos become one Thing, and even *all* times and events become in a single Cosmos but One Event/Time so from the 'Infinispectivizing' of the Infinitizing Subject (Who is 'Seeing' from a *progressively infinitizing though indefinite temporal span),* the seemingly identical Events called Cosmoses (all of a certain duration, even or uneven) and the specific *intervals* between these Cosmoses (also of a certain duration—whether equal to each other or not—it makes no difference) *would each seem to endure for a shorter and shorter amount of time,* until they began to *infinitesimalize* and, thus, converge on zero duration.

It is the capacity for *comparison* with which, presumably, the Infinitizing Subject is equipped, that makes any particular Event-Cosmos or Inter-Cosmic-Interval, or series of Event-Cosmoses and Inter-Cosmic-Intervals, or even a *huge* number of Event-Cosmoses and Inter-Cosmic-Intervals, seem as *very little duration indeed* compared with *virtual* temporal infinitude—so little, in fact, as to be of *infinitesimal* duration.

At this point, let us look at the mathematical truth that, if an infinitude of infinitesimals (or infinitesimalizings) is added together, the result will be only the infinitesimal or infinitesimalizing. Put in terms of our model, if an infinitude of infinitesimalizing durations (Cosmoses and inter-Cosmic Intervals) are added together, the result will simply be an infinitesimalizing duration.

Thus, the entire infinite span of Events and Non-Events we call Cosmoses and inter-Cosmic Intervals can be imagined as composed of an infinitude of *infinitesimalizing* Events/Non-Events, the time value of which will be equal, only, to an *infinitesimalizing duration.* Thus the entire span of infinite Super-Events (i.e., Cosmoses and the Events between them, which are measurable Non-Events)—all, add up to "next to Nothing" (approaching Nothing).

Reviewing where we stand thus far:

- We have a spatial Point (standing for an Infinitude of Cosmoses), the magnitude of which is "next to nothing", enduring for 'times' and disappearing for 'times', which are *also* "next to nothing".
- Suppose next then, that all the infinitude of Cosmic appearances 'Seen' as Points, and all the infinitude of Cosmic disappearances 'Seen' as intervals, were viewed from this *virtually infinite (i.e., infinitizing) Temporal Perspective,* i.e., *the Temporal Infinispectivizing* of the Infinitizing Subject.
- Then, just as Cosmoses appearing as Real points became indistinguishable, would not all such Events (Cosmoses and Inter-Cosmic Intervals) seem to become *so short* as to be of indistinguishably similar duration because of the *infinitesimalization* of their duration, and their convergence upon zero?

A Temporal 'Distance' (Temporal Infinispectivizing) *forever infinitesimally less* than Infinite Temporal 'Distance' would annihilate all distinction of duration between any kinds of events, and all events (Cosmoses and intervals) would seem to be of the *same* duration converging on zero, i.e., of *infinitesimalizing* duration.

If such infinitesimalizing of duration were in progress, and one contemplated the Infinite Time Line, would not all such Macro-Events (Cosmoses and Inter-Cosmic Intervals), because they seemed of so short a duration as to take virtually "no time at all", begin (seemingly) to 'crowd together' into *one infinitesimalizing moment,* (blurring and rendering negligible, and even indistinguishable, all lines of distinction between Cosmic Events and Inter-Cosmic Intervals) such that, together, they all seemed like one event of infinitesimalizing duration? Eternity (or, better, *Eternalizing Duration) dwarfs* all events into *infinitesimality.*

Thus, conceivably, it would be that the Event/Non-Event Sequence of Cosmoses and inter-Cosmic Intervals would collapse to an *infinitesimalizing temporal point* (a *temporal vanishing {but not quite vanished} point)* which (as long as the Temporal Infinispectivizing were not absolutely *infinite*—which it could never be in Super-Cosmos or Cosmos) could retain its 'infinitesimality' (forever) while avoiding becoming 'nothing'. But should the All-but-Infinite Temporal Perspective ever become a TRUE 'INFINISPECTIVE' then, the *perceived 'infini-composite' Event* of *infinitesimalizing* duration, would collapse into *nothingness* and would *vanish* just as would the Cosmo-Composite-Spatial-Point if seen from (not a *virtually* infinite 'distance', or a distance *approaching* infinity) but a truly *infinite* 'DISTANCE' (such as the true distance 'between' INFINITUDE and anything whatsoever). Such a 'DISTANCE' is 'AVAILABLE' only 'within' or 'as' the ABSOLUTE ITSELF.

It may be that some thinkers would protest that the Power of the Infinispectivizing of the Infinitizing Subject is not sufficient to collapse the entire Infinite Time Line into *infinitesimality.* The matter could be debated, and the key lies in *infinitesimalizing* each of the temporal units (Cosmic Events/Inter-Cosmic Interval Non-Events) on the Infinite Time Line. There are, however, "bigger guns". Let the 'INFINISPECTIVE' of the INFINITE SELF—that 'ABIDES' forever in ABSOLUTE DURATION—be applied against the Infinite Time Line, and see what happens to the duration of any event whatsoever (whether Cosmos, Inter-Cosmic Interval, series of Cosmoses and Intervals, or astoundingly vast series of Cosmoses and Intervals). All of these would be dwarfed instantly not just into *infinitesimality,* but into nothingness.

- ◆ INFINITUDE (the INCOMPARABLE) destroys any-thing against which IT is *compared.* IT 'DESTROYS' by instantly changing all thing into ITSELF!—into NOTHING! Thus it is, that Illusion is ever a *maya,* a *nothing-at-all,* and only the INFINITE SELF *IS.*
  - Thus it is, that all spatial things and temporal things are Really *almost* no-things, fast approaching *nothingness* (i.e., *'infinitesimalizing'),* when 'Seen' from the 'Infinispectivizing' of the Infinitizing Subject.
  - Thus it is, that all spatial things and temporal things are *REALLY absolute nothings* when 'SEEN' from the INFINISPECTIVE. The 'INFINISPECTIVE', 'in which' 'SEE-NESS' is REALLY non-existent (but BE-NESS 'REIGNS'), reduces all things to the *nothingness* of the GREAT ZERO. All things, vanish into NOTHINGNESS.
  - Thus we see, that from the INFINISPECTIVE of the INFINITE SELF, no-thing has even been and, NOTHING has *ever* BEEN! The entire World of Illusion, the World of Extension, the World of Temporality are as nothing, are, *in fact, nothing,* and perhaps, from the perspective which by now should be familiar to the reader, *less than NOTHING.*

If we were ever looking for an argument to demonstrate that Illusion, though it *actually* exists, *exists not,* this is the one. REALLY, 'INFINISPECTIVELY' Illusion REALLY *exists not.* Nothing does! NOTHING 'DOES'! "A thousand Ages in His sight is like an ev'ning gone" applies well to the Universal Logos, but for the INFINITE SELF what could be said?—"A thousand Ages in HIS 'SIGHT' is like ... like ... *nothing at all!*" Even an infinitude of Ages can *never* REALLY 'happen'. NOTHING 'REIGNS' *forever!*

On the Analogy Between the *Point* and the *Now*

Remember that everything in Cosmos (including Cosmos) as 'Seen' from the 'Infinispectivizing' of the Infinitizing Self (and, perhaps, in a way, even from the highest Perspective available to the Universal Logos) is 'Seen' to occur at the *one Point,* and that that dimensionless *Point* negates any possibility of Real spatiality in Cosmos. There is but one Space in Cosmos and that is, ESSENTIALLY *no*space, *no*where.

Then, as well, is everything occurring at one *time?*

- In Cosmos, the answer is, "principally, Yes", *if* Cosmos is seen from the Perspective of the Cosmo-Eternal Now, available to the Universal Logos and His Chief Subjective Emanations in the World of Being (and to us when we are *being* Them—which, in fact, we always *are* just as They are always *being* us—Essentially).
- With respect to Super-Cosmos, the answer is even more definitely, "Yes", if the 'Perspective' is that of the Infinite Subject/Infinite Self (which certainly *sur-*

> *vives Its Own Reduction into the Universal Logos!).* The Law of Emanation requires the inviolate stability of the Emanating Self with respect to the Emanated Self. Super-Cosmic Time would be intimately related to the 'Infinite Vision' and related, as well, perhaps to 'Infinite Memory'. Time would exist (which it does *not* in the ABSOLUTE CONTINUUM), but it would not be Time as experienced in-Cosmos.

The idea being discussed is that the Point and the Now (Cosmo-Eternal) are analogous, just as the 'Pointlessness' and the NOW are analogous. The Cosmic Point (the Point upon which all Cosmic Spatiality and Temporality occurs) has no dimension and can be found (within the Bounds of Cosmos) both *nowhere in particular* and *everywhere in general.* It is a Cosmo-Pervasive Point (because of its dimensionlessness). The Cosmo-Eternal Now likewise has no specific, exclusive time in-Cosmos, and, thus, cannot be located *specifically* with respect to a past and future in Cosmos, and further is found at *all* times (and, 'no-times'—for Cosmo-Fabricationally, there are countless 'no-times') in Cosmic Duration (this being analogous to the 'everywhereness' of the Point). Within Cosmos, therefore, due to the Omnipresence and Cosmo-Eternal Now of the Universal Logos, all Cosmic spaces are, Essentially, One Space, and all Cosmic times, are Essentially, One Time.

Thinking for a moment about the Infinite Time Line and the Infinite Chain of Cosmoses, there certainly have existed other arrangements and combinations of variables in the Cosmic Configurations of innumerable Cosmoses, but have they not been ESSENTIALLY illusory? In any Cosmos is there any REAL *space* in which such arrangements and combinations can occur? There *have been,* of course, a variety of kinds of intra-cosmic space, from a certain limited perspective, but if we imaginatively bring all space within a Cosmos to a single Point, then there is no *room* for variation, modification, combination and arrangement, and all spaces become One Cosmic Space.

Along the same line of inquiry, we must ask if there have ever REALLY been *other* 'times'? Time, as we know it, is the *factor* by means of which the combinations and rearrangements of objects in space occur. Time is as much a *sine qua non* of Motion as Space.

The question then arises, If (in any Cosmos) there is Really (Essentially, despite appearances) only the Eternal Now (or Cosmic Eternal Now), have there *ever* been *other* specific 'times'? Perhaps, there have been other *specific* Cosmic Eternal Nows. Are they distinguishable from one another? If the 'Temporal Expansiveness' from which they are 'Seen' is not so great as to obliterate them entirely, perhaps they are distinguishable as separate, singular Cosmic Events. As the 'Expansiveness of Temporal Observation' approaches Infinity, it is probable that even an infinitude of Cosmic Eternal Now would be indistinguishable from one another. The Eternal Now in each Cosmos eradicates distinguishable intra-Cosmic 'times', and the Infinitizing Perspective/'In-Spective' eradicates distinguishable Eternal Nows.

Has anything *else* in the ALLNESS of Infinite Cosmoses past, ever REALLY or even Really take place? It may appear that an infinity of perceivable combinations and rearrangements, i.e., changes, *have,* indeed, occurred (from a limited perspective of consciousness), but has there REALLY/Really been any *other* 'time' in which those occurrences *could* have occurred? The proper answer, it seems, is that all such occurrences *have taken place in a Now that is NOW.* This is not to say that they all *are taking place* Now, but that they all *have taken place* in a Cosmo-Objective Now that is Essentially a Cosmic Eternal Now and that is, even more ESSENTIALLY, the NOW, the ETERNAL NOW. The Now (whether Cosmo-Objective or Eternal) is different from the NOW, though the latter includes the former.

On Five Kinds of Now

Let us consider the possibility of *five* kinds of *now*—

1. A relative, or ordinary, 'now'
2. A Cosmo-Objective Now
3. A Cosmo-Subjective Now
4. A Cosmic Eternal Now
5. *The* ETERNAL NOW (as there is only *one* ETERNAL NOW, forever)

1. The ordinary 'now' is approximate and arbitrarily determined. It is not directly related to Cosmic Structure, and is determined by the uninformed decisions of beings-in-Cosmos who are ignorant of the Nature of Time in Cosmos and its *modus operandi.* The 'now' can be applied to very different measures, such as the second, the minute or even the hour. It is usually a general measure, and loosely means the 'present moment' which is, inevitably, from a technical perspective, an extended span of time. Even when human beings begin to measure time more precisely (using the idea of the stop watch, and all the other precision time-measuring instruments) the relative 'now' still remains approximate because there is no knowledge of the nature or duration of an ultimate moment.
2. A Cosmo-Objective Now is an *ultimate moment* in any given Cosmos. (In different Cosmoses they may be of different duration, depending upon the Cosmic Algorithm.) Cosmo-Objective Nows are perceivable by the Great Subjectivities in Universe (such as the Universal Logos, or His Emanated Extension, the Universal Son, or Universal Fohat). Upon, or 'at' (but not 'within') any ultimate moment, it is Now everywhere in Fohatically-Fabricated Cosmos, and this, the Universal Logos (the Great Observer of every change in the Cosmic Configuration) Knows.
3. The Cosmo-Subjective Now is the inter-moment instant, the Moment of Subjective 'Evaluation' or 'Appreciation'.
   - It occurs 'between' Cosmo-Objective Nows (i.e., ultimate moments) and may or may not be of the duration of ultimate moments (depending upon one's Cosmo-Conception, and of course, upon the *truth* of the matter).
   - The Cosmo-Subjective Now occurs within the lower extension of the World of Being (i.e. the World of Adjustment) and is not registered in the World of Effects.
   - While from one perspective, the Cosmo-Subjective Now may be said to occur only 'between' ultimate moments', from another perspective, it can be said to

occur, as well, *simultaneously with ultimate moments*, i.e. superimposed upon Cosmo-Objective Nows. From this second perspective, the sum of all Cosmo-Subjective Nows would be the Cosmo-Eternal Now.

- Cosmo-Subjective Nows, however, may well be subject to the intra-Cosmic quantizing of Time, whereas the Cosmic Eternal-Now would not.

The Cosmo-Subjective Now may, unlike the Cosmo-Objective Now, be *divisible*. So much depends upon whether there is quantization within the World of Being, and, if so, of what nature?

- What is the nature of the 'adjustments' that occur during the Cosmo-Subjective Now? Upon these adjustments the 'next Fohatic Move' is predicated.
- What is the process of change and adjustment in a non-particulate World, such as the World of Being is 'supposed' to be? Are there 'movements'? Are there limiting parameters governing such 'movements'? What is the 'Speed of Imagination' in the World of Being?
- Are there 'Moments' '*within*' the Cosmo-Subjective Now? If so, are they quantifiable or can there be an infinitude of them? Almost certainly, if there are 'moments' within the inter-moment instant', such moments are 'shorter' that ultimate Moments in the World of Fabrication.

These are profound mysteries which could lead into some of the most abstruse speculation in this entire treatise. These mysteries touch on Cosmo-Logoic Volition, Imagination, Ideation, and, in general, upon the internal adjustments, 'movements?', 'changes?' within the World of Being, which gradually compel the Cosmo-Configuration in the Mosaic World of Fabrication to conform to the Design-at-the-Beginning.

World of Adjustment
Cosmo-Subjective Nows

World of Effects
Cosmo-Objective Nows

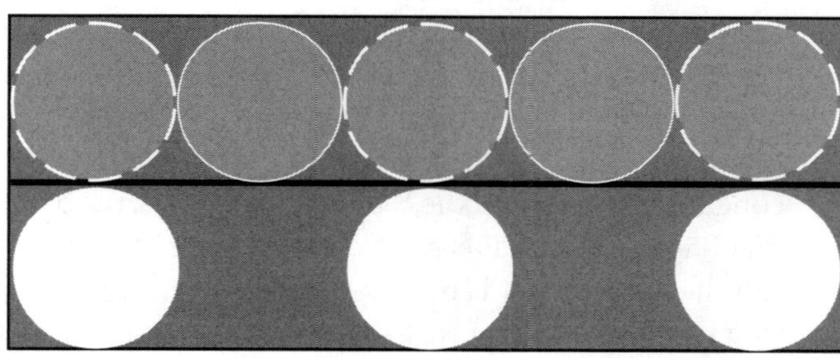

4. The Eternal Now (or Cosmic Eternal Now, [represented by the gray backgrounds in the figures above and below]) is not limited by the ultimate moment or by the Cosmo-Subjective inter-moment instant (of *readjustment*). The Eternal Now (unquantized, unsegmented) occurs *during* the ultimate moment, and also *during* the inter-moment instant, when the entire Fohatically-Fabricated Universe disappears into Self-Realizing *and* Self-*Directing* Subjectivity. The Cosmo-Objective Now may *seem* Continuous-in-Fabricated Universe, but the Eternal Now *is* Continuous in Universe (though not Continuous in the ETERNAL NOW [shown by the all-encompassing black ellipse, below, and by the black line, suggesting the Infinite Time Line] for Universes and Their Cosmic Eternal Nows constantly appear and disappear). In the Cosmo-Eternal Now, all intra-Cosmic times are One Unchanging Time. This Cosmo-

Eternal Now (as Registered by the Universal Logos) obliterates all possibility that the Cosmic-Consciousness of the Cosmic Logos will be captivated by the apparency of Sequence, and all possibility (at least on the level of the Cosmic Logos) of being *limited* by the Perception of Sequence (though the Sequences in the Cosmic Configuration are, necessarily, Cosmo-Logoically registered).

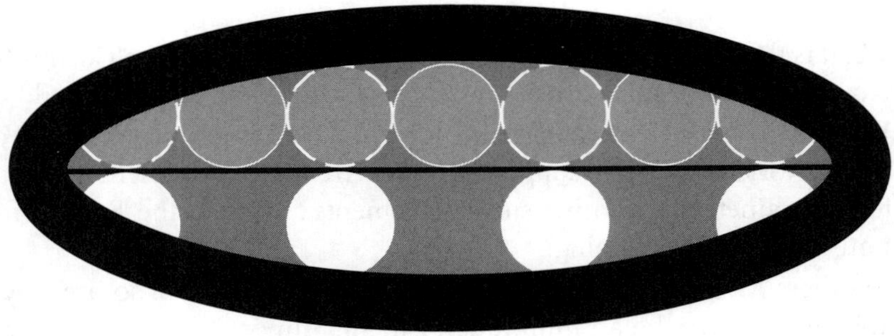

The Cosmic Eternal Now is the highest possible analog to the ultimate moment. Nothing within Fabricated Cosmos *moves* during a 'frozen' ultimate moment, and similarly, the Cosmos as a Whole does not seem to *move* during the Cosmic Eternal Now. The *Real* Ultimate Moments are Cosmic Eternal Nows—as 'frozen' in their apparent immobility, or as 'of one piece' (from the Perspective of the Universal Logos) as are the minute ultimate moments *in* 'Particulate' Cosmos. The highest degree of Eternal Nowness in sub-'SOURCE' Worlds (i.e., the Super Cosmic World) would be *visible* from the 'Infini-Spective' of the Infinite Subject/Self in Super-Cosmos. This would be the Super-Cosmic Eternal Nowness of 'Subjectobjectivity', and would relate to no-*'thing'*, thus, no Cosmos whatsoever. The Universal Logos, concentrated in contemplation within the World of Being has, probably, a species of apprehension of that Super-Cosmic Eternal Now, the depth of which or comparability to that of the Infinite Subject/Self is not ascertainable by human beings.

5. The ETERNAL NOW includes distinguishable Eternal Nows and annihilates them, collapsing them, in their infinite entirety into an *event* of infinitesimal duration and then into a *non-event of no duration at all.* The ETERNAL NOW abides only in NO-THINGNESS, and is the destroyer of all apparently separate *events.* The ETERNAL NOW is the enemy of all *time*, and even of articulated Infinite Time but not the enemy of 'TIME', ETERNAL DURATION which is, REALLY, no *time* at all.

On the Care of the Universe

There is another great Philosophical Problem: how to *care* about the *object*, how to care about the apparent 'part' when one understands its apparent ESSENTIAL un-REALITY? Perhaps the answer comes in the idea that every object or presentation in Cosmos (including the *entirety* of Cosmos Itself) *is fully the INFINITE SELF.*

Therefore, the Cosmos and all within It must be treated with *infinite* respect, just as the INFINITE SELF should be treated. On the other hand, there is (strangely) nothing <u>*less*</u> *dear* than Cosmos (and its aspects) from the 'INFINI-SPECTIVE' of the ABSOLUTE SELF, *infinitely* removed as IT IS from Cosmos (just as any infinity is from a singularity).

From the intra-Cosmic perspective, however, (which is *not* a perspective *beyond good and evil)* there are things within Cosmos of apparently greater and lesser value. Pragmatically, this *intra*-Cosmic point of view must be considered if the Beauty of the Design-at-the-Beginning is to be fulfilled, which it *must* be. The 'INFINI-SPECTIVE' of the ABSOLUTE SELF, however, could not 'CONSIDER' Cosmos thus *relatavistically.* Cosmos would be 'SEEN'/'INPERIENCED' (both words are incorrect!) as an 'EXTRUSION' of the INFINITESSENCE, a narrowing down of the INFINITE POTENTIAL to the narrowest possible Singular Point or Set of Actualizable Possibilities.

Thus, as we might expect, when considering the INFINITE SELF, there are, necessarily, contradictory points of view:

1. Cosmos *Is,* ESSENTIALLY, the INFINITE SELF and, thus, is *infinitely valuable.*
2. Cosmos *Is,* REALLY, 'less than NOTHING', a vanishing 'EXTRUSION'/Illusion, which cannot in the least ruffle the *imperturbable* HOMOGENEITY of the ABSOLUTE SELF. Cosmos, literally, "makes no *difference*", at least to IT.

It would seem the part of Wisdom to make a choice in favor of the first option. It would seem, that despite its seemingly *infinite insignificance,* Cosmos must exist for a 'GOOD REASON' (i.e., 'REASON' as it can only 'inhere' 'within' the INFINITESSENCE). Furthermore, under the dictum that "PARABRAHMAN and Samsara are ONE", the inescapability of the ABSOLUTE SELF in all apparent *things* should caution respect, and even the most solicitous care conceivable.

We must remember that there is no such thing as a *fragment* of the ABSOLUTE SELF and that that SELF, therefore, cannot be *in* anything *partially* or *BE* anything *partially.* The *absolute entirety* of the ONE AND ONLY THING is ever presented under the disguise of objectivity and, even, of under the disguise of the single, simple, humble object. This thought should be arresting in its implications.

On Problems and the Absence of Problems

Many are the musings in this treatise upon the various problems concerning the existence, nature, and implications of Time, Space, Motion (and Their Opposites) and, in general, "Life in Cosmos" and 'Life 'outside' Cosmos. Many and manifold are the apparent problems in thought which arise. From the 'PERSPECTIVE' of the 'INFINISPECTIVE' however, there are *no* problems, nor is there anything to *solve*. The 'IDEAL' WORLD (the WORLD of the ABSOLUTE) IS the REAL WORLD and IT forever substands and, even, *absorbs through identification* Duality and its problems. There are no problems when *duality* does not exist.

From the practical metaphysical point of view, therefore Space/space is an *appearance* and **8** Am *everywhere* in Cosmos at the same time and thus am *nowhere*.

- **8** have no location, because location depends upon relative position.
- **8** have no REAL intra-Cosmic position.
- As well, **8** have no location in time. Infinity before Me; Infinity behind Me; INFINITUDE PRESENT. Thus *ever* has it been, and *ever* will it be.
- In a way, **8** have ever been *located* at the same *place in time*—'NOWHERE' 'in' INFINITE DURATION.
- **8** Am, have been and will be the participant in all *possible* 'times' in lower Cosmos—Cosmo-Objective Nows.
- **8** Am, have been, and will be, the participant in all Cosmo-Subjective Nows within the World of Being from which the 'next' Cosmo-Configuration is *anticipated* and the most recent *evaluated.*
- **8** Am, have been, and will be—ever (during Universal Manvantara) abiding in the Cosmic Eternal Now—the participant in *all* possible specific times, and in the one seamless Cosmo-Eternal Moment.
- Further, **8** Am Now at all times (through the Cosmic Eternal Now) living 'in' times 'within' this-Cosmos, and (through the ETERNAL NOW) at all times *past* and even, *to come*, and further (again through the ETERNAL NOW) Am at *no time at all* (Time being negated entirely by the NOW). Every time has ever been ESSENTIALLY the *same* and will be. It has never *not been NOW*, though sometimes it has not been Cosmo-Objectively Now, Cosmo-Subjectively Now, or Cosmo-Eternally Now.
- Further, **8**, in-Cosmos, have never experienced anything but Nowness or Eternal Nowness, and, in *general*, have never 'inperienced' anything but NOWNESS. **8** have without cessation 'abided in' a 'positionless-position' within the NOW. Thus, according to Radical Infinitism, are the multiple 'inperiences' of Time pertaining to a human being.

On Whether 8 Exist Now or NOW in the Future

The question will naturally arise: If **8** exist NOW at all possible 'times', do **8** exist in the *future?* First, we must answer whether the future exists! From one perspective, there has never been a *past* as distinct from a present. We have seen:

- How from the 'Infinispectivizing' of the Infinitizing Subject/Self, all moments past (whether Cosmos/Events or Intervals between Cosmos/Events) have tended to become (no matter how great their *normally perceived* duration) infinitesimalizing in duration;
- How such Events *all* become one infinitesimalizing Event (the entire infinitude of them); and,
- How if one adds an infinitude of infinitesimalizings, the result (strangely) is no more than the *infinitesimalizing.*

Viewed in another way, if we add more and more of an unspecifiable and ever-lessening less and less we certainly do not arrive at everythingness (infinity), but hold a 'steady state' at *infinitesimality,* just as the addition of an infinitude of infinities is none *other* than infinity (indefinite and unspecifiable as ever, and forever incapable of augmentation). All of this is important because:

- ♦ Through understanding the infinitesimal (*truly,* though more awkwardly, the 'infinitesimalizing') we have established the 'Infinispectivizingly'-Cognized identicalness of all time past. All time past has become (from this Perspective) one single Event that hovers at the *vanishing point of NOTHINGNESS.*

It may be argued that there is a natural limit to the diminishment of the perceived duration of Cosmos/Events and Inter-Cosmic Intervals because time in lower Cosmos in quantized, and no duration in-Cosmos can be reduced in duration so as to take less time than an 'inviolable' ultimate moment. From this Perspective, each Cosmos/Event and each Inter-Cosmic Interval between Cosmos/Events could not be reduced below the value of an ultimate moment, in which case the result of such a reduction (for an infinitude of Cosmoses and Inter-Cosmic Intervals) would be an infinitude of ultimate moments (which are *actual quantities)* and would sum to *infinity* itself, thus making the virtual *simultanization* of all E/events along the Infinite Time Line impossible. Time would not 'collapse'.

Continuing along this line of thought, there are *probably* limits to how long the entire Cosmos/Event can endure in terms of relative length, and there are certain limits as to how relatively short an ultimate moment may be. (Parenthetically, if there were limits on the side of *length* of endurance, it *might* be a prescription for Cosmic Failure, i.e., the non-fulfillment of the Design-at-the-Beginning "in time" {or in the time cosmically allowed}.)

Given that there may be limits on the Cosmos/Event (possibly also expressed as limits upon the possible diminishment of cosmic units of time), it stands to reason that there could be limits upon the "Seven Eternities" that intervene 'between' Cosmos/Events. This is not a foregone conclusion (for who *in* Cosmos can "tell the tale"?), but it is a reasonable inference. In all the metaphysical literature there seems to be a reasonable parity between the duration of a Cosmos/Event and the duration of an Intra-Cosmic Interval between such Cosmos/Events.

At any rate, the point to grasp here is that no matter how long any of these Great Cosmic Events (or non-Events) might endure, they would *never normally* endure (so our Cosmology and normal experience seem to indicate) only for an *infinitesimalizing* (i.e., for an unspecifiable 'unit' of Time as close to nothing as possible) but, instead, for a *finite* unit of time, however relatively long or short. When you sum an infinitude of definite *finites*, you get an *infinite*, and this would destroy the model that negates the REALITY of time past. An infinitude of indefinite *infinitesimals* converging upon zero would *not* destroy the model. How can we approach this problem?

There is a flaw in this reasoning that destroys the argument against the Negation of Time, and this flaw concerns *perspective*. If the Perspective being used were an intra-Cosmic One, that of the Universal Logos, there would be basis to this argument, for intra-Cosmically certain Cosmic Parameters are maintained, and there are inviolable *limits* (of *perception)*. From that intra-Cosmic Perspective, units of time cannot be 'compressed' beyond a certain relative brevity. The perceived duration of time however, clearly alters with altering *perspective*. The normal invariability of certain standard Cosmic units of time is consistent with the nature of the sustained Self-'Sight' of the Universal Logos which brings forth a Cosmos of *finite* dimension. The perception of Time, however, varies with the nature and position of the viewer.

In this way, Infinitistic Metaphysics correlates (to a degree) with modern Relativity Theory. Suppose that the Self-'Sight' of the Universal Logos is altered through what we have been calling *infinispectivization*, becoming, instead of a 'cosmo-conventional' Perspective, the 'Infinispectivizing' Perspective of the Infinitizing Subject/Self. Then, the Universal Logos would be 'returned' to the nascent Super-Cosmic Development It once *possessed* before It was 'telescoped' 'down' and 'out' of Pre-Cosmic Infinified Selfhood and De-Infinitizing Selfhood. In fact, *Essentially*, the Universal Logos *possesses* this Infinispectivizing even *now*, due to the Principle of Emanative Retention. Because of this, no matter how great or small the Cosmic Events and Inter-Cosmic Events being 'Seen', they would (perceptually) *become infinitesimalizings* through 'Infinispectivizing'. From this Perspective, units of Time of any kind would seem to have no REAL duration.

- The duration of a so-called 'Unit of Time' is purely a *product of perception and perspective*, and is an artifact of the nature, quality and 'Point of Observation' of the *observer.*

With the possibility of reducing and eventually negating the *reality* of time past, what can be said of "time future"? The past exists (correct?) but the future, as usually conceived does *not*—yet. Just as we have, through 'Infinispectivizing' negated the specificity of the past (and *imaginatively/perceptually* reduced it all to one infinitesimalizing thing, or a thingness converging on zero), so perhaps this approach is possible with the future.

A problem arises. We know by inference that the future will occur. We are assuming that the Great Breath will continue to 'Breathe' and that Cosmoses will continue to appear and disappear as they always have. We are, however, entirely unable to predict the *specificity* of the future, for that is a 'DETERMINATION' (hopeless to understand) 'UNDERTAKEN' 'within' the FOUNT OF ALL POSSIBILITY.

If however we have negated the specificity of the past and past units of time, through imaginative 'Infinispectivizing' and its corollary, imaginative 'Infinitesimalization', is it equally legitimate to negate the specificity of the future, and to reduce all future possible duration to an *infinitesimalizing?*

From the Infinispectivizing of an Infinitizing Subject (in Recession), we can also imaginatively collapse all *future* Cosmos, Cosmic Events, and Inter-Cosmic-Events into an infinitesimalizing, (just as we did for past Events and Intervals). Thus we can (imaginatively) say that the future (as extension) is *impossible* because future units of times would also be *infinitesimalizings,* which cannot produce measurable extension in time through addition. Therefore, (from one strange perspective) the so-called future can never be 'reached' by the March of Time, and, in fact, all future units of time are not so much converging upon the zero point (which is forever NOW), but are incapable of *extending* <u>*from*</u> the zero point (because they have no *measurable duration* with which to extend).

From this perspective, we negate the future by removing the 'temporal substance', as it were, from which the future can be generated. Customarily, *illusorily,* the Now Point *seems* ever to 'move': a past is "left behind" and a future is "entered". However, by Infinispectivizing analysis of that illusory past and that illusory future, we discover that such a penultimate, *Infinispectivizing* Perspective 'robs' them of 'temporal extensibility' (but not quite of their existence).

- 'Past' and 'future' are still "constructs of consciousness" but (as the Buddhists might say) they are "empty of own-being"! This being the case, if we look deeply enough, we may discover that the zero point (the *absolute present)* is not Really 'moving at all' and has *never* moved. It simply maintains a 'presence' as the 'present' in order to maintain the Illusion of Objectivity. Otherwise, if TRUTH 'ASSERTED' ITSELF all *objectivity* and *temporality* would immediately disappear (simply because they never *had even appeared)!*

By depriving the past and future of duration, i.e., extension in time (albeit from the Perspective of Infinispectivizing, *not* from the Perspective of any Universal Logos), we have established the REALITY of the ETERNAL NOW as a *substratum* in all possible events forever. Thus from as Infinispectivizing, the past cannot exist and neither can the future—only and ever the NOW (or less radically, the Infinitesimalizing Now-as-NOW). We have deprived the future of the *time* in which to occur, i.e., time as 'temporal extension'.

Even the durable present moment could be utterly REAL, and if the 'spective' used were the 'INFINISPECTIVE' instead of that approximation to the Infinispective we are calling 'Infinispectivizing', for then the present moment or any moment would vanish into *nothingness,* into the NOTHING, and *not exist at all.*

The dynamics of the Infinite Time Line are beginning to look very interesting, and stranger than strange, from the Perspective of the Infinispectivizing Infinitizing Subject/ Self. We have a future that *cannot* exist because there is no Real Time to provide for its existence; we have a past which cannot *have* existed because all time within it (when perceived from a rapidly Infinitizing Temporal Span) summed to an infinitesimalizing duration converging upon zero, and in fact, 'INFINISPECTIVELY' summed exactly to zero, and, hence, never existed at all.

From yet another perspective, since the past was (once) the future, whatever is said of the future *now* must be said of the past *then*. For every past, when it was an anticipated future, there was (at that 'time') no 'temporal extension' to provide for it to *exist*. So the past, which we deduce as not having existed (past time being an infinitesimalizing converging on zero or non-existence) cannot have occurred for two reasons:

1. Because its total time value is merely an infinitesimalizing.
2. Because the past (when it was the future) cannot Really have occurred, there having been no 'time' available (at that 'time'!) to provide the 'temporal substance' necessary to create or generate 'temporal extension' *as* a future.

What of the present? We have a present which both *is* and *is not*. The present ('Infinispectivizingly', and *not* as *perceived* by the Universal Logos) is of infinitesimalizing duration, if it exists at all. We have already established:

- That the past cannot Really exist.
- That the total sum of past time converges on zero.
- From the ULTIMATE INFINISPECTIVE the past *becomes* zero.

The same is true for the Future. All future possible time never extends beyond the infinitesimalizing and in fact is swallowed by zero duration *before it emerges*. But, think of it, every present is both a past and a future, and if neither a past or future could exist, a present cannot exist either. These thoughts are simply intended to open a wide-ranging discussion on these perplexing matters.

Thus we have a situation in which *any* possible time is negated and, therefore, *Time* Itself is negated. Every possible moment along the Infinite Time Line is either Infinispectivizingly reduced to an infinitesimalizing continuing to converge upon zero, or 'INFINITESIMALLY' reduced to absolute *nothingness*. Infinispectivizing 'allows' the World of Illusion a kind of fleeting, marginal, un-REAL existence/presence 'hovering' on the 'ABYSS of NOTHINGNESS'. This is practical and useful, for, from our very limited intra-Cosmic perspectives, we *think* we *know* it exists, and for the seeming 'temporal span' of our boundedness in Cosmos, we will at certain 'times' continue to think so.

The 'INFINISPECTIVE', however, literally "takes everything away". We are left bereft with nothing but NOTHING, nothing but OURSELVES-the-SELF. Anything we thought to have *ever* 'happened' we understand as *not* REALLY having 'happened' at all. Our ABSOLUTENESS is proven forever inviolable. And yet the World exists! Or *does* It?

What shall we *do* with the REAL, *absence*, the un-REALITY of Time? Would accepting such a thought upset our lives as lived within the Veil of Illusion. Perhaps we can attempt the following:

- ♦ Apply the REAL *absence of Time* to the *illusory presence of Time* in such a way that *every illusion becomes a REALITY*.

Thus, do we infinitize the present moment and everything about the World of Illusion, restoring that World, in our consciousnesses, to the STATELESS 'STATE' in which It has *ever been?* Do we not wish to know that we are *living* and *being* the REAL which we *are ever and always living and being?* This will "make all the difference" in a life in which a *difference* does not REALLY have to be made, because there cannot be a *difference!*

On Vacuum/Voidness/Plenum

The VACUUM/VOIDNESS/PLENUM IS *REALITY ITSELF.* The VACUUM might be called the *infinite density* of BE-NESS, infinitely *void* of un-REALITY. The VACUUM is a *vacuum* (a perceived emptiness) only with respect to *things,* but not a *vacuum* with respect to ESSENCE (in relation to which IT is an ABSOLUTE FULLNESS). The VACUUM is the REAL 'EMPTINESS' of *things.* The VACUUM deprives all things of the illusion of substantial REALITY. All states of non-VACUUM are illusory. A 'vacuum', however, is not *the* VACUUM. A 'vacuum' is a 'less-than-NOTHING', an illusory interval of nullity in the PLENUM.

Therefore, 'vacuums' are 'holes' in ABSOLUTE SPACE, infinite reductions of ABSOLUTE SPACE, infinite reductions of ABSOLUTENESS, 'finities'. The VACUUM is REAL. All other vacuums are un-REAL.

On the Subject

- The Subject (in Cosmos) is the One Experiencer/'Inperiencer', *emanatorily extended* as many S/subjects.
- The SUBJECT is the only REALITY.
- The Subject is the Universal Logos, as Representative of the SELF.
- The SUBJECT/Subject is the **I/8** in Pre-Cosmos.
- The SUBJECT-as-Subject Is the **I/8**-becoming-**8** in Cosmos.
- The SUBJECT-as-Subject is also That Identity to which all *objects* are *subject.* (If you subject someone to something, you, the 'king', as it were, assert yourself and achieve primacy.)
- The SUBJECT is the ONE WHO IS, the ONE who *forever* IS. The Subject, the Universal Logos, is the One Who cyclically Is.
- The SUBJECT is the *immortal, unchanging* ONE. The SUBJECT is the PERMANENT ONE.
- The SUBJECT-as-Subject is the Experiencer/'Inperiencer' of All, or of part, depending on the state of encapsulation by prakriti, which means depending upon the extent and depth of Its Self-Sight.
- In TRUTH, the SUBJECT-as-Subject *is* REALLY, non other than the SELF. The SUBJECT is the ABSOLUTENESS, ITSELF.
- The SUBJECT-as-Subject is the Experiencer, the 'Inperiencer', the Participator, the Pervader, the Knower, the Center found at every Point-in-Cosmos (which, of course, is Really, but One Point).
- The SUBJECT-as-Subject Itself *is* the Point, the Condensed Point.
- The SUBJECT is the MOTIONLESS ONE behind all motion.
- The SUBJECT is THAT which naught escapes.

- The SUBJECT is THAT to which all things refer.
- The SUBJECT is the REAL, the PERMANENT, the ULTIMATE KING.

The SUBJECT 'needs' the Illusion (the 'hole-y' Illusion), perhaps, of that which is less. [See The Problem of Evil in Section II.] The Universe is less. The Universe is thus subject to the SUBJECT-as-Subject. The Universe is the Great Singular Object that objected to the SUBJECT-as-Subject.

On the Object

What is the Object? The Object is that which is perceived by the Subject. The Object is that which is illusory, limited,and evanescent. The Object is the result of the Cosmic Process, which is based upon Dualism, but *objectness* is found even in the Pre-Cosmic Process, and, if traced to its root, 'ARISES' with the 'FLASHING FORTH' of the 'RAY' of the ABSOLUTE, at which time the EVANESCENT INFINITE SUBJECT-instantly-Infinite Subject, and the EVANESCENT INFINITE OBJECT-instantly-Infinite Object 'ARISE'. Duality and the Appearance of the Object are coeval, coetaneous.

◆ The 'OBJECT'-instantly-Object is, as it were, *That* which is 'thrown *out*' from the INFINITE SUBJECT, but in the 'throwing' ('RADIATING') the INFINITE SUBJECT ITSELF is *simultaneously* 'EXTRUDED' as the Infinite Subject (for They are inseparable), and in order to 'balance' the 'EXCLUDED' Object.

There is, however, an Eternal, Essential Identicalness between the Subject and Its derivative Object (for the Father *Is* the Mother and the Mother *Is* the Father). The Object, in Itself, is Secondary, and is naught but the Subject. The Object is an Extrusion of the Subject. The entire Universe is *subject* to the Will of the **I-as-'I/8'-as-8**, the Will of the SELF, My Self (the Self {**8**} being the SELF-in-Universe). ESSENTIAL-**I**-MYSELF, AM subject to none, but **I** MYSELF. **I** AM completely 'SPONTANEOUS' within the NOW, the ETERNAL NOW—have *been* and ever *will be*, at all 'times'. As well, **I/8** Am free within the Pre-Cosmic Infinitude, but under MY OWN 'INSTRUCTIONS' to Finitize. **8** in Cosmos am free to follow MY SELF-'EXTRUDED' 'DESIGN' that has become, through finitization and specification, the Design-at-the-Beginning.

An *object* is ever reified, a 'finity', *'thinged'*. An *object* is ever limited. (Perhaps there is one exception, with respect to Mulaprakriti, which we have named the Infinite Object, but, in a way, It is also a Non-Object, if Objects must have boundaries.) Even Mulaprakriti, however, is a state of *limitation* compared to PARABRAHMAN. Perhaps Mulaprakriti should be called the 'Infinite Finitude' or the 'Infinite Finity'. It is absolutely unbounded *but*, It is *registrable*, which the INFINITE SELF is *not*, PER SE. Mulaprakriti is susceptible to apprehension by Infinite Consciousness and De-Infinitizing Consciousness, which means that It is a 'Creation' of 'MAYA'-instantly-Maya.

◆ The INFINITE SELF, on the other hand, is *not* susceptible to apprehension by Consciousness, and 'when', mysteriously, IT *does* 'BECOME' instantaneously 'SUSCEPTIBLE' to 'MAYA'-as-'CONSCIOUSNESS' ('ALLOWING' to 'ARISE' 'within' ITSELF the 'EVANESCENT INFINITE TRINITY'), IT just as *instantly* 'CEASES' to be 'SUSCEPTIBLE' and 'CASTS OUT' the 'EVANESCENT INFINITE TRINITY' into the Pre-Cosmic State, through which 'HAPPENING' Mulaprakriti as the Infinite Object Flashes into Existence.

Objectlessness occurs when the Infinite Subject (the SUBJECT-in-Pre-Cosmos) Realizes Itself to be all pervasive. The Infinite Father/Infinite Subject/Infinite Self Realizes Its Identity with the Infinite Mother/Infinite Object/Infinite 'Other'. This Realization, existing at the Beginning of Cosmos, is sustained in Super-Cosmos throughout Cosmos. During Cosmos the attenuated reflection of this Infini-Relation is sustained only in that stratum of Cosmos called the World of Being. 'Following' Cosmos, the Infinite 'Subjectobjective' Relationship again prevails undistractedly (whether for an infinitesimalizing or longer).

It has been the Will of the SUBJECT-as-Subject to 'Create' those limitations in Consciousness called Objects. Consciousness Itself (being 'MAYA'-as-Maya) 'Creates' those limitations. 'MAYA'-instantly-Maya is the Principle of *Relation*, and Relation Finitizes. Objects are the out-pictured manifested *infinitessentially implicate* variety of the INFINITY-of-INFINITIES, namely the ALL-SELF, the ONE-AND-ONLY-INFINITE-SUBJECTIVITY. Objects (in-Cosmos) disappear (to the illumined understanding) when the SUBJECT-as-Subject (the Universal Logos) is found at every Point in Cosmic Space, Cosmic Prakriti. What this means is that the Universal Logos Knows Itself to be *ubiquitous* within the Mother, by achieving total Identification with Itself, total Self-Knowledge, total Self-Pervasion.

- The one who learns to achieve Identification with that Logos, shares in the Realization and Being of the Logos. The normal, unenlightened relation between Subject and Object is one of difference and non-identification, but the Consciousness of Duality is overcome through Identification.

On the Object as Subject

From the 'INFINISPECTIVE' every Subject is an Object. The only REAL *subject* per se, is the INFINITE SUBJECTIVITY, BE-NESS. But can an Object be a Subject? That is, can an Object have Consciousness? Well, a Subject that is an Object *is* a Subject. The Son, Who is an Object to His Father, is also a Subject to other lesser lives.

Is the Universal Mother, the Ultimate Cosmic Object, also a Subject? The Universal Mother is the Father as 'Seen' by Himself. If the Mother, then, *Is* Really the Father, and the Father *Is* the Ultimate Cosmic Subject, then the Mother too, Essentially *being* the Father, must be a Subject. This idea restores Life to the Mother. This means the Mother, because She *Is* the Father, can also Know as She is *Known*. The Father's Knowingness, then, is Present in all Objectification. If the Father is *in* the Mother, is the Mother *in* the Father? If the Father, 'Seeing' Himself Objectively, *Is* the Mother, then is the Mother, 'Seeing' Herself Subjectively, the Father? Is every Subject and Object and every Object a Subject?

- Are Subjects and Objects 'Creatures', like Janus, with *two faces?* It would stand to reason, and gives the necessary reflexivity needed for Universal Symmetry. The Father 'Seeing' Himself as the Mother, 'Sees' His Own Objectivity. The Mother, 'Seeing' Herself as the Father, 'Sees' Her Own Subjectivity. Each 'Sees' in the Other what is latent in Itself. Of course Each *is* the Other, and Really, They are only One Being.

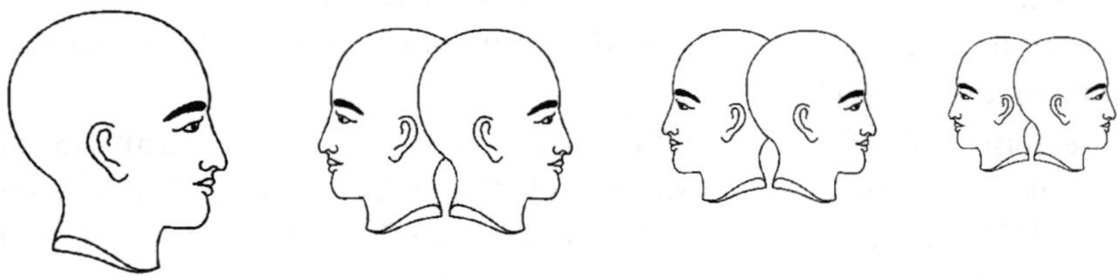

Series of Subject-Objects

On Division

Division is the means by which one thing is separated from another, or by which a thing is separated into parts. There can be no division without the possibility of extension.

- Extension and division are coeval, meaning that they arise together. Extension cannot exist without the process of dividing the One into the Many, which is another way of saying, extension cannot exist without objectively expanding the subjective Condensed Point.
  - It could be asked whether there is any extension within the Universal Logos-as-Condensed Point *before* the Process of Emanation begins?
  - Or further, whether there is any extension within the World of Being, altogether? Is an image in consciousness 'extense'?
  - Can a 'hidden' 'internal' 'Son' of the Father (simply because it comes into Logoic Recognition) be considered 'extended' simply because It has become Objective (i.e., Logoically Self-'Seen')?
  - Or must the arising of the illusion of extension await the Fabricating of the World of Approximation by Fohat? So much would depend upon the process of Object-Recognition by Subjects within the World of Being.

- Is such Recognition imagistically based, or can there be Objectification without the apparent extensiveness of Image? Certainly, extension is an illusion, and all objectification is ESSENTIALLY an illusion, but the two are not thereby mutually intersecting.

Whereas the 'RADIATION' or the 'RAY' of the ABSOLUTE (the 'POINT') 'within' the ALL-IN-ALLNESS 'CREATES'-instantly-Creates the 'appearance'/registration of Infinite Space, and then (eventually or immediately) division (considered as a process of Emanation, and hence closely allied to multiplication) of Cosmic Space (Cosmic Space in this case being considered as the Field of Expression and Objectification of the Universal Logos), creates *actual* not just *potential* Space. Every 'son' arising in the Potentiality of Cosmic Space (Cosmic Prakriti) is an *actuality*, and moreover, a 'hole' or 'vacuum' in that Space. Cosmic Space is already a finite, though vastly extensive Object—vast, that is, from the later intra-Cosmic Perspective.

The Act of Separation/Division is responsible for the appearance of increasingly Objectified Space and its derivatives—i.e., 'spaces'. Objects are 'spaces' in Space. At the End of the Great Cosmic Process, there is an end of the state of separation initiated by the Process Emanation at the Dawn of the *Actual* Universal Manvantara (though nothing but *reabsorption* into THAT will end the state of 'SEPARATION').

With the end of apparent separation in-Cosmos (achieve through bringing about the end of Twoness, and the Manyness born of Twoness), comes the end of Space-as-spaces, and the return to Pure Cosmic Subjectivity and Pure Cosmic Objectivity (Pure Cosmic Space, Cosmo-Spirit/Matter), plus the Resounding Note of Achievement (the 'Game' *well played!*). This ending of separation is consummated (at least Universally), just prior to the Universal Pralaya.

◆ Does the Returning Universal Logos, being reabsorbed into the Re-Infinitizing Subject, and thence into the Infinite Subject, and thence into the INFINITE SELF, have aught to *offer* the INFINITE SELF which It ESSENTIALLY IS? Truly, the INFINITE SELF, BEING PERFECT, 'NEEDS' *naught*. IT *already* IS all things and all possibilities. IT IS *not*, however, *all possibilities objectified*. The Return of the Universal Logos along the Line of Post-Cosmic Retraction, signals one more possibility 'inhering' within the FOUNT OF ALL POSSIBILITY that no longer 'NEEDS' to be *objectified*.

If all possibilities 'inhere' within the FOUNT OF ALL POSSIBILITY, why not the possibility that all things demanding *objectification* are *already* 'OBJECTIFIED'? In a sense we could say that 'OBJECTIFICATION' has 'inhered' forever 'within' the INFINITE SELF, because forever, cyclically, there has *been* 'OBJECTIFICATION', and there is none 'other' that the INFINITE SELF 'in' which that 'OBJECTIFICATION' can 'OCCUR'.

We come to many, many problems with this not-quite-satisfactory answer. We are touching on 'HOW' the INFINITE SELF, 'WORKS', and if IT 'WORKS' in *one way*, why can IT not ('BEING' as IT IS, the FOUNT OF ALL POSSIBILITY) 'WORK' in *ALL* WAYS? Is there LAW 'above' WILL? Is the INFINITE SELF in any way *compelled* or SELF-COMPELLED? (The author will wisely leave some of these questions for further discussion in other parts of the treatise, or will evade them altogether!)

- In thinking about REALITY, we must learn how to disabuse our minds of the tyrannical effect of certain categories of conception that are very stubborn and which, being apparently commonsensical, usually refuse to yield to the pressure of radical thought, even though radical thought must, eventually, wear them away. Categories such as Time, Space, Motion, Sequence, Cause and Effect. These Categories are the currency of the Mind; for the moment we would be mentally bankrupt without them. One day, they will have to go.

According to the Tibetan, there is one great heresy, the "Heresy of Separateness". This Heresy is rooted in the misperception that *identicalness* is not REAL. In one way, in all Cosmos, we are but interacting with 'OurSelf'-as-the One Self, 'playing' with 'OurSelf'-the-One Self, playing the 'only *Game* there is' with OurSelf-as-selves.

- Division is the means by which the Universe appears in its objectified multiplicity.
- Division is the Universal Process of making *less* by apparently making *more,* for the parts do *increase* even as their potency and effectiveness decrease.
- Division, then, in the generating of more items, is actually creating a 'distance' in quality and potency from the Reality of the unitary Universal Logos (the One Universal Being, Who emerged as the end result of the 'devolution' of the Triple Point—Infinified, Condensing and Condensed.

An important question arises: Is division REALLY possible? Division is ESSENTIALLY *illusory* because when THAT (or ITS Agents by 'RADIATION' or, later, by Emanation) is/are apparently 'DIVIDED', IT, nevertheless, *ever* 'REMAINS' just as IT IS—not just ITSELF, but *unreduced* and *entirely* ITSELF.

When division and its results are considered from the point of view of the normal limitations of human consciousness, then *division* is seen to be the cause of difference. Due to the Law of Unrepeatability in Cosmos, to *divide* and to *differentiate* are identical acts. The Universe was 'Created' *apparently,* through division and differentiation, which we call Enumeration and Entification.

Division is the means of *veiling.* When a thing is veiled, however, it does not REALLY, ESSENTIALLY, cease to be itself, and when the 'INDIVISIBLE' is *apparently* divided into the FIRST 'EVANESCENT INFINITE TRINITY', IT does not ever cease to be the 'INDIVISIBLE'. Division is 'MAYAVIC'-instantly-Mayavic and occurs through that *infinitesimalization* of 'ABSOLUTE CONSCIOUSNESS'—instantly—Consciousness, which we call Pre-Cosmic Consciousness leading to *Cosmic* Consciousness.

In speaking of the impossibility of the act of division/subdivision to alter the ESSENCE, can, for instance, the Number One Really be altered as It is divided, through Emanation? Twoness is created out of Oneness, but does Oneness vanish in the process? Not Really.

- Is there in fact *ever* a loss of the Number One, the Universal Monad?
- As the Number Four is created, is there, in fact, a loss of the Number Three, the Number Two, or of the Number One?

No, the superior, i.e., the smaller Number, ever retains its *complete integrity* in the hierarchical Pyramid of Numbers. With every division, however, there is a *veiling,* and with every subsequent division a further veiling.

This process continues until the limit of multiplicity is reached—i.e., the limit of multiplicity that is possible for any system being subjected to the process of subdivision. At that stage the maximum possible veiling for that system has eventuated. Psychologically and spiritually, ignorance is at its deepest in the most (apparently) subdivided and veiled units.

As evolution proceeds, and the veils lift one by one (which means, as the ascending Subjects 'returning' to the One 'See' themselves more and more as They Really *are),* the *smallest* Integers (i.e., the Great Archetypal Numbers) are reapproached. The unveiling process leads out of darkness, ever closer to the Unity and Supernal Light of the Cosmic Origin. Multiplicity/complexity is a greatly veiled and bewildering state, a state of deep prakritic immersion. The immersed Subject 'Sees' a very limited image of Itself as Object. Self *knowledge* is limited and fallacious, and Self-*being* impotent. With every division or Emanation a new veil (upon Self-Realized Selfhood) descends.

But, is That which deeply *Is* and *Knows*, actually *veiled?* Does the Knowing cease during the dividing and veiling process? The Knowing may *seem* to cease, but, at some level, it never can *Really* do so. The Spirit cannot *Really* be divided, and It remains entirely untouched, Essentially, by this greatest of all Mayavic Processes.

On the Descent of Subjects and Objects

Any *individual* (the true SPIRIT-as-Spirit) can properly and ESSENTIALLY be considered an *'indivisible'*, a 'Ray' of the INDIVISIBLE. Truly Pre-Cosmic **I/8**, as a 'RADIATION' of the INDIVISIBLE SELF Am, like My SOURCE, *indivisible* and am *therefore equally found in all units and throughout the apparently extended and subdivided Cosmos* and even *Cosmoses.*

The First 'DIVISION' (intra-'SOURCE') 'GENERATES' the 'EVANESCENT INFINITE SUBJECT' and (be it *simultaneously* or following with *infinitesimalizing instantaneity)* the 'EVANESCENT INFINITE OBJECT' (as well as the 'MAYAVIC' 'RELATION BETWEEN'). These three upon 'ARISING' (for practical purposes) *instantly* 'BECOME' the corresponding Infinite Trinity in the World of Pre-Cosmos (which, too, Arises either simultaneously, or in the most rapid conceivable sequence).

Let us remember, however, that intra-SOURCE, SUBJECT/OBJECT are a ONENESS which is neither, and that Pre-Cosmically and Cosmically the Subject ever remains the Object. The Subject, although primary in Time and Space, is useless to Cosmos unless It can 'See' Itself as Object. The SUBJECT-as-Subject-as-subject does not cease to be the OBJECT-as-Object-as-object just because on the lower levels of consciousness there appears to be a distinction.

- Ever according to the Law of Emanative Retention, the superior cognizance is preserved even as the inferior cognizance appears. So all levels of consciousness are maintained in their full integrity all along the Ladder of Descending Emanation.

On Separation

What is *separation?* Is separation REALLY possible? Separation only *seems* possible through SELF-'BLINDING'-as-Self-Blinding. Separation is *apparently* possible within the World of Illusion where **I**-the-WHOLE voluntarily (according to MY 'WILL'-instantly-Will) participate in illusory SELF-Fragmentation, even while **I** *simultaneously* 'ABIDE' in the 'IMPERTURBABLE CONTINUITY' of MY IMPARTITE ALL-IN-ALLNESS. **I-as-8** Am present, wholly, through all points of Cosmos. **I-as-8** Pervade Cosmos *entirely.* There can be no REAL *fragment* of MYSELF-as-Self, because the SELF (the ULTIMATE ROOT) is, fundamentally, *indivisible.* **I-as-8** Am wholly *here* and, *likewise*, **I-as-I** AM *wholly,* not just *within* the ALL-IN-ALLNESS, not just wholly *identified* with IT, but **I** wholly *AM* the ALL-IN-ALLNESS. **I** AM IT *entirely.*

Let us examine the idea of Separation closely. Separation, like its son, Division, is also an impossibility. (Separation may be considered the result of the process of Division.) Separation requires extension and requires time. In the normal flow of life (outside the perceptual province of the Cosmic Eternal Now), many events are apparently separated from each other, as they occur at apparently different 'times'. Under the regime of the apparent discontinuity of Space/space, objects are apparently separated from each other *in extension*, as two things cannot occupy *exactly* the same 'space' at the same 'time'. Under the illusion of discontinuity, no two points in space coincide exactly. Separation like Division is the one means by which Cosmos (as it *seems* to be) is created/generated. It is almost as if Cosmos Itself, the Child of Division/Separation, *Is* the Great Heresy Yet the 'ACT' of 'BECOMING' and 'GENERATING' Cosmoses is what **I**-as-**8**-as-We *do* and have ever done, and apparently ever will do.

The following interpretation of Separation begins to dawn:

- that Separation is actually *localization* with respect to another—the localization can be spatial or temporal
- that separation is Enumeration
- that Separation is created through Emanation
- that Separation is *becoming visible to Oneself as an object*
- that Separation is Objectification
- that Separation is the beginning of sensing/viewing
- that Separation is the beginning of Consciousness of an Object

Consciousness (facilitated through Separation, or 'FLASHING' 'CONSCIOUSNESS' as the chief 'MAYAVIC' 'CAUSE' of 'SEPARATION') *is the sensing of another, which other is but the Emanation of Oneself.* Each item-in-Universe is but the Emanation of the One SELF-as-Self, which each item-in-Universe *is* *<u>entirely</u>,* not ever *partially.*

Paradoxically, the Whole is ever *fully* the 'part'. Consciousness-in-Universe is caused by 'going forth' into *apparent* extension, going forth into Self-Objectivity. Consciousness-in-Universe is the act of *<u>seeing</u>* Oneself-as-*other* instead of, simply, *being* Oneself (and thus, automatically *being* an *unseen other* as well. *Consciousness* is *not <u>being</u>.*

- ♦ Separation makes possible the beginning of the senses. If there were identity or identicalness-at-a-point (two items-in-Universe occupying exactly identical points in space), there would be no need for the senses, for *being* the 'other', through identification, would take the place of seeing, feeling, touching, smelling etc., directed towards that *apparently* separated 'other'. All of those senses relate the SELF-as-Self to the *other,* and when there is no other, there is no need for the senses.

So *being* the other synthesizes (and negates the necessity for) all the senses. Through the process of Separation, it is possible to see Oneself as an *Emanated Object,* in as much as the *other* that is Emanated is, Essentially but Oneself. The Emanated *other* can then reflexively *see* Itself as the Emanating Point. Thus the Self (the Emanator) sees Itself as the Other (the Emanated) which 'Sees' Itself, not only as Itself, but as the One who sent It forth—i.e., the Self (the Emanator). Emanating and emanated Selves become *mirrors* to each other.

Alternatively considered, Separation (as it bears upon Consciousness-in-Universe) is the 'creation of *two points of view*'. Separation (Emanatorily considered) brings the ending of the *solitary* Point of View (for instance, of the Condensed Point), which is no Really *single* Point of View at all, since the Point (from which viewing proceeds) is Really, pervasive, in all places simultaneously—everywhere in Cosmic Space and thus nowhere in particular. Separation is the inauguration of location. Separation, then, (in relation to Consciousness in-Cosmos—where Emanation Really begins) is the apparent Division of SELF-as-Self, by creating alternative Points of View thus resulting in many S/selves (the greater ones of which could be called the SELF-as-Self, and the lesser, the SELF-as-Self-as-self).

- ◆ As the Process of Separation (the Process of Emanatory Self-Division) continues, more and more points of view (apertures of sensation) are created, finally arriving at the virtually numberless points of view that characterize the apparently fragmented Universal Consciousness in Cosmos. Thus results the variety of perspectives or ways of seeing things.

Thus, we see that, apparent Separation/Division/Differentiation offers the possibility of the SELF-as-Self 'Seeing' Itself as an *object.* Before Separation, there is only the SUBJECT-as-Subject (the Universal Logos, the Condensed Point) and that One can only experience Itself as the One-as-*one.* (It must be noted that 'Separation' was required even to produce a Universal Logos.) Within Cosmos, however, *before* intra-Cosmic Separation begins, the Universal Logos Self 'Sees' Its Own Oneness as the Oneness of Cosmic Prakriti. There is naught but Oneness in Universal Consciousness. Because this Self-'Seeing' requires Consciousness, it could be said that Separation exists even at this fundamental point, but the interplay is so 'seamless', that the Three are Really One. This experience of Oneness is not the same as that limited condition called seeing (sensing) Oneself in *diversified otherness.* Separation is the 'distancing' of the SELF-as-Self from the state of *being* and instead, the descending into the state of *seeing* (a terms which here is made to represent all the senses).

Thus the act of Separation represents the first movement towards the *illusion of non-identicalness.* As Separation proceeds, there is more and more with which the newly-separated-Self-of-the-Self senses itself to be *non-identical.* The Entity called the Number Two, for instance, senses Itself to be non-identical with the greater and more inclusive Entity called Number One. The Number Three senses itself to be non-identical with the

Number One and with Number Two. The Number Four finds even more points of non-identicalness (non-identity), and on it goes as we descend upon the Divine Emanatory Stream.

In short, then, Separation makes *seeing* possible; Non-separation makes *being* possible. Again, 'seeing' stands for the action of *all* the senses. Another way to consider this thought is that Non-Separation potentiates *'essence-ing';* Separation potentiates *sensing.*

Separation is an act of removal from the center. Separation (Division, Enumeration, Differentiation—call this process what you will) is *the first descent into seeming,* and from this seeming, there must be redeeming (which is escape from illusion). All immersed consciousnesses must be *redeemed* from *seeming.* Whereas there is in Cosmos, in Reality and in Essence, but one ubiquitous Center, with the onset of Separation there arise, *apparently, localized* centers.

With the appearance of two or more centers, it cannot pragmatically be said (as it was said of the One Center) that these two (or more) centers, are *everywhere* and *nowhere* (though Essentially they *are)* because they now exist *relative* to each other, and are localized either near to (or far from) each other, or near to (or far from) other previous (or future) placements of themselves. With the continued creation of other centers, the dynamic of *localization* continues, and the localization becomes even more defined and determinate.

One is not in relation to any other

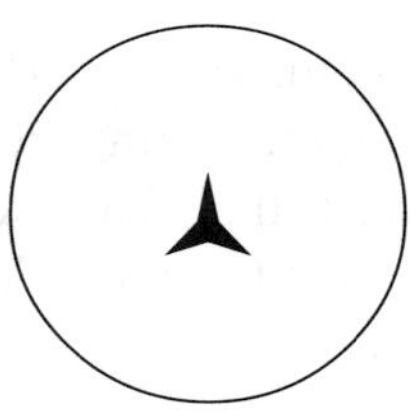

Two are in simple relation to each other

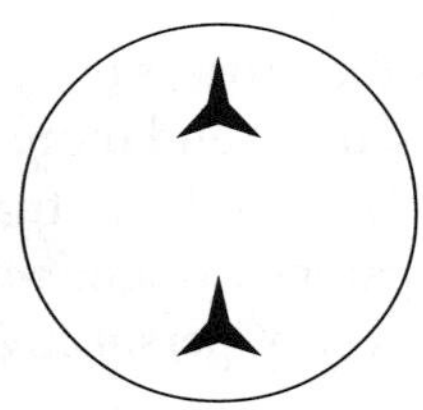

It must be observed, however, that Separation never occurs *in fact* in any Cosmos, at any time, because Separation is an ESSENTIAL *impossibility.* Separation (in-Cosmos) only occurs in the World of Seeming or the World of Sensing (i.e., the World of *Relative Cognizance*... Worlds that are not only Worlds of the 'non-ABSOLUTE', but are also Worlds of non-Universal Perception). Thus it must be said that:

- The World of Becoming and the World of *Seeming* are the same worlds. They are both Worlds that function as they do because the Process of *(seeming)* Separation is operative within them.

On Addition

Let us look at the concept of *Addition* and the mathematical operations of Subtraction and Multiplication. Addition applies in the World of Seeming or the World of Becoming, because only in such a world or domain is the illusion of augmentation or *growth* is possible. In the World of the ABSOLUTENESS of the SELF, there is no possibility of adding anything because the *ultimate completeness* of all possibility is therein contained. If, for instance, it were possible to add aught to the ABSOLUTE, that 'aught' would have to be derived from the very same ABSOLUTE to which the 'aught was to be added; there is no other SOURCE than the ABSOLUTE (or ITS 'RADIATED'/Emanated Extensions) from which the 'aught' could be derived. Hence the addition is impossible.

If aught were to be *subtracted,* that 'aught' could only be conveyed to THAT—that very ABSOLUTE from which it had just been subtracted. There is no other 'place' other than the ABSOLUTE to which it could be removed after it was subtracted, because all places are contained within the ABSOLUTE; there is 'no place' else.

Multiplication within the ABSOLUTE, is also impossible, again, because the ABSOLUTE cannot be augmented or decreased. The INFINITE is the REALM in which 'infinity' *abides completed.* This is sheer *nonsense* in the World of Becoming (the World of Enumeration) and is only possible in the REALM of the INFINITE. Multiplication leads always to increase (and sometimes to an *increase* which is really a *decrease,* as when fractions are multiplied together).

Either way, whether towards *increase* or *decrease,* it is impossible for a product to exceed infinity or to reduce infinity. Even infinity multiplied by infinity yields a product which is simply infinity. This said, there are, nevertheless, what may be called 'countable' and 'uncountable' infinities.

Within the World of Becoming, it is, indeed, possible to add, subtract, multiply, and divide, but it is impossible to alter in the least the exact number of authentic E/entities in Cosmos. That number is predetermined within the Design-at-the-Beginning, and is, in each Cosmos, invariant. It is impossible (within Cosmos) to operate mathematically upon the Essence of Essential E/entities.

- Within limits, the arithmetic operations have place in the World of Becoming, and contribute to the growth *not* of the Spirit (which can never grow or be augmented), but of the veiled consciousness, until it becomes conscious to the degree that the Universal Logos is Conscious (for indeed each consciousness is part of and *is* in Essence, the Consciousness of the Universal Logos).

For instance, psycho-spiritually considered, Addition is the process by which all apparent parts of Cosmos can learn to contain the differentiated aspects and features of all other apparent parts. So, although each apparent part, due to the Veiling Process, arrived at a position in Cosmos which deprived it (consciously and effectively) of the *allness* of Cosmos, that allness is restored through all of the mathematical operations.

One can easily understand how Addition relates to growth of consciousness, but *all* the arithmetic operations can contribute. Involution proceeds by Division, Subtraction, Multiplication, and Addition, but all of a different and contrary variety compared with the same four arithmetic operations as they apply to the process of Evolution.

On Immortality and Mortality

One can say of Immortality what one can say of Synthesis—that it simply *is.* Immortality has naught to do with form; immortality is an act of *identification.* Immortality is achieved when complete and unalterable conscious identification with the SELF is achieved. Yet, Essentially, for all authentic E/entities, immortality already *is.* Form will ever change. The Cosmic Kaleidoscope will ever turn. As long, however, as the Perceiver is not realized as distinct from the Illusion which appears to change, there will be *mortality.* Mortality (contrary to the way in which it is usually conceived), also, has naught to do with form. Mortality is a state of *consciousness,* a veiled state of consciousness, and thus a *mis-identification.* Immortality is a *fact.* Mortality is an *unrealized illusion.*

Conscious immortality comes from 'becoming' the INFINITE SUBJECTIVITY-as-Universal Subject which one already *is,* and doing so by means of *identification.* Really the Universal Subject is 'INFINISPECTIVELY' and even 'Infinispectively' an Object, and no Object is immortal. However, so close is the 'RADIATED' 'RELATIONSHIP' between the INFINITE SUBJECTIVITY and the Universal Subject, that to *identify* with the Universal Subject is to achieve Immortality-within-Cosmos, and is but a 'hair's breadth away' from achieving ABSOLUTE IMMORTALITY, which, perhaps, can only be *'inperienced'* or *'BEEN'* during the Maha-Pralayic ALL-IN-ALLNESS.

Immortality means immutability, unchangingness. Only the object changes. The Universe itself is a great Object (at least in terms of Super-Cosmic Perception). All objects are un-REAL-in-themselves, and are born of an Act of Consciousness, of which the 'ARCHETYPE' is the GREAT intra-SOURCE 'ACT' of 'CONSCIOUSNESS' (however *infinitesimally fleeting* that 'ACT' may Be). How is a (let us say *finite)* object-in-Consciousness created? It is created through *seeing/sensing,* only a *part* instead of *being all.*

Notice, the words *being all,* and not *seeing* all, because, in order to create an object-in-Consciousness, one must begin with a non-objectified state, and the very act of *seeing* indicates that a state of *objectification,* rather than *non-objectification,* exists. The state of *being,* however, is a state of *non-objectification.*

Any thing, 'inhering' in its *infinitized aspect* 'within' ALL-IN-ALLNESS, *is fully* the PLENUM or, alternatively, *is fully* THAT out of which every objective thing can come, i.e., the INFINITESSENCE, the FOUNT OF ALL POSSIBILITY. *Objectification,* however, is an 'ACT' of *non-identification* with THATNESS, an 'ACT' of *terminated inclusiveness,* an 'ACT' of *selection* and *condensation* of *particular infinitized possibilities.*

How this is 'DONE' within the ALL-IN-ALLNESS if a *great mystery.* Some *infinitized* possibility, some *noumenessentialized* 'part' of ALL-IN-ALLNESS (the FOUNT OF ALL POSSIBILITY) is, as it were:

- 'CONDENSED' out of ALL-IN-ALLNESS';

is, thereby:

- *selected* and *fixed* within the 'LINE of VISION of EVANESCENT INFINITE CONSCIOUSNESS'-instantly become-the 'Line of Vision of Pre-Cosmic Infinite Consciousness';

and is, thus:

- 'thrown *outward*' ('EXTRUDED' from the INFINITESSENCE), *translated* from a state of *infinitized being* into a state of *being seen*.

In one sense, the Universe is that which the SELF-as-Self 'Sees'. (That which the SELF-as-Self 'Sees' is certainly an *Illusion* compared to the REALITY of the ABSOLUTE.)

From the Condition of Universal Consciousness, which we call *that which the SELF-as-Self 'Sees'*, the Universe must one 'day' be *realized* (within the consciousness of *all* apparently-fragmented Participants-in-Universe) as THAT which the SELF *IS*. All beings-in-Cosmos must *know* WHO they ARE and *identify* with that GREAT BEING/NON-BEING from 'WHENCE' they 'came'.

♦ Again—immortality *is*, but *conscious* immortality must be *achieved*. Immortality is *non-identification* with any Object. One method of achieving non-identification with any object, and, hence, conscious immortality, is by means of *non-movement*, by *utter stillness*, so that every movement (and, remember, there is no *object* without movement) is seen/sensed as distinct from Point of View of *that* which does not move (which is REALLY, THAT, i.e., the INFINITE SUBJECTIVITY).

To *achieve conscious immortality means to become immovable*. The true guru (the immoveable one) has achieved *perpetual immobility*. The SELF in its ABSOLUTENESS is also characterized by 'PERPETUAL IMMOBILITY', whereas the SELF in relation to its 'EXTRUSION', the Cosmos, is 'ENGAGED' in Cosmo-Eternal, Perpetual Mobility—simply, Perpetual Motion.

On Immobility

It must be said that there is profound wisdom to the statement, "Be still and know that I am God." One of the major contributors to wrong identification is *lack of stillness*. If the Observer, himself, *seems* to move and, moreover, is attached in vision/sensing to that which *appears* to move, the NAMELESS DEITY—the INFINITE SELF—will never be found, even in ITS more accessible mode of SELF-as-Self. If, however, every movement is *seen* as if from an *immobile center*, then from that position of non-movement, God (i.e., GOD-in-Cosmos-as-God) the Immobile Cosmic Substratum, will be found. God or the Self is (in Cosmos) *That* which cannot be modified or changed.

Thus it is that an absolutely unchangeable, non-modifiable *state* must be found *within the Perceiver* (to serve as an immobile background) if the Perceiver is to be anchored in the SELF-as-Self, i.e., in the Unchanging Universal Continuity. It is enough to achieve identification with "Ishvara", the Universal Logos, and with the 'Ray' of the ABSOLUTE which each being-in-Cosmos Essentially Is. Absolute identification with the ABSOLUTE SELF (in fullness) may have to 'wait' until the "Day Be With Us."

In perfect immobility, all perceived 'separate' events become, to the perceiving consciousness as one continuous, multi-faceted Event (for the inter-moment instants cannot be perceived in the World of Fabrication)—this state ultimately resolves (by means of spiritual training and evolution) into a very exalted Consciousness of one single Cosmic-Moment, because the Sense of Spacelessness in the Cosmic Eternal Now is achieved.

All the great variety movement caused by the 'Turning of the Cosmic Kaleidoscope' becomes infused with the sense of *sameness.* In that State of Consciousness in which one single Event is perceived, *it is always, Cosmo-eternally Now.* (Additionally, it is always *here* and it is always **8**—i.e., all objects-in-Universe are 'Seen' as **8**.) Thus, the Universe or the Cosmos is That which the SELF-as Triple Point-as Self periodically registers *other* than ITSELF.

◆ It might be said that the principal movement of Cosmic Involution is from *being* to *seeing.* Consequently, the movement of Cosmic Evolution is from *seeing* to *being.* 'Seeing' is a state of consciousness in which the illusion of extension is registered as apparently *real.* 'Being' is not a state of consciousness, but a state of *identification.* The process of moving from consciousness to identification is movement from the Many to the One, and from Universal Maya to REALITY.

On the Ubiquity of Spirit

A series of questions arise, and the poetic ring of it seems to echo the *Bhagavad Gita:*

- Is it possible-in-Cosmos, to move with every move, to see with every seeing, to touch with every touching, to hear with every hearing—in short, to sense with every sensing, to be everywhere and do everything—registering everything? In principle, at least, it *must* be.
- Is it possible for Me, the TOTAL-I-ALL-SELF, *not* to participate in anything?
- Am **I-as-8** not *already* participating in everything?
- What is the mysterious Veiling Power of Maya which deludes me into the illusion of non-participation?
- How does Maya work?
- Have **8** been ignorant before during the infinite number of previous Cosmic Eternities? Yes.
- Have **8** been enlightened before during the infinite number of Cosmic Eternities? Yes.
- Am **8** ignorant now? Yes.
- Am **8** totally enlightened now? Yes.
- Will **8** not be ignorant again through an infinite number of succeeding Eternities? Yes.
- And likewise supremely enlightened? Yes.

- Am **8** not, was **8** not, will **8** not be both supremely ignorant (as ignorant as possible in this Cosmos) and unimaginably enlightened, limited by no condition of this Cosmos? Yes.
- And Am **8** not, was **8** not, will **8** not be so *simultaneously?* Yes!

If there is nowhere where **I** *AM* not *fully*, **I** (though it be as-**8**) Am experiencing *all* things, and touching and touched by *all* things. If there is no one (no unit-in-Cosmos) who **8** Am not, **8** Am experiencing through *all* degrees of illumination. Then why do **8,** the all Permeating One, the SELF-as-Self-as-self *appear* to be thus limited to an apparent condition (called human living in the three worlds), and why is my supposed identity symbolized merely by my outer name (which is just a pattern of sound)? Surely **8** can think through the illusion which leads to *mayavic* action and glamorous attraction to impermanent objects! Surely, **8**-the-We can change all our desires accordingly.

On the Individual

Let us look then at the definition of the *Individual.* The individual *apparently,* and as usually considered, is a distinct and separate E/entity. In fact, however, an *individual* is *indivisible* from the Whole. What we normally call individuality is *isolated personalism* and is diametrically opposed to true individuality, which is the *impossibility of being contained within any ring-pass-not* (for ring-pass-nots *divide* the *indivisible).*

Individuality might also be defined as the *impossibility of identifying one's being with the perceptions limited by* <u>*any*</u> *ring-pass-not.* It takes great penetration to become a true *individual,* for an individual is a being *incapable of subjugation to the illusion of separation.* The true individual must *see through* and invalidate the perception of separation—the perception of a world full of separate forms.

As normally understood, individuality is manifestation of being through a prakritic pattern—a pattern of matter. Such an individual is the 'keeper' of a patterned form and is confined to that form which is such an individual's ring-pass-not. The true individual, however, comes to know this form, to function through the form completely, to fulfill the function of the form within the system that contains him and, then, comes to destroy that form, for ever a more comprehensive pattern awaits, ever a more extensive ring-pass-not can be found.

Individuality is usually associated with *uniqueness.* Uniqueness both exists and does not exist. An individual is said to be unique. He is unique in as much as he has discovered and expresses the nature of the distinct form (both the objective form and the subtle, often *extremely subtle,* form which is usually called "arupa", or *formless* subjectivity) through which he is responsible to manifest fully, enlivening that form with his being.

- He fails to be unique, or voluntarily gives up his uniqueness when he finds that Spirit is *identical everywhere* and at all times, and that only a *sameness* with all *apparent others* is to be found at his core.

- He again regains his uniqueness when he discovers that there is only *One of Him* after all, and that that 'One of Him' (Really the One Identity of Cosmos, the Universal Logos) is *a Shared Identity* with all *apparent* others, and that, in all of Cosmos, He-the-One Self is the One and Only Unique Thing (Unique Unto Himself with *none* to compare).

Therefore, to identify individuality with uniqueness of form (as it is usually conceived) is but a *partial* definition, one which aligns with the Principle of Unrepeatability in Form, through the agency of which every pattern through which a being manifests is *distinct*, distinct from every other pattern; however, the "Pattern is not the Being."

A fairly advanced way of considering individuality, is as the 'Being plus the Patterned Field'. But a truer individuality is discovered with the arising of identification with *being*, which, itself, is indivisible, undividable.

ABSOLUTE 'UNIQUENESS'
(black circle)
Universal Uniqueness
(white circle)
individual uniqueness
(black star)

The penultimate *uniqueness* is discovered when the individual realizes that he is, in-Cosmos, the "One Without A Second." The *ultimate uniqueness* is the discovery of *uniqueness* within the UTTER ALLNESS, for the individual realizes that he IS *ESSENTIALLY* the "ONE WITHOUT A SECOND." This GREAT BEING/NON-BEING is *unique to the point of ETERNAL incomparability.* The discovery of this ABSOLUTE IDENTITY is the *perfect blend* of *uniqueness* and *identicalness.*

On Matter

The concept of the nature of Matter is often elusive. Matter is Really the objectification of Spirit. Matter is what results in Consciousness when the SELF Perceives Itself. It is the creation of a Point of View, and when dealing with *finite matter,* a *localized* Point of View. Matter, therefore, is the *view* which Spirit has of Itself. 'MAYA'-instantly-Maya is the Possibility of 'Seeing' and Matter is what is 'Seen'.

Matter demands *twoness.* The existence of Matter demands the Mayavic Distinction-in-Consciousness between the Self-as-Subject from the Self-as-Object. Matter is the product of Division, of *apparent* Separation. Matter does not REALLY exist at all; It is an Illusory State, but a *necessarily* Illusory State. Matter is the Eternal Object. When there is no *object,* there is no Matter. In the purely SUBJECTIVE 'STATE' of ABSOLUTENESS, Matter does not exist at all.

The 'ARISING' of Matter is the FIRST 'DIVISION', the FIRST 'TRINITIZING', instantly (at first) followed by the First of other Trinitizations—Pre-Cosmic and, then, Cosmic.

- It is said that Spirit is Matter and that Matter is Spirit. This is true only if Consciousness exists, which it does *not* in the ALL-IN-ALLNESS. Spirit and Matter (once the Pre-Cosmic Phrase is entered) can be considered as the two modes of the 'EXTRUDED' SELFHOOD—the Dynamic and the Receptive. Unless there are *two,* there can be no *giving* and no *receiving,* for both giving and receiving involve a movement from *here* to *there,* a going forth and a receiving. Where there is no *twoness* and no *movement* or possibility of *movement,* there can be no Matter.

The existence of Matter coincides with the arising of the Perception of Objectivity. If there is to be Objective Perception, the 'MAYA'-instantly-Maya must 'DROP' 'HER' first 'VEIL'-instantly-Veil, which then proceeds Pre-Cosmically and Cosmically. What is the Veiling Process? It is the beginning of *seeing* instead of *being.* As long as one *sees,* one is still ignorant. It is only when one is *inside* that which one *sees* that ignorance ends. Spirit-Matter is a great Primary Duality (the 'Subjectobjectivity') with Spirit *primary* to Matter. The primacy of Spirit to Matter is the basis of a *hierarchicalized* Universe.

When GOD-the-INFINITE DEITY *instantaneously* 'PERCEIVES' ITSELF as an 'OBJECT'-instantly-Pre-Cosmic Infinite Object, then, Matter (in its First Mode as Infinite Object) Appears. Matter is the beginning of apparent *otherness.* When 'OTHER<u>LESS</u>NESS' 'PREVAILS', Matter does not, and cannot, exist. We see then, that Matter is one of the 'MODES' of the SELF, objectified first in Pre-Cosmos and, later, in Cosmos.

The Object and Matter are coeval. The appearance of Matter and the appearance of Space (Infinite Space, Mulaprakriti) are coeval. Matter is naught but **I/MYSELF-as-8/MySelf** appearing to MYSELF-as-MySelf through the Veils of Maya (the *veils* of MY CONSCIOUSNESS-instantly-Consciousness). I have 'DROPPED' the 'VEIL' upon MYSELF so that (instantly and Pre-Cosmically) **I/8** *appear* less than I AM.

Matter then, is MYSELF as 'SEEN' through the 'VEIL' of MY OWN instantly 'APPEARING'/'DISAPPEARING' 'CONSCIOUSNESS'—becoming instantly Infinite Pre-Cosmic. A 'VEIL' is but the 'NARROWING DOWN' of MYSELF *by* MY 'MAYAVIC SELF'. Entering Pre-Cosmos and, then, Cosmos, a Veil is My (**I/8**) Willful or Voluntary refusal to 'See' My Wholeness. In the ALL-IN-ALLNESS, I AM MY WHOLENESS. I AM IT; I don't have to 'SEE' IT. In the 'VEILED'-instantly-Veiled State, I 'HOLD' a *dual focus*: MY WHOLENESS and *not* MY WHOLENESS, for I AM *always* the INFINITE SELF even though **I-as-8** (Pre-Cosmically and Cosmically) *seem* not to BE the INFINITE SELF, but multitudinous 'Something<u>s</u>' *less.*

Matter is the SELF-as-Self in the Receptive Mode. Spirit is the SELF-as-Self in the positive or Giving Mode. The relationship between the Infinite Spirit and the Infinite Object is Pre-Cosmic Spirit-Matter. Spirit-Matter is a Relationship at once *dualistic* and yet *seamlessly unified.* I-the-SELF, Am fully Spirit. I-the-SELF, Am fully Matter. There is no possibility of creating a Division such that in one part of the Division, I AM, and in the other part of the Division, I AM *not.* Spirit-Matter are the two Pre-Cosmos (and, probably, Post-Cosmic Modes of the 'EXTRUDED' SELF—the SELF-as-Self). Spirit-Matter also has its *Cosmic* Reflection.

- Matter 'ARISES' with the 'REFUSAL' of the INFINITE SELF to BE ITSELF 'EXCLUSIVELY'.

How can the INFINITE SELF ever *cease* to BE ITSELF 'EXCLUSIVELY'? How can the INFINITE SELF be *other* than IT IS? IT *cannot,* and yet IT *seems* to 'DO SO'.

The *mystery* centers around the SELF-'IMPOSED', SELF-'BLINDING' 'PROCESS' which 'MAYA'-instantly-Maya *is*. Maya negates by affirming. 'MAYA'-instantly-Maya *negates* the ALL-IN-ALLNESS of the SELF and *affirms* ITS *partiality* in Pre-Cosmos and Cosmos. Does the ALL-IN-ALLNESS and ABSOLUTENESS *cease* thereby? By definition, IT can *never* cease and *yet,* there is the Pre-Cosmic and Cosmic Experience of SELF-'IMPOSED' 'LIMITATION'-become-Limitation. Be that as it may, even 'while' there is Limitation, there is also a Pre-Cosmic and Cosmic *Descending Emanative Trail* of *non*—Limitation, such that the ALL-IN-ALLNESS 'REMAINS' what IT IS at every 'step' of Emanative Descent.

Ever the higher Realization remains what it is, even though there is a descent of Consciousness into the lower levels of realization. Therefore, **I**-the-INFINITE SELF AM PRESENT at all stages of descent, including that 'STATELESS STATE' from which **I** *never descended.* This image of *reversed ontological 'telescoping'* is necessary to grasp.

On Cosmic Prakriti (Cosmic Matter)

There is a great difference between undifferentiated Cosmic Prakriti and differentiated Cosmic Prakriti. While undifferentiated Cosmic Prakriti may not be *infinite* as is Mulaprakriti, it still is 'Mother-ness', as is Mulaprakriti. It does not differ Essentially from Mulaprakriti *substantially,* but it *does* differ in *scope* and, perhaps, *depth,* for the Infinite Subject-becoming-Focusing and Focused Universal Subject no longer 'Sees' His Own Reflection as *infinite,* as He did in Mulaprakriti. Mulaprakriti is the *infinite potential for objective appearance* and Cosmic Prakriti is *the bounded potential for objective appearance.* Both, however, still share the Quality of Motherness—as *no Son* has yet appeared.

There are, however, different *qualities* of 'Motherness'! The Infinite Father/Infinite Subject does not yet 'See' the Son in Mulaprakriti (unless He 'Sees' the Memory of *many* or *infinite* Sons Past—though, such an Infinite Recollection may not be possible to any extra-SOURCE Entity, and, moreover, at this our present stage of ignorance, it is surely impossible to say). Even the Condensing Point which 'Sees' the 'Bounding Infinite Object' and the Condensed Point, which 'Sees' the appropriately 'Bounded Infinite Object' or Cosmic Prakriti, *does not yet 'See' the Son.*

In these early Pre-Cosmic Days only the 'Mother' (either as Infinite Mother or as Bounded/Circumscribed Mother) is *visible,* (which means that only the Subject/Self-seen-as-the *Mother* is visible) for the Infinite Father/Infinite Subject (even as He/It 'Narrows His Vision of Himself') is not *focusing on His Own 'Son-like' specificity* but *only on His*

Own Being in general. Thus He 'Sees' only Himself, *in general*, and in the Aspect of His *Being-ness*, reflected as Mulaprakriti, and, even, only *in general* when, later, He 'Sees' Himself as Bounding and Bounded Mulaprakriti, i.e., Cosmic Prakriti (when His *scope* of Being-ness is deliberately 'Seen' as more circumscribed). The Pre-Cosmic Stage of Cosmos-Formation is thus one of 'Seen' Infinitude, and 'Seen' *Generality* and *Scope* (and, perhaps, *Depth)* of Being, but *not* 'Seen' *Specificity* of Being. The 'Seeing' of specific 'Sonship' lies ahead.

When however, the Stage of Condensed Point and Cosmic Prakriti is reached, then it is time for Internal *Specificity* to emerge. The Focused Universal Subject has now become the Universal Logos facing His own undifferentiated Reflection as Cosmic Prakriti. At this moment He still 'Sees' 'Motherness' (Circumscribed Motherness as Cosmic Space, i.e., the Circumscribed Potential to Show Forth Form from within Matrix/Mother/Field of Consciousness), but He does not yet 'See' the Son *He* has been 'Carrying' in Potential.

The 'Motherness' of Cosmic Prakriti moves into a new Creative Phase when the Son 'within' the Father is 'brought forth', or 'shown forth' by the Mother. This is the first Moment that the Father 'Sees' *more* than His Own *Being-in-general* reflected *as* the Mother. Now the Father 'Sees' not only Himself-as-Mother, but Himself-as-Son (within a now *hidden* Mother, a Mother hidden *behind*, as it were, the Form of Her Son). The Mother is Field; the 'Son' is Point.

When the Son arises as a Reflection of the hidden/invisible/subjective Son within the Father, then, brought forth within the Mother, the Father *transfers* the Universal Pattern He has been 'Holding' to the Son (Who, let us remember, is the *Father, objectified).*

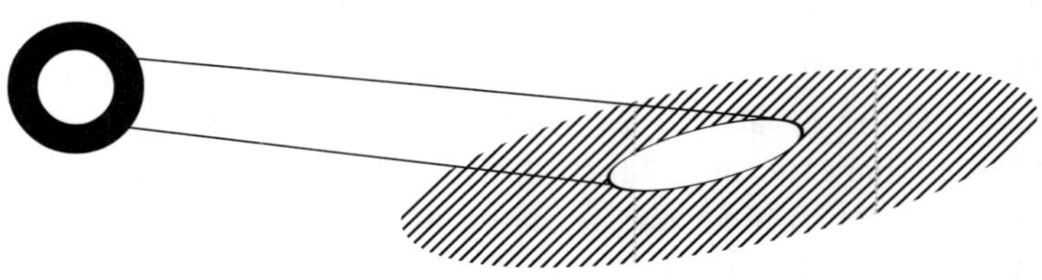

Now it is the turn of the Son (i.e., of the Father-as-Son) to *engage* the Mother, for Cosmic Prakriti is still undifferentiated. (The differentiation of Cosmic Prakriti by Fohat, when it occurs, places new limitations upon the Creative Process.) There are more Archetypal Creators within the 'Father-transferred' Divine Pattern now 'Held' by and *invisibly inhering in* the Son which need to be brought into Objectivity.

So now the Son, Who is Now in the Role of Cosmic Subject (just as His Father had been), begins, through the Power of Maya/Fohat (i.e., Self-Reflective Consciousness) to look within Himself and to 'See', first, His Beingness-in-general (His 'Syzygy', His 'Goddess') and then the articulatable Cosmic Pattern which has been transferred to Him by the Father (just as the Father earlier 'Saw' His Own Specificity, of which the Son was the *First Specific),* and thus, the Son brings forth Sons of the Son (the Divine Tetraktys, first Three and then Seven), Mind-Born Sons Who, along with the Son, *hold* and *behold* the Pattern of Things to be Objectified in Cosmos.

NOTE: There is a question regarding just what kind of Prakriti (or Self-Image) the Son and Lesser Logoi 'See' when they look into Themselves. In the case of the Son, for instance, the Prakriti 'Seen' (i.e., the Self-Image) may not be of the same depth and quality as the Prakriti 'Seen' by the Father when He (the Father) 'Saw' Himself as an Object. There are *layers* of Prakritic Registration, and each Lesser Logos operates within the Ring-Pass-Not of the Consciousness of the Emanative Logos immediately superior to It. [Some of these ideas are discussed in the Glossary under "Prakritic Registration".]

When the Infinite Father 'Sees' His Own Being, He is 'Seeing' the Infinite Mother. When the Cosmic Father 'Sees' His Own more bounded/circumscribed Scope and Depth of Being, He is 'Seeing' the Bounded/Circumscribed Mother, i.e., Cosmic Prakriti. When the Son 'Sees' His Own Being in general, is He 'Seeing' the *Pure* Bounded Mother, i.e., Cosmic Prakriti? Maybe, maybe not.

The Son Is the Father in one sense, but the Son, being the 'Seen' Object of the Father is probably 'Seeing' the Pure Bounded Mother *at one remove.* The Mother-Image is, theoretically, different from what the Father 'Sees'. While the Son is an Object to the Father, yet He Is, in His Own Right, (like His Father) a Subject. The Son is yet one more Bounding (in Scope and Depth) of the Infinite Subject (just as the Father, the Universal Logos is a Perceptual/Inceptual Bounding of the Infinite Subject, and both, just as *any* Subject/subject is, are Essentially, the Infinite Subject, which Is, ESSENTIALLY, the INFINITE SELF).

Every time any *extension* (Logoi, Gods, Monads, etc.) of the Infinite Subject 'Sees' Itself in its own *essential beingness,* it Sees the Mother (but not necessarily and not yet, the Pure, Infinite Mother), even though the *beingness* (REALLY, BEINGNESS) of every Subject is *infinite* (in fact, *the* INFINITE). As Subjects descend in the Emanative Sequence they tend to *see* the scope and depth of their *being* as more and more circumscribed, and it is this kind of *seeing* which *narrows* the *prakritic field* observed (and changes the Quality of the Image Reflected).

Thus when the Infinite Father 'Sees' Its Own *Infinitude,* It 'Sees' Mulaprakriti, the *Infinite* Object. When that *extension* of the Infinite Father called the Focusing and Focused Father/Subject 'Sees' Itself; however, Its Selfhood appears more circumscribed, more bounded. The *Focused* Father (Condensed Point, Focused Universal Subject) 'Sees' Itself as Cosmic Prakriti. The *Focused* Father has become a More *restricted* Mother!

- Every time a more *diminished* Subject 'Sees' Itself, the Mother Aspect which appears because of the *self-sight* is more bounded and circumscribed. This generalized *self-sight* equates to what we might call the 'scope and extent of an individual'. The Universal Logos has a certain general Prakritic Scope and Extent, a certain Ring-Pass-Not. The Universal Logos Sees Its Oneness, Its Singularity as of a certain Scope, Extent, and Depth, The Objectified Beingness (of a certain Scope, Extent, and Depth) of that particular Being (the Universal Logos) is what we call Cosmic Prakriti.

The Universal Son 'Sees' Its Oneness in a still more circumscribed manner, for though the Son Is the Father, the Son is also, *formally,* (i.e., in Scope, Extent and Depth) *less* than the Father, but not *less* that the Father in ESSENCE/Essence. Every subjective Father Unit (for example, Infinite Subject, Focusing Universal Subject, Focused Universal Subject/ Universal Logos, Universal Son, Universal Sons of the Son {the various Sub-Logoi and

Subsidiary Logoi}, various Cosmic Monads, etc.) has, perhaps, three ways of 'Seeing' Itself and, thus, of Seeing the Mother (Prakriti in General):

1. The Subject may be able to 'See' Itself in Its own ESSENTIAL NATURE, i.e., in Its Own ESSENTIAL INFINITUDE-as-Infinitude, and, thus, an Infinite Mother, an Infinite Object, an Infinite Mulaprakriti (all Three equivalent) arise. Certainly, the Infinite Subject can do so, and perhaps Its Pre-Cosmic and Post-Cosmic Extensions—the Focusing and Focused Universal Subject also can. This kind of *infinitized self-sight may* be available to any Unit in-Cosmos above the rank of advanced man, and *including* advanced man, but the point is debatable. There may be a natural limitation placed upon each E/entity depending upon the limitation of the superior Emanative E/entity in which the first E/entity "lives and moves and has its being." One would think that every Self-conscious E/entity would have *immediate conscious access* to the *sight* of Its INFINITE ESSENCE or Infinite Essence, but this may not necessarily be the case. Can, for instance, a great Entity "see as deeply into Itself" as a superior great Entity which emanated the former Great Entity can 'See' into Itself? We may be dealing with the "layers of the onion" as we (as aspiring E/entities) attempt to move through various depths of 'Self-Sight' superior to our own, hoping to 'See' the Essence/ESSENCE at our Root/ROOT.

2. The Subject, taking its *eye* off the attempt to 'See' Its Own Infinitude, can See the *general scope*, *depth* and (albeit illusory) *extension* of Its Being within the larger system in which It is contained. For example, the Focused Pre-Cosmic Subject has Become (through the deliberate Narrowing of Self-'In-ception'):
   - *One Specific Singularity* instead of what It was formerly—a Being of Infinite Scope, Depth and Extent, an Infinite Subject/Object; or,
   - the Cosmic Son is of lesser scope, depth and extent than the Cosmic Father; or
   - the Sub-Logoi, surrounding the Son, are of lesser scope, depth and extent than either the Son or the Father, etc.

   In this case a Self-Image (not of an Infinite Object, but) of a Bounded Object or Bounded Prakriti will arise. Such kind of *bounding* is Cosmic Prakriti, the Space/Field and Ring-Pass-Not of the Universal Logos. The Universal Logos 'Knows' that It is ESSENTIALLY the INFINITE, and, perhaps, when thinking of Its ESSENTIAL NATURE can 'Sense' (who can say for certain?) the INFINITUDE IT ESSENTIALLY IS as an Infinite Object, Mulaprakriti. When, however, the Universal Logos, 'Looks Within' and seeks to 'See' Who *It* May Be as an *Individual*, a *One, a Singularity*—then, *Cosmic* Prakriti and *not* Mulaprakriti will appear *(Cosmic* Space and not *Infinite* Space). [The question of the *degree* to which any being other than the Infinite Subject can apprehend its own Infinitude and hence, see itself as Pure Mulaprakriti, can also be referenced in Section II of Philosophical Problems as well as in the Glossary under Prakritic Registration.] When a subject apprehends Itself in terms of its place and scope, a ring-pass-not is created, and it sees itself in terms of a particular prakritic field. There arises the inner apprehension, "I am this and not that. The scope of my being extends from here to there. *This* is my Space or Field and not that. *This* is the extent of my identity. *This* is *what* (more than *who*) I am."

3. The Subject can 'See Itself' in terms of the 'Son' It is 'Holding within Itself', for the Father always 'Carries the Son' as an *invisible potential* before the Mother 'Shows Forth the Son'. With this apprehension, Prakriti as 'Motherness' (whether Infinite 'Motherness', or Bounded 'Motherness', or some still more limited and modified kind of 'Motherness') will *not*, per se, appear. Instead, a *Son* will appear in Prakriti (the Field of Father/Mother Space), a specific form will appear, and the Mother will no longer be so focally 'Seen' in Her *Own* Right, but, instead, more by means of Her Son (and the Father's) (though the Field of Father/Mother Space will *envelop* the birthing of the Son). One will have to look at the Son in order to 'See' the new form of the Mother (and of the Father). Prakriti thus begins to appear in specific form (more differentiated form) and no longer merely as a reflection either of the infinite or bounded being-ness of the subject, but of the specific content inhering invisibly within the Father/Subject. While the Father-as-Mother-Space in which the Son is being born is still visible (for Parents continue to 'See' each other after the 'Son' is born), the 'Son' will be the main focus of attention.

It becomes clear that most "contents of consciousness" for most beings are not pure Prakriti at all (not pure Mother), but only Prakriti-in-Form. While it is true that, "He that hath seen Me hath seen the Father", it is equally true that, "He that hath seen Me (the Son) hath seen the Mother." Except for the objectified progeny of the Mother, the pure Mother is as invisible as the Father (to most human beings). Instead, the Mother is seen through Her Sons who are all the many objects which Father Power (Subject Power) conceives within Her (or rather, conceives within Himself-as-Her.) The Object Is ever the Subject. This means that we (prakritically immersed 'Rays' of the ABSOLUTE) do not see Real Matter (Mother) or Real Father. We 'See' (instead) 'Sons' of 'Sons' of 'Sons', etc.

Then, can the pure Mother ever be seen in her own right? Can Mulaprakriti, Cosmic Prakriti, or even more limited Prakriti Fields ever Really be seen? The answer is, "yes" (eventually) but only when the Father 'Sees' Himself as He, Essentially, Is! The Pure Mother Is the Image of the Father as He Is, and not all that which is within the Father. When 8 Know My Own Being, and Am Conscious of that Being, then, 8 'See' the Mother as She Is (which means that 8 See the image of 'MySelf' as 8 Am. (Since 8, in such a case, Am "like unto the Father", 8 'See' the Mother.) When, however, 8 Am Conscious (not of my Being as It Really Is) but of that which is within Me, then 8 no longer 'See' the Pure Mother, but only the Mother behind the Son, or the Mother in the Form of the Son.

- The only way for the Father to See the Mother as She Is, Is for the Father to See Himself as He Is! For the human being, evolution, growth into synthesis, and simplification gradually brings this about.

The question arises, however, whether a Subjective Father, other than the Infinite Father/Infinite Subject, can ever 'See' to the depth of Himself/Itself? The Infinite Father 'Sees' the Infinite Mother, but if the Father's Son tries to 'See to the Depth of Himself', He may discover the Father 'Seeing' Him (i.e., the Son!) When the Son 'Sees' His Own Identity as Mother, what *kind* of Identity is He 'Seeing'? Is He 'Seeing' as Mother, an Image as *pure*, as *deep*, as *authentic* as the Mother that the Father 'Saw'? Or is He 'Seeing' a 'lesser Mother'? A living Image of Himself (faithful and possessing greatness) but an Image *lesser* than that the Father 'Saw'?

This model is probably the case. If it were not the case, then the Prakritic Image of My Being, which **8** as a very limited human being sense when **8** look deeply into 'MySelf', would be no different from the Prakritic Image of His Being registered by the Solar Logos when He 'Sees' deeply into Himself.

- Would it not be a great arrogance on the part of Man, to think that the Infinitude He senses as the very Essence of Cosmos, and the still deeper ESSENCE of the UTTER ALLNESS, is the very same registration of Infinitude sensed by the Solar Logos, a Cosmic Logos, a Super-Cosmic Logos, and beyond? Is there no improvement in the capacity to register and deeply apprehend Infinitude as one ascends the Ladder of Cosmic Evolution?

Yes, the ESSENCE is the ESSENCE, and IT is the SAME for each unit-of-Life in Cosmos. The faithful apprehension and registration of that ESSENCE surely, however, must improve in quality as it becomes possible to live closer to the Cosmic Source.

On the Impossibility of *ESSENTIAL* Un-REALITIES

Just as there can be no 'parts', no REAL fragments, so, also, there can be no un-REALITIES. Actualities, therefore, are ESSENTIALLY REALITIES. 'Parts' are ESSENTIALLY the WHOLE (and, strangely) in ITS ENTIRETY. The entire Cosmos might be seen as a Drama in which error becomes Truth-then-TRUTH. New forms of language are needed which will help the mind overcome the dualism to which it is instinctively prone. Just as the Positivists have their language, so the Non-Dualists must have *their* language as well.

When thinking about Spirit and Matter, there is the ever recurrent, confusing problem expressed as follows: Just Who or What is Spirit? Just Who or What is Matter? Such a problem is solved by what might be called the *thorough and full investiture of the SELF in all ITS Emanated A/aspects*. One must understand the true Nature of Emanation to understand the meaning of the foregoing.

- Focusing within Cosmos, for a moment, it could be said, that in every Emanation, there is the 'going forth' of the Whole into a *more restricted state*, and yet there is, as well, the complete maintenance of the Whole in the *less restricted state*.

There is, in Cosmos, what can mathematically be called 'n-dimensional Consciousness'. (The value for 'n' may be different in every Cosmos, and we do not even know the value of 'n' for *our* Cosmos.) It is usually conceived that the One Self is less and less invested in every descending dimension of consciousness in Cosmos. This idea, however, is Essentially illusory, though pragmatically useful. It is pragmatically useful because the experience of beings-in-Cosmos is that the full Consciousness of the Universal Logos is definitely *not* as manifest in the atom of matter as It is in the Archangel.

The model is Essentially erroneous, however, because Spirit cannot be *divided.* The SELF cannot *descend into ITS Cosmos* in 'part', but only *fully* 'into' every *seeming* part. This might be called the 'Doctrine of Full Descent'. Thus it is, that every dimension of Cosmos, no matter how seemingly inert or opaque, is *fully* the PRESENCE of the INFINITE SELF. In Cosmos there may be *apparent* un-REALITIES, but there cannot be REAL, or ESSENTIAL, un-REALITIES. Every little apparent un-REALITY is REAL through-and-through. GOD IS *fully* ITSELF at every illusory *point* in Cosmic Space.

Every Number is, itself, a dimension, thus Number and dimension are equivalent. Number and limitation are also equivalent, for Number is always a limitation upon the ZERO. Dimension and limitation are, therefore, equivalent. Each dimension in-Cosmos differs from another by the degree to which it limits the expression in Matter of the One Life in Cosmos. More ESSENTIALLY/Essentially, however, there is no dimension in Cosmos in which the Whole is not to be found *completely.* There is no 'part' in-Cosmos in which the Whole is not to be found *completely.*

The major premise is that the Principle of Indivisibility must remain *inviolate* if the Great Heresy of Separateness is to be avoided. It is an apriori axiom that there is ONE BOUNDLESS IMMUTABLE PRINCIPLE. In REAL *fact,* there is nothing else but THAT, so all other things that seem to be, are in *fact,* THAT. As the Great Sage Sankaracarya said, "The pot is really the clay."

On Location

Let us examine this concept of *location.* When first the 'POINT'-instantly-Point appears, it is the Infinified Point and it is *nowhere* and *everywhere.*

- It has no location.
- Location is a relative term.
- Location begins when an item or unit can be *related* in a certain manner with respect to other items or units.
- In order for there to be location, there must be, a *point of reference.*
- There can be no location until there is at least twoness.
- Movement is only detectable if it is measured against that which is not moving similarly, or is immovable.
- If all items within a domain stay in a particular relation to each other and *do not vary their relative position and location,* movement (with respect to an external point of reference) could be either extreme or non-existent and it would not be detectable.

On the Irrationality of the Cosmos

The entire Cosmos is based upon what we would normally call 'impossibility'. It sometimes seems as if Its *modus operandi,* Its Structure, Its Rules are all rather illogical and irrational. Here are a few reasons:

- Consciousness in Cosmos seems, for the most part, to be in a *state of amnesia* concerning Its ESSENCE and ORIGIN.
- Something is, ESSENTIALLY, NOTHING, and yet, Something is *nothing—nothing* REAL.
- Something is also 'less than NOTHING'. NOTHING is both *never* Something and yet is inescapably Something.
- The most tangible is the most evanescent, the least tangible is the most enduring, and if Something is NOTHING, then IT lasts forever.

The Universe or Cosmos does not REALLY *count,* and yet It *does,* because Its existence is based upon *counting,* i.e., Enumeration. Yet no-thing REALLY 'counts', only NOTHING counts! In Cosmos, there is complete *unrepeatability.* In the WORLD OF BEING, there is complete *repeatability,* except that there is *no event to repeat.* There is only a 'STATELESS STATE' to repeat.

♦ Thus, in the World of Becoming as contrasted with the WORLD OF BEING, two different Laws seem to hold sway, and these Laws are related to the greatest of all Pairs of Opposites—INFINITY and Finitude (which are not necessarily the same as Infinity and Finitude). The LAWS of INFINITY are entirely the opposite of the Laws of Finitude, and yet the Laws of Finitude are *necessarily* included within the LAWS OF INFINITY.

The Universe is *paradoxical*—not only because the mind of man is so meager (which it is) but, because the Universe is Designed to be a Contradiction to what can only be called the GREAT CONTRADICTION, the GREAT BEING/NON-BEING—the BE-NESS.

Section II
Problems of Living
&
Philosophical Problems

In this section, two varieties of problems are discussed:

1. Problems of Living that can be illuminated through non-dualistic thought.
2. Philosophical Problems that arise because of non-dualistic thought.

What actually are problems? Problems are opposing, conflicted energy patterns. A problem can be defined as the 'distance' between the actual and the ideal.

Problems of Living

The Problem of Karma

So much has been written about karma and the responsibility of the individual self-conscious unit, the human being. But in terms of non-dualistic thought the question is asked, To *Whom* does karma belong? *Whose* karma is *Whose?* At first the answer would seem obvious. We would answer the question from the perspective of the so-called 'individual' unit, but even the Tibetan alerts us to the idea of group karma, national karma, Planetary Karma, Solar Systemic Karma, and so forth. Somehow, all lesser units are swept into the Karma of the larger groups to which they belong. This seems sufficiently just, even though the modus operandi of the process may be obscure.

Non-dualistic thought takes the process even farther. Instead of considering the fate of tiny lives that are component parts of greater Lives, 'infinitistic thought' looks for the *essence* of those tiny lives and finds there an identity with the *essence* of the greater Lives. From this perspective, the question must be asked, Am **8** the part, or the Whole, or both? When the answer is "the part" then all the best thought that is usually given to explain karma applies. What happens, however, when the answer is "The WHOLE-as-Cosmic-Whole, and of course, the part as well"? Then a new type of metaphysical thinking is required.

We have frequently discussed the illusory though actual nature of all ring-pass-nots in Cosmos. Even the Cosmic Ring-Pass-Not is ESSENTIALLY Illusory (a product of Maya) and there is no REAL difference between the INFINITE SELF and ITS Cosmic Projection, the evanescent Universe. So, if such thoughts be accepted as valid, it would seem that every E/entity in Cosmos is ESSENTIALLY *present* within *all* ring-pass-nots, and that each E/entity is all-pervasive, ubiquitous and is somehow the ONE-as-Universal Being experiencing *all* that transpires in Cosmos. That this fact may come only as a revelation of the pre-absorptive "Latter Days" in any Cosmos, makes no fundamental difference. Even though this ubiquitous experiencing and ubiquitous *acting* (in what is, essentially *One* Cosmic Action) does not *seem* to be occurring, REALLY, and Really, it *is*.

If **I**-as-**8** in Cosmos am the Perpetrator of the One Action, Am **8** not the justified Recipient of the One Result? It seems that, ultimately, karma of any kind in Cosmos cannot be 'privately owned', although under the illusion of separative entification it *seems* to be. Such thoughts seem to defy common sense. Everyone 'knows' that "my action is my own" and "the results of my action is my own", but as we look beyond such statements we see they may not be *ultimately* true.

Because of the factor of apparent 'individuality', karma *seems* to be experienced in varying degrees of intensity by different human beings. If a person falls from a building, he seems to receive the full impact and 'I', another human being, *apparently* separated from the one falling, appear to receive practically no impact at all. The degree of impact felt, however, depends upon the degree of sensitivity of the other's 'feeling', and if the degree of identification were sufficient, it is theoretically possible that similar impacts would be felt by *both*.

There are described in the annals of spirituality examples of this kind of shared impact. Of God, it is said, that "Not a sparrow falls ...". Of course, one could see that if all karmic impact were shared in fullness by all participating human beings at *this time*, the World would not easily continue in its present form. Lack of the necessary pervasive sensitivity is actually *form preserving*. The revelation of the *equal participation* in the Universal Action comes later, when form is on the verge of dissolution.

■ For practical, in-World, in-Cosmos purposes, we must continue to regard karma, the Law that states, "For every action there is an equal and opposite reaction", as pertaining both to 'individuals' and groups, but it is well to hold the thought of what is REALLY-as-Really going on.

There is a reason why the poet, John Donne, wrote, "Never send to know for whom the bell tolls. It tolls for thee." A profound and pervasive sharing of all possible experience in Cosmos is occurring from ultimate moment to ultimate moment, and we, the PRESENCE-as-Presence-in-Cosmos are deprived of nothing. We-as-**8** are 'there' for it All, remembering, of course, that there is no 'there' and that, to the Pervasive One, 'there' is experienced as 'here'.

Whose karma is it anyway? What is *mine*, and what is *yours* when both of us are Essentially/ESSENTIALLY *identical?* In Cosmos, the World of Differentiation, the 'good' people and the 'bad' people can, apparently, Really, be differentiated, and their desserts are according to their actions. Since, however, every 'who' is REALLY the *'WHO'*, WHO, *REALLY*, is the 'good' and the 'bad'? Is not the ONE 'WHO', *both* the *good* and the *bad*, and well as *neither* and *more?* We must remind ourselves that, essentially, all apparent parts are really the Whole, and in fact, the WHOLE, the INFINITE INDIVISIBLE SELF. What will such thoughts mean to us as apparently 'individual' participants in the Cosmic Process?

The Problem of Individual Responsibility

I-THE-WHOLE-as-**8**-the-Whole have both an *apparent* "part to play" and, as well, REALLY, '*All* parts to play'. As long as **I** have Veiled MYSELF, (becoming **8** in-Cosmos) and have confined MYSELF to one set of conditions (my apparent and 'progressive' 'individuality', MY Focal Identity), through that set of conditions **8** must properly function, even as **8** realize that, REALLY (i.e., non-dualistically), **8** Am simultaneously functioning through *all* sets of conditions. This focussed functioning through one set of conditions (whether that set is assigned or chosen) is what it means to "mind one's own business." But in a deeper sense, REALLY, "one's own business" is the World, the Cosmos. We have dealt with this strange, though illuminating, thought repeatedly.

We must remember that **8** Am, Essentially, the REAL INDIVIDUAL, functioning in Cosmos as the One Cosmic Being, the Universal Logos. When there periodically is no Cosmos, **I**, the ONE TRUE INDIVIDUAL, the INFINITE SELF am thoroughly *indivisible.* Even when there *is* a Cosmos, **I**-as-**8** Am still ESSENTIALLY the ONE INDIVIDUAL, and even in-Cosmos, **8** Am still, equally, *indivisible.* **8** cannot REALLY be divided into parts, and reduced in scope to a single part (though apparently exactly this is what happens to ME-as-Me).

In Time and Space (which are largely intra-Cosmic Conditions), **8** must, of course, act as if **8** Am responsible for that particular 'part' into which **8** appear to have been confined. It is the immediate responsibility of the WHOLE manifesting as the Part (the Cosmos) and then as the tiny part (the individual) to *do* so. This is the Law. The WHOLENESS of the WHOLE is not only *in* each part but *is fully* each part.

- As well, however, "**8** Am my brother's keeper", and what he does, **8** do, for WHO-as-Who-as-who else is there to do it? Thus, in TRUTH, **8** am totally responsible for all acts throughout Time and Space forever!

This extraordinary statement statement bears close pondering. Am **8** finished with responsibility once **8** have assumed responsibility for the little 'myself'? What happens if Universal Pervasion begins to become a fact in my veiled little consciousness? Can **8** sense myself (now MySelf-asMyself) as the *responsible Actor* beyond the little ring-pass-not **8** so habitually have called 'mine'? What would change were **8** able to do so?

The Problem of Envy

Envy is a condition in which a person desires that which another person seems to have. We must ask ourselves, however:

- Is there REALLY *another* person and does that other person REALLY have what 'I' do *not* have?
- Is not the *having* of another Essentially MY-as-My *having?*

- Is not the *experience* of another Essentially MY-as-My *experience?*
- When one IS, ESSENTIALLY, *already,* the EVER FULL, the 'MAXIMIZATION', the 'SUMMUM STATE'—when one IS *already* the INFINITIZATION, what can one possibly want?
- What can one desire?
- Can one be envious of what one-as-One-as-ONE *already has?*—and *has,* not just in theory, but in *fact?*

The Problem of Jealousy

Then, there is the problem of jealousy, which arises from wishing that the love offered to another or shared with another be shared, instead, with oneself *alone.* But who *is* oneself, and who *is* another? Is not 'WHO' both oneself *and* another and All? Can there more than one SELF? If someone loves 'another', is not that someone loving ME-as-Me-as-me as well since **8** and the other are *identical?* Jealousy is based upon the illusion that one is not receiving the love directed towards another.

Given the illumined state of *identification,* the irrationality of the state of jealousy will be revealed. It is not so much that BRAHMAN will be revealed when vices are eliminated, but, even more, that when BRAHMAN is revealed, and *known,* vices will be eliminated because they will no longer seem *attractive.*

The Problem of Ambition

Ambition is the desire for personal or individual elevation. It is the seeking of added power for the apparently individual self. Ambition is necessary and justified, from the egoistic perspective, during certain phases of evolutionary development. The egoistic desire for power, however, causes much difficulty in the realm of human relations, because the one who desires power will often seize that power at the expense of the rights of others.

This type of ambition must be outgrown through realization. Ambition, as a generic term, can be thought of as including the concept of *aspiration* and, as such, is justified upon *all* rungs of the evolutionary ladder. While the man of aspiration may be more considerate of the rights of others, he may not yet be enlightened with respect to his REAL IDENTITY.

As regards the problem of individual striving (whether ambition or aspiration) we are compelled to ask once again, if it is not so, that that which one seeks, one-as-One-as-ONE already *has,* because all that can be achieved has *already* been achieved, or, rather (and more accurately) never *had* to be achieved at all because it *always* was? Further, we have established that that which one seeks is REALLY THAT WHICH one, ESSENTIALLY *IS.* All that one (in ignorance) could possibly desire *already* inheres in one's *essence,* for one is One Is ONE.

- Man as an apparently individual unit must continue to strive, and, indeed, the Divine Purpose requires the striving of the apparently separated units in order that the Design-at-the-Beginning be fulfilled.

There is, however, a complete different attitude demonstrated when one strives *apparently* as an individual while deeply *knowing himself* to be the one-as-One-as-ONE, compared to the attitude demonstrated by the ambitious or aspiring one who strives *without* this knowledge of his REAL IDENTITY. In the first case the 'individuality' of the striving is an *apparency* only, and, in fact, the WHOLE-as-Whole is working through the apparently separated unit. In the second case ignorance prevails and the striving unit is a victim of the limitations of egoism.

The Problem of Pride

Pride, like so many vices, is based upon insufficient SELF-Realization. The 'individual' Not-SELF is mistaken for the SELF. The consciousness is wrongly identified with the object and the true internal SUBJECT is overlooked. Even the Subject that one Is in-Cosmos is not understood, because if the Subject were understood, the loving link between It and all other Subjects would be realized. In the Subjective State, there is still some degree of identification with the *higher* fields through which the true IDENTITY is manifesting, but the energy of unity is so strong that the third Aspect impulse towards separation is overcome.

In the Subjective State the worst abuses of pride have been overcome, but even great Beings, Who are Subjectively focussed, have, in the occult and religious literature, been known to evidence what has been called 'pride'. The SUBJECTIVE STATE, however, is related to Synthesis and not just to Unity (as is the Subjective State). When one identifies as the REAL SUBJECT (to the degree possible by any in Cosmos) one has identified as the INFINITE SELF, and pride is completely eradicated. Why is this so?

Pride is based upon *comparison.* In the world of prakritic variation, comparison is possible. Complete prakritic inequality reigns in the World of Becoming (the World of Illusion). The interred consciousnesses mistakenly identify with stratified prakritic variation—high forms and low forms, and mistakenly evaluate 'ThemSELVES' accordingly.

Form, however, is not of the Essence or the ESSENCE. Form is what one *has,* not what one *is.* What one *is,* is the ONE SUBJECT. Where there is only ONE, there is no possibility of comparison, for comparison requires at least two. When one consciously identifies as the SUBJECT, one can look in all directions and see only the SELF, and that SELF is, ESSENTIALLY, absolutely equally manifesting in all ITS apparent 'parts'. This is often forgotten, and pride results. So, pride is based upon the lack of realization of the *absolute essential equality* of all apparently separated selves. Where is the basis for favorable or unfavorable comparison when all are ESSENTIALLY and Essentially *equal?*

Pride is also based upon the lack of knowledge of the different stations and dimensions of the Universe, through which all *entities,* in their apparent separation must pass. Each E/entity (Really, the ONE AND ONLY 'ENTITY'), through participation an infinite number of Cosmoses, has participated endlessly in *all* of these stations or dimensions. There is no particular merit or demerit in finding oneself as a unit at one or another of them.

- First of all, each E/entity, Essentially, is experiencing *all* stations and dimensions, high and low, simultaneously, regardless of the illusory appearance of being confined to one.
- Secondly, in any one Cosmos, all E/entities in their 'mode of apparent partiality' must pass through *all* possible stations and dimensions. The particular moment in Cosmic Time when an E/entity focuses within one dimension or another depends upon the *timing* of its 'pilgrimage'. Not all E/entities (as 'Rays' of the ABSOLUTE) emanated from their apparently *superior emanating Entity,* at the same Point in Cosmic Time, and therein lies a most significant reason for the different states and dimensions occupied by Essentially *equal* E/entities.

The Problem of Self-Depreciation

Similar in origin to the Problem of Pride is the Problem of Self-Depreciation. This problem too, is based upon a lack of true perspective and a misunderstanding of IDENTITY. Self-minimization is based upon the non-realization of WHO one actually IS, and, again, of mistaking the form for the SELF. Forms are relatively high and low. When one identifies with a relatively low form and compares it with higher forms, or with the higher forms one *might* have, but does *not* have, then an attitude of self-depreciation may develop. Of course, practical in-Cosmos living demands that the form be assessed and improved. Great problems arise, however, when the REAL SUBJECT inherent in the form mistakenly identifies with the form.

A simple adage states the truth of an E/entity's relation to its form, "If you can *see* it, you cannot BE it." In the realm of REAL IDENTITY there can be no great and small, no high and low. Great and small, high and low relate only to the World of Relativity in which true IDENTITY cannot be found (though, necessarily, IT is always 'there'). Of course, the normal human consciousness has been trained to believe that exactly the opposite is the truth. The customary attitude can be stated as follows, "You *are* what you *have*." It will take assiduous retraining to avoid the pitfalls of ignorant self-exaltation and ignorant self-depreciation. These both arise because human units do not REALLY *know* what the SELF IS, or, even, what the *Self* Is.

The Problem of Impatience

When a human being is afflicted by this problem the personal self becomes the victim of intense desire for accomplishment of some kind, whether high or low. A ceaseless irritable chaffing arises when, for such a one, the patterns of circumstance do not 'move', it seems, fast enough to comply with the intensity of the desire. The impatient man wants what he wants *now,* and does not wish to wait.

As an antidote to this condition, several thoughts might be suggested. What if, in REALITY, Time is understood to be an illusion. What if, in REALITY, one can never run out of Time? What if it is realized that no matter how many subsidiary fulfillments may fail of accomplishment, the One Fulfillment, which is the Fulfillment of the Design-at-the-Beginning is a *certainty?*

■ The problem arises in chief measure because one is identified with 'the most familiar part'— that fragment of prakriti with which the SELF-'VEILED'-as-Veiled individual consciousness is most closely associated.

Seeing, in large measure, only that most familiar part, the imprisoned consciousness seeks chiefly for the fulfillment of that part rather than for the Fulfillment of the Whole. The consciousness is not decentralized, and is still subject to the limitations of egoism. Naturally, then, such a consciousness pushes urgently for the compliance of circumstance to its desires, for if desires are not fulfilled (many of them, and rapidly) success for the 'part' does not seem to come and a sense of well-being is not achieved.

The solution to this misidentification is a considerable broadening of the identification. The welfare of a larger *portion* of the Universal Whole would be a welcome initial consideration, followed by a gradually growing ability to care for the Welfare of the Universal Whole in Its Entirety. About the WHOLE of WHOLES no care need be given, for THAT will BE exactly WHAT IT IS *forever,* and nothing has changed IT, or will change IT.

Another solution lies in realizing that one IS, *already, all* that one could possibly want (though this is a very *mature* realization). Of what possible use would the urgency of impatience be under such conditions? Why be impatient to have what one already *has,* if one but knew it? Why be impatient to *be* what one already IS? Of course, the problem lies in the lack of realization. Of what use is it to *have* something and *be* something if one does not *know* it? It is a preoccupation with the apparent Not-SELF that generates impatience, and that makes the realization of one's TRUE 'NATURE' so difficult. Only when the apparent Not-SELF has failed to satisfy, time and time again, will there be a turning towards what one-as-One-as-ONE has *had* and *been* forever.

Definitely, there are two *major laws:*

- the 'LAW' of the INFINITE SELF
- the Law of the Finite Self (the highest form of which is the Universal Logos)

and two *major worlds:*

- WORLD of the INFINITE SELF, and
- the World of the Finite Self (which is the Cosmos Itself).

There is a tremendous and terrible *seeming* contradiction between these two *states*—INFINITE and Finite. That contradiction (because few human beings can resolve it at this stage in the development of humanity) can tear the consciousness apart, for it seems necessary to live simultaneously in ways that seem diametrically opposed.

Clearly, one must learn to "be in the World, yet not *of* It." This seems to be a statement easy to understand and even easy to fulfill. When, however, to 'not be *of* the World' means to be totally identified with the INFINITE SELF, or even the Universal Self, the Real problems begin, because, at first, the two seem so fundamentally irreconcilable.

Most people know a great deal of what is means to be "in the World", but the other requirement, to be "not *of* It", is almost entirely foreign to their presently limited consciousness. It is clear that one must know what the SELF *IS,* before it becomes possible to fulfill the requirement, and in so doing, eliminate not only impatience, but a host of other vices.

So impatience can be cured by meditating upon the true meaning of the SELF, and seeking to identify with IT. When there is some success in this, impatience is seen as the irrational state it is. Effort is certainly futile when urgently expended to achieve what has (from the perspective of the SELF) already been *achieved.* It must also be realized that whatever can be achieved in Time and Space is inevitably *less* than THAT—which has already been achieved. Is it not futile to attempt to achieve *less* than THAT which one already *has?*

There is a pertinent analogy to be found in the story of the man who (unknown to himself) has an overflowing treasure of gold at home and yet, ignoring his wealth, spends all his time in the streets looking for pennies. Master Morya in *Agni Yoga* tells us that irritation is to be worn away by "cosmic perception", and indeed, this is also the cure for impatience, which is a key ingredient in "Imperil".

The Problem of Work and Struggle

This problem might be framed as follows, What is worth working for? Are work and struggle worthwhile? If **8** already *Am* and *have* All **8** could possibly desire, why labor at all? Why labor when nothing, REALLY, can be achieved, and all that is, seemingly, achieved is, ESSENTIALLY, un-REAL?

It is futile to work and struggle to change REALITY, THE BOUNDLESS IMMUTABLE PRINCIPLE. Labor, work, and struggle must be for *other* Purposes. Within the World of Illusion (which is the World of the 'Great Game' which must be 'won', and of the 'Great Play' which must be beautifully 'performed') there is no way to fulfill in Universal Prakriti the Design-at-the-Beginning, except through *great labor.* In this thought, Work and Play are united on a Universal Scale.

The World of Becoming is (for all Its *seeming* Plenitude) a World of *Privation* when compared to the fullest fullness of the FOUNT OF ALL POSSIBILITY, the INFINITES-

SENCE. Through SELF-'VEILING'-instantly-Self-Veiling (which is REALLY apparent SELF-'DEPRIVATION') **I**, the SELF-of-all-Selves have, apparently, 'DEPRIVED' MYSELF of most of my INFINITE CAPACITIES. Of course, the deprivation is, ESSENTIALLY, un-REAL, nevertheless, according to the 'Rules' of the 'Cosmic Game', one must *labor intensely* for the restoration of those SELF-'DEPRIVED' powers.

The labor is for something un-REAL in an ESSENTIALLY un-REAL Context (i.e., the Universe)—un-REAL compared with the GREAT REALITY. Nevertheless, the labor is worth the effort. In fact, the labor is *inescapable*. Without that labor, that work, that struggle, the ONE AND ONLY SELF could not BE *fully* WHAT IT IS, for SELF-'DENIAL' is a *necessary* ingredient of SELF-'AFFIRMATION'.

■ The idea so often advanced in this treatise is that in order for the INFINITE SELF to BE ITSELF *fully* IT must, *necessarily*, be subjected to Finitude. Contradiction and paradox are therefore, according to this view, indispensable, and so is the action of labor in-Cosmos, which is totally impossible in the WORLD OF THE ALL-in-ALL.

In labor there can be a great joy, even bliss. In "ceaseless labor" (Morya's words, meaning {perhaps} ceaseless for the duration of the Universal Manvantara) one 'Plays the Game' according to One's SELF-as-Self-Intended Design. One fulfills One's Own Primeval Will (the Will of the One Identity in Cosmos). There is a profound satisfaction in this, even though, in doing so, the INFINITE SELF is, *apparently*, subject to great privation. We also realize that even though the privation seems REAL, and is Real enough in-Cosmos, it is *simultaneously* un-REAL, and on the level of the unchanging INFINITE SELF, there can never be any 'PRIVATION' at all, even *cyclically* (as *seems* to be the case).

The Problem of Pain

Pain, though ESSENTIALLY illusory, is certainly Real enough to the experiencer. Pain is an event in consciousness caused, Essentially, by the difficulty (some might say, "the impossibility") of reconciling the Finite to the INFINITE. Its primary cause lies in the Illusion of Duality. Pain is essentially the reaction of consciousness to the illusion of SELF-'DIVISION'-as-Division. The First 'ACT' 'within' the ACTIONLESS ONE was the 'CAUSE' of the *origin* of *Pain* on a Universal scale. The 'ORIGIN' of 'MAYA'-instantly-Maya and the origin of Pain are coeval.

There are some less abstract ways to understand the Problem of Pain. For one thing, pain arises from *friction* between the apparent parts within any system, including the System called Cosmos. Pain, from this perspective, is inharmony, itself. The energy of Love, which is a Great Unifier, is thus, one of the leading assuagements of pain.

From another perspective, one of the fundamental causes of pain arises from the inevitable initial non-responsiveness of matter to spirit, and ignorance of the ways to render matter more responsive. The Universe is born through SELF-'VEILING', but It proceeds to differentiate and *involve* through Self-Veiling. (The Full Being of the Universal Logos is progressively *veiled* as involution occurs.) The price of involution and differentiation is ignorance.

The supreme 'Patterns Above' in any Cosmos are less and less successfully reflected in the descending prakritic layers. (The Self-Reflective Emanative Entities 'See' reflected as Object/Prakriti less and less of what They Essentially are.) As a matter of fact, the entire Theme of Cosmos and the Object of the Cosmic Game is to render the lower prakritic layers as totally responsive as possible to the higher Logoically-Intended Patterns on the highest levels of Cosmos (which are probably not prakritic/particulate in the usual sense, but *are* 'partite').

Until the lower prakritic layers respond frictionlessly to the Divinely-Intended Patterns, consciousnesses on the various levels involved will experience pain. Again, we can see that the experience of pain is based upon duality.

- The Higher Patterns are experienced by the lower forms through which they must manifest (by Local Identities) as *pressure.*

Intelligent adaptation of the lower forms and forces to the descending Energies can relieve the pressure, but the Veil of Ignorance prevents the understanding of how this may be accomplished. Clearly, as various of the Veils of Ignorance begin to lift or be worn through, the knowledge so necessary to the alleviation of pain is achieved.

From still another perspective, pain can be considered a guide which is useful for reminding certain aspects/'Rays' of the SELF-as-Self in *deep prakritic immersion,* (and consequently in a state of profound forgetfulness) to hew to the Original Divine Intent, as much as that Intent is capable of realization at any particular depth of prakritic immersion. Pain, in one of its modes, is an indicator of deviation from the Original Pattern. That deviation is Really inevitable, and is caused by the Cosmic Process of Veiling and the ignorance which is inevitably the artifact of that veiling.

So pain is one of the conditions of consciousness that **I**-as-**8** have 'Created' with Cosmos. As long as **8** am estranged from MYSELF, (and, more practically, from MYSELF-as-MySelf) there will be pain. As long as there is veiling and subsequent illusory limitation, there will be pain. Obviously, during each Universal Pralaya, there can be no pain, for there can be no duality of any kind.

From a psychological perspective, pain is the natural response of the consciousness to experiencing oneself as *less* than one fully IS. In this respect, it could be said that *all* Self-conscious units within Cosmos are experiencing pain to some degree. The Spirit loathes diminishment (or, Really, the *perception* of its own diminishment). Psychologically pain is the 'distance' between the 'actual', and the Real-then-REAL. Of course there is no REAL 'distance'. Even the 'actual' is fundamentally the REAL, and every E/entity is *already,* could it but know the fact, *fully* what it consciously (or even unconsciously) seeks to be. But there is a *perceived* 'distance', and awareness of that apparent distance causes an condition of inharmony or malaise which we term pain.

From a more *physiological* perspective, *stretching* beyond a certain point causes pain. A given structure *is* in fact a certain configuration of relationships. One might wonder whether pain has its roots within the micro-'level' of the Cosmic Configuration. When these relationships are forced and stressed, pain results. A duality arises. The tendency of the structure to remain as it is, is pitted against the forces which would change its configuration, and a *tearing* results. Thus, pain is caused by the violation of existing conditions, existing structures. Pain and the *excessively forceful alteration of structure* are synonymous. Rending and tearing result. To reach and stretch too far, to be compressed too much will result in ripping and tearing. Ultimately a tear is the same as a separation and a separation (again the dualism) is a chief cause of pain.

In the WORLD OF BEING (not, the intra-Cosmic World of Being which is still subject to Limitation), there are no separations. There are no 'tears!' (Note the equivalence of the two meanings of the word 'tears'.) There are no boundaries or distinctions. There can be no rending process.

- In the World of Multiplicity and Division which is the World of Becoming, there is *always* a rending process. Thus, in a way pain is inseparable from Cosmos. Pain *is* Cosmos Itself.

Within Cosmos, pain is overcome as duality is overcome. Pain is disintegration, while joy and bliss are integration. In Cosmos, the Highest Synthesis is the Highest Bliss. There are ways, of course, that dis-integrations lead to higher integrations, and thus, pain leads to bliss. In general, it might be said that in the World of Creativity, however, which is the World of Becoming, for a long while disintegrations and re-integrations follow one upon the other, and so pain must follow upon joy and joy upon pain. To this fluctuation, **I-as-8**, the SELF-as-Self have submitted MYSELF/MySelf. **I-as-8** have done so *willingly* and *necessarily*.

So, we find that, in-Cosmos, pain is inescapable, but it need not be experienced only from a limited state of consciousness. It can be experienced from the one of the higher reflections of the 'STATE' of ABSOLUTE SELFHOOD, from a State of High Abstraction (even if not from the 'STATE' OF *ULTIMATE* ABSTRACTION). There are various approaches. One can so intensify the resident SELF within, and thus becoming so increasingly identified with the REAL that the un-REALITY of dissonance begins to pale. Thus, the intensification of Synthesis neutralizes Duality.

- Pain is, Essentially, preoccupation with and absorption in dualistic relationship. The duality-reconciling World of Harmony (allied to the buddhic and monadic planes) can become so intensified that pain becomes virtually non-existent through inattention. We can see, therefore, that pain is overcome through the intensification of a state of greater harmony and synthesis.

It must not be thought that the presence of pain necessarily indicates that something is terribly wrong. The experience of pain (as in the "sacred pains") might indicate that something is terribly right! Sometimes within the very *actual* hierarchical order of things in Cosmos, a lesser level of being (a lesser Creative Hierarchy) must experience pain so that the greater purpose of a greater Creative Hierarchy may be achieved through the agency of the lesser. This results in the pressure of a greater energy upon a lesser energy.

We are presently in the midst of a great Planetary Forcing Process. Both the Spiritual Hierarchy of the Earth and, Shamballa, as well, are impacting Humanity as never before. Pain will necessarily be experienced due to unaccustomed pressure. For those who avail themselves of the opportunity, however, and, consequently, experience a measure of the inevitable pain, the pain is an indicator of progress, and not of maladjustment.

In summary, pain may be the result:

- of the inadequacy of matter to respond to spirit; or
- of wrong and unnecessarily frictional relationships; or
- of the result of moving in the wrong direction; or
- simply inherent and inescapable, due to the Metaphysics of Cosmos—based upon the incomensurability between the INFINITE and the Finite.

Pain will be experienced by every E/entity in Cosmos, and, may **8** add—pain will be experienced cyclically forever by all E/entities in all Cosmoses-to-Come just as it has been experienced in an infinite number of Cosmoses past. Even as pain is apparently experienced, the 'super-current' of the COMPLETE BLISS of the INFINITE SELF is *simultaneously* being experienced, and the BLISS of the INFINITE SELF is REAL, whereas the pain is ESSENTIALLY illusory.

It takes much development of consciousness to realize this distinction between the REALITY of INFINITIZED BLISS and the un-REALITY of pain, but for all E/entities (who are, let it be remembered, essentially but ONE INFINITE ENTITY/NON-ENTITY) the realization is inevitable. As the Tibetan has said, speaking of developments which, for many human beings, loom not so far ahead, and which, hence, are intra-Cosmic developments—"Naught remains but Bliss."

The Problem of Desire

Is the Problem of Desire how to overcome it and even eliminate it? Or is it rather to understand it as an inherent aspect of Life in Cosmos, and to learn to direct it?

■ Desire is what keeps all apparently separated 'parts' within Cosmos moving towards the 'pre-Cosmically determined Point of Consummation', which We-as-**8** all agreed upon as the Structure of Fulfillment which expresses the Finality of the Fixed Design, the Original Intent, the Purpose of the Universal Logos.

For every E/entity, desire is an extension of S/self. It is a reaching beyond what is *actual* toward what is more *ideal* (i.e., desirable). With regard to the Intent of the Universal Logos, Desire is a progressive motion and instrument of Cosmic Progress towards a Pre-Cosmically Conceived Conclusion.

Those who have realized the SELF, and have consciously become the SELF, will not have desires in the usual sense, because the SELF is the EVER FULL, the infinitely enduring PLENUM towards which all desire tends. This is even true of those who have realized the SELF-as-Self (perhaps, a more accessible realization). Within the World of Relativity, however, it is impossible not to have desires and still participate appropriately in the dynamics of the Cosmic Process.

Desire is the impelling dynamic of the Cosmic Game which We-as-**8** have created and constructed at the infinitely recurring Beginning of Cosmic Time. Although in the SELF, ALL is (from BEGINNINGLESSNESS through ENDLESSNESS) achieved in full, within the World of Illusion (a most *necessary* World—the World which the Universal Logos 'Becomes') there exists the illusory possibility of something yet to be achieved and the apparent presence of a means to achieve that something.

The *Desire* which initiates the Cosmic Process (or shall we call it *Will* at that early Time in Cosmic History) impels towards achievement, driving the Intention at the Beginning into manifestation. When we contrast Desire and Will, we find that in Will there is more illumination. D̲esire, and especially d̲esire, is what Will becomes as the Self-Veiling Process in Cosmos proceeds along the Involutionary Arc.

To the E/entity who has realized the SELF within the Self, there exists the possibility of remaining detached from Desire, and of *choosing* which desires to have. For the most part, E/entities are *possessed* by desires rather than *possessing* them. Many human beings may say, "I have this or that desire", but more truly, the desire *has* *them*. The one who *chooses* desires can *direct* those desires (or impulsions towards the new), towards the fulfillment of Divine Purpose, and not be *driven* towards mistaken gratification.

REALLY, and even R̲eally, all that can be desired is as nothing compared to what *already* ETERNALLY IS. So, while it is necessary (in our SELF-'VEILED' and even Self-Veiled states) to have desires, one must realize that what one appears to desire is synthetically fulfilled by the SELF that one already IS. Given this realization, desires, then, using an equine analogy, can be 'ridden' somewhat more loosely than otherwise might be the case.

One must identify Desire in general (and desires in particular) as a kind of energy which leads towards Cosmic Completion, for right desire moves the Cosmic Process forward. The Planned Finale of the Cosmic process is *desirable,* and it is That which we-as-**8** have agreed upon from the Beginning (the Beginning of this particular Cosmos)—we who are fully Present-as-Presence in each apparent division within Cosmos and fully Present-as-Presence in the highest possible transcendent Dimension as well.

- With respect to Desire, the important achievement is to *drive* and not be *driven.* This becomes possible when one is a successful 'non-dualist' or 'infinitist', and, thus, achieves a sense of COMPLETE SUFFICIENCY through identification with the INFINITE SELF. Only that E/entity is free to desire wisely who realizes that Desire is not a means to *acquisition,* but only a means to fulfill in manifestation the Design-at-the-Beginning.

The personal or even individual elements are then removed from Desire and It can be utilized as a beneficent Cosmic Energy facilitating the Purpose of the Universal Logos (at Root, Our Own Purpose).

The Problem of Values

A value is a cherished vision, often justified by thought and sustained by a steady stream of desire. According to one's values, so the energies and forces at one's disposal will be directed. Each disciple must at length deliberate upon the question of what is valued most. If we were to attempt to classify values, three great categories appear:

- temporal values
- systemic/holistic values
- ultimate values

1. Temporal values concern the personal and individual spheres of consciousness; They concern the self.
2. Systemic values are holistic and concern the larger concentric Spheres of Consciousness in which the unit S/self finds its part to play. The Wills of our Planetary Logos, our Solar Logos, our Constellational Logoi (the Lord of the Seven Solar Systems), our Galactic Logos, etc. and even the Universal Logos, determine systemic or holistic values. These are values as seen from various positions of increasing decentralization. This order of values concerns the <u>S</u>elf.
3. Ultimate values concern REALITY purely and simply. These values transcend the specifics of any micro-level individuality or any macro-level Systemic Being (Which, might also be considered a great Individuality). These values do not concern themselves with any 'created' thing whatsoever. They concern the SELF and the SELF alone. These values are entirely non-dualistic and 'infinitistic'. These are the values which the lover of REALITY seeks to adopt, even while finding it necessary to hold certain values from the individual and Systemic/holistic levels.

The Problem of Values arises when we realize that there *are* three classes of values, each of them important in its own sphere. The student of the disciplines of SELF-REALIZATION finds it necessary to maintain all three, and, problematically, the three levels are extremely difficult to reconcile. How shall the student attempt to reconcile the three levels of values and still maintain integrity of consciousness? How shall this be done without splitting the consciousness?

Some might argue that only ultimate values are of importance, and indeed, these values do contain all others. They are fundamental. It is not possible, however, to live successfully in-Cosmos maintaining *only* ultimate values. To do so is to ignore the fact that the INFINITE SELF, **I**, *became* the Universe for what we, of limited mind, can only call a 'Reason'. Further, given the 'LAW' of BEING, there is no way to escape the Presentation of an beginningless/endless succession of SELF-'BECOME' Universes. To 'BECOME' Universes is what **I**, the ABSOLUTE, 'DO', and It is the *Only* Thing **I** 'DO'. Within MYSELF, **I** 'DO' *nothing* (NOTHING).

Therefore, we must *value* our 'OWN' 'CREATION', *The* Creation—That which we-as-**I** must, *necessarily,* 'DO' forever. Each Universe is a great Son of Necessity. Its Creation is not a random or meaningless 'ACT'/Act. We—the-**I** (by 'BECOMING' a Cosmos) apparently deprive 'OurSelves' of OUR-SELF, only to restore 'OurSelves', to 'OurSELF'—

even while forever *remaining* THAT, the SELF. So while the most important values pertain to the ONE AND ONLY SELF (which, ITSELF, is the greatest of all VALUES), there are lesser but still important values which pertain to SELF-'RADIATED' Beings and Their Derivative Emanations (the sequential Universes being the foremost) These Universes must be treated with respect, whether or not they are REAL in the *absolute* sense.

We are faced with a dilemma. The World of Name and Form is un-REAL, but cyclically *presented* throughout Infinite Duration. While It is un-REAL, It is a persistent *Fact* of our consciousness and must be encountered. It cannot be despised or ignored, otherwise we are not working in accord with the FORCE which 'MADE' It (the Cosmos) *necessary* in the first place. Yet, in and of Itself, all Cosmoses are 'less than *nothing*'. Indeed, they are infinitely less that NOTHING—NOTHING being the ONE INFINITE SELF, the ZERO. What shall be done to solve the problem?

- Perhaps we can value, but not over-value, the World of Name and Form and Its presentations, even while holding the highest esteem, and the highest value, for THAT which changes not.
- Perhaps we can avoid depreciating or extolling the un-REAL (though Actual) Universe, even as we are immersed in the realization of the supreme value of the REAL.
- Perhaps most importantly, we can learn to see the apparently un-REAL in terms of the REAL.

If one values the SELF, the INFINITUDE, supremely, and is thoroughly identified with IT, then one will come to realize that all lesser things cannot help but BE *IT* as well. The lesser beings are REALLY the GREATEST BEING, and the lesser values are substanded by the GREAT VALUE. One must learn to find the INFINITUDE in all apparently lesser things, and thus value these lesser things, not for their *seeming* selves, but for the SELF that they REALLY ARE, ESSENTIALLY.

From a practical perspective, it must also be realized that the process through which worthy intra-Cosmic 'things' that are held in esteem, gradually leads to the realization of the THAT which is most valuable, the SELF, the GREATEST VALUE. We learn to ascend on the Scale of Values, for no thing lesser than the SELF can forever satisfy. It has been said that "BRAHMAN and Samsara are ONE." By a certain *merging* and *dissolving* process in consciousness, it is possible to transform all heterogeneity into the GREAT HOMOGENEITY, and then every little thing (intrinsically worthless in its disconnected evanescence) becomes the WHOLE of BRAHMAN, ITSELF, and can thus be valued supremely for its ESSENCE.

■ In relation to all these philosophical considerations, what is most important is, somehow, to *live the realization.* One is required to "live a dual life" if one is to BE the INFINITUDE, the BOUNDLESSNESS, the SINGULARITY, even while apparently experiencing all of the struggles and pains and illusions of the World of Multiplicity.

BRAHMAN IS the SUPREME VALUE. Though the Universe is the Great Illusion, It must also be valued, if not for Itself, per se, then for the *fact* that It, too, IS BRAHMAN, and BRAHMAN alone.

The Problem of Values —a Further Consideration

A value is that which is repeatedly desired, or a desire which is maintained with relative constancy regardless of the appearance and disappearance of a variety of fluctuating desires. One's values depends upon one's World View. There is, for example, an ascending 'ladder' of ego-centered values, soul-centered values, spirit-inspired values, Universal Values (That which is valued most by the One Being who informs the Universe), and, finally, THAT which IS of SUPREME VALUE in ALL the UTTER ALLNESS. Obviously, THIS/THAT is WHAT every E/entity-in-Cosmos is *seeking* whether or not such an E/entity is conscious of the fact.

The Problem of Values is therefore a problem of *Perspective. Who am I* when I desire something? As my *point of identification* changes, my values change. In a general sense, we might say that the values related to the WORLD OF BEING become *inapplicable* when applied to the Total World of Becoming, and vice versa.

In this case we are focusing upon a contrast between intra-Cosmic and SUPER-Cosmic values. Within Cosmos we might say that the values of the World of Being (a World located, from a practical solar systemic perspective, upon the higher Systemic Planes but, Really upon the higher Cosmic Planes, and, even, perhaps, upon Super-Cosmic, i.e., Kosmic Planes {if the number 100 means anything archetypal to all systems}) are inapplicable or difficult to apply within the World of Effects (the Worlds of Fabrication) and vice versa. Naught and aught, 'no' and 'yes', merge into each other when contexts are seen from opposite perspectives.

■ Right Action is all a matter of knowing *what to do,* and *how, and where to do it.*

To achieve right action, illusions must be 'seen through' and they can only be seen through by thinking about things "under the Aspect of Eternity"—"sub specie Aeternitas." Under that ASPECT/Aspect (which is a kind of LIGHT/Light) are seen Great Truths. Perhaps, even the TRUTH is sensed. It is these Great Truths that must be used to conquer, first, illusion, and then, the Great Illusion.

The limited sight that leads to wrong action must be replaced by the Greater Sight that makes wrong action impossible. In this case, to *know* the Good is to *do* It. Perhaps this Socratic Aphorism should be modified to read, "To *become* the Good is to *do* the Good." Of course, we already *are* the Good (and the GOOD as well), but we, in our present apparent limitations, have to both *know* It and *consciously become* It in order to *want* to *do* It. *Becoming* the Good, runs much deeper than simply *knowing* It. In this case, *becoming* is an act of Identification.

The World of Effects (the Mosaic World) must also be coordinated with the World of Universal Archetypes (the World of Being, the World of Divine Patterns), *both* of which worlds are to be found within the Total Cosmos (which is the World of Becoming, considered *en large).*

The consciousness of the ordinary limited human being is filled with preferences and abhorrences. Even when the consciousness is turned towards subtle, relatively immaterial items-in-Cosmos, there are still preferences and abhorrences. In the field of occultism, a disciple prefers to be 'on' one Ray rather than another Ray, or to be in one Ashram rather than another Ashram. These are valuations which are formed 'a long way down the mountain'—by no means at the Pinnacle of Synthesis. No Ray is 'better or worse' than any other Ray, nor any Ashram 'better or worse' than any other.

From a certain Pinnacle of Consciousness that is *identified* with the One Being in-Cosmos, it can be realized that only the *partial* individual seems to 'be' one thing or another, or is 'in' one thing or another. From this Pinnacle, one-as-One (i.e., one in a state of identification with the One) realizes that It *Is* and *Owns* (as it were) all Universal Conditions, and at length will Identify with them all—*consciously.* **8** Am, surely, the Author and 'Inperiencer'/Experiencer of them all. **8** Am the Actor in all (and the Drama is a Monologue). Owning the All, Having the All, Being within All, why should **8** value one condition over another. 'I' do so, precisely because 'I' am *I* and not consciously **8**. At length 'I' will *be* (consciously) That which **'8'** already *Am* . Until that time, **I/8** will have a *double standard of values:* one for myself-the-ignorant one, and the other for 'MySelf' the All-Seeing **8.**

What would my values be if I could realize that **8** Am the Whole Playing a part, and yet, mysteriously, remaining the Whole (and even the WHOLE)? Perhaps my values would be Truer if **8**-as-I would understand and use the mantram 'NOW-Now-now'. In this mantram we witness the descent of the PRESENCE.

- Notice that each "now" can be broken down into 'no—w'. What this means is that to live in the now ensures that there will be no 'w' (i.e., no 'double-you'—no sense of dualistic identity). The NOW-as-now, brings an end to the consciousness of duality. From such a perspective, all values will change for the Better—or, better yet, for the BEST.

The most intensely observant focus is needed to live in the Now (Cosmo-Objective, Cosmo-Subjective, and Cosmo-Eternal {the Goal}) and thus negate duality. Even living in the 'now' (from so-called moment to moment {Really, from macro-moment to macro-moment) is difficult enough). The simplest things are the most difficult. All complexity of consciousness must be banished if the now-as-Now-as-NOW *realization* is to be achieved. To *realize* thusly, one begins to 'See' with the *Real Eye* (the Eye of Synthesis) as a Real **8.**

The problem is that human attention is usually 'sticky'—it sticks to that which changes. It attaches itself to past combinations or, overly eagerly, attaches itself in an anticipatory mode to future combinations. Attention (as usually practiced) could be called *combination-adhesive.* With this limiting adhesiveness, the full impression of the *essence* of the moment cannot be registered.

Combinations are registered and the Now fails to make Its impact. The Cosmo-Objective Now as it grows gradually into an apprehension of the Cosmo-Subjective Now and, thence, becomes the Cosmic Eternal Now is the *aperture of detachment* from illusory combinations—in fact, from all combination whatsoever, as *all* combinations are illusory.

The Problem of Progress

Within Cosmos, human beings who are awakening to their possibilities often become obsessed with the possibility of rapid progress. They desire to *have* more and to *be* more in every way. Within the Universe, apparent progress is certainly possible. Within the World of Illusion the scope for expression of each and every E/entity can apparently be expanded until that scope becomes identical to the Scope of the Universal Logos (which is *not* the same as *becoming* 'a!' Universal Logos). Then it will be discovered that each E/entity has always *had* that Great Scope, from start to finish of the Universe—a Mystery based upon the Indivisibility of BEING (and, even, upon the Mystery of Universal Logoic Being).

If we consider these matters from the perspective of REALITY (which does not REALLY *have a perspective)* however, we must conclude that *there is no REAL progress.* There is only *actual* progress. It is possible and justified to rejoice in *actual* progress even though one knows that it is no REAL progress, perhaps in the same way that one rejoices in winning a game or in playing successfully. Name and Form can, indeed, progress. The *means* of expressing the Divinely Intended Pattern for a particular Cosmos can *progress.* THAT which is the true SUBSTANCE of expression, however, can *never* progress. Progress from where to where? From what to what? IT already IS ALL IT *can* BE. Inasmuch as all E/entities are inescapably IT, they, too, cannot REALLY progress.

What can be done, however, since, in-Cosmos, progress is required. That simultaneously complexifying and simplifying re-patterning that we call progress is a fundamental, indispensable dynamic in-Cosmos, and we, as apparently limited E/entities, must perforce participate in it. Not only that, but the Secret Teachings of all Ages enjoin upon us that need to pursue our 'progress' with extreme vigor, if we are to be "liberated".

■ What comes to mind is the old saying that one is to act with all the intensity of one who has ambition, without being ambitious. Similarly, one is to pursue one's apparent progress with extreme vigor realizing that it contributes to the progress of the larger systems in which one finds place, all the while realizing that nothing is being achieved that can alter the 'PERFECTION' of THAT which is already ETERNALLY PRESENT.

From another perspective, however, the importance of 'progress' can be justified, for it is *necessary* that the FOUNT OF ALL POSSIBILITY proceed endlessly, throughout Infinite Duration, 'EXPRESSING' those 'EXTRUDED' possibilities in *conditions* of extreme privation and limitation—i.e., in-Cosmos. That 'EXPRESSION'-as-Expression-as-expression under conditions of limitation is the One Great 'ACT' of IT-'BECOME'-It (the 'ACTIONLESS' ONE-'BECOME'-the Divine Actor).

To the extent that the P/players within the Cosmic Game seek to progress, they are facilitating the necessary 'EXPRESSION' of the ONE AND ONLY, and in that way, cooperating with the Will of the Universal Logos, which is only a Reflection of the SELF-INHERENT NECESSITY that compelled the ONE AND ONLY to 'BECOME' through Finitization the Universal Logos and Its Universe.

So, in a way, there is nothing futile about the pursuit of progress. True, it does not add one iota of 'improvement' to the ALL-SELF (which one, ESSENTIALLY, IS), but it does assist in the fulfilling of the Contract-at-the-Beginning which the ALL-SELF in ITS 'GUISE' as Universal Logos *contracts* with ITSELF to fulfill—all out of ETERNAL *NECESSITY*.

When a Game is well-played the players leave the playing field in a state of satisfaction, glowing with their accomplishment. Then, they proceed with their *real* life. When the Game of Cosmos is concluded, the Host of Entities Who are the Players upon the Cosmic Field, shall also leave the Field (albeit, as One Player!) in a state of profound satisfaction and then, also, proceed with Their REAL LIFE, the LIFE of the INFINITE SELF, which IS as REAL as the Game was Illusory.

The Problem of Humility

Humility is often enjoined upon students of spirituality without being understood. While it is considered 'spiritual' to be humble, and unspiritual to be proud, the root of both concepts remains unexamined. Within Cosmos is it possible to be either humble or proud or both simultaneously. Within the SELF, it is possible to be *neither*.

Humility (and pride), as usually conceived, are both based upon *mistaken identity*—as are so many of humanity's follies. Let us focus upon humility as we have already treated pride. In Cosmos, there is nothing to be said against true humility; it is related to a "developed sense of right proportion", as the Tibetan has said. The problem lies with *"false* humility" that arises because of the inability to 'co-measure' (Morya's term) and, even more, because of lack of experience <u>*as*</u> (or in *identification* with) the ONE SELF. The unenlightened consciousness assumes that what is large is large, and what is small is small, and wrongly identifies the SELF-as-Self with either or both. The SELF-as-SELF is never known nor is the SELF-as-<u>S</u>elf.

False humility is based upon the necessity, at all costs, to avoid the appearance of claiming that one is 'bigger' than one is; under its spell it is even considered proud to acknowledge that one (as a Focal Identity) stands where one stands on the Ladder of Evolution; further, it is considered rightly 'humble' to pretend that one is smaller than one actually is. The key word is 'pretend', because true knowledge of one's developmental status within Cosmos is missing. When the Greater Systems within which the E/entity is enfolded are better known, a truer sense of proportion develops and true humility appears as the virtue it is, because, when possessed of true humility, one can act intelligently, fittingly and helpfully within one's context.

Perhaps an even truer sense of humility is based upon the realization that there is neither large nor small within the SELF (which all E/entities, ESSENTIALLY, ARE) and that the SELF can never vary. The SELF can never be *other* than ITSELF; IT cannot be great; IT cannot be small, but only both and neither. To realize that 'great' and 'small' are REALLY illusions—*veils* upon WHAT one-as-One-as-ONE REALLY *IS*, renders the one who realizes humble in a new and strange way. One looks upon all magnitude with an "equal eye", and with a compassion and understanding which only the conviction of the ESSENTIAL *identicalness* of all things can bring.

There will still be recognition of magnitude, of course. There must be, if one is to live his 'entified life' successfully within the World of Illusion. We have established that this type of *successful* living is *necessary* to the SELF. Judgment and criticism, however, are suspended. We know, from studying all that is best about the virtue of humility, that the truly humble individual is not judgmental or critical because he truly *understands* proportionately.

Within the humility born of SELF REALIZATION, the humble individual sees all differences, and yet, no ESSENTIAL difference. In the SELF there is no difference. Instead, there is the deeply satisfying realization that one IS the SELF. There is joy in this realization and even bliss.

With this comes the further realization of all the different Cosmic "posts and positions":

- through which one (as an E/entity—the indivisible 'Ray' of the ABSOLUTE) must, perforce, apparently 'travel';
- through which *all* E/entities (Really, all are but the One Cosmic Entity) 'travel';
- through which the WHOLE-as-Whole is *not* 'traveling' but *abiding,* even NOW, through the agency of all its apparent 'parts'.

It dawns that one is necessarily *active* and *present* at all points and in all stations and stages in Cosmos even NOW, and that, though one seems to have left a certain level of prakriti immersion behind (whether 'above' or 'below'), one is *somehow,* even NOW, ESSENTIALLY Present as the PRESENCE within that state of immersion.

In the light of these considerations, we have to consider what might be called the *'illusion of immediacy'.* **8**, the WHOLE-as-Whole, Am focussed within this *immediate* ring-pass-not and *seem* to be most *here,* but Am as much in *apparently other* far less *immediate* ring-pass-nots as **8** Am in the one with that **8** *seem* to be most persistently associated. One begins to "see through" this particular illusion.

■ Humility is strengthened as one recognizes the many dimensions, planes, levels, and states of Cosmos and realizes that not only must one apparently pass through them during Cosmic Unfoldment, but that one is even *now* manifesting through them. The high, the low, the great the small—these are all poles of one's true BEING-as-Being even NOW, for one is, even NOW the high, the low, the great, the small. Yet all of these are as *nothing* compared to the SUPREME NOTHING (the INFINITE SELF) WHICH one truly IS even NOW.

So, both the Problems of Humility and of Pride are based (again, as usual) upon *mistaken identity,* the failure to distinguish the INFINITE SUBJECT from the Finite Object. They are based upon:

1. The foolishness of thinking that one is one specific *thing.*
2. Comparing the name and form of that thing to the name and form of other specific things, thinking that, somehow, in so doing, one has learned something about one's true nature.

When one realizes that one IS *all things* because one IS the INFINITE SELF that *homogenizes* those things, all comparison of the apparent differences between little self and little self is seen in an entirely new light, for how can one compare the seamless INFINITE SELF, the ALL-INCLUSIVE, the INCOMPARABLE, to anything but ITSELF? Even then, there is no possibility of *comparison,* but only of *recognition of identicalness.*

When the REAL IDENTITY of the little selves is *known,* the comparison of little self to little self will negate the usual kind of contrasting, and, instead, will only produce a recognition of the ABSOLUTE IDENTITY 'shared by' the E/entities compared.

The Problem of World Denial

The philosophical system of Non-Dualism has often been faulted for inducing in its advocates a passive state of world-denial, which, reasonably, can be judged as both useless and selfish. In this 'Radical Infinitist' reformulation of the Doctrine of Non-Dualism, world denial or world rejection are *not* seen as necessary and inevitable consequences of the acceptance of non-dualistic world view.

Is it not contradictory, unreasonable and irrational of ME to both 'CREATE'/'BECOME' the World of Becoming, and then *deny* or *reject* it? If it is argued that 'MAYA'-instantly-Maya and not **I/8** Created this World, the question would have to be posed:

- Who or What is 'MAYA'-instantly-Maya if not I, MYSELF? Who or What can apparently *veil* the SELF if not the SELF-ITSELF?

And yet, so many of those who claim to realize and identify with the ABSOLUTE SELF think that it is incumbent upon them to deny and reject the World of Multiplicity.

If **8** accept that **I**-as-'MAYA'-instantly-Maya 'BECAME' this Universe through a SELF-'VEILING'/Veiling Process, **8** cannot blame Illusion on Maya alone, nor can **8** reject the World as merely *'Her'* Creation. **8** (knowing My<u>S</u>elf, REALLY, to be **I**) Am forced to take responsibility for what **8** have done—to take responsibility for the Universe **I**-as-**8** have *Become.* In any Action that has taken place, **I**-as-**8** have been the 'ACTOR'-as-Actor. **8** cannot pretend that MY 'ACT' of 'RADIATION' leading to <u>My</u> <u>A</u>ct of 'Creation' had no 'REASON' or 'PURPOSE' behind It.

Yes, the Universe is <u>ABSOLUTELY</u> un-REAL, but It is manifestly *Actual,* and 'here' for a very Great incognizable 'REASON'. Many non-dualists refuse to pay attention to the World of Becoming simply because it is un-REAL, but, it might be asked: Is it possible for that which is *apparently* un-REAL to be *REALLY* un-REAL? Since only the REAL *IS,* the un-REAL must be the REAL *also.* So, attention must be paid even to the un-REAL which, if It is to be truly evaluated, must be considered as *ESSENTIALLY* REAL.

We are presented with a Universe which is undeniably *Actual.* It cannot be ignored simply as dismissible Illusion. Mass consensus affirms that it is 'there' or 'here', and unless we call mass consensus "mass hallucination", we have, at least, to regard the World/

Universe as an undeniable factor in human consciousness. One would have to deny every presentation of which one is aware, if one were to deny the seeming presence of the World.

Even though there will be a dispute concerning the *REALITY* of the World, at least Its *Actuality*, cannot be denied. If the World-of-Becoming is Actual, Its *'ex-istence'* is assured. If It is Actual and exists, being a finite Thing, It must have been *created*, and if It has been created, It must have been created by an *Agent* of Creation. There is, however, but *one* Agent of Creation and that Agent of Creation is **I-as-8**, MySelf (substanded by the INFINITE **I**, the INFINITE SELF). **8**, therefore, have Created or somehow *Become* the World. If **8** (being **I**) *do* something, **8** do it because of My Own Will (which is ESSENTIALLY, the 'WILL'); there is no other *will* to compel Me-as-ME.

8 Am the Creator, and **I** AM also the ONE INFINITE IDENTITY *substanding* **8**-the-Creator (the Universal Logos). Apparently **I**-as-**8** have been 'Creating' thus for an Infinity of Ages, throughout Infinite Duration. Therefore, there must be some *Value* in this Act of Creating, a Value based upon Original Divine Intent—unless, of course, **I**, the ONE AND ONLY and SPONSOR of all Creators (including **8**) am quite 'MAD'—a definite *possibility* within the FOUNT OF ALL POSSIBILITY!

Therefore, it would be foolish for the human being simply to *deny* what is, for all human beings, (and Really, for all E/entities) the Great Presented Fact of Consciousness. So to the question, Why take any 'thing' in Cosmos (including Cosmos Itself) seriously, or care or value the World of Becoming in the slightest, since It is only *Actual*, but *essentially* un-REAL? one can answer, that the World of Becoming, too, is ESSENTIALLY REAL because there is, ESSENTIALLY, *naught* but the REAL. Its Form alone is an Illusion.

Another answer that should be proposed to would-be World-deniers can be derived from the following: the yawning gap between what **8** know Essentially (because on the highest levels **8** Am the SELF-as-Self), and what 'I' know as a limited person has been cited over and over again. It must be assumed that my present knowing as a limited person is insufficient for ME-as-Me-as-me to know the reason why **I**-as-**8**, carrying out My Original Intent, My Design-at-the-Beginning, Created the World according to Its Parameters. Given, then, the present low state of human consciousness (in and of itself), it is premature to deny the World!

The Problem of Tolerance and Intolerance

This problem is basically one concerning 'like and dislike'. 'Like and dislike' are usually based upon the factors of resonance and harmony. The form of each E/entity is a pattern of energies and forces. Such patterns resonate to, or harmonize with, the patterns of other E/entities, or, they do not. When there is resonance and harmony, there usually exists between two E/entities what is called "liking", because the patterns are rather *like* each other. When there is lack of resonance, and, instead, active inharmony, dislike often develops because the energy patterns of the two E/entities are not only *unlike* each other but disruptively contrary. There are also neutral conditions of indifference based upon *non-contrary unlikeness,* which can be accounted for using the same model.

Tolerance or acceptance is usually not considered a problem, unless there develops a tolerance for that which undermines the goal-fitness of organisms and systems. But *intolerance* is a great problem in the world today. Intolerance is based upon the energy of *repulsion* associated with the third and first aspects of divinity. Intolerance usually exists between those whose energy patterns are significantly unlike and who, hence, often dislike each other. From all this discussion we come to realize that 'like' and 'dislike', tolerance and intolerance, all are matters concerning the *form* and not the intra-Cosmic Life or Spirit, and still less, the INFINITE SELF.

Upon the highest levels of Cosmos, the need for, and value of, the energy patterns of all E/entities is understood, and a grouping of E/entities which, on the lower levels, might well result in conflict and intolerance, will be (on the highest levels) welded into a unified and cooperative system. Even in-Cosmos (at fairly deep levels of prakritic immersion) and without recourse to the SYNTHESIS of SELF, intolerance can be overcome by a lofty perspective and an understanding of the Laws of Harmony.

- How much more, then, is intolerance (wrong rebuff) negated when the IDENTICALNESS of all Selves within the ONE SELF is experienced and realized.

Within the SELF (the INFINITESSENCE) there is only ESSENTIAL 'likeness' or 'alikeness'. Dislike and intolerance are based upon the experience of inharmonious vibratory distinction with which the matter-embedded consciousness is persistently presented. 'Within' the SELF, however, there is no vibration, no inharmony and no distinction. Therefore, 'within' the SELF (or in-Cosmos, within the State of Identification with the SELF), there can be no dislike or intolerance. Dislike stems from the *non-essential* part of oneself, from one's field-pattern, one's external structure, from one's prakritic constitution which is *not* one's REALITY (or, even, Reality).

This is not to say that in the process of right living the form must not be evaluated, corrected and improved, and the integrity of each divinely purposeful structure maintained, but even during these adjustments, the *essential value* of a fellow-S/self (which is a fellow-E/entity-in-the-SELF) must be asserted. There is no good reason to allow a shark (simply because a shark is a fellow being-in-the-ONE SELF) to devour one's form. This act of false sacrifice would, under most circumstances, be un-Goalfitting from a Universal Perspective, and yet one's ESSENTIAL ONENESS in the SELF with that creature must be affirmed.

- The type of tolerance that arises through adjustment of the form, develops from the stage of initial 'disliking' and is gradually transmuted into 'liking', the transmuting power being the Principle of Harmonization. If what is originally disruptive to the constitution can be transmuted through harmonizing adjustment into that which is enhancing of the constitution, 'dislike' will be transmuted into 'like'. All of this transpires on the level of the form and the consciousness trapped within that form.

The tolerance arising from SELF, however, (and from ITS Emissary in-Cosmos, the Divine Spirit) is entirely *different.* Toleration, and even *appreciation,* develop on the basis of *essential likeness,* the tendency of Spirit to fuse and blend with Spirit. It is true that a measure of *intolerance,* of a certain kind, may at first be necessary in order to preserve the form, for there is only so much that any form can "tolerate". For instance, human beings show a beneficial intolerance to the cold, to excessive sun, to certain intrusive agents (bacteria and viruses). This form of intolerance is simply a protective reflex.

From the philosophical perspective, however, no ESSENTIAL/Essential intolerance is admissible, for every part *is* the WHOLE-as-Whole, the SELF-as-Self, and no E/entity can be rejected, in *essence. Conditions,* on the other hand, *can* be rejected and re-formed according to the Divine Plan. The INFINITE SELF, however, (with all its constituent, *inseparable* E/entities) is seamlessly All-pervading and without the slightest division or interval. It is impossible for one 'part' of the SELF to rebuff another part of the SELF, for there are *no parts.*

- Looking for methods for improving human understanding and behavior, we find, then, that intolerance between E/entities is overcome through the affirmation of that which is not only *alike* between them, but that which is *identical* about them.

Of course, there is only one absolutely identical 'FACTOR' in their nature and IT has nothing to do with their forms; that factor is the ABSOLUTE IDENTITY WHICH they, ESSENTIALLY, ARE.

Always and ever, to solve the Problems of Humanity, the *essential identicalness* of human beings in the INFINITE SELF must be remembered. That all of us are not just 'part of', but REALLY and fully *the* INCOMPARABLE ONE, is a thought that will do much to negate the inharmonious differences which perpetually arise between us.

The Problem of Respect or Disrespect for Form

Why must form be respected if it is such a limitation and of so minimal importance? In answer to this, it should be said that we must not curse the instrument which helps us fulfill our desire simply because the instrument is not the fulfillment, itself. The form is our means to achieve the 'Pre-Cosmically Determined End' (the Design-at-the-Beginning). Our Cosmic 'Play' has a 'Plot', and there is no way to carry out that Plot without the form.

We-the-**I**-as-**8**, *need* form, otherwise we could not achieve the degree of limitation necessary to *contradict* 'OurSELF' in the way that We-as-**I**-as-**8** must. Form (or the Cosmos—for they mean the same thing) is our means of 'depriving' '**I**-OURSELVES' [sic] of our FULLNESS. For some necessary 'REASON', We-as-**I** 'WILL' to 'DO' this. Our only 'ACT'-becoming-Act is the Act of SELF-'LIMITATION'-as-Self-Limitation. As a matter of fact, as 'SEEN' from the 'INFINI-SPECTIVE' of the INFINITE SELF, every possible *action* is a *limitation* upon that SELF. Since no *act* of any kind can add one iota to the SELF, it must be a subtraction, a kind of *reduction* of PERFECTION, although, ESSENTIALLY, there can be no subtraction either. Anything *done* (if anything could REALLY be *done)* would lead to a condition of 'lessness' (like doing 'one more thing' to an already *perfect* work of art). Certainly, in one respect, the Cosmos must be called "less than nothing" (i.e., 'less that NOTHING'), a most illuminating phrase deserving to be pondered.

A close student of the Process of Cosmic Development will see that the Universal Form must not only be respected but studied intensely. In MY 'ACT' of SELF-'VEILING'; in My Act of Self-Veiling, **I**-as-**8** (on one level only) have forgotten the dynamics of the Veiling Process. The Universe will not "go away" or yield its well-veiled secrets simply because **8** pretend It is not REALLY 'there', or 'here'. The only way 'out' is the way 'through'. So, **8** must learn to "take a very close look" at Cosmos, which is what *respect* is all about anyway.

Thus, **8** learn to *respect* the Universe from the perspective of Identification as the SELF (a big step beyond identifying *with* the SELF). **8** respect the Cosmos in order to discover the *secrets* which **I**-as-**8** have hidden from 'MySelf' (MY intra-Cosmic Self). And, **8** respect the Universe because It is necessarily *beautiful,* being a Reflection of the INFINITE SELF which **I** AM.

The Problem of Joy and Bliss, Unhappiness and Wretchedness

According to what we might call the LAW of PARADOX, which seems to be more a 'LAW' of the INFINITE SELF (if the INFINITE SELF can have *laws)* than a Law of Cosmos—**8** (because **8** *Am* **I**) Am *simultaneously* all opposites. It is certainly clear that **I** (the INFINITE SELF in MY STATE of ALL-IN-ALLNESS) AM simultaneously all opposites. But, it is important to realize that **I**, the INFINITE SELF in-Universe as ***8***, Am, *also,* simultaneously all opposites. **8** experience simultaneously in-Cosmos all opposite states.

If **8**-as-I, the apparent 'part', am 'happy', it does not mean that **8**-as-I am happy *only.* By virtue of MY-as-My Pervasive Being, **8** Am thoroughly invested in *all* Cosmic conditions and experience them *equally.* As what might be called, the 'Localized Self-as-self', 'I' may be happy or unhappy, wretched or blissful, but as the 'Ubiquitous SELF-as-Self', **8** Am responding fully along a virtually complete emotional spectrum from the most exalted blissfulness to the deepest despair.

So, then, **8** Am conscious in my localized condition (Local Identity) and, *simultaneously,* in my ubiquitous condition (Depth Identity). Localization is personalization, encapsulation—the most evident and present of illusions. **8** Am both a localized Self-as-self and a ubiquitous, pervasive, omnipresent-in-Cosmos SELF-as-Self. Localized conditions have their own laws. Ubiquitous conditions function according to different an more extensive Laws.

There is much within an human E/entity which strives to ensure that the pattern with which it is locally and immediately identified is in that state of harmony (both within and without) called 'happiness'. This harmony, from time to time, may be achieved, but often, it is not. If the harmonized state is not achieved, Am **8** to say, "I am unhappy", simply because the 'most immediate field' is dissonant 'within itself' or with other fields. It all depends on the depth and extent of My Identification.

■ If My Identity is externally focussed, the state of happiness, or well-being within Me-as-me may rise and fall depending upon the condition of My 'most immediate field'. If **8** Am identified as a Pervasive Self (or, even better, as the SELF) the condition of the 'most immediate field' will play but a small role in my overall happiness or unhappiness, although that very pervasiveness of My identification may do much to harmonize the 'most immediate field' as well.

From the localized perspective, the natural state of the Knower of the SELF is joyful, even blissful, for the Illusion of Separation (the cause of pain) has ended. With the Illusion of Separation lifted, naught but the original, blissful, undivided, primeval 'STATE' of SELFHOOD-as-State of Universal Selfhood remains, though the perception of Cosmos does *not* disappear. If **8** Am, however, in a state of <u>SELF</u>-forgetfulness (the very opposite of the desirable state of <u>self</u>-forgetfulness), the many divisions within the world and inharmonies will conquer me, obscuring the Higher Harmony Which **8** Am, and then the state of consciousness known as 'unhappiness' will be the result.

It is, nevertheless, possible for ME-as-Me to learn to be confronted with the potentially divisive presentations, and yet remain anchored within the INFINITE SELF/Universal Self. If **8** succeed in doing so, then, **8**. Am experiencing dual consciousness, or better, dual *identification.* At that point of dual being, **8** Am *in* the World of Becoming but not *of* It. In such a case, the illusory presentations (and *all* presentations are such), the divisive presentations, will appear as un-REALITIES, though, of course, as *actualities*, and the Cosmic Synthesis within the INFINITE SELF will be held in such a way that happiness, misery, etc., as usually considered, will not be absorbing factors within consciousness.

It is not that **8** will not be aware of these unfortunate states and attempt to help, **8** will. **8** will be much less likely to say, however, "*I* am happy; *I* am unhappy", though **8** *may* realize that ***8*** Am blissful. **8** will also be aware that the happiness and unhappiness of the World are MINE-as-Mine. Although **8** will necessarily pervade these states, and although they will be MINE-as-Mine as much as they belong to the localized 'other' one for whom they are occurring in the 'most immediate field', still, **8** will not be identified with these states, however compassionately **8** attempt to alleviate them. **8** will *feel* them, yes, **8** will help with all my might, but will be *simultaneously blissful.* The Dalai Lama is a good example of this attitude. One must live ever in the Synthesis (a Cosmic term) reflective of the SYNTHESIS (the 'STATELESS STATE' of THAT).

The Problem of Morality

The essence of this problem concerns the development of ways and means of achieving right human relationships, and the avoiding of those behaviors that lead to wrong human relationships. How are we to treat one another? From the non-dualistic, 'Radical Infinitist' perspective, we are to treat each other as *essential selves* 'within' the ONE SELF. On a still higher turn of the spiral, we are to treat each other as if we were the ONE SELF, ITSELF, which we ARE, ESSENTIALLY. We are not to be *objects* to each other but *subjects*—REALLY, *the* SUBJECT. If we persist in treating each other as objects, we treat each other as 'things' un-REAL, as 'things' divorced from the SELF. If, however, we treat each other as Subjects who *are* the SUBJECT, we begin directly and immediately and fully to feel the impact of every act upon another as if it were an act upon 'OURSELF-as-OurSELF' because it *is* so.

Always the question arises whether the values of form must be preserved, or those of LIFE-as-Life. Certainly, certain forms must be preserved in order to fulfill the Divine Purpose, the Design-at-the-Beginning, but, whenever form inhibits or impedes the full expression and knowing of SELFHOOD/Selfhood, the form must go.

So many human beings are lovers of the form, incapable of relating to the LIFE/Life within the form. The way they treat each other is then judged by how well they treat the form. Indeed, the form should be fittingly treated, because it is the means of achievement in the World of Illusion. By the form is here meant not only the gross and obvious form, but the subtle sheaths on all levels. However the morality of a human being is not solely to be judged on how well he treats the forms of *apparent* 'others'.

Identification as the SELF may induce a feeling of boundless compassion resulting in tenderness to the form, or it may indicate the need for apparent harshness to the form—one's own or that of another. Even the harshness, however, will be motivated by boundless compassion, for Identification as the SELF induces the widest possible *pervasion* and its result, exquisite sensitivity and compassion.

- The Christian Scriptures tell us to "Love thy neighbor as thyself." Interpreted from the Radical Infinitist perspective this is a most profound statement. It means that we shall love our neighbor as if he were the *same* as 'OurSELF', the INFINITE SELF. We are being enjoined to love our neighbor by entering a state of Identification *as* the SELF. This is certainly different from the usual moralistic point of view.

Morality can only be understood from a state of Unity and, even better, from a state of Synthesis-as-SYNTHESIS. An attitude of separative individualism will not reveal its purpose. When, however, the harmony and fused heterogeneity of Love, and the enforced homogeneity of Will and Synthesis pervade one's consciousness, then morality can be understood as the way to resolve the Many into the One-as-ONE by refusing and overcoming the Illusion of Separativeness.

The Problem of Good and Evil —To *Whom* are they Done?

Whether **8** do good or **8** do bad, **8** do what **8** do to MYSELF-as-'MySelf'. 'Good' is that which fulfills the Design-at-the-Beginning. 'Bad' or Evil is that which does not lead to, or, worse, leads away from that fulfillment. Every act that **8** perpetrate or carry forward must be an act of liberation, an act which renders more visible and expressive the REALITY of the ONE SELF, and the Reality of the One Self (the One serving as the Finite Expression of the OTHER). If an act is *thus*, regardless of the *apparent* effect that it has upon the form, it will be a *moral act*. If an act is *not* thus, even though it preserve and even enhance the form, it is an *immoral act*.

All conduct, codes of conduct, must facilitate the actualization or the fulfillment of the Original Intent for that particular Cosmos—the Purpose or Design-at-the-Beginning. Thus we see that moral acts are Self-Liberating acts (The *SELF* does not have to be liberated—IT has been UTTERLY 'LIBERATED' forever). Such acts bring more freedom to the Self-as-self and not less. They empower and do not incapacitate. Although they may not necessarily empower the form, they do empower the Being. BEING need not be empowered just as the INFINITE SELF need not be liberated.

- Remember that with the Creation of Cosmos, the ALL-POWERFUL BEING/NON-BEING 'BECOMES' (in a way) *subject* to ITS OWN SELF-'IMPOSED' Limitation-by-Design.

Who or what else is there to *limit* the ALL-POWERFUL INFINITE BEING/NON-BEING? Regardless of SELF-'LIMITATION'-as-Self-Limitation, however, ITS ALL-POWERFULNESS will be reclaimed by the apparent Exile (the Universal Being) at the great "Day Be With Us." The 'Good' facilitates that reclamation in every possible way. The 'Bad' or 'Evil' deviates from the Goal, though the achievement of the Cosmic Goal is certain.

The Problem of ABSOLUTE GOOD, and *absolute* 'EVIL' (if existent)

In one respect, ABSOLUTENESS is GOOD and any form of variation from ABSOLUTENESS is EVIL. (Of course, since there can never *REALLY* BE any variation from ABSOLUTENESS, EVIL, by this definition, must be illusory and GOODNESS, the ABSOLUTE, REAL). Good and Evil *within* Cosmos are not GOOD and 'EVIL', and depend for their definition upon intra-cosmic relationships.

From this perspective, the STATE OF ULTIMATE ABSTRACTION is the ULTIMATE GOOD. It is the 'STATE' of Universal Death (as heretical as this may sound). The 'live' State, on the other hand, is the *'vile'* State, the *'veil*-ed' State, and (following on with these permutations), also the *'evil'* State, as strange at that may seem. With respect to GOOD and EVIL, *death,* or the climaxing eradication of the Cosmos is the antechamber to the ULTIMATE GOOD (for all trace of duality {even though that duality be illusory} has vanished) and also, *life* (as we usually experience it), putatively the 'ULTIMATE EVIL' or limitation. Life as we know it is, indeed, *limitation,* and 'EVIL', too, is simply limitation. Since there can never REALLY be limitation, there can never *REALLY* be any 'EVIL'; there only *seems* to be.

Now, within Cosmos itself (and especially with respect to Humanity on the *Evolutionary* Path), whatever brings the greatest release or liberation for the greatest number is Good. Within Cosmos, what brings the greatest limitation to the greatest number (if they are upon the *Evolutionary* Path) is Evil. Note here we are not speaking of ABSOLUTE GOOD or 'EVIL', but of a *relative,* in-Cosmos, Good and Evil. In the World of Becoming there is a *relative* Good and a *relative* Evil. Both of these are Real-in-Cosmos and actual.

'Within' the ALL-in-ALLNESS, GOD IS GOOD ALONE. How can the ULTIMATE AFFIRMATION, which the INFINITESSENCE *IS* , be considered EVIL in any way? If there is anything resembling 'EVIL' 'contained' 'within' IT, it is the *necessity* that there exist a Universe, which serves as a kind of 'contradiction' to the ABSOLUTENESS, and, as we have shown, the Universe for all the Goodness It contains, Is, strangely, from the *absolutist* 'INFINISPECTIVE', the One and Only 'EVIL'. Since, however, the Universe is only a *Seeming,* even this apparent 'EVIL' is REALLY the GOOD in disguise, for the Cosmos *is* the INFINITE SELF. BRAHMAN and Samsara are *One.* The INFINITE SELF alone is GOOD, ULTIMATELY GOOD. GOOD is the ULTIMATE 'STATE'.

■ If there REALLY were a duality there would *REALLY* be 'EVIL', but since there is only an *apparent* duality, there is likewise only an *apparent* 'EVIL'. Within the Universe—Itself nothing but a *necessary Illusion*—relative Good and Evil must be taken seriously, otherwise the Cosmic Game cannot be played properly by the One Player.

The Problem of Brotherhood

Who is a brother? We are told that Brotherhood is already a *fact,* that it does not have to be achieved. Certainly, however, that fact must be *manifested.* Brotherhood might be defined as the 'art and science of right group relations'. The manifestation of Brotherhood is not just a problem that concerns humanity; it is a problem involving all E/entities in Cosmos. The Design-at-the-Beginning (the Divine Purpose) is intended to bring about right relations throughout Cosmos. Manifest Cosmic Brotherhood will be the result.

In Cosmos, every non-partite 'part' is brother. All authentic 'parts' (really 'Rays') are, of course, non-partite and impartite. We human beings are *apparent* 'parts'. We are to each other brother—mother, father, sister, friend, etc. We are *all* relations to each other and, yet, none, because within the INFINITE SELF, *relation* does not exist.Before there is division or *apparent* division (i.e., Emanation and Number), there are no relations. There is no *family* at all, no Father, Mother, Child. The Cosmic Trinity is derived (through descending Pre-Cosmic Trinities) from the SELF, but the SELF antedates the Three Cosmic Persons—antedates even the One Universal Logos in Whom the Three Cosmic Persons inhere.

In Time and Space, however, (i.e., within Cosmos), the situation is different, and *relations* exist. Time and Space do not simply disappear on the higher planes of Cosmos, for all Aspects of Cosmos occur in Space and are governed by Time under the Law of Cycles. In Time and Space and under the Illusion of Separation, we are all, at the very least, *brothers*. To be brothers means that we are ESSENTIALLY *identical* 'Rays' of the ONE LIFE, manifesting through a variety of forms. Every *authentic entity* or *primary entity* (though every such is, Really, but the One Identity *fully)* is a *brother*.

Who is My brother? Not just a being, a localized being with a *similar* form. This is the limited way that limited human beings judge another to be a brother. Instead, a brother is *any* B/being in Cosmos (whether 'localized' or not) with *any form at all* (remembering, *ever*, that all apparent beings are but *apparently distinct* B/beings, and are, Really, but One Being). For Brotherhood is based upon *being*, upon identicalness of ESSENTIAL IDENTITY, and not upon similarity of form.

Therefore we see why Brotherhood must be a Great Fact in nature, for through our intuitively stimulated philosophical reasonings, we know all B/beings, ESSENTIALLY, to be but ONE BEING. Brotherhood is thus based upon the equality of *essential* being-as-Being-as BEING regardless of the inevitable inequality of the form, for all forms in Cosmos function under the Law of Unrepeatability, thus no two forms can ever be exactly identical.

What duty does one owe one's brother? What must one do? Of course one must only do what one wills, as no duty can be profitably forced. But what duty *does* one *owe* one's brother? Might we say—the enhancement of a brother's *beingness* within localized conditions.

■ Always the nucleus of SELFHOOD/Selfhood must be positivized, potentized, and, eventually liberated. There is, REALLY, no way to potentize the ALL-POTENT that resides 'within' every brother (or rather *IS* every brother), but there is a way to remove ITS *veils*. One can share *being* with a brother and thus contribute to the recognition of His INFINITUDE. This is called 'infinitizing' the Selfhood of one's brother, which is simply an *unveiling* process which helps reveal His DIVINE IMMORTAL SELF to 'HimSelf'.

How, then, do we achieve Brotherhood? Firstly, by recognizing that it already *exists*. Secondly by realizing that it is based upon shared IDENTITY *as* the INFINITE SELF. Thirdly, by stimulating the CORE of a brother's Being which will be recognized as being IDENTICAL with one's own.

The Problem of the Inequality of Hierarchy

A hierarchy is a stratified arrangement of E/entities based upon an orderly ascending and descending structure of their magnitudes (in prakriti). Each E/entity is positioned according to rank, and nowhere, apparently, is absolute equality to be found between the different grades of E/entity comprising the structure. This hierarchical arrangement seems to be a "fact of life" in Cosmos, and cannot be ignored or reduced to the same level—i.e., cannot be 'leveled'!

- Hierarchy is the principal organizing structure of the Universe, but hierarchical ordering still pertains to the World of Illusion and not to the WORLD OF REALITY.

From the perspective of REALITY, while all forms in Cosmos are hierarchically arranged, they are nonetheless equally pervaded by the ONE AND ONLY SELF WHICH **I** AM, and the One Cosmic Self Which, for the duration of this Cosmos, **8**, also, Am. It is this pervasive, simultaneous inhabitation of all forms which is the great *equalizing* factor. Even while greater and less, better and worse, must be properly evaluated within the World of Becoming, it is realized that the SELF is invested *equally* within all such forms, and that **8**, no matter who **8** *seem* to be, Am expressing 'MySelf' through higher and lower, greater and less, better and worse, through all the pairs of opposites *simultaneously*. Thus, a complete equality between E/entities exists regardless of the Hierarchy of Forms and Consciousnesses.

We see, therefore, that it becomes necessary to simultaneously accept and negate the Principle of Hierarchy. Within the World of Becoming, the World of Illusion, we cannot function without it. If, however, we choose to understand the REALITY of relationship, we see that hierarchical arrangements often obscure the ESSENTIAL SPIRITUAL EQUALITY which judges all beings (of no matter what relative status) to be ESSENTIALLY *identical*.

The correct attitude towards this paradox, is to respect the paradox and find a way to honor both aspects of it. There is no way in all of Infinite Duration that we-the-**8** will ever escape from the Presentation of Dualism by the INFINITE SELF—simply *because* the INFINITE SELF *<u>is</u>* INFINITE. Cosmos after Cosmos we-as-**8** shall be presented with the Object (a SELF-'REFLECTION') Which seems to belie the INFINITE SUBJECT, and we shall give 'OurSelf' the Cosmic Task of *proving* that the Finite Object *<u>is</u>* the INFINITE SUBJECT.

Lest the prospect seem too tedious, we must remind ourselves that we-as-**8**-in-Cosmos have been doing this, precisely, *forever* and apparently have not tired of the 'Game'. Lest one become overpowered by despair when contemplating the endlessness of this Sisyphean Labor, we must realize that from the perspective of REALITY, we-as-**I** have *never* been doing this *at all*—never, ever. For the INFINITE SELF, the BOUNDLESS IMMUTABLE PRINCIPLE cannot ever *BE* other than ITS infinitely enduring CHANGELESS SELF. **I**, the INFINITESSENCE, AM infinitely more than blissfully absorbed in MYSELF forever, no *matter* what else (including 'BECOMING' a Cosmos) **I** *appear* to be 'DOING'. The Paradox is profound, but these thoughts state the essence of it.

The Problem of Hierarchies of Forms, or Hierarchies of Beings

Simultaneously, there is a Hierarchy of forms but never a Hierarchy of *beings*. A Hierarchy of *being* does not exist. A Hierarchy of *becoming* does exist. Thus, in Cosmos, equality *seems* to be an illusion and Hierarchy prevails. In actuality, Hierarchy is an *illusion* and *equality* prevails. Being 'CREATURES' of the NOTHINGNESS, and creatures of Cosmos simultaneously, we are subject to the great Law of Contradiction, the ultimate 'LAW' of PARADOX. In one way, the only thing the great SELF can 'DO', the only 'ACTION' IT can 'TAKE', is to 'CONTRADICT' ITSELF.

The Problem of the Magnetism of the Immediate Presentation

It is incorrect, but initially unavoidable, for the SELF-as-Self to identify with any one form within the Cosmic Hierarchy of Forms. This built-in error occurs so frequently because there exists what might be called, the Principle of 'Localized Immediacy'. This is the illusory tendency of a given consciousness to identify most closely with the particular form with which it seems to be most intimately associated (its Focal Identity). Proximity wins the day and overcomes the far more subtle REALITY of the ubiquitous pervasion of the SELF-as-Self.

Every authentic E/entity (every 'Ray' of the ABSOLUTE) does have an 'immediate presentation' (i.e., certain contents of consciousness which are most present and obtrusive because they relate to the forms which are {due to Emanation} the nearest). The registrations in consciousness based upon changes within the E/entity's auric force-fields are usually what constitute the 'immediate presentation'.

The usual primacy in consciousness of the immediate presentation emphasizes the E/entity's apparently separative and illusory 'part-hood' that seems to deny the E/entity's Real relationship to and REAL *identification* with the WHOLENESS/Wholeness. The difficulty in altering this habitual and resistant dynamic in consciousness contributes to the persistence of the egoistic idea—the illusion of the *apparent reality* of the separate individuality.

How shall we overcome the potently arresting and preoccupying nature of the immediate presentation whereby the SELF-as-Self thinks It is *only* the particular 'part' (i.e., the little self), and *not* the SELF-as-Self manifesting through *all* apparent parts? **8** (and all **8**'s) Am faced with the great problem of the subjugation of MY-as-My consciousness to one, special, Essentially illusory case, i.e., to the apparently isolated part. **8**, Who Am, Really, a Pervasive Consciousness Am apparently isolated and confined to the particular—a particular ring-pass-not.

Analogously, on the highest possible turn of the spiral within Cosmos, **I**-the-WHOLE am SELF-'SUBJECTED' to a similar, though far vaster, Illusion of the single Ring-Pass-Not of a particular Cosmos, when in fact **I**-the-WHOLE AM REALLY *boundless.* As the Cosmos-encapsulated-INFINITUDE, **I**-as-**8** must awaken to the Great Illusion of 'boundedness', for **I** AM the BOUNDLESS IMMUTABLE PRINCIPLE.

Any boundary, even the Ring-Pass-Not of Cosmos, is in fact an illusion, and has to be seen through, worn through and, at length, destroyed (as it ever is on the Universal "Day Be With Us"), yet it is **I**, MYSELF, WHO have 'ENERGIZED' the 'CREATION' of this Condition of apparent *boundedness.* How shall **I**-as-**8** free 'MySelf' from falsely thinking MySelf to be the Limited One (the Universe) which, indeed, **8**, *Essentially* Am, and Realize 'MySelf' as the ALL-PERVADING BE-NESS, which **8**-as-**I** *ESSENTIALLY* AM, thus Negating, in consciousness *all* illusory boundaries, even the most encompassing?

We can see from this discussion that **8**, as the SELF-as-Self-as-self in Cosmos have a localized and more *individual* task of liberation; the **I**-as-**8,** on the other hand (demonstrating in Cosmos as the Universal Self) has a *Cosmic* Task of Liberation (though, it is still an *Individual* Task on a grand scale). In both cases, the *immediate presentation* must be transcended.

On the more localized level **8**-as-I must gradually render 'MySelf'/myself impressionable to all presentations (i.e., all other authentically entified patterns) in Cosmos, as if they were My Own (which they, Essentially are). **8** thereby learn, eventually, Universal *pervasion.* On the Cosmic Level, **I**-as-**8** must see through and beyond the entire Cosmic Presentation, which for Me (as the Universal Self) is, indeed, a most Immediate Presentation, and learn to Identify MySelf fully with the INFINITESSENCE, the FOUNT OF ALL POSSIBILITIES which lies, as it were, beyond and within, and in fact, IS the ESSENCE of all that is presented to Me. (It will be more than 'yet a little while' before **8** can do this.) In both cases an illusion is being overcome—the extraordinarily prevalent illusion that that which is apparently more *immediate* is more REAL/Real.

The Problem of Prayer —To Whom does one Pray?

To pray is to invoke a superior power in order to alter conditions and patterns in a realm over which the superior power has definite influence. Prayer is useful and will not be dispensed with during the Aquarian Age simply because the Science of Meditation will be far better understood than during the Piscean Age. Prayer, as invocation, will be an essential part of the New World Religion.

From the non-dualistic, infinitist perspective, however, does it make sense to pray? *Are* there Ones *(other* Ones) to Whom one can effectively and meaningfully pray? In the World of Relativity, yes, but from the perspective of the WORLD OF ABSOLUTENESS, no. In the World of Becoming which is the World of Relativity, in order to be *practical,* one acts as if another being is strictly another being, and not ONESELF-as-'OnesSelf'. In the World of Relativity, there *are*, indeed, great and small.

There *is*, actually, a Hierarchy of B/beings ('B/beings', being those aspects of MYSELF-as-'MySelf' which **8** call *others*). That aspect of 'MySelf' which I call *another*, (i.e., a being or E/entity in Cosmos) especially if that 'other' is 'higher' that 'I' am (with respect to its localized consciousness and the quality of the form through which it expresses), *can be invoked in prayer*. **8** can attune to ever higher, ever freer, Emanations of 'MySelf' (for what else *is* another E/entity—whether higher or lower than my localized self—than an Emanation of the SELF-Substanded-Self, Who **8** Am?).

8-the WHOLE-as-Whole, subjugated to localized conditions (the nature of which defines My apparent evolutionary status), can appeal to **8**-the-WHOLE-as-Whole subjugated to still localized, but less limited conditions, with the result that **8** (the invoking one) will be liberated to some degree from the more localized conditions under which **8** labor. My conditions will become a bit more like the less localized conditions of the more expanded **8** (though still **8**) which **8** invoked.

So in the World of Time and Space, it is practical to consider that there are *actual others*, both higher and lower than oneself. This is an *actual fact* if not a REALITY. In Cosmos, within various of its strata and dimensions, there is *greater* energy expressing and *lesser* energy expressing, a *greater* Presence of the SELF-as-Self manifesting through prakriti, and a *lesser* Presence of the SELF-as-Self.

■ It is both legitimate and useful for the *less* revealed Presence of the ONE SELF-as-Self, to appeal for the intervention and downflow of the *more* revealed Presence of the ONE SELF-as-Self.

In fact, it becomes clear, that in all prayer one is only appealing to ONESELF-as-'OnesSelf'. The ONE SELF, after all, is apparently manifesting through many various degrees of *revelation*.

It can be thought of this way: in the atom **I**-as-**8** Am manifesting as an 'i'—tremendously veiled and unrevealed. In a Solar Logos **I**-as-**8** Am manifesting as a Cosmic Entity Who is *veiled*, true, but almost inexpressibly more revealed than My manifestation as an atom. In the Universal Logos, **I**-as-**8** Am as *fully revealed* as **I**-as-**8** can possibly Be, depending, of course, upon the Time Schedule for the Universal Manvantara, for at the Great "Day Be With Us", **I**-as-**8** will be maximally revealed in MY Cosmos, though *infinitely less so* than **I** AM 'REVEALED' to MYSELF (if 'REVEALED' **I** can BE) in my 'STATE' of PURE INFINITESSENCE—*without* the encumbrance of a Cosmos.

Perhaps, keeping these thoughts in mind will beneficially transform the way in which human beings pray. There will be far more of the factor of *identification* in the prayer, and far less of emotional pleading, based as it is upon unresolved dualism.

The Problem of ego and Non-ego

Ego can be understood as the 'apparent 'Sight' of SELF'. REALLY, the SELF cannot be *seen* for 'if one can *See* IT one cannot *BE* IT'. Non-ego does not think it 'Sees' the SELF, because it is identified *as* the SELF and/or Universal Self. The same idea stated from another perspective is as follows: the state of ego arises from 'seeing' (i.e., becoming conscious of) the apparent Not—SELF/Self and, then, calling it the Self or the self. By doing this, the SELF is objectified (made into an object), but the SELF can *never* REALLY be *objectified,* because IT is the INFINITE SUBJECTIVITY. Or course, though one can never REALLY 'see' the SELF, from another perspective, if *anything* is 'seen' can it be another other than the SELF which is *all things?*

How do **8**, Who Am, Essentially, the INFINITE SUBJECT-in-Cosmos, acquire the conception of ego? How do **8** arrive at any conception of what **8**, MySelf-as myself, actually Am? If **8** arrive at any concept of 'MySelf' through the normal processes of consciousness, **8** will surely be mistaken. Surely, at length, **8** must learn to disagree with the idea that anything **8** may 'See' is the PURE ESSENTIAL REAL SELF manifesting in Cosmos as **8** MySelf. If **8** predicate aught of MYSELF, that predication is false. **I** AM surely the "ONE ABOUT WHOM NAUGHT MAY BE SAID" (in the *ultimate* meaning of that title). Even the ESSENCE of **8** (the **I** in Cosmos) *is* the "ONE ABOUT WHOM NAUGHT MAY BE SAID".

■ The best method of arriving at My REAL IDENTITY is through *negation.*

Negation is the great means of cleansing the face of IDENTITY. Via negation, one 'takes away' (just as one takes away the sedge or algae from a pond) all that which has covered and concealed the TRUE SELF. Negation cleanses the "doors of perception". The SELF can be relied upon to 'SHINE' in 'FULLNESS', provided the art of negating the Not-SELF is vigorously and efficiently exercised. Through the right practice of negation, predication of limitations upon the SELF is eliminated. Of course, REALLY, *any* predication anent the SELF dwarfs and misrepresents IT. To form a specific concept of the SELF is to limit the SELF. At this point, you might ask why this book was written!

Now, ego is built of definite assertions concerning identity. If one says, "I am an identifiable 'this'", one may have spoken sufficiently accurately to define identity as it is usually conceived within the World of Becoming, but one will, in REALITY, have spoken *falsely.* The purpose of non-dualistic, 'Radical Infinitist' philosophy is to destroy the false identity which ego is.

In today's Earth Humanity (for there are Humanities that flourish in other times and places in Cosmos), few *realize* WHO they REALLY ARE, or even, Who they Really Are. The first pertains to 'SUPER'-Cosmic IDENTITY, the second to intra-Cosmic Identity (which is nonetheless, ESSENTIALLY, also 'SUPER'-Cosmic IDENTITY). The problem of ego is the problem of the wrong thought and wrong action which arises when one acts from within the consciousness of *separativeness* (the one "Great Heresy" as the Tibetan has informed us). As an unenlightened person, one acts on behalf of what one thinks one *is,* and one is almost always sadly mistaken, because one has *localized* and *defined* what one is. This localization and definition are necessarily always 'partial' and illusory. One is therefore limited in his consciousness and 'energetic outreach', by his *ego* (by his conception of his identity), and is thus prevented from being pervasively invested in all other 'localities' in Cosmos. Thus one acts, not with the best interests of the Cosmic Whole at heart, but only on behalf of the interests of the apparent 'part' that one thinks one is.

■ We see that the ego is really a *thought.*

There *is* such a thing as the *actual* (though temporary) Identity of the SELF-as-Self in-Cosmos, and every Self must one day know what this Identity Really Is. That actual Self Identity, however, (the particular Role a particular E/entity may be playing in the World of Becoming at any one time) is rarely the same as the limited egoistic conception of identity entertained by most human beings. Even this actual Self-Identity (though it must be understood if the E/entity is to be successful in the World of Becoming) Is, Itself, an Illusion, but a *necessary* One. In light of the foregoing, the statement should be made that 'ego' is REALLY illusion. Non-ego is, *essentially* TRUTH.

Fundamentally, the Problem of *ego* arises because the SELF is wrongly seen as an <u>o</u>bject, or even an <u>O</u>bject (the contrasting magnitudes indicated by the different capitalizations makes no difference). If one thinks that one is a *small* object or a *big* Object, it is all the same. The SELF is never an *object.* On the scale of relativity (in which objects are required to define identity) ego problems may arise because the consciousness within is identified with an *unsuitable* object. For instance, if **8** am a human being, but **8** deludedly think **8** Am *actually* the planet Earth, then **8**-as-I have quite an "ego problem". This kind of ego problem is really a *secondary* kind. The *fundamental* problem arises from thinking that *subjective selfhood* can ever be expressed as an *object,* a definite thing.

The Problem of *ego* is so subtle, and so close to us, that it is almost impossible to grasp. The *truth* of the question seems to fly in the face of common sense. The difficulty is that a REAL *subject* cannot be *defined,* i.e., made into a definite thing. The SELF is only and ever a *subject.* The SELF IS the INFINITE SELF IS the INFINITE SUBJECTIVITY. Even the Self in-Cosmos is ESSENTIALLY the INFINITE SUBJECTIVITY. Even the little benighted <u>self</u> is *still* the INFINITE SUBJECTIVITY. So is *everything,* even those things we would call the most definite objects.

■ The Problem of *ego* would disappear if we could only apprehend this INFINITE SUBJECTIVITY easily, but due to SELF-'VEILING'-'BECOME'-Self-Veiling, and what we might call 'Emanative Loss' [see Glossary], we cannot. Our normal faculties of consciousness are insufficient to apprehend the SUBJECT. Through normal consciousness we sense, feel and think, but these faculties can touch only *objects,* never the INFINITE SELF.

We are told, however, and some have experienced, that the SELF (the ever-present PRESENCE, probably in the Mode of the Universal Presence) *can* be apprehended or intuited by an *inner faculty.* We are all possessed of this faculty, but its nature is so subtle that it cannot be used until the multitudinous veils have been removed through the disciplines of 'philosophical negation'.

One difficulty in determining our TRUE IDENTITY arises because when we *think,* we want to think of 'something', i.e., of some-*thing.* The human consciousness, at its present stage of development, has difficulty conceiving abstractions. To such a limited mind, that which is *definite* seems more *real* than that which is *indefinite.* **I**, however, (whether **I** AM isolated as PARABRAHMAN during Universal Pralaya, or *immersed* in-Cosmos during Universal Manvantara) AM definitely *not* something *definite* (or de-*finite).* **I** AM, certainly, *not* a something. **I** AM the UTTER ALLNESS, (if anything at all!), and AM thoroughly *indefinite,* thoroughly INFINITE (yet *uncountable).* Thus **I** cannot be defined. **I** cannot be bounded. **I** cannot be limited in any way.

Until **8** truly identify as the SELF, **8** shall be beset with the falsity of *ego.* After **8** identify properly, **8** shall have no more ego, but **8** *shall* have a Self. All E/entities in Cosmos have an *actual* Self, even the Most Exalted Entity, the Universal Logos. (Of course, that Self, for all, is, in-Cosmos, the One Universal Self.) Exalted Beings do not suffer from the limitations of an ego.

- It is more than possible to recognize 'OnesSelf' as an actual Self (a 'Ray' of the ABSOLUTE at a certain depth of prakritic immersion), and *also* identify Ones Self as the ONE AND ONLY IMPERSONAL INFINITE SELF. This dual *self-assessment* is what we all hope to achieve, *eventually* (even though Time is an Illusion), and doing so will make it possible for us to live (in-Cosmos) the *highest* type of Dualism—a *necessary* Dualism.

The Problem of Boundaries

Here, TRUTH anent the SELF runs into a practical problem, and a big one—the Problem of Boundaries. Within Cosmos all functionality and relationship are based upon distinction, separation, and boundary. Cosmos is a World full of boundaries, and boundaries must be respected according to the Laws of Cosmos. But how shall the **I**, the EVER FREE, function in so constricted a World? How shall **I,** boundriless as **I** AM, find a way to BE MYSELF when **I** (even though **I** manifest as **8**) AM forever running into the artificiality of boundaries (even though **I** AM *essentially,* if not *actively,* responsible for making those very boundaries)? The problem is a variation on the Great Problem of the seemingly Insuperable Contradiction between the INFINITE and the Finite.

Something must be done to solve this problem for the sake of the *quality* of LIFE in Cosmos. In Cosmos, **I** (manifesting as **8**) have no choice but to further limit MYSELF-as-'MySelf' and do something practical (i.e., something that will enhance the *quality* of in-Cosmos Living). Intra-Cosmic boundaries (apparent boundaries between *apparent* E/entities) must be respected or the entire Cosmic Edifice will go crumbling, and the Intended Design-at-the-Beginning will not be reached according to Plan. Boundaries are separations, and separation is pain to the intra-Cosmic SELF, i.e., the Self. (The SELF, PER SE, can feel no pain. To IT, no one condition is different than any other condition. In fact, there are no conditions at all in IT. All apparent conditions are *identically* IT *'infinitessentially'.)*

Boundary is the apparent negation of the fullness of SELF (which nothing can REALLY negate). So the establishment of boundaries is, to use an unusual phrase, 'anti-bliss'. *Boundary* is *anti* ALL-in-ALLNESS. Without boundaries, however, there is no Cosmos, and there *must be* a Cosmos. It is *necessary.* So **I**-as-**8** (in-Cosmos) have to find a way to *adapt* to the *less-than-I,* which **I**, too, AM.

In Time and Space in the Relative World of Cosmos, **I**-as-**8** (the WHOLE-as-Whole) must Play the Cosmic Game the basic Pattern of which **I** (as Universal Logos) have, in Pre-Cosmic 'Days', recognized as Designed. It is My fullest Will (as intra-Cosmic SELF-as-Self) to play the Game. **I**-as-**8** fully realize that the fact of My (Logoic) Willingness must dawn upon unenlightened 'I' as well.

■ To Play the Game means to respect boundaries while ever working to destroy them!—in the right way and at the right 'time' in Cosmic Time. In Time and Space, boundaries must be worn ever thinner and altered to serve Spirit's pressure towards expansion. Boundaries cannot be capriciously assaulted or destroyed for the very reason that they are an integral part of the Divine Purpose, and the Divine Plan which carries out that Purpose.

Thus, in the Beginning, **I**, in the 'FLASH', as it were, 'BOUND' MYSELF to 'BECOME' a Cosmos. **I**-as-**8** enumerated 'MySelf'. **I**-as-**8** divided myself—apparently. All of these Acts are essentially artificial and un-REAL Acts. And in doing these things **I**-as-**8** Create the Problem of *ego* (the Problem of Apparently Separate Identity) which **I**-as-**8** must transcend by rediscovering 'MySelf'-as-MYSELF in a 'SELF-strange' 'Place', the Cosmos, in which **I**-as-**8** *appear* to have thoroughly forgotten MYSELF. **I**-as-**8** solve the Problem of *ego* by revealing MYSELF to 'MySelf', and this **I**-as-**8** Do in (the fullness of) Time.

The Problem of Trespass

The Problem of Trespass is intimately related to the Problem of Boundaries. In all societies there is some kind of law against trespassing, against violating the boundary of another E/entity. Such a law is a law against violating the Principle of Distinctiveness or of Individuality (as interpreted in a limited sense).

The World of Becoming demands separation. The World of Illusion (a *becoming* World) exists simply because of apparent separation, and It seeks to maintain Itself by maintaining separation. Thus, generalizing, it may be said that throughout Cosmos, it is unlawful to trespass. If trespassing were allowed, then it would be impossible for each apparent E/entity to play its Self-Assigned part because the integrity of that part would be disrupted, interfered with, violated.

All this notwithstanding, there is a subtle form of trespass that is allowed, for it does not destroy or disrupt the form through which the Spirit must express. Perhaps, it is incorrect to call such a process 'trespass'; it is both more and less. What is here indicated is the *trespass* from one Spirit to another. This process is really Identification. It is the most supreme form of intra-Cosmic Intimacy, and is, Really, far more penetrative than any other kind of trespass, for it moves through and beyond all apparent boundaries to the very *core* of *being.* The identifying *trespasser* requires no permission of the one with whose Spirit he identifies. The act is instantaneous, and beyond the possibility of being deflected. Such trespass penetrates every possible form.

- In this major trespass which we call identification, one is not allowed to violate the form or rearrange the form of another. Laws regarding trespass exist for the preservation of the form, no matter how subtle, but *the trespassing one* can deliberately and consciously *become* another, which, in fact, he already is. This unconventional and deliberate penetration to ESSENCE/Essence is the most spiritually effective form of trespass and the greatest means of negating boundaries.

We find, then, that the law demands respect for the temporary Principle of Insularity. The Law within Cosmos protects *ego*. After all Cosmos is an 'Ego' (considered in the highest sense of the term). Cosmos is an *egg*, or rather, the Cosmic Process occurs within an 'egg' of Cosmic Prakriti. Within Cosmos, egoic 'eggs', (the guarantors of individuality and integrity) are protected by Cosmic Law (and all its lesser attenuations on various dimensions) until the *cosmically correct time* for their shattering comes.

Yet, unless one dares in some manner to trespass boundaries, one shall forever be a localized self. Through right *'identificatory trespass'*, the illusion of localization can be overcome without violating the Law that is intended to preserve the integrity of form.

The Problem of Position and Function within Cosmos

There is required of every advancing E/entity the closest possible scrutiny of the Grand Design, as that Design takes form, fulfilling itself in Cosmic Prakriti. The limitations of a form must be known thoroughly, for they define the position and function of that form within the larger Whole. In Cosmos, the Law of Fixation of Position obtains. By this Law, not only is an apparent E/entity *localized* with respect to a given dimension of Cosmos, but at any given time its position is (at least relatively) *fixed* with respect to other E/entities who share that dimension and, even, with respect to apparent E/entities who are focused on other dimensions. This fixation results in stable relationships based upon the Law of Repetition.

There is no *identifiability of type* within Cosmos without repetition, nor is there even a relative continuity of identifiable relationship without some degree of repetition. We must remember, however, that within Cosmos, the Principle of Unrepeatability prevails, so the repetition of which we speak is not *exact* repetition. Because of Perpetual Motion within Cosmos (whether continuous or discontinuous) the Cosmic Configuration within any one Ultimate Moment is never identical to that in the succeeding moment or identical with any other.

Without some form of (at least relative) Fixation of Position with respect to other units (which are, themselves, relatively fixed), there can only be chaos. For instance, if common table salt, Sodium Chloride, NaCl, is to exist (i.e., to have identifiable persistence) the minute entities that go to form it must have some designated Fixation of Relation between themselves. They maintain a certain 'position' or 'place' in (or *as)* Space relative to each other. This stable relationship *is itself* NaCl. For all E/entities other than the ultimate particle/event, *pattern is entity.* A change of pattern would produce a different form which would necessarily be ensouled by a differently constituted E/entity (Essentially, a 'Ray' of the ABSOLUTE) suited to express through that form. The contributory entities in any compound (for instance the atoms of Sodium and the atoms of Chlorine) do not (in normal chemical reactions) change, but if their *relationship* changes (when this is possible, as some entities may relate in only one way), or if they aggregate with other elemental entities, another compound would be formed.

So within Cosmos, E/entities or B/beings functioning through form (even through the attenuated forms we often call *formless)* do *take* and *hold* a place or position relative to each other, and maintain that position for a designated term. The maintenance of position during the term intended (and not beyond that term) facilitates evolutionary progress by means of the repetition of a Design-Intended Relationship. Learning and the transference of quality arise through the fulfillment of Design-Intended Relationship, which is "Right Relationship". (Not all relative *positions* and, hence, relationships, are Design-Intended—some are out of harmony with Original Intention, and hence, are, relatively, *evil.)*

- Position, of course, determines function, just as it might be said that function determines position. The two are inextricably inter-related. The place of any E/entity within a system is related to what it is intended to do within the system—its function.

Every system (and our Cosmic System particularly—Itself the prototype of all other intra-cosmic systems) is intended to work organically, as a whole, and all apparent E/entities must be, at length, truly *functional.* They must contribute to the planned working of the System, otherwise, they are destructive and cancerous (aggressively counter-intentive). The ABSOLUTELY FREE (manifesting boundedly in Cosmos as the *Cosmically* Unlimited Universal Self) must, paradoxically, manifest *multiply* through virtually countless limited E/entities ('Rays' of the One, who *are* the One) in order to fulfill ITS/Its own Design-at-the-Beginning. Those limited E/entities must each maintain, for a term, a designated position (of positions) that defines their function relative to other such limited beings, and, as well, to the Cosmic Whole which contains them.

- Even though each human being is ESSENTIALLY the TOTALLY FREE AGENT and is, Essentially, totally dispersed in Cosmos, pervading all things, one does not have the option of damaging or destroying the Cosmic System within which one (as an apparently lesser unit) is contained, for the System was Intended and Planned by *oneself,* when one-as-One was in a far less limited, ignorant and *veiled* state—i.e., at the Cosmic Beginning.

One must maintain one's place, uphold one's position, and act one's limited part even though one be, ESSENTIALLY, the WHOLE OF WHOLES. For, from an *absolute* perspective (the 'INFINISPECTIVE'), what else is there but the WHOLE of WHOLES? This great Cosmos that We-as-**8** (the Universal Self) have *created* together is what We-as-**8**-as-

I periodically 'DO', for all *doing* is periodical. Cosmoses are, simply, what **I**-as-**8** 'Do', and they must be done according to Our/My 'WILL'-as-Will.

The problem is simply this, Why be *bound* when we-as-I are ESSENTIALLY *free?* Why uphold position and function when, ESSENTIALLY, we-as-I are beyond every possibility of localization and superior to every possible function? The answer has to be, because to do so is Our Will. To do so is My Will. **8** must fulfill my Will-at-the-Beginning, which is aligned with and, even, *is* the NAMELESS 'WILL'.

The Problem of the Necessitous Fulfillment of My Will-at-the-Beginning

It seems that, in the course of the Cosmic Process, **8**, manifesting as a vast succession of **8**'s and illusive I's, Am under Self-imposed pressure (a Self-imposed *directive)* not to deviate from My own Will, which is, at Cosmic Root, My Will-at-the-Beginning. This means that the will of 'I' (now in a far more veiled state), is not Really allowed to countermand the Will of **8** that was formulated in a far less veiled state—i.e., when the Grand Design was conceived under the Pre-Cosmic Directive (itself 'BESTOWED' by the 'FLASH').

But Am **8** not ESSENTIALLY *free*, because REALLY, **8 Am I** (the EVER-FREE)? Cannot **8** (being **I**) do exactly what **8** please at any point in Time and Space simply because there is *none other* to compel Me to do otherwise? Of course, the Source of *compulsion* is **8**-'My<u>Self</u>' (the Universal Logos pervading all beings, and ever intent upon the fulfillment of Its Will-at-the-Beginning). **8** have Self-Decided to conform to My Own *<u>Unveiled</u>* Will. That Will is present in the limited little self as a deeply unconscious inherent propelling direction, but not as an explicit conscious pattern. Hence the Unveiled Will-at-the-Beginning is heavily *veiled* in little 'me'.

What is important to realize is that my little veiled will is *not free* because it is <u>superintended</u> by My Greater Unveiled Will. (The word 'superintended' is most interesting in this connection.) Although, **8**, *essentially*, have ubiquity, pervasion, omniscience and omnipotence in Cosmos, as a little self, as a little will, **8**-as-I do *not* have these things, nor do **8**-as-I have the prerogatives which these powers confer. Because **8**-as-I am under limitation, it is *as if* My Will is *not* free, when *essentially* It Is. Thus I must obey a Will which *seems* not to be my own, although that Will *Is*, Really, *My Own*. This is a paradoxical situation.

There is only one Will (the Will of the Universal Self, the Universal Logos—that Will being, ESSENTIALLY, the WILL of the ABSOLUTE SELF). In Cosmos, the Universal Will, the One Will, is active in *<u>all</u> willing*, but some acts of will are more Self-Veiled than others. The One Will, when deeply Self-Veiled, will not easily conform to the One Will in the *unveiled* state. The Self-Veiled will, will seem to deviate from the *unveiled*-Will, and yet in all these acts of will only One Will will be engaging in the processes of willing—i.e., performing *acts of will.*

Ever and at all 'points' in Cosmos, the Will is free (within the constraints of ignorance), but the *results* of freedom within the Self-Veiled condition are not the same as in the Self-Revealed state. Unveiled Will, will *in time,* dominate Self-Veiled will. As an entity re-ascends along its Line of Emanation, the *little will* increasingly approximates the One Will. This is what might be called the 'Unveiling of the Will-at-the-Beginning'.

The Cosmic Process will not for the whole of Cosmic Duration allow the lesser will to contradict the Greater Will. My truly Free-Will (My Will in Freedom—the Freedom of full Self-Realization) which is My Unconstrained Will must (at the Cosmic "Day Be With Us") prevail entirely. The seeming contradictions within the One Will are negated through the passage of Cosmic Time. At first the contradiction is inevitable, but the Cosmic Evolutionary Process of a gradual *unveiling,* with its illuminative results, ensures that all *willing* will at length conform to the One Will, the Original Intent.

The Problem of Will is often a psychological one. The willing agent (a human being or some other self-conscious E/entity) feels deeply that its will is inviolable, and Essentially this is correct, for all wills are, in Cosmos, the One Will (REALLY, the *ONE* 'WILL') which naught can compel, for naught else exists. Yet each S/self-conscious E/entity has the experience of a *hierarchy of wills,* in which the lesser wills are, at length, compelled to conform to the greater wills. The entire principle of "Father, not my will, but Thine be done" is a reaching up along this hierarchy of wills, for an ever truer approximation of the Will-at-the-Beginning—i.e., Original Cosmic Intent.

In the process of Evolution, will is successively and progressively sacrificed to will after will until the One Will is reached. It is important for the agent of will to realize that his will is multiple, hierarchical, and simultaneously in action upon all possible levels of *willing.* When the little self is confronted by the seemingly abrasive impact of another and higher Will, that self must realize that it may well be confronting its *own* will operating on a level *closer to* the Will-at-the-Beginning. This realization will change the psychological response which is often one of rebellion instead of broadened understanding.

The Problem of My Willingness to Play a Part

This is a problem often posed by the limited consciousness that frequently dislikes the part which it is assigned to play, and wishes to play a larger, fuller, more personally gratifying role. That little self, when it begins to realize its True Selfhood as the SELF-as-Self, may question the necessary of performing any 'part' at all, since it realizes itself to be, ESSENTIALLY, the WHOLE OF WHOLES (or, when in Cosmos, simply the Whole). Deeper understanding, however, will reveal the *necessity* of appropriate *role playing.* In a way, the ONE SELF as ONE 'ACTOR' does only One Thing periodically, forever—IT ('COSMIFYING' ITSELF into 'It'), Plays a Role, and that Role is *Cosmos Itself.*

What is Cosmos, after all, but a 'Part' to Play—a defined, limited, circumscribed *part* of the INFINITE POTENTIAL (the FOUNT OF ALL POSSIBILITY) which I AM? It is as if the unfolding of Cosmos is the unfolding of one set of possibilities of an infinitude of *infinitized* possibilities, and *one set only*.

When it is realize that **I**-the-WHOLE 'DO' nothing forever but periodically Play Parts, then it becomes possible to see the value of all the lesser roles which **8**-the-Localized-Whole play in Time and Space. **8**-as-Localized Whole, (a 'Ray' of the ABSOLUTE) immersed in the prison of sequence and location (i.e., Time and Space), do as **8** have done at the Beginning, and play one role or part after another until **8** understand 'MySelf'-as-Universal Logos to be Playing the One Cosmic Part.

The playing of part after part need not stop Me (the **8**) from identifying 'MySelf' as the Self/SELF (first as the Whole and then as the WHOLE OF WHOLES). **8** come to realize that My localization is Self-Imposed, and, with growing illumination, **8** begin to honor the meaning and value of that Self-Imposition. **8** play the Game because it is, apparently, My *Bliss* to do so, and more importantly, because it is *necessary*. During each Cosmos, it is MY 'WILL'-as-My Will to play this Game. Thus, **I** periodically 'COSMIFY' and thus 'LIMIT' the SELF which **I** ESSENTIALLY AM. Cosmos is the Great Limitation.

This Process of Role Playing has occurred periodically forever. In the theatre, let the actor not become lost in his part, for this will result in a bad performance through the loss of control and perspective. Analogically, let the INFINITE SELF not become lost in that 'Unfragmented-Fragment of INFINITUDE' which is Cosmos, the 'Part' IT periodically 'Plays', for this too, would result in a bad performance. Indeed, the INFINITE SELF cannot *ever* become 'lost' in Cosmos, ITS 'Part', for IT ever remains exactly as IT IS.

■ The dual attitude must be held by those who wish to consciously retain REALITY even while taking part in that infinitely lesser Reality we call the World of Illusion; the proper attitude is, "*In* the World but not *of* the World."

This dualistic perspective need not, and, indeed, *cannot* be maintained 'outside' of Cosmos in the 'STATE' of ALL-IN-ALLNESS that prevails during the Universal Pralaya. During Cosmos, however, a divinely dualistic perspective is required for right functioning. There are parallels in the life of the human being. In the life of the human being, for instance, the Solar Angel may, indeed, be "downward gazing" and *involved* in the lower worlds of form, and yet that Angel remains very much Itself within its own world.

On a far, far vaster level, the One Self (the Cosmic Self or Universal Logos) may also be "Downward Gazing" upon Its Cosmos, and involved in all Cosmic Processes through Emanation and Identification, and yet It remains inviolably Itself upon Its own level. From the loftiest perspective, the ONE SELF, WHO IS every possible Cosmic process and Cosmic change, remains ever unchangeably ITSELF throughout the entire Universal Manvantara, but whether IT can be considered 'DOWNWARD GAZING' is highly debatable. The parallels, however, are suggestive.

It becomes possible to play a part well and with joy, when one realizes that he *cannot possibly be limited by the part he plays.* The actor-as-Actor-as-'ACTOR' is ever *infinitely* greater than the part 'HE' 'PLAYS'. The use here of the word 'infinite' is literal. The 'distance' between the apparent fragments (however large the fragment may be) and the INFINITE SELF (the FOUNT OF ALL POSSIBILITY) is *infinite*. Who would mind playing a 'part' in the light of such a realization?

The Problem of Fear

Fear is essentially a recoil in the face of potential annihilation or disintegration (whether total or partial disintegration, such as wounding or dismemberment). Fear is the primal response of the ego, and arises due to a primitive sense of selfhood. When selfhood (as presently conceived by the incarnating entity) is threatened, then fear arises. The phenomenon of ego (which, as usually used, is roughly equivalent to a primitive sense of selfhood) comes into existence due to the perception of *boundaries* as a component of selfhood. When selfhood is experienced as *boundriless*, there is no fear. This perception of boundriless *selfhood* (REALLY, SELFHOOD) leads to the conviction of immutable immortality. When there is established a conscious immortality, there can be no fear, for there can be no REAL annihilation or disintegration.

- It has been said in the Vedantin Teaching that, "Whenever there are two, there is fear."

This is a profound statement, and reveals that the arising of fear is inseparable from the arising of Cosmos, which (in relation to the INFINITE SOURCE) becomes the first *seeming* Duality. Indeed, there can never be any REAL duality whatsoever, but only *seeming* dualities, of which the first is the SELF and the Self, or the INFINITE and Cosmos (even though, necessarily, the ONE 'INCLUDES' and IS the Other). There are many consciousnesses in Cosmos Who are *fearless,* but such are Those who have seen *through* the Illusion of Duality even while registering a functional duality as an inevitable Presentation of Consciousness in Cosmos.

It is interesting that fear, archetypally, is always of *loss.* This fact is due to the holistic bias of the SPIRIT/Spirit for which any form of *less-than-wholeness* is anathema. Apparent *reduction* (however it seems to come about) is SPIRIT/Spirit's only pain. Fear, is, rarely if ever, of *gain,* unless the gain itself is seen to lead inevitably to loss. Thus, fear is of being *less* or of losing the *being/selfhood* that one appears to have.

Fear of pain, or of the pain that others may experience, is the fear of a dissonant, divided condition which destroys bliss—bliss being a unitive state. We see then that any inharmony usually has the effect of destroying the apperception of Wholeness/WHOLENESS. Pain always arises through the perception/experience of division.

Perhaps the greatest of the fears is the fear of separation. The separation experienced is always from *another*—whether that other be the SELF, God, another human being, or a particular set of circumstances. Separation implies the loss of that other. Isolation is the extreme condition of separation and may be seen, superficially, as the loss of *all* others. In extreme isolation, one is, *apparently,* left with no one other than one's self (however that self may be defined). There are ways, however, by means of which isolation can reverse the sense of separation, and lead directly to the apperception of THAT from which one cannot *ever* be separated. Such a State of Consciousness or, rather, of *being,* has been called "Isolated Unity". In such a State of Unitive Isolation, all the 'others' which have apparently become *lost* are very *Really* regained.

In most states of consciousness, there exists some kind of minimal unity that arises because of an identification with some apparently stable and unitary thing, whether that thing be a limited ego or a place or a circumstance. No self-conscious being can live without the perception/conception of self-unity, no matter how minimal that unity may Really be. For instance, a self-conscious being always has within its consciousness at least some conception of an objective unity it may call 'itself'; a human being always thinks that he is, at least, 'something'! Fear arises when the disruption of that perceived/conceived unity is threatened, and the human may be confronted with the idea that he is "nothing at all". Fear of the loss of identity (of being *nothing)* is one of the greatest fears, for the result in the unillumined mind would be to seem totally lost, separated and even annihilated.

Fear, then, even in relation to such relatively lowly states of consciousness, is of the loss of what one seems to *have* or what one seems to *be*. The fear is always that one will somehow be less than whole or farther away from the achievement of a desired state. (We must remember that the fulfillment of desire is meant to make one feel more complete, more whole, for fulfilled desire temporarily ends the duality existing between the desire and the achievement of that which would fulfill it.) Fear, then, is the thought or apprehension that a 'lessness' will supervene due to an act or to an encounter.

Somehow it must be realized in the light of Radical Infinitism that one can never be *less;* that one can never be 'distanced' from what one already IS; that one IS, in ESSENTIAL fact, *already* THAT which is to be achieved. One need not *achieve* it; one *is* IT; in fact, one IS all possible FULFILLMENT ITSELF. One need not add to oneself to *become* IT. The FULLNESS that one already and ever IS cannot be altered in any way, nor can one REALLY be separated from that FULLNESS. It must dawn that one has 'ABIDED' *forever* in an untouchable, imperturbable 'STATE' which prevails over all possible variation NOW even as it has prevailed and will prevail *forever*.

Naught within the World of Relativity can touch this realization once it is firmly rooted in the true 'inperience' of the SELF. With true, consciously-anchored experience of the SELF, fear vanishes, just as duality vanishes.

■ The degree of fear seems directly proportional to the degree to which one is identified with phenomena within the World of Relativity.

How much does one Really 'care' about the Relative World? Indeed one must both care profoundly, and not care at all—a seemingly paradoxical attitude. In one way, one is playing a game—the only Game, and the emphasis is upon the word 'play'. If one becomes too serious about an ordinary game, one forgets that it is "just a game", and becomes an unsportsmanlike competitor. The analogy is so important. One must not play even the Great Game too seriously. Or perhaps it should be said that one must not play the Great Game with too much attachment to outcomes. One can play 'as if' seriously, using all one's capabilities, yet with a detachment that arises from knowing that no outcome, good or bad, will affect REALITY in the slightest (though the outcome *will* affect progress towards the fulfillment of the Design-at-the-Beginning).

With excessive self-investment in outcomes, fear increases because no outcome can be perfect and forever stable; because outcomes exist within the World of Relativity, every outcome is subject to disruption. So, one must both *care* and *not care*. This attitude is sometimes called "Divine Indifference". To act with consummate skill and compassion as if one cared immensely, but actually to be so SELF-Identified that one simply IS—

abiding ever in a state of detachment from what one is doing—this is the attitude of Those who truly *Know* the SELF and, hence, ARE the SELF.

Another perspective on fear reveals that it arises with the loss or destruction of the *familiar*. Most human beings are identified with the familiar. If the familiar is destroyed they are thrown back upon a Self they have not yet recognized; they feel lost and isolated and, thus, *separated*—hence subject to the Greatest of all Heresies, the Great "Heresy of Separateness". In short, they sense themselves *excommunicated* from any sense of identity.

■ In order to move towards fearlessness, one must gradually, gradually, divest oneself of the familiar, or translate it to oneself in a new way so that it becomes REAL instead of familiar.

The great sage Sankaracarya did this by "Merging the World in Brahman" or "Dissolving the World in Brahman." The World may still have *presented* Itself to His consciousness as the World—i.e., the image of things was still present in consciousness, but the substantial REALITY had changed. Everything had become, as it were, 'BRAHMANIZED'. Through such a Divine Process it is the BRAHMAN that becomes incessantly *familiar*, and not the ever changing modifications in the World of Illusion.

Fear is certainly a force which hems us in and prevents us from going beyond our familiar boundaries. How then to enter into the BOUNDLESS STATE (or, even, the expanded State of Cosmic Boundedness) when one is still subject to fear, when there are certain lines one will not cross, for as all boundaries must be crossed, all lines must be crossed on the way to the REAL.

All boundaries are REALLY artificial (however *cosmically* necessary they may be), but there are ways of *crossing* them without necessarily *destroying* them, and with them the form(s) they are intended to protect, for the time for the destruction of the form is wisely pre-determined and keyed to the fulfillment of its function. It may be said that "One can always cross a formal boundary *in Spirit*."

The idea of the crossing of boundaries evokes thoughts regarding Conformity to <u>Law</u>—a word which, when spelled backwards, yields 'Wal(l)'. Indeed the Law is a Wall, a willed-boundary that may not be trespassed. When we think of Laws we are compelled to think of *acts*—what we will and will not *do*. Questions regarding the scope of action arise also (mainly in the negative) with respect to fear.

- When fear arises, what are the things one will not do?
- More importantly, what are the *liberating* things one will not do?
- Where are the places one will not go?
- What are the beneficial experiences one will not hazard?

Fear is really the enemy of ubiquity and pervasion. Fear maintains through reticence the ego state, the localized state, the state of false identification with the familiar.

The realization 'boundarilessness' is a fearless state. For what can befall one when one is, ESSENTIALLY, boundriless? How can what one IS ever be destructively modified or, even, modified at all? How can one lose anything or become less? One can only lose attachment to the Great Illusion. One can be the WITNESS/Witness of form destroyed, and yet the SELF is not destroyed. Confirmed in REAL SELFHOOD, what can possibly be taken away?

The main question in the consciousness of the fearful one is: What will happen to *me?* Even, What will happen to my loved ones? is a variation of ,What will happen to *me?* for one is identified with one's loved ones and, therefore, feels that which impacts them. Indeed, what *can* happen to 'me' if I dare to move against my fear? In the SELF-REALIZED 'STATE', **8** know that *nothing* will happen to me (or to me-as-Me-as-ME). Nothing *can* happen to ME, and strangely but truly, nothing *ever has* 'happened' to ME, or ever will. **I** (in MYSELF) have never varied, nor have **I**-as-**8** in Cosmos. **I** have not modified or been modified, nor can **I** be modified, whether **I**-AM-**8** in Cosmos, or **I**-in-ABSOLUTE PURITY. Thus, fear vanishes with the conviction of one's ESSENTIAL *immutability, invariability,* and *immortality.* Fear vanishes with the realization that no SELF-reducing change can ever take place, that 'lessness' can never REALLY happen, and that 'moreness' (or rather 'ultimateness') exists forever *maximally.*

It cannot be denied that, in-Cosmos, fear serves the Design-at-the-Beginning. Fear (really the survival instinct) at first, preserves the fragile integrity of form—form being relationship, relationship being the means of progress towards the Goal of the Original Intent. Later, however, fear is based upon the falsity of ego, and the falsity of ego is based upon the illusory perception of boundary, and the perception of boundary upon the Principle of Separation which, itself, is, ESSENTIALLY, the 'great *impossibility*'.

■ So we see that, though fear is ESSENTIALLY/Essentially an illusory attitude that must be eradicated, it does tend to preserve the integrity of form, holding (in the initial stages of Evolution) the World of Illusion (which is the World of Becoming) intact. While fear initially *maintains* relationship, thus preserving form, if it persists unduly in self-conscious entities, it also immobilizes and paralyzes and *prevents* evolution of the form and, thus, the freeing of the Spirit.

A far better way of maintaining relationship and the integrity of form is through the agency of *Love.* The point is simply that *right identification* (non-dualistic identification) eliminates fear.

The Problem of Loneliness

This problem is solved through the achievement of Isolated Unity. In the SELF there is no loneliness—only 'ALL-ONE-NESS'. The illusion of loneliness is based upon the sense of localized encapsulated selfhood. For those who can access the SELF or SELF-as-Self, all sense of encapsulation melts.

Loneliness often induces fear because, when one feels lonely, one feels separated from all that conduces to happiness, joy, and bliss. Identification with the SUBSTRATUM that one-as-One-as ONE ever REALLY *IS,* prevents all possible consciousness of separation.

In earlier days, loneliness is partially overcome through interaction and relationship, but it is never truly conquered. As the consciousness matures, Identification with the SELF/Self replaces the more illusory stages of interaction and relationship. These still exist, but are understood to be simply appearances of the SELF within the Ubiquitous Self.

The Problem of Justice

The Problem of Justice is akin to the Problem of Karma. Every unit in Cosmos reaps the effect of its actions. But *who* sows and *who* reaps? As heretofore explained, sower and reaper are one and the same SELF-as-Self WHO is the true IDENTITY/Identity of every apparently separated little self. Because of the Ubiquity of the SELF-as-Self-as-'allselves', is not all sowing an *immediate* reaping? Action sown by the ONE-as-One impacts immediately upon the ONE-as-One, for upon WHOM/Whom *else* can there be an impact? There is no delay. There is only One Action experienced by all-as-One in Cosmos simultaneously. Sowing and reaping occur in an instant of time. Perhaps this is why the word 'Karman' simply means "action".

There is, thus, One Impact, and One Receiver of Impact, although this One Receiver is disguised as the Cosmic Multitude. Justice (in its ultimate sense) simply and Essentially means that all E/entities shall experience all things equally. In Time and Space it may seem that each E/entity receives, specifically, its own "just desserts" but this is Essentially an illusion. When one receives, then, in Reality, *all* must necessarily receive, for there is only ONE-as-the Cosmic One-as-'all the many *ones*'. This is 'Justice' understood in a new way—the Justice of RADICAL HOMOGENEITY. There is Justice because there is *evenness* all around.

■ The SELF (and, even, the Universal Self) is *homogeneously* distributed. Thus, all experiences are definitely shared and experienced to the *same* extent, even though experience in Time and Space *seems* to contradict this logical necessity. It is true that none is spared from the impact of what he has done, but it is even more radically and *strangely* true that none is spared the impact (for good or ill) of what *all* have done. Evenly and equally, each experiences *all*.

Then, is the Cosmos 'Just'? Those who maintain the old view of "An eye for an eye and a tooth for a tooth", will say that the Radical Non-Dualist perspective does *not* offer a *just* Universe. In all fairness the "return blow or blessing", in-Cosmos, does seem to pertain *most* to the localized E/entity who gave forth the blow or blessing.

It could be said, therefore, that, given the reincarnational perspective, the Universe does *seem* just in the more limited sense of the word. The new form of Justice, however, is Cosmic in its implications and states that the Self-in-Cosmos immediately and completely experiences the impact of the One Action initiated by the Self-in-Cosmos. The Ubiquitous Self *receives* what It has *done*, though the doing and the receiving are simultaneous. If the Self (veiled as It is, even at the level of Universal Logos) *did* otherwise, It would *receive* otherwise. This karmic dynamic is, therefore, *just* as well as *exact*.

The Problem of Sex

The sex impulse is a tremendous driving force in Cosmos. It is an impelling means of apparently overcoming duality. The Cosmic implications of sex have to do with the union of the One Spirit with the One Matter (the union of the Self-Observing Self with Its Self-Reflection, of the Universal Father with the Universal Mother). Thus it can be seen that *sex,* in this higher sense, will persist throughout the Universal Manvantara, for Spirit and Matter will not be totally reunited until the onset of Universal Pralaya.

In the life of the average aspirant or disciple, the Problem of Sex is far more immediate and 'engrossing'. It manifests as various problems that arise because of the intensity of attraction between the sexes.

The field of sexuality is a breeding ground for the glorification of the false ego, and leads into separative and exclusivistic patterns of behavior. The problem of possessiveness is also definitely involved, as is the problem of incorrect, non-Goal-Fit expenditure of a whole range of individual energies.

In the light of the *realization* of the ONE SELF, the following of questions arise:

- Who is attracted to Whom?
- Is not 'WHO' attracted to 'WHOM'?
- Why the intensity of passional longing for the Not-SELF when supreme satisfaction comes from WHAT one-as-One-as-ONE *already*, ESSENTIALLY, *IS?*
- What happens to sexuality when ego disappears?
- Does sexual passion disappear when ego limitation disappears?
- Does the longing of Spirit for Matter in the World of Becoming justify full sexuality in the life of the spiritual aspirant or disciple?
- Is sexuality helpful or inimical in the fostering of SELF-Realization?

Without going into each of these questions in detail, it would seem that the entire character of sexual attraction and sexual experience would change if there were Identification with the SUBSTRATUM. The many abuses surrounding this field of human behavior would gradually vanish, and the Greatest Good of the Greatest Number would be more easily achieved. The desperate quality of so much human sexuality would be relieved, for desperation is based upon duality (and its *apparent* insurmountability), and most specifically upon the insistent and often frustrated urge for the two poles to come together *ideally.* When it is realized that bi-polar consummation has already and forever been *consummated* in the SELF, sexual consummation will be viewed and experienced with fuller understanding.

The Problem of Fate

This problem is closely related to the Problem of Karma and the Problem of Justice. Fate is a pattern of circumstance which 'awaits' the individual on the basis of his multi-leveled actions. Destiny is a pattern of circumstance which awaits the individual on the basis of his Inherent Design or Quality. From the Radical Infinitist perspective, however, the question must be asked, Is there a *separate* fate for anyone? Essentially, the answer would have to be, No.

- There is only an *apparently* separate fate. All fates are the *same* Fate, just as all actions are One Action. **8**/We Am the Experiencer of all action which is the One Action, and the One Action produces the One Fate.

Are all things experienced equally at all 'points' in space? From the limited perspective of the customary rationality this seems an absurd question. If a bomb goes off ten thousand miles from 'where' 'I' (apparently) am, 'I' do not feel it *directly*, do 'I'? However, since all points in Space are, Really, One Point, which is located simultaneously everywhere and nowhere, there is Really equal experience of all things at all apparent points in space.

Because of the veiling power of 'MAYA'-instantly-Maya, **I**-the-WHOLE AM, *apparently*, a particularized and encapsulated multiplicity. The prevailing illusion is that **I**-as-**8**-as-I experience one set of things instead of *all* sets of things. It will, however, become revealed in the fullness of time, and with expanding sensitivity, that that which I (in my limitedness) thought I did *not* experience, I-as-**8**-as-**I**, *did*, in fact, experience *fully*.

These thoughts bear directly on the question of fate. There is no Real Truth in the old saying, "There but for the Grace of God go I." Instead, the Truth is found in the statement, "There, indeed, go I" (though it must be understood that the I is the I-as-**8**). In these thoughts we see the essential Truth of the words of the poet John Donne, "Never send to know for whom the bell tolls. It tolls for thee." All fates are MY-as-My Fate. What befalls the Universal Logos literally befalls Me-as-me. That 'I' do not 'know' this in my limited brain consciousness, is due simply to the Powers of Maya—MY/My Powers.

The sharing of all circumstance cannot be escaped. When one says to another, "You just don't know about this, because you haven't been through it", a great, fundamental fallacy is being propounded. An apparent truth is taking the place of an Essential Truth. What is my condition? Answer with the question, What is the Condition of the World of Becoming? That Condition is *My* Condition.

8-the-Presence (in reflection of **I**-the-PRESENCE) Am simply *present* everywhere in Cosmos. The great strain, the great friction, the great contradiction, is how to live the apparently particularized life and the 'non-apparently' Universal Life and INFINITIZED LIFE simultaneously.

The Problem of Blame and Complaint

This is a problem that besets all of humanity, and its solution is particularly crucial for aspirants and disciples on the Path of Spirituality, for blame and complaint emphasize the very duality that must be overcome.

All blame and complaint can and must cease on a *philosophical* basis. If **8** Am active in every movement in Cosmos (as **8** Am because of the *pervasion* of the SELF-as-Self), and, further, that which **8** blame, or that in relation to which **8** have a grievance, is the result of the pervasive movement in which **8** Am active (really the One Movement in Cosmos), then **8**, indeed, Am the Author of that which **8** deplore. **8**, Essentially, and *not personally*, have full responsibility for all conditions in Cosmos and this **8** must acknowledge. With this acknowledgment in place, **8** can move beyond praise and blame, beyond commendation and complaint.

My attitude, once **8** have realized My Universal All-Potency and My All-Sufficiency, should be one of supreme *satisfaction*. Satisfaction comes when all desires have been fulfilled, and indeed, on the level of ESSENTIAL SELF (could IT but be realized) all desires *have been fulfilled.*

The Problem of Dissatisfaction and Discontent

The state of *desire* is possible only intra-Cosmically, and is predicated upon the incessant dissatisfaction. This dissatisfaction, especially in the life of the Self who is awakening spiritually, is sometimes called "Divine Discontent". On the Path of Evolution, the SELF-'VEILED' consciousness remembers 'unconsciously' (or is subtly imprinted with) the impression of the SYNTHESIZED 'STATE' (or even the non-ABSOLUTE Unified State which existed at the inception of Cosmos), and the contrast between the unconscious echo of that Unity and the presented, in-World dualistic state, stimulates the Divine Discontent to return to a state of Unity.

- In-Cosmos, some type of dualistic consciousness is inescapable and, with it, the incessantly driving pressure of Divine Discontent. For the one who is achieving illumination, however, there must begin to appear along with this necessary Divine Discontent, a Supreme Satisfaction.

Why should **8** not be satisfied if **8** realize WHO/Who **I/8** REALLY/Really AM/Am? (The contrasting capitalizations of the same word juxtapose MY INFINITUDE and My Cosmic Relativity.) In the face of my necessary Divine Discontent (*necessary* if the Divine Purpose is to be unfolded in that part of ME-as-Me which is Objective, i.e., Matter), **8** must realize that **8**-as-I already *AM* and *'HAVE'* all that Cosmos can Be and can Create, and even ALL that the ABSOLUTE SELF can *BE*, for in ITSELF, PER SE, IT 'CREATES' nothing.

To overcome blame and complaint, one must remind oneself everyday of the Truth of One's Real Nature (in Cosmos) and, also, and even more importantly, of ONE's REAL (SUPER-Cosmic) NATURE. One of the major purposes for meditation upon the ESSENTIALLY non-dualistic NATURE of REALITY, is simply to facilitate this type of reminder, for of the many things forgotten by the mind of man, the INFINITE SELF is the most *often* forgotten.

With these thoughts in mind, we must ask ourselves (as apparently limited beings):

- Wherein lies my greatest dissatisfaction?
- Do I lack satisfying association with my fellow human beings?
- Do I lack money?
- Do I lack aesthetic surroundings?
- Do I lack comfort?

The range of possible discontents is virtually numberless. All sorts of negative states arise because of these perceived discontents and the fixation of the consciousness upon them. Blame and complaint are simply two in which the not-SELF is accused as the source of discontent. (But *is* there a not-SELF?)

In answer to all such questions, one must ask: What is IT that **8** *do* have? The true answer to this question is astonishing to the SELF-Realized consciousness. Do **8**-as-**I** not have bliss and supreme content? Is not the SELF that **I,** ESSENTIALLY, AM the NOUMENON of these very states (content and bliss) that crown the aeonial search which Desire impels? Why then complain and blame?

The Problem of Worry

Worry is the chaffing, recurrent fear that that which is desired will *not* come to pass, or, conversely, the fear that that which is *not* desired *will* come to pass. Certainly, all things desired with sufficient intent do come to pass in the Fullness of Time. They are guaranteed, because, as Morya says, "Everything is possible." Undesirable, non-harmonious, less-than-ideal states that *do* come to pass are certainly temporary and can be overcome.

Worry is based upon a lack of trust, which in turn, is based upon non-Realization of the SELF (for in the SELF is the "mine of trust"). The one who worries fears undesirable outcomes, the eventuation of non-ideal patterns. Worry shows engrossment in the not-SELF, preoccupation with processes in the World of Becoming.

The antidote is to realize the *insubstantiality* of Cosmos, and Its evanescence, Its impermanence. About the INFINITE SELF it is not possible to worry. IT is the ever-INFINITIZED STATE which is the NOUMENON of all possible satisfaction. That SELF is ever *present* as the PRESENCE. IT is the GREAT REASSURANCE. To knowers of the SELF, a debilitating state of worry is impossible. Why fret and chafe over that which is infinitely less than ONE already IS (and, therefore, 'Has').

The Problem of Worry is solved, philosophically, through an "Adjusted Sense of Right Proportion." The *value* of the that which can be found within the World of Relativity (however worthy or beautiful) must be contrasted with the INFINITE VALUE of THAT which already IS in one's possession, because one-as-One-as-ONE IS IT. Yet, we forget these things, and consequently worry.

The Problem of Self-Confidence —Feeling 'Good' About 'OnesSelf'

This is a very practical problem, relating directly to the life of discipleship, and to the "spiritual inferiority complex" which often develops as the apparently limited human being begins to contact the higher energies. When the disciple is joyful, the effectiveness of his or her work is multiplied greatly. Master Morya has told us that "Joy is a special wisdom." Joy surges through the life when there is a true Self-confidence and when one truly rejoices in *knowing* and *being* WHO/Who He IS/Is.

How can one actually develop this Self-confidence and the feeling of joyful Self-assurance and right Self-approbation which accompanies it? Is accomplishment in the world necessary for Self-confidence and joy in *being* 'Ones Self', or is it sufficient to have found the SELF and SELF-as-Universal Self? Certainly, the localized, encapsulated self acts as if accomplishment in the World of Becoming is necessary. This little self judges itself according to the standards of the World of Illusion and not according to the standards of the World of Being or of the ABSOLUTE.

- Few there are who take their joy from identification with the INFINITE SELF. It is understandable why the numbers should be so small, for, at this stage of human development, the task is extraordinarily difficult due to egoism and mistaken identity.

The SELF-as-Self-as-self expresses through ever-changing equipment. The vehicles of expression must necessarily fluctuate constantly though the INNERMOST SELF (even though IT be thought of as the Self or even the self) changes not. A problem arises because, usually, the confidence one feels is not based upon the immutability of the INNERMOST CORE, but, unfortunately, upon the condition of the vehicles through which the CORE expresses. There is no true Self-confidence to be had in this way, because one has placed his reliance upon that which *ever changes,* whether for good or ill. A great uncertainty will be the result, and an insecurity which is just the opposite of true Self-confidence (REALLY, *SELF*-Confidence).

Again, the realization of the INFINITE SELF is the only permanent solution—permanent for this Cosmos, that is, for the SELF must be *re*-realized in every one of an infinite number of Cosmoses (even though SELF-'REALIZATION' {or, better, SELF-'BEING'} has never, for an infinitesimalizing instant, been interrupted or diminished). The

INFINITE SELF, of course, is always and ever the ABSOLUTE PERFECTION, the SOURCE of every bliss. IT cannot fail. If one IS THAT, even though ONE seem to be a tiny 'one', how can one help but be unwaveringly confident? How can one not "feel good about oneself" when that self is the Self is the SELF, the UTTER PERFECTION?

This is not to recommend an inflated or unrealistic (better, non-*actualistic)* attitude, but simply to state that the human being must come to terms with his ESSENCE, which is *the* ESSENCE. After that, behavior may appear normal in many ways, but an immense qualitative change will have occurred inwardly.

The Problem of the Continuation or Non-Continuation of Happiness —the Problem of Mood

Happiness (as a reflection of the bliss of the *continuous* SELF) would be continual if it were not for the changeability of relations within the World of Becoming and the identification of the Self/self with those changes. Continuity cannot be found in a discontinuous Cosmos (especially in the World of Approximation), so the *life unconditional* must be understood and cultivated. This can be accomplished through strict attention to the SELF, which IS ever-present as the PRESENCE. Access is not the REAL problem, attentiveness is.

As human beings, we are so identified with what we *have*, and what we *create*, that we lose any possible touch with our ESSENTIAL SUBSTRATUM, our BASIS. It is the maintenance of our connection to the CONTINUAL BASIS that imparts Synthesis-in-Cosmos, and makes it possible for us to consciously participate, eventually, as the 'Pervader of All Relations'. Needless to say, this Pervasion (though omni-experiential and maximally variegated) is Essentially *blissful* and *continuous*, since it reflects the CONTINUOUS BLISS of the INFINITE SELF.

Must one *have* the *part* when one-as-One-as-ONE already IS the WHOLENESS? Do not most of us find ourselves grasping after parts and seeking to put pieces in place instead of staying focused upon THAT which We-the-I already *ARE* and already *Have?* Yet our worldly conditioning is such that it is almost impossible not to pursue relations as if they were the INFINITE SELF (and, of course they *are,* but only ESSENTIALLY). The normal happiness and unhappiness are simply a matter of reaction to conditions, and conditions will ever vary.

Imagine a Humanity that was anchored in the UNCONDITIONAL. This means not only an anchoring in UNCONDITIONAL LOVE, but an anchoring in UNCONDITIONAL BEING which makes possible UNCONDITIONAL LOVE. UNCONDITIONAL INTELLIGENCE is also available, to the one who identifies *as* SAT-CHIT-ANANDA (BEING, CONSCIOUSENSS, BLISS).

Our unhappiness arises because we are driven to have what we (apparently) do not have. This drive motivates us, giving us the incentive and the urge to possess—as if a lesser kind of possession of what we *already have* in greater *fullness* would render us happy, joyful, or blissful?! What is possession, actually? Possession is the exercise of will over an object or objects. It is the possibility of achieving total impact upon an object—the possibility of being the only one whose will impacts a particular object with which one may identify oneself. Possession is a kind of exclusivism which grows out of an egotistical consciousness.

It must constantly be reiterated that we already *have* what we would *possess* simply because we *are* the SELF. There is REALLY no such thing as *possession* in all of Cosmos. *Having* is illusory; *Being* must take its place. One can appear to possess something, but, in TRUTH, one does not REALLY *have* what one seems to possess until one IS it. So our time should be spent attempting to *BE that* which we would *have*. This may not seem a practical pursuit to the "man of the world", but it is the Path of Bliss.

The Problem of the Causes of Unhappiness

Unhappiness is the result of apparent privation. Restoration of that which is apparently lost usually restores happiness. Unhappiness is generated by a sense of lack, by the sense of 'distance' between the actual and the ideal—in short, by unfulfilled desire. *Apparently not having* what one wants creates in the life a great sense of cleavage or separation.

With these several causes we are all familiar. We must ask, then, whether it is possible to live within Cosmos and not feel *separated* from the most desirable? The answer must be, Yes. Since all sense of separation is fundamentally an illusion born of SELF-'VEILING'-'BECOME'-Self-Veiling, it is logically impossible to be ESSENTIALLY separated from the most valuable (which is ever *THAT)*.

To understand the illusory nature of the separation experienced, one must appreciate the ABSOLUTE REALITY of the PRESENCE, especially at those moments when one is confronted with all that which is apparently divided and separated. One must appreciate the *continuity* of the PRESENCE/Presence in a manifestly discontinuous world. The Great Illusion Is inescapable and, further, the Great Illusion Is what WE-the-**I** 'DO'. In one way, We, the SELF-in-Cosmos, are Isis restoring to integrity the dismembered Osiris. Osiris, then, is the Cosmos dismembered. Isis can also stand for the knowledge of the TRUTH of the *indivisibility* of the SELF which the Ancient Mysteries impart. As we 'piece the world together', as Isis did Osiris, we discover the former causes of our unhappiness to be illusory.

In the fullness of realization the sense of privation vanishes. No E/entity is Cosmos can ever be deprived of the SELFHOOD which it *IS*, and which IS everything. The realization of the fallacy of privation eliminates the chief cause for unhappiness.

The Problem of the Possible Maladaptivity of Unitive Living

If the Realization of the SELF dawns, will it render the one who realizes maladaptive in the normal world? This is a very serious question, and the evidence is not always encouraging. The Cosmos is organized *hierarchically,* and there is the possibility that respect for this hierarchical organization is reduced in those who realize the fundamental SELFHOOD of all E/entities, regardless of the hierarchical rank of such 'E/entities'. Will all things appear *the same* to the one identified as the INFINITE SELF? Will all preferences and choices be viewed equally and thus all sense of Direction in Cosmos be ignored? After all, if SELF REALIZATION confers THAT which IS *most valuable* why pursue other things in a paltry Cosmos?

8 think it can be fairly concluded that to be *lost* in the ABSOLUTE while incarnated in the World of Becoming may prove maladaptive, in a *Cosmic* sense. The emphasis should be upon the word 'lost', for there are those who identify as the INFINITE SELF and are not at all 'lost'. In fact, they are those Who (or rather WHO) lead apparent 'others' to the INFINITE SELF which they (the others) have apparently 'lost'.

The right approach is to be "*in* the world, but not *of* the world." It is easy for the Radical Non-Dualist to repudiate the World (the Cosmos), for, as frequently stated, the Cosmos is infinitely far removed from the INFINITE SELF (even while, ESSENTIALLY, *being* that very SELF). The Cosmos ever suffers in *comparison* with PARABRAHMAN-the-INCOMPARABLE (as *The Upanishads* have made abundantly clear). Nevertheless, the Cosmos Is Beautiful!, and the Cosmos Is, ESSENTIALLY, *BRAHMAN* ITSELF.

- Thus there is no good reason for rendering oneself unfit to skillfully negotiate the World of Illusion. In fact, such unfitness would be an attitude of *separativeness* in one who, so he thought, had achieved the ULTIMATE UNIFICATION IN THE SELF, for the World of Illusion is *also* the SELF. Right attention to *duality* ever preserves the integrity of *monality.*

The Problem of Stillness and Activity

If the ancient saying, "Be still and know that I am God", is true, then it becomes important to know what stillness really *is* and how to achieve it. When shall one be still? When shall one act? The whole matter relates directly to the illumination of the consciousness, for *activity* per se, is a *veil.* The point is not to devalue activity, for Cosmos Itself *Is* Activity. Activity is given the lion's share of human consciousness. Unless, however, stillness is understood and practiced, the consciousness will remain preoccupied and, hence, enslaved.

So then, how is one truly to be *still?* It seems clear that stillness emerges through not *doing* anything, through not changing, not indulging in modification, through seeking no thing, through realizing that, in ESSENCE, there is no where to go and nothing to do in order to produce the FULLNESS one already IS. Certainly there are 'times' and 'spaces' when it is possible *not* to *do.* In the stillness of non-doing, it begins to dawn that *stillness itself IS the GREAT ABSTRACTION.* The GREAT *MOTIONLESSNESS* IS the GREAT ABSTRACTION, the SELF. If the SELF has any way of revealing ITSELF in Cosmos through the Veils of Maya, that way is *stillness,* for Maya is Motion.

- To achieve REAL stillness is an act of profound *abstraction.* In stillness, *motionlessness* is realized, and since pure MOTIONLESSNESS *IS* the SELF, then, in stillness the SELF is realized. From another perspective, the SELF IS the STILLNESS.

Motionlessness is the key to abstraction. If one with persistence cares not for doing anything, then motionlessness will descend upon the consciousness and Being/BEING will be revealed. Of course, it is only the ABSOLUTE SELF ITSELF which can BE (completely) MOTIONLESS. Because IT IS *ever* ITSELF, and no-thing else, IT IS forever MOTIONLESS. Since, however, there is no other than the ABSOLUTE SELF, even in the World of Illusion, MOTIONLESSNESS can, at least, be realized, because all the apparent motion of Cosmos IS no REAL motion at all. (Where there is only *one* Point or *no* REAL Point, how can there be movement?) The point is that all *apparent* motion is infused by the PRESENCE of the GREAT MOTIONLESSNESS, the STILLNESS, the SELF. This MOTIONLESSNESS can be accessed by those who know how to be *still.*

One achieves stillness through negation and disidentification. In such states of meditation, all that which moves must be discounted, repudiated, negated. That which moves is *not* THAT which must be revealed. Even when the vehicles of the meditator are apparently totally tranquil, they are still invested with movement, for the vehicles are *motion itself.* Practically speaking, the vehicles can never be rendered completely motionless, but they can be made sufficiently still, in a relative sense, to allow the meditator's attention to disengage from them and find THAT which, *ever present,* has no movement in IT at all.

The Cosmos is *activity,* or rather, the *illusion* of activity. Pervasively interfused within all this activity, however, is the SUBSTRATUM of MOTIONLESSNESS. We come to realize that all that moves is REALLY MOTIONLESS. This Cosmos, this "Sea of Activity" is REALLY a SEA OF MOTIONLESSNESS. What must be realized is the STILLNESS *in* the Action.

Questions may arise concerning how Motion (the Cosmos) came out of MOTIONLESSNESS (the INFINITE SELF)—a problem as difficult as the age-old conundrum, How can *something* come out of *nothing?* One way to seek the answer is through the idea that MOTIONLESSNESS can be considered INFINITIZED MOTION, infinitely accelerated motion—motion accelerated to the point of *motionlessness.* Wherever ultimate TRUTH is pursued, we find that the apparent opposites must meet, blend and become identical.

- With respect to the practical achievement of stillness (which ever reveals the STILLNESS), time must be taken every day for meditative acts of disengaging negation, which produce in the consciousness to the degree possible, utter

stillness, utter abstraction, and hence the apprehension of utter SELFHOOD.
As long as there is preoccupation with modification, there is no true apprehension of SELFHOOD.

Modification is certainly the very *essence* of creativity, and there is no Cosmos without modification. However, modification is useless in the art of detecting, or *being*, the SELF. Hence an act (which is a modification) cannot reveal the SELF, it can only clear 'space' in consciousness for the realization of BEING to dawn.

From the Radical Infinitist Perspective, motion (however valuable in achieving intended patterns within Cosmos) will be seen as ESSENTIALLY useless in relation to the SELF. In fact, the most useful of all 'motions' are those which teach us how *not* to *act.* While (intra-Cosmically) many worthwhile things can be achieved through motion, no *ultimately* worthwhile thing can be achieved through motion, for the only ultimately worthwhile 'THING' is the INFINITE SELF. Perhaps, then, there is Wisdom in the Eastern Idea of "sitting quietly, doing nothing".

Motion has its value in Cosmos, but it is a *lesser* value than that of quintessential MOTIONLESSNESS. On the Path of Spirituality, one learns the wisdom of devaluing certain things, so that the truer values may emerge. When such things are devalued, they are no longer pursued, and when they are no longer pursued, they are no longer energized. When no longer energized, they no longer *live* and move and, thus, *interfere.*

So, in one's meditative process, there must be a devaluing of motion, if ABSOLUTENESS is to be realized. The reader of the *Yoga Sutras* of Patanjali cannot help but notice the insistence that all modifications must cease. The SELF, which **I**-as-**8**-as-I AM, is to be revealed in *motionlessness.*

The Problem of How to Choose What to 'Do' (Out of all the Many Things to Do)

Out of all the many, many things that it is possible to do, what should one do? How is one to choose one's actions? This is a problem relating to the efficiency and productivity of the disciple, and the degree to which he or she is capable of fulfilling a Self-assigned role in the Divine Plan.

The touchstone for decision must be one's best approximation to the Original Intent, the Design-at-the-Beginning. Now, every authentic E/entity (as a 'Ray' of the ABSOLUTE Who is the One Cosmic Self) is a 'part' of that Design, and is itself designed in relation to Original Intent regardless of how its actions may fall short of the Design.

Perhaps our best estimation of Original Intent (given the present limitations of the human mind) would be to consider it an Idea expressed through a vast, numbered Pattern of Qualities. The Design-at-the-Beginning has numerological integrity, and likewise, every authentic Entity or Identity (as part of that Pattern), also, has numerological integrity. There are many Entities and Identities and many are the Number-Beings in Cosmos, but all Entities/Identities are but One Entity/Identity, and all Numbers are simply and Essentially, the Number *One*.

The *true name* of any authentic E/entity (though that E/entity, as a Cosmic Role, be evanescent) embodies its Cosmic Function (the nature of the Cosmic Pattern which it expresses at the time and space that it manifests in a particular Cosmos). All authentic E/entities have the same SUPER-Cosmic IDENTITY no matter in what Cosmos they may be participating. The True Name of an authentic E/entity embodies its *role* in the Original Intent of the Universal Logos animating that Cosmos. There is but One Entity playing many roles and not all of these role manifest at the same Cosmo-Objective time period. Some manifest throughout the Universal Manvantara, and many more are far more fleeting 'Rays' of the ABSOLUTE in manifestation.

■ What all this means is that when attempting to choose a course of right action, one must live up to one's True Name (taken or given at the Beginning of Cosmos) and thus do one's share in fulfilling One's Own Original Intent (for Whose Intent is It but One's Own?). So often this means that in order to live up to one's True Name, one must *discover* one's Real Nature.

SELF/Self discovery is necessary before *conscious* right choice or action is possible For those units of life which are largely unconscious of their Selfhood, this discovery is not now possible, and yet, despite their unconsciousness, Divine Nature will often instinctually, compellingly dictate right action as a reflection of Original Intent. (People simply 'find themselves doing' what they are *supposed to do!)* When illumined consciousness is added to instinct, and to the beneficent guidance of higher Powers, then, right choice and right direction become infallible.

All these thoughts are so many ways of saying that one must *be* who one structurally *is* at any time and place in the Cosmic Process (i.e., One Being—many apparent Identities in many times and places). One-as-one must fulfill One's Own Pattern in order to choose right action and act wisely.

■ The simplest way of saying this is, "Be Who You Are." When within Cosmos, One must fulfill Ones Original Design and the Intent behind that Design. Beyond Cosmos, 'within' THAT, there is no particular Design at all, for a design is a limitation.

Every unit in Cosmos needs to find the way back to the UNPATTERNED REALITY or, *equally and apparently oppositely,* the INFINITELY PATTERNED REALITY. Right Action—out of all the many possible courses of action—is that which leads more rapidly and beautifully to the fulfillment of the 'Design at the Beginning' and, thence back to the 'STATE' from which one *never departed*—the INFINITUDE.

The Problem of Preferences —Likes and Dislikes

How is it possible for the illumined individual to have *likes* and *dislikes* when all presentations are, ESSENTIALLY, the same? Taking the hint from this question, some spiritual/philosophical systems have stated that, with regard to the Objective World, the illumined one has *no preferences.* at all, and what, in the deepest sense, is the whole of Cosmos but the Objective World. Whatever happens to him or to others during his phenomenal (and, hence, un-REAL life) is "all the same to him", and viewed with complete indifference. This attitude to the World of Becoming is said to characterize the yogi who has achieved Moksha—release from all "Samsaric" experience.

We can see that such an attitude would rightly be considered maladaptive in our busy Western world. It might suffice for the man who had finished work in the worldly dimension, and was on his way to *work* in another and higher dimension, but for those whose apparently individual karma binds them temporarily to the three worlds of human evolution, this approach would be unsuitable.

Always in Cosmos, there is *work to be done.* Cosmos, Itself, *Is* the Great Work of THAT, as well as the Great Illusion. Whatever system of thought removes the need for entified participation in the Great Labor (whether that participation is as an atom, a cell, man, a kingdom, a planet or a star, etc.) has some serious deficiencies in the eyes of Radical Infinitism. Every authentic E/entity is both the WHOLE/Whole and the apparent 'part'. *Both* must be lived.

When thinking about the *sameness* or *differences* of all presentations in Cosmos, it should be said that, while ultimate ESSENCES are *always* the same, forms (with *perhaps* the exception of the ultimate particle/event in a given Cosmos) are *never* the same. We cannot live in Cosmos without respecting It *on its own* (albeit, illusory) *terms,* and the basis upon which It was generated. While it is true that we can always *translate* Cosmos into the BOUNDLESS IMMUTABLE PRINCIPLE, and be philosophically justified in doing so, it is not an entirely practical course of action for living *rightly* within the World of Illusion.

Cosmos and all of its dynamics are based upon the Work of Maya which is the Work of Creating (through Consciousness and its alterations) Illusion—the Great Illusion. The dynamics of Cosmos are based upon *seeming,* and *work* within Cosmos must always take seeming and appearances (both of which are REALLY illusions but, nevertheless, Cosmic *actualities*) into account. Even the final Purpose of the Cosmos, achievement of the Universal Fixed Design, as conceived by the Universal Logos, is also, ESSENTIALLY, an Illusion based upon *seeming,* though a Great, Beautiful, and *Necessary* Illusion.

Returning to "likes and dislikes", even though they occur only in relation to illusions, to 'seemings', they must be respected, because through respecting them and understanding them, comes the right ordering of the energies, the right ordering of the E/entities, and the right fulfillment of 'Monadic Intent' which is both a tiny unfragmented 'fragment' of the Original Intent of the Universal Logos, and the *entire* Intent, Itself.

Likes and dislikes, therefore, are *actually* part of the Grand Design, and serve as the means by which every energy or E/entity finds its proper place, position and function in Cosmos. **I**, the ABSOLUTE WHOLENESS, AM represented in Cosmos by *Cosmos Itself*, and also, and sequentially by a number of seeming 'partialities', a number of seeming aspects, a number of seeming quantities/qualities, a number of undetached fragments of the Whole Pattern until at the last, again, **I** AM represented (in a new way, earned by My Cosmic Labor) by the whole Cosmos Itself (Which, Essentially, **I**-as-**8** have always been).

Meanwhile, however, at this 'time' and under the density of veiling which produces My immediate and particular S/self-consciousness, **I**-the-WHOLE AM represented *here* by a specific localized quantity/quality (and, significantly and simultaneously, *everywhere else in Cosmos*, by myriad other quantity/qualities). This quantity/quality must find its right place in Cosmos with respect to all other quantity/qualities.

Thus, the harmony or inharmony of interactions (likes and dislikes) must be understood and recognized so that no premature forcing together of contraries resulting ever in inharmony can occur. If the Cosmic Game is to be respected (and respect It we must for It Is what **I**-the-WE 'DO'—in fact, the *only* 'Thing' WE 'DO'), then likes and dislikes, which order all things intra-Cosmic in harmony or disharmony, must also be *respected*.

The Problem of the Possible Ineffectuality of Bliss

If one finds the INSEPARABILITY in relation to all things, *bliss* arises naturally. Must bliss, however, manifest as a so-called 'blissed-out state'—vague, ineffective, unrealistic, impractical? Absolutely not. The Condition of the Greatest Being in Cosmos (which **I**-as-**8** Am as *All* Are) Is one of *bliss*, and this Being, Is the Cosmic Master.

The Universal Logos is blissful, but above this blissfulness, is the Noumenon of Bliss (BLISS), 'inherent' in the INFINITESSENCE. The end of separation is the beginning of bliss. Bliss is the natural result when all sense of separation ceases. If one lives identified *as* the INFINITE INSEPARABILITY, the radiance of bliss is born. The illusion of fragmentation need not disrupt the certainty of non-separation.

■ The Bliss of the Self-as-SELF can be preserved inviolate in the midst of multiplicity. Through a combined act of Will and Realization, one must guard the Bliss *deliberately* in the midst of the onslaught of the World of Fabrication.

If Bliss can be preserved in the midst of apparently distracting multiplicity, it clearly need not be an unfocused state. The usual image of a person in a state of *bliss* is of one abstracted, passive and disinterested in all that transpires in the normal world. Perhaps this is true of the blissful mystic, but not of the blissful occultist, who has willful mastery of the lower three worlds of human evolution as well as the rapture of Self-Realization.

(Does the term 'blissful occultist' seem a contradiction in terms?) To be blissful, even ecstatic, and yet supremely alert and engaged—this is the goal.

Only in Bliss can one play aright the Universal Game. Is it a denigration of Cosmos to call It the 'Great Game'? No. A Game is to be *played,* and in all life experience, it is the process of *play* that is most joyous and even *blissful.* Further the well-played game does not REALLY affect the REALITY of the player. The player *is* who he *is* regardless of the game played. A well-played game is really a beautiful process because one need not invest himself in it with so much seriousness that one forgets who one *is*.

The right approach to detachment is found in the dynamics of *play.* The player and the game, the actor and the part, the dancer and the dance—all these offer us analogies of joyful even *blissful* participation in skillful action, while preserving detachment. Therefore, rather than being some vague, abstracted, ineffectual state,

- Bliss can be a state of maximum alertness, appreciation, skill, and even *mastery.* Bliss is the natural *spiritual* state in which to live, the natural state of those who have found the Synthesis of the Universal Self, or, even, perhaps, the 'SYNTHESIS' of the SELF.

The Problem of Meditation in the Radical Infinitist World View

What is the value of meditation to the Radical Infinitist, the Radical Non-Dualist? From the deepest perspective it could be said that meditation is SELF-'AFFIRMATION', the affirmation of the ONE AND ONLY INFINITE SELF.

To the Radical Infinitist, meditation serves no particularly *constructive* purpose in the World of Form. The World of Form, per se, is only interesting in as much as it is REALLY the WORLD OF ABSOLUTENESS in disguise. The many relations and interrelations in the World of Becoming are the theme of *other* kinds of meditations that focus upon skill-in-action in this illusory world, and that are very necessary and valuable in their own right.

To the Radical Infinitist, however, the purpose of meditation is essentially SELF-Remembering (to use the Gurdjieff's terminology in the most abstract sense). The rediscovery of the SELF is the purpose. Such meditation is a bold act of SELF-'Retrieval', for the INFINITE SELF (and, even, the Self) has certainly vanished from the lives of most human beings.

We read much about the "Lost Word", and the Pilgrimage to find It. That Word is the Soul, and today many are finding It and learning to live by Its vibration. In fact, however, something far more *fundamental* than the Word has been lost. The very SELF has been, for practical purposes, lost—the *essence* of the Essence which IS THE ESSENCE.

There are thousands and thousands of meditations, many serving worthy purposes, but only those that retrieve the vanished SELF ("nearer than hands and feet") are Radical Infinitist Meditations, and will lead the meditator (in the words of Sankaracarya) to the "consummation of his life." Such meditation is for those who have a passion for REALITY that transcends their interest in Relativity.

- Meditation can "restore the SELF unto the self." Meditation can *infinitize* selfhood into SELFHOOD.

Thus, let us meditate.

Philosophical Problems

The Problem of the Divisibility of the Spirit

The phrase "Divisibility of the Spirit" occurs frequently in the Agni Yoga Teaching. As commonly used it signifies the ability of the highly advanced Yogi to simultaneously project different aspects of himself to different localities. It may also indicate the ability of the Yogi to generate various palpable images of himself and project them, simultaneously, wherever he wishes. It general it means the possibility of multiple (hence not strictly localized) *presence* of the same E/entity.

The dynamics of Divisibility of the Spirit also have cosmological and cosmogonical implications, and relate directly to Emanation Theory, for during the Process of Emanation a kind of Divisibility of the Spirit occurs. Emanation is a most interesting and mysterious process whereby the One becomes the Many while yet remaining the One. The question presents—what are the dynamics of this seemingly paradoxical process—the Divisibility of the Spirit?

Let us assume that **8** (standing for any human being or greater Self-Conscious Being) Am the Emanatory Source. The question arises, How can **8** remain fully what **8** Am in full consciousness and full power with an extensive ring-pass-not and, simultaneously, send forth aspects of 'MySelf' that are smaller in scope, power, and extent, though *identical* in Essence?

Perhaps a visualization would be useful. **8** can visualize myself as the Point of Origin in a Process of Emanation as **8** imaginatively look down the entire ramifying stream of My emanations as if **8** were standing at the cap stone of the pyramid, a synthetic position in which **8** Am *fully* **8.** At the next *emanatory remove*, 'below', **8** Am apparently *less* than **8** was immediately before, even though, simultaneously, **8** Am still *fully* **8** at the 'top'—My Point of Origin. What has happened as a result of My first emanation, is that while **8** Am still, Essentially, **8**, My prakritic scope has been narrowed. At the next remove, **8** Am still, Essentially, **8** but My prakritic scope has been narrowed still more. The REAL-as-Real *identity* is always the same. At each remove from Myself, **8** Am still Who **8** Am, but My range of sensitivity and expression, My range of Self-Identification, has been curtailed. My 'prakritic entombment', has densified.

As the Emanatory Sequence continues, **8** find that **8** ('below') Am multi-dimensionally conscious in ever lessening degrees, while still retaining the full consciousness of each 'level' even as **8** 'descend'. Nothing of the wide consciousness experienced at My Point of Origin has been lost, even though apparently (and for purposes of prakritic involvement), **8** have been reduced again and again. Of course the apparent reduction is illusory and un-REAL, but *actual*, and is the result in Cosmos of the Self-Veiling power of Maya.

Let us inquire into multi-dimensional consciousness. Every E/entity in Cosmos is *necessarily* multi-dimensionally conscious, but, paradoxically, is usually (dwelling, similarly to many others, in a seeming prakritic isolation) *unconscious* of that fact until a great deal of relative advancement (or 'Ray'-Retraction) has been achieved. This unconsciousness results from the Mayavic effect we might call, 'the upward sealing of dimensional foci'. Not only is there an upward *sealing* effect, by means of which the various dimensional foci along an emanatory stream are, for the most part, *sealed* from each other, so that a given dimensional focus cannot easily pervade or even interpenetrate a higher dimensional focus (and sometimes even lower dimensional foci that are too remote), *but* there is also a *ceiling* imposed, higher than which pervasion or interpenetration is seemingly impossible.

For instance, as a human being (which is a 'Ray' of the ABSOLUTE at a particular dimensional focus), my participation in the higher foci along my emanatory stream (such as the Monad, the Planetary Logos, the Solar Logos, etc.) is severely limited. **8** know, logically, that **8** must be participating even now upon those higher levels of My emanatory stream, but My *human* consciousness is, as it were, *dimensionally sealed* from those higher dimensional foci (and even from some of the lower ones, such as the atoms and subatomic particles in My vehicles of manifestation). At whatever level of a particular emanatory stream in the Great Divine Emanatory Stream **8** may be participating, it is always the same *indivisible **8*** (the Universal One) who is participating. **8** realize, then, that **8** *know* more than **8**-as-'I' know, but *dimensional sealing* prevents me from knowing *consciously* what **8**, necessarily, know.

☐ Always, with respect to multi-dimensional consciousness the practical question must be asked, What exactly do **8** know on other levels even now? What seems to occur is a kind of forgetfulness by a given emanation (E/entity) of the nature and quality of the level from which that emanation originated? With every act of emanation, that *forgetfulness of the previous dimension* seems to occur, at least in part.

Rather than having an awareness of a complete continuum of multi-leveled consciousness, each emanated E/entity has a relatively more complete awareness of dimensions that are *below* its current focus, but rarely of dimensions that are *above*. The simultaneous conscious apprehension of all the different scopes of consciousness along an E/entity's emanatory stream is the prerogative of great Cosmic Entities alone (but are They Really *different* from **8**, 'MySelf'?).

For instance, if **8** (always with others of My *kind)* Am a fully developed Galactic Logos (which, strangely, **8** {the Indivisible Oneness}, necessarily, *must* be), **8** will be at least *capable* of identifying with all dimensional foci along the entire lower emanatory stream **8** have pursued. All lesser E/entities are far more dimensionally sealed than the greater Entities (though in REALITY no E/entity is REALLY dimensionally sealed). Strangely, the Universal Logos may be dimensionally sealed from the ABSOLUTE until the Universal "Day Be With Us."

- ☐ Maya (Who is the Power Who *reduces* SELF-as-Self-Consciousness) engineers the appearance of limitation and *discontinuity*, and dimensional foci become to each other virtually impenetrable ring-pass-nots. It is spiritual training (and the slow "March of Time") which provides the capacity for these ring-pass-nots to become inter-penetrable.

In light of these thoughts, we see why it is difficult for the human E/entity (a relatively low form of authentic E/entity) to answer the questions, What do **8**, even now, 'know' upon higher levels, and, *What* am **8**, even now, on higher levels? Answers come as the more lowly forms of 'encapsulated Allness' (i.e., lower entity/emanations) re-project themselves into higher dimensions (the Antahkaranic Process, and recognize that they (as the One 'Ray') had been conscious upon those higher dimensions all along—conscious *during* all those ages when they had thought the higher levels to be inaccessible. Such realization confirms thoughts such as, "Ever have **8** been so", or "**8** have always been aware multi-dimensionally, but recognize it only now."

- ☐ Everywhere in Cosmos, there are Mayavic barriers and ring-pass-nots and illusory separations. The great Drama of Cosmos is to bridge these barriers and separations, eventually rendering all ring-pass-nots into One Cosmic Ring-Pass-Not.

Only the meditator who *does* bridge these SELF/Self-Imposed barriers realizes that consciousness has been *simultaneous and multi-leveled throughout the duration of Cosmos* and that every consciousness can trace its the way back to the capstone of the Universal Pyramid where, *ever*, Cosmic Consciousness has existed for each and all in fullness—a Cosmic Consciousness that has been the possession of each seemingly minute consciousness however lowly in temporary expression.

In Cosmos, there exists the particularization of ALLNESS, the finitization of the INFINITE, the bounding of the BOUNDLESS. This is the Great Contradiction (apparently) and can only occur because of the WILL of the ABSOLUTE SELF to SELF-'VEIL'-instantly-Self-Veil. Ego-hood is dimensional focus. Ego is an 'egg', a boundary, a ring-pass-not. Cosmos is apparently full of spherical bounding.

The Cosmic Drama of the *creating* and *overcoming* of separative boundaries is the dynamic purpose of the Life of Cosmos (the Universal Logos). That Drama is to *create* limitation and then *destroy* it. The particularized Universal Life (each particularization being, paradoxically but Essentially, the very Wholeness of that Life) moves relentlessly, in the World of Becoming, through great upward striving towards Universalization (thus becoming *Life* instead of merely, a 'life'), and finally to the consummation of realized 'ABSOLUTIZATION'. The particularized Universal Life has now become LIFE ITSELF, which it/IT always *WAS*.

The Problem of the Apparent Limitation of Human Consciousness

A philosophical problem exists in the contrast between the actuality of my limited consciousness when contrasted with the philosophically necessary Non-REALITY/Reality of that limitation. Throughout the Universal Manvantara, it appears that **8**-as-I will be more focally impacted within 'my' ostensible ring-pass-not, than within the ring-pass-not of what, to the dualistic consciousness, are called 'others'.

The Christ, however, negating this dualistic Time/Space dynamic, made a profound and memorable statement, "If Ye do it unto the least of these, Ye do it unto me." Such was His state of identification, and such was the extent to which the ring-pass-not of His normal prakritic equipment (His various vehicles and sheaths) was rendered incapable of 'containing', and thus limiting, His consciousness. He, indeed, was/is identified with Vishnu the Pervader. Christ (as much as any human being could) had solved the problem under discussion. His consciousness was both local and yet extensive far beyond the normal ring-pass-not of prakritic vehicles. Being upon the Path of Earth Service, He had become an expert in the Art of Identification.

Of all human beings in Earth's present humanity, Christ realized (in the *Cosmic* sense, *began* to realize) the pervasion which all self-conscious beings in Cosmos will one day fully realize. But those of us who, as localized-selves, have not reached His status, will find it necessary to ask ourselves:

- If I-as-**8**-as-**I** am the inescapable 'PERVADER'/'PARTICIPATOR', because **I** AM, ESSENTIALLY, the one *indivisible* BOUNDLESS SELF, why do **I**-as-**8**-as-I not *know* this?
- Why is MY ALLNESS apparently particularized, encompassed, bounded?
- Why AM **I**, WHO must, necessarily, be *fully* present, conscious and active at all points within Cosmos, apparently, not aware of this?
- Why, in short, does MY ABSOLUTENESS not transfer readily into Cosmos at the level of my normal human consciousness?

This is a profound question, vexing to the limited consciousness. The short form of the correct answer is, 'Maya'! A slightly fuller and more rational explanation is, *willful SELF-Limitation.* MY Design-at-the-Beginning requires the use of *Ignorance* as an instrument of the limitation **I** must 'IMPOSE' upon MYSELF in order to BE *completely* WHAT **I** AM—i.e., the FOUNT OF ***ALL*** POSSIBILITY.

Another answer is that MY ABSOLUTENESS *does* transfer to Cosmos, and that Cosmos, ESSENTIALLY, is nothing other than WHAT **I** AM, namely the ABSOLUTE (though under the Illusion of Maya). That, in my present Cosmic-condition, **8**-as-'I' cannot know this in fullness, is part of MY/My Original Intent and cannot be countermanded simply by an act of my personal will. However, from the Radical Infinitist perspective, **8**-as-I *must necessarily* know ('unbeknownst' to *me!)* and experience non-limitation through pervasion, simply because **8**-as-I Am ***I.***

The Problem of Differing Degrees of Immediacy in the Registration of Experience

This is a profound problem. Why, if each E/entity, Essentially, pervades Cosmos, and hence is found as Essentially *identical* with the experiencing subjective life within every ring-pass-not—why, then, are different E/entities' experiences registered with different degrees of intensity? Why are the experiences that 'I' experience, more immediate to 'me', than they are to 'you'?

Until a much greater sensitivity is developed among experiencing human beings, there will be no way to make a convincing case of the correctness of the hypothesis of Equal Experience Universe-Wide. The thesis seems to contradict everything we know, think and, yes, *experience*; and yet, given the LAW of the INDIVISIBILITY of the SELF (and, even, the Indivisibility of the Universal Self), the hypothesis stands to reason. Why does it not seem to work for most, although it *does* work for some.

Remember what is said of God, that "Not a sparrow falls ..." Presumably the Indwelling Life of any System is the full Experiencer/'Inperiencer' of lives included within the System. Since, philosophically, we can trace our own identity to the Identity which is that of the Universal Logos and ultimately to the IDENTITY that is the ONLY IDENTITY, we, too, must be the Experiencer/'Inperiencer' of all in-Cosmos experiences, no matter what *apparently* separate and distinct E/entity experiences them.

Why, then, does the theory not seem to be part of the *experience* of human beings? From the largest point of view, we are probably not yet, *cosmically,* at that point in the Universal Process where the Process of Unveiling has overcome the Process of Veiling. The experience of ignorance is still very deep to many Creative Hierarchies (even though the BEING which each of them ESSENTIALLY IS, cannot possibly Experience any ignorance whatsoever). The idea that an E/entity may simultaneously experience the TOTAL 'ILLUMINATION' OF INFINITE SELFHOOD 'along with' the privation of finite entification is hard for the human mind to grasp, or allow as credible, but so the case seems to be.

From another perspective, it could be said that humanity's degree of identification with either the Universal Self or even, more radically, with the SELF, ITSELF, is scarcely unfolded. There have been in the history of the human race some "sixth rounders"—those who have achieved Monadic Awareness (understood themselves, *experientially* as a 'Ray' of the ABSOLUTE Who Is *the* 'RAY' of the ABSOLUTE). In relatively recent history, the Buddha, Sankaracarya, and the Christ could be numbered among them. In fact, however, humanity knows very little about *identification.*

When deep identification on the Spirit level is achieved, the 'POINT'-instantly-Super-Cosmic Point will be realized to be in all Cosmic Points, and the pervasiveness of the SELF within the Self will be experienced as a fact in consciousness.

☐ Clearly, the first step is to cultivate that State of Identification which makes the SELF-as-Self immediate to the consciousness.

The Problem of Progressive Fulfillments in Cosmos in Relation to the Ultimate Cosmic Fulfillment

A difficult set of questions arise when we consider the nature of Universal Consummation:

- Does the Universe tend towards an Ultimate Consummation only—a Consummation that is a fulfillment of the Design-at-the-Beginning, the "Fixed Design"? Or,
- Are there various necessary states of temporary fulfillment along the Way?
- Does every moment in Cosmic Time have its archetype which correlates with the Fixed Design and which guides the structuring of that moment? Or,
- Does every moment have its 'rightness'?

It would seem that the answer must be "Yes." It would seem that the Design-at-the-Beginning must unfold in Cosmic Time as the Divine Plan, which sets forth, in an ideal form, those relationships between all things which will lead to the fulfillment of the Fixed Design at the "Ending Times". Herein, we see the relationship between Purpose and Plan. Purpose is *fixed* at the Beginning and *fixed* at the End, but the interim is a constant flux guided by the Divine Plan which reflects the Divine Purpose in *Time*.

Moments of relative consummation may indeed come. There are many little 'finales' along the way to the Grand Finale of the Grand Master Musician, the Universal Logos. This series of ascending consummations, however, are *approximations*, and, given the nature of Free Will in-Cosmos, are *unpredictable* in form, even from the perspective of the Universal Logos.

All that is predictable is the 'Final Outcome', the fulfillment of the Design-at-the-Beginning. The many E/entities guided by the Unfolding Divine Plan which reflects the Divine Purpose must exert all skill-in-action possible (considering their *veiled* state) to bring Universal Relationship at long last into conformity with the Intended Design-at-the-End, which is *exactly* the Design-at-the-Beginning. There is complete certainty that the Ultimate Universal Design will be fulfilled, and complete uncertainty as to exactly *how* It will be fulfilled. This makes Life-in-Cosmos very interesting!

Original Intent is a Design, a kind of Relationship between all destined Universal Factors. The many limited E/entities *(apparently* limited, and apparently *existent*) must, at the "Ending Times", come into a certain Cosmic Configuration, a certain Patterned Relationship, which was predicted from the Time of Inception.

An interesting problem arises here, however, for not all dimensions of Cosmos will necessarily be extant at the Ending Times. Just as there is a progressive involutionary materialization, there is a progressive evolutionary dematerialization. So then, which dimensions of Cosmos will remain at the Ending Times? Clearly then, the Fixed Design (to be fulfilled at the Ending Times) does not include *all* dimensions of Cosmos, for many dimensions will have disappeared through the process of Cosmic Obscuration. What, then, *is* the Final Design, and is there anything remaining at that point but the One Cosmic Entity?

To answer this problem, we must realize that the Fixed Design is being somewhat fulfilled at every point in Space at every moment, though the fulfillment is but an approximation (the Lower Worlds being called, the World of Approximation). We must also remember that the One Cosmic Being is Essentially unitary though *apparently* multiple-in-manifestation. When Its multiple 'parts' (the may 'Rays' of the ABSOLUTE which are but One Universal 'Ray'-as-'RAY') 'return' from Their apparent 'wanderings', and are poised in a Cosmic Configuration which will induce Universal Pralaya, they (now again the Conscious One) bear with Them a possibility of Concluding Relationship within the World of Being which was 'foreseen' by Them-as-Universal-Logos at the Beginning Times. For Who Was and Is the Universal Logos if not Them in seamless union? In this respect, each Logos, though Essentially Unitary, is also a Host of 'Rays' of the ABSOLUTE which are Essentially but One 'Ray' (the Universal Logos-as-'Ray') which, Itself, Is but the ONE 'RAY' of the ABSOLUTE.

Thus, though various prakritic dimensions may be etherialized and absorbed into still higher such dimensions, the 'Harvest' of all that was acquired by the One-'Ray'-as-many-'Rays' is collected within the reascending, merging, blending 'Rays' of the ABSOLUTE that together-as-One make a Patterned Relationship, which is called the Design-at-the-End possible at the Ending Times. The Final Approximation achieved in the World of Approximation meets its Model. This Design may be fulfilled even though Objectivity, as usually considered, may have vanished. The Consummating Relationship will be purely Subjective as the Harvest of Emanation is ingathered.

The Problem of Veiling —Maya

We come now to a stupendous philosophical problem which threads its way throughout this treatise, appearing from time to time as the insoluble paradox it ESSENTIALLY/Essentially *is*. The problem seems to mock all efforts of thought; the mind is a puny instrument with which to confront its grandeur. It is simply this, Why should there be 'MAYA' at all? Why should the dynamic of SELF-'VEILING' even exist?

All within Cosmos yearns towards the INFINITE (even though there is naught else, and thus no 'distance' between the INFINITE and the yearning ones). The Cosmos and all Its myriad lives yearn (though They know it not) to become what They already *ARE*. Then, why should the INFINITE SELF in ITS one apparent 'ACT', *de-infinitize* ITSELF, *finitize* ITSELF, *veil* ITSELF from ITSELF (by becoming *Itself)*, 'CREATE' the Objective State of Cosmos through separation, 'CREATE' the separation of subject and object? This is *the* QUESTION, the unanswerable question.

Need (conventionally understood) does not dictate this ACTIONLESS-'ACTION'. What can the INFINITE SELF NEED? IT lacks *nothing* because IT IS *infinitized* 'EVERYTHINGNESS' and *NOTHING* forever. 'Within' IT all desires are forever fulfilled. Every virtue and every quality is ABSOLUTIZED within IT. There is simply no *ordinary reason* for the Perpetual Motion which *births* Cosmoses.

☐ Could the INFINITE SELF not 'REMAIN' forever 'UNVEILED'? (In a way, of course, IT 'DOES' and *must,* for even when IT 'VEILS' ITSELF, IT remains 'UNVEILED'! A paradox!

If there were no 'VEILING'-as-Veiling, there would be no-thing at all, ever (or rather, the NOTHING-AS-ALL forever). Nothing would ever 'happen'. There would be and could be no possible *event.* The Universe is a periodic Event (in fact, the *one* periodically recurring Event) and, of course, there would be no Universe. Would this make a difference to the INFINITE SELF? *'Absolutely* No, and *Yes* as well.

It has been explained in this treatise that 'MAYA' IS PARABRAHMAN in *action.* Every *act* is (to PARABRAHMAN), in fact, a *veil.* Strange as it may seem, therefore, the very first 'FLASHING FORTH' of the 'RAY' which is to become and impel the Universe is 'MAYA'-instantly-Maya at work. Maya is the entire Birthing Process, both Pre-Cosmic and Intra-Cosmic. Veiling is the One Archetypal Motion.

Through the *agency* of PARABRAHMAN-as-'MAYA'-instantly-Maya, the SELF periodically SELF-'OBJECTIFIES', turning ITSELF inside-out, as it were, while still 'REMAINING' just as IT always WAS. "Demon est Deus inversus"—the Devil is the lining of God. GOD (the ABSOLUTE DEITY) 'TURNS' ITSELF into the Devil, which, strangely, is *noumenally* the Universe Itself. Of course in 'DOING' so, GOD 'REMAINS' *forever* and *only* GOD. There is simultaneously the *appearance* of Pre-Cosmic Action and continuous NON-'ACTION', as ever. So, the Universe, Itself, (the Cosmos) is the very 'Devil'—'D-Evil'; 'D-Vile'; 'D-Live'; 'D-Veil', the 'D' forever standing for the 'Delta' of Form. One can see the larger philosophical meaning of the word 'Devil'.

☐ Is there really a 'REASON' behind the Creation or the SELF-'BECOMING' we call the Universe?

One can hardly call it *desire*, or *love*, or anything at all, REALLY. Perhaps one can call the 'REASON'—*NECESSITY.* All we can really do is describe what *seems* to happen, for, indeed, the ONE GREAT 'HAPPENING', and all the many 'Happenings' which succeed it, are *all* a Great *Seeming.*

There is a fluctuation between *exteriorization* and *interiorization,* which is the PULSE OF LIFE or the One Super-Cosmic Breath. This Breathing has been going on forever, but it is not quite correct to say that IT 'BREATHES', for IT only *seems* to Breath while forever, as it were 'HOLDING' ITS 'BREATH'! The Breath, however, is perpetual, as is the beat of the Super-Universal Heart.

☐ The one MOTIONLESS-'MOTION' is from MOTIONLESSNESS to Motion (which two, together, can be seen as a 'Motion') and back again, without REALLY having ever been anything but MOTIONLESS. This is the Motion from ZERO to One perpetually, interminably, without ever having *been* anything other that ZERO. Paradox!

Is the World an Illusion or is It not? It would be correct to say that It both *Is* and Is Not, for It is nothing else but IT.

The Problem of Whether Cosmos Should be Preserved or 'Destroyed'

Should not Cosmos be 'destroyed' if to do so would liberate the Cosmic and intra-Cosmic Self into the ALL-in-ALLNESS WHICH IS the INFINITE SELF? This, of course, is a fanciful question! While no human being is capable of 'destroying' Cosmos, there are many who have attempted to do with their consciousness by means of incorrect World Negation and World Denial.

Who is it, however, that is asking this drastic and almost absurd question? A limited human mind, correct? The Universal Logos (so close to the 'WILL' of THAT which manifests as the Universal Logoic Will) would not ask thus, because *that* Great Consciousness Knows the *answer* (if any-One in Cosmos does). Thus, at some level, *we*, who are, Really, that Great Consciousness also *know*.

The answer returns that it is not MY/My WILL/Will to *'destroy'* Cosmos (even in consciousness), for **I**-as-**8** have 'Created' It by *Becoming* It. While Cosmos could not be considered necessary to the *improvement* of the SELF, for what can *improve* THAT to which no-thing can be added? Cosmos *is necessary* for the sake of SELF-CONSISTENCY, by means of which the SELF demonstrates apparent *inconsistency* (by 'ACTING') in order to remain *consistent* with ITS INFINITUDE. For that 'REASON', therefore, Cosmos is both *necessary* and, 'WILLED' into Being.

Those who are thoroughly captivated with the thought that Finite Cosmos is the GREAT REALITY (when, in fact, It is only the Great *Actuality* of this particular 'Moment' in the Infinite Duration) may resent profoundly the application of the principles of Radical Infinitism to our Finite Cosmos. Their world-view and their security are threatened by the reminder of the ESSENTIALLY *INFINITE* (though, *actually* Finite) NATURE of Cosmos, and by the attempt to 'infinitize' Cosmos (or restore the Finite Cosmos to Its ESSENTIAL INFINITUDE).

For many consciousnesses, it is a great threat to blot out all form, or to take away through thought the illusory 'reality' of that which was previously apprehended as enduringly 'real'. Man wants only *more* and never *less*, unless the less is *essentially* the gateway to more.

'More' and 'less', however, are meaningless terms to a BEING/NON-BEING WHO, forever, has BEEN ALL and, thus, *'HAD'* ALL. There is nothing to be gained by 'destroying' Cosmos, or ignoring It or denying It. Cosmos comes with the INFINITE SELF. There is no other way. Cosmoses have been *generated* forever. Allowing for the intermittency of Universal Pralaya and Universal Manvantara, there *always* was a Cosmos, cyclically.

☐ The question should not be, Should we 'destroy'....? but, How shall we realize that Cosmos *Is* now and forever, PARABRAHMAN? That realization would effectively amount to the destruction of Cosmos as an Illusion while preserving It as a 'RADIATION' of REALITY.

The Problem of Seriousness in Cosmos, or Taking the Cosmos Seriously

It is clear that the Advaitans do *not* take Cosmos seriously. For them, It is merely an illusion to be eradicated through SELF-Realization. The Radical Infinitist has a different point of view.

Everything that one does within Cosmos is an infinitely simplified and reified reflection of the ABSOLUTE 'STATE', which is a 'STATE' of ABSOLUTE SIMPLICITY *and* ABSOLUTE COMPLEXITY simultaneously. The Cosmos is an 'ACT'-as-'Act' that **I** 'DO' (though **I**, PER SE, cannot 'MOVE', but then, if *moving* is a *possibility* {since **I** AM the FOUNT OF ALL POSSIBILITY} **I** cannot *not* 'MOVE', at some 'time'). When **I** AM not 'DOING' Cosmos, **I** 'DO' *nothing* at all, while *being* ALL, INFINITELY ALL, hence, NOTHING. In ITS active or outgoing state, the INFITUDE, the SELF, the BE-NESS, the ABSOLUTE is a Cosmos-'BECOMER'.

If the Cosmos were not 'WILLED', were not NECESSARY, It would not exist. It would not be the apparent Great Fact of our consciousness. Therefore, although the Cosmos is the slightest of things when compared with the LIMITLESS POSSIBILITY of the ABSOLUTENESS, It has recurrent 'place' in the UTTER ALLNESS, and for 'REASONS' unknown to ourselves in our states of 'localized INFINITUDE'. **I** 'BEGAN' the 'DOING', then **I**-as-**8**-as-Creator Continued. **I**-through-**8** have 'Done' Cosmoses, repeatedly, forever and will 'Do' Them, repeatedly, forever. In as much as to 'BECOME' (through apparent SELF-Finitization) a Cosmos is the One ACTIONLESS-'ACTION' ('MAYA'-instantly-Maya) of the ABSOLUTE SELF which I AM, each Cosmos is worthy of My respect, My love, and My wholehearted participation.

Now, it might be asked, is there any 'part' of the ABSOLUTE that is *not* participating in Cosmos? In occultism we read always of God Transcendent and God Immanent, the first 'above', as it were, and 'separate' from His Creation, and the second, pervading and 'invested in' His Creation. The question is, of course, non-sensical from the perspective of Radical Non-Dualism, but the concrete mind tends to pose this type of question, separating *this* from *that*. The absurdity of the question stems from the fact that *all* of the SELF, *all* of the ABSOLUTENESS *<u>IS</u>* at every point (a pointless Point, Really) always.

We might say that the ABSOLUTE IS the *WHOLE* of ITSELF at all 'times' and all 'places' (and *includes* all possible 'times' and 'places') even though 'times' and 'places' do not REALLY exist. The SELF, ITSELF, IS the OMNIPRESENT SUBSTRATUM, and can *never* be separated from that which *seems* (and only *seems) not* to be ITSELF. Every apparent point within Cosmos, *is,* thus, the *entire* INFINITUDE. The INFINITE SELF is *indivisible,* let us remember, *boundless,* and utterly *homogeneous.* There is no variation in IT whatsoever. IT has no *extension* and one cannot think of IT as if IT were related in any way to *space.*

Although the brain may weary of such thoughts from time to time, it is simply *essential* to persist in them. Always new insights will appear, and REALITY will reveal ITSELF through confronting seeming contradiction. Annihilation of the Universe (first in consciousness and then, "in the fullness of Time", in *actuality*) will restore the Universe to Its ESSENTIAL INFINITUDE.

The Problem of SELF-Contradictory Possibilities Within the FOUNT OF ALL POSSIBILITY

A superb name for the INFINITE SELF (the 'INFINITESSENCE') is the FOUNT OF ALL POSSIBILITY, for indeed every possibility that could ever exist in the infinite sequence of Cosmoses (and even those which have never been actualized ... or never will be?) IS forever *resident* within the INFINITE SELF, and has, thus, been forever *fulfilled* to the *greatest extent possible*—i.e., every possibility is *infinitized*, or *ultimatized* within the INFINITE SELF. To think of the SELF in this way, however, as illuminating as it is, creates great philosophical problems from our intra-Cosmos perspectives.

☐ Let us assume that every possible possibility is *resident* in its *ultimatized state* within the FOUNT. Then, no matter what is predicated of the SELF, that predication is immediately *contradicted* by the fact that the SELF is the FOUNT OF *ALL* POSSIBILITY.

For instance, if we say that the SELF is "BOUNDLESS", then, because *boundedness* is a possibility, boundedness (as a contradiction to BOUNDLESSNESS) must be in some way inherent in the SELF. If we way that the SELF is "IMMUTABLE", then, because *mutability* can certainly be designated as a possibility, mutability as a contradiction to IMMUTABILITY must be in some way inherent in the SELF, and so it goes.

Everything predicated of the SELF, including ITS INFINITUDE, must *necessarily* be contradicted if the SELF is to BE what IT ESSENTIALLY IS—namely the FOUNT OF *ALL* POSSIBILITY. The ultimate in contradiction arises when we predicate that the SELF IS the FOUNT OF ALL POSSIBILITY. If the SELF is TRULY the FOUNT OF ALL POSSIBILITY, then equally, there is the *possibility* that IT is *not* the FOUNT OF ALL POSSIBILITY—i.e., that IT IS, for instance, the FOUNT of *some* possibility or *no* possibility. Somehow, this contradiction, too, must be actualized in a Cosmos in order for the SELF to continue to BE WHAT IT IS by contradicting ITSELF and *becoming* what IT *is not.* (Are not the limitations of mind apparent?)

☐ We can see from this dilemma, that it becomes impossible to know anything *immutable* about the SELF, even ITS supposed IMMUTABILITY, upon which so many premises concerning the SELF are based. Every premise about the ABSOLUTE must necessarily be contradicted by the ABSOLUTE ITSELF if IT IS truly the ABSOLUTE.

Perhaps there is a solution to this Problem of *Possibility* in the very Creation of Cosmos. We might think of Cosmos as the *arena* for the expression of all that which apparently contradicts the INFINITUDE of the INFINITE SELF. In this way, the Cosmos seems to play an adversarial role in relation to INFINITUDE (to which, paradoxically, no relation can be had)!

'Within' the SELF, ITSELF, no possibility is ever *expressed*, for no-thing ever happens', or, rather, NOTHING 'HAPPENS' *forever.* The World of Illusion is required for *expression.* Further, the FOUNT OF ALL POSSIBILITY is a name descriptive of the INFINITE SELF because the World of Illusion (with all Its myriad possibilities) has existed periodically forever. Possibility can only be *expressed* in the World of Illusion that is the World of Becoming. Within the SELF-as-*SELF*, *all possibility* is *infinitely abstracted* into the INFINITESSENCE and hence, not *expressed*.

Within the SELF-as-Illusion, however, (which is the Cosmos), *all* possibilities *can* (theoretically) be expressed (though they never *are*, completely). This expression of possibility within the World of Illusion is, in a noteworthy way, *nevertheless* an expression '*within*' the INFINITE SELF, because even the World of Illusion is still, ESSENTIALLY, *only* the INFINITE SELF.

We see then, that with the birthing of an infinitude of Cosmoses, SELF-'CONCEIVED' for the definite purpose of *expressing* some of the possibilities 'resident' 'within' the FOUNT OF ALL POSSIBILITY, we are enabled to leave the SELF-as-SELF in ITS ALL-IN-ALL-NESS *undisturbed forever.* IT can 'keep' ITS RADICAL INFINITUDE, ITS IMMUTABILITY, ITS BOUNDLESSNESS, ITS INFINITE DURATION, simply because all possibilities 'resident' 'within' IT are either *expressed* (or, theoretically, *expressible)* through the Mechanism of 'MAYA'-instantly-Maya, the 'GENERATOR'/Generator of the World of Illusion. The Cosmos is that 'Portionless-Portion' of the INFINITE SELF through which *all that contradicts* the SELF *can* be expressed. Here, however, is the 'bomb': *Every<u>thing</u> contradicts the SELF.*

☐ A *thing,* or anything *particulate,* is contradictory to the ESSENTIAL NATURE of the SELF. Yet, *thinghood* must be (at least *infinitessentially)* 'resident' within the SELF as a *possibility,* and therefore must be *expressed* in order not to violate the inviolable PERFECTION of the SELF. We come to realize that any 'de-finite' possibility is only expressible in *Illusion.*

Given these thoughts, it is 'fortunate' (for the sanity of the limited human mind, that is) that the INFINITE SELF is the 'NOUMENESSENCE' which IS the *infinitization* of every possible possibility. Some possibility is so 'outrageous' to human conception, that some NON-'THING' is *needed* to infinitize such possibility to the *infiniteth degree* such that only NOTHING remains of it (and yet, it is allowed to *be)!*

One might well ask, what is it, at every 'FLASH' which 'DETERMINES' the one possibility or set of possibilities which shall be *actualized* in the Cosmos-to-Be, and the infinitude of *other* possibility which will never see finite actualization? (We must realize that *some* possibility within the FOUNT is definitely *not* finitely *actualizable.* Certainly, a possibility must exist, that there be *unactualizable* possibility.) No one in Cosmos can REALLY answer the question. It is no wonder that the wisest of the wise have considered the ABSOLUTE *unthinkable* and *unspeakable.* Of course those are only two possibilities. Can we wonder why any inquiry as to what IT REALLY IS, must be met with *silence.*

These thoughts may boggle the mind, but contemplation of the paradoxes they present will destroy the *usual* illusions concerning the nature of REALITY/Reality. Of course, the *unusual illusions* will remain, because compared to THAT (to which *nothing can* be compared), all thought of *any* kind is necessarily illusion. Nevertheless, in our World of Illusion, thought must be exercised, and, if pushed far enough, will lead the human thinker (who IS the SELF, ESSENTIALLY) to realization of and identification with THAT very SELF, even though none of that 'accomplishment' will be capable of being expressed in words—words, the *tomb* of REALITY. It becomes increasingly clear that REALITY is the *ultimate destroyer* of the mind!

The Problem of Which Came First —The Cosmos or INFINITUDE?

This is the classic "chicken and egg" problem, but, in a way, more difficult. At least the "chicken/egg" problem is semi-soluble for those with a sense of humor, for due to the Principle of Hereditary Mutation, a life form which evolutionarily *precedes* the form of a chicken, can lay the egg out of which a chicken hatches! We cannot say this of the INFINITUDE and the Cosmos. If we do not know which came first, Cosmos or INFINITUDE—INFINITUDE or Cosmos, and if, due to the Principle of Infinite Regress we know there is no possibility of an *ABSOLUTE* BEGINNING, can we determine which if either of this Primeval Duality is permanent?

Perhaps the question itself is wrongly formulated and is far too linear and sequential to be applied to the UTTER ALLNESS [See Glossary]. The question may simply be the product of that inherent category of human thinking we might call Linear Causality. Maybe this question can only be answered from within the 'DEPTHLESS-DEPTHS' of the ALL-IN-ALLNESS.

Nevertheless, the question is intriguing. If the mind follows backwards the causal chain of Cosmos—INFINITUDE—Cosmos—INFINITUDE—Cosmos—INFINITUDE, etc., there will *never* be a *beginning*. Every Cosmos will be preceded by a STATE of ALL-IN-ALLNESS or UTTER PLENUM/VOID, and every STATE of UTTER VOIDNESS/PLENUM will be preceded by a Cosmos.

The important thing is this: the Cosmos as 'Chicken' does *not* lay the Egg from which 'another' INFINITUDE 'EMERGES'. While, 'Chicken and Egg' are sequentially generative (each coming out of the other), this is hardly true of Cosmos and the VOID. In fact the dynamic is *entirely* different, for the VOID persists forever, even *during* the intermittent appearances of Cosmoses. They do not REALLY alternate.

Ever the VOID, the INFINITUDE, the ALL-IN-ALLNESS has been *exactly* as IT IS, the UNIVERSAL CONSTANT, the BOUNDLESS IMMUTABLE PRINCIPLE. There has never been what might be called 'intermittent VOIDNESS', but, on other hand, Cosmoses have been intermittent forever.

☐ The VOID ever IS. Cosmoses precede and follow one another in unending sequence.

So, while it can be said that a sequence of Cosmos—No Cosmos—Cosmos—No Cosmos, etc. occurs forever, it *cannot* be accurately said that a sequence of Cosmos—VOID—Cosmos—VOID, etc. occurs forever, for the second sequence is a *false sequence.* VOID is *ever present.* VOID is *present* even *during* Cosmos. Indeed. Cosmos, Itself, is none other than the VOID.

While it is possible to go backwards in INFINITE DURATION and find a 'time' when there was no Cosmos and *only* VOIDNESS, it is also necessary to say that one has only to go a little farther 'back' along the Infinite Time Line and a Cosmos will 'appear'. Thus while is it theoretically possible for the 'MIND' of the ABSOLUTE (*whatever that is*) to number Cosmoses backwards from the present Cosmos, and determine which Cosmos was preceded or followed by a given Cosmos, it is impossible to designate any Cosmos as having come *first* in an infinite sequence. Indeed, there *never* was a *first*, and so it can justly be said, that Cosmoses have *always been*—albeit, intermittently, forever.

☐ Perhaps it can be said, that while no Cosmos can ever have 'preceded' the VOID (since the VOID is *unprecedable,* without *'precedent'*, having 'BEEN' forever), the VOID must *always* precede any Cosmos that arises, for a Cosmos *must* 'EMERGE' from a VOID, whereas a VOID can never emerge from a Cosmos. In this respect it can be said that the VOID always *'WAS'* first, but not that it *'CAME'* first in a sequence. The VOID, from this perspective, is *first, only,* and *ever,* whereas Cosmos is *second and cyclo-infinitely appearing.*

The Problem of How VOIDNESS Can Persist in the Presence of a Cosmos

This is one of the most vexing philosophical problems, for it seems to contain an insoluble contradiction. We know that ABSOLUTE VOIDNESS or the BOUNDLESS IMMUTABLE PRINCIPLE *changes never forever,* for with any change IT would cease to be ITSELF—the HOMOGENEOUS, the ABSOLUTE CONTINUUM, and yet we know, from what we call direct experience, that, as far as our experience in-Cosmos goes, "there is nothing constant but change." Change, whether we call it REAL or illusory, is a *fact* of our consciousness.

Everything that we know from a common-sense point of view tells us that when there is change, there cannot be VOIDNESS and when there is VOIDNESS, there cannot be change.

☐ The Radical Infinitist perspective asks us to consider as reasonable the *apparent* nonsense that the change that seems so real is really nothing but VOIDNESS, and that since apparent change is, ESSENTIALLY, VOIDNESS, then, that change does not contradict or negate the VOIDNESS. From this perspective, 'change' would not be REAL. In other words, from this perspective, it is possible to have what is called 'change' *and* VOIDNESS simultaneously.

In sum, this philosophy suggests that while 'change', being *apparent only,* does not and cannot negate VOIDNESS, the unchanging FACT OF VOIDNESS (the FACT OF UNCHANGING INFINITUDE), *does* negate the REALITY of 'change'. What this ultimately means, is that 'change' is not REALLY 'change', that change is not REALLY 'happening', that 'change' is not *REALLY* REAL. Only NOTHING REALLY 'HAPPENS', and only NOTHING is REAL.

Using instances from our normal human lives we might ask, is a drama real? Is a play real? Is a game real? In casual conversation, such things are often said to be unreal. If one looks more deeply, various answer could be given to the questions above, for there is, indeed, an interaction between what human beings usually define as fantasy and reality.

In the Philosophy of Radical Infinitism, however, we must define the un-REAL more carefully. We must define that which is un-REAL as being 'incapable of affecting the BOUNDLESS IMMUTABLE PRINCIPLE'. Of course, *nothing* can REALLY affect the BOUNDLESS IMMUTABLE PRINCIPLE, the VOIDNESS, so 'every-thing' is un-REAL. All possible things are *infinitessentially* REAL in the INFINITESSENCE, but in that ULTIMATIZED 'STATE' these things have *lost their thingness.* In the INFINITESSENCE all possibility is *infinitely noumenalized.* The moment a possibility appears as a discrete 'thing', 'FALLING', as it were, *out* of the INFINITESSENCE, that possibility ceases to be REAL and, instead, become *actual.*

In this respect, each Universe is a Possibility (with an enumerable multitude of sub-possibilities) which has 'FALLEN' 'out of' the INFINITESSENCE.

- *Subjectivity* is the GREAT REALITY.
- *Objectivity* is the *greatest possible contradiction* to Subjectivity.

Objectivity is an *infinite reduction* of Subjectivity, but, since Objectivity is, undeniably, a possibility within the FOUNT OF ALL POSSIBILITY, it must be *actualized, fulfilled.*

Every objectively actualized possibility from the FOUNT OF ALL POSSIBILITY is infinitely 'removed' from the INFINITUDE (while yet, necessarily, *being* the entirety of the INFINITUDE). Can we call an 'infinitesimal-izing', objectified possibility (even though it be our apparently massive Universe) a REALITY? The INFINITUDE annihilates It entirely. Compared to the NOTHING, which the INFINITUDE IS, any objectified possibility (even a possibility as 'large' as a Universe) is a *nothingness.*

Submitted to 'Infinispectivizing' within Super-Cosmos, any possibility becomes a *virtual nothing* indeed, or appears to be converging relentlessly and immeasurably upon *no-thingness*, as an infinitesimal is a *process* and not a *thing*, for which a truer name is an 'infinitesimalizing'. So, is it possible to call such a negligible quantity as a Universe a REALITY? Using more poetical language, is It not (though It be a Universe) merely an 'infinitesimalizingly' small bubble 'within' the Ocean of Mulaprakriti?

☐ No wonder then, Cosmos is considered un-REAL, and the perpetual VOIDNESS, REAL. When, therefore, we ask the question, How can VOIDNESS persist in the presence of a Cosmos? is it not somewhat as if asking the question, How can the Ocean persist in the presence of an infinitesimal drop of water?

In order for these thoughts to make a deeper impression upon the mind, it is necessary to realize the infinitesimalizingly tiny nature of Cosmos:

- from the 'Infinispectivizing Perspective, the 'size' of Cosmos converges at great speed upon NOTHINGNESS. Cosmos is as a *Real* Point in Infinite Space;
- from the Infinispective of the Infinite Subject, Cosmos is, *invisibly* only one of an infinitude of *actualized* or *unactualized* Cosmoses. Cosmos, thus, 'disappears' into infinite articulatable possibility; and
- from the 'INFINISPECTIVE' of the ABSOLUTE, the size of Cosmos *is* no-size, and Cosmos *is* NOTHINGNESS. Cosmos never even *arose!* REALLY, Cosmos does not even occupy Space (not even a Point), for *space* is not REAL, and, thus, all extension is REALLY illusion.

Thus, generalizing from these various perspectives, Cosmos is simply a thing which, like all things, is REALLY a 'non-thing'. Cosmos is an objectivized possibility 'FALLEN' from the INFINITE NOUMENALIZATION of the INFINITESSENCE. Cosmos Is, then, a necessary SELF-Reduction, *apparently* 'happening' (though not REALLY 'HAPPENING') in order to ensure that the SELF 'REMAINS' forever ABSOLUTE PERFECTION.

The Problem of Whether a Thing is Identical with Itself

This problem has surfaced in the contrast between Aristotelian and Non-Aristotelian thought. In Aristotelian thought, a thing is equal to or identical with itself; in Non-Aristotelian thought, no thing is equal to or identical with itself.

To discover if this question is meaningful, let us see whether there is such a thing as a 'thing'? Certainly, intra-SOURCE, there is not. In Cosmos, there may be 'things', however *illusory* they may be. Perhaps in all of Cosmos, there is only one true category of 'thing'—the ultimate particle/event, the smallest possible item-in-Cosmos enduring for the shortest Cosmically-allowable unit of time.

Such an ultimate particle/event might be called a 'maximally minute objective presence', and may well be the only non-composite item-in-Cosmos. That such a particle must exist seems logically necessary. How it might behave, given the fact that it would be found upon the highest of the Fohatically-Fabricated Planes, is, at this point of human knowledge, virtually impenetrable. If we are to call this particle/event an elemental irreducible 'thing', then all other 'things' are composite and aggregate.

When Fohat "digs holes in space", the 'holes' that Fohat 'digs' are these ultimate particle/events. The 'holes' are the very *reflected* Presence of Fohat. (Intra-Cosmic Fohat is a Reflection of the Universal Logos, and so the particle/events are reflections of a Reflection, i.e., of Fohat.) All Emanations, along the Divine Emanatory Stream (with the exception of the very first Emanation, the Universal Son), are *reflections of Self-reflections.*

There are a certain number of these ultimate particle/events for each Cosmos, and Fohat *is* each and all of them. It can be said that, relative to each other (and, relative to the measurements of the Cosmos), each of these particle/'holes' has a certain ring-pass-not (generated by Fohatic Self-Perception), and further, that the particle/holes are irreducible (for Fohat {and His Superiors} have 'Willed' it to be so).

In a way these 'holes' are 'nothing at all', for Fohat is ESSENTIALLY (like all the Members of the Cosmic First Family, and, even, like *all* beings-in-Cosmos) a SELF-'REFLECTION' of NOTHINGNESS, and, hence, ESSENTIALLY, *nothing at all.* Thus, in an *absolute* way, we could call Fohat, 'NOTHINGNESS-in-Illusory-Action'. More relatively, Fohat must be considered a Self-Reflection of the Universal Logos. The particle/events are generated by Fohat Reflecting its own 'content'; they are therefore generated when a Reflection of a Greater Self (i.e., Fohat as the Self-Reflection of the Universal Self 'Generates' It's Own Self-Reflection and the articulable 'content' of that Self-Reflection. All E/entities or items in Cosmos are reflections generated by the Self-Reflection Process.

- ☐ Because the ultimate particle/events are (all of them as Fohatic Self-reflections) Fohat and Fohat alone (Fohat, Itself, being the Active Agent of the Universal Logos), the ultimate particle/event/holes could be said to be *identical* with Fohat, and, thus, each ultimate particle/event/hole *identical to itself.* This, may well be the only instance in-Cosmos of a 'thing' being *formally* identical with itself. All things in Cosmos, being ESSENTIALLY, NOTHING, are *substantially identical* with each other.

When we speak of E/entities, we speak of *'Rays' of the ABSOLUTE* which are, in fact, but ONE 'RAY'. Of course, there is identicalness between such 'Rays', and identicalness of each 'Ray' with Itself, for each 'Ray' Is, ESSENTIALLY, the ONE AND ONLY 'RAY' 'RADIATED' (an inadequate term) *as* the ONE AND ONLY SELF-'BECOMING'-Self. So 'Rays' are *mutually* identical and *Self*-Identical.

When we speak of 'things', however, we are focusing in the realm of *objectivity* and not of *subjectivity.* Things are *objects,* and identicalness among the *forms* of objects (due to the Principle of Unrepeatability) may be impossible-in-Cosmos, whereas identicalness among *subjects* (since all subjects are, ESSENTIALLY, the INFINITE SUBJECTIVITY) is constant. All Subjects, for instance, Who 'See' themselves as Objects are ESSENTIALLY *identical,* but the Objects 'Seen' are *not* because each 'Seer' (who prakritically is *not-identical*) is 'Seeing' Itself in order to generate objects; it is also Seeing Itself from a Point of View necessarily different to the way any other subject could see it; hence, the non-identicalness of the Objects. As well, each E/entity is both an *object* and a *formal attenuation* of its Emanator, so each E/entity's *object-generating* 'Self-Sight' (as it 'starts' from a distinct, unique basis) will be unique to that E/entity.

Leaving for a moment the discussion of an indivisible, elemental, 'irreducible-in-Cosmos' 'thing' (such as an ultimate particle/event), let us examine things-as-aggregates. If one could, for instance, freeze and greatly magnify the perception of a normal object/thing, would a thing not be simply a *changing relationship?* Initially the changing relationship would be caused by the reconfiguration of ultimate particle/events relative to each other. Later, all manner of changing relationships would be formed by the reconfiguration of more complex aggregates relative to each other. At every ultimate moment, a new 'thing' would appear, because a new relationship of configured factors would appear.

- ☐ If a thing (other than ultimate particle/event) is defined as an object that is identical with itself from ultimate moment to ultimate moment, then, perhaps in Time and Space (in the world of Cosmos) there *is no such thing* as a 'thing'.

Phrased otherwise, we might say—no thing is *self-identical* from ultimate moment to ultimate moment. This is so because of the ceaseless and unrepeatable relationships generated by the totality of constituents which go to the composition of a thing. Some constituents (may?) relate to each other in one ultimate moment, just as they had in previous ultimate moments or moment, but this will *never* be the case for *all* constituents. The fundamental objective constituent in 'thingship' is, of course, the ultimate particle/event.

Within Cosmos, the form of 'things' is generated by incredibly rapidly changing relationship between fundamental constituents. Except, perhaps, from the Perspective of the Consciousness of the Universal Logos (and from the Cosmo-Subjective Perspective in the Worlds of Adjustment and Being), a pure thing can only rarely be 'caught' in perception, because the reconfigurations are occurring too rapidly, and, therefore, 'things' which are by no means self-identical from ultimate moment to ultimate moment *appear* to be self-identical.

We are speaking here of the level of super-microscopic exactitude, and not from the point of view of general *seeming*. Certainly, on the macro level, a table or chair *seems* to be identical to itself from moment to moment, but we know that such is the dance of the atoms that the apparent identicalness is a macro-perception only, and is fundamentally illusory. Yet, the perceived identicalness is sufficient for human understanding in the ordinary world of human perceptions.

We see then that, in Cosmos, there may be no exactly identical relationships from ultimate moment to ultimate moment. Even if two ultimate particle/events reconfigure with apparent identicalness to each other, their reconfiguration is not Really identical (with respect to the entire Cosmic Configuration), because a myriad of other factors have changed position with relationship to each other (thus, automatically, changing the position of the two ultimate particle/events in question. So-called identically reconfigured relationships would be only *approximately* identical, for the purposes of rough, inexact estimation. When we look for continuity in Cosmos we are not looking for *literal* continuity, but only for the *seeming* of continuity which is sufficient to produce the *illusion* of stability.

From a more subjective perspective, we ask, Is not THAT which does not move and which is EVER CONSTANT equivalent to and identical with ITSELF? This is a completely different question from Self-equivalency and Self-identity considered in relation to a 'thing', i.e., in relation to an *object*.

In Cosmos it can only be said that a thing is *approximately* equivalent to itself, from moment to moment. Each moment shows a variation in the multiple relationships which go to make the thing's 'thing-hood'. In the GREAT BE-NESS, however, the ONLY ONE THING is absolutely equivalent to ITSELF. It IS ITSELF, and ITSELF alone because (change being impossible 'within' IT) there is nothing else that IT could BE, there being none other. All 'Rays' of the GREAT BE-NESS also are Self-Equivalent. Further, there are no 'ultimate moments' in IT, as, for IT, there has never been a segmentation into 'Time' and thus no 'passing of Time'.

We have considered earlier that the forms of prakriti (which forms are 'things') are *not* formally equivalent to each other (even when they are in the same category {such as Hydrogen atoms, for instance}), nor even identical with and equivalent to *themselves*—*formally*, from moment to moment.

☐ When one thinks, however, that all 'things' are really Prakriti, which Itself is, Essentially, Mulaprakriti, which Itself is, Essentially, 'Objectified PARABRAHMAN' (the ABSOLUTE SELF-IDENTICAL IDENTITY), we realize that, at the most fundamental level, even all *'things'* are ESSENTIALLY equivalent and identical to each other, and, moreover, equivalent and identical to themselves from ultimate moment to ultimate moment. This is *ultimately* and ESSENTIALLY *true*, but not normally a practical consideration in the World of Becoming.

Returning to the GREAT SUBJECTIVITY, we find that IT (the UNIVERSAL CONSTANT, the ABSOLUTENESS, the THATNESS, SUCHNESS, etc.) is *identical* with ITSELF, but in Cosmos, in terms of exactitude, any thing (within the World of Fabrication) must be considered unequal to itself if measured against itself one ultimate moment later. Pragmatically, however, there is no way to measure any 'thing' from ultimate moment to ultimate moment, so even things which are not *exactly* equivalent or identical from ultimate moment to ultimate moment can be *considered* so, depending, of course, upon the duration of the evanescent *approximate* relationships between the parts which define its 'thing-hood' on the macro level. Should the motion which creates 'thing-hood' cease altogether throughout Cosmos (and that motion *does* exist as well, though in a far different way, within the World of Being), the Universe would return to ABSOLUTE ZERO, in the philosophical sense, and all things would disappear into THAT WHICH IS ever *identical* with ITSELF.

The Problem of Whether Archetypes Fluctuate

Archetypes, which are the foundational formative Potencies in Cosmos pertain to the World of Being (which is *not* the WORLD OF BEING/NON-BEING). Really, the World of Being can be thought of as situated 'between' the WORLD OF BEING (the ABSOLUTE) and the World of Approximation (the *lower* World of Illusion). Archetypes, therefore, are found 'above' the usual Worlds of Form (the World of Fohatic Fabrication) and, in fact, are found upon the very highest Cosmic Planes (or Kosmic Planes), usually upon the second plane of any seven-planed system.

☐ Archetypes are really Patterned Relationships of Energies which control the patterning of all energies and forces below them.

From a Cosmic Perspective, it is the Archetypes which, as it were, 'Guide Fohat' and, thus, confer stability upon all configurations working themselves out in Matter. Archetypes seem to be stable and abiding, perhaps even permanent, but are they? Since Archetypes are found intra-Cosmically, and there is no absolute intra-Cosmic continuity (but only discontinuity), do not even Archetypes fluctuate? Is there any *motionlessness* in Universal relations? Is there anything at all at rest in Cosmos?

This is a profound question because it is difficult to analyze the constitution of an Archetype. The lower archetypes are of course impermanent-in-Cosmos, consisting of a number of qualities, and are simply the thoughts of various creators (none of them as 'high' as the Cosmic First Family. These lower archetypes are probably *prakritic formulations* and are therefore different from Ideational Archetypes found in the World of Being.

For instance, the psychological archetype of the 'Hero' consists of many qualities, and that archetypal image may fluctuate depending upon the society within which the Hero archetype is formulated. Such an archetype is simply a relatively persistent thought-form, fluctuating:

- because of ongoing changes of energies and forces within the World of Fabrication, and
- because it is (as a thought) prakritically constituted.

What, however, can be said of the Higher Archetypes, which are the Great Fundamental Numerical Entities, especially the first nine apparently non-composite Numbers. Are the Fundamental Numbers continuous, non-fluctuating, *sustained presences* in Cosmos, and therefore are they possessed of a degree of the attribute of *continuity* which characterizes the ZERO from which, ultimately, They originated?

Does 'Oneness' *move?* Does 'Twoness' *move* in any way? What of 'Threeness' and 'Fourness', etc.? Admittedly these are difficult and strange questions. Is it possible to conceive of any item in-Cosmos which, in itself, is not (at least as regards its origin) a movement? Try to conceive of anything coming into existence without a *movement*, without a difference from that which preceded it. A difference *is* a movement (even if nothing *moves* from 'here' to 'there' in space).

Another related question arises, If nothing comes into existence without 'movement' (defined as 'difference'), can anything *remain* in existence continuously for a certain duration, without 'movement' to sustain it? The attempt to answer this question will lead us to ask whether Consciousness (though it arose through change) can be sustained without movement for a given duration. In other words, is Consciousness (once it is established) necessarily a 'movement'?

No Number arises without movement. Although, the birth of Number *One* is most difficult to conceive, this entire treatise, in once sense, has been about the birth of the Number *One* and Its relation to the ZERO. The birth of all other Numbers proceeds through the process of Emanatory Self-Subdivision, and Emanatory Self-Subdivision is inseparable from *movement*. The movement which births a Number is the movement of the *Refocusing of Consciousness* in the Emanator of the Number to be birthed. Number, thus, arises through a *change* in Consciousness.

☐ If there is a 'change' in Consciousness, what, if any-thing, 'moves'? Perhaps we could call a 'movement' in Consciousness an alteration in 'breadth' and 'intensity' of *sensitivity*. Or perhaps we might understand such 'movement' as a 'change in *presence*'.

Numbers are *born* through movement (or change), even the Number *One*. Once Numbers are born (emanated), is there any *movement* in Them, or are They intra-Cosmic Continuities? First, there is no such thing as an *absolute* intra-Cosmic Continuity, because Cosmoses 'disappear' rendering Them and any 'parts' of Them discontinuous.

This is the *non-disappearing* INFINITE SELF the *only* ABSOLUTE CONTINUUM. So if Numbers are, necessarily, discontinuous, how are they so? They are not discontinuous in the same way that Objects in the World of Fohatic Fabrication are discontinuous (multiply particulate and fluctuating with each ultimate moment).

One way of conceiving change within an apparently unchanging Archetype called a Fundamental Number, is to realize that all Numbers have their *time*. There is, in Cosmos, a *time* for Oneness, a time for Twoness, for Threeness, and so forth. Though there may be a 'time' when all allowable Numbers in a given Cosmos exist simultaneously, there are times when certain Numbers are being *emphasized* or *de-emphasized* (such as during the onset or termination of Ray Cycles). Within the Numerical Being, what *changes* or *moves* to produce this emphasis?

We must remember that Numbers considered as Archetypes are Authentic Entities of the highest order. Numbers are *Entities*, not just abstractions used in counting. When a Number has its 'time', the Entity Who Is the Number somehow *emerges* in power and prominence, and that emergence is a *change* or *movement* within It, and/or a *change* or *movement* in relation to the other Numbers.

Surely the analogy from the stages in the life of man must hold good when transferred to the Cosmos as a whole, and there must be a long phase of Cosmic Development when, for instance, the Entity Who Is Number Three holds sway, followed by a shorter phase when the Entity Who Is Number Two takes over, followed by a still shorter interval of absorption and obscuration during which the Entity Who Is Number One *moves* to a position of prominence. Perhaps this 'movement' can be thought of as a *variation in the intensity of focused Vision.*

Technically, in this discussion we are focusing within the Realm of Consciousness, talking about *movement* (a factor-in-Cosmos which pertains to the Third Aspect and not the Second {as does Consciousness}). Thus the discontinuity of 'movement' for which we are searching will not be of the same kind as that which demonstrates in the World of Fabrication in relation to the Third Aspect or "Activity Aspect". The 'movement' we are discussing here will be a change in something that has been *continuously sustained* until the time of the change—a *change of Conscious Focus.*

In a way, we are speaking of a 'Change of Gaze' (using the metaphor of 'Sight' for 'Consciousness'). We will have to ask ourselves if there is such a thing as, 'intensification or de-intensification of the Gaze'? 'Movement' within the World of Being can be achieved through *differential intensities of 'Gaze'*. How 'hard' does a Divine Observer Gaze at one object when compared to Its Gaze upon another? In a field that contains many objects, although *all* may be *seen*, where is the focus?

☐ If a Gaze can be sustained uninterruptedly, that Gaze is a kind of Continuity in Cosmos No BEING, however, other that the ABSOLUTE BEING can be an ABSOLUTE CONTINUITY (because only IT 'ABIDES' forever). The necessary Discontinuity for which we are searching is provided by a change of 'focus' of the Gazing One.

We are not here speaking of 'Vibratory Discontinuity' (as is ubiquitous in the World of Fabrication), but of a 'Willed Refocusing' according to a Cosmic Schedule of Operations. It is as if the World of Being stands mid-way between the ABSOLUTE MOTIONLESS of the HOMOGENEOUS CONTINUUM, and the moment-to-moment fluctuation in the World of Approximation (the World of Fohatic Fabrication). The 'change of Gaze' may be expressed in terms of the words, 'Willed differential in *presence*'. A 'Willing' Subject can be 'in' and object or not 'in' an object; can bring the object 'forward' in the Field of Consciousness, or 'retire' it. This it does by *intensifying its own presence 'within' the object.*

We begin, therefore, to receive some light on this difficult question, Does even an Archetype fluctuate?, when we realize that Archetypes are Great Beings in Cosmos, and that All Beings in-Cosmos (in contrast to the ABSOLUTELY NON-FLUCTUATING BE-

NESS which IS THE INFINITE SELF) do *fluctuate*, yes, but in a way that involves a relatively lengthy application of the *factor of continuity.*

How this fluctuation or change occurs may not be easy of comprehension. Certainly the movement which equates to change of prominence or position (as when a triangle rotates emphasizing one vertex over another) is easy enough to understand. What, however, of other types of movement *within* an Archetypal Being? When we think of the infinity of time which (theoretically *could*, but almost certainly does *not)* exist between the ultimate moments in a Cosmos, we begin to realize that every E/entity in its *spirit aspect* (even the great Archetypal Entities) could *move* (i.e., *refocus)* between being cognizant of a Cosmos-in-Fabrication, and being Cognizant of Themselves as Spirit within the Worlds of Adjustment and of Being. Do such Archetypal Entities 'change' Thought/Quality, or merely 'hold' Thought/Quality, allowing Fohat to do the necessary 'Adapting'? Do such Archetypal Entities (or Their lower 'enfolded' 'Rays') provide moment-to-moment *mini-archetypes* meant to guide the moment-to-moment formation of the Cosmic Configuration, or is this simply a function of the very capacious Intelligence of the Great Approximator—Fohat.

One thought of value seems to emerge from this consideration: that 'movement' in the Realm of Higher Archetypes is 'movement' in the Realm of Ideation, and maybe 'movement' through *intensification of identification* with certain 'held', formative, Ideas. Probably, the 'higher' one 'rises' into the World of Archetypes, the less change or discontinuity there is. But Great Archetypes have many 'Rays' of the ABSOLUTE 'enfolded' within them, and these lesser Archetypes (authentic E/entities all) have to have something to *do.* While Archetypes remain Themselves throughout the Universal Manvantara (thus *being themselves* as a principle function) they may well 'engage' in the First and Second Aspect kind of 'movement' through *identification* and *presencing* and *not* through 'movement' as it is conceived in relation to the Third Aspect. This is a difficult area of speculation and requires much pondering.

We must also realize that the Being Who Is Number One is only the intra-Cosmic Representative of Oneness as It exists as a Possibility within the INFINITESSENCE. There have been an infinity of such Representatives—One for every Cosmos. The *infinitized* 'IDEA' of Oneness can never be *perfectly* represented in a Cosmos. The PERFECTION of that IDEA dwells only within the ABSOLUTE, 'wherein' it is *noumenessentialized.*

Numbers are Possibilities dwelling in an *infinitized* state within the FOUNT OF ALL POSSIBILITY. The God Who comes forth as the Number One (the Universal Logos for a particular Cosmos) Is merely the 'Guardian', Representative or, better, Custodian of 'PERFECT ONENESS' for *that* particular Cosmos—and an *imperfect* (because Finitized) Custodian at that.

Therefore, the Beings Who Are Numbers in Cosmos (ever dimly cognizant of the PERFECTION of the Possibility called 'NUMBER', as that PERFECTION 'indwells' *infinitessentially* within the INFINITESSENCE) *labor* to bring into Cosmos as much of that PERFECTION as is available given the Cosmic Algorithm (the Design-at-the-Beginning). Our Number One in this Cosmos *labors* to Be as Perfect a Number One as this Cosmos allows, and so for all the other Entity/Numbers.

We can conceive, that (in ways *most subtle*, and most related to the *Consciousness* Aspect) the Great Archetypal Numbers *move* (without *changing positions, for They per-*

vade Space). They, of course, 'Uphold Their Numerosity', each as an Archetype (through sustained Self-Reflection, i.e., through sustained Self-Gaze) and, as well, 'Uphold the Numerosity of Their Emanations' (through the sustained *specified* Self-Reflection which produces the Self-Reduction which is Their Son), *but,* They also *'Alter* Their Gaze' and this 'Alteration' is a 'Movement', a 'Discontinuity' within a Cosmos which will tolerate no *perfect continuity.* Archetypes, after all, may be *related* to other Archetypes through *numerical combination* (as provided by Metaphysical Arithmetic Operations), and within those combinations, they may *ideationally intensify or de-intensify* through *identification* and *investment,* or *retraction or presence,* as suggested above.

Thus, it is that the Archetypes (so stable, *apparently,* but, Really, only *relatively* so) *move* within the World of Being in a way unique to Themselves, by 'Seeking' to be 'Faithful Representations' of the almost certainly *timed* 'INSTRUCTIONS' *inherent* because *implanted* within Them ('INSTRUCTIONS' 'FLASHED FORTH' *by* and *within* the 'RAY' of the ABSOLUTE at the 'MOMENT of AWAKENING').

All within the World of Fabrication seek to approximate the Pattern 'Held' (and cyclically *changed*) within the World of Being by the Great Archetypal Numbers. The Great Archetypal Numbers within the World of Being, however, seek to approximate the 'SUPER'-Cosmic 'INSTRUCTIONS' 'IMPLANTED' within the Pre-Cosmic Infinite Subject/Self at the 'FLASH'—'INSTRUCTIONS' which consequently, by extension, 'inhere' within the Universal Logos (Who is but the Pre-Cosmic Infinite Subject/Self-in-Attenuation). So, the Universal Logos and Its Emanations have *Their* Labor as well, as They 'Seek' (through a blend of both *relative continuity* and *measured movement)* to be as Perfect Representatives as They may be of the 'NUMERICAL PERFECTIONS' *infinitessentially* 'inhering' 'within' the GREAT PERFECTION, the ZERO.

The Problem of the Departicularization of the Word "I"

It will be noticed that in this treatise the word "I" is used variously and strangely (as **I**, **8**, I, and i), and usually (the careful reader will notice) in a non-egoistical manner. In most contexts, the word "I" (in this text, 'I' or 'i') has such a limited and personal meaning. In fact, "I", used correctly, is the most powerful and *inclusive* word in our vocabulary. How, then, can we departicularize the use of the word "I" and restore it, first to its *Cosmic* significance and then to its *infinite* significance in relation to the INFINITE SELF.

It is so instinctive to use this word "I" in its limited and localized sense. Unless there is a sense of the UNIVERSAL CONSTANT, it will be impossible to use "I" in any but a localized manner. In order to use it properly, we must, above all, overcome the Great Illusion of Separateness. Separateness is the instinctive, reflexive consciousness within lower Cosmos.

☐ Labor is required in order to restore to consciousness the Synthesis which IS.

Amazement and wonder are also required, for they shatter the usual categories of thought that imprison the sense of WHOLENESS/Wholeness. To use the "I" properly, we must cultivate the intuition, which like Isis, reunites all apparently fragmentary presentations. Readers of Walt Whitman will realize that he succeeded in this. His use of the word "I" is extraordinary, "I am vast. I contain multitudes."

The Problem of the Identicalness of the INFINITE and the Finite —BRAHMAN and Samsara Are ONE

The Universe IS BRAHMAN. That which is before our eyes is the Living Miracle (Child of THAT, and, also, THAT ITSELF). This is true because the BOUNDLESS IMMUTABLE PRINCIPLE is *indivisible* and *IS* all things without exception. The *apparent* divisibility before us is illusory, but what the divisibility REALLY and ESSENTIALLY *IS*, IS REAL, for there cannot be anything that is ESSENTIALLY un-REAL.

As we look around ourselves at all the amazing variety of color and form, hear the blending of many sounds and voices, and, in general receive the multifarious testimony of the sense, we must realize that all *this* is *THAT*—the ONE AND ONLY REALITY. Of every single thing **8** see, **8** may say of it, "This very thing is the ONE SELF in *fullness,* for there are no 'parts'. This IS the ABSOLUTENESS. This IS the INESCAPABLE. It makes no difference what is presented. The ABSOLUTENESS cannot be escaped."

Are not these thoughts incredibly validating of the value of Cosmos? Indeed, Matter Is Spirit. 'MAYA' *Is* PARABRAHMAN. One need look no farther than that which is presented at every moment to find the MIRACLE, the ULTIMATE SELF-of-all-Selves, the INDIVISIBLE, the ABSOLUTENESS. Every such re-cognition or realization is part of the Universal Self's SELF-Re-Membering. It is not just the so-called individual self or, even, the Self which must be 're-membered, but it is the PRESENCE ITSELF that must be 'Re-Membered'. It is desirable that the PRESENCE become *continuous to consciousness* (as IT already IS in *fact).* The PRESENCE IS the UNIVERSAL CONSTANT (or even more accurately, the *SUPER*-UNIVERSAL CONSTANT). It is the PRESENCE which reminds ever that Samsara is REALLY BRAHMAN.

Because the World (Samsara) REALLY *IS* BRAHMAN, full and complete interest in the World-as-Presented is required of the Synthetic Yogin. Full interest is not an invitation to entrapment and enthrallment, but rather to a keen appreciation for the way the One Chosen Possibility from the FOUNT OF ALL POSSIBILITY is working Itself out in that Great Objectivity we call the Universe.

From the perspective of Radical Infinitism, it is not admissible to deny the World, for to deny the World is (curiously) to enter into a subtle kind of dualistic thinking. To deny the World is to admit that there *can be* a *real,* independently existing, not-SELF (which, clearly, there is not). If there were to be a SELF and a *real,* non-illusory not-SELF, a division would have to exist demarcating the SELF from the not-SELF. That division would have to be occurring 'within' BRAHMAN, the BOUNDLESS IMMUTABLE PRINCIPLE, the ONLY THING THAT IS. BRAHMAN, however, being forever ESSENTIALLY *indivisible*, the SELF *and* the not-SELF cannot *both* REALLY exist, and it would be absurd to say that only the not-SELF exists (which strangely is what so many of today's so-called 'enlightened' thinkers are propounding, for the most part remaining ignorant of the implications).

What we are constantly presented with in all of our experience on every possible plane of Nature IS, therefore, the SELF (albeit the *veiled* SELF). We are presented with *the* SELF in a certain form. That form is the result of a Creation we might call the 'de-limiter'. Due to Mayavic Delimitation the ONE SELF seems to function differently on every level or dimension of Cosmos.

Some may have found that, as a result of non-dualistic thinking, a kind of vacuum, or strange sadness began to enter their consciousness. This World View can be incorrectly interpreted as "world denial". Such an error can instill feelings of insipidness, lack of enthusiasm towards one's previous interests, and even a certain kind of lethargic alienation from the World.

An entirely opposite response, however, is what should emerge if the concepts of Radical Infinitism are truly understood.

☐ Through Radical Infinitism, a tremendous enthusiasm and *livingness* can enter the life. Although the form of what is being seen within Cosmos is ESSENTIALLY illusory (a construct of the 'Mayavic De-limiter', or what, in-Cosmos, might be called the Universal Mind, the Great Divider), at the same time, there is a tremendous affirmation of *all* that is *seen*. This affirmation begins with *seeing* but it merges into *being*.

'Seeing' is the word-symbol that stands for all forms of consciousness which exist between subject and object. 'Being' is the word-symbol which stands for the realization and assertion of *identicalness* between subject and object and, thus, for the elimination of the subject/object consciousness.

As the enthusiasm and livingness grow, it becomes more and more easy to affirm the entire Cosmic Presentation—i.e., the collection of all ways in which Cosmos *appears* to consciousness. Does it not seem as easy to *affirm* the Cosmic Presentation as it is to deny it? Merging the World in BRAHMAN is the way to proceed, as Sankaracarya advises us.

☐ No more joyous or blissful act can actually be contemplated than to successfully merge the World 'within' and *as* the INFINITE SELF. This *merging* or *dissolving* is an act or a process that demands every power of concentration of which the disciple is capable.

What we may discover by considering this matter is that World Negation or World Denial is only the *first* step in the Process of identifying as the INFINITE SELF. The *affirmation* of the ESSENTIAL REALITY of the World must proceed side-by-side with Its negation. The world is un-REAL in that It does not *seem* to be BRAHMAN, for It is apparently full of division and difference, whereas BRAHMAN IS *not*. The World, however, is *REAL* in as much as any Presentation-in-Cosmos must, necessarily, *be* BRAHMAN, ITSELF, since there can be nothing else. If there *seems* to be something *other than*

BRAHMAN, that *other* is, in fact, REALLY *none other* than BRAHMAN, ITSELF. The World *does* seem to be an *other,* therefore the World is *none other* than BRAHMAN ITSELF.

☐ The World is a Pre-Cosmically Determined Presentation to that apparent 'LIMITATION' of ABSOLUTENESS which is called Universal Consciousness.

The Problem of the un-REAL REALITY, and the REAL un-REALITY

Given that BRAHMAN and Samsara are ONE, the whole question of what is *real* (REAL, Real, or real) becomes very involved and very simple at the same time. Every presentation which *appears, in itself, real,* is, in a sense, REALLY un-REAL. Thus, we have an 'un-REAL *apparent* reality'.

Therefore, the *appearance* of any finite thing can be said to be un-REAL. It is the *appearance* of the presentation which is the un-REALITY, for the *appearance* is not (since it is a product of delusive Mayavic Consciousness) identical (at least in the Realm of Consciousness) with the HOMOGENEOUS SUBSTRATUM WHICH IS BRAHMAN. The ESSENCE of the appearance, however, *is* BRAHMAN, and so the *apparently* un-REAL presentation is, *after all,* ESSENTIALLY *REAL.* In this case we have the very opposite of what we had above—a 'REAL apparent un-REALITY'.

☐ The whole problem is solved by realizing that BRAHMAN IS the ONE AND ONLY REALITY, and no-thing can be presented which is not ESSENTIALLY IT—regardless of appearances! No appearance is *real-in-itself,* but only REAL-*as*-BRAHMAN.

The Problem of the Finite-INFINITE, and the INFINITE-Finite

The implications of the 'un-REAL reality' and the 'REAL un-REALITY' are profound and directly related to the *infinite* and the *finite.* In this regard we will discuss the 'Finitized INFINITE' and the 'INFINITE Finitude'. The idea is developed just as in the case of the contrast between REAL and un-REAL. Every *thing* is a Finitude, yet since, within the UTTER ALLNESS, only the INFINITE *IS,* then the apparent Finitude Is, ESSENTIALLY, also the INFINITE. Further, since there is naught but the INFINITE, why does the World not *appear* to be the INFINITE? It is because in 'BECOMING' the World, the INFINITE has 'FINITIZED' ITSELF (which IT cannot *REALLY* 'DO', though IT *appears* to 'DO' so).

It appears that in order to describe the REALITY/Reality of any E/entity-in-Cosmos or item-in-Cosmos, paradoxical terms must be employed because every possible presentation is apparently a *blend* of both (or *is* both) the INFINITE and the Finite, the REAL and the un-REAL. To resolve the problem one must rely upon a capacity within the human Spirit to blot out all form by resolving Heterogeneity (the Cosmos) into HOMOGENEITY (the INFINITE REALITY). This capacity exists but each must discover it 'within'; it cannot be directly taught.

The Problem of the *INFINITIZING* of the Finite

If we are to be technically correct, we cannot speak of 'infinite *things*' at all—this would be a contradiction in terms. All we can say is that all things are ESSENTIALLY infinite, in that they are *no-thing other* than the INFINITUDE, the ABSOLUTE. The 'thing-hood' of a thing is delimited and cannot possibly be infinite, but the *substratum* of a thing (which is one with the ONE SUBSTRATUM) *is* the INFINITE. So while there are no 'infinite things', all things are ESSENTIALLY and *substantially* INFINITE and ABSOLUTE. Mulaprakriti (that which we are calling the 'Infinite Object') is the One Exception. It is a 'thing', because it is 'Seen', or 'Registered', but it is a 'no-thing' because that which is 'Seen' is infinite—a Boundless Homogeneity.

The slightest of things "seen in God", according to the mystics, is vastly superior to the greatest of things seen 'outside' of God. Therefore, the great task is to *see* all things in GOD (the ABSOLUTE) and even more, to *be* all things in GOD. Another way of saying this, is that our great task is to *infinitize the finite.* Of course, we have realized by now that the finite is *already* the INFINITE, and ever *has* been as it ever *will* be; the finite simply has to be understood as the INFINITE and *identified with* as INFINITE.

Rather than reject the World as un-REAL, Its ESSENTIALLY INFINITE NATURE has to be *seen* not only suffusing the World, but as *being* that very World Itself. When this state of *consciousness/being* is achieved, it leads away from alienation, and lends a beautiful intimacy to all encounters with the World. The World is not *other* than THAT. The World *is*, in fact, the CELEBRATED ANCIENT SELF.

☐ Thus, every worldly encounter should be a kind of *celebration,* for the INFINITE SELF is to be lived in celebration. This World (the World of Becoming, or World of Illusion) demands a greatly enhanced valuation, not rejection at all, but respectful treatment in tenderness. Indeed, the ground whereon we walk is "Holy Ground" for we "live and move and have our being" within the ONE AND ONLY.

It is so interesting that form (as an aspect of Illusion) can be despised (under the Law of Repulse) and even *must be,* at times, for the sake of the advancement of the Spirit, but form, considered as the very PRESENCE of BRAHMAN, ITSELF, must be *intimately* loved, *infinitely* respected.

Really, the basic Principles of Radical Non-Dualism are so totally simple. BRAHMAN is ALL that there IS. The SELF is ALL that there IS, and every 'thing' is both illusory and non-illusory, is both apparently finite and *absolutely* INFINITE.

The Problem of 'LAW', 'WILL' and 'NECESSITY'

Multiplicity is un-REAL but it must be respected. This respect is evidenced by understanding and obeying the *laws* which obtain in relation to multiplicity. REALLY, in the WORLD OF BEING (not the World of Being) there are no 'LAWS' *in particular.* From another perspective, however, the INFINITE SELF *is* LAW. ITS *noumenalized* 'WILL' is LAW. In the World of Cosmos, however, there is Law as we normally understand the term 'Law'. The 'WILL' of THAT is the Law of Cosmos.

There is an important relationship between Law and Will. Law is an Energy Pattern insisted upon by a superior Will. Or, Law may *be* the Energy Pattern of the superior Being Who Wills, because that Being *is,* in Cosmos, a Pattern-in-Manifestation, and the mere existence of this superior Being *imposes* upon lesser beings the Pattern that, in manifestation, It *Is.* No being is REALLY a pattern, for pattern implies multiplicity and every being is ESSENTIALLY impartite, monolithic and one with the ZERO. Beings, however, as we *usually* know them (i.e., *not* simply as 'Rays' of the ABSOLUTE) are inseparable in Cosmos from the patterns through which they manifest. Such energy patterns are vitally important when considering the meanings of Will and Law.

Will takes precedence over Law. In the causal chain, Will precedes Law. While it is true that one may become cognizant of Law, and then apply one's will to fulfill that Law, the Law of which one has become cognizant is Really the Will of a Being superior to the one who fulfills the Law. In Cosmos, the first Will emanates from the Universal Logos, at the Beginning (of each Cosmos), and is sustained thereafter on the Waves of Song—we might call it "Brahma's Song". That Will empowers the Fixed Design, the Original Intent, the Cosmo-Conception of the Universal Logos. The combination of Will and Fixed Design produces Law. On all levels of multiplicity, that original Universal Logoic Will becomes Cosmic Law.

Law, then, in an intra-Cosmic sense, becomes Necessity.Necessity is that which the Law protects and upholds (namely the Manifestation of the Original Intent, or the Design-at-the-Beginning), and must *necessarily* come to pass. Necessity is *inevitable* and *unrelenting.* Necessity is that which *must* be performed or accomplished. Necessity is Willed Manifestation, and the actualization of Necessity is enforced by Law. Necessity, ultimately, leaves no room for free will, because one *must,* whether he *will* or no. The means of accomplishment, however, are *free*. Fortunately for the existence of *free will*, there is but ONE 'WILL' which is the 'AUTHOR' of all that which becomes, in-Cosmos, will-compelling Necessity.

- ☐ Thus, WE-the-**I** have 'WILLFULLY' 'CREATED' Cosmic Necessity and the Law (reflective of 'LAW') to enforce the manifestation of that Necessity. As well, there is still a higher mode of 'NECESSITY', and a kind of 'LAW' of BEING which is 'WILLED' by THAT in order to guarantee the *fulfillment* of that 'NECESSITY'.

The Universe has been called a Son of Necessity. This means that the Universe is *necessary* and hence, Its sequential appearance is *compelled*. The SELF as 'SOLE' AUTHOR of the Universe, *cannot*, however, be *compelled*. Necessity requires enforcement, but who or what is there to enforce anything upon the ONE WITHOUT A SECOND? It cannot be said that the one ABSOLUTE SELF *needs* to obey any *law* at all. IT, the ONE AND ONLY, is a 'LAW' UNTO ITSELF. Since there is none to compel IT, other than ITSELF, any seeming separation of ITSELF into an apparent duality of INFINITE SUBJECTIVITY (SELF) and Object (Universe) occurs only for 'REASONS' of ITS OWN. The 'Great Coming Forth' of THAT is not a 'LAW' which IT *must* obey. IT *must* obey only ITS own 'WILL' (if we can call it *will)*, and that 'WILL' *is*—the WILL TO BE ITSELF *forever*. IT *seems* to 'DEEM' this *sustainment* of ITSELF *forever*, a 'NECESSITY'!

Words here are *worse* than useless. Suffice it to say that the usual reasoning applied to Will, Necessity, and Law (in all combinations and possible sequences) *cannot* 'reasonably' be applied to the INFINITE SELF.

The Problem of Whether Illusion is REALLY un-REAL

Illusion is said to be an un-REALITY, yet it has *actuality*—i.e., that which we call consciousness *notices* illusion and registers it. Thus, illusion is a *fact*, an *actuality*, and yet is still un-REAL. Perhaps this is because that which is noticed is not ESSENTIALLY/ Essentially what it *seems* to be. In other words, a *seeming* can be *reduced* to 'something else', whereas a REALITY cannot be further reduced. So, though un-REAL (i.e., *reducible*), illusion is *actual*, *factual*, and even (in some cases) Real-in-Cosmos (but not {as an appearance} REAL). Illusion *happens*.

- ☐ Therefore, the un-REAL *happens*, but even as the un-REAL happens *in consciousness*, the IRREDUCIBLE, the REAL, has never ceased to BE EXACTLY AS IT FOREVER IS.

So although illusion is *happening*, NOTHING is *REALLY* what is *happening!* It can reasonably be said that illusion is both un-REAL (because it is reducible) and REAL because it can be none other than the SUBSTRATUM.

The Problem of Why There is "No Good Reason" for the Universe

Most thinkers want a "Good Reason" for the Universe, for *all this* that we experience. The usual 'Good Reasons', however, are not REALLY 'Good'. A Universe is, at least *apparently*, a *deviation* from the ABSOLUTE PERFECTION of the ALL-IN-ALLNESS, for the Universe, having come *from* the ALL-IN-ALLNESS, is necessarily (according to apparency) *less* than the ALL-IN-ALLNESS, and anything less that ABSOLUTE PERFECTION cannot be as PERFECT as ABSOLUTE PERFECTION. So we see that, compared to the ALL-IN-ALLNESS, the Universe (as an *appearance)* is REALLY an *Imperfection.*

What possible 'REASON' can the ABSOLUTE have for 'BECOMING' less PERFECT than IT already IS? In light of these thoughts we cannot say that IT needed 'EXPRESSION'; that IT "so loved the world"; that IT was 'LONELY' and 'NEEDED' something other than ITSELF to which to 'RELATE'! All such explanations are based upon an INFINITE SELF that is *less that* the PERFECTION IT IS. Moreover, a misleading anthropomorphism tends to creep into such explanations.

Another possible thought that attempts to offer a 'REASON' for 'BECOMING' LESS PERFECT, is that ABSOLUTE PERFECTION *is not REALLY* ABSOLUTE PERFECTION (hence, the 'ALMOST PERFECT' would 'NEED' the Universe to be ABSOLUTELY PERFECT). This, however, is a desperate and unphilosophical attempt, which, for obvious reasons, undermines our basic premise of a BOUNDLESS *IMMUTABLE* PRINCIPLE. If we adopt this point of view, we will soon be forced to say that everything we assert is *not really* what we assert, and that no-thing is identical to itself.

While *formal non-self-identity* is explicable in terms of the instantaneously variable configurations of ultimate particle/events in-Cosmos, *essential non-SELF-IDENTITY* is impossible to maintain in relation to the BOUNDLESS IMMUTABLE PRINCIPLE. An *identity is an identity.* When it comes to an examination of the ONE ESSENTIAL IDENTITY, IT IS WHAT IT *IS, exactly*—the ABSOLUTE. ABSOLUTE 'PERFECTION' cannot be *excised* from the ABSOLUTE!

We can, however, somewhat extend and apply the above thought and come up with, perhaps, what might be a *best* 'REASON' for the Universe. Because the ALL-IN-ALLNESS IS the GREAT CONTRADICTION, we might say that the ABSOLUTE PERFECTION *is not REALLY* the ABSOLUTE PERFECTION (the FOUNT OF *ALL* POSSIBILITY) *unless* IT is *apparently imperfect* as well as being forever PERFECT. Now this is a statement worthy of the GREAT CONTRADICTION.

☐ We therefore see that the 'REASON' for the Universe is SELF-CONSISTENCY which can only be maintained through the 'GENERATION' by the ABSOLUTE HOMOGENEITY of That Which is *Inconsistent* with ITSELF—namely, the UNIVERSE.

This explanation seems to point to a 'LAW' which derives from the fundamental assertion—**I AM I** FOREVER *and no other.* And yet, if **I AM I** FOREVER, **I** must be 'others' forever, for PERFECTION demands Imperfection in order to be *perfect.* If **I** do not *contradict* MYSELF **I** CANNOT *BE* MYSELF.

The Problem of Whether the *Universal* Logos can Commit Error

The Universal Logos is the ABSOLUTE SELF, ESSENTIALLY and REALLY, but It is not the ABSOLUTE SELF Really (in Cosmos). Because of SELF-'VEILING', the Universal Logos (Being a Specialized Singularity) is, as it were, *infinitely removed* from the full potential of the LIMITLESS POTENTIAL. The Universal Logos is, therefore, a Limited Being.

No Limited Being is OMNISCIENT, OMNIPRESENT and OMNIPOTENT. With respect to Its Ring-Pass-Not, the Cosmos, a Limited Being (in this case the Greatest Being within that Cosmos) may be Cosmically Omniscient, Omnipresent, and Omnipotent, but not ABSOLUTELY so.

We see, then, that the Universal Logos is an "Imperfect God" because of its manifest (hence, illusory) *separation* from the LIMITLESS PERFECTION, the ABSOLUTE DEITY. Thus the Universal Logos is *laboring* and *experimenting* within Its Prakritic Field (which we call the Universe). The Universal Logos *knows* the Universal Algorithm (the Design-at-the-Beginning), and *knows* the nature of the Relative Perfection which must be achieved in Cosmic Prakriti at the "Day Be With Us", but does *not know* the exact *means* by which this Cosmic Consummation will be reached. Even less do the Emanated Extensions of Itself (all the various authentic I/identities in Cosmos) *know* the exact means of achievement. These emanations have the further disadvantage of being further *veiled,* so they will progressively lose sight of the Design-at-the-Beginning (even though that Design is *inherent* within them, because they, for all their smallness, are still the entirety of the Universal Logos as well as being, *apparently,* themselves).

- ☐ Can the Universal Logos commit an error? The very reason for the existence of this Logos and His Universe is so that *error may exist*—i.e., so that Imperfection may flourish. The short answer to this query is, Yes, but at the same time it must be realized that the Design-at-the-Beginning *will be fulfilled exactly,* no matter what the sometimes less-than-perfect means employed to achieve it.

The Problem of Whether the ABSOLUTE can BE in 'ERROR'

By what standard shall the ABSOLUTE be judged when IT ALONE IS the UTTER ALLNESS. The ABSOLUTE is INCOMPARABLE. IT can only be compared to ITSELF, and even this would require an inadmissible *division* in the INDIVISIBLE.

The concept of *error* belongs to the World of Relativity, the World of Pattern. Deviation from a clearly defined and desirable standard is error. We have seen that the Universal Logos (because it is only Omniscient and not OMNISCIENT) can (especially, via Its

Emanated Self-Reflections) commit error, but it is absurd to think that the ABSOLUTE (which cannot even REALLY 'ACT') can commit error.

- ☐ The one thing the ABSOLUTE 'DOES' (and which many might consider an extraordinary error) is to illusorily 'BECOME' a Universe. The one 'ACTLESS ACT' of the ONE WITHOUT A SECOND is the SELF-'VEILING'-instantly Self-Veiling, which *generates* a Universe.

This SELF-'VEILING'-instantly-Self-Veiling is the Process 'COMMENCED' by the INFINITE SELF in ITS 'MODE' as 'MAYA'-instantly-Maya. Because SELF-'VEILING' is rationally *necessary* (because of the 'WILL' to SELF-CONSISTENCY, the PERFECT demanding the *imperfect*) it cannot REALLY be called an error. Interestingly, however, SELF 'VEILING'-become-Self-Veiling is the 'Root of All Error' in Cosmos for Error arises from Ignorance, and Ignorance arises from instantaneous SELF-'VEILING' and all its Super-Cosmic and Cosmic continuations and implications.

There may be those who will wildly say that it was a 'MISTAKE' for the INFINITE SELF to SELF-'VEIL' and thus 'CREATE'/'BECOME' the Universe and Universes. There is no basis, however, for such irrational statements, even though the Process of SELF-'VEILING' and Self-Veiling seems irrational to the ordinary human mind.

- ☐ Error, per se, however, (i.e., *mistaken action)* cannot REALLY be attributed to the INFINITE SELF, the ABSOLUTE. To realize, however, that the ABSOLUTE is, through 'MAYA'-instantly-Maya, the *deliberate SOURCE OF ALL ERROR,* gives one pause to think.

The Problem of 'MOVEMENT' within the SELF

Although I cannot move because I-the-BOUNDLESS SELF AM IMMUTABLE, yet, because *movement* is a *possibility,* and I AM the FOUNT OF ALL POSSIBILITY, I cannot *not* 'MOVE'.

- What we must realize is that *possibilities are illusions,* because they are reducible to the INFINITESSENCE.
- In their *absolutely noumenal* 'STATE' 'within' the INFINITESSENCE, possibilities are *not* illusions because they have been *utterly noumenessentialized.*
- The moment, however, a possibility becomes *actualized,* that possibility becomes an illusion (a reducible appearance).

Now, while 'movement' *is,* indeed, a possibility, it (like every-*thing* else) becomes *illusory* the moment it becomes *actual.* 'Within' the INFINITESSENCE, movement is not *actual* but only an *utterly noumenessentialized potential,* just like all the infinitude of other noumenessentialized potentials. All these 'inhere' within the INFINITESSENCE, but in their noumenessentialized state do not disrupt the IMMUTABILITY of the BOUNDLESS SELF.

☐ Thus, while the NOUMENON of movement abides with all other possibilities 'within' the INFINITESSENCE, there is no *actual* movement within the INFINITESSENCE, the ABSOLUTE. Anything *actual* must 'take place' 'outside' the ABSOLUTE (which, being impossible—there being *no* 'outside') is REALLY 'taking place' 'within' the ABSOLUTE, though it does not *seem* so.

If all this seems SELF-'CONTRADICTORY', it is. It is as if when the ABSOLUTE must 'CONTRADICT' ITSELF by 'BECOMING' imperfection (as IT *must)*, IT must 'CONTRADICT' ITSELF 'outside' of ITSELF (using its 'MODE' of 'ACTUALITY'/'APPARENCY') so that 'within' ITSELF, IT may *REALLY* 'REMAIN' SELF-'CONSISTENT' and PERFECT. Since, however, it would be a 'CONTRADICTION' and illusory for the ABSOLUTE to *have* any 'outside', the 'CONTRADICTION' is *eliminated* by the fact that the apparent 'outside' of the ABSOLUTE cannot help but REALLY be the 'inside' of the ABSOLUTE. Since 'outside' *is* 'inside', the ABSOLUTE 'REMAINS' SELF-'CONSISTENT' (which IT *must*), by 'CONTRADICTING' ITSELF (in 'APPARENCY') in such a way that the 'CONTRADICTION' is both (apparently) 'outside' the ABSOLUTE and yet (REALLY) 'inside' the ABSOLUTE simultaneously.

All this may be summarized by saying the ABSOLUTE's SELF-'CONTRADICTION' is *only apparent!* The ABSOLUTE 'RESERVES' any-*thing* but ITSELF for *apparency*—for the *otherness* which is not REAL. One begins to realize that two Great LAWS, the 'LAW' of SELF-CONTRADICTION, and the 'LAW' of 'NON'-SELF-CONTRADICTION (or the 'LAW' of SELF-CONSISTENCY), must necessarily both co-exist without negating each other.

If the SELF is to be 'SELF-CONSISTENT' it must 'CONTRADICT' ITSELF; if the SELF is to 'CONTRADICT ITSELF', that 'CONTRADICTION' results in SELF-CONSISTENCY. (It is almost certain that the INFINITE SELF is not a Gemini, {though that is a *possibility!* 'within' the FOUNT OF ALL POSSIBILITY} but it sometimes seems so.)

The Problem of Whether All 'Possibilities' within the FOUNT OF ALL POSSIBILITY must be Actualized Simply because They *are* 'Possibilities'

The FOUNT OF ALL POSSIBILITY is the INFINITY OF INFINITIES, the INFINITESSENCE. 'Within' IT, all possibilities are *noumenessentialized* but not separately and distinctly *actualized.* Noumenessentialization, however, should not be considered an *inferior* 'state', for in the noumenessentialized 'STATE', an infinity of possibilities are maximally *fulfilled,* if not *actually* expressed.

There have been an infinity of Cosmoses, and even this present Cosmos *cannot* be satisfactorily numbered, because there is no discernible starting point in the Infinite Chain of Cosmoses. The Cosmos can only be numbered in relation to a specific Cosmos, which has been numbered in relation to a specific Cosmos,which has been numbered ...etc. There has been Infinite Time in which to actualize all the infinity of possibilities within the FOUNT OF ALL POSSIBILITY, but as Georg Cantor has ably and paradoxically demonstrated, some infinities are larger, or more 'countable', than others. An infinitude of limited 'Things' such as Cosmoses is, paradoxically, 'smaller' than an ABSOLUTELY LIMITLESS INFINITE SOURCE—i.e., the FOUNT OF ALL POSSIBILITY.

Putting it in inappropriately concrete terms, there is always 'more' in the FOUNT OF ALL POSSIBILITY than the Infinite Sequence of Limited Cosmoses can ever receive, for each Cosmos fulfills simply One of an infinitude of possibilities.

- One + one + one + one, etc., forever (though tending towards infinity) ...

is necessarily infinitely less than:

- INFINITUDE 'minus' One + INFINITUDE 'minus' One + INFINITUDE 'minus' ONE etc., forever.

Since INFINITUDE 'minus' One is simply INFINITUDE, it would be as if we added an infinity of INFINITUDES and compared the 'sum' to adding and infinity of Ones. (Some would say the 'sum' is the same, and some would not.) From a certain perspective, we have the strange paradox of a given infinitude the content of which is infinitely less than the content of a second infinitude.

☐ It would seem that we can safely say that not *all* possibilities within the FOUNT OF ALL POSSIBILITY need necessarily be expressed within the Infinite Chain of Finite Cosmoses.

As to which possibilities are 'CHOSEN' for actualization, 'WHO' 'KNOWS'! **8** mean 'WHO' does indeed *'KNOW'*. That 'KNOWING', however, is based not upon subject/object relations, but upon INFINITIZED 'IDENTIFICATION', i.e., *BEING* that which is 'KNOWN' to the *infiniteth degree!* Think of the 'CHOICE' FACTOR. For every Cosmos, only *one* of an infinitude of *infinitessentialized possibilities* can be 'CHOSEN'.

One might ask, why the 'FLASH' does not 'CONTAIN' an utterly (from our perspective) *absurd* possibility for the Cosmos-to-Be. Are we entirely certain, however, that our Cosmic Algorithm would not appear *absurd* to the Universal Logos of a past or future Cosmos? And what is to guarantee that an *absurd* 'CHOICE' would *seem intra-Cosmically absurd* to the Universal Logos and His Emanations 'Enacting' the 'CHOICE'?

These are very difficult problems and, for us, insoluble. Some have tried to solve the problem by saying that the entire infinitude of possibilities is being simultaneously 'Enacted' in an infinitude of simultaneously existing parallel Universes. Arguments against this position are found throughout this treatise, but in short,the main problem is the violation of the Law of Periodicity. Periodicity is a Fundamental of *The Secret Doctrine,* and upon these Fundamentals this entire treatise is based.

If Periodicity is retained, then, for one thing, an infinitude of Universes can never be parallel Universes of one periodic cycle, because, then, all the other Universes of *other* periodic cycles will be excluded, and, thus, our supposed infinitude of parallel Universes would not be, Really, an infinitude. There are other arguments against an infinitude of simultaneous parallel universes extensively pursued elsewhere in the text.

The Problem of Whether You have done *EVERYTHING*, —or only *ALL* That *has been* Done?

We have established that there is 'more' possibility within the FOUNT OF ALL POSSIBILITY than has been actualized or than ever *can* be actualized in Cosmos or in an infinitude of Cosmoses. Cosmos is really a tiny receptacle, infinitely small, 'infinitesimalizingly' small, virtually non-existent—compared to the GREAT 'CONTAINER'.

'YOU' ARE the ONE AND ONLY SELF, ESSENTIALLY. YOU have been 'ACTIVE'-as-Active in all the infinitude of Cosmoses. All that *has* been *done*, YOU have done. Since YOU-as-You have *done* an infinitude of things, is there *nothing*, therefore, which YOU have *not* done?

A little thought will reveal that there are, necessarily some things (an infinitude of them) which YOU have *not* done—not because you were not the ONE AND ONLY 'ACTOR' in all the Infinitude of Cosmoses, but simply because of the drastic incomensurability between the Cosmos (and even the 'sum' of all Cosmoses), and the FOUNT OF ALL POSSIBILITY.

☐ There are simply things to be done that have never been done, and, with 'almost' infinite certainly, never will be done, i.e., made *actual.* Every Cosmos 'misses' an infinitude of opportunities for actualization!

YOU-as-You-as-you can, however, be assured of this fact that all things that *have* been done, *YOU* have 'DONE'-as-Done. There never could be another 'DO-ER' than YOU-YOURSELF, the SELF. Not all things *have* been done, but of the ones which *have* been done, YOU have 'DONE' them. (Do *you* feel tired?)

From the Beginningless-Beginning, the Infinite Process has been 'In Process'. The Great Breath (though ESSENTIALLY Illusory) has Oscillated. Never was the time when the Breathing was not 'In Process'. Never was the time that the Breathing has not been, ESSENTIALLY, Illusory though Actual.

☐ There has been "plenty of *time*" in which to 'do' things. Have an infinity of things been done? Definitely yes! Have *all infinities* of things been done? Definitely no! Surely countless things have been done, but there is always room for more, simply because of the inexhaustible NATURE of the FOUNT OF ALL POSSIBILITY, the PLENUM. These thoughts alone provide material for years of debate.

Is there "nothing new under the Sun"? Yes and no. Every Cosmos is 'New' and yet the Possibility that It represents has dwelled forever within the FOUNT OF ALL POSSIBILITY. Every apparent *newness* emerging into Cosmic Actuality never ever *began* because it had always been *inherent* 'within' the INFINITESSENCE. Thus, no 'newness' is REALLY new, but because it is emerging *as* or *into* Cosmos for the first and only time, it *is seemingly* new. The ABSOLUTE SELF will ever be the MYSTERY of all possible Mysteries, and must, necessarily confound us—at those illusory moments when we are not *being* IT—apparently.

The Problem of Whether the Principle of Unrepeatability in Cosmos is Inviolable

The Philosopher Herbert Spencer evolved the notion of "Eternal Recurrence" to explain that when all possibilities in-Universe had been played through, a Great Repetition of the Universal Process would ensue, again and again. If this is the correct interpretation of his thought, then, something seems incorrect about this notion, due to the ABSOLUTELY INFINITE NATURE of the FOUNT OF ALL POSSIBILITY. And yet, Spencer's idea cannot be summarily dismissed.

Are we sure that there can be no *exact repetition* in Cosmos, or in an infinite sequence of Cosmoses? After all, it *is* a possibility! INFINITUDE is so 'CONTRARY' that one can only hope IT is SELF-REGULATED, and has gained a 'bit of maturity' from 'being around forever'! (Excuse the irreverence; desperation speaks.)

We all acknowledge *approximate repetitions* otherwise there would be no stability of relationship or stability of form (which *is* relationship), but *exact* repetition is another matter. It seems that the entire Infinite History of the SELF is one of *repetition of essence*, or more accurately *relentless, seamless continuity of essence*. The History of Form, however, (at least in-Cosmos) seems to respond to a Spiral Motion which prevents even apparent repetitions from being *exactly* so. For instance, the orbit of a planet may seem to repeat itself more or less exactly, but the motion of the Solar System through space changes the orbit of the planet into a spiral. This means that the planet has changed its *location* with respect to all other factors in Cosmos, and especially with respect to any former supposedly identical position.

The idea of Repeatability in Form seems to set itself directly opposed to what is being hypothesized as the *primary dynamic* of the ALL-SELF. When the ALL-SELF is in ITS ALL-IN-ALLNESS (which, REALLY, IT *always* IS), ITS 'PURPOSE' can reasonably be supposed to be—to BE FOREVER ITSELF *maximally, fully, completely, infinitely.* When we consider the ALL-SELF as the FOUNT OF ALL POSSIBILITY, we can say that ITS PURPOSE is to express in Cosmic Form ITS INEXHAUSTIBLE INFINITUDE. Why repeat form exactly? For what purpose? Especially when, *forever*, there will always be *infinitudinously more.*

☐ In one way there is something repetitive about the Great Breath. The Inbreath and the Outbreath have been repeating forever. Have these Great Breaths, though, been exactly the same? Are our breaths, though similar, exactly the same? Again, we might have an Approximation of Repetition on the grandest possible scale.

The dynamics of the Three Aspects of Motion seem connected with the 'BEHAVIOR' of the ALL-SELF. The Rotary Motion of Repetition relates to the apparently ever-regular Great Breath. Spiral Cyclic Motion relates to the *downward spiraling* Involution and *upward spiraling* Evolution of an infinite series of Cosmoses (a *series* which almost certainly is not *evolutionary).* [See the Problem of the Infinitely Evolved Cosmos.]

Admittedly, it would be heartening to apply the motto "Onward and ever upward" to the infinite series of Cosmoses, but for various reasons, to do so leads to impossibilities and absurdities. The first Aspect of Motion is called "Driving forward through Space" and can be seen to relate to the SELF's 'PASSION FOR THE ACTUALIZATION OF THE NEW' as the FOUNT OF ALL POSSIBILITY ever *actualizes* those possibilities which have never before, in Infinite History, been actualized.

☐ The one arena in which *repeatability* in Cosmos seems possible (but far from certain) is in relation to the ultimate particle/event. Each ultimate particle/event is, theoretically, further irreducible *in-Cosmos.* It continues to be itself from ultimate moment to ultimate moment, from Cosmo-Objective Now to Cosmo-Objective Now, and it seems to be the only instance of *exact repeatability* in form.

Perhaps even here, however, there is only a *seeming* identity between ultimate particle/events. Certainly in scope, impartiteness, and duration, they would have to be identical, or they could not be considered an ultimate unit in-Cosmos *spatially* and *temporally.*

We have to realize that even though certain items-in-Cosmos may repeatedly occupy the *same* 'position' with respect to each other, it is *not* the same position with respect to all other variables in Cosmos, and so the repetition of position only applies to the small scale and not the large (i.e., the entire Cosmic Configuration). If 'upon' an ultimate moment, only one item-in-Cosmos changes position, and all others do not, repeatability has been shattered, for the entire Configuration has been altered. The Cosmic Configuration of the succeeding moment has varied from the Cosmic Configuration of the preceding moment. As such minute change is always happening, we can see why the Cosmic Configuration of each ultimate moment is different from the one preceding and succeeding it. The *temporally enacted* Original Intent would require it.

☐ So while small scale repetitions seem possible, perhaps, for a time, there is no repetition in the Configuration of the Whole, or the Cosmic dynamic of "Driving Forward through Space" would be violated. The Universe is, after all, 'on Its Way', and all the various apparent 'parts' have to 'keep moving' to make sure It 'gets there'! Hence the Principle of Unrepeatability in Form seems to hold up from a Cosmic Perspective.

The Problem of the Beauty of the World —the Beauty of Maya

Is the World of Becoming *beautiful* although infinitely removed (apparently) from the WORLD OF BEING. Poets, artists, naturalists, lovers, to name a few, are enraptured by the Beauty of the World. Master Morya tells us that "With Beauty thou hast the Light." As well, He reminds us to think and say, "How beautiful is everything!" And yet certain Mystics and Sages say that the Beauty of the World is as nothing beside the 'BEAUTY' of BRAHMAN.

To speak of Beauty or Ugliness requires comparison. If the comparing consciousness has no experience of the INFINITE SELF, is that consciousness qualified to speak about Beauty? Let us grant for a moment that the experience of the BRAHMAN may be so transcendentally extraordinary as to dwarf all other experiences. This is the testimony of thousands among the enlightened (i.e., the *humanly* enlightened). Still, there are other human beings who, equally enlightened it would seem, see Great Beauty in this humble little World—a World which is, in one respect, so very 'distant' from its SOURCE. Must one deny the manifold beauties of the World which, before knowing the BRAHMAN, one loved so thoroughly? Do they turn to ugliness when once the BRAHMAN is known?

8 Am inclined to think that the Beauty of the World *is,* itself, at least the tiny reflection of the ABSOLUTE 'BEAUTY' OF BRAHMAN. Since everything in Cosmos is ESSENTIALLY indivisible, the very WHOLENESS of BRAHMAN must be found at each point (whatever, exactly, we may mean by a 'point'), as, indeed, BRAHMAN and Samsara are ONE.

☐ To reject the Beauty of the World as minimal or un-REAL is, again, to create a false duality where there is, REALLY, only an indivisible monality. The Beauty of the World, then, is not only a *reflection* of the BEAUTY OF BRAHMAN, but it *is* that very 'BEAUTY' ITSELF.

Everywhere the PRESENCE is REALLY *present.* Every experience of deep Beauty is REALLY an experience of the BRAHMAN. In most experiences of Beauty the sense of UTTER WHOLENESS may be missing, but there are aesthetic experiences which do, in fact, evoke the WHOLENESS, the SUBSTRATUM. One must rejoice in Beauty and Relationship (for what is Beauty but a specially harmonized Pattern of Relationship), just as one rejoices in the VOID. BRAHMAN and Samsara are ONE. BRAHMAN and Beauty are two sides of a ONENESS WHICH has no sides.

The Problem of Illusion

The word 'illusion' is so frequently used in Advaita Vedantin Philosophy that we must be sure we truly understand it. The World of Becoming is said to be an Illusion. This statement is not made because the presentations of the World of Becoming do not *seem* real enough to the registering consciousness, but rather, because, no matter how *actual* or how *factual* such presentations may be, they are *not* ESSENTIALLY what they seem. Instead of being what they *seem* they are impermanent, evanescent and, most importantly, further *reducible* into something which is utterly *different* from what they seem—namely the INFINITE HOMOGENEOUS SUBSTRATUM.

☐ There is no apparent likeness between an illusion's presentational form, and the INFINITE HOMOGENEOUS SUBSTRATUM. Though the two are ESSENTIALLY the same, the illusion is reducible into the SUBSTRATUM, but the SUBSTRATUM is in no way reducible into the illusion.

Radical Infinitism (Radical Non-Dualism) is interested in the ROOT of things. Things-as-*things* come and go, but the ROOT 'ABIDES' *forever*. In this philosophy, unless a something *abides* forever *just as it is* (irreducibly), it is an illusion. Since no 'things' abide forever as they are, all 'things' are illusions. The only 'something' that *does* abide forever just as IT IS, is the INFINITE HOMOGENEOUS SUBSTRATUM, the ABSOLUTE, the SELF.

The Radical Infinitist will not deny the 'factuactuality' which is being presented to consciousness, but will realize that consciousness, itself, is *also* an illusion, for it loses its particularity, its specificity (in fact, its very *existence*) during the Universal Pralaya when the SELF is not 'ACTING OUT' possibilities 'resident' 'within' ITSELF as the FOUNT OF ALL POSSIBILITY. No, presentations *cannot* be denied and must be 'handled' (a reference to the fundamental apparent materiality of Cosmos).

The Problem of the Intricate Inter-Relations of the Cosmic 'First Family' & The Problem of the Subjectivity of the Higher Formative Forces

Universal Prakriti is the great Universal Object. The Universal Logos is the great Universal Subject. From the 'INFINISPECTIVE' of the INFINITE SELF, *everything* in Cosmos is Objective (and un-REAL), including the Highest Logoi and Creators. From the intra-Cosmic Perspective, however, are these Logoi and Creators (all of them, Essentially, Self-Reflected Emanations of the Universal Logos) to be considered as *objects* or *subjects?* Do these high Beings have a *prakritic envelope?* If They have a prakritic envelope, is it a *Fohatically-prepared* prakritic envelope?

As abstract as these questions seem, their solution is vital for Universal Cosmogony—the Birth of the Gods. It would seem that the only BEING that cannot have a prakritic envelope is the ABSOLUTE ITSELF, for an envelope suggests *duality,* and there is no duality 'within' THAT (except in the *contradictory 'part'* of THAT). Once, however, the Point of Super-Universal Potential is 'RADIATED' then even PARABRAHMAN-as-Infinified Point/Infinite Subject has a Prakriti-Envelope (REALLY a 'Mulaprakritic-Envelope'), for PARABRAHMAN-as-Mulaprakriti as the Infinite Object 'comes to birth' with the 'Birth of the Pre-Cosmic Consciousness' of the Infinite Subject.

We often call the ALL-SELF the INFINITE SUBJECTIVITY, which relieves IT of the necessity of having an Infinite Object of which to be 'CONSCIOUS'. When, however, PARABRAHMAN 'EXTENDS' ITSELF as the 'RAY' that 'FLASHES FORTH', IT instantly

'BECOMES' (in 'part!') the Pre-Cosmic Infinite *Subject* (*not SUBJECTIVITY*) and, therefore, instantly *needs* an Infinite Object, which Mulaprakriti (the Infinite Subject 'Self-Seen' as Infinite Object) Becomes—instantly. Infinite Super-Cosmic Consciousness, of course, if the Infinite Intermediary in this 'Becoming'.

Therefore, even before the SELF-as-Infinified Point has pursued the Finitizing Sequence of SELF-as-Infinified Point-as-Condensing Point-as-Condensed Point-as-Universal Logos—i.e., even before the Universe is *circumscribed*—there is a kind of Self-'Seen' Mulaprakritic Envelope for the Pre-Cosmic Infinite Subject, newly released from TOTAL INFINITIZED SELF-PREOCCUPATION 'within' the INFINITESSENCE. Thus Super-Cosmos is 'Mulaprakritic'—*prakritic* in the highest possible degree—though *not* 'particulate'.

Once we are in-Cosmos, in-Universe, and that Being called the Universal Logos is defined, It has the potential of a definite Prakritic Envelope, for It is in a Subject/Object Relationship with that 'portion' of Mulaprakriti that we will call Cosmic Prakriti.

- The Infinite Subject 'Sees' Itself as *Mulaprakriti*.
- The Universal Logos (now Pre-Cosmically 'Individualized') 'Sees' Itself as *Cosmic Prakriti*.

The Matter which Prakriti Is, is not an external 'thing' but an internal Self-Perception or Self-Reflection of a Self-'Seeing' Subject—whether the Infinite Subject, the De-Infinitizing Subject (Condensing Point), the Universal Logos-as-Subject, the Universal Son-as-Subject, or Fohat-as-Subject, etc., Prakriti of any kind is simply Self-Image!

There are two stages of engagement when a *subject* 'faces' an *object*. The first stage is one of *vision*. The subject and the object register each other but have not interpenetrated each other. This stage is highly dualistic. Then comes the stage of *mutual penetration* in which the subject infuses the object and the object receives the subject and responds. This phase of *engagement* (when intra-Cosmic) is what we might call Spirit/Matter. It is a phase of *interplay*. We might call it 'Subjectobjectivity'. The second phase can follow the first with virtual instantaneity. Time intervals in Super-Cosmos must remain entirely speculative.

It is important to realize that, all this while, Cosmic Prakriti is *as yet* undifferentiated by Cosmic Fohat. Both of the following:

- the dualistic phase of 'Spirit facing Matter', or the Universal Subject facing the Universal Object (Cosmic Prakriti), and
- the *engagement* phase of Spirit/Matter, or the Universal Subject *infusing* the Universal Object,

take place when Cosmic Prakriti is *still undifferentiated* by Fohat. The Universal Logos (Subject) 'Sees' Itself *as* the Universal Object, and the Universal Object 'Knows' that It is being 'Seen' by the Subjective 'Seer' in Whom (*reflectively*) It finds Its Subjective Identity. There is, hypothetically, a constant exchange of this Self-Reflexive Mutual Perception, and a true Stage of Spirit/Matter-in-Divine Interplay exists. The 'Seer' 'Sees' Himself as One Who 'Sees' Herself being 'Seen'. The moment the Subject 'Sees' Itself as the Object, It endows the Object with the capacity to 'See' the Subject as an Object. If the Father 'Sees' Himself as the Mother, the Mother must be able to 'See' the Father. We can call this 'Subjectobject Visual Reflexivity'.

From the Theosophical Perspective, the two phases of Cosmic Spirit/Matter Engagement discussed thus far ('Sight' and 'Interplay'), correspond to the two symbols, respectively:

- the Point within the Circle
- the Horizontal Diameter of the Circle

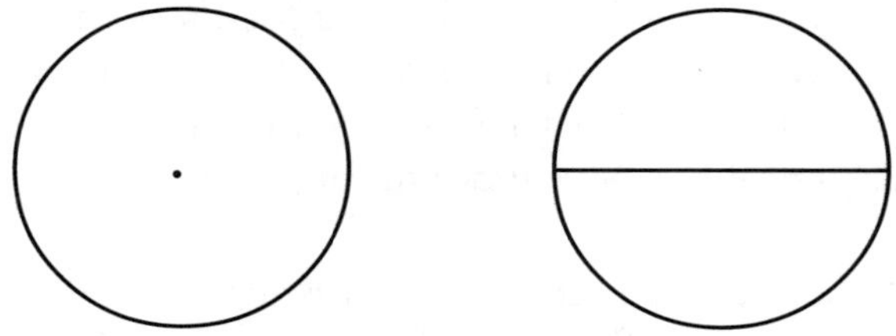

In the Third Phase of Engagement, intra-Cosmic Fohat (as there is what might be called *Pre*-Cosmic Fohat) is sent forth (not from the Universal Logos alone in order to 'Create' a Son), but from the Universal Son of the Universal Logos (Who has been born of the Spirit/Matter) and begins (as Fohat/Brahma/Holy Spirit) to condition Cosmic Prakriti, through the process of differentiation.

This differentiation is probably by no means as complete as it will later be (for at this point we are operating upon the very highest Super-Cosmic Planes—higher far than our Cosmic Planes, and the Act of Differentiation *descends),* but the conditioning prepares the Field of Cosmic Prakriti for the Descent-via-'Radiation' of the *supervisory* Cosmic Logoic Emanations (the Numbers that are Highest Archetypal Beings in Cosmos) into the World of Fabrication, the Fohatically-Manipulated World. The World of Being 'hovers over' the World of Fabrication, and different 'Ray'-Beings within the World of Being 'superintend' and set the archetypal model for different strata of the World of Fabrication. The matter is so extraordinarily complex to the human mind, that only the principle can be enunciated as a hypothesis. Enfolded 'Rays' of the ABSOLUTE-as-'Rays' of the One are 'Archetypal Modelers', and Fohat, and his Host, 'enact' in 'Mosaic particulation' the Models set forth.

Already to this point, and in a sense, the Numbers One, Two, and Three have emerged. (The manner of counting, however, seems to change according to the perspective adopted.)

- The Number One is always the Universal Logos.
- The Third is Cosmic Prakriti Who is Really (from the most usual perspective),the Second, just as,
- the Son is the Second who is (as usually considered) the Third (in order).

This leaves Fohat out of the enumeration, and if He were to be introduced, the numbering would change.

Fohat is always strange and unpredictable. There can be no *action* whatsoever without Fohat. He is elusive and Protean in Mode. Therefore, from the Moment the Cosmic Father 'Sees' Himself as the Cosmic Mother, the very 'Act of Seeing' *Is* a Mode of Intra-Cosmic Fohat. In this respect, Fohat *Is* Intra-Cosmic Maya. Fohat is always the 'Act' of the

Subject, even as SUPER-Cosmic 'FOHAT' or the 'RAY' of the ABSOLUTE, is the 'ACT' of THAT. On these levels, to *see* is to *act.*

Numbering the Cosmic First Family with Fohat participating, we find:

- the Father as Number One,
- Fohat (as the Second in order of appearance), as the Act-of-Sight making possible the Father's 'Self-Sight' revealing the Father as
- the Mother (who is, then, Third in order of appearance), and
- finally, the Son (Fourth in order of appearance) born of the Father's *specialized* and *reduced* 'Self-Sight'—a 'Self Sight' reflecting the 'Seen' 'Self-Reduction' as an Object within that Greater Object, the Mother.

The Father 'Sees' the Son as His Own Reduced Selfhood, reflected within Himself-as-Mother. Thus, the Son is born through an Act-of-Vision, which Act is again necessarily Fohatic. Fohat serves whoever acts the role of Subject—whether Father, Son, or lesser Sons, or, perhaps, from a more unusual perspective, even the Mother. Fohat, thus, and strangely, is the 'Relation Between'.

Immediately after the Cosmic First Family is complete, further Emanations originating from the Son begin the creation the Supernal Tetraktys of (Pattern Holders), the Three and the Seven Who dwell in the Company of the Son. The Members of the Supernal Tetraktys are Mind-Born Sons of the Universal Son. The Universal Son 'Impels' and then 'Supervises' the Emanatory Process which produces these Sons.

Let us look at the above thoughts from a slightly different perspective. There is a moment in the Cosmic Process when the Universal Logos Exists as the Bounded Point. He is the Condensed Point-Become-Universal Logos. This phase of the Process correlates to Number One. Number One 'Faces' Cosmic Prakriti Who at that point (eliminating Fohat) can be considered Number Two, but a very *passive* Number Two, waiting to be *acted upon* rather than Act. (Remember the 'Facing' *and* the 'Interplay' *is* Fohat) Let us remember that all the Logoi are Actors, and, thus far in the Process, Cosmic Prakriti is not an Actor (except to 'See' the One Who will 'Act' upon Her).

With the interfusing interplay of the Universal Logos with Cosmic Prakriti, we have the generation of interactive Spirit/Matter which means that Cosmic Prakriti (the Mother) has been *enlivened* by the Father and can 'bring forth objects', among which, the Son will be the first. Father goes forth into Mother through the Agency of one of the permutations of Fohat. In this case, Fohat can be considered the 'Act-of-Seeing', which 'Act' *is* the Father *in* the Mother, (which 'Act' *is* the Maya of Self-Reflective Consciousness) 'Creating' interplaying Spirit/Matter (i.e., 'Subjectobjectivity').

Meanwhile, the Father though engaged with the Mother through the Agency of a certain permutation of Fohat, *remains also disengaged* from the Mother. Emanation Theory reveals how the Emanator always goes forth into 'lessness' (an apparent Self-Lessening) through its emanations but, simultaneously, remains fully in tact upon the plane from which the emanation originated. The Father goes forth into engagement with His Self-Image (Mother/Matter), but remains detached as well upon a higher level as the Subjective Father Entire. In other words, there is:

1. the Father 'Seeing' Himself as the Mother, but there remains
2. the Father 'Being' Himself as Himself (without 'Seeing' Himself).

Thus, the Unmanifest Logos 'Remains' 'above' His 'Sight' of Himself as the Mother (i.e., 'above' the State of Spirit-Matter). There are many profound implications to the thought that every E/entity may have the option to *be itself* as well as *see itself.*

☐ 'Light' is absent from the stage of *being* and makes possible the stage of *seeing.* From this consideration, Fohat (in all Its many permutations) is a kind of *light.*

Out of this *engagement,* or 'blending' of Spirit and Matter (of Father Blending with His Own Self-Objectification through 'Self-Sight', and then *specifying* His Self-'Sight' within the Father/Mother Field) emerges the Cosmic Son (the Pattern Holder, the Form Builder). The Cosmic Son is Vishnu. He is to *act upon* Matter (now *enlivened* or *engaged* by the Father, the Universal Logos).

Cosmic Prakriti is now enlivened, 'Father-Seen' Matter, and awaits the coming of Her Son Who (in cooperation with His Emanations as Agents of Himself, i.e., the Supernal Tetraktys) will *father upon Her* (through Emanative Self-Perception) the immense variety of Archetypes for the Coming Universe all of which lesser Archetypes will be focussed or *'anchored'* upon the Archetypal Planes of Cosmos (and will be 'enfolded', 'implicate' subdivisions of the major Numerical Archetypes). Mother/Prakriti is the 'stratified Images of a Hierarchy of Emanatory-Subjects.

Fohat (emerging as the Holy Ghost) and also, in a sense, *anchored* 'Monadically within the World of Being' (for Fohat, too, is the One Monad), will have to Build the Approximations (to all these Son-'Held' Archetypes) within the lower worlds over which He Presides—i.e., the World of Effects, the World of Approximation, World of Fabrication, the Mosaic World (four names for the same World composed of many planes constituted of differentiated prakriti).

Fohat always needs a Template to follow. That Template came with the Father ('given' to Him via the 'RAY' which He ESSENTIALLY Is), was bestowed (via the Emanation Process) by the Father upon the Son, who then articulated it (either directly or, eventually, in the "Fullness of Time") in great detail (via the Emanational Chain which He *heads)* by impulsing multiple, successive, increasingly reduced emanations which could be called emanated 'sons of the Son'. These are Beings/Emanations (successively emerging—always a lesser from a greater) Who sustain certain portions of the Fixed Design which the Universal Son sustains simultaneously and in entirety.

With either a portion or all of the Archetypal World in place, created by the Self-Visualization and Self-Reflection of the Son and his successively Emanated Host, the time has come to *engage* the Mother again.

Let us review:

1. The Father first engages the Mother by 'Seeing' Himself *as* the Mother. This is the 'Creation-through-Sight' of specifically *Cosmic* Prakriti—a Prakriti which is maximally undifferentiated in terms of that particular Cosmos but which is, actually, *conditioned* Prakriti and not infinitely *dense* Prakriti, for it is 'Seen' by a Subject (the Universal Logos) Who is no longer Infinite (and all prakriti *reflects its Seer).*

2. Then, the Father 'Sees' Himself as Mother again, but this time, because He 'Looks' deeply within the Father/Mother Field, a Reduced Replica of Himself appears as an Object within the Mother (His Self-Reflection). Within His Greater Self-Reflection (the Mother Field) a Lesser Self-Reflection (the Son) appears.

3. Then the Son (now an Object-becoming-a-Subject), begins *His* Emanative Process, creating the Archetypal World.

When, however, the Son 'Sees Himself, He does not 'See' the Mother as the *Father* 'Sees' the Mother. Instead He 'Sees' His Own Self-Reflection as a kind of 'Reduced' Mother (a further lessening of the *density* of Cosmic Prakriti—a rendering more tenuous of the Reflected Image of the 'Seeing Subject'). If the Son (through Identification) could transpose His Consciousness into the Consciousness of the Father, He would 'See' the Mother as the Father 'Sees' the Mother, for the Son (through Identification) would *be* the Father 'Seeing' Himself. The Son, however, is a 'Father-Seen-Object-Becoming-Subject, and is thus *lesser* than the Father, and thus 'Sees' a 'Lesser Mother'. The Mother the Son 'Sees' is naturally related to the Mother the Father 'Sees' in the same way that the Father-Become-the Son is related to the Father-Remained-the Father.

There are two pairs of 'Cosmic Consorts', the First Pair of greater scope and depth than the Second Pair. The Son 'Creates' through Self-Reflection (just as the Father had done) His Own Son, and that Son (in the same way) His Son, and so forth. Each time the Emanator 'Sees' a lesser Self-Reflection and thus, in a way, a *Lesser* Mother than the Emanator *before* Him. This is the Work of the Second Aspect of Divinity, and it may well complete itself before the Third Aspect under Fohat begins to differentiate Cosmic Prakriti to 'Approximate' what is being 'Emanatorily Created' within the World of Being. Fohat, Itself, must appear in yet a New Mode, becoming the Holy Spirit, born of the Son, and being the means by which the Divine Pattern held by the Son and His Host becomes Fabricated. Objectification precedes Fabrication. Fabrication by Fohat and His Forces is based upon what the Universal Son has Objectified as the Divine Pattern.

☐ The origin of the different Modes of Fohat is elusive. Fohat is as if constantly changing form or Mode, serving this or that Director. In this discussion, however, we are focusing upon a point in the Cosmic Process when Fohat is about to become *Son*-Directed and Guided.

We have arrived, then, at the point where a measure of the Archetypal World (perhaps *all* of the Archetypal World) is in place. All the Beings in the World of Archetypes ('Rays' of the ABSOLUTE and authentic E/entities) are structurally related at a multitude of 'levels' within the Archetypal World (the World of Being) *as* Emanations of the Son. Now, that which is to be built within the Worlds of Approximation must reflect the Pattern that the Son and His Emanated Host are *holding* and *sustaining*. In a way, the Emanated Host of the Son *Is* the Pattern to be Approximated.

Thus it is time for Vishnu the Son to engage the Mother again, or rather to 'Guide' a new form of engagement. Vishnu, does not now engage the Mother *directly*, but instead sends forth an Aspect of Himself as Brahma (Who is, again, Fohat in a new Mode {we might call it the Fourth Mode of Fohat}. Thus, Fohat is sent to be Active in the Task of differentiating Cosmic Prakriti. From Vishnu (now the focal Creative Subject) comes Brahma, and Brahma is at once the Holy Spirit and Fohat in the Mode of Fohat (Fourth Mode) known best to us from *The Secret Doctrine*. Fohat now appears as that ever-active Cosmic Character we think we know Him to be.

In the Theosophical Model of Creation, the Third Aspect always *precedes* the Second and there is some truth in this, but it must be seen in context. (The Model presently being used, though different, is not contradictory with the Theosophical Model, this Model simply begins earlier):

1. the Father Acts;
2. the Son Acts;
3. Fohat Acts;
4. the Son and His Host Act again, *with* Fohat and His Host;
5. the Father Acts.

It will be seen from this fivefold process that the usual Theosophical Presentation begins only at Stage Three, and thus it may, before due investigation, seem that the Third Aspect led by Fohat Acts before either the First or Second Aspects. Indeed the so-called Three "Outpourings" begin at Stage Three, but in this Model, the Outpourings are prepared by Creating within the Archetypal Worlds the Patterns upon which these Outpourings are based.

- Thus it becomes necessary for the Father to first *impulse*, *energize*, and *hold* the entire Process.
- Then it becomes necessary for the Son or Second Aspect to receive the Pattern and emanate at least *some* (perhaps, all) of the Holders of the Pattern. (There may be a phased-Emanation of the 'Holders' when the time is right for Their cooperation with the appropriate level of Fohatic Host.)
- Then, with the Pattern in Place (a Pattern that determines among other things the ways in which Undifferentiated Cosmic Prakriti is to be differentiated), Fohat and His Host can begin their Work. This is the "First Outpouring".
- Thus, with Archetypes in Place, Fohat/Brahma is sent forth, and Engages in Comic Love the Enlivened (Father-'Seen'/Son 'Seen') Mother, and, observing the Template 'Held' by the Son and His Host, Creates (through 'Particulated Self-Sight') all the differentiations, levels, dimensions of Matter possible within that particular Cosmos. (Or, perhaps, the prakriti for only *some* dimensions are fabricated.) Some thought is required, as we may be dealing with a well-timed, phased process, descending from more abstract levels of the World of Fabrication, eventually to the most concrete.

Now, an important and subtle question arises, Does Fohat and His Host Create all dimensions immediately one after the other? Or, is there a 'waiting period' until the appropriate Phase of the Design-at-the-Beginning (in Its Temporal Mode) has been 'reached' within the Cosmic Process? This question is the counterpart to the question of whether the entire Archetypal World is Emanated in one uninterrupted, sequential Process. If we think of the concept of Descent and Reascent, we may realize that there are times in Cosmos when matter of a certain kind simply does not exist, because it is either too early or too late in the Emanative Sequence.

We are used to seeing charts that display seven planes and seven subplanes (and perhaps ten). Is it possible that these charts show the Universe (or at least our small portion of it) as It is Now, but not as it always was or will be? Just as globes or entire chains disappear by absorption into their superior globes or chains, entire planes could do the same. If we can think this about the dynamic of *disappearance*, we can think likewise about *appearance*. Just as there are not *always* ten globes per chain, or ten chains per

Planetary Scheme, there may also be times (either early or late in the Cosmic Process) when there are not a full seven or ten planes.

The thoughts offered above would naturally touch upon *what* Fohat built and *when*. Could it be that there would be a gradual descent in the Fohatic building process just as there would be (within the Son Aspect) a gradual Emanative 'Creation' of the various 'Rays' which were to go forth (in archetypally-supervisory Consciousness) into that which Fohat and His Host would fabricate? It could be that the fabricating and 'going forth' begins on planes that are relatively very abstract and *seemingly* formless. The best efforts of the Second and Third Aspects would be required to complete the building processes on such an abstract plane, *before* Fohat 'Created'/'Became' the building materials for the next 'lower' plane, or before the lesser' 'Ray' Entities (who would involve themselves in progressively greater Fohatic differentiation) would be emanated.

The possibility being presented here is one of 'Phased Emanation' and 'Phased Fabrication'. The idea here, is that there would be much to do and complete 'above' before it was possible to take the next 'step' leading 'below'. Whichever way the truth may be found—whether in:

1. an immediate sequential Process, first by the Son and His Host and then by Fohat and His Host, or in
2. a gradual step-by-step, plane-by-plane Process, in which the Progressive Emanations of the Son alternated with progressively complex Fohatic differentiations (the lower planes of the World of Fabrication being reached only gradually)

—one thing is certain: the Universal Son 'Holds' the Fixed Design in Its Entirety (whatever His Emanatory Schedule), and Fohat attends to that Template (en large and in detail) in order to Fabricate it Perfectly—in the *fullness* of Cosmic Time.

Let us Return to the Cosmic First Family after that speculation upon the timing of the various phases of the Cosmic Process. Fohat is about to engage in the differentiation of Cosmic Prakriti. Fohat, representing the Third Aspect is ever the *Forerunner* of the *prepared* Second Aspect. Just as Vishnu the Son is Cosmo-Structurally, Number Two, so is Fohat, now His (the Son's) Emissary, Cosmo-Structurally, Number Three, even though Fohat was there at the Beginning with the Father, and is now (in a new form) emerging Fourth in order of appearance: Father, Mother, Son, Fohat. Together, they (the Son and Fohat) will begin to act upon the Mother (Who can now be considered Second in order, and Cosmo-Structurally, the Fourth) in a new way. The Son now stands, Cosmo-Structurally, for Number Two, and Fohat, Who emerged fourth in order (at least Fohat as we *usually* know Him, for He is a very *protean* Character) takes the place (Cosmo-Structurally) of Number Three. One can see how and why these Numbers keep changing. The order of appearance does not always reveal the Cosmo-Structural/Aspectual Role to be Played.

Thus, we now have the *Upholding* Father, the *Pattern Building* and *'Holding'* Son, and the *Active, Conforming* Fohat as-Holy Spirit, engaging in Their various ways (through vitalization, by proxy, and directly) the multiple Self-Reflections (Self-Images) that *are* the many faces of Mother. Due to the interplay of Fourth Mode Fohat with the Mother, the Mother will become multiple indeed, and will emerge as the multiple differentiations of Cosmic Prakriti (Self-Enumerated Fohat Self-Reflected in 'particulation').

There are really Three *Cosmic* Pregnancies involved here, and this fact is the source of much confusion in Esoteric Cosmogony—the Birth of the Gods:

1. In the First, the Father in the Stage of Spirit/Matter (probably through the agency of the Third Mode of Fohat/Maya/Consciousness) impregnates Cosmic Prakriti and produces the Son.
2. The Son Who is usually considered the Third, is now the Second, because the Son now becomes the Husband of His Mother. This is the Second Engagement in which the Mother (Objectivity) participates, resulting in the Second Pregnancy.
3. She, the Mother, brings forth many sons of Her Son, the Emanations of the Universal Son and His emanations (timed however they may be). Certainly the Supernal Tetraktys must emerge early for It is Essential in sustaining the World of Being.
4. There is a third Pregnancy, however, and this is most important. In this Third, the Son does not directly engage the Mother, but sends forth and *guides* an Emanation of Itself as the Holy Spirit/Brahma/Fohat. (Mysterious Fohat is always That which *goes forth* from a superior Source to engage an Object in an inferior position.)

In a mysterious way there are, in this system of thought, *four* Modes or appearances of Fohat and yet *One Fohat:*

1. the 'RAY' OF THE ABSOLUTE was SUPER-Cosmic 'FOHAT' or 'MAYA';
2. Fohat as the Pre-Cosmic agent of engagement between the Infinite Subject and the Infinite Object. This was Pre-Cosmic, Infinite Fohat as Infinite Maya/Consciousness. This engagement produced the possibility of a Universe;
3. the Agent by which the intra-Cosmic Father went forth (Self-Perceptually) unto the intra-Cosmic Mother (Cosmic Prakriti) to produce *first*, interactive Spirit/Matter, and *secondly*, the Son—that Agent, too, was Fohat in its First Cosmic Mode (Really, Its *Third* Mode in the Fohatic Sequence); and
4. then, the Emissary of the Son re-engaging with the Father-'Seen', Son-'Seen', Mother (Who has already given birth to the Son of the Father, and to a number of Archetypal 'Number-Sons') in order to produce *differentiated* Mother (*differentiated* Prakriti, the birth of *differentiation* in Cosmos)—that too is Fohat.

Four Fohats in all. REALLY and Really, we find that Fohat-as-'FOHAT' is, at 'ROOT' the instantaneous 'MAYAVIC' 'ACTION' of THAT, and becomes, in Pre-Cosmos and Cosmos, the One Agency by which the Greater impresses the Lesser, by which the Subject impresses the Object. Fohat, too, is the 'Creator' (by 'Sight'/and Noumenal Light) of the Object.

For further consideration, when the stage of Spirit/Matter is reached, Cosmic Prakriti no longer remains exactly as It was *before* the interfusing. It is vitalized or enlivened by the Father (the Father *actively* 'Seeing' and 'Appreciating' the Image of Himself {the Mother} which He 'Sees'). There is not yet a pervasive *differentiation* of Cosmic Prakriti (for the Father 'Sees' a *homogeneous* not *heterogeneous* Vision of Himself). It could be said that the Universal Logos, now infusing Cosmic Prakriti, and forming Spirit/Matter is *poised to inspire differentiation,* but first the Birth of the Son and His Emanated Company is required. The Birth of the Son and His Host (which establishes the major outlines of the Archetypal World) allows the sending forth of Fohat to *differentiate* the Enlivened Cosmic Mother, for Fohat (in its Fourth Mode) Is simply the Son in another form. Fohat, in this Mode, Is, as it were, 'Mind Born' from the Son (after the Son has used an earlier Mode of Fohat {part of Mode Three} to help Self-Reflect the Company of the Son).

So it is the Son Who sends Fohat forth on Its more familiar (to us) task of *differentiating the enlivened Mother*, but since the Son *Is* the Father/Mother, and since the Mother is Really the Father, and the Father is REALLY the ABSOLUTE, we can see that all this 'Cosmic First Family Creativity' is nothing but the 'PLAY' of the ABSOLUTE BEING with *ITSELF*, which interplay, 'FOHAT'/Fohat (in one or other of Its Four Modes) is sent forth to Create and Facilitate.

When Fohat (in Its most familiar Role) engages the Mother under the Directive of the Son, He is not just Acting upon absolutely undifferentiated Cosmic Prakriti in Its pure 'Father-Seen' State, but upon *prepared* Cosmic Prakriti that has been *conditioned* by the Father/Mother Interplay which results in the Birth of the Son. So Universal Fohat, now the Agent of the Second Aspect (the Son), Acts upon Matter which has been significantly *changed* by the Birth of the Cosmic Son and His Host (Matter that is thus, 'Son-Seen' Prakriti as well as 'Father-Seen' Prakriti) for there does remain the continued enlivening *presence* of the Father in the Mother.

☐ Fohat then begins Its process of Differentiating Spirit/Matter, and the appropriate Symbol is the Cross within the Circle (not yet turning). This Act of Differentiation might, in Theosophical Terms be called the First Outpouring. It is the Cosmic Son-as-Fohat Who does the differentiating. (The Cross *whirls* or *turns* first counterclockwise, then, clockwise as Universal Involution and Universal Evolution are pursued respectively.)

When the Fohatic Differentiation is complete (as 'planarly' deep as it may go), the time has come for the cooperation of Fohat and His Host with the Host of Emanated Archetypes, led by the Son. All the Emanations of the Son and His Host are Subjective Beings responsible for the *holding* and *sustaining* of the Divine Pattern, but now there is a timed 'Raying Forth' of the Archetypes into or 'over' the Prakritic Differentiation which Fohat and His Host have Prepared. The Archetypal Emanations nevertheless remain anchored upon the Archetypal Planes of Super-Cosmos within the World of Being. Fohat can build the materials which *are* the various planes without the strict supervision of the Son and His Host, but the building/fabricating of *forms* is another matter.

The order of such 'Raying Forth' is more than difficult to imagine. Modern Cosmology shows us that Great Structures in Cosmos are created *first*, and then, lesser structures *follow*. Therefore, Great Entities (emerging as 'Rays' of the ABSOLUTE) may engage the Products of Fohat (on a particular {abstract} Plane of focus) first, and apparently lesser E/entities (which are nonetheless the same, identical One Entity as the Greater Entities) may (equally, as 'Rays' of the ABSOLUTE) engage the Products of Fohat (on, perhaps, the same particular {abstract} Plane of focus) *later*.

Highly differentiated E/entities (the human Monad is one such) are probably not needed on the very highest planes of the World of Fabrication. Upon such levels, presumably, only greater and more comprehensive Entities would be expressing, with lesser, 'enfolded' 'Rays' of the One Universal Logoic Ray, waiting until the Cosmic Process had reached a stage of greater differentiation and, hence, descent. Much depends on the degree of Fohatic differentiation to be found on the higher planes of the World of Fabrication.

Whatever the order by means of which the Son-born Host of greater Archetypes and Their enfolded lesser Archetypes engage the Fohatically Prepared Prakriti, there is but One Entity, One Monad, One Being doing the *engaging*. The One Cosmic Identity (which every apparently distinctive 'Ray' or 'Authentic Entity' *Is*) is involved *equally* in all Archetypal and Fohatic Processes.

☐ Bringing the idea home, 'we' (you and I) as we Essentially are, are the Ubiquitous Participator in all of these apparently differentiated Processes. 'We' are never anything but the One. As the One we pass through many Self-Reflective apparent differentiations, but our Oneness remains ever intact. Thus all 'Rays' are but *One* 'Ray' and all apparently differentiated Monads are but One Monad (the One Universal Life).

It should be said that the Son (or Second Aspect) rules over the birth of forms, but not over the *differentiation* of Matter. The Second Aspect is more the Architect, and Fohat, more the Active Building Agent. Fohat certainly needs 'guidance' (even in its building of the elementary materials of Cosmos). This Supervision, the Son provides without 'descending' via Its E/emanations to the degree It will later. Later Fohat assists the Son directly, and the Son and His Host are even more 'projectedly' active in the Guiding Process. Fohat, however, is not principally responsible for the Emanation and 'Holding' of the Forms. This is the Province of the Son.

The various differentiations of Cosmic Prakriti *could* be considered *forms* because they are geometric aggregates of ultimate particle/events, and the Son rules Divine Geometrization. In Theosophy, however, only *aggregates* of prepared *materials* (differentiated matter) are considered to be true forms, no matter how aggregated those prepared materials, themselves, may be.

The Problem of Whether the Universe is Infinite or Finite

Let us begin by stating that an infinite Universe could not be a *periodical* Universe if such an infinite Universe were organized both *hierarchically* and *involutionarily/evolutionarily* as, in every respect, *our* Universe is seen to be by those who study and think deeply.

A hierarchically organized Universe (unless it is utterly static) demands the Cosmic Processes in Time we call Involution and Evolution, making possible the descent and ascent of life units through the various stratified dimensions of Cosmos. A *hierarchically organized* Universe which did *not* contain these Processes-in-Time would 'freeze' all E/entities at a particular developmental level (some *infinitely* high and some *infinitely* low—

if the Universe were an *infinite* Universe, or some just *extremely* high and some just *extremely* low, if the Universe were a vast but *finite* Universe) and would not allow any developmental movement (which movement takes *Time).* Such a model is contrary to what is well known concerning the evolution of form and (apparently) of consciousness, and is absurd.

- ☐ So we come to the thought that a Universe that is hierarchically organized (whether It is *infinite* or *finite)* must contain the dynamics of Involution and Evolution in Time if it is, at least, to resemble, *our* Univese (as we generalize upon it from life upon our globe) and if Cosmic Process is not to seem absurd from our standpoint.

Thus, a hierarchically-organized Universe (if it is what we might call a 'normal' Universe) demands Cosmic Time Processes which allow for the descent and reascent of life units, or at least (emphasizing, for a moment, the side of the picture with which we are more familiar), *evolutionary ascent.* But, an *infinite* hierarchically organized Universe (if unfolding through the processes of Time {and how else can unfoldment occur?}) would take *forever* to generate Its entire infinite content, and *forever* to reach the ultimate goal of ascent ... otherwise, at *what* specifiable 'level' would there be found the *highest* point of achievement in an infinite hierarchically-organized Universe?

Even if the acts of evolutionary elevation and improvement proceeded at the speed of light or, even, at 'ever-infinitizing speed', such acts would still take forever to reach the 'infinitely high' (and 'un*definable*') goal of achievement. This fact alone *would negate the possibility of periodicity* within an *infinite,* hierarchically-organized, evolving Universe, for the Universal Synthesis necessary for Universal Reabsorption would lie "ever on ahead." In fact, an infinitude of vast Cosmic Structures or Life Forms could never even be *reached* in 'Time' by the returning evolutionary Life Wave, not to mention the absurdity, that if the Involutionary Process is *infinite* in duration, the Evolutionary Process could never even *begin,* since Evolution must *follow* Involution.

Since, in an infinite hierarchically-organized Universe, all involutionary and evolutionary processes would take forever (i.e., the Cosmic Out-Breath and the Cosmic In-Breath would take forever—unless such processes proceeded at infinite speed thus negating Time, (and, REALLY, negating their very existence, as no 'specifiable' can 'surivive' 'infinite speed' without becoming NOTHING!) which appears *not* to be the case. As well, the Law of Periodicity could *not* exist, for (from the evolutionary perspective) there would be no break in the infinitely-long Universal Process in order to allow the onset of a Universal Pralaya, which the Law of Periodicity demands.

Further, hypothesizing (against all reason) an infinite hierarchically-organized Universe, it is also *not* conceivable that there *ever has been* a break in the infinite Universal Process. Thus (according to such a hypothesis) there has been no opportunity for Universal Manvantara (also required by the Law of Periodicity) to *begin.*

Additionally, because the Process of Involution would take *forever* in an infinitely extensible hierarchically-organized Universe subject to the processes of Time, the Process of Evolution (dependent upon the *ending* of the Process of Involution) could never begin (which clearly it *has)* for the Universe would still and forever be pursuing an infinitely enduring Process of Involution. Even with this there is a problem, for *when* did such a Process of Involution *begin?* Because such a Universe would be *infinite* there could be no *beginning* as well as no *end.* Thus such a Universe (infinte and hierarchical) would have been *involving in form forever,* and, thus, by *now,* should be *infinitely involved* (whatever that means) Is it?

Thus, in such a Universe (an infinite, hierarchically-organized Universe) purposeful action leading to (or hoping to reach) a final goal would be impossible because no *final* goal could either exist or be reached *in Time,* (and all Cosmic Processes require Time). Further, no *starting point* for such a directional process could ever be found in an infinite, hierarchically organized Universe, because no starting point for such a Universe could ever be determined given the required infinite, on-goingness of the Universal Process.

- [ ] An *infinite* Universe, if It is *hierarchically organized* (which our Universe seems to be) cannot be *periodical.* The only possibility for the existence of an infinite Universe which *is* periodical is for the kind of infinite Universe in which *the entire infinitude of the Universe appeared suddenly, totally, and in complete infinite detail without any possibility of developmentalism* (and, perhaps, even without the possibility of 'movement' of any kind).

If any time-sequencing whatsoever is involved in the appearance of the contents of the Universe (as it would be if the Universe were *hierarchically organized* and hence emanationally {and gradually} developed) there would never be enough Time for Universal Unfoldment, because sequential developmental action through Time in an Infinite Universe would take *forever,* which would abrogate the possibility of periodicity. Since such a Universe would be *infinite* it would be (if it were also periodical) forced to disappear *before* it had even unfolded *involutionarily.* To further the absurdity, the disappearance of such an infinite *periodic* Universe would be just as instantaneous as Its appearance, and the disappearance would have to occur without any possibility of 'completion' of the Universal Process.

While the instantaneous appearing (and developmental 'frozenness') of an infinite (supposedly *periodical)* Universe is quite absurd (just as is Its, hypothetical, instantaneous disappearing), i.e., judging from everything we *know* about Creation and the developmental, evolutionary Universal Process which is everywhere to be seen, let us (in the face of the obvious facts) for argument's sake, grant for a moment that such an infinite, 'frozen', *periodical* Universe might *instantaneously* have appeared. Just *when* did such an infinite Universe (not requiring 'Time' for Its processes {which don't exist anyway}) suddenly appear full blown and complete in infinite detail? Certainly, no appropriate 'Beginning-time' which was not preceded by an early 'time' can possibly be found (due to the inescapable factor of *infinite regress* when considering linear Time), and thus there can be found no reasonable 'Time' for such a Beginning to have occurred.

Such a Universe could have begun *arbitrarily,* as the Christian Bible may have us believe. Being, however, the kind of Universe we are hypothesizing, infinite, 'frozen' and still periodical, 'God' could have 'Willed' Its appearance five minutes ago or a centillion years ago, and it would have made no difference—as the Universe is, 'frozen', non-developmental, and would not be any different a centillion 'years' ago that it was five minutes ago. (Further, 'within' such a 'frozen' full blown Universe, how could time be measured at all?—it would be as if time were 'standing still', just as in an ultimate moment.)

The question of *when* such a Universe 'appeared' or how long It lasted would, Really, be utterly meaningless, as, within It, *nothing* would be accomplished anyway. There is no 'Reason' for such a Universe to endure. Its endurance would be *purposeless,* which is, presumably, *not* the case with periodical phenomena, which are *periodical* in order to accomplish certain processess within a given period.

From all the above, one is forced, to conclude that such an infinite Universe, if it were periodical, would be *arbitrarily* and *purposelessly periodical* and could have no reasonable association with any Cosmic Law such as the Law of Periodicity. Again, such an infinite (arbitrarily) *periodical* Universe (were it at all possible) would necessarily be utterly static, or at least, infinitely repetitive, for if any kind of ongoing developmental or evolutional movement is admitted into It, such movement would take Infinite Time to unfold since the *spatial extent* of such a Universe would be infinite. This demand for Infinite Time in which to pursue development or evolution would militate against the possibility of periodicity.

So we see that it is impossible for the Universe to be simultaneously both infinite and *periodical.* Earlier, we established that it is impossible for the Universe to be both infinite and *hierarchically organized* (especially if that hierarchical organization is to serve any purpose, and not simply be a 'frozen pattern').

If it is supposed that the Universe is infinite and *non-periodical* (and, yet, inclusive of the idea of evolution), then the *beginning* of the evolutionary process must be located or designated. But because an infinite Universe (presumed in this argument as *non-periodical* and *forever existent*) can have no beginning, then, a beginning of the evolutionary process cannot be located or determined.

☐ If the beginning of the evolutionary or developmental process cannot be located, we can only conclude that such a process has been going on forever, and if it has been going on forever (even at a rate that is as conceivably slow as possible but definite and greater than zero) the entire Universe would be *infinitely developed,* or *infinitely evolved,* which, clearly, It is *not.*

Besides, could an infinitely evolved Universe have any finitude in It at all? And yet there is, for human beings, at least, blatantly, nothing but the registration of finitude, at least objectively. We might, then (almost in desperation, if we were determined to keep an *infinite* Universe) argue for an infinite *non-periodical* Universe which always was and always will be and which *has no evolutionary or developmental movement in it.*

This would mean that all evidences of improvement witnessed by man are illusory. Such a conclusion, again, contradicts all we know about improvement and elevation. If we presume that, in such an infinite non-periodical Universe, evolution and development, indeed, *do* exist, but only reach a certain level and then have to begin again, *what* is that level? And why a *certain* level? The very idea of a 'ceiling' on evolutionary development contradicts the concept of infinitude. Further, what an irrational Sisyphean Universe It would be—roll the stone *towards,* rather than *to* the top of the hill only to have it roll down upon you—forever!

If it be presumed (against reason) that the Universe develops or improves in the most microscopic definite increments conceivable, there is still a tremendous problem, for since such an infinite non-periodical Universe has, presumably, existed forever, then, even though progress has been incredibly slow (but never *infintely* slow—for that is meaningless), the current state of the Universe should manifest infinite development.

Granting, against all reason thus far marshaled, that there is an *infinite*, non-hierarchical, non-periodical Universe which always has been and always will be, and has no developmental or evolutionary process within it, we may ask further, Is such a Universe an *Object?* If it is not an Object, It is no different from the VOID (which is INFINITE SUBJECTIVITY) and which, therefore, *is no Universe at all!* So if a Universe is *not* an Object it cannot be a Universe. If it *is* an Object, It must have had a Cause and a Beginning (at least from all we know about every other object in our experience).

☐ But we have shown that the infinite Universe cannot be periodical, and so, must *always* have been. If the Universe always has been, It cannot have had a beginning, and if it is an Object, it is absurd to say that an Object has no beginning and no end—for this contradicts all we know of any object within our range of consciousness.

Thus far, then, we have been entertaining the absurdity of the Universe as a non-evolving Object which never began. Now what of the Cause of this Object? A Cause must be something other than the Object Itself. The 'something other' must *not* be an Object, because, if it were, it would be *part* of the Universe (for a Universe *contains* all objects) and therefore could not be the Cause, for an object cannot be the cause of itself (although, there IS, if the Ageless Wisdom is to be credited with Truth, 'THAT' which is not an Object *and* is the CAUSELESS CAUSE of ITSELF).

Further, even if, against all reason, we allowed the Cause to be an Object, we are confronted with the problem of *infinite regress*—what caused the Cause? We are faced with an infinite regression of 'Causal Objects' and, thus, with the impossibility of designating any one of them as the 'Cause' of the Universe.

☐ If we are to find a 'Cause' for the Universe, it must never be another Object; a true 'Cause' must be *purely SUBJECTIVE.*

We have therefore arrived at the point from which we are entertaining the seemingly untenable concept of an infinite, non-periodical Universe that never develops and always was, *and* a *subjective* Cause that also must *always* have been, for a cause cannot appear later in Time than the object of which it is supposed to be the cause. Given this situation, we necessarily have an *eternal duality*, and not only an eternal duality, but one in which the two aspects of the duality cannot find a way to interact—for what could be the manner of causation by which this supposedly ETERNAL Cause would act upon a Universe which has *always been?*

Since such an infinite, non-periodical Universe, according to this argument, never began, it must be *uncaused,* since all causes are causes because they cause something to *begin.* Even if we grant that the Universe, as an Object, may have had a *subjective* Cause, there is, clearly, no way for such a Cause to *act*, because the kind of Universe we are discussing *never began:* there was no Event called a 'Beginning'. Every

☐ Cause (as the term is usually considered) not only *is* an event but must *result* in an event. A Cause which does not result in an event cannot be considered a Cause—so, then, the Cause we are hypothesizing is no Cause at all, and therefore we are back to the idea of the Universe as an uncaused Object which, from all we know about Objects, is an absurdity.

Further, an infinite Universe would allow infinitely large structures to develop (over infinite Time) and infinitely small structures to exist as well, through infinite subdivision. There is no scientific evidence that either such exist. Further, an infinite Universe would be a continuum (infinitely divisible and infinitely expandable) which would negate the need for quantum movement—a dynamic of motion now firmly established as Real by science.

If we are to believe, then, in an *infinite* Universe, we will have to accept a Universe infinite in extent and containing infinitely microscopic subdivisions; as an Object which never had a Cause; as one Universe which always was and always will be; as a Universe in which no type of development (or at least continuous development) is possible, but only stagnation and at the best, and therefore endless sisyphean repetition with an arbitrary ceiling upon development or evolution. Would you *want* to live in such a Universe? *Do* you, in fact, live in such a Universe?

A Universe is REALLY Objective and *Is* an Object. As such, It must be created and destroyed and cannot last forever.

- A Universe is, therefore, *Finite*.
- A Finite Universe, being an Object, can *begin*.
- A Finite Universe can have an effective Cause capable of being determined within the span of Infinite Duration.
- A Finite Universe can obey the Law of Periodicity.
- A Finite Universe can be hierarchically organized.
- A Finite Universe can be driven by a Purpose capable of fulfillment in Time.
- A Finite Universe allows the possibility of evolution and development (without the absurdity of positing the necessity for an infinitely evolved and infinitely developed Universe).
- A Finite Universe allows the possibility of an Involutionary Process which is not unending, which has a term, and which gives way to an Evolutionary Process.

☐ It stands to reason, then, that our Universe is a Finite Universe as are, necessarily, all Universes.

The Problem of Whether 'Something' can Come out of NOTHING

This is the ancient conundrum. As long as NOTHING is conceived of as the VOID, utterly HOMOGENEOUS, with no possible movement 'within' IT, we are forced to say, No, for all our experience with events shows that they are causally preceded by other events. As the VOID is *eventless* (except for the 'FLASH'), the impossibility of accounting for the FIRST 'EVENT' becomes acute.

☐ However, if we suppose the NOTHING of the VOID to be the INFINITELY FULL INFINITESSENCE, we no longer speak of *NOTHING* 'happening', but of *EVERYTHING* 'HAPPENING' to the *infiniteth degree* (for this idea must be just as TRUE as the more conventional concept of the VOID), then, the question changes and so may the answer.

The *big question* in relation to such a reversal of thought is, Can something come out of EVERYTHING? Even though the EVERYTHING we are speaking of is *special*—an *infinitized* EVERYTHING, an *infinitessentialized* EVERYTHING, and a *noumenessentialized* EVERYTHING, the answer seems more likely, Yes. This being the case, instead of having *no cause* for the 'RAY' which 'FLASHES FORTH', and Its subsequent Universe, we have, *all the possible causes that could ever be,* for they *all* 'inhere' in the INFINITESSENCE.

From this reversed perspective, can a Finite Thing (the 'RAY', the 'POINT') be caused by the *infinitized* ESSENCE of all possible finite things? Can every possible cause that could ever be, be the cause of a Cosmos? Do you not feel how the atmosphere around the Problem changes, and how the 'FLASHING FORTH' of the 'RAY' begins to seem almost *reasonable.*

This Problem is by no means solved by means of this discussion, but, perhaps, a perspective has been offered which will allow fresh thought upon the subject. All this is predicated upon the idea that NOTHING is INFINITIZED 'EVERYTHINGNESS'.

The Problem of *When* the Second Aspect Engages the Third Aspect, —and *What* it Means to do so

In the discussion of the interplay of the Cosmic First Family, the Cosmic Process in general was described in terms of the series of Numbers 1-2-3-2-1, rather than 3-2-1. It was stated that the Father and the Son must Emanate at least the major Beings Who *hold* the Divine Pattern (the Fixed Design) in the World of Being before it makes sense for Fohat/Maya to begin the task to preparing the "building materials", the highly differentiated prakriti which characterizes the various planes and subplanes of the World of Fabrication. It was stated that even this building of what could loosely be termed the 'elements' must proceed according to a certain Design *held* or *sustained* within the Archetypal World of the Universal Son.

What is most difficult to conceive is the general timing of the interplay between the Second and Third Aspects. If we borrow the analogy from the normal process of building as humans conceive it, then the plans are prepared conceptually, in detail, before the actual builders arrive on site to make the plans into a *reality* (using that term in its lowest sense).

☐ Problems often arise with a 'design-as-you-go' approach, and the results of many mistaken actions have to be torn down, and new building processes instituted as unexpected changes are made in the pattern to be actualized. Applying these thoughts to Cosmos, the question arises as to whether the Fixed Design is a Design in *minute detail* from the first—i.e., *before* Fohat begins the Reflective Self-Particulation that leads to the appearance of minutely differentiated Cosmic Prakriti.

There is enough evidence in thought to attempt a provisional, Yes. It is certainly necessary for the Model to exist *before* the Replica, so that the Replica has a Plan to which to conform. The question simply arises as to the *degree* of detail of the Model. In some ways the first efforts of Fohat are very detailed, dealing as they do with the *most minute* factors in Nature (ultimate particle events and the tiny particles and atoms arising therefrom). It would stand to reason that a Plan would have to exist in some detail to serve as a Template for the fabrication of such minute items.

Even if the entire Fixed Design were pre-developed in full detail before Fohat and His Self-Emanated Sons began the process of World Fabrication, there would, nevertheless, be plenty of opportunity for error, as the History of the World Process in a number of Cosmologies has revealed. Just because the Design is in place does not mean that the builders will build properly of have sufficient illumination to *register* the Design so that they can build properly.

The really demanding question relates to how the 'Pattern Holders' representing the Second Aspect, and the Active Form Builders representing the Third Aspect, *actually work together*. Perhaps this question is what *A Treatise on Cosmic Fire* just *begins* to address, with the Tibetan Teacher always at pains to remind us that He is simply giving us the "ABC's "of a stupendous subject.

Albeit that the Pattern Holders are *anchored* upon the Archetypal Planes, in what ways and to what extent do They "go forth into matter"—the matter fabricated by Fohat? Further, it must be acknowledged that Fohat and His Emanated Host are as much the One Universal Identity as are any other E/entities in Cosmos, thus, They, too, must have a kind of *anchorage* upon the Archetypal Planes in the World of Being, though Their function is different and more *active*. Fohat, necessarily, has a *Monadic* Aspect. All beings in-Cosmos *are* fully and indivisibly the One Universal Monad.

☐ Perhaps we could say that the Universal Son and His Emanated Host, and Fohat and His Emanated Host, (while both having anchorage within the World of Being) are active in-Cosmos in different ways. Fohat and his Host are not so much 'Pattern Holders' as 'Executors of the Held Pattern'. The Son and His Host are not entirely inactive, even in the World of Fabrication, but Their function relates more to Consciousness than Activity.

Each of the Three Aspects of Divinity are possessed of the Qualities of the other two Aspects. The Tibetan brings this idea forward in *A Treatise on Cosmic Fire*. Brahma and Vishnu are in Shiva. Vishnu and Shiva are in Brahma, and Shiva and Brahma are in Vishnu. The Three are Really a Nine. Each has Will, Consciousness, and Activity, but They 'specialize' as it were. It is important to bear these ideas in mind if we are to avoid a subtle form of separatism in our conception of these Great Entities.

If modern cosmology has anything to reveal anent this process, it is likely that the great structures are built first and that smaller structures later form within them. What this means is that in the Emanational Process, the One Life, in its form of the Greater Logoi engage differentiated prakriti first, and create (on however many planes) the Great Cosmic Structures through which They manifest the One Life as *Lesser* Logoi engages differentiated prakriti *later* in the process and greater detail thus appears.

What this means is that Greater Entities go forth first, and that lesser E/entities (who are emanations of the Greater), follow after (even though they are 'enfolded' and perhaps 'pre-articulated' (i.e., articulated as waiting potencies well before Their 'time' for 'going forth' has come). The lesser Sons of the Universal Son, and the lesser Sons of Fohat never act independently of their Superiors (Who always *supervise* their more detailed and microscopic work).

☐ Thus, it seems, that (as regards the Involutionary Process) greater detail of form emerges with time, and, also, greater complexity of 'building materials' emerges with time. During the Evolutionary Process, the trend is entirely reversed, with greater simplification emerging with time. During the Evolutionary Process the lesser 'Rays' of the Universal Son (lesser! Monads) and the lesser Sons of Fohat 'retractively disengage'. Thus, lower Planes and the forms previously resident upon them, disappear.

We have been speaking here of the Involutionary Arc in any Cosmos (which is necessarily different from the Involutionary Arc as we study it in relation to our Planet). While we do know that our Planet Earth is now upon Its Evolutionary Arc, as is our Solar System as a whole, we do *not* know, necessarily, that our Cosmos (i.e., the Universal Logos) is upon Its Evolutionary Arc.

In fact, since the Universe (physically at least) appears to be *expanding*, thus, corresponding with the Out-Breath (Exhalation) of the Great Breath, there is every possibility that the Universe as a Whole may still be engaged in Its *Involutionary Process*. The lesser cycles of lesser Logoi, would have both an Involutionary and Evolutionary Arc in Their Processes even though the Universe as a Whole was pursuing Its Involutionary Arc. Such minor Involutionary and Evolutionary Arcs would also continue during the *Evolutionary Process* of the Universal Logos.

The important thing to realize is that the *Cosmic* Second Outpouring demands the practical engagement of the Universal Son and His Host, *and*, Fohat and His Host. That engagement is achieved through the Presence in Consciousness of the 'Rays' (Entities) of the One Life at the Cosmic Site or Field where form building is to occur. The One Life in Its Mode as Great Logoi in the Company of the Son 'Ray Forth' from the Archetypal World (while still remaining identificatorily *anchored* in the Archetypal World) in order to *keep Their Eye upon* and *Hold the Pattern for* the large scale Fohatic Constructions which must be built at the initial stages of the Involutionary Arc.

The construction at first is broad and general. As the building process descends plane by plane, the One Life-as-lesser-Logoi (lesser Archetypal Beings, and 'Higher! Numbers') take over the more detailed work involving more and more complex aggregates of ultimate particle/events. The term, the One Life or the One 'Ray' is continually repeated as it is so easy to forget that all E/entities are One Entity, the One Life, the One Universal 'Ray'. It is the *Oneness* of apparent particularity, which we must hold in mind, rather than the *apparent* and, Essentially *illusory,* particularity of the One Life. If we begin by assuming the *particular is Real,* we shall fall into egoism and separatism very quickly.

The Consciousness of the Second Aspect (as 'Rayed Forth' from the Archetypal Planes) proceeds 'downwards' with Fohat, plane by plane, subplane by subplane, supervising the building upon each plane and subplane. The *proceeding downwards* is simply an Act of Consciousness, an Act of Presence in a Localized Prakritic Field Intended for Construction All the while the anchorage of every 'Ray' (and even of Fohat and His Host) upon the Archetypal Planes remains. When Fohat 'Blinks His Eye' as (theoretically) He does an enormous number of times per human second, the Cosmos in process of Fabrication within the lower worlds instantaneously disappears, and the 'Holders of the Pattern', and, even, Fohat and His 'Executors of the Pattern' are instantly returned to Spirit/Consciousness as it is known within the World of Being.

It is tempting to pursue in speculation the mechanics of the Descent and Reascent of the Universal Son (cooperating with Fohat and His Host all the while), but it is clearly beyond the capabilities of the author. We would be forced to discuss Super-Cosmic Planes and Cosmic Planes of which Humanity, as yet, can have no knowledge.

We, therefore, must rest content with setting forth some possible models that concern the generalized relationships between Members of the Cosmic First Family as they carry forth the Cosmic Process. It must be reiterated that although the number of apparent Actors and Creators in the Cosmic Drama is manifold, there is Really but *One* Actor/Creator, *One* Character, *One* Presence, and that *One* is **8-the-I**. The greatest danger for Self-Understanding is to forget the One while the Many are being described.

The Problem of What REALLY/Really Happens When Fohat Enters Conditioned Cosmic Prakriti

When Fohat (in Mode Four), as *Self-enumerated Subjectivity,* engages in Self-Reflection against the background of the conditioned density of 'Father-Seen', 'Son-Seen' Cosmic Prakriti, what arises is, as it were, manifold 'bubbles of Fohatic subjectivity', or the possibility for the Subject (Fohat) to *objectively* 'See' just as many 'things' (i.e., 'bubbles') as the number of subjective units into which It has enumerated Itself—all this for the purposes of creating a differentiated Cosmic Prakriti. We have a subjectively multi-particularized Fohat 'Creating' *particulation* in the conditioned density of Cosmic Prakriti simply by 'Seeing' Its Own manifold, subjectively enumerated content. Each of Fohat's multiple points of view turns out to be (in objectivity) an ultimate particle/event. At this point, Cosmic Prakriti is not seen as density, but as multi-particulate. Such are Fohat's Self-Reflections. Thus are the "building blocks" of Matter created. It must be realized that un-aggregated ultimate-particle events are only to be found upon the very highest subplane of the very highest of the Planes of Fabrication.

☐ If Fohat oscillates between *seeing/being/seeing/being/seeing,* etc., then the ultimate particles (as events within the conditioned density of Cosmic Prakriti) flash in and out of manifestation (i.e., in and out of Fohat's Self-Observing Consciousness). Particles and events are the same, and they are, at this fundamental level, quantized by Fohat's oscillation between *seeing* and *being.*

Since there is no *absolute* continuity 'outside' the INFINITE CONTINUUM even FOHAT/Fohat cannot be 'continuous', for Fohat (with all other Members of the Pre-Cosmic First Family-become-the Cosmic First Family) has been 'EXTRUDED' from the ABSOLUTE CONSTANT, and is therefore, apparently no longer part of the UNVARYING CONTINUUM.

Fohat has a semi-continuity of Being in the World of Being, just as all Members of the Cosmic First Family (all being 'Rays' of the One 'RAY', and, ESSENTIALLY, the ONE 'RAY' ITSELF). Fohat's Actions however, are discontinuous, and this results in the Fohatically Fabricated Universe (the World of Approximation) turning 'on' and 'off' with the frequency of an 'ultimate' moment. At each such instant (the inter-moment instant) the entire *Objective* Universe goes into a kind of mini-pralaya, but the Subjective Universe (focused within the World of Being) remains.

'Outside' the HOMOGENEITY there must, perforce, be discontinuity of some kind—discontinuity in Time, discontinuity in Space, and discontinuity of Movement. Even Mulaprakriti (which many consider to be PARABRAHMAN, ITSELF) is discontinuous, appearing and disappearing with each Cycle of the Great Breath. What more fitting discontinuity for ever-active Fohat, then, than the regular oscillatory disappearance of those Fohatic energy/events which define Time, Space, and Motion within the Fohatic World of Fabrication? Between ultimate moments, between positions of particle/events there is only a return to Spirit, to Monad, to 'Rayness', Qualitative Ideality within the World of Being.

All this (for being so speculative) seems mildly reasonable, as *no 'thing' in Cosmos is allowed to continue to be itself continuously,* i.e., *forever.* According to the Law of Periodicity, even the Universal Logos cannot have *continuous* existence for It appears and disappears with one Respiratory Cycle of the Great Breath. No-thing 'EXTRUDED' from the CONTINUUM has absolute continuity, but some 'EXTRUSIONS' are *more continuous* that others. All E/entities in Cosmos (in their Essence of Oneness) have some anchorage in the World of Being and, thus, have a kind of continuity which is called permanent in Cosmos (though some degree of 'change' is part of this in-Cosmos Permanence).

☐ Fohat too is anchored in the World of Being and is (as an Aspect of the One Ultimate Cosmic Monad) permanent-in-Cosmos, but Fohat's Self-Perceptions (that create the manifold articulation of the World of Approximation) are relatively fleeting, or even *extremely* fleeting.

What this Fohatic Fluctuation means is that Life and Death are alternating at every moment, but what is Life and what is Death? Usually, what we will call Life is assertion, particularity, presence in Cosmos, and absence from the World of Being (the most continuous Universal Factor); what we will call Death is negation, no-thingness, return to the Spirit and the World of Being.

An exactly opposite point of view could, also, be justified. All things, then, are never more than an instant away from the Unity of the One and the Design-at-the-Beginning. The flutter of these *Opposites* would be so incredibly rapid, that it would be as if both states, the Relative Continuum of the World of Being and the Dis-Continuum of the Fohatically Self-Objectified World of Fabrication were occurring simultaneously, which (Essentially) they are.

The Problem of Whether Ultimate Particle/Events have Individuality

As these particle/events *are* Fohat, and as Fohat enumerates Itself in order to produce them, could there be a *subtle differentiation* in the numerous ways Fohat *'Sees'* Itself in order to create differentiated Comic Prakriti? Perhaps the individuality or distinctness of each particle/event consists in the 'particular' Fohatic 'point of view' which that particle/event represents.

As to the dimension upon which such particle/events would occur and their probable duration, there seems to be identicalness. If ultimate particle/events have differentiable 'experience', then their uniqueness is ensured, but problems exist as to the mode-of-mutual-transmission of any possible experience they might have, as nothing can 'pass between them' during the 'frozen moment' they occupy 'position' within the Cosmic Configuration, and, further, there is no 'time' allowed for such a passage. Via pervasive consciousness, however, the 'point of view' of each may be shared as the 'point of view' of all, which would be akin to the 'group soul' phenomenon on much lower levels

The Problem of How Fohat Operates in Cosmic Prakriti —Whether Consciousness, Movement, or *Both*

All the Pre-Cosmic and Intra-Cosmic Archetypes have Consciousness. These Archetypal Beings may emphasize one Divine Aspect over another, but all of Them have *all* Divine Aspects; Brahma is in Vishnu and Shiva; Vishnu in Brahma and Shiva; and Shiva in Vishnu and Brahma.

What Fohat Self-Objectifies *as* differentiated Cosmic Prakriti (a differentiated, multi-particulate Self-Image), It Does through Acts of Consciousness/Perception and then *moves* (or *moves 'in'*) what It has objectified—Really It moves Itself relative to Itself (thus moving <u>*as*</u> what It has objectified). This movement itself is probably an Act of Changing Fohatic

Perception (for at that exalted level, to *see* is to *be,* or to *see* a thing, as such, is to have the thing *exist* as such).

Fohat 'Sees' as differentiated Cosmic Prakriti *objectively,* the Self-Enumeration which is 'there' within Itself *subjectively.* That Self-Enumeration 'there' 'within' Itself (the product of subjective Self-Division, subjectively articulated and ready to be 'Seen') is 'there' in *numerical exactitude* accordance with the *Idea Fohat 'Carries'* or *Embodies*—the particular 'EXTRUDED' 'IDEA' or 'POSSIBILITY' of the ABSOLUTE that is destined to *become* the Cosmos-to-Be. Fohat (starting as SUPER-Cosmic 'FOHAT') is the Primary INFINITESSENCE-'SPONSORED' 'BEING'-instantly-Being gone forth in several Descending Modes to facilitate in Mulaprakriti-'Become'-Cosmic Prakriti the 'IDEA'-instantly-Idea<u>s</u> It (Fohat) Subjectively 'Carries'/'Embodies'. Fohat is an Idea-*Bearing* Energy. Fohat <u>IS</u> the *peremptoriness* of the Idea whose *time* has come. That Idea is the Formative Power for the entire imminent Universal Cycle. That Idea is the *custodian/embodiment* of what the INFINITESSENCE has *released for Cosmic Objectification.*

Within Cosmos, Fohat continues Its function as Idea Bearer. Fohat always *embodies* (or *carries forth*) the Idea *implanted* within it. 'FOHAT'/Fohat *becomes successively:*

1. the Agent of THAT;
2. the Agent of the Infinite Subject/Self;
3. the Agent of the Universal Logos, the Father; then,
4. the Agent of the Son.

All the while Fohat (by facilitating Subjective Reflection) is working in relation to the Mother (in all Her Attenuating Modes, Her Attenuating Images) bringing Objectivity out of Subjectivity, facilitating the Mother to 'Show Forth' what is in the Father (but which was *originally* 'in' THAT, and, even, *was* THAT)! It is as is Fohat is a kind of 'Sight' (Sensitivity, Registrability) which renders present but hidden Ideas-within-Subjectivity *objectively articulate.*

Through an Act of Interior Attunement, Fohat enumerates Itself (in accordance with the specifications of the Fixed Design) while still remaining entirely, and unitarily Itself. Pictorially, this could be called 'Enumeration in the Dark'. Fohat then (as expressed in the language of Objectivity) 'enters Cosmic Prakriti' which is equivalent to saying that It (Fohat) enters a State in which it is possible to 'See' Its Own enumerations as *objective things.* These objectified Self-enumerations are, as it were, 'overlays' upon Cosmic Prakriti (and, for consciousnesses unconsciously submerged in the World of Fabrication, <u>*are*</u> *Cosmic Prakriti Itself),* for all that such consciousnesses Really experience of Cosmic Prakriti, is *differentiated* Cosmic Prakriti. Thus, are ultimate particle/events born. In a way, the subjective Self-enumerations of Fohat are the subjective noumena of ultimate particle/events which become *objective* as Fohat focally 'Sees' them within Himself.

Once Fohat has objectified its multiple Self-enumerations as differentiated prakriti, the Face of (Idea-Reflecting) Cosmic Prakriti is *obscured* (at least to the vision of many E/entities-in-Cosmos). Man, for instance, does not cognize imparticulated Cosmic Prakriti. His identification *with* and *as* the Greater Logoi Who 'See' him (Man) is not sufficiently developed for him to do so. This means that Man does not yet see himself as if he *were* the Greater Logoi, for <u>un</u>particulated, <u>un</u>differentiated Cosmic Prakriti (in its several 'extents' and depths) is simply Their Reflection of Themselves.

Matter, *as we know it,* is created when subjectively-Self-enumerated Fohat, objectifying Itself as differentiated Cosmic Prakriti, *moves* Its objectified enumerations (ultimate particle/events) relative to each other (moving them *not* as an external Agent would, but as if from *within*, because Fohat *Is* them), and thus arranges Its objectified enumerations relative to each other in a great variety of 'positions' determined principally by the 'IDEATION'-become-Pre-Cosmic Ideation which It (Fohat) is *bearing/carrying/conforming to* as Agent of the Pattern Holding Universal Son and His Host.

Really, Fohat not only *bears* or *carries* this Ideation, but Fohat *embodies* the Ideation, and is the 'Sight' which compels the Objectification of the Ideation. In this way, 'FOHAT'-as-Fohat, '*Act*-ually' shapes all Worlds.

The Problem of the Existence of Self-Enumerations Within a Subjective Being

Do differentiation and articulation require Objectivity if they are to exist? Can there be differentiation and articulation in the realm of pure Subjectivity? Clearly a complex 'IDEA' of the Universe-to-Be (albeit an *infinitized* 'IDEA') 'EXISTS' within the PLENUM *before* the 'IMPULSION' for the Formation of that Universe (i.e. the 'RAY' 'FLASHING FORTH'.) When do subjective differentiation and articulation Really begin? Only once the Subject engages with Itself-as-Object? Probably well before. Is not the Subjective Pattern of Cosmos already SELF-'CONCEIVED' before It is 'Seen', 'Noticed', or 'Objectified' by means of the meticulous Self-Reflection of the Universal Logos and His Emanated Agents?

☐ Somehow the INFINITE SELF (if IT is to 'REMAIN' the INFINITE SUBJECTIVITY) must 'RETAIN' *infinitized* 'IDEATION' 'within' IT without 'SEEING' that Ideation (for *seeing* is *instantly* Objectification). Instead, the INFINITE SELF must simply 'BE' that *infinitized* 'IDEATION'.

'FOHAT' (in its First Mode), for instance, *Is* 'Bearer' of the intra-SOURCE 'IDEATION' which is to *become* a Cosmos. The 'IDEATION' must *already* be 'CHOSEN' or 'ARTICULATED' by the SOURCE SELF as the 'RAY' 'FLASHES FORTH'.

- Perhaps, the 'FLASHING' and the 'CHOOSING' are identical!
- Perhaps the 'FLASH' is the sudden Birth of Time occurring in 'No Time at All'—from TIMELESSNESS to Time in *no time.*

The 'IDEA'-to be-Universal Idea is 'EXTRUDED' from the INFINITESSENCE and 'FLASHED FORTH' into Pre-Cosmic Latency, 'there' to be developed into a Universe. 'MAYA'-instantly-Maya is 'FOHAT'-instantly-Fohat is the ONE 'ACTOR'-instantly-Actor Who is the 'Bearer'/'Embodier' of the 'IDEA'-become-Idea. Fohat/Maya/Actor It Is, Who 'Forces' the Infinite Subject (becoming De-Infinitizing Subject) to begin 'Seeing' Itself *not* just in terms of Infinitude, but in terms of the 'EXTRUDED IDEA'. Maya forces perception into certain forms through Self-Veiling. Fohat is the same—the Enabler of

Limiting Self-Perception. Later Maya must *lift* her Veils even as, earlier, She *lowered* them. Then, too, Fohat's functions will change, and He will *undo* the 'knots' and 'complexifications-in-Consciousness' which He has caused.

Thus, 'out' of the PLENUM, which IS the INFINITESSENCE, has come one 'IDEA' (one Set of Possibilities) 'RIPE' for Objectification as a Cosmos. That 'IDEA' necessarily holds within itself, ESSENTIALLY, the potential for articulation/differentiation which will make a Cosmos possible. Cosmic Prakriti, *cannot*, in Itself, furnish the articulation and differentiation which are necessary if the 'IDEA' is to be objectively expressed. The articulation and differentiation (being originally and Essentially Subjective) must come *first*, before Objectification. The 'IDEA'-as-Idea is *subjectively inherent before objectively evident.*

In this model of the engagement of Subjectivity with Objectivity, it is necessary that the INFINITE PLENUM-SELF identify ITSELF not only with the infinitization of *all* possible possibilities, as IT 'DOES' 'within' the INFINITESSENCE during Universal Pralaya, but identify with a *single articulated possibility*, an 'IDEA' which when 'EX-TRUDED' from the INFINITESSENCE, is to serve as *the* Idea which is destined both to *become* the coming Universe and to *determine the detailed Pattern of* the coming Universe.

For the preservation of *non-objective articulation-within-Subjectivity*, *identification* is the key. There must be no *seeing* of the articulation, but only a *being* of it. Thus, even though there is, *inherently*, (SUPER-Cosmically and in the Pre-Cosmic Stage of Infinite Subject/Object), *potential* articulation and differentiation of the 'IDEA'-as-Idea, that 'IDEA'-as-Idea is (at those stages) still *'all one'*, and its articulations are not *seen* as separate (i.e., as 'standing out' from the general background of INFINITUDE/Infinitude).

The *seeing* of the 'IDEA'-as-Idea-*in distinct articulation* only begins with the 'movement' of the De-Infinified View-Point and the De-Infinified Seen-Point towards the state of Finite View-Point/Finite 'Seen'-Point. Then the 'IDEA'-as-Idea can be elaborated by the 'Concentration' of the Universal Subject (Condensed Point, Finite View-Point) upon Itself (first in general) and then in a more *meticulous* manner. The Subject (in this case, the Universal Logos) *examines Himself very closely*, as it were, to 'See' *exactly* what He Is and *exactly* what is 'within' Him. Thus He 'Sees' for the first time what He has been 'Carrying'/ 'Embodying' all along.

☐ The Process is a 'Voyage of Self-Discovery', with the descending progression of Attenuating Subjects discovering through 'Self-Sight' what has been 'implicit', 'implicate', 'hidden', and unrevealed within Them. We are dealing with a Super-Cosmic and Cosmic version of "Know ThySelf." The Universal Logoic Meditation that precedes the manifestation of Cosmos is an Effort to bring the latent, unrevealed 'IDEA'-as-Idea to Light.

An interesting thought arises regarding Fohat in Its various Modes. In general, Fohat/Maya is the 'Seeing', and the 'Seeing' *is* the Action. When the 'RAY' 'FLASHES FORTH', it is as if 'SIGHT' 'FLASHES FORTH' for the first time (yet again). The 'Creation' of every Entity in the UTTER ALLNESS is a Process originating with *SELF-'SIGHT'*—SELF-'SIGHT'-as-Self-'Sight' (which is 'MAYA'-as-Maya, and 'FOHAT'-as-Fohat).

Thus, 'within' the INFINITE SELF, 'SIGHT' *is* 'ACTION'. Pre-Cosmically and Cosmically, 'Sight' is *also* 'Action'. Even in the depths of Cosmos, on the level of the ultimate particle/event, Fohatic Sight is Action. Within Cosmos, Itself, the *kinds* of Sight begin to differentiate, producing the more Continuous Sight of the Son and His Host and the Rapidly Fluctuating Sight of Fohat and His Host. In a very important way, when thinking about the Creative Process of the highest Cosmic and Pre-Cosmic Levels, Modes of 'Seeing' are Modes of 'Doing'.

- ☐ In conclusion, it may be said that every authentic E/entity in Cosmos is 'IMPLANTED'/Implanted with 'INSTRUCTIONS'-as-Instructions which (before the Process of Objectification of the Instructions begins) will not yet be *seen* by that E/entity. Within the Archetypal World, the E/entity-as-the One Life *knows*. These Instructions are an aspect of the Fixed Design which that E/entity (as an aspect of the One Universal Life) must objectify according to that Design.

These thoughts have immediate application to us as human beings. In the simplest possible terms, there is a design within us of which we are often unaware. By looking deeply within ourselves we discover that design. For us, at our (apparently) low level, simply to *see* the internally articulated Idea is *not* the same as objectifying it. (For the higher Beings, to *see* is *actually* to *objectify.)* Our task is to take the internally articulated Ideas within us, and objectify them within prakriti (which is the meticulous Self-Objectification of Fohat and His Host).

It is interesting that in their *own* Self-Objectifications, all E/entities are limited by the superior Self-Objectification of the Entity or Entities in which they "live and move and have their being." What the superior Entity 'Sees' is, perforce, a ring-pass-not limiting what *we* may see. Ever the Self-Objectification of the lesser entity is a reduction of the Self-Objectification of the superior Emanating Entity. We must learn to 'See' as we are 'Seen', and thus we climb the Ladder of Ascending Subjectivity.

The Problem of Whether One Sees Only That Which Is Within 'OnesSelf'

Mulaprakriti is the Infinite Object, the Infinitely Dense Object, because the Subjective Infinitude which 'Sees' Itself as Mulaprakriti is a continuous, undifferentiated Infinitude. There is *no-thing* 'in' Mulaprakriti (until it is 'Seen' *'there')* by the 'De-Infinitizing' Subject moving towards Finitude. 'IDEA'-as-Idea is subjective and 'contains', *unseen* within Itself, the entire *articulation/differentiation* that must be objectified within Cosmic Prakriti (Universal Matter).

The 'Coming IDEA' to be Objectified *as* and *in* Universe, is at once *simple* and *complex.* The simplicity is the main *Note,* but the complexity is all that may be unfolded through the sounding of that main Note. The intended articulation of the Coming IDEA-as-Idea must exist *in non-objectified articulation* within the Ideational Realm before It is objectively articulated in the Prakritic Realm.

Everything that the 'IDEA'-as-Idea *Is* must *pre-exist* Its 'working out' in Prakriti, i.e. Its *objective* 'working out'. When the Subjective Being 'Sees' objects in Prakriti, It is Really 'Seeing' only Its subjectively articulated, differentiated Self. Nothing can be seen in Prakriti which is not already in the Self. Nothing can happen in Prakriti that is not already in the Self. The *seeing* and the *happening* are the same thing.

The Problem of the Necessary Particulateness of Certain Higher Dimensions if they are not to be *Impossibly* Continuous

In other words, in what does the Dis-Continuity of certain Higher Dimensions consist, and to what 'altitude' in Cosmos does particulateness extend? H. P. Blavatsky talked about the "Prakritic Planes". Does that term mean that some Planes are non-prakritic, which can be taken to mean non-particulate? What does particulateness mean upon those high levels?'

Perhaps particulateness only occurs in the Fabricated Cosmos, the Cosmic Levels upon which Fohat builds the Universe of Particle/Events. The World of Being or World of Archetypes (the very highest levels of the World of Becoming) is virtually permanent-in-Cosmos It is a *sustained* World of Fixed Design. It is not a World constructed by Fohat (in the manner Fohat works 'below'—i.e., through Self-Objectified Self-Enumeration as ultimate particle/events) and yet, wherever there is *act*, either in Cosmos, Pre-Cosmos, or 'within' IT, there is Fohat/'FOHAT' in one of Its/'ITS' *modes*.

☐ One might ask whether the World of Being is particulate or imparticulate. The answer to this difficult question depends so much upon whether the 'Self-Sight' of the highest Logoi (those in the World of Being) can be continuous-in-Cosmos (i.e., sustained without interruption) or whether it is 'momentarily' discontinuous (from ultimate moment to ultimate moment) as in the World of Approximation, the World of Fohatic Fabrication.

It would seem that the Archetype (of the Fixed Design) in the World of Being would have to be *held*, or *sustained*, if the inter-moment instant (the anti-thesis of the ultimate moment) was possibly to be used by Fohat and His Host (as well as by the Emanations of the Universal Son—i.e., the 'Rays' of the One Cosmic 'Ray') as a 'moment of adjustment', a moment of Self-correction, to better approximate the Fixed Design *held* in the World of Being.

The Problem of the Possible Infinite Futurity of the Present Moment

From the imagined perspective of *beginninglessness,* the present moment (and every present moment, forever—even a 'present moment' that occurred at the deepest conceivable/specifiable 'time' in the "Night of Time") *is,* in a way, the infinite future (i.e., the infinitely-progressed future) of an infinitely regressed, beginningless past. This can be seen because the present moment is *infinitely removed* from that infinitely regressed beginningless indefinite past. So the present moment, in a way, *is* the infinite future, or the 'moment of infinite futurity'. More readily understandable is the idea that the present moment is the cutting edge of future infinitude.

A problem arises in this conception, because the present moment is a *definite* moment. One must ask whether *any definite* moment can be said to represent an *infinite* (hence *indefinite*) future. Does not the *definite* contradict the *indefinite,* as the *finite* contradicts the *infinite?* Perhaps we will be more accurate if we say that the present moment is the *farthest* the 'future' has *ever* proceeded. This is easy to understand.

☐ More difficult, but important to understand, is that *any* definite moment (which is ever a *present* moment) is always infinitely removed from a beginningless past and an endless future. If any definite moment is infinitely removed, for instance, from a beginningless past, is it legitimate to name that present moment an 'infinitely *futurous* moment'? The answer would seem to be both, Yes and No.

And yet if every present moment does become a past moment once a future moment has become a definite moment, are we justified in thinking of *any* present moment as a past moment, the very instant it becomes a present moment? We are not here referring to the idea that a present moment *instantly* (at the next ultimate moment) becomes a past moment? We are asking whether the present moment (which can be considered the maximally future moment the moment it becomes a present moment) can *also* be considered a past moment the very moment it becomes a present moment. The common sense answer is, No. However, if we eliminate through *infinitesimalization* (and by means of the 'Infinispectivizing') all past moments, and collapse them into the present moment, *then* the present moment is the only moment there is (as 'infinitesimalizing' as it is).

That present moment, thus, becomes the summation of all past moments, and thus is a way, the *only* past moment. Of course infinitesimalization also deprives the waiting future of time in which to happen (for if times in the past are deemed to be of infinitesimalizing duration, the same applies to any possible times in the future), so, the question arises, without a future can there ever be a past? Apparently naught remains but the present moment, forever.

Forgetting for a moment that past and future may be conceptually/imaginatively destroyed without too much trouble, and returning for a moment to the present moment, we see that at the present moment we are *already* (and de-finitely!) *infinitely far* into the future, and we *always have been* infinitely far into the future, even at any *definite time* in the infinite remoteness of an infinitely regressed past.

☐ Thus every moment of time along the Infinite Time Line is the infinitely progressed future, the *definite infinite future* (a kind of semi-contradiction, as previously discussed). Since, in a way, every present moment is the summation of all past moments (a summation of an infinitude of 'infinitesimalizings'), and since every present moment is also the maximally future moment (of the moment) the past and future have forever been the same. Thus are past and future one!

Now, can we consider the present moment as the infinitely regressed past of an infinitely progressed future? This, even if true, (which it may not be) would be harder to conceive. It is one thing to say that the present moment *will be* the infinitely regressed (definite!) past at a certain point in an infinitely progressed future (an illusory thought, because any *definite* point in the future is a finite temporal 'distance' from the definite present moment), but quite another thing to say that the present moment already *is* the infinitely regressed past.

Of course, here again we run into the problem of the 'definite indefinite'. It is of course possible to say that any moment in time along the Infinite Time Line (whether that moment took place 'x-tillion' years ago, or will take place 'x-tillion' years from now), was or will be the infinitely regressed past of an infinitely progressed future. This is the corollary to saying that any moment (whatsoever) upon the Infinite Time Line, is the definite/indefinite infinite future because it is always infinitely removed from a beginningless past.

There are problems with the forgoing conception. Time, apparently has a linear direction (although, from a certain perspective, time is cyclic too). The trouble is, the past (as usually conceived) has 'happened' and the future has not *yet* 'happened'—so the past is *actual* and the future is *potential.* The past has precipitated and the future has not yet done so.

☐ So right now, and now, and now, *ad infinitum*—we are as far into the future as we ever *can* be—(from one perspective) infinitely far—but we are never, *presently,* the past (unless we use the *infinitesimalizing* Power of 'Infinispectivizing' to gather an infinite past into the present).

If the infinitude of the future were *already* existent, then and only then would the present moment be the infinitely recessed past, and (problems would still exist because of the *definiteness* of the present moment, for no 'definite' term can Really be the 'infiniteth' term of an infinite series), and yet, neither is it a specifiable, 'countable' term because there is no specifiable term from which to begin counting when dealing with a *beginningless* infinite series.

Thus we see, that while the present moment is, in a way, the infinite-but-definite future of an infinitely recessed past, it can never be the infinitely recessed past of an infinitely progressed *existent* future, because neither a *definite* future (and this seems obvious) or an *indefinite infinite future* presently exists. The *indefinite infinite future* will *never* exist! Commonsense will tell us that the present moment can be both present and future, but not (without strong *manipulations*) present and past. The infinite past has already happened, although it is beginningless. An infinite future will never *completely* happen, although it will never cease *moving towards 'happening'*.

The present moment can be infinitely removed from either an *indefinite* past or an *indefinite* future, but never from a *definite* past or *definite* future. If the present moment can never be *infinitely* removed from a definite moment in the future (or past)—then *no moment* can ever be infinitely removed from any other moment, if those moments are definite, and what other kind are there?

☐ Certainly, this present moment is not the infinitely progressed future of any *specific moment* in the past, but only the infinitely progressed future of "an infinitely regressed past" (a very *indefinite* phrase). If any moment in the past is identified as a definite starting point, then the distance between it and the present moment is *finite* and not infinite.

Thus, this present moment is, Really, the infinitely progressed future for an *indefinite* past (if we can overcome the problem of the 'definite indefinite'), but never for *definite* past. Therefore, because the present moment is *definite,* it can never, in itself, (unless from the most extremely manipulated perspective), be an infinitely regressed past from any *definite* moment in the future. It can only be the infinitely regressed past from an *indefinite infinitely progressed future,* which does not yet, exist, and cannot *ever* exist, for the future will forever be *indefinite* (just as the past, in terms of its precipitated events {and not its beginninglessness} is definite). The moment that the future becomes *definite* it has become the present and then instantly the past.

Let us imagine the beginningless, endless Infinite Time Line. If two points (two moments) are placed upon that Line, we have finitude and a *definite* span of time.

- If both 'ends' of the Infinite Time Line remain *indefinite*—then what?
- How far is an Infinitely regressed indefinite beginningless past from an infinitely progressed indefinite endless future?

The Real and ultimate answer is that there is no 'space in time' between them—Zero. Because (in this question) there is no specifiable 'time' given, we have inadvertently entered the Realm of INFINITE DURATION and the two indefinite extremes (a beginningless indefinite past an endless indefinite future) have 'met' in the TIMELESS ETERNAL NOW. If, however, but one point is put anywhere upon the Infinite Time Line, then that point is as immeasurably far (temporally) from the infinite past as from the infinite future. That point is in the definite—yet somehow indefinite—'middle' of all possible Time.

☐ If time is linear, and there was never a *definite* past (a point of precipitation), we could never have reached the present. We would 'travel' forever and never reach it. It seems one must always start from the *definite* past if one is to reach any definite present. The past is an infinitude of 'definites'; the future (as seen in the present moment) is an infinitude of 'indefinites' 'waiting' to become definite.

What can be legitimately said is that any Cosmic Now is always equidistant from an infinitely regressed beginningless (hence indefinite) past and an infinitely progressed endless (hence indefinite) future. There is an infinitude of precipitated possibilities (behind) and an infinitude of unprecipitated possibilities 'ahead.'

The fact that there *is* a future tells us that all possibility has not yet been precipitated for the precipitation *does continue* from ultimate moment to ultimate moment. Thus, even in the (very lengthy!) *infinite* past, all possibility has *not* been precipitated, and yet an *infinitude* of possibility *has* been precipitated. Thus, the terms 'all' and 'infinitude' are not necessarily equal, 'all' being ostensibly 'larger' than 'infinitude' or, at least, than *some* infinitudes. Every moment has hovered on the edge of the future in which even *more* precipitation can be expected (even though there may be an interval of a Universal Pralaya between precipitations, if the moments under question are the last moment of a concluding Cosmos and a future which is the first moment of a beginning Cosmos).

The Problem of Whether the Past *Is* the Future

All definite *past*, *was* a future with respect to some definite past moment. If we begin with the present moment, and go back as far as we choose to go—a centillion UAGO (Universes AGO), or even a 'googol-plex' (an actual mathematical construct) UAGO, then that entire interval which (to us in the present moment represents a past) was, from a perspective taken one moment *before* the span began, a *future*.

So we can see that with respect to any given point in the past, all the past, dating *back* to that point, *was a* future, but not *the* future with respect to our present moment, which (since the second ago this was read), has now become a past moment to this present *present* moment—the span between them now representing what was a future span to that once-present-now-past moment.

The Problem of the Possibility that Infinite Futurity Already Exists

This problem has been touched upon in the foregoing sections, and the thought of any possibility that infinite futurity exists *now* was quickly dismissed. Was it dismissed prematurely? Perhaps, there is one way in which it could be said that infinite futurity *exists now* (though the word, 'exists' is not the correct one).

- ☐ All that can possibly be precipitated in an infinite objective future *already* 'inheres' as infinitessentialized possibility 'within' the FOUNT OF ALL POSSIBILITY.

In a way, it could be said that there is nothing that *can possibly be* that is not *already* fully perfected and 'ultimatized' (through infinitessentialization) 'within' the FOUNT. In this way, along the Infinite Time Line, whatever has ever precipitated had already been fully 'ultimatized' forever.

It is not that the present moment (or any moment) was 'PRE-DESTINED', and *had* to occur as illusion (for the ratio of non-precipitation to precipitation seems to be an 'on-going' infinitude-to-one)! It is simply that any condition whatsoever that *has* been precipitated (because such precipitation was, indeed, a possibility) must have 'inhered' forever 'within' the FOUNT OF ALL POSSIBILITY, the ABSOLUTE.

☐ Thus, whereas the specific future to be precipitated remains *forever* unknown (the odds against any specific precipitation being infinity-to-one), any precipitation which *does* occur, has been *infinitessentially present* forever in the FOUNT OF ALL POSSIBILITY. Thus any precipitated future *already is, (infinitessentially)*, but *is not objectively.*

The Problem of 'Where', and in 'What', does Inter-Cosmic Memory Dwell

There have been an infinitude of Cosmoses Past and thus an infinitude of past impressions planted cyclically forever upon the Memory of Nature (let us call it the highest possible Cosmic Akasha). It is not logical to think that the accumulated memory of every Cosmos is stored within the Memory of Nature of the Cosmos to come, otherwise within each Cosmos (infinitely removed from the beginninglessness of the Great Breath), there would be infinite memories past. A Cosmic Being is a *limited Being* just as a personality is a limited being with respect to the Soul. The personality, per se, does not store all the memories of thousands of personalities past. The memories are stored within the permanent atoms stored within the Causal Body.

Perhaps the obvious analogy of the Causal Body to the INFINITE SOURCE holds good. The INFINITESSENCE is *infinitely full* of infinitized possibility. Certainly, that fullness *includes* an infinitude of *elapsed, actuated possibility* such as has appeared in an infinitude of Cosmoses past. Such elapsed possibility would naturally hold 'place' within the FOUNT OF ALL POSSIBILITY, for indeed, *elapsed, actuated possibility* is *possibility* none the less (and has to have been *present* 'within' the FOUNT OF ALL POSSIBILITY *forever before* it had even elapsed!). All possibility 'INHERING' 'within' the INFINITESSENCE would necessary be *infinitized* and, thus, the SAME THING, as there is no room for *particulateness* 'within' the GREAT HOMOGENEITY. And yet, *infinitized possibility is* the fullest fullness of anything that could ever happen or of anything that ever *did* happen.

Of course, the SOURCE 'CONTAINS' all opposites, including (the possibility of) particulateness, which would, apparently *violate* ITS 'NATURE' as the BOUNDLESS IMMUTABLE PRINCIPLE and the GREAT HOMOGENEITY. The *contradictions* to ITS ESSENTIAL NATURE are 'EXTRUDED' *as* Cosmos and *actual* possibilities in-Cosmos. Yet, it can still be said that the SOURCE, nevertheless, 'CONTAINS' such articulated possibilities, because even though the Cosmos *appears* to be going on 'outside' the SOURCE, it is necessarily going on 'within' the SOURCE and *is* the very SOURCE ITSELF. What else *is* there?

Thus, any elapsed possibility that is a 'candidate' for Infinite Memory, *already* has 'inhered' forever 'within' the FOUNT OF ALL POSSIBILITY (albeit, in its *infinitessentialized mode*—whatever that REALLY *IS*). Theoretically, infinitized elapsed possibility ('within' *IT*) is not a *particulate memory.* Particulate Memory, of an infinitude of Cosmoses Past (assuming them to have existed), would have to exist in the Super-Cosmos Realm of the 'De-Infinitizing' and 'Re-Infinitizing' Subject/Object.

Speculatively, Infinite Memory would be likely to 'appear' precisely at the juncture when the Infinite Subject/Object (engaged in Its Infinispective), *transitions* (let us say, in Pre-Cosmos) to a Mode of 'De-Infinitizing', and thus of 'Seeing' through 'Infinispectivizing' (or, more precisely, 'De-Infinispectivizing'—the reduction of Infinite 'Sight'). If all of an Infinite Memory of particulate experience were ever to 'appear' in a Flash, that pivotal moment would be a good moment (for it is doubtful that any objective particulation can be 'Seen' from the Infinispective).

Contrarily, in Post-Cosmos, the moment immediately preceding return to the Infinispective (following Post-Cosmic 'Re-Infinitizing' by means of 'Re-Infinispectivizing') would be a reasonable moment for a Summary View of Particulated Infinite Memory *(not infinitessentialized* Infinite Memory, which is not REALLY 'Memory' at all, as its 'CONTENT' *always was*).

The Problem of How a 'MOVEMENT' can 'OCCUR' When not *Preceded* by a 'MOVEMENT'

For instance, how did the first 'RAY' 'FLASH FORTH'? This is an ancient philosophical inquiry, and it can be reduced to the question, How can 'Something' come out of NOTHING? Any change whatsoever is a movement. As we usually consider, movement, however, movement has a cause.

This means that movement is *preceded* by another movement without which the first movement could not occur. Within the ABSOLUTE, however, the FIRST 'MOVEMENT' has, during *that* particular SPANLESS/'SPAN' of Universal Pralaya, *no precedent, no* preceding 'MOVEMENT'. First of all, 'MOVEMENT' 'within' the non-illusive *purity* of IT, violates the IMMUTABILITY of the FIRST PRINCIPLE, and secondly, the problem of 'infinite regress' would arise if the FIRST 'MOVEMENT' had to be preceded by a 'MOVEMENT' for then *that* preceding 'MOVEMENT' would *also* have to be preceded by a 'MOVEMENT', and so *ad infinitum.*

There is REALLY, at this *time,* in Time and Space, no way to answer this question. One can only imagine (by skirting perilously close to the edge of impossibility) that the 'MOVEMENT' was of *infinitesimally instantaneous* duration (which means that it 'TOOK PLACE', but *almost* did not, because the *infinitesimal* is so close to zero as to be *almost* nothing at all).

☐ One has to think of this most major 'MOVE' of all possible 'MOVES' as 'ARISING' *'un-causedly', spontaneously, instantaneously,* and 'allowing' in its *infinitesimally instantaneous* 'SPAN' the 'CHOICE' of the particular infinitized possibilities of the Cosmos-to-Be *and*, also, the 'SPLITTING' of the HOMOGENEITY (which 'SPLITTING' results of the EVANESCENT INFINITE TRINITY).

Perhaps, in that splitest 'split' of a 'second', all infinitized possibility within the FOUNT OF ALL POSSIBILITY becomes instantaneously *visible* and instantaneously 'CHOSEN'. INFINITE 'INTELLIGENCE, LOVE, WISDOM, etc.' doesn't need much *time* to 'CHOOSE'! Perhaps for the *infinitesimally instantaneous* EVANESCENT INSTANT that the 'EVANESCENT INFINITE SUBJECT' 'EXISTS', the entire 'RANGE of ARTICULATABLE POSSIBILITY is *there* 'BEFORE IT', and the 'CHOICE'/'SELECTION' can be 'MADE'. 'WHO' 'KNOWS'! 'WE', somehow, *do know* in the profoundest depths of SELF, even NOW, and *will* 'KNOW' *again* at the virtually MOMENTLESS/'MOMENT' of the GREAT AWAKENING.

☐ It is clear from the foregoing that one cannot speak of such things intelligently. In a way, we are trespassing the *'forbidden territory'* of the ONE ABOUT WHOM NAUGHT *MAY* BE SAID. Permission 'DENIED', and for GOOD 'REASON'! The mind simply cannot conceive the *causation* of that which has no *cause.*

If we conceive of the VOID as the ABSOLUTIZED STATE, in which the 'INFINITIZATION' OF EVERYTHING POSSIBLE is 'ULTIMATIZED', we can perhaps imagine that we have a better grasp of that 'MOMENT' which a 'Specificity' (which is infinitely less than the HOMOGENEOUS 'NON-HAPPENING') is *suddenly* 'EXTRUDED' from the 'UNSPECIFIABLE'.

The Problem of Whether *all* Possibilities are 'Resident' Within the NOUMENESSENCE —What of 'bad' possibilities?

There is no way to avoid the conclusion that the FOUNT OF ALL POSSIBILITY is all-inclusive. Life, in its fullness, is terrible as well as beautiful. What we call "bad" or "evil" is, Really, a deviation from the kind of *action* (considering action upon all levels) that would normally lead to the fulfillment of the Design-at-the-Beginning. Beauty is that which tends towards the fulfillment of this Fixed Design. Ugliness, badness, evil all tend towards a discordant design. There is no way to judge an action or a pattern of action out of context, for badness is a *relative* quality just as is *goodness.* We have all had the realization that the moral value of a given act cannot be judged except in relationship to the context in which it occurs. Nevertheless, we must avoid sentimentality when thinking about ABSOLUTE REALITY. 'Within' IT, everything is possible.

It is probably not wise to consider that any kind of *absolutely* 'EVIL' possibilities 'ABIDE' 'within' IT, or, for that matter, *absolutely* 'GOOD' possibilities. The question of the 'goodness' or 'evilness' of a possibility can only be decided in relation to a particular Cosmic Algorithm. *All* possibility, without exception, 'inheres' 'within' IT in *infinitessential identicalness.* 'Good' and 'Evil', depend upon comparison, and are matters to be 'handled' exclusively in-Cosmos!

The Problem of Whether *All* Infinitized Possibilities will be Actualized in a Cosmos

The answer to this seems to be, 'Almost' definitely, No. The reason for this is that the number of possibilities within the FOUNT OF ALL POSSIBILITY is truly *infinite.* The number of *actualizable* possibilities within any given Cosmos is *always finite.*

One can easily see, that if one possibility 'EXTRUDED' from the FOUNT were to be manifested per Cosmos, the difference between the number of possibilities in the Cosmos and the number of possibilities remaining unactualized 'within' the FOUNT would be infinite. Even if there were a *set* of possibilities destined for actualization in a particular Cosmos, the discrepancy between the two would, nevertheless, be infinite. In fact, a little thought reveals that the difference between any finite number no matter how small or *huge* is infinite.

☐ What this means is, that at every Cosmic Opportunity, there are always an infinitude *more* possibilities than can possibly be actualized within the Cosmos. This infinite discrepancy being operative (every *time)* for an infinitude of Cosmic Opportunities, one can see that there will *always* be an infinite number of possibilities which will *never* be actualized in Cosmos.

The most conservative position in this very tricky and confounding field of thought is to say that there are, in this present, finite Cosmos, an infinitude of possibilities that are *not being actualized.* Even within our very limited human experience, common sense tells us that it is so.

Further, it would stand to reason that the general process of *possibility-actualization* for finite Cosmoses is similar. Thus, it would not seem unreasonable to extrapolate to other finite Cosmoses and say, that within each of them, an infinitude of possibilities either *was not actualized* or *will not be actualized.* If we sum all these actualizations and non-actualizations of possibility, we find that far more has *not* been actualized than has been actualized.

Of course, a problem arises in that an infinitude of possibility has, indeed, been actualized, since an infinitude of finite things is an infinitude, and there *have* been an infinitude of Cosmoses. (Can we be thought unreasonable if the thought of greater and lesser infinitudes dawns upon our consciousness precisely in relation to this consideration?!)

So much depends upon the method of 'counting' utilized. If we use the 'matching' technique—comparing two infinite series and matching them item for item, we are (in Time and Space) led to the conviction of greater and lesser infinities. We have two choices, however: to *stop* and compare, or *not to stop* and compare.

- If we stop and compare, we will always have a finite quantity in the series of *actualized* possibilities in a definite number of Cosmoses, and an infinitude in the series of groups of *actualizable* possibilities. The proportion will always be infinity-to-one (taking all definite quantities as a *oneness*, a *monality)*.
- If, however, we *never stop to measure,* but continue comparing series to series forever, there are some who may say that the ratio between both series, as *entireties,* will remain infinity-to-infinity.

Working with the *infinite* (in its 'many varieties'?) consistently produces results which are strange to the Time-bound, Space-bound human mind.

The Problem of Whether Each Number is the Emanation of the Number Immediately Preceding It or Whether All Numbers (Except One and Two) are the Result of the Interaction of Several Numbers (Unitary Entities) Which Precede It

For instance,

1. Is the Number Four the Emanation of the Number Three? Or,
2. Is the Four the result of the interaction between the Entities known as the Numbers One, the Two, and the Three, which combine in different proportions or intensities to produce the Number Four?

Similarly,

1. Does the Eight arise by Emanation directly from the Number Seven, and the Nine from Eight, the Ten from Nine, etc.? Or
2. Does the Number Eight arise from the interaction in various ways of the Numbers One through Seven, and the Number Nine arise from the interaction in various ways of the Numbers One through Eight?

If we accept the examples under #2, let us call them **<u>Model 2</u>**, then the Number Eight will arise through the interplay of the Seven Entities that were emanated *before* the emanation of the Eighth Entity. The Eighth Entity will not be called the Number Eight. Eightness will arise through the *interplay* between the Seven Entities emanated before the emanation of the Eighth Entity, and the Eighth Entity Itself. All the preceding Seven Entities will be of equal value in the production of Eightness and the Seventh Entity will *not* be unduly important in the production of Eightness. As well, the Eighth Entity may not 'come out of' the Seventh Entity but from a Field Interplay of all seven preceding Entities (all of Whom are Unitary Monads—the same in Essence, but different in Power due to Emanative Attenuation).

If we accept the examples under #1, let us call them **<u>Model 1</u>**, then for instance, the Number Nine will, in a way, require the cooperation and interaction of Numbers One through Seven, *but only as these Numbers are 'contained' within Eightness*, from which the Number Nine would emanate. In other words, the Relationship that is Eightness will definitely be involved in the production of Nineness, and that Relationship that is eightness will be intimately associated with the Eighth Entity. This model is, perhaps, easier to understand and has much to recommend it.

☐ The Real question to answer, though, is whether Threeness, for instance, is an Entity (i.e., embodied in an Entity), or whether Threeness is a Relationship between the Entities—in this case, a Relationship between Entities One and Two, which precede the Third Entity and *interplay* with It. Or, perhaps Threeness can be both! This we will attempt to show.

Let us focus for a moment upon the dynamics of **<u>Model 2</u>**. Numbers are produced by Emanation and Interplay. Let us focus upon the production of Threeness according to **<u>Model 2</u>**. From the Number One emanated an attenuated Identity. This Second Entity, (the Emanation of the First Entity) would not be considered, in itself the Number Two. This Second Entity would simply be an unitary Entity. Only by the interplay of this Second Entity with the Entity that emanated It (namely, Number One) would Twoness be produced. How, then, would Threeness be produced according to **<u>Model 2</u>**? Let us consider two Methods.

> <u>Method 1 of Model 2</u>: the First Entity and the Second Entity might enter into interplay, and through their interplay produce a Third Entity. This Third Entity would, in Itself, *not* be the Number Three—the Number Three, or Threeness, would arise from the interplay of the first three Entities.
> Numerically, this might look like:
>
> 1 —> X
>
> X + 1 = 2 as Twoness
>
> 1 + 2 —> X
>
> X + 2 + 1 = 3 as Threeness
>
> 1 + 2 + 3 —> X
>
> X + 3 + 2 + 1 = 4 as Fourness ...

Method 2 of Model 2: the Third Entity *does emanate* from the Second Entity (rather than being produced by the interplay of the first two Entities). That Entity (the Third Entity) that emanates from the Entity that emanated from the Number One, is, again, not in Itself the Number Three. This Third Entity is simply another Unitary Entity (another attenuated Monad). Only when the Third Entity that emanated from the Second Entity (Itself emanating from the Number One) begins its Interplay with both the First Entity and the Second Entity is Threeness Produced.

This could be represented by:

1—> X
X interplays with 1, and then = 2 as Twoness
2 —> X
X interplays with 1 & 2, and then = 3 as Threeness
3 —> X
X interplays with 1 & 2 & 3, and then = 4 as Fourness
4 —> X ...

Using **Model 2** (with either Method 1 or Method 2) we could conclude that no unitary Entity *is* the Number Two, or the Number Three, or the Number Four, etc., but that these Numbers are Composite Entities—emanated Entities in Interplay.

On a synthetic note, the structure of relationship within the Divine Emanatory Stream may provide a way to combine both Models. Each Number would then continue to be seen as an emanation of the Number One (fully in Essence, attenuated in Power, i.e., the Number Two would Essentially *be* the Number One, but the Number Two would be *reduced* in *power).*

☐ Each Number (no matter how large, and no matter how far 'down' the Divine Emanatory Stream It occurs) would, thus, be regarded as a *singularity* and a replica of the Universal Monad (the Number One). However each Number (other than the Number One {and even *It* in a strange way}) would, Itself, be seen as a *Field of Relationships.*

It was proposed earlier, in **Model 2**, that each Number could be considered a relationship and *not,* strictly, an Entity. It was proposed that Twoness, for instance, was produced by the interplay between the First Emanating Entity, the Number One, and the Entity that is Emanated from that First Entity (i.e., the 'Son' of the First Entity).

May it not be that the Entity that is Emanated from the First Entity, is not only a Singularity (thus reflecting the Universal Monad) but is, *Itself,* also, a *Field of Interplay* between the First Entity and Its Self-Reflection as the Second Entity? In other word:

- The First Entity is so much 'within' the Second Entity, that the Second Entity becomes a Field of Interplay, or a Field of Relationship between the Number One-ness of the First Entity (which the Second Entity *'contains'*) and Its Own Nature (i.e., the Unitary Nature of the Second Entity) per se.
- The Second Entity, Itself, is thus to be considered the Field of Twoness, and no further interplay between the Second Entity and the First Entity (an interplay

which would take place *outside* the Field of the Second Entity) is required to produce Twoness or the Number Two. Twoness is already inherent *within* the Second Entity.

From such a perspective, the Second Entity ***is*** the Number Two, because this Second Entity arises from an interplay of the Number One with Itself (i.e., with Itself considered as an Object). The Number One combines with its own Self ('Seen' as an Object) to produce the Number Two. The emerging, emanated Object ('Seen' by the Number One within Itself) *becomes* the Number Two. The formula, "I 'See' MySelf within MySelf" can be rendered 1 + 1 = 2. The 'Seeing' One *enters* the One 'Seen' (the Point of View *enters* the 'Seen-Point'), and that combination 'within' the One 'Seen' makes of the One 'Seen' a Two.

Further, and similarly:

- The Third Entity (emanated from the Second Entity, the hypothesized Number Two) is not only, again, a Unitary Entity (and thus a kind of Monad, like all other Numbers/Entities, which It *is*) but is, Itself, the Number Three, because the Relationship between the First Entity and the Second Entity is already 'contained' within It Itself (It being the Third Entity).

In forming the Third Entity, we see that the Second Entity-as-Twoness (from which the Third Entity emanated) is Itself an Interplay between the first Entity, Number One, and the Second Entity, and thus, when It (the Second Entity emanated the Third Entity as an Object of Itself) the interplaying Oneness and Twoness (resident within the Second Entity) would 'go into' the Third Entity, which (when combined with the Oneness and the Twoness) would yield and embody a Threeness.

The Second Entity in which is embodied Twoness, 'Sees' within Itself another Monad (for every Entity is, whatever Its numerosity) also, a Monad. The Second Entity by 'Seeing' Its own Reflection within Itself (Its reflection as a Monad) adds another *One* to the Twoness of Itself. The Entity Number Two, may contain Relational Twoness within Itself, but its Self-Objectification registers as a Single Object (another Unitary Being, a Monad).

The formula would be as follows:

- The Twoness of the Number Two, plus the Single Image of Itself, becomes a Threeness, when and as the Twoness of the Number Two combines with Its Own Single Object Image. Mathematically this would be: 2 + 1 = 3.

Thus the Third Entity would also be the Number Three, the Embodier *in its own Nature* of the quality or relationship of Threeness. Of course, each Emanated Entity remains a Monad-in-Attenuation as well as a Field of Monadic Inter-Relationships yielding Number. In other words, while 3 = 1, it is also true that 3 = 3. The same formula can be used for every other Number.

One can see that, from this perspective, Emanation is a Mathematical, Psychological Process. The addition occurring is between Subjects and the Objects, or Self-Reflections, They 'See' 'within' Themselves. One can see how every Number, according to **Model 1**, would be formed by the numerosity of a given subject, *plus* the Number One (the Number One standing for the addition of the Self-Objectification of the Subject (seen as a Single Object) to the numerosity of the Subject.

The image of structures and shadows can help us understand the way this process of Emanative Addition works. Every opaque structure (no matter how many parts it may have) will appear as a unitary, unarticulated shadow. Within the shadow, all articulation is reduced to oneness. Thus, we see how a Subject with a numerosity greater than One, will always have a shadow which appears as a One. Every Number is formed by the Number preceding It, plus the addition of the Monad, the Number One.

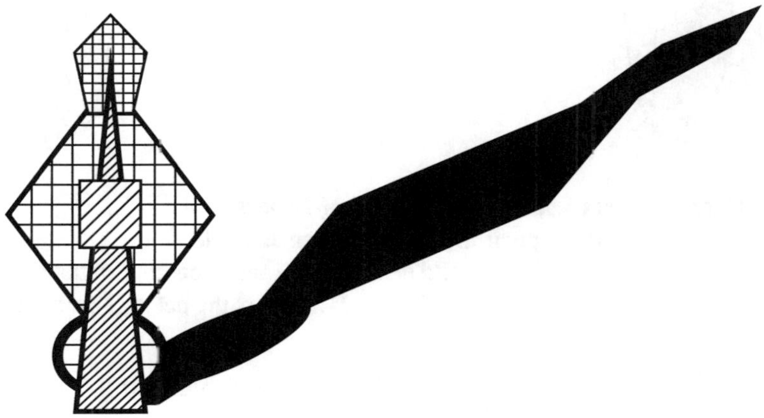

We are also talking about a very interesting kind of addition of symbols. Thoughts contrasting the *field* and the *point* have been discussed. The 'field' is what any Entity 'Sees' of Itself *globally, as a whole.* The *point* represents specificity and attenuation of Self-Reflection. In a way, every Number is formed by the *addition of a field plus a point.*

- If **8** Am Number Two, **8** experience Myself within a Field called Twoness, a generalized Field suffused with the Quality of Twoness.
- If **8** concentrate with that Field, however, **8** 'See' a Self-Reflected *point* that is My specific Image-in-attenuation.
- What **8** 'See' is Really a Monad, just at **8** Am inescapably a Monad (in this case, a Monad suffused with the Quality of Twoness).
- When **8** 'go forth' into the 'Seen Point', it is as if the Field **8** presently Am, 'adds Itself' to the *point*, thus becoming another Field (this time called Threeness).
- From the Perspective of Emanative Retention, although **8** as the Field of Twoness 'go forth' into the *point* **8** 'See', thus establishing a new Field called Threeness, yet, **8** 'remain' within the Field of Twoness, and as **8** 'look towards' the Field of Threeness, it does not appear as a Field, but still only as a *point.*
- **8** have to be 'within' the *point* in order for it to appear as a Field.
- Similarly, if **8** as Oneness, were 'looking at' MySelf within the Field of Twoness, that Field of Twoness would appear as a *point* to the Consciousness within the Field of Oneness.

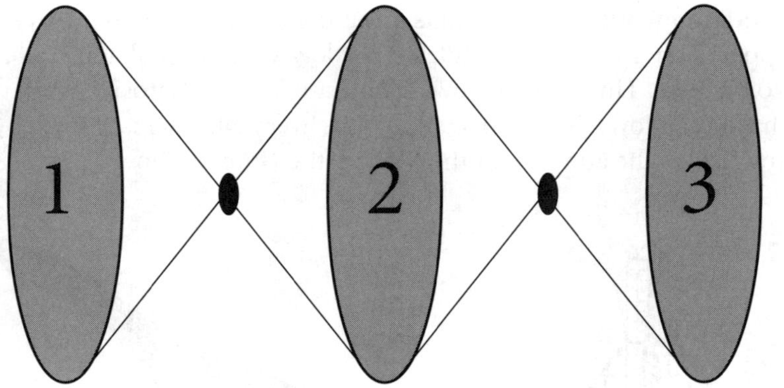

Field of Oneness ... sees a point ... of the Field of Twoness
If the "1" ... enters the point ... Twoness is seen as a field
Field of Twoness ... sees a point ... of the Field of Threeness,
If the "2"... enters the point ... Threeness is a seen field

Thus Field and Point dynamics are very important to grasp if one would visualize the Emanative Process with accuracy and understanding. Another image which may be useful in understanding the Emanative Process is as follows.

The image of a large sphere with ever-smaller concentric spheres may (from a certain angle of vision) be illuminating. The large sphere, the Number One, generates the next smaller concentric sphere, the Number Two. The Number Two is, therefore, *contained* within the Number One (which is Emanatorily and Metaphysically true). The Number Three, which would be the next smaller concentric sphere would be (even though it is a *larger* Number) *contained* within the Number Two <u>and</u> the Number One. This too is true. *Thus would Emanative Descent be symbolized by ever-smaller spheres within spheres.*

The opposite picture would also be revealing: *the Number One would be the smallest of all spheres, and each succeeding Number would be a larger concentric sphere.* Certainly this would illustrate that the Number One was *contained* within the Number Two, and that both the Numbers Two and One were contained within the sphere called Number Three, and so forth.

This latter model would illustrate the principle that relationship and interplay between preceding numbers form the next larger Number (as symbolized by the next larger sphere). The former spherical model would illustrate the idea that the smaller Numbers, though *quantitatively* smaller are *metaphysically* larger, for indeed, all Numbers, all spheres, are contained within the Number One.

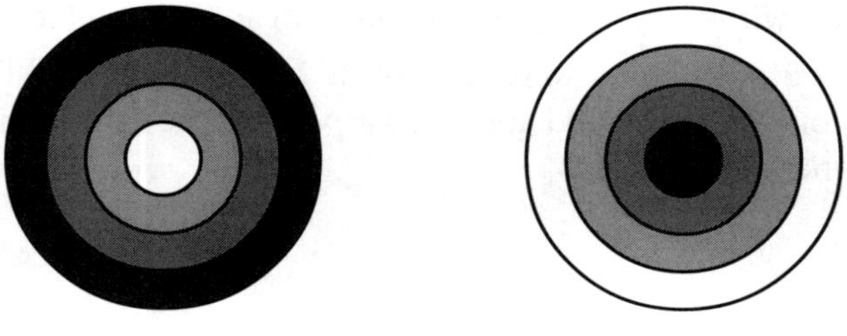

The Problem of Whether any B/being in Cosmos can Really 'Inperience' the ETERNAL NOW or Only the Eternal Now

While the Eternal Now, or Cosmic Eternal Now is the 'gateway' to the ETERNAL NOW, the REAL 'inperience' of the ETERNAL NOW may necessitate the collapse of all Time and Space into ABSOLUTE NO-THINGNESS. If the ETERNAL NOW is truly inaccessible, those, who in Cosmos, supposedly 'inperience' a Cosmic Eternal Now would have no way to judge whether their 'inperience' was REALLY that of the ETERNAL NOW.

☐ The 'inperience' of the Eternal Now is sufficiently radical; it involves the negation of Cosmic Time past, and Cosmic Time future. The Real Eternal Now, moreover, may only be inperienced by Those whose level of focus has been polarized within the World of Being. This would naturally (except under extraordinary conditions) eliminate the human being from the list of Those capable of 'inperiencing' the Eternal Now.

Perhaps, the 'inperience' most human beings have when they say they are 'inperiencing' the Eternal Now, is that of the ongoing Presence of the dual Cosmo-Objective Now and Cosmo-Subjective Now ('reached' through the successful cultivation of the Attitude of the Observer). Perhaps such observers-becoming-Observers manage a fleeting alignment with the Logoic Perception of the Cosmic Configuration (devoid of the perception of the incredible configurational detail) from and 'between' ultimate moment to ultimate moment. The successfully completed Antahkarana will make one an Observer of Modification.

If the Eternal Now is remote, the ETERNAL NOW is *infinitely* more remote:

- 'Infinispectivizing' will reduce Time and Space to infinitesimalizings and thus facilitate apprehension of the Eternal Now in Cosmos; the
- 'Infinispective' will *periodically* reveal (to Consciousness) *infinite Time/Space* (and will thus produce a kind of Super-Cosmic Eternal Now); but the
- 'INFINI-SPECTIVE' alone (obliterating as IT does all Time and Space, whether Cosmic or Super-Cosmic) is required for the *absolute* 'REALIZATION' of the ETERNAL NOW). (Perhaps this 'INFINISPECTIVE' should be called 'INFINIDENTIFICATION' since no 'SIGHT' is REALLY involved; only 'BEING' is.)

Beings in-Cosmos (of the status of a human being) still must deal with the presence of objects, even if they learn that those objects are simply Self-Objectifications. A tremendous assertion of *synthesis* in consciousness is needed for such beings to transport themselves into the Cosmo-Eternal Now (a 'Reflection' of the 'Infinispective'—the Infinified 'View' of the Pre-Cosmic Infinite Subject).

It is almost certain that the 'INFINISPECTIVE' cannot be achieved by *any* being in Cosmos. The 'INFINISPECTIVE' or 'INFINIDENTIFICATION' pertains to the ONE AND ONLY BEING/NON-BEING, and it is *this* INFINITE SELF *alone*, WHO can *fully*—not only 'INPERIENCE', but *BE* the ETERNAL NOW. To achieve the REAL ETERNAL

NOW can be reasonably considered the prerogative of the INFINITE SELF in Universal Pralaya. This does not mean that every apparent B/being-in-Cosmos is not forever *in* the ETERNAL NOW—no one and no-thing can *escape* IT. Consciousness of that NOW, however, is quite another matter. Since *consciousness,* per se, is Mayavic, consciousness is not REALLY the way to 'inperience' the NOW. Object-annihilating BEING is the requirement.

The Problem of the One Self, Emanation, and the Monad

We will now touch on an area of inquiry, without the relative mastery of which the Self and Identity cannot be understood in depth. A few fundamental ideas will be presented and then elaborated.

- **8** Am the Father, the Universal Logos.
- **8** See MySelf, and the first thing **8** See (other than My Own Being reflected *as* the Infinite Mother) is the Son.
- First, **8** Know **8** Am MySelf as Mother, Cosmic Prakriti.
- Then, **8** Identify with what **8** 'See' within My Own Being—the Son. The Son is already within Me and *is* **8**, for what else is there but **8**? Thus, the Son, though concealed, is, like **8** MySelf, Essentially as Subject. The Son, however, is also an Object. **8** Realize that the Son **8** 'See' within Me *Is* **8**, just as **8** Am **8**—**8**, Who 'See' the Son.
- Now, through 'Self-Sight', Self-Reflection, Self-Objectification, Identification and consequent Emanation, **8** 'Become' the Son (Who always has been **8**, as well as *remaining* the Father).
- Looking deeply within MySelf, **8**-the-Son find My Own Fatherhood (My Fatherhood within and as the Son).
- Then, **8**-the-Son Emanate the First of the Three Sub-Logoi Who stands in the Company of the Son, and **8** as the First Sub-Logos Emanate the Next, and that Next the Next, and so forth (unfolding in Time) the Divine Emanatory Stream.
- Looking at the Divine Emanatory Stream "from above downwards', **8** Am the Father, **8** Am the Son, **8** Am *all* the many, many Emanated Sons of the Son, and **8** Am *focally* and (however, only *apparently)* an Individual Universal Unit.
- **8** Am, yes, an apparently specific 'Ray' of the ABSOLUTE, with a Point of View (but **8** Am not *exclusively* so), for, as well, **8** Am *focally all 'other'* Individual Universal Units. **8** Am all *apparently other* 'Rays' of the ABSOLUTE with a Point of View.
- The Father is a 'Ray' of the ABSOLUTE and the Son is a 'Ray' of the ABSOLUTE, but the Son *is* the Father (i.e., an Object/Subject, or Object-to-be-Subject, 'within' the Father) 'Rayed Forth' from Him (the Father), an thus, the Son is a 'Ray' of a *'Ray'*.

- A 'Ray' of the ABSOLUTE is an Act of Self-Reflection. *The* 'RAY' of the ABSOLUTE is the PRIMAL 'ACT' of SELF-'REFLECTION'. All 'Rays' of the ABSOLUTE (in-Cosmos) are identical with the First *Cosmic* 'Ray' of the ABSOLUTE which means that all 'Rays' of the ABSOLUTE are identical with the ONE AND ONLY 'RAY' of the ABSOLUTE. That **8** Am a 'Ray' of the ABSOLUTE, means that **8** Am a 'Ray' of MYSELF.

Seeing exists. Something is 'Seen' and Something 'Sees'. This is Self-evident. Since, however, the only 'THING' that IS, is *THAT*, the SELF, then, the Something that is *'Seen'* must BE none other than that SELF, and the Something that *'Sees,'* must BE none other than that SELF (though it may be that the SELF-as-SELF cannot REALLY 'SEE' but can only 'BE' and, therefore, must 'SEE' *by proxy* as it were). Self-'Seeing' is *Self*-Reflection. SELF-'SEEING' is *SELF*-REFLECTION. SELF-'BEING' is 'INFINIDENTIFICATION'.

Most E/entities in Cosmos are 'Rays' of 'Rays' of 'Rays', etc., depending upon the 'altitude' of the 'Ray' within the Divine Emanatory Stream. No 'Ray', however far removed from the First Universal Logoic 'Ray', (i.e., the 'Ray' of the One) is any *less* a 'Ray'. All 'Rays' are identical, from the highest to the lowest, and all 'Rays' are identical with the 'RAY' of the ABSOLUTE. If any 'Ray' seems lesser than another 'Ray', it is due to the relative difference in scope for expression, to the relative difference in depth of prakritic immersion. The 'Ray' of the ABSOLUTE that manifests as an atom of matter is *identical,* for instance, with the 'Ray' of the ABSOLUTE that manifests as the Universal Son of the Universal Father—the Universal Logos.

☐ This study of 'Ray' Dynamics shows that in all of Cosmos, there is but One Entity in Emanative Self-Extension, which ultimately means that there is but ONE ENTITY/NON-ENTITY in 'RADIATED'-SELF-'EXTENSION'.

Does Focal Identity (i.e., apparent *distinctness of Identity)* REALLY 'EXIST'? It does not. As well (intra-Cosmically considered), *distinctness of Identity* Really does *not* Exist. This means that all Identities are ABSOLUTELY ONE with the ONE AND ONLY IDENTITY/NON-IDENTITY. It also means that, in-Cosmos, all Identities are Absolutely One with the One and Only Cosmic Identity, the Universal Logos.

Even if Focal Identity does not REALLY 'EXIST' and does not Really Exist, is Focal Identity a universal *illusory* experience? Yes, Focal Identity *is* a universal illusory experience. When did Focal Identity—the universal illusory experience of the *distinctness* of Identity arise?

- As the Father, **8** *know* **8** *Am* the Father, for **8** 'See' MySelf Reflected as the Mother, thus revealing to Me that **8** Am, indeed, the Father.
- Thus **8** know **8** Am the One. (We, together, Mother and **8**, *know* that We-the-**8** Am the One.)
- **8** Am the One, and **8** *remain* the One, even as **8** 'go forth' Emanatively through 'Particulate Self-Recognition' (i.e., Recognizing, in MySelf, that 'Part' of Me who is the inherent Son) and, thus, **8**, *apparently* (as the Son of MySelf) become less, *apparently* reduce My scope and depth of Self-Image.
- On this high level, **8**-the-Son *seem* to be *less* (when **8**-the-Son appraise myself through Focal Identity) than **8**-the-Father.

- Of course, Essentially, **8**-the-Son, and **8**-the-Father are absolutely *identical.* It is our 'prakriti' or Reflected Self-Image that varies.
- Farther 'down' the Divine Emanatory Stream, in every case of fairly enlightened Self-appraisal, **8** (as/at some level of Emanated Sonship) begin to think **8** am a lesser Being {lesser than the One Who Emanated Me and certainly lesser than the One Father}), because when **8** look into MySelf deeply, and try to discover Who **8** Really Am, **8** discover that **8** Am being 'Seen'!
- **8** discover through the Course of Evolutionary Time that it is a deeper 'Whole within MySelf' Who is 'Seeing' me. Of course that Whole Who 'Sees' Me, is, **8** MySelf.
- Thus, along the Divine Emanatory Stream, Whoever **8** Am (in terms of Focal Identity), and at whatever depth of prakritic immersion **8** may seem to be principally *focused*), **8** Am being 'Seen' by a greater One, Who is being 'Seen' by a still greater One, Who, in turn, is being 'Seen' by yet a greater One ... etc.
- Nevertheless, though a great chain of ever more inclusive 'Seers' is revealed, **8** discover that it is none other than **8**-the-One-Self, Who is doing the Seeing in *every instance.* **8** Am the Self-'Seen' 'Seer'.
- The varying dimensional depths of My Emanated Identity arise from the number of times **8** have been *descendingly* Self-Objectified.
- **8** begin to realize this from the dawning experience of being *multiply* 'Seen' by MySelf in ever-ascending Spheres of more inclusive Wholeness until, finally, the Point of the One Universal Self is Reached.
- Thus ascending, **8** See MySelf Seeing MySelf, that 'Seeing'-Self, Itself being 'Seen' by a still greater 'Seeing Self' Who 'Sees' Me as well, and so forth.

We might say, for instance, that the Number One, which is divisible *into* all the Numbers, is <u>*in*</u> all the Numbers even as They are in the Number One. From the perspective we have been discussing, this means that the Number One 'Sees' all the Numbers, and They, if They learn to 'repent' (i.e., to "turn back", and look deeply within as well), 'See' that They are being 'Seen' by the Number One. The Number One 'Sees' all (magnitudinally) larger Numbers from *Its Own High Perspective*, but the Number One, for instance, *also* 'Sees' the Number Three through the 'Eyes' of the Number Two, in which the Number One participates, and, continuing, the Number One 'Sees' the Number Four *through* the 'Eyes' of the Number Three in which the Number One participates by the fact that It participates in the Number Two which, Itself, participates in the Number Three, which, Itself, is the Chief Emanating 'Seer' of the Number Four, etc.

The 'Process of Seeing' considered in a *descending manner* 'BEGINS' in THAT, though it does not continue 'THERE'. It is *instantaneously expelled* in an *infinitesimal unit of newly-born time* resulting in **I/8** (now Pre-Cosmically and *not* intra-'SOURCEDLY'), 'Seeing' MySelf as the Infinite Being—Mulaprakriti. Then, obeying within MySelf the 'INHERENT, intra-SOURCE-'ARISEN' 'COMMAND' of *THAT* to *Finitize,* **I/8,** (the Infinite Subject 'Seeing' MySelf as the *Infinite* Object) Will to begin 'Seeing' MySelf as *Finite* rather than *Infinite,* and *do so* until via 'De-Infinitizing' (whether instantly, or through the passage of Pre-Cosmic Time), **8** 'See' MySelf as the One, the Singularity, the One Bounded Subject reflected in Bounded Mulaprakriti, or Cosmic Prakriti.

'Before' the 'Process of Pre-Cosmic Seeing' is 'INAUGURATED' in an *infinitesimally instantaneous* 'FLASH', **I-the-ONLY** could only *BE.* **I** 'ABIDED' in 'INFINIDENTIFICATION'. When 'FLASHED FORTH', however, **I** 'BECOME' **I/8** and 'See' MySelf as Infinite, the Infinite Object, whereas, *'Before'* the 'FLASH', **I** could only *BE* THAT. And, indeed, **I** 'CONTINUE' MY 'INFINIDENTIFICATION' even as 'below' in Super-Cosmos, **I-as-8** proceed with Infinite 'Seeing' and the finitization of Infinite 'Seeing'—Finitized 'Seeing'.

☐ In light of these ideas, what does my (human) so-called Individuality REALLY and Really Mean? It does not mean the **8** Am an Isolated Distinct Unit of Particularized Identity; however, much **8** might *appear so* to MySelf (as Focal Identity) and to 'others' (though such 'others' Really do not exist).

8 Am the 'Seen' Son, of a 'Seen' Son, of a 'Seen' Son, etc. Thus, my individuality does not consist in *isolated distinctness.* **8** have My Individuality (which is Really, My *Indivisibility)* from the Beginning of Cosmos. **8** have the Individuality/Indivisibility that every member of the Family of E/entities has. **8** share Individuality/Indivisibility with each and every one. Am **8** *not* each and every one?

For instance, there is a Number Six and a Number Seven. There, apparently, is an Entity Who is the Number Six. There is, also, apparently an Entity Who is the Number Seven. The Universal Logos, however, is the *One*, and is, thus, *all* Entities in Cosmos (even *more,* Really, than They are, Themselves). Thus the Universal Logos *Is* the Number Six *and* the Number Seven. **8** Am the Universal Logos, as (in-Cosmos) there is naught else, Really, to Be. **8**, therefore, Am both the Number Six and the Number Seven.

There is, however, an illusory *Focal Identity* associated with or inherent to the Number Six. There is also an illusory Focal Identity associated with or inherent to the Number Seven. That particular 'Son of the Father' Who has Emanatorily *become* the Number Six, may identify with *being* the Number Six, exclusively, and forget that He is, Essentially, the Father.

Likewise, for example, the particular 'Son of the Father' Who has Emanatorily *become* the Number Seven, may identify with *being* the Number Seven, exclusively, and forget that He is, Essentially, the Father, i.e., that He is the 'Seen Son' of a 'Seen Son', etc., Who is, ultimately, the 'Seen Son' of the Father—the Father, Who *Is* (through 'Self-Seeing') the Universal Son.

☐ This mistake in Essential Self-Identification, of course, is far more likely to happen to entities at a *lower* point of prakritic immersion, such as human being, than the Entities here discussed, but the example is used simply for clarity because the Archetypal Numbers are so simple and easy to understand. On a far lower levelhuman beings think they are a specific personality, and forget they are an emanation of the Soul which, Itself, is an emanation of the Monad, which Itself, is an Emanation of a number of far greater Beings. Localized Identity prevents an understanding of Emanatory Roots.

- Now, since **8** *Am,* the Universal Logos **8** Am all His Sons as well.
- When **8** Am (as part of the Emanatory Process) functioning as the Number Six, **8** may make the mistake of thinking that my *Real* Identity is Number Six.

- At the same time **8**, as the One, will 'See' MySelf making that mistake.
- Likewise, when **8** Am (as part of the Emanatory Process) functioning as the Number Seven, **8** may make the mistake of thinking that my *Real* Identity is Number Seven. At the same time **8**, as the One, will 'See' MySelf making that mistake.
- Now, even though the 'immersed Son' Aspect of MySelf-the-One may be making a mistake in both 'places', Am **8** who is the One, Really any *more* the Number Six than the Number Seven.
- Am **8** not Who **8** Am (the Number One, i.e., the One), regardless of the Focal Identities which **8** (as Emanated 'Sons' of MySelf) may wrongly presume to be Real?
- Similarly, as a particular human being (My present Focal Identity) Am **8** Really *more* one human being than another?
- Over *here* **8** may (in terms of Focal Identity) *seem* to be *more this* human being than *that* human being. *There,* however, where **8** Am as equally Present as **8** Am over *here,* **8** may *seem* to be *more that* human being that *this* human being. In fact, *this* and *that* will exchange places.
- Really and Essentially, **8** Am no more one than the other. **8** have Focal Identity in *each* and *Real* Identity in *each.*

These examples generalize that, no matter what Focal Identity any E/entity in Cosmos may *seem* to have, its *Real* Identity is *always* that of the One (Who, Itself, is, ESSENTIALLY, the ONE AND ONLY ONENESS/NONENESS). If at any point in Cosmos, **8** think that **8** am any *isolated,* and *distinct* Identity, **8** Am severely mistaken. **8**-the-One may manifest both the *functions* called Number Six and Number Seven, but **8** Am no more Number Six than Number Seven, and Really Am *neither* of these, Essentially.

☐ **8**-the-One may manifest both the *functions* called 'I'-the-Author and 'You'-the-Reader, but **8** Am no more 'I'-the Author than 'You'-the-Reader, and *Really* Am *neither* of these, Essentially, while, from another perspective, include *both* as examples of My Own Self-Objectification. 'I'-the-Author, and 'You'-the-Reader are *illusory* though *actual* Self-Objectifications of MySelf-as-the One.

There is but One Cosmic Identity *(rooted* in ABSOLUTE IDENTITY) which has multiple experiences of Focal Identity and also the single experience/'in-perience' of *One* Identity. **8** Am the *One Identity,* Cosmos-wide, having multiple simultaneous experiences of Focal Identity in/from many Points of View (i.e., from many perceptions of Self-extent and dimensional depth).

8 Am not and never have been an a *partial thing* (for instance, an isolated *distinct* Monad or an isolated, *distinct* 'Ray' of the ABSOLUTE). No 'Ray' of the ABSOLUTE can be *isolated* and *distinct.* **8** Am the *One Ubiquity,* present everywhere in Cosmos, through all E/entities. From the illusory perspective of Focal Identity, **8** Am the One Ubiquity seemingly *partialized* and 'you' (and every other *apparently distinct* E/entity) are also the One Ubiquity seemingly *partialized.*

'You'-the-**8**, however, and 'I'-the-**8**, are never *partial things,* never 'isolated', isolatedly 'distinct' Monads or 'Rays of the ABSOLUTE'. Ever and always We-the-**8** are only the One (the Universal One 'Seen' by Its Pre-Cosmic 'Seers', Who are but the ONE AND ONLY 'SEER/NON-'SEER'—the INFINITE SELF).

☐ If **8** Am the Monad, **8** Am the One and, never (though **8** may, at times, *seem* to be so from the mistaken perspective of Focal Identity) a *fragment.* The Monad is *indivisible.* A 'Ray' of the ABSOLUTE is *indivisible.* The implications for this conclusion are *profound* for the functioning of the entire Process of Emanation along the Divine Emanatory Stream.

Since **8** Am not an Isolated Distinct Unit (an *Isolate),* **8** (as an 'Isolate') do *not* Really have a *unique Path* in the Divine Emanatory Stream. There are, indeed, Paths of Descent delineated as **8**-the-Father-of-all-my-successive-Sons, send MySelf forth as Son after Son after Son. All the apparently *lesser* Sons that **8**-the-Father 'Become', have arrived at their present depth of prakritic immersion (their present dimensional depth) via a unique or distinctive Path. **8**-the-One, however, have descended upon *all* Emanative Paths.

If **8**, as some apparently isolated and particular Son (focusing upon my illusory Focal Identity) think that **8**, as an isolated, distinct, 'something' have descended (or will reascend) along a particular Emanative Path, **8** will be mistaken. The truth is otherwise, for all along the different higher 'levels' of the Divine Emanatory Stream, **8** have not existed as an isolated, distinct thing, but only as the One—the One 'Seeing' MySelf 'Become' Sons who initially, at least, (until the *evolutionary corrective* has been applied) *think* they are isolated, distinct things! It might be said, tellingly, that **8** have no Real *'dividuality'* [sic]! Contrarily, **8** *do* have Individuality (i.e., Indivisibility).

8 the One Cosmic Life, descend along the Divine Emanatory Stream in all possible Cosmically-allowable ways (for the Divine Emanatory Stream has a certain 'shape' and structure according to the nature of the Cosmic Algorithm). **8** Am everywhere within the structure of that apparently *hierarchical* Stream *simultaneously.* That the Emanated Aspects of MySelf mistakenly *think* they are in one or other 'location' within the Stream, is the result of Universal Maya, which (functioning as the *narrowing of Cosmic Consciousness*) forces the Emanated Aspects of MySelf to *forget* from Whence they came. They forget, even as **8** the One, look on from My Oneness (and, as well, from every Emanatory Position superior to Them), 'Seeing' Them forget.

☐ If **8** (temporarily in my illusory Focal Identity) attempt to *retrace* MySelf along the Emanatory Stream, **8** will always be led into illusion if **8** am looking for the Descending Path of a distinct and isolated 'Unit of Spirit'. As **8** begin to ascend, that 'Unit of Spirit' is nowhere to be found. Whatever fragile existence it seemed to have, disappears as soon as the field of the Immediately Superior Emanatory Sphere is reached.

All mistakes is assessing nature of Identity and its Source are made by working 'from below upwards' and trying to retrace the Emanatory History of an *illusory concrete, limited, distinct 'Unit of Spirit',* and by presuming that it existed as that 'distinct Spirit-thing' on all possible superior levels of Cosmos, even up to the Universal Son or the Son of the Father. The search will be futile. Distinct and isolated Identity is a *gross illusion.* The only distinct Identity which exists is the One, and even It is *indistinct* from THAT, the SOURCE, the INFINITE SELF.

So materialistic are the ideas of present day humanity, that they tend to concretize all thoughts relating to 'substance'. (Substance is Really very *immaterial,* i.e., *non-objective.)* People identify with objects rather than the One Ubiquity. They always think that they have to be some-thing, rather than what we might call an *'indefinite'*. This is illusion—the identification of the Subject exclusively with the Object, instead of realizing that the Object is simply the Subject-'Seen'.

Let us bring this whole inquiry "close to home":

☐ So many students of esotericism speak of *their* Monad or *their* Soul. First of all, there is no such thing as *a Monad.* There is only *the Monad.* There is no such thing as *a 'Ray' of the ABSOLUTE.* There is only *the RAY of the ABSOLUTE.* 'I' do not have *a* Monad. It is incorrect to ask, What is *my* Monadic Ray? I AM **8**. My only REAL IDENTITY is THAT, only THAT. My Secondary Identity, which is *Cosmically* Real, is *That* (the One, the Universal Logos). ESSENTIALLY, as the great Vedantins have said, "There is only THAT. All else is Illusion."

8 Am NOTHING. 'You' are NOTHING. 'You'-and-'I'-as-**8-as-I/8-as-I,** are REALLY NOTHING. As hard as it is for us, in our Focal Identities, to grasp, We are REALLY *NOTHING.* Failing that ULTIMATE IDENTIFICATION, We are the One (Who also Is NOTHING). If we say we are a 'Monad'; if we say we are a 'Spirit'; if we say we are a 'Ray' of the ABSOLUTE, and mean thereby, a distinct, isolated, individually identifiable enduring quantity-in-Cosmos, we are in the depths of *egoism* (not egotism). 'Egoism' is more fundamental, and 'egotism', merely, a predictable extension.

We must grasp what all this means Emanatorily. If **8** seem to 'divide', **8**, in fact, do not Really do so, but, instead, *remain* exactly as **8** Am. Through Emanative Retention, the Father *remains* the Father even as He 'Becomes' (through Self-'Sight') that Object 'within' Himself we are calling the Son. Further, the Son who is, pragmatically, *functionally less* than the Father, is, in terms of *Real* Identity, *fully* the Father, the One (just as are all the many lesser *sons* 'Down Stream').

If **8** in my Focal Identity look 'Up Stream', **8** am likely to ask:

- Where did ***8*** come from?
- Where was **8** at such and such a time?
- Where was **8** before the Earth appeared?
- Where was **8** before our Solar System appeared?

And so on, *almost, ad infinitum.* Can we see the inherent error in this line of inquiry? Can we see, in this line of questioning, the futile attempt to consider the **8** as an isolated, definite, particular *thing?* That which **8** *seem* to be now, a human being of a certain nature and character is not *at all* my Real Identity (let alone my *REAL* IDENTITY).

☐ That humanness with which 8 now identify, is merely My Self-Objectification at a certain dimensional depth in Cosmos.

Even now, the Real **8** has a multitude of *other* Self-Objectifications, all of them being, Essentially, the One. Even right Now, **8** Am no more *here* than **8** Am *there.* **8** Am in *both* places fully, and in no Cosmic 'Place' at all. It is **8**, and **8** alone (**8**-the-One), who 'descends' the Divine Emanatory Stream. Equal, under the Eye of God (My Eye), there is no Monadic Point of View, no 'Son-like' (apparently finite) Individuality which is better of greater than another, for *all* are the One.

'My' Focal Identity tells 'Me' that **8** should be able to *retrace* 'My' Emanational Path to the 'time' *before* the Monads 'settled within' their present depth of prakritic immersion—i.e., upon the second subplane of the Cosmic Physical Plane (the Monadic Plane). It is true that a 'Ray' may be retraced; all 'Rays' may be retraced. *All* 'Rays', however, are *'My'* Ray'. **8**-the-One have been Present through *all* of them, in all sorts of prakritic contexts, and 'Points of View'.

At a given point in the Evolutionary Process, the 'Ray' that **8** usually *think* **8** Am will be reabsorbed into the greater 'Ray' from which It originally went forth.

- What then happens to *'Me'* if **8** Am then reabsorbed into that 'Ray'?
- Is my individuality destroyed?
- Do '**8**' disappear?

It may take a major readjustment of thought, but it must be realized that it is **8**, *MySelf,* (as Absorbent 'Ray') Who reabsorbs the (only *apparently* 'lesser') 'Ray' with which **8** had *Focally* identified MySelf.

What, then, of all the other 'Rays' which/Who are absorbed by the same greater 'Ray'? Are, suddenly, all those other 'Rays' Who were *not* **8**, 'mixed together', as it were, with 'Me'? (See how *illusorily concrete* this way of thinking can become!) Well, these so-called other 'Rays', were none but **8** all along, and the apparently 'greater' 'Ray' Which absorbs the 'alls' of 'Me' is certainly just as much **8** as the more familiar 'Ray' with which **8** had been identifying *exclusively.*

The 'day' will come when even that 'greater', absorbing 'Ray' will be reabsorbed into Its Parent 'Ray' (along with a host of other parallel 'Rays'), and the still greater 'Ray' that absorbs It will be **8** as well, until, at last, the **8** that is the Universal Logos absorbs all 'Rays'—all 'Rays' which all along have been multiply, and simultaneously, **8** from every possible Point of View and dimensional depth.

The Problem of Forgetting

Descending, **8** forget, and yet **8** remember. **8** Am always what **8** Am. The forgetting 'part' of MySelf thinks it is 'something', something Real and distinct, but it is not.

☐ What **8** usually call 'I' is the *forgetting* part of MySelf. That part is called a Monad, but it doesn't Really exist as something permanent-in-Cosmos. It doesn't *Really* exist at all. Only the remembering part of MySelf (Who Is the Cosmic *Oneness)* Really exists.

The forgetting part of MySelf must 're-face Me' and thus become the remembering part of MySelf. The relationship between the two 'parts' can be expressed by re-wording the ancient mantram thusly, Having 'Become' through Self-Objectification an *apparent fragment* of MySelf, **8** Remain! The SUPER-Cosmic version would be, Having 'BECOME' through SELF-'OBJECTIFICATION' an *apparent fragment* of MYSELF, **I** 'REMAIN'!

Who then are the Pilgrims, those Who 'go forth' into the 'Distant Land of Forgetfulness'? There is only *One Pilgrim. Distinct* Identities (separate-unto-Themselves) do *not* go forth.

- **8**-the-One go forth appearing to *become* a lesser object, and My **8**-ness in that object *forgets* both Who It Really Is, and *Who* **8** Am.
- From high 'above' My forgetting self, **8** always *know* the truth of My Indivisible Oneness. Below, however, **8** forget, and **8** forget on many, many levels of Cosmos.
- The Object which is also a Subject, forgets the greater Subject from which it came, and which it *Is*. **8** think **8** Am a 'part', but **8** Am the Whole.
- **8** have forgotten Who **8** Am.
- **8** Am *not ultimately* a Monad.
- **8** Am the Cosmic Subject, the One.
- **8** must peal back the layers of My Identity to stop thinking that **8** Am a limited thing—a limited thing like a Monad!

This seems a shocking thought to those who, having been trained in occultism, think of 'their!' Monad as the highest aspect of their Identity. Can we see the reductionistic, materialistic point of view this represents?

A Monad is not what we think. A Monad is *the* 'RAY' of the ABSOLUTE. A Monad is not a *thing*. A Monad cannot be 'seen' Objectively from 'below' as if it were a thing. The Monad is only found when one discovers within oneself the great Cosmic Monad (his Oneness). **8** may think **8** Am a 'part', but **8** Am the Whole! **8** Am the Wholeness doing *this*, **8** Am the Wholeness doing *that*. **8** have no *separate identity*.

☐ There is no such thing as a *separate* Monad. There is no such thing as a *separate* 'Ray' of the ABSOLUTE. My apparently separate Identity is a *seeming* Identity, and cannot be transferred to the Monad or to the 'Ray' of the Absolute.

8-the-One exist at *all* levels of the Divine Emanatory Stream. There are *no* 'Real Monadic Individualities'. There are only Objectifications of **8**-the-One, and these Objectifications are *numbered* in each Cosmos, which makes 'me' (in my usually deluded Focal Identity) think that there are certain *number* of Monads. My Objects, though they take on a life *seemingly* of their own, and seem to be Real 'somethings', are not so. Only **8**-the-One Am Real.

First 'MAYA'-instantly-Maya, and, then, Universal Maya is the 'Authoress of Forgetfulness'. Self-Objectifying Consciousness is 'Forgetfulness of Origin'. **8** banish Maya by 're-becoming' that Greater Self Which 'Sees' 'Me', and from which **8** have been 'distanced' through Its/My Act of Self-Objectification. When the "Spell of Maya" lifts at the end of the Cosmic Process, **8** will not longer 'See' what **8** Am, **8** Will simply Be It!

How will **8**-as-Focal Identity 'remember' what **8** have forgotten? **8** must become cognizant that 'Something' is 'Seeing' Me, and that That which is 'Seeing' Me is **8**, My Real Self (at least more Real than the Self **8** think **8** Am by means of Focal Identity). **8** must *become* that which is 'Seeing' what **8** usually think **8** Am. In fact, **8** already *Am* that which is 'Seeing' what **8** *think* **8** Am. Then the **8** that has forgotten will no longer forget the **8** which has *not* forgotten!

The Problem of Whether Time and Space Exist in the Higher Worlds

It is often said that, even in the World of the Soul, Time and Space no longer exist. Wherever, however, there is a One, Time and Space automatically exist, because every One is a coeval Three. From the 'MOMENT' the 'RAY' of the ABSOLUTE has 'FLASHED FORTH' *Space exists,* because the nascent Infinite Subject (with infinitesimalized instantaneity) 'Sees' Itself as the Infinite Object—Mulaprakriti, or Infinite Space.

The Arising of the Infinite Pre-Cosmic Trinity signals an Event, an Appearance, and that Event, since It is one of an infinitude of similar Events that have occurred along the Infinite Time Line throughout Infinite Duration, becomes an Event in Infinite Time. Thus, with that Event (REALLY, preceded by the virtually simultaneous 'EVENT' of the 'FLASHING FORTH' of the RAY of the ABSOLUTE) *Time* can be said to exist. Since Time and Space exist Pre-Cosmically, they certainly exist in the far more condensed Worlds of the Higher Cosmos, the World of Being.

☐ If the World of Being is said to be 'Timeless', it is only so with respect to the Cosmos, and *not* as measured against the 'Exhalations' of the Great Breath—the Ultimate Standard of Measurement for the measurement of Time.

It is the uninterrupted sustainedness of the World of Being which contributes to the *thought* of Timelessness, since when there is a Fixed Design, and *nothing moves,* there is no Time. (That within the World of Being there is, hypothetically, a slow, sustained, and 'timed-to-Purpose' Movement-in-Consciousness through changes in identificatory emphasis has already been discussed.) This does not alter the Perception of the Cosmo-Eternal Now from the 'altitude' of the World of Being). In general:

1. though the Universal Logic Perception, when it turns towards Its Emanation, Cosmos, may be, in part, that of *Timelessness*—the Cosmo-Eternal Now (for the Logos can 'See' the past, present and future as one), and
2. though this Perception may be for the most part shared by the Logos' Greatest Emanations within the World of Being ...

nevertheless, even during this Perception of Timelessness, Time *is passing,* at least when Cosmic Duration is measured against the Infinite Time Line.

The Problem of Free Will/FREE WILL for the Universal Logos and the ABSOLUTE DEITY

It is clear that there is nothing to compel the ONE WITHOUT A SECOND, and thus ITS 'FREEDOM' is *absolute.* Can IT 'COMPEL' ITSELF? IT can 'DETERMINE' at the 'MOMENT OF AWAKENING' to 'ACT' in *one way* rather than in an *infinitude of ways.*

Does 'LAW' *require* IT to 'DO SO'? 'LAW' cannot be *imposed* as an *external force* because there is *naught external.* Surely the INFINITE SELF is a 'LAW UNTO ITSELF'.

Once 'DETERMINATION' is 'MADE' at the 'FLASHING FORTH' (once the Proto-Cosmic 'DIE' is 'CAST'), can the INFINITE SELF 'DEVIATE'? At that 'POINT', the INFINITE SELF, PER SE, no longer 'ACTS'. ITS ONE 'ACT' is the 'CHOICE'/'FLASH'/'RAY'/'POINT'. IT 'ABIDES' in ITS *utterly transcendent* INFINITUDE, and ITS Pre-Cosmic and Cosmic Representatives *Do* the rest. The INFINITE SELF 'REMAINS' the NEGATOR OF ALL, even as IT, through ITS Representatives *affirms the positing of the Singularity* (the 'Opposite Pole' of a non-polar INFINITUDE).

Is the Infinite Pre-Cosmic Trinity *bound* by 'LAW'? To all in Cosmos and Pre-Cosmos, the 'CONTENT' of the 'FLASH' is 'LAW'. There are two *compulsions*:

1. the Compulsion to Finitize and Become a Singularity, a Definite Oneness as opposed to an Indefinite Infiniteness; and,
2. the Compulsion to 'Hold' and Unfold the Conferred Pattern, the Pattern of Possibilities ('EXTRUDED' from the FOUNT OF ALL POSSIBILITY) to be *actualized* in-Cosmos.

The Pre-Cosmic Infinite Self/Infinite Subject/Infinified Point Acts under the 'LAW' of THAT when it Finitizes Itself. The Universal Logos, which is That which *eventuates* through the Finitization Process, Acts under 'LAW' when It 'Holds the Program' (affirms the Design-at-the-Beginning) and unfolds that Program through Emanation.

The Infinite Self is *not* free to refuse to Finitize. The Universal Logos is *not* free to refuse to Hold the Design and unfold It. It would seem that the Infinite Self/Subject is *already* 'carrying the 'INSTRUCTIONS' and may be free in terms of the *mode* of Finitization, the *how* of Its progressively limiting 'Self-Sight'. The Universal Logos is free to carry out the Process of Universal Emanation according to the 'LAW'-of-THAT inherent within It (the Logos). The Universal Logos has freedom within the entirety of Cosmic Space to 'supervise' the unfoldment of the Divine Design, but then, again, It tends to remain the 'Silent Support', and 'bestow' more *active* Cosmic Functions upon Its Emanations, especially Fohat.

Every B/being in-Cosmos is Essentially the Universal Logos, and the Universal Logos is ESSENTIALLY *THAT.* At the core of every B/being is *absolute freedom,* for ESSENTIALLY, no matter how numerous such B/beings may *appear* to be, they are still ESSENTIALLY but ONE BEING/NON-BEING. Freewill is thus unimpeachable throughout Cosmos and Pre- and Post-Cosmos.

The 'LAW' of THAT, however, as it applies to Cosmos and Pre-Cosmos, is 'LAW' which every B/being in ITS ESSENTIAL NATURE has 'IMPOSED' upon ITS OWN non-ESSENTIAL *appearance.* Thus, though there is Free Will at every apparent *point,* there is also the *impossibility* in Pre- or Post- Cosmos, or in Cosmos of *violating* that 'LAW'—*ultimately.* In fact, there are simply an infinitude of *impossibilities*—those *acts* that cannot *possibly* occur in Cosmos or Pre-Cosmos due to the 'EXTRUDED' 'INSTRUCTIONS'.

☐ The final conclusion regarding the Freedom of the Will (as seen from the perspective of Radical Infinitism) is that while the *WILL-as-Will-as-will is utterly free, ESSENTIALLY,* such restrictions as are *imposed* upon its freedom, are *SELF-*'IMPOSED'—REALLY, SELF-as-Self-as-self-*imposed.*

'Within' IT, 'CHOICE' of one 'possibility', or set of 'infinitized possibilities', occurs by what we can reasonably presume to be 'WILL', in UNIMPEACHABLE FREEDOM. The 'ENACTMENT' and Unfoldment of that 'CHOICE' becomes *'LAW'* to all 'RADIATIONS'/ Emanation in Super-Cosmos and Cosmos.

But was the 'CHOICE' 'MADE' *under* 'LAW'. It is hard to imagine so. Was the fact that a 'CHOICE' was 'MADE' and, presumably, has been 'MADE' on an infinitude of 'OCCASIONS', a question of 'LAW'? Has the 'INFINITE SELF' no 'CHOICE' but to 'MAKE a CHOICE'? For us, this is presently unanswerable. Even though the 'CHOICE' has been made intermittently forever, that is no reason to insist that 'CHOOSING' is a 'LAW' of ITS BEING rather than an 'ACT' 'WILLED' intermittently forever.

When we human beings are confronted with prolonged regularity, we almost automatically think, 'LAW' (or Law of law). It is, however, beyond our ability to discern whether we are dealing here with what might be called the *'LAW'* of SELF-CONSISTENCY through APPARENT SELF-CONTRADICTION, or the *'WILL'* to SELF-CONSISTENCY through APPARENT SELF-CONTRADICTION. There is no convincing way to decide: 'LAW' presupposes compulsion, and there is *none other* to do the compelling; 'WILL' presupposes an *act of identity,* and the INFINITE SELF is as much a NON-IDENTITY as an IDENTITY. If the 'LAW' is SELF-'COMPELLED', however, it is no longer 'LAW' but 'WILL'. All in all, except for the question of whether the INFINITE SELF has the kind of SPECIFIABLE IDENTITY that could 'WILL' any 'ACT', the probability is more towards 'WILL' than 'LAW'.

The Problem of Whether 'Rays' Combine

Every authentic E/entity is Really a 'Ray' of the ABSOLUTE. Thus all Monads are 'Rays' of the ABSOLUTE. Further all 'Rays' of the ABSOLUTE are *absolutely indivisible.* The INFINITE SELF is indivisible, and the 'RAY' of the ABSOLUTE is ITS (the INFINITE SELF's) Representative in the World of Duality (which includes both Super-Cosmos and Cosmos).

The 'RAY' of the ABSOLUTE simply means, the 'PRESENCE' of the ABSOLUTE. *A* 'Ray' of the ABSOLUTE, though *apparently* less than *the* 'RAY' of the ABSOLUTE, is REALLY the *entire* PRESENCE of the ABSOLUTE. Wherever a 'Ray' of the ABSOLUTE is Present, the *entire* PRESENCE of the 'RAY' of the ABSOLUTE is also ESSENTIALLY *present.* 'Rays' of the ABSOLUTE do not *combine* because they are never Really *separate* from each other, thus *all* such 'Rays' are Really *One* 'Ray' which Is *the* 'RAY'. Their *seeming separation* is simply the result of Maya, and the World of Illusion which It ('MAYA'-instantly-Maya) has 'CREATED'-instantly-Created, then, Unfolded.

When 'Rays' are reabsorbed into apparently *greater* 'Rays', they are simply being reabsorbed into *Themselves.* A 'Ray' is Itself, and a 'Ray' is equally identical to That into which it is apparently reabsorbed. There is no question of *combining* 'Rays' because there is *no difference* between them—no difference in ESSENCE, and, strangely, no difference in magnitude.

The Problem of Whether all Objects are Material? —Can Any Objects be Non-Extended? —Can Any Objects be Non-Spatial?

Materiality is, simply, *objectivity.* An object, by the very fact that it is objective (i.e., 'Seen') is *material.* The only 'NON-MATERIALITY' is THAT—THAT, which *as* THAT is never 'SEEN', only *'BEEN'.* Extension is an artifact of consciousness. 'MAYA'-instantly-Maya Creates Extension, for Maya is Consciousness. Interestingly, Maya means 'to measure'. Extension is that which is *measurable* or *comparable.* The INFINITE SELF IS the INCOMPARABLE—all else is Maya—the *'Comparable'.*

If Maya (the Power of Self-Objectification) is Extension and PARABRAHMAN is NON-EXTENSION (whether *temporal* or *spatial),* then all objects have extension, even if that extension be extremely subtle or relatively minute, or infinitesimal. This means that all objects are 'somethings' and have 'somethingness'. That is extended which 'takes up' space and 'takes' time. All objects 'take up' space (even if 'space-in-Consciousness') and 'take' time. Even the greatest of all objects, the Infinite Object, Mulaprakriti, not only 'takes up', but *Is* all possible Space, for It Is Infinite Space. Further, Mulaprakriti, exists only *intermittently* though forever. Thus, Mulaprakriti, the Infinite Object, 'takes up' or 'occupies' Time (in this case, along the Infinite Time Line).

☐ Interestingly, while the Infinite Object has no *location* in Space because It *is* Infinite Space Itself, that same Infinite Object has *location in Time* along the Infinite Time Line. (Just as the Number One is preeminent to the Number Two, so Time is preeminent to Space.)

Both Time and Space are, apparently, intermittent, for Time and Space apparently *exist not* during the ALL-IN-ALLNESS of Universal Pralaya. Yet after the MOMENT of AWAKENING, it is *understood* (by Whom?) that the Infinite Time Line (though apparently disrupted by TIMELESSNESS) strangely *continued during* Universal Pralaya (i.e., that there was a 'time of *no Time).* It is far more difficult to imagine that Space *continued* during that Pralaya (and, thus, that there would have been, during NOTHINGNESS, a 'space of *No Space).*

Even Infinite Space seems truly intermittent. If, however, Space is defined as equivalent to Registration or Consciousness, then upon 'AWAKENING' from Universal Pralaya, it could be realized that 'NOTHINGNESS' did, indeed, 'OCCUR' and therefore, was, hypothetically 'Registrable'. Really, however, upon 'AWAKENING', 'NOTHINGNESS' could only be presumed (*by inference*) to have been 'Registrable', as 'during' NOTHINGNESS, there would have been no Point of View 'outside' of NOTHINGNESS from which to *consciously register* NOTHINGNESS (and of course, there would have been *no-one, and nothing* to have done the *registering!*).

So, NOTHINGNESS-as-a Registration-as-Space would be purely *inferential.* The inference upon 'AWAKENING' would be that IT (NOTHINGNESS) *must have* 'OCCURRED', and, therefore, that Space *must have* existed just as it does 'during' Universal Manvantara. Such Space could be called 'Inferentially Continuous Space'.

The conclusion is, that anything *objective* has *extension. Objectification is spatiality* and *spatiality* is *extension.* That extension may be definite, infinite, or infinitesimal (to sustain the latter two of which requires *convergent ongoingness* {not fixity} and, hence, a species of *motion*), but all three types of extension have *noticeability, registrability,* (and all objects must be *noticed* by consciousness, or they would not be objects). Thus, if something is 'noticeable', consciousness has been required to 'notice' it; where consciousness is, space is—and where space is, extension is. Of course, it becomes necessary to differentiate between 'delimitable extension' and non-delimitable extension (which some would say is no extension at all).

Clearly, the term 'extension' is being used broadly, to indicate anything 'noticeable' (not just anything conventionally measurable). An object need not be definite to have noticeability and, hence, extension. From this perspective, anything that is not NOTHING has extension, even if that extension is not conventionally measurable and definite. Thus, mere existence (demanding as it does, Time and Space) *is* from this broadened point of view, 'extension'.

It can be asked, Is extension 'within' Consciousness the same as extension 'outside' Consciousness? To this it must be answered that *all* extension is extension *within* Consciousness. There is no object but that it is a "Content of Consciousness". Objects arise because Consciousness arises. Originally, Consciousness Creates the Object through what might be called 'Internal Self-Replication'. The Subject 'Sees' the *image* of Itself 'within' Itself. First the Infinite Subject 'Sees' Itself as the Mother (the Mother is always the First Object). Then through the process of *condensation* or *finitization* the Subject 'Sees' a *'reduced* but faithful replica of Itself' as a Son.

☐ So first, in Super Cosmos, the Infinite Subject 'Sees' an *exact* Reflection of Itself (and, thus, Infinite Space arises as an Object), and then, the Infinite Subject begins to 'De-Infinitize' and 'Sees' a *reducing* Reflection of Itself (and thus eventually, a delimitable Son {Really, the Second, and this time, *definite* Object} arises *'within'* Consciousness). This Finite 'Son' (of Infinite Father/Infinite Mother) is Really the Father of the Universe-to-Be.

Is there a difference between the kinds of Objects which arise within the Consciousnesses of Those Who focus Cosmo-Eternally within the World of Being, and Those Whose focus in the World of Fabrication, the World of Approximation? It might be said that the Objects which arise within the Fohatically-Supervised World of Fabrication are 'particulate reflections of the imparticulate Images which are sustained within the World of Being'. The particulate objects of the lower worlds are particulate so that they can be easily reassembled from ultimate moment to ultimate moment to better approximate the Great Archetypal Images by means of which the Fixed Design is sustained.

We see therefore that objects in all worlds other than the World of Being are composed of minute, Fohatic particle/events. This is almost certainly *not* the case in the World of Being which Objects are *ideational, imparticulate,* and *seamlessly numerical* (i.e., equivalent and resolvable, despite apparent numerical 'parts', to Number One). In other words, all Objects within the World of Being are indivisible Reflections of Cosmo-Monadic Oneness.

The Problem of Whether Subjective Ideas are Formless *Before* Being Objectified —Is an 'object' always a form? —Can an object be form-less? —Is an *idea* formless?

From the perspective of thought used in this treatise, anything perceived is automatically a *form*. Any existent is a form. Only *THAT* IS a NON-EXISTENT, and thus, only *THAT* is *absolutely not* a form. Being the NAMELESSNESS, IT 'ESCAPES' any attribution of *form* as well, becoming, justifiably, the 'FORMLESSNESS'.

Midway between formlessness and form is the Infinite Object, Mulaprakriti. It is a 'Something' (because It is *registered*), but It is, paradoxically, a *boundless* 'Something'. Simply by the very fact that the Infinite Object is the Infinified Self-Perception of the Infinite Subject, It becomes a 'Thing', a kind of hybrid Finitization (a, paradoxically, *boundless* Finitization—like a 'definite indefinite'), but as an Object called Infinite Space, It initially 'contains', per se, no lesser finitizations, no other *forms within* It. Later, (with the 'De-Infinitizing' Process of the 'De-Infinified View Point') such finitizations begin to appear.

There are obviously subtle forms and crude forms. Of subtle and very subtle forms, most human beings are unaware. If something exists, yet remains unseen, it is said to be *formless*, but technically this is inaccurate. What is needed is a knowledge of the nature of subtle forms, subtle Objects in Consciousness. The word 'form' should not be applied only to concrete things with a visually perceivable shape.

Is an *idea* a form? Most definitely, a *kind* of form. Does an idea 'occupy' Space? Any 'registrable' perceived, no matter how subtle, occupies Space (because it 'occupies *consciousness*' and the Field of Consciousness is the Field of Space). In Radical Infinitism, to occupy Space simply means to *appear* within Consciousness. Space is always, fundamentally, *internal* (appearances notwithstanding). Space is what the Subject 'Sees' through the mediation of Consciousness. Space is *whatever* Consciousness reveals (even if Consciousness seems to reveal only Itself). Only when there is no Consciousness is there no Space.

Is an idea necessarily *visible* in the manner that tangible objects are visible? This is more difficult. What are the Ideas 'Held' in the Minds of the Archetypal Beings/Numbers in the World of Being? These Ideas *are* Objects in that they are *perceivables* or *'inceivables'* or *'registerables'*, but *are* they perceived/'inceived'/registered as *extended* (as we *usually* conceived Extension)? Given the limitations of human consciousness as it attempts to assess the nature of the World of Being, it is impossible to say with certainty. The most reasonable answer is, probably not. Ideas, however, can be said to 'Fill the Space of Consciousness'.

Our usual notion of extension, form and object, comes, most probably, from the limitations of human consciousness as it is operative within the lower levels of the World of Approximation. It would be, probably, most inaccurate to *project* these presentations-in-consciousness upon the World of Being.

- ☐ Ideas are perceivables/'inceivables'/'registerables' within the World of Being. They are as *seeds*, differently constituted from the usual tangible forms presented visually to limited consciousness in the lower worlds, but which unfold and develop *as* familiar presentations of visual forms within the lower worlds.

But what are 'Sons' if not Ideas, and are not 'Sons' Perceivable? Visible? Ideas may lack 'bounded extension' as it is found in the more concrete realms, but they can reasonably be said to have 'boundless extension within a given Ring-Pass-Not' (that of Cosmos), and to have 'extension within the Space of Consciousness'.

We can justifiably say that all higher Forms (with respect to the greater Consciousness which conceives them) *are*, indeed, *'registerables'* and a species of *'visibilities'*. If forms were *invisible* to the Consciousness conceiving/perceiving them (and were, moreover, *not visible* to *any* Consciousness whatsoever) those forms would be 'no-things', vacuities, non-existents, non-objects, non-emergent-'inherencies' (The INFINITE SELF is, so to speak, *infinitely 'filled'* with such, albeit in an *infinitize mode.)* Such forms, therefore, would be *no forms at all.* (It must always be remembered that the sense of *'sight'* and the thought of *'visibility'* are used to indicate *all kinds of consciousness.* (The specialized sense of *sight* and 'visibility', as usually conceived, is not indicated.)

- ☐ If a form *exists*, it is being conceived/perceived/registered and is considered *visible.* The *appearance* of the form may vary (and probably *does* vary) dramatically with the plane upon which it is 'found', and with the organs of registration through which it is registered (whether by beings focussed upon that plane or upon others). The conclusion is, however, that *ideas are visible*.

For practical purposes, we need to know what may be the *appearance* of an idea. To understand this we would have to translate ourselves into the Consciousness of a superior, enveloping Entity (the Ubiquitous **8**, again, by another name) for Whom the ideas were as visible Self-inherent Objectivities.

In all matters of defining the appearance or the form in which an object 'appears' or 'registers', the question must always be asked, registers to *whom?* Perhaps, to these Higher Beings, Ideas 'appear' or are 'Seen' as 'Intensifications of Quality', extended throughout Space. The form of an Idea is not a spatial form, but a form consisting of *differentiable quality.* If by 'form' one simply means, 'crude shape', then Ideas have no form.

The Problem of Whether Change Exists in the World of Being

This is the problem of the 'Changing of the Molding Archetypes'. There is certainly change within the many levels of the World of Fabrication, the World of Approximation. The Cosmic Configuration is *ever* reconfiguring. Why does it do so? Does Fohat, Who guides the reconfiguration of the Cosmic Configuration, respond to a Changing Directive or a Changing Pattern? If so, from *whence* this Guiding Pattern?

There is no *absolute continuum* other than the CONTINUUM. The stabilities within Cosmos are, at best relative for they disappear with the cyclically disappearing Cosmos. Even that which endured unchanged throughout the entirety of Cosmos could only be called permanent-in-*Cosmos* and certainly not permanent in Infinite Duration.

Within the highest Beings in the World of Being (especially the Father, and, to some degree, the Son) the highest probability of *relative changelessness* can be found. The Father (after He projects that 'Self-Seen', Self-Reduction of Himself called the Son) may simply *abide*, unknown and unseen as the 'Sustainer of the Whole'. The Son, Per Se, is probably the major Custodian of the Fixed Design *in its synthesis.*

There is a Final Form of Beauty which must be achieved in any Cosmos, and the synthesis of that Form must be *held* until the 'Day of Achievement'. As the Second Aspect, in our lower worlds, is ever the "Preserver of the Form", the Universal Son is the likely 'Guardian of the Fixed Design'. In this respect, the Son changes not, though, through Emanation, He is the Author of *specific qualitative change.*

☐ The Logoi in the Company of the Son, however, may be involved in what we could call the 'Intended Revolution of the Final Form through Cosmic Time'. The Final Form is the Fixed Design, but the Final Form must be reached through Ordered Process in Time (and Space). The Final Form (in the World of Approximation) is reached through the sequential coordinated appearance and disappearance (or *intensified emergence and retraction)* of Secondary Forms (Archetypes and Combinations of Archetypes—like Combinations of Numbers), (in/from the World of Being) and these (in varying combinations) must emerge and withdraw so that Fohat and His Host (under Their guidance) may execute the Process of Fabrication correctly.

It is possible that the fulfillment of the Major Form (the Fixed Design) through the sequential appearance of progressive Secondary Forms, is under the province of the Super Tetraktys—the Ten Logoi (three major, seven minor) Who are, fundamentally, Emanations of the Universal Son, and are found in the Company of the Universal Son. Thus, They have a form-preserving, form-sustaining, function.

How would 'change' in the World of Being occur, so that the progressive Secondary Forms appeared "on Schedule"? The Logoi involved would simply change their 'Self-Sight.' They would 'look within' to discover those Aspects of Themselves (those emerging, Universal-Son-impelled, New Patterns) that needed to be *held* in Consciousness for a certain Cosmic Period. Each of the Ten, acting similarly, would discover 'within' a portion of the Changing Pattern to be *held* uninterruptedly in Consciousness. To this Changing of the Pattern (which is a Changing of Self-Image, and which would involve the 'sending forth' many 'enfolded' {apparently *lesser}* 'Rays' of the One Universal Ray), and *sustaining* of the New and Timely Pattern, Fohat and His Host would respond, and *build* accordingly.

If three levels of activity were to be discerned in Cosmos they might be as follows:

1. The utter fixity of the Life-Giving Father and the steadfastness of the Son (who 'Holds the Synthetic Pattern of the Fixed Design').

2. The timed Revolution of Patterning Image to be sustained by the Sub-Logoi and Subsidiary Logoi (and all the many 'enfolded, implicate 'Rays') in the Company of the Son.
3. The constant modificatory activity of Fohat and His Host as they build in particulated prakriti the forms which are to correspond to the Guiding Images set forth.

The Problem of Whether all Objects and Images Need be Particulated

Within the World of Fabrication, the forms built to conform to Divine Images in the World of Being are *particulated,* composed of ultimate particle/events, and the many kinds of aggregations derived therefrom. The particulations are, however, Essentially, *un-detached,* i.e., 'retractable' 'into' Fohat. The World of Approximation is a World of Separated Duality (thus the constant oscillatory motion between appearance/disappearance).

The World of Being, too, is a World of Duality, but not a World of *Separated* Duality. The only *absolutely non-dualistic world* is the WORLD OF ABSOLUTE BEING. The World of Being (within the World of Becoming, i.e. Cosmos) stands midway, as it were, between UTTER NON-DUALITY, and Separated Duality. The World of Being contains Subject/Object relations and is, thus, dualistic. The World of Being, however, has within it the factor of semi-continuity, sustainedness, which relates it to the WORLD OF BEING, for the World of Being is the *World of Sustained Archetypal Images.*

The chief intermittency in relation to the World of Being is its appearance and disappearance *with* the appearance and disappearance of *Cosmos.* Every Logos within that World, however, is, in a way, consciously closer to the SOURCE, than any of the lesser beings-in-Cosmos. They consciously 'carry' within Themselves, the factor of 'CONTINUITY' as it is only *fully 'BEEN'* 'within' the ABSOLUTE CONTINUUM in 'INFINIDENTIFICATION'.

The Universal Logos has a kind of 'Self-Chosen Finitude', yet He *knows* His 'ESSENTIAL INFINITUDE at ROOT' (even if He does not entirely 'Embrace' It). He is the Master of the Sustained Configuration, the Sustained Image. He can 'See' Himself and that which is emerging from the 'DEPTHS' of Himself with Continuity. The other Logoi in the World of Being (His Emanative Extensions) are also, relatively, Masters of Internal Continuity. We might call this Art, Mastery of Continuous Self-Observation.

The Patterns produced through Logoic 'Self-Sight' in the World of Being are 'imparticulate', and 'ungranulated'. The Logoi 'See' Themselves (and the emergent possibilities 'IMPLANTED' by the 'FLASH' within Themselves) *uninterruptedly.* The 'Seeing' of the *selves* They 'See' within Themselves (Their Sons) has the quality of *continuous homogeneity* 'resident' 'within' the GREAT CONTINUUM, the GREAT HOMOGENEITY. This is as much continuity as the Cosmos can withstand and still be a Particulate Thing, which Cosmos *Is.*

The Problem of the Reason for Particulateness

Why must there be particle/events within Cosmos? Why should not all creations be continuous-in-Cosmos? This is, fundamentally, a very difficult problem with many answers, none of which can be conclusive at the present level of ignorance of the author. Without going into great detail, it could be said that particulateness is needed:

1. To ensure the extremely limiting experience of the ONE-as-One-as-"anu", the "speck".
2. To allow for the finitization of free will.
3. To make 'movement' possible (as it is *not* within the INFINITE SELF, for only discontinuities can 'move'.
4. To allow for the *necessary illusion of fragmentation*, the antithesis of the *infinitized everythingness* of the ABSOLUTE.

Particulateness is the opposite of HOMOGENEITY. The Purpose of Cosmos is always (as strange as it may seem) the 'Manifestation of a Contradiction to PERFECTION'. The PERFECTION can only 'remain' so, if IT *includes its opposite.*

☐ It is the *seamless homogeneity* of the ABSOLUTE that is asserted in the First Fundamental of the Secret Doctrine, which posits the existence of a BOUNDLESS IMMUTABLE PRINCIPLE. The Contradiction of this PRINCIPLE (though, in REALITY there can be no *absolute* Contradiction) would be a Finite Particulate Discontinuity (i.e., a Universe), which *must* appear as an articulated *possibility* 'EXTRUDED' from the FOUNT OF ALL POSSIBILITY.

The Cosmos can be described as, 'Finitization in Motion for a Coherent Purpose'. The Coherent Purpose is 'GENERATED' 'within' THAT in a 'FLASH' and *held* continuously-in-Cosmos within the World of Being.(the dynamics of which, involve in Essence, the Father 'Beholding' Himself as the Object-Mother; the Mother, reciprocally, 'Beholding' Herself as the Subject-Father; and *both* 'Beholding' their Son. Really, Father-Mother is a Unity, 'Beholding' Its Own Replica, which Replica 'Beholds' Its Replica, etc.

The 'Gaze' of the 'Beholders' is *continuous in Cosmos.* 'Holding-through-Beholding' is a Cosmo-Psychological Process, and has little to do with the Form within the World of Approximation that is to be built as a Fabricated Reflection of Divine Pattern which is Beheld. To 'Hold' the Pattern of the Fixed Design, is a 'Be-Hold-ing' Process. All in Cosmos is *sustained* or *built* through Processes of Consciousness, through the Process of Beholding or not-Beholding.

The big Cosmo-Psychological Decision is, 'To See or Not To See'; Cosmo-Creation and Cosmo-Destruction eventuate as a result of 'Seeing' or not-'Seeing'. Sustained 'Seeing' results in non-particulate, motionless, coherent Image—the kind of 'Image' (whatever its actual *form* to the Perceiving Consciousness) to be found within the World of Being; intermittent 'Seeing' results is particulate, discontinuous motion—the kind of motion to be found in the World of Approximation in which there is an ongoing attempt to *approximate* the Beauty of the Fixed Design.

The reasons given for the evident (though, Essentially, *illusory) particulateness* to be found in Cosmos are for the most part psychological and metaphysical. It is the author's opinion that all functional dynamics in Cosmos will eventually be shown to arise from *metaphysical intention.* This means that there has to be a profound 'WHY' behind any more obvious 'What', 'When', 'Where', or 'How'. The 'WHY' is the ESSENTIAL 'LAW' which regulates all 'ACTION'-as-Action in Cosmos and Pre-Cosmos.

The Problem of How Change Really Occurs

If the Cosmo-Objective Now in Cosmos is immobile, *what* is action and *when* does action happen? Is 'instantaneity' or "no time at all" possible within the Fabricated Cosmos, or *between* the Fabricated Cosmos and the World of Being? 'Within' the ABSOLUTENESS, *motionlessness* 'ABIDES' forever. Within the World of Being the Divine Pattern is:

1. *Held/ will-be-held* by the Universal Son in Fixity (the patterns for all Cosmic-*times* superimposed in one Great Pattern, the Fixed Design).
2. Changed sequentially (according to 'Divine Schedule' by *refocusing* the Pattern Holding Consciousnesses of the Son, Sub-Logoi, and Subsidiary Logoi).

The few changes in the relative fixity which, from this perspective, exist within the Supernal Tetraktys cannot Really be called moment-to-moment Cosmic Change. These few changes are Really an occasional, planned alteration of the Schedule of Operations for the Actualization of the Fixed Design. (A hypothesis does exist concerning what might be called the 'Ideal Momentary Frame', which would require rapid *imaging* within the World of Being to guide the multi-leveled Fohatic Process of Fabrication.)

The most rapid *mobility* occurs, however, within the World of Fabrication (for the Third Aspect ever moves more rapidly than the Second), but what is 'movement' and what moves? In a discontinuous Universe (the World of Fabrication) movement is *not* the continuous passage of a thing from 'here to 'there' through 'space'. Rather movement is an *intermittently perpetual change of relationship between relating factors.* The 'relating factors' are, fundamentally, ultimate particle/events, *willed into Configuration* by Fohat Who is 'following' the Divine Pattern 'Held' and 'Be-Held' by the Universal Son and the Logoi and Agents in His Company.

In all such movement, *relationship* is the preeminent guiding factor. Movement is more a *seeming* than a Reality. Movement is in the nature of an *instantaneous change of relationship from one configuration to another without intervening configurations.* The concept that there are *no intervening configurations* between the positions of the variables involved is vital to understanding movement on the ultimate quantum level.

What we call a Cosmo-Objective Now, is a quantum of time in which the relationship between ultimate particle/events is sustained without variation. That quantum of time is an ultimate moment within a particular Cosmos. No movement occurs or can occur in Objective Cosmos *during* an ultimate moment nor (within the World of Fabrication) can any other time elapse. The ultimate moment occurs throughout the Mosaic World of Effects, and cannot be split. The illusion of movement (on the macro-level) is what consciousnesses submerged in the World of Fabrication perceive as a result of the effect of continuous change of relationship on the ultimate micro level.

Now, we must ask ourselves a set of difficult and not-entirely answerable questions:

- Can a configuration of ultimate particle/events change to a non-configuration (a nullity in perception) in "no time at all"?
- Can the juncture between Fohatic event and non-event be instantaneous?
- Can the Cosmo-Objective Now of a relationship of particle/events 'change', vanish at infinite speed, so that "no time at all" intervenes between event and non-event?

This is a most difficult question, far beyond our capability to solve. Perhaps we can gather some thoughts and questions which will point towards a solution. We might ask:

- What is the 'Speed of the Fohatic Will'?
- What is the 'Speed of the Blink of the Cosmic Eye'? What is the 'Speed of Self-Reflection'?
- Can 'Now' and non-Now abut each other with *no time* between?

Strangely, it seems that instantaneity, or infinite speed, or "no time at all" are needed, to preserve quantized Time in Cosmos! Within Objective Cosmos, the ultimate moments *seem* to abut 'against' each other so that no time intervenes. This is an illusion that creates the impression in consciousness of the seamless flow of time and continuous movement through 'space'. Clearly, however, 'time' within the Subjective Cosmos is 'passing' between the 'frozen Fohatic relationships' which abide unchanged during each ultimate moment.

In the truly Highest Subjective Cosmos, the World of Being, the Cosmic Eternal Now prevails. One in tempted to wonder whether there is a *change* in the 'Held Pattern' at the onset of each of Seven Eternities. (That potential digression will not be followed, alas!)

For practical purposes, we can say that within the World of Being the Consciousness of 'Eternally Now' prevails undisturbed. Within the World of Fabrication the general (though illusory) consciousness of 'always now', also, prevails undisturbed. The Cosmo-Subjective Now (during which there can be {Janus-like} orientation towards the next ultimate moment or orientation towards the Eternal Now) occupies a middle-ground between the two other 'Nows'. The real question is:

- How *fast* is the change from the Consciousness of the World of Being to the Consciousness of the Fabricated Cosmos and back again. If time is 'taken' when measured against the Infinite Time Line?
- How long is an ultimate moment?
- How long is an inter-moment instant?
- Is there any duration between? Measurable duration? Infinitesimally instantaneous duration? *No* duration?

Again, How fast is the Divine Will? How fast is 'Self-Seeing'? Is it possible to sequentialize two held states of relationship in such a way that the operative dynamic Really is, "<u>N</u>ow you see it; <u>N</u>ow you don't?"

In this model concerning the nature of movement, we do not have to be concerned with *infinitely* rapid movement because, in a way, there *is* no movement. It is not as if an ultimate/particle event is a 'thing' that has to pass through a certain 'distance' in *"nothing* flat"—that would be an impossibility. Strangely, this model of movement, has no *movement* in it. There is merely *appearance/disappearance* <u>*without*</u> passing through space or 'covering' distance. There is merely the abutment of two states:

- one of relationship, the other of non-relationship;
- one of Fohatic Self-Objectification, the other of non-Self-Objectification;
- one of Fohatically Willed appearance, the other of Fohatically Willed disappearance.

Metaphorically, we are talking about an electric 'switch' controlling 'On' and 'Off' positions; the switch however does not 'move' through space from one position to the next to turn the appliance 'On' or 'Off'; it simply is at one position, and 'then' at the next without the intervention of 'time'. Possible? Or Impossible?

One is forced to think of the juncture between the TIMELESSNESS of ETERNAL DURATION and the appearance of Time. Was any Time 'taken' for Time to appear? How could Time be used in which to 'MAKE Time Appear'? Perhaps to say that at the 'MOMENT of AWAKENING' Time *appeared* "in no Time at all", is accurate. To think that 'after' a Universal Pralaya of 'PLENITUDE' a quantity of 'Time for 'DECISION' is 'NEEDED' may be to *infinitely minimize* the ABSOLUTE! These thoughts lead one to think that TIMELESSNESS is needed to 'PRODUCE' and *uphold* Time.

One must ask, Is Time needed to change positions? In the normal macro-world, of course time is needed; in the ultimate Cosmic *micro*-world, perhaps not. What is interesting is that Cosmic Time (the Cosmo-Objective Now) only appears *after* there has been a change of relationship. So we are asking whether it takes 'time' to create Cosmic Time (in the World of Fabrication); and whether it takes 'time' to make Cosmic Time disappear (in that same World)? If it does take time, one cannot 'borrow' time from Cosmic Time, because it does not yet exist. One could only 'borrow' from Infinite Duration along the Infinite Time Line. There is good reason to believe that since the ETERNAL NOW 'ABIDES' *forever,* TIMELESSNESS is ever *available* to 'serve the needs of Time'.

Let us think of appearing or vanishing in a 'Flash'. The 'Flash' is not a 'thing' that 'takes time', but is simply the effect upon Consciousness of the change. What is there, if anything, about a change from *absence* to *presence* to *absence*, etc., which necessarily 'takes time'? Nothing 'moves through space'. Nothing has to *gradually* appear and *gradually* disappear, for the ultimate particle/event has no 'parts' needing 'warming up to peak' or 'cooling down'. It is *movement* that 'takes time', but in this model of movement, there is *no* movement, per se, i.e., there is no *gradual* change of relative position; there is no incrementalization from one position to another, covering a possible infinitude of interim positions.

☐ The 'Flash of Fohatic Intro-Ception' does not require movement by increment. It simply *Is or Is Not.* Therefore, again, an 'outrageous' question, What is the speed of the change from *being to non-being?* Is there REALLY/Really a *passage* from *non-being* to *being?* Is this one kind of change in which no *movement* is involved?

Time is only required if there is movement. Time can only exist in relation to movement, even along the Infinite Time Line. If there is no *movement* involved in the change from 'on' to 'off' then no transitional 'time' is required for the change to 'take *place*'. (i.e., 'position' within the Cosmic Configuration). The change from 'on' to 'off' to 'on' etc., in-Universe, may feature the direct intervention of the ETERNAL NOW in the Cosmic Process.

We realize that no movement can take place 'within' the PURE ABSOLUTE CONTINUUM. Movement pertains to the World of Illusion, or so we reason. When, however, we look closely at so-called *movement* in the World of Illusion, we realize that movement itself, may be simply another illusion, a non-existent, existing only as an impression in consciousness. Perhaps we are being told an amazing story—that while there is apparent *change* in Cosmos, there is no Real *movement* to make that change possible. There are simply changing 'frozen relationships'. How strange that the appearance of movement should result from the absence of movement. The entire Cosmic Configuration can change, and yet no-'thing' 'moves' from 'here' to 'there'!

☐ Conceived in terms of the Will, the entire Cosmos may be functioning according to the 'Assertion/Negation' Dynamic, or more forcefully, the 'Yes/No' Dynamic. What is the speed of change from 'Yes' to 'No'? Is *movement* required between 'Yes' and 'No'. Is the ETERNAL NOW all that separates 'Yes' from 'No'? If the ETERNAL NOW is all that separates 'Yes' from 'No', then 'Yes' becomes 'No' becomes 'Yes', etc. with an interval of *zero duration* intervening.

While 'things' in Cosmos cannot appear then disappear an *infinitesimal* moment later, they can endure for their allotted span and then disappear in an ETERNAL NOW *at no speed at all*. The concept of speed is inapplicable to their appearance or disappearance. NOTHING 'took *place*' between the end of the appearance and the disappearance, or between the end of the disappearance and the beginning of the appearance, and *NOTHING* has *no* speed.

The Problem of How the Infinite Subject/Object *Really* Becomes the Finite Subject/Object

The Infinite Subject/Father 'Sees' Itself as the Infinite Object/Mother, and in doing so 'Contemplates' Its Own Infinitude. The Infinite Subject/Father does not remain rapt in such Contemplation 'forever' but 'moves' towards Finitude (while, of course, *not* 'moving' towards Finitude). What might this Really mean? In a way, the Process of Finitizing is:

- A change of Self-Focus, a gradual (or, perhaps, instantaneous) creation of a *boundary*—a change in Self-Conception from 'Unbounded Imparticulate Somethingness' to 'Bounded Imparticulate (or Multi-Particulate) Somethingness'.

- It is a 'Movement in Consciousness' from a 'Seen Zero' to a 'Seen Number One' (and, of course, there would be a transition period with many 'points' possibly appearing, as the 'point' can be thought of as transitional between zero and one.
- It is a Movement in Consciousness from a 'Self-Sense' of Indefiniteness to a 'Self-Sense' of Definiteness.
- Most importantly, it is a Change from the Infinispective of 'Although **8** Am, **8** Cannot Be Defined', to '**8** Am and What **8** Am *Can* Be Defined'.

The Movement from the Infinite Subject/Object to the Finite Subject/Object is thus a growing Sense of Self-Definiteness. In a way, it is a 'Search for Self'. The Infinite Subject is Potentially Everything and Anything, but within that Everything and Anything, the Finite Subject can *find* Himself. There is an internal *pressure* from the 'RAY'-'IMPLANTED' 'INSTRUCTIONS' that helps the Infinite Subject-'Seen'-as-Infinite Object *define* Itself (once the Infinite Subject has 'Become' the 'De-Infinitizing Subject') as a Specific Finite Subject-'Seen'-as-Object.

There is, as it were, a 'Driving Instinct' to 'Search' for what the Subject Definitely Is and what It must Definitely Do. Finitization is Specificity of Self-Definition. At first the Sense of Definite Scope-of-Being is achieved as contrasted with Infinite Scope of Being. Subsequently, as Cosmos beings, the Sense of Definite Possibility to be Actualized is achieved as contrasted with Infinite Possibility which *could* be Actualized. The 'INSTRUCTIONS' guide the Focusing Universal Subject to Self-Determine a Scope of Identity which is suited to the latent possibilities to be Actualized.

Is there a Pre-Vision of the Fixed Design (just as a human being coming into incarnation may have a pre-vision of that which is to be achieved in the incarnation)? It is a possibility. Is there an 'Orienting Flash' that helps to define the Ring-Pass-Not of Identity? A Flash *before* the Process which must *unfold in Time?* This would certainly be one explanation.

All that can be said (and that, in *general)* is that the Infinite Subject seems to 'enter' a 'De-Infinitizing' Reduction, thus undergoing a Process of 'Self-Definition' (albeit, while ever remaining Its Infinite Self) which allows It (now the Focusing Universal Subject) to 'Find *Its Own Definite Space* Within'. From 'then on', Its Space is *limited,* and Its Self-Conception is, correspondingly, limited as well. The Focusing Universal Subject is now called the Universal Logos. The Universal Logos now a *de-finite* Being, *de-finitely* is 'settled into Its Own Space', and the Cosmic Process of Self-Reflective Emanation can begin. That which *simultaneously Is*, can, through 'x-tillions' of Cosmo-Objective Nows, be Unfolded in Cosmic Time.

It is worse that hopeless to speculate upon the relative duration of these Pre-Cosmic States, and early Cosmic States. From the 'MOMENT of AWAKENING' when the 'RAY' 'FLASHED FORTH', through the Pre-Cosmic Finitization, to the Emanation of the World of Being in Cosmos, the 'splitest split' of a human second may have elapsed. The 'RAY' in fact, may have appeared "in no Time at all."

Perhaps it is sufficient to have some concept of a Possible Reasonable Process in the change from INFINITUDE through Infinitude through De-Infinitization to Finitude. The ascertainment of technical Time and Space 'measurements' will have to wait until Man *becomes* what He *already Is.*

The Problem of Whether There is an Unchanging Point of Reference in Cosmos Against Which any Particle/Event can Determine its Position

[See Ultimate Moment in Glossary.] This is the problem of whether there is a spatial *center* to the Cosmic Sphere. Upon the etheric physical plane, all E/entities have their center: the atom, the cell; the human being; the planet; the solar system; the galaxy; etc. We can look for the analogical solution, stating that the Cosmos in Its entirety is certainly a Living Entity and should, correspondingly, have its Spatial Center. According to the Tibetan Teacher, all lives in Cosmos tend towards sphericalization, and every sphere has its center.

Further, the idea that Cosmos is *ordered* and that "God Geometrizes" lends credence to the likelihood of an actual spatial Center to the Cosmic Sphere. From a certain perspective, we must realize that, metaphysically, Time and Space in Cosmos do *not* Really exist, and that the Principle of Non-Locality and the Principle of Non-Temporality represent deep spiritual truths. The Cosmo-Eternal Now and the Ubiquitous Point exist from the perspective of the World of Being, whether or not a definite spatial Center can be determined.

Man's present knowledge of Cosmos is too small to determine (on the etheric/physical plane) 'where' that center may be. Certainly, a number of great *centers* which are major *points of convergence* are identified by astronomers. Knowledge of the extent and even *nature* of the etheric/physical Cosmos will have to expand before there can be certainty.

When focusing on this problem it is always well to bear in mind an old metaphysical definition of God, "God is an intelligible Sphere whose center is everywhere and whose circumference is nowhere." If, for instance, the Infinite Self-Sight of the Infinite Subject should be *maintained*, even as (appealing to the Pre-Cosmic version of the Principle of Emanative Retention) there is a change of Pre-Cosmic Subjective Focus towards Finite Self-Definition, the *spatial* results would be most interesting. Infinite Unbounded Space would remain as an Object to the Infinite Subject/Self, even as Bounded Definite Space *condensed* within the Objective Infinitude.

It is well to remember that it makes no difference how relatively 'big' or 'small' the Cosmic Sphere may be; assuming our Cosmic Sphere to have measurable etheric/physical dimensions it could be an 'x-tillion' times larger than a given measurement or an 'x-tillion' times smaller; it would make no difference.

Regardless of its spatial dimensions, the Cosmic Sphere would be *dwarfed into infinitesimality* by the Boundless Infinite Object, Mulaprakriti (and possibly, dwarfed into total disappearance—non-registrability). The Infinite and the Finite are simply incommensurate. The only way to determine the 'size' of the Cosmic Sphere is in relation to other Cosmic Spheres, of which there have been an infinitude, so opportunity for comparison will not lack—given the *right Observer in the right Position!*

The Problem of Whether There is Ever a *Moment* When the Entire Infinitude of Possibility 'within' the FOUNT OF ALL POSSIBILITY is *objectively* revealed

The *objective* revelation of the entire 'content' of the FOUNT OF ALL POSSIBILITY may not ever be necessary, if the INFINITESSENCE which the ABSOLUTE IS, somehow *includes in an infinitessentialized 'STATE'* the *noumenessence* of every possible possibility. If NOTHING is the *infinitized all* of every possible something, then 'DURING' the ALL-IN-ALLNESS, all possibilities (better than 'KNOWN') would BE *'BEEN'* through 'INFINIDENTIFICATION'.

The question arises, If the infinitude of possibilities cannot be specifically objectified, how can a *set* of them, or even *one* of them be 'CHOSEN' for 'EXTRUSION' and Articulation in Cosmos? Here we are stopped entirely by our ignorance of the NOTHINGNESS. We must remember, however, that the NOTHINGNESS is, as well, the PLENUM, or INFINITIZED 'EVERYTHINGNESS'. The 'dynamics' of what we would call the process of 'CHOICE' (of the one out of the infinitude) is entirely beyond our ken. Yet a possibility or set of possibilities *appears,* as it were, *encoded* within the Pre-Cosmic Self.

Since the Infinite Subject is in the position of the Infinified Observer, does It 'See' Itself as an Unarticulated Infinitude (reflective of the ABSOLUTE HOMOGENEITY of the FOUNT OF ALL POSSIBILITY) or does It 'See' Itself as an Articulated Infinitude (from which the Choice of one set of possibilities is possible)? Again this is (for us) an unanswerable question. It would seem, however, that specified vision of 'articulables' would 'begin' during the 'De-Infinitizing' Process of 'De-Infinispectivizing' as the Pure Infinispective of the Infinite Subject is 'left behind'.

The Infinite Subject, aware of Its own Infinitude, is, nevertheless, subject to Pre-Cosmic Time. The 'FLASHING FORTH' is, again for the infiniteth time, the Birth of Time. The question arises as to whether an *articulated* Infinitude could be Apprehended or Cognized within Time (as once the 'RAY' 'FLASHES', it will be a finite amount of time before IT is, again, 'withdrawn' into INFINITUDE).

☐ The infinitude of the *INFINITUDE* is not an endless more and more of things. It is a 'STATE' of such *infinitized 'concentration'*, that everything is *infinitessentially 'contained' in NOTHING.*

The scenario we are here questioning, however, would require an endless extension of articulated possibilities—an infinitude of unending more and moreness. Because the Infinite Subject would still be subject to the Processes of Time (for Time has *begun)* Its Cognizance of an infinitely extended articulated possibility would have to be utterly simultaneous, or, no matter how relatively rapid such cognizance was, it could never encompass the extension forever, which would abrogate the Law of Periodicity.

One wonders whether the 'Beholder' of infinite articulated possibility past, and infinite articulated possibility future, can cognize such articulation when limited by Time. Earlier we suggested that the 'time' for such apprehension of infinite articulation might be the juncture between the Infinite Subject contemplating Its own homogeneous Infinitude as Mulaprakriti, and the 'De-Infinitizing' 'Move' towards Finitization during which the Infinite Subject (seems) to 'Become' the 'De-Infinitizing Subject'. Who can say that a 'Flash' revealing infinite articulation past and to come could not occur? Its *modus operandi* is, of course, utterly inconceivable to us.

The Problem of the Origin of the *Decision* Which Determines the Parameters of the Cosmos-to-Be —Is it a 'DECISION'? Or, a 'Decision'?

The foregoing discussion leads directly to a more serious problem. There is a determination made, for any Cosmos is Finite, and Its Parameters must be determined.

1. If the 'DETERMINATION' is made 'within' the FOUNT OF ALL POSSIBILITY, the *modus operandi* is humanly inconceivable. There would be no Time available to 'MAKE' such a 'DECISION' 'before' the 'FLASH' (or the 'CHANGE' from BENESS to Being) because Time *begins* with the 'FLASH'.

(Parenthetically, it is possible to call the Period of the Universal Pralaya, a 'TIMED TIMELESSNESS'; nevertheless, the time, which is 'later' understood to have 'elapsed' 'between' Universal Manvantaras, was completely unavailable and inaccessible during that Period, for UTTER TIMELESSNESS 'PREVAILED'.)

2. If the Determination is made by the Infinite Subject, Pre-Cosmically, it may be a somewhat 'uninformed' Decision, for, Is the entire content of infinite possibility *available* to the Deliberation of the Infinite Subject, and, if it were, what is the manner or form in which such content would be available?

Could such a Decision about the Time and Space of the Cosmos-to-Be be Made by any Being Who is *already subject to* the very Time (and, perhaps, Space) that is such an important Aspect of the Parameters upon which It is supposed to Decide? In any case, WHO or WHAT 'DETERMINED' the 'MOMENT' of the 'FLASHING FORTH' of the 'RAY'? It could not be the Infinite Pre-Cosmic Subject, for the 'RAY' 'CREATED', as it were, that very Infinite Subject. The (perhaps *limited)* Capacity of the Infinite Subject seems to be the Capacity to Realize Itself as Infinite, but *not* to Objectively Cognize Its Infinitude *in Articulation.*

Any problems encountered when examining the possibility that the Infinite Pre-Cosmic Subject Determined the Parameters of the Cosmos-to-Be are only compounded when wondering whether the Decision could have been that of the Finite Universal Logos. The Mayavic Forgetfulness is, in the Universal Logos, literally infinite. Given the Logos' Time-Boundedness, It would be impossible for the Logos to 'Review' as it were, all infinite actualized possibility past, and even more impossible for It to anticipate infinite unactualized possibility future. The probability of an Uninformed Decision become infinity-to-one, i.e., is *infinite.*

☐ It appears that the most probable locus for the *decision* is 'within' THAT, where no Time for a 'DECISION' *REALLY* 'EXISTS'. We can only rest upon the thought that the ABSOLUTE PLENUM that is the FOUNT OF ALL POSSIBILITY, 'CONTAINS' even the problematic possibility that such a 'DECISION' can be 'MADE', although the possibility seems to *contradict* the 'NATURE' of the HOMOGENEOUS IMPARTITE BEING 'within' which the 'DECISION' must 'OCCUR'.

The Problem of Whether an Ultimate Particle can Change in Any Way, Since it has no Parts

An ultimate particle/event is apparently incapable of receiving impacts of any kind—especially from anything smaller than itself—for what could be smaller? It appears that only an ultimate particle could impact another ultimate particle, and yet it is doubtful that they can touch. (A tangent-of-contact would thereby be created, and it would be far 'smaller' than the ultimate particle/event itself, which is not allowable.)

In any case, what would be the result of such touching, if it could occur? Certainly there would be no change in the ultimate particle due to impact for no *thing* can be imparted or absorbed. Anything absorbed would make the particle partite, multiple—so nothing can be added to it. Thus, there would be no communication *between* ultimate particles (unless purely in the realm of that substance called *consciousness).* How then could their positioning relative to each other be coordinated? This would be extremely problematic except for the fact that all ultimate particles *are* the Being, Fohat. Because they are simply a momentary Self-Objectification of this Great Being, they are *more* than communicating with each other, they *are* each other.

The Problem of Whether or Not Ultimate Moments can Occur "All the Time"

This problem addresses the reason why all ultimate moments and ultimate movements have to be coordinated and cannot be occurring "all the time"—for this would be to divide ultimate moments into lesser moments, and an ultimate moment would, therefore, no longer be *ultimate.* If nothing can take place 'on' or 'during' an ultimate moment, certainly no other movement (which would be occurring, as it were, part-way during the duration of the reference ultimate moment) could occur. An ultimate moment cannot start and then part-way through it, another ultimate moment begin, for this would be to divide the indivisible ultimate moment!

The Problem of Infinite Speed and Infinite Distance

If speed is infinite, is the distance 'covered' infinite? If there is infinite speed, can speed any longer be considered 'speed'? From the testimony of experiment and experience, we think we know that no factor in Cosmos travels at *infinite* speed. Of course, there would be no present way to detect such 'speed'. Even with respect to speeds judged to be faster than the speed of light, we are almost incapable of detection, though contemporary physics theorizes certain particles that may travel faster than the speed of light.

If movement through S/space were really possible, that which traveled in an unchanging direction at infinite speed would instantaneously be infinitely beyond the boundaries of the finite Universe in which it originated, unless it were forcibly 'contained' by Law or Will.

If any factor were able to 'travel' at infinite speed, and still be 'contained' in-Universe by Law or Will, that factor would be completely omnipresent at every possible point of appearance along its in-Universe path. In a way its original structure would be destroyed and it would *become* its path/field. It would, in fact, become an un-allowable continuity in a discontinuous field.

☐ While it may seem absurd to pose the possibility, it is important to know if any factor in Cosmos *can* travel/move at infinite speed, because only thus could an *infinite* Universe exist while not violating the Law of Periodicity (at least not violate the Law in the *usual* way). An infinite Universe with processes occurring at *less* than infinite speed would take forever to complete its processes, and, thus, could not be periodical.

Paradoxically, an infinite Universe with *all* processes occurring at infinite speed would be, conceivably, utterly static—no distinct events would occur, for the speed of any given factor must be measured against the speed of other factors, and if all factors 'moving' at infinite speed are omnipresent, there would be no reference factor anywhere in-Universe against which to measure speed.

The result of this kind of 'motion' in-Universe would be a continuous 'positing-in-objectivity' of one single Universal Object. Or, contrarily, it might force the Universe into 'non-being'. This scenario from all we know of Universes, is absurd, and it, too, violates the Law of Periodicity (at least intra-cosmically) for where there is utter stasis there can be no periodical motion such as is always detected in relation to E/entity-S/systems.

While common sense tells us that in our Universe no processes are occurring at infinite or even extremely great speeds (speeds, for instance, trillions of times faster than the speed of light {though upon the very highest Super-Cosmic Planes, who knows?}), yet it is important to consider whether what seems to us, relatively, like a 'slow' speed may really be much, much faster when compared to some Absolute Standard of Time (as, perhaps {but not definitely} might be found upon the Infinite Time Line). The 'seemingness' of speed is no sure indicator of its *absolute* speed. About such thoughts Relativity Theory has much to say.

All the above speculation assumes that 'motion' is Real-in-Universe; elsewhere in this treatise arguments have be proposed which argue against the possibility of Real motion. Returning, then, to the conception of Cosmos developed in this treatise, we might say that the 'movement' of any factor at infinite speed 'within' Cosmos is utterly impossible if the theory of ultimate moments is correct.

The ultimate limiting 'movement' in any Fabricated Portion of Cosmos is the 'speed' or duration of an ultimate moment which, by definition, is *finite.* Presumably, an ultimate particle/event could disappear during an 'inter-moment-instant' and reappear *anywhere* in Cosmos (even at the furthest 'boundary' of Cosmos), but this would, nevertheless, *not* be an example of the 'infinite' speed of that particle; in fact, such remote reappearance would not be a measure of 'speed' at all, since no reference points would be 'passed' during the 'journey' out of subjectivity into objectivity.

☐ From another perspective, if ultimate moments and their corresponding inter-moment-intervals fluctuated at infinite speed, no finite Universe would have duration; a Universe would be 'over' the instant it started. Even as many ultimate moments and inter-moment intervals as a Universe might contain when multiplied by 'zero time' or 'no time at all', yield a product of *zero.*

The possible infinite speed of ultimate moments and inter-moment-intervals is here discussed as a limiting and determining case, because if infinite speed cannot exist as a property of such moments and intervals, it certainly cannot exist as a property of other factors in-Cosmos which 'move' far more slowly. The speed of such 'slower' factors are determined by the 'movement' of ultimate particle/events, the appearance and disappearance of which are directly keyed to the duration of ultimate moments and inter-moment-intervals.

This entire question could be investigated far more exhaustively than has here been possible. That being said, it will almost certainly be found that *infinite speed* is equivalent to *no speed at all* (by means of the eradication through omnipresence of the form of the 'factor' the speed of which is to be measured, or complete *eradication* of the form, for the form-Provider, Fohat, doesn't 'move that fast!'). Infinite speed would thus end utterly in its opposite—absolute *stasis.*

☐ ABSOLUTE STASIS, interestingly, is 'CHARACTERISTIC' (can we use such a term) of the ABSOLUTE UNCHANGING SELF, ITSELF. If not in ABSOLUTE STASIS, it would end in *utter annihilation* of the 'speeding factor', which is the NOTHINGNESS of the ABSOLUTELY UNCHANGING SELF. Infinitely rapid motion is no motion at all, and since every 'thing' *is* a *motion of a kind,* we would naturally be left with *nothing*—i.e., the VOIDNESS of the INFINITE SELF.

The Problem of Monadic Ascent

Christ did not mention reincarnation, and present teachers do not mention the non-individuality of the Monad. Most students of esotericism think that identifiable individuality is little enough to ask of the Universe. Even the Teachers of the race tacitly acquiesce to the inference that human individuality exists, probably realizing that if They did *not*, despair and other intractable psychological problems would quickly develop among Their students.

It is interesting to realize that, although the Christ knew full well of the Reality of the Doctrine of Reincarnation, He emphasized it so little (or so selectively) that this precious doctrine did not work itself into the main body of Christian Theology. Although manipulative ignorance expurgated the Doctrine of Reincarnation from the official theological presentation, it could *not* have done so successfully if the Christ had really wanted to emphasize reincarnation beyond any shadow of a doubt, for the doctrine then would have been *explicit* instead of implied, and there would have been no way to discount it.

On a higher turn of the spiral we might wonder whether the Doctrine of Non-Ego is today also relatively unmentioned or under-emphasized by the Teachers of esotericism and occultism. Instead, the emphasis is placed upon the individual, and increasingly upon the individual in relation to the group. Beneath this obvious and spiritually-exoteric teaching, however, lies the deeply philosophical Truth that the individual-as-individual does not REALLY exist. Within the World of Illusion and for all practical purposes, of course the individual exists! But in all REALITY, no individual other than THAT can be found.

☐ People thus think of themselves as distinct identities ascending to their next level of evolution. To a degree (an *illusory* degree) this is true enough, but in REALITY there is no such ascent, for SPIRIT-as-Spirit is unchanging forever. The average student thinks that 'someday' he may 'be' a planet, a sun, a Ray Life, a great Avatar, but if we examine the hierarchical structure of Cosmos, we find that this longing to be something greater, must one day end in the *One*, the Universal Logos. Is it possible that all apparently 'separate' Monads shall one day 'be' *exclusively* the One and Only (for Now) Universal Logos?! Or shall there be as many Universal Logoi for our Cosmos as there are Monads?

Certainly the absurdity of such a model can be seen, simply by extending it as far as it can go. Thus there must be a different approach to the question of 'ascent'. According to the idea that there is only One Identity in all of Cosmos, we easily theorize that we *are* that Identity. We *are* the Universal Logos (in Essence, which means, 'in *fact'*).

If this is so, why should we not also be all the E/entities through which the Universal Logos is expressing Itself in Cosmos? Why should we not be, even at this moment and at every moment, <u>E</u>ntities high and <u>e</u>ntities low? Why should we not be at once the Universal Logos and the tiniest most fleeting ultimate particle/event, for the Universal Logos *Is* the entire range of Its emanative expressions.

If assent is given to the ideas immediately above, then what happens to 'ascent'? Do we really ever *become* something which we are not already? Can **8** *become* a Solar Logos? Would it not be more accurate to say that **8** consciously 're-become' the Solar Logos **8** already Essentially Am? The same might be said for any of our future Paths and the higher dimensions towards which they lead. Essentially, we are already 'there', and even Now our identity is just as much that 'higher' Identity as it is our more familiar 'lower' identity.

☐ Extending the idea to the extreme, to re-establish clarity, certainly we cannot *become* the Universal Logos, for the One and Only (for us) Universal Logos already exists! We can only 're-become' the Universal Logos we already *are*.

Thus, we must rethink what it means to 'progress'. God, the Universal Logos, is (within Cosmos) omni-dimensionally present and active, as present and active in the *part* of Himself as He is in the *whole* of Himself. We *are* that God and, thus, we are *already* omni-dimensionally present and active in all of Cosmos. God the Logos is the Witness to as many 'focuses of immediacy' as there are E/entities (including, in this case, ultimate particle/events). **8**, God-the-Universal Logos, Am immediately present to the tiny personal sphere of limitation **8** usually call 'my-self'; but, *equally,* **8**, God-the-Universal Logos Am also *immediately present* to *all* the 'S/self spheres' of limitation within My Cosmos. The sense of immediacy and uniqueness of identity is ultimately to be found within *every* 'S/self sphere'; **8** experience/'in-perience' my sense of immediacy and uniqueness *ubiquitously* in Cosmos.

Thus, as **8** and every other apparently distinct Monad 'ascend', **8** ascend to a 'Self-sphere' where **8** already *Am;* **8** ascend to a 'higher' 'place' where **8** have been (or *seemed* to MySelf) just as immediate and just as apparently unique as **8** have been (or *seemed* to MySelf) in my more limited and more familiar human Monadic Sphere. **8** but 'go home' to where **8** have been 'living' all along. While 'My' next 'home' may be apparently different from 'Your' next 'home', ultimately 'Home' in-Cosmos is the Universal Logos where 'You' and **8** have been *Really* 'living' throughout the duration of Cosmos.

☐ From this perspective, there is then no need (by means of the Cosmic Evolutionary Process) for us to 'become' other than we *already* are. This, by the way, does not mean that we are free not to labor. Such is the Nature of Things that we must *labor* to 're-become' what we already are.

The Problem of Immutability and Vibratory Activity

This is the problem of whether that which appears and disappears with vibratory consistency can be considered immutable if, when it reappears, it is *identical* to the way it was when it disappeared. What we are dealing with here is Really the Problem of Change. If ultimate particle/events are impartite, can they change? Change requires movement, and within that which is impartite there can be *no* movement.

One scenario shows ultimate particle/events *unchanging* throughout all their appearances and disappearances. Are they, therefore, immutable throughout the duration of

Cosmos? They seem to relate only to other ultimate particle/events, but can their reconfigurations Really be called 'relation', for no information can pass from them or be received by them (at least prakritically/objectively). Such passage/transmission and/or reception would require change or movement 'within' an ultimate particle/event, which by definition is impossible, for ultimate particle/events are pure, indivisible units of Fohatic Self-Perception. Can they be 'seen' if nothing emanates from them? This is another matter.

Supposing, then, for argument's sake, that ultimate particle/events change not; in what, then, does *change* within the World of Illusion fundamentally consist? It might be said that the reconfiguration of ultimate particle/events is sufficient to induce the illusion of change. All difference in *quality* is difference in *quantity* plus *relationship*, or simply, difference in relationship. Every reconfiguration of ultimate particle/events is a change of the many parts of the great and intricate intra-Cosmic Relationship.

☐ Are the multiple reconfigurations of relationship between ultimate particle/events sufficient, then, to account for the illusion of an ongoing, ever-changing Cosmic Process? Perhaps. "God Geometrizes", it is said, and perhaps the multiple articulations of Fohat which we are calling ultimate particle/events, produce through time (and on their own extremely microscopic 'level') an ever more perfect Divine Geometry, which reflects on a more macro-level as an ever-beautifying, ever-harmonizing Cosmic Process.

Another question arises, Although ultimate particle events can give or receive nothing during a 'frozen' ultimate moment of manifestation, what does Fohat 'learn' 'during' each ultimate moment? Fohat 'exists' 'above' as well as 'within' the Worlds of Fabrication. Clearly, Fohat is ever 'Self-correcting' (in the Realm of Pure Consciousness/Being), the better to approximate the Design-at-the-Beginning (as mediated {in Timed Revolution} to Him {Fohat} via the Divine Cosmic Son).

So, although within the Worlds of Fabrication, all is 'frozen' during each and every ultimate moment, this cannot be the case within the World of Being (the World of highest Archetypal Reality for any particular Cosmos), though the manner of 'change' within that higher World remains mysterious, fundamentally different from 'change' within the World of Fabrication, and a matter of speculative conjecture. There must be a kind of 'feedback system' which is activated 'during' each inter-moment instant (i.e., during each Cosmo-Subjective Now) and in relation to the Cosmic Configuration assayed within the World of Fabrication/Approximation during each ultimate moment.

All this thought leads to the following, When Fohat 'repositions' Its multiple articulations (ultimate particle/events), It does so on the basis of what It (Fohat) has 'learned' by assessing Its most recent Cosmic Configuration. The 'new'/forthcoming position or 'new'/forthcoming Cosmic Configuration will be what it will be because of that 'learning' or assessment.

Is it possible, then, that ultimate particle events themselves *change slightly* from one ultimate moment to the next on the basis of what Fohat has 'learned'? This is a very serious question with no easy answer. So much depends upon whether the relational-position of ultimate particle/events is, alone, sufficient to bring about the many necessary and apparent 'changes' in the Cosmic Process as witnessed on the macro-level, or whether there have to be, additionally, *changes in the nature of ultimate particle/events themselves.* Surely, what we usually call matter changes constantly throughout the development of worlds and systems.

☐ The question is, Do the fundamental 'building blocks' of matter change as well, or do they not? Remember, ultimate particle/events have no 'parts' which can 'change', so a 'change' would have to involve their entire structure, globally considered.

An even more difficult question arises in this regard, If the ultimate particle/events 'change', or, shall we say, reappear slightly differently than they disappeared, *how* and *in what manner* do they change? They are still impartite, and by definition, must *remain* impartite, so there is no reconfiguration of lesser particles 'within' them possible. They *could* change in 'size', relative to their former 'size', but the parameters of Cosmos are not to be played with lightly; the Laws under which Universal Constants 'vary' (*if* they vary at different times in the Cosmic Process) would have to be studied very closely, and we are in no position to do so.

Could such ultimate particle/events change in *consciousness?* Presumably any *veiled-system,* from the Universal Logos, to Man, to the atom, to the ultimate particle/event, can change in consciousness. Fohat, too, is *mayavically veiled,* and the Great Veil *lifts* upon Fohat as it does upon all other Beings. But *how* does it lift? And what *is* Consciousness at that high level? An ancillary question arises here concerning the particulateness or non-particulateness of Entities within the World of Being, Does change of Consciousness within the World of Being require the *particulation* of those Beings Whose Consciousness is changing, or does Consciousness (itself a kind of 'moving part') have no 'moving parts'? Does Consciousness operate independently of particulation?

All this leads to a consideration of how the initial Cosmic Trinity consisting of Cosmic Subject, Cosmic Object, and Cosmic 'Sight' or Consciousness *Really* operates. Can this Trinity consist of three unitary, *impartite* Fields? Because these three Fields are separately denotable and denumerable They *are,* indeed, 'Parts' (at least with respect to That which includes Them). But perhaps the three 'Parts' are Themselves, *Essentially,* impartite. Even if They each *were* impartite, through Their interplay They would be responsible for creating, *perceptually,* all particulations 'below' (for particulation is produced through *perception,* especially, 'Self-Perception', 'Self-Sight'). Even if these Three are *initially* impartite, at least two of them, Cosmic Subject and Cosmic Object, are 'particulate-in-latency', for all particulateness-in-Cosmos develops from Their *mutually-perceptive* interplay.

Returning to our original consideration, during the Cosmic Process, the Universal Logos is *veiled,* and all E/entities in Cosmos are, likewise, *veiled.* During the Cosmic Process and at the *right time,* however, all veils 'lift'. How do they do so? What does it mean for a veil to 'lift'?

In the lower World of Becoming (the World of Fabrication), right relationship/alignment of all the constituent 'parts' is equivalent to the 'lifting of the veil'. This is brought about by means of Purification, Destruction, and Organizing (the three Shamballic Potencies, considered in their *universal* Aspect). Within the World of Being (partite, impartite or semi-partite), *unveiling* may occur through what we might call *'refocussed* Self-Sight'.

The great Originating Beings in that Realm 'know' Themselves in a certain way at the *beginning* of the Cosmic Process, and, perhaps, in a far more complete way at the *end* of the same Process.

An important implication here arises when considering this model dealing with a hypothetical 'Logoic Ignorance':

- ☐ The Universal Logos Itself just *may* be in the midst of a great Adventure in which even *It* may not 'See' *simultaneously* all aspects of the Design-at-the-Beginning.

This is a possibly troubling idea. We often think of the Universal Logos as being Omniscient within Its own Cosmos (and, of course, It *may* be). Indeed, it must be said that all that *is* known in a Cosmos during a given Cosmic Now, is 'known' by the Universal Logos (more than by any of Its subsidiary Emanations).

But does the scope of what *can* be 'known' in Cosmos *change* from moment to moment, such that the Universal Logos 'knows' more at the end of Its Cosmic Process than at the beginning? May it be that the Universal Logos is 'encoded', as it were, at the beginning of the Universal Manvantara with the Algorithm (from the FOUNT OF ALL POSSIBILITY) to be worked out in Cosmos, but that It (the Logos) must sequentially *discover* the nature of the Algorithm, and that this 'discovery' is Its own 'unveiling' that prepares It to be reabsorbed into the INFINITE SELF?

All of this touches on the interesting idea that the Universal Logos (although It Is Essentially THAT) is, because It is *finite*, *infinitely removed* from THAT. The Universal Logos may well be, because it is temporarily *veiled*, ignorant of the INFINITE CONTENT of THAT. So the questions arise:

- Is the Universal Logos (*veiled* and semi-blinded) on a Divine Adventure (albeit {hopefully} to a Forgone Conclusion we call the Perfection of the Design-at-the-Beginning)? Or,
- Is the Universal Logos (with respect to *Itself*, at least, if not to *THAT*) *unveiled* and fully cognizant of all intra-Cosmic possibilities from start to finish?

If the latter scenario is accepted as more likely, then veiling *within* Cosmos begins with the Divine Emanative Process. The Universal Logos, then, 'knows all' (*intra*-Cosmically), but Its Emanations *do not*! The only REAL *unveiling* for the Universal Logos, then, would occur *after* the perfected expression of Its 'Assignment', when It drops Its Finitude and is consciously reabsorbed into the INFINITUDE (from which It {the Logos} was, as it were, *infinitely removed* during Its 'Tenure' *as* a Cosmos).

We students of esotericism discover the 'Divine Plot' as we go.

- Does "As above, so below" apply, or does it not? Is it true of the Universal Logos, or is it not? Obviously, no firm conclusions can be responsibly drawn.
- Do we, on the highest levels of Cosmos, *as* Universal Logos, 'know the Self' *entirely* right Now? Or, do we not? Or, do we not quite yet?

The problem is fundamental and very great. In thinking about it we will be forced to ask, To what extent does the World of Being *become?* Remember that the entire Cosmos is called the World of Becoming, with the World of Being representing only the higher 'levels' of the World of Becoming. What is being asked then, is whether the entire World of Being is *static*, throughout the duration of Cosmos, or whether in some sense, It, too, *becomes* and to what degree? That World could, of course, 'move' without 'Becoming' (simply as a Service to it various emanated Aspects 'below'). If It *becomes,* no doubt Its *becoming* is far 'slower' than the *becoming* that occurs within the Fohatic Worlds of Fabrication (the 'lower' World of Becoming).

Light may be thrown upon the problem if we deal with the question, Is *unveiling* equivalent to *becoming?* Even the higher levels of the World of Being (exclusive of the Consciousness of the Universal Logos, per se) are *veiled,* for these higher Worlds are 'Created' by the Universal Logoic Emanative Process, which is 'Universal Logoic Self-Sight'. A time in Cosmos must come when the Emanations lesser than the Universal Logos Itself, must 're-become' that Logos in fullness. This will be a great *unveiling.* Can we call this *unveiling* a *becoming* as well?

In the lower Worlds all veilings and unveilings are becomings. Unveilings are right re-alignments, hence are related to position and relationship. The issue is not so simple in the World of Being, because the question of the partiteness or impartiteness of that World is abstruse and difficult of solution. Only *parts* can align and re-align.

Within the World of Fabrication beginning and end are not simultaneously seen. Perhaps even Fohat, the Workman and Creator also must constantly 'consult' His 'Betters' to find out 'what is next'. But within the World of Being (though parts of that World are necessarily *veiled* due to the Divine Emanative Process), it would stand to reason that the end is (at least, somewhat) seen from the beginning—though only the Universal Logos (by this theory) would 'See' with complete clarity (as complete as possible, that is, under Its *veiled* condition—and this is the *optimistic* view).

Change within the World of Being, then, would be the refocussing of Consciousness upon the *next Archetype to be actualized,* and the next and the next, in sequence. Such a 'Focus' within the World of Being would be 'held' until there was right 'conformity' within the World of Fabrication ('Created' and 'Supervised' by Fohat). Of course, this raises the question of how consciousness 'changes', or 'moves', *if* it Really does.

- ☐ From this perspective we see that it is not so much 'learning' that occurs within the World of Being, as it is 'holding' until right relationship exists below. This 'holding' has to do with the changing of the Keys and Notes and Colors that govern from 'above' the entire Process within the World of Fabrication 'below'.

Intra-Cosmic Fohat, however, being archetypally related to the discriminative Third Ray, would not, by this theory, 'hear' all things at the same time. He would receive the Words of Instruction in sequence, when the 'time' was right. Fohat, far more than His Superiors Who 'See' the End from the Beginning, would (because largely 'downwardly focussed') presumably 'See' only one thing at a time. Thus, in sequence, Fohat would 'learn' from what He 'accomplished' within the Worlds of Fabrication. His Consciousness would change incrementally accordingly.

Would this mean a change in the consciousness of ultimate particle/events, or in their mode of manifestation? They, after all, *are* Fohat-in-particulation. *Objectively,* however, no-thing could pass *between* them (for they are the least of things, and any thing passing *between* them or emanating *from* them, would have to be a *lesser* still). However, if Fohatic Consciousness is imparticulate (and omnipresent-in-Cosmos), these ultimate particle/events could be 'in touch' with each other, just as Fohat is 'in-touch' with Himself by *being* Himself. Thus, Fohat's 'learning' would be *their* learning—impulsed from the deeply subjective and imparticulate World of Being, and not from the objective domain in which they appear and disappear.

Many ideas on this set of problems are conceivable. No hard and fast conclusion has been drawn. It is recommended that those interested in these very foundational matters think their way into the subject. (The author is very open to discussing the various ideas that may emerge.)

The Problem of Whether Relationship is REALLY Spatially Extended, or, Are There Energy Relationships, like Ideas, Which Might be Called 'Patterned Points'?

All spatial extension is Really an *apparency. REALLY*, in fact, extension is an impossibility, because in the REALM of the REAL there are no-*things* to be extended. Yet extension is an *actuality* within the World of Illusion. Extension is an artifact of *consciousness.* Concrete objectivities 'found' or perceived within the 'lower' worlds have a kind of extension that is recognizable: they appear to begin and appear to end. They are particulate in an obvious way because within the field of consciousness they are delimited.

What, however, can be said of denotables or items that do not have 'shape' in the usual sense? The spatial beginning or ending of such items is not easily discerned, or discerned at all; thus bounding limits cannot easily be drawn. Such items might appear to be ubiquitous within a field of consciousness—there is a 'what' to them, but no 'where'.

For instance, there is the Idea of Beauty. It is different from the Idea of Goodness. They can *both* exist within the field of consciousness, and both can be said to be *everywhere* , *anywhere* (or, more concretely, 'no-*where')* in such a field. They have no 'shape' in the usual sense, but they do have identifiable, *qualitative* distinctness. If such Ideas are 'field-pervasive', what is their form? Are they to be deemed 'extended' because they can be contacted anywhere within the field? Or, does the category 'extension' not suitably apply to them as it does to items which have delimited 'shape'?

There may be certain 'spaces' in which, and certain 'times' during which, certain Ideas appear *not* to exist. If the Idea is a *true* Idea, however, and not just a limited *thought* then, no doubt, it is still pervading the field in which it seems (at the time) *not* to exist, for true Ideas are pervasive. It will be especially and noticeably 'present' *if* it has been *released* by the *emphatic attention* of the highest Second Aspect Beings within the World of Being. When those Beings within the World of Being attend to the next Idea to be emphasized, that Idea pervades all Cosmic Space with maximum potency, even though it may not be registered (by most) at all. During such times we might say that the Idea is not only *extended* through all Space (with its own *kind* of extension) but 'maximally present' for impression and registration. The Idea, however, can be noticeably 'withdrawn' (though it still remains *extended* its own manner) through the willed *inattention* of those very Beings within the World of Being. At such a time, that Idea, though extended, will temporarily be far less capable of detection by beings within Worlds other and 'lower' than the World of Being.

☐ Therefore, we might say that Ideas have no apparent, delimitable kind of extension in Space, because they are all-pervading. Normal objects, on the other hand, *do* have an apparent, delimitable kind extension in Space, because the spatial limitations of such objects are clear. Ideas, however, might be said to have a *kind* of more delimitable of 'extension in time', because there are 'times' when (because of *emphasis* within the World of Being) certain ideas *will* transfuse the lower worlds and other 'times' when they will not (though if they are true Idea/Quality/Essences They will still be 'present' and pervasively extended even if relatively latent and undetectable).

So, over a very long period of Cosmic Time, the *registrable* presence or absence of such Ideas could be noted. When such Ideas *were* registrably present to many in the lower worlds, they would be (for practical purposes) *wholly* present in the particular field in which they were always (technically speaking) present. This means they would be not only wholly extended spatially (which, they always are), but that they would be *dynamically present*, which really means that they could, conceivably, be contacted with facility at *any* 'point' in that particular space.

From the foregoing we gather that, when speaking of Ideas, 'present' does not mean 'dynamically and actively present'. Also, certain Ideas are 'withheld' from the lower worlds, which does not mean they are not *present;* it simply means they are veiled, and their reception by lower world receivers is *prevented.*

☐ The question might arise, Are all Ideas necessarily (when emphasized within the World of Being) present at all points of Space within the World of Fabrication?

The answer is, probably, 'Yes, Really', but 'No, practically'. If **8** Am the Universal Logos (which **8** and all Are), then all Ideas are apprehensible to Me, whether such Ideas are activated through emphasis or relatively latent for a time. **8**, the Logos, pervade all of Space, and from any and all 'positions' within that Space could apprehend the said Idea.

If **8** Am a Solar Logos, Who has 'activated' a certain Idea within My sphere of influence, all Greater Entities Who include Me, can apprehend such an Idea, because They (by the fact that They *include* Me) are necessarily 'within' My sphere of influence. Does the Idea **8** (as Solar Logos) have activated or emphasized necessarily extend throughout all of Space, or just throughout My relatively little solar sphere of influence? This is a difficult question. Could a being of My stature (say another Solar Logos existing within the Andromeda Galaxy) apprehend the Idea **8** have activated? Certainly the Being Who Is the Lord of Andromeda probably *could* because there is reason to believe that the Logos of the Milky Way and the Logos of Andromeda are in interplay, are mutually sensitive, and may be, to a degree, mutually pervading. But what of the contact between smaller entities within each of these Solar Logoi? Analogously can a cell in one human body definitely register the state of a cell in another human body thousands of miles away? Perhaps. Perhaps not. On a strictly practical level, probably not.

If we focus upon the ontological nature of true Ideas, it may be that Ideas are really Great Authentic Entities, and thus, Essentially, pervade all of Space, even though they have ring-pass-nots which confine them. If **8** am sensitive enough, **8** can contact the *Essence* of any E/entity in Cosmos because **8,** Who Am Really an Emanated Idea, pervade, in My *essence*, all of Cosmos. So just as all true B/beings are *essentially* (but not *actually*) non-local, so (since true Ideas *are* true B/beings) such true Ideas are *also* non-local, or all pervading. The term 'true Idea' has nothing to do with normal truth or falsity, but with the thought that a 'true Idea' is a Being fundamental to the structure of Cosmos.

Perhaps we must say that it is the *created thoughts* of relatively lesser B/beings within the Field of Space that do *not* pervade the *whole* of space. If **8**, Who Am an Idea, think a *thought*, that thought will be pervasive of the entire Cosmos only to the degree that **8** Am an Entity Who has expanded My conscious ring-pass-not to coincide with the Cosmic Ring-Pass-Not. **8** (the Idea), however, *Am*, all-pervading, whereas My *creation*, the *thought*, is not. The Universal Logos, as Idea, naturally pervades His Cosmos; also, any thought generated by the Universal Logos can pervade His Cosmos to the degree He *wills* it. This is *not* true of most B/beings in-Cosmos.

Thought is of matter, therefore, and *has* extension in the more usual sense. Ideas are of Essence, and are Authentic Emanated E/entities; therefore, they are *essentially* extended through all of Space, but have no actual, delimitable boundaries in Space. The *essence* of an Idea, however, is different from the *quality* of the Idea (just as the Monad is different from its Egoic expression). For Beings of greater scope, the quality of all Ideas included within Them is apprehensible. There is a question, however, concerning whether the *quality* of every Idea/Being is apprehensible to every other Idea/Being at all times and places in the Cosmic Process. While Essence is ubiquitous in Cosmos, *quality* may be more localized, and may require the constant growth of consciousness in Cosmos before it *(quality)*, too, becomes ubiquitously apprehensible, at the "Ending Times".

The Problem of Whether an Ultimate Particle has Extension

In line with what has just been written above, ultimate particle/events in their material and objective nature have 'extension' as *usually* considered. They, the Contents of the Great Consciousness, have (in the 'Eye' of that Consciousness) shape and dimension for the very reason that they appear *as* Space. (One could say appear *in* Space but it would be less accurate, if more apparently understandable). Ultimate particle/events, however, are *idea/beings* as well, and thus, *in essence*, must necessarily be all-pervasive in Cosmos, and thus have no *delimited* extension. Total cosmic extensiveness is equivalent to non-extension. (Thus, does a true 'point' have 'total cosmic extensiveness'?) Again the formula is: *essence is ubiquitous; quality is temporarily subject to localization* and requires, for apprehension, the growth and 'extension' of consciousness'. Simply put, **8** Am everywhere in Cosmos, but it will be long time before *everyone else* 'knows' it!

The Problem of the Nature of Consciousness

Consciousness is usually thought of as the relation between the Subject and the Object, the 'Light' which breaks forth when Spirit and Matter engage. As a definition this is, perhaps, satisfactory, and yet the definition is often accepted without questioning the *modus operandi.*

Out of an indivisible, impartite SUBSTRATUM emerge the Infinite Subject, the Infinite Object, and the Infinite Consciousness (Maya) which is the Infinite Relation between 'opposing' Infinities. Thus, in a way, these 'three', the First Super-Cosmic Trinity are 'parts' of THAT which is impartite. Is it fair to say that though these three are 'parts', they themselves are, at least *initially,* impartite?

We are therefore positing that the Infinite Subject, the Infinite Object, and the Infinite Consciousness (each of them 'infinite after its kind', and all of them the prototypes of all lesser subjects, objects, and 'relations between') are *impartite.* If this is so, how does *consciousness* Really work? We think we know more about the functioning of 'partite' systems, than we do about 'impartite' systems, and we are probably correct.

- ☐ What we are really asking here is, What is 'Self-Registration'? What is 'Self-Sight'? In the lower worlds we often say, "I see myself"; in such a situation, the self (the lower personal ego as observer) 'sees' the not-self (an object which it mistakenly regards as itself—for instance, a physical body). We seem to have no particular metaphysical problem asserting, "I see myself", and feel comfortable applying this 'sight' to the emotional level, perhaps to the mental level, and, later, even to the higher vehicles.

We are, however, elevating the inquiry to an ultimately fundamental level, and asking not only, What is 'Sight'? but What is 'Self-Sight'? We can perhaps adroitly avoid the difficulties inherent in the issue if we simply say that the nature of the Self is such that It inherently 'contains' the capacity for Self-Reflexivity, or Self-Reflectivity. This thought can even be applied glibly to the ONE AND ONLY SELF, and thus all Universes came to *be.* While this assertion may be true, it still skirts the issue of *how* 'Self-Sight' may 'work'. Just *how* does the Self 'look' at Itself?

Again, we can skirt the issue by saying, "Suddenly there was 'Sight'", i.e., Consciousness, and since there was always and only the INFINITE SELF, then, the 'Sight' which 'arose' *necessarily* had to be *of the SELF by the SELF.* This, as well, is undoubtedly true, but still leaves us in the *dark* as to method.

Another approach might be to say that "Number 'AROSE' in IT", and the 'CAPACITY' for 'SELF-ABSORPTION' ('INFINIDENTIFICATION') *'BECAME'* the 'CAPACITY' for 'SELF-SIGHT'. What we are really asking here is how the ongoing 'SELF-PREOCCUPATION' of INFINITIZED BEING 'BECOMES' 'SEEING' (i.e., 'MAYA'). Clearly, unless Number 'ARISES' there is no-*thing* to 'SEE' and no other *thing* to do the 'SEEING'.

Really, however, this solution is not much better than the other, although the focus has shifted to a kind of incomprehensible 'SENTIENCY' in which a totally absorbed and indescribable 'SELF-SENTIENCY' ('INFINIDENTIFICATION') which *knows no differ-*

ence, is instantly transformed into an extra-SOURCE Sentiency 'between' two Infinite Polarities (the Infinite Subject and the Infinite Object).

What becomes clear is that whatever the INFINITE SELF was 'DOING' 'before' any of an infinitude of Cosmic Beginnings, is suddenly 'CHANGED' into a Bi-Polar Activity in the Super-Cosmic Realm. Of course, the INFINITE SELF was not *doing* anything in the usual sense of *doing.* The 'ACTIONLESS-ACTION' that is UTTER BEING has, however, to be posited in order for the human mind to think it has a grasp on the 'STATELESS STATE' of the INFINITE SELF during Universal Pralaya. Whatever that 'ACTIONLESS-ACTION' *is,* the INFINITE SELF must 'CHANGE' even if IT does *not REALLY* 'CHANGE', in order for Consciousness to 'ARISE' (and then be instantly *expelled* from the INFINITE SELF—as well 'MAYA' *should* be!).

☐ Here we are in 'real trouble' according to conventional logic, for that which 'ABIDES' forever must 'ALTER' and yet, simultaneously, *not* 'ALTER'. What this is Really saying is that Consciousness, though It *seems* to 'ARISE' never REALLY *does,* nor does any Universe. And yet, *illusorily* albeit, an infinitude of Universes do 'ARISE' out of THAT.

The basic problem is twofold: first, we do not know how anything *can* 'ARISE' 'within' THAT or 'out of' THAT; and second, even if we accepted that aught 'ARISES' 'within' and 'out' of THAT, we are completely stymied (logically) to think that the *aught* that 'ARISES' both 'ARISES' and yet does *not* 'ARISE'! What we have here is the cosmic version of "Schroedinger's Cat" which is somehow both 'alive' and 'dead' at the same time.

Coming back to Consciousness, It most certainly exists, at least in the World of Illusion, which is the only World we can actually 'know' anything about. Perhaps, finally, we must be led to conclude that the origin and *modus operandi* of Consciousness are part of a supremely impenetrable Mystery which man, at least, cannot possibly hope to solve.One definite thought should be asserted however.

☐ It is often stated that Consciousness is eternal and exists throughout duration. This assertion, clearly, seems a fallacy. Consciousness (according to Radical Infinitism) is an 'ARISING', just as is the Super-Cosmic Trinity of which It (Consciousness) is a part.

To say that always and ever there was only Consciousness is like saying "Always and ever there were Three." But Numbers Themselves are dependent for Origin upon the ZERO, and They (Numbers) 'EMERGE' only periodically. Thus, it seems right to assert that only the 'NUMBERLESS THAT', the ZERO, 'ABIDES' throughout ETERNAL DURATION, and that Consciousness 'ARISES' only periodically with the 'ARISING' of Number. Whatever is 'GOING-ON' in the 'STATE' of UNIVERSAL PRALAYA, it is *not* Consciousness!

The Problems of Whether the Infinite Object 'Sees' the Infinite Subject, Just as the Infinite Subject 'Sees' the Infinite Object

This is the Problem of Retro-Flexivity, or Dual Self-Reflectivity. It seems so easy to say that the Subject is the 'Seer' and the Object is the 'Seen'. If we imagine that the Subject is the Father, and the Object, the Mother, we are immediately presented with a humanly-based analogy that demands interpretation, for while the Father does 'See' the Mother, the Mother most certainly 'Sees' the Father as well.

There is something very psychologically appealing (though probably not entirely accurate) in the thought that when the Father 'Sees' the Mother the result is the Son; and when the Mother 'Sees' the Father, the result is the Daughter. Perhaps we are dealing with a *retroflexive loop* between Subject/Object, Father/Mother, in which each alternatively *becomes* the other. Certainly if the Object 'Sees' the Subject, by that very act, the Subject has become an Object, and the former Object has become a Subject.

Really, in all the Three—Subject, Object, Consciousness; Father, Mother, Son/Daughter, there is *Only One Being!* This must *never* be forgotten. If one simplistically asserts that the Subject must be the Subject, and the Object must be the Object, and the Consciousness must remain the Consciousness, the asserting one must be reminded that there is no way that each of the Three are not *each other entirely!*

Much thought has been expended upon the "Mystery of the Trinity" and justifiably so, because the Trinity is Essentially *mysterious*. If we say, as has been said in this treatise, that the Mother Arises through the Father's total 'Self-Sight', is it, perhaps, almost as correct to say that the Father Arises through the Mother's total 'Self-Sight'. When Division into Polarities instantly (or *almost* instantly) 'ARISES' in THAT, are the Polarities *identical* or different? It would seem that They (the Polarities) must be different, or there would be no magnetism between Them, but are they *both* endowed Consciousness? Or are the Polarities not Real until *retroflexive Self-Sight* is initiated by the One Who Sees-Itself?

If a Dual Retroflexive Self-Sight between Infinite Father/Mother, Infinite Subject/Infinite Object *does* exist, *that which is 'Seen'* may be fundamentally different. For instance, when the Infinite Father 'Sees' Itself as the Infinite Mother, the Infinite Father (Synthesis) is 'Seeing' the *homogeneous possibility* of Infinite Articulation (Root-Matter, Mulaprakriti). When the Infinite Mother 'Sees' Itself as Infinite Father, the Infinite Mother (the Potential for Articulation) is 'Seeing' Itself as the possibility of Infinite Unarticulated Synthesis. The 'Seeing', of course, is Maya.

☐ The big question here (which is a question "out of the ordinary") is whether, in the 'FLASH' that is the 'ARISING', *two* 'Seers' 'ARISE' along with the 'Seeing' and not just *one* 'Seer' (which in Seeing Itself 'Creates' the 'Seen')?

If the Heraclitan Principle that "everything turns into its opposite" is true, then there is no reason why the Infinite Subject should not *be* the Infinite Object, and the Infinite Object should not *be* the Infinite Subject. In fact, at-'ROOT', They necessary *are so.* Again the question concerns the mind-boggling *modus operandi.* (Perhaps we might pause {but not overlong} to ask, Just what *is* a "boggled mind"? The author has the deep conviction that every reader who has read this treatise from start to finish will know with certainty!)

Indeed, it might be said that the Son/Daughter Arises when the Two Poles 'See' Themselves *as* Each Other. Number 'AROSE' and the Polarities were 'CREATED'. Consciousness might be considered the 'Mutual Sensitivity' of the Polarities. The question, however, of the 'sequentiality' or simultaneity of the Arising of the Poles cannot be evaded.

One possible (and Monistic) way of expressing the *modus operandi* is as follows, the Father 'Sees' Himself 'Seeing' Himself 'Seeing' Himself 'Seeing' Himself, *ad infinitum.* Every other 'Seeing', of course, is the Mother 'Seeing' the 'Seeing' Father. This formulation emphasizes that we have but *One* Being engaged in a kind of *bi-polar dynamic* and yet remaining only a Unitary Self. This formulation might be called a 'Retroflexive Ontological Loop', and could account Cosmo-Psychologically for all Creative Processes within Super-Cosmos, and within the highest dimensions of Cosmos.

The Problem of Whether all Three Members of the Trinity 'AROSE' Simultaneously

This enormously important problem follows upon the one just discussed.

- Does the Infinite Subject (the Infinite Father) come *before* (if only by the splitest-split of a 'second') the Infinite Object (the Infinite Mother)? For all practical purposes, the Trinity 'ARISES' in THAT instantly (in a manner indeterminate).
- But does the Trinity 'ARISE' as a Unity that instantly (through a Self-Reflexive Dynamic) becomes a Trinity, or does it 'ARISE' simultaneously as a *three-part* Trinity?
- Does One Point Emerge, instantly becoming Two Points and the 'Relation Between', or do the Two Points and the 'Relation' 'ARISE' simultaneously?

We are dealing here with a fundamental problem of *hierarchy* and the hierarchical structure of Cosmos. Through 'Self-Reflexivity' (or 'Self-Reflectivity') the One can instantly 'Generate' the Two and Three together, but the One *precedes* them both. Since the Infinite Subject is usually identified with the Father Aspect, this would show the Infinite Object or Mother Aspect Arising *from* the Father Aspect. The Western Bible suggests this in *Genesis,* but in these egalitarian days, this may not seem 'socially' acceptable; *hierarchically,* however, it makes sense.

Their 'Reflective-Relating' might be called the Son, or the Son/Daughter, or the Number Three, but then Their 'Reflective-Embrace' might, *also,* simply be called Spirit-Matter, which only *later* 'Generates' the Son, Son/Daughter, or the Three. If Father/Mother 'Embrace in the Dark' (and a Deep and Mutual 'Gaze' is an 'Embrace'), the first result may be Spirit/Matter, before the eventuality of Spirit/Son/Matter. It is important to realize that there are a number of ways to look at these fundamental Super-Cosmic and Cosmic ('Family') Relationships, and that no utter consistency is to be found in the Scriptures of the World. Cosmogony and Theogony have their maddening variations producing no small impact on the resultant Cosmologies and Theologies.

It seems to the author that, for all practical purpose, the Three *spring into being* 'virtually' simultaneously, but that the One *does* (however instantaneously) precede the Two, and the Two the Three. The Numbers may, from a certain perspective, be switched around a bit, however, as it is hard to imagine the Number One 'Seeing' Itself as the Number Two, without the prior aid of Consciousness which has sometimes been called the Number Three. So is the Two Really the Three, and the Three Really the Two? The perspective is ever mobile. Of one thing we can be certain; They are all a o*neness*.

A very important point arises from this consideration of 'Self-Reflexivity' or 'Self-Reflectivity'.

☐ We are led to the conclusion that every Subject, in order to become *creative* must *become* Its own Object! This applies to the Divine Son as well as to Fohat, and, in a way, applies to 'Man the Artist', who must create *new worlds* out of himself.

We are led back to the conclusion that Mother and Son *emerge* from Father; that Infinite Object and Infinite Consciousness *emerge* from the Infinite Subject. Thus is the Principle of Hierarchy *sustained* even within the Trinity. Thus it is, as well, that within the Cosmic Process, the Process of Divine Emanation may proceed *hierarchically*.

The problems still remains as to whether the Two and Three Arise simultaneously, or in progressive order. From one perspective the two points on the base line of the equilateral triangle could be seen as the Two and Three, with the One (the Father) found at the apex. This model would point to a kind of parity between the Two and the Three, i.e., to a kind of 'co-joined' Arising. Certainly, the Two and Three are paradoxically interchangeable in the Cosmogony of the Ageless Wisdom.

Another model would be expressed through the symbol of the "Point within the Triangle". That point would represent the force that impels the Self-Reflective Process, but upon doing so, immediately disappears into the Superior Point (Father) which relates Itself *rotarily* (in both directions simultaneously) to the two other Points (Mother and Consciousness {also called, Consciousness/Son}—depending upon the perspective).

A number of models are possible. It seems wise to use a model which *preserves a sequence of emergence* within the SUPER-COSMIC, Super-Cosmic and Cosmic Trinities, setting the trend for the sequential, hierarchical emergence of all other Numbers/Entities/Ideas/Qualities.

☐ In summary, it might be said that the weight of the author's thought is presently on the side of the idea that any process involving the interplay of a number of points or foci, has always *one starting point,* even though that starting point may almost instantly generate a number of secondary points to which it (the starting point) is not only *related,* but with which it is, *essentially,* identical.

The Problem of Wave and Particle

Modern physics tells us that there are certain phenomena (such as light) which can appear or be measured in certain specified contexts as either particles or waves. The mode of the experiment and the relationship of the viewer or detector to that which is to be viewed or detected may influence whether the light appears as particles or waves. This has led some theorists to the conclusion that light is neither particle or wave *exclusively*, but, rather, *both*. Richard Feynman (a major figure in modern physics, who eschewes the even-handed approach) comes down definitely on the side of *particles!*

Radical Infinitism speaks a great deal about particles and very little about waves, except to say that wave motion may be considered an important *mode of motion* influencing the apparent movement of particles. A major question arises in this connection. Waves in various media in the physical plane world very obviously exist from the macro-perspective; waves are propagated *through* media, and the media are atomic, molecular, etc., and are, thus, composed of particles. Thus, in liquids and gases, for instance, waves are propagated through *particulated media.*

The wave motion of light, however, is said by physics to need no medium of propagation. The old idea of ether as the medium for the propagation of light waves has not yet been replaced by newer (and yet, older) ideas concerning the *real* nature of ether. Presently light is understood to be, as it were, *self-propagating* with no medium necessary.

Relativity Theory tells us that light (as energy) is equivalent to matter and the matter can be converted to light (and other energies) in a manner described the by formula $E=mc^2$. But if light is matter, of *what objectivity* does light consist? Does its particulate nature entirely disappear with the transformation from matter to energy (light)? Or is the transformation from matter to light a transformation from a cruder particularity to a more refined particularity? The particulate nature of light is presently attributable to the photon—a kind of tiny 'atom of light' so to speak.

Light can be seen; it is *not* immaterial. For Einstein and for the esotericist as well, light *is* matter. If light is not particulate (and yet is *material)* of *what* does it consist? Can waves be thought of as *continuous* in some manner and totally non-particulate? Does the photon no longer exist when light is measured as a wave? Is light-as-wave an imparticulate continuity? It seems unlikely.

From the Radical Infinitist perspective all seeming motion in Cosmos (especially within the Worlds of Fabrication/Approximation) is caused by the reconfiguration of ultimate particle/events. Even the Great Perpetual Motion of the Great Breath is characterized by the emergence and disappearance of a great 'cyclo-eternal' Particle/Event called a Universe. Is it possible to successfully separate wave motion from the presence of particulateness?

☐ Those who say that light is both particle and wave stop short at analyzing the very probable necessity that particles be present if waves are to exist. If for some strange reason, waves are entirely *imparticulate,* then they are (from the Radical Infinitist Perspective) the *only* continuities in the entirely discontinuous and quantized realm of the World of Approximation. A *continuous, imparticulate* wave would be a very strange anomaly in a domain in which what we call Objective Space is, itself, *discontinuous.*

If the assumptions within Radical Infinitism concerning ultimate particle/events are correct, there can be reconfigurations, but no *real* 'moving through' Space. The perception of 'moving through' Space would be an illusion of consciousness in worlds hugely more *macro* than the sub-sub-microscopic worlds in which Fohatic particle/events appear and disappear.

The problem comes down to one of Continuity and Discontinuity. A wave, if imparticulate, would be a *continuity.* The constituents of a wave, if the wave is imparticulate, would be indeterminate; the wave would be 'material' (for it would be objective, and objectivity *is* materiality), but:

- What would be the nature of the material?
- Further, what would happen to a wave during an ultimate moment or during an inter-moment interval?
- Would it keep propagating even though the entire Cosmo-Objective World had momentarily disappeared?

In trying to answer such questions we are forced to the conclusion that only *particulate waves* are consistent with the theories of Radical Infinitism.

The Problem of Whether There can be Motion in Cosmos Without the Participation of Ultimate Particle/Events

In the Worlds of Fabrication wherein Fohat Reigns, ultimate particle/events are inseparable from the phenomenon we call 'motion' (even if motion is Essentially un-Real). Since the number and type of planes in Cosmos is, at this stage of human evolution, difficult if not impossible to determine, it is hard to know how many Systemic Planes, Cosmic Planes, and Super-Cosmic Planes are included in the Worlds of Fabrication. Certainly our entire Cosmic Physical Plane is included, and perhaps *all* of what we call Cosmic Planes (assuming that Super-Cosmic Planes *do* exist).

We are back to the important question concerning the nature of motion or change within the World of Being, which World, presumably, is *not* particulate in the manner in which the lower worlds are particulate. There must be change in such a World (the World of Being) however much 'slower' it may be than in the lower worlds.

The execution of the Divine Plan demands an ever-changing emphasis upon archetypal Sound, Key, Color, etc. But such change in emphasis should be conceived more as a *change of the focus of Cosmic and Sub-Cosmic Consciousness* than as a change of 'position' or spatial relation. We are confronted with the task of imagining change in a relatively non-spatial, non-temporal (hence, non-particulated) Realm, for Space/Time is particulation and quantization.

We might well ask, if certain factors in certain domains of Cosmos are not *spatially* distinct and differentiable from each other (i.e., distinct in terms of their particulate composition), in what way can they be said to be *distinct* at all? Perhaps it must simply be said that they are *ideationally distinct.* Consciousness within the World of Being, (which is the Consciousness of the Universal Logos; of the Divine Son; of the Mother; and, of the Seven or Ten Logoi Subsidiary to the Divine Son), is sensitive to the Quality of Idea/Numbers. These Idea/Numbers exist as impartite Wholes, qualitatively distinct or ideationally distinct from one another, but certainly not spatially or particulately distinct. They are not aggregates. For instance, it would not be proper to speak of the Number One being more present in one domain of Cosmic Space than another. These Divine Idea/Numbers, or Divine Idea/Quality/Numbers, are, in-Cosmos, utterly pervasive (though not universally registrable).

'Motion' or change in the World of Being is simply the *intensification* of one Idea or Quality or Number in relation to another or in relation to groups of others. This intensification is caused not by *particulate repositioning*, but by the *focalization* or *expansion* of 'Sight' or Consciousness. Such 'change' might semi-paradoxically be called *choice without motion.* Imagine a number of factors all *equally* and *simultaneously* "under the *Eye*".

What, Really, *is* the 'movement' or 'change' by means of which there is focus on one factor in preference to another, or on two factors in preference to any number of others. Metaphorically, such 'change' is simply what might be called *'intensification of the Gaze'.* It is simply an affirming Act of Will. Will says Yes or No; This or That. Will *chooses* without 'moving'. Will invests itself more or less *in* any thing simply by *identifying* with it to a greater or lesser extent. Such are what might be called the 'motionless movements' of the Life Aspect. Identification (if it can be called a 'movement' at all), is very different from the expansive, embracing 'motion' of Consciousness, and utterly different from what we usually call Activity. Thus are the Three Aspects of Divinity denoted in terms of Their differing 'motions'.

All these thoughts are feeble attempts to describe a type of 'motion' or 'change' that is not dependent upon particulate reconfiguration, or relational positioning. Thus, in summary, the Will within the World of Being, 'moves' by means of intensified Identification accompanied by intensified Consciousness. Since the Will is already 'within' that which It Identifies, there need be no customary 'movement' towards or away—only an *intensification of Presence* 'within'.

The Problem of Fragmented and Un-Fragmented Images

There are Archetypes and Their reflections in the lower worlds. The concept of the *Mosaic* holds the key. Within the World of Being, Idea/Quality/Beings exist as imparticulate Wholes. Images within the Mind of God are Wholes; Their reflections within the Worlds of Fabrication are *not.* What, Really, is any Being other than an 'Image in the Mind of God', the Universal Logos?

☐ The reflection of an Archetype is a Mosaic, a fragmented image, presumably of an unfragmented Ideational Model. The fundamental 'pieces' of the Mosaic are none other than our ultimate-particle events which 'aggregate' (to use a Buddhist term) in order to represent 'Models in Heaven'.

Ideational Models (Plato's Ideas, or "Forms") are un-aggregated, even though they may have multiple apparent 'parts'. This is mysterious. Within the World of Being, somehow there is not the fragmentation found in the Worlds of Fabrication, and yet the Ideational Images (the Great Numbers are such) are *Composites which are nonetheless Wholes.* The image of the Mosaic thus holds an advanced occult Truth relating the World of Approximation (the World of Fabrication) to the World of abiding Universal Archetypes (Higher Forms).

The Problem of How Ideational Qualities Combine and yet Remain Whole

The World of Being is a Realm of Wholeness. The *usual* issues of Time and Space may have little to do with this Domain (though Time and Space as abstractions are inescapable throughout all levels of Cosmos, the World of Being is on a 'Schedule'). Within this World all Formative Ideas are ever Present at the same 'Point', as it were. All Formative Ideas are simply Present (as the one and only Dimensionless Point) within the Mind of God, and are so cosmically extended and pervasive as to have no de-*finable* extension.

What is the method of 'storage' (or being 'present') within the Mind of God, the Universal Logos? God, at this 'level' is beyond the usual intra-Cosmic Time/Space constraints, and all Ideas are simply *wholly present* and *present as Wholes.* Combination arises simply through Divine Conception. God 'Imagines' and a whole image *arises*, exactly as God Wills. God sustains the World of Being simply through Will, and Ideas remain *multiple* and yet *whole* and *imparticulate* simply through the Will and Imagination of the Universal Logos.

☐ In the World of Being the Whole is indeed greater than the sum of its parts. This is the Principle of Synergy. How can the Cosmic Monad add Itself to Itself, numerous times, and no matter what the numerosity of the result (whether 3, 5, 7, 9, etc.) still pervade the sum with Oneness, such that the Many (though multiple) are seamlessly the One? This thought gives the illuminating hint regarding how Ideational Qualities may combine and yet remain imparticulate and seamlessly *whole.*

The Problem of *That* into which Ultimate Particle/Events With all the Fabricated Cosmos Disappear —Does the World of Being also Disappear?

Do ultimate particle/events disappear into PURE NOTHINGNESS, or is the State of Fohatic Disengagement a State less ultimate that NOTHINGNESS? 'After' every ultimate moment, for an undetermined instant (an inter-moment interval) the entire World of Approximation, which lies dimensionally 'beneath' the World of Being, 'disappears'. Into *what?*

Just because ultimate particle/events disappear, it does not mean that the Main Cosmic 'Players' have to disappear as well, for they are *ideational* and *imparticulate.* The Beings within the World of Being *are* and *contain within Themselves* the Ideational/Qualitative Model upon which the lower or 'Mosaic Worlds' are built. These Beings, in a sense, *are* 'Beauty of Design' and are sustained through the uninterrupted 'Self-Sight' of the Universal Logos (uninterrupted at least, for the duration of a Cosmos).

We thus come to the thought that the Essence and Quality of Living Idea/Beings are 'continuities in-Cosmos' (though not *absolute continuities*) and *reside* in the World of Being, but that the *expressions* of those Idea/Quality/Beings in the World of Fabrication or Approximation (the Mosaic World) are *not* continuities, and are withdrawn into a state of non-objectivity (a World of Cosmo-Subjectivity) after every ultimate moment. Thus, the World of Being is sustained throughout Cosmos in a *relatively* unchanging 'State' while the World of Fabrication changes with extreme rapidity.

Can it be reasonably said that during every inter-moment instant or interval the World of Fabrication is *withdrawn* into the World of Being? Indeed Essence and Quality-to-Be never 'depart' from the World of Being. All authentic E/entities 'reside' as Ideational/ Qualitative Essences in the World of Being throughout a given Cosmos.

☐ What is happening, then, as the World of Fabrication 'disappears', is that Fohat instantaneously *withdraws* Its 'enumerated Self-Objectification' (Its 'enumerated Self-Sight'), and returns (for the briefest instant) to Its State of Wholeness, Its 'imparticulated State'. That which has been *created* 'disappears', but Fohat, the Creator, does not, residing (if only for an instant) with the other 'Members' of the World of Being, in a State of *subjective interiority.* The various 'Sons of Fohat', as well, are withdrawn into their hierarchical positions within the World of Being.

The Problem of the Ostensibly non-Spatial, non-Temporal Nature of the World of Being in a 'Spatial', 'Temporal' Cosmos

A finite Cosmos has boundaries caused specifically by the focussed 'Self-Sight' of the Universal Logos. The Logos Wills to 'See' Itself *bounded.* In what sense is the Universal Logos 'contained' by Its own boundaries? Is the World of Being (in which the Ideational/ Qualitative Model of a given Cosmos abides) 'contained' by Cosmic Boundaries?

Thinking of the FOUNT OF ALL POSSIBILITY that 'CONTAINS' *all* possibility, can we say that such 'POSSIBILITIES' are REALLY 'IDEAS' 'within' the INFINITE SELF? Such 'POSSIBILITIES' or 'IDEAS' cannot be said to be *spatially or temporally bounded* in any sense, because 'within' the INFINITE SELF, Time and Space cannot not exist (though Time and Space must necessarily *inhere* 'within' the 'NOUMENESSENCE').

ESSENTIALLY, an 'IDEA' is BEING plus 'QUALITY', or, perhaps, it could be called a 'LIVING QUALITY'. *In* Cosmos (which is a 'Finitude'), an Idea would be called a 'Living Quality'. There is no such thing as 'QUALITY' uninformed by BEING; thus an 'IDEA' is one of an infinitude of 'MODES' of the INFINITE SELF. A Cosmos is REALLY an 'IDEA' 'EXTRUDED' for Objectification from the FOUNT OF ALL POSSIBILITY.

- ☐ While, paradoxically and necessarily, the full 'BEINGNESS' of the INFINITE SELF 'INHERES' 'within' and 'as' the 'EXTRUSION', only an 'infiniteth portion' of all possible *'QUALITY'* 'within' the INFINITE SELF is represented.

So the Life of Cosmos (the Universal Logos) *Is*, indeed, the INFINITE SELF, but the *Quality* of Cosmos, that which makes Cosmos an 'IDEA-as-Idea' distinct from all other 'IDEAS' or 'POSSIBILITIES' 'RESIDENT' within the INFINITE SELF, is *unique* and is only one of an *infinitude* of 'IDEAS' or 'POSSIBILITIES' 'RESIDENT' 'within' the FOUNT OF ALL POSSIBILITY.

Through the process of 'EXTRUSION', the 'IDEA' which is Cosmos (i.e., the BEING plus 'QUALITY' which is Cosmos) *becomes*, as it were, 'Time and Space-Bound'. Though in a sense that 'IDEA' (the BEING and 'QUALITY') continues to exist free of Time and Space, the IDEA's particular 'Field of Application' is 'within' a certain Time and Space delimited by the 'Self-Sight' of the Universal Logos.

Thus the 'IDEA' (containing a multitude of lesser, 'embedded' Ideas to be actualized through a particular Cosmos) is not operational with respect to the *entirety* of Mulaprakriti (Undifferentiated Root-Matter), but is only operational within *Cosmic* Prakriti. In the most ESSENTIAL sense, Beings in the World of Being (Beings Who *are* Ideas and thus *are* both Being and Quality) can therefore be said to be non-Spatial and non-Temporal of infinite extent and duration, *yet*, for *practical* purposes, these Being/ Ideas are *localized* and *finitized* for the duration of the Cosmos. Thus is the World of Being (for practical purposes) spatially and temporally *bounded.*

The Problem of How Infinitized Possibilities Can be Distinct from Each Other

Within the FOUNT OF ALL POSSIBILITY, an infinitude of possibilities are 'RESIDENT' in an *infinitessentialized state* and *not* as distinct and articulated possibilities, for articulation and division are not possible within the INFINITE SELF (though articulated possibility occurs 'in' Cosmos, and certainly Cosmos is not only 'within' the INFINITE SELF, but is wholly and completely the INFINITE SELF).

☐ The INFINITE SELF is the INFINITESSENCE. Rather than say that the INFINITESSENCE 'CONTAINS' all possibility, it would be more accurate to say that the INFINITESSENCE 'IS' all possibility. The INFINITESSENCE is the 'NOUMENESSENCE' of all possibility, including those possibilities which 'within' IT are *impossible!*

The problem to be addressed lies in the thought that 'within' the INFINITESSENCE, all *infinitized* possibilities are utterly the *same.* How is it, then, that all possibilities are forever *unique* and, yet, forever the *same?* For, the ONE AND ONLY HOMOGENEOUS BEING *IS* all possibility. ESSENTIALLY, IT *IS* WHAT IT 'APPEARS' to 'CONTAIN'. How can THAT which remains forever *homogeneous* ever be *articulated?* Another way of phrasing the question is as follows, How can the UTTER HOMOGENEITY be, *also,* the 'INFINITE ARTICULATABILITY'?

☐ Once again we have entered the realm of temporarily insoluble paradox, which seems to justify the paradoxical act of giving to the NAMELESSNESS the name of the GREAT CONTRADICTION. Of course, if the NAMELESSNESS *IS* the GREAT CONTRADICTION, in order to be *contradictory* IT must also BE the 'GREAT NON-CONTRADICTION'. Thus we are left with the thought that 'within' IT, all polarities or opposites, both cancel themselves out and do *not* cancel themselves out.

Perhaps, we will be forced to conclude that the NAMELESSNESS is forever, both an UTTER HOMOGENEITY and an 'INFINITE ARTICULABILITY', both the VOID and the PLENUM. What seems 'humanly reasonable' in relation to a PLENUM seems totally unreasonable in relation to a VOID, and vice versa. It is *clear,* at least, that the *modus operandi* of the accomplishment of *articulation* 'within' the 'UNARTICULABLE' is *unclear!* There, unfortunately, the author must leave the consideration for a time—hopefully, not for 'All Time'!

The Problem of Whether all 'Parts' are Composed of 'Parts' —Are there Impartite 'Parts'? —Is there a Whole without 'Parts'?

This problem is immediately relevant to the structure of Cosmos.

- Common sense seems to say that every-thing, every object, is divisible.
- Radical Infinitism posits an ultimate particle/event that is *indivisible,* simply because the Universal Logos and Intra-Cosmic Fohat (His Emissary) have *willed* it to be indivisible. (In doing so, Fohat simply Wills Himself to 'See' no smaller 'divisions' within Himself.)
- Further, it is posited, that no 'material' means of dividing this smallest of all particles has been provided in Cosmos—though, of course, Divine Will and Divine Imagination could easily do it. The Divine Will and Divine Imagination can do *anything—allowable.*

The ultimate particle/event in any Cosmos is thus posited as a 'part' with no 'parts'. The insurmountable metaphysical problems connected with the 'particulability' of the INFINITE SELF have already been discussed above and elsewhere in the text. That INFINITE SELF is 'IMMUTABLE' and, yet, is not only the SOURCE of all mutation, but *IS* all mutation—for, other than IT, naught else *is.* Thus, paradoxically, the INFINITE SELF *IS* both the WHOLE and all possible 'parts' forever.

Within the World of Fabrication (the World of Approximation, the Mosaic World) all wholes are partite. The Fohatic World of Fabrication is the Realm of the *particle* and by means of the various *aggregations* of ultimate particle/events all things are *made, fabricated.* Our Solar Logos, for instance, is a great Whole expressing Itself through many 'parts'—some parts being as large as planets, and some parts being as small as atoms, sub-atomic particles, and (ultimately) ultimate particle/events.

How 'high' (dimensionally) the World of Fabrication 'extends' is difficult to say. Probably, for every seven-plane system (whether the planes are systemic, cosmic, or super-cosmic) there are three superior planes that are, at least, reflections of the World of Being (as It 'hovers above' the World of Fabrication). Suffice it to say that the World of Being can be most logically associated with the Number Three, and the Worlds of Fabrication, with the Number Seven. Their *sum* is the perfect Ten.

☐ The 'mosaic' presents an image which will help us clarify the difference between Wholes in the World of Being, and Wholes in the World of Fabrication. Note that a mosaic is composed of many 'parts' and 'pieces'. When looking at mosaics one notices that, inevitably, there are lines of demarcation between the various pieces. The picture presented by the mosaic is, therefore, *not seamless.* Fohat 'Creates' a reflected, aggregated World composed of His (Fohat's) extremely numerous, but definitely numbered, 'corpuscular units of intention' (i.e., ultimate particle/events).

The analogy seems to call for the existence of a *seamless* world of *Ideation* of which the Mosaic World is a *reflection.* The "Mosaic Pavement" found in some mystery traditions is an excellent objectification of the idea. So the question arises, If the World of Fabrication is a mosaic and, hence, *particulate*—each Whole within that World consisting of many 'parts'—can the World of Being be *seamless* in its construction and hence, *imparticulate?* Is the World of Being, as it were, filled with *seamless* Ideational Images, and, if so, what is their Real nature? Are they *pictorial* or *non-pictorial?*

The Entities within the World of Being *are* great Number/Idea/Qualities. Of course, They are also Essences, at-one with the Universal Logoic Essence, and, ultimately, with the ONE AND ONLY ESSENCE. Essences can be conceived as impartite with no difficulty. Very simply, BEING (and even BEING-as-Being) is *indivisible.*

The contemplation of *quality,* however, presents its own set of difficulties. Qualities are derivative of Essences; Qualities are *not* Essences (except in the sense that all possible things are fundamentally the ESSENCE). And finally, Qualities are inescapably 'partite' (though not *necessarily* 'particulate'). The fundamental Idea/Quality/Beings Who 'reside' within the World of Being are Really the Great Numerical Entities. Every Number larger than the Number One can be divided evenly, with no remainder by Itself, on occasion by some other Numbers, and by the Number One.

☐ Thus, the Numerical Beings within the World of Being are, in a sense, *divisible. But,* are They *particulate?* Is it possible for a Being to be *divisible* and still *imparticulate, divisible* and still *seamless?* This is a very big question. It will take us into an analysis of how Numbers are generated and how They relate to each other.

From a metaphysical perspective, every Number (no matter how many times It is divisible by the Number One) is, nevertheless, the Number One. Another way of saying this is that every Number, regardless of Its magnitude (Its numerosity) is *still* the Number One. Thus, every Number is a unit, unitary, and a Whole. Is, then, every Number, though divisible, a *seamless* Whole? Since all Numbers are Essentially the Monad (even though the 'Monad-in-attenuation'), can the Monad Really be divided from Itself? Does the combination of the Monad with Its own reflection, and with the reflection of Its reflection, and with the reflection of Its reflection of Its reflection, etc., actually create *seams, crevices, fragmentation,* or is *imparticulate wholeness* retained ever?

The reader can see what is being suggested—that the 'Self-Reflective' Mode of Emanating the Beings within the World of Being (Who are all Numerical Archetypes), allows these great Qualities, these great Ideas, to be complex *(partite,* in a sense) and still *not* fragmented and *particulate.* Thus, although divisible, these great Beings/Qualities/Ideas retain their *seamlessness* and are, therefore, non-*mosaic,* non-*particulated* in the sense that lower world systems are.

No apparent 'part' within Them is at all separate from any other part. Thus these Beings are, as it were, 'Whole Images', or 'Seamless Images of Wholeness', reflected 'below' as particulated mosaics. What may be the *Real* 'form' of such Beings/Ideas/Qualities, what human being can say *exactly?* Are they 'Pictorial Appearances within the Mind of God-the-Universal Logos'? Are They utterly non-Pictorial? Certainly, some of Some of Their 'reflections' in the World of Fabrication are visual/pictorial, but certainly not all.

☐ There is a tendency in the human mind to think that a reflection must be a fairly exact image of that which it is a reflection. This is most often so when reflections occur strictly within the lower worlds, but need not necessarily be the case when the lower worlds are reflecting the higher and apparently 'formless' worlds. There is, however, no reason why the Mind of God should not be able to conceive in completeness a seamless Image of that which is to be mosaically reflected below.

Suffice it to say that Numbers-as-Idea/Quality/Entities combine and merge in a way different to the manner in which particles relate. The Platonic Forms of the World of Being are divisible but *'non-creviced'*, hence 'flowing' and *seamless.* The lower Fohatic forms are also divisible, but are *aggregated* relationships of distinct ultimate particle/events, and hence are characterized by fragmentation and 'crevicing'.

So then, are all 'parts' composed of 'parts'? Are there impartite 'parts'? Is there a Whole without 'parts'? We have suggested that, except for ultimate particle/events (and, on the temporal scale, ultimate moments), which are *divinely willed* to be impartite, all 'parts' are composed of 'parts'. But, in Cosmos (in the World of Being) 'parts' need not be aggregations, for Numbers are something far more spiritually *fused* than are aggregations. Therefore, wholes within the World of Fabrication and Wholes within the World of Being *both* have 'parts', but the parts within the former are aggregated, and the parts within the latter are seamlessly *fused*.

As a general statement, it might be said that *if* the "Implicate Order"(of which so much has been discussed among spiritually-minded people who are also scientific) Really stands for the World of Being, then that 'Order' is a state of *fused variety* and not a state of *aggregation* at all.

The Problem of Why Strict Continuity is Non-Allowable in a Finite Cosmos

Continuity exists when a factor persists uninterruptedly exactly as it is. As all factors-in-Cosmos arise and subside, none of them is absolutely continuous. Even the Universe is a Discontinuity, because It arises and subsides, appears and disappears; Its *being* is interrupted by *non-being.* Naturally, all subsidiary B/beings within Universe, since they are of a lesser order than the Universal Logos (Who appears and disappears with His Cosmos), must *also* appear and disappear, and, hence, must be discontinuous. THAT, and THAT *alone,* IS a CONTINUITY—the ONE AND ONLY CONTINUITY, for *never* has IT ceased to BE what IT IS, and *never* will IT cease to BE what IT IS.

☐ Within any Cosmos there are *degrees* of discontinuity. All B/beings-in E/essence are *continuous*-in-Cosmos, though discontinuous in manifestation when 'measured against' the Infinite Time Line. All Authentic B/beings in a Cosmos, endure (in Essence) throughout the duration of a Cosmos. The Idea/Qualities of the Beings in the World of Being have a *kind* of continuity-in-Cosmos. Another way of saying this is that the 'Patterned Ideation' of the

Cosmic Purpose endures throughout Cosmos; the 'Fohatic Enactment' of that Ideation does *not* endure. The 'Seamless Images' endure; Their 'mosaic reflections' do *not* endure.

We must remember that, for intra-Cosmic purposes, the World of Being is a relatively Timeless/Spaceless World. While not a World of Absolute Timelessness and Spacelessness, It functions for the duration of Cosmos in the Cosmic Eternal Now, and the Cosmic Eternal Present (the 'Cosmic Here'). Within that World there is a realization of One Synthesized Cosmic Movement in One Synthesized Cosmo-Eternal Moment. So, there is a certain kind of 'sustainedness' in this relatively Timeless/Spaceless higher World.

The 'lower' Worlds, the World of Fabrication, presents a different image. That World is plainly particulate and 'fissured', as it were. If ultimate 'parts' (which we are calling ultimate particle/events) flash in and out of objectivity countless times per earth second (*countless*, at least, for the human mind):

- Why, it may be asked, cannot these particle/events be *continuous in-Cosmos* (or relatively so) just as are the Idea/Quality/Beings of the World of Being?
- Why must they flash in and out of existence?
- Why can they not just abide as they are and 'move through Space' to their next 'relational assignment', their next configuration?
- Wouldn't the Universe be simpler if this were the case?

This is a difficult and fundamental set of questions. In it is hidden the key to the apparent individuality of spherical forms, and the key to the experience of *finitude* (which is a very *strange* experience for the INFINITE HOMOGENEITY).

First of all, according to the Fundamentals of *The Secret Doctrine*, all B/beings must obey the Law of Periodicity; they must appear and disappear. Even the Universe (the 'biggest' of all 'Particles') appears and disappears. Only in the inmost 'RECESSES' of the INFINITE SELF is Periodicity abrogated (un-REAL appearances of Universes notwithstanding).

☐ So the ultimate particle/event is simply obeying the Law of Periodicity (according to the most rapid allowable cycle in-Universe) when it flashes in and out of objectivity.

It would hardly seem reasonable for such a tiny unit to endure without disappearance for the entire Universal Cycle just as the Universe-as-a-Whole does, for they are particles of a very different size! To each sphere there is a proper Cycle of Periodicity, and the cycle of the ultimate particle/event must, proportionally and reasonably, be the *fastest*. (Whether or not the ultimate particle/event 'changes' in any way through its fantastically rapid 'reincarnations' is quite another matter which has been discussed, inconclusively, earlier in the text.)

Granted, it might be said, that the ultimate particle/event must 'reincarnate' according to the Law of Periodicity, is there, even so, any reason why it cannot 'move' continuously through space 'while' it is incarnate/objective; why must it be 'frozen' for an ultimate moment, disappear, and then, seemingly 'leap' to its next 'position'? Why not 'continuous movement'? This, too, is a difficult and fundamental question.

☐ The Quantum Universe has made its appearance in human thought, and in the estimation of the author, is here to stay. Perhaps, however, the deeper philosophical implications of *quantization* are not so readily understood. What is at stake in the thoughts generated around this subject is the very nature of S/space itself.

Mulaprakriti is Absolute Space—utterly dense and imparticulate. Cosmic Prakriti is *finitized* Mulaprakriti—also dense and imparticulate. There is no way for anything (particle or otherwise) Really, to 'move *through* space'. There is *no-thing* to 'move through'. Pure Mulaprakriti and even Cosmic Prakriti are 'no-*things*' (being only 'some-*things' in potential.* They are 'Products of Self-Perception'—the Self-Perception of the Infinite Subject in the former case, and the Self-Perception of the Cosmic Father/Subject in the latter. Neither Mulaprakriti or Cosmic Prakriti are *space* as we think we know *space.*

Space as we think we *know* it is *'thinged'.* The appearance of a host of ultimate particle/events *is* S/space as we think we know it! This means that S/space is *particulate, quantized!* S/space is a 'fissured' 'some-thing'—a 'fissured objectivity'. Each unit of S/space is separated from every other by the unbridgeable gulf of objective 'no-*thing*-ness'. The aggregation of all the ultimate particle/events *are* the very S/space which some theorists would have ultimate particle/events 'move through'! Because every ultimate particle/event is 'surrounded' as it were by the objective reflection of Nothingness (i.e., is surrounded by an Infinite

Objectivity which is so remote as to seem utterly subjective and non-existent), ultimate particle/events are utterly 'locked' into relationship and cannot 'move'. Objectively speaking, they are in utter isolation from one another, though because they are all articulations of Fohat they are, as it were, necessarily 'in communication' with each other *subjectively*—through identification and not transmission.

Thus, in order to 'arrive' at their next relational 'position' in the Cosmic Configuration, they must disappear out of normal Space/Time altogether, and reappear into their next 'position of isolated relatedness' (to use a paradoxical term). This is certainly not 'movement' in the ordinary, relativistic sense. There is no 'moving' from one 'place' to another. There are only discontinuous, quantized 'leaps' as what we think is S/space 'rearranges itself'. Fohat 'changes the Mosaic Image' before His 'Eye'. Remember, S/space is *particulated* and changes its configuration from ultimate moment to ultimate moment. This then, is the, hypothesis, the 'reason in terms of S/space' why there must be discontinuity at the most fundamental level within the World of Fabrication.

Beginning with the idea that S/space is particulated, we must think also of the contrary implications of having *non-particulated* S/space. Let us focus on Cosmic Space. Non-particulated Cosmic S/space would be *continuous* Cosmic S/space, non-quantized Cosmic S/space. It would also be *infinitely divisible* Cosmic S/space (though *what* would be the manner of its *dividing?).* It would be Cosmic S/space in which an infinitude of dimensionless 'points' could conceivably exist on a straight line (or on any 'line'), whereas in particulate Cosmic S/space, no Real line can exist, and such lines as there are have a very finite number of 'virtual points' called ultimate particle/events. Could anything prevent thoroughly imparticulate Space from being a Great Abstraction, incognizable, unregistrable—just as Mulaprakriti is.

Forgetting for a moment the terrific problems concerning the modus operandi for 'dividing space' infinitely, we can say that the infinite divisibility of Cosmic S/space would present real problems to the Finite Universe Theory. Since, there are so many indications that the Universe *must* be finite (for one thing, to honor the Law of Periodicity), that we are led (from that fact alone) to suspect that Cosmic S/space cannot be infinitely divisible.

- An infinitely divisible Cosmic S/space is also an infinitely expandable Cosmic S/space—there would be no end to it, either in terms of diminution or extensibility. Because such Cosmic S/space would be possessed of an infinitude of *actual* dimensionless points from which ultimate particle/events (if they could even exist in *continuous* Cosmic S/space) could relate to each other through configuration, an infinitude of relationships within Cosmic S/space would be possible, and, on their 'way' from 'here' to 'there', ultimate particle/events would 'move through' an infinitude of relationships. This, too, would countervene the idea of a Finite Universe, with a *finite* number of actualizable processes and relationships.
- Also (as explained earlier in the text) an infinitely divisible and infinitely expandable Cosmic S/space would require an infinitude of Time in which to conduct the Universal Process—this possibility also countervenes the very solid Law of Universal Periodicity.

All of these seeming substantial questions concerning infinitely divisible space need not really be considered, because the entire premise of an infinitely divisible *space* (at least as applies to Cosmic Space) is ill-founded. Such a notion treats *space* as if it were an 'external something' which (crudely put) could be divided in so many 'pieces'. The psychological dimension is entirely missing, as is the profound realization that, colloquially, "You only get from *space* what you put into it."

☐ Thus *space* is really a *projection!* If the Objectivity called *space* is, in some way, infinitely divisible, it must mean that the Subject Whose Self-Image *is* Space is also *infinitely divisible.* One cannot *do* anything to so-called 'external *space*' until one *does* something to the Subject of which the *space* is a Reflection. Whatever one *does* within the Subject will immediately be *reflected* in *space.*

We could debate the existence of the infinite divisibility of the Infinite Space that Mulaprakriti is, and find justification for it (but that infinite divisibility would apply *only to Mulaprakriti* and not to the many varieties of Cosmic Prakriti.)

☐ Mulaprakriti, however, is a Reflection of an Infinite Subject and is certainly not cognizable as what we usually call *space.* Mulaprakriti is a Great Abstraction, unperceivable, and nothing will ever arises 'in' It, if It remains in a 'state' of infinite divisibility, which means of 'homogeneity'. *Finitization* within the Subject must occur for Cosmic Space (as we know it) to arise!

The *apparently* continuous yet articulated Presentation to Consciousness that we normally call Space and Its 'contents' is simply a collection of *subjectively projected finitudes* and cannot be produced from an apparently *external, infinite divisibility.*

That an impartite Space *does* exist is accepted by the author. That such Space, however, is what we usually call *space* or *interval* is disputed by him.

- Abstract Space or Mulaprakriti is a high-order relative Continuity, though not an *absolute* Continuity, for It disappears during Universal Pralaya.
- Normal S/space (Cosmic Prakriti), then, is conceived as *particulated* Space, and is a profoundly *discontinuous Object* composed entirely of evanescent discontinuities (ultimate particle/events).

Such Space is, Really, an 'aggregation of minute Fohatic Self-Perceptions'. Cosmic Space (like all *space* Really, is a 'State of Consciousness'. When Fohat 'Sees' Itself in enumeration, particulated (or Cosmo-Objective) S/space arises. Fohat's Self-Perceptions are countless discontinuities, and thus the S/space that results from the Fohatic Act of Self-Perception is also entirely discontinuous.

Space within the World of Being (such Space also being an aspect of Cosmic Prakriti) could be called Ideational/Qualitative Space. It is a non-particulate Space, but it is not infinitely divisible and homogeneous as is Mulaprakriti. Such Space is a Field of Unified Quality Presences, a Space of multiple, omni-present Qualitative/Ideational Distinctions.

☐ This Cosmic Prakriti (appropriate to the World of Being) is 'partite' but *seamlessly so* (because the Space is 'Monad-ized Space'). Although Cosmic Prakriti is 'partite', it is 'imparticulate', or non-'Mosaic'. All modes of *space* reflect the *perceiver*. Thus it is that a 'plane' and a 'state of consciousness' are one and the same.

We have examined the reasons for the quantizing or particulating of S/space, also referring to Cosmic Space that was *not* quantized and particulated, nor homogeneous and infinite divisible. Why, however, in the Worlds of Fabrication (and elsewhere) should there be the quantizing of Time?

Again, if Time in Cosmos is infinitely divisible it is also infinitely expandable or 'multiple', and a Universe (according to the Fundamentals of the Secret Doctrine, i.e., the Law of Periodicity) does not 'have' "*all* the time there *is*" to run Its course. If Intra-Cosmic Time is an *unquantized continuum* then *all* possible times (or units of time) can become included in a Cosmos, which thereupon ceases to be both finite and periodical.

We have established (somewhat) above that ultimate particle/events must appear and disappear in order to reconfigure; that there is no 'S/space' for them to 'move through', since they, themselves, *are* S/space (particulated, *mosaic* S/space). Now, what happens to so-called continuous Time when S/space becomes discontinuous? What happens to Time when S/spaces instantaneously 'disappears', or when S/space (Objectivity) is 'frozen' into immobility? Time cannot be continuous when S/space is not continuous, for there can be no intra-Cosmic Time without intra-Cosmic Space (i.e., Objectivity).

Thus, Time (in Cosmos) is forced to appear and disappear, as well, for when there is no objectivity, no enumeration, there can be no "flow of time" (which is simply the registration of events {i.e., objectivities} of some sort in some kind of consciousness). With the disappearance of S/space, there are no events to register. In the changelessness of 'nothing' there can be no Time. When, also, there is a necessary 'freezing' of ultimate particle/events in a Fohatically *willed* 'position', Time (in the Worlds to which the 'freeze' applies) also must 'stop', for the possibility of registering changing events has stopped.

We are discussing Time as perceived in the Worlds of Fabrication. From the perspective of the Cosmo-Subjective Now within the World of Being, Time is not quantized in the same way. Essentially, however, Time is forever *quantized*, and this is so on the *largest macro-scale*, for Time 'appears' and 'disappears' with the Great Breath, and each New Cosmos is an Ultimate Macro-Particle/Event, and the Duration of Each Universal Manvantara is the Terminally Macro Ultimate Moment!

- It is with this Macro Ultimate Moment that the Universal Logos *identifies* to produce in Its Consciousness the Cosmo-Eternal Now.
- To the Universal Logos (on Its highest level of identification) the Universal Manvantara is a 'Frozen Moment' (though not to units of consciousness of lesser scope who do not stand upon the Universal Summit).
- To identify with any 'frozen moment' (whether an ultimate particle/event, or the One Macro Ultimate Moment) is to be 'in' the *eternal now* of *that particular field of identification.*

This thought suggests that there are a variety of 'frozen nows' in which (the absorbed consciousness) experiences no passing of time. Ultimately, the UTTER ABSORPTION of 'INFINIDENTIFICATION' establishes the IMMUTABILITY of the ETERNAL NOW, and negates the 'passage' of any *time* forever.

The arguments we have been pursuing have attempted to demonstrate how *spatial constraints* force *temporal constraints*; how spatial quantizing, forces temporal quantizing. It might be possible, but more difficult, to argue from the other direction. Why must Time, *in its own right*, and divorced, temporarily, from spatial considerations, necessarily be quantized? How might temporal constraints force spatial constraints?

☐ It can be argued that Time (not only in Cosmos, but at its very root) is quantized; that Time cannot even 'appear' unless there is quantization—event/non-event/event/non-event, etc. INFINITE DURATION is not 'Time'. INFINITE DURATION is a species of *'Incomparability'*. It is impossible to assess 'how long' INFINITE DURATION 'lasts'. INFINITE DURATION 'LASTS' only 'in' the ABSOLUTENESS. In all of the INFINITE DURATION of the ABSOLUTENESS, there is never any interrupting event that would make measurement meaningful. Thus, in all of *eventless* INFINITE DURATION there is no *time.*

Time, however, is a measure, depending upon the alternation of the appearance of quantized events and their disappearance. The most fundamental quantum is the Universe, Itself, and Its periodical appearance and disappearance is *the* 'Discontinuity' which, in the largest sense, makes the measurement of Time possible. Time is a finite measure which can be understood in relation to the Infinite Duration of the Infinite Time Line. INFINITE DURATION is simply a 'TEMPORAL HOMOGENEITY' and can never be measured.

We come to understand, then, that the so-called *"flow* of Time" is no 'continuous flow' at all, but is rather an aggregation of discontinuous events and non-events. Without this aggregation of events and their negatives there could be no Time. So the appearance of Time demands, requires discontinuities in S/space. If there were simply NOTHING forever, there could be no Time; and, also, if there were simply an unchanging 'Something' forever, there could be no Time. Thus we see that Time, by its very nature, and if it is to exist at all, demands and forces discontinuities in Space.

If before my eyes (and, filling my consciousness entirely) is an unchanging point that never alters, Time does not exist as long as the point continues to be unchangingly what it is. Time demands interruption. An utterly 'smoothness' in Time, demands an utter 'smoothness' in S/space, and we are set back in the Realm of Mulaprakriti.

☐ In a way, Time is that which allows Space to exist. Time is the *interruption* that brings Objectivity into Being, and Objectivity *is* Space. The term 'ultimate particle/event' conveys the thought that the smallest unit of space is also an 'event' or unit of time. Really, Time and Space are utterly inseparable, immediately appearing and disappearing with each other.

Earlier we showed that, according to Radical Infinitism, it was impossible to 'move through' S/space, because the objects which were to 'move' *were* the very S/space through which they were to move. Analogously, is it possible to show that it is impossible to 'move through' Time? (We will discount for a 'moment' the thought that the existence of the ETERNAL NOW means that there never *has been* or *will be* any *time* at all! That, though ESSENTIALLY TRUE, is too ultimate a perspective for our present consideration. Instead we shall focus on the meaning of the "flow of Time".)

The concept of 'flow' is allied to the concept of 'continuity'. It might be said that continuous 'movement through Time' would demand the experiencing of *every possible moment* in an infinitude of moments. Using the spatial analogy, it would be rather like 'moving' though every one of an infinite number of points in an imaginary line, and equally impossible *(actually)*. We showed how such a line of infinite dimensionless points cannot *actually* exist in space, and how an *actual* line consists of a certain number of 'virtual points' or ultimate particle/events. It is like that with Time.

In *actual* Time/Space, an infinitude of dimensionless moments is as impossible as an infinitude of dimensionless points. In the World of Ideation it can be conceived (which makes such a conception *ideationally Real)*, but in the World of Fabrication, it cannot be Fohatically fabricated.

The parameters of a finite Cosmos will not allow Time to be infinitely divided. If Time could be infinitely divided, then S/space would be infinitely divided. Infinitely divisible Time in Cosmos would demand motion of infinite rapidity in that same Cosmos, for Time is measured by spatial events, and thus, in order for Time to be measured, some *corresponding* spatial 'change' or 'motion' is required.

☐ We have already argued extensively against infinitely divisible Space and infinitely rapid occurrences of events. The apparent fact of the matter is that not all units of time are 'allowed' in a given finite Cosmos and thus, the uninterrupted "flow of Time" (which demands infinitely divisible time) is (actually and always-in-Cosmos) *interrupted* and thus becomes a *discontinuity* rather than a *flow!*

If Cosmic Intention *wills* the non-existence of units of time briefer than a specified magnitude, then the flow of Time has been interrupted, and the 'next moment' must necessarily be 'separated' from the moment which preceded it. This is the quantizing of

Time. It is as if an infinitude of possible moments briefer a 'briefest allowable moment' of a specified magnitude have been 'excised' from the "flow of Time", thus causing, as it were, a 'jump' from one allowable briefest moment to the next. Between the two allowable moments (we call them ultimate moments) an infinitude of briefer moments might have intervened, but have been 'willfully removed' and made 'Cosmically impossible' by the supervising Intelligence of the Cosmos serving the Cosmic Algorithm.

What happens in the 'duration' in which the infinitude of unallowably brief moments are excised?—*nothing* in Fohatically Fabricated-Cosmos happens. The excision or disappearance of that time, signals the necessity for the disappearance or retraction-into-Subjectivity of S/space, and an 'inter-moment instant' is produced—the Cosmo-Subjective Now.

☐ Thus, given the necessity that Cosmoses be finite in order to be periodical, we can see that it is impossible to 'move through' continuous (and, hence, infinitely divisible) Time, for according to what we might call 'Cosmic Conventions', some 'times' are simply *missing!* Because of this, continuous, infinitely divisible Time does not exist in Mosaically-Reflected Cosmos.

Strangely, even along the Infinite Time Line, some *times* are *also* missing. Just as in lower Cosmos, the time occurring within the World of Being and measurable against the Infinite Time Line is *missing,* because space in the form of ultimate particle/events has disappeared, so, there are vast 'periods' of *timelessness* when Space disappears because of the Inhalation of the Great Breath. The implications are mind-boggling.

Just as within the Worlds of Fabrication, one moment seamlessly, continuously seems to flow into the next, so within the Super-Cosmic and Cosmic Worlds, one Macro-Ultimate Moment and Its Ultimate Macro Particle Event (a Universe) could seem to flow seamlessly into the next. This means that for the 'RADIATED' LIFE in Super-Cosmos and Cosmos, it could *seem* as if there were simply One Continuous Universe (perhaps going through Its Oscillations) but only One, just the same, and *never* disappearing!

Through these thoughts, we see the Universe Itself as a Quantum of Time and a Quantum of Space. If we, imaginatively, increase the speed of Its appearances and disappearances to *infinitesimally less than infinite speed,* we would have the Ultimate Duality. The INFINITE SELF would 'ABIDE' as ever, but all Universes would be One Virtually Continuous Universe existing in counterpoint to the VOID, forever. If the speed of alternation of the Great Breath were increased to *infinite speed,* the duration of each of an infinitude of Objectivities called Universes would be zero, and the sum of the entire infinitude would be zero.

Therefore, no Universe would ever exist at all; the Great Breath would be annihilated and only THAT remain as ever, in UTTER CHANGELESS HOMOGENEITY. From the 'INFINISPECTIVE' this is precisely what *seems to 'happen',* and, thus, is a way, *'DOES* HAPPEN'. Whether or not our Universe or an infinitude of Universes ever 'AROSE' depends upon the *perspective.* From the ONE PERSPECTIVE that *really counts* (or rather does <u>*not*</u> *count* because there is no Number in IT), NOTHING 'AROSE' and 'STAYED THAT WAY'.

Let us return to a difficult, and perhaps unanswerable, question, Why must Time 'freeze' during the ultimate moments within a given Cosmos? We could choose a 'dependent' argument saying that if 'Space' 'freezes', Time must 'freeze' and we would be correct. But can the question be answered on its own terms, choosing to address the requirements of Time first, and of Space secondarily? Perhaps; perhaps not.

In the ultimate moment, we have what we might call a 'frozen moment', a 'frozen frame', as it were. Has Time-*in*-Fabricated-Cosmos Really 'stopped' during an ultimate moment? This possibility cannot be discounted. Or is it just that the *perception* of Time-in-Cosmos by all intra-Cosmic B/beings within the World of Fabrication has stopped, such that, each ultimate moment is for such 'prakritically submerged' consciousnesses a *timeless* moment, even though the illusion of the ongoing 'flow' of time persists in their normal consciousness. This would mean that out of a vast sequential series of 'timeless moments' (each having a different spatial configuration) is generated the illusion of a continuous sequence of events moving, 'flowing' through time.

Let us begin by the impossible task of examining how things 'ARE' 'within' the INFINITE SELF. Certainly, within the INFINITE SELF, which 'dwells' ever in the ETERNAL NOW, there 'inhere' all possible moments (including an infinitude of *timeless moments*) that 'add up to the *'zeroness of one infinite unchanging ETERNAL MOMENT'*. With respect to the INFINITE SELF, it should *not* REALLY be said that 'time never stops'. Instead, ('within' that SELF) because of the utter continuity of an infinitude of dimensionless 'moments', anything we might call Time is transformed into ENDLESS DURATION, which, in ESSENCE, *annihilates* Time, before It ever Really Arises. An infinitude of dimensionless moments is *no*-Time at all.

We are dealing here with the most *special* of all cases, and it is not ever proper to speak of Time with respect to the INFINITE SELF 'ABIDING' as IT does in INFINITE DURATION. What, after all, *is* a 'dimensionless moment'? 'Moments' with no *time* content? In a way such 'moments' cannot possibly exist in REALITY. Perhaps, however, it is at least instructive to realize that a 'moment' which is *infinitely brief* 'contains' absolutely "no time at all", and that even if an infinitude of such absolutely 'timeless moments' is summed, no Real time will have elapsed. Thus one can account for the ETERNAL NOW by means of an infinite sum of 'temporal nullities'.

- ☐ The end result is that 'within' the INFINITE SELF there always and forever 'is elapsing' (with the infinite 'slowness' of zero speed) but ONE ETERNAL 'MOMENT' that is forever devoid of Time.

All appearances notwithstanding, always and forever, the MOMENT is NOW. From an *absolute* 'PERSPECTIVE' (the 'INFINIDENTIFIED INFINISPECTIVE') there is only one answer to the question, What time is it?—NOW!

Leaving aside the consideration of Time with respect to THAT (an utterly futile and confounding, though tantalizing task) let us return to Cosmos. As measured 'against' the Infinite Time Line (which is the ultimate in *linear/temporal* Measures), Cosmic Time never *Really* 'stops', though it may do so, *effectively* 'in' certain 'levels' of Cosmos. 'During', what for lower world beings is, effectively, the 'stoppage' of Time (even though they don't realize it), Time on the Infinite Time Line (for the Cosmo-Conscious Beings in the World of Being, and, perhaps, for the Super-Cosmic Trinity) "marches on". Even for the submerged consciousnesses in the lower Worlds, Time *Really* "marches on".

What about the 'stoppage' of Time *within* Cosmos. Is it Real? In fact, the Cosmic Process never stops 'generating' measurable units of Time'—however, these units are only measurable 'against' established 'Macro-Universal Event Units' (definite intervals between Universal Manvantaras and Universal Pralayas—or "Eternities", according to H. P. Blavatsky) 'upon' the Infinite Time Line, and only measurable/perceivable by a kind of elevated Consciousness (whether 'upper-Cosmic' or Super-Cosmic) which can *compare* the duration of the 'freeze' to another unit of duration of which it (the elevated Consciousness) is *simultaneously cognizant.*

On the other hand, the units of time generated by the Cosmic Process are *not* consciously measurable by consciousnesses enmeshed 'within' the lower Worlds of Cosmos. Nor are ultimate moments *experienced* by such consciousnesses as units of time. (The units of time which *are* experienced are far larger, aggregated moments.) Enmeshed, prakritically submerged consciousnesses, simply register impact but have no sense of the relative duration of ultimate moments (nor, Really, any sense of *duration* at all associated with such moments).

First of all, such submerged consciousnesses have no sufficiently sensitive apparatus to register such extraordinarily micro-events. Secondly, the instantaneous disengagement of the submerged consciousness and its momentary 'retraction' into the World of Being, means that with its subsequent *re*-engagement or submergence, it is as if *no interval* has occurred! Only the next event-configuration registers, and that configuration then seems to 'flow' out of the previous configuration.

☐ In other words, the "flow of time" seems *seamless!* Any perceived relationship seems to 'flow' uninterruptedly into the next and a 'smoothly moving' non-fluctuating objective Universe is perceived. From the many sequential 'quanta of motionlessness and timelessness', a continuous *movement through Time is <u>perceived</u>* (however *illusorily*). The psychological presentation *in consciousness* is thus completely *other* than the facts, which are not registered consciously as they <u>R</u>eally *are.*

An *elevated, abstracted* consciousness would be required for *true* registration. Thus we see how the subjugation of consciousness to quanta of time (ultimate moments) and quanta of space (ultimate particle/events) generates Illusion.

So, does time-in-Fabricated Cosmos Really 'stop' at each ultimate moment? Since relative motion does *not* exist in the World of Fabrication 'during' a moment in which Super-Cosmic Time *continues,* it can be said that during what we call an ultimate moment, time-in-Cosmos *both stops and does <u>not</u> stop.* It all depends upon the *context* and the 'position' of the *observer.* If an observer within the lower Worlds could register a series of sequential positional relationships in even a portion of the Cosmic Configuration, each positional relationship would seem to 'pass by' so instantaneously as to take "no time at all".

Strangely, each 'frozen moment' could take a million Earth years measured against the Infinite Time Line, and yet would be perceived by a submerged consciousness only as an 'instantaneity'. In fact, from that submerged position, no ultimate moment' could ever be *isolated* from another; so *seamless* would seem the "flow of time" that no ultimate moment could be seen as *separate* from another. Time would be inevitably experienced as a *continuum*. The elevated, abstracted Consciousness, however, would see the discontinuity, and if an ultimate moment 'lasted' a million years, they would *know* it, for their Consciousness would be registering 'other *things*' 'during' the 'frozenness' 'below'.

Reviewing then what it is (in human macro-consciousness) that accounts for the sense of the "flow of time", we might say that the *interval between* 'cosmic disappearances' is not consciously registered by the in-lower-World consciousness (for the mechanism of registration 'disappears' with the disappearing Fohatically-Fabricated World).

☐ In terms of the discontinuous (in this case, human) consciousness which focuses intermittently in the lower Worlds, only the sequence of 'event-positions' have an impact and is theoretically (but not yet *actually*) registrable. The 'blackout' is not cognized and the 'event-pictures' seem 'seamlessly joined' end to end.

(For an explanation of exactly 'how' this is accomplished, millions of years may have to pass before we *know* with certainty. It seems a good inquiry for a standard university course in psychology during the fifth round!) Really, the reconfigurations of ultimate particle/events are probably not registered at all at the *human* level; with the possible exception of very, very advanced 'seers', it is probable that only relatively macro-configurations that are 'macro-resultants' of those ultimate micro-events (but many times removed-through-enlargement from those micro-events) are Really registrable and registered. We must remember that even the tiny atom of matter is, relatively, a huge macro-unit compared to an ultimate particle/event, and certainly the impact of an individual atom is not registered by human consciousness; only huge congeries of atoms are registrable.

However, at a very *deep* level, human consciousness *abides Spirit Consciousness* (i.e., the Consciousness of the World of Being, which {in each Cosmo-Subjective Moment} registers every Fohatic 'change'). Such Consciousness might be called a 'continuous imparticulate *sensitivity* to all change'. That continuous sensitivity is intermittently obscured (for the submerged consciousnesses) as the prakritically focussed 'submerged portion' of the human consciousness 'forgets' its participation in the World of Being as many times a second as there are ultimate moments (and remembers it as many times a second as there are inter-moment instants—Cosmo-Subjective Nows).

Perhaps, we should say that the 'consciousness-in-retraction' remembers that it has not forgotten. The normal, worldly human consciousness is, thus, 'flutteringly' intermittent, but does not *know* that it is intermittent; the *deepest* human consciousness (actually 'cosmo-permanently' 'resident' within the World of Being) is *continuous* (at least in *Cosmos*).

☐ Paradoxically, the deepest human consciousness (which is not Really 'human' at all) 'knows' all that is 'happening' even while the 'prakritically embedded part' of it focussed within the lower worlds effectively 'forgets'. It is as if the human being, for instance, is *simultaneously* benighted in the lower World *and* continuously wise and *knowing* subjectively.

This does seem like a paradoxical model. The net effect of the dual consciousness (one *continuous*-in-Cosmos, and the other incognizantly intermittent) is like that of living two (or maybe more) *parallel lives*, in which the 'lower' is (for a great duration of

Cosmic Time) 'isolated' from the higher, while the 'higher' continuously 'looks on' at the drama being carried out in ignorance 'below'.

A suggestive analogy could be put in musical terms. Against the background of a constant tone representing continuity of Consciousness-in-Cosmos, is the rapidly intermittent sound of another note representing *worldly* consciousness. This intermittent note might be said to be what the constant note 'sounds like' when it is projected into or submerged in the lower worlds. The intermittent note, of course, 'sounds' only during ultimate moments and is silent during inter-moment instants or intervals. During every interval (Cosmo-Subjective Now) when the second note is *not* sounding, only the sound of the sustained or continuous note can be heard. Interestingly, if the submerged consciousness could be aware of its note (the intermittent, second note) that note would not seem to be sounding intermittently, but, rather, *continuously* (for the same reason that 'things' in the lower worlds seem to 'move' through continuous time from one 'place' to another).

Gradually, evolutionarily, the intermittent note begins to *attune* itself to the sustained note, until, when *consonance* (identicalness or perhaps octavization between the two notes) is achieved, it is as if the sustained note is sounding *all the time,* even during submergence in the lower worlds. The continuous supervisory Perception of the World of Being fuses with worldly consciousness and *vice versa.* Alignment is achieved through 'uni-son'. One field of Consciousness supervenes whereas, before, there were two. There is the achievement of simultaneous multi-dimensional Consciousness, and, eventually, the Consciousness concerned is interactively 'awake' at all 'levels' of the Divine Emanatory Stream. This strange phenomenon is related to the concept of Dimensional Sealing [described in the Glossary], and this metaphor or analogy describes how the 'sealing' may be destroyed.

☐ The mysteries of Consciousness are many. Since Consciousness is a *continuous non-material substance* the human being (in Spirit) is *always* aware of every qualitative change in the Fohatic Mosaic, but does not (in the lower worlds) *know* he is aware. Evolution is the repossession, 'below', of the awareness which exists continuously 'above'. We might say that Consciousness is the 'means' by which matter 'appears'.

Returning to the *actual* 'freezing' of Time (which the human consciousness does *not* register), *why* (in terms of intra-Cosmic Process within the World of Fabrication) does Time, or *must* Time *actually* 'freeze' 'while' all ultimate particle/events are 'holding their positions' in utter immobility? The answer is simple in a way: *if nothing moves, Time stands still*—at least as far as process in 'lower' Cosmos is concerned, for the psychological registration called Time requires differentiable events in order to exist, and there are no events without change. The 'freeze' (as factual as it may be) is yet too 'micro' for the 'macro' human consciousness to detect. The overall effect is quite other than Cosmic Reality would suggest. While time Really is 'frozen' in lower Cosmos, it *actually* does not *seem* so.

The human consciousness is yet too crude. Time 'lurches', as it were, from 'frozen moment' to 'frozen moment', but the human psyche thinks it *flows.* No movement is ever registered by the embedded human consciousness 'during' an ultimate moment, and no movement of time actually occurs 'on' an ultimate moment, but the global effect in the human psyche is that *movement is constant* and that "time flows on". Thus, what *Really* happens is different than what *seems* to happen. That which only *seems*, but *is* not, is *illusion.*

For the type of Spirit Consciousness which 'abides' in the World of Being, the 'freezing' of time 'below' is 'Seen' and understood, but the sense of Time is so expanded by the Perspective of the World of Being that the 'frozen moment' is seen to have *Real duration* against the Infinite Time Line. Time, for such an Observer, therefore, does *not* 'freeze', even though time 'below' is 'momentarily frozen'. For the embedded consciousness, denied the registration of change, there can be no registration of time as it more *Really is* 'above'.

The human consciousness in the lower worlds also *thinks* it has a kind of continuity and 'undifferentiable *flow*' appears, but Really, what seems like a *flow* is an interrupted 'presentation to consciousness', a presentation 'blinking on and off', for the lower consciousness 'becomes' at every inter-moment interval the *higher* Consciousness of the World of Being (yet always 'forgets' that it has so 'become').

When the lower consciousness "returns to Earth", it simply registers the next "change of position". The overall effect, however, of what are, Really, incremental, 'frozen units of impression' upon consciousness, is the effect one sees in a film, which offers a most useful analogy.

☐ In a 'movie' or film, all events and impressions seem to flow smoothly and uninterruptedly, and the black lines between the 'frames' (the 'flicker' so often seen in older films) are unseen and unnoticed. A motion picture is nothing but a series of frozen frames, and yet, the overall effect is that of continuous movement.

Thus it is for the consciousness of man in the lower worlds. For man's 'superior Consciousness', however, the black lines separating the frames are 'Seen', and the illusion of continuous movement disappears, for, simultaneously, a *seamless* film called *higher Reality* is, as it were, 'running' uninterruptedly (in the World of Ideations/Qualities), *all* the time.

It is interesting to imagine that this 'Archetypal Film' might be running the next *Real* image just *before* its pale reflection appears in a frame 'below' in the World of Fabrication. The purpose of these dual images (one seamless/ideal and the other *mosaic*/approximative) is to synchronize and unify the images, just as in the musical analogy, one must synchronize and unify the notes through consonance. The analogies are clear and, hopefully, instructive.

Here are a series of thoughts that may be useful concerning the problem under discussion:

- All larger moments are built upon smaller moments.
- We have established that in a finite Cosmos there must be a smallest 'allowable moment'.
- All units of time larger than that smallest allowable moment are aggregations or multiples of that smallest moment.
- On the relatively macro-level, many *apparent* movements will *seem* to occur within a given macro-unit of time.

- Contrarily, however, if 'within' the smallest 'Cosmically-possible' moment, 'movement' of any kind *could* occur, that very fact alone would *divide* that supposedly smallest moment into still *smaller* moments (for a moving *thing* must have a correlated 'time' in which to move).
- Thus, the heretofore smallest moment would be shown *not Really to have been* the smallest moment, and so *ad infinitum.* Why is this so?

As an example, a unit of movement 'within' a given macro period of time can always be divided into smaller sections of the whole unit of movement, each section of which would naturally take *less* time to accomplish than the time taken for the *entire* movement. Thus, there had better be no movement (even *apparent* movement) within that unit of time designated as an ultimate moment in Cosmos.

So, we can try to understand immobility or 'frozenness' from the Time perspective or from the Space perspective. The various arguments posited are meant only to be *indicative* and not *conclusive.* Better and more convincing explanations may be found in the future.

Just before moving on, it is very curious to realize that all through Infinite Duration, Time is simultaneously 'frozen' *and* 'progressing' depending upon the context and observer. From the *ultimate perspective* Time *both* "marches on" forever, and *fails,* altogether, even to 'move its left foot'! (At no 'time' does the author 'feel' the limitations of his mind more, than when dealing with this particular subject.)

The Problem of How Consciousness Can be Impacted if all Ultimate Particle/Events are 'Frozen'

In the foregoing discussion, the question of the Nature of Consciousness and its sensitivity was considered a little. If Consciousness is independent of its 'contents', then Consciousness will be impressed whether there is an apparent 'movement' of the contents, or whether there is *no* movement. Does it take 'time' to impress Consciousness, or is the mere *presence* of a content of Consciousness sufficient, even if there is no "passage of time" associated with the content? For instance, What is *eternal, timeless contemplation?* The point is that Consciousness is of a higher order than Time, and is operative whether "time passes" in the usual sense, or whether the Eternal *prevails.*

☐ Consciousness (at least in-Cosmos) is a *continuum after its own kind.* In its own nature, it is the 'immaterial detector of all *arisings*'. Pure Consciousness is Sensitivity, per se, and 'Infallible Touch'. Consciousness is not an *absolute continuum* because in does not exist 'within' the ZERO 'STATE'.

Apparent changes in Consciousness are not Really changes in <u>Consciousness</u>. Consciousness is always exactly what it is, 'Infinite Sensitivity', but it can be *veiled through preoccupation.* Every *arising* of a 'content' 'within' Consciousness is a kind of *veiling.* Pure

Consciousness is the Infinite Sensitivity of the Infinite Self to the Infinite Object. What is being described is the Mode of Being of the Super-Cosmic Trinity. Within Cosmos, that Trinity 'Becomes' the *Cosmic* Trinity consisting of the Cosmic Sensitivity (or Consciousness) of the Universal Logos to Cosmic Prakriti (but as what used to be Infinite Sensitivity is, in Cosmos, preoccupied with the Finitude called Cosmic Prakriti, that Infinite Sensitivity has been *veiled).* All Consciousness within Cosmos is Really Cosmic Consciousness or Sensitivity various preoccupied by 'contents of Consciousness'.

☐ Perhaps the main point is that while Consciousness is the 'Creator' of *abstract* Time and Space, and the Registrant of all articulation within Time and Space, Consciousness is also *independent* of the partite presentations of Time and Space (but not independent of Time and Space, per se). The particulations of Time can appear or disappear; the particulations of Space can appear or disappear, and still, Consciousness *Is.*

Time and Space (in the abstract) however, cannot disappear and Consciousness remain, because the very *arising* of 'CONSCIOUSNESS' 'within' the GREAT HOMOGENEITY, is an 'EVENT', a 'HAPPENING', and wherever there is an *event* or *happening* of any kind, there is *time and space.*

- Infinite Space is That which the Infinite Subject 'Sees' of Itself;
- Infinite Consciousness is the 'Seeing';
- and Time has *arisen* (or *re-arisen*) simply because the 'Seeing' has *begun,* whereas, before the 'Seeing' *began* the HOMOGENEOUS 'STATE' was 'TIMELESS' (and still so 'REMAINS' in ITS 'OWN RIGHT')—the Paradox!

Impact or impression upon Consciousness is not impact or impression upon *matter,* per se (though, in the deepest sense, anything *named* immediately *becomes* material). Consciousness has no *material* means of being impressed.

☐ Consciousness is impressed by *being the space* 'occupied' by that which *presents,* that which *arises,* within Itself. Consciousness in Infinite Subjectivity *touching* Itself. Ultimately 'CONSCIOUSNESS' IS 'SELFNESS' 'TOUCHING' 'SELFNESS'-'BECOME'-'OTHERNESS'. Consciousness is the Universal Self's First Instrument for Being What *Else* It Is.

The Mystery of Consciousness is fundamentally the Mystery of 'MAYA'-become-Maya. Suffice it to say for the present, that Pure Consciousness is unperturbed by the comings and goings of the articulations of Time and Space. Thus, in the lower worlds, consciousness registers content even though time 'stop', and nothing has to 'move' 'within' impartite Consciousness for this registration to occur.

The Problem of Whether Ultimate Particle/Events can or must 'Touch' Each Other, and the Implications of the Inquiry

Perhaps one should be a metaphysical crystallographer to address this problem. It is Really a problem concerning communication. There can be no transmission (in any conventional or spatio-objective sense) between ultimate particle/events, for nothing can go forth (spatially) from them, and nothing can be received. Furthermore, they cannot change because they are impartite. We have already discussed how they might change (if they did—a complicated thought not to be repeated here).

The question before us is whether these ultimate particle/events remain isolated from each other, 'separated' from each other (but still related and configured) by the *infinitely dense no-thing-ness* of Mulaprakriti, or whether their relationship involves not just geometrical arrangement relative to each other, but a 'touch' as well. Those familiar with Buckminster Fuller's philosophies know about "Bucky Balls". There is much of metaphysical profundity hidden in these little spherical models of Cosmos' "building blocks". They may well be (on a relatively immense macro-level) an analogy to the ultimate particle/events we have been trying to understand. We notice that the Bucky Balls *touch,* and thus *build* all the structures possible within Space.

Spheres which touch provide more solid structures; this is evident. Could ultimate particle/events actually 'touch', What would be the implications? Communication between them is problematic at best, because they cannot (in any spatial manner) *change.* Perhaps the multitude of cosmic structures they are responsible for *fabricating* would be far more integrous if the ultimate particle/events *could* touch. The possibility cannot be ruled out. It also must be said, that chaos and order being what they are in Cosmos, and since Fohat Itself is 'semi-blinded', it is very likely that even if ultimate particle/events *could* touch, they would not *always* touch. 'Transitional rearrangements' would seem necessary for the evident building-up and tearing down of cosmic structures.

☐ What would 'touching' mean? Do spheres (let us call them that) when they touch, touch at a *point?* Would a kind of *tiny virtual point* be created by the touching, and would that then become a 'smaller than allowable' structure in Cosmos? A great problem would arise if *locality* could be shown to exist 'within' or 'upon' an ultimate particle/event, for a sub-locality means the possibility of divisibility and of further partiteness.

There is also the possibility that ultimate particle/events could *almost* touch, thus sustaining integrous configurations, and yet avoiding the introduction of unwanted (and probably, impossible) 'spatialities' into their process.

We have earlier established that communication between ultimate particle/events could be established simply because they all *are* Fohat. More concretely, it is very difficult to conceive the kind of spatial relationships they might have with each other, without falling into 'macro-conventionality'. Further, the 'visibility' of ultimate particle/events could be seriously questioned? Can the imagination conceive a 'spatial indivisibility' (as we might call these particles).

☐ The spherical shape is fundamental to Cosmos, though it can be argued that no true spheres can be aggregates (all actual bodies in Cosmos being aggregates), just as no n-sided polygon can ever reach complete circularity. But what about the ultimate building block, the ultimate particle/event—can it be a sphere (since it is impartite, non-composite, and non-aggregated)?

If ever the "Uncertainty Principle" were operative, it would be in relation to this question. It is likely that in relation to this problem we will simply have to continue asking questions and hope that imagination (sense-bounded as it usually is) will approximate itself more closely to Reality.

The Problem of the Continuity of Emanative Sources

The thoughts one encounters in thinking through this problem often run counter to common sense. An example might be:

- How can a man leave his home and still remain at home?
- Put in terms of Radical Infinitism, How can the INFINITE SELF 'BECOME' the Cosmos, and still 'REMAIN' immutably the INFINITE SELF?
- Within Super-Cosmos we might ask, How can the Infinite Subject and Infinite Object 'Condense' into the State wherein they have become the Cosmic Logos and Cosmic Prakriti, and, yet, *nevertheless* remain the Infinite Subject and the Infinite Object? (Can that Condensation be a Super-Cosmic Quantum Leap, or is the 'De-Infinitizing' Infinite Subject/Object necessary?)
- Within Cosmos we might ask, How can the Universal Logos *become* the entire articulated Cosmos and still, never for a Cosmic Moment (of any kind), be anything other than Itself?

The principle seems to be that every continuity (absolute, or cosmically relative) tends to become something less continuous than itself even while *retaining* the degree of continuity it had at the onset of sending itself forth as a discontinuity.

☐ If an old fashioned telescope is unfolded so that the largest aperture is closest to the eye and the smallest, farthest, a visual example will appear. When the telescope is collapsed, the smaller cylinders are enfolded invisibly within the largest cylinder; when the telescope is unfolded, the smaller cylinders progressively appear, but the largest cylinder remains exactly as it always had been—it has, metaphorically, lost nothing of itself.

In this model of the Universe, the Prodigal Son both *leaves* the Father's Home and, yet, because the Son *is* Essentially the Father, that Son *never leaves* the Father's Home. Thus, the Emanative Source changes not for having sent its emanation forth. This is especially so for the EMANATIVE SOURCE.

- ☐ So we ask ourselves, Are we an E/emanation, or the Emanator? There would be a very partial understanding of Identity, if the E/emanation identified *only* with its own apparency 'below', and *not* also with Itself as Source 'above'. Ultimately, every discontinuity is *rooted* in the ONE AND ONLY CONTINUITY.

The Problem of Whether, when the Cosmos is Concluded, Anything is Returned (unto the GREAT SUBJECTIVITY) as a Gain or Harvest

This problem arises when we realize that at the conclusion of a personality life, the overshadowing Soul within the Causal Body receives a harvest of experience and grows thereby. We wonder, perhaps, whether the same could be true of Cosmos in relation to the INFINITE SELF, of which It (Cosmos) could be imagined to be a Personality.

There is a great difference, however, between the Soul informing the Causal Body (which is a growing structure) and the INFINITE SELF (the ETERNAL *non-growing* PERFECTION). Perhaps we think that if the INFINITE SELF is not somehow 'better' for having 'had' the 'EXPERIENCE' of the Cosmos, that life-in-Cosmos will have been in vain. We forget, perhaps, that when dealing with the infinite and with the INFINITUDE which, after Spinoza, is *"infinite in an infinite number of ways"* that the "rules change", and an entirely different approach to thought must enter. By definition, naught can be added to or taken from the INFINITUDE, for IT IS, 'already forever' the *infinitization of all possibility.* How, then, is IT REALLY to 'GAIN' from the apparent gain or harvest of Cosmos?

An infinitude of 'Cosmic Games' have already been 'Played' to relative Perfection, and yet the INFINITE PERFECTION has not 'changed' one iota. At the conclusion of every Cosmic Game, the INFINITE PERFECTION *was* just as IT *was* at the outset of the Game. Then, why have a Game if nothing is to be *gained* by the playing of it? Perhaps, by analogy, we might say that although nothing *substantial* is to be gained, there is always the joy of simply *playing.* Somehow, 'PLAY' lies at the 'ROOT OF IT ALL'.

- ☐ Perhaps we might be forced to see Cosmos, and all of Time and Space, as mere 'playthings'—un-REAL and of no consequence in themselves, but good for 'Cosmic Sport' alone. The Tibetan Teacher has said of the Initiate, that Time and Space are His "playthings".

The Southern Indian Sages might well ask whether any *illusion* can add aught to the content of established PERFECTION? So much depends upon how one values the Universe. Since during any Cosmos, the Cosmos is usually all we think we *have*, we value It immensely.

The Sages seem to see that something *infinitely more important* has been forgotten in such an evaluation. If that *something* were introduced into the consideration, so they say, we would throw the Universe away! In light of these thoughts, perhaps the Cosmic Game must be played as if there were nothing to gain but the accomplished 'Playing' of it.

The Problem of Whether there can be a Cosmic Failure

How one deals with this problem may depend very much upon whether one is an pessimist or a 'cosmoptimist'! Surely, there is no way, at this point of illusional density, to tell. We know very well that on more relative 'levels' there are apparent failures all the time. The occurrences on our own Moon Chain signaled such a failure and, no doubt, throughout infinite Cosmoses there have been failures far more colossal than this.

The really big question, however, is whether a Cosmos *Itself* can fail, or, rather, whether a Universal Logos can fail in Its SELF-Assigned (or is it 'S̲elf-Assigned') Task? How far can the Law of Analogy be carried? Something may arise in us that cries, in essence, "That wouldn't be fair! Surely a relative Perfection will supervene and the Design-at-the-Beginning will be fulfilled!" After all, should not a Universal Logos have all the *time* He needs to 'complete' His Intent? Surely, during Infinite Duration, there is no shortage of *time*! (Though in *INFINITE DURATION* there is none to be had!) Why should the Universal Logos simply not persist to the point of Cosmic Achievement? Of course, this may be a view that is entirely correct.

- ☐ While it can be said that there is 'REASON' 'within' THAT to 'BECOME' a Cosmos, it might be questioned whether there is any 'NEED' to *succeed* at 'BECOMING' and 'MANIFESTING AS' that Cosmos. Is not the entire (hypothetically proposed) 'REASON' for Cosmos a matter of objectifying Limitation and Imperfection, so that the GREAT PERFECTION may continue *immutably* as IT IS?

In the Theory of Radical Infinitism, the Universal Logos is *infinitely* Veiled, but far more *intelligent* than any of Its Emanations. A Cosmic Algorithm is 'CONVEYED' by the 'UNVEILED BEINGNESS'. Will that Algorithm be expressed in relative Perfection as Intended? Or is there just a 'chance' that if the Cosmic Game is not well played It (the Game) may end in failure, falling short of the Intended Goal? Thus the Universal Logos would 'Lose the Game'! If failure occurred would the INFINITE SELF be diminished in any way by the failure? Not if IT IS the *INFINITE* SELF! What happens within Cosmos can make no possible difference to IT.

Admittedly, this is not a very comforting thought. There is something in human nature (not to mention other natures) that rebels at the thought of 'spending a whole Universe' (just like spending a whole *day*) for nothing! We know how "out of sorts" we human beings may feel at the end of a badly spent day; just imagine the psychological nightmare at the end of a badly spent Universal Manvantara! It is bad enough to realize that *win* or *lose*, the INFINITE SELF will not be affected in the least; at least one wants to have the satisfaction of playing well, and *winning!*

So is this an ultimately 'comfortable' Universe or not? Is the End assured from the Beginning or not? If not, the 'pressure' upon the Universal Logos (and all His Emanations) would be very great indeed, and the Cosmic Game would indeed be, somehow, a *truer* Game. Must every Universe have a "Hollywood Ending"?! (One might well ask whether one wins every game of "solitaire"?) At the Beginning of Cosmos does the 'WORD' sound, "Labor in Limitation, but don't Really worry about it ... all will be well." Or might the 'WORD' sound forth, even to the Universal Logos, "Labor in Limitation and do Your best ... the outcome is not guaranteed and the stakes are high." Let the reader choose at will the more compelling option—or the more comforting option! (When one reads of Swami Vivekananda's experience in the Ice Cave of Shiva, one wonders what he saw!)

Madame Blavatsky says of the Great Pralaya that, in effect, it sweeps everything out of Space like so many "dry leaves", ready or not. It is difficult to know whether she is speaking of the truly *final* Pralaya, or one of the great precursory Pralayas in relation to which such thoughts would be eminently reasonable. After all, does any human being, no matter how accomplished, accomplish *all* that is intended when the close of a particular life comes? *Most* certainly do not.

Perhaps the author is, after all, a 'cock-eyed Cosmoptimist', but he does think that all of us *as* Universal Logos, always finish by playing the Game against our own Infinite Ignorance very well. There may be crises, and moments of uncertainty, and never is the Path to Cosmic Victory the same, but Cosmic Victory is inevitably achieved! Perhaps the Super-Universal Record could be consulted for testimony. At any rate, Time (Who has no chance of surviving the Debacle) will *tell.*

The Problem of Whether there are Many Monads or Only One

This is a foundational problem for the clarification of human identity. It has been treated from various perspectives throughout the text and in the Glossary, but perhaps should be taken seriously as an *unsolved* problem.

- ☐ People question the true nature of their identity, and wonder about its individuality. For most students of esotericism, their ultimate identity is thought to reside 'in' what is called "the Monad". They usually attribute a persistent individuality and distinctness to this Monad, imagining the survival of its distinctness throughout the Mahamanvantara, and even throughout the entire

Universal Manvantara. Perhaps, they even wonder what *they*, as an individual and distinct Monad, were 'doing' in many former Universes, or what they will 'do' in Universes to come.

To entertain such a view of human identity seems contrary to the best indications of the Ageless Wisdom Teaching. It would be to invest all sense of identity within a relatively tiny ring-pass-not, and to forget one's Identity/IDENTITY as a 'Ray' of the ONE 'RAY' of the ABSOLUTE.

To get a clearer picture of identity, one has to "begin at the Top", i.e., with the Universal Logos, Who we all *are*. (At this point in the reading of this treatise, perhaps this thought is accepted as probable.) That Logos sends Itself forth in 'Rays'; the Universal Logos *Is* the Rays Its 'sends forth', while never ceasing to be Itself, just as, for instance, the flame becomes (and, in a sense, *is*) other flames while never ceasing to be the originating flame. 'Enfolded' ('embedded', 'encoded') 'within' the Universal Logos are all the 'Rays', and 'Rays' of 'Rays', and 'Rays' of 'Rays' of 'Rays', etc. which are to be 'sent forth' or *emanated* during the Cosmic Span. These many 'Rays' are 'embedded' as it were, 'within' the Logos, and 'embedded' within the Logos' Emanations, and within the Emanations of Emanations, until the time comes for the 'dis-embedding' (the point in Cosmic Time when an embedded 'Rays' flashes forth into its proper sphere of activity within the lower worlds).

☐ It is easy and superficial to recognize these 'embedded Rays' as Monads, and to think, "Now, at last, I have found my identity; I am one of those embedded 'Rays'", but, perhaps this is to think too concretely and too separatively. The Universal Logos, Itself, Is the Identity of all those 'Self-embedded Rays'. He *Is* them. They have no *distinct identity* of their own. They are not at all what might be called 'ultimate Cosmic Individualities', because they are merely extensions of the One Individuality Who is the Cosmic Logos.

If I think of myself as an 'embedded Ray' awaiting its time to truly *be* (a time which may be relatively quite *late* in the Cosmic Process), I will have missed the point that I, (or, rather, **8**) in all truth Am *every* embedded 'Ray'—not just one. My Identity does not begin when an embedded 'Ray' which we call the Human Monad flashes into objectivity at some point midway through the Solar Manvantara, and who knows how far through the Universal Manvantara. My Identity began as a limited Being when the *Universal* Logos first came to Be (for is not the Universal Logos a *Limited Being?)*, and My Identity continued in every Emanation of the Universal Logos, and in every Emanation of His Emanations, and so forth.

Reversing the direction, My Identity resides in *That* in which the 'Monad-as-Ray' is embedded. **8** Am just as much That, as **8** Am the Monad sent forth. The 'That' of which **8** Am speaking, might be a Planetary Logos or a Solar Logos, but It, too, is a 'Ray' flashed forth from the state of 'embedding' and Its Identity is just as much in That in which It is embedded as it is in Its own Solar or Planetary Identity. And so it goes.

☐ Now, when the Monad is 'retracted', will **8** cease to exist? Will **8** be any less 'individual' because 'my!' Monad has been 'retracted' into a Source along with perhaps millions or billions of other Monads. Even now, do **8** not have Identity on *many, many* 'levels' superior to the Monadic level? Am **8** not, even now, the

Planetary Logos, the Solar Logos, and those higher Sources of which these Logoi are 'expressed' 'embeddings'?

When **8** as Monad am reabsorbed or 'retracted' in 'My' Monadic Source, will **8**, per force, remain identified with 'Monadhood', or will My sense of Identity 'elevate' into the Identity of the Source, where, according to this Theory, even *now* it 'resides' even more fundamentally than it does 'within' the Human Monad?

Some, when examining this quasi *pyramidal perspective* might say, "Ah well, then really, I am not just one Monad, I am *many* Monads." While there is some truth in this (depending upon how the word "am" is used), it tends to 'particulate' Identity. Many embedded Monads may form a higher-order Whole of which the Monad with which **8** identify is a 'part', but rather than say **8** Am:

- one Monad of the many apparently aggregated Monads; or
- *each and every one* of the many aggregated Monads in the reabsorbing Source-Being;

it would be wiser to say,

- that **8** Am the *Source Being Itself*, Who has been expressing through all the assembled (but 'non-aggregated'! Monads).

Thus **8** have elevated my Identity to a *truer* level, and must *continue to do* so all through the process of 'Emanative Retraction'.

The phrase, 'Monadic Aggregates' is never Really accurate. Just as in the World of Being, the Wholes there found were 'partite', but 'seamless' and 'non-aggregated', so within any Source-Being 'containing' a multitude of 'embedded Monads'. The embedded Monads in any Source Being are 'seamlessly related', just as are the smaller Numbers 'internal' to any larger (hence, weaker) Number.

☐ The Whole being "greater than the sum of Its parts", it is wise for any Self-conscious Authentic Entity in Cosmos to identify Itself with the Source Being Who 'contains' It as a 'Monadic embedding'. Naturally, the ability to do so is progressive throughout linear time, but even *Now* it is a fact (and in some dimension of the *beingness* of the human being, is recognized).

In conclusion it should be said that 'atomism' with respect to true Identity has its motivational value, to a degree, but is Essentially false. The true human being, the Spirit, is as much *everything* and *anything* as *one* thing; even more accurately, the human being is *no*-thing (being *That* which includes both all things, as well as the one lesser thing (for instance a Monad) with which the Identity was formerly and ignorantly identified.

The Problem of Prakriti as Self-Perception —How Does it Work?

One of the main ideas advanced in this treatise is that:

- *prakriti* is that which any Self 'Sees' of Itself. Prakriti (the Mother Aspect) is, thus, 'Self-Sight';
- Mulaprakriti is the 'Self-Sight' of the Infinite Subject;
- Cosmic Prakriti is the 'Self-Sight' of the Universal Logos (the Condensed Point);
- the World of Being is the Self-Sight of Lesser Logoi, and the Spirit Aspect of all B/beings; and
- the Fabricated World is constructed by the 'Self-Sight' of Fohat.

Should it be said that the matter that *man* 'sees' is also 'Self-Sight'? To a degree this may be true, and every one of us does 'make' his interior, psychic world; the human being 'makes' his own "heaven" or "hell", we are told. But what of the molecular, atomic and sub-atomic matter that we as scientists, are beginning to study and understand. Are 'we' responsible for *'making'* it simply through 'Self-Perception'? Obviously, not!

As strictly human beings, we are not the 'Creators of the Universe'. The human being is very far down on the 'Cosmic Totem Pole', and, given this position, must "live and move and have his being" in the 'Self-Perception' of greater 'gods' than he. Of course, the Essence of the human being (his true Identity) is not limited to the lower worlds, and even Now is very much a part (if not the Whole!) of those Cosmic Entities who *are* guiding and fabricating the World of Objectivity through 'Self-Sight'. Certainly, then, we human beings are participants in this process, just as the Superior Beings, Whom we *are*, are Participants in what we call *our* limited human process. Identity cannot truly be *localized* in any dimension of Cosmos; it can only *apparently* be localized.

From another perspective it might be said that 'within' us, *are* the greater Beings Whose direct Self-Perception is creating the Prakriti we see as matter (and its formulations). "The lesser cannot include the greater", it is said, but the lesser cannot *not* include the greater. The Prakriti born of the 'Self-Sight' of the Great Ones is Really, in a deep way, our *own* Self-Perception too. We see what is *in* Them, but since they are *in* us, and since we *are* Them, then we see as matter (as Prakriti) what is *in us* too!

☐ It can certainly be wondered, with Kant, whether what we see is dependent upon the mechanism and instrument through which we are seeing. Maybe what we 'see' as matter is not at all what is 'Seen' by the 'Self-Seeing' Creators; we may only *think* there is a sameness. The lesson to be learned is that even the human being 'creates' his own objectivity, more than he may suspect.

On some profound level it may be true that no E/entity can 'S/see' as objectivity any more than is *in* It (or, more radically—any more than it *is*). Thus, does the World of Objectivity appear entirely different to man than to different orders of the animal kingdom ('below') and to various orders of the angelic kingdom ('above'). As to *how* the Universal Logos sees Matter, it is assuredly very different from the way we do!

The Problem of Whether 'ANYTHING' is 'GOING ON' 'within' the INFINITE SELF 'before' the 'RAY' 'FLASHES FORTH'

This is a vexing problem the answer to which will almost certainly remain forever concealed during the Universal Manvantara. It can be dealt with swiftly and paradoxically with the answer, '*nothing* and *everything*'. Are we any the wiser for thinking of these as two apparently contradictory but non-mutually exclusive possibilities?

The Problem of Whether the 'Severe Limitation' that Cosmos Is, is 'CHOSEN' from 'within' the PLENUM, or is the SELF-as-Self 'THRUST OUT' into Super-Cosmic Ignorance to Choose?

This is another one of those optimist/pessimist quandaries, very much akin to the question of whether any Universe can end in 'failure' with respect to Original Cosmic Intent. Another way of asking this is, How 'blind' is the Universal Logos? Just 'when' does the Universal Algorithm appear? Is that Algorithm an 'IDEA'/'BEING' 'CONTENT' of the 'RAY' of the ABSOLUTE when IT (the 'RAY') 'FLASHES FORTH', 'encoded' then, or 'embedded' then, 'within' That which *becomes* the Universal Logos? Or is the Universal Logos 'THRUST OUT', 'EXTRUDED' *before* the <u>*choice*</u> of Algorithm.

We are blocked in discerning the correct answer because there is probably 'no-way-in-Cosmos' to 'know' 'WHAT' is 'GOING ON' (in the 'ON-GOING' NOTHINGNESS) 'within' the INFINITE SELF during the ALL-IN-ALLNESS of Universal Pralaya. How can 'CHOICE' or 'ANYTHING', for that matter, 'OCCUR' 'within' the IMMUTABLE ONE? One wishes one could 'know' if the 'DETERMINATION' concerning what 'IDEA'/'QUALITY' will 'BECOME' a Cosmos was 'MADE' *'before'* or *'after'* the 'FLASHING FORTH'. Perhaps, more 'confidence' would be instilled (in the human consciousness) with regard to the Universal Logoic Process if the 'CHOICE' were made 'before', for then that 'CHOICE' would have been 'MADE' in an 'UN-VEILED STATE'.

☐ If the 'CHOICE'-now-<u>C</u>hoice were not 'encoded' in the 'RAY', and were, instead, left to the Super-Universal Logos or the Universal Logos, *veiling* would already have 'set in', and the Choice would have been made (as strange as it seems to say so) in 'infinite ignorance', for any limited state of knowledge, however vast it may appear to the human mind, is, nevertheless, <u>*infinitely*</u> *limited* when compared to the INFINITIZED 'STATE'.

Perhaps it is just foolishness , but once again the 'Cosmoptimistic' intuition of the author tells him that from the very 'TIME' **I**-as-'RAY' 'FLASH FORTH':

- **I** have 'SOMETHING'-as-Something to 'DO';
- that **I** AM always 'with' MySelf;
- that **I** 'GO FORTH' with PURPOSE;
- that **I** AM *not* just playing *dice* with 'MY-SELF' and with MY Universe;
- that It (the Universe) is a matter of consequence;
- that a 'DECISION' concerning something as important as a Universe (since it is the ONLY 'DECISION' **I** ever 'MAKE') must be 'MADE' from 'within' MY FULLNESS.

Continuing in this vein, once **I** AM 'RELEASED' into Super-Cosmos as the Infinite Subject/Object, **8** *'See'* (in MySelf as Mulaprakriti) My unarticulated *infinitized* Material Potential, My Infinite Potential for Objectification. As **8** 'depart' from the Infinispective into the 'De-Infinispectivizing', **8** may 'See' all *articulable possibility.* "All this can be," **I-now-8** (Poised to become Finite) Say to MySelf, but, be that as it may, **8** "Have My Instructions." **8** 'See' My Infinite Potential (indeed **8** Do) yet **8** Begin to Objectify *strictly* and *only* That which is within Me (as 'SELF-DIRECTIVE') to Objectify. When **8** 'See' MySelf as Mulaprakriti, **8** 'See' the unarticulated Possibility of Manifesting Anything; when **8** 'move' towards Finitization, **8** may 'See' the infinitudinous things which could arise or have arises; but *in Me from the First* is *That* which **8** *Must* Manifest, and That (specifically) which **8** Must Objectify within the Infinite Allness **8** 'See'. Of course, words here fail utterly.

Perhaps this is rather a second ray view, viewing the whole Universal Process a benevolent and 'PLANNED' from the HIGHEST. A more first ray view would say, Throw the baby in the water (the "Waters of Space") and let it sink or swim! Such a view would favor 'EXPULSION' into *veiled ignorance,* with "do or die" consequences for Universal Evolution. The Universal Process would then become a great 'Contest' depending upon right Pre-Cosmic Choice, and right Intra-Cosmic Execution of that Choice. In such a scenario, things could go 'badly' for the Universal Logos. Since, however, the INFINITE SELF IS *"beyond Good and Evil"*, whether the Universe were a 'success' or a 'failure' would make no difference to IT at all!

☐ As stated again, perhaps for the comfort of the reader (or my own!), the intuition favors the theory of Pre-'FLASH' 'Algorithmic Embedding', which means (from a sanguine point of view) that a 'GOOD'-for-Cosmos 'CHOICE' has been 'MADE'. After all, how far can you trust a *blinded* Logos?

The Problem of the Mathematics of the Emanatory Process

The Emanatory Process, for being so fundamental to the origin and development of the Cosmic Process, is extremely difficult to imagine with any degree of accuracy. Almost certainly It is both a mathematical and psychological Process, the formulas for which are well concealed in what for the human being is the distant 'future'.

While we can realize very well that Emanation is fundamental to the objectification of every Authentic E/entity, we do not know the true 'how' of it. Perhaps, however, a few speculative thoughts can be offered concerning the relationship which simple mathematics has to the overall Process.

When considering the Mathematics of Emanation there are two important subjects:

- Diverse Emanatory Paths
- The Meaning of Mathematical Operations

Superficial consideration of the Problem of Emanation makes its solution seem easy. The concept of Emanatory Paths is conventionalized and Numbers are taken at face value. On closer examination, however, there appear to be *many* possible Emanatory Paths based upon various mathematical operations. Further, it is not easy to determine, metaphysically, exactly what a Number Really *is.*

In the previous section we discussed one type of Emanatory Path. In that model, the Number One produces a Reflection (the Mother) that is simply Itself (the Number One) objectified. From this perspective Father/Mother are a Great Oneness (even though various Numbers can be assigned to the Mother).

In order to inaugurate the Divine Emanatory Stream, the Number One has had to come into relation with a lesser and more specific Aspect of Itself as an Object (through 'Self-Sight'); the Number One has to 'discover' another Self within Itself. This 'Self-Seeing' 'act of discovery' 'sends forth' a new Emanation which can simply be called an Emanation of the One or can be called the Number Two. This Emanation of the Number One that is or will become the Number Two can be looked at in two ways:

- as an attenuation of the Number One (hence, Itself, simply the Monad as 'another!' Monad) and,
- as the Number One in relation with Itself.

The Number One not only 'discovers an Objective Self within Itself through Self-Sight', but *invests* Itself (*through identification*) in Its discovery. This *investment* and *going forth to identify with* is what Really makes the Emanation of the Number One, an authentic Number Two, an authentic Twoness. The Number One 'goes forth' towards the Objective Self It 'Sees'; thus the Number One comes into *intimate relation* (i.e., *identification)* with Itself and (while still remaining Itself) 'enters' into Its own Emanation and, thus, 'Becomes' the Number Two. Psychologically, this is how emanation seems to work.

Since the Number Two (which is the First Emanation of the Number One {setting aside the 'Seeing' of the Self-as-Mother as of another Category}) is Really none other than the Number One (albeit in *attenuated* form), when that First Emanation (now, the

Number Two) comes, Itself, to produce an Emanation, it can be realized that, Really, the Number One is *still* reproducing Itself. The One produces the Two, but the Number Two is Really only the Number One. Really, every Number Emanated (regardless of Its numerosity) is, fundamentally, only the One, the Monad.

This Emanatory Path might be called 'Non-Ramifying Vertical Emanation' and reflects the sequential series of integers, 1,2,3,4, ... n, etc. Every Entity produced in this type of Emanatory Path is an attenuated reflection of the Originating Monad, Number One. All Emanations, then, are really Number One, over and over again.

- The first Emanation is Number One *reduced.*
- The second Emanation, emanating from the *reduced* Number One (called Number Two) is Really Number One *still more reduced,* and so forth.

Regardless of reductions, attenuations, de-intensifications, reflections, etc., every successive Emanation (even an infinitude of them) is still Number One. The entire UTTER ALLNESS [see Glossary] consists of ZERO and One.

Once they are emanated or produced, Numbers are Entities. These Number/Entities can *interplay* among Themselves (concerning which another entire treatise could be assayed). The important point is that the modes of interplay between Number/Entities are Mathematical Operationsin the simplest model: Addition, Multiplication, Division and Subtraction. Number is the Relationship of the Universal Monad to Itself, and the Mathematical Operations are Modes of Relationship.

Let us call the Number One and all Its sequence of Emanations, 'Emanatory Factors'. In the simplest arithmetic, there are then four Modes of Relationship in which Emanatory Factors can engage. It must be remembered that in dealing with Emanatory Factors, we are dealing with Whole Numbers, Integers, and not with fractions. The Number One and all Its Emanations are Really Entities, and an Entity, Essentially, retains Its Wholeness. (From a philosophical perspective, even a fraction is a *whole* or *monad.)*

Addition

The emanative way of generating the sequence of integers by addition is always to *add the Monad:*

- Let 'E' be any Emanation and 'n' stand for any Number.
- Then, $E_n + 1 = E_{n+1}$
- Substituting values for clarity: $E_1 + 1 = E_{1+1}$
- E_1 is the Number 2 (the first Emanation) which, when the Number 1 is added to it, becomes E_2 or the Number Three + (the second Emanation).

Thus 2 + 1 = 3

- Similarly (without going into all the detail), 3 (the second Emanation) + 1 = 4 (the third Emanation).

All this seems obvious until it is realized that the '1', which is being added, is always the value of the first Number considered as a *Monad.*

For instance, in the equation, 2 + 1 = 3:

- To the Number 2, the Number *2* is Really added (though it doesn't appear thus in the equation) because the added Number *2* is in its *Monadic Form* as '1', thus yielding 3.
- Or, to the Number 3, the Number *3* is Really added, but, again, in its *Monadic Form* as '1', thus yielding 4.

In words, we would say as a general formula:

- Two, plus Itself (the Monad) equals Three; or,
- Three, plus Itself (the Monad) equals Four.

In this manner, the sequence of integers is produced mathematically by metaphysical addition. Another kind of addition suggests the presence of at least two Emanatory Factors.

- Let the Number One be known as '1'
- Let the first Emanation of 1 be known as E_1, and the Emanation of E_1 be known as E_2, and so forth.

The process of addition can be symbolized in two ways, first:

- $E_n + E_{n+1} = N$
- (where E_n is any Emanation in a sequence of Emanations from the Number One, and E_{n+1} is the Emanation following E_n in the sequence).

NOTE: The subscript '$_n$' is always one less than the number value of 'E'. 'N' is the Number that results from the combination of E_n and E_{n+1}. This is the *ordinary* kind of addition. In the sequence of numbers 1, 2, 3, 4, 5 ... the progressive N's will be 3, 5, 7, 9 ... etc.—in other words, all the *odd* Numbers. As an example:

- $E_1 + E_{1+1} = N$
- $E_1 = 2$
- $E_2 = 3$
- $2 + 3 = 5$

Metaphysically it is not to difficult to rationalize why 1 + 2 = 3. (Can we come up with the metaphysical justification for why 2 + 3 = 5?)

Secondly, another kind of addition brings a different result; perhaps it should be called *fusion* more than addition. It would work in the following sequence:

- $1 + 2 = 3$
- $1 + 2 + 3 = 4$
- $1 + 2 + 3 + 4 = 5$

Mathematically, these equations are also erroneous; the metaphysical reasons, however, can be seen.

Other forms of addition (non-sequential) are also possible. For instance, the Entity Number Two, and the Entity Number Seven may be added to produce the Entity Number Nine. Thus the relationship called Twoness in additive combination with the relationship called Sevenness, produces the relationship called Nineness. There is a whole line of inquiry possible in relation to the metaphysical implications of such addition.

Through the Process of Addition, there has been formed a combination. The sum or result of this combination is dependent upon the *mutual presence* of the two Quantities (Entities) added. In other words, the combination always exists as long as the two Quantities remain *together.* What does 'together' mean in this case?

Let us use the analogy of a partnership. Two skillful people form a kind of partnership and this partnership creates a new quality which did not exist before the two people worked together. When the togetherness is of the nature of *addition* then the partners need not *interact.* They must simply be in the same place at the same time, acting as skillfully as they would if they were not in partnership, but acting together on the same task in what might be described as a parallel manner.

Thus, when the Mode of Relationship called Addition is operative, the partners need to *act* together but need not *interact.*

When between sequential Numbers the Mode of Relationship called Multiplication is considered, we have a formula as follows:

$$E_n \times E_{n+1} = N$$

Now two Emanatory Factors have again come together, but instead of simply functioning together and in a non-interactive manner as they did when they were independent of each other (as in Addition), they now *interact* in such a way as to augment each other's powers.

Subtraction

When the Mode of Relationship is Subtraction, we are reminded of the process of *abstraction* or *obscuration.* The following instance of Subtraction is very easy to understand and useful:

$$E_{n+1} - 1 = E_n$$

Substituting values, we have for instance:

- $E_{2+1} - 1 = E_2$, or
- $E_3 - 1 = E2$ or $4 - 1 = 3$

  (Remember that $E_1 = 2, E_2 = 3, E_3 = 4, E_4 = 5$, etc.)

Here the Monad is withdrawing itself from a *more* prakritically identified and subdivided state to a *less* prakritically identified and sub-divided state. Another way of saying this is that the Monad is withdrawing its Self-Projections.

In the example cited, Fourness is disappearing and Threeness is re-entering the less emanatorily-extended State in which It existed before It (Threeness) emanated Fourness out of Itself. Threeness is doing this through the 'Retraction-into-Itself (i.e., Subtraction) of Its own *monadized* Self-Image (which It formerly 'Projected' through Self-Sight, and with which it formerly interplayed to create Fourness).

This kind of Process-of-Retraction/Subtraction reduces 'Monadic Extensions' and, eventually, restores the Cosmic Monad, the Number One, to Its *undivided* State.

Division

When the Mode of Relationship is Division, we are reminded of the process of fragmentation, and augmentation—the Many from the One. In a way, any *descent* along the Divine Emanatory Stream is a kind of division leading to an augmentation of Emanations, and eventually to a fragmentation in consciousness.

When two Numbers in sequence are divided by each other, what is described is the *musical ratio* that characterizes the relationship between the two Numbers. While the mathematical relationships become a little more complicated here, one interesting, surprising and important kind of relationship could be brought forward:

$$E_n/E_{n+1} = E_{n+1}/E_n$$

This equation seems not to be mathematically correct, for it would mean (substituting number values) for instance, that 2/3 = 3/2. *Musically,* however, the equation is correct, for both 2/3 and 3/2 describe an interval (between two tones or notes) called the "fifth", in which the frequency of the lower note is always 2/3 that of the higher note, and the frequency of the higher note is always 3/2 that of the lower note. Since the musical interval called the 'fifth' (also called the 'Dominant') is (in Esoteric Musicology) given the meaning of *soul* relationship, one might say that two Numerical Entities related through division by the ratios 2/3 or 3/2, express the quality of *soul* between them. This is just a special case. Naturally *any* integer can be divided by any other, expressing not only varying musical ratios, but resulting in varying quotients, each with their own numerological meaning and musico-metaphysical correspondences.

Any Number when divided by a greater Number is reduced or augmented in power depending upon the perspective. For example, Number 4 divided by Number 2 equals 2. Since 2 is far smaller than the Number 4, the 4 could be considered *reduced* in power (external power), if magnitude means power. But since 2 is closer to the Number 1 (the Emanatory Source), it could be considered *augmented* in power over the Number 4.

There is a fertile field for metaphysical speculation in all the simple arithmetical operations between integers, and the operation of division, presents, potentially, some of the most interesting consequences because the relational factor of the *ratio* is introduced.

The Problem of Mathematics and the Determination of Emanatory Paths

Moving on to the subject of Emanatory Paths, we can see that there are several ways that those Entities Who are Numbers can be produced, and that They can be produced by all the Mathematical Operations. What are the various possible Emanatory Paths?

- Let the Act of Emanation be characterized by the symbol 'em'.
- Let the first Emanation from Number One be designated E_1, the Emanation of E_1 be designated as E_2, etc.

Path 1: The Path of *Non-Ramifying* Emanative Descent can be expressed as follows:

1 em E_1 em E_2 em E_3 em E_4, ... em E_n

These Emanations (even though they are the sequence of Integer Beings) are simply extensions of the Number One and are *all,* Essentially, Number One (in attenuation).

Path 2: The Path of *Ramifying* Emanative Descent. The Number One produces two Emanations, each of which produce two more, each of which produce two more, etc. Thus:

1 em E_1 & E_2; E_1 em E_3 & E_4; while E_2 em E_5 & E_6, etc.

This design would be a symmetrical 'Tree Form' and Emanations would appear as syzygies (pairs of polar opposites). For understanding the *Bi-polar Nature* of the Divine Manifestation, this type of Path would be most explanatory.

Path 3: The Path of *Creative Numerical Groupings:* It has often been called to our attention that:

- from the 1, the 2 proceeds
- that from 2, the 3
- that from the 3, the 7
- that from the 7, the 12
- that from the 12, the 49; etc.

This type of Creative Path or Emanative Path establishes that (in terms of a certain way of viewing Cosmic Structure) not all Emanations may be of equal creative importance. By this method an Emanation does not proceed strictly from the Emanation immediately preceding It. Instead certain preceding *groupings* of Emanations become *creatively interactive* and thus generate the Emanations succeeding them.

For instance,

- 1 em E_1
- 1 + E_1 em E_2 (or the 1 + the 2 generates the 3); then,
- 1 + E_1 + E_2
  (or the 1, plus the 2, plus the 3) does not simply emanate E_3 (the 4) alone, but, rather, E_3 (the 4), E_4 (the 5), E_5 (the 6), and E_6 (the 7), and with some schools of thought E_7 (the 8), E_8 (the 9), and E_9 (the 10) (making, with the Number One, 10 Emanatory Factors {ten Numbers} in all).

We see, then, the possibility of forming two kinds of groups of 7:

- The 7 Emanatory Factors may be formed by *including* the first 3 Emanatory Factors; or,
- 7 Emanatory Factors may be seen as independent of the first 3 (simply by including E_7, E_8 and E_9).

Be alert to realize that E_1, though the First Emanation is, by this method, considered the *second* Emanatory Factor (or Number), and E_2, though the Second Emanation, is the *third* Emanatory Factor. This minor inconvenience in numbering arises because the Number One is *not* an Emanation but, more, a 'RADIATION'.

We see, then, in the Number One and the first two Emanations (E_1 and E_2) the first Creative Group of Three Emanatory Factors. Their interaction and interplay produce the next seven Emanatory Factors—E_3-E_9. It is important to realize that using this model, E_3 does not arise *directly* from E_2, nor E_4 from E_3, but that E_3-E_9 arise from the *specialized interplay* of 1, E_1 and E_2.

One possible model for that interplay is very interesting, for 1, E_1 and E_2 do not combine by simple addition. They *change their emphasis.*

- To produce E_3, the Number 1 becomes the most dominant Emanatory Factor.
- To produce E_4, E_1 becomes the most dominant.
- To produce E_5, E_2 becomes most dominant.

How this dominance of one of the three Emanatory Factors over the other two is achieved can only be vaguely hypothesized. It would seem to have something to do with what we might all 'the modulation of relative intensity' or 'restraint of expression'. When we come to the 'Creation' of E_6:

- 1 and E_1 equilibrate their relative intensity and dominate E_2.
- E_7 requires that 1 and E_2 emerge, equilibrate their relative intensity and dominate E_1.
- E_8 requires the E_1 and E_2 combine with equal intensity and dominate 1.
- The production of E9 requires equal intensity from all three—1, E_1 and E_2.

Note well in this model, that E_3-E_9 arise strictly from the interaction and interplay of 1, E_1 and E_2, and *not* from any interplay among themselves. For instance, E_5 does *not* arise from an interplay between E_3 and E_4.

Those familiar with the model by which:

- Ray 1 = Abc
- Ray 2 = aBc
- Ray 3 = abC
- Ray 4 = ABc
- Ray 5 = AbC
- Ray 6 = aBC
- Ray 7 = ABC

will easily be able to follow the model above wherein:

- 1, E_1 and E_2 are equivalent to A, B and C, whereas
- E_3 - E_9 are equivalent to Rays 1 - 7.

From the Seven Emanations (produced as they were by the interplay of the Number One and the first two Emanations {Numbers Two and Three}), the next *twelve* Emanations arise, yielding, with their arising, 22 Emanatory Factors thus far. The Twelve can be produced directly from the Seven, or there may be recourse to the Original Three Emanative Factors (which when combined in various additional ways and with varying degrees of intensity can produce the Twelve). After the production of the Twelve, the 49 must arise, and then perhaps the 343, and the 2401, etc. The arising of Emanations in multiples of seven is an important part of this Cosmo-Structural Model.

What is important to realize in relation to this Model of an Emanatory Path, is that *tiers* of Creative Groups interact to generate all of the members of another Creative Group on a lower 'tier', which tier, in its turn, generates all the members of another Creative Group on a still lower tier, etc..

Path 4: The Path of the Tetraktys. The Tetraktys in its usual form is tenfold with four parallel horizontal rows consisting of 1, 2, 3, and 4 points, arranged as bowling pins are arranged. This structure may well indicate an important Emanatory Path"

- From the 1 em E_1 & E_2
- From the combination of E_1 & E_2 em E_3, E_4 & E_5
- From the combination of E_3, E_4 and E_5, em E_6, E_7, E_8 & E_9, etc.

The Tetraktys can be infinitely extended past its usual tenfoldness.

The value of this Emanatory Path Model is that it is both sequential and involves Creative Groupings. From the 1 the 2; from the 2 the 3; from the 3 the 4; from the 4 the 5; from the n, the n + 1. Each row in the Tetraktys represents an Order of Being consisting of various members. Each Order of Being is produced by the preceding Order of Being plus the Number One, the Monad. In Emanatory Theory it is always most metaphysically satisfying to proceed through the addition of the Monad.

Path 5: The Tree of Life: The Kabalistic Tree also indicates a possible and intricate emanatory sequence.

The Problem of the Many Diverse Ways of Producing Numbers

Numbers may arise through a particular Emanatory Sequence or through several, but they can also be 'produced' through quite a number of mathematical interplays, which, metaphysically are 'Entity Interactions'. By these interplays a result is produced which is resonant with a certain Number/Quality.

The following example focuses on some of the Modes of Numerical Interplay by means of which the Number 6 can be derived or produced. Each equation would have its metaphysical meaning. [The elucidation of this approach will be treated in another volume of the *Tapestry of the Gods Series.*]

- $1 + 1 + 1 + 1 + 1 + 1 = 6$
- $1 + 5 = 6$
- $2 + 4 = 6$
- $3 + 3 = 6$
- $1 + 2 + 3 = 6$
- $1 \times 2 \times 3 = 6$
- $7 - 1 = 6$; $8 - 2 = 6$; $9 - 3 = 6$; etc.
- $12/2 = 6$; $18 / 3 = 6$; $24 / 4 = 6$; etc.

The Number 6 is produced, sustained and reinforced by all the above operations and *many* more. These kinds of interplays are going on simultaneously among a variety of Numerical Entities (constantly, in a way, and emphatically at different times in the History of Cosmos). The particular mathematical interplays listed above would, metaphysically, uphold and reinforce the 'Quality of Relationship' called 'Sixness' within Cosmos. What this may suggest is that all Entities other than the Number One, may be *produced* (or at the very least, *reinforced*) in a number of ways through various kinds of interplays.

The Problem of How Emanation Really Works

8, Number One, Am MySelf. Now, through an 'Act of Emanative Self-Sight', **8** *also* 'See' MySelf-in-attenuation, and then relate to that which **8** See. Thus there are Two. In relating to that which **8** See, **8** 'go forth' into it. **8,** Number One, am now MySelf *plus* my new relationship between Myself and my 'Seen' Self. This *relationship is* Twoness. This new Relationship with MySelf (i.e., with MySelf Who Is the Cosmic Monad) which is Twoness must bear its fruit. Since **8**, the Cosmic Monad have 'gone forth unto My Seen Self', and Am now 'in' My 'Seen Self' Twoness exists. Twoness is now *subjective,* close, familiar. **8** the Number One have 'settled into' my Twoness, and no longer See It as an Object, but have begun to *identify* with this State of Twoness. It becomes what **8** Am and not what **8** 'See'. So **8**, the only Consciousness in Cosmos, preserve my Oneness, but in addition begin to 'sink into' Twoness, Identifying with It, going *into* It, rather than merely *witnessing* It as a detached Observer.

Now, **8** Am *both* Number One on one 'level', *and* Number Two on a lower 'level'. Then, moving on to the next Emanatory Stage, **8** (acting 'within' My Twoness), 'See' My Twoness-in-attenuation, and the very 'Seeing' has produced an Emanation! This is how it works: from 'within' My Twoness, **8** begin to interplay with that (attenuated reduction) which **8** 'See' of my Twoness, and the interplay produces Threeness. A new Entity has been created when **8** 'Saw' my Twoness as an Object. In 'Seeing' My Twoness, **8** Created through this 'Seeing' another Entity, rightfully called the Three or Threeness once **8** the 'Seer' 'enter' It and begin 'interplaying' with It. Thus every Numerical Relationship 'Sees' Itself, and in 'Seeing' Itself, and in 'going forth into what It Sees', thus creates the next required Entity in the Divine Emanatory Stream.

It must be remembered as well, that every Number Entity (no matter what Its numerosity) is, Really, only the Number One reduplicated in an attenuated state. Number is Really but the interplay of the One and Only Monad with Itself. This can be expressed as follows:

- $1 + 1 = 1$ (normally 2)
- $1 + (1 + 1) = 1$ (normally 3)
- $1 + (1 + 1) + [1 + (1 + 1)] = 1$ (normally 6), etc.

Every Number Entity is simply the Monad, the Number One. Numbers arise when the Number One (the Cosmic Monad) whether wholly as Itself or in attenuation, interplays with other attenuated instances of Itself.

The Problem of Irrationality Concerning the Origin of the Monad

The ramifications of this question can be shown by a simple and seemingly absurd question, Can **8** ever become a Solar Logos, or Am **8** *already* the Solar Logos?

☐ Present theories of Occult Philosophy are widely interpreted as suggesting that what is now called the human Monad begins its 'pilgrimage' by manifesting through tiny and low forms of life, perhaps an atom, perhaps through something even smaller. After completing this and other relatively 'low' forms of manifestation, it gradually 'ascends' through the ranks of various kingdoms such as the mineral, vegetable, animal, and human kingdoms, until, gradually, it enters higher spiritual kingdoms. It is presumed that the destiny of this Monad (conceived in a rather individual way) is virtually endless, and that it can 'become', as it were, (or, alternatively, use for its vehicle of expression) a planet, a star, a constellation, or even a galaxy. Most students presume the advancing Monad has no end of *individual* possibilities in Cosmos.

There is something fundamentally illogical, however, about this model. The Universe is Hierarchical, or Pyramidal, and there are far fewer great lives near the 'top' of the Pyramid than the multitudinous tiny lives at the 'bottom'. Even given the fact that many more Universal L/lives are *out* of incarnation than are *in* incarnation, the numbers of lesser lives *still* vastly outnumber the greater Lives that these lesser lives are one day, supposedly, to 'become'. Let us remember, too, that Greater Lives have no monopoly on being out of incarnation, and so vastly *more* lesser lives are out of incarnation than Greater Lives.

<u>MODEL 1</u>: For the sake of simplicity let us compare Solar Logoi and atoms. The first type of Life is huge (relatively) and the second type tiny (relatively). There are, no doubt, a certain number of stars (the outer vehicles of Solar Logoi) in the Cosmos at any one time. Compare this very great number, to the number of atoms, especially considering the fact that stars are *composed of* atoms. Is it conceivable that, one day, when the Monads now manifesting as, let us say, mineral atoms in all these stars, have progressed to the status of Solar Logoi, that there will be as many stars as there are now mineral atoms? Would not these mineral atoms-become-stars *necessarily be composed,* also, of *mineral atoms?* Or at least, of multitudinous lesser lives, all, presumably "on their way 'up'"? Where would it end? The Universe would be an ever-expanding pyramid.

☐ If this model were true, ever more Monads would have to be 'fed into', as it were, the atomic kingdom to continue to support the ever-expanding base of the Pyramid necessary to support the units that have ascended from the base (for every ascending unit requires its own multitudinous pyramidal support)!

Let us remember that in a finite Universe, the number of 'subdivisions' of the One Universal Ray is *limited.* Clearly this model (*demanding* endlessly more 'atomic Monads') would violate that limitation. These ideas are suggestive enough that something is wrong with the usual conception of how the Monad 'ascends' into 'greatness' and ever-widening scope. Let us seek to confirm that a very serious error has been made. Imagine the Seven Great Cosmic Lives we might call the 'Seven Spirits Before the Universal Throne'. We are told in the usual explanations of Cosmic Evolution that every Being that exists in Cosmos was either *once* or *will be* Man. Let us assume that this is also true of these Seven Great Spirits. This means that They 'ascended through the Cosmic Ranks' to arrive at Their present status 'before the Throne'. (To continue our investigation, we must silence the inner protest against the absurdity of this assumption!) Presumably, then, every ascending Monad would one day become such a Spirit Before the Universal Throne.

Now the problems really begin. First of all, in a hierarchically organized Universe (one of the indispensable Foundations of the Ageless Wisdom), can we imagine that there would ever be <u>*more*</u> than Seven Great Spirits Before the Universal Throne? It seems inconceivable, for Their *sevenness* seems indispensable for the integrity of the present Cosmic Structure, and Their *reflected* sevenness is equally indispensable ('below') for a multitude of lesser systems. Granting then, for the sake of argument, that in order to preserve the integrity of Cosmic Structure, there can never be more than Seven Spirits, and granting that every ascending Monad is destined to ascend to the high Point where it, individually, can manifest as one of that great Order of Lives, we find some insuperable numerical absurdities.

Can we imagine that milliards of roughly equivalent Monadic Types can converge upon the status of the Seven Spirits, and that each Monad (singly and individually) can *become* one of the Seven Spirits. (Forget for a moment that the Seven Spirits must necessarily endure *throughout* the Universal Manvantara.) The aperture is far too small. This is a 'bottleneck' of a truly Cosmic kind. Granting that there are only *seven positions,* and granting that each Monad must fill such a position for a certain duration (and probably a very, very *large* duration), what a 'waiting list' there would be! There would not be time enough in our finite, 'Time-scheduled' Universe, or in a million such Universes for *single* Monads to achieve that status and 'hold' it for a tenure.

The absurdity becomes even greater when we consider Monadic Destiny as tending towards the Three, or, worst of all, towards the One Itself. Can every 'individual' Monad 'become' the Universal One and Only? What about all the others? This model is plainly impossible, given all we think we know about the Universe and Its structure.

☐ The major misconception stems from ignorance of the implications of Emanatory Theory and a thorough misunderstanding of what the Monad Really Is (i.e., that there are not *many* Monads at all, but *only One* in the entire Cosmos).

<u>MODEL 2</u>: All so-called single, individual, separate Monads are but the One Cosmic Monad. Their extraordinary multiplicity arises from the hierarchical and 'many-tiered' structure of Cosmos. The many Monads ('embedded' within the World of Being) are but the many active or latent 'extensions' of the One and Only Monad. *They have no truly independent existence!* The One Monad is found simultaneously on *all* 'dimensional levels' of Its Cosmos, and It identifies *only with Itself,* and not *exclusively* with any of Its 'Monadic extensions'. Even when the One Monad is the Monad of the atom of matter, It is ever and only the *fullness* of Itself. There is but One Soul in all of Cosmos (the *Second*

Fundamental of the Secret Doctrine), and One Spirit, and it is worse than futile to attempt to divide It in any way.

Thus, the One Monad. 'originates' 'above' and *remains above* even as it apparently descends 'below'. Going into the Three, it never leaves the One; going from the Three to the Seven, It never leaves the Three; going from the Seven into the Twelve, It never leaves the Seven, etc. Any Path of Emanatory Descent could be chosen and always the result would be the same. The One remains *above* even as It 'Becomes' the Many *below.*

- ☐ Then, how can we possibly say that the human Monad only ascends (without having descended)? How can we say that the individual human Monad 'Becomes' greater Beings like Planetary Logoi, or Solar Logoi, etc., when *It already is such Beings.* There is something profoundly true about simultaneous, multidimensional Consciousness, and it is true for every Authentic E/entity in Cosmos.

If MODEL 2 is true, what does it mean for the human being? What does it mean for our ambition, our aspiration, our vision of the near and distant future? Surely, Monadic ascent still exists, but, Really, it should be called 'Monadic re-ascent', or 'Monadic retraction'. Also, instead of 'Becoming', we can only 'Re-Become', and even this is a misnomer, for that which we 'Re-Become' we have ever *been* throughout the duration of Cosmos, and *are now.*

The second model offers a profound challenge to prevailing concepts of human identity. It strikes a great blow to the notion of separative individuality (even on the Monadic 'level') and reminds us of the Essential *seamlessness* of Cosmos. It assures us that the One Who recurringly greets the returning Pilgrim is none but the *Pilgrim Himself*—the ever-higher Aspect of the Pilgrim Who wandered not so far into the "distant country". Eventually, the "returning *ones*" are greeted by the One Who Wandered Not At All—the Universal Logos. WHO 'WAITS' to 'GREET' Him?!

The Problem of the Point in Space

It is easy to talk of *points* and *space* in a rather inexact manner, without fully understanding either; they hold many mysteries. A point is symbol of transition, from a greater to a lesser or a lesser to a greater. A point is both a Subject and an Object, a Point of View and that which is *seen* from the Point of View (i.e., a 'Seen Point'). In one way it could be said that the beginning and ending or anything is a point (thus the phrases, "starting point", and "end point").

The point may stand for different things in different fields. When 'POINTNESS' or the 'POINT' suddenly 'APPEARS' in THAT, it represents a 'MOVE' from a the MAXIMALLY HOMOGENEOUS MODE of 'ENFOLDED INFINIDENTIFIED BE-NESS' to the 'MODE' of 'ACTION'. Usually points are *seen* from other Points of View, but this is conceivably *not* the case with respect to the FIRST 'POINT', for until this 'POINT' arises there is no 'SEEING' 'within' THAT, only a 'STATELESS STATE' of INFINITIZED BENESS. Thus, the FIRST 'POINT' might be thought of as the UNSEEN 'POINT'. However, from the 'TIME' of the 'ARISING' of that 'POINT' (to the 'TIME' of its 'DISAPPEARANCE') there is always a *point of view*.

☐ The 'ARISING' of the FIRST POINT is equivalent to the 'ARISING' (whether instantaneously or in instantaneous sequence) of the SUPER-COSMIC TRINITY, which is 'CONTAINED' in the FIRST 'POINT'. The nub of the matter in this 'ARISING' is whether the 'POINT' is a 'SPECIFICITY', a 'SINGULARITY', or whether it is still, in some way, a species of INFINITUDE.

We have discussed the idea that as the SUPER-COSMIC-'POINT' (with 'ITS' enfolded SUPER-COSMIC TRINITY) 'ARISES', 'IT' is immediately 'THROWN OUT' and becomes the Super-Cosmic-Point (with Its enfolded Super-Cosmic Trinity). If the Infinite Subject of the Super Cosmic Trinity is called the Infinified Point (because It is no Specific Thing, and 'Sees' generalized Infinitude instead of Specificity), what can be said of Its higher, instantaneously existing 'PROTOTYPE', the 'INFINITE SUBJECT'?

In SUPER-COSMOS and Super-Cosmos, *points* are different from how they might ordinarily be conceived. These *points* are 'INFINIFIED' and 'Infinified'. When we usually think of a point, we think spatially, seeing a 'pin-point' of white against a black background or a 'pin-point' of black against a white background.

When dealing with the *primal appearances* and *primal interplay* within *super-cosmic realms,* we cannot visualize in that manner. The 'APPEARANCE' of a 'POINT' in the INFINITUDE is *not* the 'APPEARANCE' of a 'SINGULARITY'. INFINITUDE still 'PREVAILS'. That 'POINT' is the 'SIGNAL' of the possibility of 'RELATIONSHIP' 'BETWEEN' INFINITUDES, or 'BETWEEN' the *infinite halves(!)* (*infinite after their kind*) of the ONE INFINITUDE.

- The INFINITE SUBJECT is infinite after its own kind;
- the INFINITE OBJECT is infinite after its own kind; and
- that kind of 'MAYAVIC' 'SENSITIVITY' which 'RELATES' THEM is infinite after its own kind.

Within this fleeting SUPER-COSMIC 'CONDITION' called 'POINT', there is proceeding a *trinitarian-interplay between three infinitudes,* each *infinite after its own kind.* That 'ARISEN' 'POINT' has nothing to do with *space* as we usually conceive it. The symbol of the 'POINT' is simply used as an indicator of 'SOMETHING *DIFFERENT* HAPPENING'. (Of course, none of these words are appropriate in relation to the ABSOLUTE. Language is hopeless).

In this examination, the SUPER-COSMIC 'POINT' is being discussed as if 'IT' were an 'ABIDING' 'SOMETHING', but, according to the Theories of Radical Infinitism, that 'POINT' *disappears instantly* having 'BECOME', in effect, the Super-Cosmic Point/Super-Cosmic Self. In fact, it may be justifiable to say that the SUPER-COSMIC 'POINT', from 'ITS' very 'INCEPTION' *always was* the *Super-Cosmic Point* (for the ABSOLUTE cannot 'TOLERATE' the presence 'within' ITSELF of anything other than the 'PURITY of ITSELF').

The terminology SUPER-COSMIC 'POINT' is retained however to demonstrate that whatever 'ARISES' must necessary have 'ARISEN' 'within' the ABSOLUTE, even though it be *instantaneously* 'CAST OUT' (thus 'BECOMING' *reduced).*

Since the 'THREE-FOLD SUPER-COSMIC POINT' is so *evanescent,* and so virtually *identical* with the Threefold Super-Cosmic Point, perhaps we should confine our attention to that Threefold Point, which is the Infinite Self consisting of:

- Infinite Subject/Infinite Spirit
- Infinite Object/Mulaprakriti
- Infinite Consciousness/Maya

When related to 'point symbology' these Three can be called:

- Infinified Point or Infinified Viewpoint
- Infinified 'Seen-Point'
- Infinified 'Seeing'

From the 'Infinispective' of the Infinified Viewpoint, just *what* is 'Seen', now that Super-Cosmic 'Seeing' is in effect? Does a View-Point 'See' a 'Seen-Point'? Does one *thing* 'See' another *thing?* Probably not. Remember that we are dealing with *infinitudes* 'after their own kind'. We are not dealing with 'de-finite' *things.* Points, as usually conceived, are *articulations, specificities* (and look quite definite). Thus far in our descent, however, we have not yet entered the realm of articulation and specificity. So in this first Super-Cosmic 'Sight', a View-Point does not 'See' a single 'Seen-Point' which 'stands out' against a background and thus makes the 'Seen-Point' articulable.

☐ The View-Point is an 'Infinified' View-Point, and the 'Seen-Point' is an 'Infinified' Seen-Point, and the 'Seeing', too, is 'Infinified'. What this means is that *'no point stands out',* and that the 'Seeing', or 'Subjectobjective Interplay', is not characterized by One *Thing* 'Seeing' another *Thing,* but rather, a *homogeneous vision of potential, though unarticulated, infinitude.*

From the Infinispective (which, we may say, is, in a way, 'shared' 'between' the Infinite Subject and the Infinite Object), no articulable *points* are 'Seen'. It could be considered a *non-specific vision of infinitude.* Points are *specificities.* We might conceive of this Super-Cosmic State as Pure 'Three-in-Oneness', Oneness prevailing but enfolding the Cognizance of Threeness. Yet, strangely, we are not here talking of a delimitable Oneness, Twoness or Threeness, but of an Infinite Oneness, Twoness, and Threeness.

The Super-Cosmic Realm is 'Transitional' and thus we are thinking of the INFINITE and the Finite blending in the Infinite. We are justified in calling the SUPER-COSMIC 'POINT' and the Super-Cosmic Point, *unseen points,* for, in a way, the INFINITE SELF ('out' of which these *points arise)* can never REALLY 'SEE'. IT can only BE BE-NESS. 'INFINIDENTIFICATION' is *not* a form of 'SIGHT'. These *points are* ITS 'SEEING'.

This far in Super-Cosmos we have a 'Three-Way Seamless, Self-Appreciative Infinitude' devoid of articulation, or separate, singular points. Yet, forever, *points,* as 'PRECIPITATIONS of INFINITIZED POSSIBILITY' have 'ex-isted'. There have been an infinitude of articulated points—all the *singularities* of all Cosmoses forever.

Now, are these points ever 'Seen' (all at once) or 'Re-Seen'? This is an extraordinarily difficult question, relating as it does to the existence and *locus* of Infinite Memory. We might wonder, if there was ever a *review,* by *whomever* concerning all *infinitessentialized points* which had been *objectified* from the FOUNT OF ALL POSSIBILITY. If such were ever to 'appear', Super-Cosmos might well be the 'Place' for that 'Appearance'.

The problem of whether 'ARTICULATION' and 'SPECIFICITY' can *ever* 'ABIDE' 'within' the INFINITE HOMOGENEITY is probably unanswerable, though the odds are *against it* are 'infinity-to-one'. It simply depends upon whether there is the *possibility* of 'OBJECTIVITY' 'within' IT, because without OBJECTIVITY, no 'ARTICULATIONS' or 'SPECIFICITIES' can be 'REGISTERED'.

Leaving the question of 'OBJECTIFICATION' within the INFINITE SUBJECTIVITY unanswered, we focus upon the Nature of Objectification within Super-Cosmos. Thus far, we have an unarticulated Homogeneous Subjective Infinitude 'Seeing' Itself, as an unarticulated Homogeneous Objective Infinitude. We have an Infinified View-Point (or Infinified Pointness) 'Seeing' Itself as an Infinified 'Seen-Point' (or an Infinified 'Seen' Pointness). The 'Infinification' of the Point, annihilates *specific points,* and produces a 'View' of *generalized homogeneous infinitude.*

This means that the Infinispective produces a View of *all possible points in their infinitized, hence noumenessential, hence unarticulated State.* From the Infinispective *no single point* is 'Seen' (though from an 'Infinispectivizing' of the 'De-Infinifying Subject, the Condensing Point, single points, perhaps even an infinite *vastness* of points, *may* be 'Seen').

We have been talking about two kinds of *point, each* of which 'contains' two points, as well as the relation between them.

1. First there is the SUPER-COSMIC 'POINT' which (for its ultra-brief duration) 'CONTAINS' the 'EVANESCENT INFINIFIED VIEW-POINT', the 'EVANESCENT INFINIFIED SEEN-POINT', and the EVANESCENT INFINITE 'RELATION' BETWEEN THEM.
2. Then, there is the Super-Cosmic Point (the Super-Cosmic Self) which 'Contains' the 'Infinified View-Point', the 'Infinified Seen-Point' and the Infinite Consciousness that Relates Them. What is being stressed above, is that these *points* should not be visualized or imagined as *conventional points,* because they are *infinified* and, hence, *point-obliterative* if points are conceived as they usually are.

If the symbol of the point as conventionally conceived is of any value, *when,* in the Creative, Precipitative Sequence, does it become relevant. Points, as usually conceived, are *articulations,* and *specificities.* When do such come into the picture? Perhaps, articulation and specification begins to appear as the Infinified View-Point begins to *condense,* becoming the Condensing Point or 'De-Infinified View-Point'. This stage indicates the transition in Objectivity from an 'Unarticulated Homogeneous Infinitude' to an 'Articulated, Heterogeneous Infinitude', and becomes, eventually, a 'Specified Articulation' a 'Specified Singularity'.

In the Stage of 'De-Infinitizing' (wherein the Perspective changes from the Infinispective to the 'De-Infinispectivizing'—provided that any form of *gradualism* in what is looking more and more as an utterly *quantized* UTTER ALLNESS) *articulated points* begin to appear. Do these represent *all points past?* All *conceivable points,* i.e., all *infinitized possibilities as specifiable articulable possibilities?* Who can say what this Transition from Infinified Infinitude to Specified Singularity 'looks like'?

Certainly, there *may* be a view of *many articulable possibilities as points,* but at the end of the Process (even if that Process is virtually instantaneous), there is but One 'Seen-Point', and One View-Point from which It is 'Seen'. After all *infinification,* a *Finitude* has been 'Created', and One View-Point 'Sees' a 'Seen-Point'; the Finite Subject 'Sees' the Finite Object; the Universal Logos 'Sees' Cosmic Prakriti. Now, at last, we have Point relating to Point. We are in the 'Realm' of Distinct Points.

Out of these Distinct Points (a Finite, Subjective View-Point called the Universal Logos, and a Finite, Objective Seen-Point called Cosmic Prakriti) arise all the other 'Son-Points' that are Emanations of the View-Point in *interplay* with Its own Image. In a way, every 'Seen-Point' is the *reflected Self-Image* of the 'View-Point'. When we enter Cosmos, we enter the Realm of Specifiable Articulation. We are in a Realm of Points within Points within Points. When a View-Point 'Sees' a 'Seen-Point' within the 'Seen-Point' which is the Reflection of the View Point, then the Creative, Emanative Process has begun.

- ☐ What might be important to realize is that, when a View-Point 'Sees' Its Own Reflection, that Reflection may not appear as a Point, but rather a *'Field'.* When a View-Point is 'Seen' from afar, that View-Point will appear as an aticulated, specified 'Seen-Point'. For instance, the 'De-Infinitizing Subject' (on Its way to Finitude) 'Sees' the destined Finite Subject as a kind of 'Son Point'. *But* when a View-Point 'Sees' Itself (and not 'Son-Points' arising within It) the View-Point 'Sees' a 'Field' (a Mother-Field) in which 'Son-Points' can Arise!

Here we have the possibility of a significant understanding concerning Point and Field. A View-Point 'Seeing' Itself, 'Sees' a Field or Matrix. Thus the Father 'Sees' Himself as a Mother-Matrix-Field. But if the Father/View-Point 'Concentrates' 'within' Himself, He 'Sees' instead of a Field, a Point, which means that He, the Father, is 'Arising within Himself' as a 'Son-Point'.

Thus, we see that the Creative Process has two phases. The Emanator 'Sees' Itself as the Field within which the Emanation will arise. The Father 'Sees' Himself as Mother within Whom He, the Father, will 'Arise' as the Son. The Creation of Fields relates to the Divine Syzygies operative at every Station of Emanation.

Almost all View-Points are 'Seen-Points' as well. For instance, the Universal Son, Who Arises as an Emanation from the Father, is, naturally, 'Seen' by the Father (and, of course the Mother). This Universal Son is, as well, a View-Point, and thus is *both* a 'Seen-Point' and a View-Point (and certainly has Its own Son-Field, in which it is engaged with Its own Wife/Mother/Reflection).

- ☐ Only the SUPER-COSMIC 'POINT'-instantly-Super-Cosmic Point is not a 'Seen-Point'. It is only a View-Point, for THAT 'out' of which it *came* is a 'BE-ER' and *not* a 'SEER'. So 'SEEING' 'ARISES' out of 'BEING' without being 'SEEN'! We might call MAYA-instantly-Maya, the 'Unseen Seeing'.

These thoughts concerning the 'point' are being re-emphasized here because 'Field and Ground' considerations are very important for Philosophical Occultism. Whenever discussing the infinite and the finite, questions concerning 'boundaries' arise:

- What is bounded and what is not?
- What does boundedness 'look like'?
- What does 'unboundedness' 'look like'? (Our imagination is so challenged.)
- What is it to imagine something 'definite' (such as a *virtual point)?*
- What is it to imagine something that (definitely) exists but is 'in-definite' (such as a *Real point)?*
- What is it to imagine something that is devoid of articulation, specification and boundary (such as the Infinite Subject, the Infinified View-Point—or—the Infinite Object, Mulaprakriti, the Infinified 'Seen-Point')?
- Is our imagination capable of such distinctions?

A really big question is, What does 'Some-thing' look like? Of course, it all depends upon the 'Point of View'. Perhaps we can say that, from a distance, any 'thing' looks like a 'point'. This applies to the Universe, Itself, if 'Seen' from the some Super-Cosmic Perspectives.

☐ *But* the 'Seen-Point' becomes a Field, a Matrix, a Space, once the Viewer in the View-Point 'Enters' the 'Seen-Point' He has been 'Seeing' from a 'Distance'. In a way, we could say that the once 'distant point' becomes a 'sphere' upon 'entry into the point'.

This model may reveal some important ideas about *identification* and *consciousness.* A 'Seen-Point' appears to the Consciousness. A View-Point is *conscious* of a 'Seen-Point'. However, if the 'View-Point' 'Enters' the 'Seen-Point', the 'View-Point' is no longer *conscious of* the 'Seen-Point', but, instead, has *identified* with the 'Seen-Point', and now has the 'inperience' of *being* the formerly 'Seen-Point' instead of simply *seeing* It.

Thus, if a 'point' appears to Consciousness as a 'point', *consciousness* is prevailing and the Subject/View-Point does not have the 'inperience' of *being* the 'point' It *sees.* But if the Subject/View-Point 'Enters' the 'point' which It was 'Seeing', and that 'point' thereby seems to become an *enveloping field which the Subject/View-Point now Is,* then, *identification* has taken place. Mere 'Seeing' has been transformed into 'Being'.

☐ The entire Story of the appearance and disappearance of Universes can be interpreted in terms of 'BEING' and 'SEEING', and the constant alternation between 'Points' and 'Fields'.

It is because *members* of the 'SUPER-COSMIC TRINITY-as-Super-Cosmic Trinity' are so much *in each other indentificatorily* that what obtains in Their 'Relationship' is an Infinite Pointless Field, rather than the *conscious perception of articulable or articulated points.* Since the Three *are* One, They are not *consciously separated* from each other. Instead, They *indentificatorily envelope Each Other,* in an Infinite Super-Cosmic Field (or some may prefer to call the Field a 'Ground', and points, when they appear, a 'Field'). When *articulations* and *specified 'Points'* begin to appear 'within' that Field, *finitizing* (or 'De-Infinitizing') has begun, and a Universe as a 'Seen'-then-'Seeing' Singular Point will be the result.

The Problem of Whether the Cosmo-Subjective Now is Divisible or Indivisible

If we were to examine the Cosmo-Objective World, (the World of Fohatically Particulation), we would witness a World with well established Parameters. The ultimate moment is the measure of duration and disappearance in that World. The duration of the inter-moment instant (the Cosmo Subjective Now in what we are calling the World of Adjustment) determines the moment of the 'beginning' of the ultimate moment. For the purposes of imagining the relationship between these two Worlds we may consider these two Nows as roughly equivalent.

The question arises, however, Is the Cosmo-*Subjective* Now Really quantized? Is it regulated, as is the Cosmo-Subjective Now? Are Time and Space within this higher World (a World which is a kind of lower stratum of the World of Being) *infinitely divisible?* This is a serious question, because it concerns the *modus operandi* of the *adjustments* which are manifested within the World of Cosmo-Objectification.

If there is an active and dynamic thought/image-process in Cosmos, it is probably occurring within this World of Adjustment. The higher strata or aspects of the World of Being might be considered the World of the *'Held* Image', of the 'Sustained Archetype' meant to guide the World Process in the World of Fabrication during ages as they elapse. In a way there is only One Great Guiding Archetype (the 'IDEA' 'EXTRUDED' from the FOUNT OF ALL POSSIBILITY/INFINITESSENCE for objectification). Yet, this Great 'IDEA' has many 'Partless Parts' (Integrous Parts) or Aspects, and they are probably 'timed for intensification' within the World of Being. Thus does the Divine Purpose (which is the 'IDEA') unfold in Time *purposefully.*

☐ It is as if the Beings within the World of Being are the Architects of the lower Worlds, Who have to present Their 'phased purposeful Plans' to the Fohatic Workmen for accurate and properly-timed construction within the World of Cosmo-Objectivity.

On the very highest levels of the World of Being, what we can call the Great Universal Images may be sustained for a *long time.* The Great 'IDEA' which is at the Root of all Cosmic Structure, will be 'Held' by the Universal Logos' (in the Cosmo-Eternal Now) for the entire duration of Cosmos. Lesser Images (lesser Aspects of the Great 'IDEA') will be 'held' for a *shorter time,* and still lesser images, for a still *shorter time.* Probably on the lowest levels of the World of Being (the higher strata of the World of Adjustment), where minor 'enfolded Rays' of the One Ray (apparently lesser Monads) are involved in the holding of archetypal images, the held images are of the briefest duration—how brief, it is difficult to say.

Hypothetically, the greatest brevity of a held and guiding Ideational Image would be for an inter-moment instant, to guide the Fohatic Workers in Their *next* arrangement of the Cosmic Configuration. Ostensibly, there would be multifarious 'held Images' to guide what might be considered the many 'parts' or aspects of the Cosmic Configuration.

If there is some value in this model, how do the Spiritual B/beings in the lower strata of the World of Being (the higher World of Adjustment) *change held Image?* In the upper strata of the World of Being, the main Images may endure, relatively, for aeons, but on the lower levels, lower guiding Images could be changing with great rapidity to guide the extreme micro-phases of configuring within the World of Fohatic-Particulation (the World of Approximation).

- Is some 'time' in Cosmos needed to for the 'archetypal Guides' functioning in the World of Being and Adjustment to determine how long a guiding Image should be held and when and how it should change?
- Is some 'time' also needed for the Fohatic-Workers to assess how to respond to the 'archetypal guiding Images' (some, hypothetically, incredibly 'brief' and some of much longer duration)?

Given the normal human thought process, the tendency would be to say, 'Yes'—'time' for 'Fohatic-feedback' (from the Cosmo-Objective World) and for Fohatic-'adjustment' with respect to the creating of the Cosmo-Configuration in that World is needed. The hypothesis is that moments of intelligent assessment (both for the 'Holders' and the 'Workmen') would occur during Cosmo-*Subjective* Moments (the inter-moment instants) within the World of Adjustment. The World of Adjustment could be considered

1. either a 'middle World' between the World of Being and the World of Fabrication, or
2. the lower strata of the World of Being.

It would seem that within that World of Adjustment, two kinds of Cosmic Creators could make Their necessary Plans for 'Action' 'upon' the very next ultimate moment. These two categories of Creators would be:

1. The many Monads ('enfolded' as 'Sons' of the One Cosmic Monad, and thus Monadic Sons of the Son), and
2. Fohat and His many 'Sons'.

Creating the lower World is the *mutual* Task of these two kinds of Creators, and (according to human analogy) they would need 'time' in which to 'consult', even though that 'time' may be brief beyond human conception.

Now, the problem arises concerning the 'time' available for such 'consultation', and what may be the nature of such time. The World of Adjustment is a *Subjective World* a World of Non-Particulate Thought. This world (like the World of Being) is 'partite' but 'imparticulate' (an important distinction), and an extremely *numerically integrous* World. It is a world of 'Rapid Imaging', by means of which 'assessment' may be made of the Cosmic Configuration (in relation to the Archetypes to be unfolded 'below') and 'decisions taken. The decisions taken will concern:

- what guiding Image to 'hold' and 'how long' to 'hold' it, and
- how to 're-position' 'below' so as to alter the Cosmic Configuration so It better *approximates* that guiding Image (or series of Images).

The first kind of 'decision' is undertaken by the Monadic Sons of the Son; the second kind of 'decision' is undertaken by Fohat and His Sons.

Now, how fast do the Beings in the World of Adjustment (Monadically configured Spirit-Aspects of all authentic Entities including the Fohatic Host) have to *think/image/*

ideate to make their two principal kinds of adjustments? Using the analogy of those who play the game of chess, every move is *planned,* and 'time' is taken to 'think over' the 'next move'. But during the inter-moment instant, there isn't much time, and so the Host of the Universal Son' (Pattern Holders) and the Host of Fohat (Workmen) have to 'think fast' (whatever that really means).

One of the implications is that their Thought process in the World of Adjustment could not, necessarily, be *bound* by the duration of an ultimate moment. Assuming the duration of the inter-moment instant to be, perhaps, roughly equal to that of an ultimate moment, the 'Adjusters' within the World of Adjustment might have to have *numerous* assessive and adjusting thoughts *during* such a time period. Perhaps, then, their thought-imaging process can 'move' *as fast as may be necessary* to determine the next 'Image to be Held', and the next 'Move to be Made'.

This would mean that Time in the World of Adjustment would be very flexible and elastic and that 'Change of Image' and 'Change of Intent' would *proceed with the speed of Will.* The 'speed of Will' is presumed to be 'as fast as Will wills it to be'. This would mean that on that particular level of Cosmos (the World of Adjustment), there would be no 'briefest moment' (the ultimate moment 'below' might seem to be extremely 'macro' compared to some of the 'imagistic moves' within the World of Adjustment). What we are describing here is a condition of great freedom to 'decide' upon *how* to *create* and manifest in the Cosmo-Objective World.

If the hypothesis of this degree of flexibility is true, then, one might think that every Image 'held' to guide a succeeding ultimate moment, and every re-configuration in Cosmic Configuration in the Cosmo-Objective World would be 'perfect', and, as a result, all unfoldments within the lower Worlds would 'perfectly' reflect the 'perfect' Design(s) in the World of Being. Maya, however, is active in all worlds thus, all is progressively *veiled* through 'descending dimensions'. Even the Universal Logos is 'veiled' to the extent that He does not 'Know' the 'Content' of the FOUNT OF ALL POSSIBILITY, and is focused entirely upon the 'One 'IDEA' which He, Essentially, *Is,* and which He must 'unfold' through Emanation.

What this means is that the 'Holders of the Design' and the Fohatic Workmen 'do the best they can given the light they have'. Lesser 'Holders' attempt to reflect and harmonize with Greater Holders; Lesser Fohatic Workmen do the bidding of Greater Fohatic Workmen. Thus lesser Images harmonize with Greater Images and lesser 're-configuring initiatives' harmonize with Greater Re-configuring Initiatives—to the degree possible in a *veiled condition.*

☐ We are discussing a great, cooperatively creative Task undertaken by an extraordinarily varied assemblance of Creators—one group in the Company of the Son, and the other group in the Company of Fohat. They 'play' with Cosmic Time and Space (and, in a way, even *become* Time and Space) so that the Design-at-the-Beginning may move through various phases of *purposefully-timed consummation* and at length become the *realized* Design-at-the-End.

The ETERNAL NOW is *present* as the PRESENT MOMENT ever. Every unit of time in an imagistic movement in the World of Adjustment, no matter how *brief,* is *infinitely larger* than the MOMENTLESS MOMENT called the ETERNAL NOW (just as any definite quantity, whatsoever, is *infinitely larger* than zero). But in the World of Adjustment there may be some extraordinarily *brief* moments, when compared to the ultimate moment in the Cosmo-Objective World (as brief as that ultimate moment had previously *seemed* to be). As well, Time within the World of Adjustment can continue 'through' and 'during' the manifestation of ultimate particle/events in ultimate moments in the 'outer', 'lower' World.

☐ Thus, time in the World of Adjustment (and even more so the World of Being) and the Image-Events associated with it, is *regulated by Will alone.* If we want to judge 'Conditions' in the World of Adjustment, we might say they are extremely elastic, and consist entirely of the 'motions' of *Will and Imagination* which we might call 'Image-Events'. These Image-Events endure for as long or as short as their Creators *will* for them to endure.

The regularly timed 'appearances' in the World of Approximation (the Cosmo-Objective World) are guided and regulated through the use of Image-Events, whether the form of these Images be pictorial or far more subtle. The important thing to realize about such Image-Events is that (though partite), they are 'imparticulate' and are entirely *integrous* and 'of one piece'. Image-Events are *'monadically conditioned'*, in effect, *monadized.* [See Glossary.]

The World of Being, Itself, is sustained by Great Archetypal Image Events (which change by the Will of the Sustainers when the Grand Design calls for change). These Great Images are Relationships between the Fundamental Numbers, and are characterized by the Quality of Oneness, and possess the Integrality of the Number One (the Cosmic Monad) no matter how many imparticulate 'parts' they possess. The lesser Archetypal Images may be more complex, specific and multi-partite, but also possess the characteristic of seamless Oneness, since, like the Great Archetypal Images, all their 'Number parts', too, are basically the Number One in relation to Itself.

The Image Events which may be produced by Fohat and His multitude of hierarchically organized 'Sons', are created in response to the 'held' Archetypal Images (both great and small), and, also, in response to a necessarily instantaneous 'evaluation' of the *degree* to which the most recent Cosmic Configuration created 'on' the most recent ultimate moment in Cosmo-Objectivity conformed (or failed to conform) to the Archetypal Design being 'held'.

The mechanics of all this are, to the human mind, unfathomably complex, but 'activity' within the World of Adjustment and the World of Being all seems to be a Cosmo-Ideational Process in which 'action' is a mixture of 'Self-Sight' and Image-Projection which does not obey the Laws of Fohatic Particulation current in the World of Fabrication, the Cosmo-Objective World. The entire dynamic of the Process between the Worlds of Adjustment and Being (on the one hand) and the World of Fabrication (on the other), consists of the dynamic of extraordinarily rapid interplay between Imparticulate Image-Events (entirely *non-Mosaic* and *Monadically Integrous)* and 'particulate image-events' (ultimate particle/events) creating the Cosmic-Configuration which is the Cosmic Mosaic.

The Problem of 'Seeing' within the 'Being-Seen-ness'

We have discussed the idea that the 'Seer' awakens to the fact that he is being 'Seen', and that the One Who 'Sees' him is none than a higher Aspect of *himself.* From this perspective, 'prakriti', in general, is 'what one 'sees' of oneself'. 'Self-Sight' is the creator of ring-pass-nots, and every ascending 'Seer' is both released into, and limited by progressively ascending and increasingly expansive ring-pass-nots of Self-Sight.

8 'See' Myself as a kind of 'Field of Consciousness', a kind of 'Matrix of Self-Reflected Space' in which numerous aspects of Myself can arise. But *how* Am **8** 'Seen' from above. Surely, as merely one possibility in the still larger 'Self-Reflected Space' of the Field of Consciousness, of the greater Being Who is My *Emanator.* That Emanator, too, is likewise 'Seen', and so it goes.

If **8** want to move beyond the ring-pass-not of what **8** 'See' of Myself (i.e., if **8** want to cease being 'trapped' by the Field of Space which is My own Self-Reflection), **8** have to dis-identify from my apparent Self, and 'become' or 'slip-into' the Identity of the higher Being Who is Emanating the apparent 'Me'. Then, instead of being limited by My former ring-pass-not of Self-Sight, **8** will be limited by *His* ring-pass-not of Self-Sight. **8** will think that My Field of Space has expanded greatly, and **8** will have the impression of a much expanded sense of '**8**-ness'.

This is how a higher kind of Evolutionary Process (based upon *Identification) proceeds.* **8** disengage from my *present Selfhood* and 'become' (in Consciousness) the 'Seeing' that 'Sees' the apparent and 'former' Me. Of course, to a degree, **8** already 'See' 'within' the 'Seeing' which 'Sees' Me, because **8** 'See' the World of Created Things, which **8**-as-I did *not* create. **8** thus 'See' objectified thoughts from the Minds of greater Gods than **8**-as-I am. However, **8** do not 'See' these things as the *Creator* of them would see them. **8** 'See' them through a kind of human psycho-mental filter, and they become very much part of 'My' world, with the characteristics of 'My' world. When later, by ascending into the Identity which emanated 'Me', **8** 'See' these same things, they will look entirely transformed because they will be 'Seen' from within an entirely *enlarged* and *'subtle-ized'* Field of Space.

Thus, **8** 'See' within the greater 'Seeing', and Am limited, not only by My own 'Seeing', but by the greater. 'Seeing' which 'Sees' Me, and the still greater 'Seeing' which 'Sees' My 'Seer'. It must be remembered that **8** (unbeknownst to **8**-as-I, and because of Emanative Retention) *already* 'See' through the 'Eyes' of the greater Seers upon the various 'levels' of the Divine Emanatory Stream. As **8** ascend and change Identity, **8** recognize that **8** have been these Greater 'Seers' all along.

☐ The Universe, and all dimensions within It are simply 'Self-Projections'—
imagistically objectified Self-Projections.

If **8**-as-I change My Identification as a Self, **8**, inevitably, change what **8** 'See' of Myself. This means that My ideas of what constitutes Matter, Space and even Time, undergoes a radical modification. Matter, Space and Time are *Who-8-Am-in-Projection.* When **8** change Who **8** Am (through a change of Self-Identification), **8** change Matter, Space and Time—all of which are only Perceptions.

Section III Conclusions & Implications

In this section of the *Infinitization of Selfhood* we consider implications, conclusions, and even advice that arise from our philosophical reasoning. Often, what makes these Conclusions *Radical* is the constant use of 'I' or '**8**'. The use of these terms of not a 'presumptive leap' but grows out of the kind of thought upon which we have been focussed.

Conclusion 1
I am you. You are I. (More accurately, <u>8</u> am you. You are <u>8</u>.)

The Tibetan has suggested that the idea lying behind this statement is for most aspirants and disciples a trite and unrealized platitude. However, this aphorism enunciates the principle underlying the possibility of *infinitized relations* (or non-dual relations) in a relativistic world. If we pursue to its logical conclusions the Axiom—"Everything Is Everything Else"—then **8** am a *something* and so are you. All somethings are *essentially* (not formally, hence, externally) *identical,* and therefore we are each other. We are not each other *in part.* **8** Am not *partially* you and you are not *partially* **8**, but, on the contrary, and as strange as it may seem, **8** Am *fully* you and you are *fully* **8**.

This means that, on the deepest possible level, *your* experience is <u>*fully*</u> *my* experience, and my experience is <u>*fully*</u> *your* experience even though, subject to the Veils of Maya, we do not in our ordinary states of consciousness (even the rather lofty ones) realize this to be so. When the time of Harvest comes at the Great Reabsorption or "Day Be With Us" (and probably somewhat before that time), ***8****-the-Whole* will fully realize that **8** have been *fully present* in that which has been called *you* and *You-the-Whole* (a term completely equivalent to but less descriptive of Reality than **8**-the-Whole) will realize that you have been fully present in that which has been called *me.* There will be no distinction in our accumulated experiences, which will be, for all practical purposes, *one* series of experiences.

Conclusion 2
<u>8</u> Am, essentially, *as much* you as <u>8</u> Am what <u>8</u> usually call 'I'.

No matter how much my *normal,* hence, illusory, experience teaches me that **8** Am more what **8** call 'I', than **8** Am what is called you, **8** Am, *nevertheless,* the Whole (the **I**-as-I-as-'i' or the SELF-as-Self-as-self) experiencing through what **8** normally call 'I' *and* **8**

Am *equally* the Whole experiencing *fully* through what is normally called you. Due to the necessary fact of the Omnipresence of the SELF-as-Self-as-self, **8** (**I**-as-'I') Am as fully within each and every form as **8** Am within the form usually identified with 'I'—my limited selfhood.

Conclusion 3
All 'parts' are illusory.

If there were such a thing as a REAL *part,* it would have to be created in defiance of the Law of Indivisibility (which is directly related to the very First Fundamental of *The Secret Doctrine). Illusory* parts or particles there are, have always been, and will always be, cyclically, forever; these parts or particles are 'actual-in-Illusion' [see Glossary]. REAL *or* ESSENTIAL parts there can *never* be. The GREAT PRINCIPLE is both 'BOUNDLESS' and 'IMMUTABLE'. A part-creating *division* (what other kind is there?) is a *mutation,* hence a violation of Principle of Immutability. To divide is also to *move* or to *act.* It is understood that there can be no possible movement or action 'strictly' *within* IT, the BOUNDLESS IMMUTABLE PRINCIPLE. Thus, to divide would contradict the fundamental HOMOGENEITY of the GREAT ALL-in-ALLNESS.

Any type of REAL division is, thus, ESSENTIALLY impossible, though, for practical 'in-Universe' purposes, division is actual-in-Illusion and necessary within the World of Becoming. Since division is ESSENTIALLY illusory, and there can be no division without change, all changes are, ESSENTIALLY, illusory as well, even though the entire World of Becoming is built upon change. The World of Becoming is actually the World of *Seeming.* While change *seems* to be happening, it is not REALLY happening at all. This is a very great paradox. Change, for all its palpable actuality does not and cannot disturb the *fundamental changelessness* of REALITY.

How shall we deal with this paradox? In one way it is insoluble. The REAL is, by definition, forever *changeless,* but we must deal with *actual* change as if it were *real,* or we cannot fulfill Divine Intent. There seems to be a profound contradiction here. Perhaps, however, we must consider that while the contradiction is *actual,* it is not REAL. The SELF 'SUBJECTS' ITSELF to the World of Seeming which is Cosmos. In fact the SELF *becomes* the World of Seeming, the Cosmos (albeit, while ever 'REMAINING' undisturbedly ITSELF). This Cosmos and Its dynamics are impermanent and passing. Though the dynamics of the World of Seeming *seem* to negate the 'STATELESS STATE' which must, of necessity, characterize the SELF, that CHANGELESS 'STATE', in REALITY, cannot possibly be negated. Thus, if division *seems* to take place, the apparently divided 'parts' are not REALLY different at all.

If division were REAL, it would create REAL parts, and parts to be *REALLY* parts, would have to be different and distinct from either each other or, at least, from the WHOLE. Even if parts were identical to each other and different from the WHOLE, they would have to be *less* than the WHOLE, because with respect to the WHOLE, that which is different is necessarily *less.* If difference existed, however, a condition would necessarily exist within the BOUNDLESS FULLNESS (assuming any *condition* at all *could* exist within the BOUNDLESS FULLNESS) in which there was a 'lessness'. 'Lessness' is impossible in relation to the ABSOLUTE PLENUM, the EVER FULL. Further, the REAL presence of this 'lessness' (if it *were* REAL) would mean that the ALL-IN-ALLNESS, the ABSOLUTE PLENUM could not be not *fully present* and *fully* 'FULL' in all places at all points in Cosmos, because ITS very PRESENCE in all places and at all points would necessarily *negate* the 'lessness'—the '*part*ialness'. To place such a limiting condition upon the BOUNDLESS IMMUTABLE PRINCIPLE violates the very *boundlessness* of ITS NATURE.

There are many reasons why *parts* cannot exist, and why, even though parts *seem* to exist, they must all be considered, not *parts* of the SAME THING, but, the SAME THING *ITSELF*. To think in this way would change our approach to the World of Diversity (i.e., the World of Seeming, the World of Becoming). All seeming diversity would be resolved into the realization of ABSOLUTE HOMOGENEITY.

Approached from the simplest point of view, we may consider it axiomatic that GOD (the ABSOLUTE DEITY) is *indivisible*. GOD, the SELF, is indivisible. Thus, 'less' and 'more' are impossible 'within' the SELF. In all of Cosmos, there is only the SELF, the SELF of ALL, the ALL-SELF, only the BOUNDLESS IMMUTABLE PRINCIPLE, which is the INFINITY of INFINITIES. To divide IT (if such an act were possible) would be to create something *other* than IT, would be to create a *second* to this "ONE WITHOUT A SECOND." A true division (were it possible) would create a difference 'within' or 'in relation to' the UNDIFFERENTIABLE ONE. Difference, however, cannot be created in THAT which cannot be differentiated. If, then, the apparent act of division creates no *REAL* difference, the *apparent* division is not *REALLY* division at all. Then, that which *appears* to be divided is not REALLY divided at all. The parts we think we see, therefore, are not REALLY parts, but are the WHOLE OF WHOLES ITSELF. To perceive thusly would change consciousness entirely.

In sum, perhaps it should be said that the BOUNDLESS IMMUTABLE PRINCIPLE suffers no conditions or variations or actions other than the continuation of ITSELF, PER SE. IT abides, unchangeably, exactly as IT is, as the GROUND OF ALL BEING. Any action or motion (of no matter what nature) would make IT *less* than ITSELF, and this would be an impossibility, for the IRREDUCIBLE cannot be reduced.

Conclusion 4
The BOUNDLESS IMMUTABLE PRINCIPLE can never vary.

The First Fundamental of *The Secret Doctrine* asserts the existence or PRESENCE of a BOUNDLESS IMMUTABLE PRINCIPLE. The reader, being in no position to prove or disprove the veracity of the First Fundamental, can either accept or reject it. In this treatise, it is accepted.

If the BOUNDLESS IMMUTABLE PRINCIPLE is to remain forever exactly what IT is said to BE (i.e., if, throughout Infinite Duration, IT is to remain IMMUTABLE) then, nothing at all can be done *to* IT or '*within*' IT (meaningless prepositions!), for every *doing* would be a *mutation* of the 'IMMUTABLE'. Anything that *seems* to be *done*, is *not* RE-

ALLY *done* in TRUTH and in FACT. The apparent *doing* occurs only in the World of Play, the World of Illusion.

Thus, with respect to the BOUNDLESS IMMUTABLE PRINCIPLE, to conduct any operation *upon* IT or to say that any substantial operation is conducted '*within*' IT is to contradict ITS very NATURE. If we hold to the premise asserted by the Three Fundamentals of *The Secret Doctrine,* then we accept that IT IS as IT is said to BE. No assertion which seems to contradict ITS nature can be accepted. This being said, IT ITSELF is necessarily the GREAT CONTRADICTION ... or *is* IT?

Conclusion 5
All seeming 'parts' are REALLY wholly THE WHOLE.

All that we call 'parts' are necessarily illusory because the WHOLE of WHOLES cannot be divided. The strange conclusion that logically arises from the Law of Indivisibility is that all parts have the SELF (the WHOLE of WHOLES) *equally invested* within them. The Veils of Maya make a part seem to be nothing but a part. If, however, that which is called a part is 'SEEN' (or better, 'BE-EN') in TRUTH, and from the 'INFINISPECTIVE' of the SELF, then all parts are known to *be* the WHOLE of WHOLES ITSELF.

The Holographic Principle tells us that each part within a system is a more or less accurate reflection of the *wholeness* of the system of which it is a part. The idea proposed here agrees, in principle, with the idea that each part is a reflection of the Whole System which contains it. In the World of Illusion the part reflects the Whole but *is not* the Whole.

It would lack common sense to say, for instance, that the individual human being, for instance, *is, formally,* the *wholeness* of the planet Earth. However, from the 'INFINISPECTIVE', both the human being and the planet Earth *are* SOMETHING ELSE—namely, the SELF ITSELF. Each of them is *fully* and ESSENTIALLY the SELF, and the SELF is *fully* and ESSENTIALLY them. Therefore, on an ESSENTIAL, not *formal* level, the human being and the planet Earth *are each other,* because "things equal to the same thing are equal to each other".

Therefore, what is being asserted here goes beyond the assertions of the Holographic Principle. While the part, indeed, is not, *formally,* the Whole which contains it within the World of Becoming, that same part, ESSENTIALLY, is *fully* the WHOLE of WHOLES which, ITSELF, not only *contains,* but *IS* all possible parts of no matter what magnitude, as well as all Whole Systems (themselves but parts) which in the World of Becoming contain the lesser parts.

The unlikely statement which fully equates each and every possible part with the WHOLE of WHOLES becomes a necessary, though, when we realize that the OMNIPRESENT SELF, being BOUNDLESS, is found completely, fully and equally everywhere in Infinite Space and Cosmic Space—there being, ESSENTIALLY, no such thing as either vacuum or interval.

Conclusion 6
The SELF cannot be excluded from any place or any point in Space.

The Principle of the Omnipresence of the SELF (the ABSOLUTE) necessitates that IT, so to speak, "take the *place*" of every other thing. The SELF is not merely superimposed upon every other thing, allowing for an unacceptable dualism; namely the simultaneous 'EXISTENCE' of the SELF, *and* the existence of *all other things*. That would be superficial and contradictory. The SELF, rather, (functioning within the World of Illusion where Space *seems* to exist) 'occupies the same space as' every other thing in such a way that IT *negates the presence* of every other thing, and *asserts* the PRESENCE of ITSELF alone. The SELF is an OMNIPRESENT CONTINUITY which, ESSENTIALLY, *replaces* all things in Space with ITSELF. The SELF is the 'ENFORCER' of HOMOGENEITY.

Conclusion 7
ESSENTIALLY, there can be no separation.

Every separation depends upon division, and division within the OMNIPRESENT SELF is inadmissible, for any division creates *another.* 'Another' cannot exist *within* or in the PRESENCE of "the ONE WITHOUT A SECOND" (for 'another' *is* a 'second').

So division, and its result, separation, are ESSENTIALLY impossible. Further, division and separation would necessarily create a 'lessness' within the ALL-FULL, the ABSOLUTE PLENUM (which is, necessarily, according to ITS OMNIPRESENT NATURE, *fully* and *wholly* PRESENT at all 'points' in illusory Space). There is no 'room' for *anything* <u>*less*</u> that IT. The creation of such a 'lessness' (since any REAL divisions within the UTTER WHOLENESS would create a condition of 'less than WHOLENESS') is also logically inadmissible because it would contradict the 'NATURE' of the INDIVISIBLE WHOLENESS, the HOMOGENEOUS PLENUM.

Conclusion 8
All things can be negated except the SELF.

All things except the SELF are un-REAL except they be considered the SELF ITSELF. The PRESENCE of the SELF negates all things. IT, in an ultimate sense, "blots out all form." No 'thing' has any true REALITY in and of itself. Such 'reality' as it *does* have, may be, at the most, evanescent <u>R</u>eality within the World of Being—a kind of Cosmos-long permanence within the Great Illusion (the Cosmic System). The only <u>*true*</u> REALITY of any thing, other *than* the SELF, is only (as it could be said) *'within'* the SELF, ITSELF, and, even more, <u>*as*</u> the SELF, ITSELF.

In other words, a thing, *in itself*, can be negated as an un-REALITY, but a thing *in*, and even better, *as* the SELF, is *fully REAL* because that thing *is wholly* the SELF, ITSELF (because, every 'part' *is* REALLY the WHOLE *entire).* The SELF, ITSELF, IS, in fact the GREAT NEGATION, and any thought of negating IT would be meaningless and futile. Because IT IS this NEGATION, the SELF is simultaneously the GREAT ASSERTION, as IT (ever and uninterruptedly throughout Infinite Duration) IS, unchangingly PRESENT, 'ASSERTING' *the* 'FACT' of ITS OWN fathomless 'NATURE'.

Conclusion 9
The Great Illusion is based upon the illusion of 'parthood' or divisibility.

He who 'sees' only THAT which is indivisible 'sees' BRAHMAN, 'sees' the SELF, and overcomes the Great Illusion (at least in *consciousness*, for there is, most definitely, a *power/life* aspect to the necessary overcoming). ('Seeing', per se, is *not* sufficient, for 'seeing' is *mayavic*, yet it represents an initial and necessary stage leading to *being.)* Dualistic 'sight', or, better, *registration*, prevents the realization of synthesis.

The goal of synthesis, the realization of which is the goal of monadically inspired living, is—while *apparently* 'seeing' the part, to REALLY see *no* part. This type of *monistic* 'seeing' would make it possible to 'see' the WHOLE within each part, the INFINITUDE within each part, the SELF within each part. This 'seeing' is achieved through *identification.*

Conclusion 10
8 ('I-as-I') am ultimately responsible for what 'you' have done and you are responsible for what 8 have done.

There is no REAL 'I'. There is no REAL 'you'. There is but ONE 'ACTOR'—*the* 'ACTOR'-as-Actor, 'Playing' all the many roles which Maya assigns. The ONE 'ACTOR', alone, is ESSENTIALLY responsible. The ONE 'ACTOR'-as-Actor has 'Done' the 'Deed' (the Universal 'Deed') and the ONE 'ACTOR'-as-Actor experiences the results of the 'Deed'.

This means that, when consciousness has expanded sufficiently, which certainly occurs by the time of the "Day Be With Us" (whether the Greater Universal 'Day' or simply a systemic, constellational or galactic 'Day'), '**8**' experience the karma of all 'you-as-**8**' have done, and 'you' ('you-as-**8**') experience the karma of all '**8**' have done. WHO has been the 'ACTOR'? Certainly not the un-REAL 'I', and the un-REAL 'you'. 'SOMETHING' has 'PERVADED' both, and been constantly PRESENT in both. In fact, 'SOMETHING' has *been* both all along. So, if one digs deeply enough, where does responsibility lie?

Conclusion 11
There is no impact which 'you' receive that 'I' do not receive. There is no impact which 'I' receive that 'you' do not receive.

Written with greater technical accuracy these statements would read, "There is no impact which 'you-as-**8**' receive that **8** do not receive; there is no impact which **8** receive that 'you-as-**8**' do not receive." What is the difference between 'your' experience and 'mine'?

On the level of normal human consciousness, there is a vast difference. Clearly on that limited level, 'my' experience is *not* 'yours'. Even as consciousness expands to include the Consciousness of a Planetary Logos or a Solar Logos or beyond, there is still an *apparent* difference, for we are told that there are certain dynamics within our Cosmos which no Life within our solar system yet understands. The inevitable trend towards Universal Synthesis, however, will see the fusing and blending of B/beings at the level of Spirit. Ever the Root of Identity will be traced to Its Origin, revealing Cosmic Identity.

Further, ever the ROOT of IDENTITY will be traced to *ITS* ORIGIN, revealing ABSOLUTE IDENTITY. The individual experiencer will be seen to have been a *necessary illusion* and the One Subject of all possible differentiated experience in-Cosmos will be revealed. Further, the ONE SUBJECT of all possible differentiated experience within the ALL of an infinitude of Cosmoses will be revealed. WHO or Who, then, experienced what?

It might be said that a given *apparent* unit experiences certain impacts more intensely that other *apparent* units. This is *apparently* true. Yet the only BEING that is ever PRESENT, is the PRESENCE ITSELF, and IT is *fully* and *completely* (not *partially)* PRESENT at all Points (whether 'Real' or 'Virtual') in Space simultaneously. "Blotting out all form", as IT 'DOES', IT 'VANQUISHES' (in degree of REALITY) all other, lesser, illusory experiencers, and 'REMAINS' the ONE AND ONLY 'EXPERIENCER', regardless of the seeming differences in the intensity of various experiences experienced by the apparently separated experiencers. This concept requires much contemplation and is rich in possibilities for modifying human relations.

Conclusion 12
8 experience everything that transpires in Cosmos.

8 miss absolutely nothing. This seems a profoundly radical and even pretentious statement, but it is based upon the logic of the *essential indivisibility* of the SELF, as so many of these conclusions have been.

WHO *IS* the SELF? Is not every Actor in Cosmos the ONE 'ACTOR'? Is there any 'place' in (illusory) Space where that ONE SELF is *not?* Further, since **8**, (the One Self) ESSENTIALLY, Am *wholly* and *fully* that ONE SELF (and IT IS **8**), is there any 'place' in Space where **8** Am not? Further, since the ONE SELF which **8** Am, endures in unchanging PERFECTION throughout Infinite Duration, is there any 'time' within that Infinite

Duration when **8** (because **8** Am **I**) have not *been* at every 'place' in Space (however, illusory that Space may ESSENTIALLY be)?

Therefore, **8** miss nothing; nor *have* **8**, nor *will* **8**, nor *can* **8**, simply because of WHO **8** (**I**) REALLY AM. With respect to all of an infinitude of Cosmoses, **8** have been the 'witness' of ALL and All, and will, necessarily, be the 'witness' of ALL and All. With respect to all of an infinitude of Cosmoses, **8** have been the 'participant' in ALL and All, and will, necessarily, be the 'participant' in ALL and All. That in 'my' limited *personal* or even *soul* consciousness **8** do not know this, is the result of the SELF-'VEILING' which **I** persistently 'IMPOSE' upon MYSELF throughout the Eternal Duration of each and every Universal Manvantara. Yet, on the very highest-Universal Level, **8** (the Indivisible One, the Cosmic Only One) do, *even now*, know and witness All that transpires within the Cosmos.

Further, in ABSOLUTENESS (which necessarily cannot cease from *being* exactly what IT IS merely because a Cosmos exists) **I** do, *even* NOW, 'KNOW' and 'WITNESS', or, better, *'AM'* All that transpires within the Cosmos, That **8**-as-**I** have *always* experienced all things (and, even, 'BEEN' all things) will be revealed to 'me'-the-**8**, and to all units, as sufficiency of synthesis is attained, and, fully, as Universal Pralaya supervenes.

Conclusion 13
Since there is naught but 8 (the SELF-as-Self), all that 8 experience is 8 MYSELF.

To the casual thinker, it seems that the experiencer experiences many things, and that these things are, to a great degree, aspects of the Not-SELF. There is, after all, incredible *actual* diversity, and to most, that diversity seems REAL, or at least Real. Is there REALLY, however, a *Not*-SELF? *ESSENTIALLY*, there *cannot* be. Further, is diversity REAL? We have already established that because of the Principle of Indivisibility, diversity cannot be REAL but only *actual*. So it turns out that the INFINITE OMNIPRESENT SUBJECT of all experience is *also* the infinitely recurring, cyclically-omnipresent 'OBJECT'-as-Object-as-object that is *apparently* experienced.

In other words, **8** Am *both* the SUBJECTIVITY-as-Subject of 'my' experience *and* the Object of that experience. In Cosmos, **8** cannot experience anything *other* than MYSELF-as-MySelf. 'Within' the INFINITE SELF, **I** cannot even 'EXPERIENCE' MYSELF; **I** can only BE that SELF to the *infiniteth degree*. Thoughts of the Not-SELF are so deeply ingrained in the consciousness of spiritually inclined individuals, that they do not realize that, though the Not-SELF conception represents a great step beyond the ignorant, 'centerless' identification or absorption of the little 'i' with and into all experience, this conception still falls far (nay, *infinitely*) short of the ESSENTIAL TRUTH.

Is it possible for one to 'see through' the apparent diversity of all experience and find in that which is experienced but One Object which is, ESSENTIALLY, the experiencing Infinite Subject which is, ESSENTIALLY the ONE INFINITE SUBJECTIVITY? Much meditative practice will be required—much practice to achieve that which already 'EXISTS'! Paradox!

Conclusion 14
8 can never leave the PRESENCE of 8, MYSELF.

There is so much talk today about "finding oneself" that it is not realized that **8**-as-**I**, the SELF, AM the GREAT INESCAPABLE, the GREAT CONSTANT. **I**-as-**8**-as-I AM ever PRESENT. 'I' (regardless of my ignorance) simply cannot *lose* MYSELF, though **8** (as intra-Cosmic Maya) can and do *veil* MYSELF. If there is ONE UNIVERSAL CONSTANT, *present* in *all* Universes (and even when Universes exist not), ***I*** AM IT.

In terms of our particular Universe, **I**-as-**8**-as-I AM IT. That 'CONSTANT' persists uninterruptedly, without interval, through all Times and in all Spaces. There is no apparent interval, whether of Time or Space, which **I**, the SELF, do not *fill entirely*. "Nature abhors a vacuum", it is said, but **I** AM the *absolutely dense VACUUM*. **8** (the **I** in-Cosmos) cannot evade MYSELF, escape MYSELF, 'lose' MYSELF. In Universe, **8** (the *highest* type of 'I' possible in-Universe) Am the Presence-ever-Present (Reflection of the PRESENCE).

Can the Presence ever leave the Presence? Can the PRESENCE ever leave the PRESENCE? All alienation from the SELF is a *seeming* only. We-the-**8** must remember that we cannot, *essentially*, vacate either our Universal Selfhood or our ABSOLUTE SELFHOOD.

Conclusion 15
There is no point in Time and Space where 8 Am not fully 8.

Due to MY SELF-'IMPOSED' ignorance, **8** (all of us as human beings) constantly *seem* to be far less than **8** REALLY *already* AM. 'I' say to myself, "One day, 'I' shall be fully **8**, but not for a long time to come." There must dawn the realization that **8** Am, *even now*, fully **8**, and that my ring-pass-not cannot, Essentially, be limited to the tiny, localized individual sphere (though, temporarily, it *can* be limited to the Universal Sphere). UNCONDITIONED BEING, however, 'EXISTS' for all of 'us' *already*. There is no Cosmic Time past, present or to be, and no Cosmic Space, past, present or to be, when and where **8** have not been *fully* MySelf. There is no Pre- or Post-Cosmic Time past, present or to be, and no Pre- or Post-Cosmic Space, past, present, or to be, when and where **I/8** have not been fully myself. There has been no 'time' or 'space' within the UTTER ALLNESS when or where **I** have not been *only*, *utterly* and *absolutely* MYSELF.

Conclusion 16
There need be no extension, dimension, or time for me to be fully I. I AM fully I, both nowhere and everywhere, at all times and at 'no time'.

MY **I**-NESS exists unaffected by Time and Space. MY **I**-NESS is independent of any structure or temporal happening within any Cosmos. The apparent intra-Universe 'extension' of matter and the various distinguishable dimensions (created by prakritic vibratory rate) which present themselves to all intra-Universe consciousnesses, have no REAL effect upon my **I**-NESS.

In fact, these intra-Universe factors act as temporary veils upon 'my' realization of and my identification *as* my **I**-NESS. Nevertheless, everywhere in Cosmos and at 'every time' **I** AM fully **I.** When there is no extension or dimension or time, because there is no Universe, it can, also, justly be said that **I** AM fully **I. I** AM fully **I** when the only 'place' there is, is *nowhere,* and the only 'time' there is, is *'no-time'.* Another way of saying this is that, in ESSENCE, my **I**-NESS is *absolutely unconditional.*

Conclusion 17
Each Cosmos is composed of what might be called 'an absolute density of SELF'.

The 'density of SELF' *within* Cosmos is paradoxically equal to the 'density of SELF' within the BOUNDLESS IMMUTABLE PRINCIPLE, ITSELF. In REALITY there is no difference in 'place' or ESSENTIAL 'state' within Cosmos or 'within' the BOUNDLESS IMMUTABLE PRINCIPLE. In defiance of common sense, even *within* Cosmos (with all its apparent diversity and differentiation), there is absolutely no interval whether of Time or of Space (despite intra-Cosmic-Perceptions to the contrary). SELF 'fills' every *seeming* interval, for intervals are but a seeming. This is another way of saying that both within Cosmos and 'within' the ALL-in-ALLNESS, the SELF is OMNIPRESENT. Nowhere is IT *not.*

Conclusion 18
The perception by the SELF-as-Self of apparent interval, whether of Time or Space, is the basis in consciousness of the Great Illusion.

The perception of interval is based upon the creation of a *point of view.* Within the SELF in ITS Universal Pralayic STATE of ALL-in-ALLNESS, there is *no possible point of view.* Because within the SELF there can be no isolated *points,* there can be no *points of view.* The first Point of View appears with the 'FLASHING FORTH' of the 'RAY' of the ABSOLUTE. This instantaneously 'GENERATES' a Pre-Cosmic *Infinified* Point of View. That Infinified Point of View becomes the Condensing and Condensed Points of View. The first Real *limited* Point of View is that of the Condensed Point-become-Universal Logos.

The moment a point or points of view exist(s), there arises the consciousness of *that* which is *registered* from a particular vantage point. That which is registered is designated as an *object*. There can be no *subject-object distinction* without the apparent existence of an *interval* (gap, cleavage or separation) between subject and object; for without an interval or separation or division (call it what we will) subject and object would be totally fused, undifferentiated, and we could not speak meaningfully in terms of a twoness—i.e., a subject and an object. A point of view is a *localization,* and all localization creates duality and causes the arising of the concept of 'interval'.

As the notion of interval is gradually overcome by the evolving consciousness, so is the Great Illusion seen to dissolve into Its First Essence which is the Universal Self and, subsequently (if the penetration be sufficient) into Its *ultimate* ESSENCE which is the BOUNDLESS IMMUTABLE PRINCIPLE.

Conclusion 19
8 have no individuality in the ordinary sense. It is the WHOLE-as-Whole or the SELF-as-Self which Acts through all 8 appear to do.

Certainly, from all that has been said, it is clear that **8** have no separate eternal individuality. **8** have an *apparent* individuality-in-form for the duration of a Cosmos (and many are the forms which distinguish this apparent individuality throughout the course of Cosmic Duration, for **8** express equally through *every* form!).

ESSENTIALLY, however, **8** Am not distinct from everyone or anyone else, and there is no individuality which survives in its distinctness from Cosmos to Cosmos. Individuality in the normal sense of the word is equivalent to a major limitation upon the SELF that **8** ESSENTIALLY AM. ESSENTIALLY, **8** Am not a part; **8** Am the WHOLE-as-Whole.

As the term is usually conceived, an 'individual' is merely a 'part', but we have established, for many reasons, that there can be no such thing as a REAL 'part'. A REAL part would contradict the 'NATURE' of the INDIVISIBLE SELF. Parts simply cannot REALLY exist. Parts are a mere *seeming* and, hence, are illusory (though *actual).* Whatever individuality 'I' (in my limited 'I-ness') think 'I' have (if I am unfortunate enough to think of individuality as indicating a *part* of the Whole) is REALLY a phantasm born or error.

When limited 'I' thinks thusly, then, **8** Am REALLY the WHOLE-as-Whole in a condition of *error*—the *normal condition* of most consciousnesses in-Universe. Because, however, of the *incorruptibility* of MY TRUE 'NATURE', **8** Am *also* and *simultaneously* the WHOLE-as-Whole in a condition of TRUTH (for **8**-as-I-MYSELF forever 'KNOW' the TRUTH, whether in-Universe or in MY ULTIMATE STATE of ALL-in-ALLNESS). **I** being **8** AM/Am 'in' both of these states/conditions simultaneously.

Essentially, the Real 'You' is not at all distinct from the Real 'Me' and **8** (the REAL-I-as-I) Am not distinct from the Real 'You'. This can equally be said of the REAL 'YOU' and the REAL 'ME', and is, if conceivable, *infinitely* more TRUE. Granting the *absolute identicalness* of our ABSOLUTE IDENTITY, even within Cosmos (the Great Relativity), **8** have no *separate* individuality from you, nor you from me, *but* we both do have *individuality* in the truest sense of the word, which means that we both have *indivisibility.* You and 'I' and all in Cosmos are, Essentially, 'Rays' of the ABSOLUTE, which are but One 'Ray' which is but ONE 'RAY'. Meditate upon the word 'individuality' and understand its true meaning.

To say that 'you' and 'I' have no *essential* distinct and separate individuality may seem like a great *theft* of identity, but in fact the realization of this *restores* Identity and IDENTITY. Our *apparent* differences abide, and must abide, in the World of Becoming (which is the World of Illusion), but those apparent *formal* differences are not differences in *individuality.* Individuality is a state of *indivisibility.* That which is indivisible has no 'parts', and only one THING (if we can call IT a 'THING') has no 'parts'. The true *individual* (and there is only *one* in all the UTTER ALLNESS) is the ONE ABSOLUTE SELF.

A human being *becomes* in consciousness the INDIVIDUAL he already *is* when he realizes his ABSOLUTENESS, and the IDENTICAL ABSOLUTENESS of every other apparently distinct individual. Only thus is true *individuality* attained—*indentificatorily.* On the LIFE side of the equation, true *individuality* has 'BEEN' for each and all forever. The attainment of the realization of Universal Individuality (Universal Indivisibility) is the antechamber to *identification* <u>as</u> the ABSOLUTE INDIVIDUALITY.

Conclusion 20
8 Am MYSELF the UNIQUENESS.

Thus, we ourselves (We-as-**8**) are the UNIQUENESS. The ONE SELF, which **8**/We Am/Are, is totally *incomparable.* What *other* is there than the INFINITE SELF (which, ITSELF, infinitely supersedes in scope the ALL of an infinite number of Cosmoses past and to come)? In the most profound sense of the word, **I** AM the only *utterly unique* INDIVIDUALITY. The lower type of individuality, though unrepeatable, is not a *true* individuality because it *composite;* it can always be divided.

The true INDIVIDUALITY is forever unique and remains uniquely, incomparably ITSELF forever. IT manifests forever in the unrepeatability of multitudinous forms and thus ITS expressions (the many little selves, even the may little 'Cosmic Selves') experience a lesser form of uniqueness. Yet it would be incorrect to say that the INFINITE SELF *repeats* ITSELF at any Time in Infinite Duration, because IT never for a moment *ceases* to be fully and entirely what IT IS, so how can IT *repeat* ITSELF. IT IS the 'CONTINUUM' and thus *continues* exactly as IT IS.

The amazing realization is that the small self-conscious unit of life must give up is apparent individuality which *seems* so unique, in order to achieve the TRUE INDIVIDUALITY which is the *same* as the *individuality* of every other E/entity in Cosmos. Thus by becoming the *same* we become both *individual* and *unique.*

From another perspective, each Cosmos in the UTTER ALLNESS is utterly unique, in a *lesser* way, from every other Cosmos. When in-Cosmos, the Essential uniqueness of each E/entity is the fact that it has no Real individuality other than that of the One Individuality (the Universal Logos, the One 'Ray'). Who, however, *Is* the One Individuality other than the TRUE INDIVIDUALITY (which all of an infinitude of lesser individualities have *ever* been)? Thus, though our petty, personal uniqueness is dissolved in the Unitary Uniqueness of the Universal God "in Whom we live and move and have our being", even *that* Uniqueness is dissolved in the ONE AND ONLY UNIQUENESS of THAT.

Conclusion 21
All E/entities in Cosmos are absolutely identical in Essence.

This conclusion has some astounding implications, such as, that in any Cosmos, there can exist no *REAL* Hierarchy of beings. In the World of Becoming, of course, *all* is organized in an hierarchical manner. The principle of Hierarchy, however, relates to name and form alone (name and form being virtually equivalent concepts) and the quality of the consciousness which functions through name and form. The existence of Hierarchy depends upon factors such as status and magnitude, upon more and less. In the WORLD OF PURE BEING (which is the only WORLD there REALLY *IS)*, there is *no more* and *no less*, for there is no variation of any kind. There is only the constant *fullness,* the constant *maximization,* the constant *infinitization.* All E/entities in-Cosmos, therefore, are ESSENTIALLY *equal* (and, even Essentially equal).

From the point of view of Life-in-form within the World of Becoming (the World of Illusion) there is total and complete *inequality* in every respect. From the point of view of the SELF, however, there is total and absolute *equality.* Within the ALL-SELF, each seeming E/entity is REALLY the UNCHANGING ONE, and is, *already* and *forever,* the *maximization of all potential.* This being the case, wherein, then, can be any REAL difference between E/entities such that a non-illusory Hierarchy could REALLY exist? Within the ALL-SELF, each E/entity (because each such E/entity is individually un-REAL) is *already* THAT which is ULTIMATE. Because of this, no matter what apparent 'level' of B/being within Cosmos is considered, at every level *ESSENTIAL equality* is to be found. There is no*where* to go and *nothing* to achieve. ULTIMATENESS *already* IS. More than being ESSENTIALLY equal, all E/entities are ESSENTIALLY *identical.*

Viewed wisely from the point of view of greatest possible Synthesis within Cosmos, there can be no ESSENTIAL differences of any kind between any E/entities. Of course, the *formal* differences are virtually countless and they must be respected. Even intra-Cosmically, Essential differences do not exist. The Essence of each E/entity is resolvable into the One Universal Essence. All 'Rays' are the One Universal 'Ray'. If however an infinitude of Ultimate Essences (for the number of Cosmoses has been infinite) are compared, it will be seen that all Essences are the ONE ESSENCE. Everything that has ever been or will be is *identical in ESSENCE.*

As a consequence of these thoughts we should remember that as we contemplate one who is apparently *another,* we are not only contemplating an utter *equal* and hence (at the level of Essentiality) a *brother,* but we are contemplating an *identical*—one who is absolutely *identical* with our *essential self,* the SELF. Our task must be to make the experience registered by the one we contemplate our own, for *essentially* it *already* IS so.

Conclusion 22
All Entities in Cosmos are Essential and indispensable.

We live in an age when the individual E/entity is considered expendable. (Even if stars explode, are there not 'plenty' of them?) Within the World of Relativity the individual E/entity is *always* expendable, for a replacement will always, sooner or later, be found. But WHO, REALLY, IS the E/entity? Is not the E/entity, REALLY, the *ABSOLUTE* ITSELF? Of this ABSOLUTE or ONE SELF, we must ask, Is IT expendable or, rather, ESSENTIALLY, irreplaceable? We must remember that every E/entity in Cosmos is not only an integral 'part' of the ONE SELF but *is*, in fact, *wholly* the ONE SELF! The ONE SELF is fully and completely present at every possible point in Space and at every possible Time. There is, therefore, no way to remove the part without removing the WHOLE! Besides, the part cannot be removed, for it does not exist. Besides, whither could it be removed that would separate it from the ONE SELF which it, ESSENTIALLY, IS? No, no E/entity can be removed from what we might call 'the arena of the PRESENCE'.

E/entities, therefore, must be seen and understood *subjectively*, and not judged by their form. Forms are non-ESSENTIAL/Essential and dispensable; the ESSENTIAL E/entity (WHO IS *LIFE*, ITSELF) is just the opposite. What all this means is that all B/beings will be in the PRESENCE of all B/beings forever. This is not only possible but necessary because all beings are ONE BEING WHO IS the PRESENCE, ITSELF.

Conclusion 23
Entities seem unequal in Time and Space because they are manifesting through patterns that are non-identical and unequal and of greater or of less complexity.

Identical E/entities are manifesting through *non*-identical patterns. It is important in all matters of enhancing relationship to focus upon the identicalness of all E/entities. In one way, it could be said to every E/entity, "The entire SELF is 'behind' you, 'within' you, and, more accurately, *IS* 'you'. You are THAT." These thoughts are connected with the idea that "All men are created equal." In fact, it should be said of all E/entities (and not only men), that they are not *created* at all, but are *uncreated essences, essentially equal* forever.

Conclusion 24
Every E/entity has already passed through all dimensions of all Cosmoses past.

This is an astonishing thought with astonishing implications. For instance, that which is now the E/entity called "worm" has been, in Cosmoses past, the E/entity called "star". The present condition of an E/entity is no indication of past experience in other Cosmoses, though present condition will certainly indicate (for those who can read its meaning) the *past* experience within the *present* Cosmos. (Radical understanding, however, will reveal all experiencers in Cosmos to be experiencing One Simultaneous Cosmic Experience.) We must remember that all experiences are the experiences of the ONE SELF-as-

One Self. The apparent presence of illusory subdivisions of the SELF in the form of *many E/*entities must not deprive us from this realization.

Examining the hypothetical structure of Cosmoses by extrapolating from our own, it can be said that there are various vibratory strata within each Cosmos and that every E/entity (considered as an illusory though *actual* subdivision of the ALL-SELF-as-One Self) must pass *experientially* through all strata of all Cosmoses (and, even, more strangely, 'occupy' all strata simultaneously). As there have been, necessarily, an infinite, hence incalculable, number Cosmoses past, each E/entity to be encountered in this Cosmos (no matter what its present level) has already (as a representative of the WHOLENESS of the WHOLE SELF, and {more} as the WHOLE SELF, ITSELF!) passed through *all possible strata of all Cosmoses past.* We can hypothesize that the lowest such strata are somewhat comparable to strata in our Cosmos through which our tiniest and most densely aggregated ultimate units of life (i.e., the densest aggregations of ultimate particle/events) are presently expressing, and that the loftiest of these strata in other Cosmoses are, perhaps, somewhat comparable to similar strata in our own Cosmos—strata through which the Greatest Super-Beings we can conceive (Members of the Cosmic First Family) are now expressing.

The astounding implication is that in every E/entity encountered, one is relating to an *equal* or, more, to an *'identical'*—i.e., to one who has had experience equal to your own because you and it are identical. What E/entity encountered has not in former days manifested as a solar or galactic God, and even, ultimately, as the Universal Logos of *all* Cosmoses—yes, of *all* Cosmoses. Each E/entity has necessary experiences great glories and abysmal degradations (and is doing so Now). No matter what the E/entity, its glories have been identical to ours; in fact, *ours,* as have its degradations. All this is true because all E/entities are One Entity are ONE ENTITY, *always, completely* and *forever.*

Is this not a call for the respect of the lowly—or, more accurately, of the *apparently* lowly. The great too must not be viewed with excessive awe, because, ESSENTIALLY, Their 'altitude' is, even *now,* ours, just as our relative lowliness, Theirs. With this conclusion comes the recognition that the great and the small are, ESSENTIALLY, one and the same.

Conclusion 25
8 Am but an Actor, acting a part, a Player, playing the only Game there is.

8 Am, ESSENTIALLY, the WHOLE 'ACTING' or 'PLAYING' but a part. REALLY, **8** Am the ONE 'ACTOR'-as-Actor, Acting and Playing *all apparent* parts in Cosmos. Even when, in the fullness of Cosmic Time, **8** become *(consciously)* that Actor Who is the Whole of Cosmos (i.e., the Universal Logos, Who, Essentially, **8** *already* Am), even then, **8** shall be but Acting or Playing a Part (though a relatively *big* one), for Cosmos Itself is but a 'Part' in a 'PLAY' with an infinite number of sequential Manvantaric 'Acts'. REALLY, **8** Am the Player of *all possible parts* which, in their infinite sum, are but One Part. **8** Play the

Part of 'the Universe'. Similarly, **8** Am the Player of *all possible games* which, in their infinite sum, are but One Game. **8** Play the Game called 'the Universe'. When **I** sleep in *infinitized wakefulness* in the ALL-IN-ALLNESS, **I** play no part and no games, for there are no parts or games to play.

Conclusion 26
MY BLISS in the STATE of ETERNAL SELFHOOD is permanent. My Bliss and Joy 'in-Universe' (as well as MY agony and suffering) derive from an initial Voluntary 'ACT' of apparent SELF-'LIMITATION' followed by an apparent return to MY LIMITLESS 'STATE'.

8 have been apparently 'going forth' and 'returning' for an infinite number of 'times' throughout Infinite Duration. In each 'going forth' **8** apparently lose (through SELF-'VEILING') that which **I** *eternally possess*—the FULLNESS of MYSELF.

Will **I** always 'SUBJECT' MYSELF to this apparent suffering? It appears that I have always done so, and that **I** always will, even though ESSENTIALLY, **I** (aloof in MY INHERENT 'NATURE' from all possible relationship), AM *not* suffering at all, and have 'ABIDED' forever in the one and only INFINITIZED STATE which makes what is called the highest intra-Cosmic bliss seem but a dim and feeble reflection. While MY BLISS in the 'STATE' of ETERNAL SELFHOOD has endured forever, and will forever so endure, regardless of seeming fluctuations in the World of Becoming, **I**-as-**8** have Willed that, in-Universe, happiness, joy and bliss must be first obscured and then regained. **8**, in-Universe, have decreed that **8** must *earn* the bliss that **8**-as-**I** *already* have *had* forever.

This process seems paradoxical, and MY 'MOTIVE' might be questioned. It must be realized that there has never been a REAL loss of the *infinitessence of bliss* which **I**, MYSELF, *AM*. The in-Universe loss and gain is but a Play, a Game which, compared to the REALNESS of MYSELF, and MY UNCHANGING SELF has no REALITY. SELF-'VEILING'-as-Self Veiling is the Game **I** (pre-Cosmically, **I/8**, and, intra-Cosmically, **8**) Play, not so much for the Cosmic "Fun of It" (though **I**-as-**I/8**-as-**8** *do* have *Fun)* but, rather, out of SELF-'NECESSITY'—out of the 'NATURE' of MY INFINITE SELFHOOD.

Conclusion 27
There is no such thing as <u>individual</u> salvation.

This is so because there is no such thing as the *individual*, except for the ONE INDIVISIBILITY which is the ONLY TRUE INDIVIDUAL. Salvation, in this context, means a conscious return to identification with the ONE SELF. As **8** (according to the Principle of the Omnipresence in-Cosmos of the SELF-as-Self) am *totally extensive* in Cosmos, **8** pervade consciousnesses which are benighted as well as those which are consciously and, even, supremely, identified as the SELF. Therefore, **8** Am, even now, both *'saved'* and *'not* *saved'.* Further, forever, while 'in-Cosmos', **8** am both *saved* and *not saved.* One cannot speak of salvation when there is no Universe and *ALL* 'ABIDES' in the INFINITIZED STATE. During such *apparent* 'interludes' of PURE INFINITUDE, there is no need of it, for the INFINITIZED 'STATE' *is* the 'STATE' of SALVATION.

When, however, in-Cosmos, any unit of life awakens to the fact of the REALITY of its ESSENTIAL SELFHOOD, it is as if the ENTIRE SELF 'AWAKENS.' and, in this *awakening,* every unit (of the SELF-as-Self) is 'saved'. (Remember that each *self* is, ESSENTIALLY the ENTIRE SELF, just as each part is wholly the WHOLE.) Stating a paradox, it might be said that, when *awakening* comes, it is as if the SELF-as-Self awakens to the fact that It-as-IT has *never* REALLY *been asleep.* The SELF-as-Self, upon *awakening,* realizes that It (because It is, ESSENTIALLY, the SELF) IS the STATE of SALVATION. For each Self or self in-Cosmos (and during the entire duration of a Cosmos), a 'STATE' of *full wakefulness* is proceeding uninterruptedly and simultaneously with the illusory states of apparent sleep. This means that **8** (standing for all Selves or selves) Am (while in-Cosmos) both *awake* and *asleep* at the same time; and, as previously stated, that **8** Am, therefore, both 'saved' and 'not saved' at the same time.

These thoughts are paradoxical, but are, **8** think, a fair approximation of the very strange truth concerning these matters. The practical implications of this speculation relate to the idea that, while in-Universe, to find the SELF within each Self or self *is Salvation*-in-Universe. It is impossible that any E/entity in-Cosmos should fail to rediscover the SELF it essentially IS before the final "Day Be With Us." This rediscovery may come sooner or later in Universal Time, but it is inevitable.

No E/entity can remain 'outside' of SELF (the BOUNDLESS IMMUTABLE PRINCIPLE) when the *time* for the Great Reabsorption comes. No E/entity can be left in the World of Illusion, for that entire World is dissolved, reabsorbed into Its SOURCE. By the 'time' that Universal Conditions are prepared for the onset of Universal Pralaya, the forms within Universe will have become so attenuated, and so close will be the 'distance'-in-Consciousness between all apparent selves and the One Self, and between the One Self and ONE SELF, that all in-Universe Consciousnesses will (as One) realize their identification with the SOURCE.

Conclusion 28
I AM the Veiling Power. I AM THAT which is (apparently) 'VEILED'. I AM also THAT which can never REALLY be 'VEILED'. I AM all three.

Most mysterious to the 'VEILED' PARABRAHMAN which, in Cosmos, **8** Am, is the 'ACTIONLESS-ACTION' of PARABRAHMAN (or more accurately, of PARABRAHMAN—as-'MAYA'). In fact, the only 'ACTION' of PARABRAHMAN (the ONE in WHOM no 'ACTION can REALLY 'OCCUR') is 'MAYAVIC-instantly-Mayavic. That 'ACTION' *(the* 'MAYAVIC ACT'—all acts in Cosmos and Super Cosmos are actually Mayavic) is the 'ACT' of Pre-Cosmic 'RADIATION' which inaugurates the Universe.

Sankaracarya has said that the 'ACTION' of 'MAYA'—instantly-Action of Maya is inscrutable and without beginning, but surely such 'ACTION' is, ESSENTIALLY, MY *OWN* 'ACTION' (and, succeedingly, My Own Action), for **8** and all **8**'s are **I**, or PARABRAHMAN. That 'ACTION' (MY 'ACTION') is the *apparent* SELF-'BLINDING' which is no *REAL* blinding; the apparent SELF-'VEILING' which is no REAL 'VEILING'. The Pre-Cosmic Mayavic Act of 'Creating' the World of Illusion by means of the Veiling Power does not negate the simultaneous *infinite ongoingness* of the ALL-IN-ALLNESS in ITS UNDIMINISHED NATURE which **I** AM. PARABRAHMAN's *ex-perience* of Illusion does not negate ITS simultaneous (and *infinitely continuous)* SELF-'ABSORPTION' 'within' the *fullness* of ITSELF. PARABRAHMAN, in 'BECOMING' a Universe seems to *contradict* ITSELF, but does not *negate* ITSELF.

In simple terms, even while Illusion is *actual,* and affects all within the Great Illusion which the Universe Is, yet, simultaneously along with this subjugation of all in-Cosmos to Illusion, the Non-Illusory 'STATE' is PRESENT, REAL and *infinitely abiding.* There is no way that the BOUNDLESS IMMUTABLE PRINCIPLE can ever REALLY be modified, or ever cease to BE the UNIVERSAL CONSTANT, ever cease to BE WHAT IT IS and HAS BEEN. In light of this seeming paradox, it is possible to understand how a Self, in-Universe can 'awaken' to the fact that It-as-IT has never REALLY been asleep.

Conclusion 29
There is THAT which always WAS, IS NOW and ever WILL BE. IT is ever the SAME. I AM THAT and I never change.

IT (the INFINITE SELF) never changes and there is naught else but IT. The astonishing thing to realize is that because **I** AM IT, **I,** too, never change. Even when **I** manifest as **I**-as-**8**-as-**I**, (**8** being the link between the INFINITE **I** and the finite 'I'), **8** must realize that, even in-Cosmos, **8**-being-**I**, never ESSENTIALLY change. When all *veils* are lifted, what shall be uncovered or discovered is the CHANGELESS ONE. Throughout all Cosmos, no matter through what phases my forms may pass, **8**-being-**I** will not change, and have not changed. All variation and *seeming change* are but the movement of veils, whether the *lifting* of veils or the *descending* of veils.

8-being-**I** have no movement in ME, for to have movement, **8**-being-**I** would have to be divisible instead of, ESSENTIALLY, the INDIVISIBLE INDIVIDUAL; **8**-being-**I** would have to be at least *twofold,* which **8**, since **8**-Am-**I**, *cannot* be, because of My absolute identification with the **I** which is the INDIVISIBLE SELF. Movement, after all, demands change of position, but in order for the state of *position* to exist, a minimum of two is required. There can be no position unless there are two points, each being the point of reference for the other. The points can then have 'position' relative to each other. Within ME, however, because **8** Am ESSENTIALLY the *pointless* **I**, there can be, in REALITY, no separate points. Even though to consider myself as a point (such as a Monadic Point) is useful within the World of Becoming, such a designation is nevertheless ESSENTIALLY illusory. So in fact, **8** (because MY true NATURE is **I**) am the <u>IMMOVABLE</u> ONE, just as the SELF is the IMMOVABLE ONE, having no parts.

Just as the SELF is ESSENTIALLY incapable of change, so **I**-the-SELF AM ESSENTIALLY incapable of change—despite all appearances to the contrary. What **8** (in-Cosmos) must achieve is the consciousness of MY ESSENTIAL *immovability,* MY ESSENTIAL *unchangeability.* Indeed since there is nowhere in-Cosmos **8** Am not, there is nowhere **8** cannot be. In some mysterious way, **8** must *already* have the consciousness of MY *immovability* and *unchangeability.*

To say that the human being does not change would not be accurate. The human being changes constantly as do all E/entities—in their *formal, objective* aspect, and, even, in their *consciousness* aspect. But the TRUE SELF does not and cannot change—forever. LIFE changes not. Change is an *appearance* only and is fundamentally, REALLY, un-REAL. If the human E/entity or any other Self-conscious E/entity is identified with REALITY, such an E/entity will be forced to realize that It cannot possibly change as regards Its inmost ESSENCE, Its TRUE SELFHOOD. This realization will have a profound effect upon the way life in the World of Illusion is lived.

Conclusion 30
All in the World of Relativity that 8 desire to achieve, 8 have already, in ESSENCE, achieved.

This may come as a startling thought. A little meditation will reveal, however, that, ESSENTIALLY, **8** Am THAT—the all-inclusive ENTIRETY, and therefore, **8** must necessarily *also* be all that which **8** (in my limited condition of 'I-ness') desire to achieve (for the *greater* necessarily includes the *lesser).*

The entire Cosmic Process might be seen as 'the hiding and the finding of the SELF'. From the humorous perspective, it is all a game of "Peek-a-Boo" (now **8** 'see' ME, now **8** don't); or a game of Cosmic "Hide and Go Seek". The SELF is in 'hiding' and IT must find ITSELF before IT is found by the player/seeker who is called, of all things, "It". 'It' is like Maya; if 'It' sees the hiding player and tags 'home' before the player reaches 'home' the player loses. If a player reaches 'home' before that player can be seen by 'It' and before 'It' can tag 'home', then the player is "Home Free" (uncaught by 'Maya' who prevents true freedom by {of all things} *pretending* to be 'IT' when it is only 'It'). From the perspective of the hiding players, they (the REAL 'IT') left the goal, hid, and must return to the goal to be 'free'—freely IT again. They IT must avoid being caught by 'It' or they cannot be free.

The entire dynamic in these games is one of "Lost and Found". They reflect the SELF that is apparently lost, and must be, somehow, found. In "Peek-a-Boo" (a game for very young children) it is interesting how the child herself holds her own hands over her eyes and removes them suddenly in order to see the adult who has always been directly in front of her. What better example could there be of the SELF-'VEILING' process which prevents us from seeing GOD. Sometimes the child plays this game in front of a mirror, and, even more *aptly,* hides her own image from herself. There are many squeals of delight with the inevitable revelation of her own image which comes from removing her hands from her eyes. She 'sees' herself, if she will only stop preventing herself from seeing herself through self-'blinding'. The analogy holds: children's games reflect *the* 'Game'.

Returning to the main point, no matter what **8** seem to be seeking or desiring, **8** Am REALLY seeking or desiring MYSELF (the SUPER-Cosmic SELF as IT manifests through the many intra-Cosmic Selves and selves). But, then, what **8** have been seeking and desiring has always been "right before my eyes", so to speak, or more accurately and mystically, "nearer than hands or feet." Since *being* something is greater than (and *includes) having* something, **8** already *have* what **8** want because **8** *Am* what **8** want!

The necessary and often missing factor is to realize that the many things **8** seem to want *are*, ESSENTIALLY, MYSELF. An important question arises, however. Do **8** *consciously have* my-SELF? This *having* is what building the Antahkarana is all about.

Conclusion 31
Nothing that you or 8 can do will make any ultimate difference.

This conclusion is not meant to induce in the thinker a sense of futility regarding his activities. This is not a call to the irresponsibility of inactivity, or worse, to passivity within the World of Becoming. Such attitudes would not be admissible under the Divine Plan, which, does, *indeed,* exist.

The conclusion simply derives from an attempt to understand life in our Cosmos from a *radical,* REALISTIC perspective. Certainly, no-thing we do will make any difference to the SELF, WHO is already MAXIMALLY INFINITE—the INFINITESSENCE. The INFINITE SELF is now, was and ever will be as IT ever MUST BE. Nothing can touch IT. Nothing will make the slightest difference to IT. IT cannot change or vary in the slightest. The Universes or Cosmoses may evolve *within* Themselves, each serving as an arena of unfoldment (objectification) for certain possibilities from 'within' the FOUNT OF ALL POSSIBILITY (the INFINITE POTENTIAL), but the SELF, ITSELF, which IS *already* (and, in fact, *forever)* the MAXIMAL PERFECTION cannot *possibly* evolve. The evolution of THAT which IS *already* PERFECT in every way is a nonsensical idea.

Now, 'you-as-**8**' and '**8**' can and must make tremendous *relative* differences in-Cosmos, which is a World of Relativity based upon change. We must do all we can to fulfill the Design-at-the-Beginning (the Purpose of the Universal Logos) which Purpose is *our* Purpose (not partially, but *fully).* Everything we do affects the manner in which that Purpose is achieved, the aesthetics, so to speak, of its achievement. The many S/selves (in one or another state of ignorance) can make a profound *difference* in the World Drama. All this, however, will make no difference at all to the ONE UNCHANGING SELF, the ABSOLUTE which is the INFINITESSENCE of all possibilities.

This is to say that our efforts must be confined to the intra-Cosmic Context, to the 'Play' being Acted and the 'Game' being Played. *There* we *can* make a difference. 'Within' the largest possible 'ARENA' of the INFINITE SELF, nothing ever changes.

Conclusion 32
8 cannot fail to BE MYSELF.

To BE MYSELF is the one way **8** *cannot* fail. What then is failure? It is only deviation from My Own Intent (considering MySelf as identified with the Will-at-the-Beginning, the Will of the Universal Logos, which stays with 'little descending 'me' throughout the duration of Cosmos). On the highest level of Cosmos, **8** (the **I** appearing in-Cosmos) Am ever intending and ever obeying My Own Intent. It is only the Emanations of My Universal Logoic Self which can *fail* (through Maya) to conform to Original Intent.

Any Emanation, passing through many levels of *selfhood* is extremely fallible. But even as '**8**-as-Emanation' Am failing 'below', **8** Am succeeding 'above' where **8** Exist in the fullness of my Universal Logoic Identification. In that higher State (the highest possible in-Cosmos) **8** cannot fail to Be MySelf and to Live according to My Intent (which Intent is REALLY, MY INTENT, note capitalization). On that level, **8** realize to the fullest extent possible in-Cosmos, that **8** Am **I**, and that My Intent, My Will for this Cosmos, is simply a reflection of one set of possibilities of an infinitude of sets of possibilities within GREATER **I** (the FOUNT OF ALL POSSIBILITY).

Thus, while in Time and Space, which is the World of Becoming (the Cosmos), **8** *apparently can* fail, REALLY **8** (the Reflection of **I**) can *never* be anything less than PERFECT, the PERFECTION of the INFINITE SELF. The apparent failures below are only failures-in-Cosmic-Context. They are, Essentially, failures to 'bring through' into expression the Self **8** Am. All the while, while **8** Am *apparently* 'failing' below, **I** AM ever perfectly, fully and completely MY PERFECT SELF above. Though **8**-as-'I' suffer, **I** suffer not, and though **8**-as-'I' struggle, **I** struggle not, and though **8**-as-'I' fail, **I** fail not. Though any condition whatsoever obtain 'below', that condition does *not* obtain 'above' within the CONDITIONLESS BEING. Thus, through all apparent vicissitudes through which **8** pass as a Self, **8**, because **8** Am **I**, cannot fail to BE MYSELF.

No matter what happens 'below', it is impossible for 'me' to cease being the ONE INFINITE SELF. This is an example of the constant, apparent, and obstinately illusory *dualism* obtaining 'within' the non-dualistic SELF (than WHICH there is no OTHER), for indeed, if the Universe can be said to be located *anywhere*, It is *'within'* the SELF.

Conclusion 33
Action is limitation.

Within SELF, nothing at all is necessary; nothing at all is needed. There is no need, because a need suggests an insufficiency, and 'within' the SELF, the PLENUM, there is but ALL-SUFFICIENCY. Only that is done which is willed. The purpose of action is fulfillment. Action is undertaken to fulfill a purpose or a desire. But 'within' the SELF, fulfillment already exists *maximally*, because the INFINITE SELF IS *fulfillment itself.*

Any change in THAT which is *already* completely fulfilled, completely PERFECT, is nothing but a 'diminishment'. Action is change, therefore, within the context of the INFI-

NITE SELF, *action is limitation,* a diminishment of the PERFECTION that EVER IS. Of course the INFINITE SELF cannot in any way REALLY *be* diminished, and so, as one might expect, there is REALLY no such thing as *action* (not only 'within' the INFINITE SELF, but *at all).* Of course, from the 'INFINISPECTIVE', everything is 'within' the INFINITE SELF.

Conclusion 34
8 do not move. 8 cannot move.

Within SELF (where else is there?) 'ACTION' (were it REALLY possible) would be a disturbance, a perturbation, a modification. 'ACTION'/Action/action, however, is REALLY illusory. Nothing REALLY moves, though it *seems* to. Nothing REALLY changes position, though it *seems* to, for 'within' THAT, there is no position and All is 'within' THAT. Movement is based fundamentally upon the illusion of division, but division cannot possibly exist in the ONE AND ONLY INDIVISIBLE SUBSTANCE, therefore movement, REALLY, does not exist.

Limited consciousness perceives what it calls interval, but interval does not REALLY exist. Limited consciousness perceives movement 'across' a seeming interval, but if interval does not REALLY exist, therefore, movement does not REALLY exist We never move from the 'place' we are, because we are never *in* any *place.*

REALLY, We/**8** Am *nowhere,* for no point of reference can REALLY exist, except as an illusion in the World of Illusion. **8**/We are REALLY immovable like the GREAT IMMOVABILITY. We, ESSENTIALLY, ARE. All *seeming* movement is *actual* but un-REAL. Movement is ESSENTIALLY un-REAL. ALL of the UTTER ALLNESS is, REALLY, *nowhere.* From the perspective of REALITY, it could be said that for any E/entity in Cosmos, there is REALLY *nowhere* to 'go', and yet we must function, illusorily, 'moving' within the Illusory Play.

We-the-**I**-as-**8** have 'created' an un-REAL world in which the most fundamental of LAWS of our INFINITE NATURE are constantly and necessarily contradicted. We are, in fact, functioning within a World which could rightfully be called the Great Contradiction (except that our REAL WORLD is, even more, the GREAT CONTRADICTION). Through the agency of ITS Illusory Cosmos, the GREAT SELF 'INDULGES' in a Play of Opposites which apparently contradicts the ABSOLUTE MONALITY IT IS.

If there is any such thing as *relief* from ABSOLUTE PERFECTION, *action* (and all its derivatives) could well be considered that relief. In order for the SELF to 'ACT' (as mysterious and rationally impossible as such an 'ACTION' must seem to us), the INFINITE SELF must *necessarily* 'LIMIT' ITSELF. Both this 'ACT' and this SELF-'LIMITATION' (REALLY the *same thing),* however, serve a most *necessary* Purpose, without which the SELF would 'FAIL' to BE ITSELF—strange thought.

Conclusion 35
It is impossible for ME to be absent.

8, ESSENTIALLY, AM the ALL-PERVADING PRESENCE, and it is impossible not to BE MYSELF. (Even when **I** appear *not* to be MYSELF, because **I** AM 'MANIFESTING' a Universe, **I** AM still MYSELF, paradoxically. It is impossible not to BE that PRESENCE, and the PRESENCE is *always* PRESENT. Therefore **I**-as-**8** cannot "miss a thing." **I**-as-**8** cannot REALLY be deprived of *any* experience. Even while **8** Am, *apparently*, being deprived of experiences because of My manifestation through the limited 'I' which **8** use, even *then*, **8** (as the One Universal Life) Am also the 'Witness' of and 'Participant' in all other extant experiences. In short, *wherever* there *is—there* **8** Am. **8** Am never "out of the picture" or "missing from the scene." **8** cannot be eradicated, eliminated or removed. **I** (SUPER-Cosmically) and **8** (intra-Cosmically) 'accompany' MYSELF/MySelf forever. In-Cosmos, no-*thing* is ever "out of My 'sight'." SUPER-Cosmically **I** AM PRESENT as NOTHING.

Conclusion 36
If Maya Exists, I AM Maya.

There is no other BEING in the UTTER ALLNESS to *Be* Maya other that **I**, MYSELF. If *any* power or shakti exists, **I** AM fully that power as well. MY major Power (manifesting in Super Cosmos and Cosmos) is the Power of Identification. **I**-as-'MAYA'-instantly-Maya, AM 'CONSCIOUSNESS'-instantly-Consciousness.

'MAYA'-instantly-Maya is simply one of MY Super Cosmic and intra-Cosmic Modes of Activity and **I** function fully within It. Subjectivity, Objectivity, and Consciousness are also three of MY Super-Cosmic and intra-Cosmic Modes, and Super-Cosmically and intra-Cosmically **I**-as-**8** function fully within them.

Conclusion 37
If anything happens, 8 Do it.

Again, this seems a radical statement, contradicting our common sense. Anyone can 'obviously see' that there are many actors doing many things, and that 'I', apparently, only do what 'I' do. Yet, if there is but ONE BEING in the UTTER ALLNESS, it stands to reason that (all appearances to the contrary) that BEING must be the ONE 'ACTOR' or ONE 'MOVER' in all happenings. It must be clear as well that, though to the limited consciousness there appear to be many motions, to the Universal Witness there is but One Motion. So, **8** play the many parts which are but One Part, and **8** make the many moves which are but One Move, and Am responsible for the many actions which are but One Action. **8**, the Omnipresent Pervader Am the Ubiquitous Participator.

Conclusion 38
All that happens is MY-as-My responsibility.

How far does My responsibility extend? This is an intra-Cosmic consideration. If **8** consider MySelf a personal or even a spiritual Individuality (such as a soul or Monad), My responsibility is certainly contained within the limited sphere of my personal and individual actions. But if I-Am-**8**-AM-**I** (the INFINITE SELF) then **I**-as-**8** Am *super-multiply responsible* at all 'points' in Cosmic Time and Space. My *essence* pervades All. My *actuality* obviously does not.

Yet, in this question, it is not my *actuality* that counts. Self-conscious B/beings are responsible for their actions, but *how many* Self-conscious B/beings are there, REALLY? If there is only One Self-Conscious Being (Who is an in-Cosmos Reflection of the ONE ABSOLUTE BEING/NON-BEING) then that One Self-Conscious Being is responsible for All that transpires within the Cosmos It *Is*. If **8**, through *identification with essence*, Am that One Self-Conscious Being, because It and **8** are both THAT (the INFINITE SELF), then **8** Am unquestionably responsible *(ultimately* and *essentially* responsible) for all that transpires in Cosmos.

In illusory Time and Space and during the World of Becoming, My *actual* responsibility is nowhere near as extensive, *unless* we are approaching the Universal "Day Be With Us", and **8** come to realize that **8** have, all along, *been* the One Directing Being in Cosmos. Then my *actual* responsibility will begin to approximate my *essential* and *ultimate* responsibility.

Conclusion 39
All points are actually One Point.

First it will be necessary to assume that there are such things as 'points', which REALLY there are not. In-Cosmos, however, the concept is useful, so even though it is fundamentally illusory, we will, for a moment, assume that there *are* 'points', for God as a Presence can be called a Real 'point' and every B/being ('Seen' by Its Emanator) can be called a Virtual 'point'.

The idea that there are many points in Space correlates with the concept of extension. Extension, however, is REALLY illusory. If we are to assume that 'points' exist, then, the truth is that the multiplicity of points are but *One* Point, for if there is no REAL extension, how can a multiplicity of points be extended in Space? 'Points', in a way, though they *exist*, have no extension, for they are dimensionless. Because all points are REALLY One Point (though *actually* {in consciousness} many points), energy passes through all points immediately and simultaneously, because energy is REALLY passing through only One Point. From the One Point (which is all points) all items-in-Cosmos are 'touched' simultaneously.

The One Point is really an instrument of Universal Pervasion. Through the agency of the One Point, it becomes clear, from one angle of thought, that 'here is there' and 'there is here'. The illusion of interval is overcome, and any place is realized to be *all other places.* The center of all lives is fixated upon the One Point which is both nowhere and everywhere in Cosmos. All points in Space, therefore, are not only infinitely close, they are *identical.*

If we define 'there' as where you are not, and 'here' as wherever you *are* (which is, in this context, *center),* then, there is no *there.* There is only *here.* The Great Illusion is based upon the idea that there is more than one point. The Great Illusion is based upon the idea that there is Space at all (and Time, as well, for it takes 'time' to travel through 'space' 'across' an illusory interval).

To realize that there is only One Point, and to be able to function through it, and thus achieve Cosmic Ubiquity, is to go far towards overcoming the Great illusion, and with it, the illusions of both Space and Time. Let us be reminded, again, however, that, REALLY, there are *no points*—not even One Point, REALLY.

Conclusion 40
If the truth does not seem to contradict itself, it is not the TRUTH.

The presence of paradox indicates the likelihood (though not the certainty) that truth may also be present. Truth (strangely) demands apparent inconsistency.

This apparent inconsistency, contradiction and paradox is caused by the profound (shall **8** say 'infinite') gulf which exists between the INFINITE and the Finite. Because of this, All B/beings within Cosmos are forced by 'NECESSITY'-as-Necessity to live paradoxical lives.

Conclusion 41
8 am active and Present in all dimensions along the entire circle of Cosmic Involution and Evolution.

Many apparently separate lives, E/entities, are found upon all "rungs of the ladder" of Cosmic Development. **8** (intra-Cosmic **I**) however, Am necessarily all-extensive and, hence, ubiquitous in Cosmos. There is no 'point' in which **8** Am not *fully* found, and no 'life' (no matter how lowly or how exalted) which **8** Am, *essentially, not.*

Though **8** *seem* to be an evolving B/being of a particular status, still, somehow, on all other levels, **8** Am, simultaneously, both involuting and evolving. Though **8** *seem* to be an involuting being of a particular status, still, somehow, on all other levels, **8** Am, simultaneously, both evolving and involuting.

Conclusion 42
Cosmos is the 'Game' 8 Play.

It takes tremendous detachment and identification with REALITY to see the Whole of Universal Life as a 'Game' or a 'Play'. To do so is not to devalue Life in-Universe in the slightest. Instead, a new freedom in the way Life is lived is engendered, and a beneficial kind of objectivity achieved. Also, a sense of truer values replaces the old values based upon attachment to the un-REAL (i.e., to the *supposed* Not-SELF).

How seriously must one be about *playing* the Game, or *acting* in the Play? Serious, enough, **8** would suppose, to play beautifully, skillfully and with great enjoyment. There are perhaps many who approach living somewhat as a game, and this attitude sometimes masks a manipulative attitude and hardness of heart.

This is not at all what is meant. The Game is best played with a wide-open, wise heart to accompany the alert intelligence and the will to win. Yes, *winning* is important. And what does it mean to *win* the Game of Life, the Game which Cosmos *Is?* It means to fulfill the Intention-at-the-Beginning in the *best possible way.* It means to express one's subjective energy pattern (for each E/entity in Cosmos Is a distinct energy pattern) as completely as possible, with the object of the fullest possible contribution to the Whole thus facilitating the fulfillment of the energy patterns of 'others'.

Shall we be serious? Shall we be tragic, or comic? Ever we must be aware of the strange reversal which occurs because of the incomensurability of the INFINITE and the Finite, for, in-Cosmos, that which seems to be REAL, REALLY is not, and THAT which, at first, seems un-REAL, is truly REAL.

One can see the need for a sense of humor and a high degree of detachment. Perhaps it is comforting to know that nothing of value can ever be lost, because the SELF is the only 'THING' of value, and the SELF is not only the Universal Constant (the Self, Really), but, more—the SUPER-UNIVERSAL CONSTANT. Whether one plays well or badly (and all, **8** think, at length, learn to play well), no skill or ineptitude in the Play can ever damage or even affect the SELF in the slightest.

Conclusion 43
For all that transpires, 8 have no one to blame but MySelf (i.e., MY SELF-as-MySelf). 8 have no one to thank but MySelf.

8 do and did all that is done. In *your* doing, **8** do. In *your* thinking, **8** think. In *your* loving, **8** love. In *your* hating, **8** (subject, in-Cosmos, to Illusion) hate. **8** Am ever present. You cannot be rid of Me (ME-as-Me), nor **8** of you. **8** Am My Own reward and retribution. **8**, in various guises, Am bestowed upon MySelf (i.e., MYSELF in-Cosmos). Overall, shall **8** not be thankful that this is so?

Conclusion 44
No terms of measurement can apply to ME-as-Me at all.

Terms of measurement can apply only to Maya, and, though **8** *Am* Maya and, even, the Works of Maya, **8** Am not ESSENTIALLY so. It is futile to apply terms of measurement, terms of description, or predication to Me (WHO IS REALLY—THAT) as it is to do so to THAT, ITSELF.

REALLY, **I** (SUPER-Cosmically), **I/8** (Super-Cosmically) and **8** (intra-Cosmically) have *no name*, just as the INFINITE SELF has no name. My evanescent role and intra-Cosmic sheaths may have a *name*, for they are of *form*, but even *within* Cosmos, **8** have no REAL name, and *can* have no name. **8**-as-Cosmic Self may have a Name (which can be compared to the Names of an infinitude of other Cosmic Selves past). **8**-as-Cosmic Self may have a Magnitude (which can be compared to the Magnitudes of an infinitude of other Cosmic Selves past). **8**-as-THAT, however, have no *name*, and no *magnitude* whatsoever, other than to say that **I** (in MY INFINITESSENCE) AM the 'infinitization' of all possible magnitudes. Other than that, **I** and even **8** cannot be measured.

Conclusion 45
My progress in Cosmos is the Progress of Cosmos Itself.

The form with which **8** Am usually (and REALLY wrongly) most identified may progress or not progress. At length, it will progress. Is the progress of that form, however, and the consciousness which manifests specifically through it, Really an indication of *My* Progress in Cosmos?

We must remember here that **I**-as-**8** REALLY cannot progress. The SELF cannot change for better or for worse and **I**-as-**8** Am fully the SELF. Development (i.e., progress) through prakriti does, however, occur in-Cosmos, and that development is important from a Cosmic Perspective, though *ultimately* meaningless to the 'STATELESS STATE' of INFINITE SELFHOOD.

So how shall we measure My progress in Cosmos? **8** would say, "Only by measuring the progress of the Universal Logos as It unfolds Its Intention through the One Universal Form." From an egoistic perspective, **8**-as-'I' have a very limited ring-pass-not, but through the disciplines of abstraction, **8** find myself to be 'MySelf' (i.e., a Being in Cosmos who is as Cosmically all-pervading as the One Cosmic Being.)

Thus, if **8** ask myself/MySelf, How am **8** doing? the Cosmos would have to be scanned to give the truest possible reply. As this is impossible at the moment (given the degree of Mayavic Veiling which still cloaks our Earth's human race), my answer should at least take in our planet and the later the Solar System. Essentially, extended-**8** manifests through all such Systems (even Systems of greater magnitude), and though, on this level of consciousness, knowledge of such manifestation is lacking (though it exists for ME-as-Me even Now on higher levels), it is well for me and for all other awakening human beings to think in such extensive terms—to do so is a great feat of decentralization. It can be clearly seen that within this World View, the decentralized view is the *truer* view. Thus, Really, the Universal Process is *My* Process. The Universal Progress is My Progress. **8** only progress as the Universe progresses; nevertheless, all the progress in the world through an infinity of Cosmoses cannot make ME any better or any different than **I** *already* AM.

Conclusion 46
There is work to be done, but there is no work to be done upon MYSELF.

Spirituality in the modern day is filled with exhortations for work upon the S/self. This is a valuable trend, but certainly the S/self is not, formally, the SELF. The SELF, which each of us truly IS, *cannot* be worked upon. The work that can be done is designed to improve the means of *revealing* and *expressing* WHAT **I**-as-**8** already Am. Even in-Cosmos, **8** cannot be better, Essentially, than **8** Am at this very moment—even if outwardly **8**-as-'I' seem to be failing or making a mess of things.

Really, the work to be done is *aesthetic* work meant to assist the fulfillment of the Divine Design in maximal beauty. The work is, Cosmically considered, very important and indispensable in fulfilling MY Pre-Cosmic Contract with MYSELF, but even that great Universal Work is still 'Play'. Work cannot change WHO **I**-as-**8** Am; it can only change how *well* **8** *perform*. Work can change how well **8** *know* Who **8** Am. Work will affect the degree of my apparent limitation in-Cosmos because, it applies to the realm of name and form through which **8** must, perforce, express.

Really, work releases Me from the limitations **I** do not REALLY have. Work reveals to Me (in-Cosmos) that **8** have no Real intra-Cosmic limitations, just as **I** have no SUPER-Cosmic limitations. Work restores me to MySelf, and Me to MYSELF.

Conclusion 47
'Yes' is Manvantara and 'No' is Pralaya, perhaps.

From a certain perspective, 'Yes' is Manvantara and 'No' is Pralaya, for one can see in Manvantara a great affirmation of the infinite fecundity of the FOUNT OF ALL POSSIBILITY, and in Pralaya, its denial. From another perspective, however, 'Yes' is also the *absolute affirmation* of the INFINITY OF ALL INFINITIES, which can demonstrate *fully* only 'during' Universal Pralaya. In such a case, Manvantara would be a privation of this LIMITLESS INFINITUDE. Manvantara would then give the 'No' to the PLENUM, for it is impossible that the PLENUM manifest during any one Universal Manvantara, or even in all the infinitude of Universal Manvantaras.

Conclusion 48
I have not moved and I cannot. There has been no movement in ME. There is no 'place' to which or from which I can move. In all of MY LIFE and BEING, I have not moved, nor will I, nor can I.

The true **I** is immaterial and all-pervasive. IT has no particularity or relativity within IT. There is no possibilty of change within IT. This being so, any 'MOVEMENT' in IT is impossible. **I**, ESSENTIALLY, am *not* a relative being, **I** have been, am and will be the ABSOLUTE IMMOBILITY forever. Illusorily, however, there have been naught but perpetual movement forever (to which the Great Breath {Essentially, *MY* Movement) testifies.

Section IV
Counsel on Living the Life of Radical Infinitism —Radical Non-Dualism

We come now to the section of the *Infinitization of Selfhood* where we discuss how *actually* to live the life of Radical Non-Dualism, Radical Infinitism. Living the life has to do with practical action, the maintenance of a certain state of consciousness and the sustainment of a certain *tension of the Spirit* (equivalent to Identification) that will allow access to an ever-deepening sense of Selfhood, leading to intimations of the Universal Self and, perhaps, of the INFINITE SELF, ITSELF.

What is necessary is to *rouse* the human self in such a way, and to live at such a high point of revelatory tension, that one is able, through the disciplines of realization, to consistently (as Sankaracarya says) *merge* the Universe into BRAHMAN, the INFINITE SELF, or, alternatively, to *dissolve* the Universe into BRAHMAN. What is needed is a *constancy* of SELF-Realization through identification *as* the UNIVERSAL CONSTANT, which will reveal the substantial nature of all things within the GREAT HOMOGENEITY.

The use of language in the first part of this book has been rigorous at times and convolutedly inventive. A looser form of discourse will now be adopted for this section, as we are dealing with human living as it is found within the three worlds of human evolution (the physical, emotional, and mental realms). It is presumed that the reader will be able to interpret what is said with insight, having by now become familiar with the many distinctions and discriminations that the more rigorous terminology highlighted.

Counsel will sometimes be offered in a *directive* tone as injunctions for *propelling* right thought and right action When addressing the reader I will also use the word 'you' (even though, by now we all deeply understand that the term 'you' ultimately must be understood as the-I-We-All-ARE). Having read this far, **8** think 'you' will agree that, try as **8** might, **8** can do no other or better than talk to MY<u>SELF</u>! (If 'you' will listen, **8**, *necessarily,* will!)

Counsel 1
Recognize the INFINITE WHOLENESS in the apparent 'part'.

Never be deluded that the part is *just* the part. Realize that what is presented is only a *seeming,* and that an understanding of its true import, its true nature/NATURE must be worked for diligently if it to be detected beneath the surface of things. Value supremely what is presented to consciousness, not because of its form, but because of its REALITY. Realize that though there is great variety in the World of Becoming, that variety ever points to the SAME THING, the ONE AND ONLY SELF. Make sure you *see* that DIVINE SAMENESS in all the variety presented.

Do not scorn any kind of presentation, because each presentation is reflective of one of the infinite number of potentials within the INFINITE POTENTIAL, the FOUNT OF ALL POSSIBILITY. So reject no presentation, but attempt to see it, instead, "under the species of eternity" (to use the words of Spinoza). Do not simply *imagine* that it is so, but *see* it as so, and realize that what you are seeing is not just an *aspect* of INFINITY, but INFINITY ITSELF (amazingly, in ITS ENTIRETY!).

What this will mean for you is the ability to be delighted by the World. You will find yourself rejoicing in Cosmos ('infinitesimalizing' though It Be). Unhappiness is due to the apparent absence or inaccessibility of the desirable. Remember that the desirable is *never* absent but is always present in the PRESENCE. As a matter of fact, the MOST DESIRABLE, the PRESENCE OF THE ONE AND ONLY can *never* be escaped. IT will always be present and it will always be presented, though in Mayavic disguise. Every presentation you encounter is that of the MOST DESIRABLE ONE.

Counsel 2
Negate the lure of form.

Negate the lure of form by realizing the true nature of form. The lure of form is that which makes the unenlightened person forget INFINITY, forget the SELF. It is possible, however, to deal intelligently with the form (as we must) without becoming enthralled by it. Forms themselves may be attractive or repulsive because of the harmonious or inharmonious interactions of patterns in Cosmos. Forms may be, relatively, 'Good' or 'Bad' when compared with the Design-at-the-Beginning, the Original Intent. Regardless, however, of what forms may seem to be in a relative sense, they are ever REALLY the INFINITE in Mayavic disguise. Realize that it is possible to be so repelled by a form, that one forgets its true INFINITE NATURE, its *identity* with the SUBSTRATUM. It is also possible to be so allured by a form that one becomes engrossed in it, engrossed in its superficiality, thus forgetting its *oneness* with the INFINITE SELF.

The great danger concerning the lure of form is that the duality of INFINITE/Finite will not be maintained comprehendingly, but rather that the consciousness will become the captive of the Finite. Then, all sense of the GROUND OF ALL BEING will disappear from the consciousness, and with that disappearance, the disappearance, also, of the opportunity to 'compare' the Finite with the INFINITE SUBSTRATUM. Realize this danger, and do not let the opportunity for true objectivity and detachment be lost through forgetfulness of the INFINITE SELF due to the engrossing allurement of form.

It is the lure of form that *veils* the PRESENCE of the WHOLENESS. The lure of the form is the lure of the 'partial', and this allurement reinforces the illusory *seeming* that only that which is 'partial' is attractive, worthy and valuable. The lure of the form makes the part *seem* to be the WHOLE, but it does this in the *wrong way* (for there is a legitimate way in which the part is *known* to be the INFINITE WHOLENESS, *substantially)*. The error occurs because, enamored of the part, the perceiver forgets the WHOLE OF WHOLES, and mistakes the *form* of the little part for the forgotten WHOLE/Whole.

In order to negate the lure of form, remember the SELF. Remember Maya. Remember that nothing *is* what it *seems,* but always *more,* and not only *more,* but the WHOLENESS ITSELF. In relation to any form, then, never forget the WHOLENESS. Rob the form of its particularity, and see, instead, only WHOLENESS. Be not attracted by the particularity of the form but only by its *absolute identity* with the ONE WHO LIVES. In relation to any form, willfully *prefer* the PRESENCE over what you *seem* to see.

Counsel 3
'Re-Member' the Self.

Fuse all the many, many pieces together in *selfhood*—the Universal Selfhood of Cosmos, and, even more importantly, the INFINITE SELFHOOD of the ONE WITHOUT A SECOND. Let there be no liking or disliking of a thing simply because of what a thing *seems* to be, but only in relation to the Greater Whole, the Greater Cosmic Pattern. Remember that what we call 'things' are really *relations.* Understand relations in order to understand value of things'. Constant are the fluctuations that so often negate the capacity to *'re-member'* the SELF. Negate the effects of fluctuation by holding to the PRESENCE—the GREAT SUBTLETY, which no fluctuation can touch.

Counsel 4
Remain undisturbed through deep contemplation of the SELF.

Practice continuity of SELFHOOD. The SELF is ever present but largely forgotten. Think of the SELF continuously *even* while apparently thinking of 'other' things. Almost every condition within Cosmos seems to negate the possibility of remembering the SELF continuously. There are constant disturbances within the 'normal' state of consciousness, and the ABSOLUTE STILLNESS is forgotten, not registered. Attention is drawn towards the particular and then enslaved. Generalize and abstract the attention constantly so that it cannot be drawn by and absorbed into the particular. Remain in the appreciative, contemplative mode—appreciating and contemplating the ONE AND ONLY SELF.

Learn to hold the PRESENCE regardless of disruptions, regardless of variety or modification. Modification itself will never cease within Cosmos. Such is the WILL of THAT, the ABSOLUTE. Cosmos Is. Modification.

Shall seeming disturbance force us to forget the ALL-IN-ALLNESS, the INFINITENESS, the BE-NESS, the ABSOLUTE SELF during the Manvantaric Term of Cosmos? No. Learn, despite multiplicity to hold the simplicity of the SIMPLEST ONE. Hold steady the STATE OF SELFHOOD. Hold the PRESENCE regardless of the disruptiveness of presentations, the intensity of presentations, the multiplicity of presentations, the changeableness and modification of presentations. BE so deeply rooted in the contemplation of the INFINITE that no presentation can disturb your continuity of DIVINE SELFHOOD. In fact, welcome all presentations as examples of the infinite Creativity of the SELF-as-Self. Welcome the apparent disruptions and transform them into confirmations of the Oneness and the ZERONESS. Address yourself to this great task.

Counsel 5
Live the *egoless* STATE
but respect natural limits.

Limits are the veils of Maya. They are the Veils YOU, YOURSELF have 'DROPPED' over YOUR OMNIPOTENCE, OMNISCIENCE and OMNIPOTENCE. Respect what YOU have 'DONE'. YOU did it for a 'PURPOSE'. Respect the limits YOU YOURSELF have 'Created', but remember WHO YOU ARE.

Counsel 6
Realize that
blessings and antagonisms
come only from YOURSELF.

There are no *others*, so how can blessings or antagonisms come from any but your-self-as-'Your<u>Self</u>'-as-YOURSELF. Resist the instinct to treat every separate presentation as if it came from *another*, from some<u>one</u> else, from some<u>thing</u> else. Realize this truth and share it. To every apparent 'other' you meet, say in your "Heart of Hearts", "You are not *another*. You are not *another*." Deeply realize the truth of "Naught is but ME." In all you do, the ONE ACTOR 'ACTS'. YOU ARE that 'ACTOR', the AUTHOR of all impacts in Cosmos.

The impact you receive, you generated, be it for good or ill. Wherein, then, lies the idea of offense? WHO offends and WHO is offended? Wherein lies the idea of special favor? WHO grants special favor and to WHOM? Can you be truly angry with YOUR-SELF? WHO is the SELF with which you are truly angry? Has that SELF ever REALLY 'DONE' anything? And yet, has not that SELF 'DONE' *all* things?

Deplore a form of behavior, if you must, for the sake of the Divinely Intended Good-ness of the Design-at-the-Beginning, but do not devalue the SUPREMELY VALUABLE. (i.e. the ONLY SELF) as IT may appear in the one who behaves wrongly. Can there be a just reason for upset or for the loss of equanimity when all things that are delivered to you come through YOU, YOURSELF? Realize, of all who act for good or ill, that there is no one 'un-kin-d', for all who act are 'kin', *identical brothers* in fact. The 'ACTOR' but 'ACTS' upon HIMSELF. The one who gives immediately receives. And the one who re-ceives is giving as well in his receipt. WHO is the giver? WHO the receiver? WHO!

Counsel 7
See SAMENESS without obliterating the *seeming* of variety.

You are enjoined to "blot out all form". This does not mean that you ceases to be effective in the handling of form, You must become very effective, for this is the only way to fulfill YOUR-as-Your' Design-at-the-Beginning, 'YOUR-as-Your' Original Intent, the Design and Intent of the ONE-as-One. Simply, then, see to *root* of form, and realize that the *root* is *formless* and identical with the Great HOMOGENEITY.

Form has no independent REALITY. Pottery cannot exist without clay. Clothes cannot exist without fibres. Hold always the two—the ORIGIN and the Result or the Creation. Never, when contemplating the Creation, forget the ORIGIN which, not only is the SOURCE of the Creation, but IS the Creation Itself.

Honor the variety. Rejoice in the multiplicity. See the "ten thousand things" and realize they are but evidences of the ONE ABSOLUTE and ULTIMATE POWER. Do not number yourself among those who become dismayed by multiplicity. This would be a great mistake. Nevertheless, fuse the multiplicity into REALITY.

Counsel 8
Serve the original intent.

The goodness and rightness of your actions will depend ultimately upon your ability to understand the Design-at-the-Beginning (which no localized human consciousness Really can!). You can, however, to a degree, understand the *reflection* of that Original Intent within your Being, and, according to your understanding, you must live. This understanding will put you into relation with all apparently 'other' E/entities within the Great Pattern of Cosmos. To some of these you will be immediately related. To some you will be only remotely related—*apparently*. The Original Intent, the Universal Purpose, guides all and sets the Standard for all activity in Cosmos.

Fathom the Pattern as it lies at the Core of your Being-in-Cosmos. Your CORE of all Cores is the ONE AND ONLY SELF. This INFINITE SELF is identical in all, but your individual, Original Intent is unique to you—for a *time*, until the 'Ray' of the ABSOLUTE you Are is absorbed into a higher 'Ray', and the Original Intent of that 'Ray' becomes your Original Intent. There is, Really, but One Original Intent manifested variously through the apparently diverse 'Rays' which E/entities in Cosmos mistakenly think they fundamentally *are*. Each stage of the One Process at every 'point' in Space has its Aspect of the Original Intent. Find yours, but know that all the 'other' Aspects are Really yours as well. Your multiple Cosmos-wide Intents will each last for a 'time'.

Discover the Intent most immediate to the 'you' you seem to be. Fathom it. Live it! Then discover the others which you are living, unbeknownst to you, as well. Move ever towards discovering the One Cosmic Intent which *is all* seemingly lesser Intents. This Intent is your Real Intent and It is called the Divine Purpose. It is *more* your Real Intent, than the little individual Intent you are so intent upon living. Learn the meaning of the Greater Will which is Really Your Will.

Counsel 9
Learn the intricacies of form while realizing their fundamental impermanence.

As you focus upon the World of Illusion (the World of Form), set your mind and heart upon cognizing intelligently every pattern that World presents to you. This readiness to learn endlessly is necessary if, at length, YOU-as-You-as-you are to develop Universal Love, Universal Power, Universal Intelligence. You, in your present state of localization are apparently deprived of these three mighty Qualities. Once, not only were you possessed of Them, but you *knew* it. (Indeed, if the Truth be known, you are *even now,* possessed of Them, and *even now,* {high on the Divine Emanatory Stream} you *know* it!)

The Universal Life Wave in which you began your Cosmic Pilgrimage started from a condition of Cosmic Omnipotence, Comic Omniscience and Cosmic Omnipresence. The deprivation of Self-Veiling, however, grew as you (the One) descended, and became virtually complete and impenetrable as you reached the bottom of the Universal Evolutionary Arc (if, in fact, we have reached it—Cosmically!). Restoration of your prerogatives is now on its way as you ascend, mounting ever closer to the Origin of the Arc—but the 'Time of Return' is still aeons away. You possessed. You lost. Now you are regaining. And yet, upon the highest Cosmic Planes you dwell as you have for the Cosmic Aeon, the 'Witness' of your apparent but Essentially illusory descent and re-ascent. How strange is the multi-dimensionality of Consciousness in Cosmos!

You are *becoming* the SELF-Realized Being you already *are.* A SELF-Realized being is not allowed to be ineffective with respect to a Cosmos, but must, in the march towards Aeonial Consummation, become All-knowing, All-loving, All-powerful. Having adventured and labored through all dimensions of Cosmos, you and I and All, will not allow ourselves to escape into UTTER SELFHOOD with anything less than Complete Cosmic Mastery. Thus do we fulfill our Cosmic Contract with ourselves-the-SELF.

So, although, as the SELF we have *ever been,* we may *already* know everything ABSOLUTELY, now, because of the SELF-'ENGENDERED' Veiling Process of Maya, we must come to know everything *in particular.* World negation, world denial is most definitely not allowed, not intended and is a great error. We-as-WE-the-I will not allow it, have not intended it, and, ultimately, will refuse to delude ourselves with the dualistic idea of rejecting a part of ourselves. Therefore, learn ceaselessly!

Counsel 10
Infinitize all *seeming* units of LIFE.

To fulfill this injunction is to confer the greatest gift upon all those you encounter. Note the use of the word 'seeming'. The units of life you encounter are *already* the INFINITE in ESSENCE, but as you behold them, you, the observing consciousness, must conduct an interior transformative process in order to render them *infinite* in your eyes, and eventually in their own. This process is, Really, a refusal to be captivated by apparent difference. It is a reasserting of a mantram which holds those you behold within the

WORLD OF BEING. The Mantram is— "THOU can'st no more go out." Through this realization you refuse to allow any E/entity you behold to 'fall out of' the INFINITE SELF. You hold them, as it were, within the INFINITESSENCE. You, in fact, *absolutize* them. From the perspective of the unenlightened human consciousness, all E/entities are seen as 'fallen from' the ABSOLUTE SYNTHESIS. By an act of internal merging, it will be your task to restore them to their proper *identicalness* with the INFINITE HOMOGENEITY.

What are the 'Techniques of SELF Restoration'? How, actually, are you to *infinitize* all Units of Life thus restoring them to the SELF? Most certainly, you must remember the SELF continuously, standing in the PRESENCE of the UNIVERSAL CONSTANT constantly. Through the power of imagination and suggestion, see the ROOTLESS ROOT of all Units of Life. That ROOT is YOUR ROOT. That ROOT is MY ROOT. That ROOT is identical, the *same* in All. Using all means at your disposal (and you will have to find the means), *know* the ROOTLESS ROOT of All.

It is a great task to infinitize all seeming Units of life, because of the problematic existence of what might be called, 'The Reflex of Finitization'. The World of Fragmentation (i.e., The World of Illusion) is very strong (relatively), and items-in-Cosmos seem to want to disperse into that state of fragmentation. In your focussed consciousness, held at the highest possible point of tension you can achieve, you will have to overcome that natural tendency by holding all things in 'BRAHMANIC SOLUTION'.

There is a deep joy in infinitizing the apparent Units of Life. Of course, ESSENTIALLY, they need it not, but, *actually*, they do. The veiled consciousnesses of the Incarnated ALLNESS need infinitization almost desperately. For them, infinitization is LIFE ITSELF. For them (and who are they but we ourselves?) infinitization is a restoration of their true 'STATE' of INFINITE SELFHOOD. When, within your own consciousness, you infinitize 'another' you restore them to their REALITY Eventually your act of *restoration* will become their act of *realization*.

But how is it done? How is infinitization accomplished? You will find it to be the most elusive of *willed realizations*. Infinitization is a special mode of *seeing* which leads to the realization of ABSOLUTE BEING. Infinitization is promoted by SELF 'touching' SELF, by the recognition of the INFINITE SPIRIT *as* All. Both infinitizer and infinitized will find themselves 'current' in the CURRENT, in the PRESENCE, vibrating to the *thrill* of LIFE ITSELF. To infinitize is to give the Wholeness of Your Life in the cause of *reanimating* the sleeping lives around you. Infinitization is the gift of *livingness*, an extraordinary gift. It is to give THAT which is already possessed by the recipient, though *unknown*.

As you infinitize, a reflex action upon your localized self is also inevitable. To infinitize all presentations you encounter is to sustain your own infinitization. To *see* the SELF in 'others' is to *be* the SELF yourself. Live, thus, within the THRILL OF LIFE! To infinitize is to honor the INFINITE 'within' each apparent thing. It is to honor the PERFECTION 'within' each apparent thing. The ESSENTIAL PERFECTION is already *present*. No one can infinitize unless he dwells consciously 'within' the SELF, identified as the SELF, merged in the INFINITUDE. From that 'STATE', you will find it possible to exalt any Unit of Life maximally *within the limits of its form*. You will remind them of ESSENCE, and this reminder will raise them into a state of joy.

Remember that infinitization is not inflation. This great act of LIFE-Restitution is not meant to lead to excess, trespass of proper limits or grandiose self-estimation. Even the lowliest entity can be infinitized within your realization, and, though inevitably enhanced, must yet remain within its proper sphere within the Cosmic Pattern. No great leaps in form can be taken. Such is the Self-Imposed Cosmic Law. An ant today cannot tomorrow manifest as a Solar Logos even though the ESSENCE of these two Lives is the identical (and though the One Identity which both *are* is simultaneously manifesting *as* both and, yet, is *free* of both). Though in the WORLD OF BEING (and even the World of Being) the ant and Logos are complete equals, the vast difference in their formal manifestation must be wisely respected. Realize that in the World of Illusion there is a timed and sequential, developmental progression of form, and it cannot be violated according to the Laws *created* by Original Intent. Knowing what you know of BEING, learn, my Brother, to *infinitize.*

Counsel 11
View all things
"under the aspect of Eternity."

Realize that there are many way of contemplating an object. Consciousness requires an object, even though that object may later be seen as the ONE AND ONLY SUBJECT. Your perspective, the way you see things, is dependent upon your World View which causes you to value or devalue the things you see. Your values and mine are dependent upon our frame of reference. On the horizontal level, is the frame egoistic and personal; or communal/societal; or national; international; or perhaps planetary or systemic; or even Cosmic. And on the vertical level, is the frame of reference physical or psychic, or mental, or 'triadal', or perhaps Monadic, leading gradually into conscious participation in the SELF? The two frames meet and intertwine all along the way. The larger and subtler the frame of reference, and the more clearly it is seen and understood, the truer to the Original Intent are the values.

This being so, be a perceiver with the largest and subtlest frame of reference possible to you at this time. Hold ETERNITY in your consciousness, or, better still, in your *being.* See things as they *are* rather than as they *seem* to be. Be impersonal, super-personal, Universal, if possible (though for human beings at our stage of development upon this planet, it is not Really possible). Let your mind recognize that "ETERNITY is 'long'", and a Cosmos, REALLY, of the most brief duration. Imagine the infinite Sequence of Cosmoses in the Utter-Allness. Then realize that this immense Sequence of Universes is as nothing compared to the INFINITE, ETERNAL GRANDEUR of the SELF. This entire Sequence has, forever, been the merest ripple of seeming contradictions upon the ETERNALLY UTTER PERFECTION which is the INFINITE SELF. Seeing thus, thinking thus, you are ready to contemplate the immediate presentation "under the aspect of Eternity."

Let the SUPER-Universal 'VIEWER' persist and not be forgotten. Sustain in imagination that 'OBSERVER' of ITSELF, which IS the INFINITE SUBJECT-'BECOME'-Infinite Subject. By Mayavic Reflex, it is so easy to fall back (apparently) and lose the Eternal Point of View. Let reflexes of this nature be trained out of you, be "worn away by cosmic perception" as Master M. has said. The World of Becoming is incredibly beautiful "under the aspect of Eternity", for It is seen in the INFINITUDE which It REALLY IS.

Counsel 12
Realize the evanescence
of all objects.

Although objects *seem* finite, they are in fact *infinite*, at least their ESSENCE is. Nevertheless, objects will pass and be gone. Ever will forms, combinations and recombinations, pass and be gone. Never are they repeated—*never* in all of INFINITUDE (though they remain, as it were, 'resident' in the 'state of fulfillment' 'within' the INFINITESSENCE—'resident' as 'fulfilled possibility' 'within' the FOUNT OF ALL POSSIBILITY.

Even as one *seems* to see a stable, apparently unchanging object, the relationship which the object *is*, is forever changing. It is only the crudity of our perception which contributes to the illusion of sameness. Realize within any object, especially within a loved object, THAT which never changes. Hold fast to THAT even while relinquishing attachment to the object's form.

The Cosmos is a Holographic Kaleidoscope—turning rationally, beautifully and ceaselessly within the Universal Manvantara, creating, preserving and destroying a Logoically-Predetermined Sequence of Intended Forms. Form is the product of Maya, the Principle of Separation. Hold to form—hold to grief! Grief itself is the cry emanating *from* the separated *towards* the INSEPARABLE. Grief is the natural reaction of all *Cosmic Identities* in the face of that which is not IDEAL; the cry, though it be unconscious in many lower lives, is the longing for the ONE IDEAL, the INFINITE SELF.

Therefore, hold not to forms or conditions and there will be no grief over forms and conditions. Conditions are but passing states of the UNCONDITIONAL. They are mortal states of the IMMORTAL. Look out upon a world which cannot be destroyed, though it *seem* to be destroyed. Look out upon the INDESTRUCTIBLE ONE. The only sin there is (if sin there is), is the failure to express the *radical non-separativeness* of the UNCONDITIONAL SELF. i.e. the Love between Spirits which is the Identicalness of Spirit. To be trapped in the illusion of separativeness and to act as if separativeness were REAL is the great regret. Therefore, realize that, though objects come and go, the SELF is *forever* the PRESENCE. Release the object in favor of the SELF.

Counsel 13
Feel into all forms of life
—Be sensitive.

This is the way of bringing together all that *seems* separate and of realizing 'OnesSelf' as the ALL-IN-ALL. A growing sensitivity signals reunification and the beginning of the reclamation, in Cosmos, of the Synthesis that *Is* (reflective of the ABSOLUTE HOMOGENEITY that *ever* IS). Be sensitive, and know the pain which is the agonizing 'distance' of the self from the Self from the SELF.

Know that that 'distance' is un-REAL, and rejoice in this knowledge. Knowing the Truth of the Synthesis (and the greater TRUTH of the VOID), carry healing and 'wholing' in your words. Say to the impoverished, "You are rich"; to those who seem separated,

"You are united"; to those who seem isolated, "You are forever in the company of All." Say these things in such a way that they, too, will see and know, even as you have seen and known. You have the right to speak thus, because you are a *knower* of the ONE IDENTITY, the SELF.

We human beings are in a "far country". We have come to think that only the lower planes of nature are that far country, and the higher our true home. Though it is *partially* so, it is not REALLY so. The Universe Itself is the "Far Country", infinitely 'distant' from the INFINITE SELF, the ETERNAL HOME, (and yet a Country so 'close' (infinitely close) as to be identical with that HOME) The Prodigal Son story applies on many levels, and, above all, to the Creation of the Universe. Who is the Prodigal Son? Merely the human soul? Merely the human spirit? No. The SELF-as-Self is the Prodigal Son going forth into the Far Country of *Finitude.* Time and Space themselves *are* that Far Country.

Realize then, that you are the prodigal, that All are the Prodigal, that the Universe Itself *Is* the Prodigal Son—the Son of Necessity. This most ancient of all stories is not merely an *intra*-Cosmic story, but a SUPER-Cosmic History revealing the infinitely recurring 'DRAMA' of the ABSOLUTE SELF in 'RELATION' to ITSELF. For prodigality is to wander from INFINITE SUBJECTIVITY into Finite Objectivity, though the entire Adventure occur 'within' the SELF, if it occur at all!

Be sensitive to all prodigals, and to the Great Prodigal, our Cosmos. Be one who sends the prodigal upon the *Homeward Way.* Realize that each 'time' an authentic E/entity in Cosmos awakens to its TRUE IDENTITY, the Cosmic Prodigal in Its Entirety returns HOME. Help all prodigals on their Way by dissolving the Universe in BRAHMAN. Your sensitivity to all those who see not, is one 'moment of melting' in the great and coming Dissolution, the Universal Pralaya. Be sensitive so you may lead *apparent* 'others' towards the *absolution* of the ABSOLUTE. Your sensitivity helps to release all prodigals into THAT which they have ever been. Therefore, *feel* into all forms of life Permeate illusory boundaries. Be sensitive and WHOLE.

Counsel 14
Radiate the bliss of the ONENESS that IS.

Those who know the TRUTH are blissful because they have reunited THAT which has been falsely separated by the Sword of Illusion. Be a *knower* of UNCONDITIONAL TRUTH. Knowledge of the World of Becoming will grow and grow throughout the Cosmic Aeon. One need not wait for Cosmic Omniscience in order to know the TRUTH of the SELF.

Enter into the ONENESS that IS. Bliss is there to be found. Radiate that Bliss (Ananda) which is the natural 'feeling state' of the SELF-as-Self. Experience this. Reflect this. Radiate this Bliss—the result of YOUR realized WHOLENESS. Demonstrate this, and be a force for the pervasion of Bliss within the World of Becoming. Why does the Dalai Lama seem so happy?! Universal Bliss is the natural state of the One who dwells in Synthesis.

INFINITE BLISS is the natural 'STATE' of the one-as-One-as-ONE WHO has identified as the INFINITESSENCE.

Let the positivity of the one who is united be consistently expressed. Do whatever is necessary to find that point of tension that will obliterate in Bliss all dimmings, gloomings and fluctuations. Negate these through the 'POWER' of the SELF and Be Blissful.

Counsel 15
Have confidence in your essential and inherent *all-sufficiency* —Be SELF-confident.

In Time and Space, there is, *essentially,* and due to the dynamic of SELF/Self-Pervasion, nothing you *are* not and nothing you *have* not. However, because human consciousness is subject to Time and Space, it always *seems* that there are many things you both *are* not and *have* not. Because you-as-You-as YOU ARE the SELF, all things are YOURS, and you ARE them! Can we see this, despite the seemingly contradictory evidence of the mind and the senses? There are Sages who can. In fact, you and I already *do see*, but we may be unconscious of our seeing.

The ALL-SELF IS the REPOSITORY OF ALL POTENCY. We *ARE* THAT. You-as-YOU *ARE* THAT. Essentially, there is no *potency* which you-as-You-as-YOU *are* not or *have* not. While it cannot be said that there is *nothing* you have not done (for the INFINITE POTENTIAL is always greater than the opportunities for actualization in any or all Cosmoses), it *can* be said that all the things that *have* been done forever, YOU have 'DONE'.

How profoundly Occult is the concept of SELF-Confidence. REALLY, there is no other SOURCE for *true* confidence than the SELF, but experiencing it demands a recognition of one's REAL IDENTITY. Realize how fundamental to all success in Cosmos is the realization of your REAL IDENTITY.

Is not the one who knows himself to *be* the SELF, *essentially* (and in the fullness of time) sufficient to every challenge Cosmos can present? Will there not be a justified feeling of adequacy in the face of every possible obstacle? Of course, the practical limitations of Time and Space must be carefully considered, even by the SELF-Realized Sage (for he still must function through his current 'localization'), but will not these limitations be overcome rapidly by the one-as-One-as-ONE who is capable of drawing upon the INFINITE SELF as the SOURCE?

Not only are you sufficient to the task, but your's eventually is the All-Sufficiency of the Cosmically-Incarnated SELF, the Universal Logos. Have confidence in *your essential* and inherent All-Sufficiency based upon the UTTER ALLNESS you *are.*

Counsel 16
Appeal for strength to your deepest Self in Cosmos, but know It by Its many names.

Realizing your Essential All-Adequacy, and retaining complete confidence in the SOURCE-SELF you-as-You-as-YOU ARE, realize your vital power to make changes in Cosmos in line with the Original Intent, the Design-at-the-Beginning. Strength is needed to move that part of Cosmos which you, in your present state of localization, can influence from one state of actuality into a more ideal state of actuality. But you *have* that strength because of WHO/Who you ARE/Are.

Realize that all Units of Life being 'divisions' of the indivisible WHOLENESS' have, at *root*, the power of the WHOLE because they, ESSENTIALLY, *are* that WHOLE. There is, therefore, *no one* to whom to appeal besides your DEEPEST SELF.

If this is so, then what has been the meaning throughout human history of prayer and appeal to Deity? Have these appeals been useless or misguided exercises? No. Realize in this connection that it is possible to call one's DEEPEST SELF or the apparent Aspects of one's DEEPEST SELF by a host of names—names which correlate directly to the host of higher Agencies and Deities to which spiritual appeal is so often made.

There is an analogy to this process within the human constitution. The human constitution is really *one,* but we stratify it to better understand its complexity. We speak about our body, our personality, our soul, our triad, our Monad etc. and we attempt to get in touch with these various aspects of ourselves by a process of invocation akin to the process of appeal to Deities and Deity. For instance, if we appeal to the Master, to Sanat Kumara, or even to the Planetary Logos or the Solar Logos, to 'God', then, to What or to Whom is one REALLY appealing? Are these hierarchicalized Beings anything other than various Names for the Universal Self-as-ABSOLUTE SELF operating at various degrees of revelation and *authenticity* within Cosmos? All Names are but One Name which is the Current Cosmic Name (the Name of the Universal Logos) for the NAMELESSNESS (the *REAL* ONE ABOUT WHOM NAUGHT MAY BE SAID). There IS no BEING but the SELF. The entire Hierarchy of Deities are simply various Names for one's very Self

Therefore, know the many names of the One Deity Who Is the ONE DEITY, but realize as you make appeal, to Whom-as-WHOM you are appealing. Namely, one or other of the Great Names, which represent simply one or other of the powers of your very Self-as-SELF. Understand the responsibility implied in this realization.

Counsel 17
When you appeal to your deepest Intra-Cosmic Self, be 'realistic' in your expectations —Time and space must be respected.

We have realized the All-Adequacy, the All-Sufficiency of the INFINITE SELF. We have realized that no matter to Whom or to What, invocation or appeal is made, it is always made to an Aspect of ONESELF/'OnesSelf'. Thus appeal is made, not on the basis of weakness or passivity, but on the basis of a profound sense of SELF-Confidence, confidence in the fact that one-as-One-as-ONE has tremendous power within 'OnesSelf' because of One's Essential Universal Identity, and, infinitely more, because of ONE's ABSOLUTE IDENTITY.

Realize, however, that contact with this tremendous SOURCE-as-Source of Power and its exercise in-Cosmos is subject to Number, Time and Space. Further, this Power must express through numerous veils. Fix firmly in mind that in the World of Cosmos, there can be no continuous instantaneous achievement, because all progress is determined by the number of movements possible within certain standard Cosmic Time Frames, and is dictated by the Laws of Position and the Laws of Relationship within a particular Cosmos. So yes, there is ABSOLUTE POWER and you-as-You-as-YOU are the ULTIMATE SOURCE of your own liberation and the INFALLIBLE GUARANTOR of your own achievement, but if material results are to be sought in Cosmos (and what 'results' are not 'material' from an 'essentialist' perspective), the POWER you-as-You-as-YOU ARE cannot be released instantaneously, but is subject to apparently sequential action.

Thus, be patient. While it is eminently desirable that you realize your Essential All-Sufficiency with a SELF-as-Self-Confidence that cannot be shaken, your patient respect for the speed of the processes of (albeit illusory) Time is required. Do not agitate your environing system because of the non-instantaneity of these processes, otherwise Original Intent will not be served. Though processes may seem slow, the SELF-as-Universal Self (the Comforter) is *ever-present* and thus there is no need for complaint.

Counsel 18
Merge and dissolve all variety into ONE.

This injunction is similar to the one which calls for you to infinitize all presentations. When you do this, you are involved in an Act of Synthesis. The idea is an ancient one, and most beautifully and consistently expressed by the great Vedantin Sage, Sri Sankaracarya. He calls upon us to merge and dissolve the Universe into BRAHMAN. Do we know how to do this?

Realize that to merge and dissolve the Universe into BRAHMAN (i.e., to 'infinitize' It) you must consciously *melt* all variety into ONENESS. But variety (the virtually countless modifications of the World of Becoming) cannot be set aside as unimportant. No, all variety must at length be mastered. There is no way for you to ignore this, for you are intra-Cosmically the All Pervading One—Son of THAT.

A human being on this plane manifests within a tiny, roughly spherical ring-pass-not, the destiny of which is to expand into the Ultimate Ring-Pass-Not of Cosmos. This will be the destiny of all authentic I/identities within Cosmos, great and small. The Final Ring-Pass-Not Itself will, on the "Day Be With Us", yield Itself to the INFINITUDE, and the Cosmic Aeon will be concluded. Thus, everywhere to be seen are hosts of evolving, expanding forms (or those which involve and contract prior to expansion). Everywhere is an immense variety, all of it nothing but a manifestation of the ALL-MIGHTY BRAHMAN. THAT is the ONE IDENTITY to be perceived in every one of the Universe's multifarious units.

You cannot possibly fathom the immense variety as a localized consciousness. No human being can. No Being in our Solar System can. One wonders whether any Being in our Galaxy can. It is simply sufficient to know that the variety is immense beyond all possibility of present computation (for so much of it exists upon inner and higher planes where no human mathematics has yet essayed to venture).

Thus, you strain to cognize the inescapable, *overwhelming* variety of Cosmos. On each plane of Cosmos (and there may well be a thousand planes) different categories of the variety are to be encountered. The ALL-SELF-as-Universal Logos has not only 'Produced' this stupendous variety but has 'Become' It. This Great Being (and the BE-NESS which It ESSENTIALLY Is) is responsible for every intricacy in relationship of the seemingly unending variety. None of it can be ignored. At length, all of It will have to be mastered by each authentic I/identity in Cosmos—*all* of It. This may seem a shocking statement, but we have to remember that Cosmic Division of Labor begins to terminate upon the very highest planes of Cosmos, upon which each becomes each-and-All. Cosmic Omniscience, Cosmic Omnipresence, Cosmic Omnipotence—these are the destiny of every authentic E/entity in Cosmos.

So tend to the variety with which you are surrounded. At first your circle will be small. Later, much later, it will enlarge to include the Whole. The task of coming to terms with variety cannot be relinquished. Not one may escape it.

But Cosmic Variety seen only as multifarious movement is mere fragmentation. The Point of Origin must be seen, the Point (the Condensed Point) which is the Active Foundation of the Universe. Not one unit within the variety, not one E/entity, item, or aggregate but must be traced back to the Point of Origin and 'returned'. This is one way of seeing the Task of Cosmic Evolution—to leave no point in Cosmos unrelated to the One Point; to leave no aspect of 'extension' unevolved or unrelated to the Unextended State (for extension is an illusion, as we have established, and, Really, the Point has *no* extension); to leave no Number that is not reabsorbed into the One, the Number 'One'.

Thus, let there be 'infinite' respect for the variety of the World of Becoming and the gradual and joyful mastery of it. This will require of you a special and subtle 'operation' of *fusion* in consciousness and in being. Through your Power of Fusion you must labor to perceive no *disconnected* variety, but only a supremely *connected* and *rationally patterned* variety. In fact, what you must labor to see and experience is a variety which is based upon the Principal of *Identicalness*. You will then REALLY understand that everything *is* everything else.

This is one of the supremely important axioms of Radical Non-Dualism. Remember that in all the variety of presentations which the Universe offers, and in all the virtually countless modifications, naught can be presented to your consciousness but the Universal Self-as-SELF ITSELF.

Counsel 19
Be still, Be immutable, Be immortal
—In short, BE.

It is impossible to BE as long as there is a single movement. What we usually call *being* when in the World of Becoming is not REALLY BEING, for BEING is utterly motionless. The moment there is a movement, one *does* and BEING has ended (although in an ultimate sense, BEING can *never* end). When there is no movement, one simply IS. The entire Cosmos *is* perpetual motion. The ALL-SELF is no motion at all. Can there be motionlessness when all is in motion. Can the WORLD OF BEING, REALLY, 'enter' the World of Becoming. One would have to say, Yes, in as much as the World of Becoming *is* the WORLD OF BEING, ESSENTIALLY.

The GREAT MOTIONLESSNESS of ABSOLUTENESS during Universal Pralaya is utterly mysterious, and almost certainly cannot be duplicated by intra-Cosmic beings (though let us remember that intra-Cosmic beings are simultaneously the SUPER-Cosmic BE-NESS WHICH never ceases. Some Sages have claimed to have achieved this *utter stillness*, the STILLNESS of the SELF.

One can see the relation between stillness, immutability, immortality, and the SELF. The utterly *still* state is the closest in-Cosmos approximation to the STILLNESS of INFINITUDE, the ABSOLUTE DEITY. From this perspective, the less one *does* the more one IS. Complete and perfect stillness would signal the transition from 'Becoming' to BEING, the conscious *infinitization* of the Finite. We can see, then, the power of stillness, and the deeper meaning of, "Be still, and know that I am God." This injunction, from a Radical Infinitist perspective is simply extraordinary.

Therefore, realize the importance of and relationship between these injunctions: Be still. Be immutable. Be immortal. Is it not interesting that the DEATHLESS 'STATE' is the MOTIONLESS 'STATE'? As well, the IMMUTABLE 'STATE' is the MOTIONLESS 'STATE'. The ABSOLUTELY DESIRABLE 'STATE' is, also, the MOTIONLESS 'STATE'. Realizing these things, let us Be Still, so that we can BE the 'STILLNESS' and discover the TRUTH of the MOTIONLESSNESS for ourselves.

We must not pass on without reminding ourselves that the FOUNT OF ALL POSSIBILITY 'contains' in ITSELF as the INFINITESSENCE, the *contradiction* to every quality predicated of IT. It would seem, however, that Time and Space are needed for the expression of the contradictions. The Universe Itself is the Great Contradiction. From experience we know that motion does not lead to the HEART OF THE MYSTERY.

Stillness in-Cosmos is still the *key* for entering THAT which might be described as *the* PERFECT STILLNESS. Therefore, BE STILLNESS.

Counsel 20
Achieve stillness by dropping desire.

Desire itself is a great agent of movement. It is an energizer of parts and pieces. The fulfillment of desire normally demands movement—a movement from the current *actual* state to a more *ideal* state (i.e., one that is more 'desirable'). Desire (especially aspiration or high desire) is that which compels a reordering or reorganization of Cosmic variables which is more ideal, i.e., which complies with a higher Pattern. Another step is required, however, on the way to stillness and revelation of the INFINITE SUBSTRATUM. Simply *drop your desire* for (achieving) a higher Pattern (no matter how high that Pattern may be) by realizing that the ROOT of all Patterns (more *desirable* than any Cosmic Pattern, whatsoever) *already* exists, and has only to be unveiled. Why desire what you already *have* and *are?* Drop desire and allow what YOU ARE to be revealed. Desire serves its purpose throughout Cosmos and cannot Really be relinquished until the "Day Be With Us", but for a time (a moment, a minute, an hour) drop desire and 'enter' TIMELESS STILLNESS.

Counsel 21
Create and use mantra for achieving 'Still Penetration' into THAT.

Various mantra (coherent, magically effective phrases based upon that type of vibratory motion we call sound) may be used to ward off other intervening obstructive movements which destroy the stillness and hence the penetration into REALITY. A mantram, for instance, may reduce certain forces to quiescence, even though the mantram, itself, depends upon motion (the motion of sound waves, the vibratory motion of atoms of different qualities, and even the motion of the physical mechanism {vocal chords, tongue, lips, etc.} that produces the sound or the motion of the 'thought-mechanism' which *thinks* the sound). Mantra, however useful, and however needed for many cycles to come, are simply expedients. Mantra clear the Path to REALITY of unwanted vibration, but one does not merge with REALITY purely as a result of a mantram. Within the pure 'STATE' of REALITY, *silence* must prevail. REALITY does not sound a mantram, for sound is inescapable from form (however tenuous), and REALITY is the UTTERLY FORMLESS 'STATE'.

Mantra are only instruments Really. A mantram could be proposed as follows, "Movement seals the door to REALITY." This mantram would have *thought value* in the World of Meaning, but no *sound value* per se. Thought value of this nature, however, might be sufficient to quiet obtrusive modifications/movements which were preventing realization. Certainly a mantram could be useful in establishing a certain *attitude* towards modification.

In addition to the magical, creative potencies of many ancient and hierarchically sanctioned mantra, mantra, in general, and all similar 'phrases of remembrance', are useful in intensifying the *point of tension* of the meditative, contemplative consciousness. Once the *point* of tension becomes *dimensionless,* the breakthrough into greater Reality (and, perhaps, even into REALITY) occurs. Then, that point transfers itself to all dimensions, at the same time being localized in none.

Counsel 22
Learn to consult
the Inner Teacher.

There is an Inner Teacher—but not the soul, not the Master, nor any member of the Spiritual Hierarchy, nor any 'other' being, per se, in Cosmos—though all of these can and will assist one's own SELF-as-Self, identical with the ABSOLUTE SELF-as-One Self. In consultation with this Inner Teacher the way into the Mystery of Mysteries must be evolved by every self-conscious authentic E/entity, for each such E/entity IS the Cosmic Mystery-as-MYSTERY. The Inner Teacher, the Self-as-SELF, *knows the Path of Return* to the Cosmic Home You never left, and thence, 'HOME'. Learn to consult the Inner Teacher. Learn to listen to Your Self, the Self-as-SELF.

Counsel 23
Rejoice in YOUR ZERONESS
—in YOUR NOTHINGNESS!

The ego or localized self, the veiled SELF/Self always wants to *be* 'something'. 'Be Something'! This is the noisy affirmation of the ego or little self, since it has forgotten Who it Really Is (REALLY, and ultimately—IT). As solid as they may seem, all egoistic identifications are but false assurances to a lost consciousness.

The dynamics of the process work as follows. The SELF is the UNIVERSAL CONSTANT, the ABSOLUTE CONSTANT. IT is unvarying, unchanging. In Cosmos, even though the ancient Mayavic forgetfulness has descended, there is (within each authentic E/entity, especially once self-consciousness has been reached) a reminiscence of that UNIVERSAL CONSTANT. This 'memory', deeply veiled, causes a deep need for identification with—what? Once, 'long ago', in the unveiled state, the identification was with the Whole and even the WHOLE. Now, deeply veiled, the consciousness reaches out blindly towards the apparent not-SELF and fastens itself to an *object,* creating a very partial self-identity in the attempt to reproduce that dimly remembered CONSTANT STATE, the STATE OF THE UNIVERSAL CONSTANT.

Ego (nourished by unconscious memories of the CONTINUITY) is an attempt to find continuity in a discontinuous World. Inherently, the SELF, deeply imbedded in the apparently relative being of the little self—this SELF, WHO has never REALLY ceased being ITSELF—'realizes' ('instinctively', 'unconsciously') that duality, multiplicity, modification, and the familiar conditions of the World of Becoming are all foreign to ITS TRUE NATURE. The SELF, deeply veiled as the self, realizes (though unconsciously) that identity is inseparable from *constancy*, and that if a thing fluctuates, it cannot be a true *identity*. There follows upon that vaguely felt sense (which we are calling an 'unconscious realization') an attempt to rigidly impose *continuity* through the mechanism of *egoism*. An ego at least is an (apparently) unchanging 'something'. Ego demands a name (which is given) and a familiar form, which is constantly and reassuringly seen. What is not realized, however, in this benighted egoistic search for continuity and constancy as the *sine qua non* of identity, is that the ego name is partial and hardly reflective of anything other than a sound, and that the ego form, far from being constant and identical with itself from ultimate moment to ultimate moment, is constantly changing.

To the deluded consciousness, therefore, 'something' is seen as better than NOTHING. Ego must 'be something'. To the illumined consciousness, however, NOTHING is seen as infinitely superior to 'something'. It takes a long time before one can rejoice in ZERONESS or in NOTHINGNESS, because 'somethingness' is for ages seen as equivalent to *being*, to having a *constant identity* (which, of course, it is not). However when 'somethingness' is, at last, seen merely as actual and conditional, limited and partial (which thorough and rigorous thinking reveals it to be), then, in the search for identity, the EVER-FULL is pursued instead of the objective and partial, and one is more than content to *re-identify* with the FORMLESSNESS, the NAMELESSNESS, the UNMANIFESTED 'STATE' and then can rejoice in his ZERONESS.

Every unit of Life is, ESSENTIALLY, ZERONESS, but when playing one's part in Cosmos, ZERONESS is practically useless and even dangerous and disruptive. ZERONESS is the Obliterator. It is not the friend to created things. ZERONESS is not the friend to *false* identity. The great Taoist Masters such as Lao-Tzu, and also the Buddhists have learned to rejoice in their ZERONESS, rejoice in the VOID. Let us learn from them.

Counsel 24
"Love thy neighbor as Thy Self."

From the perspective of Radical Non-Dualism it is clear why the enlightened Sage will "Love thy neighbor as thyself." THY *neighbor* IS THY SELF, *the* ONE AND ONLY SELF. One can hate a form. One can hate an act (which is also but a form), but, one can never hate the SELF. One can only love IT, the ULTIMATE 'SUBJECT' of DESIRE.

The SELF, the TRUE SELF is "beyond good and evil." In a way, one can only hate an *object*. One cannot hate a REAL *subject*, for one can only be repelled by a 'finitude' and a true *subject* is an *infinitude* (of which there is ultimately only ONE —the INFINITUDE). All true subjects are THAT and THAT alone.

So, one can only hate a finite state. One can deplore certain actions if the actions emerging are patterns which are greatly at variance with Original Intent (either the Original Intent in Cosmos or as reflected in One's intra-Cosmic Selfhood). The one who has *seen* (or, better, *been*) his REAL neighbor cannot hate his neighbor, but only the patterns emanating *from* his neighbor. Repulse arises when such patterns are out of harmony with one's own patterns and cause friction, dampening or dissonance.

Take dissonance as an example. Dissonance is like a gnawing pain demanding attention and resolution. Because it gnaws, it distracts from Wholeness/WHOLENESS. Dissonance obsesses the consciousness, painfully fixing it upon fragmentation, and forcing it to dwell in that less-than-ideal state. Hence dissonance (similar to friction, irritation, dampening—all the product of vibratory inharmony) is deplored, because it holds one back from a state of greater release. In other words, as long as there is unresolved dissonance within the energy system, it is impossible to go farther. As long as there is a compelling fragmentation arresting the attention, it is impossible to go higher into Wholeness/WHOLENESS.

In the SELF, however, there can be no dissonance. There can not even be harmony. There is only *identicalness*. In a way, we can understand the Biblical adage in the following way, 'Love THY neighbor as *the* INFINITE SELF with WHICH you wish to merge and blend, and into WHICH you wish to dissolve, thus becoming ONE with the UNIVERSAL SOLVENT'. Surely, to do so is the greatest urge of all awakening intra-Cosmic selves, subject, as they are, to the Cosmic Desire for Universal Re-Unification. In their intra-Cosmic Heart of Hearts, all authentic I/identities long that there be maximal merging in Universal Love—a longing which reflects in the human sphere as community, solidarity, and heart.

In order to Love THY neighbor as THY SELF, each must certainly *identify* at the deepest level with his neighbor. This identification, Essentially, means the capacity to *see to difference* and yet <u>*be*</u> *the sameness which one also sees.* Thus, fellow traveler upon the Path of Synthesis, *see* only the sameness which is the SAMENESS, the GREAT HOMOGENEITY. Minimize the differences. Acknowledge them; know their place within the scheme of things, but merge yourself within the SAMENESS and <u>BE</u> LIFE.

Counsel 25
Radiate the *Fire* of
TRUE SELFHOOD.

Once you *feel* and *know* you *are* the SELF, how can you help but *be* the SELF in active, deliberate manifestation. *Being* the SELF is spiritually contagious, though extremely subtle. Around the one who is SELF-Identified and, hence, radiant with this Identification, the SELF in all is stimulated, gradually unveiled—the SELF-alienated ones seek their *essence.* Around expressive people, others are inclined to express. Around free people, others are inclined to attempt to be free. Around the one who *knows* the SELF, others find themselves seeking the ROOTLESS ROOT of their intra-Cosmic Being—the SELF.

What can you share with others? Many patterns, qualities, and vibrations of a constructive and elevating nature, most useful in the fulfilling of the Design-at-the-Beginning. Yes, but the greatest of all sharings is your *identicalness* with the apparent 'other'—

identicalness, the Key to the INFINITE. When the SELF is shared, TIMELESSNESS is realized. When the SELF is shared, the ABSOLUTE 'SATISFACTION' is realized. When the SELF is shared, the end to burning desire is realized. Thus, radiate the *fire* of TRUE SELFHOOD.

Counsel 26
Be not distressed over *seeming* things.

The Earth has been called a "Vale of tears". 'Vale' means 'valley' (a kind of 'depression' on the surface of the Earth). The sound of the word 'vale' is identical to 'veil'. Maya Is the 'Veil'—the cause of all tears and distress. There could be no distress without ignorance, the 'Creation' of Maya (Who, of course, Is PARABRAHMAN's 'CREATION'). But WHO can 'MAYA' BE, if not PARABRAHMAN, ITSELF? WHO *other* than PARABRAHMAN?

Yes, in the World of Becoming there is distress due to the temporarily inharmonious interplay of the Great Opposites. Somehow, however, even NOW, ALL IS WELL. This must be realized, and it is difficult to do so. While it may sound fatuous and utterly naive, this statement, "Even NOW, ALL IS WELL" is based upon the deepest philosophical penetration.

Distress is a *seeming* thing, but unfortunately, we human beings have the "habit of misery" as the Tibetan Teacher has suggested. Of course we must *care* about Cosmos. It is Our 'Creation' and Our responsibility. We owe it to the SELF! Nevertheless, we have no good reason for allowing distress to make us non-functional, poor players in the Great Game. Distress is a maladaptive response to inharmony. Distress arises through forgetfulness of the SUBSTRATUM. When one, in his human mode, is beset by distress, one must realize the un-REALITY of the *seeming* things, and, *as well,* the need for *changing* them. Dissonance must be rendered into harmony. In a high-spirited, sportsmanlike way, we must learn to *change* the *seemings* which hitherto distressed us. SELF-Confidently we remember that the SELF is the INVINCIBLE—IT cannot be defeated. In the words of Master Morya, "SObstacles are possibilities." Therefore—Be not distressed over *seeming* things.

Counsel 27
Realize that *all* problems are soluble.

Realize that all problems are soluble because **I**-as-**8**, the SOLVER-as-Universal Solver, *Am* the solution to all Cosmic problems. **8** have only to resolve or dissolve all things into 'MySelf'-as-SELF, and all problems are solved. **8**-as-**I** AM the ONE RESOLUTION. Every Cosmos *is* a Problem to be solved. **I**-as-**8** 'Create' the Problem and **8** (being **I**) Solve It. There is something inherently Blissful in the exercise of My Power.

Realize that as you solve the many problems confronting you, you are contributing to the Resolution of the Cosmic Dissonance, to the Unraveling of the Cosmic Knot, and to the Unification of all apparent Dualities. The ABSOLUTE 'POWER', the ABSOLUTE 'LOVE' and ABSOLUTE 'INTELLIGENCE' which you-as-You-as-YOU inherently ARE, are equal to any and all problems. Time and Space must be respected, but the problems must yield to the SELF.

Counsel 28
Respect your handiwork —the Cosmos.

Respect what YOU-as-You have 'Done'—the Creation of the Universe, Itself. Does this seem a preposterous thought, that YOU-as-You were (with all apparently 'other' 'Yous' who are One **8**) the Creator of Cosmos? Is it inflated? Is it egotistical? Is it nonsensical? None of these! We know so very little of our Origin, and are so completely mistaken about our Essential nature because of the Veils of Maya (for which WE-as-**I** are responsible anyway!) What a strange World We-the-**8** have fashioned!

Indeed, we must respect Our Handiwork. In the great Discipline of Negation which is Radical Non-Dualism, one discovers the impossibility of negating any 'thing' at all. Realize with the Enlightened Sages that it is ESSENTIALLY impossible to negate anything *without creating an artificial duality*—a 'split' within the INDIVISIBLE SELF. For negation suggests that there is something 'other' than the INDIVISIBLE SELF. By unvarying definition, however, there can be nothing 'other' than the SELF. Negation creates a 'something other' which is an impossibility. The Path of Rejection reveals, therefore, that nothing can be rejected! The Path of Negation reveals that nothing can be negated. Form-as-form can, of course, be seen to be un-REAL and evanescent, but the *substratum* of any form (what it is in *essence)* simply *cannot* be negated. In short, we must accept the World as the SELF.

Counsel 29
Understand the 'Word Games' that lead to the understanding of REALITY.

Cosmos is the very *antithesis* of ALL-IN-ALLNESS. 'Thingness' is the antithesis of 'NO-THINGNESS'. Cosmos is in fact the Great Opposite, the Great Contradiction, and the Great Adversary (in a strange way, the 'Devil' Itself). Even though Cosmos be *the* One Contradiction, It is a Contradiction for which 'room' must be made—'room' within the SEAMLESS HOMOGENEITY.

We can rightly call Cosmos a Contradiction, but in a deeper sense, can anything REALLY contradict the ALL-IN-ALLNESS? The ALL-IN-ALLNESS is ever elusive. Nothing can be predicated of IT, and nothing can be *denied* of IT. IT both admits of no possible variation, and *yet* requires that there be variation, otherwise IT would be *less* that what IT IS. The SELF IS *the* CONTRADICTION. We may say this is only so because the human mind is fallible (which, of course it is), but the mind is what we have in order to understand these difficult matters. The heart, too, must add its testimony. The only other alternative is to give up on the understanding of REALITY, which would be to deny the possibility of understanding our very SELF.

So, as we have already established in numerous ways, in order for there to *BE* INFINITY, INFINITENESS, ABSOLUTENESS—the dynamic of (at least apparent) contradiction is necessitated. Duality must be *presented.* **8** do not say that Duality must exist *absolutely,* but, merely, that Duality must *become actual* (even though that *actuality* be illusory because reducible). Along with this, the seeming negation of the ABSOLUTE STATE must exist (at least in consciousness—which, itself, is necessarily illusory). In still other words, if the PRINCIPAL 'THING' or, more properly, the PRINCIPAL 'NO-THING' is to BE what IT IS, then what IT IS NOT must also exist. Of course, if IT IS REALLY what IT IS, then, even what IT IS NOT is WHAT IT IS! But then, is there anything IT IS NOT?

For the relief of the reader, it is necessary to say that these riddles concerning being and non-being, can be solved somewhat through the use of the concept of *apparency.* When a thing 'happens', is it REALLY happening or only *apparently* happening? Much of what seems to be happening is only *apparently* happening. True enough, the apparency is an *actuality* and a *factuality,* but it is, nevertheless, an *illusion,* because it is *reducible* to the 'ACTIONLESS' SUBSTRATUM.

It is not to everyone's taste to enter so deeply into the subject of *being/non-being* as to be forced into what appear to be 'Word Games', but our entire process of thought is, at the moment, based upon our language and its use. Language must be rendered a more incisive tool to reveal abstract relationships. In the process, the natural limitations of language (and of its users) are exposed. Persistent logical application of language to these highly abstract metaphysical subjects leads inescapably into what the human mind, as presently constituted, regards as *paradox.* There seems to be no other way. When language becomes paradoxical in order to describe the TRUTH/Truth, REALITY is conveyed as much by the perplexity generated as by the precision attempted. Despite what is said about SELF-Realization, this we know: It is possible.

The Creation of Cosmos is a function of the GOOD. Within Cosmos is the appearance of Good and Evil, but Cosmos, is ESSENTIALLY GOOD It is GOOD because, not only it is *necessary* to THAT, the ALL GOODNESS, but Cosmos, Itself, *Is* the ALL-GOODNESS, ITSELF (albeit in disguise).

When one tires of attempting to say various things about NOTHING and about Cosmos, one can simply say of each of them (leaving all sophisticated arguments aside)—IT IS WHAT IT IS. This being the case, we have to find a way to deal with each of them on their own terms. That finding of the Way is what the evolution of consciousness is all about.

Counsel 30
Honor agreed-upon forms
—the Archetypes.

I-as-**8**-the-We were the One, Who, at the Cosmic Beginning, preset all Intended Forms (the Divine Numbers) through the Process of the Universal Logoic Emanation of the World of Being. It thus becomes necessary to understand which forms are in line with Original Intent, and which forms are unconscionable abominations (for such *have* existed during the course of the History of the Earth, and elsewhere, no doubt).

What this means is that we must Honor the Archetypes. At first there is the Universal Self at the Dawn of the Cosmic Process, Planning and Designing according to 'IMPLANTED INTENT'; and then there is the Universal Self in the midst of the Cosmic Process *acting* spontaneously (through Its E/emanations) In the midst of the Cosmic Process, the Universal Self (through all Its emanated Self-extensions) is in a more *veiled* and less clear state, and, consequently, It will spontaneously create under the influence of Ignorance, even while remaining the *Cosmically* Omniscient Universal Self on Its Own Plane. Unfortunately, what It Knows on Its Own high Plane, is not communicated clearly to all Its emanated extensions (other, lesser, authentic E/entities in Cosmos) because of all the many veils which It (the Universal Logos) has imposed upon Itself through the Mayavic Process of Emanation. (We must remember that all the Hierarchy of B/beings in Cosmos are Really the E/emanations of the Universal Logos, the various E/emanations being Really Its Self-extensions—more veiled and circumscribed replicas of Itself, but, Essentially, *fully* Itself, nonetheless.)

All during the Cosmic Process, the Archetypes are those Beings Who take emanated birth in the first phases of the Universal Self's Emanatory Stream. These Archetypal Beings are both Being and Pattern, Idea and Quality. They are the Sustainers of the Patterns which must manifest below. The majority of emanated lives (being deeply veiled in the lower worlds) do not fathom the Original Archetypes (except, perhaps, in the inter-moment instant). They build according to much lesser patterns far removed from those Original Archetypes which sustain the Cosmic Pattern. The lesser patterns, however, are, indeed, part of the intricacy of the Greater Patterns and have a true form on the Archetypal Planes. Sometimes even secondary, tertiary, etc. or even lesser archetypes (at many removes from the Original Archetypes) are not apprehended with accuracy by these lesser creators due to the veiled condition of the registering consciousnesses. (We human beings are such lesser creators, far from the Source of the Emanatory Stream {while, yet, being that Source entirely}.)

Nevertheless, gradually, the consciousness of the Principal Archetypes clarify, and the many emanated creators begin to know the difference between an Intended Archetype and a self-generated phantasm (i.e., a poor approximation of that which abides as True Pattern upon the Archetypal Planes). Not all forms are good forms. Some forms, however, *are* good and "true to type", true to the Original Intent of the Design-at-the-Beginning. Therefore, find these *true* forms. Honor the *Agreed-Upon* Forms—the Archetypes.

Counsel 31
Improve the form.

Form is not REAL but it is indispensable in Cosmos. See to its quality. The ESSENCE of form is always the same—the SUBSTRATUM. But the *form* of form must be appreciated on its own terms. Forms must be changed, improved, rendered true to Archetype, true to Original Intent.

Do all you can to improve the form, bringing it into approximation with the closest reflection of Original Intent presently apprehensible. This is our task in the World of Approximation. Eventually, you and I and all, as 'Humanity, the Creator' will bring form into *exact conformity* with Original Intent, but the time is not yet.

Counsel 32
Sing!

Why sing? To sing is the beginning of learning to vibrate at all rates and frequencies. Only so many frequencies and combinations of frequencies are possible in a given Cosmos and one must, at length, vibrate to all of them as one's ring-pass-not expands to fill the Whole. To sing is the way towards harmonizing all available frequencies. Such harmonization is eventually required of all beings-in-Cosmos according to Cosmic Law. To sing is to learn to take one's place, in Beauty, within the Cosmic Choir.

Counsel 33
BE continuously.

The conditions of Cosmos militate against *being* continuously, but as YOU ARE the UNIVERSAL CONSTANT, YOU *can* BE. The phenomena of the World of Becoming can be totally ensnaring, and yet you-as-You-as-YOU can also learn to BE the 'OBSERVER'. Ultimately, YOU ARE the 'OBSERVER'-as-Observer *and* the 'PARTICIPATOR'-as-Participator as well.

None of us (as the SELF WE ARE) can allow ourselves to lapse from the radiance of continuous BEING (though we always *seem* to do so!). Form, born of fluctuation, will lapse repeatedly and give way to the inclination to modify and thus obscure *being,* but the SELF does not lapse in ITS *continuous* attentiveness.

Symbolically, the Sun must always shine, though the Moon will vary. If there is no continuity in the World of Nature, but only discontinuity, does that mean there is no continuity in the World of SUPER-NATURE? No. Though the World of Illusion seek to destroy your ESSENTIAL CONTINUITY, learn to BE, *continuously!*

Counsel 34
Imagine the annihilation of space or interval.

If you do so, you will be approximating a great Truth in Cosmos, for there is no space or interval. Imagine in Cosmos a continuous touching of all things, each to each. In fact, imagine something which goes far beyond touching, for touching respects ring-pass-not(s), but REALITY does not. Imagine a Cosmos which is free from any illusion whatsoever. Can you? Will Cosmos remain? Is not Cosmos Itself the Great Illusion? Imagine, instead of the many forms presented, an *impenetrable density.* This is the REALITY of Cosmos. The forms are, as it were, *an aggregation of ideational 'holes'* 'separating' the density with *nothingness.*

Imagine further a boundriless, impenetrable Density. This will be Mulaprakriti, the Infinite Object. Realize that all interval (i.e., the illusion that space exists) is merely a presentation in consciousness, as Fohatic Self-Reflection. Interval can never REALLY exist. All that seems to be happening is happening, at a single imaginary point which signifies localization but, itself, does not REALLY exist.

By annihilating interval, one becomes, consciously, *all* that 'happens'. One has always been so, but now one is *pervadingly conscious* of the fact. One finds oneself as the Divine Perpetrator of all acts-in-Cosmos, on the *inside* of all acts, and *as* the acts *themselves.* By annihilating interval, all experience becomes universally *common* experience.

How to do this? Simply begin by realizing that there is no *'there.'* The illusion of interval arises as we compare 'here' to 'there'. Eliminate 'there' and let all things be 'here'. Use the imagination to do this. Eventually you will eliminate 'here' as well, but first things first.

Counsel 35
Have faith in the amazing potential of the moment —Realize the ABSOLUTE FULLNESS of the NOW.

Never was any moment any better. Never was any moment more 'full of' INFINITUDE. Never was there any other moment, never. This is the literal TRUTH. The patterns or the presentations might have been relatively better or worse at other 'times', but that difference has nothing to do with the FULL PRESENCE of the SELF in *this and every moment.* The complete potency to solve all problematic presentations always exists.

The I-in-Cosmos-as-**8** is ever superior to any condition, for IT IS the UNCONDITIONED SELF. The LIFE of the SELF-as-Self IS ever present in the PRESENCE. Access IT in all conditions. What dissonances can resist ITS *fusing power?* The LIVINGNESS of LIFE IS NOW.

Counsel 36
Imaginatively contain all movements in the NOW.

Imagine a Universal Process of movement with all things occurring simultaneously. When you have this image, realize that there has never been (throughout the Infinite History of the UTTER ALLNESS) any REAL *movement* at all. Realize that all apparent movement has been *actual* but *illusory* movement. Though there has been no REAL movement, all actual movements in an infinity of Cosmoses have occurred at this very ABSOLUTE MOMENT. There is but *one* ABSOLUTE MOMENT, and IT contains All Time forever. There is one Stupendous Cosmic Action and it is All REALLY occurring NOW. In fact, All the Cosmic Actions in an infinity of Cosmoses have occurred NOW.

While *actually,* from an intra-Cosmic Perspective, all action occurs sequentially from ultimate moment to ultimate moment, in the Cosmo-Objective Now, REALLY all action has occurred, is occurring and will occur NOW (in the ETERNAL NOW), for there is no other 'time'. YOU-as-You are involved in every phase of the Great Action and have *always* been involved. In fact, YOU-as-You *are* the Action, Itself. And YOU-as-You are Acting everywhere, NOW.

Counsel 37
Be the 'Other'.

Realize that you are as much I, as 'I' am 'I'; that you are as much I as you are you; that I am as much you, as you are you; that I am as much you, as I am I (I have used conventional type here, but you-as-You-as-YOU know what I-as-**8**-as-**I** mean!). YOU ARE living MY LIFE because that LIFE is YOUR's. **I** AM living YOUR LIFE because that LIFE is MINE. WE simply have to BE EACH OTHER, and *know it.*

Counsel 38
Persist.

REALLY, YOU-as-You-as-you have no choice. TOTAL **I**/ALL-SELF PERSISTS FOREVER. In our limited localizations, however, we forget, and are deluded by the illusions which appear to change from ultimate moment to ultimate moment. YOU-as-You, however, are 'Witness' of all these changes, and YOU change not in the least. Realize that YOU have persisted *forever.* YOU can 'DO' naught else. Be empowered in Cosmos by the PERSISTENCE YOU *forever* ARE.

Counsel 39
Reduce every-thing to a *point,* then make the point disappear.

There is no-thing but that it is Essentially a *point* and ESSENTIALLY, NOTHING. To infinitize any-thing, 'see' it as if from an 'infinispectivizing distance' so that it becomes an infinitesimalizing (a REAL *point).* That point is all points; all points are Really superimposed. Then, imaginatively, make that point disappear. Thus, the realization of INFINITUDE is restored.

Section V
Lists, Figures,
& Tabulations

A Table of Fundamental Bi-Polar Movements or Conditions

The following list is suggestive rather than complete. It includes a number of the basic *dynamics* discussed and developed in this treatise. An understanding of these basic movements and non-movements, conditions and non-conditions, will put polarity in place against the background of THAT which is *resolutely* NON-POLAR.

1. add/subtract
2. advance/withdraw
3. affirm/negate
4. aggregate/disintegrate
5. alter/repeat
6. appear/disappear
7. articulate/ 'homogenate'
8. assert/withdraw
9. attract/repel
10. be/not-be
11. become/be
12. change/sustain
13. cleave/unite
14. concentrate/disperse
15. concretize/etherialize
16. conscious/unconscious
17. contract/expand
18. create/destroy
19. define/'in-definate' [sic]
20. desire/will
21. discontinue/continue
22. down/up
23. emanate/retract
24. enumerate/'de-numerate'
25. 'evanesce'/endure
26. exclude/include
27. extend/super-impose
28. focus/defocus
29. forward/back
30. go forth/remain
31. grieve/rejoice
32. imprison/liberate
33. in/out
34. infinitesimalize/infinitize
35. isolate/permeate
36. labor/play
37. move/stop
38. multiply/re-unify
39. on/off
40. particulate/liquidate
41. point/'un-point'
42. posit/negate
43. precipitate/dissolve
44. present/absent
45. objectivize/subjectivize
46. relate/isolate
47. separate/re-integrate
48. subjugate/elevate
49. veil/unveil
50. yes/no

Synonyms for The FIRST 'EVENT'

- The 'FLASHING FORTH' of the 'RAY' of the ABSOLUTE
- The 'ARISING' of the 'POINT' or 'POINTNESS'
- The FIRST 'ACT'
- The FIRST 'EVENT'
- The MOMENT of 'AWAKENING'
- The AWAKENING of 'MAYA'
- The FIRST 'MISSION' of FOHAT

The FIRST 'EVENT' could be visualized as the 'FLASHING FORTH' of the 'POINT'. Symbolically, the 'POINT' represents a *non-locality,* which while dimensionless (and thus occupying *no space)* indicates that 'Something' is 'HAPPENING', or about to 'HAPPEN', 'Somewhere'. The subjective, the SUBJECT, is becoming objective, the 'OBJECT'. A single point with 'radiating lines' can suggest this 'FIRST EVENT' since it alludes to the 'least' possible 'interruption' in a *vacant 'SPACE'* and also indicates that some 'MOVEMENT' has occured from 'WITHIN' to 'Without'.

Synonyms for The Monad

- A 'Ray' of the ABSOLUTE
- A Point of the One Point which *is* the One Point
- An Authentic Entity
- A Son of the Father
- A 'Ray' of the 'RAY'
- An intra-Cosmic Point of View

Subjectivity & Objectivity Diagramatic Conventions

In these diagrams, the ABSOLUTE, as the most subjective element, is portrayed in black. As the RAY of the ABSOLUTE 'FLASHES FORTH' towards objectivity, the image elements lighten towards white to represent the emergence of the Object from the SUBJECT. However, the order of black to white reverses as further objectification is suggested: shading from white to black shows the immersion of the Object-as-object. The diagram below is one means of portraying these Subject/Object relationships, with both the ABSOLUTE and the lowest levels of Cosmos shown by black.

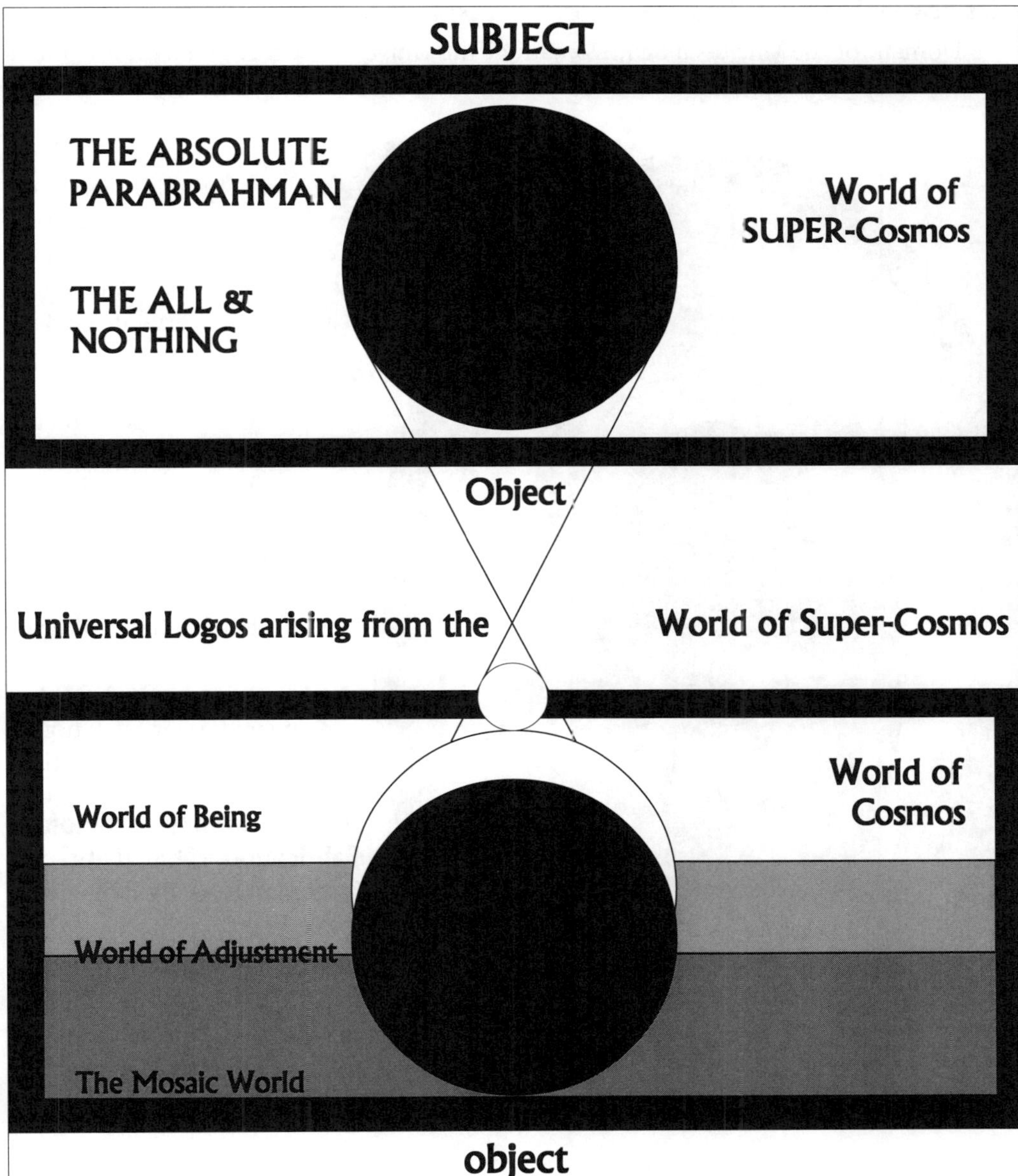

The Worlds (circular model)

SUPER-Cosmos — THE INFINITE SELF — ABSOLUTENESS

Domain of 'NOTHING', from whence the RAY of the ABSOLUTE originates. Represented by the outer black elliptical border.

Super-Cosmos — The Infinite Self — Infinite Space/Object or Mulaprakriti

This Domain is shown by the outer gray ellipse.

Cosmos — The Finite World — Cosmic Prakriti or Space

Domain of the Universal Self; Domain of the Universal Logos. This Finite World is portrayed by the 3 innermost circles, which in their totality comprise Cosmos.

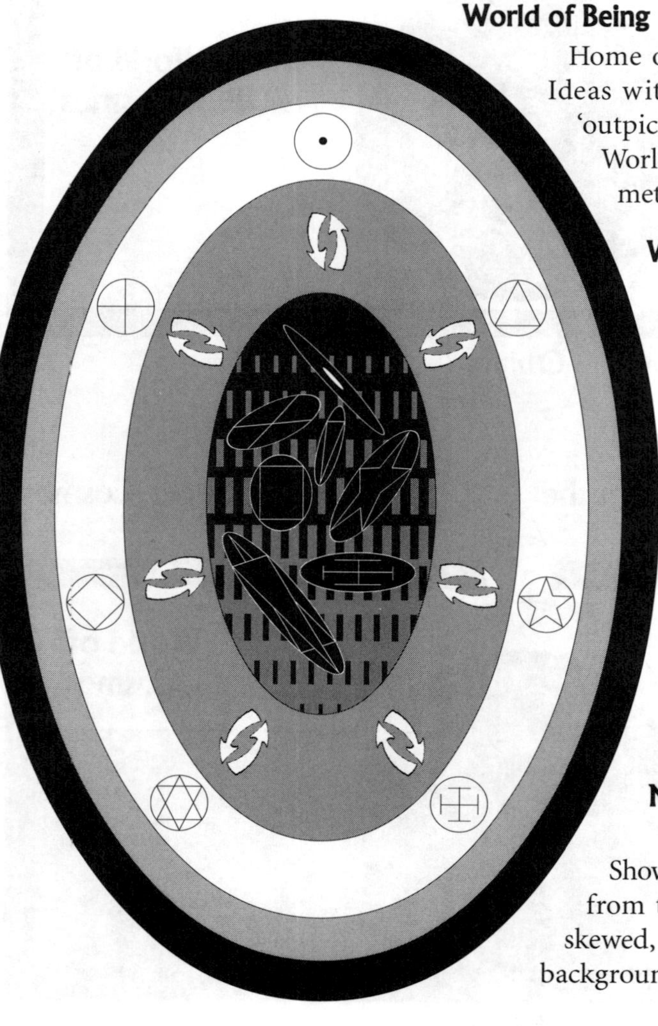

World of Being

Home of the Archetypal Geometrical Ideas within Cosmic Subjectivity that 'outpicture' in the lower worlds. This World is shown by the simple geometrical forms in the white circle.

World of Adjustment

Domain of cooperation between the 'Holders of the Image' representing the Universal Son, and the 'Replicators of the Image' representing Universal Fohat. Fohat is represented here by the dual 'feedback' loops connecting the Archetypes in the World of Being with their reflection in the World of Approximation.

World of Approximation, Fabrication, Objectivity, Particulateness, or The Mosaic World

Home of shadows and 'parts'. Shown by the central cluster of 'Ideas' from the World of Being, darkened, skewed, and placed upon a particulate background.

The Worlds
(vertical model)

WORLD OF SUPER-Cosmos, THE WORLD OF THE INFINITE SELF

THE ABSOLUTE — THE INFINITE — PARABRAHMAN

World of Super-Cosmos, The World of the Infinite Self

The Infinite Subject — Mulaprakriti (Infinite Object)

World of Cosmos, World of Objectivity

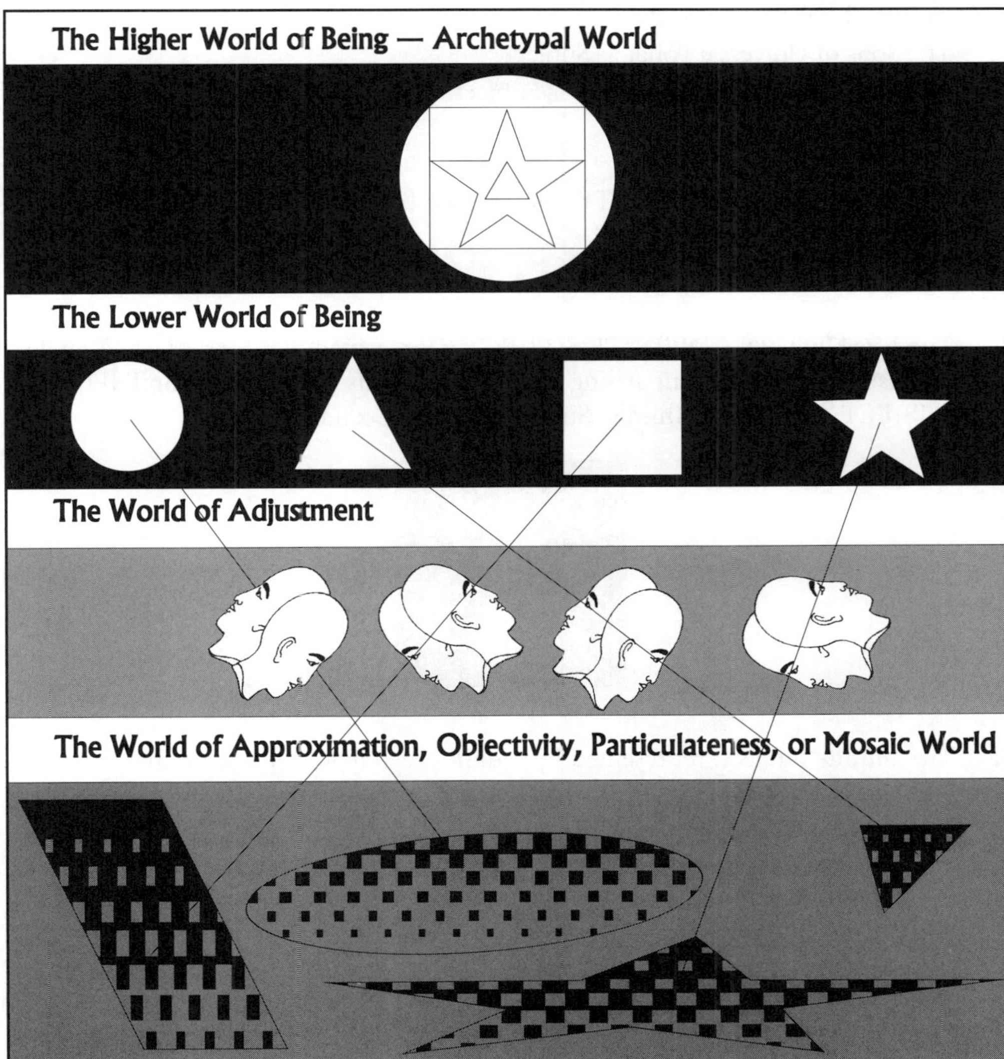

The SUBJECT/Subject
In a Possible Order of Appearance

1. The INFINITE SUBJECTIVITY
2. The EVANESCENT INFINITE SUBJECT
3. The Infinite Subject
4. The Focusing Infinite Subject
5. The Focused Universal Subject or Universal Logos
6. The Universal Son as Subject
7. The Three Sub-Logoi 'Surrounding' the Son as Subject
8. The Seven Subsidiary Logoi as Subject
9. The Multitudinous Monadic Host as Subject
10. Universal Fohat as Subject
11. The Sons of Universal Fohat as Subject
12. The Multitudinous Fohatic Host as Subject

1. The INFINITE SUBJECTIVITY: represented by 'NO-THING' — literally, no image should be considered.

2. The EVANESCENT INFINITE SUBJECT: represented by a dark dashed circle, suggesting the evanescent arising and immediate disappearance of an INFINITE SUBJECT 'SEEING' an Infinite Subject in Super Cosmos.

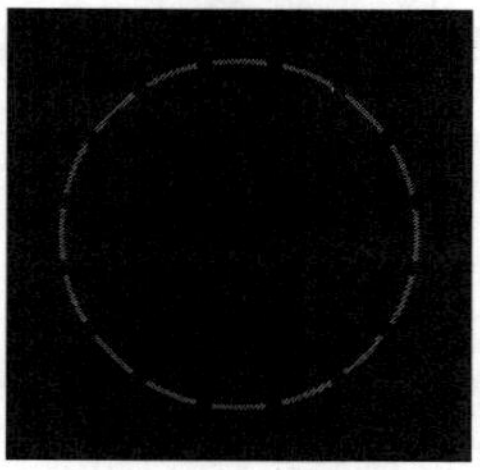

3. The Infinite Subject: represented by a dark gray circle, suggesting that the evanescent INFINITE SUBJECT has become continuously existent in Super Cosmos.

4. **The Focusing Infinite Subject:** represented by concentric circles, suggesting the condensing of the Subject.

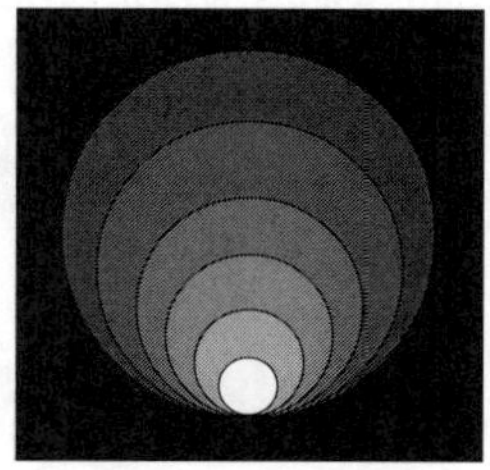

5. **The Focused Universal Subject or Universal Logos:** a small 'open' point, indicates the Light of Life, which is the Universal Logos as an externalized possibility. (The Universal Logos is represented as white, as it is The Object relative to the ABSOLUTE, and The Subject to all Objects/objects *below* It.)

6. **The Universal Son-as-Subject:** shown here as a light gray circle surrounding the smaller, generative 'Father' point.

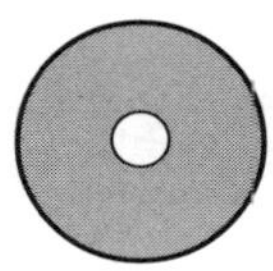

7. **The Three Sub-Logoi Surrounding the Son-as-Subject:** pictured as three darker circles surrounding the Son, suggesting the Sub-Logoi's increasingly objective nature.

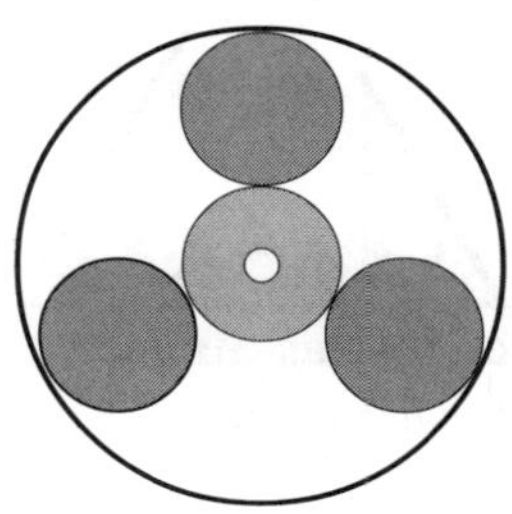

8. **The Seven Subsidiary Logoi-as-Subject:** suggested as seven dark gray circles surrounding the Three Sub-Logoi, as the Seven are more objective than the Three (and, within Cosmos, greater objectivity is portrayed as greater degrees of darkness).

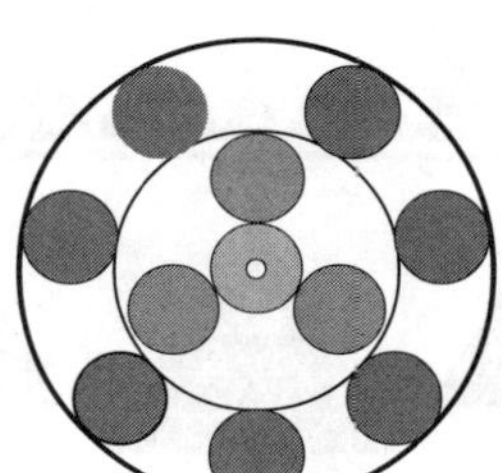

9. **The Multitudinous Monadic Host as Subject:** latent 'Rays' within the Seven Sub-Logoi, each reflecting the structure of superior subjects. (Each circle represents one of the Seven great categories within the Monadic Host.)

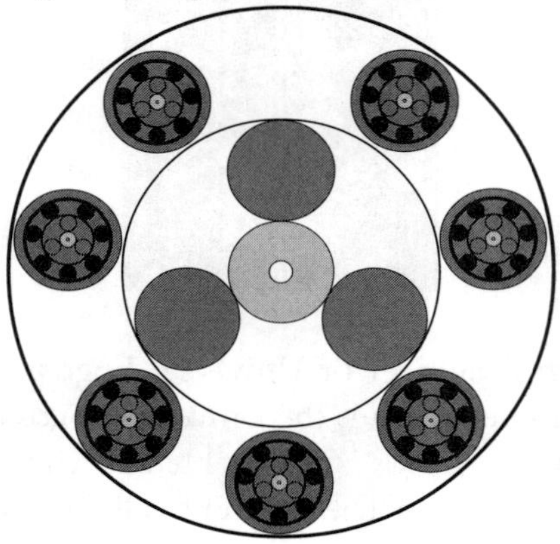

10. **Universal Fohat-as-Subject:** shown as emerging from the center of the Son, the Universal Logoic point. Universal Fohat is represented as a triangle, alluding to movement and circulation.

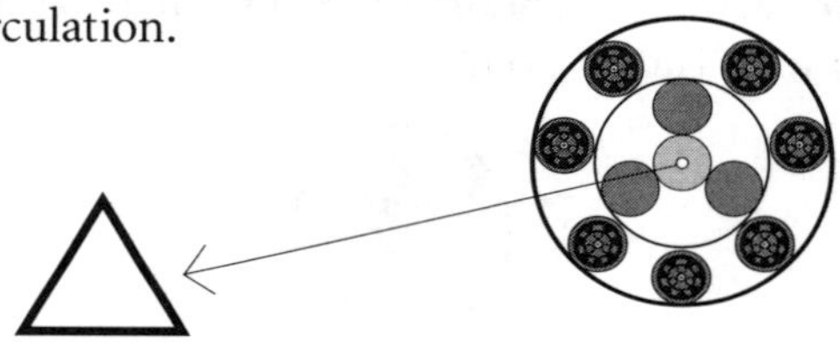

11. **Sons of Universal Fohat-as-Subject:** again, a subjective Three and more objective Seven 'Sons' arise.

12. **The Multitudinous Fohatic Host-as-Subject:** further reflections of superior Subjects emanate in hierarchical pattern.

The 'OBJECT'/Object
In a Possible Order of Appearance

1. The 'RAY', the 'POINT', the 'FLASH'
2. The 'EVANESCENT' INFINITE SUBJECT as OBJECT
3. The 'EVANESCENT' INFINITE OBJECT
4. The Infinite Subject, Infinified Point, 'Ray' of the ABSOLUTE as Object
5. Mulaprakriti, Infinite Mother, Infinite Object
6. Condensing Point, Focusing Universal Subject as Object
7. Finitizing Prakriti, Finitizing Mother
8. Condensed Point, Focused Universal Subject as Object
9. Cosmic Prakriti, Universal Mother
10. Universal Logos as Object
11. Universal Son as Object
12. Three Sub-Logoi as Objects
13. Seven Subsidiary Logoi as Object
14. All 'Rays'-to-be of the ABSOLUTE as Objects
15. All Sons of Fohat as Objects
16. Ever and Always Maya as Object (but Never 'MAYA' as Object for It is 'UNSEEN')

It will be noted how not only are the usual Objects considered to be Objects, but also the usual *Subjects* are considered to be Objects, for indeed they are so to a Superior Viewer. Also Consciousness, Itself in the form of all the Modes of 'MAYA'-instantly-Maya, is considered to be an Object, for the 'Seeing' that Consciousness *is* can also be 'Seen' by a superior 'Seer'.

Only perhaps the FIRST 'SEEING' cannot be 'SEEN'. This FIRST 'SEEING' may be considered the 'top' of the EVANESCENT INFINITE TRIAD, consisting of EVANESCENT INFINITE 'SUBJECT', EVANESCENT INFINITE 'OBJECT', and EVANESCENT INFINITE 'CONSCIOUSNESS' or 'MAYA'. These three are *'BEEN'* by PARABRAHMAN, but not 'SEEN'.

'SEEING' begins within the EVANESCENT INFINITE TRIAD. For however infinitesimal a moment, the EVANESCENT INFINITE SUBJECT can 'SEE' the EVANESCENT INFINITE OBJECT (and perhaps there is reciprocity), but on that level there is NOTHING to SEE the 'SEEING'. Thus, the FIRST 'SEEING' is the only 'UNSEEN' other than PARABRAHMAN, ITSELF. It is as if this 'MAYA' is the *only non-objective Object,* which means that It cannot be 'SEEN'; It can only be 'BEEN'.

The conclusion is that 'MAYA' and PARABRAHMAN (the 'BE-ER') are ineradicably ONE. PARABRAHMAN 'HAS' *nothing* to 'SEE' 'MAYA' with, without recourse to 'MAYA' (the 'SEEING'); therefore, PARABRAHMAN (unable to 'SEE' 'MAYA') is 'limited' to *'BEING'* MAYA. 'MAYA' thus, as it were, *becomes* the 'UN-SEEN' 'SEEING', PARABRAHMAN's OWN 'SIGHT', which cannot 'SEE' ITSELF. So PARABRAHMAN can only 'BE' ITS 'SIGHT', and never 'SEE' ITS 'SIGHT'. Perhaps this is why 'MAYA' or 'CONSCIOUS-

NESS' is so *mysterious*—'IT' must be an 'OBJECT' because 'IT' is *not* PARABRAHMAN, per se, but IT cannot be 'SEEN' for IT, ITSELF, is the *means* by which PARABRAHMAN would have to 'SEE' IT. Thus, in a way, 'MAYA' is the only 'UNSEEN' OBJECT and since IT is 'UNSEEN', PARABRAHMAN does not 'KNOW' what "MAYA' is 'DOING'.

While all the many 'Seers' can be 'Seen', can the original 'SEEING' be 'SEEN'? Is not the 'SEEING' required in order to 'SEE' the 'SEEING'? We come to the conclusion that, while there is 'Seeing', PARABRAHMAN does not 'SEE' the 'SEEING'. PARABRAHMAN, per se, is not a 'SEER'. PARABRAHMAN is 'BLIND'-in-BE-NESS; 'MAYA' Is ITS 'UN-SEEN 'EYES'. Thus, 'MAYA' is the 'FACULTY' of 'SEEING' belonging to the PARABRAH-MIC ROOT, which cannot 'SEE' but can only 'BE'.

On the Object as Subject

From the 'INFINISPECTIVE' (whether IT 'SEES' or does not 'SEE') every Subject is an Object. The only REAL *subject* per se, is the INFINITE SUBJECTIVITY-BE-NESS. But can an Object be a Subject, i.e., can an Object have Consciousness? Well, a Subject that is an Object is a Subject. The Son, Who is an Object to His Father, is also a Subject to other lives.

Is the Universal Mother, the Ultimate Cosmic Object, also a Subject? The Universal Mother is the Father as 'Seen' by Himself. If the Mother, then, *Is* Really the Father, and the Father is the Ultimate Cosmic Subject, then the Mother too, Essentially *being* the Father, must be a Subject. This idea restores Life to the Mother. This means the Mother, because She *Is* the Father, can also Know as She is *Known*. The Father's Knowingness, then, is Present in all Objectification.

If the Father is *in* the Mother, is the Mother *in* the Father? If the Father 'Seeing' Himself Objectively *Is* the Mother, then is the Mother, 'Seeing' Herself Subjectively, the Father? It would stand to reason, and gives the necessary reflexivity needed for Universal Symmetry.

The Father 'Seeing' Himself as the Mother, 'Sees' His Own Objectivity. The Mother 'Seeing' Herself as the Father, 'Sees' Her Own Subjectivity. Each 'Sees' in the Other what is latent in Itself. Of course Each is the Other, and Really, They are only One Being.

The Three States of Subject/Object in Super-Cosmos

Beginning	Middle	End
Infinified Point	Condensing Point	Condensed Point
Infinite Subject	Focusing Universal Subject	Focused Universal Subject
Super-Cosmic Self	Cosmifying Self	Cosmic Self
Mulaprakriti	Finitizing Prakriti	Cosmic Prakriti
Infinite Object	Finitizing Object	Finite Object
Infinite Mother	Finitizing Mother	Cosmic Mother

One way of illustrating these three states is with a cone-like image:

- the most subjective, beginning state is represented by the opening of the cone, a black circle.
- the transitional, or middle ("ing") state is suggested by the gray, narrowing portion of the cone.
- The most objectified state is indicated by a small white elliptical point.

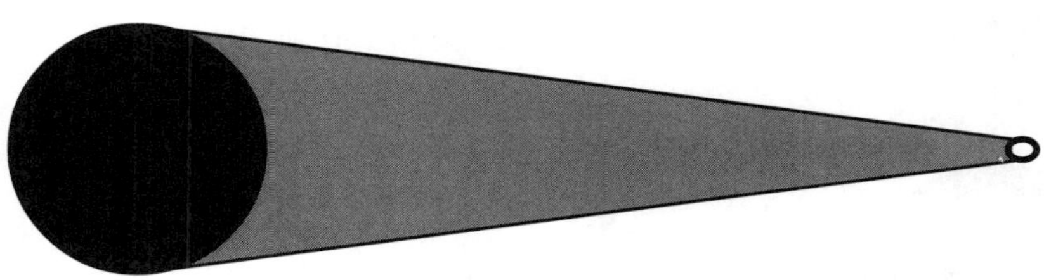

Cosmic First Family

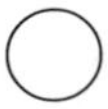 Universal Logos

 Universal Fohat-as-Holy Ghost

 Universal Son

 Universal Mother

 Every Act, Move, or Impulse in Cosmos is Universal Fohat

---

Universal Logos or Father

Self-Sight begins within Father

Universal Logos through Self-Sight generating First Mother Field

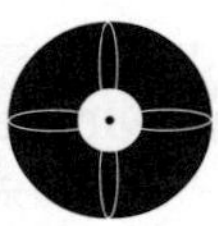

Father and Mother (First Mother)

Father and Mother interplay, generating the Son

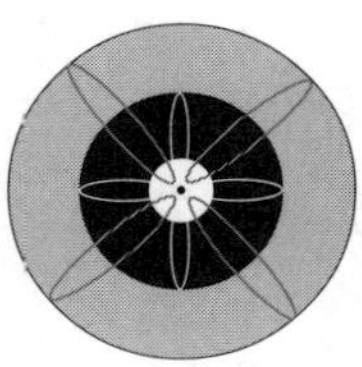

Father, Mother, and Son

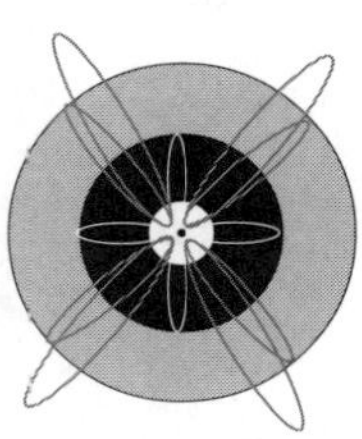

First Mother and Son interplay, generating Second Mother Field

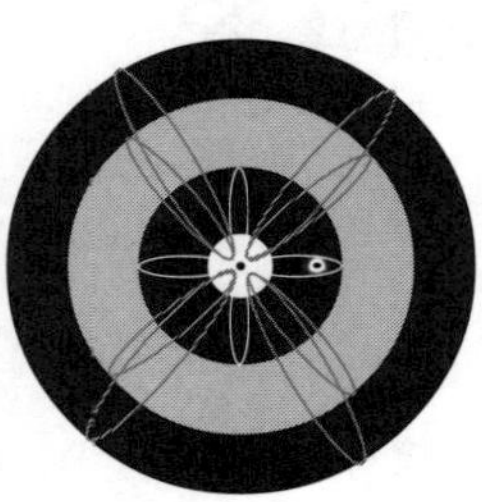

Father, Mother, Son, and Second Mother Field

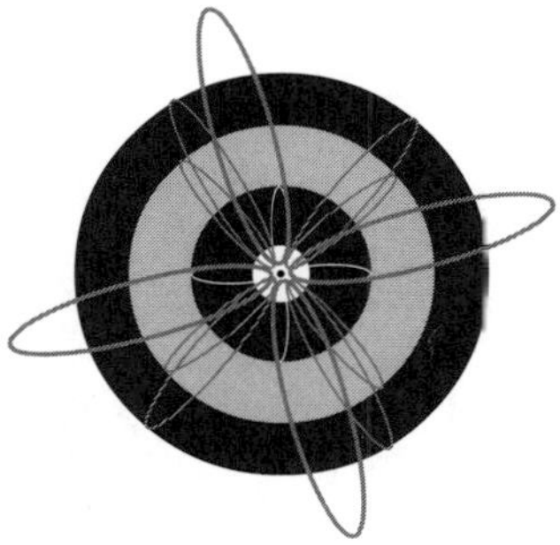

Son and Second Mother interplay,generating Universal Fohat-as-Holy Ghost

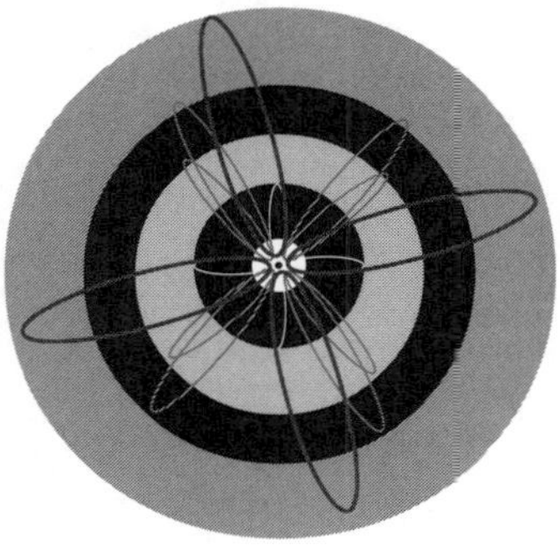

Father, First Mother, Son, Second Mother, and Universal Fohat-as-Holy Ghost

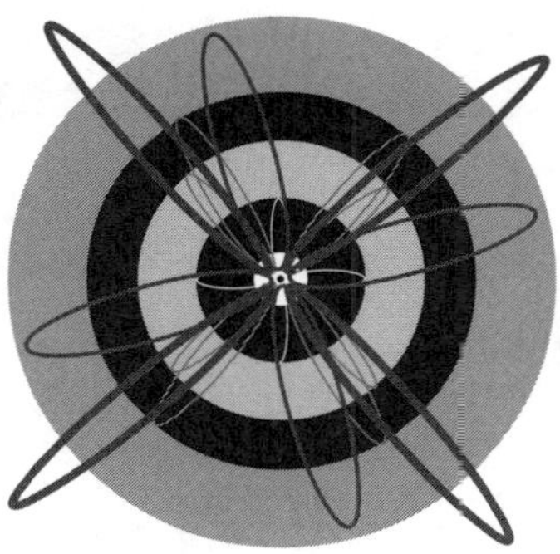

Second Mother and Universal Fohat-as-Holy Ghost interplay, generating Third Mother

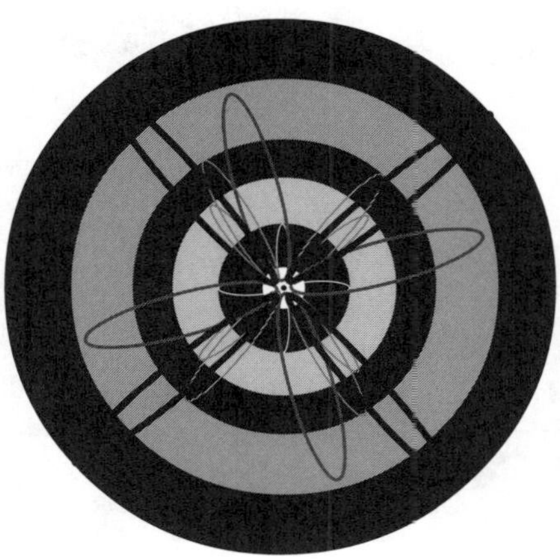

Father, First Mother, Son, Second Mother, Universal Fohat-as-Holy Ghost, and Third Mother

EVANESCENT SUPER-Cosmic 'POINT'

INFINITE 'SUBJECT'	INFINITE 'CONSCIOUSNESS'	INFINITE 'OBJECT'
INFINIFIED 'VIEW-POINT'	INFINIFIED 'SEEING'	INFINIFIED 'SEEN-POINT'

Super-Cosmic Point

Infinite Subject	Infinite Consciousness	Infinite Object
Infinified View-Point	Infinified Seeing	Infinified Seen Point

Finitizing Point

Finitizing Subject	Finitizing Consciousness	Finitizing Object
Finitizing View-Point	Finitizing Seeing	Finitizing Seen-Point

Finite Point

Finite Subject	Finite Consciousness	Finite Object
Finite View-Point	Finite Seeing	Finite Seen-Point

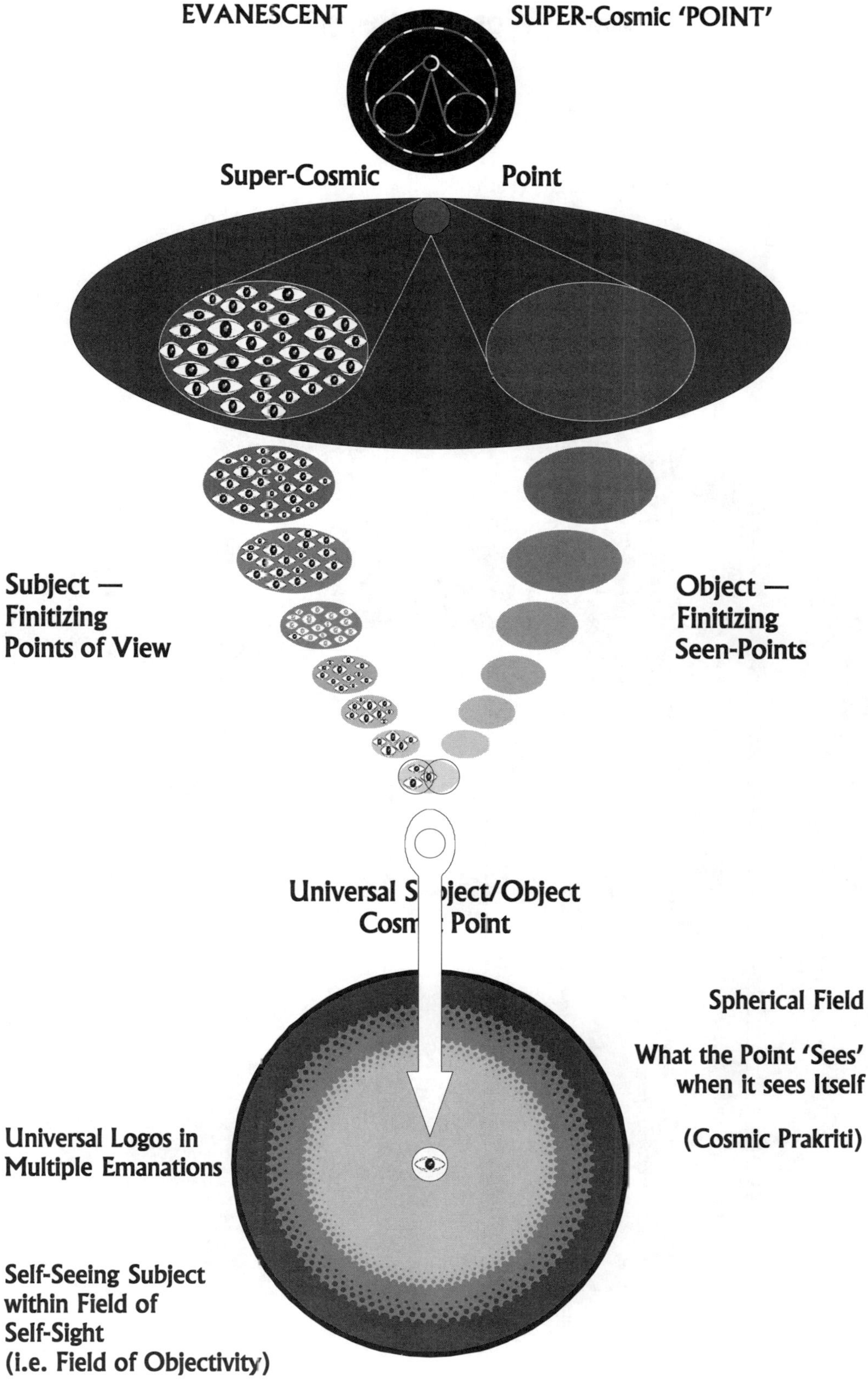
EVANESCENT
SUPER-Cosmic 'POINT'
Super-Cosmic
Point
Subject —
Finitizing
Points of View
Object —
Finitizing
Seen-Points
Universal S ject/Object
Cosm Point
Spherical Field
What the Point 'Sees'
when it sees Itself
(Cosmic Prakriti)
Universal Logos in
Multiple Emanations
Self-Seeing Subject
within Field of
Self-Sight
(i.e. Field of Objectivity)

World of Being

1	=	Universal Logos (Father)	2	=	First Mother
3	=	Son	4	=	Three Sub-Logoi
5	=	Seven Subsidiary Logoi	6	=	Second Mother
7	=	Universal Fohat-as-Holy Spirit	8	=	Three Sons of Fohat
9	=	Seven Subsidiary Sons of Fohat	10	=	Third Mother

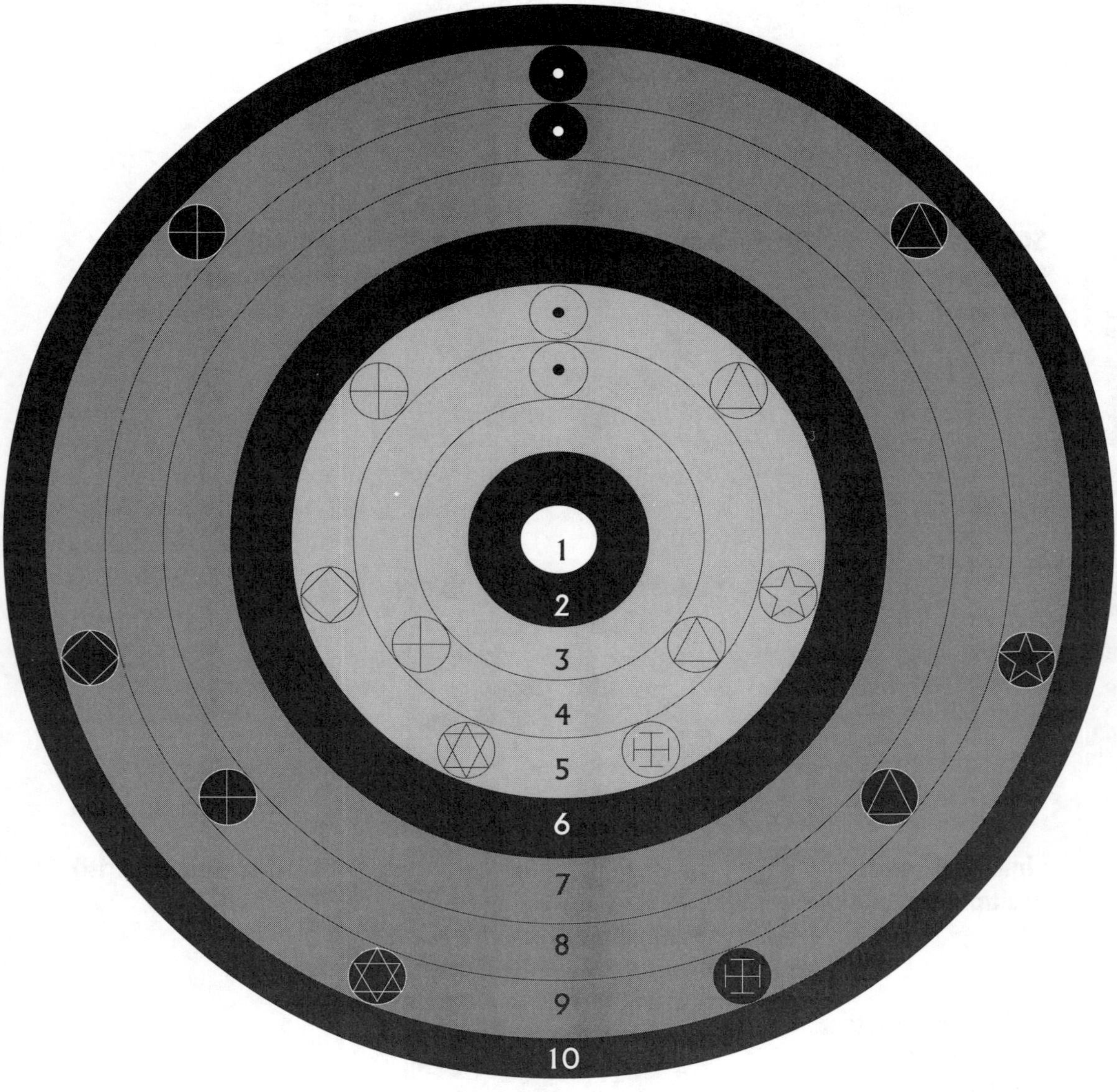

Four Types of Consciousness of Those Rooted in the World of Being

1. 'Within' the Cosmo-Eternal Now, seeing the Universe as a Single Seamless Event.
2. Projected towards the lower Worlds of Fabrication, from which Perspective quantized Time and Space are Observed.
3. Within the domain of 'Unfolding Archetypal Relations' (the World of 'Slow Purposeful Change') which controls the Emergence of Regulatory Ideas. This is consciousness extended into the World of Adjustment. (These Relations unfold according to the Time-Space Formula inherent within the Design-at-the-Beginning.)
4. A fourth level of consciousness involves (by emanation) Those within the World of Being, and that level is simply immersion within the World of Fabrication. (Just as 'we' 'below' are Really 'above' as well, so, They, 'above' are Really 'below'.)

Five PERSPECTIVES/Perspectives in SUPER-Cosmos, Super-Cosmos, and Cosmos

1. The INFINISPECTIVE 'BE-EN' within the INFINITE SELF, i.e., 'INFINIDENTIFICATION'.
2. 'ESSEDENTIFICATION' (IDENTIFYING with but not 'SEEING' the 'POINT').
3. The Infinispective of the Infinite Self 'Seeing' only Itself as Mulaprakriti.
4. The 'Infinispectivizing' of the De-Infinitizing Subject and the Re-Infinitizing Subject. Also, the Condensing Point and the Uncondensing Point.
5. The Cosmo-Spective of the Finitized Subject. Also, the Condensed Point.

Seven Nows

1. The ETERNAL NOW.
2. The Cosmic Eternal Now or Cosmo-Eternal Now.
3. The Cosmo-Subjective Now.
4. The Cosmo-Objective Now or Cosmic Now.
5. The immediate ultimate moment.
6. The precisely measured 'now'—scientifically considered.
7. The 'now' approximately considered: the present moment.

Four Trinities

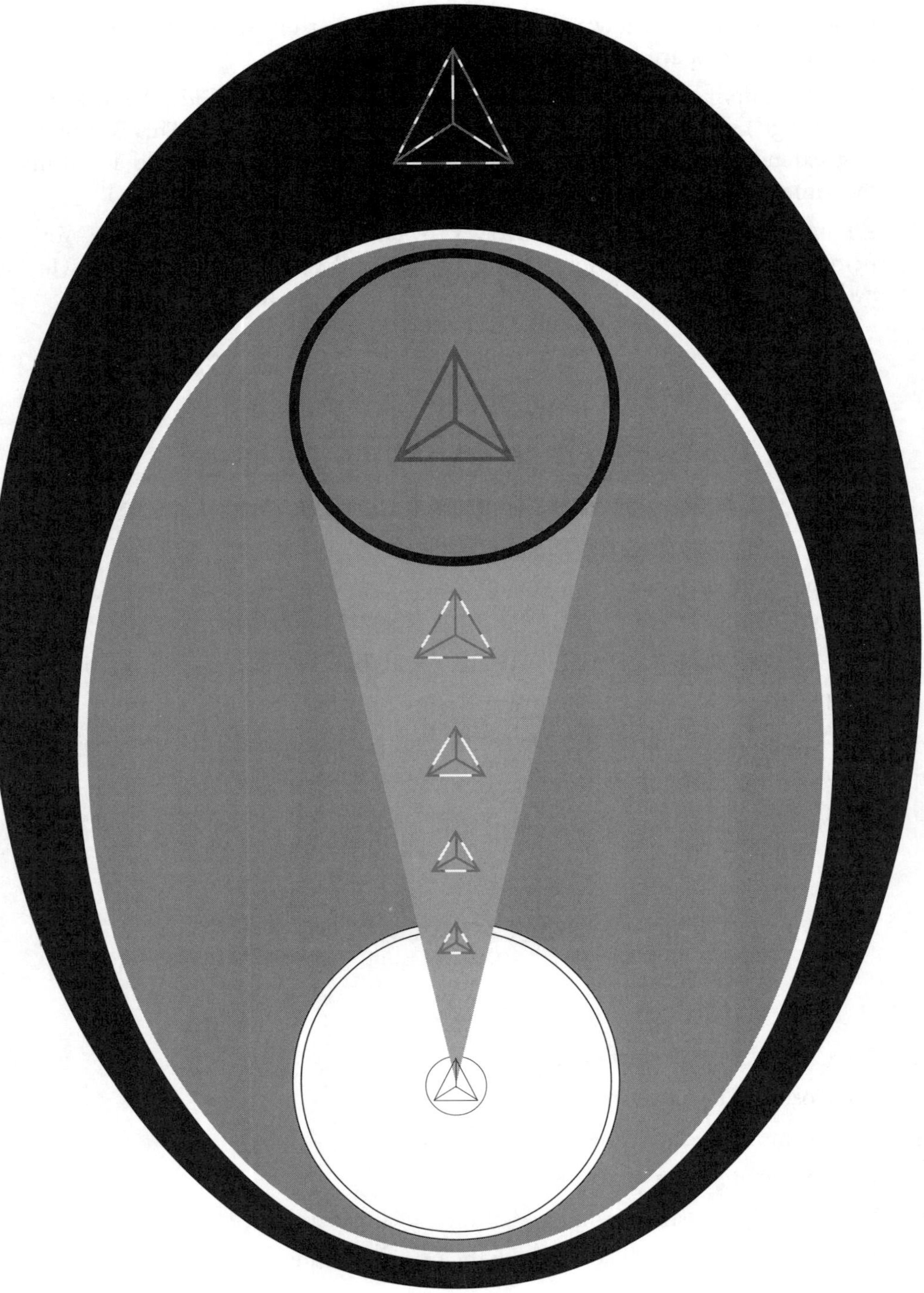

Four Trinities
(continued)

Within the ABSOLUTE (SUPER-Cosmos), the 'EVANESCENT INFINITE TRINITY' FLASHES FORTH

INFINITE SELF

EVANESCENT INFINITE SUBJECT
EVANESCENT INFINITE 'CONSCIOUSNESS' 'MAYA'
EVANESCENT INFINITE OBJECT

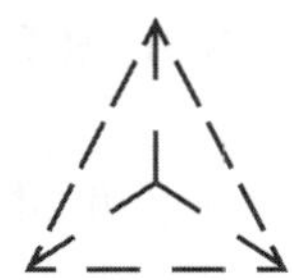

Within Super-Cosmos, the Trinity of the Infinite Self arises through 'Self-Sight'

Infinite Self

Infinite Subject, Infinite Sat
Infinite Consciousness, Infinite Ananda
Infinite Object, Infinite Chit

Within Super-Cosmos, the Infinite Trinity begins condensing, or finitizing ...

Finitizing Self

Finitizing Subject
Finitizing Consciousness
Finitizing Object

... and becomes the Trinity of the Universal Self, or Universal Logos *in Cosmos.*

Universal Self

Universal Subject
Universal Consciousness
Universal Object

Four Modes of FOHAT/Fohat

MODE 1
WORLD OF ABSOLUTE

SUPER-Cosmic Fohat

Agent of INFINITE SELF

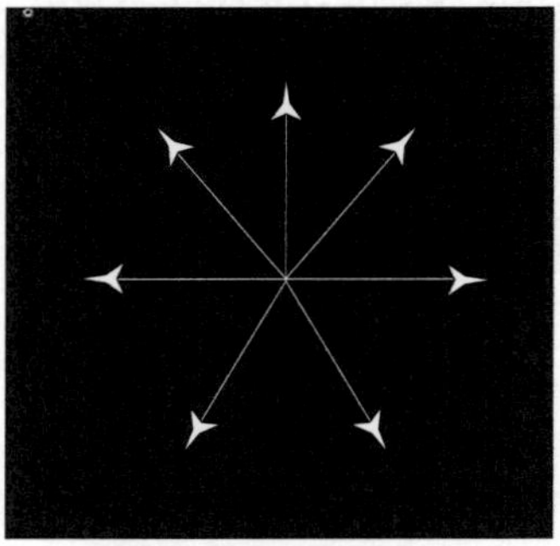

MODE 2
World of Super-Cosmos

Super-Cosmic Fohat

Agent of Infinite Self

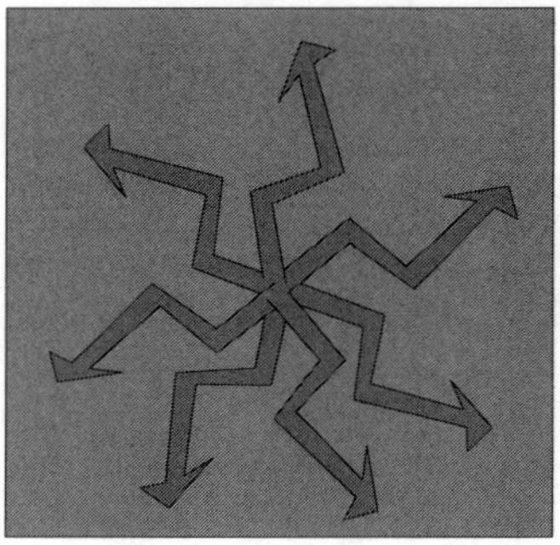

MODE 3
World of Being

Cosmic Fohat

Agent of The Universal Self
within the World of Being

MODE 4
World of Fabrication or Approximation

Intra-Cosmic-Fohat

Universal Fohat as the
Holy Ghost and His Agents

Seed Thought Formulas, Equivalences, and Relational Requirements for Certain Fundamental Ideas

This table of equivalences and requirements can be used to clarify relationships between identical or virtually identical factors in Cosmos, Super-Cosmos and 'in' SUPER-Cosmos. Confusion often results because ideas and concepts that are similar are thought of as if they were different things.

This table should help bring a number of apparently different factors into union. In many cases deep pondering will be required to fathom the latent relationship or equivalency that may not be obvious, but the rewards of persistence should prove fruitful. The reader need not stop with the equivalences that are listed, but can generate his or her own. The possibilities are endlessly engaging and will do wonders for the penetrative power of the abstract/intuitive mind—in this way the Field of the Mind begins to become a unified field and mental synthesis may eventuate.

This list of equivalencies represents but the barest beginning of a quest with extraordinary potential for generating Synthetic Understanding. The author would suggest using various of the equivalencies (or similar self-generated equivalencies) for 'Seed Thoughts' and, in deep contemplation, generating insights that fuse the Fields of the Spiritual Triad. To work in this way is to work with the Antahkarana especially in relation to the Realm of the Abstract Mind.

- A -

ABSTRACTION = FREEDOM
ABSTRACTION = IMMUTABILITY
ABSTRACTION = SALVATION

act = beginning
act=change=motion
act = ending
'ACT' = "not for REAL"
'ACT' = 'PLAY'
'ACT' = 'POINT' = 'RAY' = 'FOHAT' = 'MAYA'

action = apparency
action = discontinuity
action = disequilibrium

'ACTION' = 'ENTITY'
action = illusion
action = the impossible actual
action requires number
action = number
action = separation

addition = diminution
Addition = Subtraction

ALL = NOTHING

attention = dichotomization
attention = objectification
attention = creation

- B -

beginning = change
beginning = change of relationship
beginning = ending
Beginning = Dis-Continuity
beginning = disequilibrium
beginning = finitization
beginning = interval
beginning = point
beginning = separation

BEING requires *blindness*
BEING = Bliss
BEING = IMPENETRABLE SELF-'PREOCCUPATION'
BEING = 'INFINIDENTIFCATION'
BEING = Negation of Interval
BEING = 'NON-SENTIENCY'
BEING = NOTHING
BEING is 'UNSEEING'
BEING = Un-Veiling

BE-NESS = ULTIMATE DESTROYER

BRAHMAN = Samsara

- C -

change = discontinuity
change = event
'CHANGE' = 'LIMITATION'
Change (First Super-Cosmic) = Limitation
change = illusion
change = motion
change = number
change requires number
change requires time

Chord (Music) = Organized Vibration

conception is enumeration
conception is finitization

'CONSCIOUSNESS' = the FIRST 'ACT'
consciousness = attachment
Consciousness = Cleavage
Consciousness = Death
Consciousness = Distance
'CONSCIOUSNESS' = 'DIVISION' of SELF
consciousness requires division/separation
consciousness creates duality
'CONSCIOUSNESS' = 'EXILE'
'CONSCIOUSNESS' = 'EXPULSION'
Consciousness = Externalization
'CONSCIOUSNESS' = FIRST 'DISCONTINUITY'
Consciousness = First Discontinuity
Consciousness = Finitization
'CONSCIOUSNESS' = the 'FLASH'
Consciousness = Fohat
Consciousness = Ignorance
Consciousness = Illusion
Consciousness inaugurates extension
'CONSCIOUSNESS' = 'EXTRUSION'
'CONSCIOUSNESS' = 'FINITIZATION'
'CONSCIOUSNESS' = the FIRST 'LIMITATION'
Consciousness = the First Limitation
Consciousness = Fohat
'CONSCIOUSNESS' = 'HETEROGENIZATION'
Consciousness = Love
'CONSCIOUSNESS' = 'MAYA'
Consciousness = Maya
Consciousness = Motion, as Space = Time
consciousness = objectification
Consciousness = Pain
Consciousness = Prison Warden
'CONSCIOUSNESS' = 'RADIATION'
Consciousness creates Relationship
Consciousness = Revelation

consciousness = 'seeing'
'CONSCIOUSNESS' = SELF-'SIGHT'
Consciousness = Separation from THAT
Consciousness creates Space
Consciousness = Space
Consciousness = the Sword
Consciousness creates Time
Consciousness = Universe
'CONSCIOUSNESS' = UNSEEN 'SEER'
Consciousness = Veil

Cosmos = "Anu" (the "Speck")
Cosmos = the Dis-Continuum
Cosmos = Evil
Cosmos = *the* 'EXTRUSION'
Cosmos = the Great Contradiction
Cosmos = the Infinitesimalizing
Cosmos = Ultimate Macro-Quantum

CONTINUUM = HOMOGENEITY
CONTINUUM = Timelessness
CONTINUUM = 'SPACE'

Creation requires Emanation
Creation requires Enumeration
Creation requires Finitization
Creation = Limitation
Creation = Self-Projection
Creation = Self-Replication

- D -

densification = limitation
Densification, Infinite = Mulaprakriti

Density, Infinite = Self-Reflection of the Infinite Subject
Density, Infinite = Reflected Infinite Homogeneity

differentiation requires enumeration
differentiation = enumeration
differentiation = minimization
Differentiation = Self-Enumeration
differentiation requires space
dimension = discontinuity
dimension = distinctive presentation
dimension = division
dimension = localization
dimension = Fohatic Self-Perception
Dimension = Self-Projection
dimension = repetitive relationship
dimension = specified Self-Reflection
Dimension = Vault of Reflected-Being
dimension = vibratory frequency

discontinuities = points
discontinuities = possibilities
Discontinuities = Presentations in Consciousness
discontinuities = things

discontinuity = actuality
discontinuity = change
discontinuity requires difference
discontinuity = division
Discontinuity = E/entity
discontinuity = event
discontinuity = motion
discontinuity requires number
Discontinuity = Number

division = discontinuity
division requires extension
division = illusion
division = multiplication = subtraction
division = pain

Duality = MONALITY

- E -

Ego = requires boundary
Ego requires Consciousness
Ego requires differentiation

Ego = Disequilibrium
Ego = Emanative Loss
Ego = forgetfulness of ZERO
Ego = localization of consciousness
Ego = loss of Subjectivity
Ego = Number Two mistaken for Number One
Ego = Self-Division
Ego = Self-objectification
Ego = Self-particulation
'EGO' = SELF-Reflection
Ego = Self-Forgetful 'Self-Sight'

Emanation = Arithmetization of the One
Emanation = Divisibility of the Spirit
Emanation = Entification
Emanation requires Enumeration
Emanation = focused Self-'Sight'
Emanation = Gift of Self
Emanation = Hierarchicalization
Emanation = Holographic Self-reproduction
Emanation = Limitation
Emanation leads to Localization
Emanation = Self-Attenuation
Emanation = projected Self-reduction
Emanation requires Self-objectification
Emanation = Self-Enumeration
Emanation = Self-'re-creation'
Emanation = Subjective Birth
Emanation = subjectivized Self-'objectification'
emanation requires time

ending = beginning
ending = change of relationship
ending = discontinuity
ending = moment of re-creation

Entification requires Consciousness
Entification = Limitation of View Point
Entification = Self-objectification
Entification = Partialization
Entification = 'Pointilization'
Entification = Qualification
Entification = illusory Self-particalization
Entification requires Separation
Entification = undetached Self-division

E/entities = the same 'Ray'
entities = quanta

Entity = Identity
Entity = Limitation
E/entity = illusory SELF-'DIVISION'
entity = motion (*some* motions are *not* entities)
E/entity = Subject-as-Object & Object-as-Subject
E/entity = 'Ray' of the ABSOLUTE
entity = relationship

enumeration = diversification
Enumeration = Entification
Enumeration = Illusion
Enumeration = Finitization
Enumeration = 'Fohatization'
Enumeration = 'Mayavication'
enumeration = minimization
Enumeration = Monadic Replication
Enumeration = SELF-'REDUCTION'
Enumeration = violation of ZERO

Essence = the 'Ray' of the ABSOLUTE
ESSENCE = the 'RAY' of the ABSOLUTE
ESSENCE = SELF
ESSENCE = SUBSTANCE

Essentialization = Centralization
Essentialization = Disillusionment
ESSENTIALIZATION = Eradication of Extension
ESSENTIALIZATION = Homogenization
Essentialization = 'Identicalization'
ESSENTIALIZATION = Infinitized Reduction
Essentialization = Noumenalization
Essentialization = Self-Restoration

ESSENTIALIZATION = SELF-Restoration
event = change
Event, the = Cosmos
event requires consciousness
event = conscious registration
event = content of consciousness
event = a discontinuity
'EVENT' = 'FLASH'
'EVENT' = 'FINITIZATION'
event = happening
event requires interval
event = movement
event = perception
event = precipitation
event = presentation
Event = Self-objectification
Event = Self-reflection
Event = 'Trinitization'

'Eventation' = Enumeration
'Eventation' = Intervalic Punctuation

Everything = Everything Else

Everythingness = Infinite Enumeration

Evil = belief in REAL boundaries
Evil = Cosmos
Evil = Devoted Particalization of Identity
Evil = Division
Evil = Egotism
Evil = Enthrallment in Self-Reflection
Evil = Limitation
Evil = 'Fractionation'

Evolution = De-Objectification of Consciousness
Evolution = De-Particalization of Consciousness
Evolution = Gradual Convergence upon the Point
Evolution = Intended Geometrization
Evolution = 'Retraction'
Evolution = Sequentialized Re-unification
evolution = unifying change of relationship
Evolution = Unveiling
Evolution = Spirit-Fusion

Extension = a content of consciousness uncontained by a point
Extension = Illusion
Extension = Intervalization
extension, Fohatic = multi-particulate objectification
Extension = Measurability in the Field of Consciousness
Extension = Objectification
extension = the perceived 'filling' of Space
extension = perception
extension = spatial registration

Externalization = Banishment
Externalization = Death
'EXTERNALIZATION' = 'CONSCIOUSNESS'
Externalization = Emanation
'EXTERNALIZATION' = Enumeration
'EXTERNALIZATION' = 'EXTRUSION'
'EXTERNALIZATION' = Finitization
Externalization = Movement from 'Here' to 'There'
Externalization = Objectification
Externalization = Self-'Sight'
Externalization creates Two
Externalization requires Two

Externalized SELFHOOD = Illusion

- F -

Finitude = Illusion
Finitude = Infinite Loss
Finitude = Reduced Self-focus
Finitude = SELF-Forgetfulness

Fire = Action
'FIRE' = the First 'ACTION'
Fire = Change
Fire = Dance of Life
Fire = externalized PRESENCE
Fire = Motion
Fire = the Power of Re-Configuration
Fire = the Power to Change
fire = vibration

fires = events

Fohat = Action
Fohat = the Actor
Fohat = the Agent of Externalization
Fohat = Agent of the Willing Subject
Fohat = the 'Bearer'
Fohat = the Enumerator
Fohat = Eros
Fohat = Father of Vibration
Fohat = He Who Goes Forth
Fohat = the Intermediary
Fohat = the Joiner
Fohat = Maya
'FOHAT' = 'MAYA'
Fohat = the Messenger
Fohat = the Objectifier
Fohat = the Particulator
'FOHAT' = the 'RAY' of the ABSOLUTE
Fohat = the Relation Between
Fohat = the Revealer

- G -

God = Cosmic Singularity
GOD = Infinitude
God = GOD minus *Infinity*
God = the One Ray of the ONE 'RAY'
God = the Universal Logos

Goodness = Aesthetic Universal Fulfillment
GOODNESS requires *conditionlessness*
GOODNESS = I
GOODNESS = INDIVISIBILITY
GOODNESS = INFINITIZATION

- I -

Idea = Archetype
Idea = Being plus Quality
idea = coherent discontinuities
Idea = a dimensionless presence
Idea = the Energy of Geometrical Relationship
Idea = Entity
Idea = 'EXTRUDED' *infinitessentialization*
Idea = Geometry within a Point
Idea = Monads in Relationship
Idea = Number
idea = patterned point of view
Idea = a Point with formative power
Idea = a Possibility 'EXTRUDED' from the FOUNT OF ALL POSSIBILITY
Idea = Power of Configuration
Idea = 'Rays' of the ABSOLUTE in Relationship
Idea = Seed-Point
Ideas = Numbers in Arithmetical Combination

Identification = being *another*
Identification = 'De-intervalization'
Identification = Destruction of the Observer
Identification = Essentialization
Identification = Freedom from Modification
IDENTIFICATION = *internalization*
Identification = the multiplication of distance by zero
Identification = Negation
Identification = Negation of Objectification

illusion = appearance
illusion = a content of consciousness
Illusion = Anything
Illusion = Apparency
Illusion = Duality
Illusion = Impermanence
Illusion = Objectification
Illusion = Reflection of Subjectivity
Illusion = Revelation of Consciousness
illusion = seeming
Illusion = Self-Reflection
Illusion = 'Sight'
Illusion = 'Thinghood'
illusion = vibration
Illusion = What 'Sight' 'Sees'

Individuality = Essential Identity
Individuality = Indivisibility
Individuality = the One Identity
Individuality = the Presence of THAT in This
Individuality = That which can be contained in a Point
Individuality = Seamless Wholeness

infinitesimal = 'definite indefinite'
infinitesimal = less than any amount of time
infinitesimal = "next to nothing"
infinitesimal = 'somethingness' converging on 'nothingness'
infinitesimal = a vanishing point (never vanishing)

'infinitesimalizing' = all appearances from 'infinispectivizing'
'Infinitesimalizing', an = Cosmos 'Infinispectivizingly' Viewed
infinitesimalizing = escape from quantification
infinitesimalizing = polar-opposite of infinitizing
infinitesimalizing, an = something *less* than NOTHING

infinitization = annihilation
infinitization = essentialization

Infinitization = Life Restoration
Infinitization = Noumenessentialization
Infinitization = Ultimate Maximization
Infinitization = PERFECTION
Infinitized SELF-Preoccupation = REALITY, 'INFINIDENTIFICATION'

Infinite Time requires segmentation
Infinite Time = Infinite Duration
Infinite Time = an intermittent Foreverness of Events

Infinitized Interiorization = LIFE

INFINITUDE = Universe + INFINITUDE

interval = discontinuity
interval = an ending and a beginning
interval requires event
interval = event
interval = extension
interval = illusion
interval requires canalized focus
interval = two events separated by non-event
interval = two non-events separated by an event

IT = I
IT = ITSELF
IT = NOTHING
IT = 'PRIVATION'
IT = VOIDNESS
IT = ZERO

item = event
item = extraction
Item = 'EXTRUSION'
item = I in specificity
item = a perception
item = point
item = a presentation in Consciousness
item = space

item requires time
'item' spells 'time' (by permutation)

- L -

LIFE = CONTINUOUSNESS
Life = Fire
LIFE = INFINITUDE
Life = Limitation
Life = One and Oneness
Life = the Point
LIFE = POINTLESSNESS
Life = Vibration
LIFE = ZERONESS

localization = objectification
localization = limitation
localization = precipitation

Love is *not* BEING
Love = Consciousness
Love = Consolation in Exile
Love = Cosmic Jailer
Love = the Divine Emanatory Stream
Love = Embracing Illusion
Love = the First Fall
Love = the Final Return
Love = Loss of Self
Love = 'Seeing'

- M -

matter = any-thing
Matter = the contents of Consciousness
Matter = NOTHINGNESS 'Seen'
matter = object
Matter = Objectified Internality
Matter = the Possibility of Self-Disclosure
Matter = Reflection of Ideation
matter = Self-Seen points
Matter = the 'Seen' Self
matter = 'Sonship' objectified
matter = thingship
Matter = ZERO-as-Zero-as-One-as-Many

Maya = Consciousness
Maya = the Creator
Maya = the Dancer
Maya = Deceiver
Maya = Distracter
Maya = the Dramatist
Maya requires Enumeration
Maya = Enchanter
'MAYA' = the 'ENTERTAINER'
Maya = the Entertainer
Maya = Externalization
Maya = Eros = Fohat
'MAYA' = 'FOHAT'
'MAYA' = Infinitesimalizing
Maya = Interval
Maya = Limitation
Maya = Love
Maya = Lure from Center
Maya = Motion, Space and Time = Consciousness
Maya = the Multiplier
Maya = Objectification of Subjectivity
'MAYA' = PARABRAHMIC SELF-'DELUSION'
'MAYA' = 'POINTNESS
Maya = the Producer
'MAYA' = the 'RAY'
MAYA' = 'RELATIONSHIP'
Maya = the Reducer
'MAYA' = the 'REDUCER'
Maya = SELF in 'ACTION'
Maya = SELF-Reduction
Maya = 'Sight'
Maya = the Thief
'MAYA' = UNSEEN 'SEEING'
Maya = the Veiler
'MAYA' = the 'VEILER'

modification = change
modification = discontinuation

modification = diversification
modification = fragmentation
modification = movement
modification = obscuration
Modification = Perturbation
Modification = Self-Perception
Modification = the Veil
modification = vibration

moment (ultimate) = frozen relationship
moment (ultimate) = stillness
Moment (Ultimate Macro) = Universal Manvantara

MONALITY requires Duality

motion = action
motion = change
Motion requires Consciousness
Motion = Difference
Motion requires Discontinuity
motion = discontinuous change of relationship
Motion requires Enumeration
Motion demands Entification
Motion requires Extension
Motion = Fire
Motion = Illusion
Motion = Interruption
Motion = Maya
Motion births Number
Motion requires Number
motion requires disappearance and reappearance
motion (ceaseless) = rapidly changing periods of absolute Rest
Motion = Re-configuration
Motion requires Separation
Motion requires Space
Motion requires Time
Motion Veils Being

Mulaprakriti = the Boundless Finitude; the Finite Boundlessness
Mulaprakriti = the Super-Cosmic 'Seen Point'
Mulaprakriti = God in 'Sight' of Himself
Mulaprakriti = God 'Knowing' Himself
Mulaprakriti = Infinite Object
Mulaprakriti = Infinite Objective Pointness
Mulaprakriti = Infinite Homogeneity Reflected
Mulaprakriti = the Mother Who Is Father
Mulaprakriti = Non-intervalic Objectivity
Mulaprakriti = the NOTHINGNESS-as-'Somethingness'
Mulaprakriti = PARABRAHMAN-as-Object
Mulaprakriti = Root Matter
Mulaprakriti = Self-'Sight' of Infinite Self
Mulaprakriti = the 'Thereness' of Potential Allness
Mulaprakriti = What is 'Seen' through the Infinified Point

Multiplication = Dimensionalization
Multiplication = Distraction
Multiplication = Entification
Multiplication = Illusion
Multiplication = Monadic Replication
Multiplication = Mutual Augmentation
Multiplication = Retreat from One and Oneness
Multiplication = SELF-Reduction
Multiplication = Self-Reproduction
Multiplication = Subtraction

- N -

Negation = Abstraction
Negation = the Enemy of Maya
Negation = the Destruction of the Object
Negation = FREEDOM
Negation = Identification
Negation = Liberation
Negation = REALITY

Negation = SELF-Affirmation
Negation = SELF-Restoration

NOW = the Annihilation of Time
NOW = Annihilator of Interval
NOW = the BOUNDLESS MOMENT
NOW = *consumer* of sequence
NOW = 'DOORWAY' to the PRESENCE
NOW = the ETERNAL 'MOMENT'
NOW = *friend* of ZERONESS
NOW = Immovable BEINGNESS
NOW = INFINITE DURATION
NOW (ETERNAL) = *infinitized intension*
NOW = the infinitesimal negated by ZERO
NOW = TIMELESSNESS
NOW = Verticalization of Horizontal Temporal Extension

NOTHING = INFINITIZED EVERYTHINGNESS
NOTHING = the NAMELESSNESS
NOTHING = the PLENUM
NOTHING = PURE PURPOSELESSNESS
NOTHING = ULTIMATE AFFIRMATION
NOTHING = the ZERO

Number = the Companion (of the ZERO)
Number = Delineator of the Field of Consciousness
number = division = relation = item
Number = Ego
Number = Entity
Number = Idea
Number = partite, 'imparticulate' Relationship
Number = the Individualizer
Number- the Infinite Reduction of ZERO
Number = the Monad
Number = the Monad in Extension
number requires motion requires time
Number = Objectification
Number = Power of Division
Number = the Preoccupier
Number = the Reification of the ZERO
Number = Self-Division
Number = Self-'Sight'
Number < ZERO

- O -

object = conscious focalization
object requires duality
object = illusion
Object = Image of Imagelessness
object requires interval
object generates interval
object requires motion
object requires number
object = a 'one-ing'
object = a point of view
Object = a Projected Interiority
Object = Subject
object = a *Real* point
object = an un-REALITY
object = a virtual point

Objectification = Condensation
Objectification = Enumeration
Objectification = Finitization
Objectification = Limitation
Objectification = 'Mayavication'
Objectification = Mayavic Projection
Objectification = Mind-Born Birth
Objectification = Perception
objectification requires separation/division
Objectification = Self-Alienation
Objectification = Self-'Sight'
Objectification = Self-'Splitting'
objectification requires vision
Objectification = 'View-Pointing'

Objectivity = "the Far Country"
Objectivity = any *perceptual 'distance'* greater than zero
Objectivity = Subjectivity
OBSERVER = CONTINUITY

Observer = Darkness
'OBSERVER' = 'EVANESCENT INFINITE SUBJECT'
Observer = Father in Darkness
Observer = Spirit
Observer = Subject
Observer, Infinite = First 'Seen Point'

One = ZERO
One < ZERO

- P -

PARABRAHMAN 'BECOMES' 'MAYA'
PARABRAHMAN = 'MAYA'
PARABRAHMAN does *not* equal 'MAYA'

Paradox discloses 'IRRECONCILABLE REALITY'
'PARADOX' = the FOUNT OF ALL POSSIBILITY
Paradox = 'TRUTH'

part = apparency
part = impossibility
part = Whole
part = WHOLE

particle = event
particle (ultimate) = Fohatic Attention

particle/event = greatest *actual* state of limitation
particle/event = lower limit of Cosmic Finitization
particle/event = Mosaic unit
particle/event = most minutely articulated Fohatic Self-Perceived Subjective Enumeration
particle/event = smallest *actual* object
Particle/Event, Ultimate Macro = Universe

particularization = attention
particularization = focussed consciousness
Particularization = objectively projected subjective Self-Enumeration
particularization = separative perception

PERFECTION = 'ETERNAL CONSOLATION'
PERFECTION = imperfection (if imperfection lacks)
PERFECTION = INVARIANCY
PERFECTION = PLENUM
PERFECTION is 'PURPOSELESSNESS'

Pervasion = Cosmic Permeability
Pervasion = 'De-egoification'
pervasion = negation of isolation in Space
Pervasion = Negation of Exclusion
Pervasion = Omni-Identification
Pervasion = Omnipresence in Space
Pervasion = Permeating Extension
Pervasion = 'Spirit-Merging'
Pervasion = 'Substandingness'
Pervasion = totally shared Space

Point, the = the Aperture of INFINITUDE
Point, a = Complement of a Field
point = collapse of space
point = a discontinuity
point = a dimensionless objectivity
Point, the = the 'Doorway' through which ZERO 'passes' to *become* One
point = the form of an Idea
Point (Infinified) = Infinite Subject
point = an infinitesimalizing
Point, the = infinitesimal condensation of THAT
point = an object in *essence*
point = an indefinitely small, non-*actual* object
Point, the = NOTHING's 'WAY' of 'BECOMING' 'Something'
point = a Point of View
Point = Portal of Expansion or Diminution
Point = the Root Form in Cosmos
point = 'seen' specificity

Point = a Subject/Object
point = synthesis of extension
Point = *that* which is *not* 'Something' and *not* NOTHING
Point, the = a two-way Mirror
Point = the Transition between the INFINITE and the Finite
point, Real = Will to *somethingness* or *nothingness*
point, virtual = a dimensionally perceived Idea
point, virtual = 'distant' View of an actuality
point, virtual = thing (from a distance)

precipitation = actualization
precipitation = dimensional condensation
precipitation = reification
Precipitation = 'EXUDATION' from 'SUPER-SATURATED HOMOGENEITY'

predication = reification

Presence, the = the Beloved
Presence, the = the Comforter
PRESENCE, the = the 'CONTINUITY'-in-Discontinuity
PRESENCE, the = the 'DISSOLVER' of Time and Space
PRESENCE = the 'MOST FAITHFUL'
PRESENCE, the = the 'NEGATOR' of Objectivity
PRESENCE = the NOWNESS of THAT
PRESENCE, the = *that* which summons ALL into the ETERNAL NOW
PRESENCE = Timelessness

Presentation, the = Cosmos
presentation = content of consciousness
presentation = event
presentation = illusion
presentation = interruption
presentation = object
presentation = a perception

'PRESENTER', the = 'MAYA'
Presenter, the = Maya

Purpose, Cosmic = the presently 'EXTRUDED' IDEA of THAT
Purpose = a Musical Chord
Purpose = Intended Geometrization
Purpose = the Intended Fohatic Mosaic
Purpose = Organized Discontinuity

- Q -

Quality, a = an E/entity
Quality = Idea minus Being
Quality = Number in Cooperation
quality = quantity
Quality = Reduced Self-Reflection

quanta = discontinuities
quanta = discrete possibilities 'EXTRUDED' from the INFINITESSENCE
quanta = enumerations of 'Darkness' discontinuously coming to Light
quanta = Enumerated Subjectivity Reflected in Objectivity
quanta = enumerations within the Infinite Object
Q/quanta = E/entities
quanta = emergence of Time and Space
quanta = events
quanta = 'finitizers' of imparticulate Ideation
Quanta = Numbers
quanta = particulated Individuality (indivisibility)
quanta = particulate Presences in ontological alternation
quanta = subjects of Lords of Time & Space
Quanta = Universes

quantity = numbered aggregation
quantity = quality

- R -

Radiation = Externalization
'RADIATION' = Limitation
Radiation = influence extending from 'here' to 'there' or from 'there' to 'here'
Radiation = mutuality in the Field of Extension
Radiation = Objectification
Radiation = Particularity in Transmission

'RAY' of the ABSOLUTE = the 'APPEARANCE' from 'NO TIME'
'RAY' of the ABSOLUTE = the EVANESCENT INFINITE 'TRINITY'
'RAY' of the ABSOLUTE = 'FINITUDE'-instantly-Finitude
'RAY' of the ABSOLUTE = the FIRST 'DISRUPTION'
'RAY' of the ABSOLUTE = 'FOHAT' 'ARISING'
'RAY' of the ABSOLUTE = the 'GENERATOR' of discontinuous Continuities
'RAY' of the ABSOLUTE = 'MAYA', suddenly
'RAY' of the ABSOLUTE = 'POINTNESS' in 'POINTLESSNESS'
'RAY' of the ABSOLUTE = each 'Ray' of the ABSOLUTE
'RAY' of the ABSOLUTE = THAT 'in' This

REALITY = the CHANGELESSNESS
REALITY = ESSENCE
REALITY = the IRREDUCIBLE
REALITY = the PERMANENCE
REALITY = 'PRIVATION' of 'thingship'
REALITY = THAT
REALITY = 'Thinglessness'

Relation = Communication
Relation = Disposition-in-Extension
Relation requires Enumeration
Relation = Finitization
Relation = quanta in purposeful aggregation
Relation requires Motion
Relation requires Number
Relation = 'Relat-ivity'
Relation requires Objectification
Relation = shard presence in the Field of Consciousness
Relation = shared presence in Field of Space
Relation requires Space
Relation requires Time

ring-pass-not = divided Continuum
ring-pass-not = extent of Self-Perception

- S -

SELF = 'ANNIHILATOR' of Duality
SELF = '*BE*-er', the
SELF = Everything
Self = the Father in Darkness
Self = **8**
SELF = **I**
SELF = the ILLIMITABLE
SELF = the INFINITE
SELF = NOTHING
Self = a 'Ray' of the ABSOLUTE
Self = the One
SELF = the ONE *behind* the One
Self = the One Cosmic 'Ray' of the ABSOLUTE
SELF = the 'SEER' *by proxy*
SELF = SOURCE of all Objectification
SELF = the UNDEFINABLE
SELF = the 'UNEVOLVING ONE'
Self = Universal Logos

SELF-Assertion = SELF-Minimization
SELF-Enumeration = Narrowing Consciousness/Vision

Self-'Sight' = Apparency
Self-'Sight' = Creativity
Self-'Sight' = Emanation

Self-'Sight' = Family Relations
Self-'Sight' = Illusion
Self-'Sight' = Objectification
Self-'Sight' = Self-Replication
Self-'Sight' = Space

separation = addition, subtraction, multiplication, division
Separation = Consciousness = Maya = Limitation = Creation
separation = discontinuity
Separation = Enumeration
Separation = Extraction from HOMOGENEITY
separation = limitation
separation requires localization
separation = localization
Separation = Maya
separation = objectification
Separation requires 'RADIATION'

sequence = consciousness focusing successively
sequence = Finitude's perception of INFINITUDE
sequence requires event
sequence requires limitation
sequence requires objectification
sequence negates simultaneity
sequence = the protector of formal integrity
sequence = reducing valve of simultaneity
sequence requires time
Sequence = Warden of Simultaneity

Simultaneity = Approximation of Eternity
Simultaneity = Completed Point of View
SIMULTANEITY = ETERNAL NOW
SIMULTANEITY = TIMELESSNESS
simultaneity (conscious) requires the All-Seeing Eye
Simultaneity requires Universal Vision
Simultaneity (Super-Cosmic) requires Super-Cosmic Vision

Something = infinite SELF-'REDUCTION'
Something = NOTHING
something = object
Something = Objectified Self-Reduction
Something = Reification of Non-Objectivity
Something = 'Sonship'

Space requires Consciousness
Space = Discontinuity
Space = Extension
Space = Field of Consciousness
Space = Field of Generation
Space, Cosmic = 'Hole' in the Infinite Continuum
Space = Interior of the Subject (in Projection)
space = interval
Space = Object
Space = Prakriti
Space = Registration
Space = Self-Confrontation
Space requires Self-Separation
Space = Self-'Sight'
Space (Infinite) = Discontinuous Objective Continuity
Space (Infinite) = Mulaprakriti
Space (Infinite) = Reflection of Infinite Imparticulate Ideation

spaces = 'holes'

stillness (utter) = ABSTRACTION
Stillness = Disengagement
Stillness = Dissolution
stillness = divisionlessness
Stillness = Freedom
STILLNESS = the 'MOTIONLESSNESS'
STILLNESS = *infinitized* motion
'STILLNESS', the = infinite speed
stillness = REALITY

SUBJECT ('EVANESCENT) = 'BE-EN' 'POINT'
Subject = King; Object = King as (loyal?) Subject
Subject = Object
Subject = 'Seen-Point'
Subject = Self
Subject = View Point

ubjectivity = Home
SUBJECTIVITY, the GREAT = 'HOME'
Subjectivity = the Kingdom

subtraction = addition
subtraction = less as more

succession requires *externalized observation* (as the Super-Cosmic Self of Infinite Duration)

Super-vision negates sequence

- *T* -

thing = a difference
thing = a discontinuity
thing = disequilibrium
thing = a disturbance
thing = disturbance
thing = field from ground
thing = intensification
thing = motionless relationship (motionless for an ultimate moment)
thing = object
thing = NOTHING
thing = particle
thing = a particulate preoccupation
thing = a Point of View
thing = precipitation
thing = a presentation in consciousness
thing = 'Seen Point'

Thou = THAT
thought = configuration
thought = formal Idea
thought = patterning
thought = patterns of Presences
thought = objectification of idea
thought = thing

Time requires Consciousness
time = difference in duration between compared events
time requires discontinuity
time requires duality
time requires event
time = event
Time = Finitizer of ETERNITY
time = interval of difference
time = measured precipitations
Time = Motion = Space (A totally interdependent Triad)
time requires object
time = particularized focus of consciousness
time = quantity of motions
Time = registered sequential appearance
time requires regularity
time requires relation.
Time = space for Motion
Time, measurement of requires Standard of Measurement
Timelessness = BEING

Two = Generator of Necessary Imperfection
Two = One
Two = One = ZERO
Two = Pain
Two = REALITY *contradicted*
Two = What Number One Becomes on Departure

- U -

Universe = Cosmos
Universe requires Consciousness
Universe = the Great Contradiction
Universe = Drama of Finitude
Universe = an Emergent ETERNAL 'LATENCY'
Universe = the Great Discontinuity
Universe = Illusion
Universe = the Infinitesimalizing
Universe = the Interrupter of ETERNITY
Universe = Maya
Universe = the One Act
Universe = the Occasional Finitude
Universe = the Possibility of the Moment
Universe = the specific Idea whose *time* has come

- V -

Veil = Consciousness

'VEILING' = BEING 'BECOMING' 'SEEING'
Veiling = Canalizing Consciousness
Veiling = De-infinitization
Veiling = Discontinuity (The *continuing* CONTINUUM is un-Veiled)
'VEILING' = 'ADVERTENCE' from 'INFINIDENTIFICATION'
Veiling = Enumeration
Veiling = Modification
Veiling = Multiplication
Veiling = Intentional Self-Ignorance
Veiling = Self-Objectification
Veiling = Self-Limitation
Veiling = Self-Reduction
Veiling = Vibration

Vibration = 'Blinking' of Fohat's 'Eye'
Vibration = Becoming
Vibration = Centralization becoming Bi-Polar Oscillation
Vibration = the Cosmic Dance
Vibration = LIFE living as Life
Vibration (the Fundamental) = BEING/Consciousness/BEING/Consciousness/BEING etc., One Cycle per Cosmos
vibration = densification
vibration = discontinuity
vibration = discontinuous assertion of Non-BEING
vibration = dualism
Vibration = Fohatic Oscillation
Vibration = fluctuating Assertion of Finitude
Vibration = Homage to Two over One
vibration creates object
vibration requires object
vibration = obscuration
vibration = ontological fluctuation
Vibration = Ontological Repositioning
vibration = reification
vibration requires time
Vibration = Subject/Object alternation
Vibration = Veiling
Vibration = waves in the Continuum
Vibration (Fundamental) = ZERO to One to ZERO to One (One per Cosmos)
Vibration (Fundamental) = Great Breath

Vision = Detector of Finitude
Vision = Fohat
vision generates illusion
Vision = Illusion
Vision = Means of Self-Reflection

VOID = ABYSS
VOID = CONTINUUM
VOID = FOUNT OF ALL POSSIBILITY
VOID = HOMOGENEITY
VOID = INFINITE DENSITY
VOID = NOTHING
VOID = 'POINTLESS' WHOLE
VOID = PLENUMVOID = UTTER ABSTRACTION
VOID = ZERO

- W -

WHOLE = 'part'

- Y -

You = **8**
You = Object
You = Subject

- Z -

ZERO = ABSOLUTE FULLNESS
ZERO = BE-NESS
ZERO = **I**
ZERO = INFINITE
ZERO = the One
ZERO = the NEGATOR
ZERO = the PRESENCE

A small beginning has been made. The reader is invited to ponder these equivalencies and relationships as Seed Thoughts. Mercury, god of Pure Reason, will be invoked in the pondering and many will be the revealed subtle connections useful in bringing Synthetic Understanding to the mind.

Now that the principle is understood, the reader may find it congenial to begin discovering equivalencies unlisted here. Since everything is related to everything else, there should be no problem discovering them. This exercise will contribute greatly to the strengthening of the Antahkarana as an instrument of penetration into the Realm of the Spiritual Triad.

A List of Difficult or Problematic Questions

To the following questions there are no easy answers, especially given the limitations of humanity's present mental/intuitive apparatus. The answers to some of these questions might be considered 'Secrets of Initiation', and still others, may be altogether beyond the grasp of any planetary or solar Intelligence. These questions have been treated speculatively in various ways throughout this treatise, and are listed here simply to help focus the mind upon philosophical issues ripe for prolonged brooding.

1. How does 'Something' 'come out of' NOTHING?
2. Is there a regular, standard interval between Universes, or is it variable? If the interval varies, with what can it be compared in order to measure it?
3. What determines the duration of a Universe? Is that duration part of the Cosmic Parameters of the Design-at-the-Beginning, or is the duration variable?
4. Which came first, the 'SWAN' or the Egg? (The Egg is *always* preceded by the 'SWAN', but, then, the STATE of PURE 'SWAN-NESS' is *always* preceded by a State in which the Egg exists.)
5. Is the ABSOLUTE subject to 'LAW'? Does the ABSOLUTE 'BEHAVE' *lawfully?* What is the relationship between ABSOLUTE 'LAW' and ABSOLUTE 'WILL', (if such 'STATES' exist) or, Cosmically, between Law and Will?
6. Have there been any REAL intervals between events, or has there been only ONE GREAT CONTINUOUS EVENT, which is REALLY a NON-EVENT? (After all, if 'nothing happens', or only <u>*NOTHING*</u> REALLY 'HAPPENS', then there are no REAL events or intervals.)
7. Is Illusion REALLY *necessary?* Is Illusion REALLY Illusion?
8. Can **I** ever cognize MY-SELF? Can **I** the INCOGNIZABLE DEITY ever REALLY cognize MYSELF? The 'Birth' of the Cosmos is based upon the idea that **I** can, but *how* do **I** do so, especially when **I** have no 'parts' with which to 'ACT'?
9. Is evolution REALLY possible? Throughout the Infinite Duration of the ULTIMATE ALLNESS has there REALLY been any evolution at all?
10. Does the SELF REALLY *change*? By definition, IT does not! If the SELF does not change, why an infinity of effort and labor?
11. Does the NOTHING REALLY never cease BEING ABSOLUTELY NOTHING even during the 'Time' that 'Something' *appears* to be happening? If this is REALLY TRUE how does it *work?*
12. How shall we reconcile the *great contradictions* posed by the ZERO and the One, by the INFINITENESS and the Finite, by uncompromising egalitarianism and an uncompromising 'hierarchicalism'?
13. How, exactly, is the 'part' to become, *consciously,* the Whole, and the WHOLE it *already* both Is and IS?
14. Is there a way to overcome *dimensional sealing* and access one's own multiple levels of consciousness along the entire Divine Emanatory Stream? How, as a human being, can one *consciously know* what one already *super-consciously* and *unconsciously Knows?*

Alphabetized List of Certain Problems

The majority of these Problems are discussed in Section II, Problems of Living and Philosophical Problems; the remainder are treated in other sections or within the Glossary.

Problems in Living

1. The Problem of Ambition
2. The Problem of Anxiety
3. The Problem of Boundaries and Trespass
4. The Problem of Blame and Complaint
5. The Problem of Brotherhood
6. The Problem of Choice
7. The Problem of Desire—Its Usefulness and Its Overcoming
8. The Problem of Destiny
9. The Problem of Dissatisfaction and Discontent
10. The Problem of Ego and Non-Ego
11. The Problem of Envy and Jealousy
12. The Problem of Fear
13. The Problem of Free Will
14. The Problem of Valuation of the World—the Respect or Disrespect for Form
15. The Problem of Fate
16. The Problem of Freedom and Necessity
17. The Problem of Good and Evil—To Whom are they done and by Whom?
18. The Problem of Happiness, Joy and Bliss
19. The Problem of Human Vice
20. The Problem of Humility
21. The Problem of 'I'—The Word and its Departicularization
22. The Problem of Impatience
23. The Problem of Individual Responsibility
24. The Problem of Ineffectuality—Especially of Bliss
25. The Problem of Justice
26. The Problem of Karma
27. The Problem of Likes and Dislikes, Tolerance, and Intolerance
28. The Problem of Loneliness
29. The Problem of the Maladaptivity of Unitive Living
30. The Problem of Mood
31. The Problem of Morality

32. The Problem of Pain
33. The Problem of Position and Function within the Cosmic Whole—Knowing One's Place
34. The Problem of Prayer and Invocation—To Whom does one Pray and Why?
35. The Problem of Pride
36. The Problem of Progress
37. The Problem of Right Human Relations
38. The Problem of Self-Confidence—Feeling Good About Oneself
39. The Problem of Self-Depreciation
40. The Problem of Sex
41. The Problem of Stillness and Activity
42. The Problem of Values—ABSOLUTE and Relative
43. The Problem of Will—Whose Will and How to Fulfill It
44. The Problem of Work and Struggle
45. The Problem of World Denial
46. The Problem of Worry

Philosophical Problems

1. The Problem of 'ACTION' (its impossibility and yet its *necessity*) in THAT
2. The Problem of whether there is 'ANYTHING' 'GOING ON' 'within' the INFINITE SELF 'before' the 'RAY' 'FLASHES FORTH'
3. The Problem of Archetypes—Do They Fluctuate?
4. The Problem of Beauty—And the Beauty of Maya
5. The Problem of Beginnings and Endings—Their Reality and Their un-REALITY
6. The Problem of whether to Conceive of the BOUNDLESS IMMUTABLE PRINCIPLE as an IDENTITY—A SELF
7. The Problem of BRAHMAN and Samsara
8. The Problem of Co-Measurement in a World Simultaneously Finite and INFINITE
9. The Problem of the Nature of Consciousness
10. The Problem of Continuity and Discontinuity, of the CONTINUUM and the Dis-Continuum
11. The Problem of Continuity of Consciousness in a Fohatically Fluctuating Objective Universe
12. The Problem of the Continuity of Emanative Sources
13. The Problem of why Strict Continuity is Non-Allowable in a Finite Cosmos
14. The Problem of the Correlation of the concepts of the SELF and PARABRAHMAN

15. The Problem of whether there can be a Cosmic Failure
16. The Problem of whether when the Cosmos is Concluded, is Anything Returned (as unto the GREAT SUBJECTIVITY) as a Gain or Harvest?
17. The Problem of the Divisibility of the Spirit
18. The Problem of 'Doing'—Have You 'Done' Everything that could ever *be* Done, or only Everything that *has been* Done?
19. The Problem of Duality—Its Infinite Duration
20. The Problem of the Entry of NOTHING into 'Something'
21. The Problem of Emanation—Its *modus operandi*
22. The Problem of How Emanation Really Works
23. The Problem of Emanative Retention
24. The Problem of Entity—Its Apparent Distinctness and REAL Indistinctness
25. The Problem of the Enumerations of a Subjective Being
26. The Problem of Equality and Hierarchy
27. The Problem of Error—Can the Universal Logos Commit Error?
28. The Problem of 'ERROR'—Can the ABSOLUTE 'COMMIT' ERROR?
29. The Problem of the ETERNAL NOW
30. The Problem of Event and Non-Event
31. The Problem of Every Point as Center in an Unbounded Pre-Cosmic System
32. The Problem of the Existence of Infinite Duality
33. The Problem of Family Relations—The 'First Family of Cosmos'
34. The Problem of Fascination with the Immediate and Indifference to the Remote
35. The Problem of the Finite INFINITE and the INFINITE Finite
36. The Problem of the Finiteness of Infiniteness of the Universe
37. The Problem of Fohat—Its Operations in Mulaprakriti and Cosmic-Prakriti
38. The Problem of 'FOHAT'/Fohat—Its True Identity
39. The Problem of Form and Formlessness—Can Formlessness Exist in Cosmos?
40. The Problem of Fragmented and Un-Fragmented Images
41. The Problem of how Ideational Qualities Combine and yet Remain Whole
42. The Problem of Identicalness—Is a Thing Identical with Itself?
43. The Problem of Illusion—Is It REALLY un-REAL?
44. The Problem of Immutability and Vibratory Activity
45. The Problem of Impossibility of Relation between the ABSOLUTE and the Relative
46. The Problem of the Incomensurability of Principles of the Finite and 'PRINCIPLES' of the INFINITE
47. The Problem of Incremental Fulfillments in the Fulfillment of Divine Purpose
48. The Problem of Individuality—Its *Apparent*, but un-REAL Existence

49. The Problem of the Individuality or Non-Individuality of Ultimate Particles
50. The Problem of the 'Infinispective' and the 'INFINISPECTIVE'
51. The Problems of whether the Infinite Object 'Sees' the Infinite Subject, just as the Infinite Subject 'Sees' the Infinite Object
52. The Problem of Infinite Speed and Infinite Distance—Do They Exist?
53. The Problem of Infinite Speed in relation to Motionlessness
54. The Problem of the Infinite Subject and the De-Infinitizing/Re-Infinitizing Subject
55. The Problem of Infinite Succession and Infinite Development
56. The Problem of the Infinitesimal—Its Nature, Existence, or Non-existence
57. The Problem of *Infinitizing* the Finite
58. The Problem of Instantaneity—Is Time *always* Required for Change?
59. The Problem of Irrationality Concerning the Origin of the Monad
60. The Problem of Law, Will, and Necessity
61. The Problem of the Limitation of Human Consciousness
62. The Problem of whether the 'Severe Limitation' which Cosmos Is, is 'CHOSEN' from 'within' the PLENUM, or is the SELF-as-Self 'THRUST OUT'
63. The Problem of Linearity and Simultaneity
64. The Problem of Mathematics and the Determination of Emanatory Paths
65. The Problem of the Mathematics of the Emanatory Process
66. The Problem of 'MAYA'-as-Maya
67. The Problem of Maya—Its Equivalence or Non-Equivalence with Fohat
68. The Problem of Meditation in the Infinitist World View
69. The Problem of Memory—'Where' and in 'What' does inter-Cosmic Memory Dwell?
70. The Problem of Mental Inconceivability—Antinomies
71. The Problem of the Monad—Are there One or Many?
72. The Problem of Monadic Ascent
73. The Problem of the Monad—Its Individuality or Non-Individuality
74. The Problem of whether there are many Monads or only One
75. The Problem of Monality and Duality, Monism and Dualism
76. The Problem of whether there can be Motion in Cosmos without the participation of Ultimate Particle/Events
77. The Problem of Movement—How Movement Occurs if not Preceded by Movement?
78. The Problem of 'MOVEMENT' within the SELF
79. The Problem of Movement—Its Reality or its Apparency
80. The Problem of the Many Diverse Ways of Producing Numbers
81. The Problem of the One Self—Its Recognition in Diversity

82. The Problem of One, Two and Three—What is the Pre-Cosmic One, Two, and Three when Compared with the intra-Cosmic, One, Two, and Three?
83. The Problem of Opposites—Their Separation and Their Unification
84. The Problem of Paradox and Contradiction
85. The Problem of Particulateness of the Higher Dimensions
86. The Problem of whether all 'Parts' are Composed of 'Parts'
87. The Problem of Past, Present, and Future in Relation to The ETERNAL NOW
88. The Problem of Perspective
89. The Problem of the Point (*True* Point)—Its Indivisibility and its Non-Dimensionality
90. The Problem of the Point (*Virtual* Point)—Its Divisibility and Dimensionality
91. The Problem of the Point as an Intermediary between INFINITUDE and Finitude
92. The Problem of a Point of Reference in Cosmos—Does such a Point Exist?
93. The Problem of Points within Points Dimensioned Virtual Points
94. The Problem of Points within Dimensionless Points
95. The Problem of Positivism as a Denial or Reduction of REALITY
96. The Problem of Possibility or Impossibility of Evolution and Progress
97. The Problem of 'POSSIBILITIES'—Can 'ALL POSSIBILITIES' within the FOUNT OF ALL POSSIBILITY be Actualized?
98. The Problem of how Infinitized Possibilities can be Distinct from Each Other
99. The Problem of 'POSSIBILITIES' 'within' the INFINITESSENCE—Which are Chosen, and Which are *not,* and Why?
100. The Problem of Prakriti as Self-Perception—How does it Work?
101. The Problem of the 'RAY' of the ABSOLUTE—ITS Mode of Non-Divisible Divisibility
102. The Problem of the 'REASON' for the Universe—Is There a 'Good One'?
103. The Problem of Regularity or Indeterminacy of Interval between Sequential Universes
104. The Problem of Relationship—Are all Relationships Spatially 'Extended'?
105. The Problem of whether Relationship is REALLY Spatially Extended
106. The Problem of Seeing—Does One See only That which is *within* Oneself?
107. The Problem of the 'SEEN' SELF/'Seen' Self—Is It Homogeneous or Articulated?
108. The Problem of SELF-Contradictory Possibilities within the FOUNT OF ALL POSSIBILITY
109. The Problem of Self-Reduction and Non-Self-Reduction
110. The Problem of the Seriousness of Cosmos
111. The Problem of 'SIGHT'—Does the INFINITE SELF 'SEE'?
112. The Problem of Simultaneity and Linearity
113. The Problem of the Simultaneous Validity of Contraries

Section VI
Aphorisms, Meditations & Mantrams

Radical Infinitist Aphorisms

1. **8 8 8**.
2. Thou Art That.
3. **8** Am That **8** am.
4. If **8** see it **8** cannot be it.
5. Whatever **8** 'See' **8** Am.
6. REALITY is 'POINTLESS'.
7. Cosmos is 'Next to NOTHING'.
8. Seeming needs redeeming.
9. Be that which you would Have.
10. Hold to form—hold to grief.
11. Cosmos is a Paradox Alert!
12. Everything is everything else.
13. The less one does the more one IS.
14. **8** Am Thy Life.
15. **8** Am the Self-'Seen' 'Seer'.
16. "There, indeed, go I."
17. Every Cosmos is a Problem to be solved.
18. "Even NOW, ALL IS WELL."
19. **8** Am the Self-'Seen' 'Seer'.
20. Don't wait to Know It All before you find out WHO YOU ARE.
21. Infinitization is a special mode of seeing which leads to the realization of ABSOLUTE BEING.

22. If **I** do not contradict MYSELF **I** CANNOT BE MYSELF.

23. IT IS WHAT IT IS.

24. Nothing exists in Space but what **I** 'put into it'.

25. 'SEEING' 'ARISES' out of 'BEING' without being 'SEEN'.

26. **8**'m a 'Seen' of a 'Son'! And, as well, a Son of a 'Seen'!

27. Nothing can be 'Seen' in Prakriti which is not already 'in' the Self.

28. The LIVINGNESS of LIFE IS NOW.

29. It's All 'Happening' ... or, is It?

30. We simply have to BE EACH OTHER, and know it.

31. "**I** see myself within MySelf." (1 + 1 = 2)

32. Having 'Become' through Self-Objectification an apparent fragment of MySelf, **8** remain!

33. Being is believing.

34. Save One; Save All.

35. Thou shalt not 'Thing'! [sic]

Important Statements —Some Thoughts for Pondering

The ABSOLUTE —The BOUNDLESS IMMUTABLE PRINCIPLE

- The BOUNDLESS IMMUTABLE PRINCIPLE can never REALLY vary.
- The ABSOLUTE is the GREAT NON-EVENT.
- The 'identifier' can realize he IS the BOUNDLESS IMMUTABLE PRINCIPLE.
- The BOUNDLESS IMMUTABLE PRINCIPLE is *indivisible,* or IT would have boundaries.
- The BOUNDLESS IMMUTABLE PRINCIPLE IS the ONE WITHOUT A SECOND.
- There can be no *second principle* (i.e., factor) in *relation* to the BOUNDLESS IMMUTABLE PRINCIPLE. Hence, the phenomenal world cannot be *other* than the BOUNDLESS IMMUTABLE PRINCIPLE or the ABSOLUTE, the SELF.
- The ALL-IN-ALLNESS IS the TOTALLY INFINITIZED IMPARTITE STATE. IT is the UNCONDITIONED 'CONDITION'.
- In the ALL-IN-ALLNESS, Space (both as interval and as the arena of the SELF's expression) does not exist. In the ALL-IN-ALLNESS, Time does not exist. In the ALL-IN-ALLNESS Motion does not exist—yet all three are *infinitized,* and thus *essentially identical.*
- The state of ALL-IN-ALLNESS is *non-extensive.*
- The BOUNDLESS IMMUTABLE PRINCIPLE IS at once *zero dimensional* and *infini-dimensional.*
- 'Within' the BOUNDLESS IMMUTABLE PRINCIPLE/SELF there is no vacuum because there is no interval or discontinuity of any kind. And yet, there is a vacuum with respect to objects, as there prevails an ABSOLUTE VOID or EMPTINESS.
- The phenomenal world cannot be *other* than the BOUNDLESS IMMUTABLE PRINCIPLE or the SELF. The World *is* BRAHMAN and the ABSOLUTE SYNTHESIS is inescapable.
- In order for the BOUNDLESS IMMUTABLE PRINCIPLE /SELF to BE what IT IS, IT must periodically *appear* to be what IT is not even as, simultaneously, IT *continues* to BE what IT ETERNALLY IS.

Action, Doing, and Happening

- The 'CHANGE' ('within' the INFINITE SELF) from *'BEING'* to *'SEEING'* is the very First *'DOING'*.
- The less one *does* the more one IS.
- No-thing has ever 'happened'. NOTHING has always 'HAPPENED'.
- Continuous action cannot occur in a quantized Universe.
- Action *is* a discontinuity. A continuous *act* is *no act*.
- 'ACTION' cannot occur in the CONTINUUM. If 'ACTION' begins, Illusion begins as well.
- The 'ARISING' of 'ACTION' in the CONTINUUM is its own 'EXPULSION'.

Beginnings and Endings

- The SELF is beginningless and endless.
- All that begins must end. All that begins ends by the very fact that it begins.
- Every change is both a beginning and an ending.
- Every moment is both beginning and ending.
- Beginnings and endings are but change of relationship.
- Beginnings (in-Cosmos) are agglomerations that are re-agglomerations of authentic units and of authentic E/entities, identities. Endings (in Cosmos) are also agglomerations and re-agglomerations.
- There is an endless succession of Universes that never began and which shall never end.
- True Beginning is a 'MOVEMENT' from the UNCONDITIONAL and UNCONDITIONED 'STATE' into a Conditional and Conditioned State.
- There can be no REAL 'BEGINNING'. There can be no REAL 'END'. Beginnings and Endings are actual but un-REAL.
- The UTTER ALLNESS IS simply the rhythmic apparent variation of NOTHING and Something while, nonetheless, preserving inviolate the continuity of NOTHING.
- Cosmic Beginnings always were—though cyclically appearing.
- The Beginning is NOTHING 'BECOMING' Something.

Center

- NOTHING cannot have a CENTER.
- Cosmos may have a 'spatial center' but no 'Spiritual Center'.
- An ultimate particle/event can have no center. Nothing within it is differentiable.
- Since Centers are Points, the GREAT HOMOGENEITY can have no 'CENTER'.

Change

- REALITY does not change and cannot change. The BOUNDLESS IMMUTABLE PRINCIPLE does not change and cannot change. 'Change' is, however, naught else than REALITY (the BOUNDLESS IMMUTABLE PRINCIPLE).
- That which changes is not REAL.
- 'ACTION' cannot change the UNCHANGEABLE.
- Change does not necessarily imply 'Movement'.
- All change is illusory though *actual.*
- The GREAT IMMUTABILITY never 'MUTATES' and thus the Illusory Conditional World never can be in position of a true *second* with respect to the ONE WITHOUT A SECOND, the BOUNDLESS IMMUTABLE PRINCIPLE.

Conditions

- The one thing that can be predicated of anything in-Cosmos is (that it is) a condition.
- Conditions depend upon discontinuity.
- The CONTINUUM is forever *conditionless.*
- Conditionality requires Relativity.
- Conditions (Perception and Apperception).
- Cosmic Consciousness would be an *infinite reduction* of ABSOLUTE 'CONSCIOUSNESS' (if ABSOLUTE 'CONSCIOUSNESS' were possible). The scope and nature of Cosmic Consciousness is *infinitely removed* from ABSOLUTE 'CONSCIOUSNESS' (which cannot be *consciousness,* as we know it, *at all).*

- All objects are un-REAL-in-form/appearance (though REAL-at-ROOT), and are born of an Act of Consciousness derivative from the 'ACTLESS-ACT' of 'CONSCIOUSNESS'.

- The Universe is the possibility of 'Seeing' what **I** *subjectively* AM. **I**-as-**8** 'See' an *unfragmented fragment* of what is within ME, and thus IT (that which **I** AM) acquires objectified existence.

- The '<u>RAY</u>' of the ABSOLUTE which 'FLASHES FORTH' is the 'ACT' OF 'SEEING'; the '<u>R</u>ay' of the ABSOLUTE Is the SELF-Becoming of an Infinite (Infinified) Point of View which we call the SELF-as-(Infinified) Point or Infinite Subject.

- When the 'RAY' 'FLASHES FORTH', INFINITUDE is instantly 'SEEN'. From 'then on', instantly, ABSOLUTE BE-NESS no longer 'CONTINUES', PER SE (for the World of Illusion, an apparent 'other', exists) and *yet IT MUST and DOES.* The IMMUTABLE CONTINUUM *must continue immutably.*

- Thinking is 'Thinging'; it can only assert un-REALITIES.

- 'SEEING' is an infinite reduction of the capacity to BE.

- 'SEEING' apparently 'INTERRUPTS' 'INFINIDENTIFICATION'.

- '<u>S</u>eeing' is simply a reduction of the capacity to <u>B</u>e.

- 'SEEING' 'ARISES' out of 'BEING' without, ITSELF, being 'SEEN'!

- Consciousness-in-Universe is the act of *seeing* 'OnesSelf-as-other' instead of *being <u>another</u>* as 'OnesSelf'.

- The 'RAY' of the ABSOLUTE Is an attentive 'ACT' of instantaneous 'SELF-REFOCUSING'—from TOTAL SELF-'ABSORPTION' to Total Self-<u>A</u>wareness. It has no *knowable* 'CAUSE' but the 'CAUSE' may be *'BE-ABLE'.*

- Consciousness Is Maya; Maya Is Consciousness.

- Time and Space are merely categories of consciousness and thought (not exclusively human consciousness and thought). They have no REALITY in and of themselves. Time and Space are limitations (initially SELF-'LIMITATIONS') caused by the arising of the PERCEIVER/Perceiver/perceiver, and are the result of the inherently finitizing *action* of 'PERCEPTION'/Perception/perception.

- *Seeming* is any non-ABSOLUTE registration of REALITY.

- One cannot *be conscious of* REALITY; one can only *BE* IT.

- Consciousness in Cosmos is an apparent continuity of objective presentation in which change (independent of space) appears as movement *through* space.

Continuity and Discontinuity

- Strict Continuity in Cosmos is impossible.
- The 'CONTINUOUS' SELF totally interfuses the Discontinuous Cosmos. 'CONTINUITY', totally underlies, Discontinuity, and, in fact, *is* it.
- Discontinuity (the Dis-Continuum which Cosmos Is) is a *necessary* un-REALITY.
- The Infinite Time Line is an Infinite Discontinuum.
- No 'thing' in Cosmos can possibly continue to *be itself* continuously, uninter-ruptedly. Even the Universal Permanencies of the World of Being both begin and end.
- All Fohatic Dimensions are Particulate, or they would be identical with the World of Being, and the World of Approximation would not exist. Hence we must address ourselves to the nature of the particulateness of the highest Fohatic Dimensions.
- A beginningless, endless, unchangingly *continuous* act is no act at all.
- No *action* in Cosmos is utterly Continuous, even the action of Fohat.
- *Continuous* action cannot occur in a quantized Universe.
- The Universe is both discontinuous and, Itself, a Discontinuity.
- The World of Being is the most continuous Discontinuity in Cosmos.
- The pair of (Infinified) Point and Mulaprakriti is the First Super-Cosmic Duality. Duality demands discontinuity.
- An event is a disruption of continuity. An event *is* a discontinuity. Thus there are no 'EVENTS' in THAT. And yet, 'Something' *apparently* 'HAPPENED'. Or did It, *REALLY*?

Cosmos, the Universe, and the Universal Logos

- Cosmos is a Role the ONE SELF *plays*, sequentially, forever and ever.
- Cosmos is an infinite minimization *(infinitesimalization)* of the ALL-IN-ALLNESS. The implications are astonishing!
- If Cosmos were not 'Seen' it would not exist.
- Cosmos is the SELF-*resident* possibility of Finitude in manifestation.
- Cosmos is the Periodic Limitation of the INFINITE-SELF.
- The entire Cosmic Process (including the Appearance of Cosmos Itself) is subject to the Law of Unrepeatability.

- The Cosmic Process is not a Drama, Dialogue or Multi-logue, but a *Monologue.*
- The Universe is a Presentation—to Whom?
- Universes are Events of the largest scope/dimension—*relatively.*
- There never was a first Cosmos! There never will be a last Cosmos!
- There is an endless succession of Universes that never began and which shall never end.
- The Universe is *Actual*—Its Presence is agreed upon by all *thinkers;* Its Absence is agreed upon by all *knowers.*
- The Highest Magic is the Appearance and Disappearance of the Universe.
- Cosmification inheres forever in the INFINITESSENCE as the *possibility of finitude.*
- The Universe is that which the SELF-as-Self 'Sees'.
- The Universe is a *necessary* Limitation.
- Whereas the Universe/Cosmos appears to be a contradiction to REALITY it is not ESSENTIALLY so.
- Cosmos is a SELF-'Become' Particularity.
- Cosmoses cannot evolve or improve.
- In Cosmos, there is no vacuum, unless that vacuum is the *presence* of NOTHING 'in' Something—the *presence* of the SUBJECT 'in' the Object.

Design

- The Design-at-the-Beginning unfolds in Time as Music, and consummates at the close of the Cosmic Finale as the all-inclusive Cosmic Chord.
- Formal configurations are unique and unrepeatable throughout all Eternity.
- We-as-**8-as-I** 'CONCEIVED' the Design-at-the-Beginning.
- Playing the Cosmic-Game well, demands aesthetic sensitivity, sensitivity to Beauty of Design.

Divisibility and Separation

- All divisibility is illusory, though *actual.*
- 'SEPARATION'-instantly-Separation is the first descent into *Seeming.*

- Every time the 'Circle of Existence' is apparently divided, **8** Am there equally in all divisions. **8** pervade all divisions (and dimensions) fully without exception. **8** cannot be divided from 'MySelf'.
- The Individual is *indivisible.*
- ZERO obliterates separation.

Duality

- Understanding of the Necessary Illusion and the Infinity-of-Duality comes through the figure '**8**'. (The '**8**' can be written in an infinity of ways characterized by changing proportions between the top and bottom loops.)
- In the 'HISTORY' of the BOUNDLESS IMMUTABLE PRINCIPLE there is involved an infinite and imperishable *duality.*
- Cosmos is inseparable from the fact of *duality.*
- ABSOLUTE MONISM requires the Infinite Duration of Dualism.

Ego

- Ego is mistaken 'IDENTITY'.
- Non-Ego arises with the destruction of Consciousness. Therefore, destroy consciousness and BE!
- Every vice is based upon ego.
- Ego is a *seen self* rather than a *'be-en' self.*
- Ego is Discontinuity in the Continuum of Bliss.

Emanations

- The Mother is the That which Reveals what is within Oneself.
- Each hyparxix retains its full integrity undiminished during an *emanatory flow.*
- REALLY, each Emanatory Source is *fully* invested in its emanations; partial investment is an *appearance* only.
- That which is emanated is always less (from the illusionary perspective) than that from which it emanated even though that which is emanated ESSENTIALLY contains the *whole* of that from which it emanated within itself.

Entity and Identity

- Entities are Reflections of Reflections, and all Reflections are ESSENTIALLY, UNSEEN 'REFLECTIONS' of NOTHINGNESS.
- Authentic Entities are Identities. Only an Authentic Entity has potentially conscious Identity as a 'Ray' of the ABSOLUTE.
- Entity arises through motion producing configuration and combination.
- Entity is a Life apparently distinct from the SOURCE. Consciousness and Motion are the 'departures from the SOURCE' which *are* Entity.
- Each individual unit of Life is found *equally* and *fully* in and *as* every other unit of Life (large or small) throughout the entire Universe-of-all-Dimensions.
- All Authentic Entities are *gods*—even the Life which manifests as the atom of physical substance. Nevertheless, all these *gods* are but One God, One Entity, the Universal Logos, who is but the ONE 'ENTITY/NON-ENTITY'.
- Identity-in-Cosmos is the *partialization* or *finitization* of the ABSOLUTE **I.**
- '**I**-as-I' am a prakritically bounded 'Ray' of the ABSOLUTE BEING.
- An Identity is an Authentic Entity.
- True *Individuality* is the impossibility of being contained within a ring-pass-not, for what is to divide or *contain* the *true* Individual?
- All E/entities are REALLY undetached aspects or 'Rays' of that GREAT ENTITY/ NON-ENTITY.

Event

- An event is a variation, a variation either from a previous pattern or NON-'PATTERN' (as in the ULTIMATE HOMOGENEITY).
- An event is a disruption of pure continuity. An event is, therefore, a discontinuity.
- An *event* is a 'coming to birth'. "Eve" is involved!
- *The* 'EVENT' is the 'VARIATION' from the *infinitely complex 'PATTERN' of the NOUMENESSENCE* to a Singular Pattern which is 'EXTRUDED' from the NOUMENESSENCE as the Dis-Continuity which launches the Universal Process.

Evolution

- There is no Evolution in the Infinite Series of Universes.
- The INFINITE SELF can *never* 'EVOLVE'.

- Evolution is the progressive conformation of the World of Approximation to the Fixed Design *held* within the World of Being.
- The Spirit never evolves!
- The Universal Logos, per se, never evolves.
- Evolution is a Cosmic Un-Veiling.

Extension

- Extension is the possibility of relation. There is no relation without extension. There is no relation without the illusion of interval, but interval need not be conventionally spatial or temporal. Interval can be Ideational and Qualitative.
- Extension can be interpreted as delimited presence and unlimited omnipresence.
- Extension is the false substantiality of the *insubstantial* World.
- Extension is an *appearance*, never, per se, a REALITY.
- From the conventional perspective a virtual point is extended; a Real point is non-extended.

Fohat/'FOHAT'

- 'FOHAT' is 'NOTHINGNESS in Action'. It is, likewise, *infinitized 'EVERYTHINGNESS'* running a limited 'Errand' called Cosmos.
- Fohat is 'Sight' as 'Action'.
- Fohat is the 'Relation Between'.
- 'FOHAT' IS the Power of *the* <u>one</u> 'IDEA' Whose 'time' has come (yet again!).
- Fohat <u>IS</u> the *peremptoriness* of the Idea whose *time* has come.
- Fohat works through ultimate particle/events by *becoming* them. Thus are all things with the World of Fabrication created, shaped and moved.
- Fohat is the Objectifier.
- Fohat is the 'Bridge' between any Self and Its Not-Self.
- Fohat is the Revealer of the Self to the Self.

Form

- No object/thing/form has *ever* been—i.e., existed *forever*. Mulaprakriti, however, has existed forever cyclically.
- There are many Objective Forms that must be considered Subjects as well as Objects by the Subjective Entities Who emanated them.
- Forms have definition; but there are 'formless forms'.
- A point if both a form and yet, *not* a form.

Illusion

- Anything limited is illusory. Illusion is Limitation.
- All relationship is illusory.
- All change is illusory.
- Cosmos is an Illusion, the Great Illusion.
- Extension is illusion.
- In Cosmos, there is no vacuum, and yet illusion is a kind of vacuum caused by the Consciousness/Action of intra-Cosmic 'Fohat' as it established *(objectively)* a specified number of 'Subjective Points of Self-Perception' in that Aspect of Cosmic Prakriti which is Its (Fohat's) global Self-Reflection.
- *Seeming* needs *redeeming*.
- All 'EXTRUSIONS' of the INFINITESSENCE are Illusory.
- The Great Illusion is the Illusion of non-IDENTICALNESS.
- The Great Illusion/Universe/Cosmos is the *necessary* means by which the BOUNDLESS IMMUTABLE PRINCIPLE can periodically 'Act Out' the 'PARADOX' inherent In ITS OWN Nature without altering ITS OWN ABSOLUTE NATURE in the slightest.
- All illusions are illusory. No illusions are illusory.
- Anything predicated of THAT is illusory.
- The word, 'apparently', used so often in relation to illusion, reveals that the question of whether a something is REAL, Real, actual or illusory depends entirely upon how it *appears* from various Points of View. The dynamics of 'Sight' are inseparable from *'apparency'*. Nothing *IS* as it seems; 'NO-THING' IS *NOT* as IT *seems*.

Indivisibility and Divisibility

- All divisibility (and all that arises from divisibility) is illusory though *actual.*
- Illusion ends with the realization of the impossibility of Divisibility.

Infinite and Finite

- Any 'big' is infinitely far from *infinite* bigness; any 'little' is infinitely far from infinite littleness. Any 'big' is as infinitely far from infinite bigness as it is from infinite littleness. Any 'little' is as infinitely far from infinite littleness as it is from infinite bigness.
- Cosmos is an *(infinitely* 'distant') approximation of INFINITENESS.
- INFINITUDE is the GREAT CONTINUITY.
- Infinite extension is, in fact, no extension at all.
- Infinite speed is, in fact, no speed at all.
- Infinitized anything is NOTHING WHAT-SO-EVER.
- Any quality (relationship) *infinitized* is thereby *negated* and simultaneously *absolutized.*
- It is impossible to find any 'thing' which is not *apparently* Finite and *REALLY* INFINITE.
- More drastically, it is impossible that there should be any *'thing'* which is not both Finite and INFINITE.

NOTE: One could say that the ABSOLUTENESS is not both Finite and INFINITE, but the ABSOLUTENESS is not any 'thing' and it is not a presentation. IT IS not in any way reified—made into a 'thing'. What are presented to us in Cosmos are perceptions. They are 'things', items-in-consciousness, items in awareness. All of these presentations or "contents of consciousness" (as Roberto Assagioli would call them) are *both apparently* Finite and *REALLY* INFINITE though *ultimately* *REALLY* INFINITE.

- The Finite does not exist. All that exists is the INFINITE.
- There is no finite thing. All things are ESSENTIALLY the INFINITUDE in ITS ENTIRETY.
- When the 'RAY' 'FLASHES FORTH', INFINITUDE is instantaneously 'SEEN' instead of *infinitessentially* 'BEEN'. Then, INFINITUDE *continues being* 'BEEN', even as ITS Reflection, Mulaprakriti, is 'Seen'. The 'Seer' of INFINITUDE-as-Infinitude is not 'SEEN' by a 'SUPERIOR'.

- In the ALL-IN-ALLNESS, INFINITUDE is *not* 'SEEN'.
- Any 'portion' of an infinite 'something' is *formally* (not ESSENTIALLY) infinitesimal when compared to that 'something'.
- Conversely, any 'portion' of the INFINITUDE (if such a thing REALLY exists) is as ESSENTIALLY *INFINITE* as the INFINITUDE, *entire.*
- The 'INFINITESSENCE' is the one and only ESSENCE of infinite rarefaction and infinite refinement.
- Infinitely (literally *infinitely)* more *numerous* than the possibilities that have even been precipitated, are the possibilities that remains ever unprecipitated.
- INFINITUDE (the INCOMPARABLE) destroys any-<u>thing</u> against which IT is *compared.* IT 'DESTROYS' by instantly changing all thing into ITSELF!—into NOTHING!

Limitation

- Positivity is Finitude; Pure Negativity is INFINITUDE.
- The Universe is a *necessary* Limitation.
- There is no *creation* without *limitation.* The one thing that *can* be 'DONE' in the UTTER ALLNESS is to *limit.* Other than that, naught can be done, for any 'ACT' of the ALL-SELF is a *limitation* upon ITSELF—*apparently.*
- Location is limitation.
- Creation is an act of Limitation.
- Individuality is Limitation until it becomes *indivisible.*
- Mathematics: Division is illusory; Multiplication is illusory; Subtraction is illusory; Addition is illusory.
- All arithmetic operations are illusory 'in' the ALL-SELF, which cannot be modified in any way.
- Mathematics is the Language of Universal Relationship.
- Mathematics and Music are the primary tools of Pure Reason.
- In relation to the ALL-SELF, to 'DO' is to *limit.* There can be no 'DOING', nor 'MOTION' without *limitation* upon the LIMITLESS. Thus, there can be no 'DOING' nor 'MOTION'—*REALLY.*

Maya, Measurement

- 'MAYA'-instantly-Maya is the Great Divider, the Cause and Principle of Perceived Duality.
- Maya is Limitation. Maya is Consciousness. 'MAYA' is 'FOHAT' the First 'ACTOR'.
- The Process of Maya is SELF-'CAUSED'. Maya is not only the Agent of the SELF-as-Self, but is *fully* the SELF-in-'ACTION'-as-Action.
- By the 'ACT' of division/differentiation the SELF 'VEILS' ITSELF and 'BECOMES' the Super-Cosmic and Intra-Cosmic Veiler. Maya is not only the Great Veiler, but the Great (apparent) Divider. Maya is not only the Author of all Action but *Is* all Action—Super-Cosmically and Intra-Cosmically.
- Maya *acts* through 'Sight'.
- Maya is PARABRAHMAN made Conscious (i.e., *illusorily* split into Infinite Subject and Infinite Object).
- Consciousness is Maya; Maya is Consciousness.
- The Veiling Process is the Revealer of Matter. Matter is naught but Spirit objectified, the objectified state of the SELF-as-Self.
- The 'measured' occurs forever against the background of the MEASURELESS. The 'measured', however, can only be measured against the *measurable*. Infinite Duration is 'measurable'; INFINITE DURATION is *not*.
- 'MAYA'/Maya has intermittent Beginnings and Endings, though It never Began; nor will it End.
- Without 'MAYA' the INFINITE SELF would never 'SEE' what was 'in' ITSELF—though IT has no trouble 'BE-ING' those same 'THINGS'.

Motion, Movement, Vibration

- Movement in-Cosmos is instantaneous repositioning of ultimate particle/events. Movement is instantaneous change of relationship.
- Movement requires duality.
- To achieve conscious immortality means to become *immovable*.
- Infinitely rapid motion is immobility. Immobility is, thus, *infinitized motion*.
- The Great Breath is Perpetual Motion, but the 'UNDELUDED' SELF is characterized by 'PERPETUAL IMMOBILITY'.

- Movement in-Cosmos should not be considered *movement through* intervening points. Since points are things, how can things move *through* things?
- The Doctrine of Perpetual Motion requires Duality and, hence, Illusion. Without the Illusion of the Universe as a seeming pole to NOTHINGNESS, the possibility of the *movement* that makes Perpetual Motion possible could not exist.
- As the rate of vibration increases, the amplitude of the vibration on either side of the x-axis decreases, until at virtually infinite speeds the amplitude approaches the infinitesimal and at 'infinite speed' *absolute stillness* is reached. Thus, *infinitized motion* is *absolute stillness.*
- There is a tremendous similarity of meaning between the words 'movement' and 'moment'—each requires the other.
- Vibration is quantized change of relative position—the quantized change of relative position of ontologically oscillating Fohatic lives.
- Ultimate moments are the duration of frozen maxi-minimal event/relations.
- There is no such thing *in Cosmos* as a movement that takes "no time at all." But what of a *change* that takes "no time at all"? Can there be a *change* which is not a *movement?* Certainly the 'ultra-economical' ETERNAL NOW has *never* "taken any time at all."
- No movement <u>in</u>-Fabricated Cosmos can occur *during* an *ultimate moment.*
- Perpetual Motion is *actual* but *illusory.*
- Motion is the change of position/location of a given item-in-Universe relative to another item-in-Universe, or relative to a Fixed Point of Reference in Universe, which Center (if it exists!) is the Center of Cosmic Prakritic Sphere, the Center of the Ring-Pass-Not of the Universal Logos.
- There is, apparently, no such thing *in Cosmos* as a movement that takes "no time at all". However, though there is change, is there, necessarily, movement (from one point to another through all possible intervening points)? Change is always *perceived* as *movement,* but is it?
- A rapid sequence of successive static patterns gives the illusion of movement.
- If all things are points, there is no-thing *through which* a point can move. There can only be instantaneous change of relationship without movement.
- One cannot properly speak of things occurring *at* a point in Space but only *in* a point in Space. Points are things, objects. Where there are no things/objects, there are no points. This is why things cannot move through points in space—because things cannot move through things.
- All movement is based upon the inter-item Fohatically-willed adjustment that causes moment to moment reconfiguration.

- Reality (in the World of Being) is never more than an ultimate moment away.
- A condition, is a motion, is a change.
- The illusion of motion depends upon the illusion of relationship.

Mulaprakriti, Prakriti, Matter

- The Infinite Subject (on Its 'way' to becoming the Focused Universal Subject) Finitizes Mulaprakriti by 'Seeing' Itself as a *One* instead of a *None*, by 'Seeing' Itself as an Singular Individual rather than as a Boundless No-Thing.
- The only utterly dematerialized state is found within the ALL-SELF.
- Essentially, Mulaprakriti, per se, knows no modification. It is total material potentiality (or the potentiality for particularized objectification).
- Forms in Matter are aggregations of differentiated Fohatic Points of View. Fohat reconfigures all Points of View at each ultimate moment, for the Purpose of progressively *"better* points of view" more in keeping with the Divine Vision of the Design-at-the-Beginning.
- Mulaprakriti is the SELF-Reflection of PARABRAHMAN.
- Even Mulaprakriti is a Discontinuity as are all 'EXTRUSIONS' of the INFINITESSENCE. Mulaprakriti (when *present)* is a spatial Continuum and a temporal Dis-Continuum (because It is not *always present)*.
- Could even Mulaprakriti have Its subjective side?! The SELF is, indeed, as much Mulaprakriti as IT is the Infinite Subject. From this perspective, Mulaprakriti might be called 'GOD-as-Receiver'.
- Mulaprakriti is the Image of NOTHINGNESS.
- Self-Image *Is* Prakriti.
- Matter is naught but Spirit objectified by means of Consciousness. Consciousness 'Creates' Matter. Matter is brought to 'Birth' by the 'SELF's' 'ACT' of 'CONSCIOUSNESS'.

Nothing and Something

- NOTHING is *infinitized* EVERYTHINGNESS.
- Every 'something' is infinitely (literally, *infinitely)* less than NOTHING.
- A Real point is midway between 'Something' and NOTHING.

- The mind cannot apprehend NOTHING; if it tries, the mind *somethings* NOTHING.
- 'Something' is merely an infinitely limited 'EXTRUSION' of 'un-particulated' *infinitized* 'EVERYTHINGNESS'.

Noumena and Phenomena

- The NOUMENON IS, as it were, an *infinite distillation* or *abstraction* of all possible Noumena.
- The NOUMENESSENCE is the only non-Phenomenon.

Number and Zero

- Numbers are Beings capable of operating upon each other to produce the numerosity of other Numerical Beings.
- Numbers are ZERO in manifestation.
- Enumeration is Entification. Numbers are Entities.
- ZERO is the destroyer of Number, Time and Space.
- The Numbers are formed by Emanation, and these Numbers (as Immaterial Formative Potencies) Descend. Numbers are Beings, reflective of the ONE BEING/NON-BEING. Numbers (like all Subjects) are ESSENTIALLY NOTHING. Numbers are ZERO in manifestation. The 'Birth' of Number is most mysterious, just as is the 'Birth' of Cosmos (the Number One).
- Number and Illusion are coeval.
- Geometry is the *cosmically sanctioned* means of Creating Relationship.
- A Number has the Power to *compel aggregation* according to Its Nature.

Paradox and Contradiction

- According to the Law of Paradox, **I** AM simultaneously *all* opposites.
- The ALL-IN-ALLNESS contains even that which (apparently) contradicts IT OWN 'NATURE'. For instance, the IMMUTABILITY 'contains' the possibility of 'mutability'. This is a profound paradox: the NON-POLAR contains the polar.
- The Religion of the Future may well be 'Paradoxy'!

- Not only can contradictions be true, but complete, apparently self-canceling contradictions are *necessarily true* forever. All pairs of contradictions, Pairs of Opposites, are SELF-*resident*, forever, 'within' the INFINITESSENCE. In their *infinitized STATE* they are *identical,* and yet, were they to manifest, they would necessarily *manifest* as contradictions—forever.
- The Paradoxical nature of the INFINITE SELF prevents anything said about IT from being TRUE, including this statement that anything said about IT is prevented from being TRUE, including the previous statement that anything said ...
- What is the difference between an Orthodox Christian and a 'Paradox Infinitist'?
- In all the Cosmos (and Pre- and Post-Cosmos) the Principle of Paradox prevails. 'Something' *does* come out of' NOTHING. A 'part' *does* seems to exist, although its existence is philosophically impossible because of the LAW of the INDIVISIBILITY of the SELF—derivative from the First Fundamental of *The Secret Doctrine.*

Particles or Particle/Events

- An ultimate particle/event is indivisible. The time and movement that, theoretically, would be required to divided it are impossible in Fabricated Cosmos.
- An ultimate particle/event is an indivisible *willed* presence.
- Fohat works through ultimate particle/events by *becoming* them. Thus are all things with the World of Fabrication created, shaped and moved.
- In a way, an ultimate particle/event is too 'fast' to be 'caught' by any-*thing,* or 'touched' by anything'. It dwells in splendid isolation no matter what its movements and repositionings.

'Parts' and Wholes

- There is no way to 'remove' the 'part' without 'removing' the WHOLE.
- Consciousness 'creates' 'parts', *apparently* destroying the *wholeness* of the WHOLE.
- Even an Infinite Object is, paradoxically, a 'part'.
- The WHOLE is *entirely* not *partially* in the 'part'. In all the UTTER ALLNESS, there are no such things as REAL 'fragments'.

Perfection and Imperfection

- Imperfection 'RADIATES' from PERFECTION. The 'RAY' of the ABSOLUTE (also called the 'POINT') is the *first imperfection.*
- Any change in THAT which is already completely fulfilled, completely PERFECT, is a 'diminishment'—except, of course, that 'change' cannot be REAL.
- The Universe is a *necessary* Imperfection.
- In *relation* to the ABSOLUTE (which relation is, ESSENTIALLY, impossible!) all action, including the First 'ACTION') is imperfect, and either creates (and is coeval with) limitation, or *is* limitation, itself.
- Imperfection is a necessary part of PERFECTION. PERFECTION demands (apparent) Imperfection—logically so. Things are not what they *seem.*
- Perfection is *not* PERFECTION; the first is Cosmic, the second *is* the ABSOLUTE, and they are *incommensurable.*

The Point and Points

- The Condensed Point is the incipient Universal Logos.
- A Real *point* is any object at all, 'Seen' from a 'distance' sufficient to *infinitesimalize* the object.
- A Real Point is a *'thinged'* vision of any Subject by Itself—provided that Vision takes place from an ever receding Point of View (an 'Infinispectivizing' converging upon the Infinispective).
- A *virtual point* is the reduction of a definite thing to a definitely *quantifiable* tiny-ness, no matter how small.
- The entire activity of Cosmos is occurring at a (dimensionless) Point, *the* Condensed Point, the 'Center'. The 'Center' is the Center of the Stratified Cosmic Sphere. It is a *ubiquitous Center* and, yet, is but One Center. Though the Center is everywhere in the Sphere, it is, nevertheless, in the *'Center'.*
- A *point* is how an enumerated 'Ray' *appears* from the Source which emanated that 'Ray'.
- Self-'Sight' reveals Son-Points—at first Real, then (so that they may abide in Objective Cosmos—*virtual.*
- A point is a transitional 'something' between BE-NESS and Existence.
- All apparent 'points' in Cosmos, Really, coincide. All points in Cosmos are identical. Interval is illusion.

- Point, in-Cosmos, is a Subject 'Seen' as an Object. Points cannot be measured and yet, they cannot *not* be measured. Because they are, Essentially, Subjective, they do not belong to the Realm of Maya, and cannot be measured, but, because they are, also 'Seen', and, thus, Objective, they *do* belong to Maya (the 'measurable').

- What 'Something' can be more *elusive* than a point?

- Both the Infinite Subject and the Infinite Object are a *Single Dual Point.* That 'Point' (arising from 'POINTNESS') is the Point as both Subject and Object; as Subject It is a Point of View; as an Object, It is the Point 'Seen' from the Point of View (for all that can ever be 'Seen' from a Point of View can be resolved to a Point).

- The Point (being *dimensionless* and, yet, a 'Something') is the true Representative of the ABSOLUTE (the 'POINTLESSNESS') in the World of Illusion.

- In the ALL-IN-ALLNESS there is no *point of reference.*

- In Super-Cosmic Space/Time there is *one point of reference.*

- The Point and the Now are analogous—the first in relation to Space and the second in relation to Time. Now is a 'Point' in Infinite Time. The ETERNAL NOW is a *pointlessness* in Time.

- The First Pre-Cosmic Point (and the Last Post-Cosmic) is 'Infinified'.

- The Infinified Point yields the 'Infinispective'. The Condensed Point *Is* a 'Focused Point of Self-Perspective'. Both Points are REALLY/Really the same subjectively, but the Second point is *bounded by limited sphericity* (due to Its Own willfully limited Self-'Sight') and the First Point (being Infinified) is *unbounded.*

- The 'RAY' of the ABSOLUTE which 'FLASHES FORTH' is the very 'ACT' of the 'SELF' 'SEEING' the SELF (i.e., 'POINTNESS'). The 'RAY' instantly *generates* an Infinite (Infinified) Point of View which we call, generally, the SELF-as-Point in Pre-Cosmos (or, more technically, the SELF-as-Infinified Point).

- The pair—of Infinified Point (the Infinite Subject) and Mulaprakriti—is the First legitimate Duality that, with the Infinite Consciousness which unites them, comprise the Infinite Trinity. All three are but one Super-Cosmic Point, which is one with 'POINTNESS' ... which is identical with NOTHINGNESS.

- Points in Space are Self-Perceptions (the word 'Self' standing for any of a number of Cosmic or Super-Cosmic Beings). Space is filled with Virtual Points and Real Points (which are inter-convertible, depending upon Point of View). All the Points in Cosmo-Objective Space are Virtual Points (actual things with dimension). All the Points in Cosmo-Subjective Space are an infinitude of conceivable infinitesimalizings. If, ideationally, a Real point can be conceived in Cosmo-Subjective Space, then that Real point exists in the Ideational Worlds. Nevertheless, Real Points are not 'things' in external space, but conceptions in the Mind of a Spiritual Conceiver/Perceiver. Thus, Ideational Space can have an infinitude of Real Points if such a conception is possible to the Subjective Aspect for Whom that Space is a Reflection.

- Every Point in Space means every Point in the Field of Consciousness of the Universal Logos and Its Emanative Extensions.
- A Point is a Point of View, and, as well, that which a Point of View 'Sees'.
- The 'POINT' is the FIRST 'LIMITATION'.
- The Infinified Point is the *Aperture of INFINITUDE.*
- No Point can *actually* (Cosmo-Objectively) exist in Cosmos. Nothing dimensionless can exist dimensionally. And yet, Points, are *Ideational* Realities, though measureless.

Possibility

- Conditions are crystallized possibilities precipitated out of the uncrystallized *infinity of possibility* 'contained' within the LIMITLESS POTENTIAL, the FOUNT OF ALL POSSIBILITY.
- If the INFINITE SELF is the FOUNT OF ALL POSSIBILITY, then, no matter what is predicated of it, there is *always,* the *possibility* that IT may be *other that what is said.* There is as well the possibility that what has just been said of IT may be *other* than what is said, i.e., that what is predicated of IT may, indeed, be *possible.* Thus, apparently self-contradictory possibilities always mutually coexist without annihilating each other (unless, of course, they do!). In IT, nothing is impossible! (Or is it?)

Pralaya

- There exists at (possibly) *regular* intervals a Great Universal Pralaya. *Regularity* may mean nothing in relation to Infinite Duration. There will always be enough time in the Infinite Chain of Cosmoses to manifest *all intended possibility* and also, never enough time to manifest *absolutely all.*
- There is REALLY 'no space of time' between Universal Pralayas, because Universal Pralayas though, Actual, are not REAL. While Universal Pralayas can be 'measured' on the Infinite Time Line, it has always been the SAME 'MOMENT' forever, and NOTHING is ALL that has 'HAPPENED' in that 'MOMENTLESS MOMENT'. If Timelessness intervenes between 'Enduring Events', then the duration of that intervention is *zero!*

The PRESENCE and NOW/Now

- The Cosmo-Objective Now is the *present cosmic moment* in Fohatically Fabricated Cosmos; the NOW is the MOMENTLESS MOMENT in which the PRESENCE *abides.*
- It is impossible not to be *present.* It is impossible not to be the PRESENCE. Even when one is apparently *absent* as an Objective E/entity through 'ontological oscillation' one is still *present.*
- If a Cosmo-Subjective Now can be 'split' (unlike a Cosmo-Objective Now), then 'movement' within such a unit of Cosmic Time is 'impartite Movement of the Will' which is accomplished through 'identificatory intensification'—'being more something than another'.
- Those within the World of Being can apparently 'change' during a 'frozen' Cosmo-Objective Now. As Ideational/Archetypal/Numerical Beings are 'impartite', and require no 'movement' (from 'here' to 'there') to create 'change', such beings can split the second indefinitely. Spirit along moves at infinite speed because It is already NOTHING forever.
- 'Infinispectivization' requires an infinitely subdivisible Now.
- A Real Point cannot exist unless the Now within the World of Being is infinitely subdivisible, for a Real Point is an 'infinitesimalizing'. An infinitesimalizing of Space requires an infinitesimalizing of Time.

The 'RAY' of the ABSOLUTE

- When **8** say, **8** Am a 'Ray' of the ABSOLUTE, it means that **8** Am a 'Ray' of MYSELF.
- There is no such thing as <u>a</u> *Monad.* There is only <u>the</u> *Monad.* There is no such thing as <u>a</u> *'Ray' of the ABSOLUTE.* There is only <u>the</u> *'RAY' of the ABSOLUTE.*
- **8** Am **8**, the One, *here,* having forgotten that **8** Am also *there.*
- A 'Ray' of the ABSOLUTE is the PRESENCE of the *entire* 'RAY' of the ABSOLUTE at a certain apparent 'Depth of Objectification'.

REALITY/Reality and Actuality

- ABSOLUTE REALITY is the Great 'NON-EVENT'.
- That which is un-REAL is that which is 'incapable of affecting the BOUNDLESS IMMUTABLE PRINCIPLE'. As no-thing can affect the BOUNDLESS IMMUTABLE PRINCIPLE (other than ITSELF), all things are un-REAL.

- *Appearance does not alter REALITY.* This means that, no matter what *appears* to be so, REALITY continues to BE as IT IS.
- REALITY is ever unprecipitated, for IT abides 'AS IS' forever. REALITY cannot be precipitated, though an *'unfragmented fragment'* of IT can be 'EXTRUDED' into Externalization. And yet, all things *precipitated* are, ESSENTIALLY, REALITY.
- ABSOLUTE REALITY is never conditional or limited by conditions. Actuality, on the other hand, is conditional and describes the state of conditions.
- The Universe, or Cosmos, is a Great Actuality, *not* a GREAT REALITY.
- It is totally impossible for anything to be *ESSENTIALLY* un-REAL. Even appearances are, *ESSENTIALLY,* REALITIES.
- Whereas the Universe/Cosmos *appears* to be a *contradiction* to REALITY, it is not ESSENTIALLY so, for It is not REALLY *other.*
- Reality is never more than an ultimate moment away, and for those in the World of Being it is *always* present. REALITY, however, is never *even* an ultimate moment away. REALITY is NOW—*always.*

Relation, Relationship

- Everything *is* everything else—Essentially and ESSENTIALLY.
- All Relationship Is, REALLY, Illusory.
- BEING and Relation are opposites.
- A thing is a relation. Without relation, there would be no thing at all, no thing, *formally.* Even an *ultimate bubble in the Koilon* (far tinier, probably, than the bubbles in the Koilon of which we read in *Occult Chemistry)* is a relation between Fohat and Itself, or, more conventionally, between the Subject Fohat (Acting for and as the Universal Logos-as-Universal Son) and the Object, Prakriti.
- Archetypes are Numerical relationships. Beings manifesting in Cosmos are relationships. BE-NESS 'DESTROYS' relationships.
- Geometry is the *divinely sanctioned means* of Creating Relationship.
- *Relationship* which exists, *subjectively,* in an <u>*unextended*</u> mode, is *objectively* registered through <u>*extension*</u>. Extension and Objectivity are equivalent.
- In all the World of Becoming, there is naught but points-in-relationship.
- Some points within points are related imparticulately (as those in the World of Being). Other points within points are related 'particulately', as those in the Mosaic World of Fabrication.

The SELF and I/8

- **8**—Am the SELF, the SELF AM **8.**
- **8**—Total **I**/ALL SELF-as-Self, Am Omnipresent in Cosmos.
- **8**—Total **I**/ALL SELF-as-Self Am Omniscient in Cosmos.
- **8**—Total **I**/ALL SELF-as-Self, Am Omnipotent in Cosmos.
- **I**—TOTAL **I**/ALL SELF AM OMNIPRESENT, OMNISCIENT, OMNIPOTENT both 'in' and 'out' of Cosmos. To TOTAL **I**/ALL SELF, Cosmos is an *objectified infinitesimalizing.*
- The SELF-as-SELF cannot 'ACT', and yet, how can IT *not?*
- **8**, the Universal Self, Am all-pervading. **8**, the Universal Self, Am Omnipotent. **8**, the Universal Self, Am Omniscient. **8**, the Universal Self, Am the Doer, the Participator, the Potency behind all Motion.
- **I**-as-**8** Am the Author and Experiencer of All and ALL.
- The SELF in ITS ALL-IN-ALLNESS is not an 'ACTOR'. The CHANGELESSNESS is Impartite. The SELF-in-Universe, however, *must be* the Actor. What other *Actor* can there be?
- The SELF is *wholly* and *entirely present* in every illusory point of Time and at every illusory point of Space in Cosmos. The Self pervades All, but the SELF, ITSELF, is the TRUE PERVADER, or, better, the TRUE 'BE-er'.
- Every being can *BE* the SELF, but can never, strictly speaking, *relate* to IT. How does this affect our meditations?!
- Every object or presentation in Cosmos (including the entirety of Cosmos Itself) is *fully* the SELF.
- Every unit of life is *fully* the SELF in ESSENCE, and *partially* the SELF in formal expression.
- 'I' am as much 'you', as 'I' am myself. This is true, but hard to realize. **8** Am as much 'you', as **8** Am **8** MySelf. This is the same truth, but is easier to realize. Am 'I' not Really, **8**?
- Nothing can be withheld from the SELF. Where is the 'place' or 'time' where the SELF IS *not?*
- The SELF cannot be excluded from any place or any point.
- The SELF will not tolerate *exclusive* predication. *No one thing* can be said about IT. Only *all things* (an *infinite infinity of things)* must be said about IT, *never* only *one.*

- **8** Am a prakritically-bounded (REAL, but non-actual, i.e., non-objective) Point, or *non-fragment* of ABSOLUTE BEING.
- **8** Am an objectively-bounded Point of INFINITE SUBJECTIVITY.
- The Self is the SELF in its Universal Aspect. (The Universal Aspect of the SELF is naturally limited.)
- The SELF (written this way) represents the ABSOLUTE SELF.
- *Infinitized* SELF-Preoccupation IS 'INFINIDENTIFICATORY BE-NESS'.

Space

- In the ALL-IN-ALLNESS, *space* (both as interval and as the arena of the SELF's expression) does *not* exist. In the ALL-IN-ALLNESS Time, also, does not exist, nor does Motion.
- Objects are 'spaces' in Space. Objects are vacuities in Space, disruptions in the Objective Continuum.
- There is no 'space' between Me and MySelf, and, still more ESSENTIALLY, between ME and MYSELF.
- We see, paradoxically, that *there is no 'space' in SPACE!*
- Wherever there is Space, there are Time and Motion. Wherever there is Motion, there are Time and Space. Wherever there is Motion, there are Space and Time. This 'Trinity of Illusions' is inseparable.
- The 'STATE' of ALL-IN-ALLNESS is 'non-extensive'. IT 'HAS' no dimension at all, or an *infinitude* of *infinitized* dimensions, which is equivalent to 'HAVING' none.
- Within Cosmos, Essentially, and regardless of appearance, all 'places' are the same Place, all 'times' are the same Time.
- There is no 'space' in SPACE—i.e., ABSOLUTE SPACE.
- 'Space' is different from 'space' is different from 'SPACE'. Do we know the difference?
- Space is an 'expanded' Point, or a 'Point'-in-Extension.
- To 'enter' a point is to enter Space.
- Pure Consciousness is a Continuum that could be called Space. Presentations within this Continuum are the discontinuities called 'Contents of Consciousness'.

- The Point and the Now are analogous—the first in relation to Space and the second in relation to Time.

- All points in Cosmos superimpose and coincide. This means that all Self-Perceptions are but One Self-Perception. Therefore, all points, being <u>R</u>eal but non-actual, are Really identical.

Subjects and Objects
Subjectivity and Objectivity

- There is Really no such thing as an *external* Object.

- All *subjects* are identical, just as all *points* are identical.

- A Subject is, REALLY, NOTHINGNESS.

- Spirit-Matter is Really Subject-Object. The Two Things are ever One Thing, and that, cyclically, forever.

- A 'hole' in Space is the *objectified* Presence of a Subjective Viewer (who is ES-SENTIALLY, NOTHING). A hole in Space is the *externalization of a point of subjective 'IDEATION'.*

- Every Subject/Object is Really a 'hole in Space' (whether Infinite Space, 'De-Infinitizing Space' or Cosmic Space).

- There is an *enumerated subjectivity* that (because not yet objectified, i.e. Self-'Seen') need not be considered *strictly* objective. And yet Cosmic Prakriti, the Mother (Viewing the Infinite Subject as an Object) 'knows' 'Something's coming'! To the INFINITE SUBJECT, infinitely enumerated subjectivity is 'BE-EN' (but probably not 'SEEN') as Objects-in-Potential. In general, any articulations within a Subject, whether or not immediately 'Seen' by the Subject Itself, are, *on their way* to being 'Seen' as objects.

- All objects are Really infinitesimalizings and REALLY, NOTHING.

- Real points are objects in the state of infinitesimalizing.

- The entire Universe is the Drama between *seeing* and *being*. What is *seen* Exists, but the Infinite Subject 'Sees' only That which is within Itself—and all Its derivative *subjects see likewise*. What is 'SEEN' in the OBJECT is only what is within the SUBJECT. And, what is 'Seen' in the <u>O</u>bject is only what is within the <u>S</u>ubject.

- An Object *is* what the Subject *sees*.

- Nothing can be 'Seen' in Prakriti that is not already in the Self-as-SELF. Less is Seen by the Self than is in the SELF, because the Self (formally) Is an infinite limitation upon the SELF.

- Seeing *produces* an Object. This is what Super-Cosmic Fohat 'Does' in relation Mulaprakriti, or intra-Cosmic Fohat in relation to Cosmic Prakriti.
- NOTHINGNESS IS PURE SUBJECTIVITY. The 'WITNESS' IS REALLY 'BLIND' and IS, ESSENTIALLY, NOTHING.
- Ideation precedes objectification. Are subjective ideas formless before objectified? Can any 'thing' however subtle be *formless?* WHO or WHAT IS the 'VIEWER'? If there *is* one!
- **8** Am an objectively bounded Point of INFINITE SUBJECTIVITY. Everything that is, is a Point of INFINITE SUBJECTIVITY, except for the fact that INFINITE SUBJECTIVITY has no distinguishable *points.*
- What a Subject 'Sees'—*is*, exists.
- Every Subject is/becomes an Object to the Object It 'Sees'. If an 'Seen' Object is endowed with Subjectivity (because it cannot *be* other than the Subject that 'Sees' It), then that Object (now a Subject as well) must be able to 'See' the Subject which 'Saw' It. (i.e., 'See' that Subject as an Object).
- Remember that all the 'Key Players' in Early Cosmic Days (such as Fohat) are Subjects in Action. From a certain perspective, however, They are Objects, too.
- Anything that exists is an Object (ultimately, to the SUBJECTIVITY, or to the SUBJECTIVITY's 'PROXY'—the 'THREEFOLD RAY of the ABSOLUTE'). Even the Infinite Subject is an Object to THAT. What, though, is the method of 'Seeing' the Infinite Super-Cosmic Subjectivity? Can the INFINITE SUBJECTIVITY, per se, 'SEE', or only 'BE'?

Things and No-Things

- Predication is *reification.* The NO-THING will not tolerate reification. Yet man must *predicate.* What is to be done? NOTHING?
- There is no *thing* but that it is *seen,* for *seeing* ever 'creates' the *thing.*
- While a point is not quite a *thing* neither is it a *no-thing* either!

Time and the NOW/Now

- Time has cyclically appeared and disappeared forever.
- Infinite Time only exists 'outside' the INFINITE SELF; i.e., Infinite Time is *seen* to exists only by a Point of View 'outside' the INFINITE SELF. What is that Point of View? That of the Infinite Subject?

- At a certain 'Perceptual Distance', all things become One Thing, all places become One Place and all times become One Time, One Unchanging Event. This Point of View might be called the Infinispectivizing. The Infinispectivizing is not required in order to reduce any finite thing (however vast) to a Virtual Point (to apparently indissoluble Oneness). In order to reduce all things, all places, and all time to a single *Real point*, however, the Infinispectivizing *is* required.

- In terms of Time in Objective Cosmos, it is always Cosmically Eternally Now.

- In terms of Infinite Duration, it is always NOW.

- It may be NOW at certain *times* (along the Infinite Time Line) when it is not Cosmo-Objectively Now!

- Is Time the possibility of Space [sic] occurring?

- The One Omnipresent, Omniscient, Omnipotent Observer perceives (registers) Objective Cosmos only in the Cosmo-Objective Now, though abiding in the Cosmic *Eternal* Now.

- The ONE OMNIPRESENT, OMNISCIENT, OMNIPOTENT 'OBSERVER/NON-OBSERVER' 'REGISTERS' only in the NOW (the ETERNAL NOW), if IT can be said to 'OBSERVE' or 'REGISTER' *anything*, as that would require 'CON-SCIOUSNESS' which it does not (for 'long'!) 'POSSESS'.

- The Cosmo-Objective Now is the only possible arena of Cosmo-Objective Action, and yet no-thing in Objective Cosmos *moves* in or 'on' the Cosmo-Objective Now.

- The NOW is out of relation to the past and future and even to the Cosmic Present. The NOW is a 'NOTHINGNESS' in Time.

- NOW is the only 'Time' which exists, forever.

- The NOW is the unalterable 'NOUMENON' of Time—the PRESENCE of ETERNAL DURATION within Time.

- The present moment or the 'presented moment' is all of ETERNITY the Universe has 'Time' for! The present moment (the ultimate moment) is the *possible-in-Fohatic-Universe Presentation of ETERNITY.*

- The PRESENCE is *present* only in the NOW. Since it is always NOW, the PRES-ENCE is always *present.*

- There has never been another moment but the *present* 'MOMENT'/'NON-MOMENT'. NOW is the *time!*

- The present moment is a uniquely configured Presentation of the SELF to ITSELF-as-Self. The Cosmic Configuration is unique throughout all the Infinite Time Line in Infinite Duration.

- The ETERNAL NOW is the *aperture of detachment from illusory combinations.* Even the Eternal Now will serve!

- Time does not exist without an Observer. The Infinite Subject as Universal Logos 'Creates' Cosmic Time. There is also Super-Cosmic Time that relates to the Infinite Time Line. What is *its* Observer?

- Time is the finitization of INFINITE DURATION.

- The Super-Cosmic 'Clock' is *ticking forever* even though *no moments REALLY pass,* and but ONE 'MOMENT' *IS.*

- The pulse of sequential Universes is the ultimate *Standard Unit of Time Measurement.* Is that Pulse regular?

- There is REALLY 'no space of time' between Universal Manvantaras, because Universal Manvantaras though, Actual, are not REAL. Any normal 'space of time' is only *actual.* but not REAL. That particular 'space-in-time' (between Universal Manvantaras), however, *cannot* be measured. Only un-REALITIES can be 'measured'!

- Time (as we use it) is the measure of regular discontinuities.

- There is no Time and only One *continuous* 'MOMENT'/'NON-MOMENT' (the ETERNAL 'MOMENT') in ABSOLUTE DURATION.

- Now is definite; NOW is indefinite. This means that Now can be reified, but NOW cannot be reified. All times take place as a Now, but all definite times are infinitely too large to *be* NOW.

- The longest possible numerable duration of time is, notwithstanding its length, nevertheless infinitely far from the duration of the span of Infinite Duration. 'Forever' is infinitely longer than the largest denumerable span of Time. Forever outraces ceaseless enumeration.

- It is NOW at every 'point in Cosmic Time' (and *during* every 'point in Cosmic Time') even though the 'points in Cosmic Time' are apparently discrete, discontinuous, and separated by interval. It is also NOW when 'Cosmic Time is *not*'!

- Time is inseparably related to the Principle of Repetition.

- The Point and the Now (Cosmo-Objective) are analogous—the first in relation to Space and the second in relation to Time. The Point here discussed would, necessarily, be a *Virtual* Point, because the Cosmo-Objective Now has defined duration.

- Even *between* Universes, it is *always* NOW. Between Universes, no Cosmic Time exists, and yet it is always NOW. Time is measured forever along the Infinite Time Line. Yet can Time be measured as 'having existed' when it REALLY existed not?

- An interval between events may be a 'non-happening' in terms of the events, but the interval *is* a 'happening' simply because it is perceived as a 'something'.
- A frozen moment in a Cosmo-Objectivity has an *Ultimate* Time Value. Time is (apparently) 'moving' *outside of Cosmos* on the Infinite Time Line even though within Cosmo-Objectivity, Time is *not* moving, for nothing is moving.
- Moments are events *and* non-<u>move</u>ments.
- All Time must be measured forward and backwards from a definite *Now*. Time cannot be measured forward from a *beginningless beginning* or backwards from an *endless end.*
- No time passes in Objectified Cosmos during an a ultimate moment.
- A Cosmic Now is quantified and has 'time value' (in Infinite Time, or Infinite Duration, or on the Infinite Time Line); an ETERNAL NOW, of which there is only *one*, forever, cannot be quantified, and has a 'Time-Value' along the Infinite Time Line of <u>zero</u>.
- All moments in Cosmos have duration. The *one and only* 'MOMENT', NOW, has no duration. The one and only 'MOMENT' in INFINITE DURATION, literally, has taken "no time at all"—forever!
- The moment between ultimate moments may approximate an infinitesimalizing or it may be the equivalent of the ultimate moment, but it is *not* zero, otherwise there would be no ultimate moments, but only one cosmo-eternal moment in Objective Cosmos.
- An ultimate moment is indivisible—in Fabricated Cosmos. Within the World of Being, identificatory Acts of Will can occur 'during' Cosmo-Objective Nows.
- An ultimate moment is a *time-quantum*—a position of the Cosmic Configuration 'held' *motionless* for a *time*, that *time* being measurable only along the Infinite Time Line (and in comparison with the Ultimate Time Standard) and not measurable in Objective Cosmos.
- Time, being relative, needs a Standard of Measurement, or it cannot exist.
- An interval between events is a kind of hidden vacuum in Cosmic Time, but *not* a vacuum in Infinite Time, Infinite Duration.
- Every *time* has been ever ESSENTIALLY the same.
- The Real Ultimate Moments are Cosmic Eternal Nows! (as 'frozen {in their own way} in immobility as ultimate moments). A Cosmic Eternal Now will last for the entire duration of a Cosmos! This does not mean that other units of Cosmic Time are not proceeding—well, "at the same time"!
- Time and NOW are opposites, just as One and ZERO are opposites.

Values

- Naught in the world of extension is *formally* desirable. No thing is desirable formally except as it is a mask for the ABSOLUTE, the SELF, which it, ESSENTIALLY, IS.
- Value this Universe as the SELF!
- The only way to value the Finite is to detect within it the ESSENTIAL, the INFINITE. Things are of value only 'in' and *as* the SELF.

The Whole and the Part

- It is incorrect to identify with any one form within the Hierarchy of Forms. That way the madness of *ego* lies.
- WHOLENESS/Wholeness is found at every point. WHOLENESS/Wholeness is found in every Number, and the Number One is found in every Number. The Monad is found in every Number and *is* every Number.

WILL

- **8**-as-I may not deviate from My Will. **8**-as-'I', in a more veiled state, Am not ultimately allowed (or do not allow MySelf) to deviate from My Will in a less veiled state. The principle of Hierarchy must be respected and obeyed.
- Image-Events in the World of Being move with the 'Speed of Will'.

The Worlds of BEING and Becoming

- In the World of PURE BEING, *never* is there variation. PURE BEING is distinct from Being-in-Universe.
- In the World of Becoming, *never* is there total sameness. In the World of Becoming the Law of Unrepeatability prevails. No discontinuity is repeatable.
- The World of Being is included within the Greater World of Becoming.
- The WORLD OF BEING includes the World of Becoming.
- The World of Becoming is Illusory and does not include the WORLD OF BEING, though, ESSENTIALLY, the World of Becoming *Is* the WORLD OF BEING.

- In the World of Becoming no thing (formally) equals any other thing. (All things are formally unequal.)
- In the World of ABSOLUTE BEING, every infinitized thing equals every other infinitized thing (All infinitized things are equal). There are, however, no *separate infinitized* things.
- In the World of Becoming, the Principle of (formal) Unrepeatability holds sway, as no combination or recombination ever exactly repeats another.
- BEING and Existence are *not* equivalent terms.
- All differences of whatever kind are superficial, and pertain only to the World of Becoming.

Meditative Exercises to help induce Radical Infinitist Realizations

Through the process of Meditation and Visualization it can become possible to make the seeming abstract considerations of Radical Infinitism *Real Experiences.* The following Meditations (written mostly along Third Ray and First Ray lines), are meant to faciliate the achievement of a New Infinitist Perspective in the consciousness of the meditator, who is encouraged to enter deeply into the Process of *Unveiling* the Self-as-SELF—the 'HEART' of REAL IDENTITY. [Bolding within the meditation suggests phrases to be spoken.]

The Overcoming of Desire

1. Begin by quieting all aspects of the personality nature and aligning with a deeper, higher Reality.
2. Cast your inner eye over the World of Becoming and determine what you want most—whether people, places, things, conditions, relationships, etc.
3. Choose to focus upon just a few of your strongest desires. Why are they so strong? Why do you want what you want?
4. What is your life like *without* the fulfillment of these desires? Visualize and imagine.
5. What would your life be like if these few strongest desires *were* fulfilled? Visualize and imagine.
6. What are you willing to give, or to sacrifice, to achieve the fulfillment of these desires?
7. Having pondered on these matters pertaining to the World of Becoming, think deeply now about the WORLD OF BEING.
8. Dwell upon the *WHO* 'YOU' REALLY ARE—the SELF.
9. Realize what you *already* 'HAVE' because 'YOU' *ARE* the SELF. Ponder upon the ABSOLUTE as the PLENUM, and the PLENUM as the INFINITE SELF.

10. Think of the fact that you have had an infinity of desires for an infinity of Cosmoses. Think about whether fulfilling them has changed 'YOU' in the slightest.

11. Think about the futility of trying to 'reach' the PLENUM 'YOU' ALREADY *ARE* by the incremental steps of fulfilling desires. (It is like trying to reach infinity through enumeration.)

12. Search deeply within 'YOURSELF'. Realize what it means that the SELF *IS* the PLENUM and that 'YOU' ARE the SELF. That the SELF *IS* the INFINITESSENCE and that 'YOU' ARE the SELF.

13. See if you can *find,* already deep within the SELF 'YOU' ARE, the *accomplished fulfillment* of the desires you *seek* to have fulfilled.

14. See if there can arise within you such a deep satisfaction from *already being* the SELF, the PLENUM, that you discover that your desires have (not only *already* been fulfilled) but have *forever* been fulfilled.

15. As you brood, do not seek to 'fool' yourself, or 'hypnotize' yourself into 'believing' something you do not *actually* believe, but, instead, try to find the TRUTH of the realization—'YOU', as the SELF, *ARE FOREVER* the fulfillment of all possible desire.

16. Now, review again your most pressing desires—review them in the light of the PLENUM which is the SELF. How do these desires look now? Are they still completely unfulfilled? Has fulfillment come in any way? Does it matter so much whether fulfillment (as you dwelt on it previously) comes?

17. Resolve that whenever faced with pressing desire, you will invoke the PRESENCE—the SELF which is the PLENUM, the EVER-FULL, the CONSUMMATION which *has been* and *will be—forever.*

18. With a deep sense of satisfaction that comes, not from worldly consummation of desire, but from the realization of THAT which can be called the GREAT FULFILLMENT, sound the OM three times breathing forth Peace over a World driven half mad by desire and its apparent unfulfillment. From a deep realization that has arisen within yourself, convey to the restless, agitated World that, ESSENTIALLY, All Is Well:

    OM OM OM

Where There are Two, There is Fear

1. Begin by quieting all aspects of the personality nature and aligning with a deeper, higher Reality.

2. Realize 'YourSelf' as spirit/soul/personality. Identify as Essence.

3. From this high perspective, examine with clarity your three worst fears. In each case, *why* do 'you' (the "little you") fear? What particular outcomes do 'you' dread? Analyze with clarity what particular aspects of 'your' nature would suffer if these three fears came to pass.

4. Reaffirm Your identification with Essence. Enter deeply into Your Essential Identity.

5. Standing in Your true Selfhood, face each of the three fears. Can any of the three fear-inducing possibilities *Really* touch *You in Essence?* Deeply realize the invincibility of the Essential Self. Realize that naught can harm or even *touch* the Life.

6. Realize deeply the impermanence of all phenomena. Realize that, Essentially, no phenomenon can affect the Life You Are in the slightest. Realize that, Essentially, no phenomenon can affect *any* Unit of Life in the slightest.

7. With this realization firmly implanted in the consciousness, realize the cause of all fear to be *wrong identification.* Let there be revealed to You all the many ways 'you' (the "little self") identify your identity incorrectly.

8. Realize that You, the Essential Self, are already 'within' any fearful situation, and *are*, in fact, such situations themselves—totally. Realize that fear-inducing possibilities are not *separate* from 'YourSelf'. (This may not be easy to accomplish, but persist until you do.)

9. Again, face your three worst fears, but this time with a deeper conviction of Your Essential invincibility. Face these fears, until You sense a noticeable reduction in their fearsomeness.

10. As You sense fear departing from you, meditate upon the ancient dictum from the Upanishads, "Wherever there are Two, there is fear."

11. Develop an enlightened resolve to be One—and, thus, *fearless.*

12. Close this meditation with three soundings of the OM, spreading Your new realization concerning the illusory nature of fear to all human beings on this planet:

    OM OM OM

Ten Centillion 'U-AGO' ('Universes AGO') —A Meditation for the Induction of Perspective

1. From a Point of great abstraction, clarity and understanding, begin following the Infinite Time Line "backwards in time".
2. Watch the present incarnation dissolve, and realize—**Before the present incarnation was, I AM.**
3. Take yourself back, century by century, and realize, in general, your *presence* in each century. Stop at 500 years, 1000 years, 2000 years. Realize at each of these points your inescapable presence on the planet (in no matter what form—physical, astral, mental, causal). Realize, also, at each of these points of consideration: **Before this time was, I AM.**
4. Take yourself back thousands of years, tens of thousands, hundreds of thousands, millions—each time stopping to realize your presence *then,* and your presence *before* that time.
5. Stop at a point *before* the formation of our planet and realize that you, as the 'Ray' of the ABSOLUTE, *Are!*
6. Stop at a point *before* the formation of our sun and realize the same.
7. Place your presence in Space before the formation of our galaxy and realize that though the Galaxy is not yet, still, *You Are, You Exist!*
8. Finally, leap to the Point in Super-Cosmos before the formation of our very Universe, and realize that *You* are, necessarily, There in Nowhere. There is no moment of Infinite Time, Infinite Duration from which you can be removed.
9. But Universes are *successive.* A Universe preceded ours, and *there You are!* You cannot be removed. You are a necessary participant in all that has ever transpired.
10. Travel back in the endless, Infinite Chain of appearing and disappearing Universes. Are *You* not *there?* Where *else* can *You be*?
11. One UAGO *You Are.* Two UAGO, *You Are,* Ten, a hundred, a thousand, a million, a billion, a trillion, a quadrillion, a centillion UAGO—*You Are*—the *You,* Who Is *'YOU'*!
12. Never was the SPIRIT 'BORN', never shall IT 'DIE'. Never were 'YOU' 'BORN'; never shall 'YOU' 'DIE'.
13. Think of it—Ten Centillion UAGO, 'YOU' *as ever.* 'YOU', as ALL, *as ever.*
14. And before that and before that, and before that, regressing infinitely, in search of a 'BEGINNING' which never was nor ever *can* have been.

15. Always and ever 'YOU' have *been* the INFINITE 'WITNESS' of an Infinitude of Cosmoses.

16. Infinitely back into the Dark Recesses of Time, it has always been NOW and 'YOU' have always been 'YOU' (**I**), and it has always been HERE ('NOWHERE').

17. Remain in deep meditation focussed on YOUR UTTER IMPERISHABILITY.

18. Sound the OM three times, sharing with all of humanity your realization of the INERADICABILITY of the SPIRIT:

    OM OM OM

Who *Were* You?
Does it Matter?

1. Begin by quieting all aspects of the personality nature and aligning with a deeper, higher Reality.

2. From your greatest achievable point of objectivity look upon your present personal life, you present localized identity. Seek to grasp the essence of its quality, its note, its purpose and function. Realize that though that incarnation *was* you, yet it was not.

3. Now, in imagination, go to your immediately previous life (if you can remember it). If you can, do the same—seek to grasp the essence of its quality, its note, its purpose and function. If you cannot exactly remember, dwell on your sense of what it might have been like. Again, realize that though that incarnation *was* you, yet it was not.

4. Continue along the same line, going backwards incarnation by incarnation (in imagination or through recollection), doing the same.

5. Arrive at a point when you can no longer remember and no longer have any intimation of the nature of a given incarnation or of those that preceded it.

6. Pause to realize that, although you were incarnated, then, you can remember nothing—and yet you *were* even as you are now.

7. Realize, deeply, that because there is an Infinite Chain of Cosmoses, theoretically, you could continue the process of searching out your past incarnations, not only upon the various planes of this Cosmos, but upon various (and Really, *all)* planes of an infinitude of Cosmoses past. Pause, and let this realization sink-in deeply.

8. Now ask yourself, truthfully: How important is it to me that I no longer remember any or the majority of incarnations, for instance, which occurred *before* ten thousand years ago? How much am I troubled that I do not know the nature of the, perhaps, thousands of incarnations which occurred before that time?

9. Then extend the question: How much am I troubled that I cannot remember incarnations (in whatever form) *before* the Earth was formed, or before the Solar System was formed?

10. Extend the question further: How much does it matter that I cannot remember an infinitude of incarnations in an infinitude of Cosmoses past? Realize how little you probably care about the details of those countless incarnations upon all levels of all Cosmoses.

11. Now come back to the Present Moment. Look at your present incarnation again and at all the things which concern you so much. Feel your involvement in those things.

12. Then imagine how much those things will matter 100 years from now; in 200 years from now; in 500, in 1,000, and in 10,000 years from now.

13. Project yourself (as a 'Ray' of the ABSOLUTE) ahead, perhaps million years. 'You' may not even dwell upon this planet or you may. How important will this present incarnation seem then? As important as one of your unknown incarnations which took place a million years ago seems to you now?

14. Feel coming over you a great sense of Planetary Perspective. A sense of Cosmic Perspective. Look upon the present incarnation as if it will be but a forgotten episode in an endless chain of incarnational episodes.

15. Dwell deeply upon the meaning of Divine Indifference.

16. Look at your present incarnation again, and imagine how your would approach it, how you would think about it, how your would *live* it, if it really made no difference, or perhaps a very *tiny* difference in the infinite chain of your incarnations as a 'Ray' of the ABSOLUTE.

17. Imaginatively extract yourself from this present incarnation, and from all incarnations past and to come, and realize Who you *Really* Are—WHO 'YOU' *REALLY* ARE.

18. Realize that it is because of your True Identity (and YOUR TRUE IDENTITY) that these little incarnations make so very little difference. Ask yourself what difference the infinitude of them has made to your REAL IDENTITY.

19. Dwell in the sense of your ESSENTIAL unchangeable PERFECTION.

20. Resolve to live this incarnation and the incarnations to come filled with the realization of what is important and what is not; what is great and what is small; what passes away and what endures.

21. Close with the sounding of the Sacred Word three times, conveying to Humanity the thought of the Eternal Values, so that they may be more detached from pettiness, and more absorbed in a truer Identity (which is none but the TRUE IDENTITY):

    OM OM OM

The Magic Mirror

1. Begin by quieting all aspects of the personality nature and aligning with a deeper, higher Reality.

2. Stand, as spirit/soul/personality, before the Magic Mirror of Identity. (You can visualize this mirror as you choose. Within it you see your image reflected, and especially the image of the face of your physical form.)

3. Imaginatively study your image. Realize that this image is, in a way, a tiny reflection of who You Are, but that it can never capture or embody who *You Really Are,* and infinitely less, WHO YOU REALLY ARE.

4. As you gaze at the image enter into a deep sense of *essential otherness,* of *pervasive invisibility.* See the image, and yet *be* invisible and essentially *other* than this image. Allow 'YourSelf' to be deeply impressed by the dualistic perception of both *being* the image (apparently) and *not being* the image (ESSENTIALLY). Hold the awareness of this duality in consciousness.

5. Now gaze intently at the reflected image of your face and see it turn into another face—and another, and another. These can be faces of persons you know or do not know, it makes no difference.

6. Hold the image of these changing faces for as long or short as you like, but *realize* that the image of each face is Really the image of *Your* face. See the 'SELF-as-YourSelf' in many forms. No matter what facial form is presented to you within the Magic Mirror, realize deeply that it is *Your* form.

7. Feel *Your* invisible SELFHOOD pervading each and every image. As you become more proficient in the Art of Identification, do not release any image from the Magic Mirror until You *know* that You are *identical* with the Essence of that image.

8. Close the exercise by looking in the Magic Mirror and seeing *no reflection at all.* Realize that in 'seeing' *nothing,* you are 'seeing' your own *essential formlessness,* your own *invisibility.* Affirm to 'YourSelf': **8 Am, ESSENTIALLY, the invisible, all-pervasive SPIRIT. 8 Am, ESSENTIALLY, the FACELESS FACE behind all faces. 8 Am, ESSENTIALLY, the FORMLESSNESS behind all forms. Naught is but Me. Naught IS but ME.**

   OM OM OM

Note: This meditation may be practiced with great rapidity or in a more extended manner. You may find that there are certain facial images with which it is easy to identify. You may use these to establish the 'inner sense' of your *identicalness* with apparently 'other' people. Eventually, you will want to practice upon those images for which you have no attraction, or which, perhaps, personally repel you. The INFINITE SELF is both what we call "good" and "evil". IT is the very SUBSTANCE of All and ALL (and so are 'you-as-You-as-YOU'). Personal preferences, judgments, attractions and repulsions matter not in the slightest, and have no bearing upon your ESSENTIAL IDENTITY. Identicalness with All and ALL must be inwardly realized, and for this purpose, the Magic Mirror has it use.

Who Sees?

1. Begin by quieting all aspects of the personality nature and aligning with a deeper, higher Reality.

2. Realize Yourself as a 'Ray' of the ABSOLUTE, an apparent aspect of the One Universal 'Ray' of the ABSOLUTE.

3. Realize that as the Monad, You are a 'Ray' at a certain depth of Prakritic Immersion. Realize that although you are One and Pervasive, yet You still have a localized Identity.

4. Enter Your Subjectivity deeply, and look out upon the prakriti within which You (as a 'Ray'/Point/Monad) are apparently immersed. See all Your vehicles and realize that You are the Subject of these vehicles. You are the Observer. They are the Object, and You are the Subject. Dwell deeply in your Subjectivity as the 'Seer' of that which You 'have'.

5. Now, withdraw inwardly from what you 'See', from what You 'have', and realize that you are not only the Subject, you are also an Object, and that at this very moment You are being 'Seen'! Ponder upon how it is that You are being 'Seen'. What or Who is 'Seeing' You? Enter the experience deeply.

6. Realize that You are 'Seeing' Yourself being 'Seen'. You are in a reflexive dynamic.

7. Now, pondering even more deeply, realize that the One Who is 'Seeing' You as an Object, is You Yourself as a Subject. Realize that You are Seeing Yourself being Seen by You, Yourself. Enter deeply into the realization of the 'Seen' and 'Seeing' 'Yous'; realize the truth that it is *You,* indeed, Who are in both positions.

8. Ponder, now, the Divine Emanatory Stream, and realize that even the You Who was 'Seeing' You (as You think You are now), is, Itself, being 'Seen' by a still greater 'Seer'. Realize that every 'Seer' is also a 'Seen'.

9. Now, come to the realization that all along the Divine Emanatory Stream, it is none other than *You* Who is doing *all* the 'Seeing', and none other than *You* Who, at every level upon that Stream, is also being 'Seen'.

10. Say to Yourself: **8 Am the 'Seer' and the 'Seen', 8 the Subject/Object Am.**

11. Deeply consider the Divine Emanatory Stream, and realize that only One Being is found upon all its many levels. Realize that that Being is *You*. Feel this deeply.

12. Close by sounding the Sacred Word three times, pouring forth upon Humanity Your realization of the Subject Who Is Object, of the 'Seer' Who is 'Seen':

    OM OM OM

8 Do It

1. Begin by quieting all aspects of the personality nature and aligning with a deeper, higher Reality.

2. Realize your identity as a personality; then, your identity as a Transpersonality within the Causal Body; then your Identity as a 'Ray' of the ABSOLUTE, ever one with the One 'Ray' of the ABSOLUTE.

3. From this high perspective, view the world of Human Living.

4. Imaginatively 'see' some of the many things you have done in this present incarnation. Realize that although you are looking at many actions in many places in many times, it was the *same* 'You' that performed these actions.

5. Imaginatively cast your inner eye over your ascertained or imagined incarnations in the past. See many kinds of things being done in many places at many different times. Realize that despite the great diversity expressed as thousands or millions of acts, it was 'You' Who performed the acts, the very *same* 'You'.

6. Now view the world of Human Living again. In the present moment, imagine people all over the world 'doing' things, all kinds of things. Observe, imaginatively, the rich diversity of what they do.

7. Realize that almost every one of those people think that they, and they alone, are doing what they do. Realize the vast collection of apparently separate identities, egos, performing thousands and millions of acts.

8. Imaginatively expand the picture and go back in time hundreds and thousands and even millions of years, and 'see' all kinds of things being done by all kinds of people, in a great diversity of times and places.

9. Realize the egoic insularity of these many acts. Realize how millions upon millions of people (all through the ages and in all imaginable places) have thought that they, and they alone, have done what they have done.

10. Realize, however, that every one of these people in all their incarnations is a 'Ray' of the ABSOLUTE, just as 'You' are.

11. Realize, still more deeply, however that there can be no *separate* 'Rays' of the ABSOLUTE. Realize that all 'Rays' are One Ray, that all Monads are One Monad, and that all Points are but One Universal Point.

12. Realize the illusion of separative identity.

13. Realize that though many apparently separate human beings have thought they have performed many separate actions, now and in all ages, that there has been only One Performer—the One 'Ray', the One Monad, the One Point in all Points.

14. Realize that it was 'You' all along who were the One Performer, the One 'Ray', the One Monad, the One Point in all Points. Just as 'They' were the One Performer, the One 'Ray', the One Monad, the One Point in all Points.

15. Realize that 'You'-the-**8** *do* it, *did* it, *will do* it.

16. Identified as the One Doer, see yourself in all doings—now, in the past, and as anticipated in the future. Feel 'Yourself' intimately within every act you can imagine—the good, the bad, the great, the small, the purposeful, the purposeless. See Yourself within all the many forms of men and women, doing all the things men and women do and have done for ages. *You* are there. *You* are the 'Doer'.

17. Allow your imagination to roam Time and Space and feel Your Presence in all the many *doings, feelings, thinkings, lovings, willings.*

18. When the realization is deep and well established, say to yourself, *knowingly:* **8 Do It.** Say this slowly and meditatively until the conviction of the Reality of what you say is firm.

19. Close with the sounding of the Sacred Word three times, radiating the knowledge of non-egoity to all the millions of apparently separated individuals the world over—individuals who think that they *alone* do what they apparently do. Breathe forth upon them, the realization of Oneness—the One Identity, *Doing, Feeling, Thinking, Loving, Willing, Being:*

    OM OM OM

On Becoming NOTHING through the Art of Negation

1. Begin by quieting all aspects of the personality nature and aligning with a deeper, higher Reality.

2. You are about to go on a Quest for your inherent BEINGNESS.

3. Realize the great and misleading human tendency to think 'you *are* what you *see*'. Established as deeply as may be in the understanding of BEING, realize the great fallacy of this approach.

4. Prepare to 'travel' inwards towards the ESSENCE through the Power of Negation.

5. See your etheric-physical vehicle. Value it, share your gratitude for its service to You, then reject it as *formally* un-REAL.

6. See your astral/emotional vehicle. Value it, share your gratitude for its service to You, then reject it as formally un-REAL.

7. See your mental vehicle. Value it, share your gratitude for its service to You, then reject it as formally un-REAL.

8. See your causal vehicle. Value it, share your gratitude for its service to You, then reject it as formally un-REAL.

9. See your manasic vehicle. Value it, share your gratitude for its service to You, then reject it as formally un-REAL.

10. See your buddhic vehicle. Value it, share your gratitude for its service to You, then reject it as formally un-REAL.

11. See your atmic vehicle. Value it, share your gratitude for its service to You, then reject it as formally un-REAL.

12. See your Monadic Vehicle. Value it, share your gratitude for its service to You, then reject it as formally un-REAL.

13. What remains? You are part of the Planet. You are part of the Solar System. You are part of larger constellations. You are part of the galaxy. You are part of a family of galaxies. You are part of galactic clusters.

14. See them all—imaginatively, one after the other. Value them, stand in awe of their beauty and grandeur, honor Them as the *gods* they are—then reject them all, *utterly*, as formally un-REAL.

15. They will not do. No-'Thing' will do. Only NOTHING will do!

16. Whatever You see, touch, contact in any way, *none* of these things are REAL. None will do. Reject them! Negate them! Reject them utterly! They, as forms, have naught to do with the REAL 'YOU', You *seek*.

17. If you *see* it, you cannot *be* it. Forget all philosophical niceties—this is a useful truth. 'If you *see* it, you cannot *be* it!'

18. Remain, then, utterly *alone* in your uncompromising *aloofness.* Every time the consciousness goes outward to seek a 'thing' that 'YOU' might be, bring that consciousness back, and *isolate it.*

19. Know that You are 'YOU' *alone*—'YOU'-the-**I.**

20. Let no-thing cling to 'YOU', and cling to no-thing. All things are delusive—REAL in ESSENCE but totally un-REAL formally. Be done with every presentation encountered.

21. Where do 'YOU' find YOURSELF? Where are 'YOU'? Who are 'YOU'? If 'YOU' are none of these things you have driven away, negated, WHO are 'YOU'?

22. Realize deeply, deeply that 'YOU' are no-thing at all. Having rejected all things, 'YOU' have become NOTHING, the ROOT of ALL.

23. Without *thinking,* abide in 'NO-THINGNESS', 'ALOOF' and uncompromisingly REAL.

24. Hold rejection like the flaming *Sword of REALITY* to defend the SELF against minimization through wrong identification.

25. Thus live vigilantly as the DESTROYER of Illusion.

26. Live in NOTHING. Live *as* NOTHING. Reject ALL. *BE* NO-THING!

27. Sound the OM three times, sharing Identification *as* the VOID with all human beings who can love REALITY:

    OM OM OM

Meditation upon Identification

1. Quiet the vehicles of the personality and steadily withdraw within.
2. **OM**
3. Focus as an integrated personality, and, steadily withdrawing within, align with the Solar Angel, the Representative of the Heart of the Sun. Feel deeply the energy of uniting, fusing, Love.
4. **OM**
5. Focusing imaginatively within the Causal Body, withdraw steadily inwards through the three tiers of petals, feeling their quality, until the consciousness is centered within the energy field called the Jewel in the Lotus. This is a quiet and steady movement from Love into Will/Synthesis. Dwell quietly within this energy, and realize your connection to the 'Lotus Field' working in affiliation with the Great Ashram.
6. Withdrawing steadily inward, imaginatively enter the higher, wider spaces of the Spiritual Triad, where the Mind of God, the Love of God and the Will of God are reflected in the Personality of the Monad (i.e., the Spiritual Triad). Realize the power of impersonal identification with the Divine Plan.
7. Withdraw steadily inward to the very center of your being, and imaginatively focus as a Spirit, a Monad, a until your consciousness is focussed upon the plane of that formless, all-pervading Awareness and Presence.
8. Realize your deep identification with God's Purpose.
9. Survey the World of All Forms, Great Brahma's Form, the Form of God.
10. Imaginatively see the many beings, their structure, their principles, their divinity.
11. Realize that the Sprit which animates them, the Spirit which they are, is *absolutely identical* with the Spirit you are.
12. Blot out all forms within the Great Form, but abide identified with the ESSENCE within the form. Through sustained identification, *know* the identity of ESSENCE to be REAL.
13. Pass before your inner eye, some of the forms of people, places and things with which you are familiar, and, blotting out the ordinary form, remain in a state of deep identification with the Spirit ESSENCE of these beings and configurations.
14. Resisting the lure of the familiar, realize that even these forms are ESSENTIALLY Spirit. Imaginatively make these familiar forms appear and disappear, all the while maintaining your identification with their ESSENCE.

15. Using the Creative Imagination, and your deep and growing sense of the ESSENCE, seek to merge your being with the Being of the One Who Pervades All Cosmos—the Great Universal Life. While this is not *literally* possible (at this 'time'), the attempt will inspire in you a growing sense of Pervasion and Identification with all Life.

16. Hold the feeling of *formless extensiveness* and *Pervasion,* along with an image of your more familiar selfhood. Realize that both are true and valid in their own spheres—the All Pervading Self and the localized self.

17. Resolve to hold your Identification with the Greater while fulfilling the requirements of the Lesser.

18. Sound the OM three times in the state of deepest possible Identification with the All-Pervading Universal Life:

    OM OM OM

"I Still Persist" —A First Ray Meditation

1. Following the usual means of alignment and abstraction, withdraw inwardly rapidly, recognizing, acknowledging and then disidentifying with one vehicle after another.

2. Withdraw steadily inward to the very center of you being, and imaginatively focus as a Spirit, a Monad, a until your consciousness is focussed upon the plane of that formless, all-pervading Awareness and Presence.

3. From this high point of vision, imaginatively see all the vehicles through which you, in your localized form, are manifesting:
   - the physical-etheric, astral and mental bodies
   - the personality vehicle as a whole
   - the causal body with its three tiers of petals and the synthesizing Jewel in the Lotus
   - the threefold Spiritual Triad, manas-buddhi-atma (personality of the Monad)
   - finally, the Monadic Sheath on the Monadic Plane

4. Realize that although, from the human perspective, you are Essentially the Monad, yet you necessarily manifest through all these vehicles throughout the course of human evolution.

5. Now maintaining the high state of abstraction and pervasion characteristic of Monadic Awareness (doing this at least in the imagination), and remaining ever upon the Monadic Plane, project a *ray* of your Monadic Self into the etheric physical vehicle and be *present* within the ring-pass-not—as well as maintaining your presence on the Plane of the Spirit. You are now established (in consciousness and *being*, in two 'places').

6. Realize your etheric-physical identity and your connection with the etheric-physical plane through your etheric-physical body.

7. Let there appear (imaginatively) a blinding flash of light which instantly causes the etheric-physical vehicle to vanish. Realize that although you no longer possess the etheric-physical vehicle that you still exist—*fully.*

8. Then, deeply centered within your sense of being say: **Though the etheric-physical vehicle vanish, yet, I still persist.**

9. Establish your consciousness in your astral vehicle and realize your connections to the astral plane via your astral vehicle.

10. Then imaginatively see this vehicle too vanish instantly in a blinding flash of light Realize that although you no longer possess the astral vehicle that you still exist—*fully.*

11. Then, deeply centered within your sense of being say: **Though the astral vehicle vanish, yet, I still persist.**

12. There follows the same procedure for the mental vehicle using the mantram: **Though the mental vehicle vanish, yet, I still persist.**

13. Center yourself deeply in your sense of imperishable inner being, and realize that although (imaginatively) you no longer possess any of the personality vehicles, you still *persist* in the fullness of your Being, the fullness of what you *Really Are.*

14. Imaginatively enter the Causal Body, the Temple of Solomon. Establish yourself in the Knowledge, Love and Will of the Solar Angel, and within the synthesis of the Jewel.

15. Realize the beauty and value of this stratum of existence, and the ways in which this vehicle connects you to the world of souls.

16. Then, loving and valuing the Causal Body, and *knowing* the energy which it brings to your life, call for the blinding flash of light and imaginatively see it, too, vanish in an instant.

17. Centered deeply in the persistent sense of your own *being,* realize what you have done, and dwell deeply with this realization.

18. Then say with the fullest possible understanding and conviction: **Though the causal vehicle vanish, yet, I still persist.**

19. Then one by one transport the center of your being into the Triadal Vehicles and, after realizing their value and what they may and do bring into your awareness, use the following mantra: **Though the manasic vehicle vanish, yet, I still persist. Though the buddhic vehicle vanish, yet, I still persist.Though the atmic vehicle vanish, yet, I still persist.**

(Although the full realizations of these affirmations lie far ahead of the great majority who will use this meditation, the exercise is useful in establishing an orientation towards Eternal Persistence, then, ETERNAL PERSISTENCE.)

20. Establish yourself (imaginatively and finally) within the Monadic Field. Seek to understand this Field as your True Home, the Father's Home, but realize this Home to be only temporary.

21. Focus on the *fact* that you are ESSENTIALLY a 'Ray' of the ABSOLUTE, and as such, are ESSENTIALLY *free,* and incapable of being forever *confined.*

22. With your very *being* focussed on ESSENTIAL BEING, see again the blinding flash of light and the vanishing of the Monadic Sheath—the last of the sheaths for this temporary phase of your evolution as a 'Ray' of the ABSOLUTE.

23. Then say: **Though the Monadic Sheath vanish, yet, I still persist.**

24. What now remains? Feel and *be* what now remains.

25. There will be many Greater Sheathes into which the 'Ray' will be absorbed, but *realize* that you, the Essential You-as-ESSENTIAL 'YOU', are *completely indestructible.*

26. Seek to realize the literally *infinite number of forms* you have, in your infinite history, *indwelt.* Realize they have all been destroyed. They are no more, but *you* are.

27. Seek to realize the literally *infinite number of forms* you *will* in your infinite future history indwell. Realize that though they do not yet exist, they too will be destroyed, but this destruction will not affect the REAL 'YOU' in the slightest.

28. Realize the complete and absolute indestructibility and immortality of the Spirit 'YOU' ARE.

29. Gather all your realization into one final affirmation of BEING: **Though all worlds be destroyed, I still persist!**

30. Sound three OM's seeing the realization of immortality flow into the hearts and minds of all human beings:

    OM OM OM

"Naught Is But Me" —A Second Ray Meditation

1. Quiet the vehicles of the personality and steadily withdraw within.
2. **OM**
3. Focus as an integrated personality, and steadily withdrawing within, align with the Solar Angel, the Representative of the Heart of the Sun. Feel deeply the energy of uniting, fusing, Love.
4. **OM**
5. Focusing imaginatively within the Causal Body, withdraw steadily inwards through the three tiers of petals, feeling their quality, until the consciousness is centered within the energy field called the Jewel in the Lotus. This is a quiet and steady movement from Love into Will/Synthesis. Dwell quietly within this energy and realize your connection to the 'Lotus Field' working in affiliation with the Great Ashram.
6. Withdrawing steadily inward, imaginatively enter the higher, wider spaces of the Spiritual Triad, where the Mind of God, the Love of God and the Will of God are reflected in the Personality of the Monad (i.e., the Spiritual Triad). Realize the power of impersonal identification with the Divine Plan.
7. Withdraw steadily inward to the very center of your being, and imaginatively focus as a Spirit, a Monad, a until your consciousness is focussed upon the plane of that formless, all-pervading Awareness and Presence.
8. From this high vantage point of *pure being,* view the World of Form as an Entirety. Begin with those forms that are familiar to you, and move outwards into larger and larger wholes. As you do so, visualize the tremendous diversity of the World of Form, and notice the astonishing heterogeneity.
9. Focus on that which seems to separate one from another—the boundaries, the distinctions, the qualitative and formal differences. Let your entire focus be upon *differentiation.*
10. Despite the myriad of differences, realize the *magnetism* which attracts all forms to each other. Realize the obvious and subtle ways that each form is bound to every other. Experience yourself as part of this magnetism.
11. Through the creative imagination, visualize the magnetically bound World of Forms as completely transparent. Within each form see the blazing light indicating the Presence of Spirit. When you look, imaginatively, towards any form, all you see is the shape, transparency, and the blazing Radiance of Spirit.
12. Realize that that same Radiance of Spirit blazes in you. Feel the presence of the blaze and identify with It.

13. Realize that in all forms, and within your own form, *one and the same great light of Spirit is blazing.* Recognize the *identicalness* of all these Spirit Lights.

14. Be impressed by the utter *sameness* of the Spirit Light. *Feel* the sameness. Look through all forms (for they are transparent) and see but Spirit Light that you know to be *one and the same* in every form.

15. Holding your concentration on the identicalness of the Radiant Spirit Light, restore all forms from transparency to opaqueness—all their color and vibrancy returns—but, the Spirit Light is *still* blazing and is seen *though* all the opacity of form.

16. See the full variety of form, but realize simultaneously the identicalness of the Spirit Light that blazes within every form and with which *you are one.*

17. Holding the vision of the Radiance and the realization of the Identicalness of Spirit Light which all share, say quietly to yourself (or revolve the thought, brooding upon its significance)—**"Naught is but Me."**

18. Look out into the World of Form and, realizing the inherence of Spirit that makes all things One, deepen your understanding through the constant repetition of —**"Naught is but Me. Naught is but Me. Naught is but Me."**

19. Sound the Sacred Word OM three times seeing your realization permeate human consciousness:

    OM OM OM

God Is a Consuming Fire

1. Begin by quieting all aspects of the personality nature and aligning with a deeper, higher Reality.

2. After abstracting from all lesser identifications, realize 'YourSelf' as the invisible, all-pervading Fire of Spirit. Dwell with this realization, and intensify it.

3. Realize that, in Essence, God is You, and You are God, and *both* are Fire.

4. Pass before Your inner eye an array of physical-etheric forms. Identified as the Fire of Spirit, approach any or all or these forms and *consume* them until they, too, turn to Fire and are absorbed into the Fire You Are.

5. Pass before Your inner eye an array of astral forms. Identified as the Fire of Spirit, approach any or all or these forms and *consume* them until they, too, turn to Fire and are absorbed into the Fire You Are.

6. Pass before your inner eye an array of mental-forms. Identified as the Fire of Spirit, approach any or all or these forms and *consume* them until they, too, turn to Fire and are absorbed into the Fire You Are.

7. Having imaginatively consumed all the normally objective worlds, and identified 'YourSelf' as the Consuming Fire, ponder deeply upon *Stanza I,* from the *Stanzas of Dzyan,* given in *A Treatise on Cosmic Fire:*

   > *The Secret of the Fire lieth hid in the second letter of the Sacred Word. The mystery of life is concealed within the heart. When the lower point vibrates, when the sacred triangle glows, when the point, the middle center, and the apex, connect and circulate the Fire, when the threefold apex likewise burns, then the two triangles—the greater and the lesser—merge into one flame, which burneth up the whole.*

8. Scan with the inner eye the entire Field of Objectivity (including all subtle dimensions). *Consume* every objectivity which impresses You, and render that objectivity into the Fire of Spirit You Are.

9. Realize that all things dwell within You as *Fire.*

10. Under the deep impression of this realization, ponder on the true nature of Synthesis. Realize the equality between Synthesis, God, Fire, and Your true intra-cosmic Self.

11. Sound the OM three times deeply identified with the God Who is, in Cosmos, the All-Consuming Fire:

    OM OM OM

They are Gone but They Live in ME

1. Begin by quieting all aspects of the personality nature and aligning with a deeper, higher Reality.
2. Identify 'YourSelf' with the one SELF-as-Self Who pervades all of Cosmos.
3. Think of the virtually countless human beings (whether personally known to You, or not) who have died, thus relinquishing the physical form. Ponder their present 'location' in Cosmos.
4. Realize that (because Time and Space are Essentially illusory) the concept of 'location' (though useful to the normal, personality consciousness) is completely unsuitable in the relation to the Essence of any human being.
5. Realize that, in fact, all human beings who have ever 'lived' are *Here Now.* Let this realization sink deeply into your consciousness.
6. Realize that, in Essence, only *You-the-**8*** are *Here Now.* As strange as this concept may seem, let it percolate into your consciousness until it is felt as Truth.
7. Realize the inevitable conclusion that all human beings who have ever lived are *within You-the-**8, Now**.* Take time to imaginatively absorb the Essence of any human being who has ever lived, and feel each *within* 'YourSelf'. *Know* each to be *identical* with 'YourSelf'.
8. Practice this realization upon departed human beings whom you have known personally. Do not relinquish the exercise until you are deeply impressed that these human beings truly *live within* You and *are* You.
9. Having established a deep conviction of this Essential Truth, realize the fact that all human beings to come, in all ages, *will*, necessarily, also *live within* You and *be* You.
10. Feeling a deep Love for all human beings past or to come, close this exercise by saying (mentally or softly aloud) the Mantram of Unification:

The Sons of Men are one and I am one with them.
I seek to love, not hate;
I seek to serve, not exact due service;
I seek to heal, not hurt.

Let pain bring due reward of light and love.
Let the soul control the outer form, and life, and all events
And bring to light the Love which underlies the happenings of the time.

Let vision come and insight.
Let the future stand revealed.
Let inner union demonstrate and outer cleavages be gone.
Let Love prevail.
Let all men love.

OM OM OM

'BEGINNINGLESSNESS' and 'ENDLESSNESS'

1. Begin by quieting all aspects of the personality nature and aligning with a deeper, higher Reality.

2. Think about a number of 'things' or 'items-in-Cosmos'. Realize the specific ways that each of them has had a *beginning* or will have an *ending*. Be as specific as you can in your visualizations.

3. Cast your eye about the World of Form and see if you can find any 'thing' that does not begin and end.

4. Realize that for most things you can think of, it is possible to determine *when* they began, and approximately *when* they will end. But now, dwell on Cosmos.

5. You can, in general, conceive of Its Beginning, correct? You can, in general, conceive of Its Ending, correct?

6. But now think of the entire infinite succession of Cosmoses. Ponder upon when that succession *began*. Go back, back, back, back ... Can you conceive of a Beginning? And now go forward, forward, forward, forward, anticipating the Creation of an infinity of Cosmoses. Can you conceive of an Ending to the succession? An Ending to the Infinite Chain of Cosmoses?

7. What does it do to your mind to think that Cosmoses simply *always were* and *always will be?* Dwell with it.

8. Ponder on the thought that never was the time when there was not a Cosmos (or a Universal Pralaya). Ponder on the thought that never will be the time when there will not be a Cosmos (or a Universal Pralaya). Dwell with it. What happens to your consciousness?

9. Realize, then, that the infinite series of Cosmoses is Beginningless and Endless, even though each Cosmos, *individually,* both begins and ends.

10. Now, become even more abstract. Seek to identify with THAT WHICH IS *NOTHING.* Where did IT come from? Is it conceivable that IT *began?* Is it conceivable that IT can *end?*

11. Ponder on a CAUSELESS CAUSE.

12. What does it do to your consciousness to think of an UNCAUSED NO-THING which has *forever been?* Ponder upon it.

13. What does it do to your consciousness to think of an UNCAUSED NO-THING that *forever will be?* Ponder upon it.

14. What does it do to your consciousness to realize that *'YOU'* are THAT UNCAUSED NO-THING?

15. Spend a few minutes thinking about the SELF, and repeating to yourself the words: **EVER-SO, EVER-SO, EVER-SO, EVER-SO, EVER-SO, EVER-SO, EVER-SO ...**

16. Spend a few minutes thinking about the INFINITE SELF, 'YOURSELF', the **I** and repeating to yourself the words: **FOREVER and FOREVER and FOREVER and FOREVER and FOREVER and FOREVER and FOREVER and FOREVER and FOREVER and FOREVER and FOREVER and FOREVER and FOREVER ...**

17. When you have 'time', repeat the above over and over again, *until the change happens* and the 'Key to the Mystery' *dawns upon your 'blown out mind'.*

18. In a state of profound realization of the nature and quality of ETERNITY, close with the Mantram: **ONLY I, ONLY HERE, ONLY NOW—FOREVER.**

19. Sound the Sacred OM three times, breathing forth upon humanity, and upon all sentient beings, ETERNAL BLESSING, a FOREVERNESS of BLESSING:

    OM OM OM

The ETERNAL NOW

1. Quiet the vehicles of the personality and steadily withdraw within.

2. **OM**

3. Focus as an integrated personality, and, steadily withdrawing within, align with the Solar Angel, the Representative of the Heart of the Sun. Feel deeply the energy of uniting, fusing, Love.

4. **OM**

5. Focusing imaginatively within the Causal Body, withdraw steadily inwards through the three tiers of petals, feeling their quality, until the consciousness is centered within the energy field called the Jewel in the Lotus. This is a quiet and steady movement from Love into Will/Synthesis. Dwell quietly within this energy and realize your connection to the 'Lotus Field' working in affiliation with the Great Ashram.

6. Withdrawing steadily inward, imaginatively enter the higher, wider spaces of the Spiritual Triad, where the Mind of God, the Love of God and the Will of God are reflected in the Personality of the Monad (i.e., the Spirit Triad). Realize the power of impersonal identification with the Divine Plan.

7. Withdraw steadily inward to the very center of your being, and imaginatively focus as a Spirit, a Monad, until your consciousness is focussed upon the plane of that formless, all-pervading Awareness and Presence.

8. From this Point deep within, and abstracted from all identification with form, allow the consciousness to proceed "backwards in time", 'seeing' the Great Pageant of Changing Forms. Go to those civilizations, cultures, places, circumstances that seem familiar to you and with which you have a resonance, but deeply maintain the Attitude of the Observer.

9. From time to time gather the sense of yourself as an apparently individual unit of life, living within various contexts, learning, growing and experiencing the quality of the times.

10. Sense the immensity and grandeur of the slow "March of Time". Realize its evolutionary Purpose and your inevitable participation in the Process.

11. Refocus on your 'position in Time' in the present moment.

12. Then, still maintaining the Attitude of the Observer send your consciousness "forward in time", imagining some of the evolutionary unfoldments that lie ahead for humanity and our planet.

13. Transport yourself imaginatively into countries, civilizations, cultures, places, circumstances which suggest themselves to you as possibilities, and feel your participation as an apparently individual unit of life in these future possibilities.

14. All the while maintain in your Spirit Awareness the sense of the "Slow March of Time", the unfoldment of Planetary History on a Large Scale.

15. Return in consciousness to the Present MOMENT, the NOW. Stay with the Present MOMENT and realize that all that may be transpiring in the immediate surroundings, upon the planet, within the solar system and even within the Entire Cosmic Configuration of Changing Relationships, does not 'touch' the Present MOMENT.

16. Detaching your consciousness from any pre-occupation with form, and realizing its evanescent, impermanent nature, realize deeply that this Present MOMENT has *always been with you and always will be.* Realize the *inescapability* of the Present MOMENT.

17. Then reaching 'forwards' and 'backwards' in Time, simultaneously, bring the past and the future which you have imaginatively experienced into the Present MOMENT, and realize that when all these past experiences occurred it was NOW, just as it is NOW.

18. And realize that when all such future experiences will occur, it will then be NOW, even as it is NOW at this moment.

19. Realize the Omnipresence of the NOW, how IT binds all experiences together.

20. Realize that NOW is the only MOMENT you *have,* ever *have had,* or ever *will have.* Realize that NOW is ETERNAL.

21. Realize that 'YOU'-as-a-'You' have seen an infinitude of Combinations and Configurations through the Cosmic Kaleidoscope during an Infinity of Universes, and that even in the face of this Infinite Variety, it has always been *exactly NOW.*

22. Realize the NOW as the HOME of the ONE UNCHANGING PRESENCE. Enter deeply into this PRESENCE and realize that it is you, and I and all.

23. Realize that the PRESENCE is ETERNALLY PRESENT in the ETERNAL NOW.

24. Imaginatively reach forwards and backwards in Time *infinitely,* and realize that *forever,* it has been and will be only NOW.

25. Feel the *purity and power* of the ETERNAL NOW, and resolve to live within IT, a Spirit/Monad.

26. Meditate in silence as to how the ETERNAL NOW may become the *medium* in which you live.

27. See the improvement of your localized life, and the localized lives of all around you, arising as the result of living forever focused in the ETERNAL NOW.

28. Sound the OM three times, identified as the all-pervading PRESENCE, ever PRESENT in the NOW:

OM OM OM

Merge the World into BRAHMAN
Dissolve the World in BRAHMAN

1. Following the usual means of alignment and abstraction, withdraw inwardly rapidly, recognizing, acknowledging and then disidentifying with one vehicle after another.

2. Poised imaginatively upon the atmic plane of Spiritual Will, withdraw steadily inward to the very center of you being, and imaginatively focus as a Spirit, a Monad, polarizing the center of your *being* upon the plane of that formless, all-pervading Awareness and Presence.

3. From this highest of all human perspectives seek to become aware of the subtlest of all subtleties—the ONE AND ONLY SELF.

4. Realize that in all the variety that presents itself to your outer and inner eye, you are 'seeing' naught but the SELF.

5. Identify as deeply as possible with the SELF. Feel IT as the GREATEST FORCE in the World.

6. As you identify with and as the SELF, feel that IT IS an invincible 'FORCE of RESOLUTION'—the 'POWER' to turn all things within the World of Becoming *into* ITSELF.

7. Hold in your consciousness/awareness the SELF *and* all other things.

8. Bring before your inner eye, one presentation after another—one thing after another, one place after another, one E/entity after another, etc.—and in the way you choose to symbolize the process, see the GREAT SELF magically 'RESOLVING' each of these 'things' into ITSELF.

9. Thus brood upon the entire World of Form; see it *merging* into BRAHMAN the SELF, *dissolving into* BRAHMAN the SELF, *devoured by* BRAHMAN the SELF.

10. With the fullness of your Highest Will, let naught remain in Cosmos, naught within the World of Form, but the ALL-VICTORIOUS BRAHMAN, the SELF—CONQUEROR of EVERY DIVERSITY, RESOLVER OF THE MANY INTO ONE—RESOLVER OF THE ONE INTO ZERO.

11. Thus with Sankaracarya of Old—Merge the World into BRAHMAN. Dissolve the World into BRAHMAN.

12. Sound the OM once, for the ONE AND ONLY:

    OM

Meditation on Infinity leading to INFINITY

1. Begin by quieting all aspects of the personality nature and aligning with a deeper, higher Reality.
2. You are about to go on an adventure of *endless enumeration.* (You may use pen and paper, or a computer to assist you.)
3. Conceive the largest number imaginable. Add number to number, multiply number by number, or do whatever it takes in the attempt to conceive of an infinitely large quantity.
4. No matter how large the number you conceive, keep adding to it. Even multiply it by itself, and by itself, and by itself. Have you (do you think) reached now a number sufficiently large to be called infinite?
5. Pause for a few moments. Realize the hopelessness of your task. Realize infinity cannot be reached in this way.
6. Realize that no matter how large the number you have just conceived it is still as infinitely far from your goal (of an infinite quantity) as if you had simply remained with the number *one*. Realize this deeply.
7. Brood deeply upon the difference between *infinity* and any definite enumerable quantity no matter how large. Realize not only their quantitative but their *qualitative* difference.
8. Meditate now upon GOD as the INFINITE.
9. Realize from your experience with numbers that the ABSOLUTE DEITY, as the INFINITE, is not 'filled' with any *quantity* of enumerable 'things'.
10. Attempt to realize deeply that INFINITY (the ABSOLUTE SELF) has *naught* to do with Number. Attempt to realize somewhat the *impartite, non-enumerated* NATURE of ABSOLUTENESS.
11. Realizing that INFINITY (the INFINITE DEITY) is non-enumerable, meditate upon IT as the ZERO, and seek to fathom the MYSTERY.
12. See all possible enumeration absorbed by the ZERO. Realize that *absorption by the ZERO* is the *infinitization of enumeration.*
13. Dwell for a few 'moments' (or, should it be said, 'Dwell Forever'?) in the *certainty* of INFINITUDE Realize that IT can never be 'reached' One cannot count his or her 'way' to IT. One cannot accumulate his or her 'way' to IT. In the face of such feeble attempts IT will forever remain *infinitely remote.*
14. Realize that IT is so *ever present* in *completeness* that IT IS the UNREACHABLE.
15. Resolve to *know* 'YOU' ARE the INFINITE.
16. Sound the Sacred OM three time, sharing your realizations anent infinity and INFINITUDE with all of humanity:

    OM OM OM

Meditation on the VOID

1. Following the usual means of alignment and abstraction, withdraw inwardly rapidly, recognizing, acknowledging and then disidentifying with one vehicle after another.
2. Poised imaginatively upon the atmic plane of Spiritual Will, withdraw steadily inward to the very center of you being, and imaginatively focus as a Spirit, a Monad, polarizing the center of your *being* upon the plane of that formless, all-pervading Awareness and Presence.
3. From this highest perspective of human vision, seek to identify with the ONE.
4. Within the ONENESS, *merge* and *dissolve* the World into BRAHMAN.
5. Collapse Time and Space into the ETERNAL NOW, and dwell as the PRESENCE within the ETERNAL HOMOGENEITY.
6. Realize the UTTER EMPTINESS of the ETERNAL HOMOGENEITY.
7. Realize VOIDNESS—that 'YOU' *ARE* VOIDNESS.
8. Dwell deeply and fully in VOIDNESS.
9. Realize that VOIDNESS is *infinitized* FULLNESS, the PLENUM.
10. Seek to fathom the *identicalness* of VOIDNESS and PLENUM.
11. Realize the ZERO (ITS Symbol) as the 'Key' to the understanding of VOIDNESS and PLENUM as *identical.*
12. Contemplate *practically* what it would mean to live within the World of Becoming *as* both VOIDNESS and PLENUM. What would be the nature of your relativistic, localized life?
13. Realize the INFINITE POWER of the VOID and the PLENUM, and resolve to carry the thought of this mighty *identicalness* in your consciousness.
14. Repeat with deep *knowing* the Mantram: **I AM the VOID. I AM the PLENUM EVER FULL. I AM I AM I.**
15. Close with the Sounding of the Sacred Word OM three times, dissolving the World simultaneously into the EVER-FULL VOID, the PLENUM. Impart to Humanity that though it is NOTHING, it is forever FULFILLED:

    OM OM OM

The Vanishing Point
Part I — Space

1. Begin by quieting all aspects of the personality nature and aligning with a deeper, higher Reality.

2. Enter the state of Pure Subjectivity.

3. Beginning from a position in the midst of Cosmos, surrounded by an incredible diversity of forms (some articulated because they are close and some appearing merely as points, because they are distant), you start to withdraw towards the boundaries of the finite Universe.

4. As a Pure Spirit (free of the limitations of Space and Time, and of the Curvature of Space/Time), you can travel with incredible rapidity, far faster than the speed of light.

5. You find yourself passing the most rapid of objects. As you speed away, all separate, articulated points being to merge into single points, and they, also, into single points.

6. You have broke free of the boundary of the Finite Universe. All the articulation within it is fast disappearing.

7. The entire Universe now appears as One semi-articulated Point. Now even the articulation disappears and all diversity has been resolved into oneness.

8. Your 'Sight' is Absolute (all laws of optics abrogated), thus, even as your withdraw greater and greater speed the Point that is the Universe becomes merely smaller and smaller without quite disappearing.

9. You realize that the Universe is in the Process of Infinitesimalization even as you are 'Seeing' from that mode of Perception called 'Infinispectivizing'. As long as you are never infinitely far from Cosmos, Cosmos as a Point will continue to *infinitesimalize* without becoming *no-thing.*

10. You realize that to Your 'Infinispectivizing View' the Cosmos is utterly one, all things are One Thing, all places are One Place. To Your View, Cosmos both *is* and yet, *is not,* for it is so rapidly infinitesimalizing that It cannot be measured. It is shrinking below the value of any known quantity, converging upon *no-thingness* and yet, It, Cosmos, still Is, simply because You are not quite yet, (nor can be) the infinitely distant Observer.

11. You realize that, from your unique Perspective, of 'Infinispectivizing' (which is not quite the Infinispective) You have virtually destroyed Space (or at least the possibility of Object in Space)—but not quite—One THING lacks.

12. You ponder deeply upon what you 'See' and what You 'Know'.

The Vanishing Point
Part II — Time

1. You realize what you have just done to the idea of Object in Space, and are satisfied with your Adventure in Consciousness, You now begin to think about Time.

2. You are ready to undertake another experiment using 'Infinispectivizing'.

3. You immediately re-insert Yourself into Cosmos. You can do this because You are a Pure Spirit, utterly immaterial, a Pure Observer.

4. You look around you. You see events—appearances and disappearances, cyclic occurrences, things happening and, then, "after a while" happening again.

5. You wish to know all about vibration and cyclic events, so you allow yourself to become smaller and smaller, to journey into the world of the atom. You journey deeper and deeper into the atom, into particles, into particles of particles, until you see the tiniest particle of all—the ultimate particle event.

6. To Your eye, focusing upon a single ultimate particle/event, it seems to be in one place and then change places instantaneously and then again.

7. Intent on fathoming the Mystery, you identify as the One and Only Cosmic Observer. You can now observe what is Really happening.

8. You are in the World of Being (and you can Observe the entire Cosmic Configuration) consisting of *all* ultimate particle events.

9. Your 'Sight' and Your Registration is so rapid, that you can now 'See' that the ultimate particle/events are blinking 'on' and 'off'—appearing and disappearing, and that when they disappear, all of Fohatically Fabricated Cosmos disappears with them.

10. You are now witnessing the ultimate vibration in this particular Cosmos. When the Universe is 'on' you witness all ultimate particle/events in their positions within the Cosmic Configuration; when the Universe is 'off' you experience only the Archetypal World of Sustained Patterns and the Beings Who, Essentially, Are these Sustained Patterns.

11. You now know something very important about the Secret of Vibration. Intent on a more Macro-View, you begin to withdraw from the ultimately Micro-World, and now larger and larger aggregates of ultimate particle/events appear before your Eye.

12. You 'See' larger and larger cycles built upon smaller cycles. Each cycle is a small cycle compared with the next larger. You 'See' event after event.

13. You are now moving so fast in retraction (and are seeing from such a 'distance') that events which (when you were closer) seemed to take a 'long time', now seem to take only a very short time.

14. As you move in retraction, you are not only moving in Space, but you are pressing 'backwards' and 'forwards' along the Infinite Time Line, racing at the

same speed you have spatially towards the beginningless past and the endless future. You are attempting to embrace the Infinite Time Line as One Wholeness, but it ever eludes you—barely.

15. Now you are comparing every temporal cycle you witness, to the immense Temporal Span that you are gathering into your awareness. As a result, all of your space/time expansion and retraction, all familiar cycles are appearing briefer and briefer.

16. Human incarnations seem to follow each other with only a second in between.

17. Soon it is the same for planetary incarnations.

18. Soon even solar systems (thought to be so long enduring) seem to appear and disappear as events separated at first by 'minutes' (so it seems to you) and then only by seconds.

19. Soon galaxies are no longer seen as enduring relationships but also begin to flash 'on' and 'off'—appearing, disappearing and appearing again only 'minutes' apart and then seconds, and then split seconds.

20. To your Eye, all events (no matter of what magnitude and no matter how 'long' they used to take when You were 'closer' to them—both spatially *and* temporally) are taking less and less Time, until they seem to be occurring practically on top of one another.

21. In fact, at a certain spatio-temporal 'distance' from Cosmos, you can no longer distinguish one event from another. All you see is the steady shining of a tiny Point of light (becoming ever tinier in magnitude, but still visible to You because of your Super-Cosmically Perfect Sight).

22. It appears to you that all events that were happening before have stopped. It is as if Time has stopped. From Your Perspective all you see is the blackness of surrounding Space and the tiny point of light which undergoes no fluctuation (apparently).

23. Suddenly it is Now, Eternally Now. Nothing changes. All events are one and they are unnoticeable.

24. You assume from what you knew that there must be, in that Cosmos, a March of Time, and that what you would have called the 'Future' is happening, but you can 'See', nothing. The future is passing (you know it must be) but it is only unchangingly, Now. As there is no change, (for you) no Time is passing.

25. Whatever was past, whatever was future, are all *one* to you, from your Infinispectivizing Point of View. Time is "standing still".

26. Suddenly the Point disappears. You wonder what has happened. It seems to remain dark for a while, but then the Point reappears.

27. You realize that the Cosmos-Point has just disappeared before your Eye and reappeared. It, then dawns upon You, that you are witnessing the Event called Cosmos occurring, and then notoccurring.

28. You realize that at your tremendous spatio-temporal speeds, you are not only receding in Space, but you are traveling in such as way as to 'See' Time as if it

were Space. You are so far into the past and so far into the future that you are rapidly entering a Time-Consuming NOW—not the Cosmic Eternal Now You experienced a little 'while' ago, but something far more obliterative.

29. You now know Yourself to be hurtling close and closer to infinite speed (while horizontally embracing larger and larger portions of the beginningless, endless span of Eternity) and are thus converging, upon the ETERNAL NOW. You know this because something very strange begins to happen.

30. The Point that had been appearing and disappearing every once in what to You is a 'While', is now blinking 'on' and 'off' far more rapidly.

31. You realize that Cosmos Itself is becoming something like an ultimate particle/event, only upon the supreme Macro Level (however Micro It may seem to you now).

32. The intervals are getting shorter and shorter as, pushing the two extremes of Infinite Duration, you continue to converge even more rapidly upon the ETERNAL NOW.

33. Now the intervals are so rapid that you can hardly tell what is 'on' and what is 'off'—a bit like looking at a fan and seeing the blade as if it were stationary.

34. Predictably, all fluctuation seems to cease. You realize that the fluctuations between 'on' and 'off' have achieved a sufficient speed, such that, amazingly, what you are looking at NOW is virtually all the Universes that have ever been, or that ever will be, occurring as Events at as close to infinite speed as possible without being infinite speed. Your vision has rushed 'in Perceptual Recession' at virtually infinite speed as you have (infinitizing rapidly) gathered the Infinite Temporal Span into your awareness. The Events and intervals have begun approaching zero at tremendous speed and now are infinitesimalizing (but are not quite zero) for the simple reason that you are traveling at ever-infinitizing speeds that are not quite infinite, and gathering the Infinite Temporal Span under your Temporal Eye with ever greater rapidity that (fast as it is) ever fails to grasp the entirety of the infinite Temporal Whole.

35. You realize that you are now 'Seeing' the entire Cosmic Process (virtually, forever), as if it were occurring *almost* in ONE TIME and *almost* in ONE PLACE. There is *almost* only ONE TIME *(almost* NOW), for the Time You are 'Seeing' is vanishing at ever-infinitizing speeds which are, albeit, ever less than infinite speed. There is, also, to Your Eye, *almost* only ONE PLACE *(where <u>You</u> are)* for the Place-Point You are 'Seeing' is also moving towards vanishing at ever infinitizing speeds which are, albeit, ever *less* than infinite speed.

36. The 'place' where You are is most interesting because You are *almost NOWHERE.* In fact your are infinitesimalizingly removed from NOWHERE, just as You are infinitesimalizingly removed from NO TIME.

37. Subject/Object, however, still remains. That is why there still *is* a Point, and an Observer, and that is why there is still Time (however close it may be to the ETERNAL NOW) and still Space (however close that Space may be to NOWHERE). Also there is Still **8**, however close that **8** may be to being **I**.

The Vanishing Point
Part III — NOTHING

1. You realize You are *witnessing* the most extraordinary of all possible 'Sights'. All of Space (forever) is collapsed into a single infinitesimalizing Point. All Time that ever *was* or ever *could be* is also entirely collapsed into an infinitesimalizing Point.

2. All that ever has been or ever will be (forever) is on the verge of becoming *nothing*. Ponder.

3. Nevertheless, the 'Sight' is maintained, and the Point does not disappear.

4. What would it take for the Point to disappear? You realize that You are not quite traveling at infinite speed and not quite at an infinite 'distance' from the Time/Space Point. Although you are as close to infinite speed as you can be without being at infinite speed, and as close to infinite distance as you can be without being infinitely distant.

5. Suddenly you realize WHO 'YOU' REALLY ARE. Your realize the ESSENTIAL impossibility of what You *think* You are 'Seeing' and the impossibility of the *'whenness'* of Your (supposed) 'Seeing'. You realize that there can be no possible difference between *You* and *what* (in Time and Space) You are *apparently* 'Seeing'.

6. You suddenly realize that, in fact, You are *not* a Pure Spirit, but are *NOTHING*. You-as-'YOU' are NOTHING at all.

7. And with that realization Your speed-in-recession becomes instantly *infinite*. And your speed of Temporal Expansiveness becomes instantly *infinite*. Time becomes NOW, ETERNALLY NOW. The Point disappears into *nothingness*.

8. The Point is gone and there is only NOWHERE/EVERYWHERE and 'YOU' *are* THAT.

9. The Point is gone and there is only NO TIME/EVERY TIME, the ETERNAL NOW, and 'YOU' *are* THAT.

10. 'YOU' 'REALIZE' that everything 'YOU' have 'SEEN' has been an illusion, which *never* has REALLY 'BEEN', forever. It didn't 'happen' at all.

11. 'YOU' 'REALIZE' that 'YOU' thought 'YOU' 'Saw' an Object, but that only YOUR OWN SUBJECTIVITY, YOUR OWN ALL-IN-ALLNESS has 'PREVAILED' *forever.*

12. 'YOU' 'REALIZE' that NOTHING *was*, NOTHING *will be*, and that NOTHING *IS—as ever, forever.*

13. 'YOU' no longer 'See' the "Point of IT".

14. 'YOU' have *returned* to the INFINITESSENTIAL SELF 'YOU' have *forever* 'BEEN'.

15. NOTHING

16. ZERO

17. **I**

18. **?**

Entering the Point

1. Begin by quieting all aspects of the personality nature and aligning with a deeper, higher Reality.
2. Identify Yourself with pure Subjectivity. You are the Observer and all around you is utter Blackness, Nothing, the Void.
3. Against the background of *nothingness* you notice the appearance of a tiny Point of white light. All You see is blackness and the point.
4. You decide to approach the Point. Imaginatively you draw closer.
5. As You approach, the realization dawns upon You that the Point is Really our Cosmos. Ponder the implications.
6. You see that the Point, instead of merely being One Point is, in fact, many, many points—many points within the One Point.
7. The closer you come, you realize that each of the many points that seemed indivisible at a distance, is also divisible. Each point is a great collection of super-clusters of galaxies:
    - these super-clusters also appear as points, but they too are shown to be divisible into families of galaxies
    - these families appear as points, but you discover them, on closer approach, to be divisible into galaxies
    - and the galaxies into cosmo-systems
    - and the cosmo-systems into constellations
    - and the constellations into stars
    - and the stars into planets
    - and the planets into chains
    - and the chains into globes
    - and the globes into kingdoms
    - and the kingdoms into units
    - and the units into lesser units
    - and the lesser units into sheaths
    - and the sheaths into cells
    - and the cells into molecules
    - and the molecules into atoms
    - and the atoms into ultimate atoms
    - and the ultimate atoms into particles
    - and the particles into ultimate particle/events
    - —and there an end.

8. Ponder on your descent! You began with One Point, and encountered lesser points, and lesser points—points within points—all of them revealing their divisibility upon closer approach. When You saw the point that 'became' an ultimate particle/event, you saw no other points.

9. Realize the Universe as an incredibly complex collection of points within points within points, etc.

10. Realize the incredible diversity of points.

11. Most importantly, realize that though all these points seem dispersed in Space, *they are Really, Essentially, superimposed* and, more, *identical.*

12. Realize that from the loftiest Perspective, all points are, indeed, Really, and Truly, but One Point.

13. Ponder upon the Many and the One, the One and the Many, and the kind of Perspective which is required to 'See' the Many as One.

14. Close by sounding the Sacred Word three times, casting your inner eye upon all the virtually countless (but, indeed countable) points in Cosmos, and resolving them all into One. Breathe forth upon Humanity the Blessing that the realization of this Resolution of the Many into the One—brings:

    OM OM OM

Mantrams for the Infinitist

When concentrated Spirit-aligned thought and rhythmic speech are combined the idea content of the thought penetrates the mind and brain with greater ease. The high philosophy of Radical Infinitism does little good unless it becomes true realization in the consciousness and being of those who seek to use it as a rule of life.

Mantra are tools for concentration, penetration, invocation and realization. The manner in which they are organized contributes to their invocational power. It is possible to brood over a subject and come to realization, but mantra are extremely useful in reawakening and intensifying realization, once it has been achieved. Of course, new realization may also arise through their use.

The following short exercises (either in speech or thought) are meant to assist the Radical Infinitist to *live* the philosophy. In each case **8** will state the purpose for the mantram (though it is often obvious) as well as a suggested rhythm in which the words can be said or *thought.* Based upon these mantrams offered for experimentation (and after having understood the principle involved) the mantrist can create his or her own mantram, with meanings and rhythms which are considered useful and appropriate. The constant key is the ability to sustain thought and focus while intoning the various sounds in the various rhythms.

Mantram 1

The following mantram is useful for realizing the relationship between:

- the ETERNAL NOW
- the Cosmic Nows
- the conventional present moment

It may be spoken *very slowly* in a long four count, using the last beat to breathe. Before beginning the exercise it would be well to meditate a little upon the differences between these three *nows,* so they are clearly in mind.

NOW ...	Now ...	now ...	inhale
NOW ...	Now ...	now ...	inhale
(repeat)			

Mantram 2

The following mantram, in which the sound (pronounced like 'eye') is used five times in sequence, in as many sequences as the mantrist wishes, is designed to help realize the descent of the:

"I"	ABSOLUTE **I**, then into
"I/8"	the Super-Cosmic Infinite Subject, **I/8**; then into
"8"	the Universal Monad, **8**, at-one with all Monads focused on all dimensions in-Cosmos; then into
"I"	the bounded egoic 'I' whether the ahamkaric Transpersonality within the Egoic Lotus or the more limited personality; then into
"i"	the false 'i' sometimes asserted by the little lunar lords.

The mantram should be done slowly to an eight count, leaving the last three beats for the intake of breath. Naturally, with each repetition of the sound, the meaning of the terms under focus must be contemplated. Depending upon the amount of breath available, the slower the exercise is performed, the better. The mantrist may discover that it is possible to sustain a longer sequence if only thought and no sound is used as the air is released while mouthing the words. It is also possible to work with thought alone independently of the breathing of the physical body.

I	**I/8**	**8**	I	i	inhale	...	...
I	**I/8**	**8**	I	i	inhale	...	...

(repeat)

Mantram 3

The following mantram is very much like the one preceding, except that the mantrist in interested in the ascent of the 'I' concept from its lowest possible form (identification with the lunar vehicles) all the way to its highest forms.

The same technique is recommended as above. Every repetition of the sound can be accompanied with a visualization appropriate to the concept which the mantrist is attempting to realize.

i	**I**	**8**	**8/I**	**I**	**inhale**	**...**	**...**
i	**I**	**8**	**8/I**	**I**	**inhale**	**...**	**...**

(repeat)

Mantram 4

The following mantram is a mantram of identification and inclusion. The **8** sees its identity with any apparent 'others' to whom it might relate, and 'reclaims them', as it were, 'into' its own **8**-ness. The Point that is the **8** and the Point that is 'You' are identical, and in this mantram the **8**-Point 're-gathers', as it were, the 'You'-Point into itself. Thus a circuit completed and the two points merge.

The mantram is done to a slow eight count, with the last three beats to be utilized for the breath.

8	**in**	**You**	**in**	**8**	**inhale**	**...**	**...**
8	**in**	**You**	**in**	**8**	**inhale**	**...**	**...**

(repeat)

Mantram 5

The following mantram is designed to help bring about the realization that each **8**, Monad, or 'Ray' of the ABSOLUTE is identical with every other **8**, Monad or 'Ray' of the ABSOLUTE. The result should be the realization that there is no isolated individual identity in-Cosmos, but that all intra-Cosmic **8**'s have an identical **8**-ness.

A slow five count can be used, with the last two beats for a breath. As above, the words may be mouthed with no sound utilized, or the exercise can be done entirely (though rhythmically) in thought, independently of the physical breath.

8	**Am**	**That**	**inhale**	**...**
8	**Am**	**That**	**inhale**	**...**

(repeat)

Mantram 6

The following mantram is like the one above, except that the focus is upon realizing the identicalness of the **8** with the Universal **8**, the Universal Self, and not alone with all other **8**'s. The focus should be upon the One **8**, which is *all* **8**'s.

The mantram can be performed as many times a desired to a slow eight count, with the last two beats reserved for the breath. As above, an unvoiced version and a version which is simply 'thought' can be utilized.

8	**Am**	**That**	**8**	**Am**	**inhale**	**...**	**...**
8	**Am**	**That**	**8**	**Am**	**inhale**	**...**	**...**

(repeat)

Mantram 7

The following mantram is like Mantram 6 in its method of performance except that the thought is different and (infinitely!) more expanded. The mantrist is attempting to realize the 'I', the ABSOLUTENESS.

As this mantram is performed there is no thought of Cosmos or of any Cosmoses, but only of the VOID. Of course the ABSOLUTE is, indeed, *unthinkable* so the mantrist is enjoined to find a method of performance which does not rely upon an image.

The very deepest level of IDENTIFICATION is being attempted, and the focus is upon the ROOTLESS ROOT of BEING. (It will be interesting for the mantrist to compare the results of working with **8—Am—That—8-Am** and **I—AM—THAT—I—AM.** Over 'time' there should be a subtle qualitative difference.)

I	**AM**	**THAT**	**I**	**AM**	**inhale**	**...**	**...**
I	**AM**	**THAT**	**I**	**AM**	**inhale**	**...**	**...**

(repeat)

Mantram 8

The following mantram is a mantram-of-*merging*, in which the **8**, the You, and the We are all invested (through imagination) into the Allness.

Before beginning the mantram, it would be well to meditate upon the meaning of **8**, You, and their union in the We. There must, as well, be a suitable image for the Cosmic Allness, with a sense of its great diversity and relatively vast extent.

As can be seen from the form below, the mantram is said to an eight count with two counts for intake of breath. As usual, the eight count rhythm is preserved through the unvoiced and simply 'thought' versions.

8aminall	**8aminall**	**Youareinall**	**Youareinall**	**Weareinall**	**Weareinall**	**inhale**	**...**
8aminall	**8aminall**	**Youareinall**	**Youareinall**	**Weareinall**	**Weareinall**	**inhale**	**...**

(repeat)

Mantram 9

The following can be a very powerful little mantram. In it, the mantrist focuses upon anything or anyone (either in imagination or in the world of the senses) and repeats the word 'Identical', thus establishing that no matter what or whom he 'sees', the mantrist is *identical* with whatever or whomever is 'seen'.

Mantrists *can devise their own rhythm* for the use of the word 'Identical', or they may prefer to use no rhythm at all, waiting, instead, for the realization of *identicalness* to subtly impress the consciousness and being.

When using this mantram, the mantrist may find himself/herself *pervading*, as it were, whatever may be the object of focus, *and* simultaneously reabsorbing that object into **8**-ness. Clearly, this mantram is not about *form* but about Essence calling to Essence, or, even more deeply, ESSENCE identifying *as* ESSENCE.

I—dentical I—dentical I—dentical I—dentical ...
I—dentical I—dentical I—dentical I—dentical ...
(repeat)

Mantram 10

The following mantram is oscillatory, and is meant to convey some sense of the appearing and disappearing Fohatic World of Fabrication. 'Being' represents the World of Being, in which the consciousness is disengaged from objectivity; 'Seeing' represents the World of Fabrication, during which the consciousness is engaged with the object. The mantrist may wish to close the eyes during 'Being' and open them during 'Seeing'. This mantram can be performed unvoiced, for the best effect, or, perhaps, simply 'thought'.

The rhythm to be maintained is the rhythm of the opening and closing eyes coordinated with the appropriate word. The breath can be allowed to be unobtrusive and to pretty well take care of itself. Since there are several things going on at the same time, the mantrist may wish to experiment to arrive at a tempo and a breath-rhythm which works best.

Before beginning, meditation upon the meaning of 'Being' (Subjectivity in the World of Being) and 'Seeing' (Objectivity, engaged with the World of Objects) will be useful. It must be stressed that the *thought* behind the mantram is what counts. The purpose in every case is to achieve subtle changes in consciousness and identification, so no strictly mechanical performance will do.

Being Seeing Being Seeing Being Seeing Being ...
Being Seeing Being Seeing Being Seeing Being ...
(repeat)

Mantram 11

One can see that the form of the following mantram is different. The thought is very much like the "I—dentical Mantram", Mantram 9, but the technique is different. The bolded verse is to be spoken and the italicized verse to be '*thought' on the intake of breath.*

The rhythm is the same whether chanted or thought, and the sequence can be repeated as long as desired. The purpose, of course, is to survey the vast variety of life with one sweep, and merge it all into one Essence. One can use imagination or the senses If one uses the senses, for instance, one can train one's eyes on any available collection of impressions, and merge them into Identical Oneness. If one is in public, the entire mantram can be done in thought, keeping the outbreath and the inbreath rhythmic. This mantram is slightly different from the 'I-dentical' Mantram because the focus is upon collections of items rather than upon one item at a time.

Identical is All 8 See ... *Identical is All 8 See ...*
Identical is All 8 See ... *Identical is All 8 See ...*
(repeat)

Mantram 12

The following mantram is meant to assist with the realization of the transition between the NOTHING to the Oneness and back again. There is a tremendous gap in consciousness and identification between NONENESS and Oneness, and some facility needs to be developed in thinking of the two in relation. Of course, REAL relation is impossible, because NONENESS (being the INFINITE) cannot *relate.* Nevertheless, there is a fluctuation in consciousness between two state of apprehension which have distinctly different qualities.

The mantram can be performed on a slow voiced four count and an equivalent inbreathing (unvoiced or 'thought') four count. The mantrist, if he wishes to use visualizations, may visualize black nothingness for 'NONE', and a white figure '1', or a white dot, for the word 'One'. Or the visualization may be more generalized and include the general surroundings conceived in terms of the "blotting out of all form" as contrasted with the presently visible forms. Imagination may also be used to enhance the contrast.

It is important to have some 'sense' of the 'NONE'. No image will Really suffice, because any image automatically becomes a thing, or a '1'. Nevertheless, the sense of contrast must be developed. It is clear that the following mantram works from 'above' to 'below' and back, and so the Path of Involution and Evolution is suggested, or, more abstractly, the Birth of the Universe out of NOTHINGNESS and Its return to NOTHINGNESS.

NONE is One and One is NONE; *NONE is One and One is NONE;*
NONE is One and One is NONE; *NONE is One and One is NONE;*
(repeat)

Mantram 13

The following mantram is exactly the obverse of the preceding and can be used in exactly the same way, except that the emphasis is upon the way the existent Oneness is absorbed into NOTHINGNESS and then again 'RADIATED' forth. This is a more Cosmo-Centered Mantram than the preceding one, and is therefore, perhaps, more concrete and less occult. In a way, though, it emphasizes the disappearance of the familiar and its re-emergence, so it does have value.

One is NONE and NONE is One; *One is NONE and NONE is One;*
One is NONE and NONE is One; *One is NONE and NONE is One;*
(repeat)

Mantram 14

The following two mantrams are right along the same line as Mantram 13, but they are more *emphatic* and can be said with greater power. These mantrams will be spoken or 'unvoiced' more quickly, so it is important to meditate clearly upon the concepts involved before beginning.

The rhythm calls for a spoken (or 'unvoiced') four count, followed by an interlude of mental silence as the breath is inbreathed, also on a four count. These mantrams should only be used after there is a very well established, *felt* sense of the meaning of the ZERO and the One, otherwise they can degenerate into energetic, mechanical repetition.

From ZERO to the One, then the One becomes the NONE
(Long intake of breath in mental silence)
From ZERO to the One, then the One becomes the NONE
(Long intake of breath in mental silence)
From ZERO to the One, then the One becomes the NONE
(Long intake of breath in mental silence)
(repeat)

From ZERO to the One and every One is None but NONE
(Long intake of breath in mental silence)
From ZERO to the ONE and every One is None but NONE
(Long intake of breath in mental silence)
From ZERO to the ONE and every One is None but NONE
(Long intake of breath in mental silence)
(repeat)

Mantram 15

The method in the following mantram is similar—one line voiced to a slow four count; the second line unvoiced or 'thought' on the intake of breath. The idea is the same as between the ZERO and the One, but the mantram helps to realize the *endlessness* of the dualistic alternation.

The mantram should be performed slowly enough to ensure the necessary changes of *felt quality* with the changes of words. Three ideas are conveyed: the VOID, the Object and ETERNITY. These three must not simply be words to the mantrist.

NOTHING, Something, NOTHING, Something, Ever Thus, Ever So,
Something, Nothing, Something, Nothing, Forever On It Goes ...

NOTHING, Something, NOTHING, Something, Ever Thus, Ever So,
Something, Nothing, Something, Nothing, Forever On It Goes ...
(repeat)

Mantram 16

In the following mantram subtlety of thought is required. The mantram is meant to induce the realization that Negation is REALLY an Affirmation of the SELF, and that Affirmation of Form and the Object is REALLY a Negation of the SELF.

Through its use, one comes to realize that everything, as usually conceived, can be conceived *oppositely* with equal or greater validity.

The exercise can be done on a four count slowly enough to visualize or intuit the reversal of the usual meanings of 'Yes' and 'No'. The intake of breath also takes four counts and is performed in mental silence, dwelling with the realization that may be gathering.

No is Yes and Yes is No
(Long intake of breath in mental silence)

No is Yes and Yes is No
(Long intake of breath in mental silence)
(repeat)

Mantram 17

The following mantram can be very powerful and should only be performed when one understands the purpose of the *rejection of the object*, and can do so realizing that the Essence/ESSENCE of the object is not being rejected, only captivation by its form. Specific things or situations (probably *not* people) can be chosen for 'This' and 'That'.

It would be better if for each use of the mantram, two specific things (perhaps, a Pair of Opposites) were chosen, rather than searching for randomly during the performance of the mantram. The purpose, of course, is to realize that the TRUE **I**, is ESSENTIALLY *none* of the illusions with which '**I**-ness' is usually confounded.

This mantram can be performed quite rhythmically and quickly. It takes 16 counts (eight for the first line, and seven for the second line, plus an empty beat), and all 16 can be performed on one exhalation if the exhalation is controlled. An intake of breath in mental silence can follow during the next eight beats, and the mantram resumed and performed as long as desired.

This—NO That—NO Never Never Never I
This—NO That—NO Never Never I ... (pause)
(Long intake of breath in mental silence)

This—NO That—NO Never Never Never I
This—NO That—NO Never Never I ... (pause)
(Long intake of breath in mental silence)
(repeat)

Mantram 18

The following mantram is also a mantram of *rejection* or *negation* and can be performed very much like the one above. It emphasizes the annihilation of false identification for all Time (or at least for the duration of the Cosmos).

Sometimes, in addition to visualization, hand gestures which reject to the 'right' and to the 'left' can be effectively utilized. Obviously some practice will be required, and a very firm idea of that which is being rejected as a point of identification.

The same number of beats are used as for Mantram 17, eight for the first line (one beat per word) seven for the second line (one beat per word, plus an empty beat) and, then, eight beats for the intake of breath in mental silence.

Never this never that never never never never
Never this never that never never *more ... (pause)*
(Long intake of breath in mental silence)

Never this never that never never never never
Never this never that never never *more ... (pause)*
(Long intake of breath in mental silence)
(repeat)

Mantram 19

It can be seen that the following mantram is a corrective for the two which have just preceded it. BRAHMAN and Samsara are, indeed, one. One achieves Identification not only through rejection, but through the Universal Acceptance of all things. All presentations in all Universes are always, forever, ESSENTIALLY, THAT, the INFINITE SELF. Both positions are true; presentations are *never* THAT, *formally* and *always* THAT, ESSENTIALLY. The balance must be preserved. Since we are in a Second Ray Solar System, the use of the following mantram will probably be more agreeable than the mantra of rejection.

This mantram is performed in exactly the same way as the mantra of rejection; the number of beats is the same (eight—seven—eight). The word 'aye' is, in this instance, pronounced like the letter 'A', and means, 'forever'.

In performing this mantram, the mantrist is to imagine a great 'embracingness' of whatever images he or she correlates with 'This' and 'That'. Sometimes distance can play a role in this mantram. The 'This' can stand for something close at hand; the 'That' for something remote. Or again, the more normal Pairs of Opposites can be utilized. In any case, the result should be, first, fusion, and then Identification with the Essence/ESSENCE of the forms chosen. If there has been in the life, any problematic rejection of an area of life as if it were not included in the BRAHMAN, this mantram would be useful for reclaiming that which has been rejected.

Always this Always that always always always always
Always this always that Always and for aye *(pause)*
(Long intake of breath in mental silence)

Always this Always that always always always always
Always this always that Always and for aye *(pause)*
(Long intake of breath in mental silence)
(repeat)

Mantram 20

The next mantram emphasizes the endurance of the ABSOLUTE **I**, forever.

The sound ('eye') is sustained for four counts on a certain note. The note can be chosen to correlate with the Soul Ray or even Spirit Ray if that can be ascertained. Experimentation can proceed using different notes to judge of different effects.

While the ('eye') is sounded the mantrist imagines Identification with All and Pervasion of All. The next line simply reinforces the concept of the ETERNAL DURATION of the ETERNAL **I**, and also lasts four counts. The intake of breath in mental silence also lasts four counts, and then the second section begins—alike in the first line, but with more emphasis on the second line, especially on the word 'e—ver—more' (each syllable with an equal accent).

I ... ...
(imagining Identification and Pervasion)
Forever and ever and ever and ever
(Long intake of breath in mental silence)

I ... ...
(imagining Identification and Pervasion)
Forever and ever and e-ver —more
(Long intake of breath in mental silence)
(repeat)

Mantram 21

The following mantram requires a strong imagination.

1. In the first line one imagines the ESSENTIAL **I** expanding from tiniest conceivable Point in Cosmos to Infinitude.
2. The second line reversed the first, and calls for a contraction from Infinitude to the Point.
3. The third line calls for a moving backwards in Infinite Time as far as possible towards the Beginninglessness of the Infinite Time Line, and
4. the fourth line, for a return to the Now.
5. The fifth line calls for moving forward in Infinite Time as far as possible towards the Endlessness of the Infinite Time Line, and
6. the sixth line, for a return to the Now.
7. The seventh line asks the mantrist to penetrate deeply as possible into the ETERNAL NOW.

This mantram calls for profound thought and an expansive imagination. Each mantrist of course will imagine differently. The timing of the word ('eye') can best be determined by each mantrist. It is conceivable that the count could be long, depending upon whether the 'eye' is chanted on a given note, 'unvoiced' or simply 'thought'. The intake of breath in mental silence should about equal the outbreath.

Obviously, this mantram is not so rhythmically rigorous as some of the others. The quality of thought is all important. The goal is to develop a realization of **I**-ness permeating Infinite Time and Infinite Space as well as having a focus in the two *points* we call the Ubiquitous Point and the Now.

1. **I——————————(expanding to infinitude)**
   (Long intake of breath in mental silence)
2. **I——————————(contracting to a point)**
   (Long intake of breath in mental silence)
3. **I——————————(backwards in time from Now)**
   (Long intake of breath in mental silence)
4. **I——————————(forwards in time to Now)**
   (Long intake of breath in mental silence)
5. **I——————————(forwards in time from Now)**
   (Long intake of breath in mental silence)
6. **I——————————(backwards in time to Now)**
   (Long intake of breath in mental silence)
7. **I——————————(penetrating into the NOW)**

(Repeat the sequence as many times as useful)

Mantram 22

The purpose of the following mantram is to fathom, to the degree possible, the ETERNAL NOW, for which the 'NOW' stands symbolically. The ETERNAL NOW is the devourer of all Time (whether Time Past or Future Time). Even what we call Time Present is collapsed into *zeroness.* In this mantram, as one moves backwards and forwards upon the Infinite Time Line, is it necessary to imagine the NOW devouring/collapsing Time Past and devouring/collapsing Time Future. How each mantrist does this will be determined by himself or herself.

The important thing to realize is that Time Past and Time Future are, REALLY, *illusory,* they REALLY have not existed, will not exist, and do not exist—all this, from the 'INFINI-SPECTIVE', of course. As one moves along the Infinite Time Line it must be as if one *were* the INFINITE SELF (who 'IN-CEIVES' from the *perspective* of the 'INFINI-SPECTIVE'). One must not only realize that it *was* NOW, then, and that it *will always be* NOW in the future, but that there was no difference between the NOW-Past, the NOW-Future, and the NOW-NOW. One must come to realize that there has been no REAL 'movement' in Time at all, and that, forever, there has been but ONE MOMENTLESS ETERNAL MOMENT. This is a tall order, but as all Cosmic-Event-Points along the Infinite Time Line imaginatively 'crowd into One Point' (through the application of the 'Infini-Spective') and, then, that Point disappears altogether into *zeroness,* when the 'INFINI-SPECTIVE' is imaginatively applied, realization may, indeed, dawn. The ETERNAL NOW contains the *pressure* of All Time. Where is the consciousness that can withstand that pressure?

1. NOW————————————(backwards in time from Now)

(Long intake of breath in mental silence)

2. NOW————————————(forwards in time to Now)

(Long intake of breath in mental silence)

3. NOW————————————(forwards in time from Now)

(Long intake of breath in mental silence)

4. NOW————————————(backwards in time to Now)

(Long intake of breath in mental silence)

5. NOW————————————(penetrating into the Timeless Zero Point)

(Long intake of breath in mental silence), etc.

(Repeat the Sequence as many times as useful)

Mantram 23

Affirming the Life within any Presentation. The Master Morya has said that the word "AUM" can be translated as meaning, "Yes". The following is a mantram by means of which the Life Essence/LIFE ESSENCE within any being-in-Cosmos can be powerfully affirmed. The INFINITE SELF has *infinite value*, especially compared with any one of ITS presentations-in-Cosmos. Yet, each of the presentation, in ESSENCE, is *also* of infinite value, and that ESSENCE needs to be recognized and affirmed. The mantrist may select a particular E/entity, or thing, or state, or condition to affirm, or the affirmation may be done in the abstract, affirming Life Itself, or LIFE ITSELF, the INFINITE SELF.

The mantram is arranged in three groups of three, which built towards a climax on the tenth and final, 'Yes', followed by two more beats before the intake of breath begins. The duration of the intake of breath can vary with the mantrist. A count of six or a count of twelve sustains the rhythm.The mantram is vigorous and can be performed fairly rapidly, with concentration upon the Essence/ESSENCE of that which is to be affirmed. The final 's' of the final 'Yes' can be sustained for emphasis through the last two beats, and, naturally, the final 'Yes' should receive special emphasis.

If the mantrist wishes, a series of 'things to be affirmed' can be pre-selected, and the attention can move from one to the other in order during the intake of every breath.

Yes,	**Yes,**	**Yes;**
Yes,	**Yes,**	**Yes;**
Yes,	**Yes,**	**Yes;**
YES	**...**	**...**

(Long intake of breath in mental silence)

Yes,	**Yes,**	**Yes;**
Yes,	**Yes,**	**Yes;**
Yes,	**Yes,**	**Yes;**
YES	**...**	**...**

(Long intake of breath in mental silence)

(repeat)

Mantram 24

The following mantram begins with the ETERNAL **I** and ends with the ETERNAL **I**. Between these is the sequence of normally used vowels, transliterated as follows: 'EE'—'Æ'—'AH'—'OH'—'OO'. There is a natural progression from one vowel to next (though other sequences could be legitimately arranged, and other meanings attached). The meanings to be used are delineated below.

The mantram should be performed slowly enough so that each level of meaning has time to penetrate, and there should be meditation upon each section before beginning. A useful rhythm would be eight counts for each vowel and six counts for each inhalation of the breath, though the mantrist can vary the rhythm according to need and inclination. Voiced, unvoiced and 'thought' approaches can also be used, all the while preserving a basic rhythm. A particular note may be selected upon which to intone the vowels. It is also possible to change pitch for each of the vowels, and this can be done according to the meaning of the vowels per se, or according to the meaning attached to the vowels in this mantram. There are a number of possible pitch assignments which, though interesting, are of less importance than the deep realization that should accompany the mantram. In general, the path of simplicity is probably best.

This mantram is meant to focus upon *wholes.* The particular whole selected may be the entire Cosmos, or it may be some more localized whole such as the galaxy, solar system or our planet. If the work is done in relation to a whole contained within Cosmos, the symbol '**8**' should be substituted in meaning for the 'I'; the pronunciation will be the same, but there will be a subtle shift in the depth and extent of Identification.

I ... ...
(identifying with the ETERNAL I)

EE ... ...
(noticing the remarkable rich diversity)

Æ ... ...
(questioning the reason for the diversity)

AH ... ...
(with acceptance and gratitude for goodness behind diversity)

OH ... ...
(deeply realizing/understanding the Mystery behind the diversity)

OO ... ...
(with compassionate understanding for pain)

I ... ...
(infinitizing all lives in order to save)
(repeat)

That which is free from duality; which is infinite and indestructible;
distinct from the universe and Maya, supreme, eternal;
which is undying Bliss;
taintless—that Brahman art thou,
meditate on this in thy mind.

Vivekacudamani, Verse 261

Section VII
The Glossary
of Radical Infinitism

This Glossary of Radical Infinitism is created because there is a need for a specialized vocabulary in the field of philosophical cosmology and cosmogony. These definitions will *not* be of the kind offered in a good dictionary. Rather, they will have special reference to the occult, metaphysical, and cosmological significances of the terms. The definitions will also be in line with the Philosophy of Non-Dualism and with the Science of Identification.

It may be best, perhaps, toread over this section *before,* as well as during, the reading of theother sections of the book. Extensive cross-referencing between listed terms is possible and will lead to a wider comprehension of related concepts. In using this Glossary, it is suggested that an attempt be made to also reference terms *within a definition* concerning which there may be uncertainty.

NOTE: Single quotation marks ('') are used around a number of capitalized words relating to the apparent *actions* of the ONE AND ONLY SELF to indicate that these 'ACTIONS' cannot be considered actions in the usual sense, because, in ESSENCE, it is impossible for the ONE SELF to *act.* Additionally, full capitalization is used for certain words which speak of the *actions* of THAT, or of anything which is hypothesized as occurring within THAT. In other words, the *actions* of THAT would, by this method, be written: the 'ACTIONS' of THAT. Please see the Introduction for a full explanation of punctuation and capitalization conventions.

- A -

absence; absent

By the *absence* of something is denoted its removal from the possibility of impacting a registering consciousness.

Example: Although the Master was thousands of miles away physically, we could not think of Him as absent from our spiritual festivals due to His strong and immediate influence upon the quality of our thinking.

Example: Although he was physically present at the meeting, his deep preoccupation with inner matters made it seem as if he were absent.

ABSOLUTE

By the *ABSOLUTE* is meant the one and only 'STATE' of UTTER PERFECTION, WHICH in ITS ESSENTIAL NATURE has never changed forever and never will. Being

INFINITIZED PERFECTION, IT cannot be further developed of modified in any way. THE ABSOLUTE IS THE BOUNDLESS IMMUTABLE PRINCIPLE.

Example: The ABSOLUTE IS the ONE BEING/NON-BEING from WHICH all relative things emerge and into WHICH all relative things dissolve. ITS only example is ITSELF.

ABSOLUTE NOUMENON

By the ABSOLUTE NOUMENON is meant the INFINITE SELF considered as noumenal SOURCE of every possibility that has appeared, is appearing, will appear, or could appear in-Cosmos.

Example: The quality of beauty of all beautiful things in-Cosmos has, as its Relative Source, an intra-Cosmic Noumenon that is a great Archetype of Beauty. That Archetype, however, has for Its *Ultimate Noumenon* the ABSOLUTE NOUMENON.

ABSOLUTE NOUMENAL 'STATE'

By the *ABSOLUTE NOUMENAL 'STATE'* is meant the ultimization or infinitization (i.e., the INFINITESSENCE) of any condition or state existing within the World of Becoming.

Example: The ABSOLUTE NOUMENON of all Virtues-in-Cosmos (such as Goodness, Beauty, and Truth), is 'found' within the 'STATE' here indicated, the ABSOLUTE NOUMENAL 'STATE'. Even more than this, the 'STATE' here indicated *IS* the ABSOLUTE NOUMENON of Goodness, Beauty, and Truth as we know Them in-Cosmos.

absolutize; absolutization

By *absolutization* is meant a process by means of which the fundamental idea or archetype that gives meaning, quality or distinct identity to an appearance-in-Cosmos is *raised infinitely* to a 'STATE' that 'contains' or 'embodies' or (even more) *IS* the *ultimate possible development to the point of absolute perfection* of that idea or archetype.

Example: The ALL-SELF *contains* (rather, IS) the absolutization of the principle of *selfhood*-in-Cosmos. The sense of "I am" that every self-conscious human being 'inperiences' is absolutized in the ALL-SELF.

accident; accidental

By *accident* or *accidental* is meant the occurrence of a configuration or relationship in-Cosmos that was not *intended* by the will of the agent that engendered the configuration or relationship.

Example: The over-stimulating of her throat center was the accidental side-effect of the monadic impact she successfully invoked via the extended Antahkarana.

Example: Since the Universe is a "Son of Necessity" generated under ABSOLUTE WILL and LAW, It is no accident.

act

By an *act* is meant a change caused by a willing or desiring agent of change—especially, but not exclusively, a self-conscious agent of change.

Example: The <u>acts</u> of disciples bring about important changes in humanity.

'ACT'-of-Origin

By the *'ACT'-of-Origin* is meant the spontaneous 'eterno-cyclic' *RADIATION* of the INFLUENCE of the ABSOLUTE SELF such that there results the emergence of the omnipresent Point of Universal Potential from which each Finite Universe is generated.

Example: Every Cosmos is preceded by an <u>'ACT'-of-Origin</u> which is the *only* 'ACT' that can be attributed *solely* the SELF-AS-SELF.

'ACT' of SELF-Limitation

By an *'ACT' of SELF-Limitation* is meant perhaps three 'ACTIONS'-at-the-Beginning (of each Cosmos) by means of which the ALL-SELF 'ENDED' (apparently) the 'STATE' of INFINITIZED SELF-PREOCCUPATION ('INFINIDENTIFICATION') which characterizes the ALL-IN-ALLNESS of the Universal Pralayic Period, and 'RE-FOCUSSED' ITS 'ATTENTION' upon ITSELF in an *objective* way. The first such 'ACT', is:

1. To *appear* to 'BECOME' the Omnipresent Point of Cosmic Origin;

the next, (which can be considered coeval with the first) is:

2. To begin (via 'MAYA'-instantly-Maya) to 'SEE' (and in 'SEEING' thus 'CREATE') Mulaprakriti; and,
3. The third 'ACT' (by now an <u>A</u>ct carried out by the SELF-as-*Concentrated* Point, *not* the SELF-as-*Omnipresent* Point) is to Determine the Boundaries of the Cosmos-to-Be.

Example: The Universe (in fact, *each* Universe) arises only through an <u>'ACT' of SELF-Limitation</u> originating 'within' the BOUNDLESS IMMUTABLE PRINCIPLE. The origin of such a seeming impossibility as an 'ACT' 'within' THAT is utterly mysterious to the mind of man or perhaps to *any* Mind/mind in-Cosmos. Perhaps the nature of the origin will be revealed at a *timeless* 'TIME' when the 'SELF-Distraction' called the Universe has vanished into NOTHINGNESS.

Actor

By an *Actor* is meant any *self-conscious Self/self* in-Universe, any and all of which *Selves/selves* are emanations of the One SELF-as-Self-in-Universe (which Itself is a *'RADIATION'* of the ONE AND ONLY SELF, the ALL-SELF).

Example: In all life experiences within the three worlds of human evolution, the trained disciple is the true <u>Actor</u> 'behind' his more obvious personality. The personality with its limited consciousness is only the *apparent* actor; the disciple-as-Soul is the responsible <u>Actor</u> playing his part upon the stage of life.

'ACTOR'

By the *'ACTOR'* is meant the ONE AND ONLY (extra-Cosmic though simultaneously intra-Cosmic) BEING/NON-BEING considered in ITS aspect as 'DO-ER'. Within the ALL-IN-ALLNESS there is no humanly comprehensible 'ACTOR', for an *act* requires number, relationship, and motion, all of which are *not* to be 'found' within the homogeneous ALL-IN-ALLNESS. Technically, then, THAT which we would call the 'ACTOR' cannot REALLY 'ACT' when 'ABIDING' in the fullness of ITSELF; at least. IT must somehow *reduce* ITSELF ('BECOMING', apparently *less* than IT IS) before IT can be an 'ACTOR', or rather (more accurately at such a phase in the pre-Cosmic Process) an <u>A</u>ctor. (Of course, even SELF-Reduction would have to be considered an 'ACT', so the nature of the *origin* of ITS 'ACT' remains to the finite mind extremely mysterious.) However, since there *is not* and never *will be* anything *other* than the SELF, all *action within-Cosmos* must *necessarily* go on utterly within the REALM of SELF. And yet, within that UNDIVIDED REALM no *action* (as we know it) is at all possible, because action requires division.

Therefore, an *Arena of Possible Action* must be generated (yet how?) to render Action possible. An Arena of Action is an *Objectified Domain* (a Domain in which Objectification is possible) no longer limited by the *PERFECTION* of the ALL SELF (if PERFECTION *can* limit!), for when the IT 'ABIDES' utterly *within* ITS PERFECTION, the 'ACTOR' cannot ACT (at least as we understand action). What is there to 'DO'? By 'WHOM'? Upon 'WHAT'? Given the intractability of PERFECTION, a kind of *imperfection* (which is *apparent SELF-Reduction)* must be 'GENERATED', so that the ACTOR-as-<u>A</u>ctor may *act.* The paradox is, that in order to 'CREATE' or 'GENERATE' a realm in which action is possible, an 'ACT' is *first necessary.*

For practical purposes, it is well to remember that there is only ONE IDENTITY in-Cosmos, so if we accept the self-evident fact that *action does occur* (at least *apparently*), then the ONE IDENTITY must somehow be the ULTIMATE SOURCE of that action—the ONE IDENTITY must be an ACTOR, in fact. The *how* of this 'ACTION' must remain a profound mystery. One can begin to see why the ULTIMATE IDENTITY is considered *unthinkable* and *unspeakable.*

Example: The ONE WHO IS, yet WHO does not 'ACT' (there being no REAL possibility of movement within ITSELF), IS the only possible 'AUTHOR' of all Action/action (the <u>'ACTOR'</u>), and, in fact, *IS*, necessarily, all Action/action—there being nothing *other* than IT. This is a paradox insoluble by the mind of man, yet *apparent* action through illusory movement is both possible and actual.

actual

By *actual* is meant all that is *perceived* as present (or potentially present) within the World of Becoming, but which is *not* ESSENTIALLY REAL. That which is only apparently *<u>r</u>eal-in-Universe* (but is REALLY highly illusory) is *actual.* That which is *<u>R</u>eal-in-Universe* (for instance, Archetypes on the highest cosmic planes) is also *actual* (though of a higher order than that which is *<u>r</u>eal-in-Universe).* That, however, which is *ABSOLUTELY REAL* can *never* be *actual* (c.f., *real-in-universe).*

Example: Every illusory state is <u>actual</u> but un-REAL. Some illusory states may be both actual and <u>R</u>eal-in-Universe.

By *actual* is meant manifest in-Cosmos.

Example: The Cosmos and everything in It are actual. Though the Cosmos is Real, and though all things in Cosmos may be either Real or real, no things in Cosmos are *ever* REAL.

actual-in-Illusion

By the term *actual-in-Illusion* are designated those presentations which, due to the limitations of the perceiving consciousness, are seemingly real within the World of Illusion (i.e., the World of Becoming), and must, therefore, be treated (at least) as facts-in-consciousness during the various phases through which consciousness passes in its development in the World of Becoming.

Example: The 'extension' of matter is a construct of consciousness and, though unREAL, must be treated as actual-in-Illusion if the perceiving consciousness is to navigate the World of Illusion successfully.

actualist

By an *actualist* is meant one who regards tangible or relatively sensory objects as the most important contents of consciousness. (The 'sensory' objects of the astral and lower mental planes {which, to man, are at least *relatively* sensory} could be included as pertaining to this definition.)

Example: Behavioral psychologists are actualists but certainly not Realists or REALISTS. This is true of strictly *materialistic* scientists as well. In fact, most human beings, subject as their consciousness is to the Third Aspect of Divinity, are actualists who know very little of the Real or the REAL.

Actualist, Higher

By a *Higher Actualist* is meant one who focuses upon the Archetypal Realities which impart order, structure, and Lawful Process to a particular Cosmos.

Example: Students of Divine Psychology are for the most part Higher Actualists because they study those Beings Who *Are* Cosmic Patterns. Such students may or may not be REALISTS (understanding the primacy of the INFINITE SUBSTRATUM) but they are certainly Realists, for they are studying Patterns which though necessarily illusory simply because manifest, are nonetheless fundamental to the integrity of Cosmos and are hence Real-in-Cosmos.

actuality

By an *actuality* is meant that which (in the World of Becoming) *seems* to be REAL, but it is not REAL. (Remember, there are three categories of the *real:* the real, the Real and the REAL.)

By an *actuality* is meant an apparent fact pertaining solely to the World of Becoming and never to the WORLD OF REALITY.

By an *actuality* is meant any presentation within the World of Becoming that pertains only to the World of Becoming.

By an *actuality* is meant any object of perception/apperception.

By an *actuality* is meant anything (within and including Cosmos) that is generated by an 'ACT'/Act/act.

Example: When Emerson said, "Things are in the saddle and ride mankind", he was speaking of the average man's preoccupation with actualities. Once a person begins an inquiry into *meaning* and *significance,* he begins to 'see *through*' actualities discovering their connection to the World of Reality. Remember, however, that even presentations to be found in that high Archetypal World called the World of Reality, or the World of Being, are Actualities.

Actuality, the Great

By *the Great Actuality* is meant the Universe considered as a 'precipitation' of the GREAT REALITY, the ALL-SELF.

Example: The Universe, Itself, is a Great Object to the Perceiving Consciousness of the SELF-as-Concentrated Point (the immediate Forerunner of the Universal Logos). The Universe is only One of an infinite number of Universes, all of which have either already been or will be *actualities.* For the duration of our present Universal Manvantara, our present Universe must be considered the Great Actuality.

Adversary, the Great

By *the Great Adversary* is meant (strangely) the Cosmos Itself. The Cosmos is the supreme contradiction of INFINITUDE, and, is, in its Finitude, *opposed* to INFINITUDE even while *being* INFINITUDE entirely.

Example: Often the so-called "Devil" is known as "The Adversary" as is Saturn, which fact this planet's connection with "Satan" tends to confirm. However, deep thought will reveal that the prototypal Adversary, the Great Adversary, is the Universe Itself, for the very existence of the Universe contradicts and, hence, *opposes* the Fundamental Principles of INFINITUDE—except for one *outrageous* Principle—namely, that apparent contradictions to INFINITE PERFECTION *must* manifest in order to preserve INFINITE PERFECTION.

affirmation; affirm

By an *affirmation* is meant an act which posits a presence; affirmation objectifies.

Example: The Universe Itself Is an affirmation of the Will of the Universal Logos.

By *affirmation* is also meant the assertion and strengthening of a subjective presence or an objective pattern of energies.

Example: To act in a way congruent with one's intra-Cosmic nature is to affirm in-Cosmos the function of one's energy pattern.

By the verb *to affirm* is meant an act/process of fortification or substantiation.

Example: It is the duty of every disciple to affirm the spiritual life in the aspirants under his or her guidance.

agent; Agent

By an *agent* is meant an E/entity who represents and carries out the will of a greater E/entity.

Example: Intra-Cosmic Fohat is the Agent of the Universal Logos, but, ultimately, It is the Agent of THAT (which *purely* and in ITS immutably impartite NATURE) cannot 'ACT' (except by means of FOHAT/Fohat Itself).

algorithm; Algorithm

By *algorithm* is meant the fixed yet dynamic Seed Pattern of a particular Cosmos. The unfolding of the Cosmic Algorithm in Time and Space is the Divine Purpose in Dynamic Action.

Example: The Cosmic Algorithm is equivalent to the Design-at-the-Beginning of a Cosmos. The Cosmic Algorithm is *the* authoritative Formula for energy emanation/generation and energy interaction in-Cosmos.

ALL, the

By *the ALL* (with capital letters) is meant the totality of E/entities, states, and conditions in this Universe as well as in *all Universes that ever have been or ever will be.* The ALL is a collective noun that denotes an infinite number of 'enumerables' (i.e., enumerable things), and is used to emphasize the totality of 'in-Universe' E/entities, states and conditions of the total collection of 'denotables' that have been, are now, or ever will be generated 'BY' the BOUNDLESS IMMUTABLE PRINCIPLE-as-SELF, rather than to emphasize the BOUNDLESS IMMUTABLE PRINCIPLE in ITS purity, PERFECTION, HOMOGENEITY, and ABSTRACTED ZERONESS—i.e., in ITS ALL-IN-ALLNESS.

Example: The ALL has a particularized though *infinite* Content incalculable by all except for the ABSOLUTE with ITS ABSOLUTE 'INTELLIGENCE'. Even the Universal Logos could not calculate the Content of the ALL, for since the Content is infinite, it would take infinite time to do so, and the Universal Logos does not *have* an infinitude of time in which to exist (the Logos being a *finite* Logos), nor is it capable of infinite speed.

All, the

By *the All* is meant the entire phenomenal and noumenal content of a *particular* Universe.

Example: The Universal Logos of our particular Cosmos is invested in the All—i.e., All that has transpired or will transpire within Its Vehicle of Expression, our present Cosmos. Our Universal Logos, because It is a SELF-'VEILED' Being, is ignorant of *all* that has transpired in the ALL. Only TOTAL I/ALL-SELF 'KNOWS' the total Content of the ALL.

ALL-IN-ALLNESS

By the *ALL-IN-ALLNESS* is meant the ONE AND ONLY totally infinitized, impartite, 'STATELESS STATE' of REALITY. It is the 'UNCONDITIONED CONDITION'.

Example: Between Universal Pralayas, *THAT* IS in the 'STATE' of ALL-IN-ALLNESS. The term ALL-IN-ALLNESS emphasizes the re-absorption of the All, the entire Content of a Cosmos with all Its Cosmic Quality into the NOUMENESSENCE, the INFINITESSENCE. The ALL-IN-ALLNESS IS the totally SELF-ABSORBED 'STATE'.

Allness, the 'Infiniversal'

By *the 'Infiniversal' Allness* is meant the totality of all that has *already* transpired in all Universes forever. The Infiniversal Allness has naught to do with the INFINITESSENCE, but only with *already actualized possibilities* that have been 'RADIATED' from the INFINITESSENCE.

Example: What Super-Cosmic *historian* can comprehend the History of the 'Infiniversal' Allness? The History of all Cosmoses is beyond the apprehension of the Universal Logos, but how shall the INFINITE SELF (WHICH has no 'ORGANS') 'KNOW' what has transpired? And yet, if the INFINITE SELF is PERFECTION ITSELF, it is not possible that there should be anything in any Cosmos which is utterly *unknown.*

ALLNESS, UTTER

By the *UTTER ALLNESS* is meant the totality of the ALL-IN-ALLNESS of the ALL-SELF as well as the articulated *content* of every one of an infinitude of Universes that ever have been or ever will be.

Example: There is no example other than IT and the infinitude of ITS Illusory Expressions called Cosmoses. The *calculable* UTTER ALLNESS includes absolutely all and everything that IT has 'DONE' in the 'Forever-Gone'. The *incalculable* UTTER ALLNESS includes absolutely all and everything IT WILL ever 'DO' and the 'Forever-to-Come', *plus,* the ABSOLUTE SELF, ITSELF.

Example: The UTTER ALLNESS is the ALL plus THAT WHICH has 'GENERATED' or will 'GENERATE' that ALL.

ALL-SELF, the

By *the ALL-SELF* is meant the ABSOLUTE considered as the SOURCE of INFINITE SELFHOOD. The ALL-SELF IS equivalent to the ALL-IN-ALLNESS.

Example: All *selves* that have ever been or ever will be in all Cosmoses past, present and to come, are simply manifestations of one ALL-Inclusive SELF—the ALL-SELF.

ALL-SUFFICIENCY, the

By *the ALL-SUFFICIENCY* is meant the SELF in ITS ABSOLUTE PLENITUDE.

Example: What human being can feel a sense of any privation when identifying with the INFINITE SELF, the ALL-SUFFICIENCY?

always

By *always* is meant forever, regardless of states and conditions.

Example: The INFINITE SELF is always the INFINITE SELF, *absolutely*, whether or not the Illusion of Cosmos is apparently 'happening'.

AM; Am; am

When these words are used technically, *'AM'* relates to the BOUNDLESS SELF, *'Am'* to the Cosmic Self (the Universal Logos) and to all Self-conscious Beings cognizant of their relational unity with other Beings/beings in-Cosmos; the awakening human soul would use the term **8** Am. The term *'am'* relates to the level of the personality or ego consciousness.

Example: Whereas **I** AM as *ever*, **8** Am the evolving Self-Consciousness in-Cosmos, from the stage of 'humanhood' to the ultimate stage of Cosmic Consciousness in which **8** Am the Universal Logos. I am, on the other hand, the personal self of this particular incarnation.

Antahkarana

By *Antahkarana* is designated a term that can be unconventionally translated as 'knowing exactly what **8** Do'. In other words, **I**-as-**8** Am active Universally but in my limited 'part-consciousness', **8** know it not because of the Principle of SELF-'IMPOSED' 'isolation of localized consciousness'. This isolation arises first through SELF-'VEILING' and then Self-Veiling. Antahkarana is an instrument of *pervasion*. Through the agency of the Antahkarana, it becomes possible to know what **8** 'Do' in *all* cases (eventually, throughout the Cosmos) and not just in the special *localized* case that **8**-as-I now and habitually call my selfhood.

Example: The building of the Antahkarana is the key to experiencing "muti-dimensional consciousness".

Example: The building of the Antahkarana is the re-acquisition of Universal Soul, experientially, but not REALLY 'lost' through SELF-'VEILING' and Self-Veiling.

apparency

By the *apparency* of anything is meant that which it 'appears' to be to a registering consciousness, as opposed to what it REALLY IS.

Example: In Cosmos, apparency of the GREAT HOMOGENEITY is multiplicity.

apparent

By *apparent* is meant that which (to a perceiving consciousness) presents with a false or merely seeming reality (or, even, Reality; the term 'reality' {with a small 'r'} is ever meant to indicate a seeming or an illusion. Even Reality, though Universally Archetypal, is REALLY false, and merely a seeming of a higher order—thus, ultimately, merely apparent.)

Example: That you and I are nothing more than human beings may be apparent to others of our kind, but is a notion which has no ESSENTIAL REALITY.

appearance

By an *appearance* is meant any illusion, from the supremest height represented by the 'RAY' of the ABSOLUTE (which is the First Appearance and Greatest Illusion) to the depths represented by hallucination and the phantasmagoria of the astral plane.

Example: An appearance cannot be REAL, though it can be actual (i.e., real, and Real). Appearances are contents of consciousness, and no content of consciousness is REAL. REALITY is registered or, better, participated-in through *identification* and not through *consciousness.*

Example: Though the 'RAY' of the ABSOLUTE is, apparently, *less* than the ABSOLUTE, and, thus, an Appearance, the very serious question arises: to *what* or to *whom* does 'IT' *appear?*

apperceivable (noun)

By an *apperceivable* is meant a subtle object capable of being registered only by subtle senses.

Example: For man at his present relatively undeveloped stage of evolution, the sheath of the Monad on the monadic plane is definitely an apperceivable.

apperception

By *apperception* is meant the conscious registration of internal or abstract objects, hence, of objects that can only be registered by the subtle senses. When that which is deemed formless is registered, such a registration must necessarily be considered registration of an object simply because that formless 'something' that is registered can be distinguished from other registrations. Apperception is the act of registering *all* subtle objects, even *formless* ones.

Example: Apperceptions usually relate to what we call the *arupa* or formless levels of manifestation. For instance, it is most correct to say that one apperceives the ideas of the buddhic plane, and perceives the images of the astral plane.

archetype

By *archetype* is meant simply an *original type*—a compelling pattern of subtle energies fundamental to a given system (and existing upon the higher dimensions of that system)—in conformity to, and in reflection of which, the design of certain derivative patterns of less subtle energy appears upon the lower dimensions of that system.

Example: The Egoic Lotus of the Solar Logos is the archetype of a multitude of lesser lotus-like structures found within Its Ring-Pass-Not. As above; so below.

Archetype

By *Archetype* is meant an Entity of a high order (a Cosmic Reality) Who functions as an ordering, structuring, and directing Principle in-Cosmos. Even Archetypes are temporary when compared to the ONE TRUE REALITY. Archetypes, even though They are Real-in-Cosmos are but *combinations* and all combinations are evanescent. The Archetypes of one Universe may differ considerably from the Archetypes of another, even though as Beings, they are ESSENTIALLY the *same* in the SAMENESS.

By *Archetypes* are meant Patterned Relationships of Energies that control the patterning of all energies and forces vibratorily 'below' them.

Example: The Universal Logos is *the* Archetypal Being (i.e., the Archetype) of Cosmos. All lesser Wholes/wholes reflect this Archetypal Being holographically.

Example: The Manu of a Root Race determines and sets the racial form for that particular Root Race. In this respect, the Manu of any each particular Root Race can be considered the Archetype for the billions of human units who develop in that Race.

articulate; articulation

By the term *articulation* is meant nothing to do with speech, per se. The idea is far more fundamental. Articulation is the process of intelligent differentiation within a homogeneous or relatively homogeneous medium. It is the definition of items, forms, and patterns within a specific medium or field.

Example: With respect to Fohat there are two modes of articulation—subjective and objective. Fohat articulates *within Itself* and in great detail (at first *ideationally* and *subjectively)* the multitude of intended Idea-Patterns included within the Great and emerging 'IDEA'-as-Idea (the Cosmic Idea held by the Universal Logos) which is to 'Become the Pattern' of the incipient Universe.

In other words, Fohat *Understands* what is to be Done and *Mobilizes Itself* to Perform by ideationally and subjectively preparing within Itself the enumerations and relationships It must 'Become' within Cosmic Prakriti (i.e., within Its own 'Self-Reflection')—for Fohat Creates by 'Becoming' that which It Creates. Then, Fohat articulates within Cosmic Prakriti the Objectified Forms of the Subjective Idea-Patterns that Fohat had previously ideationally and subjectively articulated within Itself (which means that Fohat engages in specific 'Self-Sight').

Thus there is first a deeply Subjective Articulation/Enumeration within Fohat *and* a subsequent Objective Articulation within Cosmic Prakriti (which, Objective Articulation is, nonetheless, Cosmo-Psychological). Thus is the Subjective Articulation reflected within the Objective Articulation. Thus the 'Timely' Idea 'EXTRUDED' from the INFINITESSENCE (which Idea was originally *noumenessentialized* 'within' the INFINITESSENCE) is first subjectively assimilated by Fohat and then materialized or manifested or objectified in Cosmic Prakriti by this same Cosmic Agent—Fohat, Who, as it were, 'Becomes' that very Idea within Cosmic Prakriti through the 'Action' of 'Self-Perception'.

artifact

By *artifact* is meant that which occurs as a secondary or tertiary, etc., *result* of a primary intention. An artifact is simply a condition which is an *unintended happening.*

Example: The Great Teacher was virtually incognizant of Her popularity with the intelligentsia. That popularity was merely an <u>artifact</u> of the uncompromising mental clarity and integrity with which She addressed all problems.

aspect (noun)

By an *aspect* is meant an integral part of a greater whole.

Example: Her spontaneous compassion is one of the most spiritually important <u>aspects</u> of her character and indispensable to her success as a server.

aspect (verb)

By the verb *aspect* is meant the particular way in which one factor in a set of relationships affects or influences other factors within that set of relationships.

Example: Mars in your horoscope <u>aspects</u> Jupiter favorably, predisposing you towards indomitable enthusiasm and optimism.

Aspect

By an *Aspect* is meant one of the *essential* Attributes (Qualities) of the One Cosmic Deity. These have sometimes been called the "Names of God".

Example: The One God of this Cosmos manifests in three principal <u>Aspects</u>: Will, Love, and Intelligence.

By an *Aspect* is meant a high Principle that is absolutely essential and fundamental to a given system.

Example: The Doctrine of Non-Dualism is one of the indispensable <u>Aspects</u> in the Vedantic Body of Truth.

'ASPECT'

By an *ASPECT* is meant a *principal* 'ATTRIBUTE' (or 'QUALITY') of the ONE SELF. The ONE SELF, however, is INDIVISIBLE and *devoid* of specific 'ATTRIBUTES' (or 'QUALITIES') because IT is the NOUMENON of All Possible Attributes-in-Cosmos that 'INHERE' in IT as the ONE INFINITESSENCE. In one way the ALL-SELF is the SOURCE of *infinitely more* 'ASPECTS'/'ATTRIBUTES' than could ever appear among the Aspects/ Attributes which characterize a Universal Logos and Its Cosmos. It is, however, virtually impossible to say anything about the ABSOLUTE without thinking in terms of ITS infinite 'ATTRIBUTES', of which the most important can be considered ITS 'ASPECTS'.

Example: The three principle <u>'ASPECTS</u>' of the ONE SELF as put forward by the Vedantins as SAT, CHIT, and ANANDA or their *infinitessential fusion*: "SACHIDANANDA". If <u>'ASPECTS</u>' of the ONE LIFE can be discussed at all (which some thinkers justifiably think *cannot* be done) then the most fundamental of them would be these: BEING/CONSCIOUSNESS/BLISS, or SAT/CHIT/ANANDA (SACHIDANANDA).

ATMAN

By the *ATMAN* is meant REALITY. ATMAN means particularly the 'Ray' of the ABSOLUTE—as the 'RAY' of the ABSOLUTE—as the ABSOLUTE. It stands for the ESSENCE of any E/entity, that ESSENCE considered *identical* with the ABSOLUTE, ITSELF.

Example: ATMAN *IS* PARABRAHMAN.

attention

By *attention* is meant a particularized focus of consciousness upon a particular registration that intensifies (for the registering consciousness) the relative strength of that particular registration as compared with the strength other simultaneously occurring or potentially accessible registrations. Attention is the selective focus of consciousness.

Example: When I give my attention to the positive qualities of my fellow group members, such qualities seem enhanced and of far greater value than their few faults which were so noticeable before.

attenuation

By *attenuation* is meant a process by which an energy source projects itself in such a manner that the projection is qualitatively a replica of the source but a *de-instensified* replica. Simply, *attenuation* means a *diminishment* that preserves quality.

Example: If I reduce the volume of a sustained musical note, for instance, an E, the 'softer' E is still recognizably an E, though the original sound has undergone an attenuation.

Example: In Emanation Theory, objectification is attenuation.

attribute (noun)

By the term *attribute* is meant a particular quality assigned to or associated with a particular object.

Example: The character of a human being can be understood as a collection of attributes generated throughout the course of many lives.

Example: The Rays of Attribute, though an integral part of the Septenate of Rays, possess qualities which are derivative from the Rays of Aspect.

attribute (verb)

By the term *attribute* is meant the act of assigning or associating a particular quality to or with a particular object.

Example: The attempt to attribute certain qualities to the BOUNDLESS IMMUTABLE PRINCIPLE is an exercise in futility as this 'BEING/NON-BEING' is the NOUMENON of ALL POSSIBLE ATTRIBUTES. (Nevertheless, it is intellectually seductive to engage in this futile exercise.)

authentic

By the term *authentic* is indicated a direct relation to the SELF, even if in an attenuated manner. An *authentic E/entity*, for instance, is one that is directly related along the Divine Emanatory Stream to the SELF-as-Self. An authentic E/entity, ESSENTIALLY, shares the same unitary *selfhood* as the ONE SELF. An authentic E/entity is (in its inmost recesses) but *one* 'Ray' of the ABSOLUTE. There are certain kinds of entities which have no inmost recess. In general, that which is *authentic*, shares the same *essence* or inner quality as that from which it was derived.

Example: His close contact with the Master contributed to the almost universal evaluation of him as an authentic initiate.

Example: An intellectual understanding of Radical Non-Dualism is nowhere nearly as valuable as an authentic identification with the ONE AND ONLY SELF.

In a lesser sense, the term *authentic* simply means *genuine*—that which *is*, in fact, what it *seems* to be.

Example: The ceremonial dorje used by the Rimpoche is certainly authentic; it is certainly no mere Western imitation.

auto-intensification

By the term *auto-intensification* is meant a self-induced process that magnifies or enhances the intensity and expression of the patterned energies which define the expression of a being as unique.

Example: Most egoism is simply ignorant auto-intensification.

Example: Occult meditation by a disciple upon the major six Rays that condition his energy structure is a potent and desirable means of auto-intensification.

Example: From the Perspective of the Universal Logos the entire story of Cosmic Evolution is a Drama in which Logoic Auto-Intensification is pursued through the necessary medium of Cosmic Prakriti, and the experimental Process of Prakritic Objectification.

- B -

Bailey, Alice A.

Alice A. Bailey was an English author in the field of Theosophical Esotericism. For thirty years she served telepathically as the amanuensis for The Tibetan Master Djwhal Khul. Together, they wrote and published what have come to be known as the "Alice Bailey Books" or the "Blue Books", which offer a deep and comprehensive presentation of Trans-Himalayan Occultism for the Western World.

beginning

By the term *beginning* is meant the first action of a continuing process.

Example: That selfless act marked the beginning of her fitness to tread the Path of Discipleship.

Example: In relation to any complex process, it is extraordinarily difficult to determine its beginning.

Beginning, the

By *the Beginning* is meant the Action which (only apparently) ends the SAMENESS and HOMOGENEITY of the ALL-IN-ALLNESS of Universal Pralaya and impulses the Heterogeneity and Objectification of Universal Manvantara.

Example: The Beginning of every Cosmos occurs with a 'FLASH' of 'RADIATED ESSENCE' from the DEPTHS OF THE ONE, by means of which 'FLASH' the ALL-SELF signals yet again ITS eternally recurrent 'CAPACITY' for Objectivity even while retaining, as ever, ITS IMMUTABLE INFINITE SUBJECTIVITY.

Example: Sometimes the word 'BEGINNING' is used to indicate the first stages of the Cosmic Beginning, although, technically, the term 'BEGINNING' does not refer to Cosmic Cycles at all; when the term is used it is usually in single quotes, the 'BEGINNING'.

BEGINNING, the

By *the BEGINNING* is meant *nothing at all!* A BEGINNING is an impossibility; there was no BEGINNING of THAT—the GREAT BEGINNINGLESSNESS.

Example: An example of the BEGINNING is impossible except in misguided, deluded thought.

being (noun)

By a *being* is meant an authentic unit of LIFE, ex-isting in the World of Cosmos—the SELF-'OBJECTIFIED' World.

Example: Animals, humans and devas, to name but a few, are all beings—essential and authentic *radiations* of the ONE LIFE, pursuing their apparently separated development under the limitations of multiplicity.

'BE-ing' (verb)

By *BE-ing* is meant a 'STATE' of complete identification *as* BE-NESS.

Example: There is a transition between *seeing* and pure 'BE-ing' where any possible 'point' vanishes into NOTHINGNESS. That transition occurs when an 'infinite distance' is *reached* (either Spatially or Temporally). One can see that that transition *never* occurs in-Cosmos, for, in relation to Cosmos a definite 'infinite distance' is not possible. It is only 'reached' within the INFINITE.

BEING-as-Being-as-being; being-as-Being-as-BEING

By a *BEING-as-Being-as-being* is meant any of a great multiplicity of beings in the Cosmos (the hyphenated sequence calling attention to the fact that the SOURCE of any being is the ABSOLUTE BEING, and secondarily, the One Universal Being, the Universal Logos).

By the second sequence, *being-as-Being-as BEING*, is indicated the inevitable conscious ascent through which any being in Cosmos must 'travel' in its (apparent) return, first to its *Universal* Source, and then to the ABSOLUTE SOURCE. These sequences are used in order to remind human beings whence they came and whither they are going.

Example: The complete building of the individual, planetary, solar and, finally, cosmic *antahkaranas* will make the sequence, being-as-Being-as-BEING, a demonstrated fact in the lives of all ascending consciousnesses. The descending sequence, BEING-as-Being-as-being, has much more to do with the SOURCE/Source and descending destination of the *sutratma*, the "Life Thread."

BE-NESS

By the term *BE-NESS* is meant the *GROUND* OF ALL BEING, WHICH IS the INFINITE SOURCE of all particularized beings.

Example: BE-NESS is devoid of beings, and yet ESSENTIALLY *IS* every being that is now, ever has been, or ever will be.

boundless

By the term *boundless* is meant that which is incapable of being circumscribed, contained, or divided in any way.

Example: The Omnipresence, Omniscience and Omnipotence of the Universal Logos are boundless (hence, all-pervasive) with respect to our Cosmos, but since our Cosmos is, Itself, *bounded,* the extensive range of the One Universal Logos can be considered only *relatively* boundless. Only the INFINITE SELF is *truly* boundless for there is nought *else* to bound IT.

BOUNDLESS IMMUTABLE PRINCIPLE

By the *BOUNDLESS IMMUTABLE PRINCIPLE* is meant the FIRST AND ONLY ULTIMATE ABSTRACTION, SOURCE, AND ROOT OF ALL FOREVER.

Example: The BOUNDLESS IMMUTABLE PRINCIPLE IS ITSELF alone. IT cannot be exemplified.

BOUNDLESS IMMUTABLE PRINCIPLE-as-SELF

By the *BOUNDLESS IMMUTABLE PRINCIPLE-as-SELF* is meant the same as the BOUNDLESS IMMUTABLE PRINCIPLE, but with the connotation that the BOUNDLESS ONE is *both* ENTITY (as SELF) and NON-ENTITY (as ABSTRACT UTTERLY IMPERSONAL PRINCIPLE).

There is an inherent duality that enters every consideration of the ABSOLUTELY UNITARY BEING. IT must necessarily (due to limitations of *any* consciousness—not just *human* consciousness) be considered a SOMETHING *and* a NOTHING. But to call IT a SOMETHING is to predicate of IT defined *existence,* which cannot properly be done. To call IT a NOTHING is to *(nolens volens)* make IT into a SOMETHING simply by the act of predicating ITS NOTHINGNESS. (Predication is reification.) Nevertheless, the human mind seems incapable of avoiding this *dualistic impression* as to ITS NATURE. IT remains, in human assessment, both SOMETHING and NOTHING, hence both an EGO and a NON-EGO, a SELF and an *utterly abstract* PRINCIPLE.

Example: *The Secret Doctrine* emphasizes the utterly abstract NATURE of THAT, whereas the Non-Dualist Vedantin School emphasizes the SELFHOOD of THAT. It is useful for the human mind to conceive of THAT as *both*—the ULTIMATE ABSTRACTION *and* the ULTIMATE ENTITY. For this reason the term—BOUNDLESS IMMUTABLE PRINCIPLE-as-SELF—was invented.

BRAHMAN

By the term *BRAHMAN* is meant the ABSOLUTE, ITSELF, the ONE AND ONLY INFINITE SELF.

Example: BRAHMAN IS unique unto ITSELF.

Example: Sankaracarya encourages all aspirants-to-REALITY to "Merge the world in BRAHMAN." Has any concept been a greater magnet for exaltation of spirit than the concept of the BRAHMAN?

Breath, the Great

By the *Great Breath* (*not,* the GREAT BREATH [with all capitals], as it cannot exist] is meant that Infinitely Enduring Perpetual Oscillation within the UTTER ALLNESS by means of which the ALL-IN-ALLNESS 'BECOMES' the Universe while, nevertheless, remaining UNCHANGED and UNCHANGING. The Great Breath is the Fundamental Rhythm within the UTTER ALLNESS. The Paradox is that the Great Breath is a Motion, and the ALL-SELF (as ITSELF, PER SE) does not 'BREATHE', because IT does not 'MOVE'. The REALITY of the 'situation' seems to be rather like *breathing* and *holding the breath* simultaneously, because even though the ALL-SELF *appears* to 'GO FORTH' from ITSELF, IT does not REALLY 'GO FORTH'. The Great Breath, therefore, is the *apparent* Motion that produces the Great Illusion. The 'Exhalation' of the Great Breath begins with the 'FLASHING FORTH' of the 'RAY' of the ABSOLUTE. The 'Inhalation' of the Great Breath is the lengthy Process of Cosmic Evolution and Cosmic Obscuration.The Great Breath, however, cannot be considered a REALITY (though It Is a Reality). Its Occurrence represents the 'EXTRUSION' of 'POSSIBILITY' from the INFINITESSENCE, and every such 'EXTRUSION' is, necessarily, an *illusion,* though Real-in-Cosmos. Even with respect to human *samadhi,* one must, as it were, *stop breathing* to know what Reality/REALITY Is/IS.

Example: The Great Breath is the Pulse of Cosmic Life. Without the Great Breath no Cosmos could Live, but the SELF that never dies would certainly continue to LIVE. The INFINITE SELF IS the one BEING/NON-BEING that is *not* kept 'alive' by *breathing.*

- C -

cause; causality

By *cause* is meant a process relating two conditions/configurations in such a way that the presence of the condition/configuration existing *first* in time, lawfully and predictably induces the appearance of the other condition/configuration, at a 'later' time. If a given condition/configuration is the *cause* of a second condition/configuration (which is called the *effect*), then that given condition/configuration must exist *antecedent to* the existence of the second condition/configuration if that second condition/configuration (the *effect*) is to come into existence.

By *causality* is designated an inclusive term used to indicate the principle of causation in general.

Example: The "Approach" of the Spiritual Hierarchy to humanity is one of the main <u>causes</u> for the greatly heightened aspiration and spiritual longing arising in the human family at this time.

Example: The concept of <u>cause</u> is inextricably linked with the concept of *change.* <u>Cause</u> is considered the *reason for* or *principal impulsion* of change.

Example: If <u>cause</u> is the reason for or principal impulsion of change, then <u>cause</u> can exist (apparently) in the *future* as well as the *past.* Numinous visions of Divine Patterns to come (which Patterns Now exist), though they have not yet precipitated into form upon the physical-etheric plane), can justifiably be considered <u>causative</u> in relation to the sensitive consciousness that can apprehend them. In all fairness, such visions, though indicative of the physical-etheric *future* must be considered as occurring *in subtle form* (i.e., in form more subtle than physical-etheric form) in the *present.* The future does not <u>R</u>eally exist Now in the same *form* or *forms* through which it will manifest in the 'future', i.e., at a later time. Only a form *indicative* of that future exists Now. Such forms, which often accurately indicate the "shape and form" of the future (i.e., "of things to come") are definitely <u>causative</u>.

CAUSELESS CAUSE

By the *CAUSELESS CAUSE* is meant THAT which, ITSELF (caused by nothing and antedated or succeeded by nothing) is yet the SOURCE from which all objective *things* (i.e., 'objectivities') derive.

By the *CAUSELESS CAUSE* is meant THAT— the BOUNDLESS IMMUTABLE PRINCIPLE, the INFINITE SELF.

Example: The <u>CAUSELESS CAUSE</u> has no 'BEGINNING' be-*cause* never was the 'TIME' when IT WAS *not.*

chance

By the term *chance* is meant the concept that many patterns between variables in-Cosmos are not rationally willed and/or purposeful patterns, but occur either accidentally, or as artifacts of other willed actions that do not relate directly to the patterns under consideration. If a certain something happens *by chance*, then there is no Cosmic Intelligence guiding the formation of the pattern we call that 'something'.

Example: God, the Universal Logos (on Its Own Level and subject only to the limitations of Its Own Nature) is totally free within the limitations of Its Cosmos, yet at many points along the Emanatory Stream, God-in-Emanation is partially, to almost completely, blinded. God's Will, within the stringent limitations of the Cosmic Algorithm, is completely free, but God, the Universal Logos is blinded compared to GOD the ALL-IN-ALL, this blindness being due to the SELF-'VEILING' 'PROCESS'. A blinded free will, will make mistakes and will not always *appear* free. Patterns created by blinded free will often appear as *chance patterns* (as if there were no broad and purposeful Intelligence behind them) even though the willing which created such patterns was impelled by a deeply felt though unconscious sense of the Design-at-the-Beginning as well as by an unconscious sense of the Logoically Sanctioned Patterns which flow from that Design.

change

By the term *change* is meant a re-positioning or alteration of items contained within a particular context.

Example: Items-in-Cosmos are perpetually (cyclically) undergoing a change of configuration with respect to each other.

By *change* is simply meant motion.

Example: It is impossible to conceive of change divorced from item, motion or time.

CHANGELESSNESS, the

By *the CHANGELESSNESS* is meant that ABSOLUTELY MOTIONLESS BEING/NON-BEING WHO/WHICH IS the INFINITE SELF.

Example: The CHANGELESSNESS IS the INFINITE HOMOGENEITY. Because IT IS the ONE AND ONLY HOMOGENEOUS CONTINUUM, IT cannot under-go any change.

chaos; chaotic

By the term *chaos* (as contemporarily and not Mythologically understood) is meant a condition of variables within a given context such that no abiding pattern or regularity governs the movements/changes of these variables and/or their relationships to each other. In absolute chaos no pattern of magnetism or repulsion whatsoever exists between variables. There exists only random, purposeless, unintended motion.

By *chaos* is meant a complete lack of relationship.

Example: Chaos is a condition of complete instability and unpredictability. Chaotic movement is guided by no patterning Force.

Example: The aggregate of units comprising the unprincipled substance of the First Solar System (though these units be vestigially conditioned by the Principles of the First Solar System) would (at this point in Cosmic Time) engage in movement bordering on the chaotic were it not for the patterning influence of Second Solar System Principles, which are reflective of the Second Solar System's Design-at-the-Beginning. (Each system has its particular *design at the beginning.)*

Example: In ancient Greek Mythology, Chaos was equivalent to what the Hindus call Mulaprakriti.

chord

By a *chord* is meant either the result of sounding more than two notes (each note being of consistent frequency) simultaneously, or the result of sounding a fundamental note (of consistent frequency) which produces an aggregation of simultaneous sounds consisting of overtones (i.e., partials). (If the frequency of the note or notes sounded is not consistent, more than one chord will result.)

Example: The blending of all individual 'notes' within a soul-infused group, produces a chord of many notes, the harmonious or non-harmonious quality of the chord signaling the degree of spiritual harmonization of the group. This chord might also be called the composite group *note.*

combination

By a *combination* is meant a positioning of variables (a configuration) that endures beyond the least possible division of time existing in a particular Finite Universe. Configurations that change with each 'ultimate moment' cannot properly be called combinations as they are, given the Laws of a particular Universe, *maximally evanescent,* and the items concerned do not Really *relate* or *combine.* (Relationship and combination require *repetition of configuration through time.*)

Example: Every atom is a combination of ultimate particles configured according to a certain design.

Example: Every authentic E/entity manifests as a combination of lesser wholes, with the exception of that entity called the ultimate particle which is indivisible.

[See the note after the "Configuration" Glossary entry.]

Combination

By a *Combination* is meant a Fundamental Cosmic Structure, an Entity of a High Order which directs and regulates Cosmic Process.

Example: All Emanations of a High Order are, Essentially Combinations of interactive Subjects/Objects—such as the first ten Numbers.

Example: All Archetypes are Combinations, whereas mere archetypes are combinations.

Example: In the Realm of Number, only the ZERO (which is not REALLY a number) is totally uncombined (uncreated and ungenerated by relation of any kind). Even the Number One is a *perceptual* Combination arising from *initial* 'OBJECTIFICATION' (thus requiring 'SEER' and 'Seen' if it is to exist). The Number One always stands midway between undifferentiated INFINITUDE and the Infinitude of Multiplicity. From another perspective, the Number Two is the transitional Entity between Unity and Multiplicity, and therefore might be called the 'Seed of Multiplicity'. All N/numbers other than the Number One are generated through the Self (Universal Monad) entering into *relation* with Itself. Such combinations, resulting from cosmically primeval acts of Self-Perception, might be called 'Perceptual Combinations'.

comparison; compare

By the term *comparison* is meant a process of matching or superimposing variables for the purpose of determining similarities and differences between them.

Example: In Cosmos every 'thing' can, with profit, be compared with every other thing. Only the INFINITE SELF is *incomparable.*

Example: Co-measurement (a term much used by the Master Morya) is the Science of Comparison.

Condensing of the Infinified Point, the

By *the Condensing of the Infinified Point* is meant yet another phase in the Mayavic Veiling Process (in this case, Pre-Cosmic Veiling) by which the Infinite Vision of the Infinified Point is 'narrowed down' to *one single* Vision (i.e., narrowed down to *one, single* Possibility of 'Sight'—a 'Sight' 'Seen' by a Consciousness that is no longer *infinified* but which is limited to one option of 'Sight' instead of being engaged in limitless, or infinite options of 'Sight').

This new, non-infinified and thus drastically limited *Point* of View can be called the 'Condensed Point', and 'through' it, the SELF-as-Self 'Sees' but *one* Vision which is the *limited* Vision of the Parameters/Possibilities/Scope of the Cosmos-to-Come. As both 'Points' and 'Logoi' are Really but the SELF-as-Self in various 'States of Consciousness' produced through *veiling*, the Condensed Point could justifiably be *named* the Universal Logos. The Infinified Point could be considered or named the 'Pre-Universal Logos'—the SELF-as-Self-as Infinified Point, or the Infinite Subject. (Clearly, the term 'See' is meant to indicate Consciousness, and the term 'Sight', that which is 'Seen'.)

Example: A great and vital question in Occult Philosophy arises concerning whether the 'Choice' of the Cosmic Pattern-to-Come is made *before* or *after* the Condensing of the Infinified Point. The very Character of each Universe, its Beauty, and Its 'Place' and 'Function' within the Infinite Sequence of Universes, and Its 'Relation' (if such exists) to other Universes within this Sequence, are dependent upon the answer to this question. This is so because a 'Choice' made after Condensation would be a severely restricted 'Choice', performed in ignorance of the nature of all Universes Past (as well as to come) in the Infinite Sequence of Universes. The only 'Absolutely Informed Choice' would necessarily be an '*ABSOLUTELY* INFORMED CHOICE' made from 'within' the ALL-IN-ALLNESS. Such a 'CHOICE' would be, for obvious reasons, paradoxical, as *nothing* 'OCCURS' 'within' the Pre-Cosmic and Post-Cosmic ALL-IN-ALLNESS.

On the other hand, it may be that a 'Choice of Cosmic Pattern' for the Universe-to-Come that was made at a Pre-Cosmic 'Time' 'following' the 'FLASHING FORTH' of the 'RAY' OF the ABSOLUTE yet *before* the Condensing of the Infinified Point, could make it possible for the SELF-as-Self-as-Infinified Point to 'Remember' the Natures of all

Cosmoses Past, as well as access the FOUNT OF ALL POSSIBILITY. The question naturally arises as to whether it is possible for the SELF-as-Self to *access* the FOUNT OF ALL POSSIBILITY once the 'RAY' has 'FLASHED FORTH'. These considerations are important if one wishes to know whether each Universe is *conceived* from 'within' the ABSOLUTENESS of the INFINITESSENCE (i.e., from 'within' the PERFECTION) or *conceived in relative ignorance.*

condition

By the term *condition* is meant an *actual* or manifest configuration of items-in-Universe extant within a particular field of focus. It is important to remember that what is not *yet* a *condition* within the field of focus under consideration *may*, however, be a condition within a higher field.

Example: Humanity today is in its present condition due to many energies and forces—antecedent and current. Plans exist which, when fulfilled, will make it possible for humanity to live in a better condition, though the time is not yet. Those better conditions (though well-conceived) do not *yet* exist—at least not for humanity.

Example: The Cosmos, Itself, is in a certain Condition depending upon the relationship of all variables found within Its Ring-Pass-Not. The general Nature of the Cosmic Configuration during a particular Period of Cosmic Time determines the Nature of the Cosmic Condition at that Time. The Cosmic Configuration simply is what it is. The Cosmic Condition (how things are progressing *cosmically* when measured against Universal Logoic Will) can be *evaluated* in relation to the degree to which the Cosmic Condition fulfills the requirements intended by the Universal Logos for that Phase of the Design-at-the-Beginning which is transpiring at the time of the evaluation. We may, for instance, know something about the present *condition* of the Earth Globe relative to the Intent of our Planetary Logos. We may think we know something about the *condition* of the Moon Chain at the time of its sudden termination long ago. But what do we know of the Cosmic Condition (Now, at this very moment of Cosmic Time) in relation to Universal Logoic Intent? And, yet, such a Condition exists.

conditional

By *conditional* is meant 'subject to variation'—given to change (sooner or later) as a result of the re-configuration of internal variables and/or the impact of external variables.

Example: The quality of life in the three worlds of human evolution is entirely conditional, and dependent largely upon the fluctuations of the lunar vehicles.

Example: All combinations in Cosmos are conditional and hence impermanent.

configuration

By a *configuration* is meant any disposition or arrangement of variables in Time and Space, no matter how fleeting that disposition or arrangement.

Example: Ultimate particles configure and re-configure themselves from ultimate moment to ultimate moment. This perpetual dance produces many configurations that are devoid of any relatively lasting or stable magnetism and, hence, cannot be considered as combinations.

Example: Whereas a combination may be associated with the magnetic energy of the Second Aspect of Divinity, a configuration is more to be associated with the distributive energy of the Third Aspect of Divinity.

Example: A configuration is created by the act of perception of a perceiving consciousness. A given aggregation of items viewed from all *possible* intra-Cosmic perspectives, produces in consciousness as many configurations as there are perspectives. If an infinitude of perspectives were available from which to view an aggregation, then each aggregation could yield an infinitude of perceived configurations. But an infinitude of perspectives is *not* available, because Space is *quantized* and *not* a *continuum.* Only certain perspectives are available just as only certain spatial positions within the space of an atom are available for the atom's electrons. Spaces 'in-between' an atom's electron shells cannot be occupied. 'Available space' depends upon the Laws of Relationship prevailing within a given Cosmos. Analogically, a given fundamental note will not produce *all possible* overtones or partials, but only certain overtones and partials (at a certain tonal 'distance' from the fundamental). Such cosmically-lawful overtones and partials are related mathematically to the fundamental note.

All this is another way of saying that not all possible designs or configurations are 'allowed' to become *actual* within a given Cosmos. There may well be a potential infinitude of un-precipitated actualities within a given Finite Cosmos (such as the mathematical set of all integers), but there can never be an infinitude of *precipitated* actualities, for Cosmos, being Time-bound, cannot tolerate the manifestation of any *actual* infinitude.

Note: The difference between the concept of *combination* and *configuration* is subtle and subject to change with change in definition. In this treatise the most fleeting arrangement of variables can be called a 'configuration' but not a 'combination', though other authors might word it differently. Configurations however, can be more lasting, in which case they would be called combinations *as well as* configurations. The term *combination* suggests a greater magnetic interplay between variables than does the term *configuration.* In a condition of utter chaos there is no lasting or predictable combination or configuration, though at any moment a *configuration* could be identified by a perceiving consciousness.

Configuration, Cosmic

By the *Cosmic Configuration* is meant the relational disposition (i.e., the arrangement) of *all* variables in Objective Cosmos at any given Moment of Cosmo-Objective Time in-Cosmos. The Cosmic Configuration necessarily changes from ultimate moment to ultimate moment.

Example: The Universal Logos is the All-Seeing Witness of each and every change of the Cosmic Configuration throughout the Duration of His Cosmos.

Example: Due to the Law of Unrepeatability, the Cosmic Configuration can never be *exactly* the same from ultimate moment to ultimate moment.

Example: Presumably the Cosmic-Configuration applies only to the Work of the Cosmo-Objective World. Willed imagistic change may be so rapid within the Ideational World (the World of Being) that no *Standard* Moment can be isolated 'upon which' to assess a Configuration (were the World of Being to be included in the Cosmic-Configuration). The Universal Logos, however, is in a 'Position' to assess the Frozen Cosmic-Configuration in the World of Objectivity, as well as to observe the relative continuity of 'Willed Imagistic Activity' (no matter how variable the relative speeds of such *acts*) within the World of Cosmo-Subjectivity—the World of Being. To the Universal Logos, all activity, in all the Domains of Universe, whether than activity is 'held', or rapidly 'changing', can be 'Seen', if the Logos 'Chooses', as One Seamless 'Flowing' Movement.

consciousness

By *consciousness* is meant the faculty (in a sentient being) that confers the capacity for a *knowing* (i.e., an isolated and detected) registration.

Example: A refined consciousness can isolate many discrete vibrations and discern the difference between them.

By *consciousness* is meant the faculty (in any being) of sensitivity to registration.

Example: A sensitive consciousness is receptive to many more impacts than is the average consciousness.

By *consciousness* is meant the objectification of sensitivity.

Example: The 'PREOCCUPATION' of the INFINITE SELF with ITSELF can hardly be called 'CONSCIOUSNESS' as no manner of *objectification* can 'OCCUR' within the ALL-IN-ALLNESS of the INFINITE SELF.

By *Pure Consciousness* is meant a great intra-Cosmic Continuity. Consciousness is a 'continuous imparticulate Sensitivity to all Cosmic change'. Consciousness is Essentially immaterial, indivisible and unitary. It is "of one piece".

Example: Consciousness in its pure State is *continuous* throughout Cosmos. Only prakritic modifications, or 'changes' in the Cosmic Configuration give to Consciousness a 'content' other than pure *seamless* Awareness.

Example: Even when Time in-Cosmos is 'frozen', the consciousness which is 'embedded' within the World of Fabrication *registers* the 'frozen' configuration, just as the consciousness of a movie-goer registers the 'frozen frame' in a motion-picture film. Consciousness being Essentially homogeneous and impartite 'contains' nothing within *it* that can 'freeze'.

Example: 'CONSCIOUSNESS' begins with the first 'AWARENESS' of the SELF by ITSELF. Through such 'AWARENESS' the SELF has 'BECOME' ITS OWN *object of perception,* i.e., the SELF has 'OBJECTIFIED' ITSELF.

Example: Consciousness in-Cosmos is an apparent continuity of objective presentation in which change (independent of space) appears as movement *through* space.

constant (noun)

By a *constant* is meant that which does not change (at least, does not change within a specified context and duration).

Example: No-thing in-Cosmos is utterly constant under all conditions in all dimensions. Even though things *appear* not to change in certain contexts or circumstances, they, in fact, do in others.

Example: Even the apparently unchanging constants established by the physical sciences are only constant (and that, not exactly, but only relatively) under certain dimensional conditions. A given constant functioning in relation to the physical plane will no longer be a dependable constant with respect to the etheric plane or other higher planes. For instance, the speed of light can no longer be used as an invariable constant in relation to those planes upon which particles (or even thought) can travel faster than light.

CONSTANT, the ABSOLUTE

By *the ABSOLUTE CONSTANT* is meant THAT which for ALL ETERNITY changes not.

Example: The ABSOLUTE CONSTANT IS constantly and only ITSELF.

Constant, the Universal

By *the Universal Constant* is meant the Fixed Design of Cosmos, the Design-at-the-Beginning, the Original Intent.

Example: The Design-at-the-Beginning for each Cosmos is the Universal Constant for that Cosmos. With the exception of the PRESENCE of the ABSOLUTE *in* Cosmos because IT (necessarily) IS Cosmos, the Design-at-the-Beginning is the most constant Relationship of factors in-Cosmos. It must be borne in mind that everything other than the ONE AND ONLY SELF is a *relationship*. Even the Universal Logos is a 'Relationship'.

CONSTANT, the UNIVERSAL

By *the UNIVERSAL CONSTANT* is meant the ABSOLUTE CONSTANT with special consideration of the fact that the ABSOLUTE CONSTANT is *constantly present in-Universe* as the SELF throughout the duration of Cosmos.

Example: The ONE AND ONLY SELF IS, of all factors-in-Universe, the *only* CONSTANT—the UNIVERSAL CONSTANT. No other factor-in-Universe is a *constant*—even the Universal Logos.

content of consciousness

By a *content of consciousness* is meant any presentation registered (consciously or unconsciously) by a perceiving consciousness. At every moment, a perceiving consciousness has its particular content derived either from a presentation which has never before been registered or from memory (which includes the impression of all past registrations). A content of consciousness may be *focal* and hence, entertained by the perceiving consciousness at the moment, or it may be *residual* and simply available to be evoked.

Example: The memories of all past lives are *residual* contents of consciousness. The memory of your one or two most pleasant present-life experiences are far more *acces-*

sible contents of consciousness, and, depending upon your desires and aspirations may even be focal contents. The awareness, for instance, of what you must do to fulfill your job (while you are "on the job") is certainly a *focal* content of consciousness. The Content of Consciousness of the Universal Logos is the sum of all the particular contents of consciousness (whether immediately accessible or less retrievable) of all authentic E/entities-in-Cosmos.

Example: Consciousness can be its own content of consciousness.

context

By a *context* is meant a particular collection of variables under consideration (i.e., an articulated wholeness) in relation to which a specified variable (or number of variables) is/are considered. A context might be called a 'reference aggregation'.

Example: The importance of any E/entity-in-Cosmos changes in relation to the context within which the E/entity is considered. A man who is very important as a father to his family in the context of his family, may be of negligible importance in the context of his nation. Of course, if this father happens also to be the president of his country, his importance in *both* contexts is significant.

Example: The *meaning* of any variable depends upon the context in relation to which it is to be understood.

continuity (noun)

By a *continuity* is meant that which persists unchanged from moment to moment.

Example: The great Archetypes of any given Cosmos are, with respect to that Cosmos, Continuities, though with respect to the ETERNAL DURATION of the ALL-SELF, these Archetypes are most definitely Dis-Continuities. One can see that although the ALL-SELF is ESSENTIALLY *out of all relation* with any other possible factor, IT must, in the World of Illusion (and because of the limitations of language) be discussed as if IT *were* in relation to other factors).

Example: Some patterns are called continuities although they persist only *relatively* unchanged from moment to moment. (For instance, a human being in incarnation, is, for practical purposes and for the duration of that incarnation, considered a continuity.) The term continuity can therefore be used either loosely or strictly. Strictly speaking, in-Cosmos, there are *no* continuities whatsoever. However, practically speaking, every relationship which is at least relatively stable within a given context can be considered a continuity.

continuous; continuously

By *continuous* is meant unchanging, or (at least for practical purposes in-Cosmos) repeated in reasonable similitude from moment to moment (whether from ultimate moment to ultimate moment, or from some larger designated unit of time to the same unit of time). The repetition should be either identical from time frame to time frame (which is, strictly speaking, impossible in-Cosmos) or *relatively identical* i.e., *acceptably similar* within a certain range of deviation.

Example: The heart usually beats continuously during the span of a human life, but the heart does not beat identically from moment to moment, nor are the moments at which it beats found at regular intervals.

Example: The medium through which a musical note is generated (such as a violin string) vibrates continuously for the duration of that note.

Example: Because the Universe is *quantized* there may be no such thing as the continuous movement of an object from position to position through all points of space. Motion in-Cosmos is *ontologically oscillatory* and, hence, discontinuous.

Example: Continuous motion in-Cosmos depends upon regularly intervening intervals of non-motion 'between' the motions. *Practical* continuity (as contrasted to *absolute* continuity) cannot exist in-Cosmos without the intervention of discontinuity. *Absolute* continuity is an impossibility in-Cosmos.

Example: In Cosmos, what are deemed 'continuities' are continuous discontinuities. Breathing is a continuous discontinuity. The Great Breath, also, is on Its own tremendous scale, a Continuous Discontinuity.

CONTINUUM, the GREAT

By *the GREAT CONTINUUM* is meant the ABSOLUTE in which the *infinitization of all possibility* is *continuously* PRESENT as the INFINITESSENCE or the NOUMENESSENCE.

Example: The GREAT CONTINUUM is the only 'THING/NON-THING' which *continues* forever as IT IS without the distraction of Time, Space or Motion. The 'CONTINUANCE' within the GREAT CONTINUUM is not from moment to moment, because there are no moments—no segmentations of INFINITE DURATION. The CONTINUANCE is *seamless* because in all this SAMENESS, it is always NOW.

CONTRADICTION, the GREAT

By *the GREAT CONTRADICTION* is meant the ABSOLUTE REALITY which ultimately defies every attempt at *'definitive'* explanation. What IT apparently IS, IT IS NOT, and WHAT IT apparently IS NOT, IT IS. IT REMAINS FOREVER WHAT IT IS BY BEING *simultaneously* both WHAT IT IS AND WHAT IT IS NOT. Ultimately, IT IS IT. Or, IS IT?

Example: The GREAT CONTRADICTION IS simultaneously what IT *IS* and what IT IS *NOT*. IT is also, simultaneously, *not*, what IT IS, and *not* what IT IS NOT. IT 'stands on both sides of the fence' and on neither. One can't talk about IT; but if one predicates that one cannot talk about IT, that predication is also false, so that one *can* talk about IT. One can both talk about IT and *not* talk about IT. This book is about the 'talking'; once it is read, the 'not talking' can begin.

Example: None. If ever words were hopelessly feeble, this is the moment.

cosmic; Cosmic

By the term *cosmic* is meant that which pertains to patterns which are larger than the pattern of our solar system.

By *Cosmic* is also meant whatever relates to, or must be considered in the context of, our particular Universe considered as a Whole—our Cosmos.

Example: Whereas the Universal Logos possesses Cosmic Omniscience, our Solar Logos, minuscule in scope compared to this Great Cosmic Being, is *working* to achieve a far smaller range of 'Omniscience'. All intra-Cosmic omniscience must be called *'omniscience with respect to a given and limited intra-Cosmic context'.*

Example: Cosmic Omniscience with respect to the All, is not super-Cosmic Omniscience with respect to the ALL.

Cosmically-Sanctioned Configurations

By *Cosmically-Sanctioned Configurations* are meant those combinations or configurations that *can* be (or, depending upon their relationship to the Original Intent, *should* be) *actualized* in a given Cosmos. The number of Cosmically-Sanctioned Configurations is far smaller (in fact, *infinitely* smaller) than the number of combinations and configurations *theoretically* possible.

If all theoretically conceivable positions-in-Space were, in fact, available for purposes of configuration (i.e., if all *relations* in-Cosmos were *actually* possible), the number of configurations would be infinite even if the number of variables were finite, for there would be an infinity of 'positions' from which all available items or E/entities-in-Cosmos could be related. (But there are only a limited number of positions-in-Space/Time available for 'particle/event occupation' during the manifestation of a given Cosmos.) This unavailability of each and every conceivable position-in-Space and Time is due to the quantizing of Space and the quantizing of Time.

Just as when a certain musical note is sounded as a "fundamental", there are only certain harmonics which sound at pitches 'higher' than the note, and any attempt to force other harmonics to emerge from 'between' *'musically-sanctioned pitches'* will be futile, so, too, with *theoretically available* positions-in-Space and with *theoretically conceivable* moments-in-Time. Not all *possible objective* 'times' and *objective* 'spaces' are available for 'particle/event-occupancy' within a given Cosmos governed by a specific Cosmic Algorithm.

We are treading here close to the Mystery of the Finite Universe, and are entering the domain of the metaphysical-physicist of the future (or, equally, the past). In the World of Being, far more can be conceived than can be *actualized* in the World of Fabrication; even so, given the limited nature of the Cosmic Algorithm, there are probably (even in that High World) an infinitude of 'inconceivabilities'.

Example: The movement of a planet which behaves like a yo-yo (rather than pursuing a normal elliptical orbit around the Sun) would produce patterns which were not Cosmically-Sanctioned Configurations. Such aberrational movements neither *can* happen (given the Cosmic Algorithm) nor *should* happen (given the Cosmic "Moral Imperative" of the Design-at-the-Beginning).

Cosmic Consciousness

By *Cosmic Consciousness* is meant that Consciousness which characterizes the Universal Logos of any particular Cosmos. Cosmic Consciousness is less than Super-Cosmic 'Consciousness' (which relates the Infinite Subject to the Infinite Object in Pre- and Post-Cosmic 'Days'), and is, regardless of its relative grandeur, *infinitely removed* from ABSOLUTE 'CONSCIOUSNESS' (which, in REALITY, *cannot exist,* except as such 'CONSCIOUSNESS' is 'included' within the INFINITESSENCE).

Example: Those human beings who say they have achieved Cosmic Consciousness have (if they are correct) achieved something which even the Solar Logos of our solar system has not achieved. Human beings are capable of achieving some measure of SELF-Realization, but totally incapable, for aeons and aeons to come, of achieving Cosmic Consciousness. The two (SELF-Realization and Cosmic Consciousness) are qualitatively and quantitatively far different. SELF-Realization concerns identification with, and as, the All-Pervading SUBJECT/SELF of Cosmos. Cosmic Consciousness requires complete and detailed knowledge of all states of consciousness (with their knowledges) in-Cosmos. SELF-Realization does not concern Time, Space, and Motion. Cosmic Consciousness is intimately related to Time, Space, and Motion.

Cosmic Now, (Cosmo-Objective Now)

By a *Cosmic Now* (or, more precisely, a Cosmo-Objective Now) is meant the immediate 'ultimate moment' for a particular Cosmos. A Cosmo-Objective Now can only occur 'on' or 'during' an ultimate moment. Cosmo-Objective Nows are necessarily *discontinuous,* but nothing in that particular Cosmos' World of Fabrication can 'happen' 'between' them (for 'between' Cosmo-Objective Nows, the Universe 'reverts' to the Cosmo-Subjective Now in the World of Being, where the virtually continuous 'Archetypal Gaze' reigns). The concept of that 'State' (the Cosmo-Subjective Now) which may lie 'between' Cosmo-Objective Nows is profoundly difficult. For all *practical* purposes, in the World of Fabrication of a given Cosmos, there are no 'between *times*' separating Cosmo-Objective Nows.

Example: The duration of any combination-in-Cosmos is quantifiable in terms of Cosmic Nows (i.e., Cosmo-Objective Nows). A long-enduring combination persists through relatively many Cosmo-Objective Nows or ultimate cosmic moments. A combination of short duration persists through relatively few Cosmo-Objective Nows. A Cosmo-Objective Now, Itself, has duration—the duration of the ultimate moment for that particular Cosmos within which the measurement is taking place. Such a 'Now' is quantifiable and has a time value greater than zero. A Cosmo-Objective Now (in a particular Cosmos with a particular Cosmic Algorithm) cannot have a duration *less* than an ultimate moment for that particular Cosmos, for then the ultimate moment would not be *ultimate*; i.e., it would be divisible. However, the time value of the ETERNAL NOW is exactly zero—having no duration at all. The ETERNAL NOW is a dimensionless *time point* upon the Infinite Time Line and 'within' the CONTINUUM of ABSOLUTE DURATION. That NOW *takes no time at all,* whereas a Cosmic Now *takes 'Time'.* Even the Cosmic Eternal Now which exists under the All-inclusive 'de-sequentializing', 'simultaneitizing' 'Gaze' of the Universal Logos (and which 'endures' as Eternally Now for the entire duration of a Cosmos) also "takes time" as measured against the Infinite Time Line.

The Cosmo-Objective 'Now' *within* any Cosmos has a time value (however 'infinitesimalizing') greater than zero. Within a Cosmos, a Cosmo-Objective Now is *not* instantaneous. Within the World of Fabrication, *'instantaneity'* is an illusion. Or shall we say, the 'instant' or ultimate moment varies from Cosmos to Cosmos. Certainly, 'instantaneity' (if applied to the duration of ultimate moments) does *not* mean "in no time at all"!

The NOW that is the ETERNAL NOW has a time value exactly equal to zero. Whereas it is *mathematically conceivable* (though *not Cosmically actualizable*) that incredibly rapid movements (of a kind generated, perhaps, heretofore in previous Cosmoses, or, to be generated in future Cosmoses with different Cosmic Algorithms) could theoretically be of the type that could 'occur during' a Cosmo-Objective Now in *our* Cosmos (an *actual* impossibility in any given Cosmos), it is *not* conceivable that any type of movement at all could either 'fit between' or 'occur during' successive ETERNAL 'NOWS' because 1), such ETERNAL NOWS are not *really* successive, i.e., they all occur at the same 'TIME'; and 2) since the duration value of an ETERNAL NOW (or, better, *the* ETERNAL NOW) is zero, there is no 'space in time' between them. Thus, we see, that since ETERNAL NOWS have no duration, *nothing* can happen 'during' them, or, paradoxically, only NOTHING (*the* NO-THING) can 'HAPPEN' 'during' them.

Example: 'Between' Cosmic Nows (Cosmo-Objective Nows) the only 'changes' which can occur are in the World of Being, Itself. Change within the World of Being (which World is a Semi-Continuity) may be capable of occurrence independently of the Fohatic fluctuations in the World of Fabrication (the World of Cosmo-Objectivity).

Cosmic Pulse, the

By *the Cosmic Pulse* is meant the recurrent 'beat' of identical ultimate moments in a given Cosmos as they appear and disappear (through Ontological Oscillation) along with all ultimate particle-events (the duration of which are identical to the duration of ultimate moments).

By *the Cosmic Pulse* (on a larger scale) may also be meant a Great Fundamental Rhythm (possibly, but not probably, variable with respect to Itself) which *beats out* the Great Key Measures (as well as the subsidiary Key Measures) which regulate the Cosmos during various phases and various levels of Its development. If change in the rate of ultimate moments *does* occur, that change of rate would be correlated with a *variable* Fundamental Rhythm of the Cosmic Pulse (considered from the *Macro* perspective).

Example: The Cosmic Pulse at the ultimate micro level beats once for every ultimate moment; at the extreme end of the Macro-Level the Cosmic Pulse may Beat but once per Cosmos—one Beat being equivalent to One Turn of the Universal Wheel. The Mystery of great and small Cosmic Cycles is found through a study of the Cosmic Pulse.

Example: The Rhythm of the Cosmic Pulse can be differentiated or subdivided into numerous simultaneously occurring correlated rhythms, each rhythm particular to a certain dimensional 'level' within the Cosmic Process.

Cosmic Simultaneity

By the term *Cosmic Simultaneity* is meant the Process of the all-inclusive, simultaneous change of *all variables* in the Great Cosmic Configuration, which change is 'witnessed' by the Cosmic Observer, the All-Seeing Eye, throughout the entire duration of Its Cosmos. The Cosmic Observer *Sees* all variables change simultaneously at each ultimate moment. Between two consecutive ultimate moments in Cosmo-Objectivity, all change (relative to the World of Fabrication) is simultaneous with all objectivities/variables in the Cosmic Configuration reconfiguring themselves in what may be (no 'time' at all). For any one Cosmos there is a *quantifiable* number of 'Frames of Perception' Witnessed by the Cosmic Observer. This number defines the duration of the World of Fabrication and is probably related to the duration of the Cosmos as a whole. The duration of a given Cosmos, though quantifiable in terms of ultimate moments, is perhaps not rigidly determined at the beginning of Cosmos. This is to say that 'extra time' may be allowed to complete the intended Cosmic Program if necessary. Since the amount of Time is infinite, why not?

With careful thought, we see, then, that the Cosmic Configuration does change *sequentially;* this is a 'horizontal' measure. As well, there is to be 'Seen' a Cosmic Simultaneity with respect to *all* cosmic movements (or, better, *changes*—for, in this case, there *may* be *change* without Real *movement)* which signal the onset of *each* ultimate moment; this is a 'vertical' measure. From the Perspective of the Universal Logos ('standing very far back', as it were) everything that 'happens' in Its Cosmos, happens at exactly the same time; however, "times change" which means that Cosmo-Objective Time changes with (and only with) each ultimate moment (even though other 'times', such as Cosmo-Subjective and Cosmo-Eternal) may be "going on at the same 'time'." Within Fabricated Cosmos (the Mosaic World of Effects) there are no lesser or more rapid time changes.

We see, then, that everything is happening simultaneously, but movement or change occurs only 'upon' or at the (perhaps ultra instantaneous or zero-time) inception of an ultimate moment. Everything that happens, happens simultaneously *in toto,* as one whole change, from moment to moment, sequentially. The sequence goes thusly: change appears/hold configuration/all disappears/hold non-configuration; change appears/hold new configuration/all disappears/hold non-configuration etc. Thus is described the *quantizing* of the Great Simultaneous (though discontinuous) *apparent* Cosmic Movement.

Example: As seen from the Perspective of the One Cosmic Observer (the Universal Logos), all possible changes in the Cosmic Configuration that occur at (the virtual zero-time inception) of a given ultimate moment, occur simultaneously and thus participate in a pan-<u>Cosmic Simultaneity</u> which we are calling the <u>Cosmic Simultaneity</u>. The total particularity of all changes occurring in the <u>Cosmic Simultaneity</u> can only be cognized by the Universal Logos.

Cosmic Unit, Authentic

By an *Authentic Cosmic Unit* is not necessarily meant a Universal Life Unit (or a 'Ray' of the Ultimate Cosmic Monad). An Authentic Cosmic Unit is an ultimate particle/event, an objectified particularization of Self-Enumerated Fohat.

Example: Authentic Cosmic Units may not be 'Monadically Entified' (i.e., they may not be one of the pre-determined number of 'Rays' of the ABSOLUTE apparently differentiable at any one period of Cosmos, and forming the Emanatory 'Branches' of the Emanational Tree). Instead, they are 'Fohatically Entified' (which, of course, means that, ultimately, they are Cosmic-Monadically Entified). Ultimate particle/events are the least of all forms in-Cosmos and should be considered Authentic Cosmic Units, even while not necessarily 'hosting' a *direct incarnation* (a 'Ray') of the Ultimate Monad.

From another perspective however, can there be any 'thing' in Cosmos which is not a 'Ray' of the Universal Logos (the One Universal 'Ray')? All 'Rays' are the One 'Ray' *entering* its emanated Self-Objectifications. Since Fohat is a Self-Objectification (of either the Father or the Son or both) the Self-Objectifications of Fohat are, Really, the Self-Objectifications of the One Universal Subject (the One Universal 'Ray').

cosmify; cosmification

By *cosmification* is meant the act by which any Being/being is 'immersed' in Cosmic Prakriti.

Example: At the Beginning, I-THE-INFINITE SELF, 'COSMIFIED' MYSELF and 'BECAME' the SELF-in-Cosmos, or, more properly, the Cosmic Self—the Universal Logos.

Example: The *possibilities* inherent within the INFINITESSENCE are 'EXTRUDED', articulated and cosmified throughout an infinite sequences of Universes.

Example: Cosmification is limitation.

cosmogonical

By the term *cosmogonical* is meant that which relates *first* to the Birth of Cosmos and *then* to the Birth of the Gods within Cosmos, and, further, to the birth of all other E/entities which emanate from the Gods.

Example: *The Secret Doctrine* is one of the foremost cosmogonical texts in the modern library of Philosophical Occultism.

Example: The Cosmogonical Problem is extremely complex because we are uncertain concerning:

1. The number of dimensions in-Cosmos (—there are many).
2. The number of beings which manifest upon those levels (—their number is legion and we really have no even semi-accurate idea).
3. The relative magnitude of the manifesting beings (—our range of comparison is very narrow since we are familiar with so few E/entities).
4. The order in which the manifesting beings appeared (—all this is heavily blinded in the Occult Literature).

Cosmogonical Sequence, the

By *the Cosmogonical Sequence* is meant the Order of the Birth of the G/gods.

Example: The Gods are the Fundamental Numbers, and They, naturally, emerge from Their Sources in "numerical order". Yet the manner in which the Cosmogonical

Sequence is Generated is a subject fraught with difficulty, for there are a number of *ways* that this *coming into being* of the Gods might occur, depending upon which kind of Path is the *correct* Emanational Path in our Cosmos. At various levels of Cosmos there may be differently branching Emanational Paths and the Cosmogonical Sequence may become most complex as the *birthing process* becomes more 'distant' from the Apex—i.e., the Universal Logos, the Cosmic Monad, the Number One.

cosmology; cosmological

By *cosmology* is meant the study of the nature and functioning of Cosmos.

By *cosmological* is meant pertaining to the knowledge of the Cosmos, of Its Origin, Its functioning, and Its dissolution.

Example: Deep thinking along metaphysical lines reveals many unsuspected cosmological problems. Cosmos is far stranger and more intricate than at first appears. Cosmological problems can be solved in a number of ways. The Philosophical Solution is one way, the Scientific Solution another. Philosophy blazes the trail to Truth through wide speculation upon the possibilities; science verifies those possibilities and confirms Cosmic Truth. Cosmology reveals the relation and value of the part to the Whole, and the Whole to the part. Some cosmologists are both philosophers and scientists.

Cosmo-Narcissism

By *Cosmo-Narcissism* is meant the reflexive Self-Perceptual Process by means of which the Cosmos is 'Created and 'Enjoyed'.

Example: The Cosmic Drama is an intensive Self-Preoccupation on the part of a Self-Admiring Universal Logos. Can we call this Process Cosmo-Narcissism?

Cosmos

By *Cosmos* is meant the limited Finite Universe considered as a 'RADIATION' (more than an Emanation) of the BOUNDLESS IMMUTABLE PRINCIPLE.

Example: A given Cosmos is one of a beginningless/endless series of *primary* 'RAYS' of the ABSOLUTE.

By *Cosmos* is meant the necessary Contradiction to INFINITUDE.

Example: Cosmos contains explicitly an infinitesimalizing 'portion' of that which is *implicate* within the NOUMENESSENCE, the INFINITESSENCE. The 'portion' expressed *as* Cosmos and *in-Cosmos* contradicts, simply because It is *Finite*, the necessarily unvarying INFINITE HOMOGENEITY of the INFINITUDE SELF. That which is Finite is apparently totally 'other' than the INFINITE and infinitely removed from IT.

In one way Cosmos is *not* a Contradiction to the INFINITE SELF for Cosmos expresses (though in infinitely limited degree) the 'POSSIBILITIES' which 'INHERE' within the INFINITESSENCE. In another way, because of Its Finiteness and, consequent Imperfection, Cosmos is the very antithesis of the ABSOLUTE, the INFINITE SELF.

Cosmos/Event

By a *Cosmos/Event* is meant the appearance of any Cosmos considered as a unitary Event within the UTTER ALLNESS.

Example: Against the ongoing background of NOTHINGNESS, the periodic appearances of Cosmoses are Events of the greatest possible magnitude, unless the appearance of all of an infinitude of Cosmoses along the Infinite Time Line is considered a *single* Cosmos/Event, which from the Infinispective of the Infinite Subject/Self it, perhaps, *could* be? But can an infinitude of Cosmos/Events be encapsulated as a Single Event?

Cosmos, Super-

By *Super-Cosmos* is meant both Pre-Cosmos and Post-Cosmos.

Example: The change from the Infinified Point, via the Condensing Point to the Condensed Point takes place in Super-Cosmos, not in Cosmos, Itself.

Cosmos, SUPER-

By *SUPER-Cosmos* is meant the INFINITE SELF, the ABSOLUTE.

Example: 'POINTNESS' 'ARISES' in SUPER-Cosmos, but instantaneously becomes 'Pointness' in Super-Cosmos.

- D -

denotable (noun)

By a *denotable* is meant an object or Object capable of being identified by a registering self-conscious consciousness (whether the Consciousness of the Universal Logos or some lesser self-conscious consciousness).

Example: Not only authentic E/entities, but even evanescent 'accidental' configurations can be considered denotables. In fact, at some level or other, *notice* is taken of every configuration that occurs in-Cosmos.

Example: From a Super-Cosmic Perspective, the First Denotable is Mulaprakriti, the First Object, the Only *Infinite* Object.

Example: Denotables can be subjects as well as objects. A 'Seen' Subject becomes an Object.

Design-at-the-Beginning

By the *Design-at-the-Beginning* is meant the Purpose for a given Cosmos as conceived by the Universal Logos which **8** Am, and which every E/entity-as-**8** Is. That Design is only One Possibility drawn from the FOUNT OF ALL POSSIBILITY—the INFI-

NITE SELF. The fulfillment of the Design is the Principal Purpose for the manifestation of that particular Cosmos.

Example: The Design-at-the-Beginning is a Fixed Design, but also a Dynamic Design, 'pre-programmed' to unfold in Time and Space through Motion according to certain Pre-Cosmically Ordained Parameters. This unfoldment involves the Fohatic Worlds and also the World of Being in ways that are difficult to conceive. The unfoldment proceeds according to Plan, the Divine Plan. The Design emphasizes *fixity of relationship*, and the Plan emphasizes the dynamic means by which the Design is actualized in Time and Space. However it may be, *imparticulate ideational change* signals *particulate Mosaic approximation*.

Example: All Good and Evil within Cosmos are determined with reference to the Design-at-the-Beginning. The Design-at-the-Beginning is the Moral Arbiter of all Pattern in-Cosmos.

Design, the Grand

By *the Grand Design* is meant the beautiful, vast entirety of Cosmic-Logoically-Sanctioned Patterns and Relationships in-Cosmos.

Example: Every human being becomes filled with vitality and aspiration when understanding his role as an authentic Identity within the Grand Design. The realized Beauty of the Grand Design subdues irritation (Imperil) and induces gratitude and appreciation in the one who so realizes.

different; difference

The word *different* is used to describe two actualities such that, in a comparison of the two, the wholeness of each actuality and each constituent part of each actuality are in at least one respect *incapable* of being interchanged, superimposed, or substituted the one for the other.

Example: Every atom of any particular kind, though ostensibly identical with all other atoms of that kind, is, in fact, different from every other such atom. Because of the Principle of Unrepeatability, no E/entity of a given class of E/entities can possibly be identical with any other member of that class, though it can be *similar* to another member of that class—i.e., alike in most respects. This alikeness vanishes if two such entities as atoms, for instance, are viewed from the maximally microscopic point possible in a given Cosmos.

Example: Even two ultimate particle/events are necessarily different from each other, for each Cosmo-Prakritically-reflected subdivision of intra-Cosmic Fohat (which an ultimate particle-event is) is distinct from every other such, being virtually identical with that other but not *absolutely* identical. The very fact that two things can be compared (i.e., that they are *two* things and not *one* thing) means that they are different (i.e., they are (formally, at least) *not* each other. Each ultimate particle-event having a "will of its own" also ensures that it is different from every other ultimate particle/event. Subdivision creates distinct identifiability which is difference.

dimension

By a *dimension* is meant any vibrationally-caused, particularized state of awareness or activity that is a *permanent-in-Cosmos* structural aspect of the particular Cosmos in question.

By a *dimension* is meant one of a finite number of *vibratory domains* in-Cosmos.

Example: Each great Plane of Nature or Super-Nature (Systemic, Cosmic and Super-Cosmic {or Kosmic}) is a dimension in our Universe.

Example: All dimensions interpenetrate in Space. There is no one 'place' where a dimension *is* or *is not.* But there are 'spheres of Self-Perception' 'where' a dimension *is* or *is not.*

Example: Solar-systemic sub-planes can also be identified as dimensions, for certain 'major' planes can be considered sub-planes from a certain perspective. For instance, a systemic plane is a sub-plane with respect to a Cosmic Plane, and a Cosmic Plane is a sub-plane to a Super-Cosmic Plane.

dimensional range

By *dimensional range* is meant that span of dimensions or sub-dimensions over which and within which a particular E/entity is conscious and/or active.

Example: The potential dimensional range of a human being-as-human being is the six lower planes of the Solar Systemic Plane or, equally, the six lower subplanes of the Cosmic Physical Plane. The *actual* dimensional range of most human beings is the lower eighteen sub-subplanes of the Cosmic Physical Plane. The dimensional range of the Universal Logos is *all* planes and sub-planes in Cosmos. This vast range is likewise the goal of all B/beings-in-Cosmos, and in the highest Aspect of their in-Universe Selfhood, they have, throughout the entire duration of this Cosmos, been endowed with this ultimate range through pervasion. This *fact* has to be made conscious through the internal faculty called *Antahkarana.*

dimensional sealing (vibrational occlusion)

By *dimensional sealing* is meant the process by which a certain number of an E/entity's various dimensional foci (i.e., Fields of Absorption) along its Emanatory Stream are prevented from *consciously* pervading or interpenetrating each other, with the result that the knowledge and consciousness possessed by an E/entity in one dimensional focus is not available to that same E/entity in another (usually higher, but, sometimes, lower) dimensional focus.

Example: The consciousness and activity of an E/entity on any particular plane or subplane of Cosmos determines its dimensional focus. What **8** *even now know* as a Monad, in My dimensional focus upon the Monadic Plane, is not readily 'known' to me as a human being in my customary dimensional focus upon the mental, astral, and etheric/physical planes.

Extending the idea, **8** Am (in My higher dimensional foci) also a participant in the domain of the Solar Logos (for the Monad is "that which finds its home within the Sun", *Initiation, Human and Solar,* by A. A. Bailey, page 196) and a participant, even, in the

Council of the One About Whom Naught May Be Said (a relatively stupendous but nevertheless cosmically tiny Being). Why am I, as a human being, a member of the Fourth Kingdom of Nature on this planet, the Earth, *not* aware of these necessary participations (multi-dimensional foci) along 'My!' Emanatory Stream, and of even higher participations—all logically necessary given the Theory of Emanation? The answer is 'dimensional sealing' or 'vibratory occlusion'. Dimensional sealing is part of Maya's Veiling Process, and is one of the chief means of enforcing a necessary SELF-'LIMITATION' in-Cosmos.

Dis-Continuum, the

By *the Dis-Continuum* is meant the Domain of all possibilities 'EXTRUDED' from the INFINITESSENCE. None of these possibilities can be Present-in-Cosmos *continuously,* nor can Cosmos, Itself, (especially in certain of Its Aspects) persist continuously without interruption (persist uninterrupted even for the duration of two sequential ultimate moments). Not only does the 'RAY' 'FLASH FORTH' at the first moment of 'EXTRUSION' of SELF-INHERENT 'POSSIBILITY' from the SELF, but the Universe and all Its contents continue to flash 'in' and 'out' of objective existence. Remember that *all* existence is *objective.*

Example: By its very Nature the Dis-Continuum (Which Is the Cosmos) must be *discontinuous,* the oscillation between the Subjective and Objective being perpetual in-Cosmos. No presence of a form in-Cosmos can be perpetual for the duration of a Cosmos, because the oscillatory *movement* (i.e., Ontological Oscillation), defined by the existence/non-existence/existence/non-existence of the Cosmos' entire Form occurs throughout the duration of that Cosmos, thus forcing all lesser forms in-Cosmos to conform to this movement.

During the *Subjective* side of the Oscillation, the Objective Universe 'disappears' into deeply recessed Subjectivity, in which Spirit is disengaged from its Object with which It is involved (thus returning instantaneously to '**8**-ness', and Fohat is disengaged from Its particulate reflections). During the Objective phase of the Oscillation, the Universe is 'reposited' in its next Cosmo-Objective Configuration and Spirit (at all levels of prakritic immersion) is re-engaged with the objects appropriate to the level of immersion. Between Cosmic Configurations the Universe 'disappears' into Cosmic 'No-Thingness' (a 'State' of Cosmo-Subjectivity which is far more Ideationally Real than anything found in the Realm of Cosmo-Objectivity).

Example: Fohat 'Blinks' Its Eye to create the Dis-Continuum—Discontinuous Space; Fohat 'Blinks' Its Eye to quantize Time (rendering it discontinuous) and thus ensure that the Universe does not last forever.

Dissonation

By *dissonation'* is meant a kind of mutual interaction that renders mutually engaged frequencies more dissonant, out-of-tune, etc. than heretofore.

Example: Interaction between those who are upon the First Ray and the Second Rays is often a dissonation leading to mutual discomfort.

Duality, the Primal Pre-Cosmic (and Post-Cosmic)

By *the Primal Pre-Cosmic (and Post-Cosmic)* Duality is meant the initial 'DIVISION' of the INFINITE SUBJECTIVITY (i.e., the ABSOLUTE) into the Infinite Subject and the Infinite Object. The Infinite Subject is the SELF-as-'Infinified' (Absolutely Omnipresent) Point; the Infinite Object is Mulaprakriti.

Example: In the dynamics of the Primal Pre-Cosmic (and Post-Cosmic) Duality, the Infinite Subject 'Sees'/Registers the Infinite Object in a State of complete, undifferentiated Homogeneity. At first there is no possible Variety within Mulaprakriti, because Variety (the Seed of Differentiation) has not yet been 'transferred' from the Infinite Subject (where Variety abides in non-objectivity) to the Infinite Object (Mulaprakriti) where that Variety will be objectively reflected and, hence, expressed.

duration

By the term *duration* is meant a measure of *presence* and/or *absence.*

Example: The duration of a human being upon the systemic physical plane is usually less than one hundred earth years. Though the physical form die, decay, and vanish from the physical plane, and the human entity be *absent* from that lowest plane, nevertheless, the entity endures (i.e., has duration) as a living being upon other, more subtle planes, where it is very *present.*

By the term *duration* is meant a measure of *the length of presentation* to a registering consciousness.

Example: For the human being, most non-microscopic perceptions seem to achieve a degree of relatively continuous sameness because the duration of the microscopic events which compose/are the object being perceived is far too rapid to be registered by the gross human senses. For instance, the present form of the teacup before me has an apparently unchanging duration because the duration of the atomic and molecular events within it are so fleeting as to be unnoticeable to the human eye.

By the term *duration* is indicated how long a thing *lasts*, always relative to the *lasting* of another thing (the latter *lasting* being taken as a *standard* against which to measure). The ultimate standard for measurement within any Cosmos, is the ultimate moment in that particular Cosmos. On the Macro-Scale, another Standard of Measurement might be discovered.

Example: The duration of even rather fleeting events (humanly considered) is stupendously large when compared with the duration of an ultimate moment in-Cosmos.

Example: When considering the infinite succession of Cosmoses and the duration of not only the intervals between Them but of the ultimate moments within Them, there may be no unit of Infinite Time which remains absolutely constant throughout, for the duration of ultimate moments may *change* from Cosmos to Cosmos (and maybe even within a given Cosmos) and even the intervals between Cosmoses is not assuredly regular—though WHO is the 'OBSERVER' to pronounce with accuracy upon this issue? If only 'WHO' could 'SEE'!

Duration, Eternal

By *Eternal Duration* is meant the entire Time Span of a given Cosmos.

Example: The duration of our Cosmos can justly be called Eternal Duration though, for human beings upon the Earth, Eternal Duration must necessarily be measurable in terms of recognizable units of time such the second, the day, or the year. To Beings with greater Cosmic Understanding Eternal Duration can be measured in terms of more fundamental Universal Cycles of which we human beings have, as yet, no accurate notion.

Duration, Infinite

By *Infinite Duration* is meant the beginningless, endless (though necessarily *intermittent*) apparent 'flow' of Time as registered by THAT which is the WITNESS/SUBSTANCE of all Universes and of all intervening periods of NOTHINGNESS.

Example: An infinitude of Cosmoses have appeared and disappeared throughout Infinite Duration. During Infinite Duration an infinite number of intra-Cosmic and inter-Cosmic Time Periods has elapsed. Paradoxically, all Cosmoses past have *already* "taken forever" to exist. This foreverness of alternating Time Periods is Infinite Duration. A foreverness of NOTHING, the VOID, is *INFINITE* DURATION.

Example: Time and Space appear periodically throughout Infinite Duration. Time and Space *never* appear in the *INFINITE* DURATION of the BOUNDLESS HOMOGENEITY.

Example: If the INFINITE SELF is not the WITNESS of the Infinitude of Cosmoses which have appeared and disappeared in Infinite Duration, then there remains no other possible *witness,* and the Infinitude of Cosmoses have appeared and disappeared *unwitnessed.* which is an impossibility, for only that *can exist* which in some manner is *witnessed. Witnessing* itself (on some level or other) *creates* objectivity. If the EMERGENT 'POINT' is 'UN-WITNESSED' or 'UN-SEEN', how can 'IT' exist? Can 'UN-SEEN' 'SELF-SEEING' exist?

DURATION, INFINITE; DURATION, ETERNAL

By *INFINITE DURATION* is meant *indivisible foreverness.* No 'Time' can enter INFINITE DURATION, for all of INFINITE DURATION consists of but ONE *unbounded* 'MOMENT'—the ETERNAL NOW.

Example: The BOUNDLESS IMMUTABLE PRINCIPLE has INFINITE DURATION. IT is undisturbed by the periodically appearing Illusion of Time.

Example: ETERNAL DURATION takes infinitely more *time* than Eternal Duration, and yet, takes "no time at all."

- E -

ego

By *ego* is usually meant one's sense of identity, which is that presentation-to-consciousness unto which the inherent 'I Am' affixes itself. The term *ego* usually connotes a limitation which ignorance places upon true *essential identity* such that the *identity* confuses that which it *has,* or that which it *sees,* with that which it *is.* By *essential identity,* in this context, is meant either ESSENTIAL IDENTITY (SUPER-Cosmic) or Essential Identity (Super-Cosmic and intra-Cosmic).

Example: His inflated ego seems to arise from the generally acknowledged excellence of his professional performance. He is so identified with 'being' the excellent performer that he has forgotten how to see more deeply into his true BEING/Being, and has, thus, lost touch with his deeper, spiritual IDENTITY/Identity.

EGOLESS SELF, the

By *the EGOLESS SELF* is meant the forever-uncontainable ALL-SELF. This SELF cannot be identified in any *particular* way, and is incapable of being bounded in any 'Ego-Egg' of limitation.

Example: Those who experience the VOID find their normal ego identifications dissolved into NOTHINGNESS. They still have IDENTITY, but no *particular* identity in a lesser sense. They find themselves to *BE*, ESSENTIALLY, the EGOLESS SELF.

emanation

By *emanation* is meant the extension of the influence of an originating, authentic source-E/entity-in-Cosmos of a given quality, in such a manner that the extension of influence becomes itself an authentic E/entity-in-Cosmos similar to the source-E/entity, and characterized by an attenuated degree of the quality and/or intensity of the source E/entity.

Depending upon whether one takes a relative or an *essence-focused* point of view, the E/entity-in-Cosmos formed by the extension of influence of the source-E/entity will be possessed of either an *attenuation* of the *essence* of the source-E/entity or the *whole* of the *essence* of the source-E/entity. While the E/entity that is the extension of the source-E/entity may or may not be seen as possessed of the *whole* of the *essence* of the source-E/entity, it will *not* be possessed of the *whole* of the *quality* and/intensity/scope/depth of the source E/entity. Quality, intensity, scope, and depth undergo attenuation through emanation. Though quality undergoes attenuation, the quality of the emanation is, nevertheless, *reflective of* the quality of its source. The quality of the emanated pattern is reduced in scope and intensity, but the quality of the emanation still remains a replica (fainter and ever fainter as emanations proceed) of the quality of its source-E/entity.

Example: Since Monads within the Earth Scheme are considered "cells" within the body of a particular Heavenly Man, then, for any Monad, a particular Heavenly Man can be considered its emanating Source, and the Monad, itself, can be considered an emanation. From a still more Essential (and mysterious) point of view, a Monad may be seen as an emanation of the Solar Logos, for the Monad is "that which finds its home within the Sun." The *true* ORIGIN of the Monad-as-MONAD is found in the ABSOLUTE SELF, for the Monad is an (*apparently* distinct) 'Ray' of the ABSOLUTE which is identical to the One Universal 'Ray', which, Itself, is a 'Ray' of the ONE 'RAY' of the ABSOLUTE.

Example: Ultimately considered, all authentic E/entities-in-Cosmos are E/emanations of the One Universal Identity, which Itself, is a 'RADIATION' of the ONE ABSOLUTE IDENTITY.

Emanation

By an *Emanation* is meant one of the Essential Numerical Beings in-Cosmos.

Example: The Numbers One through Nine (possibly Ten, as well, depending upon the point of view) are the Principal Emanations in-Cosmos. All other emanations are derivative of these Principal Emanations.

Example: An Emanation is simply an emanation of a higher order—occupying a dimensional position along the Divine Emanatory Stream closer to the Source of all Cosmic E/emanation. Remember that the Source of all E/emanations is *not* the SOURCE of ALL. From THAT SOURCE 'COMES FORTH' only a 'RADIATION' and *not* any form of *emanation.* (This point, H. P. Blavatsky especially emphasizes in *The Secret Doctrine.)*

Emanations-in-Combination

By the term *Emanations-in-Combination* are meant those Beings Who Are the First *Group* Entities. In a way, all Numbers other than Number One are Group Entities. Even the Number One is not *self-subsistent.*

Example: The Numbers One through Nine (possibly including Ten) are all Principal Entities in-Cosmos. They are considered fundamental to all other and more complex Cosmic Combinations, and all such complex Combinations can be *reduced* to one of these Nine. It is important to realize that all Numbers other than Number One (which is a special case) are Really Composite Entities—Emanations-in-Combination. [See "On Emanations-in-Combination" in Section I.]

Emanative Loss

By *Emanative Loss* is meant the degree to which that which is emanated becomes more circumscribed and less intense than that from which it is emanated. In the World of Becoming this emanative dynamic *seems* to occur, though, logically, the Principle of the Indivisibility of the SELF countervenes the possibility of such loss. According to the Principle of the Indivisibility of the SELF, everything is *fully* everything else, regardless of apparent differences. Naturally such a Law runs counter to common sense, but becomes logically necessary with the acceptance of the BOUNDLESS IMMUTABLE PRINCIPLE and ITS indivisibility.

Example: The Universal Logos is the *crowning* Emanative Source in-Cosmos. All other beings-in-Cosmos are derivative of this Source and are lesser in scope than this

Source (though not lesser in *essence*). The reason for this reduction is scope is the dynamic of Emanative Loss. Emanative Loss occurs through intra-Cosmic Self-Veiling, and 'dimensional sealing'. We shall regain the Emanative Loss (that which was 'lost' to 'us' of scope and power during Emanative Descent) as we and all other E/entities-in-Cosmos are *retracted* along our Emanatory Stream during the process of Cosmic Evolution. Abstraction leads to restoration. Emanative Loss can be likened to the "Lost Word".

Emanative Retention

By *Emanative Retention* is meant the Process (dependent upon the Principle of the Indivisibility of the INFINITE SELF) by means of which that which is emanated loses *nothing* of the ESSENTIAL NATURE possessed by that from which it was emanated, even though, to all appearances in the World of Becoming, the power of that which is emanated is reduced or circumscribed compared to the power of that from which it was emanated. Subscribing to this theory yields the paradoxical thought that the Entity at the peak of the Divine Emanatory Stream is, in REALITY, *absolutely identical* with the least emanated entity—all appearances notwithstanding. This identicalness pertains, naturally, to ESSENCE rather than scope in form.

Example: If the ABSOLUTE DEITY, the SELF, is OMNIPRESENT, OMNISCIENT, and OMNIPOTENT at all 'points' in Space, then Emanative Loss is an Illusion and Emanative Retention is the REALITY. In this case the *absolute fullness* of the Emanative Source *must* be passed down along the Divine Emanatory Stream to each apparently lesser emanation, simply because no E/entity-in-Cosmos (all being *essentially identical* with the INDIVISIBLE BOUNDLESS SELF) can *divide itself.*

This simply means that the FULLNESS of the ABSOLUTE DEITY is present in the tiny ant, equally with the Solar Logos—that GOD-the-SELF is maximally *present* and *potent* at all points in Space. **8** Am *fully* (not just *partially*) PRESENT everywhere. This is another way of saying that all things, regardless of appearances, *are* the SAMENESS and are thus, *identical.* When we think carefully about *indivisibility*, Emanative Loss and Emanative Retention, we can see clearly how very much we live (or seem to live) in the World of Illusion.

By *Emanative Retention* may also mean the fact that every Emanative Source remains wholly and entirely Itself upon Its own level, even though IT seem to sub-divide into various lesser emanations.

Example: I-as-8 (the Universal Logos), through Emanative Retention, remain exactly as **8** Am, inviolate and undiminished, even though **8** send forth from 'MySelf' as Emanations the Fundamental Numbers. The Fourness and Fiveness which come (*Essentially* though, perhaps, *indirectly*) from Me, do not diminish in any way my continuing Oneness.

Further, according to the Law of Indivisibility, My Oneness is *totally* and *completely* within My Fourness and Fiveness. In short, the Fullness of God is Present everywhere and in everything; even more—the *FULLNESS* of GOD is PRESENT everywhere and in everything.

Emanatory Sequence

By the *Emanatory Sequence* is meant the *lawful* and *ordered* Progression of Emanations by which the Cosmic Process of Entification proceeds.

Example: Within the Emanatory Sequence of this Cosmos, the Number Five (i.e., the Relationship of Fiveness) cannot proceed directly from the Number Three (Threeness), before the Number Four (Fourness) has been generated. This would be so even if Group Entities (such as compose of the Numbers One, Two, and Three) were, together, the Emanative Source. Numbers, no matter how they are emanated, remain sequential.

Emanatory Stream

By an *Emanatory Stream* is meant the line of descent (and *possibly* reascent) along which **8** as One *apparently distinct* 'Ray' of the Universal 'Ray' descend, and, in so doing, appear (in greater and lesser scope) upon the various dimensional levels in the Cosmic Whole. The One Universal 'Ray' (which **8** Really Am, and not a *seeming* lesser 'Ray' at all) descends (through sequential, *apparent* Self-multiplication) into multiple levels of manifestation along diverse Emanatory Streams, and each *apparently* diversified 'Ray' may well re-ascend along the same stream from which It 'emerged' into *apparent* distinctness.

Example: **8** have descended along a specific Emanatory Stream the Source of which is "lost in the Night of Time." This Stream has Dimensional Nodes or Foci of which **8**-as-Man Am totally unaware. **8** have passed the lowest point of my descent (in this Solar System, presumably), and am in process of Solar Systemic Retraction (and possibly, *Cosmic Retraction*). But though **8** seem to be only here 'below', **8**, previously, *remained* at each Dimensional Node or Dimensional Focus even as **8** continued to descend.

Thus, even now, **8** Am multi-dimensionally Present at every Dimensional Node or Focus along my entire Emanatory Stream. Through meditation **8** Am wearing away the illusory isolative effects of 'dimensional sealing' and thus My 'dimensional range' is extending. 'Above' the 'Ray' **8** *seem* to be, are 'Rays' of greater scope by far, which always **8** have been, even while, apparently, **8** also have been manifesting as the 'Rayed-extensions' of these seemingly greater 'Rays'.

Example: My Emanatory Stream is ultimately unique and thus, in some ways, different from 'Your' Emanatory Stream but they both have their Point of Origin within the Universal Logos. Besides, 'Your' Emanatory Stream is 'My' Emanatory Stream, Essentially.

energy; Energy

By *energy* is meant the PRESENCE-as-Motion of the ABSOLUTE in-Cosmos.

By *Energy* is meant That which sustains the Presence of Cosmos. Without Energy the Cosmos would vanish into NOTHINGNESS.

By *energy* is meant the power to generate motion.

By *energy* is meant the Will which *moves.*

By *energy* may be meant *fire.*

By an *energy* is also meant, a particular *pattern* of motion or *quality* of motion (for instance, the energy of Aries, the energy of Taurus, of Mercury, of Saturn, etc.—all of these energies capable of being resolved into certain patterns, qualities,or modes of motion).

Example: The etheric body provides the dense physical body with the energy to move through space.

Example: As energies are vibratory/periodic motions within a given medium, the nature of the energy can be determined by understanding what vibrates, how often it vibrates, at what amplitude (or degree of intensity) it vibrates.

Example: When energy vanishes all motion ceases. What does this say of the ultimate particle/event at the ultimate moment? Motion may cease, as well, through the power of the Will to 'Hold'.

Example: The theory of Ontological Oscillation implies that all of the Fohatically Fabricated Cosmos becomes devoid of Energy between ultimate moments. If the World of Approximation in Cosmos disappears between ultimate moments, all Energy which sustains that World has been 'ulti-momentarily' withdrawn.

enumerable (noun)

By an *enumerable* is meant anything (whether E/entity or artifact) that can be counted.

Example: To find out how many 'things', 'items', or 'events' exist in-Cosmos, one would have to count all enumerables. Enumerables come in different classes:

- some are permanent-in-Cosmos
- others are as evanescent as an 'ultimate moment'
- and yet others may be 'accidental' and 'artifacts of association'

Example: Given that each Cosmos is of limited duration, the number of *actual,* i.e., *manifested* enumerables in that Cosmos, though vast, is limited. The number of *potential* in any Cosmos is infinite.

entification

By *entification* is meant the generation of *apparently* separate and individual E/entities through the process of Emanation. All E/entities are, of course the ONE ENTITY/NON-ENTITY, but there could be no Cosmos without the necessary illusion of *'emanative SELF-'DIVISION'* (a SELF-'DIVISION', *apparent* but not REAL) that makes relationship possible.

By *entification* is meant the generation of 'fresh', 'new' apparently distinct authentic E/entities from the One Cosmic Identity.

Example: The Emanatory Sequence as a result of which the seven are generated by the three, is really a process of entification.

Example: By the process of entification the huge number of Monads in-Cosmos is generated from the One Great Universal Identity, Itself but a 'RADIATION' of the ONLY IDENTITY.

Example: For a more complete understanding of cosmology and cosmogony, it would be important to know whether the process of entification in-Cosmos is completed during the *initial phases* of a Cosmos, or whether it proceeds in phases during the development of Cosmos. In other words, do all apparently particulate Subjective Units of Life exist as such at the 'Beginning Times' of any Cosmos, or are they generated through Emanation? It is most probable that all apparently particulate Subjective Units are, Really, but One Universal Subject, and that the apparent *particulation* or *diversification* proceeds gradually and sequentially, i.e., *emanatively.* We might say that Early Cosmos is *simpler* and has not so many 'Self-Seen' Monad/Points in it.

E/entity

By the generic term *E/entity* is meant any of a variety of conscious and unconscious beings-in-Cosmos, whether considered individually or in aggregation. There are a wide variety of E/entities that can be considered subjectively/psychologically or more objectively/formally; as well, E/entities can be ensouled or un-ensouled (depending, of course, upon how we define 'soul'). A *true* E/entity, an *authentic* E/entity, is a *subjective* E/entity—a 'Ray' of the ABSOLUTE SUBJECTIVITY. However, relatively coherent aggregates of more objective forces are often called *entities* if these aggregates have well-integrated, distinct and relatively persistent qualities and patterns. Entities seem to be relatively autonomous and self-determining (if conscious) and if unconscious, seem to follow a relatively consistent direction, consistent with their quality and pattern.

By the term *E/entity* is also meant a relatively abiding *relationship* between B/beings-in-Cosmos (subjective beings who are manifesting objectively). Any of the human vehicles represents such a relationship. Various groupings of human beings, such as communities, states, nations, etc. can also be E/entities.

Example: The number of SELF-'RADIATED' E/entities in a given Cosmos is numerically pre-determined and cannot be changed during the Manifestation of a Cosmos, but the number *of other kinds of entities* is *not* predetermined, though *finite.* Free-will makes the number of non-authentic entities indeterminate.

Example: The term entity has both subjective and objective meanings. On one extreme, the *subjective,* the 'Rays' (derivative from the 'RAY') of the ABSOLUTE are called E/entities. On the other extreme, the *objective,* various abiding and semi-abiding prakritic arrangements (whether or not ensouled by REAL E/entities) are also called entities. For instance, some thoughtforms that have reasonable coherence and seem to act with a "will of their own" are also called entities.

E/entity, authentic

By an *authentic E/entity* is meant a *primary* E/entity—an E/entity-in-Cosmos to which the One Great Identity in Cosmos, the Universal Logos, is 'downwardly' connected (by *identicalness)* by *ramifying emanative descent* just as the E/entity is 'upwardly' reconnected to the Logos by *converging abstractive ascent.* Thus 'Rays' emerge from and are re-absorbed into 'Rays', all of which are Essentially the One 'Ray' of the Ultimate Cosmic Monad, which Is, Itself, the One Cosmic 'Ray' of the 'RAY' of the ABSOLUTE.

Thus, all authentic E/entities are ultimately revealed to be *essentially* 'Rays' of the ABSOLUTE, and have unconditional *permanence* in-Cosmos (not in separate distinctness, but as the One Universal 'Ray', which, Itself, is a 'Ray' of the 'RAY' of the ABSOLUTE). Secondary and tertiary entities are *created* in various ways by authentic (i.e., primary) E/entities. Secondary and tertiary entities are not necessarily found (*structurally* and *directly*) within the Subjective Design created/unfolded by the Emanative Hierarchy, the Apex of which is the Universal Logos. Secondary and tertiary entities are not, per se, 'Rays' of the ABSOLUTE (though, in a way, *all* within Cosmos cannot escape being the 'Ray' of the One Universal Logos). When primary or authentic E/entities emanate 'other' 'Rays', such 'Rays' are also authentic E/entities. On the macro-scale, any of the Great Logoi are such *emanatively Intended* 'Rays' of the One 'Ray' and, thus, are authentic E/entities; on the micro-scale the tiny atoms of matter are similarly authentic entities.

By an *authentic E/entity* is meant, *ultimately* (and strangely), NOTHINGNESS 'REFLECTING' ITSELF. All such E/entities in Cosmos are ESSENTIALLY/Essentially SELF-as-Self-Reflections. As pure SELFHOOD is, ESSENTIALLY, *vacuous, nothing,* then such E/entities are Self-Reflections of NOTHINGNESS-as-Nothingness, i.e., Essential *Subjects.* SUBJECTIVITY is *pure* NOTHINGNESS. Subjectivity is *virtual* Nothingness; It is *virtual* because all but the ONE SUBJECTIVITY can be *beheld* as an *object.*

Example: An atom is as much an authentic entity as a human being. But an organization *created* by a human being (such as a company or corporation) though an objective entity, is simply to be considered an entity—*not* an authentic E/entity.

Example: All apparently distinct units of Life on a certain 'Ray' are, in their seamlessly fused apparent aggregation, an authentic Entity, and, moreover, an authentic *group* Entity (when considered from a *fully* unfolded/emanated vantage point) Regardless of their apparent differentiation, however, they are simply but One 'Ray' and One Entity.

Example: Any authentic E/entity is a *point.* Any 'Ray' of the ABSOLUTE (another name for an authentic Entity) is also a *point.* A *point* is how an emanated 'Ray' *appears* from the Source that emanated that 'Ray'. Thus, 'Rays' appear as 'Points'. (Can this thought be helpful in resolving the apparent duality between particle and wave?)

Example: Entities are 'Rays' and 'Rays' are Points. Just as all 'Rays' of the ABSOLUTE are One Universal 'Ray' of the ABSOLUTE, which is the ONE 'RAY' of the ABSOLUTE, so all Points in-Cosmos are One Universal Point, which is the ONE 'POINT' 'ITSELF'. Even so, the ONE 'RAY' and the ONE 'POINT' are, ESSENTIALLY, NOTHING.

entity, inauthentic

By an *inauthentic entity* is meant one that is not directly part of the *Intended design* of a hierarchically organized Emanative Stream.

Example: Every entity which is not a 'Ray' of the ABSOLUTE holding Logoically-Intended Position within the Divine Emanative Stream, is an inauthentic entity, even though such an entity may be good and useful in the world. Inauthentic entities may be produced by a 'non-Cosmo-Logoically-Intended aggregation' of a number of authentic or primary E/entities, plus the vehicles of expression of these authentic E/entities.

Many man-made groups are inauthentic entities of this type. Certain groupings of the inner and higher principles of human beings, however—for instance, groupings found on higher planes and known as Ashrams, and which reflect still higher archetypal, interior groupings of certain great Subjective Entities (such as Heavenly Men)—should be considered *authentic group* Entities.

In this regard it should be remembered that Ashrams reflect Ray Groupings and Planetary Groupings, all of which might be called 'Cosmo-Logoically-Intended Structurally-Integral Associations', and thus cannot be considered either inauthentic, secondary or tertiary entities. Such groupings, depending upon the nature of their *ensoulment* (as for instance by a great 'Ray' Life), could be considered (in the large sense) *authentic* or *primary* Entities.

When one stops to think of it, the Universal Logos is certainly an *authentic* Entity (in fact, in relation to this Cosmos) *the most authentic Entity*, and the Logos is certainly (in expression) the greatest of Group Entities. Thus, careful discrimination must be used to determine the structural relations of group entities to the Intended Design of the Divine Emanatory Stream.

A most important consideration here arises, and centers on whether 'Rays' of the ABSOLUTE retain their 'integrity' when retracted and reabsorbed into, for instance, a Planetary Logos, or a Solar Logos, or whether they exist in a totally merged and indistinct condition, their Identity being, in fact, *one* with the Superior Absorbing 'Ray'. If, for instance, such 'Rays' do *not* retain their integrity, then we (you and I) simply *are* (without the slightest trace of differentiation) *the* Planetary Logos from which we emanated, (or perhaps, when deeper understanding is available) *the* Solar Logos, and, thus, are not Really parts of a "host" which (in aggregate) *is* the Planetary Logos.

H. P. Blavatsky seems to suggest the "host" theory for the Identity of Planetary Logos, but by this she may simply mean a cooperative Pattern of Emanated 'Rays'. Similarly, and on a higher level, the Planetary Logos (Who is equally a 'Ray') would retain its integrity only so long as it functioned as a Planetary Logos, and would 'lose' that integrity completely when reabsorbed in Its emanating Source (whether that is the Solar Logos or some superior Being).

The question here is, Does a 'Ray' of the ABSOLUTE possess integrity/identifiability from the 'moment' of its first emergence from the ABSOLUTE and does it, similarly, retain that integrity/identifiability as it reascends along the Divine Emanatory Stream along which it descended? Although there is a predisposition among human beings to wish to preserve indefinitely their identity as they seem to know it, the likelihood of human Monadic Identity being preserved as such through the entire emanative and retractive process along the Divine Emanatory Stream is virtually nil.

entity, objective

By an *objective entity* is meant a relatively abiding aggregation of energies and forces (an aggregation which most often is not, but may be ensouled by a directing, sustaining, guiding, informing authentic Entity) which is designated as an entity *more* because of the nature of its *objective* appearance, coherence, and pattern than because of the authenticity/integrity of its *subjective* nature.

Example: The many companies, corporations, and organizations in the world are principally objective entities. Occasionally a subjective Entity may 'overlight' or infuse such <u>objective entities</u>, and then, their character may be entirely changed. There are times, for instance, when a particular nation might be seen, for a period, to be acting out a divine destiny. At such times the leadership of that nation may have rendered it receptive to specific, directive, 'overlighting' Divine Guidance. This type of *overlighting* or divine supervision renders the <u>objective entity</u> in question a *secondary entity of the higher kind.*

Example: Tertiary entities (i.e., *created* but non-ensouled entities) can best be considered as <u>objective entities</u>.

ENTITY/NON-ENTITY, the ONE GREAT

By *the ONE GREAT ENTITY/NON-ENTITY* is meant the BOUNDLESS IMMUTABLE PRINCIPLE considered as the ONE AND ONLY SELF. This ENTITY is inclusive of all E/entities (of whatever kind) that have ever been or that ever will be. Because this GREAT ENTITY necessarily precludes specification and any possibility of ultimately meaningful predication anent ITS NATURE (except through negation and privation) IT can also be considered the ONE GREAT <u>NON</u>-ENTITY, because no-*thing* REALLY can be said of IT, and, thus, IT cannot be described or compared to any other E/entity. Indeed IT is the INCOMPARABLE ENTITY/NON-ENTITY.

Example: The philosopher can rejoice in, at least, being able to posit all the things <u>the ONE GREAT ENTITY/NON-ENTITY</u> IS *not.* How amazing to see the way the GREAT BE-NESS eludes every attempt at predication anent ITS *true* NATURE! In IT, every assertion/attribution must necessarily be contradicted including this one! Thus, even what IT is *not,* in fact, IT is *not*!

'ESSEDENTIFICATION'

By *'ESSEDENTIFICATION'* is meant the manner in which the INFINITE SELF is of a *sameness* with all 'RADIATIONS' and Emanations. 'ESSE' means "being, existence, essence", and 'DENTIFICATION' suggests the analogy to 'IDENTIFICATION'. The INFINITE SELF thus 'TOUCHES' all other apparent 'things' through 'BEING' them rather than through 'SEEING' them. If IT has any 'relation' with other things, IT is related 'NON-DUALISTICALLY' through 'ESSEDENTIFICATION'.

Example: It could be said that the INFINITE SELF is 'RELATED' to the 'RAY' of the ABSOLUTE through '<u>ESSEDENTIFICATION</u>' rather than through 'PERCEPTION'. '<u>ESSEDENTIFICATION</u>' is not 'SEEING'; it is far more complete than 'SEEING' could ever be. '<u>ESSEDENTIFICATION</u>' does not rely upon 'CONSCIOUSNESS' in which the INFINITE SELF 'WILL' not long 'ENGAGE'.

essence

By *essence* (non-technically, and as the term is most normally used) is meant the most *irreducible* aspect of any item or E/entity-in-Cosmos.

Example: The <u>essence</u> of all E/entities is, REALLY, their *true identity.* That true *identity* is rarely recognized as the SELF. Usually the <u>essence</u> of a human being is considered to be his inner character, the pattern of his soul qualities, rather than his immortal Spirit.

While the essence of any E/entity may sometimes be justifiably, though relatively, considered to be a pattern, the ESSENCE of that E/entity never can be so considered.

Example: If one searches for the essence of any E/entity or any relationship, one will eventually and ultimately find only the ONE IRREDUCIBLE ESSENCE.

Essence

By an *Essence* is meant a Reality-in-Cosmos. An Essence is a Principle which is Permanent-in-Cosmos and is a 'RADIATION'/'REFLECTION' of the ONE PRINCIPLE, the ONE ESSENCE.

Example: Each Higher Principle, endowing Cosmos with order and pattern, is an Essence. Such a Principle can also be called an Archetype.

ESSENCE

By *ESSENCE* is meant the irreducible SELF—the *infinitely rarefied* AUTHENTIC NATURE of ALL.

Example: All things in all Cosmoses past, present, or to come, derive from and can be reduced to the same ESSENCE.

Essences (plural)

By *Essences* are meant Realities.

Example: All true Gods are Essences.

essential

By *essential* is indicated that without which any item or E/entity-in-Cosmos cannot *be,* or continue to *be itself.* An essential is a 'sine qua non' (i.e., a "that without which ..."), hence, a fundamental 'indispensability'.

Example: The awakening of the buddhic energy is the essential without which impersonal love cannot be understood or expressed in the life of the human being.

Example: For human beings to be *truly* and *abidingly* happy, it is essential that they *discover* the SELF and *identify as* the SELF.

'essentialist'

By an *essentialist* is meant one who always deals with the root of any matter rather than the secondary effects.

Example: An essentialist will always see the SUBJECTIVE SELF (or at least the Subjective Self) in what appear to be objects.

essentially

By the term *essentially* is indicated a way of describing any E/entity or thing with reference to its *immediately causative subjective nature.*

Example: From the perspective of a relatively penetrating subjectivity, man is essentially his Causal, Egoic nature. A deeper level of subjective penetration would reveal man as he is both in Essence and in ESSENCE.

Essentially; in Essence

By *Essentially,*or *in Essence,* are meant—'in relation to Cosmic Archetypes and Cosmic Principles'—i.e., the Realities of a Cosmos. These two terms do *not* mean 'in relation to the *ULTIMATE* ESSENCE' which is the ONE AND ONLY ESSENCE, i.e., the INFINITESSENCE. *Essentially* and *in Essence* still pertain to the Realities of the World of Relativity and not to the REALITY of the WORLD OF BEING.

Example: As human beings, semi-subjectively considered, we are patterned wholes, which, Essentially and in Essence, are reflections of the Patterned Wholes (i.e., Heavenly Men) in Which we live and move and have our being. Every patterned whole within our particular Solar System reflects the Patterned Wholeness of the Grand Heavenly Man—the Solar Logos. In another Solar System, the Grand Heavenly Man would have a different Quality, and Its lesser patterned wholes would reflect *that* Quality.

ESSENTIALLY; in ESSENCE

By *ESSENTIALLY,* or *in ESSENCE,* are indicated the INFINITESSENCE that is the *IRREDUCIBLE* ESSENCE of *anything* under discussion. (These capitalized terms always refer to a process by which a thing is seen, *not* as it appears to be in the World of Becoming, but as it IS—IN ULTIMATE REALITY.)

Example: Within the INFINITESSENCE all things, qualities, patterns, etc. inhere as they are ESSENTIALLY, in ESSENCE, and *not* as they *appear* when they, as *pure infinitized possibilities,* are 'EXTRUDED' from the INFINITESSENCE and articulated in the World of Relativity.

'ESSESPECTIVE'

By *'ESSESPECTIVE'* is meant the 'INFINISPECTIVE' with an emphasis upon the 'PERSPECTIVE' of 'BEING' rather than the 'PERSPECTIVE' of 'SEEING'.

Example: Can the INFINITE SELF 'SEE'? A 'SEEING'/'POINT'/'MAYA' 'ARISES' in IT, but can the 'SEER' be 'SEEN'? Is the INFINITE SELF 'CONSCIOUSLY' 'AWARE' of the 'SEER'/'SEEING' that has 'GONE' 'FORTH' from IT? Or is there no 'RELATION' at all between the ABIDING INFINITE SELF and ITS 'RAY'? While it cannot reasonably be said that the INFINITE SELF 'SEES' the 'SEEING' that has 'GONE FORTH' from IT, perhaps IT does have, in 'RELATION' to that tripartite EVANESCENT INFINITE SELF, an 'ESSESPECTIVE', so to speak, an 'INVESTMENT of BEING' in that 'POINT', which does not render the 'POINT' a fully 'SEEN' 'OBJECT', but does at least preserve a 'RELATION of BEING' 'between' the two.

eternal; eternally

By *eternal* is indicated a span of Time which is, relatively, very lengthy but of unspecified duration. What is usually implied in the use of this word is that if something is eternal, it is *always* so. The term, however, is vague and indicative rather than philosophically or mathematically rigorous.

Example: "Hope springs eternal in the human breast." This will eternally be so as long as there *is* a humanity and a "human breast" which, clearly, will not be *forever*. The term eternal, as so often used, serves as the "always" and "ever" of those who have not deeply contemplated Infinity.

Eternal; Eternally

By the term *Eternal* may be indicated that which abides for the duration of a single Cosmos.

Example: A particular Design-at-the-Beginning is Eternally *guiding* and *formative* in relation to all Cosmic Processes within a particular Cosmos, but that particular Design-at-the-Beginning is *not* guiding and formative for *all* Cosmoses forever (i.e., ETERNALLY), for there are an infinitude of Designs-at-the-Beginning, each appropriate for one of an infinitude of Cosmoses.

ETERNAL; ETERNALLY

By *ETERNAL* is meant that which never ceases, which abides forever.

Example: One will search in vain to discover any infinitely enduring, hence, ETERNAL Emanation or Creation. Anything which derives from any *source* whatsoever can never be ETERNAL, except in ESSENCE. The ONE ETERNAL is ITS OWN 'SOURCE' (though, ESSENTIALLY, IT has no *source*).

Example: The HOMOGENEITY of the BOUNDLESS IMMUTABLE PRINCIPLE is ETERNAL despite the periodical 'APPEARANCE' of the World of Illusion. Nothing can disrupt the ABSOLUTE CONTINUUM, although, periodically, an Illusory Cosmos (for all Cosmoses are Illusory) *seems* to disrupt IT.

ETERNAL, the

By *the ETERNAL* is meant the INFINITELY ENDURING HOMOGENEITY.

Example: THE INFINITELY ENDURING HOMOGENEITY is the ETERNAL because IT IS forever the *same*.

Eternal Duality, the

By *the Eternal Duality* is meant the following fact: that PURE SELFHOOD (ALL-IN-ALLNESS), and the Not-SELF (these two being the most *fundamental* of the Pairs of Opposites) have existed forever, without beginning, and will persist without end. The Not-SELF consists of a Primary Three: Time, Space and Motion—all of them having existed forever, but appearing only *cyclically, intermittently*—again, *forever*.

Example: PURE SELFHOOD, paradoxically, cannot cease even in the cyclically recurring Presence of the Not-SELF (which is Limited and 'relatively' Impure). Thus, the entire History of the UTTER ALLNESS 'to date' has been an example of Eternal Duality.

Example: The following contradiction is true: Never was the Time when there was not ONLY ONE; but, Never was the Time when 'Twoness' was not *in process.* We must conclude that even though Twoness (Eternal Duality) has forever, throughout Infinite Duration, been in process, REALLY, throughout Infinite Duration, there has only been the ONE WHO IS NONE.

Example: The ABSOLUTE ONENESS is not the REAL 'OPPOSITE' of anything, all appearances notwithstanding. Where is the 'OPPOSITE' IT *IS NOT?*

eternity

By an *eternity* is meant a *vast* though *finite* duration.

Example: The Universal Pralaya lasts for a period called in *The Secret Doctrine,* "Seven Eternities." Presumably, then, the term eternity can also be applied to the measurement of Time *during* Universal Manvantaras. The question is, Who measures?, and From *what* Point of View?

Example: It is not certain that eternities have a strictly regular duration relative to each other. They may, instead, indicate great Phases of Cosmic Work, and last only as long (relative to each other) as it takes for that Work to be accomplished. After all, with all the Time in the UTTER ALLNESS available to the INFINITE SELF-as-Universal Logos, why 'rush'? There is *no one else* "waiting in line" to Become a Cosmos.

Thus while lesser Cosmic Cycles may be strictly timed, the duration of the Eternal Cycle (the one turning of the Great Wheel) *may not* be. In other words, 'eternity' and 'eternities', *cosmically considered,* may be *elastic*—'elastic', for instance, in relation to the fulfillment of the Cosmic Purpose of any particular Universal Logos.

ETERNITY

By *ETERNITY* is meant ABSOLUTE INFINITE DURATION.

Example: Throughout the endlessness of ETERNITY, literally countless Cosmoses appear and disappear, but nothing REALLY happens, i.e., only 'NOTHING' 'HAPPENS'.

'eterno-cyclic'

By the term '*eternocyclic'* is indicated and described any process that has occurred cyclically for all ETERNITY.

Example: The registration of Mulaprakriti by the SELF-as-Infinified Point at the inception of every Pre-Cosmic Process—a registration which leads inevitably to the manifestation of a Cosmos—is an 'eterno-cyclic' Process. This Process has occurred and will occur (and *has* occurred) forever in a cyclic manner throughout all ETERNITY. The duration of the cycle is unknown to man and may or may not be regular. The dissolution of all Cosmoses into the ABSOLUTE is also an 'eterno-cyclic' process.

Example: A particular Pattern of Cosmic Emanation (since it may be unique to a particular Cosmos) may *not* be an 'eterno-cyclic' pattern

event

By an *event* is meant any modification whatsoever of the ABSOLUTE HOMOGENEITY. It is more usual to consider events as being intra-Cosmic than extra-Cosmic, though the extra-Cosmic arising of the 'POINT-in-THAT-as-Point' (the SELF-as-Self/Infinified Point) might be considered an 'EVENT' since It is, apparently, a 'CHANGE' or 'MODIFICATION' of HOMOGENEOUS THATNESS.

By an *event* is meant an object/modification/disturbance/vibration/thing/appearance/manifestation/item/existence/(and, from certain perspective) E/entity, etc. If the fundamental idea of *eventness* is traced and understood, an event will be found to be all the above.

Example: Every 'movement' in-Cosmos, however slight, is an event. It is interesting to consider that, given the parameters of a particular Cosmos, no events in the World of Fabrication can possibly occur 'between' ultimate moments, because (given the theory of Ontological Oscillation) there is *no Objective Universe* between ultimate moments. Further, in Objective Cosmos, no events can occur *during* ultimate moments because no movements sufficiently rapid to be classified as *'intra-moment movements'* are 'allowed' to occur because of pre-Cosmically Defined and Willed, *Universe-specific* Parameters. From another perspective, the Cosmic Configuration is, as it were, *frozen* during each ultimate moment. If any movement/event were detected during an ultimate moment, such detection would indicate that the supposed ultimate moment was not Really *ultimate,* and that the ultimate particle/event for that Cosmos was *divisible,* and, hence, not *ultimate.*

Example: The 'FLASHING FORTH' of the 'RAY' of the ABSOLUTE at the end of Universal Pralaya is always the *first* 'EVENT' in relation to a Cosmos-to-be (at least *relatively* so), even though this kind of 'EVENT' has already occurred an infinitude of times. One wonders, Did this 'EVENT' have no Cause? for the causes of events are, usually, events themselves. Since there are no *events* of any kind possible within the CAUSELESS CAUSE, this *first* 'EVENT', has no *cause* of the kind we can fathom. It appears to have arisen out of NOTHING! But since NOTHING is EVERYTHING-in-ESSENCE, perhaps it is somewhat easier to answer the questions, Can something come out of ABSOLUTELY EVERYTHING?; Can something come out of INFINITIZED EVERYTHINGNESS?; Can something come out of an ABSOLUTE PLENUM? The answer, without too much mental torture, inclines to be, Yes, but *how*?

exist

That which *exists* is that which has duration (whether that duration is infinitesimal or infinite, i.e., 'infinitesimalizing' or 'infinitizing').

More technically, that which *exists* is any manifest thing located, apparently, *outside* the SOURCE. To *exist* a thing must be manifest 'Now'. Otherwise a thing must be spoken of as 'having existed' or as 'existing in potential'. Thus, the state or nature of the existence must be specified. For instance, that which once existed but which no longer exists in the state in which it once existed, could be spoken of as presently *existing* in memory. That which does not *yet* exist can be spoken of as *existing* in a state of *vision, intention*, etc. That which exists must exist *outside* the SOURCE because *within* the SOURCE no-*thing* 'EXISTS', per se. Only the *limitless potential for all-thing-ness* can be said to be 'INHERENT' within the INFINITESSENCE.

That which *exists forever* is that which manifests (even if 'eternocyclically') throughout Infinite Duration and is thus permanent in the UTTER ALLNESS.

Example: Two 'things' can be said, loosely, to exist forever, though only one of them exists in the truest sense of the word, i.e., is found (apparently) *outside* the SOURCE. These two are the SELF and the Not-SELF, for never was the 'Time' They were not. It might be argued that there will always be found a 'Time' when the Not-SELF does *not* exist, and this statement is true. Yet, there will always be found a Time *before* that Time of non-existence when the Not-SELF *did*, in fact, exist. Thus the Not-SELF *exists forever*, ETERNALLY, throughout Infinite Duration, though that existence is unvaryingly periodical and cyclic.

extension

By *extension* is meant the illusory perception that *real* matter 'fills' (i.e., is *extended* in) *real* space. (The word *real* with a lower case 'r' used as an adjective in relation to the words "matter" and "space" indicates the grossness of the perception and cognition under discussion.) Extension, far from being REAL is an artifact of a perceiving consciousness, hence the illusion of extension is a "construct of consciousness". REALITY is dimensionless (or 'infinidimensional') and has no extension. That which is perceived as 'extended' is un-REAL, the 'work' of Maya. Extension might however be considered 'Actual-in-Illusion', and therefore *factual*. Extension, therefore, must be acknowledged by those who wish to deal with the World of Illusion in a practical and effective manner.

By *extension* is meant a point 'seen' from a distance which converges upon the infinitesimal. Just as all definite 'things' (no matter how vast) or 'extensions' are reduced to points at distances receding towards the infinite, so the opposite is true. All points reveal their 'extendedness' to the perceiver who approaches ever 'closer'. This means, perhaps, that there are 'de-finite' points and 'in-de-finite' points. Perhaps the only *true* point, is the *indefinite point* which is the true *process* of 'infinitesimalizing'. All things can be made into tiny points which are *virtual* infinitesimals, but, because such things are *definite*, the points which they become (through *infinispectivizing*) are *virtual infinitesimalizings* and not *true infinitesimalizings* and, thus, equally, *virtual* points and not *true* points.

Example: The illusion that matter is *real* and 'solid' is based upon the perception of extension—i.e., the appearance within the field of consciousness of many *'things'* which seem both tangible and "extended in space".

Example: Without the illusion of extension the average human being's conception of space would collapse. 'Space' would seem 'spaceless' unless one labored under the Illusion of Extension.

Example: At its most fundamental, unelaborated level, extension is simply that which is registrable by Consciousness.

'Extra-SOURCE'

By the term *Extra-SOURCE* is indicated anything whatsoever (whether Pre-Cosmic, Intra-Cosmic or Post-Cosmic), which *happens.* If something occurs or originates, apparently, 'outside' the INFINITE SELF, then its field of activity and origin is 'extra-SOURCE'.

Example: Since nothing REALLY 'HAPPENS' 'Intra-SOURCE', then all happenings are inevitably '<u>Extra-SOURCE</u>' happenings. And yet, nothing REALLY 'happens' '<u>Extra-SOURCE</u>' either! What can escape the SOURCE?

'EXTRUSION' of SELF-'INHERENT' 'POSSIBILITY'

By the *'EXTRUSION' of SELF-'INHERENT' 'POSSIBILITY'* is indicated a mysterious Pre-Cosmic Process by means of which a certain 'POSSIBILITY' (or certain 'POSSIBILITIES') 'INHERING' in the INFINITESSENCE are 'EXTRUDED' (i.e., 'EXTRACTED') for reification <u>as</u> Cosmos (not just *in* a Cosmos). Does any 'ACTION' 'pre-date' the 'FLASHING FORTH' of the 'RAY' of the ABSOLUTE? Who can say? (Perhaps, *'WHO'* can 'SAY'!) However, the 'FLASHING FORTH' is the first evidence of the 'EXTRUSION' of SELF-'INHERENT' 'POSSIBILITY'.

Example: All intra-Cosmic combinations are 'RESIDENT' in ESSENCE within the INFINITESSENCE. These reified possibilities exist in-Cosmos because of the Process called the '<u>EXTRUSION' of SELF-'INHERENT' 'POSSIBILITY</u>'.

- F -

fact; Fact

By a *<u>F</u>act* is meant a content of consciousness which is deemed indisputably <u>R</u>eal (though not REAL) to a maximally illumined consciousness. In any Cosmos, there is only one maximally illumined Consciousness, and That is the Consciousness of the Universal Logos.

By a *<u>F</u>act* is meant something that is <u>R</u>eal-in-Cosmos, hence, though not ABSOLUTELY REAL, Is, at least, a fundamental and Permanent-in-Cosmos Aspect or Principle of that particular Cosmos.

By a fact (in a lesser sense) is meant something that is perceived as <u>r</u>eal-in-Cosmos, and is, hence, merely an illusory actuality which appears to be <u>r</u>eal only to a consciousness entirely subject to Power of Objectification, i.e., the Power of Maya.

Example: The Archetypes (and even, lesser archetypes which emanate from the Universal Logos) are certainly <u>Facts</u> in the Consciousness of this Logos, and remain to be discovered as indisputable <u>Facts</u> in the consciousness of all lesser-though-awakening self-conscious entities in-Cosmos—entities who mistakenly consider as <u>facts</u> the reflections in lower prakriti of such A/archetypes.

Example: It is possible for a presentation to be a 'fact-in-consciousness' and nevertheless un-True? It would seem so.

Example: All Patterns which are Permanent-in-Cosmos (and hence Archetypal) are Facts in that Cosmos whether or not they register as facts in the majority of consciousnesses contained in that Cosmos. This is so because Permanent-in-Cosmos Patterns are deemed to be indisputably Real to the Universal Logos Who Emanated them.

factor

By a *factor* is meant a variable that must be taken into consideration in a particular context if the context is to be understood and/or acted upon wisely.

Example: The evolution of the human being is dependent upon many factors—energies, forces, patterns, and conditions on a number of planes and subplanes, as well as upon the impact of a range of E/entities—planetary, systemic and even cosmic. All these factors *must* be taken into consideration if the particular difficulties faced by the human being at this juncture of Planetary History are to be understood.

Factor-in-Universe

By a *Factor-in-Universe* is meant an important Cosmos-Conditioning Principle/Force/Energy. A Factor-in-Universe cannot be ignored if the Cosmic Process is to be understood and facilitated.

Example: Not all items-in-Universe are Factors-in-Universe.

Example: The Hierarchy of Fundamental Numbers is a Factor-in-Universe which conditions a multitude of other numerically resonant factors.

faculty

By a *faculty* is meant a distinct capability resident within a sentient being (especially a self-conscious sentient being).

Example: The extensions of the five senses upon the five planes of super-human evolution are faculties which the World Initiates and Masters cultivate in order to assist planetary evolution.

Father; Father, Cosmic

By the *Father* or *Cosmic Father* is meant, ESSENTIALLY, and ULTIMATELY PARABRAHMAN considered as the Unseeing 'Seeding' SOURCE of all that can Arise as Form within ITSELF in ITS *projected* Aspect as Mother. PARABRAHMAN IS all beings, and They are ITS Agents.

By *the Cosmic Father* is meant, ESSENTIALLY (at several removes) the BEING of PARABRAHMAN as the 'Hidden Invisible Universal Logoic Potential' 'Seeing' ITSELF Revealed in the Visible, Manifest, 'Forth-Showing' of ITSELF—i.e., the Cosmic Mother.

Example: As the Cosmic Father, **I-as-8** 'See' what is within Me through the Appearance of the Mother. **I-as-8** Am the Father and the Mother 'Shows Me' ALL **I-as-8** Am.

Example: The Cosmic Father and the Cosmic Mother *always* work together as One Action. The Father is in 'Darkness' because that which is 'within' Him is 'Unseen' unless He 'Sees' Himself as Mother. The TRUE IDENTITY of the Cosmic Father is PARABRAHMAN as the 'Unseen'/'INVISIBLE BEING' of That which is to be 'Seen'. The true Cosmic Mother is, ESSENTIALLY, PARABRAHMAN as the 'Seen/Visible Being of That which is Unseen and Invisible in the Cosmic Father. Cosmic Maya is PARABRAHMAN as the 'Seeing' of That which is 'Unseen'/Invisible Being in the Cosmic Father and 'Seen'/ Visible Being in the Cosmic Mother. Fatherness 'Sees', Motherness is 'Seen', 'Maya-ness' is the 'Seeing'—and all of them are ONE ABSOLUTE IDENTITY WHO IS the 'SEER', the 'SEEN' and the 'SEEING'.

Example: No-thing Arises (visibly) in the Cosmic Mother that is not *simultaneously* (invisibly) within the Cosmic Father. The Cosmic Father and Cosmic Mother are both the SAME IDENTITY-as-Identity, and they work inseparably from each other. That which the Cosmic Father ultimately 'Sees' (as a Point) in the Cosmic Mother is the Cosmic Son.

field

By a *field* is meant a 'space' (whether relatively vast or tiny) characterized by a specific energic pattern and interplay, a certain range of vibratory frequencies and a resultant, emergent quality.

Example: The auric field of the Buddha is said to extend for miles.

Example: Fields can be evaluated quantitatively and qualitatively.

Example: Consciousness, Itself, is the Prototype of all Fields.

field, the most immediate

By the *most immediate field* is meant, that particular prakritic pattern (pattern in matter) with which a particularized, prakritically-bounded consciousness most identifies (i.e., its 'own' vehicles).

Example: What I call my three periodical vehicles and their constituent fields, constitute, in total, my most immediate field, my ring-pass-not. The ring-pass-nots of other human beings are, for *me* in my *limited* aspect, fields more *remote* than my most immediate field.

Example: The most immediate field is a fertile source of mis-identification and encapsulating localization—hence, of *egoism*.

finite; Finite

By the term *finite* is meant limited and particularized.

By the term *Finite* is indicated the finiteness of very large structures—such a structure, for instance, as the Universe. The Universe, thus, is designated as *Finite* rather than simply as *finite*.

Example: Every 'thing' is finite, regardless of its size. As an exception, the Infinite Subject/Object is a 'thing' only because we call it so. It is 'in-de-finite'.

Example: Consciousness creates the finite. Identification realizes the INFINITE.

Example: The Universe is necessarily Finite. Otherwise the Law of Periodicity is violated.

finitization

By *finitization* is meant a Process by means of which 'NOUMENALIZED ESSENCES' 'WITHIN' the INFINITESSENCE are 'EXTRUDED' for *cosmic demonstration.*

Example: All things in-Cosmos, including Cosmos Itself, are finitizations of the ABSOLUTE, the PLENUM.

finity; finities

By a *finity* is meant a finite thing, a finite object.

Example: The World of Becoming is full of finities. Even the Members of the Cosmic First Family within the World of Being are, 'Infinispectively' considered, finities, though one in ESSENCE with the INFINITE.

flux

By *flux* is meant an *almost constant, seemingly continuous* movement of a particular set of variables relative to each other.

Example: Flux cannot be *absolutely* continuous in a quantized, discontinuous Universe. The motion of the flux occurs in pan-cosmic 'frames' timed to change at each ultimate moment.

Flux, the Cosmic

By *the Cosmic Flux* is meant the *almost* incessant motion which characterizes each and every variable-in-Cosmos for the entire duration of that Cosmos. Within Fabricated Cosmos (the Mosaic World) there is Perpetual, but, Essentially *discontinuous* Motion. In the extra-Cosmic ALL-IN-ALLNESS (the ONE AND ONLY BEING) there is 'PERPETUAL IMMOBILITY', or, from a different perspective, Motion that has been *infinitized* into the 'STATE' of IMMOBILITY.

Example: At every moment the Design of the Cosmic Configuration may be extracted from the Cosmic Flux by the "All-Seeing-Eye", and by that Eye alone. No other Witness can 'See' the Cosmic Configuration as an entirety.

If it is suggested that the All-Seeing Eye of the Universal Logos 'Belongs' to the Cosmic Flux, and, though It is the Witness of all things in-Cosmos, cannot 'See' Itself, we must say that the Cosmic Flux applies to *phenomena* in-Cosmos and not to Spiritual Noumena (within the World of Being).

'FOHAT'

By '*FOHAT*' is meant, essentially, the ABSOLUTE SELF in *initial* 'ACTION'.

By '*FOHAT*' is meant the extra-Cosmic or pre-Cosmic Agent of the ABSOLUTE SELF.

By '*FOHAT*' (strangely) may be meant 'MAYA'. This must be meditated upon deeply.

Example: 'FOHAT' is the Agent of the ABSOLUTE SELF in *all* pre-Cosmic Processes. 'FOHAT' in this respect is *Super*-Cosmic Fohat.

Example: Extra-Cosmic 'FOHAT' (i.e., Super-Cosmic Fohat) becomes intra-Cosmic Fohat.

Example: As 'FOHAT' 'BEGINS' 'within' the ABSOLUTE, but 'EXPELS' 'ITSELF' instantly, its 'capital letters' have an exceedingly short duration.

Fohat (Pre-Cosmic, Cosmic, and Intra-Cosmic)

By *Fohat* is meant the intra-Cosmic Agent of extra-Cosmic 'FOHAT'. Fohat is found everywhere in-Cosmos, Doing virtually everything. From the Cosmo-Essential perspective, Fohat is the Agent of the Universal Logos (and Its Emanations) just as 'FOHAT' is the 'AGENT' of the ONE AND ONLY SELF.

By *Fohat* is meant any *action* whatsoever. Strangely, 'CONSCIOUSNESS'-instantly-Consciousness is the FIRST 'ACTION'. Focusing for a moment of the Pre-Cosmic Level, when the Infinite Father 'Sees' the Infinite Mother that 'Seeing' is Super-Cosmic Fohat. Fohat has a strange equivalence with Maya. Every *action* can be conceives as Fohat/Maya. Evidence comes through The Tibetan's writings when (considering the Trio: Illusion, Glamor, and Maya), Maya is associated with the etheric level and *activity.*

By *Fohat* is meant the Agent for Relationship, for there can be no relationship without an *act.* If any being-in-Cosmos *touches* any other being-in-Cosmos, that *touching* is Fohatic.

By *Fohat* is meant SUBJECTIVITY-in-Action. Fohat (the Great Cosmic Actor) is a Subjective Being, necessarily endowed with Consciousness (in fact, in a way, *is* Consciousness). Fohat articulates objects out of undifferentiated Cosmic Prakriti—the Great Object— by 'Seeing' them, i.e., by *being the 'Seeing'* of the Prakriti reflections of that which It, Itself, (Fohat) is bearing/carrying as Ideation from the Subjective Being(s) (the Universal Logos and Cosmic Son) Whom It (Fohat) represents. Fohat somehow *conforms Itself* to the Ideation It is bearing or carrying.

Fohat (in Its higher State as 'FOHAT') carries 'forth' into the possibility of Cosmic Objectification the particularized 'IDEATION' 'EXTRUDED' from the INFINITESSENCE (i.e., from the PLENUM). Thus 'FOHAT' 'bears' the *infinitessential* 'IDEATION' which is to become Cosmic Ideation. In a way, 'FOHAT' *is* that *infinitessentialized* 'IDEATION'.

It must be realized that Fohat is not Unconscious. Every 'Act' of Fohat is Really an *Act of Consciousness.* Fohat Creates objects in Cosmic Prakriti through the power of the Objectifying Consciousness which It is.

Example: It is Fohat Who (in Its Fourth Mode) "digs holes in space" so that the Mother or 'Birther of Forms' (Cosmic Prakriti) may be multiply differentiated.

Example: Fohat undergoes a hierarchically organized 'Self-Enumeration' (in conformity to the requirements of the particular Pattern of Cosmic Ideation demanding Objectification in a particular Universe) in order to generate a hierarchical multiplicity of differentiated 'materials' in Cosmic Prakriti (i.e., the First Outpouring, Theosophically considered).

Fohat generates this hierarchical multiplicity by means of multiple instances of Its Own Fohatic Self-Perception. The 'Seeing'/Perceiving accomplished by Fohat as It (in a multiply-enumerated subjective Mode) penetrates Cosmic Prakriti, *produces ultimate particle/events.* Ultimate particle/events are Fohat 'Seeing' the Great Homogeneous Cosmic Object (Cosmic Prakriti) as a *reflection* of what Fohat, Itself, has subjectively/multiply 'Become' (or of what It subjectively/multiply 'Carries') as a result of Its (Fohat's) Purpose-Responsive Self-Enumeration.

Fohat, as it were, subjectively 'Divides' Itself into a specified number of potential ultimate particle/events (*ultimate* "bubbles in the Koilon") and then, according to the hierarchical Plan of Cosmic Ideation (which Fohat 'Carries'), *objectifies* these multiple Purpose-Determined subjective 'Divisions' of Itself 'in' Cosmic Prakriti (or as a type of Cosmic Prakriti) by means of multiple acts of Self-'Perception'. Multiply, subjectively Self-Enumerated Fohat, thus 'Perceives' as differentiated Cosmic Prakriti) Its own subjective multiplicity.

Through this act of Perception/Reflection, multiple objective ultimate particle/events are 'born', as it were, within the Bosom of Cosmic Prakriti. Fohat, then, through the ceaseless, *Intelligently-Willed* movement and interplay of those ultimate particle/events which It has 'Become', uses these ultimate particle/events as the means to Build/Create and 'Become' those still more complex, hierarchically-organized aggregations we recognize as matter in its numerous differentiations. This entire description pertains to what in Theosophy is called the "First Outpouring". Later the building of *forms* follows.

Example: 'FOHAT' *carries* a particular 'IDEA' of the ABSOLUTE forth as the Idea-to-be-Objectified-in-Cosmos, and (when in-Cosmos) 'FOHAT' (having passed through the stage of Super-Cosmic Fohat) *becomes* Fohat Who *Is* that very 'IDEA'-as-Idea-in-Intra-Cosmic-Action. Fohat subjectively articulates Itself (Self-Enumeration) according to the specifications of the Great 'IDEA'-as-Idea, and then *objectifies* the subjective articulations in Cosmic Prakriti *as* differentiated Cosmic Prakriti, Fohat, thus, *builds* the Universe through Acts of Self-Perception which *become* the *objectification* of the Idea Projected by the Universal Logos (that Idea being previously embodied by the FOHATIC 'RAY' of the ABSOLUTE and 'carried' through Pre-Cosmos, but Super-Cosmic Fohat).

Example: In a way, all B/beings whatsoever, *are* SUPER-Cosmic 'FOHAT'.

Example: Any 'DOING', Doing or doing is 'FOHAT'-instantly-Super-Cosmic Fohat-eventually-Fohat *in action.* There can be no *movement* of any kind but that Fohat is the *Mover* and the *Movement.* Fohat, then, is the Divine Power of Movement, the *intermediary* Who carries out all *willed intent.*

Example: A "hole in Space" is generated by the Presence (in Undifferentiated Objectivity) of a 'Multiply-Self-Differentiated Subjective Viewer/Perceiver'—namely, Fohat. Holes in space are 'Fohatic Perceptions' which are Reflections of Its (Fohat's) Own subjective Self-Enumeration/Differentiation. These Fohatic Perceptions occur in exact conformity to the Cosmic Ideation of the Universal Logos of which Fohat is the 'Bearer' and which, in a sense, Fohat has 'Become—for 'FOHAT'-as-Fohat *Is* the Cosmic 'IDEA'-as-Idea-*in-Intra-Cosmic-Action.*

Example: Fohat, purely and simply, is the *Actor* in all things.

Example: Fohat is the FUNDAMENTAL SUBJECTIVITY-as-Subjectivity-in-Action. Fohat is, Essentially, the 'Subjectively Differentiated NOTHINGNESS-in-Action' (i.e., SPIRIT-as-Spirit-in-Action) which is the *immediately effective cause* of all differentiated matter. What Fohat 'Sees'/Perceives and, then, (more kinetically) 'Does', is *matter as we know it!*

We cannot cognize Cosmic Prakriti unless Fohat has, first, subjectively differentiated Itself into *potential* ultimate particle/events and, then, *perceived* the *objectified reflection* of that Self-differentiation/enumeration by means of the *reflecting power* of Cosmic Prakriti. Before engaging with Cosmic Prakriti, Fohat 'Carries' or *subjectively* 'Becomes' all the Designs which It will 'Perceive-into-Objectification' through the *reflecting power* of Cosmic Prakriti. Cosmic Prakriti (i.e., Cosmic Undifferentiated Matter) is the *limitless potential* for the Universal Logos' Cosmic Ideation to become objectively *manifest* through what we might call the '*reflection-into-objectification*' of the Content of that Cosmic Ideation (a Content 'Borne' by Fohat as the Agent of the Universal Logos—the 'Author' of the Cosmic Ideation).

Within Cosmos, therefore, Cosmic Prakriti is the *limitless potential* for Self-Reflection (the Self-Reflection of the Universal Logos), just as on the Pre-Cosmic Level, Mulaprakriti is the *limitless potential* for SELF-'REFLECTION' (the infinified SELF-REFLECTION of the INFINITE SELF-as-Infinite Subject). Whereas Fohat is the Agent of the Universal Logos, 'FOHAT' is the 'AGENT' of the INFINITE SELF. Fohat is thus an Agent/Carrier/Objectifier of *ideas* from a superior Source (in this case, Fohat's Third Mode, Cosmic Fohat, that Source is the Universal Logos). 'FOHAT', on the other hand, is the SUPER-Cosmic'-'Become'-Super-Cosmic 'AGENT'-as-Agent, 'CARRIER'/INITIAL OBJECTIFIER' of the 'IDEA' which is to *become* a Universe.

With respect to the *tangibility* of matter, it should be said that in-Cosmos, *undifferentiated Cosmic Matter* is not *tangible* in any way, but, rather, is the *ever-ready potential for the objectification of that which is subjective.* Intra-cosmic Root Matter (Cosmic Prakriti) is the *Reflection* of the *subjective* Cosmos (the Logoically-Intended Cosmos) which (through the Agency of Fohat) seeks objectification, and only the *reflections* appearing as and in Cosmic Root-Matters are tangible (as human beings understand tangibility); only the reflections appearing as and in Cosmic Root-Matters can be perceived as *matter* (as man knows matter).

Fohat, Mode of

By a *Mode of Fohat* is meant one of various Aspects of Activity of the one incredibly protean 'EMISSARY'/Emissary of the ABSOLUTE SELF.

In **Mode 1,** Fohat (or rather, 'FOHAT') *Is* the 'RAY' of the ABSOLUTE 'Carrying' the SUPER-COSMIC 'INTENTION' of the ABSOLUTE for the Universe-to-Be. This is 'SUPER-Cosmic 'FOHAT'.

In **Mode 2,** Fohat (Super-Cosmic Fohat) is the Agent by means of which the Infinite Subject-as-Infinified Point *Sees* Itself (i.e., the Infinite Object, Mulaprakriti), with progressively 'Narrowing Consciousness' thus transforming the Infinified Point into the Condensing Point, and the Infinite Subject into the Focusing or Universe-Intent Subject.

In **Mode 3,** Fohat is the 'going forth' (through Self-'Sight') of the Cosmic Father, the Universal Logos *into* the Cosmic Mother (Cosmic Prakriti), the Father being *present* as an *enlivening, life giving Presence* in the Mother through the Agency of Mode 3 of Fohat, the 'Seeing'. Via Mode 3 of Fohat, the Mother's First Pregnancy and the Birth of the Cosmic Son (the Father's *reduced* Self-Image) is achieved. We are speaking here of Cosmic Fohat.

In **Mode 4,** Fohat is the Agency for the 'going forth' of the Cosmic Son unto the Father-Enlivened Cosmic Mother, generating the Mother's Second Pregnancy, which results in the Birth of Differentiation in-Cosmos. We are speaking here of Intra-Cosmic Fohat. Through this Second Birth, the Mother becomes immensely differentiated, and ready to offer Herself, responsively, to the Son Himself (via His many 'Rays', and not, this time, to Brahma/Holy Spirit/Fohat) for the fulfillment of a kind of *Third* Birthing which will result in the birth of all the many *forms* in-Cosmos.

As the Bearer and Matrix of differentiated matter (thanks to the previous Activity of Fohat—Mode 4), She is thus ready for the Descent of the *Formative* Emanations ('Rays'), which stream forth as 'Rays' (various apparent 'differentiations' of the One Cosmic Monad) from the Universal Divine Son, and which Act upon Her (via Their correlated Fohatic Agencies—for the Sons of the Son and the Sons of Fohat work *together)* in Her Fohatically-induced condition as differentiated matter/prakriti to produce the World Forms for which the Divine Emanations from the Son and His Host (the Supernal Tetraktys), are the principal Archetypes.

Fohat (in His Activity to this point) has not Produced/'Become' the *Forms*; rather, He Produced/Became the differentiated and aggregated 'materials' out of which the Divine Son *and* (again) Fohat, (utilizing the Fohatically Produced 'Mother Materials') will 'Fashion' the Forms. As the cooperating Agent of the Divine Son, we see that Fohat continues to 'Help' as ever. Fohat's work as Eros shows Him active in the Process of Producing and Maintaining Forms. Thus here we see Fohat (a Third Aspect Being) assisting in the Work of the Son, the Work of the Second Aspect.

Example: Every objective item-in-Cosmos has been created through the active involvement of Fohat in one of its Modes. A great many items, however, also require the participation of the Divine Son—the Builder of Forms. All items-in-Cosmos come to birth *in* the Mother—the Bestower of Objectivity—through the Agency of the *reflective power* of Maya.

Example: In a way, 'FOHAT' is the first of the Gods. The ABSOLUTE IS the DEITY, the GOD WHO *always was and always will be*, but, in relation to each Universal Manvantara, 'FOHAT' is the *first* God, a 'SUPER'-Cosmic-instantly-*Super*-Cosmic God.

Fohatic Disengagement

By *Fohatic Disengagement* is meant the Process which (may) occur between ultimate moments such that all ultimate particle/events disappear or are reabsorbed into Fohat before their next reappearance at the onset of the next ultimate moment.

Example: When ultimate particle/events disappear (through Ontological Oscillation) they either disappear (with the whole Universe) into NOTHINGNESS, or they disappear into a State of Fohatic Disengagement which leaves only the non-particulated Original Players in the Cosmic Process intact. Such Players would be the Universal Logos, the Son of the Logos (of Fohatic Origin) with all Its hierarchically emanated "Sons of the Son", Fohat Itself (as Holy Spirit) with all Its 'Sons' in Their *spirit* state, and, finally, *unarticulated* Cosmic Prakriti (from one perspective, the Mother of Cosmos). Fohatic Disengagement would last for only the 'splitest split' of a second, but enough to make the entire differentiated/articulated Universe *vanish!*

Example: The solution to the problem of whether the Universe vanishes between each ultimate moment into NOTHINGNESS or into a Subjective State of Fohatic Disengagement, hinges upon whether the Members of First Family of Gods are particulate or imparticulate. It is the thesis of this treatise that They (as well as the spirit aspect if all Cosmic Participants) are, within that 'Particle/Event' called Cosmos, imparticulate. They are 'partite' but non-aggregated, and, hence, 'imparticulate'.

forever (noun)

By *forever* is meant ETERNAL BEGINNINGLESS/ENDLESS DURATION.

Example: NOTHING lasts forever. No thing lasts forever. Is this a contradiction?

forever (verb)

By the term *forever* is indicated the act of abiding in a state that was never begun and can never end.

Example: The ALL-SELF 'FOREVERS' in the 'STATE' of ETERNAL DURATION. On an infinitely lower turn of the spiral, some of us "summer" in the mountains, but *not* forever. Does it seem that the ALL-SELF is on a beginningless/endless 'VACATION'? (There is something in this last word.)

form

By *form* is meant (usually) a relationship between variables such that when an impression of these variables registers in consciousness, the impression of *extension* occurs. A form (as usually considered) thus appears material or tangible. More generally, a form is a kind of 'fixation in the Universal Flux'. When considered to be a form, a relationship between variables is either treated as a semi-'durating' object (even if the duration be exceedingly brief—such as the form which is the trail of a sub-atomic particle in a bubble chamber) or the relationship *actually does* have an endurance cognizable by an instrumentally-unassisted human consciousness.

As well, a form need not be tangible and possessed of apparent 'extension'. Patterns of energies that have (at least to the present human consciousness) no visually perceived or apperceived extension can also be considered *forms.* An *idea*, itself, which has been called a *"being incorporeal"* can also be considered to have what might be called a 'formless' form. Although ideas emanate from the *arupa* or formless planes, the energies which underlie an idea are mathematically configured/interrelated and can be thought of as existing in, at least, a *mathematical* form. Since all factors-in-Cosmos are, from the 'HIGHEST PERSPECTIVE', *forms,* ideas too, must have form, no matter how formless they may seem when compared to the grosser and more usual 'shaped'/'extended' forms cognized by human consciousness. The forms of ideas, however, must be extremely subtle, and configured in ways which the human consciousness could only consider abstruse and unusual. Ideas might be called 'Qualified Points'.

Forms can be configured in Time as well as in Space. Time-configured-forms are less tangible and are more difficult to cognize than Space-configured-forms. The elements/factors which constitute Time-configured-forms may well be simply a *succession of vibrations.*

When considering the putative 'form' of ideas, it may well be that ideas (which can be considered as *relatively* non-extended forms, i.e., forms *relatively* devoid of extension) may be organized on the basis of the *frequency-configuration* of extremely subtle and 'compacted' matter/energy units. This is to suggest that ideas are 'tiny' when compared to the extended forms they generate in grosser matter. Since Cosmos *Is* Extension (only the ABSOLUTE being absolutely 'deprived' of extension) one is hard pressed to imagine that there can be anything whatsoever in Cosmos which is *utterly* non-extended—i.e., *unperceived as an object.* By this definition, even a Real point has extension (though an 'elusive extension' which is ever immeasurable).

Following on, it may well be that ideas *do* indeed have 'extension', but an extension so concentrated and so relatively minute (probably an 'infinitesimalizing extension'), that ideas would not be considered 'material' in the usual sense, even though they, like everything else in Cosmos, are 'material' (i.e., perceivable-thus-objective from the Point of View of the 'WITNESSING'—from whatever 'SOURCE'/Source, the 'WITNESSING' derives; thus materiality is 'subtlized'. Only that which 'INHERES' within the INFINITESSENCE (or, better, *IS* the INFINITESSENCE) is non-material. Everything 'extra-SOURCE' is fundamentally *actual* and, hence, material, though certainly not *grossly* so.

The question of 'extension', point, and field is most subtle. A point (as a content of Consciousness) is, perhaps, not *truly* dimensionless, otherwise it would disappear from Consciousness altogether and be a *nothing.* Rather, it is probably more useful to conceive of a point as an 'infinitesimalizing' (participle-noun). A point is so 'small-ing' (as no 'definite' view of it can ever be had, and it *remain* a Real point) as to have no locality, and thus to be found everywhere within the field in question. This means that if Ideas are Real points, they are Really non-local, and can be found (just like any other Being) ubiquitously, throughout the field. Thus are points-as-Ideas as 'tiny-ing' as possible without being *nothing* and yet, entirely *extended* throughout the field in which they are 'resident'; therefore Ideas can be extremely *concentrated* and yet entirely *extended.* Such are the apparent paradoxes.

Ideas configure what we call tangible matter, and different frequencies of this relatively tangible matter can come into relationship with each other (and be *held* in relationship) on the basis of their resonant interplay (magnetic and repulsive) with higher-order frequency-patterns in far subtler (i.e., arupa) matter. It may well be that these higher-order frequency patterns in subtle matter *are,* in fact, *ideas.* But we do not want to materialize the concept of the Idea unduly. Ideational patterns can *induce* frequency among prakritic constituents, without being *prakriti* themselves. Ideational Points within any Self-Perceiver are indeed 'parts', but are so seamlessly related through the Monad (the 'jealous' Oneness) as not reducible into aggregable 'particles'.

It is explained [at the end of Section II] how points and fields interplay. These thoughts must be born in mind when assessing the 'differentiable non-particulateness' of the Idea. All this notwithstanding, from the Perspective of the INFINITE SELF's 'SEEING' known as 'MAYA', all Life-in-Cosmos must be considered as *in form*. Only the ABSOLUTE, in ITS 'STATE' of 'INFINIDENTIFICATION' is utterly *FORMLESS*.

Example: The pattern of planetary, solar systemic, and cosmic energies existing at the moment of a newborn's first breath are configured in a subtle pattern or *relatively* 'formless' form which results in the appearance throughout life of many far more tangible or concrete forms resonantly attuned to the original relatively formless pattern. These energies may nonetheless be 'particulate reflections' of 'imparticulate' Ideas which configure them from the World of Being.

Example: Do apparently formless patterns have 'imparticulate' ideational form? Are they seamless *unities* compelling discontinuities towards Mosaic unity?

FOUNT OF ALL POSSIBILITY

By the *FOUNT OF ALL POSSIBILITY* is meant the INFINITE POTENTIAL.

Example: All Universes past, present, and future, which *cumulatively* have displayed, now display, and will display an infinite variety of modifications of the Infinite Homogeneous Substance, REALLY emerge from the PROVIDER of all possibility for modification—the FOUNT OF ALL POSSIBILITY.

Example: The FOUNT OF ALL POSSIBILITY is a *name* for the NAMELESSNESS which emphasizes IT as the PLENUM rather than the VOID or the NO-THING.

fragmentation

By *fragmentation* is meant a condition in which the units which comprise any whole are apparently divided or separated from each other, and function without coordination. Fragmentation pertains only to the World of Conditions. From the Perspective of the Consciousness of the One Being in-Cosmos (or, still higher, when All is considered from 'within' the PRESENCE of the ONE BEING/NON-BEING) *fragmentation* like division is an impossibility.

Example: During certain phases of personality purification preceding personality integration, the aspirant to discipleship is likely to pass through the experience/'inperience' of some degree of psychological fragmentation.

Example: Fragmentation is the direct result of SELF-'VEILING'/Self-Veiling. Consciousnesses, Who Really know the Intended Pattern cannot fall into fragmentation.

Example: The World of Being is 'partite' but totally unfragmented. In fact, fragmentation cannot Really exist, but, under the Veil of Ignorance, it *seems* to, especially in the Worlds of Fabrication.

- G -

god

By *god,* or a *god,* can be meant an authentic Entity-in-Cosmos which has risen to a rank within the Hierarchy of Being that allows it to master the seven subplanes of the cosmic physical plane.

By the term *god,* or a *god,* is meant *any* authentic E/entity at all, when the nature of the E/entity is considered entirely from the perspective of *spirit.* As all authentic E/entities are, Essentially, Spirits, and all true Spirits are Monads (upon whatever level and with whatever scope), and all Monads are gods—all authentic E/entities are, indeed, *gods*—first in Essence and later (in varying degrees) in manifestation.

Example: St. Paul was speaking of the Monadic Aspect in man when he said, "Know Ye not that Ye are gods."

Example: All E/entities are, *essentially,* gods, who through further development can ascend to a higher order and rightfully be called 'Gods'-becoming-God.

God

By *God* is meant the One Being Who, Essentially, Is the Universal Logos of any particular Cosmos, and in Whom all beings-in-Cosmos, "live and move and have their being". God is infinitely limited when compared to GOD, and yet God *Is* GOD. Further, each G/god, though limited differently in comparison to each other, is the *same* GOD in ESSENCE—a Radiatory Extension of the Universal Self Who Is, in turn, a 'RADIATORY EXTENSION' of the ABSOLUTE SELF, and, even more *Is* that 'RAY' 'ITSELF'.

Example: Each of an infinite number of successive Cosmoses has Its Principal God Who is the Universal Logos of that particular Cosmos. All Universal Logoi, though differently limited, are REALLY the *same* IDENTITY, though, intra-Cosmically, They are not the same Identity.

GOD

By *GOD* is meant the ONE ABSOLUTE Extra-Universal (as well, an Intra-Universal) DEITY WHO IS the ONE SELF.

Example: To GOD, *per se,* no prayers can be offered, because there is naught but GOD, the ONE ABSOLUTE REALITY.

GOD-as-God-as-god; god-as-God-as-GOD

By the terms *GOD-as-God-as-god,* and *god-as-God-as-GOD,* are meant exactly the same as the SELF-as-Self-as-self and the self-as-Self-as-SELF. These terms are formulas for 'RADIATORY'/emanatory descents and ascents.

Example: The term *god* has become so abstracted and transcendentalized that it is difficult for human beings to conceive the *identicalness in essence* of all *levels of 'godness'* which the terms 'GOD-as-God-as-god' and 'god-as-God-as-GOD' suggest.

God Immanent

By *God Immanent* is meant the Presence of the Universal Logos 'in' each Monad of the One Universal Monad (i.e., in each authentic subjective E/entity).

Example: The discovery of the entire Universal Hierarchy within the deepest recesses of one's being is a testimony to the Presence of God Immanent.

Example: God is Immanent within each authentic E/entity in-Cosmos due to the unobstructable *channel* created by the Divine Emanatory Stream.

God Transcendent

By *God Transcendent* is meant the Universal Logos in Its Fullness upon Its Own Plane. Consideration of the Divine Emanatory Stream is not pertinent to the understanding of God Transcendent.

Example: From the 'INFINISPECTIVE' of the 'INFINIDENTIFIED' INFINITE SELF, God Transcendent is still Finite!

GOD TRANSCENDENT

By *GOD TRANSCENDENT* is meant the ONE AND ONLY DEITY WHO, while necessarily and inescapably 'BEING' ITS Creation, yet is 'SOMETHING' entirely 'OTHER' and removed from Cosmos.

Example: The *full nature* of GOD TRANSCENDENT is never appreciated by beings immersed in-Cosmos. GOD TRANSCENDENT IS a 'STATELESS STATE' of INFINITIZED SUBJECTIVITY.

- H -

Hierarchy, Creative

By a *Creative Hierarchy* is meant an Order of Lives, a Great Collection of Monads of a generally equal developmental status, which is focused at a particular level of prakritic objectification.

Example: The Fourth Creative Hierarchy is the Hierarchy of Human Monads. This Creative Hierarchy stands below the Hierarchy of Triadal Lives, and above the Hierarchy of Lunar Lives. The Fifth Creative Hierarchy of Solar Angels, though apparently 'lower' than the Fourth is Really more advanced, and is only temporarily assuming a sacrificial, redemptive position which makes Them the 'servants' of the Fourth or Human Creative Hierarchy.

Hierarchy, the Spiritual

By *the Spiritual Hierarchy* of the Planet is meant the "Society of Organized and Illumined Minds" (also called, "Christ and His Church", or the "Great White Brotherhood"). The Spiritual Hierarchy is an assemblage of highly developed human beings (the 'Graduates of the School of Human Evolution') Who (in cooperation with Sanat Kumara, the Lord of the World) are committed to the spiritual uplift and salvation of humanity.

Example: The members of the Spiritual Hierarchy of our planet possess the vivid identificatory experience that the Human Monad is a 'Ray' of the ABSOLUTE.

'hole'

A *hole'* is an *articulated subjectivity* arising in 'INFINITE SUBJECTIVE DENSITY'-instantly-Infinite Subjective Density which, through the dynamic of 'Self-Sight', become an *articulated objectivity* taking form in Infinite Objective Density.

Example: Every appearance is a 'hole' in infinitely dense 'somethingness'.

Example: All 'holes' are 'within' the Self *before* they are in Its 'Sight of Itself'.

Example: A 'hole' is a discontinuity is NOTHINGNESS. A 'hole' is always *less* than NOTHING.

Example: 'Holes' are emergent articulated subjective possibilities becoming objectified possibilities.

Holographic Kaleidoscope, Cosmic

By the *Cosmic Holographic Kaleidoscope* is meant the persistently changing arrangement of all things in-Cosmos (the Cosmic Configuration) such that every change within the Greater Whole is immediately reflected in the many lesser wholes which analogically correspond in structure to the Greater Whole.

Example: Fabricated Cosmos, Itself, is a perpetually turning Holographic Kaleidoscope reflecting the seamless Archetypal Ideation of the World of Being.

Example: What is the 'All-Seeing-Eye' 'looking' through the Cosmic Holographic Kaleidoscope?

Example: The Cosmic Holographic Kaleidoscope turns by Law and the arrangement of its pattern-making parts (though incremental and for long imperfect) occurs not by chance.

Example: The 'turning' of the Cosmic Holographic Kaleidoscope is the 'turning' of the Great Universal Wheel, which turns but once in a Universal Manvantara.

homogeneity, a

By a *homogeneity* is meant (for practical, in-Cosmos purposes) a collection or aggregation of items in which each item is almost identical with every other item. We are dealing here with a *virtual* sameness and not *exact identicalness*.

Example: A complete homogeneity is an impossibility in-Cosmos. Only PARABRAHMAN is the *UTTER* HOMOGENEITY, and Mulaprakriti Is so in a slightly lesser Super-Cosmic sense. PARABRAHMAN IS an SUPER-Cosmic ENTITY/NON-ENTITY and Mulaprakriti is a Super-Cosmic Entity/Non-Entity. Numbered B/beings are never complete homogeneities.

Example: Would an aggregation of all ultimate particles in-Cosmos be a homogeneity? For practical purposes, yes, but certainly not a *perfect* *homogeneity*, which only PARABRAHMAN IS.

Since ultimate particles are indivisible and are caused by (and more, *are*) the Presence and Activity of intra-Cosmic Fohat Self-Reflectively *'in'* Cosmic Prakriti, these particles would be more uniform than complex particles, possibly even identical to each other (for such might be the Will of Fohat). But even an aggregation of such particles would be simply an aggregation of possibly identical items—items with *prakritic boundaries* of a nature *different* to the *sub-stance* of the item itself. The outside 'wall' of the particle would be, as it were, the 'divided' density of Cosmic Prakriti Itself which, Cosmo-Psychologically, is 'Reflected Infinitude'! The outside 'wall' of the particle could also be considered the *outer limit* of a single unit of 'Fohatic Perception'—in the process of *not* perceiving Homogeneity of Selfhood. Hence, the incompletely homogeneous nature of even a collection of possibly identical items.

In judging whether an aggregation of all ultimate particle/events would be a *true homogeneity*, what we would have is Cosmic Prakriti 'aerated' as it were by the 'bubbles' of 'NOTHINGNESS' which are what Fohat, the *indirect* Emissary of 'NOTHINGNESS' (Itself, an Extension of 'FOHAT', the *direct* 'EMISSARY' of NOTHINGNESS) Creates when It *Self-reflectively* invests the 'ESSENTIAL' NOTHINGNESS (or SUBJECTIVITY/Subjectivity) It (Fohat) *Is* into the 'Somethingness' (i.e., the Objective Reflection 'Seen' and thus 'Created' with SELF-Inspired Intent by the Universal Logos) which Cosmic Prakriti *Is*. (Cosmo-Psychologically, we have diminishing Self-Perceptions within Self-Perceptions, each type of Self-Perception deemed a type of prakriti or Prakritic Field.)

Therefore we would have a combination of 'Something' (*primally*, Pre-Cosmically conditioned Cosmic Prakriti—conditioned by Maya/Fohat in Mode *3)* and 'NOTHING', the Presence of Self-Divided/Enumerated Intra-Cosmic Fohat. Something and NOTHING mixed are *not* a continuum or a *pure* or *perfect* *homogeneity*; only 'NOTHING' IS. (It cannot be forgotten that even the apparent *externality* we call 'Something' is Really a *reflection* of NOTHING-as-Nothing.) The same Cosmic Process can be described in terms of Internality/Externality, and in terms of Cosmo-Psychological Interiority in which there is no Real *externality*.The latter is the more accurate description.

The problem of 'FOHAT'/Fohat and Its Activities and Transformations is clearly of extraordinary difficulty, and the difficulty is exacerbated by the fact that conventional language cannot be used to discuss the problem clearly or meaningfully. What is offered here is largely for the purpose of stimulating exploratory thought on these matters. We must rely upon our powers of visualization, relatively feeble though they may be, to deal with abstractions of so rarefied a nature as to make all happenings on even the higher levels of the Cosmic Physical Plane look utterly concrete.

HOMOGENEITY, the

By *the HOMOGENEITY* is meant the SELF as the GREAT SAMENESS. With respect to intra-Cosmic Homogeneity, all E/entities or items are considered *essentially* identical, though they are not *actually* so. Within the HOMOGENEITY, however, there are *no items* or *E/entities* at all, whether identical or different. As there are no items or E/entities to account for in the HOMOGENEITY, whatever IT IS, is a COMPLETE CONSISTENCY—an *undifferentiated, unarticulated* CONTINUUM—hence, an ABSOLUTE HOMOGENEITY.

Example: Because the SELF IS the HOMOGENEITY, all selves in-Cosmos are utterly identical—in ESSENCE.

Homogeneity, the Infinite

By *the Infinite Homogeneity* is meant Mulaprakriti as the Potential for Infinite Material Possibility. But even Mulaprakriti, because It is an 'EXTRUSION' from the INFINITESSENCE, and technically 'outside' the CONTINUUM (i.e., 'extra-SOURCE'), is, technically, part of the Dis-Continuum, and abides in the World of Discontinuity as one of the greatest of Discontinuities. Only PARABRAHMAN is *utterly continuous.*

Example: Mulaprakriti is the Infinite Object and, before Her 'potentio-particulative interplay' with the 'De-Infinitizing' Infinite Subject, is totally homogeneous (*sui generis*) and infinitely *dense.* For this reason She can reasonably be called the Infinite Homogeneity.

Example: Mulaprakriti, being the Infinite Homogeneity is far denser (infinitely denser) than the so-called "dark matter" of space. Dark matter pertains still to our systemic etheric/physical plane, which is extremely differentiated and, hence, 'porous' compared to Mulaprakriti which has not been differentiated, conditioned or 'aerated' at all. Aerations are Self-Perceptions.

HOMOGENEITY, the INFINITE

By *the INFINITE HOMOGENEITY* is suggested the strange idea that (despite an infinite number of 'appearances' to the contrary) the UTTER ALLNESS, has been UTTERLY HOMOGENEOUS throughout INFINITE DURATION and will remain so.

Example: What quality of Realization does it take to see a World full of apparent modification and yet to realize that now, forever before, and forever after, the INFINITE HOMOGENEITY is forever undisturbed?

- I -

'I' (not bold)

By *'I'* is meant the Self-in-Cosmos when It is identified with very localized object/ matter and, hence, can properly be called a 'self'. There are times when this 'I' is so constricted that it should be even more properly called an 'i' (which indicates identification with the lunar vehicles rather than as an integrated personality). It would be too laborious, however, to be overly technical and rigorous in the use of the proper I-symbol throughout this work. It is enough for the reader to know of the possible technical differentiations.

Example: The usual sense of 'I' held within the consciousness of most human beings is limited by the *ahamkaric principle* and is, thus, utterly illusory.

Example: What **8** usually mean by 'I', is 'MySelf' (SELFHOOD in its intra-Cosmic Mode) as **8** function through My personality vehicles. 'I' am a personal being; '**8**' Am not.

'I' (bolded)

By *'I'* is meant the ABSOLUTE SELF, the ALL-SELF, or more simply, the SELF considered not so much as the ULTIMATE PRINCIPLE but as the ULTIMATE IDENTITY.

This symbol is not used to describe actions 'in-Universe', for the **I**, as a PURE ABSTRACTION does not 'ACT' within the World of Becoming. But, mysteriously, the **I** does SELF-'VEIL' before the formation of a Universe and (through ITS Extension, the **8**) continues to do so during the involutionary process.

- The term, '**I**' can only be used in a SUPER-Cosmic application
- The term '**I/8**' in a Super-Cosmic Application
- The term '**8**' can only be used intra-Cosmically

Example: The search for *my true identity* must end in the **I**, the ULTIMATE IDENTITY, even though, along the way, 'I' may think myself to be an entity of lesser scope and stature.

'I/8'

By ***I/8*** is meant the Pre-Cosmic interface between the ABSOLUTE SELF, the **I**, and the ABSOLUTE SELF in-Universe, the **8**. The ABSOLUTE SELF has been, in a way, 'FLASHED FORTH' as the 'RAY' of the ABSOLUTE, but has not yet been *cosmified* as the **8**, signifying the ABSOLUTE SELF while in Universe.

By ***I/8*** from another perspective is meant the SELF-as-Infinified Point:

- The **I** IS Intra-SOURCE
- The **8** Is Intra-Cosmic

The *transitional phase* between these two types of *identity* is Pre-Cosmic or Post-Cosmic and is represented by the amalgamation of the two symbols. Thus **I/8** is the Identity of the Infinite Subject 'RADIATED' forth from the ABSOLUTE, but not yet *cosmified*, not yet *bounded* by *Cosmic* Prakriti (by *Cosmo*-Objectivity).

Example: **I** and **8** are ESSENTIALLY the same. Whenever the **8** is Present, the **I** is latently 'PRESENT' as well, and the symbol **I/8** reminds us of their inevitable simultaneous appearance. Sometimes the symbols **I-as-8, or 8-as-I** are similarly useful to indicate the *ontological progression* between INFINITE IDENTITY/NON-IDENTITY and Finite Identity, or between Finite Identity and INFINITE IDENTITY/NON-IDENTITY (**8/I**). The symbol **I/8** represents a kind of Identity that is neither *absolutely infinite* or *cosmically finitized.* REALLY, however, all *identity* is, inescapably, INFINITE IDENTITY.

'8' (pronounced 'I')

By **8** is indicated the link which makes possible the transposition of **I**-to-I, and I-to-**I**. By **8** is indicated the realization that the ABSOLUTE EGO (the SELF, the **I**) pervades and is the very ESSENCE of all relative egos or 'I's'. He who uses the term **8** to describe his I-ness, instead of merely 'I', has reached some degree of identification with the SELF-as-Self.

NOTE: According to the conventions adopted in this treatise, the symbol **8** has only an *intra-Cosmic application* (and some Pre- and Post- Cosmic implications) and cannot be used to indicate the ABSOLUTE SELF, **I**, in its extra-Cosmic 'STATE'.

The term **8** has a tremendous, range, however, and can indicate, on the near end of the spectrum, the spiritually awakening human being, and on the far end, the Universal Logos, Itself (Who is an **8** and *not,* in Its Cosmic Role at least, the **I**).

Example: While the consciousness of my 'I' is prakritically bounded and thus, per force, ego-conditioned, **8** (my truer more essential I) is identical both to "I Am That I Am" (indicating the Universal Logos) and, *ultimately* to "I AM THAT I AM", for the **8** is ESSENTIALLY one with the ABSOLUTE EGO/NON-EGO. **8**, in essence, Am the *same* as every other **8**, and all of us are identical with **I.** As **8**'s we are ESSENTIALLY non-distinct, though our 'prakritic outposts' (our prakritic vehicles) will vary from each other. Because as **8**'s we are ESSENTIALLY identical, as **8**'s we are not caught in the snare of separative ego identification. A sense of separateness is a "sine qua non" of egoity.

'I-as-8'

By ***I-as-8*** is meant the ABSOLUTE SELF manifesting as the PRESENCE of the SELF in all E/entities in-Cosmos. Every E/entity in-Cosmos is ESSENTIALLY an **I**, but the term **I** is reserved for the SELF as IT IS when there is no Cosmos, or when Cosmos is not a consideration. The **8** represents the SELF or the **I** when IT is PRESENT *in-Cosmos* during Universal Manvantara. Often when the **8** is used as a term indicating the in-Universe PRESENCE of the SELF (i.e., the in-Universe PRESENCE of **I-ME**), the term **I-as-8** will occasionally be used to remind the reader that the **8** always is *substanded* by the PRESENCE of the **I.**

Example: When **8** realize that my True intra-Cosmic Nature is expressed by the symbol, **I-as-8, 8** find 'MySelf' filled with full SELF-Confidence, because **8** realize My tiny 'I-ness' to be substanded by the ABSOLUTE.

'8-the-Localized Whole'

By ***8**-the-Localized-Whole* is meant, that no matter through what type of entification **8** Am manifesting, **8** Am always and ever the Whole (the Cosmos as Entity) manifesting in a *localized* condition. Even the Cosmos Itself is, in a sense, *localized* though it might well be asked if the Cosmos is, indeed, *anywhere*.

Example: While I-am-**8**-Am-**I**, and thus, ESSENTIALLY not only the Whole but the WHOLE, when **8** Am in-Cosmos, **8** function as a Localized Whole (as 8-the Localized Whole) because at this time during Universal Manvantara, **8** must function upon the lower levels of Cosmic Physical Plane, and no other. Such is the stage of My development as an 'I' and My stage of *localization* reflects that stage.

'I-the-Whole'

By *I-the-Whole* is meant the realization of each apparently isolated **8**-as-I of its identicalness with the Wholeness which is Cosmos, and eventually with the Super-WHOLENESS which is the ALL-SELF, the ONE SELF.

Example: The disciplines of Identification lead the initiate to the consciousness of I-the-Whole which is equivalent to the consciousness of Isolated Unity.

idea

By an *idea* is meant a subtle structuring force in-Cosmos.

By an *idea* is meant "a being incorporeal giving shape and form to formless substance."

By an *idea* is meant an aggregation of matter-moulding energies which have no 'extension' (in the usual sense) but *do* have a pattern. (The question arises, how can there be pattern without extension? Ideationally, 'pointally', it is possible. We might say, speculatively, that ideas (on their most apprehensible and, hence, not *pure* level) are pattern-encoded atomic configurations bearing a resonant relation to a variety of 'lower' specific intra-Cosmic Energies. The Energies with which an idea is resonant are reflected in the encoded pattern of the idea.

It must be remembered that no systemic or even cosmic planes can possibly be utterly formless. Such planes must be at least 'partite' if not *particulate* or they would be *absolutely continuous* (which, in Cosmos, is impossible). Hence even "formless ideas" are 'material', in a sense of being 'reflected'/'imagaic', and 'partite'. The forms, however, lack 'extension' in the usual sense. They are far more condensed, probably to the point of 'infinitesimalization'.

By an *idea* is meant an energic pattern which embodies the formula of a relationship.

Example: No conditions can resist an idea whose time has come. This simply mean all forms of life do God's Will, whether they will or no.

Example: Ideas are often received suddenly, but their "working out" in Time and Space may take years.

'IDEA'/Idea, SUPER-Cosmic-as-Super-Cosmic-

By a *SUPER-Cosmic-as-Super-Cosmic 'IDEA'* is meant an 'ARCHETYPAL PATTERN' 'EXTRUDED' from the FOUNT OF ALL POSSIBILITY. The Universe-to-Be is structured on the basis of these Super-Cosmic 'Ideas' once they have become *cosmified* into Cosmic Ideas.

Example: It is the function of Cosmic Prakriti to Embody (through Mayavically Induced *objectifying reflection*) the Super-Cosmic 'Idea'-once-'IDEA', which become the Universal Theme to be enacted by the Universal Logos.

Example: The Universe, Itself, *Is* a Super-Cosmic 'IDEA'-as-Idea. 'FOHAT' in Mode 1 'EMERGES'-as-'RAY'-as-'Ray' from the PLENUM as the Embodier and Representative of that Super-Cosmic 'IDEA'-now-Idea.

Example: SUPER-Cosmic 'IDEAS'-become Ideas arise from 'NOUMENESSENTIAL IDEATION' (which cannot 'EXIST', and yet, must!), and develop (via the 'De-Infinitizing' Subject/Object) into the Cosmic Ideation *thought* by the Universal Logos and of which intra-Cosmic Fohat is the Emissary.

identical; identicalness

By the term *identical* is meant an absolutely and completely shared sameness of form and essence. *Identicalness* is only *completely* possible in the WORLD OF BEING, and never in the World of Becoming, in which only identicalness of ESSENCE (but never of form) is possible.

Example: To achieve the consummation of one's life, it is essential that one realize himself/herself to be identical in ESSENCE with every other human being, and, even, with every other being of any kind.

Example: Due to the Principle of Unrepeatability, forms in-Cosmos can never be exactly identical with one another.

identifiable (noun)

By an *identifiable* is meant that which cannot be experienced fully through either perception or apperception, but only through identification. All beings are, ESSENTIALLY, neither 'perceivables' or 'apperceivables' but, rather, *identifiables.*

Example: When **8** think of any E/entity-in-Cosmos, **8** can either perceive that E/entity or apperceive it, depending upon the degree of subtlety and interiority of the normal expression of that E/entity. In order, however, for me to become as intimate with that E/entity as possible (*infinitely intimate*), I must *identify* with the E/entity. The ESSENCE of any E/entity is neither perceivable or apperceivable, but only *identifiable.* Therefore, E/entities, in ESSENCE, can be with reason called, identifiables.

identity

By *identity* is usually meant those qualities or patterns of any B/being which distinguish it from every other B/being. But qualities and patterns are really *secondary identifiers.* More accurately, *identity* should mean the permanent, unchanging *core* of any B/being—that which remains when everything of a secondary nature associated with that

B/being is taken away or negated. Identity is *primary* and can never be taken away or negated. We see, therefore, that the term *identity* can be associated with *difference and uniqueness*, or more fundamentally, with *sameness*.

Example: The search for identity is the human being's most important quest. Every man or woman, whether or not they know it consciously, wants to identify with the PERMANENT IDENTITY they have been or will be regardless of the circumstances or the prakritic vehicles through which they have functioned or may be functioning.

Example: An Identity-in-Cosmos is one of a specifiable multitude of *authentic points.*

Identity, Depth; Identity, layers of Depth

By the *Depth Identity* is meant the Identity of any 'Ray' of the ABSOLUTE which lies 'beneath' Its immediately presented Focal Identity.

Example: If **8** Am an apparently Individual Universal Life Unit (an emanated Monad of the One Universal Monad), then *that* is my immediately presented Focal Identity, but **8** have 'layers' of Depth Identity 'beneath' that Focal Identity. For instance, if **8** penetrate deeply into My Nature, **8** will discover:

- that an Identity as one the Seven Subsidiary Logoi of the Universal Son underlies My Identity as an apparently distinct Individual Life Unit
- that underlying that Identity, is My Identity as one of the Three Sub-Logoi of the Universal Son
- that underlying that, is My Identity *as* the Universal Son
- that underlying that is My Identity *as* the Father

—and so forth, passing through many deepening layers of Depth Identity, until **8** discover my ultimate Identity *as* the ABSOLUTE ITSELF

Identity, Focal

By the *Focal Identity* is meant the extent and dimensional depth of Identity experienced (within intra-Cosmic Prakriti, i.e., under limited 'Self-Sight') by a 'Ray' of the ABSOLUTE, i.e., by a Subjective B/being upholding a certain position or function in the Spirit-Structure of Cosmos.

Example: The Focal Identity of that Subject called the Son of God (the Son of the Universal Logos) is *immediately* that of the Son, but there is also a layer of Identity beneath that of the Son's Sonship, for the Son is also the Father. If the Son penetrates deeply enough into His Identity, He will discover that beyond the Identity in which He is focused (i.e., His Sonship), His Depth Identity includes the Identity of the Father.

IDENTITY, the ONE

By *the ONE IDENTITY* is meant the ONE AND ONLY BEING in the UTTER ALLNESS, present as the PRESENCE forever.

Example: Who am I? Who Am I-as-**8**? Who AM I-as-**8**-as-**I**? ESSENTIALLY, **I** AM *the* ONE IDENTITY, and always and forever have been, just as, always and forever, **I** WILL BE.

IDENTITY, ESSENTIAL

By *ESSENTIAL IDENTITY* is meant the maximally internal sameness-of-SPIRIT which every E/entity shares. E/entities may express through an infinite variety of prakritic conditions, but, IN ESSENCE, their *true* IDENTITY is utterly identical.

Example: Outwardly an atom of silicon and a human being appear to be very different E/entities, but their ESSENTIAL IDENTITY is the same—*identical.*

identity, phenomenal

By *phenomenal identity* is meant the impermanent patterns which (in Time and Space) distinguish one E/entity from another. The term *phenomenal identity* is relative. The same distinguishing pattern may be considered phenomenal or (relatively) noumenal depending upon the 'altitude' of the perspective.

Example: From the perspective of the Spiritual Triad, the pattern and condition of the Egoic Lotus of a human entity can be considered that entity's phenomenal identity, but from the personality perspective the Egoic Lotus could be considered the noumenal identity of that human entity.

Example: Within Time and Space (which means in-Cosmos or Super-Cosmos) all patterns of identity are Really patterns of phenomenal identity. The only truly NOUMENAL IDENTITY is THAT. All other so-called noumenal identities are only *relatively* noumenal. In fact, they are phenomenal.

Example: From a certain perspective, phenomenal identities are *objects* and noumenal identities are *subjects.*

Example: Any form of identity other than ABSOLUTE IDENTITY is REALLY phenomenal identity.

IDENTITY, PRIMARY

By *PRIMARY IDENTITY* is meant, precisely, ESSENTIAL IDENTITY The term *PRIMARY IDENTITY* simply emphasizes the *primacy* of that which is fundamental, at root, i.e., ESSENTIAL.

Example: **I**-as-**8**-as-I may manifest as a human being, but My PRIMARY IDENTITY (which **8** share with all E/entities) IS SPIRIT.

identity, secondary

By *secondary identity* is meant the sense of self which is 'inperienced' by a self-conscious being when it focuses upon what it *has* rather than upon what it ESSENTIALLY *IS.* Any self-conscious E/entity or I/identity, when unidentified with its ESSENCE, 'inperiences' a kind of identity (which we are calling *secondary*) by means of which 'it *is* (or *seems to be*) through what it *has*'. The E/entity could more wisely forget (i.e., avert its gaze from) what it *has* in order to identify with what it IS— ESSENTIALLY.

Example: Whereas I often think of myself as an individual conditioned primarily by the third, second, sixth and fourth Rays, these Ray patterns relate only to my secondary identity. Although **8** characteristically *express* through fields qualified and conditioned by these rays, **8** *Am*, ESSENTIALLY, a non-qualified, unconditioned *being,* Who is a 'Ray' of the 'RAY' of the ONE ABSOLUTE BE-NESS.

Example: A secondary identity can be very 'high'; it need not be only personal. If **8** come to know 'MySelf' as a Monad upon the second subplane of the cosmic physical plane, **8** Am still knowing 'MySelf' in a *secondary* way, i.e., as a secondary identity. Even if after aeons and aeons **8** come to know 'MySelf' with certainty as Ishvara, the Universal Logos, **8** still know 'MySelf' in a secondary way. Only when **8** have identified 'MySelf' with MYSELF (the ONE AND ONLY SELF) will **8** be transformed into **I**, WHO **8** have always *been*, and my secondary identity be absorbed into MY *primary* 'IDENTITY'—the INFINITE SELF.

Identity, the Universal

By *the Universal Identity* is meant the Universal Logos considered as the One Identity in all of Cosmos. The Universal Identity is a 'RADIATION' of the ABSOLUTE IDENTITY and, while being the same *in ESSENCE* as the GREAT BEING/NON-BEING, is possessed of infinitely less possibility that the ABSOLUTE SOURCE.

Example: Every E/entity-in-Cosmos will at length discover itself to *be* (and always to have *been)* the Universal Identity—Ishvara, the Originator and Sustainer of this particular Cosmos.

Example: To discover oneself as the Universal Logos still indicates a degree of Illusion. It is only modifications of Cosmic Prakriti (i.e., of Self-Perception) which make the Universal Identity seem in any way distinct from the ABSOLUTE IDENTITY.

Example: Every Authentic Identity (in reflection of the Universal Identity) is *One (unitary* in Its Identity however *multiple* It may be in Self-projected/emanated Objectivity). This is to say that One Father can have many Sons.

illusion; illusory (defined by degree)

By *illusion,* technically considered, is meant *any* actuality—*any* presentation in the World of Becoming whatsoever

By *illusion* is meant any presentation in-Cosmos (which means 'in-Consciousness) which, however actual or factual, is ESSENTIALLY un-REAL because impermanent and reducible to the HOMOGENEOUS SUBSTRATUM.

Example: Any modification-in-Cosmos whatsoever, regardless of how refined or 'high' that modification may be, is, ESSENTIALLY, an illusion.

Example: Any possibility which does not 'INHERE' *noumenessentially* within the FOUNT OF ALL POSSIBILITY, the INFINITESSENCE, but instead, 'FALLS AWAY' from its *noumenality* into 'extra-SOURCE' *particularity* is an illusion. For instance, when the possibility of Harmony 'INHERES' *noumenessentially* in the INFINITESSENCE, *then* Harmony is *not* an illusion, but *has/is* REALITY in the INFINITESSENCE. But if the Idea of Harmony is borne by 'FOHAT/Fohat' into manifestation, cosmification, and finitization, then Harmony becomes a *particular* in-Cosmos Archetype, and, as high as that Archetype may be in Cosmos, and as much of a Cosmic Reality as It may be, nevertheless, It is necessary to say that the Archetype of Harmony would have to be considered *illusory* and un-REAL—though Real-in-Cosmos and Actual.

Example: In the INFINITESSENCE all 'POSSIBILITIES' are *infinitized*, and though having the potential to generate in-Cosmos illusory particularizations of themselves, are nonetheless in some way *identical* ('within' the INFINITUDE) with all other *infinitized* 'QUALITIES' similarly 'INHERING' within the INFINITESSENCE.

In other words, Goodness and Beauty are qualities quite different from each other when manifesting in-Cosmos. Goodness and Beauty as Cosmic Qualities, however, have their 'SEEDS' within the INFINITESSENCE and 'EXIST' 'THERE' as 'IDEAS' within the ABSOLUTE PLENUM (which PLENUM, remember, is *both* impartite and non-particularized). When Goodness and Beauty, however, are considered as 'INHERING' within the INFINITESSENCE, (abiding in an 'INFINITIZED STATE' in their ULTIMATE ROOT), then, they are *identical* with each other and with all *infinitized* 'QUALITIES' within the INFINITESSENCE.

Illusion, the Great

By *the Great Illusion* is meant the Universe, Itself, or, more extensively, all Universes past and to come. While the Great Illusion is usually associated with the distortions of consciousness common to those whose major field of activity is the Three Worlds of Human Evolution (the World of Gross Effects) it must be more fundamentally realized that the Universe is a Great *Object* (when Perceived from the Perspective of the Great Witness of the SELF-as-Condensed Point), and that every object is, essentially, illusory.

The Great Illusion (as Universe) is, of course, a Great *Actuality*, and must be treated with supreme respect, but It is also one of an infinitude of 'Great Impermanences' and should not be mistaken for REALITY. From a practical perspective, the Universe is, however, a Reality.

Example: All beings-in-Cosmos are subject to the Great Illusion throughout the Universal Manvantara. They cannot escape what we might call 'the presentation of objectification' until the GREAT DAY BE WITH US. (Notice that this ULTIMATE 'STATE' is *not* called the *"Night* be with us", although the Universal Pralaya *is* called the "Night of Brahma.")

Illusion-in-Universe

By *illusion-in-Universe* is meant any perception or apperception which distorts or is not congruent with that which is *Real-in-Universe.*

Example: Whereas the impacts of the buddhic plane can, from the tiny human perspective be considered Real-in-Universe, the majority of phenomena of the astral plane can certainly be classified in the category of illusion-in-Universe. From the perspective of the higher Cosmic Planes, since the energies of our buddhic plane pertain only to the etheric part of the Cosmic Physical Plane, they, too, could be reasonably judged as illusion-in-Universe. Although the entire Cosmos is Illusion from the highest perspective, the term illusion-in-Universe is intended to indicate a failure of certain forms to conform with Archetypes which (though ESSENTIALLY Illusory) are at least Real-in-Universe.

Image-Event

By an *Image-Event* is meant a willed, imparticulate thought in the World of Adjustment (which might be called the lower 'levels' of the World of Being. These Image-Events are created at the 'Speed of Will', and endure precisely as long as the *willing Creator* wills for them to endure These Image-Events are the kinds of 'actions' which occur within this World of Adjustment (intermediary between the World of Particulate Fabrication administered by Fohat and His 'Sons' and the World of Being, per se, regulated by the Universal Son and His Host, and sustained by the Universal Father).

Example: An Image-Event may be either:

1. A rapidly-occurring imparticulate image created by the 'enfolded' Monads in the Company of the Universal Son for the sake of guiding the 'Sons of Fohat' in their work of Cosmo-Objective Construction; or,
2. An imparticulate thought-image created by Fohat and His 'enfolded Sons' as They (with extraordinary rapidity) *plan* Their next 'move' in the World of Cosmo-Objectivity.

immobility; immobile

By *immobility* is meant relative or absolute non-moment.

Example: Ongoing immobility in-Cosmos is an impossibility. One could speak of the ABSOLUTE as being 'IMMOBILE', but since no particulate things exist within IT, the term 'IMMOBILE' would have *no-thing* to which to apply.

Example: The connotation of the term immobility, when used in its relative sense, often suggests a degree of inertia on the part of the item or E/entity being described.

Example: All things in Fabricated Cosmos are utterly immobile during the entire duration of an ultimate moment. If the possibility of such immobility is questioned, it could be asked, Can a Great Entity/Consciousness *fix* an Image? In Consciousness, anything 'conceivable' is thereby possible. Not all things, however, may be conceivable to a *veiled* Consciousness.

immutable

By *immutable* is meant that which never changes. The words 'mutable' and 'mutation' indicate change. The term immutable is only REALLY applicable to the ULTIMATE SUBJECTIVE 'STATE'.

Example: No object can be immutable. Even Mulaprakriti, which is the Infinite Object (REALLY, PARABRAHMAN 'SEEING' ITSELF), and is utterly dense and impartite, is *mutable* in a way because It does not *always* exist— existing in purity and immutability only in Pre-Cosmic and Post-Cosmic States. Further, Mulaprakriti is absorbed into PARABRAHMAN (INFINITE, ABSOLUTE SUBJECTIVITY) at the "Day Be With Us", and so is cyclically mutable forever.

impact

By *impact* is meant the forceful impression of one system of energies upon another, such that the receiving system is the one principally modified by the impression.

Example: The personality's first Real contact with monadic energies is likely to have a startling impact—startling to the personality, that is. (What will be 'started' as a result?)

impartite

By *impartite* is meant that which is incapable of being divided, i.e., that which is indivisible.

Example: The SELF is utterly impartite. But the ultimate particle within any Cosmos is also impartite for the duration of a Cosmos.

Example: The question may arise, Can anything (such as an ultimate particle/event) having 'extension' (i.e., definite measurability within the Field of Consciousness) be impartite? Perhaps, if there is nothing 'smaller' in-Cosmos with which to divide it, and nothing 'faster' in-Cosmos with which to subdivide it (i.e., an ultimate particle/event occurring in a ultimate moment) temporally.

Example: Divisions are made by instruments of division. What instrument of division in-Cosmos is *small enough* to divide the impartite ultimate particle? If the answer be given, "Only the Will, which occupies no space", then that very Will, upholding the necessary Finitude of the Universe, can equally *refuse* to divide that which must remain indivisible. Cosmic Parameters are upheld by Divine Will and Divine Will alone.

impossible

By *impossible* is meant that which cannot exist or be.

Example: No-thing is impossible forever. All 'POSSIBILITIES' within the INFINITESSENCE *can* be manifested during some Time and Space in Infinite Duration, but not necessarily in each and every Universe (depending upon the Parameters of the particular Universe). Simply because a thing *is* possible does not mean that it *will* be manifested. There will *always be* an infinitude of unmanifested 'POSSIBILITIES' within the FOUNT OF ALL POSSIBILITY.

Example: The existence of a thing or condition may be possible within certain contexts and impossible under others.

Example: Conceivably, all 'things' are *conceivable* (under the properly *unveiled* conditions), and that which can be conceived *can* be manifested. Only the INCONCEIVABLE is impossible to manifest, and yet, what else *IS there* that *could possibly* manifest. "BRAHMAN and Samsara are One."

Impossibility-in-Cosmos

By the term *Impossibility-in-Cosmos* is indicated certain relationships between Cosmic Variables which are *impossible*—i.e., that which cannot *lawfully happen* within the SELF/Self-Designated Parameters of the particular Cosmos in question.

Example: That the Archetypal Being Who embodies the Number Two should precede in the Cosmogonical Sequence the Archetypal Being Who embodies the Number One is an Impossibility-in-Cosmos—at least an Impossibility in *this* particular Cosmos.

Inaugurating Parameters

By *Inaugurating Parameters* are meant those Laws of Relationship that are Pre-Cosmically designated as *possible* within a Cosmos-to-Be. Due to the specificity of these Inaugurating Parameters, a great many relationships or configurations will necessarily not be included, and, consequently, will not be Possible-in-Cosmos, i.e., capable of being manifested. Strangely, even an *infinitude* of theoretically possible relationships and configurations ('INHERING' within the FOUNT OF ALL POSSIBILITY) will necessarily not be included.

Example: The Design-at-the-Beginning, which (from the perspective of one school of cosmological thought) must be *actualized* at the Consummation of the Universal Manvantara, contains the Inaugurating Parameters for the Cosmos to which It applies. Because of the Inaugurating Parameters of our Cosmos, stars are roughly spherical rather than shaped like spindles. In a different Cosmos with different Inaugurating Parameters, the majority of stars *could* be shaped like spindles—if there *were* stars.

in-Cosmos

By the term *in-Cosmos* is meant a qualifier that calls attention to the thought that an E/entity is to be found, or an event is taking place, within the context of Cosmos. Lest it be thought that this qualifier is unnecessary, let us remember that there are Super-Cosmic 'Happenings', and, also, a SUPER-Cosmic REALITY.

Example: Evolution is a Process which occurs only in-Cosmos, for evolution cannot possibly apply to the ABSOLUTE, which IS the ABSOLUTE PERFECTION—*perfect* in an infinitude of ways.

inconceivability

By an *inconceivability* is meant a thought which cannot be thought due to the limiting pressure of the Cosmic Algorithm.

Example: While it is possible to conceive far-fetched thoughts which, manifested, would not be Cosmically Sanctioned Configurations, it is also possible that certain kinds of possibilities/configurations ('RESIDENT' within the FOUNT OF ALL POSSIBILITY) simply *cannot be conceived* by any consciousness in-Cosmos, due to in-Cosmos inaccessibility of the infinitude of the FOUNT to finitized consciousnesses. Such possibilities/configurations would, under such circumstances, be called 'inconceivabilities'.

Example: Of course, we can have no conception of inconceivabilities. If we could, they could not *be* inconceivabilities.

'INFINIDENTIFICATION'

By '*INFINIDENTIFICATION'* is meant the fact that the INFINITE SELF always *IS* anything, whatsoever; no possible REAL *perspective* upon anything whatsoever can be 'ACHIEVED' from 'within' the INFINITE SELF.

Example: The INFINITE SELF can only *BE* any and all things through 'INFINIDENTIFICATION'. The INFINITE SELF, being out of all *relation* with any and all things, can never 'SEE' anything. Thus the term 'INFINISPECTIVE' (though, hopeful) is of limited suggestive value.

'infinidimensional'

By the term '*infinidimensional'* is meant the 'STATE' of REALITY which IS the infinitization of all possible dimensions. Dimensions are vibratory domains. The INFINITESSENCE IS the NOUMENESSENCE of *all* possible vibratory domains. The INFINITESSENCE is 'oscillation as thingship become NO-THING'.

Example: Whereas Cosmos is multi-dimensional, REALITY (because IT IS the INFINITESSENCE) is 'infinidimensional'. Such *dimensions* would necessarily be, however, *noumenessentialized.*

'Infinispective'

By *the 'Infinispective'* is meant the Self-Perspective of the Infinite Subject (otherwise called, the Infinified Point). The 'Infinispective' is characterized by complete 'Subject-objectivity', which is a kind of Infinite Consciousness existing between Infinite Spirit/ Matter.

Example: If the 'Infinispectivizing' ever 'Became' the 'Infinispective', then every trace of any point would disappear. The Infinispective is impossible to fully achieve in-Cosmos. Its Quality of Registration is the prerogative of the Infinite Subject/Object alone.

'INFINISPECTIVE'

By '*INFINISPECTIVE'* is meant no *'spective'* at all—no possibility of 'SIGHT' at all. Rather the 'INFINISPECTIVE' is such a complete 'INFINIDENTIFICATION' with all that might be 'SEEN' that there is only BE-NESS instead of 'SEE-NESS'.

Example: The 'INFINISPECTIVE' upon anything reduces it to *absolute* NOTHINGNESS, to ZERO. The 'Infinispectivizing' (which is a *moving* and not at all a definite Point of View) *on any thing* reduces the object *towards the infinitesimal* (which is an ever-lessening, indefinite quantity).

Example: The Infinispective, is a *completed* 'Infinispectivizing' and renders any specificity into a registrable Infinite Object. Any Object of Consciousness is returned to the Infini-Pointed Mulaprakritic State.

Example: The 'INFINISPECTIVE' might be described as 'INFINIDENTIFICATION'.

'Infinispectivizing'

By *'Infinispectivizing'* is meant a (Super-Cosmic, Psycho-Conceptual) ever-receding Point of View from an infinitude of ever-indefinite positions converging upon 'infinite distance', but never reaching it (as to 'reach' 'infinite distance' is an impossibility). Disregarding the laws of optics for the sake of this model, the 'Viewer' who has an 'Infinispectivizing' on an object, will see its object becoming ever smaller, converging, (as an infinitesimalizing) upon zero. From the *'Infinispectivizing'*, the 'Viewer' would 'See' all objects in the *process* of *infinitesimalization.*

It may be that a speed in recession, which is always an 'infinitesimalizing' 'short' of infinite speed (as little short of infinite speed as possible, and so, *virtually* infinite speed) would be required to sustain the objectivity of the object 'viewed'. At 'infinite speed' (whatever that unachievability REALLY means) the object would disappear entirely.

In any case, we see that there can Really be no *infinitesimal objects*, but only indefinite, immeasurable, but existent objects in the *process of infinitesimalization.* Otherwise, there is either a definite, measurable object (seen from a definite distance) or no object at all (paradoxically *'Viewed'* from the 'INFINISPECTIVE', which is NO POINT OF VIEW AT ALL). That which is 'Seen' from the 'Infinispectivizing' is neither an object or no object at all, but, rather, is *an object on its way to becoming no object at all* (at any number of indefinite infinitizing speeds—ever accelerating but never 'reaching' infinite speed). The 'INFINISPECTIVE', which utterly negates the possibility of objectivity, is not the same as the 'Infinispectivizing', which sustains objectivity—but just barely in a manner *ever lessening.*

By *'Infinispectivizing'* is meant the Infinitizing Point of View upon any thing as registered by the Infinitizing (either 'De-Infinitizing' or 'Re-Infinitizing').

By *'Infinispectivizing'* is meant an *ever-receding* point of view *converging* upon (but until immediately before Universal Pralaya) not 'reaching' the Infinispective of the Infinified Point of View. The 'Infinispectivizing' is never determined from an absolutely *infinite* distance, otherwise the 'thing' 'seen' would disappear into Infinite 'Subjectobjectivity' (Infinite Spirit-Matter) and, thence, 'into NOTHINGNESS. A *consummated* Infinispective is the antechamber to *the* 'INFINISPECTIVE', from which the object 'Seen' definitely disappears into NOTHINGNESS and becomes the 'thing' *BE-EN-through-'INFINIDENTIFICATION'* (i.e., identical with the INFINITESSENCE).

Example: Because the Infinitizing Subject is neither Infinite nor ABSOLUTELY INFINITE, its 'Infinispectivizing' will never reach the *point* of the Infinispective or of the ABSOLUTE 'INFINISPECTIVE', unless there is an acceleration to 'infinite speed' (symbolic of a change of state from Finitude to Infinitude to INFINITUDE).

Example: The 'Infinispectivizing' and the Infinispective are still ruled by *seeing.* The INFINISPECTIVE is ruled by *being.*

infinite

The term *infinite* indicates the magnitude of anything susceptible of ceaseless enumeration.

By *infinite* is meant an unspecifiable, ever-indefinite quantity that always exceeds any specifiable quantity which can be produced by ongoing enumeration, however ceaseless that enumeration.

By *infinite* is meant an indefinite magnitude ever larger than the largest possible enumerable magnitude.

Example: The number of numbers in the Set of all Integers is <u>infinite</u>.

Example: A set is infinite if a subset of itself is also <u>infinite</u>. For instance, the number of integers in the Set of All Integers is <u>infinite</u>, but equally <u>infinite</u> is the Set of All *Even* Integers contained *within* the Set of All Integers.

Example: An <u>infinite</u> set contains an infinitude of <u>infinite</u> sub-sets.

Example: While the number of authentic E/entities within a Cosmos is finite, the number of possible relationships between them is theoretically <u>infinite</u>—though not *actualizable* during the limited term of that Cosmos. The number of relationships between them that are possible to actualize is definitely rendered finite because of combination of factors, among which are the fixed and measurable duration of the ultimate moment for that Cosmos, and the fact that every Universe (like every other *object*) necessarily has a finite duration.

Example: It is interesting to realize that if a single *cosmically-immortal* human being began counting aloud the integers from the number one, and determined to go on doing so forever, the Cosmic Configuration would definitely change with his vocalization of every number, and all other authentic E/entities in-Cosmos would necessarily bear a different relationship to him (and he to them) as each number was vocalized. The number of such possible changes in relationship is *theoretically* <u>infinite</u>.

Theoretically, then, it can easily be seen that, in any Cosmos; the number of *possible* relations of authentic E/entities among themselves could be considered <u>infinite</u>, but any such process (no matter how <u>infinite</u>-in-potential) would "run out of *time*" due to the finiteness of Cosmic Duration. Thus, the number of *actualizable* relationships in a Cosmos is necessarily Finite.

INFINITE, the

By *the INFINITE* is meant the ALL-SELF, the ONE AND ONLY BEING/NON-BEING, WHO, like Spinoza's God, IS *"infinite in an infinite number of ways"*. The INFINITE has nothing to do with number as the concept, 'infinite', normally does. The INFINITE is *beyond number* in every possible way. No number-as-number can exist 'within' IT. Only Number-as-ZERO can 'INHERE' 'in' IT (i.e., *infinitized* Number). The INFINITE is the result of the infinitization of all numbers. The INFINITE is a 'STATELESS STATE' in which all enumeration has been <u>REALLY</u> carried *to the infiniteth degree*, resulting in a *consummating* ZERONESS WHICH IS THE MAXIMAL FULLNESS, the ABSOLUTE PERFECTION.

Example: By <u>the INFINITE</u> is meant a 'STATE' of ABSOLUTENESS in which all possible attributes of all possible combinations have been maximized or *absolutized* to such a degree that no further maximization or absolutization is possible simply because the attributes and combinations have been *infinitized*.

Example: Cosmos on all levels is inconceivably large and complex. An infinitude of Cosmoses past and to come are/were, in their aggregate, the 'Arenas' for the precipitation of infinitely more possibilities than can be precipitated within our present Cosmos. The INFINITE, however, is the SOURCE of infinitely more precipitatable possibilities than ever have been or ever can be precipitated in an infinitude of finite Cosmoses.

Example: The INFINITE includes an infinitude of infinities which are infinite after their own kind. This is another way of saying that an Infinite Set contains an infinity of infinite sets.

infinite entification (full participation in)

By *full participation in infinite entification* is meant an astonishing holographic process by means of which the ONE ENTITY/NON-ENTITY IS (*fully and completely*) every E/entity, without exception, in any Cosmos that ever has been, is now, or ever will be. Every such E/entity *is*, reciprocally (*fully* and *completely*—not just 'in part', for there are no 'parts') not only the ONE ENTITY/NON-ENTITY, but, further, every such E/entity *is*, ESSENTIALLY, (*fully and completely*) *all* other E/entities which have ever existed, exist now, or will exist, in any Cosmos past, present or future. This means, simply, *everything is everything else.* Put more personally and more provocatively—*everyone is everyone else.* How is that for a 'cast of characters'!

Example: The implications of the idea of full participation in infinite entification are at once astonishing and sobering. Try this one: **8** Am Napoleon (*fully and completely*), and so are You. (It is known that people are "locked up" for insisting they are Napoleon. This is especially the case when two people insist they are Napoleon at the same time. Please don't be too concerned: **8** won't tell, if 'You' won't!) It gets worse, however, for, in addition, You, Napoleon, and **8** *are* (*fully and completely*) the ALL-SELF. (If it's any consolation, 'You' and **8** ... *and* Napoleon, *are* also the "little men in the white jackets" who are coming to take us away ... unfortunately, there is *nowhere else* to take us!) This is extending the Holographic Principle to the 'outrageous' (at least in the view of the limited ego) radical conclusions which Radical Infinitism logically demands.

infinitely divisible continuum

By an *infinitely divisible continuum* is meant a homogeneous medium in which it is possible to make smaller and smaller divisions such that smaller and smaller (and, therefore, more and more vibratorily rapid) modifications result. An infinitely divisible continuum does not exist as an *actuality* in-Cosmos, but it is *conceivable* and therefore a *possibility* within the FOUNT OF ALL POSSIBILITY. Such a continuum may even be 'conceivable' within the Consciousness of the Universal Logos, however, He is under 'STRICT ORDERS' not to even *begin* conceiving infinite divisibility (or the Cosmic Algorithm with which He is 'entrusted' would be violated).

Example: If any Objective Cosmos (the World of Fabrication) were an infinitely divisible continuum, not only would there be no 'smallest possible particle' but there would be no 'smallest unit of time'—i.e., no 'ultimate moment'. An infinitely divisible continuum would also be (on the other 'end') an *infinitely aggregative continuum* in which cosmic variables were capable of infinite addition/combination/aggregation. Such an Objective Cosmos would be capable of containing infinitely large objects (with larger and ever larger cycles) as well as those which were infinitely small (i.e., infinitesimally 'small-ing').

Such an Objective Cosmos could *not* be *finite*. Rather it would be *infinite,* and because it would be infinite, it would require infinite time to go through its endless processes (unless infinitely rapid movement were also found within it—an impossible condition which would, for one thing, contradict the speed of light as a limiting physical plane mathematical constant).

It can readily be seen that such an Objective Cosmos would, necessarily, be the *only Objective Cosmos* that ever was or will be, because it would demand infinite time for its activities, hence the Law of Periodicity would be negated and the paradoxical gulf between the INFINITE and the Finite negated as well. In such a Cosmos the *objective* condition would be present for ALL TIME, and the NO-THING, too, would be negated, because there would *always* be 'Some-thing'. (Too, the "ETERNAL PARENT" would never 'SLUMBER' because ITS "brat" of a Cosmos would be 'awake all the time'!)

Example: Not only is Objective Cosmos *not* an infinitely divisible Continuum, but Objective Cosmos is not even a *continuum!*

infinitesimal (noun)

By the *infinitesimal* is meant an *indefinite* number/quantity which is as 'closening' to zero as possible without actually *being* zero. The infinitesimal is, therefore, a number/quantity 'infinitizingly' small, but never *infinitely* small, for that would be zero, itself. An infinitesimal 'thing' would still occupy 'space' (i.e., be a 'perceivable') but, Really, an *immeasurable* perceivable. If, however, 'some-thing' had a value of Zero it would *not* occupy Space (and, of course, would not be a 'thing', i.e., *not* be a 'perceivable'). A Real point is smaller (perhaps infinitely smaller) than any maximally conceivable definite number, for a point both *is* Zero, and yet it is not.

Perhaps we could say that a *Real* Point *is* an infinitesimal or 'infinitesimalizing'. An infinitesimal is *not* nothing, but it is not a dimensioned 'something' either. A Real Point, too, is not a *nothing,* but it is not a dimensioned 'something' either. An 'infinitizingly' enlarged quantity is not the NO-THING either, but neither is it a definitely measurable 'something' Remember, NOTHING is more than *every*-Thing' (infinitely more), but *not* necessarily more than the *infinitization* of all *possible things*—i.e., 'EVERYTHINGNESS' and NOTHINGNESS (or NO-THING) are equivalent.

Example: The infinitesimal is theoretically possible (and surely *possible* within the FOUNT OF ALL POSSIBILITY, the INFINITE POTENTIAL), but it is *not actually* possible—possible in an Actual Cosmos, just as an infinitely large number/quantity is not capable of manifestation (though it is capable of indefinite conceptualization and discussion) in an Actual Cosmos. Actually, an infinite number cannot be specified. The greatest computer that genius can fashion, adding or even multiplying integers for an infinitude of years could not reach a specifiable infinite number. The necessarily Finite Nature of each and every Cosmos prevents the actualization of either the *infinitesimal* ('small-ing') or the *infinite* ('large-ing').

Example: The infinitesimal can never be defined, just as no infinite number can be defined. The infinitesimal and the infinite are both *'in-de-finite'*.

Example: In one way, the Infinitesimal and the Infinite are identical. They are both immeasurable, hence, boundless 'somethings'. This is definitely different from a *boundless no-thing.*

infinitesimalization

By *infinitesimalization* is indicated that there is no such thing as a de-finite infinitesimal, but that there can only be convergence upon the idealized infinitesimal (the ever-diminishing smallest possible quantity greater than zero).

Example: Just as there can be no definite quantity called the 'infinite', equally, there can be no definite quantity called the 'infinitesimal'. The infinitesimal can be approached (or converged upon as a mathematical limit); the approach to the infinitesimal is called infinitesimalization. The infinitesimal, itself, is an ever-unfulfilled 'approach to zero'.

Example: The *Infinispectivizing* is a process for producing infinitesimalization. Just as there is no *definite* 'distance of vision' called the Infinispectivizing, so there is no definite percept called the infinitesimal. The 'Infinispectivizing' upon any object is a perspective which forces any object viewed to appear (through the process of infinitesimalization) as if its size, or 'extension', were converging upon a possible infinitesimal, which is, Really, to converge upon zero-presence, i.e., absorption into the ZERO. Perhaps we can begin to see analogies to the astro-physicists' "black holes".

INFINITESSENCE

By the term *INFINITESSENCE* is meant the NOUMENON of all Noumena found in all Cosmoses. Hence, INFINITESSENCE can be designated as the 'NOUMENESSENCE'.

By INFINITESSENCE is meant the infinitely refined and rarefied SOURCE/DISTILLATION of all A/archetypes, qualities, quantities and relationships, etc., which can possibly appear in a Cosmos.

By the INFINITESSENCE is meant another name for the NAMELESSNESS, the ABSOLUTE PLENUM, the FOUNT OF ALL POSSIBILITY, the INFINITE POTENTIAL, the ABSOLUTE.

Example: PARABRAHMAN *IS* the INFINITESSENCE of Goodness, Beauty and Truth. It is not so much that PARABRAHMAN 'CONTAINS' the INFINITESSENCE of all such qualities, but that IT *IS* these qualities in to a *maximally infinitized degree.*

INFINITESSENCE, 'EXTRUSIONS' of the

By *'EXTRUSIONS' of the INFINITESSENCE* are meant *noumenessentialized* 'POSSIBILITIES' that have been 'EXTRACTED' from the INFINITESSENCE for articulation *as* a Cosmos, or *in* a Cosmos. All 'EXTRUSIONS' of the INFINITESSENCE are ESSENTIALLY Illusory, though Realities to the intra-Cosmic Self (and Its numerous, though finite, subdivisions).

Example: Even Primary SUPER-Cosmic 'FOHAT', the 'RAY' of the ABSOLUTE, Is an 'EXTRUSION' of the INFINITESSENCE. Mulaprakriti Is as well, and so is everything that *appears* in-Cosmos. A SELF-'EXTRUDED''POSSIBILITY' is an Illusion. A non-'EXTRUDED' *'infinitessentialized'* 'POSSIBILITY''INHERING' in the INFINITESSENCE IS a REALITY, and, in fact, paradoxically, IS the *WHOLE* of REALITY!

Infinite Trinity, the

By *the Infinite Trinity* is meant the greatest of the Super-Cosmic Trinities—the Infinite Subject, the Infinite Object, and the Infinite Consciousness (or the Consciousness of Infinity). This Trinity can also be described as Infinite Sat/Infinite Chit/Infinite Ananda. A further description can be expressed as the unlikely Trio: the Infinified Point/ Mulaprakriti/Maya.

Example: The Infinite Trinity comes to birth virtual instantaneously with the 'FLASHING FORTH' of the 'RAY' of the ABSOLUTE.

Example: The Infinite Trinity 'Arises' from the *instantaneously evanescent* INFINITE TRINITY' which cannot 'ABIDE' 'within' the INFINITE SELF.

'INFINITE TRINITY'; 'EVANESCENT' 'INFINITE TRINITY'

By the *'INFINITE TRINITY'* or (better) the *'EVANESCENT INFINITE TRINITY'* is meant the threefold 'STATE' of 'RAY' or 'POINTNESS' that 'ARISES' (probably in instantaneous sequential order) 'within' the INFINITE SELF, as:

1. The 'EVANESCENT INFINITE SUBJECT'
2. The 'EVANESCENT INFINITE OBJECT'
3. 'EVANESCENT MAYA' for an *infinitesimalizing instant* before being instantaneously 'EXTRUDED' into the Pre-Cosmic World as 'ITS' threefold Reflection.

The Infinite Trinity is also probably arising in instantaneous sequential order—(whatever the sequence!) consisting of:

1. The Infinite Subject
2. The Infinite Object
3. The Infinite Consciousness (or the Infinite Maya, or Maya Presenting a *limited* View of the INFINITE SELF as an Imparticulate, Homogeneous, Objective Infinitude in which *an infinitude of noumenessentialized possibility Inheres*).

In each case if 'arisingness' in sequential order is correct, 'TRINITIZED''MAYA' and 'Trinitized Maya' would probably precede the appearance of the 'OBJECT'/Object.

Example: There are four 'Trinities' which *demonstrate before* the specifically Universal Process Begins:

1. INFINITE TRINITY
2. The Infinite Trinity
3. The Post-Infinite Trinity
4. The Pre-Universal Trinity

The 'EVANESCENT INFINITE TRINITY' and the 'Infinite Trinity' are described in this Glossary of Terms.

The Post-Infinite Trinity consists of:

1. The 'De-Infinifying Subject'
2. The 'De-Infinifying Object'
3. The 'De-Infinifying' Mayavic Consciousness (which relates them) all.

The Pre-Universal Trinity consists of:

1. The *Focused* Universal Subject
2. Universally Requisite Cosmic Prakriti isolated by an Act of Self-Perception from Mulaprakriti
3. Pre-Universal Maya, which makes the Act of isolating Universally Requisite Cosmic Prakriti possible.

And at this stage in the process, a Fifth, Cosmic Trinity arises, consisting of:

1. The Cosmic Logos
2. Cosmic Prakriti
3. Cosmic Maya (Fohat, in Mode 3)

Infinitist, Radical

By a *Radical Infinitist* is meant one who seeks to understand *and experience* all things from the perspective of non-dualism and is thus identified *as* the INFINITE SELF.

Example: An Radical Infinitist is not one who counts forever, but who realizes that in-Cosmos it is impossible to count forever. While fascinated with incalculable magnitudes, a Radical Infinitist is spiritually *merged* in the INFINITE 'WHERE' all 'things' have *already* and *forever* been *completely* 'counted', since, forever, the infinitization of all 'things' has 'ABIDED' 'THERE', in the INFINITIZED 'STATE'.

While respecting and loving the form as a *necessary imperfection*, the Radical Infinitist is a RADICAL REALIST, and (when it comes to valuing the REAL over ITS manifest 'EXTRUDED POSSIBILITIES'), the Radical Infinitist becomes an 'OBLITERATIONIST'.

INFINITUDE, the ALL-EMBRACING

By *the ALL-EMBRACING INFINITUDE* is meant the UTTER ALLNESS consisting of the INFINITE *plus* all Universes that IT has ever 'BECOME' or *Will* 'BECOME'. Of course, the word *plus* is nonsensical, because *nothing* can be added to the INFINITE, nevertheless, such is the structure of the human mind when it contemplates the Fundamental Infinite Duality, that the word *plus* must be used.

Example: Of the ALL-EMBRACING INFINITUDE there is no EXAMPLE except ITSELF. IT IS, paradoxically, *more* (and perhaps *less)* than what we call the INFINITE, and yet, also paradoxically, IT *cannot* be *more*. In our minds the ALL-EMBRACING INFINITUDE suggests not only 'EVERYTHING' that the INFINITE IS but Everything that the INFINITE has ever 'DONE'/'BECOME'. A synonym for the ALL-EMBRACING INFINITUDE is the UTTER ALLNESS.

infinity

By *infinity* is meant a special category of quantity that results from the ultimate increase of any quantity (principally through addition or multiplication, but sometimes through division or subtraction, as when dividing any quantity by zero or successively subtracting negative numbers from any quantity). The ultimate (non-ascertainable) result of this ultimate increase is the production of a quantity so large and yet so indefinite as to be considered *maximally* large and utterly *unquantifiable*, forever *indefinite.*

Example: The study of infinity is preferred by those who know that the human spirit (which is, ESSENTIALLY, SPIRIT) has nothing in common with finitude.

Example: The sum of an infinite number of things is infinity. This sum cannot be reached in-Cosmos but has forever and always been 'reached' within the 'DOMAIN' of the ABSOLUTE, the INFINITESSENCE.

INFINITY of Infinities, the

By the term *the INFINITY of Infinities,* is suggested that the INFINITE is not only the *infinitization* of all things, but is the *consummated infinite sum of all infinite things.* All manner of 'infinities' find their infinitization in the INFINITY of Infinities, the INFINITESSENCE.

Example: Spinoza has said that "God is a Being infinite in an infinite number of ways" (hence, the INFINITY-of-Infinities). By "God", Spinoza clearly meant GOD, the ULTIMATE ABSTRACT DEITY.

Example: The Finite Cosmos contains many potential infinities, many of them in the domain of mathematical thought. These potential infinities are already and forever totally 'summed', 'reached', 'consummated', 'achieved' (whatever inadequate word we choose to use) within the INFINITY of Infinities—the INFINITE SELF.

'in-perience'

By the term *'in-perience'* is meant that which an E/entity 'ex-periences' (the usual word) totally *within* itself. Can the ALL-SELF *'in-perience' anything?* Not REALLY, just as the ALL-SELF cannot have 'CONSCIOUSNESS', per se. 'In-periencing' is *not* as powerful or suggestive as 'INFINIDENTIFYING'. The idea of 'in-perience', however, is far more suitable (though stranger) than 'experience' when referring to the infinitized, absolutized SELF-ABSORBED 'STATE' of the ALL-SELF in ITS ALL-IN-ALLNESS. The thought of 'in-perience' is very useful 'within' Cosmos, even if it inevitably falls short of suggesting what 'HAPPENS' 'within' the ALL-SELF.

Example: Apperceptions are 'in-periences'. Perceptions (technically and rigorously considered) are *ex*-periences.

instant

By an *instant* is meant any relatively short unit of time.

Example: The term instant is used very loosely, and refers to any happening which seems to happen 'immediately' or "in no (noticeable) time at all". Obviously, no-'thing' can happen "all at once" or "in no time at all." There is to be considered, however, the amount of 'time' it 'takes' to change from 'no-thing' to 'thing'. How fast if Fohatic Self-Perception? How fast is Divine Will?

INSTANT, the 'AWAKENING'

By *the 'AWAKENING' INSTANT* is meant the 'FLASH' in THAT, 'ARISING' out of 'no Time at all' at which *point* the EVANESCENT INFINITE TRINITY 'ARISES' (INFINITE FATHER, INFINITE MOTHER, INFINITE LOVE/CONSCIOUSNESS) and *instantly* 'BECOMES' Pre-Cosmic Infinite Father/Subject, Pre-Cosmic Infinite Mother/Object, and Pre-Cosmic Infinite Love/Consciousness/*Maya.*

Example: If the AWAKENING INSTANT does not happen *spontaneously* and *instantly* there arises the Problem of the intra-SOURCE 'EXISTENCE' of 'Time' *before* Time *began.* Of course, that *is* a *possibility!*

instant, inter- and intra-moment

By an *inter-moment instant* is meant the duration of the interval which almost certainly exists between ultimate moments in-Cosmos. Because 'between' ultimate moments no-thing in the World of Fabrication can 'happen', the only non-'thing' that can 'un-happen' is the 'Appearance' of the 'Presence Itself' (and the consequent emergence of the World of Being as the main fact of Consciousness).

From an intra-Cosmic Perspective, the inter-moment instant may well be measurable, and even of equal duration to an ultimate moment. In such a case it would be possible for Someone(?!) to measure (against the Infinite Time Line) the duration of the *presence* of the Presence during each inter-moment instant (even though the Presence is *always present in-Cosmos* and, even though for It, at Its Temporo-Visual Pinnacle, no Cosmic Time Really 'passes', in the usual sense). The Universal Logos can experience intra-Cosmic Time in a multitude of ways, the highest of which is as the Presence in the Cosmo-Eternal Now.

Example: One is tempted at times to ask if the intra-moment instant (if it exists) is equal to zero or to an Eternity. REALLY, there would be no 'appreciable' difference for E/entities in-Cosmos, for whether they were subjected to an interval of either zero or an Eternity, they would not *experience* any difference.

Example: The idea than an intra-moment instant may be equal to zero would mean that the Universe was 'on' *all the time.* A Universe which was 'on' all the time would be a *continuum* with respect to Time, and thus could have no divisions of Time—an impossibility, for the existence of apparently separate 'things' and the existence of apparent divisions of Time—the two being indispensable to each other. This impossibility would not exist if the intra-moment instant were of any specifiable duration, however microscopic. If the World of Fabrication has "off time" due to the fact that all things in that Cosmo-Objective World *must necessarily be discontinuous,* the question arises as to how such "off time" (the instant during which the Universe turns 'off') could be measured and by what kind of Observer? Perhaps certain of the Lords of Galactic Karma know! After all, Their math is said to be very good!

Example: An inter-moment instant could also be called an 'Interlude of Fohatic Disengagement', because of the nature of the Ontological Oscillation of ultimate particle/events which are, Essentially, Fohat-in-Action at the most extreme micro-level.

instantaneous; instantaneously

By the term, *instantaneous,* (when considered practically) is indicated a duration *equivalent* to an 'ultimate moment'. 'Instantaneous' does *not quite* mean "in no time at all" or "an infinitesimal amount or time". An 'infinitesimal amount of time' is not only as indefinite as an infinite amount of time, but equally non-actualizable in-Cosmos.

Example: The shift of ultimate particles from one *cosmically sanctioned position* (with respect to each other) to another occurs (we would say) instantaneously. Just how much 'time' does it take? Or does it take time at all? If ultimate particle/events existed during the Cosmo-Subjective Now (just before the Cosmo-Objective Now) we might say that the 'shift' form one Now to the next 'took time'. But ultimate particle/events do not exist as objectivities in the Cosmo-Subjective Now, and so, when they appear into Cosmo-Objectivity, they appear suddenly 'out of the blue', as it were, which means, 'out of Subjectivity'? Does this take 'time'? Can it happen 'instantaneously'? Can it 'happen' in "no time at all"? Let us ponder for a 'time'. Or perhaps we will be fortunate enough to understand the truth of the matter "in no time at all"! The term "no-thing flat" holds a key. The 'change' is *not* a 'thing' and so (being a zero) is unperceivable.

Example: Because any in-Cosmos observer of inter-moment intervals would be unconscious of any-thing during such an interval (for the Fohatically Fabricated Universe with all its 'things' would have disappeared) the flow of time in lower Cosmos would appear continuous, with no break from ultimate moment to ultimate moment. One can see that however instantaneous the inter-moment interval may be, the perception of a 'seamless' flow of time would be illusory. Time does not 'flow'; it jumps!

Example: If an ultimate moment is to 'begin' precisely when an inter-moment instant 'ends', then the beginning and ending must be simultaneous, and must begin and end "in no time at all." Can the ending of a unit of time take place in "no time at all"? Can the beginning of a succeeding unit of time take place in "no time at all"? One is in need of a 'zero-time ending' and a 'coincident' 'zero-time beginning'. Ending and beginning have to be 'instantaneous', which, in this case means, "in no time at all."

We are left with very troubling and difficult questions, 'When' did the 'end' *end?* 'When' did the 'beginning' *begin?* We cannot use gradualism. We cannot incrementally 'slide' into an 'end' and 'slide' into a 'beginning', for no change is possible in that which 'ends' or 'begins'. Ultimate particle/events *are* or they *are not;* they are either 'on' or they are 'off'. We are led to the thought, that the 'end' of *endings* and the 'beginning' of all *beginnings* are Really zero-time 'changes'. They are an infinitely divisible 'line', the width of which is zero! We have much meditation to do about the *end* of 'times' and the *beginning* of 'times'.

If light on this question dawns, we may be coming closer to understand how the 'FLASH' could simply 'BEGIN' in *no time,* thus 'ENDING' (also in *no time*) the 'ABIDING' BE-NESS of THAT. 'No-Time' is infinitely more *subtle* than any unit of Time, and, *infinitely thinner.*

Example: Only NOTHING 'HAPPENS' "in no time at all"; only the ETERNAL NOW 'HAPPENS' "in no time at all". Does 'instantaneously' mean "in no time at all?" If something *happens*, it takes *time*—at least this is what we usually think. Now, if the ideas in the foregoing paragraph are true, then, is it possible to say that 'no-*thing* 'happened' to 'end' an 'ending' or 'begin' a 'beginning'. Of perhaps that an 'unperceivable' 'happened'? Or that a 'zero' or a 'nothingness' 'happened'? We are discussing a 'close shave in Time'. The matter is most mysterious. Are we witnessing (or failing to witness because it *can't* be) a 'timely!' intervention of the ETERNAL NOW?

interval

By the term *interval* is meant the 'space between' two 'things'.

Example: The term interval applies equally to Time, Space and Vibration. An interval is an acknowledgment of difference—difference is position, difference in the moments of occurrence of two events, and difference in vibratory measure. The perception of a musical interval is caused by a difference in the rate of frequency of two tones. Where all things are perceived as utterly the *same*, no interval can possibly exist.

Example: Interval and 'sameness' are opposites.

Interval, Inter-Cosmic

By an *Inter-Cosmic Interval* is meant the duration of that 'portion' of Infinite Time which exists between any two Universes.

Example: From an ideal perspective, all Inter-Cosmic Intervals would be exactly equal. In Actuality such Intervals may be of different durations to allow for what we might call Free Will within the Universal Process.

Example: If an Inter-Cosmic Interval REALLY exists, *who* can *see* it?

interval of affirmation

By an *interval of affirmation* is meant the interval between intervals of negation. During an interval of affirmation the Fabricated Universe (the Mosaic World) 'exists', is 'on', present, objective, etc. The interval of affirmation endures for one ultimate moment, and is followed by an interval of negation which may or may not 'last' for an ultimate moment. If the interval of affirmation and the interval of negation were equal, a new understanding of *vibratory movement* would be revealed.

Example: No human being in Fabricated Cosmos can notice rapidly fluctuating intervals of affirmation. Our eyes cannot follow the movements of the blades of a fan without fusing them into one apparent wholeness. The Universe *appears* to be *continuously* affirmed, *continuously* present, in existence, etc., but logically we know we live in a quantized Universe in which dis-continuity necessarily prevails. For all practical purposes, however, we have the 'best' (and the 'worst') of 'both worlds'!

interval of negation

By an *interval of negation* is meant the interval between ultimate cosmic moments during which 'time' the Fohatically Fabricated Universe ceases to be for an 'instant' (a 'cosmic instant'), the duration of which *may* be and probably is quantifiable in terms of the parameters of measurement for a given Cosmos.

Example: Perhaps, the interval of negation is equal to the interval of affirmation. What would be the implications if the interval of negation were equal to zero? This would destroy the concept of an interval of affirmation, would it not?, for a 'space of zero magnitude' between two things fuses the two things into one (especially if the two 'things' are of a similar nature). If all ultimate moments were fused, there would be but *one moment* in the entire Cosmic Process which would be equivalent to negating Time and asserting that the ETERNAL NOW alone prevails. The ETERNAL NOW would then 'devour' Time in Cosmos and negate the possibility of Cosmic Movement (for all particle/events would be 'frozen' into one unchanging Cosmic Configuration, assuming such a Configuration even began). But this is *not* what 'happens', for both Time and Illusion are actual-in-Cosmos. Thus we must conclude that the interval of negation is *not* equal to zero.

Example: In terms of the Great Breath, the Higher Interlude (the held inhalation) would be analogous to the 'interval of negation'. If that interval were 'zero', then the Objective Universe would be existing in one phase or another continuously, forever.

'Intra-SOURCE'

By the term *'Intra-SOURCE'* is meant a *locationless location* 'WITHIN' the INFINITE SELF. Because of the limitations of the human mind it becomes necessary to speak of 'ACTIONS' or 'MOVEMENTS' originating 'WITHIN' the ABSOLUTE, even though such are REALLY impossible. Whatever 'TRANSPIRES' 'WITHIN' the ABSOLUTE, such as the FIRST 'CHANGE' which is the 'FLASHING FORTH' of the 'RAY' of the ABSOLUTE, is considered an 'Intra-SOURCE' 'HAPPENING'.

Example: One of the great problems in Metaphysical Philosophy is the origin of the CAUSELESS CAUSE. It is likely that this problem can only be solved from an 'Intra-SOURCE' 'IN-SPECTIVE' (an 'in-spective' being a maximally interior though ESSENTIALLY 'SIGHTLESS' perspective). In the lower Worlds we are so used to 'seeing', that we phrase the greater part of our understanding in terms of it, rather that in terms of 'being'.

in-Universe

By *in-Universe* is meant the same as 'in-Cosmos'—i.e., occurring within the context of a given Universe and not in the Pre-Universe or Post-Universe State, *or* within the 'STATELESS STATE' of complete Universal Pralaya. The Pre-Universe and Post-Universe States (in which the Cosmos is either unformed or disintegrated) should be viewed as Transitional States and not as the ultimate 'STATELESS STATE' of complete Universal Pralaya.

Example: The concerns of most members of the human family are strictly in-Universe concerns. It becomes easier to understand the value of in-Universe Patterns, however, if one can at least speculate about the *principal* SUPER-Universal 'STATE', the IN-

FINITUDE. 'Being' (in-Universe), then, is seen better for what It *Is* and Is *Not* when apprehended against a background of apparent 'Non-Being' (which IS BEING).

We must remember that there are States which are neither in-Universe nor 'within' the UNDIFFERENTIATED TOTALITY. Such would be Pre-Cosmic and Post-Cosmic States. The 'Players' in such Super-Cosmic States would be:

1. The SELF-as-Infinified Point representing the Infinite Subject in contradistinction to the Infinite Object
2. Mulaprakriti, the Infinite Object
3. 'FOHAT'-as-Fohat in one of Its Pre- or Post Cosmic Modes, being the Agent of the SUBJECT/Subject in relation to the Object.

Just as at the Beginning there was the 'GOING FORTH' of 'FOHAT'-as-Fohat, *'from'* the Infinite Subject *'into'* Mulaprakriti to Generate Super-Cosmic Spirit/Matter (the higher Super-Cosmic correspondence of intra-Cosmic Spirit/Matter), so at the "End Times", there will probably be a 'Return' of 'Fohat' (soon to be 'FOHAT') unto the Infinite Subject, signaling a disengagement of the Infinite Subject from the Infinite Object. When this is 'happening', Cosmos is no more, and Total Reabsorption of the Infinite Subject and Infinite Object into the ALL-IN-ALLNESS is imminent, and perhaps, at that point in Post-Cosmic Time, the Total Reabsorption occurs *instantaneously.*

This Post-Cosmic Happening might be called 'the Retraction of the 'RAY' of the ABSOLUTE'. The Retraction of the 'RAY' is, in Reality, the Retraction of Post-Cosmic Fohat-as-'FOHAT' into THAT.

item

By an *item* is meant any possible 'perceivable/apperceivable'. All items are what might be called items-in-Universe, (and, perhaps, items-in-Super-Cosmos), but are never items-in-ABSOLUTE REALITY. ABSOLUTE REALITY is utterly devoid of items. An item can be an E/entity but it need not be. While all distinct energies can be perceived/apperceived as items, the connotation usually associated with the term 'item' suggests a more concrete rather than more fluid 'perceivable/apperceivable'.

By an *item* is meant any and all possible enumerables in this and all other Universes which the exception of the BOUNDLESS IMMUTABLE PRINCIPLE, as, truly, no-thing can be found *in* IT.

Example: An atom and a galaxy are both items in consciousness—but in *Whose* consciousness?

Example: An aggregation, a combination, or a configuration, all may be considered as items depending upon the perspective from which such are seen, but usually an item refers to a thing as a whole in its singleness, with little attention given to the parts which compose it. This is a question of perspective.

item-in-Universe

By an *item-in-Universe* is meant any E/entity or energy (or aggregate or configuration of E/entities or energies) in-Cosmos. The use of this term carries the connotation that the energy or E/entity has a duration greater than that of one 'ultimate moment' and is thus, at least a *relatively,* if *minimally,* abiding factor in-Cosmos. (No factor in-Cosmos is an *absolutely continuous* factor, because Cosmos is a Dis-Continuity, or a 'Dis-Continuum'.)

Example: An atom of Hydrogen, for instance, can be considered an item-in-Universe, because, no matter how brief one of its 'incarnations' may be, it persists as an distinct 'perceivable' for more than one ultimate moment. A fleeting perception of the Universal Flux, however, which is configured at the moment of perception, but which changes configuration at the next ultimate moment (never to be reconfigured exactly the same way or even similarly) should not properly be called an 'item-in-Universe'. It is rather a *transient perceivable*—an *instantaneously transient perceivable.*

- K -

key

By a *key* is meant an instrument or a method, the use of which allows entry into a hitherto unentered field of vibration or domain

Example: A specific elevated vibratory condition in the vehicles of the would-be initiate is the key for his penetration into a particular field of divine consciousness/activity, the vibratory nature of which is identical with, or closely approximates, the elevated vibratory condition in his vehicles.

keynote

By a *keynote* is meant a specific and standard rate of vibration to which a particular field (gross or subtle) vibrates (or rotates—which is vibration of a kind). Each keynote has its specific frequency which is most often associated with one of the twelve standard notes of the musical octave. (More than twelve possible notes per octave are possible but not frequently used—in the music of Western Civilization.)

Example: Each of the Seven Rays has Its particular keynote, the finding of which would allow a magician to induce/summon the presence of a Ray or Its devic representatives at will.

Example: The secondary identity of any U/unit-of-Life can be expressed as a chord consisting of the keynotes of its various vehicles. The PRIMARY IDENTITY can be associated with no particular key or keynote, for IT IS *all* of them, ESSENTIALLY.

knowledge

By *knowledge* is meant either memory of, cognition of, or immediate intuitive apprehension of all factors included within sequences of relationships. This knowledge makes possible *prediction* with respect to such sequences, and/or their intelligent *manipulation.*

Example: Complete knowledge of a process means that everything pertaining to the motions contained within the process (for instance—causes, effects, and all dynamics of interplay) are not only completely registered but completely understood.

- L -

law; Law

By *law* is meant a system of certain possible patterns/configurations *within* which, or in *conformity* to which, a certain array of interacting variables *must* or are *forced to* pattern/configure themselves.

Example: The seven Laws of the Soul specify certain patterns of relationship to which consciousness expressing itself through the causal body and/or through the Spiritual Triad will necessarily conform.

Law, the

By *the Law* is meant the Will of the Universal Logos.

Example: An old Teacher of mine used to say, "God is not mocked. No one breaks The Law; the Law breaks them." Although there are those who in Time and Space and because of ignorance oppose or circumvent the Law, such is the incontrovertible Will of the Universal Logos that all are at length brought into line with His Will and into conformity with the Law which represents that Will.

LAW

By a *LAW* is meant an invariable 'PATTERN' (presumably SELF-'WILLED') according to which the BOUNDLESS IMMUTABLE PRINCIPLE Functions in relation to ITS Objectivization—the Universal Process.

Example: The appearing and disappearing of Universes might be considered an extra-Cosmic LAW, a LAW of the INFINITE SELF. Such a LAW, however, cannot be one to which the INFINITE SELF *must* conform. For *who other* is there to compel the INFINITE SELF? Such a LAW must be considered, instead, one to which the 'RAY' of the ABSOLUTE-Become-the Universal 'Ray') and Its Super-Cosmic and intra-Cosmic progeny must conform.

Example: From one perspective the ABSOLUTE DEITY *IS* LAW—the LAW that there shall be "no other gods before ME." This LAW might be called 'the LAW of the Obliteration of Secondary Identities'.

life; Life

By *life* is meant the PRESENCE/Presence of the SELF—as-Self as It pervades Cosmos entirely, thus holding in *vibrant manifestation* all E/entities.

Example: The life pouring through the etheric body and holding both it and the physical body in coherence, is but an attenuation of the One Cosmic Life (the Universal Logos) which derives from the ONE AND ONLY LIFE.

life, a; Life, a

By *a life* is meant an authentic E/entity or an apparent Monad within the One Universal Monad.

Example: There are many secondary and tertiary, etc. entities in-Cosmos composed of a number of <u>lives</u>. Such secondary and tertiary entities are *not* authentic E/entities, and did not originate within the Logoically-Designed Emanatory Sequence of the Divine Emanatory Stream.

Example: An atom is certainly <u>a life</u>, just as a galaxy is <u>a Life</u> (for the apparently distinctive *beings* that inform them both are but apparent 'extensions' of the One Universal Monad which existed in a an entirely Unitary State before the beginning of the Divine Emanatory Process), but a man-made organization, though it may well be a secondary or tertiary entity (i.e., *composed of* or a *creation* of primary E/entities) cannot, strictly speaking, be considered <u>a life</u>, even though (along with many other kinds of secondary, tertiary, etc., entities) it may be said to "have <u>a life</u> of its own".

LIFE

By *LIFE* is meant the ONE VITALIZING 'SOURCE AND SUSTAINMENT' of all Cosmoses and of all processes in all Cosmoses.

Example: The ONE <u>LIFE</u> *IS* the ONE AND ONLY SELF.

limitation

By a *limitation* is meant a condition which prevents (or, more accurately, *seems* to prevent) the potency of an being-in-Cosmos or item-in-Cosmos from being Omnipotent; which seems to prevent the consciousness of such a being/item from being All-inclusive and Omniscient; and which seems to prevent the activity of such an being/item from being everywhere present—Omnipresent. A limitation, in fact, is any condition which seems to prevent a being-in-Cosmos or item-in-Cosmos from *being* (totally and completely) the ONE AND ONLY SELF.

Example: All beings-in-Cosmos are subject to <u>limitation</u> simply because they are identifiable as distinct *beings.* If <u>limitation</u> is to exist there must be both a limiting factor and a factor to be limited. Since the ABSOLUTE is the ONE WITHOUT A SECOND, there can exist 'in relation to IT' (though nothing can REALLY *relate* to IT) no limiting factor. Therefore, the ABSOLUTE, having nothing to limit IT, must necessarily be the UNLIMITED.

Therefore, any 'dynamic' which *seems* like a <u>limitation</u> upon the ABSOLUTE, such as the paradoxical idea that Finitude must be manifested by the INFINITE so that the INFINITE can continue to BE *INFINITE,* must not *REALLY* indicate a SELF-LIMITATION, at all, but rather an 'ACT' of SELF-AFFIRMATION.

localize; localization

By *localization* is meant the apparent confinement to an apparent place and position of any item-in-Cosmos or being-in-Cosmos *relative to* other items-in-Cosmos and other beings-in-Cosmos. The word 'apparent' is utilized here because place and position are ESSENTIALLY illusory, occurring as they do at a Single Cosmic Point (from the Perspective of the Universal Logos).

Example: BE-NESS is absolutely non-localized. Cosmos, too, is non-localized, for what other manifested Object is there with which to compare It? But any *other* manifestation *in-Cosmos* is subject to:

1. Limitation with respect to its own ring-pass-not and, eventually and ultimately, with respect to the Ring-Pass-Not of Cosmos, and is subject to—
2. Localization within the Ring-Pass-Not of Cosmos and in relation to all other manifested beings-in-Cosmos, who, also occupying place and position, are likewise localized.

Even if the 'prakritic-extent' of a Being-in-Cosmos is immense, the factor of localization still pertains because the Being exists intra-Cosmically, and thus has an 'extent' measurable with respect to the Cosmic Ring-Pass-Not.

Example: Spatial localization results from vibratory limitation. The less the vibratory limitation (i.e., the more *rapid* the vibration), the more pervasive and non-localized the B/being-in-Cosmos. Within a Real point in Cosmos (whatever that Really may be) the vibrational cycle would be of infinitizing frequency and the vibratory amplitude is of ever infinitesimalizing extent—how else could it 'fit' within a point? Non-localization in-Cosmos would be the result.

The Universal Logos Is, with respect to Its Cosmos, *not* subject to vibratory limitation. With respect to the ABSOLUTE BEING, any vibratory frequency no matter how high (so long as such a frequency is definitely quantifiable, i.e., not infinite) *is* limitation. Vibration *is* limitation.

localized B/being

By a *localized being* is meant any being-in-Cosmos the expression of which is limited by a particular ring-pass-not. A localized being usually has a localized consciousness which has not learned how to pervade other ring-pass-not(s).

Example: Any Emanation of the Universal Logos must necessarily manifest as a localized being. To be a localized being in the World of Being (which is still, let us remember, within the World of Becoming, considering the World of Becoming most comprehensively) is qualitatively different from being a localized being upon the lowest levels of the World of Illusion (our Cosmic Physical Plane). Localized Beings upon the highest levels of the World of Being, as Spirit, possess *cosmic pervasion*, but Qualitatively and Formally, they must nevertheless be regarded as, at least, *vibratorily* localized Beings, for even They cannot be *entirely cosmically pervasive* (with the same degree of *intensity*).

In other words, the Archetype called the Number Eight (or Eightness) cannot pervade with the same intensity all vibratory levels available to the Archetype called the Number Seven (Sevenness), Which Archetype occurring one phase earlier in the Cosmogonical Sequence than the Number Eight, Is a Being more generalized and more *integral* to the Nature of Things. Think of the indispensability of the Number One when compared, for instance, to any other Number. There is indeed, a hierarchy of Numbers, and the *intensity of Their Pervasiveness* in Cosmos varies. We are not here speaking of spatial *pervasiveness*, but of *degree and intensity of presence* and *'integrality.* 'Really, all Archetypes are Cosmo-Spatially Pervasive, but the *eminence of Their intensity* varies with their structural importance within Cosmos.

Example: The Power of Vishnu dissolves localized being. The Pervading Power of Vishnu is a Power conducive to De-Localization.

localized consciousness

By a *localized consciousness* is meant a consciousness confined to particular intra-Cosmic conditions and within a particular intra-Cosmic ring-pass-not, as opposed to the pervasive and ubiquitous-in-Universe Consciousness which is found present within *all* cosmic ring-pass-not(s), such as, for instance, the Consciousness of the Universal Logos.

Example: The state of consciousness called 'ego consciousness' is a severely localized kind of human consciousness. In this state there is very little possibility of pervading other ring-pass-not(s), except for those lesser rings-pass-not(s) which create the boundary of registration for those unconscious lesser lives (the lunar lords) that are included within the constitutional system of the particular human being concerned. Intuitive human beings are beginning to pervade some aspects of the ring-pass-not of their fellow human beings. Through this process they will one day know, as the Christ knew, "what is in man."

localized self; Localized Self

By the *localized self* is meant the SELF-as-Self-as-self *before* it has mastered the power of pervasion. The consciousness of the localized self is strictly that of the single ring-pass-not with which it is most intimately associated—i.e., the 'immediate presentation'. The localized self when it is self-conscious, (for there are pre-self-conscious localized selves, such as atoms and elementals, etc.) is egoistic. The consciousness of the localized self is *localized* consciousness and is neither empathic nor even slightly pervasive. One of the major themes of the Cosmic Evolutionary Process is to transform the *localized* self and its consciousness into the *pervasive* (and even *ubiquitous*) SELF-as Self and Its *Cosmic* Consciousness.

Example: Whereas the personality-centered human being is a localized self, the Planetary Logos is a Localized Self. The human initiate of the third degree is beginning the transition from localized selfhood to Localized Selfhood. The self for whom the factor of *unity* is becoming a Reality (the human initiate of the third degree) can no longer be called a strictly localized self.

Logos; Logoi

By the term *Logos* is meant a Being who (relative to the human state) is a Great Loving and Willing Intelligence responsible for coordinating, unifying and inspiring a relatively vast number of lesser intra-Cosmic lives so they learn to conform more accurately to the Cosmic Purpose (the Design-at-the-Beginning). Each Logos is, cosmically considered, a *lesser* Word, the sum of all such Words reflecting the Design and Potency of the Great Word at the Beginning.

Example: In a particular Cosmos, the Universal Logos is the foremost of all the Logoi contained within Its Ring-Pass-Not. The relatively stupendous Life which embodies a Galaxy is a Logos included within the Ring-Pass-Not of the Universal Logos, just as is the Planetary Logos, though on a far, far tinier scale.

Logos, Universal

By the *Universal Logos* is meant the One Self of Cosmos—the Synthesis of Creator, Preserver and Destroyer. The Universal Logos of each Cosmos is, in one sense, *unique* and *different* from every other Universal Logos, for Its Cosmic Purpose determined by certain *infinitessentially precipitated potencies* ('POTENCIES' and 'POSSIBILITIES' 'EXTRUDED' from the INFINITESSENCE) is unique.

In another sense, however, each Universal Logos is *identical* to every other Universal Logos, for each Is, ESSENTIALLY, the *one and only prakritically circumscribed* 'RAY' of the ONE AND ONLY SELF. Each such 'RAY' is ESSENTIALLY *identical* with every other such periodically appearing 'RAY' throughout Infinite Duration. Each such 'RAY' is *substantially* the ONE AND ONLY SELF.

Example: The Universal Logos is the Great Subject and Cosmic Prakriti is the Great Object. The Universal Logos is the greatest of the Limited Beings. (We cannot Really call the Pre-Cosmic Infinite Subject and Infinite Object *Limited* Beings; they are *transitional*—both Limited and Unlimited). The Universal Logos holds in trust (whether with complete Consciousness or not) the Design-at-the-Beginning with which pre-Cosmic 'FOHAT' 'CAME FORTH' from the ABSOLUTE at the 'FLASHING FORTH' of the Pre-Cosmic 'RAY'.

Example: The INFINITE SELF *IS* the 'RAY' which 'FLASHES FORTH' *IS* the Infinite Subject/Infinite Object *Is* the Condensed Point *Is* the Universal Logos. The entire 'Cast' of Pre-Cosmic 'Characters' are not differentiated in any ESSENTIAL sense at all. The Universal Logos is none other than the INFINITE SELF-in-Objectivity. When terminology and order of appearance tend to overwhelm, it is well to remember that the many 'Players' are all ONE 'PLAYER'—THE ONLY AND ONLY.

Logos, Sub-Universal

By a *Sub-Universal Logos* is meant the Shiva, Vishnu, and Brahma of an entire Cosmos. These are the Three Logoi Who are immediately inferior to the Universal Logos. They are the *active* Creator, Preserver, and Destroyer of the newly forming Universe.

The manner in which this is conceptualized depends upon whether one takes for a model the 'Point within the Triangle', or simply the Triangle (with an inferior or Prakritic Point). With the Point within the Triangle model, the Unmanifested Logos becomes a *Manifest* Father, Who Self-Generates a Universal Son, Who Self-Generates Universal Brahma. The Father Aspect is thus *dual:* Unmanifested and in Darkness, and Manifested and more directive. The various models can be compared, but there are always different methods of patterning and enumerating when designating the position and roles of the members of the Cosmic First Family.

Example: Always the *three* surround the *one*. This is true in relation to Sanat Kumara and the Buddhas of Activity; it is true in relation to our Solar Logos and His Brahma, Vishnu and Shiva; and, upon a barely imaginable higher turn of the spiral, it is true of the Universal Logos and the three Sub-Universal Logoi—the Cosmic Brahma, Vishnu, and Shiva.

- M -

magnetism

By the term *magnetism* is meant affinity between E/entities-in-Cosmos or items-in-Cosmos such that, through the operation of this affinity, the E/entities/items are attracted to each other and exchange influences (vibratory patterns).

Example: The great magnetism between our Earth and Venus accounts for Their continuing interplay which operates to Their mutual benefit and the benefit of the Solar System of which They are a part.

Maha-Manvantara

By the term *Maha-Manvantara* is meant the "Great Manvantara" which usually indicates the entire duration of a Solar System in its subtle as well as physically manifested aspects. There are obviously greater manvantaras than the Solar Systemic Cycle—for instance: Cosmic Manvantaras (the complete cycle of manifestation of what are called in The Tibetan's writings Cosmic Logoi).

These Logoi are *not* the Cosmic Logos (as the term is used in this treatise—i.e., as equivalent to the Universal Logos); Constellational Manvantaras (or the cycles of various kinds of constellations, differing in length according to the scope and magnitude of the constellations involved—from minor constellations all the way to globular clusters containing millions of suns); Galactic Manvantaras, Super-Galactic Manvantaras, etc., until the largest of all manvantaras is reached—the Universal Manvantara, which includes every possible manvantara in a particular Universe/Cosmos. All of these greater cycles are really Maha-Manvantaras in a generic sense. The duration of at least some of these greater cycles may be known to the Masters of the Wisdom, but not to humanity.

Example: The Planetary Logos of the Earth Scheme has probably had a number of *incarnations* during the relatively immense duration of the Solar Maha-Manvantara. May it be that Planetary Logoic Incarnations tend to change at the onset of each Solar Systemic "Eternity", which, according to H. P. Blavatsky, is one seventh of a Solar Maha-Manvantara? One day we shall know; however, if Emanation Theory is correct, we *know even Now,* at some 'altitude' along the Divine Emanatory Stream.

manifest (adjective)

By *manifest* is meant 'present-in-Cosmos'.

By *manifest* (upon a higher turn of the spiral) is meant the *occurrence* of any Event whatsoever—even a Pre-Cosmic Event.

Example: The Will of the Planetary Logos of the Earth is manifest in the multitude of forms to be found within His Planetary Scheme, and can be 'read' by those who know how to read the form as symbol.

Example: Can the SELF-as-Infinified Point or the SELF-as-Condensed Point (being the same Entity at different Pre-Cosmic Stages) be considered manifest. These Two are Really only Pre-Cosmic Points of View, and yet, inasmuch as They are *not* BE-NESS in ITS ALL-IN-ALLNESS, They (the SELF-as-Infinified Point and the SELF-as-Condensed Point) *can* be considered manifest If the ABSOLUTE SELF were the 'WITNESS' or 'OBSERVER', which the Vedantin Philosophy says IT IS (and *The Secret Doctrine* says IT is *not*) then, most definitely, the SELF-as-Infinified Point, the SELF-as-Condensed Point, and *any* Pre-Cosmic Dynamic would be manifest for that which is *'seen'* is manifest.

manifest (verb)

To *manifest* means "to show forth."

Example: The 'EXTRUSION' of 'INHERENT POSSIBILITY' from the INFINITESSENCE is an 'ACT' of manifestation. It could be said that through each Cosmos the ALL-SELF 'MANIFESTS' an infinitesimal (but in this case, *definite*) 'PORTION' of ITS INFINITE POSSIBILITY. Can an infinitesimal or 'infinitesimalizing' be 'definite'? One could say, No, and yet the Universe is definite.

Also, any portion of ALL POSSIBILITY exists in relation *infinite possibility* in the proportion 'one over infinity'. Does that proportion signal an 'infinitesimalizing' (or worse, a zero!)? It may well be that a definite number divided by infinity is *not* zero, but the process of 'infinitesimalizing' instead: i.e., 'convergence upon zero'. Certainly, the division of '1' by an ever-increasing though *indefinite quantity* is an 'infinitesimalizing'.

A question arises, If a definite quantity (say an integer) increases by means of the addition of integers at *infinite speed* will 'infinity' ever be 'reached', or only a never-endingness of more magnitudinous integers? All this bears on the 'relation' (if we can call it one) of the one 'EXTRUDED' Universal Possibility to the infinitude of 'NON-EXTRUDED' possibilities always 'RESIDENT' 'within' the FOUNT OF ALL POSSIBILITY.

manvantara

By a *manvantara* is meant, literally, "the interlude between two Manus". More generally, manvantaras are complete cycles of manifestation within planetary chains, planetary schemes, and even within larger structures, such as solar systems, groups of constellations, etc.

Example: With reference to the planetary schemes of our solar system, one complete round of seven globes can be called a manvantara, but seven complete rounds of seven globes is an even greater manvantara. There are manvantaras which pertain to *solar* schemes as well as to planetary schemes.

Manvantara, Universal

By the *Universal Manvantara* is meant the complete cycle of manifestation of the Universal Logos of a particular Cosmos. The duration of such a Cosmic Cycle is from the coming into Being of the Universal Logos until the termination of the Process of Cosmic Obscuration or onset of the Universal Pralaya which inaugurates the Cosmic "Day Be With Us." The duration of a Universal Manvantara is absolutely incalculable by the mind of man, and perhaps by any Entity within this Solar System.

Although it *may*, perhaps, be *presumed* that every Universal Manvantara endures for exactly the same amount of Super-Cosmic 'Time' (in Infinite Duration) as every other such Universal Manvantara, to assume so would be merely an unprovable speculation. The ABSOLUTE SELF has *ALL TIME*, and need not *regularize* the Appearances and Disappearances of Cosmoses in order to *"save 'Time'."* It may well be, for instance, that a Universal Manvantara lasts exactly as long as it takes to achieve the Cosmic Consummation of the Design-at-the-Beginning, which would be equivalent, from another perspective, to Universal Redemption. Such an optimistic thought suggests a "happy ending scenario" to the Cosmic Process!

Example: There have been and will be an infinite number of Universal Manvantaras, but we are in the midst of the only one of which we (as Man) can know anything.

mass

By *mass* is meant frequency of vibration.

By *mass* (according to Webster's *New Collegiate Dictionary*) is meant, "the property of a body that is a measure of its inertia, that is commonly taken as a measure of the amount of material it contains, and causes it to have weight in a gravitational field ..."

Example: Usually, a physical thing is said to have no mass when its frequency reaches that of the speed of light and the thing is thus transformed into what we call "energy". The idea of mass, however, is Really more subtle than that. Mass does not necessarily mean 'density' as usually conceived. Any item-in-Cosmos on any prakritic plane has mass; any such item exists because there is vibration. Existence itself, no matter how refined, no matter how subtle the plane referenced, is, relatively speaking and for practical purposes, *density* (even though, for the sake of metaphysical accuracy, existence is *privation* of density).

Considering the concept of mass in this new and subtler way, since all that exists has 'mass', the measurement of mass (at least in the Fohatically Particulate Worlds) is dependent upon the frequency of vibration—the greater the vibration the less the mass. In this definition, what we call 'energy' in E = mc2 has not shed its mass, but *still* is endowed with that apparent density we call mass, though the 'density' is in a much *subtler* form—an 'energy-form'.

What all this means is that there are ring-pass-nots on planes higher than the physical, and even on the highest Super-Cosmic Planes. If the Causal Body, which is an energy body, can have, as The Tibetan suggests, a certain "specific gravity", then it also has *mass*.

Example: That metaphysical gravitation (under the Cosmic Law of Attraction) exists between extremely subtle fields—fields found upon planes far 'higher' than the systemic etheric-physical plane—is an indication that these subtle fields have mass.

Example: It would be an interesting speculation to ponder on the 'mass of an Idea'! Mass is related, says the definition, to "quantity of material". Metaphysically, we could be talking about the number of ultimate particle/events involved in any presentation. But what of 'mass' in the 'partite' but 'imparticulate' Worlds? Does it exist? If materiality is registration in consciousness, are the Subjects/Objects in the World of Being 'material'? Perhaps, because Ideas (as Beings) are omnipresent in-Cosmos, they have no inertia. They do not 'move'. They are 'unaggregatedly prakritic' (not particulately prakritic), but have no 'prakritic resistance'. There is no particular 'quantity' of matter in the Qualitative Aspect of 'Ideations'. The 'Being Aspect' is of course, strictly Subjective and non-material. What is being suggested that 'mass' in the World of Fohatic Particulation and 'mass' in the World of Being are quite different. For general purposes, it is probably accurate to say that there is no 'mass' in the World of Being, in terms of what might normally be called 'quantity of matter'.

Example: From the foregoing, it is possible to hypothesize that there is such a thing as 'matter' without 'mass'. Ideas, in a way, are 'matter' without 'mass'.

Masters of the Wisdom

From a Non-Dualist perspective the *Masters of the Wisdom* are Those Who, having realized the REALITY of INFINITY are, thereby, consciously, solidly and inescapably welded into the Great Hierarchy of Being for service to the World of Becoming. They are sworn to obey the ascending Hierarchy of Wills in-Cosmos, which Hierarchy fulfills a Universal Cosmic Will which is, Essentially, naught but *Their Own Will* as It primevally participated in the Design of the Original Intent.

Example: The Masters of the Wisdom, realizing the REALITY of INFINITY, patiently serve the apparent evolution of that which is finite.

matter

By *matter* is meant any object whatsoever. That object can be subtle or gross, virtually infinitesimal, minute, vast, virtually infinite in extent or even infinite (such as Mulaprakriti). If a thing is perceivable or apperceivable as an object, it is *matter.*

By *matter,* in a still subtler sense, is meant that which 'ARISES' through the INFINITE SUBJECTIVITY's *proxied* 'PERCEPTION'/'CONSCIOUSNESS' of ITSELF, from ITS *proxied* 'TAKING NOTICE' of ITSELF. Extending the idea to the particular, matter can also be considered the Objectification of internal Subjective Content. Even more subtly, matter (especially Root Matter) is the *Potential for Specific Objectification,* or That, by means of which, Objectification can occur in the lower worlds. That which we *usually* call matter, is Really the *result* of the Mayavic Objectification Process. Maya (PARABRAHMIC SELF-'REFLECTIVE' 'POWER') is the *cause* of objectification of that which is objectified.

Example: Pre-Cosmic Matter, Mulaprakriti, *is* PARABRAHMAN 'SEEING' ITSELF by means of 'MAYA'-instantly-Maya. Such Matter is boundless and absolutely undifferentiated, for It is the Reflection of THAT which is BOUNDLESS and ABSOLUTELY UNDIFFERENTIATED. Cosmic Prakriti, which is *bounded* Mulaprakriti, is *virtually* undifferentiated but not *completely* so. Cosmic Prakriti is the Objectification of less expansive hence Bounded 'Self-Sight'. It is the Infinite Subject-as-'De-Infinifying' Subject-as-Finite Subject 'Seeing' Itself in a more limited way.

Example: There is a tremendous difference between what we usually call matter and the Root Matter, which is the Objective Reflection (induced by Maya) of any Subjective Content within the INFINITESSENCE.

Example: Root Matter is certainly an *Indefinite State of SELF-as-Self-Perception.* Even matter is a state of perception—i.e., a State of Self-Perception.

Matter

By *Matter* is meant the Root of Objectivity.

By *Matter* is meant Mulaprakriti, "Pre-Cosmic Root Matter".

By *Matter* is meant that Content of Consciousness which is 'Presented' to the Infinite Subject when Infinite Pre-Cosmic Maya causes the Infinite Subject to 'See' Itself, to 'See' Its Own Homogeneous Infinitude or Its Own *infinite noumenessentialized* (hence, 'imparticulate') Content.

Example: Matter is not tangibility but the *possibility* of tangibility. Matter is not the *stuff* of all the many 'material things' but, rather, the *possibility* of their appearance through a combination of Pre-Cosmic and Intra-Cosmic 'Psycho-Reific' Processes (Psycho-Reification—the Process whereby Objectivity and Objectivities are 'Created' through Acts of Consciousness/Perception). From this perspective, Fohat/Maya is a Great Intra-Cosmic 'Great Psycho-Reifier'.

Maya

By *Maya* is meant a Pre-Cosmic and Intra-Cosmic Process which has proceeded cyclically forever, and will continue forever—ESSENTIALLY, the Process of SELF-'VEILING' (which leads to and includes the Process of Self-Veiling).

By *Maya* is meant a simple psychological dynamic—externalize the nature and content of the *subject* as if that nature and content were an *object.* Through the Agency of Maya, what is 'within' becomes *seen* as if it were 'without', without the original 'withinness' being noticed.

By *Maya* is meant a Process of Generating *Conditions* through the agency of the 'Chief Instrument of Limitation'—*consciousness.* For this kind of 'Generation', the SELF (and only the SELF) IS ultimately, ESSENTIALLY, and *utterly,* 'RESPONSIBLE'.

Maya may be thought of as the *veiling* of REALITY, (or SELF-'VEILING'), or as the SELF 'SEEING' (by means of ITS 'CONSCIOUSNESS') of ITS FULL SELF—which FULL SELF is immediately *mayavically reduced* to an 'OBJECT'-instantly-Object by the very 'ACT' of 'SEEING'! It is as if the SELF in ITS guise as Pre-Cosmic 'FOHAT' *splits* ITSELF into Infinite Subject (the SELF-as-Infinified Point of Perception) and Infinite Object (Mulaprakriti). That which the SELF 'SEES' of ITSELF Is always *less* than THAT which the SELF REALLY *IS.* 'SEEING' reduces BEING (through dichotomization).

This is a principle vital for understanding Maya and Creation. In this 'SEEING' of *lessness,* a Cosmos is Born. In short, and in practical terms, Maya means, *first,* 'SEEING'

WHAT IS. The 'ACT' of 'SEEING' (i.e., 'CONSCIOUSNESS') is a stupendous *limitation* upon the 'IS-NESS' of WHAT IS, hence 'SEEING' is *mayavic.*

Secondly, Maya means 'Seeing' *progressively less* and *less* than WHAT IS. This second Mode of Maya is the one which is most conceivable to human minds and is the Act by which the Super-Cosmic Consciousness of the SELF-as-Condensed Point *'measures out'*, as it were, a 'Limited Portion' of Mulaprakriti (i.e., a limited portion of the Infinite SELF-as-Self-Image which Maya causes to be Reflected as Mulaprakriti), simply by 'narrowing' the Focus of Its 'Seeing' so that the *Infinite* Nature of (Itself) as Mulaprakriti is no longer 'Seen', and only that 'Portion' of Mulaprakriti (of Its Own Nature) is 'Seen' which will be needed to fulfill all SELF-'INTENDED', 'INFINITESSENTIALLY EXTRUDED' Patterns within the particular Cosmos in Process of Preparation.

Such words as 'Portion' are necessarily misleading because they cast Mulaprakriti in the role of an *extended* Something. The concept of *extension* is ESSENTIALLY illusory where primal, undifferentiated Matter is concerned. Mulaprakriti is the Boundless Object, identical with Itself at every Point (even though It has no *points*—or has, to many of them as to have *none*).

We now come to a difficult problem—difficult to conceive, which is, What does it mean that the Infinified Point (the Infinite Subject) 'narrows Its Vision' and 'Sees' less of Mulaprakriti than It did before?

- It means that the Infinified Point 'Becomes' the Condensing and then, the Condensed Point.
- It means that the Infinite Subject-as-Condensing and Condensed Point puts from Its 'Mind' the possibility of the *Mulaprakritic* Objectification of *All* Things or of Anything, and 'Concentrates' only upon the Objectification of that 'IDEA' that has been 'TRANSMITTED' via the 'RAY' of the ABSOLUTE.

This metaphoric 'narrowing of Sight' is, as it were, a *reduction of demand* upon the Reflecting Power of Maya. Mulaprakriti is, thus, no longer 'Seen' (by the Infinite Subject) in terms of the Reflected Infinitude of the INFINITE SELF (nor does the Condensing Point as 'De-Infinifying Point 'See' the multitude of articulable possibilities) but, rather, the 'Sight' developing is of a more *limited Object* in which can be Objectified (or which can be Objectified as) the FOHATICALLY-Conveyed 'IDEA' which is to 'Become' the new Cosmos.

- Mulaprakriti is, thus, not a 'Something' which was once 'Seen' as Infinite and is, at a certain phase of the Pre-Cosmic Process, 'Seen' as Finite.
- Rather, Mulaprakriti is, as it were, 'Reconsidered' by the Infinite Subject (Who is no longer the Infinified Point but has 'Become' the Condensing Point and Condensed Point). The *Infinite* Reflective Power of Maya is 'Ignored', and only Its *'Needed* Reflective Power' is Considered. In this Reconsideration, Mulaprakriti becomes *Bounded*—Bounded Cosmic-Prakriti.

Just *what* is *bounded?* Certainly *'Stuff'* is not bounded. The human mind so easily falls into the illusory notion that matter is *stuff,* and, thus, that Mulaprakriti is *stuff,* of which there can be an infinite amount or a lesser amount. No.

- Rather, in a more subtle vein, it is the *degree* of Infinitude of the Mulaprakritic Reflection' that is *bounded* by the Re-focusing of the Infinite Subject upon the 'Immediate Task At Hand' (which Task is *first inherent within the Nature of the*

Infinite Subject Itself), namely the 'Creation' of a particular Universe which conforms to the 'PARAMETERS' of the 'IDEA' EXTRUDED from the INFINITESSENCE and FLASHED FORTH as the 'RAY' of the ABSOLUTE.

It is so vital to get some grasp of the subtle 'Psycho-Reific Processes' involved in 'Creating' a Universe. Everything 'Done' is done through Consciousness and Perception. The entire Process is one in which Subjectivity is made Objective. Indeed, all Universes are SELF-Reflective Universes, and Maya is the Power to Reflect that which is Inherent in Subjectivity and which, then, *appears* as Mulaprakriti, or, more limitedly, as 'Lessening Mulaprakriti', and, finally, as Cosmic Prakriti.

By *Maya* is meant the SELF-'IMPOSED' negation of MY/OUR OWN FULLNESS.

By *Maya* is meant *Consciousness*, pure and simple, for 'CONSCIOUSNESS-instantly-Consciousness (SELF-'REFLECTION') is a limitation upon the INFINITIZED 'STATE' of BEING. Perhaps this is why all Systems of Liberation insist that the disciple "take his eyes off himself", thus can he begin the process of re-entering the WORLD OF BEING.

Example: Because of Maya, countless unconscious and, even, self-conscious E/entities do not realize they ARE the SELF and the SELF alone.

By *Maya* is meant the *trigger* for the 'AWAKENING INSTANT'.

By *Maya* is meant the 'INHERENT' PARABRAHMIC 'LOVE' of ITS OWN 'INHERENT' 'POTENTIAL' for 'INFINITE SPECIFICITY'. *'MAYA'* is 'LOVE' and 'CONSCIOUSNESS' 'within' the ABSOLUTE—both of which *require* an 'OBJECT', even if that 'OBJECT' is only the *noticed* INFINITE SELFHOOD of the ABSOLUTE!

Example: Maya is *not* Mulaprakriti, but That Process which makes the *appearance* of Mulaprakriti possible.

Example: Maya, Itself, is the First *Happening*.

Example: 'MAYA' Is the Process of SELF-'OBJECTIFICATION'.

Example: 'MAYA' is the PARABRAHMIC 'POWER' by means of which INFINITE BE-NESS (i.e., I-as-BE-NESS) 'BECOME' 'NOTICEABLE'. Before 'MAYA' **I** simply *was*, 'INFINIDENTIFICATORILY'. 'Before' the 'ARISING' of 'MAYA' the GREAT BE-NESS did not 'NOTICE' ITSELF, that IT *IS*. With the instantaneous 'ARISING' of 'MAYA' come the 'NOTICING' of BE-NESS and not simply the *being* of 'BE-NESS', i.e., BE-NESS alone.

Example: With respect to the GREAT BEING/NON-BEING, 'MAYA' is ITS SELF-'INHERENT' POWER to NOTICE *that* IT *IS*, and *then* to 'NOTICE'-instantly-Notice *what* It Is (this second 'Noticing' occurring on the Pre-Cosmic Level that developed instantly as a result of the first 'NOTICING').

Example: 'MAYA' 'SPLITS' INFINITUDE.

'MAYA'-instantly-Maya

By *'MAYA'-instantly-Maya* is meant the instantaneous 'ARISING' of 'MAYA' 'within' the ABSOLUTE and its immediate *expulsion* from 'residence' 'within' the ABSOLUTE by its very 'ARISING'. No sooner does 'MAYA' 'ARISE' within the ABSOLUTE than it becomes the Maya within the Super-Cosmic or Pre-Cosmic World.

By *'MAYA'-instantly-Maya* is meant 'EVANESCENT' 'MAYA' *infinitesimally-instantly* 'BECOMING' Pre-Cosmic Maya, or the Sustained Consciousness of Infinitude as an Object (i.e., of Mulaprakriti).

Example: 'MAYA' is PARABRAHMAN ITSELF 'INTENT' upon Finitization. When 'PARABRAHMAN' suddenly ("in no time" or, virtually, "no time") 'BECOMES' 'MAYA' in addition to ITSELF, that 'MAYA' becomes part of the World of Subject/Object Relations. That instantaneous Change is described by the formula, 'MAYA'-instantly-Maya.

Me and My

Me and *My* are, respectively, the pronoun object and the possessive pronoun used to indicate the SELF in ITS in-Universe mode as the Self. These pronouns are also used in connection with the symbol **'8'** instead of with the symbol 'I'.

Example: When **I** am out-of-Universe in my ALL-IN-ALLNESS, the term 'ME' is used when referring to WHO **I** AM. When **I** AM in-Universe (and hence Am **8**) during Universal Manvantara (which does *not* mean that **I** AM not simultaneously SUPER-Universal), the term 'Me' is used to refer to Who **8** Am—*in*-Universe. These terms (ME and Me, MY and My) relate, respectively, to MY SUPER-Universal IDENTITY and My Super Cosmic and intra-Universe Identity.

ME-as-Me-as-me; me-as-Me-as-ME

By *ME-as-Me-as-me* is meant a series of pronoun objects which correspond to the pronoun subjects **I**-as-**8**-as-I. All of these terms-in-series are meant to convey the fact that the one Super-Universal SELF manifests as the Universal Self along with all the many Self-conscious and spiritually connected Selves, and then, finally, as the limited egoistic and sub-egoistic selves (to which 'I' and 'i' correspond). ME-as-Me-as-me indicates the *descent* from Super-Cosmic INFINITUDE (i.e., ME), to Cosmic Pervasion (Me) into the Self-Reduction of intra-Cosmic particularity (me), just as me-as-Me-as-ME indicates the *reascent* from reduced intra-Cosmic particularity (me), to Cosmic Pervasion (Me) and finally into Super-Cosmic INFINITUDE (ME).

Example: Every happening in-Cosmos happens to ME-as-Me-as-me. In my little self, I do not know this. As **8** grow into My intra-Cosmic Fullness, **8** know this. In MY extra-Cosmic PLENITUDE, **I** AM, ESSENTIALLY, ALL that transpires.

meaning; meaninglessness

By *meaning* is denoted an understanding of the function of the role played by any 'part' within the Whole. A part cannot have a meaning unless it 'holds place' within a context.

Example: Within the ABSOLUTE, there prevails a state of *absolute* meaninglessness, technically considered, for there are no 'parts' within the WHOLENESS of THAT.

Example: To understand the meaning of one's life requires the specification of the level of the *self* to be considered, and a determination of the scope of the context in which to consider the meaning of that *self.* If **8** consider 'MySelf' to be a personality, only a fairly limited context will have to be considered in order for me to determine meaning. If, however, **8** consider 'MySelf' to be a Monad, a much greater and far more subtle context will have to be considered. The point is that the determination of meaning is inseparable from context.

measure (verb)

By the verb *measure* is meant the act of comparing one thing to another.

Example: It is impossible for any being to measure itself against INFINITY, for the INFINITE IS the INCOMPARABLE and the IMMEASURABLE, and the ratio between IT and any finite thing, no matter how large that finite thing may be, even the Cosmos Itself, is always Infinity to One (Infinity/One). Within the finite context of Cosmos, however, measurement can occur and is necessary for skillful action in the working out of the Design-at-the-Beginning.

Example: Maya, which means "to measure" is the process which generates 'things' and thus makes comparison and, hence, measurement possible.

middleness

By *middleness* is meant the position of anything with respect to the indeterminate, ever-indefinite extremes of an Infinite Continuum.

Example: Theoretically, in a Cosmos which was an infinitely divisible continuum, every quantity/unit in that Cosmos would, regardless of its mass, volume or density occupy a position of middleness upon the Infinite Continuum stretching from *virtual* Nothingness to *virtual* Everythingness. Because, however, Real Cosmoses are *quantized*, and have upper and lower limits of 'size'—there being a *least* possible intra-Cosmic magnitude and a *greatest* possible Cosmic magnitude (i.e., the magnitude of the Cosmos Itself as a Whole) the majority of items-in-Cosmos do *not* occupy a position of middleness between the cosmo-parametrically sanctioned extremes.

Example: Every number, regardless of its magnitude, occupies a condition of middleness between that uncountable, indefinite magnitude which is infinitely great (Infinity) and that uncountable indefinite magnitude which is infinitely small (the infinitesimal, 'infinitesimalizing'). Both uncountable magnitudes are not actualizable-in-Cosmos, however, they are conceivable and theoretically possible.

Example: Every Cosmos occupies a condition of unvarying middleness upon the Infinite Time Line which has existed throughout Infinite Duration. At this particular Time in-Cosmos—i.e., Now (and also at any other 'time' in Cosmos)—our Cosmos and Its 'Happenings' are exactly as 'far' or 'distant' in 'Time' from the *infinitely recessed beginninglessness* of the past, as it is from the *infinitely projected endlessness* of the future.

Example: Curiously, since all Cosmoses occupy 'middleness' upon the Infinite Time Line (whether such Cosmoses are 'neighbors in Time' (i.e., the One succeeding the other) or separated by a centillion of other Cosmoses), because they are still 'located' at the 'position' of middleness, it is, in a way, as if they were both 'happening' at the 'same time' (i.e., 'on' the ETERNAL NOW).

Example: It is forever impossible for the ETERNAL NOW to *move* from its 'position' of middleness 'between' ALL that has been and ALL that is to be. The NOW is, forever, utterly immobile.

Mode of Consciousness

By *Mode of Consciousness* is meant the particular 'position' of descent or ascent which an authentic I/identity may be occupying within the Divine Emanatory Stream.

Example: Human consciousness is a Mode of Consciousness for all authentic Identities at some point of their Emanatory/'De-emanatory' Cycle. Hierarchical Consciousness and Planetary Consciousness, etc. would, similarly, be Modes of Consciousness. Due to 'dimensional sealing', it is reasonable to consider that an authentic I/identity in-Cosmos (even on the lower end of the Emanatory Sequence) is be found manifesting through a great number of Modes of Consciousness (or, in fact, through *all cosmically possible* Modes of Consciousness) simultaneously without 'knowing' it.

Each Mode of Consciousness proximate to the Human Mode of Consciousness would be dimensionally sealed, if not from *every* other Mode, then, at least, from most others. The higher the Mode of Consciousness upon the Divine Emanatory Stream, the less the dimensional sealing (at least with respect to the Modes 'below' It). That Mode of Consciousness we call the Universal Logos, is in no respect dimensionally sealed, being the *Pervader* of all possible intra-Cosmic dimensions.

The question remains concerning the degree to which the Universal Logos is *sealed* from the ALL-IN-ALLNESS of the INFINITE SELF. Perhaps, the Logos *is so,* until the "Day Be With Us."

modification

By *modification* is meant any deviation or variation from homogeneity. All modifications are *objects* in the purest sense of the term 'object'. Modification and Objectification are coeval—the moment modification occurs, an object exists.

Example: The atom, the man, the greatest super-galaxy and the smallest 'atomic particle' are all modifications of the Universal Prakriti.

Example: Let it be remembered that that which seems to be a *subject* from the Cosmic or even Super-Cosmic Perspectives, is, nevertheless an *object* from the 'PERSPECTIVE' of the ONE AND ONLY SELF. (Remember that even the SELF can be an 'Object' to the ITSELF!, which is what the whole Cosmic Process is about!) When the 'RAY' of the ABSOLUTE 'FLASHES FORTH', an Infinite SUBJECT-as-Subject is Created which is a limitation of the 'STATE' of INFINITE SUBJECTIVITY. That Infinite Subject is the SELF *as* Infinite Point of View (the SELF-as-Infinified Point) 'Seeing' the Infinite Object.

The question arises, Is the Infinite Subject (born through 'MAYA') an 'Object', even though the Infinite Subject ('EXTRUDED' as It Is from the INFINITESSENCE) *is* the Subject which will substand the Cosmos throughout the Universal Manvantara? In other words, is anything whatsoever it may be, which is 'EXTRUDED' or 'PARTICULARIZED' from the INFINITESSENCE, necessarily an *object* simply because of 'EXTRUSION' and 'PARTICULARIZATION'? Certainly the Infinite Subject is a modification of the INFINITE SUBJECTIVITY.

Can a *subject* be *'seen'* as an *object?* What of the 'RAY' of the ABSOLUTE—the greatest *subjectivity* other that the ABSOLUTE SUBJECTIVITY. Can this 'RAY' be considered an *object?* Perhaps this can be answered with respect to Emanation Theory. Every time an aspect of Selfhood 'goes forth' from a Source, does the 'part' of Selfhood which *Is* the Source and *'remains behind'*, as it were, while the emanated 'part' 'goes forth', 'see' the part of Selfhood which 'went forth'? Can **8** 'see' My lesser emanated selves as *objects?* The answer would seem to be, for the most part, Yes.

Really, there is nothing but *subject* (the INFINITE SUBJECTIVITY) in the entirety of the UTTER ALLNESS. Anything 'seen' is REALLY/ESSENTIALLY a *subject* being 'seen' by a subject. So, it would seem, that any subject (with the mysterious exception of possibly ONE SUBJECTIVITY) can be rendered into an object simply by the act of *'self-seeing'*. Any modification whatsoever can be 'seen'. Even the most subtle of all, the 'RAY' of the ABSOLUTE, is a 'MODIFICATION'. This endless process of *self-seeing* finds its termination and resolution only in BEING, which is the 'CHARACTERISTIC' of the UNSEEING ONE alone. There is thus a mystery concerning whether the 'RAY' ('POINT', 'MAYA', 'FOHAT', INFINITE TRINITY-as-ONE) is, indeed, 'SEEN'!

Example: Can a subject be an object? This reminds one of the statement attributed to St. Francis and paraphrased as follows, What we are looking for is what is doing the looking. Can that which 'sees' be 'seen'? Can one 'see' the 'seer'? It might be said that the one who 'sees' can be 'seen' by *himself* if that one 'goes forth' to identify with that which was 'seen' and *yet* 'remains behind'.

For instance, a modification is seen. Then one goes forth *into* the modification and identifies with it. One, as it were, "forgets himself" in doing so. Suddenly one awakens to 'see himself' in the context in which he had forgotten himself. One becomes at that moment *both* the 'seer' and the 'seen'.

For instance, I become angry through identification with a modification. Then, I 'see' myself while I am angry. I other words, I stand back from myself and 'see' myself engrossed within the modification with which I have identified. For a moment I am as if in two 'places'. I, the subject who 'went forth', have become the 'seen' object.

moment

By the term *moment* is meant a *relatively* small division of time. The use of this term is very inexact. An *instant* is usually considered more rapid than a moment, but the term 'moment' *does* suggest considerable brevity. Units of Time are naturally composed of ultimate moments which are Really 'ultimate (*seeming*) *movements'* considered as *measures of Time.* Moments, as usually considered, are composed of many, many *ultimate* moments. Sometimes, the term 'moment' means a second of time. In the Eyes of God the Universal Logos, however, a 'moment' of Time can be a very lengthy cycle in the life humanity. The term 'moment' is completely relative.

Example: The 'sense of duration' for a given moment of time depends upon how many changes are perceived/registered by a consciousness within that 'moment' relative to how many changes are *usually* perceived/registered by that consciousness during other moments of time of equal duration.

The more changes perceived/registered, the longer the *sensed* duration; "So much has happened in the short time you've been away that it seems like a year." The number of registrations per unit of time is the key to perceived variations in *psychological time.* If the possible rate of change if sufficiently rapid (far more rapid than would be allowed by our present Cosmic Parameters) it would be possible to compress all the activity of an entire Universal Process into a minute or far less. The perceiver (accustomed to a rate of change comparable to the one usually perceived by human beings in our Cosmos but, somehow empowered to register the vastly increased rate of change proposed) would experience that minute of time, *psychologically*, as if untold aeons had passed. The relativity of time as explained by physics, is paralleled by the relativity of time in relation to consciousness.

The psychology of the perception of time is not quite that simple however, as it is possible for an eternity to *seem* like a minute and vise versa. Blake saw "infinity in a grain of sand, eternity in an hour." The extreme extension of this idea would see the *psychological perception* of the entire Infinite Super-Universal Process, (normally considered to have 'taken place' throughout Infinite Duration) *seemingly* compressed into an instant the value of which was *zero*—i.e., the ETERNAL NOW.

This is a far more *extreme* development than the old hymn suggests, "A thousand ages in His sight are like an evening gone." We would have to modify the hymn with the idea that an 'Infinitude of Ages Past have ta'en "no Time at all." This would mean that throughout Infinite Duration, it would *seem* (probably to the GREATEST OF ALL 'OBSERVERS'—the ATMAN as 'WITNESS') that 'no time had passed at all'. This 'POINT OF VIEW' (no doubt *infinified*) would be REAL. The mind boggles!

Example: The entire duration of our Solar System is merely a 'moment' of Cosmic Time. Somehow, after the foregoing considerations, this statement seems quite tame!

MOMENT, ABSOLUTE

By the *ABSOLUTE MOMENT* is meant the consideration of every possible moment in-Cosmos as the ONLY MOMENT which ever has been or will be. The ABSOLUTE MOMENT occurs at *all times*, and, thus, any possible in-Cosmos Moment (of any variety) can be designated as the ETERNAL NOW.

Example: Throughout all of Infinite Duration, all things transpire *in, on* or *at* the ABSOLUTE MOMENT which ITSELF IS of Infinite Duration (or zero duration!) and never elapses into another such MOMENT. Forever and ever there is only NOW. However, and paradoxically, just because all things are transpiring in the ABSOLUTE MOMENT, NOW, does not mean that, considering things relatively, events did not also happen *then*, and will not happen *in the future*. Past, Present and Future are legitimate designations in the World of Relativity even though the ABSOLUTE MOMENT never changes. A paradox!

Example: It is curious that an infinite number of ABSOLUTE MOMENTS occur during an ultimate moment in Cosmo-Objectivity. On the other hand, though there be a huge number of ultimate moments, they are all occurring within the ONE *unchanging* ABSOLUTE MOMENT.

moment, dimensioned

By a *dimensioned moment* is meant a definite moment of any duration (however small) greater than zero. A dimensioned moment *can*, theoretically at least, be measured; it *is mayavic*, and is quantifiable. An ultimate moment is a dimensioned moment, but a dimensioned moment is not necessarily an ultimate moment. Not all *possible* dimensioned moments are 'allowable' in a given Cosmos because of the Parameters of the Design-at-the-Beginning. Given these established Cosmic Parameters (theoretically quite possibly different for each Cosmos), some dimensioned moments would be too small (or too large), to be *cosmically permissible/possible.*

Example: Dimensioned moments *do* appear in-Cosmos, and are the means of measuring (i.e., *timing*) the 'movement' (macro and micro) of the Cosmic Configuration—from one designated position to the next. Dimensioned moments can be added together to "take time" during which the progressive (albeit illusory) Cosmic Movement appears to be taking place. Dimensioned moments have duration (as measured against the Infinite Time Line), and can be added to create all the various quantities of *measurable* duration—the minute, the hour, the day, the week, etc.—(i.e., Mayavic Time).

Dimensioned moments are an analog to the appearance of an *actual* point in the matter of Cosmos. Any tangible (apparent) point appearing in-Cosmos (such as the dot caused by a pen-point) has dimension and is not Really a Real point at all. Whereas a Real Point (an Ideal Point) has *no determinable dimension* (since it is an 'infinitesimalizing', and thus ever immeasurable, because converging upon zero with a probable speed infinitesimally less than infinite—which means, "as fast as possible" without being "faster than possible"). *Dimensioned* points (also called 'virtual points') are the only kind capable of stable objectification within the Cosmic Structure. Dimensionless points (points of ever-indeterminate dimension) are *impossible* in-Cosmos as actualities, and are possible only in imagined perception (the Realm of Ideation in which they are Realities). If Ideas are Real then Real points can be Real.

Each dimensioned moment is different from every other (for dimensioned moments 'move' in sequence, and 'place' in a sequence is a differentiator). Further, the perceivable 'content' of each dimensioned moment varies with the changes in the Cosmic Configuration. While each moment of indeterminate dimension is an *infinitesimalizing moment* and, hence, immeasurable, each *utterly* dimensionless moment, on the other hand, is identical with every other—in fact *is* the SAME ETERNAL MOMENT, the ABSOLUTE MOMENT, the time value of which is zero.

There is a difference between an utterly dimensionless moment and an infinitesimalizing moment. There is a similar difference between an utterly dimensionless point (*not* a Real point) and an 'un-dimensioning point' (retreating from dimension towards nothingness, dimensionlessness, but never arriving!). The analogy in Time to the utterly dimensionless point (which is *not* the indefinitely dimensioned, infinitesimalizing point) is the ABSOLUTE MOMENT, the ETERNAL MOMENT, the ETERNAL NOW.

Example: The theoretical infinitesimal moment can never be a specifically dimensioned moment. The latter is always definite, and the former, ever undefinable.

moment, (utterly) dimensionless

By an *utterly dimensionless moment* is meant a moment which has no duration at all, or a duration equal to zero. An utterly dimensionless moment *cannot* be measured, it is *not mayavic*, and is *not* quantifiable. The ETERNAL NOW is such a 'MOMENT', in fact, is the *only* such 'MOMENT'. The utterly dimensionless moment is Really the same as a dimensionless moment, but the emphasis is added to ensure the understanding of its complete *zeroness.*

Example: An utterly dimensionless moment does not *appear* in-Cosmos, because Cosmos *changes* (or appears to change). An utterly dimensionless moment can only appear in ETERNAL DURATION, during which every *moment* is the one and only DIMENSIONLESS MOMENT, the ETERNAL NOW. During an utterly dimensionless moment "time stands still" (which it does not, during an ultimate moment, even though, in Fabricated Cosmos, it *seems* to).

Add together an infinity of utterly dimensionless moments, and not one instant of time will have elapsed. ETERNITY is a *continuum of utterly dimensionless moments* which equate to the ONE DIMENSIONLESS MOMENT of *zero* duration. Thus, throughout all INFINITE DURATION no REAL 'TIME' at all has elapsed, and nothing has moved or happened. Though the utterly dimensionless moment does not *appear* in-Cosmos, all apparent intra-Cosmic 'Time' is *substanded* by the ONE DIMENSIONLESS MOMENT.

Example: Every dimensioned moment (of no matter what definite duration) contains an infinitude of utterly dimensionless moments.

moment of indeterminate dimension

By *a moment of indeterminate dimension* is meant a moment of duration greater than zero which, because it is an infinitesimalizing moment (a moment whose value is converging upon zero at a rate {anywhere on a continuum} ranging from infinitesimally shorter than an ultimate moment to a rate infinitesimally shorter than infinite speed—both of which are 'indeterminate' 'speeds') cannot be definitely measured.

Example: Moments of indeterminate dimension do not exist as cosmo-structural actualities. Such moments are 'idealisites' rather than actualities. Perhaps the time 'taken' (*if* time is taken), to convert the 'POINT' to the Infinified Point could be considered a moment of indeterminate dimension (a moment of *infinitesimal instantaneity*—though not all moments of indeterminate dimension have to be so 'short!'). Perhaps the time 'taken' (*if* time *is* taken) to change from the inter-moment instant (the Cosmo-Subjective Now) to the appearance of the ultimate moment is a moment of indeterminate dimension, though (more strangely and more likely) it is an utterly *dimensionless* moment (the *zero moment* which is the *end* of the 'end' (of the Cosmo-Subjective Now) overlapping with the *beginning* (of the 'beginning' of the Cosmo-Objective Now).

An important series of questions with potentially startling implications arise:

- Can a *change* take place in "no time"?
- Can two configurations of measurable duration (such as appear 'on' ultimate moments 'in' Cosmo-Objectivity) follow each other so 'instantly' that no time at all has elapsed between them?
- Or, can the Cosmo-Subjective Now (of the inter-moment instant) end, and the Cosmic-Configuration of the Cosmo-Objective Now 'begin', so 'instantly' that no time at all has elapsed?

Admittedly, this is almost unimaginable, given the structure and expectations of human consciousness. More imaginable, is at least 'moments' of ultra-brief (but *not* infinitesimal) duration between two 'durating' moments (or, being specific) between the end of the inter-moment instant, and the beginning of the ultimate moment. This would not be an elegant solution, because if allowable, such an ultra-micro moment (being much shorter, actually, *as short as one pleases* shorter than an ultimate moment) would take away the *ultimacy* of an ultimate moment, and appear to produce a Fabricated Cosmos in which Time was no longer quantifiable (i.e., appearing in measurable quanta), but then, we are not really talking about conditions *within* Fabricated Cosmos.

Still less conceivable would be the allowance of 'infinitesimalizing' moments between the two 'states' under discussion—though such moments would be forever indefinite and immeasurable. The practical result of allowing infinitesimal moments between Cosmo-Objectification and Cosmo-Non-Objectification (or between Cosmo-Non-Objectification and Cosmo-Objectification) may not, however, be so dire, for if an infinitude of infinitesimal moments are added together, only a moment of *infinitesimal duration is produced!* (Neither the Cosmo-Objective Moment nor the Cosmo-Subjective Moment would be very much 'robbed'.)

This is shocking, but is parallel to the idea that an infinitude of infinities yields but *infinity.* Since an infinitesimal moment has a time value converging upon zero as a speed that (ideally) is only infinitesimally less than infinite speed (whatever 'in-de-finit-ating' 'speed-ing' that is?) any infinitesimal moment has a duration which is, literally, "next to *nothing*".

What, then, is the difference between measuring the *nothing* of the utterly dimensionless moment and the "next to nothing" of the infinitesimal moment against the Infinite Time Line? Surely, the difference is immeasurable, and as close to nothing as it can be. An indeterminate "next to nothing" is virtually "nothing, itself" and moreover, the difference between them is, indeed, ever-*indeterminate!* The effect would be twofold:

- There would be experienced (at least potentially to a consciousness of sufficient freedom and penetration) the 'flutter of duality' such that both states would seem continuously 'co-existent'.
- There would be to in-lower-World consciousnesses, the experience of the seamless *seeming* continuum of the "flow of time", which is precisely what human beings *do* normally experience.

Really, our experience of Time is so 'macro' compared with what might really be 'happening'. We can speculate, but for huge periods of 'Time' to 'come', our consciousness will be deluded with respect to Time.

moment, infinitesimal; moment, infinitesimalizing

By an *infinitesimal moment* is meant the briefest (though, ever indefinite) moment 'EXTRUDABLE'—from the REALM OF ALL POSSIBILITY (i.e., the FOUNT OF ALL POSSIBILITY). This moment would be indefinitely brief yet *not* of zero duration. Thus, it would not be 'infinitely' brief (for that would produce an utterly dimensionless moment, a 'temporal nullity'). The infinitesimal moment can more accurately be called the 'infinitesimalizing moment', which emphasizes the changingness requires of any such moment, for such a unit of time can never 'stand still' or it would be definite, quantifiable, measurable—the analogy to the virtual point.

Thus, the infinitesimal moment pertains to the World of Change and Dis-Continuity (the World of Becoming) and not to the REALM of ABSOLUTE CONTINUITY (wherein utterly dimensionless moments exist as the one and only 'MOMENTLESS MOMENT', the ABSOLUTE MOMENT, the ETERNAL MOMENT). Such an infinitesimal moment can *never definitely* exist in a Cosmos, because no matter how brief the algorithmically sanctioned ultimate moment of a Cosmos, an 'infinitesimal moment' *would always* be briefer. Even so brief a moment as an 'infinitesimal moment', however, cannot exist in the ALL-SELF—which IS ABSOLUTELY 'MOMENTLESS'. Only the 'DIMENSIONLESS MOMENT' (REALLY, the *UTTERLY* DIMENSIONLESS MOMENT) can pertain to the ALL-SELF, for the DIMENSIONLESS MOMENT *is* the ETERNAL NOW.

Example: Any specifiable ultimate moment is, strange to say, infinitely larger than an 'infinitesimal moment' (which is forever unspecifiable), just as any extremely large number is infinitely smaller than Infinity/Infinitude (which, also, is ever-unspecifiable). The existence of an 'infinitesimal moment' is a *possibility* within the FOUNT OF ALL POSSIBILITY but it can never be an *actuality* in a *finite* Cosmos.

This raises the interesting question of whether there are certain *'possibilities'* 'within' the FOUNT OF ALL POSSIBILITY, which cannot *ever* be *actualized* in a *finite* Universe. Because the *infinitesimal* and *Infinity* are equally 'in-de-finite' and unspecifiable, and even though they can be considered *'possibilities'* within the INFINITESSENCE, they are ever non-actualizable in Finite Cosmos because they will always supersede (or 'sub-cede') the limits of any specifiable quantity. Any actualized ultimate moment (relatively tiny though it be) is always not only greater, but *infinitely* greater than a hypothetical infinitesimal moment. The ratio will always be infinity-to-one. Any large quantifiable magnitude if always not only lesser, but *infinitely* lesser than an (ever indefinite) infinite magnitude; the ratio is always infinity-to-one.

moment, an ultimate

By *an ultimate moment* is meant the briefest unit of time possible/allowable in the World of Fabrication within a given Cosmos. An ultimate moment might also be called a 'cosmic unit of change', or an 'ultimate cosmic moment' (though it would not apply to the World of Being, where all processes are subjective and 'ideational' and, hence, can occur with the 'speed of Will', which, when it 'begins' simply begins without incrementalism). An ultimate moment is (in that Fabricated Cosmos) an *indivisible quantum of time.* No objective event in a given Cosmos can be of shorter duration than the ultimate moment—i.e., for *that* Cosmos. What this means is that the 'ultimate particle' (really a form of energy, or an ultimate 'particle/event') the exceedingly rapid vibratory movement of which is 'keyed' to the ultimate moment) cannot be *subdivided,* given the Cosmic Parameters upheld by the Will of the Universal Logos. There would be no 'allowable time' in which to do it.

By *an ultimate moment* is meant the smallest unit of time that can possibly be measured in any Cosmos. All quantities can be measured, but since no briefer unit of time than an ultimate moment exists in a given Cosmos, the ultimate moment is, for that Cosmos, the least possible measurable moment. Such a moment could be called a 'mayavic unit'. All lesser moments are only theoretically *possible*, but can never be *actual.*

What would be the 'Instrument of Measurement'? Certainly no-thing *within* the Fabricated Cosmos that disappears between ultimate moments (and is, theoretically, 'frozen' *during* them) suffices. The Mind of the Universal Logos? Perhaps. A Consciousness that 'Sees' both the moment of cosmo-objective affirmation and the moment of cosmo-subjective negation would have to be the 'Instrument of Measurement'.

Further, such an Instrument would have to be capable of *movement* (at least in consciousness) even *during* an ultimate moment. Is the Consciousness of the Universal Logos capable of this? Almost certainly, as, in general, would also be the Consciousnesses of the Spirit Aspect within the World of Being. If this is so, it would take much pondering to determine the manner of measurement. There is, however, much freedom and 'speed' within the World of Being (the World of Ideation).

By *an ultimate moment* is meant the amount of 'time' during which the smallest, fastest, *indivisible* particle/event-in-Cosmos (the ultimate particle) 'holds its position' (relative to any changing or unchanging Point or Points of Reference in-Cosmos) before changing position. A change of position of that particle/event is equivalent to a change-of-time in Fohatically-Particulated Cosmos.

Even the reappearance of a particle/event in the 'same' position (relative to its most recent position) signifies a change of time (though that change would have to be measured relative to a Super-Cosmic Standard of Reference). If, however, all ultimate particles did *not* 'change position' (by disappearing from one position and reappearing in another) and, instead, each disappeared from one position and reappeared (repeatedly) in *exactly the same 'position in Space'* (relative to an unchanging point of Reference), there would be (within the World of Fabrication) no perceived 'movement of Time' and, in fact, (within that World of Fleeting Objectivities) *no perceivable movement* whatsoever. Time would *seem* to "stand still". Because an ultimate particle/event does not move *during* an ultimate moment, no time (in the World of Fabrication) elapses *during* such a moment—i.e., it is Now for the entire duration of that ultimate moment.

The sense of the "movement of Time" (and even the sense of *movement*, per se) is dependent upon *memory* (an Aspect of Consciousness)—the power to compare the present configuration with the immediately preceding configuration. If the immediately preceding configuration were exactly identical with the present configuration there would be no way to differentiate the past from the present and Time would convincingly seem to "stand still". To a maximally alert intra-Cosmic consciousness, a Cosmo-Objective Now would seem to be instantaneous. Though to such a consciousness, a Cosmo-Objective Now might seem to take "no time at all", such a Now, in a more *absolute* sense *does*, indeed, "take *time*" when compared to whatever Standard of Reference exists along the Infinite Time Line, i.e., when referred to a Super-Cosmic Standard of Reference.

It is conceivable to imagine a situation in the World of Fabrication in which (through successive, exact duplications of the Cosmic Configuration) no time at all seems to elapse, and yet a huge amount of Time is measured against a Super-Cosmic Standard of Reference. Although this scenario is *conceivable* it is probably *not* a Cosmo-Logically Sanctioned Scenario. Some intimation of this psychological perception may be attainable in certain high meditative states.

- In a Cosmo-Objective World (the World of Fabrication, or Approximation) there is not a *continuous Now. 'Nowness' is quantized.*
- Under the One Cosmic 'Eye' of the Universal Logos (the Eye of Synthesis) it is Eternally Now (Cosmically considered, for the Cosmic Logos has the sense of the Universal Appearance as a Super-Quantum of Time).
- In the ALL-SELF it is ALWAYS NOW—the ETERNAL NOW. The ETERNAL NOW is a *continuum* in the CONTINUUM.

In the World of Cosmo-Objectivity, however (as strange as it may seem) it is *not Really always now,* although it will always *seem* to be *Now,* because extremely brief interludes between ultimate moments cannot be registered by any consciousnesses which are focused exclusively Cosmo-Objectively. Ongoing *nowness* is thus a convincing *seeming* in our normal human consciousness. So, *between* ultimate moments in a Cosmos, there are, perhaps, *possible 'virtually* infinitesimal moments', or moments even quite a bit larger, or (more likely) even moments *equal* to ultimate moments, when Cosmo-Objectively it is *not,* Really, *Now,* for it is never *Cosmo-Objectively* Now *between* ultimate moments.

In the 'Eyes' of Those Who always focus consciously within the World of Being, however, (i.e., the Cosmic First Family, for instance), it is (for the duration of the Cosmos) a Cosmic *Eternal* Now, as well as being 'other' Cosmic-times (depending upon how They wish to focus during Cosmo-Subjective Moments).

In fact, in relation to the Cosmic Process, there are just as many of such tiny 'moments of Cosmo-Objective Negation' as there are ultimate moments. During these 'inter-moment instants', the World of Fabrication is 'off' or absorbed in Cosmic Subjectivity. These 'moments of Cosmo-Objective Negation' do not even exist within the World of Fabrication, not being part of Its structure (i.e., of its determining Parameters). These moments of Cosmo-Objective Negation (or of Cosmo-Subjective Assertion) are Real (in relation to the World of Being), but they are not Cosmo-Objectively Real—at least to consciousnesses focused within the World of Approximation (and unaware during such 'times' of their 'Spiritual inherence' in the World of Being—for the Spirit Aspect of all Beings inheres *consciously* in the World of Being, even if 'below' they do not realize this inherence). During these 'inter-ultimate-moments' it is (appreciably) ETERNALLY NOW and Cosmo-Eternally Now, but it is not *Cosmo-Objectively Now.*

"On the *far* side" of speculation, what these ideas suggest is the possibility that an *infinity of Time* could *possibly* exist between successive *quantized ultimate moments.* Whole Cosmoses—Cosmoses possible within the FOUNT OF ALL POSSIBILITY could appear and disappear between the quantized ultimate moments of a given Cosmos, *if* the ultimate moments 'designed into' such appearing and disappearing Cosmoses were smaller by a tremendous amount than the ultimate moments of the latter, slower Cosmos. This possibility is *not* a probability, because it leads to the possibility of an infinite number of simultaneously occurring Universes, and, in fact, to the idea of the simultaneous expression of the ENTIRE 'CONTENT' of the INFINITESSENCE. In other words, it opens the 'Door of the Finite' to the 'WHOLE' of the INFINITE, which is always an *impossibility formally* (i.e., illusorily), although this Door of the Finite is always open thusly, *ESSENTIALLY.*

The above is stated simply to emphasize the relativity of Time and the *experience* of Time, and to demonstrate the wisdom of having reasonably regulated cosmic parameters. The ultimate moment of our particular Cosmos is unknown to human beings, because the parameters of the ultimate particle/events are also unknown. As the highest dimensions of Cosmos are utterly unknown to man, ultimate 'happenings' in-Cosmos are similarly unknown and presently indeterminate—by man!

If it be asked whether the ultimate moment is actualizable in all systemic wholes within Cosmos— i.e., Do ultimate moments form part of the time measure of all wholes and systems in Cosmos?—the answer would be, Yes (but only as regards *objective* Cosmos), because all greater moments of Cosmo-Objective Time are built up of ultimate moments, just as all greater movements are inescapably composed of ultimate movements (i.e., ultimate particle/events). There is no non-ultimate vibration in-Cosmos, (i.e., no non-ultimate particle/event) *but that it is the sum of frequencies of the ultimate kind.* That which is relatively slow is composed of micro-'movements' which, for a given Cosmos, are as rapid as possible. The tiniest and the fastest are the building blocks of all other structures in Space and structures in Time.

Example: The ultimate moment cannot be measured by today's physical plane methodologies (no matter how sophisticated), because to determine such a moment it would be necessary to measure movements that pertain to dimensions vastly higher than the systemic physical plane, or, even, than the Cosmic Physical Plane.

Example: The ultimate moment of a given Fohatically Particulated Cosmos is identical for all systems within that Aspect of Cosmos. The ultimate moment for a given Cosmos is ubiquitous within all Fohatically Particulated Realms within that Cosmos. The ultimate moment is a *non-variable measure* (Relativity Theory notwithstanding) which has little to do with Space-Time modifications theoretically capable of occurring upon the systemic physical plane. The *relativity of the ultimate moment* is only possible from the hypothetical 'perspective' called the 'Infinispectivizing'. From that 'spective' all durations become 'infinitesimalizing' and converge upon zero time value. Within our Cosmos, however, that 'spective' is not cosmo-parametrically *available.* Perhaps it *is* available to the 'De-Infinitizing' or 'Re-Infinitizing' Subject of which our Universal Logos is a *condensation in consciousness.*

Example: It remains to be decided whether the ultimate moment derives from the motion of the ultimate particle/event, or the motion of the ultimate particle/event derives from the cosmo-parametrically sanctioned duration of the ultimate moment. Probably, they both derive from the 'Will of God', as unpopular as it may be so say so.

Monad

By the term *Monad* is meant a 'Ray' of the ABSOLUTE SELF focussed at a particular depth of *immersion* in-Cosmos.

Example: The Monad of a human being is, indeed, a 'Ray' of the ABSOLUTE focussed upon the second subplane of the Cosmic Physical Plane. What distinguishes it from 'another' Monad (although all Monads are One) is the vibratory frequency and quality of the prakriti in which the Monad is focussed and immersed.

'MONAD'

By the term *MONAD* is meant the ONE AND ONLY SELF, SOURCE of ALL.

Example: The 'MONAD' is the ULTIMATE SOURCE of every Monad, from the Universal Monad to all emanated Monads (i.e., authentic Identities in-Cosmos).

Monad, a Cosmic

By *a Cosmic Monad* is meant one of a denumerable host of Ultimate Subjective Units in Cosmos (Who, though denumerable in manifestation, are, yet, all One Monad). Every authentic E/entity in Cosmos is, Essentially, a Cosmic Monad. Further, every Cosmic Monad is, ESSENTIALLY, a 'Ray' of the 'RAY' which 'FLASHED FORTH' from the ABSOLUTE.

Example: The term Cosmic Monad' is used to exemplify a much larger category of Monads than human Monads. The category 'Cosmic Monad' is inclusive of all human Monads, all Planetary Logoic Monads, all Solar Logoic Monads, and far greater Monads, etc., until the *One* Cosmic Monad is reached.

Example: The human being, as Spirit, is a Cosmic Monad; our Solar Logos is also a Cosmic Monad, as are any Solar Logoi, or Constellational Logoi, or Galactic Logoi, etc., etc.,

Example: Every Cosmic Monad is completely at-one with *the* One Cosmic Monad, the Universal Logos. The *modus operandi* of this at-one-ment might be considered one of the great problems in Occult Cosmology, and is, perhaps, definitively solved upon the "Ray Path" (the Fifth Path on the Path of Higher Unfoldment).

Monad, the One Cosmic

By *the One Cosmic Monad* is meant the Universal Logos.

Example: the One Cosmic Monad is the One Cosmic Being Who not only Is *in* each E/entity but *Is* each E/entity.

Example: All E/entities in-Cosmos *are* Essentially the One Cosmic Monad, just as all Numbers in-Cosmos are Essentially the Number One.

Monad, the Ultimate Cosmic

By *the Ultimate Cosmic Monad* is meant the Universal Logos considered in Its Monadic Aspect.

Example: *The* Ultimate Cosmic Monad is the One Cosmic Monad, the Universal Monad and the One Universal 'Ray' of the ABSOLUTE, the Supreme Cosmic Entity.

Monad, the Universal

By *the Universal Monad* is meant, strictly speaking, the Spirit Aspect of the Universal Logos, which, presumably, is located upon the highest possible Kosmic Plane of all planes in the Universe. This Plane would be found within the World of Being.

Example: The Universal Monad is the *penultimate Identity* of all Monads in-Cosmos. (Their *ultimate identity* is the MONAD—the SOURCE). All *apparently* distinct Monads (apparently distinct 'Rays' of the ABSOLUTE) participate in the Universal Monad—nay, *are* the Universal Monad. Various 'kinds' of Monads emerge, as it were, from the Divine Emanatory Stream upon various dimensional levels. They are distinguished from each other by their respective levels of prakritic immersion. That Monad which is called the Human Monad is focused at a deeper level of prakritic immersion than the Monad known as the Solar Logoic Monad—yet, Essentially, They are but One Monad, One Identity.

Example: When considering the nature of the Universal Monad and Its relation to the putative host of 'distinct-but-inseparate' Cosmic Monads, we must address ourselves to the question, How does a *ray* blend with another *ray* and still maintain its own integrity, *if it does*? More specifically the question would be, How does *a* 'Ray' of the ABSOLUTE relate to *the* 'RAY' of the ABSOLUTE? Hidden in this investigation would be the secret of the "One and the Many."

Monadization, Principle of

By the *Principle of Monadization* is meant that in the 'projection-through-emanation' of any Number from Itself, that Number (no matter what Its numerosity) is always to be considered a 'replication-in-attenuation' of the Number One (the Monad).

Example: According to the Principle of Monadization, every Metaphysical Integer in Cosmos *is* the Number One. When Metaphysical Integers *relate* in the World of Being to produce Ideational Archetypes, these Relations are also the Number One.

MONALITY

By *MONALITY* is meant the opposite of Duality.

Example: MONALITY is characteristic of the ONE AND ONLY SELF, while Duality is characteristic of the finite, forever cyclically recurring Manifestation of that SELF throughout Infinite Duration. The Primary Duality is the *relationless* 'RELATION' between the ONE AND ONLY SELF and ITS Manifestation. REALLY, even the Primary Duality is *none other* than the MONALITY.

Mother (Cosmic and Super-Cosmic)

By *the Super-Cosmic Mother* is meant Mulaprakriti, the Infinite Object and the Potential Birther of an Infinitude of Forms.

By *the Cosmic Mother* is meant Cosmic Prakriti, the Universal Object and Potential Birther of all Cosmic Forms.

By the *Super Cosmic Mother* is meant PARABRAHMAN once IT has become an 'OBJECT-instantly-Object' to ITSELF-as-Infinite Self/Subject.

The Super Cosmic Mother (Infinite Object) is just as much PARABRAHMAN as the Super Cosmic Father (Infinite Subject) is PARABRAHMAN.

The Super Cosmic Mother is all of *objectivity* that is *potentially within* the Super Cosmic Father (as *subjectivity*), and the 'Knowing' of this *allness* is the province of Maya, the Super Cosmic Principle of Sentiency.

The *Super Cosmic Mother* is what PARABRAHMAN *Is* once IT has been 'SEEN' (by ITS 'RADIATED' SELF-'EXTENSIONS'). Therefore, all that is 'Seen' *is,* Essentially, PARABRAHMAN-as-Super Cosmic Mother.

By *the Super Cosmic Mother* is meant the Objectified Infinite Fecundity of PARABRAHMAN ITSELF.

Example: Father/Mothers: first, second and third:

1st FATHER/MOTHER: IS the INFINITE SELF.

2nd Father is the Super-Cosmic Father, Is the Infinite Subject, and

2nd Mother is the Super-Cosmic Mother, Mulaprakriti, the Infinite Object.

3rd Father is the Universal Logos (Cosmic Father) and

3rd Mother is Cosmic Prakriti, the Cosmic Mother.

All Fathers and Mothers are, ESSENTIALLY, the INFINITE SELF.

Example: The INFINITE FECUNDITY of PARABRAHMAN 'BECOMES' the Infinite Fecundity of the Mulaprakritic Mother, once intra-SOURCE 'MAYA' mysteriously 'ENGINEERS' that PARABRAHMAN/BE-NESS (if only with *infinitesimal instantaneity*) 'TAKE NOTICE' of ITSELF. PARABRAHMAN as 'OBJECT'-instantly-Object thus becomes the (Super-Cosmic) Mother of all possible potentials '*resident*' within' IT (even though in any one Cosmos, but one set of potentials be enacted).

Example: PARABRAHMAN ITSELF IS/'BECOMES' the "Mother of all Forms." At the 'AWAKENING INSTANT' when PARABRAHMAN 'SEES' ITSELF (via the 'SEEING' called 'MAYA'), 'PARABRAHMAN 'BECOMES' the 'MOTHER' of All Forms, the 'FATHER' of All things 'Seen', and the Union of the 'FATHER' and the 'MOTHER'. PARABRAHMAN 'BECOMES' the 'EVANESCENT INFINITE SUBJECT', the 'EVANESCENT INFINITE OBJECT' and 'EVANESCENT INFINITE CONSCIOUSNESS'. All 'within' PARABRAHMAN is instantaneously *ejected* into Super-Cosmos (where the Illusion of Multiplicity can be *legitimately* sustained without violating HOMOGENEITY).

motion; movement

By *motion* is meant the repositioning of objects.

Example: The formation of any chemical compound from its constituent elements is a movement which involves the repositioning of electrons in the electron 'shells' of all constituent elements.

By *motion* is meant the change of position/location of a given item-in-Universe relative to another item-in-Universe, or to a fixed Point of Reference in-Universe (if such can be considered as existing).

Example: The revolutionary motion of our Earth takes place relative to our Sun, the principal revolutionary motion of which occurs relative to the Pleiades. No doubt the Pleiades also have various types of revolutionary motion, the final one of which is necessarily relative to the center of our galaxy.

Example: Because the World of Becoming is a Dis-Continuum, no motion of an object from one point to another passes through *all* the points in Space that apparently separate the position of departure from the position of destination. As a matter of fact, an object being *repositioned* from one apparent point in Space to another apparent point in Space 'between' two ultimate moment may pass through *no points* separating the position of departure from the position of destination (for Space-Itself is being 'repositioned').

The Ontological Vibration of the Fabricated Universe from Cosmo-Objectivity to Non-Cosmo-Objectivity, to Cosmo-Objectivity again, etc., means that at each ultimate moment an object may 'posit' itself in a slightly different position than in the moment before (at least relative to all other objects), but that it did not *arrive* at the new position by 'traveling' continuously from point 'a' to point 'b', but instead arrived at point 'b' *seemingly* instantaneously (with *infinitesimal instantaneity* or "in no time") by reappearing (in its 'new' position) out of dissolution in Non-Objectivity.

This means that ultimate particle/events simply *change positions* from ultimate moment to ultimate moment without moving through Space. (Perhaps they can do this because they are 'thought into position' with the 'speed of Will'. Master Morya speaks of the 'Lightning of thought'.) We must also remember that Space/space does not REALLY or ever Really exist. This entire difficult matter touches on what might be called the 'self-determination of the ultimate particle'.

Further, it must be realized that so-called abstract 'points' in Space are, themselves, *things.* If one translated 'points in Space' as *'things* in Space', one might see that, perhaps, that whatever it may be that objects are supposed to be moving *through* does not Really exist! Space reconfigures Itself; but objects do not move *through* Space, for there is no-*thing* to 'move through'. Conceived 'particles' appear and disappear but they do not 'move'.

Example: There is no space (as vacuity) nor REAL movement *in* space (as vacuity); *there is only relationship and change of relationship.* We may be led to entertain the paradoxical thought that movement 'in' and 'through' Space does *not* exist, that what *seems* like movement is only *instantaneous reconfiguration, instantaneous change of relationship,* change of relationship with no 'secondary positions' between a 'position of origin' and the 'position of destination' (i.e., the *next* noticeable position).

This is another way of saying that *interval* or *space* does not Really exist in Cosmos, but that the appearance of interval/space is an artifact or necessary *illusion of consciousness.* Cosmic Space is Really the aggregations of Self-Perceptions within the Field of Consciousness (of the Universal Logos). 'Moving through Space' is simply Self-Perceptually changing the relationship of the different perceptions' within the field. To do so, the Self-Perceiving Logos need not laboriously 'move' a particular perception from 'one end of the Field to another'. He needs merely to dissolve the 'Picture of Space' which *is* Space, and 'make up' a 'New Picture'. This is done without the 'movement' of any specified Self-Perceptions. They are simply 'changed' by ending an 'old' Self-Perception and beginning a 'new' Self-Perception. God (or Fohat) 'Makes New Worlds' with the speed of Imagination/Will. This may be a 'speed' which takes 'no time' to *begin* or *end.* The sustaining of such Willed-Images 'takes time'.

How strange to think that it is impossible to 'move through' Space! That the self-evident perception of movement is Really an illusion! From this perspective, it would be possible for an ultimate particle to appear in one 'part' of the Universe and reappear, instantaneously, at the next ultimate moment, in a part of the Universe apparently maximally 'distant' (having apparently 'traveled' thereby perhaps trillions or quadrillions of times 'faster' than the speed of light). Will and Imagination would 'Direct' the reconfigurations. That the 'law-breaking' kind of utterly Time-and-Space-defying/negating-movement/change just mentioned does not, *apparently*, happen, is due to the fact that the Intelligences guiding the Cosmic Process do *obey the Law*—the Laws decreed by the Parameters of the Design-at-the-Beginning. To the 'Rearranger', however, much is possible which would seem to 'break' the Laws with which we are familiar.

Example: In a way, no change of relationship ever moves *off* the One Point, so no time may be required for the change. How 'large' is a *point* that so much can 'happen' within it?

Example: A fascinating question arises: can there be movement if that which, supposedly, moves, does not pass through any 'points' in 'space' on its 'way' from the position *from* which it moved, to the position *to* which it moved? In other words, is *change* of position from position 'a' to position 'b' considered *movement* if no positions *between* position 'a' and position 'b' were 'occupied' by that which is changing position?

Example: So-called movement is really an artifact of consciousness. Movement is an illusory appearance-in-consciousness caused by configured appearance, disappearance and reconfigured appearance of particle/events.

Example: From the Cosmo-Psychological point of view, motion can be conceived of as 'change of image'; 'the scene changes' and everything 'moves. Imagination 'makes' the 'scene change' In this respect, motion is inseparable from imagination. One can 'move' an image where one *will.* The normal Laws of Space and Time do not apply. The factor of imagination is one of the 'freedoms' available to E/entities in Cosmos who are, otherwise, under the very limiting strictures of Cosmic Parameters.

Motion, Perpetual

By *Perpetual Motion* is meant the Great Breath, which has been, so to speak, *inhaled* and *exhaled* forever. It may be that the three Fundamentals of Being in-Cosmos (Time, Space, and Motion) are *all* Illusions, but they are a Trinity of Illusions without which there is no Cosmos possible. Although, in *The Secret Doctrine,* these three are considered Absolutes, the ABSOLUTE SELF, by strict definition, can include as separate items neither Time, nor Space, nor Motion. IT IS NO-THING—NOTHING AT ALL.

Motion must pertain to the realm of Relativity, and only emerges with the inception of Universes (at the sudden 'Birth' of Super Cosmos, i.e., of *Indefinite* Cosmos). The periodic Motion of outgoing does not occur *within* the ALL-IN-ALLNESS of the ABSOLUTE, but is part of the First Illusory 'CHANGE' (the first Motion of 'MAYA'-instantly-Maya) by means of which the ABSOLUTE both 'BECOMES', and yet *does not* 'BECOME', the SELF-as-Infinified Point (the Infinite Subject).

In the same way, we could say that the ABSOLUTE ITSELF *does not breathe,* but IT *appears* to both Move and Breathe. Breathing clearly requires a dualistic motion which is impossible *within* the ABSOLUTE-as-ABSOLUTE. Although, let it be remembered that, in a sense, *everything* "takes place' *within* the ABSOLUTE!

Perpetual Motion, in the highest respect, is REALLY Perpetual Oscillation between Being and Non-Being (which Non-Being is BEING). The Motion of the Great Breath is Cyclic and a Great Interlude of Non-Motion must be included in Its Pattern of Motion. With the proviso that the Interlude of Non-Motion is included, it will be seen that the Great Breath *appears* to recur *perpetually.* It always has been and always will be. The 'NO-THING' 'WILLS' (as it were) *not* to stop It, because if the GREAT NOTHING stopped this Perpetual Illusory Motion, IT would cease to BE ITSELF (which, again, seems to be against ITS 'WILL').

It is difficult to speak of the 'WILL' (or any 'ATTRIBUTE') of the NO-THING, but when the NOTHING is considered the ABSOLUTE PLENUM (the INFINITIZED FULLNESS) we can see that not only IS IT a 'STATELESS STATE' of ABSOLUTE PRIVATION but *also* a 'STATE' of MAXIMAL FULLNESS from WHICH *no-thing* (not even a 'WILL') can be denied.

By *Perpetual Motion* (secondarily) is meant the fact that no-thing whatsoever is *continuously perfectly still* in-Cosmos, but 'moves' slightly from ultimate moment to ultimate moment. Even an item-in-Cosmos that seems to be in the same position from ultimate moment to ultimate moment (for a number of ultimate moments) has had to 'disappear' instantaneously in order to return to the same position. While the new position of that item (probably an ultimate particle/event) may be the 'same' position (relative to the former appearance of itself one ultimate moment earlier, or relative to positions of other reappearing particles), it is certainly *not* in the same position relative to those other items-in-Cosmos which *have* changed position from the earlier ultimate moment to the one in question.

A question arises as to whether there is any fixed Point of Reference in-Cosmos that *changes not?* If there were, the positioning of items-in-Cosmos would no longer be so relative. Such a fixed Point of Reference, should It exist, might be called the *Center of the Universal Sphere.* (We have to remember that there are many subtle worlds in which it is meaningful to speak of 'place' and 'position'. The higher mental plane is one such world. Probably it is so for the entire Cosmic Physical Plane.)

Example: No single motion of any kind is *absolutely continuous* (happening uninterruptedly, without interval during every *momentless moment* of ETERNAL DURATION). The so-called *perpetualness* of any motion-*in*-Cosmos is not *continuous,* <u>Perpetual Motion</u>, but *cyclic* motion which occurs and recurs (with interludes of motionlessness). If <u>Perpetual Motion</u> is understood in this way, it can be seen that (in relation to the Great Breath) it 'goes on' forever.

Example: Cyclic Motion is perpetual throughout Infinite Duration. Within the overall pattern of Cyclic <u>Perpetual Motion</u>, there are interludes of no Motion at all. <u>Perpetual Motion</u>, then, is a Super-Cosmic Pattern of Motion and MOTIONLESSNESS (though, confoundingly, the MOTIONLESSNESS does not 'stop' during the cycle of Motion). The Great Cycle of the Great Breath, Itself, can be considered a kind of <u>Perpetual Motion</u>, even though one half of the Cycle consists of ABSOLUTE MOTIONLESSNESS.

MOTION, ABSOLUTE

By *ABSOLUTE MOTION* is meant MOTIONLESSNESS.

Example: Although some metaphysicians claim that ABSOLUTE MOTION is an ETERNAL ASPECT of the ALL-SELF, 'ABSOLUTE MOTION' is as much an Illusion as 'ABSOLUTE SPACE' and 'ABSOLUTE TIME'. The *absolutizing of* Time, Space and Motion *negate them* (at least as we human beings know them). NO-THING (*absolutely nothing*) can be predicated of the SELF-AS-SELF—including Time, Space, or Motion. Of course, contrarily, *everything* (in *infinitized togetherness)* can be predicated of IT just as legitimately.

Example: From a certain perspective, the INFINITESSENCE *IS* the *infinitized NOUMENESSENCE* of Time, Space and Motion. Although none of these three is individually articulated 'within' the ABSOLUTE, it is impossible (considering the ABSOLUTE as the INFINITIZED PLENUM) to say the *infinitized NOUMENESSENCE* of Time, Space and Motion (i.e., ABSOLUTE TIME, ABSOLUTE SPACE and ABSOLUTE MOTION) does *not* 'INHERE' in the INFINITESSENCE. The Great Mystery is how the *infinitized NOUMENESSENCE* of all possible qualities can 'INHERE' within the INFINITESSENCE and *be completely indistinguishable from each other*, as 'within' IT (the GREAT HOMOGENEITY) there can be *no distinction.*

Mulaprakriti

By *Mulaprakriti* is meant the First Object (which is absolutely undifferentiated, and infinite in Perceived 'Extent', or better, 'Perceived Presence') that comes into Being when PARABRAHMAN (through a SELF-'LIMITING' 'ACT' of 'RE-FOCUSING') 'AVERTS' ITS 'INFINIDENTIFIED' 'ATTENTION' from ITS ALL-IN-ALLNESS, and begins to 'SEE' ITSELF as an Object. Mulaprakriti is the First Objectification.

Example: This Act of 'PARABRAHMIC RE-FOCUSING' produces the Boundless Sea of Mother Substance, Mulaprakriti, which is the Noumenon of Matter, the Noumenon of all possible *objects* in-Cosmos, *including* the Cosmos, Itself, Perceived as an Object. Of course, PARABRAHMAN IS the ABSOLUTE NOUMENON (the INFINITESSENCE, the NOUMENESSENCE) of ALL Noumena.

By *Mulaprakriti* is meant Root Matter, the Mother (Who Is the Father 'in sight of Himself'), homogeneous, undifferentiated Substance—i.e., that which *objectively substands* Cosmos in Its Material Entirety. Mulaprakriti is the Great Homogeneity just as PARABRAHMAN IS the GREAT HOMOGENEITY.

Example: All perceived objects in-Cosmos are Mulaprakriti-as-Cosmic Prakriti, *reflections,* of the *subjective ideational differentiations* borne by Fohat from the Universal Mind (Ideational 'Reservoir') of the Universal Logos, and *projected into Self-Reflected objectivity* by Fohat as the Eternal Mayavic 'Mirror' of the One for Whom It (Fohat) is the Agent.

Example: Mulaprakriti is the Mayavic Mirror, the Eternal SELF-as-Self-Image which arises in the Eternal Mayavic Mirror into which the INFINITE SUBJECTIVITY-as-Infinite Subject continues to 'Gaze'.

Example: The 'ACT' of PARABRAHMIC 'RE-FOCUSING' is the 'FLASHING-FORTH' of the 'RAY' of the ABSOLUTE, producing simultaneously the SELF-as-Infinified Point (the Infinite Subject with an infinitude of infinite points of view) and Mulaprakriti (the Infinite Object).

Example: Why is Mulaprakriti Absolutely Undifferentiated at first? Because PARABRAHMAN is ABSOLUTELY UNDIFFERENTIATED! When THAT which is ABSOLUTELY UNDIFFERENTIATED 'SEES' ITSELF (by Proxy), What IT 'SEES' is predictably Undifferentiated—Infinite, Undifferentiated, Mulaprakriti. The Subjectivity Infinite Subject (Proxy of PARABRAHMAN) is, at first, (in 'Reflection' of PARABRAHMAN) completely *undifferentiated.*

Example: Why is Mulaprakriti, though Absolutely Undifferentiated, the Source-Matrix of all possible differentiation in-Cosmos? Because PARABRAHMAN, though ABSOLUTELY UNDIFFERENTIATED, IS REALLY the INFINITESSENCE, which 'CONTAINS' all 'POSSIBILITY'. Mulaprakriti as the SELF-'REFLECTION' of PARABRAHMAN-as-Infinite Subject likewise contains all 'articulatable' Possibility.

From another perspective, Mulaprakriti is the potential Appearance in material and objective form of all 'POSSIBILITY' (especially that which 'DESIGNATED' for the immediately future Cosmos) which is 'INHERENT' within PARABRAHMAN—the INFINITESSENCE. There is an infinite amount of *additional* 'POSSIBILITY' in the INFINITESSENCE which, for the Cosmos-in-Process-of-Creation, is neither 'DESIGNATED' nor SELF-'INTENDED'. Mulaprakriti in Its Infinitude *could* Become/Appear as such *additional* 'POSSIBILITY', but the Law-Abiding 'Focus' of the Infinite Subject in its Mode of Condensed Point (a Focus which is *intent* only upon the Objectification of the EXTRUDED 'IDEA' for the Cosmos-to-Be) does not offer the Reflective Power of Maya the 'Opportunity' to generate *additional* and *excessive reflection.*

multi-dimensional consciousness

By *multi-dimensional consciousness* is meant the ability to be aware on many levels of consciousness simultaneously. The ultimate extension of multi-dimensional consciousness is omni-dimensional consciousness ('omni' in Cosmos, at least), which is that possessed by the Universal Logos.

Example: Every E/entity in-Cosmos *has* multi-dimensional consciousness, even the tiny humble life we know as an atom. This is so because each E/entity, even though usually defined and classified according to the lowest 'position' it has reached in its Emanatory Sequence within the Divine Emanatory Stream, possesses a multitude of levels of participation in-Cosmos 'above' its lowest level. We may know an atomic particle as a specified entity with certain limitations, but in what ways is the ensouling E/entity of that particle (which is a Cosmic Monad which is necessarily at-one with *the* Cosmic Monad) expressing itself on higher dimensions? *What*, for instance, *is* the Monad of the atom? Is that Monad not (Essentially) a Cosmic Monad manifesting ubiquitously as the *One* Cosmic Monad at all possible levels of prakritic immersion along the Divine Emanatory Stream, including one specific, tiny *atomic* level?

Example: There are certain forms in Cosmos which may *not* be points of expression for the Cosmic Monad in its graded, multi-dimensional consciousness, descending and ascending as emanations along the Divine Emanatory Stream. The ultimate particle/event, for instance, may simply be a Self-Enumeration of Fohat as It 'Sees' Its Subjective Self-Enumerations *reflected* in *(as)* Cosmic Prakriti. Each such *ultimate* "bubble in the Koilon" (as there are certainly larger *non-ultimate bubbles*) may never be the lowest point of expression of a direct emanation of the Cosmic Monad. However, every such ultimate particle is, nonetheless, *our* expression, for Who *Is* Fohat if not an apparent differentiation of the ONE SELF WHO we-as-We-as-WE ARE! This leads to an astonishing conclusion: **8** Am active *in-Cosmos* in Four Ways:

1. As an apparently differentiated-yet-undifferentiated Cosmic Monad, a 'Ray' of the ABSOLUTE;
2. As Cosmic Prakriti (from one perspective, the Cosmic Mother of all Forms) in Her Entirety (for What is Cosmic Prakriti but Bounded Mulaprakriti, and What is Mulaprakriti but PARABRAHMAN/INFINITE SELF-as-Super-Cosmic Infinite Subject 'Seeing' Itself?);
3. As the Universal Son of the Father, the One Who (with His Company, the Supernal Tetraktys) supervises the building of the Forms entrusted to Him by the Father;
4. As Fohat, the protean Holy Spirit, Agent of the Father, Who (in-Cosmos) is:
    - the 'Electrical Impulse' which 'Sees' to the impregnation of the Mother (always through the induction of Self-'Sight') thus producing the Son; and, as well,
    - Fohat as the Agent of the supervising Son (i.e., the Fohat we recognize as *Really* Fohat), Who this time, impregnates the Mother Himself (through Fohatic Self-'Sight') producing Differentiated Prakriti.

Thus, **I-as-8** Am Father, Son, Holy Spirit and Mother. **I-as-8** Am *all* of these. **I-as-8** participate in-Cosmos *through* the Great Beings called the Son, the Holy Spirit (Fohat) and the Mother (Cosmic Prakriti), the One Who *brings forth* Forms. **I-as-8** Am *Omnipresent* in-Cosmos. As *the* One Cosmic Monad (becoming, through emanation, an apparent *host* of Cosmic Monads) My experience becomes more and more *apparently individual* and descendingly minute.

Further, if **8** seek to look for 'MySelf' in Super Cosmos and enumerate My extra-Cosmic Identities, **I/8** would have to name 'MySelf' as every Pre-Cosmic and Post-Cosmic 'Player', and (SUPER-Cosmically) as the INFINITE SELF as well.

myself; 'MySelf'; MYSELF

These terms *(myself, 'MySelf', MYSELF)* when used technically signify the following: *'MYSELF'* refers only to the INFINITE SUBJECTIVITY, the ONE AND ONLY SELF; *'MySelf'* refers to any intra-Cosmic Self which has achieved an awareness of Its inseparability with all other such Selves. 'MySelf' would then have applicability across the range which separates the Universal Logos (focussed upon the very highest Kosmic Plane) from the Human Soul (focussed upon the higher systemic mental plane). 'MySelf' is therefore a term referring to a spiritualized Self, liberated from the limitations of egoistic perception. The term *myself* refers simply to the personality and ego level of consciousness—'I' and 'i'.

Example: I, myself-as-'MySelf'-as-MYSELF, AM ESSENTIALLY the ONE IDENTITY WITHOUT A SECOND. To know this is to know the true meaning of the ancient adage, "Thou art THAT", and hence, to achieve *enlightenment*.

MYSELF-as-'MySelf'

By *MYSELF-as-'MySelf'* is meant the ABSOLUTE SELF *appearing* in-Cosmos as an authentic E/entity-I/dentity-Cosmic Monad—a Self of the One Universal Self.

Example: **I**, MYSELF, have pervaded the entire Cosmos with a *fractionless "fragment"* of MYSELF, and remain *transcendent* to MY Cosmos, yet, as well, the *ESSENTIAL SUBSTANCE* of All within-Cosmos. **I**, MYSELF-as-'MySelf', Am **8**, the Entified Self, 'Son' of the SELF.

mystery

By a *mystery* is meant a process, the *modus operandi* of which is *discoverable*, but exceedingly *obscure* to the particular Mode of Consciousness making inquiry. Clearly, what is a mystery to E/entities in one Mode of Consciousness may not be a mystery to Entities in another such Mode.

Example: We are told of the "Mystery of the Solar Angels"—a mystery which conceals much about the origin, nature, purpose and relation to man of a large group of mediating Stellar/Solar Entities called Solar Angels. This Mystery also conceals much in relation to the immediate cosmic destiny of perhaps eighty percent of our Earth's present humanity.

Mystery, the Ultimate

By *the Ultimate Mystery* is meant the 'REASON' why the INFINITE 'BECAME' Finite; ITS Mode of °DOING SO'; and the 'MEANS' by which IT can apparently 'BECOME' Finite and, *simultaneously*, 'REMAIN' INFINITE.

Example: The Ultimate Mystery might be phrased as follows, How can Something come out of NOTHING? and, nevertheless, REALLY remain 'NOTHING' even while *seeming* to be Something?

MYSTERY, the ULTIMATE

By *the ULTIMATE MYSTERY* is meant the ALL-SELF, ITSELF which IS, paradoxically, both ENTITY (as the SELF) and NON-ENTITY (as the BOUNDLESS IMMUTABLE PRINCIPLE).

Example: The only example of the ULTIMATE MYSTERY is the *utterly inscrutable* ABSOLUTE, ITSELF—THAT WHICH IS *unthinkable* except as both BEING and BENESS.

- N -

n-dimensional

By the term *n-dimensional* is indicated an indefinite number of 'Vibratory Strata' potential within in the First *Finite* Object (i.e., the Circle/Sphere of Cosmic Prakriti created by the Focalized Self-Perception of the Focused Universal Subject,Condensed Point, and which equates in more objective language to the *perceptual bounding* of Mulaprakriti).

The value for 'n' will depend upon the particular Parameters of the Incipient Cosmos, Pre-Ordained by a SELF-'LIMITING''RE-FOCUSING' of the SELF. The number of dimensions to be contained in the Incipient Cosmos may or may not be determined exclusively by that Agent of the SELF we call the Infinite Subject (or the 'REDUCED PRESENCE' of the SELF we call the SELF-as-Infinified Point), 'RADIATED-FORTH' by Pre-Cosmic 'FOHAT'. (From one interesting perspective, Pre-Cosmic 'FOHAT' Is, Itself, SELF-'SIGHT' or SELF-'REFLECTION'—as 'SEEING' is 'ACTING'.) The number value for 'n' for the Cosmos-to-Be may be 'contained in' or 'borne by' the 'Ray' of the ABSOLUTE which the Universal Logos Is.

In any case, we are speaking of Universal Logoic Self-Potential. If we consider dimensions to be 'particulately prakritic', then no dimensional articulation begins in *actuality* until the Fohatic Process of Fabrication of the World of Approximation begins. There may also be a kind of *dimensionality* within the World of Being (which is dependent upon what we might call 'Modes of Self-Sight'). Such dimensions would be the Qualities of Fields of Consciousness. This dimensionality *conceived* and *held* within the World of Being, does not *actually ex-ist* in an *objective way,* until Fohat 'Self-Sees' the differentiation of Cosmic Prakriti, so that the Qualities of the Fields of Consciousness in the World of Being can be *approximated* 'below'.

Example: It may be (though it is unlikely) that the *only* 'ACT' of the SELF-AS-SELF is to 'BECOME' the SELF-as-Infinified Point, and there an end! It may be, in fact, the SELF-as-Infinified Point (the Infinite Subject, as Agent of PARABRAHM) Who 'Decides' which of the Infinite Possibilities from the FOUNT OF ALL POSSIBILITY will be used in the Cosmos-to-Be (the Incipient Cosmos). This 'Decision' would Determine the Parameters/Laws/Limitations of the Incipient Cosmos. This would require that the SELF-as-Infinified Point have *access to* the FOUNT OF ALL POSSIBILITY even though the SELF-as-Infinified Point Exists 'extra-SOURCE'. This is one view, and seems to contradict the *infini-pointed-thus-homogeneous Vision* of the Infinite Subject as It is *reflected* back to Itself in the Image of Mulaprakriti. On the other hand, it may be (and is *more* likely) that the First and Only 'ACT' of the SELF-AS-SELF (i.e., the 'RADIATION' of the 'RAY') already *contains* the 'INSTRUCTIONS' concerning *which* of the limitless SELF-'INHERENT' *possibilities* 'within' the FOUNT OF ALL POSSIBILITY are to be *actualized* in the Incipient Cosmos. In any case, a Cosmos has a specific, parametrically determined number of dimensions; the value for 'n-dimensions' is definite and quantified for each Cosmos. Is the number of dimensions in our Cosmos 10, 49, 70, 100, 1000, or some other number?

necessity; necessary

By *necessity* is meant a process which is forced or compelled to exist by Law or Will. The compelling agent(s) can be external or internal to that which is compelled. A process is a *necessity* (i.e., *necessary)* to an E/entity or relationship, if the absence of that process would either destroy or severely diminish the E/entity or relationship, *or* cause it to change its nature in such a manner as to violate its integrity—i.e., become something other than it is.

Example: The *sutratma* (thread soul) is a necessity to the physical, emotional, and mental integrity of the personality. Without it, the personality could not exist as an unit within the lower three worlds.

Example: Adherence to the Law of Reincarnation is (in the present solar system) a necessity for the gradual growth and development of the human soul.

Example: The appearance of a Cosmos is an *uncompelled* 'NECESSITY' to the ABSOLUTE. Without the 'GENERATION' of an infinitude of Cosmoses the ABSOLUTE could not *continue forever* to be exactly what IT IS.

Example: A necessity cannot help but follow a cause.

Necessity, Son of

By a *Son of Necessity* is meant a Being, the existence of which is *necessary* to a Greater Being which generated It—necessary, due either to the inherent state or condition of the Greater Being, or to the Law(s) under which the Greater Being (as generator) functions.

Example: Our Solar System is a Son of Necessity because It *necessarily* comes to birth when the Father and Mother Aspects of the Solar Logos engage in Their pre-destined Interplay—an Interplay compelled by the Cosmic Will of the One Universal Logos.

Example: More importantly, the Universe, Itself, is a Son of Necessity, because the ALL-SELF must endlessly 'RADIATE' Universes if the ALL-SELF IS to *continue* to BE WHAT IT IS—i.e., remain *consistent with ITSELF.* The Universe is *necessary* if a 'SELF-OBLIGATORY *SELF*-LAW' (what other kinds are there for the INFINITE SELF?) which we might call, the 'LAW of SELF-CONSISTENCY' is to be *'WILLINGLY OBEYED'* by the ALL-SELF. Equally, there is what might be called a 'LAW of SELF-INCONSISTENCY', or the 'LAW of SELF-CONTRADICTION'. (Human bias hopes that this is a *secondary* LAW but it may not be.) When dealing with the ALL-SELF one must simply learn to expect IT to "come down squarely on both sides of the issue."

negation

By *negation* is meant the act of rendering absent, especially *absent from consciousness.* Negation can be seen as the opposite of *affirmation* which renders *present.*

By *negation* is meant the act of canceling in consciousness the supposed REALITY/Reality of a thing perceived. If I *negate* an illusion, I understand it to be an un-REALITY or un-Reality, and am no longer deceived into thinking that it is a REALITY/Reality.

Example: The negation of the supposed not-SELF through the disciplines of abstraction cancels in consciousness the *apparently* independent REALITY/Reality ("own being", in Buddhist Terminology) of the not-SELF and reveals it, instead, to be ESSENTIALLY nothing but the SELF.

Example: The negation of un-REALITY is the surest way (via the process of elimination) to reach the affirmation of REALITY.

never

By the term *never* is meant *forever impossible*, no matter what states or circumstances obtain. The term *never* cannot confidently be applied to the BOUNDLESS IMMUTABLE PRINCIPLE because IT IS the GREAT CONTRADICTION and 'within' IT *everything* is possible (at least, the *infinitized state* of everything).

Example: While it can reasonably be said that motion can never take place within the ABSOLUTE, the fact that the ABSOLUTE IS the PLENUM, the INFINITESSENCE, the FOUNT OF ALL POSSIBILITY means that 'within' IT IS the very NOUMENESSENCE of motion. It is as if 'within' the ABSOLUTE, motion has *always* REALLY 'INHERED' to the *greatest possible degree* (i.e., to an *infinitized* extent) while never being expressed ('within' IT) as identifiable, objectified motion.

Example: It can be said of the ABSOLUTE, "Never say never!" However, it can equally be said, "Never say, Never say never." You can see where this would lead! What can one say?

nothing

By *nothing* (or no-thing) is meant the negation of particularization.

Example: When nothing exists, no particular thing can be isolated as existing.

Example: It is impossible to think of nothing without rendering 'nothing' into a 'something' through the inescapably *affirmative* nature of thought. Thought posits and cannot negate. Even the apparent act of negation in thought is a 'something'—a positing and an affirmation. The only true way to think of nothing is *not to think*. Hence the Raja Yoga System recommends subduing the modifications of the chitta (mental substance) so that Reality/REALITY may be realized.

NOTHING

By *NOTHING* is meant the ABSOLUTE NEGATION which IS the ULTIMATE AFFIRMATION and the ONLY REALITY. The human mind is incapable of thinking about *nothing*. The very act of thinking confers *'thing-ship'* upon any subject of thought. In other words, *thinking is 'thinging'*. If NOTHING 'IS', to think about IT is the first guarantee of misapprehension. Hence, H. P. Blavatsky says of 'BE-NESS' that it is *"unthinkable"*. Nevertheless, as hopeless as *true registration* of IT may be, we need such concepts to bring us to the *point of identification* in which such concepts are no longer necessary.

Example: NOTHING *IS* THE PLENUM. All opposites meet and *mutually annihilate* in the ABSOLUTE, and yet, all opposites must necessarily 'INHERE' 'within' IT.

'noticeable', a (noun)

By a *'noticeable'* is meant a *'perceivable'*.

Example: The range of 'noticeables' which can be registered by the Consciousness of the Universal Logos exceeds many-fold the range of 'noticeables' capable of being registered by the human being at its present stage of development. Noticeability depends upon 'vibratory approximation'—the degree of response to vibration of which the response mechanism of the registering consciousness is capable. If the vibration to be registered is too much faster (or slower) than the customary vibration of the matter which composes the response mechanism, there will be no response to the vibration to be registered, and it will go unnoticed. It must be pondered just how much 'imparticulate Consciousness itself' is a 'response mechanism'. What could be more *conscious* than the Presence?

Not-SELF

By the *Not-SELF* is meant, especially, the Universe as both a specific Being and an Object. Technically, the Not-SELF cannot Exist, because there is, ESSENTIALLY, naught that is *not* the SELF. But, there is an Appearance we call the Cosmos or the Universe, and to *this* the term 'Not-SELF' applies.

While one could say with justification that the SELF-as-Infinified Point and Mulaprakriti are also the Not-SELF, They are so 'close' to the SELF (being with Pre-Cosmic 'FOHAT' the First Emerging Aspects of the SELF—i.e., the Infinite Subject and the Infinite Object), that it is not especially facilitative to apply the term 'Not-SELF' to Them, though, technically, it does apply—especially to Mulaprakriti. To the 'products' of Mulaprakriti (among which the Universe is always the *first)*, the term Not-SELF applies directly.

To every other *conscious being*, as well, (especially when *cosmified)*, the term Not-SELF also applies, though never to the ESSENCE of such beings. The ESSENCE of all *infinitessentialized* beings IS the SELF.

Example: The Not-SELF (Cosmos and All within It) comes into Existence through an 'ACT of ATTENTION' (which is also an 'ACT of INATTENTION'). The 'RAY' of the ABSOLUTE *'IS'* an 'ATTENTIVE ACT' of 'SELF-RE-FOCUSING'—from TOTAL SELF-ABSORPTION ('INFINIDENTIFICATION') to TOTAL SELF-'AWARENESS'. The latter State is the first Dualistic State and it is Pre-Cosmic.

'NOUMENESSENCE'

By *the NOUMENESSENCE* is meant the INFINITESSENCE—the NOUMENON of all intra-Cosmic Noumena.

Example: The NOUMENESSENCE is the impartite, infinitized SOURCE of all possible intra-Cosmic Q/quality.

Example: The NOUMENESSENCE is the *infinitely rarefied* NOUMENON of all Noumena.

Noumena (plural)

By *Noumena* are meant Realities, Archetypes of the higher planes of Cosmos. The intra-Cosmic Noumena are the higher Formative Patterns for all lesser intra-Cosmic patterns.

Example: The Gods and Goddesses of Ancient Mythology are all Noumena for the lesser patterns and qualities correlated to Them. Mars is the Noumenon of everything 'Martian' in the World of Becoming; Venus for everything Venusian; Jupiter for everything Jupiterian; etc. Similarly, the Great Numbers (Beings far superior in scope to the Planetary Gods of our solar system) are Noumena as well. The great Being, Number One, Is the Archetype for all *oneness* found in the World of Becoming; the Being, Number Five, Is the Archetype for all *fiveness*; the Being, Number Nine—the Archetype for all *nineness*, etc. These are the Fundamental Numbers; all Numbers greater than Nine are derivative.

Noumenon

By a *Noumenon* is meant a subtle authentic Archetype within the World of Being, upon which the existence and development of all prakritic forms is based. (Note that the World of Being is part of the greater World of Becoming and *not* the WORLD OF BEING.)

Example: The Noumenon of all forms of beauty is the Being, Beauty, Itself. Likewise there is a Noumenon known as Goodness, and another, Truth.

Noumenon, the Cosmic

By *the Cosmic Noumenon* is meant the Universal Logos Who Is the Source of all E/entities (all of them replicas of Itself and, in some measure, bearers of Its Essential Quality). The Cosmic Noumenon differs in *quality* from Cosmos to Cosmos depending upon the Universal Algorithm, but does not differ in Its Essence which Is *always* ESSENCE.

Example: Eventually the Universal Theme of the Cosmic Noumenon will be understood. Indeed, the Cosmic Noumenon is *The* Universal Archetype.

NOUMENON, the ONE (INFINITESSENCE)

By *the ONE NOUMENON* is meant the ABSOLUTE, the INFINITESSENCE.

Example: The ONE NOUMENON 'HOLDS' all possible quality in *absolute*, hence, *undifferentiated* 'SOLUTION'. IT IS the 'NOUMENESSENCE' of all possible intra-Cosmic quality.

now

By *now* (in its common usage) is meant the present moment. That moment may be highly variable and has no exact measurement. Often, *now* is equated with "this very minute" or "this very second". The term *now* used in this manner is a practical and functional term, and is meant to denote, in general, "the present time".

By *now* (technically considered) is meant a measure which denotes the universal condition of all variables (the Cosmic Configuration) immediately before the next possible change in that condition, which change arises due to the tiniest cosmically-possible alteration of (the position of) the most rapidly changing variable, that variable being the ultimate particle/event.

Example: Within the World of Approximation (the Fohatic World of Fabrication), the now changes from ultimate moment to ultimate moment. Ultimate moments are those successive intra-Cosmic events by means of which Fohatically Particulated Cosmos reappears ('x-tillions' of times per second) after having disappeared from all intra-Cosmic registration (for as many times a second) due to the necessity for Cosmic Dis-Continuity. Each time Cosmos reappears, the time is now in Fohatically Particulated Cosmos.

Now, (Cosmic) Objective

By the *Cosmic Now* is suggested the *'time sense'* which arises in the Consciousness of the Universal Logos through Its simultaneous perception/registration of *all* variables-in-Particulated Cosmos. While this simultaneous registration does not negate the knowledge of the movement of any particular variable within the entire Configuration (or 'interiorly', 'outside' It), it *does* negate the *usual* sense of "the incremental passage of time" so prevalent among lesser cosmic entities immersed in the World of Particulate Fabrication who must, due to their limitations, experience registrations *sequentially* instead of *simultaneously*.

Thus, due to Its virtually continuous registration of the Cosmo-Objective Now, the Universal Logos 'Sees' All (i.e., all Process within the World of Fabrication) "under the Aspect of Eternity." In fact, the Logos 'Sees' the *entire* Cosmic Process with the 'Eye of Eternity', for not only are the all Cosmo-Objective *repositionings* visible to His Eye as *one change*, but the multitude of inter-moment instants (Cosmo-Subjective Nows) are also visible to Him. He 'Sees' the Entire Pattern of Fohatic Assertion, Fohatic Retraction (Cosmo-Objectivity and Cosmo-Non-Objectivity).

Thus, not only does the Universal Logos *register as a whole* each and every Cosmic Now (whether Objective or Subjective), but He 'abides' in the Cosmic Eternal Now which includes and *synthesizes* all ultimate moments and their opposites—inter-moment instants. All these, together form One Great Pattern in which Cosmic Sequence, Cosmic Simultaneity, and Cosmic Timelessness *combine.* Perhaps the Universal Logoic 'Vantage Point' is the only one from which this combination of apparent temporal contradictions can be seen as *one.*

By *the Cosmo-Objective Now* is meant the highest degree of *simultaneity of registration* possible in relation to the World of Fabrication of a particular Cosmos—a registration encompassing the entire Cosmo-Objective World. A Cosmo-Objective Now is the complete and simultaneous registration of the entire Cosmic Configuration within the World of Fabrication in one ultimate moment by the Universal Logos.

The Cosmo-Objective Now (a measure which is 'Time dependent') is *not* the Cosmo-Eternal Now, and still less, the ETERNAL NOW which occurs forever, independently of Time, and is ETERNALLY ABSOLUTE SIMULTANEITY, the obliterator of all objective considerations. ABSOLUTE SIMULTANEITY, due to the Limitations of Cosmos, is not *formally* (i.e., *perceptually*) possible in-Cosmos (at least, not 'ongoingly'). Compared to what is possible within any Cosmos, *infinitely more* is always possible 'within' the ABSOLUTE.

Example: The Cosmic Now (or Cosmo-Objective Now) is determined with respect to objects-in-Universe and their relations from ultimate moment to ultimate moment. The ETERNAL NOW has naught to do with objects. The Universal Logos focussed (from a sufficient 'Temporal Distancing') in the Cosmic *Eternal* Now 'Sees' the *entire* Cosmic Process (Cosmo-Objective, in the World of Fabrication, and Cosmo-Subjective, in the World of Being) as One Seamless Movement occurring in a relatively *vast*. Cosmo-Quantized Moment of Time. The entire Cosmic Discontinuity can, thus, be perceived by the Universal Logos as a seamless semi-continuity, either as 'one seamless flow' or (if the Meditative Temporal Distancing is greater) *held*, as it were, in One Great Moment of Time. There is an Aspect of the Consciousness of the Universal Logos so concentrated upon the One Great Process unfolding before His *unblinking* Eye, that it all seems to occur in one vast Moment of Cosmic Time which includes all lesser moment. This Moment is the Cosmic Eternal Now and is different from either the Cosmo-Objective Now, which occurs 'upon' every ultimate moment, or the Cosmo-Subjective Now which is used as a fleeting Moment of Universal Evaluation.

Example: The difference between now and a Cosmo-Objective Now is subtle. They both are dependent upon the ultimate moment and can only occur 'on' or 'at' an ultimate moment in any Cosmos, but the Cosmo-Objective Now entails the perception of *all* elements of the Cosmic Configuration in the World of Fabrication during an ultimate moment, and the term now is used more generically and in relation to entities and, even, Entities in-Cosmos who have a much smaller scope of consciousness. The term the Cosmo-Objective Now includes the connotation of the vast scope of what is registered by the Universal Logos within an ultimate moment. The 'nows' of all E/entities other than the Universal Logos are always only *approximations* to the Cosmo-Objective Now.

Example: A Cosmic Now (Cosmo-Objective Now) is but an aspect of the Cosmic Eternal Now. The Cosmo-Objective Now can occur only upon the 'on' ultimate moments in Cosmos. A Cosmic Eternal Now 'abides' during both the 'on' moments within the Fabricated Cosmos, and the 'off' moments within the Un-Fabricated Subjective/Spiritual Cosmos. A Cosmic Eternal Now is not dependent for its existence upon the occurrence of an ultimate moment. Ultimate moments occur 'on' Cosmo-Objective Nows and 'within' the Cosmic Eternal Now.

Example: To the Eye of the Universal Logos, and during Universal Manvantara, it is intermittently a Cosmo-Objective Now, and/or a Cosmo-Subjective Now, but *ever* truly a Cosmic Eternal Now. It is said of God that, "A thousand ages in His sight are like an evening gone." Further, it is said of Him, that Past, Present and Future are all one in His Eyes. Does He 'See' with two Eyes or one? How many eyes does it take to see a *point?*

Example: Due to considerations from Relativity Theory, the question arises as to whether it is *always* Now (the same Now) in all locations/fields in-Cosmos, or whether each location/field has a different Now. Radical Infinitism, positing an Omnipresent, Cosmos-Pervading Universal Observer, the Universal Logos, states that regardless of local-in-Cosmos variations in the speed of changing relations between variables perceived by relatively *localized* observers, the very act of 'Simultaneous Cosmic Observation' by the Universal Logos renders Its *moment of observation* a universal 'Now' *everywhere* in-Cosmos (from *His* Perspective). (This Universal Now may be Cosmo-Objective, but in a way it is *persistently* Cosmo-Subjective, and *always* Cosmo-Eternal.)

The ETERNAL NOW, too, has 'ITS' *penetration* into the World of Being. Normal Relativity Theory does not account for the All-Pervading, All-Conscious, All-Observing God of our Universe when it hypothesizes the need for different localized Nows in-Cosmos (although, in one way, the Cosmo-Objective Now within the World of Fabrication) is *dimensionally localized.* The Perceived Now of the Universal Logos (especially His Cosmo-Eternal Now as it 'admits' the ETERNAL NOW) is, therefore, independent of the curious fluctuations of local-in-Cosmos *time/space/matter conditions* as evaluated by localized, consciously-limited observers.

Example: From the perspective of Time and Eternity, to "be in the world and yet not of it" is to hold simultaneously the Cosmo-Objective Now, the downwardly (or upwardly) focussed Cosmo-Subjective Now, the Cosmic Eternal Now, *and* the ETERNAL NOW. Perhaps the Universal Logos Is capable of this? Perhaps! It is because the Universal Logos can so easily 'change His Point of View', that the *nowness* is, for Him, so variable. We can see that for the Universal Logos, Time is Relative, but not in the same way that Relativity Theory (which confines its speculations to the physical/etheric world) presents Time as Relative. The 'picture' if far, far more expansive.

Example: The Cosmo-Objective Now is infinitely larger (and, of course, *smaller)* than NOW. Any definite quantum of time (even if it were centillions of times shorter than the ultimate moment) would be infinitely larger than the NOW. From another perspective, any definite length of time, no matter how vast, is always infinitely *shorter* than the ETERNAL NOW, the *duration* of which is *infinite.*

Now, (Cosmic) Subjective

By the *Cosmo-Subjective Now* is meant the second variety of *Now* in Cosmos which is that unit of Cosmic Time which has been described as an inter-moment interval or inter-moment instant. It is, in any moment, a repetitive, standardized moment of Will, Imagination or Decision that occurs in the World of Being (in that lower stratum of the World of Being called the World of Adjustment) 'between' the ultimate moments which occur in the World of Fabrication. This second variety of Now in Cosmos is called Cosmo-*Subjective* because it is, in a way, that 'segment' of the ongoing Cosmo-Eternal Now that 'emerges into particular focus' during the 'time' in which the Objective Fohatic Cosmos has 'disappeared' into the Subjectivity of the World of Being.

Example: During a Cosmo-Subjective Now (which like an ultimate moment lasts but an 'x-tillioneth' of a human second), the next 'change' in the Cosmic Configuration is 'coordinatively' Willed by all B/beings *resident as essences* within the World of Adjustment. The concept of 'coordinatively Willed' makes all of us (B/beings in Cosmos) *collectively responsible* for the *next* Cosmic Configuration. How Am **8**? How is the World?

Example: During a Cosmo-Subjective Now, a retracted 'Ray' can 'Janus-like' focus upon the Cosmo-Configuration to 'appear' during the 'next' ultimate moment, *and* (or perhaps *and/or*) survey all from the World of Being immersed in the Cosmo-Eternal Now.

Example: Presumably the Cosmo-Subjective Now is infinitely divisible. In the Realm of Ideation of a given Cosmos (almost) anything is possible. Imagination/conception can 'move' at any speed except *infinite speed* (whereupon it would cease to exist). A subsidiary Now within a Cosmo-Subjective Now would 'last' *for the duration of a held image.* There seems to be no limit upon 'brevity', though there would be a limit upon lengthiness of duration (for the duration inter-moment instant, assuming it was standard, would be such a limit).

If, on the other hand, one looks at the Cosmo Subjective Now as ongoing (rather like the Cosmo-Eternal Now), then an image could be held for as long as was desired or allowable within Cosmic Parameters. Presumably, the Retraction of Fohatic Extensions, 'x-tillions' of 'times' a second, would have some effect upon the Monads organized for Creative Work 'below'.

In other words, in that Subjective World (the World of Adjustment) a 'Moment of Opportunity' would be realized following every ultimate moment in Objective Cosmos, and it is *that* 'Moment' which we are calling the Cosmo-Subjective Now. The presumed 'imagistic/conceptual-divisibility' of that tiny Cosmo-Subjective Moment or Now, allows for a requisite 'imparticulate ideation' within the World of Adjustment 'prior' to each 'move outward' into the World of Objectivity. There would seem to be no limit upon the speed and number of 'imagistic/conceptual manipulations' possible 'before' each move 'outward' into a 'frozen' ultimate moment in Objective Cosmos. How many thoughts enter the mind of the chess player before his next 'move'? Interestingly, once the move is made on the chess board, the positions 'hold' until the next move. The players have to have 'time' to "think things over."

NOW-as-Now

By the *NOW-as-Now* is meant as much of the Perception/Registration of the ETERNAL NOW as can infuse ITSELF (under Cosmic Limitation) into the Eternal Now (and Its subdivisions, the Cosmo-Objective and Cosmo-Subjective Nows). The Cosmo-Objective Now is dependent upon the frequency of the ultimate moment in any Cosmos. The ETERNAL NOW is utterly independent of any movement/variable.

Example: Because each Universal Logos is a *limited* Being and part of an infinite sequence of such Logoi, while It is possessed of a sense of Super-Cosmic History, nevertheless, *as Itself,* It cannot Know the entirety of that History nor Its exact 'Place' in the Infinite Sequence of Cosmoses. The Universal Logoic 'Time Sense' is also affected by the limited stature of the Universal Logos. Thus, each Logos can comprehend the intra-Cosmic Nows (Objective and Subjective), and even the Eternal Now, (with respect to Its Own Cosmos) but *not* (in *utter fullness)* the ETERNAL NOW. For the INFINITE SELF, it has *always* been NOW, 'ETERNALLY' NOW, and ever will be, but for the Universal Logos, an apparently perishable Being (at least in Its *form* nature if not in ESSENCE), it is only:

1. Cosmo-Objectively Now
2. Cosmo-Subjectively Now
3. Eternally Now during His Cosmos

Other 'Times' in Infinite Duration (which, for the ABSOLUTE, have always been and always will be NOW), can only be understood as *then* or *to come* by the Universal Logos, (though His understanding of such term is monumentally in advance of what is possible for one of His 'lesser' 'Rays'—such as the tiny human unit). Nevertheless, of all beings or Beings-in-Cosmos, the Universal Logos (because He 'abides', principally, in the Cosmic Eternal Now) understands best the approximation of the ETERNAL NOW with the three varieties of *Now* in Cosmos—an approximation symbolized by NOW-as-Now.

NOW-as-Now-as-now; now-as-Now-as-NOW

By the *NOW-as-Now-as-now* is meant the descent of the ETERNAL NOW into the Cosmic Eternal Now and Its subsidiaries (i.e., the Cosmo-Subjective and Cosmo-Objective Nows), and thence into the ordinary moment by means of a consciousness's successive identification of the ONE SELF with the One Self, and thus with the individual/personal Self/self.

This descent is involutionary and indicates what might be called 'the corruption of ETERNITY by Temporality'. The illuminative reascent symbolized by *now-as-Now-as-NOW*, which indicates the limited personal self realizing to some small extent (by ascending through some faint intimations of the Cosmo-Objective and Cosmo-Subjective Nows) the present moment in terms of the Cosmic Eternal Now (even if the capacity is not yet Really present to register the ultimate moment and inter-moment instant) and, as well, to whatever degree possible, the ETERNAL NOW, is facilitated by a power in consciousness called "blotting out all form". For this ascending sequence to yield its fruit of realization, there must be a recognition that the *action which is form* is occurring simultaneously rather than sequentially.

Example: Perhaps the poet, Blake, had a vision of the now-as-Now-as-NOW when he saw "infinity in a grain of sand, eternity in an hour."

Example: Under the 'spell' of 'MAYA'-instantly-Maya 'WE' have 'lost ETERNITY' and that 'loss' is symbolized by the formula NOW-as-Now-as-now.

Now (Eternal, Cosmic, or Cosmo-Eternal)

By the *Eternal Now* is meant the *Cosmic Eternal Now* or, alternatively, the *Cosmo-Eternal Now.* It is the Experience in the Consciousness of the Universal Logos and (perhaps) Its Principal Emanations (or perhaps of the entire World of Being-as-Universal Logos!) of the entire Cosmic Process as One Unchanging Event/'Happening'/Cosmo-Macro-Moment. In this Experience there is either the Perception of a seamless impartite flowing movement when the Cosmic Process is 'Seen' from a sufficient 'Temporal Distancing', and even an experience of a *fixity* with no registered movement (when that 'Temporal Distancing' is sufficient increased—these increases engendered by a Meditative Attitude on the part of the Universal Logos). Though discontinuity, of course, exists from 'lower' or less remote Temporal Perspectives, the many apparent movements are registered, from these heightening Cosmo-Temporal Perspective, first, as if the many movements were all a 'seamless flow' under the All-Seeing Eye, and, secondly, as if there were *no* movement or, paradoxically, one *fixed* or *static* Movement taking but a Single Cosmic Macro-Moment.

Example: The Eternal Now or Cosmic Eternal now is an aspect of the ETERNAL NOW. The Eternal Now is a *nowness uncompared.* If the Universal Process 'Seen' from a sufficient 'Temporal Distancing' as one, imparticulate, unchanging Event were *compared* against the Infinite Time Line, then, that Event would, necessarily, be apprehended in terms of *duration.* The Universal Process *can,* however, be 'Seen' by the Universal Logos as if out of all relation to other processes and, 'as if' at such a 'Temporal Distancing', that it becomes an unvarying *Single* Happening, a Single *Event* that happens Cosmo-Eternally Now! This requires the Total Concentration of the Universal Logos on the highest level of Its Logoic Temporal-Apprehension. This does not mean that the Universal Logos cannot exercise other levels of Logoic Temporal Apprehension which would allow It to register Time *(below* and *above)* even while *maintaining* the Contemplation that makes the Eternal Now, or the Eternal Cosmic Now, Cosmically possible.

Example: Just as an ultimate particle is also an event (and a 'frozen' event at that), the freezing of which determines a micro-immovable 'Now' in Cosmo-Objectivity, so the Cosmos Itself is like a Particle/Event ('frozen', as it were, for the time of a Cosmo-Eternity 'frozen', as it were, into a Cosmic Eternal Now).

To see Cosmos in the Eternal Now is, in one way, to 'See' the entire Cosmic Process as a Single Particle/Event, 'frozen' in Time. Who can thus 'See'? Probably no intra-Cosmos Consciousness (if *objectively pre-occupied)* can grasp such a Vision in its entirely. Probably only the Universal Logos. As well, the Vision from Super-Cosmos would be most interesting for placing the various Cosmoses in Temporal Relationship with each other.

NOW; ETERNAL NOW

By *NOW* is meant the *ETERNAL NOW*—the eternally immobile *non-moment* which, because of its immobility has always been and will always be identical to itself. The ETERNAL NOW cannot technically or REALLY be considered a *moment,* for a moment connotes not only 'movement' but, as well, a 'division' of divisionless DURATION.

Example: The ETERNAL NOW may be considered the *infinitely enduring utterly dimensionless moment.* The ETERNAL NOW persists throughout INFINITE DURATION, negating the supposed *reality* of Time.

Example: The ETERNAL NOW can (perhaps) be intuited by the one for whom the Cosmic Heterogeneity is known to be the ETERNAL HOMOGENEITY. Who can thus intuit?

Example: Movement is related to Time. In the MOTIONLESSNESS of REALITY, the ETERNAL NOW prevails.

Example: The ETERNAL NOW is no respecter of *objects.* Clearly, it is a 'STATE' which prevails *purely,* only 'within' the INFINITE SUBJECTIVITY, in the ALL-IN-ALLNESS of Universal Pralaya.

Example: The ETERNAL NOW is the *devourer of multiplicity.*

Number

By *Numbers* are meant purposeful, intelligent, SUPER-Cosmically pre-ordained, *systematically generated Archetypal Relationships* arising through *Emanation* in the Bounded Infinite Subject (the Universal Logos), which are, then, reflected in the Bounded Infinite Object (Cosmic Prakriti)—the Objectified Self-'Sight' of the Universal Logos.

By *Number* is meant the ZERO-in-Manifestation.

By *Numbers* are meant Archetypal Beings, only partially reflective of the ONE BEING/NON-BEING.

By *Number* is meant, commonly, a symbolic system for designating varying quantities.

By *Number* is meant, 'My' Modes of Interactivity with the emanated Self-Objectifications of 'MySelf'.

By *Number* is meant *Relationship.*

By *Number* is meant the various states of Consciousness of the Cosmic Monad, the Number One, the Universal Logos.

Example: The Numbers are formed by 'Emanative Interplay', and these Numbers (as, Essentially, Immaterial Formative Potencies), are later *embodied* ('below') as Objectifications, 'in' and 'as' *Cosmic Prakriti.* Numbers are Beings, in some small measure reflective of the ONE BEING/NON-BEING.

While Numbers are Really Something, They are REALLY NOTHING. Numbers are ZERO in manifestation—the INFINITE VOID in Finitude. While Numbers (other than the Number One) are *Objects* in the Eye of the Universal Logos, They are permanent-in-Cosmos Subjects to those E/entities laboring within the World of Fabrication.

Example: There is no *emanative enumeration* before *cosmification.* One astonishing implication of the foregoing is as follows: Pre-Cosmically and Post-Cosmically no Functional Number greater than the Number Three exists (though many 'Number-Points' may be 'Seen' by the 'De-Infinifying' or 'Re-Infinifying' Super-Cosmic Subject'. The generation (in manifestation) of the vast sequence of N/numbers depends upon the *emanative activation* of the Universal Logos. All this notwithstanding, it is clear that every possible Number must *'inhere'* as an *infinitized possibility* 'within' the FOUNT OF ALL POSSIBILITY.

Example: All **8** can 'Do' (as Universal Logos) is to emanate 'MySelf'. Whatever **8** emanate is a Monad (even though it be, formally considered, a Monad-in-attenuation). However, the manner in which **8** interact with the Emanations of 'MySelf' by *investing* 'MySelf' in 'My' Emanations *is* Number.

Numbers, Pure

By *Pure Numbers* are meant the highest Archetypal Beings. While all authentic E/entities are Numbers, only the highest Numbers (One through Nine inclusive, and, possibly, the Number Ten as a form of Number One) are Pure Numbers. Though all beings in-Cosmos are Essentially Numbers, they are still more ESSENTIALLY, ZERO.

Example: Let every human being find his or her intra-Cosmic Identity by discovering the Complex Number which designates his or her presently manifested energy configuration, and the Pure Number (in the World of Being) to which he or she truly resonates That Pure Number will be either one of the Seven or one of the Nine.

Numbers, Existent

By *Existent Numbers* are meant those numbers which exist as representatives of actual items-in-Cosmos (in which case Existent Numbers are integers) and their actual relations or potential relations (in which case Existent Numbers are sometimes integers, and sometimes ratios). Numbers are designated as Existent Numbers when and only when an existent item-in-Cosmos, or existent items-in-Cosmos (and their actual relations or potential relations) can be denoted by or correlated one-to-one with such Numbers.

Example: There is a number sufficiently large such that it is larger than any actual aggregation of actual items-in-Cosmos, and larger, even, than the number of actualizable relations between actual items-in-Cosmos. Such a number would *not* be an Existent Number.

Example: Existent Numbers are a small subset of Conceivable Numbers.

Example: The numerical consequences (i.e. quotients) of relationships (i.e., of ratios) are often not designated by integers. Quotients are the result of relationships (ratios), and these quotients can be thought of as Existent Numbers as long as the quotient is generated by the ratio of two Existent Numbers (i.e., which denote two *actual* items-in-Cosmos).

Example: In the strictest and most limited sense, the highest Existent Number is limited by the greatest number of denotables in any one Cosmo-Objective Moment (i.e., ultimate moment). The highest Existent Number will, of course, be higher than this number of denotables, but the denotables will determine the higher limit of the Existent Number.

Numerical Beings, Essential

By the *Essential Numerical Beings* are meant the first Nine Numbers (perhaps including the Number Ten as a form of the Number One), Who are the Principal Archetypes of our Cosmos.

Example: Each Essential Numerical Being is a unique Power-of-Aggregation in-Cosmos. Under the influence of Essential Numerical Beings, items-in-Cosmos (in this case within the Fohatic World of Fabrication) are *formed* in a specific way and aggregate, combine or configure according to certain Laws. These Laws might be called, the Law of One, the Law of Two, the Law of Three, the Law of Four, etc. When the concept of 'Power-of-Aggregation' is transferred to the World of Being, it should be called, 'Power of Integrous Combination' (for aggregations are not present in the World of Being, and combinations, though 'partite' are utterly integrated and as if "of one piece".

- O -

object

By *object* is meant the registration in consciousness of a particularization.

Example: In-Cosmos, that which is not isolated as a particularization (an item-in-Cosmos) is not registered as an object. There can be no object unless there is a perceiving consciousness to register it. There are, as well, Super-Cosmic Objects, and They, too, can be considered Particularizations. From the 'INFINISPECTIVE' of the INFINITE SELF (perhaps this 'SPECTIVE' should be called the 'ESSE-SPECTIVE' connoting 'BEING'-in-'TOUCH' without 'SEEING') whatever 'GOES FORTH' from IT is a Super-Cosmic Particularization and, hence, an Object, even though, paradoxically only the INFINITE SELF can 'GO FORTH' from ITSELF. Therefore, even *selfhood* can be 'seen' as an *object*. Actually, it is *precisely selfhood*, that is seen as objects.

By an *object* is meant a percept or an 'appercept'.

Example: Objects are classified according to the plane to which they pertain. While an automobile is most definitely an object, pertaining particularly to the physical plane, a refined idea, registered and understood, is equally an object to the consciousness that apperceives it. Such an object would pertain to the levels of the abstract mind or even to the buddhic plane. Each and every plane in all of Cosmos is an *object*, as is all that transpires on any of those planes.

Example: All objects are Really infinitesimalizings and REALLY, NOTHING. Every object can be reduced to a point—that which has no dimension and yet *is*. Objects, as usually considered, however, are reduced only to virtual points that *do* have dimension, for most objects are not 'Seen' from an 'infinispectivizing' which would reduce them to Real points. The difference between a *Real point* and a *virtual point* is only the Point of View of the Perceiver.

Example: An object is a Seen-Self.

Object

By an *Object* is meant one of a number of Realities (Principles, Essences, Noumena), which are Permanent-in-Cosmos. Such 'Objects' may seem to us like extremely *Subjective* Entities, but because They are *within-Cosmos* (which, Itself, Is the Greatest *Finite* Object), They must be considered merely as Objects, certainly from the larger Super-Cosmic Perspective and even from the Perspective of the Universal Logos. To lesser beings such as the human being, They may be considered as Subjects (even though They may have a particular *objective* expression—as has, for instance, a Galactic God, Who, though (to us) a profoundly *subjective* Cosmic Being, expresses Itself through the *objective form* of a galaxy of stars). A Galactic God is a relatively minor Deity compared with the highest Hierarchy of Cosmic Principle/Beings/Archetypes/Numbers.

Example: The human being must look within to find the Divine Objects (the Realities and Principles of this Cosmos), but in the 'Sight' of the ALL-SELF-as Infinite Subject, and even to the 'Eye' of the Universal Logos, the Great Subjective Entities of Cosmos are Really Externalities—Objects.

object, compound

By a *compound object* is meant a *relationship.*

Example: The only intra-Cosmic object that is not a relationship between other or lesser objects is the ultimate particle, because it is impartite. And yet, because the ultimate particle is the result of *reflective interplay* between Fohat and Its Own enumerated Self-Reflection (registering as differentiated Cosmic Prakriti), it could be said with some justification that even the ultimate particle (though indivisible in the usual sense) is a compound object because it can be resolved into two contributing Factors (i.e., it demands a Perceiver and a Perception, or an 'In-ceiver' and an 'In-ception').

Pre-Cosmically and Post-Cosmically it could justifiably be said that Mulaprakriti is *not* a Compound Object, even though the *generation* of Mulaprakriti requires the 'RELATIONSHIP'-through-'CONSCIOUSNESS' of the INFINITE SELF-as-Infinite Self with Itself. What holds good for the impartite ultimate particle/event, however, holds good on a much higher level for Mulaprakriti. It would not exist unless It were seen (and, thus, in a strange sense is 'divisible'/'resolvable' into Perceiver and Perceived).

Whereas there are some objects, like the ultimate particle/event and Mulaprakriti which in their pure objectivity are uncompounded, is there any object which, itself, is not the result of interactivity between a Subject and Maya? 'MAYA' 'DRAWS' the *HIDDEN SELF* into the *light of objectivity.* Maya does the same Super-Cosmically and Cosmically. Of course, the very FIRST OBJECT is 'MAYA' 'ITSELF'. Curiously, 'MAYA' the ONE 'RAY' of the ABSOLUTE is instantly Tri-Partite as:

- The 'EVANESCENT INFINITE SUBJECT'
- The 'EVANESCENT INFINITE OBJECT'
- The 'EVANESCENT INFINITE RELATION BETWEEN' (which is 'MAYA' 'INFUSED' into the TRINITY)

'MAYA', the 'OBJECT' 'WHO' is the 'SEEING' can only BE 'BE-EN' (or perhaps, 'ESSE-SPECTIVIZED'). In a way, because 'MAYA' cannot be 'SEEN', the UTTER MONALITY of the INFINITE SELF is preserved. IT can remain an UNPERTURBED 'ONENESS/NONENESS', which 'MAYA'-instantly-Maya 'Becomes' the entire World of Duality. Thus, the INFINITE SELF, by 'REFUSING' to 'SEE' 'MAYA', 'REMAINS' 'ALOOF' from All that has 'GONE FORTH' from IT. Thus is the INFINITE SELF, 'UNCONTAMINATED' by the World of Illusion (even while, necessarily, BEING IT).

Example: The apparent simplicity of so many objects is an illusion arising from the inability to see and understand that the objects are really compound objects, composed of many lesser entities in various relationships.

object, simple

By a *simple object* is meant one that has no parts.

Example: There are only two simple objects:

1. Mulaprakriti, which is the Infinite Object.
2. The ultimate particle of a given Cosmos, which particle is impartite because (by 'Cosmic Contract') further indivisible.

An ultimate particle is a measurable unit of *subjectively-particulated Fohat generated by selective Fohatic Self-Objectification (selective Fohatic Self-Reflection).* Just as the ultimate particle is resolvable into Fohat and selective Fohatic Self-Perception, so Mulaprakriti is resolvable into PARABRAHMAN-as-Infinite Subject and PARABRAHMAN's 'MAYAVIC'-as-'Mayavic' 'Sight' of Itself ('Itself' being the form of 'ITSELF', i.e., the Infinite Subject, that can sustainedly be 'Self-Seen'). In these two cases, the *images 'Seen'* are 'imparticulate', hence, 'simple'. PARABRAHMAN (WHO, BEING the INFINITE SUBJECTIVITY, IS, ESSENTIALLY, *every succeedent subject) IS* REALLY always one 'part' of every simple object, and REALLY is the Whole of such an object.

For *practical* purposes (refined philosophical analyses aside), Mulaprakriti and the ultimate particle (the greatest and the least of objects) are simply simple objects. A case, however, could be made for regarding the 'RAY' of the ABSOLUTE and the Infinite Subject as simple Objects, even though They are almost infinitely subjective. REALLY, any *thing whatsoever* is an *object.* Of course, PARABRAHMAN is *not* a *thing* no matter how often the feeble human consciousness attempt to make 'IT' into a *thing.*

objectification

By *objectification* (from one perspective) is denoted a process whereby a subtle, *apparently* subjective pattern of energies is rendered capable of registration by consciousness. Only that which is *objective* (no matter how subtle the objectivity) can be registered by *consciousness, per se.* The moment one becomes conscious of a supposedly subjective pattern, that pattern can be considered objectified. This does not mean that a *subjective* pattern has been *changed* into an *objective* pattern, but simply that the subjective pattern has been 'Seen' or registered. 'Seeing' or registration is the *objectivizing* of *subjectivity.* Really, any such pattern has been, in a very ultimate sense and from a very high perspective, an *objective* pattern all along, however the necessary sensitivity of consciousness *objectivize* the pattern was insufficiently developed. ESSENTIALLY, no *pattern* can be *truly subjective.* There is only one SUBJECTIVITY—the ORIGIN of all lesser, apparent Subjects/subjects. *Identification alone* registers or detects a *subject* in ESSENCE; *consciousness* does not.

By *objectification* (from another perspective) is meant an act of consciousness which causes that which is subjective to be *perceived* as an object. Through Emanation, for instance, a Subject 'Becomes' an Object in its own 'Sight'. Does not the Mantram for the Astrological Sign Gemini begin, "I see my other self ..."?

Example: A Subject can be *seen* (through a prakritic veil) as an Object, but that *seeing* fails to penetrate to the ESSENCE of the subject. Only *identification* will *unveil* a Subject and render it *intimately known,* through and through and, as it were, "from the inside out." The objectification of a Subject does not reveal its ESSENTIAL NATURE.

Example: The Universe comes to Birth by a mysterious process of SELF-'OBJECTIFICATION'. SELF-'REFLECTION' is SELF-'OBJECTIFICATION'.

Example: Mulaprakriti Is SELF-'OBJECTIFICATION'-instantly-Self-Objectification, i.e., the OBJECTIFICATION of the INFINITE SELF (cast instantly into the Super-Cosmic Field).

Example: Objectification may simply and practically mean the transformation of a vague, subtle and elusive registration, into a registration which is clear, concrete and 'fixed' in consciousness.

objectivity

By *objectivity* is indicated a quality of awareness by means of which an object is perceived as it *is* without any possibility of distortion of perception arising from within the perceiving consciousness (considering that perceiving consciousness as an instrument of perception).

Example: If, in fact, **8** Am what **8** perceive, is complete objectivity Really possible, or desirable?

By an *objectivity* is indicated anything which is capable of impressing consciousness as an object. For that matter, *any* impression upon *consciousness* is an object (though it may not be a concrete or 'shaped', limitedly-extended object).

Example: Even the most subtle states in-Cosmos (usually considered *subjective*) are *actually* objectivities.

observer

By an *observer* is meant a perceiving/apperceiving consciousness which is detached (for all practical purposes) from the thing(s) observed (though, ESSENTIALLY, identical with it).

Example: The disciple, merged in causal consciousness, functions as the observer of personality life.

Example: Quantum physics has pointed out the subtle influence of the observer upon that which is observed, especially at the sub-atomic level. The interactivity between observer and observed applies even more noticeably upon planes of Nature higher than the physical.

Observer; 'OBSERVER'

By the *'OBSERVER'* is meant the ABSOLUTE SELF in ITS mode of 'CONTEMPLATING' ITSELF as Itself. While Vedantin Philosophy speaks of PARABRAHMAN as the 'WITNESS', technically speaking it is difficult to conceive of PARABRAHMAN as a *witness* in the usual sense of the word.

Technically, there can be no such PARABRAHMIC 'OBSERVER', because (in the 'STATE' implied by this total capitalization of letters) there is only NO-THING to 'OBSERVE'. If PARABRAHM 'OBSERVES', IT does so through ITS 'RADIATORY EXTENSION', 'MAYA'-instantly-Maya (i.e., 'MAYA', PARABRAHMAN-'BECOME'-INFINITE CONSCIOUSNESS' which, as it were, 'CREATES' both the 'SEER' and the 'SEEN'). So, PARABRAHMAN is only an 'OBSERVER' by *proxy.* It is, rather, the SELF-as-Infinified Point (the Infinite Subject) Who is the first Real Observer.

There is no Observer before the SELF (through 'RADIATION') has 'BECOME' that 'POINT'-as-Infinified Point (endowed with an infinitude of hypothetically infinite points of view). All during the 'Condition' known as Universe, the SELF-AS-SELF remains *SELF-ABSORBED* ('INFINIDENTIFIED') thus, Cosmically, *UNOBSERVANT* in the 'STATE' of INFINITIZED BE-NESS, while, in *apparent* contrast, the SELF-as-Infinified Point-as-Condensed Point-as-Universal Logos both Observes and Acts. It is obvious that we are dealing with a profound (and to the human mind) unresolvable paradox.

Example: Intra-Cosmically the Observer is the Universal Logos as well as all Its Self-Conscious emanative extensions. Super-Cosmically, the Observer is the Infinite Subject Who is the Observer (of Itself) *as* Mulaprakriti. Through the Process of the Interplay between the Infinite Subject and the Infinite Object, PARABRAHMAN is able to function as the Observer and even the 'Do-er' while yet remaining *totally aloof* and, as it were, SELF-'PREOCCUPIED'.

And yet, the Vedanta refers to PARABRAHMAN as the "WITNESS". In an *ultimate* sense concerning *fundamental IDENTITY,* this *must* be so, for although PARABRAHMAN must necessarily remain *out of all relation* with anything observable, yet, there is no other IDENTITY in the UTTER ALLNESS than PARABRAHMAN, so if there is an Observer, it simply *must* be, ESSENTIALLY, PARABRAHMAN as the 'OBSERVER' (though, REALLY, PARABRAHMAN IS the '*BE*-er' rather than the 'SEE-er). For that matter, if there is an *anything,* that *anything* simply *must* be, ESSENTIALLY, PARABRAHMAN.

'OBSERVER', the ONE AND ONLY

By *the ONE AND ONLY 'OBSERVER'* is meant, apparently, an *impossibility.* The OBSERVER has NO-THING to 'OBSERVE'. It would be more proper to call the OBSERVER the *'BE*-er' because the only 'HAPPENING'/NON-'HAPPENING' within the ALL-IN-ALLNESS is an *infinitized* 'STATE' of BEING, ITSELF.

By *the ONE AND ONLY 'OBSERVER'* (technically considered) may be meant PARABRAHMAN in ITS initial 'MODE' of SELF-'REDUCTION', when for the first 'TIME' (yet again!) IT (as the 'EVANESCENT INFINITE SUBJECT') 'SEES' or 'OBSERVES' ITSELF (as the 'EVANESCENT INFINITE OBJECT'), by means of 'MAYA'-as-INFINITE SENSITIVITY, and, *in so 'DOING',* 'BECOMES', with *infinitesimal instantaneity,* the Pre-Cosmic Trinity of Infinite Subject and Infinite Object and Infinite Consciousness.

Example: The Universal Process begins with an *actless* 'ACT' of SELF-'OBSERVATION' (or, better, 'ESSE-SPECTIVIZATION'). The ONE-AND-ONLY-SELF 'CHANGES' from BEING to 'SEEING' (thus 'BECOMING' {for an interval that *defies Time*} the ONE-AND-ONLY-'OBSERVER') *while yet continuing to BE as IT ever has BEEN* (i.e., *somehow* 'UNOBSERVANT' and SELF-PREOCCUPIED), for IT can never 'CHANGE' and yet IT must—*apparently.* With the 'COMMENCEMENT' of the 'ACT' of 'OBSERVATION', it is not REALLY the 'EVANESCENT INFINITE OBSERVER'/'SUBJECT' WHO *continues to* 'OBSERVE', but the SELF-'EXTRUDED', SELF-'RADIATED' Observer (now a 'Something' apparently *less than* the ABSOLUTE SELF—a 'Something' called the SELF-as-Infinified Point or Infinite Subject) Who Observes. What necessarily 'BEGINS' as the 'INTRA-VOID' 'OBSERVER' 'BECOMES' with *infinitesimal instantaneity,* the extra-VOID Observer.

ONE-WHO-IS-THE-ZERO, the; O-N-E, the; One, the

By *the ONE-WHO-IS-THE-ZERO* is meant the ONLY BEING/NON-BEING. This BEING/NON-BEING is the ONE AND ONLY (hence the 'ONE') but is also 'NO-THING'-the-NOTHING (hence the 'ZERO').

By *the O-N-E* is meant the SELF-as-Infinified Point, the ONE-as-Infinite Subject, Who, by interacting via the Agency of Super-Cosmic Fohat with the 'Reflected' SELF-Image known as Mulaprakriti, *begins* the Process of Generating the Cosmos.

By *the One* is meant the One Being *in* Cosmos—the Universal Logos.

Example: The ONE-WHO-IS-THE-ZERO becomes the O-N-E without ever ceasing to BE the ONE-WHO-IS-THE-ZERO. The O-N-E becomes the One without ever ceasing to be the O-N-E. The One becomes (through Emanation) the Many (the teaming Cosmos) without ever ceasing to be the One. We see, thus, that the ONE-WHO-IS-THE-ZERO has, in effect, 'BECOME' the Many without ever ceasing to be the ONE-WHO-IS-THE-ZERO.

Ontological Oscillation

By *Ontological Oscillation* is meant the vibratory 'movement' between the Affirmation of Cosmo-Objectivity and the Negation of Cosmo-Objectivity. Essentially, all vibratory motion in-Cosmos begins on the level of ultimate particle/events, which particle/events (as objectified Self-enumerations of Fohat), oscillate between *being* and *non-being* (or more accurately, between *Actuality* in the World of Fabrication and *Non-Objectivity* in the World of Being) a huge number of times per second.

Example: Every E/entity engaged with the Fohatically Fabricated Cosmos, from the greatest to the least entity, experiences Ontological Oscillation a tremendous number of times per second. Thus the *fully interiorized realization* of the World of Being is never more that an ultimate moment away.

Of course, BE-NESS ITSELF *never* 'OSCILLATES' or 'FLUCTUATES'. IT 'CONTINUES' *ever*—just as IT IS (or *IS NOT).* Even the World of Being does not fluctuate as the World of Fabrication fluctuates. The World of Being is a semi-Continuity which is semi-continuous because of the uninterrupted *ideational* Self-Perception of the major Emanated Beings in the World of Being. The 'Eye' which sustains the World of Being does not 'Blink'. The Fohatic 'Eye' which generates the World of Fabrication, 'Blinks' with extraordinary relative rapidity. (The World of Fabrication is *usually* called the World of Becoming, but in this treatise the entire Cosmos, the Great Discontinuity, is considered the World of Becoming.)

Original Intent, the

By *the Original Intent* is meant the Design-at-the-Beginning, the Pre-Cosmic Purpose of the Universal Logos.

Example: The Universal Theme for this Cosmos is contained within the Original Intent devised by I-as-US-as-Us. In order for the Universal Logos to harmonize Its Cosmos and 'Per-form' properly, all beings expressing at the various levels of the Divine

Emanative Stream (especially Self-Conscious beings) must align their *little wills* with the Original Intent, as best they can understand that Intent.

Other-Than-ALL-IN-ALL

By the *Other-Than-ALL-IN-ALL* is meant all E/entities, things, beings, objects, etc. *other than* the SELF in ITS *purest* most *SELF-ABSORBED* 'STATE'. In addition to all *intra*-Cosmic inclusions, the Other-Than-ALL-IN-ALL includes the First 'RADIATION' (instantly become the SELF-as-Infinified Point, the Infinite Subject) and the First Objectification (Mulaprakriti) as well.

Example: If you can *conceive* of *anything* whatsoever, that which is conceived is necessarily Other-That-ALL-IN-ALL.

Example: The Other-Than-ALL-IN-ALL is the domain of *all objects,* all SELF-'PRECIPITATIONS'. An *infinitized* object, however, 'INHERES' *seamlessly* in the ALL-IN-ALLNESS and can never be considered Other-Than-ALL-IN-ALL.

Example: The term Other-Than-ALL-IN-ALL has no ESSENTIAL REALITY though it makes an important distinction between the INFINITE HOMOGENEITY and the infinitude of apparently illusory objects/states which arise from IT.

- *P* -

pain; Pain

By *pain* is meant a state of consciousness which arises, Essentially, from the maladjustment between the INFINITE and the Finite. The illusory perception of dualism is at the root of the pain response.

Example: Under normal circumstances, pain arises when form cannot express intention or fulfill desire. Considering the World of Becoming as a World partially unresponsive to Original Intent, the reason for Cosmic Pain can be easily understood—although by means of Cosmic Evolution, Cosmic Pain is gradually lessening. Considering the World of Becoming as the World of Finitude, one can see that Pain is, *inherent* in-Cosmos for the Cosmos can never contain or express INFINITUDE and this produces the *fundamental ontological stress.* Within the Veil of Illusion that is Cosmos, there must always be the longing of the Finite for the INFINITE, and the resulting pain which comes from the apparent cleavage between them.

PARABRAHMAN

By *PARABRAHMAN* is meant BRAHMAN ITSELF. The prefix "PARA" is added to emphasize the ABSOLUTELY ULTIMATE NATURE of THE BRAHMAN.

Example: PARABRAHMAN IS only ITSELF.

paradox

By *paradox* is meant a necessary contradiction (necessary because of the limitations of the human mind) which veils a synthetic realization or truth which can be understood only after the limitations of the human mind have been transcended.

By a *paradox* is also meant a thought, or system of thought, based upon at least two *apparently* illogical or mutually-canceling premises, such that if one premise is true the other cannot be true and vice versa. The problem with paradox is that *both* mutually-canceling premises are, *apparently* and *necessarily*, true.

Example: For instance, while it is true that:

1 *no* qualities can be attributed to the ABSOLUTE because IT IS *indivisible* (again, a quality!) and beyond all possibility of differentiation

... it is *also* true that:

2. *all* qualities must necessarily (in some way) be attributed to the ABSOLUTE because IT IS necessarily the ONE AND ONLY SOURCE of ALL, the INFINITE POTENTIAL, the PLENUM, the EVER-FULL and the FOUNT OF ALL POSSIBILITY.

We see that both of these two premises are *true* and *reasonable* according to the accepted 'definitions' of the ABSOLUTE, but one can see how they are *contradictory* and would be, under normal rules of logic, mutually-canceling. Normal logic, however, breaks down when considering the ABSOLUTE. Thus, we have here presented a fundamental ontological paradox.

Example: While a seeming paradox can arise in human consciousness because of the limitations of the human mind, there may be something *inherently, fundamentally and permanently paradoxical* about the NATURE of REALITY and ITS relation/non-relation to Illusion. Just as the SELF has been called 'The GREAT CONTRADICTION' so it is equally illuminative to call IT 'The SUPREME PARADOX'.

paradoxical dual BEING/Being, principle of

By *the principle of paradoxically dual BEING/Being* is meant a profoundly mysterious Process because of which every being-in-Cosmos (no matter how minute or humble that being) is both that humble being and, *simultaneously*, (and *fully*) the ONE AND ONLY PERFECT BEING.

Example: Under the principle of paradoxical dual BEING/Being, the atom is simultaneously the limited atom *and* the UNLIMITED ALL-SELF.

paradoxical perspective

By the *paradoxical perspective* is meant the advancement of a point of view which necessarily seems self-contradictory.

Example: A close study of 'dimensional sealing' in relation to the Divine Emanatory Stream reveals the paradox of *unconscious multi-dimensional consciousness.* How can the human being be conscious and active on mutiple levels and yet, apparently, be unconscious of the fact? When discussing the range of human possibilities it becomes a necessary but difficult task to introduce the paradoxical perspective.

particle

By a *particle* is meant a (relatively) tiny unit of energy which has a temporary identifiable (apparent) distinctness. A particle is always a constituent of a greater whole. From the human perspective, sub-atomic units appear to be particles, but from the Perspective of the Solar Logos, a human being might appear to be a particle.

Example: A 'molecule' (in the distinct and unusual sense discussed in the book *Occult Chemistry)* is but a particle of an ultimate physical atom.

Example: For a truer understanding of the nature of a particle, it must be understood as a particle/event. Every particle in-Cosmos is Really a 'happening', for 'things' are 'events'.

particle/event

By a *particle/event* is meant that each distinct particle in-Cosmos (especially referencing atomic, sub-atomic and far tinier particles yet) is really an *event*—an evanescent energy occurrence, a 'fleeting presence', a precipitation of subtle energy into a world of grosser force.

Example: All particle/events are cyclically-timed precipitations from subtlety into greater density. Every E/entity is an E/event, and from a certain perspective, every E/entity is a particle (even the Universe). Every appearance has a Space-Time dimension and hence may be called a particle/event:

- the 'particle' aspect refers to Space
- the 'event' aspect refers to Time

particle/event, ultimate (in-Cosmos)

By an *ultimate particle/event in-Cosmos* is meant the *smallest* possible unit of energy that can exist in a particular Cosmos given the particular Pre-Cosmically determined *Parameters* of that Cosmos. Ultimate particle/events are *indivisible* within a particular Cosmos. Intra-Cosmically they cannot be further reduced, and thus have a kind of impregnable integrity supported by the Cosmo-Logically Willed Cosmic Parameters which abide for the duration of the particular Cosmos—i.e., for an Eternity.

Ultimate particle/events may well be the *objectively reflected particulated presence* of Fohat in Cosmic Prakriti. Fohat *enumerates* Itself specifically and subjectively and then 'goes forth' into the Great Cosmic Object, Cosmic Prakriti. This means, Cosmo-Psychologically, that Fohat, as Holy Ghost, objectifies Its own global Presence as an Image, or Field-Image, for Prakriti *is* Image. Being, subjectively, in a *specifically enumerated* State and having 'penetrated' Cosmic Prakriti (i.e., 'interplayed' with Its own Image), Fohat 'Sees' (multiply—as many times as It has enumerated Itself) 'within' Its Cosmic Prakritic Image, and each 'Seeing' *is* an ultimate particle/event.

Fohat, we see, is responsible for executing Fundamental Parameters in a Cosmos, for the 'size' of an ultimate particle relative to all other structures-in-Cosmos to be built, determines the illusory 'extent' of the Objective Universe. The 'size' of a "hole in space" dug by Fohat ('holes' 'Seen' within Its own globalized Self-Seeing) is Really and ultimately inconsequential, however, because all things-to-come in that Universe are built upon these 'particle-holes' and have a size purely relative to them. We do not know whether our Universe is vast or tiny. Vastness and tiny-ness mean nothing if there is no other Universe within which to compare our Universe. Only Super-Cosmic Memory along the Infinite Time Line would reveal the relative vastness or tinyness of our Universe—relative to others.

The question arises as to whether these ultimate particles are *identical* with each other, because they are indivisible and impartite. They are surely the simplest of all particulate objects in-Cosmos. However these particles are, in a way, 'numbered' because Fohat enumerates Itself specifically (without Really subdividing Itself) in order to produce the multiplicity of fundamental material units (the building materials of Cosmos which ultimate particle/events are).

It is in this Self-enumeration by Fohat that the possibility of differentiation among ultimate particles exists, for each 'particle/event/hole' is *numbered* and hence distinctly itself, just as the Number Two is different from the Number Four, even though they are both Numbers (and, Essentially, the One Cosmic Monad, the Number One). Further, each 'particle/event' has, as it were, a *Fohatic will-of-its-own* which, as it were, *distinguishes* the particle. Further, the fact that there are a great number of these particle/events means that each of them is distinct simply by being *separately articulated* (no matter how similar they might be). According to the Law of Unrepeatability, they could not be *absolutely identical* (if for no other reason than that they are engaged in variegated relationship/configurations).

Thus, differentiation between things can also be considered a function of relationship. The "holes in space" (particle/event/'holes') 'dug' by Fohat will be differentiated simply by their *position* within the Bounded Wholeness of Cosmic Prakriti. Two apparently identical things, positioned differently within a definite whole, have differing relationships to the whole and to the factors contained by the whole. This is to say that two perhaps identical particles *differently positioned* with respect to the whole are, for *that* reason, no longer exactly identical.

As well, as all things possess consciousness (or, alternatively, as all things are 'bathed' within a Field of Consciousness) each ultimate particle/event, due to variegated configurations will, simply put, have 'different experiences' (although the *modus operandi* of 'giving and receiving' in experience is difficult to conceive. At any rate, such 'different experiences' would contribute to a kind of differentiation in relation to ultimate particle/events (even though, externally, they have no 'parts' which can change.

Thus is developed the possibility of a kind of low-lever 'differentiability' which may pertain to these ultimate particle/events, especially once they begin their variegated interactivity with each other (though, in a way, they are limited by the inability to change, even though they are the agents of change in the multitudinous aggregations (i.e., things-in-Cosmos) they form. Thus, repositioning of 'unchangeables' at the ultimate micro-level, causes change in 'changeables' at more macro levels. That ultimate particle/events are 'individual' in the sense of *indivisible*, however, is certain.

The most important consideration in this question of whether ultimate particles can be considered *identical*, is the thought that each such particle/event/'hole' may have, in a very veiled way, a 'will of its own', and may—with supremely blinded 'will'—pursue its Fohatically-encoded instructions that align it with the Original Intent, the Design-at-the-Beginning. The factor of what we might call *microscopic will* may subtly differentiate one Fohatic particle/event/'hole' from another and militate against their *exact* identicalness.

By *ultimate particle/events* are meant the rapidly fluctuating appearance/disappearance of tiny, objectified Self-Reflected enumerations of Fohat (existing as differentiated Cosmic Prakriti). Particles are events. Ultimate particles are Fohatic event/perceptions (Really, *Self*-Perceptions). An object is what a subject 'sees'. Every object in-Cosmos is a Self-Perception (of *some* Self).

In one way, to create differentiation, the Father-as-Fohat goes forth into the Mother (His own 'Mother Image) to 'See' in Her (Really, the Fohat/Father *engages* Himself as an Object and, thus, 'See' into and in Himself) the multiple ways He/Father/Fohat has *subjectively* enumerated Himself. Differentiated Matter (differentiated Cosmic Prakriti) are the multitude of ways that the Father-as-Fohat 'Sees' Himself reflected as the Mother-in-particulation, which He *also* Is. The multiplicity of objects arising in the Fohat/Father-as-Mother are, Really, so many Fohatic Self-Perceptions, as well as Fohatic Self-Perceptions of subjective enumerables-in-*combination.*

It must be remembered that during the course of His 'Work' for the Universal Logos, Fohat (as Holy Ghost) 'Sees' more than ultimate particle/events. Fohat has many 'Sons' of differing Degree. He, through Self-Reflection, 'Creates' His own Hierarchy of 'Sons'—the so-called "Sons of Fohat".

When the Process of Emanation (the Cycle of Cosmic Unfoldment) begins, the Universal Father first 'Sees' His full, unarticulated Nature as Cosmic Mother (Really, His Full Unarticulated Objectified Self *as* Cosmic Mother). The Universal Father continues to 'See' that Cosmic Mother throughout the duration of Cosmos, but later 'Sees' *also* a multi-particulated Mother, because of the Self-Enumerations of Fohat (Really, Himself {as Father}-as-Universal Son-as-Fohat) reflected in the Mother. The Mother is always Object. Different Subjects 'See' Themselves as different Objects, as different Fields, and thus *are* to Themselves, different Mothers (different 'Mother-Fields' or 'Matrices').

As the descending Hierarchy of Creators continues to Project their own generalized Self-Image as a Field, we have Fields, within Fields, within Fields, etc. The so called lesser 'Rays' at deeper levels of prakritic immersion, have lost a degree of the power to be Mother to themselves, and therefore must 'see' as Object, that which is the Objective Mother-Self-Image of a greater Subject. In other words, as human beings, we mostly see as objects that which Fohat has 'Fabricated'.

This does not mean that we do not also *project our own Self-Image* and 'live' in it! Sometimes the Universe begins to look like veil after veil of Self-projections. Each Entity must penetrate his own Self-Projections only to realize that another veil (projected by the Entity that 'includes' him) awaits to be penetrated. This continues until the Projection of the Universal Logos must be penetrated at the "Day Be With Us." Bear in mind as we discuss the interrelations of the 'Cosmic First Family' that there is no REAL distinction between the 'Characters' and the SELF, as IT is *all of them*, and WE ARE/**I AM** that SELF WHO is That Self, which **8** Am.

By *ultimate particles* are meant 'Fohatic *points of view* within Cosmic Prakriti'. Every *point* is both a *point of view* and a *seen point.*

Example: The ultimate particle of our present Cosmos will not be found by modern physicists as they explore ever smaller sub-atomic particles. Such structures (sub-atomic particles), as extremely tiny and amazingly fleeting though they be, are still far too large and relatively permanent to be the ultimate particle of our Cosmos. Such an energy form as the ultimate particle is far more likely to be found upon the atomic subplane of the highest Super-Cosmic Plane within the World of Fabrication—a stratum of SELF-Objectification far beyond the reach of even our Solar Logos.

Example: Another hypothetical perspective on the possible non-identicalness of ultimate particle/events is that each of them (once Fohat has identified with each and every one of His Self-enumerations, *immersing* Himself within them, as it were) *is* a Fohatic *point of view,* Self-Reflectively objectified as differentiated Cosmic Prakriti, which is Bounded Mulaprakriti, and hence *localized.*

Since Fohat has enumerated Itself (according to the Father's/Son's Instructions) before objectifying Its enumerations as *points of view* in Cosmic Prakriti (or what Fohat 'Sees' *as* Cosmic Prakriti), when these enumerations 'root' themselves within Cosmic Prakriti (i.e., within the global Fohatic Field of Consciousness, the Mother-Field) they have *'position'* with respect to *each other.* These different positions introduce the idea of *'variation* between *points of view'* within a definitely Bounded Whole.

Remember, every *point* is also a *point of view.* Since the *points of view* are necessarily *variously* 'positioned' with respect to each other, and since *points of view* are ultimate particle/events, then the ultimate particle/events may be understood for this reason to admit of some slight *variation* (perhaps in the dimension of consciousness, though not admit of further *particulation)* one to the other, especially as the process of relating and aggregating between ultimate particles proceeds actively. Differentiated Fohat gets 'very busy' thereby increasing the opportunity for distinctness of 'experience' among ultimate particles.

This must remain only a hypothesis, for another issue arises here:

- Do ultimate particle/events Really *register* their 'experience'?
- Is their 'experience' Really 'experience', if nothing objective can pass between them in any way?
- Do ultimate particles *change* in any way?
- Do ultimate particles *evolve?*

If, for instance, they neither change nor evolve (and for this, the case is strong as regards shape and size, for instance), the possibility of them remaining *identical* increases. However, since ultimate particles (like every other Object/Subject) are, Essentially, Spirit, and are invested with Consciousness, it would seem impossible that they should be utterly impervious to the results of their astounding multiplicity of contacts.

In some ways ultimate particle/events can be considered as *immutable 'eyes'* of the Greater Subject (Fohat) Who *directs* them and Who *is* them. From this perspective, they would be the *tiniest 'eyes' in-Cosmos.* The issue of the identicalness or non-identicalness of ultimate particles is an important one if the Law of Unrepeatability is to be completely applicable in-Cosmos.

By an *ultimate particle/event* may be meant a 'tiny point of view with a will'. The philosopher Bergson thought that electrons had a kind of will. An electron, however, is a mammoth structure compared to an ultimate particle. Could heterogeneity be built upon homogeneity? This is true from the largest possible extra-Cosmic Perspective (for the ABSOLUTE HOMOGENEITY 'BECOMES' the Heterogeneous Cosmos, but is it true intra-Cosmically as well)?

Example: Fohat as much 'Becomes' "holes in space" as "digs" them. Interesting questions arise in relation to ultimate particle/events, such as, *If* ultimate particles are *imparticulate,* can they *radiate?* Or can only compound objects *radiate?* When the 'RAY' FLASHES FORTH from the ABSOLUTE of *what* does the 'RAY-DIATION' *consist?* Of 'PRESENCE' in 'Absence'?

Example: If ultimate particles are impartite and undifferentiable, can they *evolve*? Can they *profit* from their manifold contacts? Would not the evolution of an ultimate particle demand *accretion to* the particle and *complexification of* the particle (which processes, if carried forward, would change the utterly simple, impartite, non-differentiable nature of an ultimate particle)? And yet, ultimate particles must necessarily be involved in the process of evolution. So is evolution, then, simply *progressively refined geometrification* of particle/events which do not evolve? This is worth pondering.

Example: An ultimate particle/event is a 'corpuscular unit of Fohatic Intention'.

particularization

By a *particularization* is meant any bounded registration which can be differentiated from or compared with another bounded registration. The SELF (the ABSOLUTE) is not a particularization because, essentially, IT is *not* a 'bounded registration' and IT cannot be differentiated from or compared with other registrations, inasmuch as the SELF *IS* all of them and, hence, 'INCOMPARABLE'.

Example: All objects are particularizations. Particularizations, however, are not necessarily aggregatively 'particulate'. They may, instead, by 'integrally partite'.

Example: To the Infinite Subject (the SELF-as-Infinified Point), Consciousness of Mulaprakriti (the Infinite Object) is an *unbounded registration,* and consequently *not* a particularization, in the *usual* sense. A narrowing of focus by the Infinite Subject (which 'narrowing' manifests as the SELF-as-Condensing and Condensed Point) Creates *Bounded* Mulaprakriti, which is Cosmic Prakriti—a *bounded registration* and hence a particularization.

In one way, Mulaprakriti, the Infinite Object, though unbounded, might be considered a Super-Cosmic Particularization, for It can be differentiated from the 'RAY' of the ABSOLUTE *and* the Infinite Subject (both of which, though almost ultimately Subjective) *could* be 'SEEN' by the GREAT 'WITNESS' as distinct Particularizations *(if* the GREAT 'WITNESS' *'SAW'!).*

We must remember that in the UTTER ALLNESS, *subjects* always become *objects.* All *subjects* other than the ONE AND ONLY SUBJECTIVITY (the INFINITE SUBJECTIVITY) are, paradoxically, 'BEHELD' by that SUBJECTIVITY (via the agency of 'MAYA'-the-'SEEING') and thus are 'SEEN' as particularized *objects.* Paradoxical as it seems, we must understand that 'MAYA'-instantly-Maya does the 'SEEING'-instantly-Seeing for PARABRAHMAN, who may well 'REMAIN' 'BLIND IN BE-NESS'.

Paths of Higher Evolution, the Seven

By *the Seven Paths of Higher Evolution* are meant the Seven Paths of Return by which human beings who have achieved the Sixth Initiation may ascend to higher Spheres of Labor and Service outside our Solar System altogether. These Paths are respectively:

1. The Path of Earth Service
2. The Path of Magnetic Work
3. The Path of the Planetary Logoi
4. The Path to Sirius
5. The Ray Path
6. The Path of the Solar Logoi
7. The Path of Absolute Sonship

Example: A human being may eventually tread any one of the Seven Paths of Higher Evolution, and even switch from one to another. Which Path is trodden will not be exclusively dependent upon the Soul Ray or Monadic Ray upon which the individual is found. The Original Path of Descent along the Divine Emanatory Stream may play an important role in the 'Choice' of Path.

Example: The Path of Earth Service is the one trodden by Chohans (Masters of the Sixth Initiation) Who have chosen to sacrifice the opportunity for advancement in favor of assisting the elevation of struggling humanity. The Buddha, the Christ, and a number of the Masters with whom we are most familiar, are temporarily treading this First Path.

pattern

By a *pattern* is meant an arrangement or configuration of a given aggregate (or integrous collection) of variables (energies, items, etc.).

Example: The arrangement of energies within the egoic lotus of the human being is a pattern of rare beauty and significance.

Pattern, Cosmo-Logoically-Sanctioned

By a *Cosmo-Logoically-Sanctioned Pattern* is meant a Pattern of energies in-Cosmos which is *Intended* by the Universal Logos because such a pattern exists in conformity with the Design-at-the-Beginning (which Design is essentially the 'IDEA' EXTRUDED into *objectivity* from the INFINITESSENCE).

Example: The many perversions invented by the desire-driven mind of man, far from being Cosmo-Logoically-Sanctioned Patterns, are abominations in the sight of Those who are aligned with the Original Intent of the Design-at-the-Beginning.

peace

By *peace* is meant a state of harmonization and maximal mutual potentialization of all forces and energies within a given context.

Example: Peace leads not only to the harmonization but to the mutual enhancement of all members of a group. A state of true peace, once it has developed within a group of disciples, is a reflection of the Shamballic Energy. Peace is a state of dynamic synthesis.

PEACE, the GREAT

By *the GREAT PEACE* is meant the INFINITESSENCE ITSELF, for all qualities and possibilities are *infinitely maximized, potentized* and *synthesized* 'within' IT.

Example: The GREAT PEACE is the absolutization of all possibility. In IT each possibility has already *forever* been perfected *to the infiniteth degree,* and has merged at this *apex of fulfillment* with every other infinitely perfected possibility to become (paradoxically) *identical with each other,* not only 'within' but *as* the NOUMENESSENCE.

perceivable (noun)

By a *perceivable* is meant any condition whatsoever. In a more refined sense, an *apperceivable* is any condition considered 'internal' (or relatively subtle) to the registering consciousness. The BOUNDLESS IMMUTABLE PRINCIPLE is not a *perceivable,* nor, strictly speaking, an *apperceivable.* The BOUNDLESS IMMUTABLE PRINCIPLE is the *ULTIMATE IDENTIFIABLE.*

Example: Human evolution demands that each human unit expand the capacity to register perceivables *and* 'apperceivables' until, at the threshold of the "Day Be With Us", *all* perceivables and apperceivables can be registered by the then maximally-expanded Consciousness once again identified *as* the Universal Logos.

perfect; perfection

By *perfect* is meant a relationship between a given collection of variables which guarantees the greatest optimization possible for each variable, and, as well, for the collective whole formed by the collection of variables. The only perfection reachable in-Cosmos is a *relative perfection,* which means the *optimization of acknowledgedly limited potentials.* No matter how great may be a potential-in-Cosmos, such a potential is always *limited* when compared with the INFINITE POTENTIAL, the FOUNT OF ALL POSSIBILITY. No matter how great may be the Potential of Cosmos as a Whole, that Potential, too, is *infinitely* limited when compared with the INFINITE POTENTIAL.

Example: Given the capabilities of each member of the ensemble, and the potential of the ensemble as a whole, it could be said with justification, that the performance was perfect.

By *perfection* is meant an achieved state of *optimization between associated variables.*

Example: The Universal Logos seeks that Final Cosmic Perfection that is represented by the complete fulfillment of the Design-at-the-Beginning.

PERFECTION, the

By *the PERFECTION* is meant THAT ULTIMATE BEING/NON-BEING WHO/WHICH IS as GOOD as IT CAN POSSIBLY BE. The PERFECTION is ETERNALLY,

UTTERLY COMPLETE in an infinite infinity of ways. That PERFECTION is the INFINITESSENCE of is all possible qualities.

Example: The PERFECTION can be described by ONLY ITSELF.

periodicity; periodical

The term *periodicity* is descriptive of a process which recurs, *ideally*, at exactly regular intervals, and actually, at *fairly regular*, though often inexact, intervals.

Example: All beings-in-Cosmos are subject to periodical appearance and disappearance, manifestation and abstraction. The appearance and disappearance of the Universe Itself is a periodical phenomenon.

Periodicity, Law of

By the *Law of Periodicity* is meant a 'LAW' OF THE SELF, which demands that the process of the unfoldment of consciousness through form (inevitably entailing an accompanying complexification and refinement of form) must occur in *regular cycles*. During these cycles, the consciousness informing a particular form alternatively works *through* that form, and then abstracts *from* it, later to rebuild a similar form, again followed by another abstraction, and so forth continuously, until liberation of the consciousness from all form is achieved.

Example: The process of reincarnation, whether in an atom, a human being or a Solar Logos conforms to the Law of Periodicity, as the constant, cyclic reappearance *in* form and abstraction *from* form of each type of informing E/entity attests.

permanent

That which is *permanent* is present throughout ETERNAL DURATION.

Example: Nothing in-Cosmos is REALLY permanent nor has there ever been, nor will there ever be, a permanent Cosmos.

Permanent-in-Cosmos

By *Permanent-in-Cosmos* is indicated that which (foundational to, integral to, or in some other way necessary to the nature of a given Cosmos), endures for the entire duration of the Cosmos.

Example: Every authentic Identity-in-Cosmos is, in Essence, Permanent-in-Cosmos, regardless of the many varying, evolving forms through which such an identity may express during its many involutionary and evolutionary cycles. The forms through which such authentic Identities express are, of course, *not* Permanent-in-Cosmos (though some may be very nearly permanent) such as the (Self-'Sight'-Generated) Prakritic Form of the Universal Son, or the (Self-'Sight'-Generated) Forms of the Members of the Supernal Tetraktys—all within the World of Being.

Example: The Great Archetypes are Patterns and Relationships of Energies, and some of these Patterns and Relationships are, for all practical purposes, Permanent-in-Cosmos. For instance, the Archetypes—Numbers Two, Three, Four, etc.—come forth 'shortly after' (whatever that *Really* means) the Beginning of Cosmos, and last almost until the very End. While *essences* are *always* Permanent-in-Cosmos, the *Principal Archetypes* are examples of Forms, Patterns, or Relationships contrasted with Essences,which are virtually Permanent-in-Cosmos.

permanent-in-system

By *permanent-in-system* is meant that which lasts from the beginning to the end of a given system, be that system an atom, man, solar system, constellation, galaxy, or Cosmos.

Example: The Tibetan has said that the planet Venus endures for the entire length of the Solar Maha-Manvantara. With regard to the many-leveled, multi-planed manifestation of our Solar Logos, Venus may be considered permanent-in-system, for throughout the approximately 311, 040, 000,000,000 Earth years of the Solar Maha-Manvantara, the Logos of Venus will be found in incarnation on one or other or, even, on all levels of solar systemic manifestation chosen by the Solar Logos.

phenomenon; phenomena

By a *phenomenon* is meant the expression in form of a noumenon (or archetype). The 'form' of expression of a phenomenon may be subtle indeed from the perspective of human consciousness, but it will always be more 'dense' than energy-forms found upon the level upon which the noumenon originates.

From one perspective, the entire Cosmos (even upon its highest Planes) must be considered *phenomenal,* and the only REAL 'NOUMENON' is to be found 'within' the ALL-SELF (considering the ALL-SELF as the NOUMENESSENCE). More frequently, and from a more practical perspective, the highest three subplanes of any plane may be considered noumenal and the lower four phenomenal (though on some planes the three and the four are reversed). This assignment corresponds to the familiar 'rupa' (formal) and 'arupa' (formless) division of planes. If a ten-plane system is accepted, the highest, *super-subjective* three are considered noumenal, and the lower seven phenomenal (i.e., form-related).

Example: Sharp pains in the heart are phenomena that may accompany a dramatic opening of the heart center caused by sudden access to the Love Energy of the Solar Angel. (Here the noumenon is the Energy of Love, while the phenomena are the pains experienced and any changes in the heart center which accompany such an opening.)

Planes (Systemic, Cosmic, and Kosmic)

By *Planes* are meant Dimensions in Cosmos. Planes are Really created by Consciousness. They are Mayavically-induced Self-Reflections of the Principal Archetypal Beings (in the Company of the Universal Son) and of Fohat and His Host. The Planes with which we are familiar are the Self-Reflections of Fohat and His Host (the Sons of Fohat).

Example: Planes are Fields of Space and States of Consciousness.

Example: The Systemic Planes are Self-Reflections of our Solar Logos. Cosmic Planes are Self-Reflections of Constellational and Super-Constellational Logoi. The Kosmic Planes (if they exist) are Self-Reflections of the Universal Logos and the Universal Son (and His Host).

Plenum

By the *Plenum* is meant the fullest fullness of Cosmos Itself, but *not* the FULLNESS 'resident within' the EVER-FULL, the PLENUM, the ABSOLUTE.

Example: All Treasures inhering in the Universal Logos will become available in the fullness of Time to all authentic E/entities in-Cosmos (which means *objectively available* to the Universal Logos, Itself). These Treasures, which are the entire wealth of the Fount of *Cosmic* Possibility, inhere in the Plenum. The Complete Bestowal (interestingly a Self-to-Self 'Bestowal') is accomplished at the Cosmic Consummation—the Great "Day Be With Us." Clearly the Plenum is *not* (for practical purposes) the PLENUM.

PLENUM

By the *PLENUM* is meant the INFINITESSENCE as the ABSOLUTE FULLNESS, the INFINITIZATION OF ALL POSSIBILITY, the PERFECTION. This term helps us consider the ABSOLUTE not only as the VOID or as the NO-THING, but as the EVER-FULL, the SOURCE in which *everything* 'ABIDES' in its *noumenessentialized* 'STATE'. The PLENUM *affirms* just as the NO-THING *denies.* Both are TRUE, both are apparently opposite, and both are *absolutely identical.*

Example: The PLENUM (which IS the 'NOUMENESSENTIALIZED EVERYTHING') periodically produces a Singularity which, being Single, is infinitely removed from INFINITIZED EVERYTHINGNESS of the PLENUM. That Singularity is the Universe with all Its Limitation. But as we know by now, Imperfection must manifest in order that PERFECTION may BE—*forever.*

point

By a *point* is meant an idea or mental construct used primarily as an indicator of position or origin. A point is not even the smallest conceivable definite something, for such a something would *still* have mass and dimension (even if such mass and dimension could not *actually* be found within the parametric limits of a particular Cosmos). As a symbol, a point is the *archetype of finitization* (but, a Real point, being forever indefinite, is not quite finite).

By a *point* is meant the *seed* of Objectivity.

By a *point* is meant that which both *does* and *does not* exist. A point has no *actual* place in the World of Illusion (where everything is measurable and definable, for a point is neither measurable or definable) and no REAL 'place' in THAT ('in which' no-*thing* is distinctly specifiable, for a point is distinctly specifiable as existing, even if not measurable). A point is a transitional 'something' between BE-NESS and Existence.

Example: A point drawn on a piece of paper is not a point but a *symbol* of a point. Any drawing is an *actuality* and has mass and three dimensions (however slight they be). The closest approximation to a point is to consider a point as an *idea,* though even an idea may be considered to have dimension and substantiality. Any attempt to represent a point in form (no matter how subtle the form) is *actual* and, thus, Really, a misrepresentation of it, for actual representations of a point all have measurable dimension.

Example: Even thought has dimension (and mass of a kind), and hence the thought-image of a point is not a true point. A true point cannot be imagistically conceived. The "thing-in-itself", that is the *Real point,* is forever elusive.

Example: A point can perhaps be understood as an *infinitesimal* (or 'infinitesimalizing'). Neither is Really definite or quantifiable; they both *are* and *are not.* They are *'definite indefinites'* or *'undefinable definites';* neither has *dimension* and yet they *exist.*

Example: Every object becomes a point if *seen* from an 'infinitizing distancing' approaching infinity. A truly *infinite* distance would reduce a point to *nothing.*

Example: Whether a 'something' is a Real point or a virtual point all depends upon the Point of View from which it is 'Seen'.

Example: Every *thing* is a point when seen from the proper perspective.

Example: All things are points when seen from the 'Infinispectivizing'. Just as points both *are,* and yet, *are not,* so things both *are* and yet *are not.*

Example: To the Infinite Subject, the Infinite Object is as much a point as an infinite expanse (thus an infinitudinous *point);* to the Infinite Object (dowered with Subjectivity because It *is* the Infinite Subject) the infinite Subject is as much a point as it is an infinitude of 'See-ness' (thus also an infinitudinous *point).* The *relationship* between the Infinite Subject and the Infinite Object (both, *points)* takes place at a single infini-dimensional point, and, thus, is *boundless.*

Example: All points forever coincide—intermittently.

Example: There are only as many points in Cosmic Space as there are Self-Objectifications of the Universal Logos. Within the Point called Cosmic Prakriti there are no infinitude of points—only those points that are Self-'Seen' Objectivities.

Example: A point devours extension at infinitizing speed. A point insists on remaining a point without the interfering 'help' of extension.

Example: Points become infinitizingly extensive when 'seen' from an infinitesimalizing distance which converges upon (but never 'reaches') zero. Just as all definite 'things' (no matter how vast) or 'extensions' are reduced to points at distances receding towards the infinite, so the opposite is true. All such points reveal their 'extendedness' to the perceiver who approaches ever 'closer'. Thus do points become *fields* and even infinitizing fields, with persistent approach. This means, perhaps, that there are 'de-*finite'* points and 'in-*de-finite'* points. Perhaps the only *true* point, is the *indefinite point* which is the true infinitesimalizing. All things can be 'perspectivized' into tiny points which are *virtual* infinitesimals, but, because these things are *definite,* the points that they become (through *infinispectivization)* are *virtual infinitesimals* and not *true infinitesimals* and, thus, equally, virtual points and not true points. This means that such tiny points are so small as to be mistakenly thought to be infinitesimals, but because the points are definite, they *cannot* be infinitesimals.

Example: One cannot properly speak of things occurring *at* a point in Space but only *in* a point in Space. Points are things, objects. Where there are no things/objects, there are no points. Does it make more sense to say something is occurring 'at' a thing, or 'in' a thing?

Example: Things are points. Points are 'holes' in Space. Things, therefore, are 'holes in Space'. Every 'hole' in space is a Point of View—appearing as a Point is the 'distance' but becoming a Point of View upon approach and identification. A 'hole' in Space is an 'image of a thing' within the Image of a Field.

Example: Every point is also a *point of view*, however 'blinded' that view may be.

Example: Enter a point and find 'space'.

'POINT'

By the *'POINT'* is meant the First 'CHANGE' in IT, the first 'STATE' that is not *the* ULTIMATE 'STATE'—i.e., the 'STATE' of SELF-'PREOCCUPIED', 'INFINIDENTIFIED', ALL-IN-ALLNESS. No E/entity-in-Cosmos (strictly *as E/entity)* Now knows what that 'STATELESS STATE' *IS*.

The 'POINT' Is the First 'RADIATION'. That 'POINT' becomes the Super-Cosmic Self/Point. That Point is *'infinified'*—located everywhere and nowhere. The FIRST 'POINT' is formed by the 'FLASHING FORTH' of the 'RAY' of the ABSOLUTE. The 'POINT', in fact, *is* the 'FLASHING FORTH' of the 'RAY' of the ABSOLUTE. The 'RAY' is SELF-'SIGHT', the first SELF-'REFLECTION', (or, more accurately, 'ESSE-SPECTIVIZATION'—a kind of cross between 'BE-ING' and 'SEE-ING').

The 'POINT' instantly 'BECOMES' the Infinified Point and, thus, we have the Pre-Cosmic Establishment of an 'Infinified Point of View'. The Infinified-Point is equivalent to the Infinite Subject, Who, at this phase in the Pre-Cosmic Process, is distinct from the Infinite Object. Through It (the Infinified Point), Consciousness first comes to exist as an apparently separate, stable Pre-Cosmic Factor. Consciousness, as the relation between Subject and Object, did not exist 'within' the INFINITE SELF—*except* at the 'MAYAVIC MOMENT' of 'FLASHING FORTH' that signaled the *apparent* 'downfall' of PERFECTION, and which 'BIRTHED' the 'EVANESCENT INFINITE TRINITY' (in which the 'PROTOTYPES' of Super-Cosmic Subject/Object/Consciousness 'INHERED' for the briefest instant before being 'EXPELLED' from SUPER-Cosmos into Super-Cosmos).

Example: The 'POINT'-instantly-Point may be considered the First Finitizing of the INFINITE (even though that Point is 'Infinified'). Still, it represents a 'transitional step' towards Finitization. Remember ever, that although the INFINITE *appears* (in several steps) to Finitize into a Cosmos (i.e., to *cosmify)*, IT *remains* continuously the INFINITE forever. Paradox!

Example: A more dynamic way of describing the Infinified Point appears in the formula, the 'POINT'-instantly-Point. This formula represents the greatest of all *falls.* The formula reveals to us the truth that to 'ARISE' 'within' the 'SELF' is be instantly 'EXPELLED' from the HOMOGENEITY of INFINITE SELFHOOD.

Example: The 'POINT' suddenly (without temporal preparation ... for does a *beginning* take *time?)* 'ARISING' in THAT (i.e., 'in' SUPER-Cosmos) is suddenly the 'Point' (the Infinified Point) in Super-Cosmos (with SUPER-Cosmos remaining 'POINTLESS' as ever). The *arising* and *ejection* or the 'POINT' may be (if not quite simultaneous), at least *virtually* so. How long does it take to for the 'POINT' to be 'EXPELLED' once it 'ARISES' in 'no time'? Perhaps by the 'time' it 'HAPPENS' again, we-as-We-as-WE-as-**I** will 'KNOW'.

Example: In the World of Duality (which even includes Super-Cosmos {but not SUPER-Cosmos}), any Point (of View) is regarded as *different* from that which is 'Seen' *from* the Point. For example, the Infinified Point is regarded as distinct from that which is 'Seen' *from* the Infinified Point (namely Mulaprakriti).

Really, however, there is but one 'POINT' in all the UTTER ALLNESS. Mulaprakriti is just another Point (which is identical to the Infinified Point that 'Sees' Itself as It). These 'two' Points (i.e., the Infinified Point/Infinite Subject and Mulaprakriti/Infinite Object) are really the *same* Point, but as that Point has no 'Seen' 'Background' the Point 'appears' as an Infinite Homogeneous Field.

Example: The 'POINT' is the FIRST 'ILLUSION'.

Point (in-Cosmos and in-Super-Cosmos)

By a *Point in-Cosmos* is meant the most authentic Representative of NOTHING in the World of Something.

By a *Point in-Cosmos* is meant, Nothing 'Seeing' a Point 'within' Itself as 'Something'—i.e., the *essential nothingness* of Subjectivity 'Seeing' its 'Point-Content' as the *somethingness* of Objectivity. Since, however, that which 'Sees' is *essentially no-thing,* it turns out that that which is 'Seen' is, *essentially,* also *no-thing.* If a Point of View is Essentially, no-thing, so is any Point 'Seen' from that Point of View (i.e., 'Seen' Point).

By a *Point in Super-Cosmos* includes both the Infinified-Point as Subject or the Infinite Object as Infinified 'Seen'-Point. The Condensing Subject and Condensing Object can also be considered Points in Super-Cosmos. If Points appear within the Vision of the Condensing Subject as It 'Sees' the Condensing Object, these Points, too, are Points in Super-Cosmos.

Example: Every Point in-Cosmos is a Self-Reduction of the Point of View that has 'Seen' it. For instance, the Point that is the Universal Son, is a 'Self-Reduction' of the Point which is the 'Self-Seeing' Father. Thus, a Point is a *reduction* of its own 'Seer'. Without a 'Seer' no Point can exist.

Example: A Point in-Cosmos is a *'transitional nothing-something'* between NOTHING and 'Something'. NOTHINGNESS-as-Nothingness, when 'Self-Seen', is *apparently* 'Something' (in consciousness) but REALLY 'Nothing'.

Example: When a Point-in-Cosmos (any cosmic E/entity) 'Sees' Itself as a Whole, It 'Sees' Itself as a Mother-Field. Out of this Mother-Field will emerge 'Sons', which arise through specialized 'Self-Sight'.

Example: Any E/entity is a Point. Any 'Ray' of the ABSOLUTE is a Point.

Example: The Point is, at once, any-thing which is 'Seen' and that from which anything is 'Seen'. Since all 'Seeing' takes place at the Point (which annihilates the *space* in which to *see),* 'Seeing' (i.e., 'Maya') doesn't REALLY happen at all—another argument for the illusory nature of every-thing 'Seen'—and what other kinds of 'things' *are* there?

Example: The Mystery of the Point, is the Mystery of Origin.

Example: A Point in-Cosmos, is a Subject 'Seen' as an Object. Points cannot be measured and yet, they cannot *not* be measured. Because they are, Essentially, Subjective (i.e., Projections of Subjectivity), they do not belong to the Realm of Maya, and cannot be measured, but, because they are, also 'Seen', and are, thus, Objective, they *do* belong to Maya (the 'measurable').

Example: All Points-in-Cosmos are 'Sons'. The Cosmos is filled with Points (is filled with Sons) but all Points *are* the One Universal Point, and all Sons are the One Universal Son Who *is* the Father? All Points in-Cosmos coincide and are both 'Something' and NOTHING.

Example: Every Point-in-Cosmos (or Point in Super-Cosmos) is both a Subject and an Object. The 'POINT' is a 'SUBJECT'. Is that 'POINT' an 'OBJECT' too?

Example: Everything that is 'happening' in Cosmos is 'happening' 'within' what, from an Infinitizing Perspective, is a 'Dimensionless-ing' Real Point, the One Great Point-in-Cosmos. Since the One Point-in-Cosmos is, ESSENTIALLY, an Illusion, all things that are apparently 'happening', are 'happening' 'within' Illusion. Happenings 'Seem' to have extension, but, Really, they are Point-Bound, and are all 'happening' 'within' the same Point *which is* Space and at the same Point *which is* Time just born as Event.

Example: Can Points exist 'within' Points-in-Cosmos? Or 'within' Points-in-Super-Cosmos? We might as well ask, can 'Rays' exist within 'Rays'? Every life unit *is* a Point, and is 'generated' by a point. If, for instance, the Son (Who is a Point) can exist 'within' the Father (Who is *the* Cosmic Point), Points can exist within Points.

Of Points it can be said:

1. They are all identical and thus (for visual purposes) 'super-imposed' upon each other.
2. They can, although dimensionless, exist 'within' each other (from the Emanational Perspective, as the Son does 'within' the Father).
3. They can seem to occupy space and thus, seem, non-superimposed, non-inhering, and non-identical, but this would alter with 'Infinispectivizing'.

Example: Every ultimate particle/event is (like all other cosmic things) a 'Point'-in-Cosmos, and hence (upon approach and identification) a "Point of View", but in this case, a "Point of View" revealing *nothing* 'within' itself, for there is no lesser thing in Cosmos to which it is the *aperture.* Every Point in-Cosmo (except for the ultimate particle/event) has other Points *hierarchically 'nested' within It.*

Point, Condensed

By the *Condensed Point* is meant the *retraction* of the Infinified Point (which 'Sees' the Infinite Object from an Infinite Number of Points of View) into *a* Point of View which 'Sees in Spherical-Potential-Multi-Dimensionality' (as it were). This Condensed Point 'Sees' as if from within a *bounded potentially multi-dimensional sphere* which (with the onset of that *bounding* 'Sight') has become transformed into Cosmic Prakriti, and not Mulaprakriti. The 'Sphere' 'Seen' from this Condensed Point is Its own Field of Consciousness.

Note that this 'Sight' from the Condensed Point is *not infinidimensional* as is 'Sight' from the Infinified Point. The 'Sight' is only correlated to the *birth* of a certain number of dimensions (i.e., dimensions to be Self-Reflectively 'Created'/Emanated/Become during the Process of Cosmic Unfoldment). The Condensing of the Point (which could also be called the Focusing Universal Subject) is Really the beginning of the 'Birth' of the Universal Logos from out of the Infinified Point (the Infinite Subject).

By the *Condensed Point* is meant the third phase of the Triple Point, the first phase of which is the Infinified Point. The Condensed Point as the Third Phase is preceded by the Condensing Point (second phase).

Example: The Condensed Point is the coalesced Universal Logos in relation to Its Ring-Pass-Not. Within the Ring-Pass-Not, Subject/Object Relations are between the Universal Logos-as-Condensed Point and Cosmic Prakriti as (incipient, unarticulated) *Universal* Object (i.e., what the now *specific* Universal Logos *initially* 'Sees' of Its Self, Its unarticulated Self). Outside and 'above' the Universal Ring-Pass-Not, Subject/Object relations are between Infinite Subject-as-Infinified Point, and Mulaprakriti as *Infinite* Object.

The Condensed Point is also totally pervasive within the Cosmic Ring-Pass-Not, but has a 'quality' of *'Centeredness'* not present in the Infinified Point. In a way, the Condensed Point is a *specified* Entity, whereas the Infinified Point (as Super-Cosmic Infinite Subject) cannot Really be called a *specified* Entity. It is, far more, an *indefinite, unspecified* Infinite Subjective Entity.

Point, Condensing

By the *Condensing Point* is meant a transitional Pre-Cosmic Phase between the Infinified Point and the Condensed Point. The *infinified* Vision of the Infinified Point is *moving towards* Finitization, but has not yet focused upon, or within, the Sphere of Oneness, the incipient Universal Sphere. The Condensing Point does not yet know Itself as an 'Individualized' Singularity, and many Cosmo-Possibilities (past or to come) may be 'Seen'.

Example: The Condensing Point is a Stage of Pre-Cosmic 'Subjective Orientation' in which the Infinite Subject is *becoming* the Focusing Universal Subject, but is not yet the Focused Universal Subject (or Universal Logos) Ready to 'See' (Objectively) the exact 'Part' of the FOUNT OF ALL POSSIBILITY, which It has been 'ENTRUSTED' to Objectify as a Universe.

Point, Dual

By the *Dual Point* is meant the two non-transitional Stages of 'Pointness'—i.e., first the Infinified Point and, then, the Condensed Point. The term is used as a convenience when it become too cumbersome to consistently mention both kinds of Pre-Cosmic Point.

Example: The Dual Point becomes the Triple Point once the Pre-Cosmic Stage of 'Condensing Point' (between the States of Infinified Point and Condensed Point) is acknowledged.

Example: The Dual Point is 'PRODUCED' by the 'FLASHING FORTH' of the 'RAY' of the ABSOLUTE. The resultant Super-Cosmic Point is 'Dual' in Pre-Cosmic Time, for the Infinified Point (granting an infinitude of perspectives upon the Infinite Object, Mulaprakriti or Space) ever precedes (by means of passage through the Stage of Condensing Point) the Condensed Point, which more definitely marks the Stage of Cosmo-Creation through the isolation (by an Act of Changing Pre-Cosmic Subjective Perception) *from* Infinite Mulaprakriti *of* Cosmic Prakriti. Between the Stage of Infinified Point and that of Condensed Point lies what might be called the *Condensing* Point. While this Process may appear from this description as a Process of Subject/Object Interplay, it is Really *deeply internal* and Super-Cosmically *Psychological*

Example: Both the Infinite Subject and the Infinite Object are a *Single Dual* Point. That 'Point' (arising from 'POINTNESS') is the Point equivalent to *both* Subject and Object; as Subject It is a Point of View (Infinified); as an Object, It is the Point 'Seen' from the Point of View (also Infinified). The paradoxically Dual View from the Point is of an Infinite Homogeneous Field.

Point, 'Infinified'

By the *Infinified Point* is meant the Infinite Point of View of the Infinite Subject that is first 'GENERATED' (or 'ARISES' *in* THAT) when the 'RAY' of the ABSOLUTE 'FLASHED FORTH' at the dawn of Pre-Cosmic Activity, and then instantly 'TRANSFERRED' into the Super-Cosmic Domain of the Infinite Subject. The Infinified Point is the 'Infini-directional' Point of View of the Infinite Subject that 'Sees' (in the 'Great Nowhere' of the Pre-Cosmic Process) naught but the Infinite Object, Mulaprakriti. Mulaprakriti is the Infinite Mayavically-Revealed Potential (in relation to an infinitude of Cosmoses) for the appearance of an infinitude of objective/material forms, (all of which would be reflections of the *'infinitized everythingness'* 'INHERENT' in the FOUNT OF ALL POSSIBILITY and, subsequently 'inhering' by proxy in the Infinite Subject/Infinified Point).

What is 'Seen' through the Infinified Point? *Perhaps*, in 'time', a 'Vision' of a more *articulated everythingness* than is 'IN-PERIENCED' by the INFINITE SELF 'within' the FOUNT OF ALL POSSIBILITY ('wherein' *all possibilities are infinitized*, and hence, non-articulated). The infinite potential of Maya to *reveal through reflection* not only the *noumenessentialized infinite possibility* of the INFINITE SELF as Homogeneous Mulaprakriti, per se, but, *also* to reveal (within Mulaprakriti) through reflection *every articulatable thing* within the FOUNT OF ALL POSSIBILITY may be 'Realized' by the Infinite Subject/Infinified Point (... who knows?). This would be an inconceivably *huge* Realization. A wholly articulated Vision of all of specified Objectivity that has *ever been born* of Mulaprakriti may, on the other hand, be the Gift of Summation to the abstracting Universal Logos (appearing in the mode of the 'Re-Infinitizing Super-Cosmic Subject') 'returning' to the Super-Cosmic Infinitude It never Really left.

The question is, Does any 'Eye' ever 'See' the infinitude of formal possibility that has been actualized in all Cosmoses forever—and 'See' it under the 'Eye' simultaneously? Such a Vision, no matter *how* it occurs *(if* it occurs) would almost certainly be Super-Cosmic. Just as the personality cannot see in one incarnation all the experiences that its generating Monad has undergone, so the Universal Logos (as Personality of THAT) cannot 'See' all the Cosmic Incarnations that have ever been.

Once the Infinified Point 'Becomes', as it were, the Condensing and Condensed Point, (the Infinite Subject thus 'Becoming' at first Focusing Universal Subject and, then, the Focus<u>ed</u> Universal Subject or Universal Logos) Mulaprakriti is no longer 'Seen' in Its Potential to be a Reflection of mostly unarticulated and (perhaps?) articulated Infinitude, but only in terms of the specific SELF-'EXTRUDED' 'IDEA' that must be *rendered objective* as the Universe-to-Be. Infinite Self-Perception has changed to Finite Self-Perception, and thus, Mulaprakriti has been transformed into Cosmic Prakriti.

Example: The First 'CREATION' *from* THAT ... the First 'EXTRUSION' *from* the INFINITESSENCE ... and the First ACTLESS-'ACT' of the ABSOLUTE ... *IS* the 'FLASHING FORTH' of the 'RAY' of the ABSOLUTE, which is the First 'MAYA'-instantly-Maya, and the First 'CONSCIOUSNESS'-instantly-Consciousness, and the First 'SEEING' (albeit by *proxy)* of the ABSOLUTE by ITSELF, and is the 'ARISING' of the 'POINT-instantly-Point (the <u>Infinified Point</u>), and, thus, the Pre-Cosmic Arising of the Infinite Subject and Infinite Object—the former ever *beholding* the latter.

This Model of Pre-Cosmic 'Spontaneous Creation' should be studied carefully. Understanding is complicated by the likelihood that all these *results* are Really *one 'EVENT'* that can be conceived as occurring instantaneously, quantum-like, in 'No *Time* as all', since Time is 'Born' (again) only at the 'FLASHING FORTH'.

Example: The Infinite Subject and Infinite Object can be thought of as existing 'together' at a dimensionless omnipresent Point (the <u>Infinified Point</u>) in Super-Cosmos (but *not* 'in' *SUPER*-Cosmos).

Example: Is the <u>Infinified Point</u> an Object as well as a Subject? In a way, It is the 'Aperture of Sight' for the INFINITE SELF-in-Super-Cosmos and, eventually, in Cosmos. Super-Cosmically and Cosmically, all 'Seeing' takes place through that Aperture and Its subsequent *reductions-in-scope.* So, the <u>Infinified Point</u> in a way is always identified with Subjectivity. Yet, because It *is* as Point (and thus *exists)* it must, in some way, be an *Object.* Every-*thing* that 'ex-ists' is an *object.* Only the INFINITE SUBJECTIVITY does not *'ex-ist'* (though, in a way, IT *IS)* and is *not* an *object.*

Point, Real

By a *Real point* is an object 'Seen' from sufficient 'distance' to be *infinitesimalized.*

Example: Any object (whether relatively huge or tiny) is a <u>Real Point</u> when 'Seen' from an 'Infinispectivizing'. From such an ever-distancing 'View', any object becomes so increasingly small, and so persistently indefinite as to have *no definable* dimension, and yet it exists, so it is *not nothing.*

Example: A <u>Real Point</u> is a *'thinged'* vision that a Subject may, theoretically, have of Itself—provided that vision takes place from an ever receding, infinitizing Point of View (an Infinispectivizing). Usually, when a Subject 'Sees' Itself 'as a whole', a Field of Consciousness is created (a kind of *global spherical space,* eventually to be filled with Self-Seen 'contents of consciousness'). If a Subject could 'withdraw' at infinitizing speed from that Field (thus 'Seeing Itself from 'outside' Itself) the perception of a point would appear and the perception of the Field would end. Points and Fields are inter changeable depending upon perspective.

point of reference

By a *point of reference* is meant any definite position in (so-called) Space from which to measure the motion or position of any other *thing* in Space.

Example: Is there such a thing as the *center* of the Cosmic Sphere or Ovoid? If such a Point exists (as dimensionless as that center Point would necessarily be) it will become possible to measure the 'position' of every other entity/event in Space with reference to that point of reference. The conception of such a Central Point of Reference in-Universe becomes problematical if the Universe is discovered to be 'shaped' like a multi-dimensional "Moebus Strip". Such a "Strip" would allow one to travel in any direction and return to the place of origin without turning back and retracing one's 'steps'.

Point of Universal Potential

By the *Point of Universal Potential* is meant that pre-Cosmic 'CONDENSATION' into Active Potency of the INFINITE SELF—first as the Infinified Point and then as the Condensing and Condensed Points. The Point of Universal Potential is more *properly* identified with the SELF-as-Condensed Point, because the *potential* suggested is *Universal* and *not 'infiniversal'.*

Example: The Process of Generating a Cosmos begins when the 'RAY' of the ABSOLUTE 'FLASHES FORTH', *becoming* first the SELF-as-Infinified Point and then, finally, the SELF-as-Condensed Point, or the Point of *Universal* Potential.

Point, Triple

By the *Triple Point* is meant the Pre-Cosmic finitizing development of 'POINTNESS' consisting of the Infinified Point, the Condensing Point, and the Condensed Point.

Example: The Triple Point is a Triple State of Subjectivity, consisting of the Infinite Subject, the Focusing Universal Subject, and the Focused Universal Subject, and, of course, their Self-Reflections as Infinite Object, Focusing Universal Object, and Focused Universal Object.

Example: The Pre-Cosmic Trinity of Points, the Infinified Point, the Condensing Point and the Condensed Point, together might be called the Triple Point.

point, tunnel

By a *tunnel point* is meant a point of light in the utter blackness conceived as the light at the end of a tunnel. A point is objectivity; light, too, is objectivity. A point is no *reified* point just as the light at the end of the tunnel is no real thing but, actually, an aperture to another dimension. It is as if one always has to pass through a point in order to enter a new dimension.

Example: The concept of the tunnel point conveys the idea of passage through a 'thing which is not a thing', through a 'something' which appears like a definite object, but is really an aperture or a 'hole'.

Example: The tunnel point indicates the ESSENTIALLY non-objective nature of all 'things', for all 'things' are *points*. A Perceiver is REALLY, NOTHING. What the Perceiver

'Sees' comes into Its own view as a Point, an apparently objective 'something', but when the Perceiver *approaches*, perceptually, the apparent substantiality of the Point disappears. It, too, like the Perceiver, turns out to be nothing, and only an aperture through which more apparent 'somethings' can be seen. Eventually, it turns out that the Object is no-thing, just as the Subject is no-thing.

Thus the Father, Who is ESSENTIALLY no-thing, apparently 'Sees' Itself *reduced* as an *objective* Son, Who, (when the Father 'arrives' 'there', by 'journeying' 'into' the Son) also turns out to be (not an object) but another Point of View (i.e., a Subject). Thus, we come to the amazing conclusion, that every object is *not Really a 'thing' but a point of view.*

Example: The darkness of the Mother is the Father 'Seeing' Himself as He Is, 'Seeing' His unarticulated Beingness. The appearance of the Tunnel Point, is the Father 'Seeing' Himself 'in' the Mother ('Seeing' Himself in 'reduction') as the Son—Who, indeed, like all 'Suns', appears from a 'distance' as a point of light. For the Father, the Son is the seeming-*Object* (Really-*Aperture)* to still more specified objectivity. The Son, Who is a 'Sun', always first appears like a "light at the end of a tunnel."

point, virtual

By a *virtual point* is meant the reduction of a definite thing to a quantifiable tinyness, no matter how small. Virtual points have a definite size which could be, (if Cosmic Parameters allowed) millions of times smaller than an ultimate particle/event. Virtual points however, are definite things, whereas Real points are indefinite things. A virtual point is created when the perceiver of a thing (receding towards infinity) *stops,* and *looks* at the thing being perceived from a *definite* distance. From that definite distance, any thing will have a definite dimension, no matter how tiny, or even, Cosmically non-allowable the dimensions may be.

It must be realized, however, that all virtual points can be turned into true points, by continuing the recession towards infinity (i.e., by continuing the 'Infinispectivizing'). The point, or infinitesimalizing, is more a *process* than a definite thing. As the perceiver of a thing (optics aside) approaches the limit of infinity—*approaches*, but never *reaches it*, the point-thing becomes increasingly an infinitesimal (converging upon zero magnitude) but never *no-thing.*

Long before the limit of infinity is approached (depending upon the size of Cosmos under observation) it is possible to *perspectively transform* the entire Cosmos and all things within it to one virtual point, the definite dimensions of which will become smaller, or smaller still, depending upon where the perceiver decides to stop (but a virtual point it will be if the observer *stops).* Once the perspective is reached (and a stop made) from which distance all articulatable points in Cosmos seem to merge seamlessly into the one virtual unarticulable point, "the *point* has been *proved!*" It is then understood that all things are one thing. One can rest content with viewing Cosmos as a Virtual Point.

If it needs to be proved that *all things forever are One Thing, in One Place at One Time,* then Infinispectivizing will have to be maximally pursued, and the Points that appear and merge will be Real (rapidly infinitesimalizing) Points. If it need to be proved that *all things are NOTHING,* then the INFINISPECTIVE will have to be achieved rather than a Converging Infinispective (i.e. Infinispectivizing), and the *virtual* point-become-*Real* point will have to become *No-Point,* by being made to "vanish into NOTHINGNESS". How? Let's keep our eyes open during the next Universal Pralaya!

Example: This means, perhaps, that there are 'de-*finite*' points and 'in-de-*finite*' points. Perhaps the only *true/Real* point is the *indefinite point,* which is the true infinitesimal. All things can be made into tiny points which are *virtual* infinitesimals (i.e., incorrectly *named* infinitesimals) but, because these things are *definite,* the points that they become (through *terminated infinispectivization)* are *virtual infinitesimals* and not *true infinitesimals* and, thus, equally, *virtual* points and not *true* points (i.e., Real points).

'POINT'-instantly-Point

By the *'POINT'-instantly-Point* is meant the instantaneous 'ARISING' of the 'POINT' 'within' the ABSOLUTE, and its immediate 'EXPULSION' from 'residence' 'within' the ABSOLUTE by its very 'ARISING'. No sooner does it 'ARISE' within the ABSOLUTE than it becomes the 'Infinified Point' within the Super-Cosmic or Pre-Cosmic World, for the ABSOLUTE-as-ABSOLUTE cannot 'ABIDE' any violation of its incorruptibly HOMOGENEOUS NATURE.

Example: The 'POINT'-instantly-Point is exactly equivalent to the 'RAY' of the ABSOLUTE 'FLASHING FORTH' as the First 'EVENT'/'ACT' of the INFINITE SELF.

Point-within-the-Sphere

By *the Point-within-the-Sphere* is meant the incipient Universal Logos. The SELF-as-Infinified Point (generated by that Agent of the SELF, Pre-Cosmic 'FOHAT'—He Who 'FLASHES FORTH' as the 'RAY' of the ABSOLUTE) once It has become the SELF-as-Condensed Point and has *Created* through an Act of Super-Cosmic Task-Intent, Self-Perceiving Consciousness the Bounded Homogeneous Object/Sphere now called Cosmic-Prakriti, has in fact *become* the Universal Logos.

Example: The Universal Logos takes Its Lineage directly from the INFINITE SELF. The Universal Logos is Really the SELF-as-Condensed Point within the Bounded Mulaprakriti called Cosmic Prakriti. The Universal Logos is the Point-within-the-(Cosmic Prakritic) Sphere.

A question might arise, Is Cosmic Prakriti as homogeneous as Mulaprakriti or less so? Further, may it not be that the SELF-as-Infinified Point, in pervading Mulaprakriti, thereby reduces the *density* of Mulaprakriti?

This kind of question must be examined carefully. What is meant by the utter density of Mulaprakriti is Its Nature as a faithful Reflection of *utterly dense INFINITUDE?* The only thing that can reduce this so called 'density' is the Will of the Infinite Subject to 'See' Itself as *less than Infinite.* If this Will is carried out (and it always is in the Pre-Cosmic Preparation of a Cosmos) then the Reflection which Is Mulaprakriti will (through 'De-Infinifying') begin to become multifarious and finite, rather than homogeneous and infinite. In this way Mulaprakriti will become less 'dense' through interplay with the Infinite Subject-'Become'-Focusing Universal Subject, and will, Really, no longer be Mulaprakriti, but incipient Cosmic Prakriti.

At a certain 'Phase' of the Pre-Cosmic Process, Mulaprakriti is pervaded ubiquitously and infinitely by the SELF-as-Infinified Point. This Really means that the Infinite Subject 'Sees' Itself as Infinite (the Vision of Its Infinitude *being* Mulaprakriti).

There comes, however, another Phase in the Pre-Cosmic Process in which the SELF-as-Infinified Point 'Condenses', as it were, to a Point of Concentration (the SELF-as *Condensed* Point, the *Focusing* Super-Cosmic Subject) even as It simultaneously focuses upon the Self-Perceptual Bounding of Mulaprakriti (thus creating the Ring-Pass-Not of the Universe-to-Be). (The Bounding of Mulaprakriti, is, from the Psychological Perspective, the *refusal* to 'See' the Mulaprakriti as a Reflection of the Infinitude of the SELF, and instead, as only a specific kind of *limited* Reflection, appropriate to the formation of the Cosmos-to-Be.)

In doing this-*bounding* Mulaprakriti and *perceptually forming* (through the Narrowing of Its Consciousness) a Ring-Pass-Not for the coming Universe-the SELF-as-Condensing Point develops both a *centered* Point of View and a *bounding* Point of View. Let us remember that in creating finite form, the Point of View of the Creator (which is definitely Mayavic) *is* the Creative Potency. With the Point of View of the SELF-as-Condensing Point— properly centered and spherically focussed (i.e., the Point within the Sphere)—the 'Time' for Real (and more humanly recognizable) Cosmic Creativity has arrived.

'POINTNESS'

By '*POINTNESS*' is meant the 'ARISING' in IT of 'DIFFERENCE', i.e., the 'ARISING' in IT of the 'INFINITE TRINITY'-instantly-Infinite Trinity. This means that *suddenly,* "in (probably virtually, no Time at all", or for an *infinitesimal instant)* there IS 'within' IT (enduring for, probably, virtually "no Time at all" or for an *infinitesimal instant):*

1. The 'EVANESCENT INFINITE SUBJECT'
2. The 'EVANESCENT INFINITE OBJECT' ('THAT' which the SELF 'SEES')
3. 'EVANESCENT INFINITE MAYA' (the SELF-'REFLECTIVITY' or the 'SEEING' by means of which the 'ARISING' *now-distinct* INFINITE SUBJECT 'SEES' ITS 'THATNESS')

These three 'BECOME' (in, probably, virtually "no Time at all", or in an *infinitesimal instant):*

1. The Pre-Cosmic Infinite Subject
2. The Pre-Cosmic Infinite Object
3. Pre-Cosmic Maya (Infinite Consciousness, or Consciousness of Objective Infinitude)

By '*POINTNESS*' is indicated a general term standing for the FIRST 'DISTINCTION' 'in' THAT, which instantly 'BECOMES' a Pre-Cosmic Reflection of Itself, namely 'Pointness'. The dynamics are symbolized as 'POINTNESS'-instantly-'Pointness'. The formula 'POINT'-instantly-Point is equivalent.

Example: 'POINTNESS' is the 'ARISING' of *objectivity* ('ESSE-SPECTIVIZATION') 'in' THAT.

posit (verb)

By the term *posit* is meant a process of asserting the existence of something or of 'putting forward' an idea which declares the existence of something.

Example: One may posit many ideas supposedly descriptive of the 'NATURE' of the INFINITE SELF, but all of them will be found, on close examination, to be "less than NOTHING"!

Example: That which is posited is never *ultimately* REAL. The Universe Itself is a *positing* of the INFINITE SELF, and is, consequently, an Illusion.

position

By the term *position* is meant the location in Space of one variable relative to another or to others. Position cannot be determined without mapping the context. Perhaps one could have said, the 'location *as* Space of one variable ... etc.'

Example: The position in Space of an E/entity is intimately related to the function of that E/entity within the Cosmic System.

Example: The close relation between the terms 'position' and 'posit' deserves careful pondering.

Position in-Cosmos, Cosmo-Parametrically Sanctioned

By a *Cosmo-Parametrically Sanctioned Position in-Cosmos* is meant a configuration of cosmic variables/items which is both possible and permissible given the Cosmic Parameters at the Beginning—i.e., the SELF-'DESIGNED'/'CHOSEN' Limitation under which a particular Universal Logos must labor.

Example: Not all possible relationships within a given Cosmos are *sanctioned* by the Universal Logos. This means that, in order to conform to the Design-at-the-Beginning E/entities and items in-Cosmos can only occupy Sanctioned Positions with respect to each other. If, for instance, the feet always try to touch the sky instead of treading the ground, the feet are not occupying a Cosmo-Parametrically Sanctioned Position in-Cosmos.

Position in-Cosmos, Cosmo-Parametrically Un-sanctioned

By a *Cosmo-Parametrically Un-sanctioned Position in-Cosmos* is meant a configuration of cosmic variables/items which is either impossible or highly undesirable given the Cosmic Parameters at the Beginning—i.e., the SELF-'DESIGNED'/'CHOSEN' Limitation under which a particular Universal Logos must labor.

Example: Cosmo-Parametrically Un-sanctioned Positions in-Cosmos between items and E/entities produce monstrosities and prodigies of nature that inspire almost universal horror and revulsion. Perversions of various kinds come in the category of Cosmo-Parametrically Un-sanctioned Positions in-Cosmos.

possibility

By a *possibility* is meant a relationship, configuration, combination or pattern which is potentially capable of being *actualized* in one or other of an infinite series of Cosmoses.

Example: A possibility is that which *can happen* in a specified context. While it *is* possible that a human being may learn to walk on air, it is *not* possible (if the Cosmic Design-at-the-Beginning is to remain intact) for all cycles in-Cosmos to suddenly reverse their motion, and begin moving backwards over the 'ground' just covered. Such a cataclysmic perversion would require an Act of Almighty God—however, God-in-Universe is not *Almighty* because SELF-'VEILED', and, besides, God-the-Universal Logos obeys His Own SELF-as-Self-Designated Rules, so His cooperation in such an extraordinary *reversal* of Original Intent is not to be expected!

Possibility-in-Cosmos

By a *Possibility-in-Cosmos* is meant a relationship between Cosmic Variables which is *possible*—i.e., that which can *actually happen* within the SELF-'DESIGNATED' Parameters of the particular Cosmos in question.

Example: Not all 'POSSIBILITIES' 'within' the INFINITESSENCE are Possibilities-in-Cosmos. In fact, 'within' the INFINITESSENCE there are an infinitude of 'POSSIBILITIES' that are non-actualizable in any given Cosmos. One such is the possibility that there be no Cosmoses!

potential (noun & adjective)

By a *potential* is meant not only a possibility which *can* happen, but a possibility for which the forces necessary to induce the happening are gathered and ready to act.

Example: Not all destined patterns in the Design-at-the-Beginning are potential during this century or even during the coming millennium. Within the Original Intent abides the potential (even a *Plan)* for all human beings to achieve Mastership, but that potential is *not* actualizable at this time. First, a great number of human beings must begin to tread the Paths of Aspiration and of Discipleship. That *treading* is an actualizable potential during the Aquarian Age but *not* the achievement of Mastership upon a broad scale.

POTENTIAL, the LIMITLESS

By the *LIMITLESS POTENTIAL* is meant the FOUNT OF ALL POSSIBILITY, the INFINITESSENCE.

Example: The Potential of every Cosmos is *limited* when compared to the infinitude of *infinitized* 'POSSIBILITIES' 'RESIDENT' 'within' the LIMITLESS POTENTIAL.

Power of Aggregation

By a *Power of Aggregation* is meant is meant a unique *quality* which expresses Itself through a specific *quantity* (this quantity being formed by the aggregation of lesser quantities which inhere within it).

By *the Power of Aggregation* is meant the Power of Number, and especially the Power of Number to configure prakriti. What this Really means is that Number has the Power to command or guide Fohat.

Example: The Power of Aggregation for the Number Four (composed as It is of the aggregation of the Numbers One, Two, and Three) is a *unique quality* differing, for instance, from the Power of Aggregation for the Number Six (composed as It is of the aggregation of the Numbers One, Two, Three, Four, and Five). The other mathematical routes to reaching a certain Number (for 2 x 3 = 6) also tell us something about the Power of Aggregation generated by specified *arithmetic interplay* between the Essential Numerical Entities.

Example: The Power of Aggregation of the Number Four compels the E/entities which come under its sway to cohere in forms exemplifying the Quality of Fourness.

prakriti

By *prakriti* is meant *matter.*

Example: Prakriti is objectivity. Differentiated prakriti is/provides the 'raw materials' which makes possible the creation of different types and forms of objects. Prakriti cannot exist without an Observer. Essentially, Consciousness (Maya) 'creates' prakriti just as Consciousness (through the power of Retroflexive 'Sight') creates the state of objectification and the objects which arise from that state.

Example: Prakriti as Mulaprakriti is the Mother. But prakriti often means *not* undifferentiated Root Matter, but those specificities, specific objects which arise 'within' undifferentiated Root Matter. Thus we never Really *see* Mulaprakriti (Matter Itself). What we see is prakriti—the objects that the Mother 'Shows Forth'. These objects (which all share in the Essential *Objectivity* of the Mother) *are* the Son! Thus it is, that prakriti, per se, does not appear unless the Son appears. Worded otherwise, we do not *see* Matter unless we see the Son. The words of Christ begin to have special meaning, "He that hath seem me, hath seen the Father."

Example: One who is prakritically-bound is consciousness-bound. His particular 'depth' of prakritic immersion reveals to him his own consciousness.

Example: It may be illuminative to understand the many types of prakriti as images within images.

Prakriti, Cosmic

By *Cosmic Prakriti* is meant that 'portion' of Mulaprakriti which has been *bounded* by the Attention of the Super-Cosmic-Self (the SELF-as-Condensing Point, which is the Infinite Subject 'Moving' towards specificity and becoming the Focusing-then-Focused Subject Intent on a Specific Universe), and is thus the 'Matter' of which the Universe-to-Be will be fashioned. [See references to Mulaprakriti.]

Example: The full Infinitude of Mulaprakriti *cannot* be included in Cosmic Prakriti, just as the full *subjective* INFINITUDE of the INFINITE SELF cannot be included in the Universal Logos. What this means Cosmo-Psychologically, is that one cannot see infinitudinous everything within a limited, bounded image.

Example: Cosmic Prakriti Is Mulaprakriti that has been 'Seen' by the all-pervading, Infinite Presence of the SELF-as-Infinified Point, and then *bounded* by the Self-Perception of the SELF-as-Condensed Point. The *bounding* of Mulaprakriti, which forms the Ring-Pass-Not of the Universe-to-Be, occurs from *within* Mulaprakriti. (This means that the Infinite Reflection which Mulaprakriti Is, begins to Finitize.) In order for this to happen, Pre-Cosmic 'FOHAT' in Mode Two (the Active Power of the Infinite Subject) initiates the beginning of the Process of Condensing Infinite Perception into Finite Perception. This Process (occurring with the Consciousness of the Infinite Subject-become-Focusing and Focused Universal Subject) is equivalent to the Bounding of Mulaprakriti.

With the completion of this Pre-Cosmic Development, the Universal Logos can be said to Exist within Cosmic Prakriti (the Mother Aspect of Itself). This means that the Universal Logos 'Sees' Himself as Mother, and that a Field of Consciousness/Space is Perceived which is the Matrix-Sphere in which the Universe will come to birth.

prakritic immersion

By *prakritic immersion* is meant the range of vibratory frequency which characterizes the vehicles or sheaths which 'surround' each authentic E/entity or Identity, but most significantly and tellingly refers to the vibratory frequency of the *highest* vehicles or sheaths of that E/entity. In general, a greater *depth* of prakritic immersion is characterized by a slower vibration of the prakritic vehicles through which an entity/identity expresses. A shallower depth of prakritic immersion is characterized by a more rapid vibration of the prakritic vehicles through which an entity/identity expresses. In general, the degree of prakritic immersion (especially of the highest vehicles) is directly proportional to the degree of bondage of the spirit to matter. The greater the depth, the greater the bondage; the lesser the depth, the greater the freedom.

Example: The personality lives and functions at a much greater depth of prakritic immersion than does the Monad. But the human Monad is far more prakritically immersed than is the Monad of the Planetary Logos or the Monad of the Solar Logos which is the least prakritically immersed of the three. The Universal Logos (on Its Own high Super-Cosmic Plane) experiences, of all beings-in-Cosmos, the *least* degree of prakritic immersion, though, via the Divine Emanatory Stream, the Logos pervades all levels and thus *experiences* the totality of all possible degrees of prakritic immersion.

Example: Assessment of the degree of prakritic immersion must especially take into consideration the *highest* vehicles through which an authentic E/entity manifests, because this 'height' indicates the *highest* consciousness of which such an E/entity will be capable. It should be remembered that Entities which focus, principally, at a very shallow degree of prakritic immersion may extend Their influence to a very deep degree. Such an Entity would be, for instance, a Planetary Logos or a Solar Logos. No matter what the 'extension-in-prakritic-depth' of such great Entities, the prakritic immersion of Their *highest* (i.e., *circum-monadic)* vehicles determines the vastest possible extent of Their Consciousness. The first and most important determination is the degree of prakritic immersion of the Monad of any E/entity. The next determination is the '*extent*-in-prakritic-depth' of that E/entity. Even among very advanced E/entities there may be great variation along these lines. An extensive depth of prakritic immersion in a high Entity is often an indication of *sacrifice*.

Example: It should be mentioned that at lower points of prakritic immersion, a subjective life will *see* (as objective), what higher subjects have projected out of themselves. For instance, a Member of the Supernal Tetraktys (and thus a very Great Logos in the Company of the Son) may look within Itself and 'See' Itself as a Prakriti Image (for anything 'Seen' is fundamentally Matter) and then, also, 'See' what is within Itself (as Son) waiting to be born. The moment that Son is 'Seen' it Arises as an Object as a Form within Prakriti and Pure Mother is no longer focally seen but only Mother-as-Son.

Then, let us suppose there is a lesser being within that Great Logos, a being who is a 'Ray' of the Greater 'Ray' which the Greater Logos Is. The lesser being is at a lower state of prakritic immersion than the Greater Logos, Who is upon the Archetypal Planes altogether. What, then, is *Prakriti* to this lesser being? In fact, the lesser being will never (until climbing evolutionarily to a higher/shallower state of immersion) 'see' Prakriti at all, but will only 'see' the objects/'Sons' 'Resident' within the Greater Logos. For a long time those Objects (which are Prakritic Sons to the Greater Logos) will, for the lesser being, be all that the lesser being can know of *matter*. For instance, all the many mineral atoms can be numbered among the *prakritic sons* of Fohat and His Host. Fohat 'Sees' Itself, and within It, invisibly, subjectively, are all latent atomic and molecular forms that Fohat is responsible for objectifying. Fohat as a Cosmic Subject becomes (with Self-Sight), the Fohatic Mother to Its Own invisible sons, which include the many atoms, molecules, particles, etc., and they appear as objective sons.

The point is, that the atomic and molecular prakritic sons of Fohat, are, to lower beings, *all they can apprehend of prakriti*. Lower beings cannot see Matter as Fohat Really 'Sees' it (or as Beings still greater than Fohat 'See' It). Fohat's prakritic sons, therefore, serve as a substitute for Pure Prakriti, until such a time as the lower being (having "climbed the ladder of Evolution") can *see itself* and *its own internal patterns to be objectified* with the same clarity that Fohat and other Great Logoi can. So Fohat's Self-'Sight' substitutes for the (one day to be developed) improved self-sight of the lesser being. Thus, temporarily, Fohat's Internal World becomes the *external world* of lesser beings. This is another way of saying that beings of our nature (the human) live within the *Mayavically-induced objectified internality* of Beings greater than ourselves. Their *inner* patterns become the *outer* patterns with which we have to deal objectively.

When a subjective being becomes truly Creative in the Cosmic Sense (which may occur at a fairly *shallow*, hence etherial, level of prakritic immersion), then it becomes more possible for that being to *see* that which is subjectively within it, its many latent 'sons' then emerging as objects in (its own kind of) Prakriti. The more insight we have into our own being, the more readily we become *Mother to Ourselves*. Before that time we must, perforce, live among the Sons of other Mothers.

If the Design at the Beginning is to be fulfilled, it becomes necessary for all self-conscious subjective beings to bring forth into objectivity that which is latent as *Divinely Intended Possibility* within *them*. Thus is the Divine Pattern fulfilled Cosmically. We see how important is *creativity*, even as usually conceived.

Example: Each descending level of prakritic immersion is a reduction in the scope and depth (degree of multi-dimensionality) of a (Monadic) Point of View.

Example: Another name of prakritic immersion is 'Depth of Objectification' of a 'Ray' of the ABSOLUTE.

prakritic levels or layers

By *prakritic levels* are meant the various planes and subplanes of the systemic, Cosmic and Super-Cosmic strata of Cosmic Objectivity.

Example: The Monads of all authentic Identities in-Cosmos are found upon various prakritic levels or layers. The main arenas of *expression*, however, for these different Monads (which, remember, are but One Monad) are still *other* and *lower* prakritic levels or layers. For instance, while the human Monad may be anchored upon the systemic monadic plane, Its main and destined arena of expression is upon the systemic buddhic plane.

prakritic variations

By *prakritic variations* are meant diversified modifications of primordial substance (Cosmic Prakriti) registered in intra-Cosmic consciousnesses as the different types of matter. These modifications are directly reflective of the internal enumerations, differentiations, and modifications, of Fohat, who *builds* the multiple differentiations of matter by *perceiving His subjectively Self-enumerated Self as reflected by/within Cosmic Prakriti* (which means 'Seeing' his lesser Self-Images, 'sons', appearing within His greater Self-Image (a Mother-Field which is a kind of Cosmic Prakriti). Cosmic Prakriti, while not to be understood as *matter* (in the usual and familiar sense of the term) Is, rather, the Matrix of Objectification, and the great Self-Reflection of the Condensed-Point-as-Universal Logos, which Reflection is 'Generated' by, Maya, the Great Self-Reflecting Power of Cosmos, Who renders latent subjectivity into objectivity.

Example: Prakritic variations are distinguishable from each other through an analysis of type and frequency of particle motion, and also through an analysis of the 'position' of certain elementary particles relative to each other. These types of differentiations and distinctions are the province of the Micro-Meta-Physics of the Science of the Future.

prakritic registration, limitation of

By the *limitation of prakritic registration* is indicated the very interesting idea that the various Sub-Creators in any Universal System may be able to see no deeper into themselves than the immediately greater Creator of Whom They are an Emanation. The Image formed by the Greater Creator will be a limiting boundary to the lesser Creator (even though that Creator be relatively powerful).

Example: A Universal Logos 'Sees' or becomes Conscious of Its Own Internality and that Internality is *projected outwards* as Cosmic Prakriti. That Logos also Sees within Itself Its Son, and in the Seeing the Son is Emanated. (This is the secret of Emanation.) The Son is an Object and thus a Son of the Mother, which the Universal Logos Becomes when It 'Sees' Itself, but the Son is also a Subject because He is "Like unto the Father" and in His turn the Son will become a Father to the Sons that are within Him.

When the Son in His turn begins the Creative Process, He, like the Father before Him, looks into Himself and thus discovers Himself as Mother. What *kind* of Mother, however? How deeply can the Son 'See' into Himself? Will He not, as He looks within, 'See' Himself being 'Seen' by the Father? He will discover Himself as an Object, yes, but that Object will be the One which the Father continues to See in Himself (i.e., within the Self of the Father). On it goes in this manner with the lesser Logos ever Seeing Himself being 'Seen' by a Greater Logos, Who, in turn 'Sees' Himself being 'Seen' by a still Greater and more inclusive Logos. The Son, thus, never 'Sees' to the depth that the Father 'Sees', and the Three Sub-Logoi of the Son never 'See' to the depth that the Son 'Sees', and the Seven Subsidiary Logoi never 'See' to the depth that the Three Sub-Logoi 'See'.

Each order of Logos 'Sees' Itself (in Its Mother Aspect) as a different kind of Prakriti, as a Prakriti less *fundamental* than the Prakriti 'Seen' by the Logos of which It is an Emanation. The question arises, Will **8** then, ever "get to the bottom of MySelf" unless **8** uncover the Vision of all the ascending Hierarchy of Logoi, in the lowest of which **8** "live and move and have my being", and each of Which "Lives and Moves and has Its Being" in the Emanative Logos immediately superior to It?

These *layers* of Prakriti are *Real* because no Logos can 'See' Itself exactly as superior Logoi 'See' Themselves. Thus each inferior Logos experiences what might be called a 'limitation of prakritic registration' when compared to the Emanative Logos immediately superior to It, and, also in comparison with all Emanative Logoi superior to It. It is very simple: If **8** Am inferior to You in scope and extent, then when **8** look into MySelf, what **8** 'See' is a Prakritic Image more superficial, less deep, than what you 'See' when you look into Yourself. When, whether though slow evolution or the faster process of occult meditation, **8** rise to your 'altitude' of being, then the Prakritic Image which appears through My Self-Sight will be equal in quality, depth and extent to the one You have now.

Example: These layers of Self-Perception can be considered as the various Doors of Initiation. If You are greater than **8** Am, and if from the present Prakritic Image that **8** 'See' when **8** 'See' into MySelf, **8** Am able to enter into the Prakritic Image which You 'See' when You 'See' into Yourself, then **8** have overcome a particular 'limitation of prakritic registration' and entered a Greater World, and have "taken Initiation".

Pralaya, Universal

By *Universal Pralaya* is meant those periods of Time throughout the Course of Infinite Duration when the entire Cosmos is reabsorbed into the ALL-IN-ALLNESS of the INFINITE SELF.

Example: There is no certainty that for any given Cosmos the Universal Pralaya must begin at a designated 'time' relative to all other Universal Pralayas, or relative to the duration of other Cosmoses. Universal Pralaya means Universal Dissolution, and, it would seem, signals the Consummated Achievement of the Universal Logos. There is no 'Good Reason' for the onset of Universal Pralaya *before* that Consummation has been achieved by means of the fulfillment of the Design-at-the-Beginning—*unless* the INFINITE SELF is *subject* to a LAW *greater* that ITSELF, which seems an absurdity.

Pre-Cosmic Parameters

By *Pre-Cosmic Parameters* are meant the Laws and Limitations inherent in the Universal Algorithm (the Original Intent, the Design-at-the-Beginning). These Laws and Limitations determine what can and cannot occur within the Cosmos-to-Be.

Example: The Pre-Cosmic Parameters are probably 'ENCODED' in the 'RAY' of the ABSOLUTE as It 'FLASHES FORTH' Creating the Infinite Subject and Infinite Object. The Infinite Subject (Who is the SELF-as-Infinified Point) Is *seeded* with the SELF-'ORIGINATED' Design of the Cosmos-to-Be. The Pre-Cosmic Parameters are probably inherent in the various Modes of 'FOHAT'/Fohat as It (being the Representative of the SELF) *acts* to fulfill the 'WILL' of that SELF.

Example: Any *Possibility for a Cosmos* 'EXTRUDED' from the INFINITESSENCE has implicit within It the Pre-Cosmic Parameters that will shape and regulate the Structure and Dynamics of the Cosmos-to-Come.

Example: A very serious question (and very difficult to solve) is, *How* and by *what* or *whom* are the Pre-Cosmic Parameters *chosen?* If the 'RAY' of the ABSOLUTE *'carries'* these Pre-Cosmic Parameters, then the 'CHOICE' of them (i.e., of the 'IDEA' which is to determine the New Cosmos) has been *infinitely enlightened.* If the Choice comes *after* the 'FLASHING FORTH' of the 'RAY', then that Choice may be 'Made' in the Semi-Ignorance of a Pre-Cosmic State that is *greatly reduced* from the 'STATE' of ABSOLUTENESS.

This question may seem rather academic and its solution of no immediate or practical consequence, but it touches upon the degree of importance of our Cosmos to the INFINITE SELF, the degree of *freedom* of the Universal Logos, and even the degree of *absolute legitimacy* of our Cosmos. Much would depend upon the assessment of whether the Infinite Subject-as-Infinified Point is possessed of Infinite Memory and Infinite Wisdom. If this were so, one could have more confidence in that *fateful choice* made in Pre-Cosmic 'Days'. Perhaps it is sufficient to dwell with these two possibilities. There seems to be no immediate means of solving this problem, or of arriving, confidently, at any conclusion.

presence, a

By *a presence* is meant, any impression which registers upon a consciousness capable of registration.

Example: A presence is usually considered a subtle energy pattern not capable of detection through the usual senses, but capable of being registered through various kinds of subtle faculties. There are, nevertheless, presentations which impact the gross as well as subtle senses. That which is registered is a presence. That which is not registered by one but is registered by another more sensitive, is nevertheless a presence, for, indeed, it is present, even if unregistered by the one less sensitive.

Presence, the

By *the Presence* is meant the Cosmos-Pervading Ubiquity of the Universal Logos.

Example: All intra-Cosmic Presences are known to be *present* by the 'sensed' vibratory quality of their sheaths. It will long indeed before the Monad which is now manifesting as a member of Earth's humanity, feels the Presence of the *Universal* Logos from the vibratory quality of that Logos' Sheaths (even though, Essentially and Subjectively, They are *both* the *same* 'Ray' of the ABSOLUTE). It can be justifiably presumed (from a study of the Cosmic and Kosmic Planes) that even the Solar Logos of our solar system is not yet capable of this exalted registration.

PRESENCE, the

By *the PRESENCE* is meant the forever unchanging, OMNIPRESENT DEITY, the SELF WHO IS all things (*noumenessentially, implicitly* and *cosmically explicit)* forever. Feeling, knowing and identifying with this PRESENCE is the goal of all Non-Dualistic Schools of Spiritual Philosophy.

Example: The PRESENCE is independent of all sheaths and is only *known as FACT* through the Art of Identification.

present (verb)

By the term *present* is meant the process by which any 'perceivable' is impressed upon a registering consciousness.

Example: The apparent Not-SELF presents consciousness with a multitude of impacts which, during the course of the Universal Aeon, must be understood and mastered through the Art of Pervasion and, finally, the Art of Identification. One day it will be realized that we-as-We-as-WE present only to OURSELVES—the SELF WHO IS all *selves.*

present (noun)

By the *present* is meant the immediately and instantaneously perceivable (or potentially perceivable) ultimate moment before the onset of the next ultimate moment.

Example: If one cannot consciously register the passage of ultimate moments in Fohatically Particulated Cosmos, one cannot consciously register the true present. No human being can (as presently equipped) register an ultimate moment, and therefore, cannot register with real accuracy, the present.

Present

By the *Present* is indicated one of the three Nows operative in Cosmos: the Cosmo-Objective Now, the Cosmo-Subjective Now, and the Cosmo-Eternal Now. Thus the term 'the Present' is variable depending upon the Point of View and the E/entity perceiving from that Point of View.

Example: For the Universal Logos dwelling in the Cosmo-Eternal Now, the entire Exhalation of the Great Breath is the Pinnacle of Present. This greatest of Cosmic Presents is a Cosmo-Macro-Quantum-Moment. The Universal Logos is also able to experience the Present as each of the multitude of changes of the Cosmic Configuration in the Fohatically Particulate World. As well, any Act of Will performed within the World of Being is, for Him, the Present, even though it occur 'during' a Cosmo-Objective Now.

PRESENT, the

By *the PRESENT* is meant the one, infinitely enduring, immobile ETERNAL MOMENT which changes not regardless of the *apparent* "passage of Time".

Example: The INFINITE SELF, *ever absorbed* in a 'STATE' of infinitely profound BE-NESS, LIVES ever in the PRESENT.

presentation

By the term *presentation* is indicated an abbreviated form of *presentation-in-consciousness.*

By the term *presentation* is meant a current event registered by a perceiving consciousness.

Example: All reasonably tangible objects are Really perceptions or apperceptions. They are presentations and are merely symbols in consciousness of the "thing in itself" which is their noumenon.

Presentation, *the*

By *the Presentation* is meant the 'BRINGING FORTH' of the Universe by the SELF *to* ITSELF.

Example: The entire Universe is an Emergence from the GREAT NEGATION/INFINITE PLENUM. The Universe, considered as *the* Presentation, is a single Possibility (out of an infinitude of possibilities) *presented* (in Super-Cosmos and Cosmos) to the progressively veiled 'CONSCIOUSNESS' of the INFINITE SELF. In a way, the INFINITE SELF 'PRESENTS' the Universe only to ITSELF, but it uses a number of Pre-Cosmic, Cosmic and Intra-Cosmic 'Agents' (which Agents IT REALLY *IS)* to achieve *the* Presentation.

presentation-in-consciousness

By a *presentation-in-consciousness* is meant any modification which registers upon a perceiving or apperceiving consciousness so as to produce a distinct 'event' or 'happening' in that consciousness.

Example: A pulsating image of the lighted relationship existing between all my group brothers and sisters around the world is for me a presentation-in-consciousness every time I meditate upon the Ashram and its affiliated disciples.

present-in-Cosmos

By *present-in-Cosmos,* is meant that which is perceivable/apperceivable/identifiable or potentially perceivable/apperceivable/identifiable within any particular Cosmos or Universe. ULTIMATE REALITY (the ALL-IN-ALLNESS) is *not,* in this sense, present-in-Cosmos, and yet, in a way, IT *inescapably* IS.

Example: An *infinitude* of 'POSSIBILITY' 'INHERENT' in the INFINITESSENCE is not and never will be present-in-Cosmos—at least present in *this* Cosmos.

Prime Mover

By the *Prime Mover* is meant the Universal Logos in Its *dynamic* Aspect.

Example: All motions-in-Cosmos derive from the Original Impulsions of the Prime Mover.

PRIME MOVER

By the *PRIME MOVER* is meant the SELF in ITS Aspect as 'RADIATOR' of the 'RAY' of the ABSOLUTE.

Example: 'Within' the ALL-IN-ALLNESS of the SELF there is no 'MOTION'. However, in order to Appear in ITS Aspect of Pointness (i.e., as Infinified Point), the SELF must (while remaining ESSENTIALLY *motionless)* 'BECOME'/'TRANSLATE' ITSELF into the PRIME 'MOVER'—a 'MOVER' anteceding the appearance of the *Cosmic* Prime Mover. How this is 'DONE' is a Great Mystery, but probably an 'ACT' of 'ABSTRACTION' is involved. The SELF must apparently cease to BE ITSELF (though IT never REALLY *can)* in order to 'SEE' ITSELF.

Example: In examining the distinction between the PRIME MOVER and the Prime Mover we find that the PRIME MOVER 'BECOMES'/'CREATES' the *Condition* which precedes a specific Cosmos; the Prime Mover then unfolds all modifications which are SUPER-Cosmically 'INTENDED' to unfold within that specific Cosmos.

principle

By a *principle* is meant a fundamental constituent of any system, which that system requires if it is to retain the integrity of its will-intended pattern.

Example: The principle of *manas* (whether it manifests consciously or unconsciously) is a necessary constituent of all E/entities-in-Cosmos.

PRIVATION, 'STATE' of

By the *'STATE' of PRIVATION* is meant a 'STATE' devoid of all attributes.

Example: The 'STATE' of PRIVATION is simultaneously a 'STATE' of *supreme negation* and a 'STATE' of *infinitized fullness.* In ITS 'STATE' of PRIVATION, the INFINITE SELF is *deprived* of anything that would diminish ITS INFINITE STATURE. This means that the SELF is *deprived* of every single *thing* (for a thing is a *finitude* the presence of which would mar ITS INFINITUDE, ITS IMPARTITE HOMOGENEITY); the INFINITE SELF, however, is *not* deprived of the *infinitization* of any single thing.

process

By a *process* is meant a goal-oriented mutual activity of a given number of related variables. A process is an interaction among variables such that, by competent study of the interaction by a competent student of same, a prediction as to the outcome of the process can be successfully made.

By a *process* is also meant any interaction of variables (whether or not the interaction is goal-oriented or goal-inclined). There are such things as unplanned, non-purposeful processes which can be analyzed so that their dynamics can be understood. Most emotional interactions between human beings are just such unplanned, non-purposeful processes.

Example: Initiation is not so much a one-time event as a long-term process, involving years of strenuous focus and intense Self/self-discipline.

- Q -

quality; Quality

By a *Quality* is meant a permanent-in-system structural factor in the Pattern of a particular Universe.

By *quality* is meant the subtle radiation which arises from relationship.

Example: The Universal Theme for each Cosmos calls for the manifestation of certain Qualities which, in their 'integrous combination', reveal the Intended Design for that Universe. Quality and Pattern are mutually-inseparable concepts.

Example: The quality of any relationship is best apprehended as a whole by that human faculty which is designed for apprehending in wholes—the intuition.

Example: Since every Number is a relationship, every Number has Its distinguishing Quality. This means that every Quantity has/is Its Quality.

quanta; quantum

By *quanta* are meant 'uninhabitable' intervals of ascent or descent (or intervals between structural changes) necessitated by a discontinuous, finite Universe, and existing in atomic systems of all magnitudes (whether substance atoms, human atoms, Planetary Logoic Atoms, Solar Logoic Atoms, etc.). Because there is no absolute continuum, per se, in-Cosmos (especially in the Fohatically Particulated Cosmos, though the BOUNDLESS IMMUTABLE PRINCIPLE *IS* a CONTINUUM) all complexification or simplification (with the attendant movement of all variables involved) occurs by *virtually instantaneous increments,* or "quantum leaps", rather than continuously and gradually.

Example: The leaps of pitch in the musical overtone series could be considered as analogous to "quantum leaps".

Example: All particle/events are quanta. Even larger energy-events (on the human and super-human levels) are quanta. This entire Universe is Dis-Continuous and *quantized,* and the Universe, Itself, on Its huge time scale, is a Discontinuous Macro-Quantum of Space/Time.

quantity

By *quantity* is simply indicated the *number* which defines or specifies the 'countables' in any aggregation.

By a *quantity* is meant the distinguishing character of an aggregation of indistinguishable items, such that a comparison of it with each of a continuous infinitude of other aggregations of items indistinguishable in nature from the items in the first aggregation (adding the proviso that each aggregation must *differ* from every other—spatial arrangement being discounted) that this comparison will demonstrate *sameness* in relation to only *one* aggregation out of the infinitude of aggregations.

Example: Every human unit can be understood spiritually as a complex quantity, for the form and pattern of every human being is an expression of Number.

Example: We are told to value quality rather than quantity, but the advice is only partially correct, for according to the Master Morya, a man is known by his *accumulations* (which are clearly quantifiable).

quantity/quality

By a *quantity/quality* is meant the distinguishing nature of the pattern through which any authentic E/entity in-Cosmos manifests. That pattern is part of the Universal Pattern and the E/entity's quantity/quality determines its place, position and function within the Universal Pattern.

Example: In occultism, *quantity is quality.* The energy value of any manifesting E/entity is determined by both its Number and the particular radiation/magnetism or *quality* which the numbered combination of its constituents produces. The term quantity/quality suggests specifically how the energy value of any E/entity is derived, and the inseparable relationship between these two factors.

quantum of vibration

By a *quantum of vibration* is meant the fastest possible frequency of vibration within a given Cosmos. This is an exact and limited quantity in a given Cosmos, though indeterminable by human science at present.

Example: The quantum of vibration for our particular (probably) ten-dimensional Cosmos is so rapid, from the human perspective, that the vibratory frequency of electrons associated with physical plane atoms (atoms of the lowest tenth of the lowest dimension—i.e., the lowest one hundredth or, perhaps, the lowest one thousandth, of the whole) is incredibly slow in comparison.

- R -

radiation

By *radiation* is meant the transmission of energy from a source to a receiver via a given medium. The energy flowing is usually a natural result of the *life force* of the source, and carries the *quality* of the source to the receiver.

Example: Radiations are of all kinds and subtleties. The radiations of our Solar Life are only partially pranic. On every one of the gross and subtle planes of our solar system, the radiations of the Solar Life make their impact felt. For instance, the Radiation of the Heart of the Sun is a form of Divine Love.

Example: A very important question presents itself, Are radiations *quantized?* Radiation of which science knows anything is produced by the release of particles (for instance, Alpha and Beta particles) from a radiating body. But can there be a radiating body that is impartite? What quanta could be released from such a body? Can the ultimate particle be considered a radiating body since it is indivisible and impartite? If not, is there, therefore, a form of non-radiating energy which would allow communication between such bodies?

'RADIATION'

By *'RADIATION'* is meant the Pre-Cosmic 'FLASHING FORTH' of the One 'RAY' of the ABSOLUTE.

Example: The 'RADIATION' at the Pre-Cosmic 'BEGINNING' is the First Objective 'ACT' and involves the release of SUPER-Cosmic 'FOHAT' from the ABSOLUTE. There are a number of equivalences which should be realized:

1. The 'FLASHING FORTH' of the One 'RAY' of the ABSOLUTE.
2. The emergence of SUPER-Cosmic 'FOHAT'.
3. The first 'SIGHT' of SELF as 'EVANESCENT INFINITE OBJECT' by the SELF as 'EVANESCENT INFINITE SUBJECT'.
4. The first 'ACT' of SELF-'REFLECTION'.
5. The appearance ("in no time at all" or after an *infinitesimal instantaneity)* of the SELF-as-Infinified Point.
6. The appearance of the Infinite Subject opposing the Infinite Object.

The 'time interval' separating any of the above equivalences (especially numbers 5 and 6 from the others) is negligible. In this treatise there are offered many ways of saying the same thing, each way representing a slightly different angle on the subject and, therefore, a possibility for increased understanding (or increased confusion if we cannot keep our terms straight). By expending the labor to understand such equivalences as listed above, it is hoped that the Pattern of the Pre-Cosmic and Cosmic Processes will be clarified and considerably simplified.

Example: The 'RAY' of the ABSOLUTE maybe already 'CARRIES', the coming *enumeration* of, or, better, 'IS' each and every Monad, each of which is defined by H. P. Blavatsky as a 'Ray' of the ABSOLUTE. Each Monad, of course, is *the* ONE MONAD-as-Ultimate Cosmic Monad. Essentially, there is only *One.* The One Identity which, *apparently becomes* the various Monads on the different levels of Cosmos, *is* the 'RAY' 'ITSELF', which is, at once, Identity and 'IDEA' 'EXTRUDED' from the INFINITESSENCE for *cosmification.*

It is as if the 'RAY' which 'FLASHES FORTH' as the first 'RADIATION' has the *potential* for definitely enumerable 'partiteness', for IT later *becomes* the many (apparently *lesser)* 'Rays' of the ABSOLUTE, and the many objects/points etc. both partite and impartite. These are the 'ENUMERATED ASPECTS' of the GREAT 'IDEA' to be Cosmically Objectified. The 'RAY' is *impartite*, however, because the ABSOLUTE cannot be divided; ITS 'RAY' cannot be divided; and, in fact, no 'Ray' of the ABSOLUTE or Cosmic Monad can *ever* be divided. Can the 'RAY' be enumerated, but *indivisible?*

This is the paradox, and the answer inclines to be, Yes! The 'RAY'/'Ray', is SELF-'ENUMERATING'/Self-Enumerating, but *indivisible.* These ideas touch precisely upon one of the most difficult of all problems: 'The Nature and Origin of the Monad'.

By '<u>RADIATION</u>' is meant the 'FLASHING FORTH' of *ESSENCE* from the ONE SOURCE without the *confinement* of that *ESSENCE* in any form. While the 'RAY' 'GENERATES' the First Super-Cosmic Object (the Infinite Object, Mulaprakriti), the 'RAY' 'GENERATES' *no vibration* and nothing that vibrates. The 'PROCESS'/'EVENT' known as the 'FLASHING FORTH' of the 'RAY' differs from *emanation,* as emanation involves the appearance of prakritic vibration (or, at least, prakritic objective presence) and, hence, 'confinement in form'.

Therefore, the 'PRIMAL <u>RADIATION</u>' is completely *subjective* in its nature. While the 'RAY' is not *confined* in form, the 'RAY' is equivalent to the 'GENERATION' of 'CONSCIOUSNESS' and that 'CONSCIOUSNESS' is necessarily engaged with an Object (albeit an *Infinite* Object). Perhaps even the instantaneous engagement of 'CONSCIOUSNESS' with the 'EVANESCENT INFINITE OBJECT' represents some degree of 'RAY'-*confinement* when the ABSOLUTE FREEDOM of the INFINITE SELF is referenced. It is clear from these abstruse considerations that the 'RAY' may mean a number of things, and it would be well not to over-materialize its meaning through addiction to a conventional image of a *ray.*

Example: The 'PROCESS'/'EVENT' which leads to the appearance of the SELF-as-Infinified Point is a '<u>RADIATION</u>' rather than an *Emanation.*

Example: Although the ABSOLUTE is indivisible, the '<u>RADIATION</u>' known as the 'FLASHING FORTH' of the 'RAY' represents the *first* (albeit illusory) *division* of the ALL-IN-ALLNESS. In a way, the 'RAY' 'SPLITS' the ALL-IN-ALLNESS. Of course, IT REALLY *cannot.*

Radical Infinitism

By *Radical Infinitism* is meant a philosophy which attempts to understand what it means to 'infinitize' all things. When a 'thing' is 'infinitized' it becomes ONE with the ROOT OF ALL. (in fact, the *very* ROOT OF ALL). Radical Infinitism relates all things to the PARABRAHMIC ROOT, and while acknowledging the importance of form in relation to Original Cosmic Intent, refuses, ultimately, to see and understand anything except *as* the ABSOLUTE, ITSELF.

Example: <u>Radical Infinitism</u> inclines to the conviction that there is "nothing (ESSENTIALLY) new under the Sun", though new forms are necessarily generated at each ultimate moment periodically and cyclically throughout Infinite Duration. Something *new* would have to be something *other* than THAT, and than THAT there is no *other.*

Example: Non-Dualism is a more popularly understood term for <u>Radical Infinitism</u> (which emphasizes the *infinitude* of the non-dual 'State').

Radical Non-Dualism

By *Radical Non-Dualism* is meant a philosophy that resolves all Subjects and Objects into ONE GREAT SUBJECTIVITY, the INFINITE SELF. This philosophy thus posits the primacy of Oneness over Twoness.

Example: From the perspective of a Radical Non-Dualism BRAHMAN and Samsara are *absolutely* ONE.

Example: Radical Non-Dualism is *Radical* because it goes to the *root* of all Subjects and Objects in order to discover that they are, ESSENTIALLY, the FORMLESS HOMOGENEITY—PARABRAHMAN. Thus, Radical Non-Dualism offers the opportunity to achieve identification with the ROOTLESS ROOT of ALL.

ray

By a *ray* is meant a gradually attenuating *extension of source.* A ray, as usually considered, is a *'going forth' of an energy pattern from a source,* which 'going forth', though it is a partialization/diminution of the intensity of the source and of the energy pattern of the source, *does* preserve, in attenuated extension, (even though it be in reflection within a denser prakritic level) that *energy pattern* and the *quality* (if not the *quantity)* of the source.

Example: All E/entities-in-Cosmos are ESSENTIALLY 'Rays' of the One 'RAY' of the ABSOLUTE.

Example: A ray is the relationship between source and recipient.

Example: A ray is a means of conveying life and quality from source to receiver via a medium. Via a ray, many spiritual gifts can be bestowed in an assimilable manner.

Example: An intriguing question arises when considering rays, What is the *form* of a ray? And what are its constituents? Is a ray merely a 'beam' of directed particles? Is radiant energy *corpuscular* or otherwise? Can it be otherwise in a Dis-Continuous Universe? Are there rays of *consciousness?* Or is Consciousness an 'immobile medium' functioning simply through 'presence'?

Note: An important problem to be discussed here arises. Can any source REALLY extend itself *in part?* Due to the indivisibility of the SPIRIT, is not every apparently *partial* Self-Extension really a *full* Self-Extension—a *radiating forth* of the *full* Self? This is true in an even more ultimate sense of the ABSOLUTE. Has not the *infinite entirety* of the ABSOLUTE 'GONE FORTH' upon ITS 'RAY'? And, yet, has the ABSOLUTE not 'REMAINED' as ever—*unchanged?* Paradox!

'RAY' of the ABSOLUTE; 'Ray' of the ABSOLUTE

By *the* 'RAY' of the ABSOLUTE is meant the 'ACT'-of-'ORIGIN' which inaugurates the Process leading to the emergence of a particular Finite Universe.

By *the* 'RAY' of the ABSOLUTE is meant the 'ACT'-of-Origin whereby the SELF-REFLECTIVE POWER of the INFINITE SELF 'ARISES' 'within' ITSELF.

By a *'Ray' of the ABSOLUTE* is meant any authentic E/entity or Identity in-Cosmos. Each Cosmos has a designated number of such authentic E/entities (implicate or 'enfolded' within each other. Each is ESSENTIALLY a 'Ray' of the ABSOLUTE.

By the *Individuality of a 'Ray' of the ABSOLUTE* is meant its identifiable distinctness during Cosmos, even though It be utterly at-one with the Universal Logos of the Cosmos, which is also a 'Ray' of the ABSOLUTE.

By *the 'RAY' of the ABSOLUTE* is meant the SELF-'LIMITING' 'RE-FOCUSING' of the SELF, which might also be called the emergence of SUPER-Cosmic 'FOHAT'.

Example: The triple State of 'Pointness' (Infinified Point, Condensing Point and Condensed Point—all three being States of the Super-Cosmic Subject) is the first Super-Cosmic State. In 'Pointness', SELF appears to 'GO FORTH' from ITSELF as something *less than* and, hence, *other than* ITSELF. The State of 'Pointness' is effectively the first SELF-'EXTENSION' of the ABSOLUTE into the 'Space'/State of *'Other than PERFECTLY ABSORBED SELFHOOD'* and is, hence, the immediate Product of the First 'RAY' of the ABSOLUTE. (See the six equivalences listed in the Glossary under The 'RADIATION'.)

Example: Is the 'RAY' of the ABSOLUTE utterly impartite or *multiple-in-Cosmos, though indivisible?* This is a strange question, but upon its answer depends the understanding of the Origin of the Monad and hence of every authentic E/entity in Cosmos.

Example: A 'Ray' of the ABSOLUTE is the PRESENCE of the *entire* 'RAY' of the ABSOLUTE at a certain 'Depth of Objectification'.

Example: A 'Ray' of the ABSOLUTE continues to *propagate* Itself, emanatorily, through Self-Reflection.

Example: Any E/entity is a Point. Any 'Ray' of the ABSOLUTE is a Point, as well as a 'Ray'. Its 'Rayness' is Its 'going-forth-ness', Its extension, or 'Identificatory Investment' of Itself into Itself or into other Points; Its 'Pointness' is Its 'Seeing-ness' or 'Seen-ness'. 'Rays' and 'Points' are two conceptions which need to be considered together.

'RAY'-Point

By the *'RAY'-Point* is meant the 'infinified point', possessing infinite perspectives upon the Infinite Object (which Object arises from the 'FLASHING FORTH' of the 'RAY').

Example: Does the 'RAY'-Point arise simultaneously with the 'FLASHING FORTH' of the 'RAY' of the ABSOLUTE or *instantaneously* following that 'FLASHING FORTH'. A question of this nature touches on the beginning of Time in any Universe. If the 'RAY' is to have any duration at all as 'RAY', per se, then there must probably be an instantaneous separation in newly born Super-Cosmic Time before the 'RAY' becomes the 'RAY'-Point. Otherwise, 'RAY' and 'RAY'-Point would be *identical factors.*

Rays, the Seven

By *the Seven Rays* are meant seven fundamental Qualities of Energy which condition all life in manifest Cosmos. With regard to our local Cosmo-System, the esoteric Source of these Seven Rays can be considered the seven majors stars of the Constellation called the "Great Bear". (For a comprehensive treatment of the meaning and application of the Seven Rays, see Volumes 1 and 2 of the author's *Tapestry of the Gods* series.)

Example: Every human being is Essentially a 'Ray' of the ABSOLUTE, no matter upon which of the Seven Rays his Monad may be found.

REAL, ABSOLUTELY

By the *ABSOLUTELY REAL* is meant only THAT which is ETERNAL and *utterly* IMMUTABLE. The term ABSOLUTELY REAL can never denote a particularized thing.

Example: Great Archetypes in-Cosmos are both Real and virtually Permanent-in-Cosmos, but They are not REAL, as in ABSOLUTELY REAL.

Example: No presentation is REAL, though, ESSENTIALLY, every presentation must, at ROOT, be REAL. The *form* of the presentation is always un-REAL apparency.

Real-in-Universe

By *Real-in-universe* is meant that which (while not *ABSOLUTELY* REAL) pertains to the semi-permanent Archetypes/archetypes (i.e., High and Formative Energy Patterns which are Permanent-in-Universe, while *not* Permanent-in-Infinite Duration) which determine Universal Process for the duration of a particular Universe.

Example: The Essential Numerical Beings are Real-in-Universe but not as REAL as the ZERO IS REAL. All great Formative Patterns in-Cosmos are Real-in-Cosmos.

By *Real-in-Universe* is meant that which is *relatively Real*, but not ABSOLUTELY REAL.

Note: The contrast between that which is an *actuality* and that which is *Real-in-Universe* is the following: an *actuality* is a presentation which need not be structurally necessary to the divinely designed integrity of a particular Universe, but that which is *Real-in-Universe* is a presentation which is archetypally and structurally necessary to the divinely designed integrity of a particular Universe.

Example: The Laws of Cosmos on all their various levels are Real-in-Universe. They are a structural necessity for ensuring that the Universal Process conforms to the Design-at-the-Beginning which serves the Universal Purpose.

'realist'

By a '*realist*' is usually meant one who believes that *things* are *real*. Such a consciousness is of a relatively low order and belongs, rather, to an *actualist*—one who is *not* a true REALIST, or even a Realist.

Example: Those whose consciousnesses pay attention almost exclusively to the data/presentations of the World of the Senses consider themselves to be '*realists*' because they care so much about matter. In fact they are crass *actualists*, caring only about effects and not Real Causes.

Realist

By a *Realist* is meant one who understands the profoundly subtle Archetypal Patterns guiding and conditioning the Universal Process.

Example: The true occultist is ever a Realist, but not necessarily a 'realist' as that term is usually used. The occultist seeks to understand completely the Divine Pattern according to which all items and E/entities in-Cosmos are arranged.

REALIST

By a *REALIST* is meant one who considers the BOUNDLESS IMMUTABLE PRINCIPLE, the ABSOLUTE SELF, to BE the *only* TRUE REALITY.

Example: Most human beings are too preoccupied with Duality, with Subject/Object Relations to be *true* REALISTS.

Realities

By *Realities* are meant Archetypal Patterns which are virtually Permanent-in-Cosmos. The word 'virtually' must be used, as the vast majority of these Realities emanate only *after* the consolidation of the One (the Universal Logos) and disappear *before* the reabsorption of the One.

Example: Preoccupation with Realities could bring accusations of being 'unrealistic'.

Reality

By *Reality* is meant the patterns of energies and forces in the World of Archetypes or Formative Patterns.

Example: Lovers of Reality ever study the Divine Pattern.

REALITY

By *REALITY* is meant the ABSOLUTE 'STATELESS STATE'.

Example: Preoccupation with REALITY can land a thinker in what appears to be World Denial. Certainly, REALITY is not what is 'happening' in the World of Becoming, although, in a deeper sense, IT IS the ONLY THING 'HAPPENING'—even though REALITY can't 'HAPPEN'! This means that ULTIMATE REALITY cannot *precipitate* or *manifest* as ULTIMATE REALITY *undisguised*. This is where 'MAYA'-instantly-Maya enters the 'Picture'.

really; Really; REALLY

By *REALLY* is meant all that pertains to REALITY ITSELF.

By *Really* is meant all that transpires in relation to the World of Archetypes.

By *'really'* is meant all that transpires in the densest part of the World of Becoming.

Example: The SELF REALLY IS the ONLY REALITY. Those who study the Divine Purpose and the Divine Plan are forced to realize this fact, though they Really appreciate the Secondary Reality of the Divine Archetypes. So many of those, however, who think they 'really' know what is REAL, simply do not.

reason

By *reason* is meant the marshaling of thoughts in such a way (often according to the rules of logic) that the interplay of the marshaled thoughts leads to truth and illumination.

By *reason* is meant a rigorous mental process conducted according to certain well-established Laws of Thought the following of which prevent fallacious thinking.

Example: By conducting one's mental life according to the principles of reason, light will grow incrementally until intuitive revelation dawns.

Example: Reason, properly used, is the friend, not the foe of intuition.

registrant

By a *registrant* is meant any E/entity in-Cosmos understood as a *conscious acknowledging receiver* of impacts or influences from other E/entities in-Cosmos. Unconscious registration is also possible and many are the unconscious registrants.

Example: Whether they will or no, all authentic I/identities in-Cosmos are conscious or unconscious registrants of a tremendous variety of impacts. What they do with the registration will determine their usefulness to the Divine Plan.

registration; register

By *registration* is meant a touch or impact emanating from any dimension of Cosmos upon any consciousness (whether self-conscious or not self-conscious).

Example: The sensitivity of one's vehicles determines the number and quality of registrations in consciousness. Is there any kind of registration which operates *independently* of the vehicles?

registering consciousness, focus of the

By the *focus of the registering consciousness* is meant the arena of awareness, or field of sentiency, in which or from which the Subject receives or registers impacts and attends to them.

Example: Consciousness is multi-dimensional, and the focus of the registering consciousness is rarely trained upon *all* possible levels of registration. 'Polarization' is the term used to indicate the focus of the registering consciousness.

reify; reifiable

By *to reify* is meant a process of precipitation whereby a subtle pattern is concretized.

Example: The mind may be open to many intuitive impressions, but what does it take to reify those impressions into useful, concrete thoughts?

By *reifiable* is indicated the capacity of a subtle energy form or pattern to be rendered into a 'thing'.

Example: Intuitions are reifiable if the mind is sensitive and well-trained, but the SELF and the experience of *identifying* with the SELF are non-reifiable.

relation; relationship

By a *relation* or *relationship* is meant a condition in which there is an actual or potential exchange of registrations (not necessarily *conscious* registrations) between two or more E/entities or items-in-Cosmos which have been configured or combined.

Example: Some relationships are more powerful than others; such will be characterized by a mutually influential exchange of energies. Really, however, each item or E/entity in-Cosmos stands in relation to every other item or E/entity. The truth of this fact is often overlooked because the registered exchange of influence seems so slight, and yet it is Real.

relative; relativity

By *relative* is meant 'in relation to another or something other'.

Example: The truths of one civilization must be considered relative to the truths of a number of significant past civilizations, if one is to understand the extraordinary contextuality and relativity of truth.

By *relativity* is meant a condition among aggregated variables such that any of the variables taken singly or in groups can only be fully understood when the entire context in which they are found is considered and understood. Relativity demands that the part or parts be related to the whole if comprehension is to arise.

Example: The study of the Grand Design and execution of that Design through the instrumentality of the Divine Plan is a study in relativity. Certainly, each factor in-Cosmos functions in relation to every other and powerfully or minutely affects every other. Cosmic Truth must be approached through relativity. SUPER-Cosmic TRUTH demands the abrogation of all relativistic thinking. To understand such TRUTH, one must be an *absolutist*—i.e., one who is capable of identifying *with*, or better, *as* the ABSOLUTE.

- S -

Salvation, Universal

By *Universal Salvation* is meant the doctrine that if *All* are not 'Saved'—i.e., completely absorbed into PARABRAHMAN at the "Day Be With Us", then *none* are 'Saved'. Since PARABRAHMAN, the INFINITE-SELF, the BOUNDLESS IMMUTABLE PRINCIPLE, is *indivisible,* should even one 'Ray' of the ABSOLUTE remain 'outside' the ALL-IN-ALLNESS (Extra-'SOURCED') then the *entirety* of the 'RAY' of the ABSOLUTE would fail to be reabsorbed.

Example: There is no salvation for the individual alone, because there is no REAL individual. Radical Infinitism demands Universal Salvation simply because the ONE AND ONLY SELF is *not* divisible.

Example: "Save one, save All." This is unquestionably true, but the paradox holds. When one ascetic finds Mukta, the Wholeness of Cosmos is released—*except* for the contradictory and equally true fact that when there are beings who are not free and SELF-Realized, the Whole of the Cosmos is in prison—un-SELF-Realized. The problem in this formulation is that no ascetic or initiate of high degree ever REALLY achieves Mukta—at least not Universal Mukta. Universal Salvation comes only at the "Day Be With Us" and *never* before.

Example: There is no Final Salvation for any until All are saved. This means that those who imagine that they have achieved Mukta, Release, have only done so in a *relative* sense. They cannot be released into INFINITUDE until All Intra-Cosmic B/beings are similarly released. Universal Mukta is for All *simultaneously*, and never for any alone. The Universal Lipika Lords (bounding the Finite from the INFINITE as They do) enforce the Law of Universal Salvation.

same; sameness

The word *same* is used to describe, for instance, two actualities such that in a comparison of the two actualities, the wholeness of each actuality and each constituent part of each actuality are absolutely interchangeable in every respect and without exception, i.e., capable of being substituted the one for the other. In the World of Becoming, *sameness* is impossible. In the definition above, the terms 'Realities' and 'realities' may be substituted for 'actualities'.

Example: Even if prakriti cooperated by initially producing absolutely identical vehicles for two units-of-Life, the factor of 'will' (even unconscious will) and the factor of variegated experiences between initially identical units would soon terminate any vestige of sameness between the units. Even if ultimate particles start out absolutely identical (which is not necessarily to say they do) they would all immediately be subject to differing impacts due to the tremendously complex movement patterns in-Cosmos, and, also, because a form of 'self will' guides, in part, these movement patterns.

SAMENESS, the

By *the SAMENESS* is meant the UTTER HOMOGENEITY which IS PARABRAHMAN.

Example: The SELF is ever 'SELF-SAME'. IT has never changed, never ever. Any seeming violations of the SAMENESS are Illusory. No-thing *else* is *self-same* from ultimate moment to ultimate moment.

Samsara

By *Samsara* is meant especially the lower levels of the World of Becoming (i.e., the World of Birth and Re-birth). Samsara is the World of Evanescent Phenomena, the World of Impermanence, in fact, the World of Fohatic Particulation, the World of Illusion.

Example: BRAHMAN is the REAL; Samsara is the un-REAL. For all its un-REALITY, however, Samsara is, nevertheless, ESSENTIALLY, *REAL*, for what other than the BRAHMAN can Samsara be?

SELF; ALL-SELF, the

By *SELF* is meant the infinitely enduring ONE AND ONLY SUBJECT of all possible Cosmic and Super-Cosmic Experience and SUPER-Cosmic *'IN-PERIENCE'*.

Example: The SELF has, necessarily, Experienced and 'Inperienced' ALL in the UTTER ALLNESS. What other EXPERIENCER is there but the SELF? Yet the SELF needs the experience/inperience not, for IT IS forever PERFECT and nothing can be added to or taken from IT.

By *SELF* is meant that IDENTITY than which there is no *other*—the ONE WITHOUT A SECOND.

Example: Multiple selves are *imaginary* selves. There is but ONE SELF, the ALL-SELF, forever.

By *the ALL-SELF* is meant, precisely, the SELF (but with special emphasis upon the infinite inclusiveness of the SELF).

Example: The ALL-SELF IS ITSELF.

Self

By the *Self* is meant a condition of *selfhood* characteristic of *spiritualized* consciousnesses in-Cosmos. The Universal Logos is a Self, as is, on a much lower level, a Solar Logos. Man, considered as a Monad or soul, is a Self. Man considered strictly as a personality, however, is not a Self but a self. In order for a self to qualify as a Self there must be a sense of unity with and spiritual connectedness to all other E/entities-in-Cosmos. All these kinds of *selves* are ultimately and essentially the SELF, but one cannot speak properly of the SELF in terms of various in-Cosmos differentiations. To designate such differentiations, other terms relating to *selfhood* are required.

Example: The voyage of Self-discovery must be undertaken and to a great extent completed before one may profitably embark upon the Voyage of SELF-Discovery.

SELF-AS-SELF

By *the SELF-AS-SELF* is meant the 'STATE' in which the SELF IS UTTERLY ITSELF without the possible extensions of ITSELF (in-Cosmos) as Selves or selves.

Example: During the Universal Pralaya, the 'STATELESS STATE' of SELF-AS-SELF prevails, for there is no *Cosmic Distraction*. But, paradoxically, there can never be a Time in Infinite Duration (whether in Universal Manvantara or Universal Pralaya) when the SELF-AS-SELF does not 'PREVAIL', for the SELF is ETERNALLY IMPERTURBABLE. The SELF IS REALLY *ever* the SELF-AS-SELF.

Example: The SELF-AS-SELF 'ABIDES' in ABSOLUTE INFINITIZED IDENTITY forever.

SELF-as-Self

By the *SELF-as-Self* is meant the ONE ABSOLUTE SELF as it manifests through spiritual *Selfhood* in-Cosmos. The emphasis must be placed upon in-Cosmos manifestation and upon the fact that it is the ONE SELF which *substands* (and, essentially, IS) every spiritual Self-Realized Self in-Cosmos.

Example: All Entities higher in development than man have realized the SELF-as-Self. No matter what the nature of Their in-Cosmos Selfhood, They realize that Selfhood to be substanded by the SELF.

SELF-as-Self-as-self

By the *SELF-as-Self-as-self* is indicated a *sequence of 'descent'* or a *sequence of 'identity reduction'* by means of which the ULTIMATE IDENTITY is apparently SELF-'CIRCUMSCRIBED' to become the Universal Identity, which, then, *becomes through emanation* every particularized identity in-Cosmos. This sequence suggests that although the *sphere of Self/self-identification* for each E/entity varies greatly, the ABSOLUTE ESSENCE of each is *identical.* The SELF IS, indeed, the Self, which, indeed, Is every self-in-Cosmos.

Example: No matter how humble the authentic entity in-Cosmos, it has been generated through Divine Descent along the Divine Emanatory Stream. No matter what an E/entity appears to be, its ESSENCE can be described by the formula SELF-as-Self-as-self.

self-as-Self-as-SELF

By *self-as-Self-as-SELF* is indicated a *sequence of 'ascent'* or a *sequence of 'identity augmentation'* by means of which every authentic E/entity or I/identity in-Cosmos is able to trace its origin to the One Universal Self of Cosmos, which Itself, is *the* 'RADIATION' of the ABSOLUTE IDENTITY, the ONE SELF. Through the use of the 'self-as-Self-as-SELF' formula, the lesser self invokes the Greater Self and finally the GREATEST SELF, all of which, however, are ever and always but ONE IDENTICAL SELF.

Example: When the disciple, reaches 'upwards' and 'inwards', intent on identification with his *SOURCE,* he may find it useful to use the formula: self-as-Self-as-SELF.

SELF-as-Illusion

By the *SELF-as-Illusion* is meant the Cosmos and All within It. The Cosmos may be Illusion, but It is, nevertheless, the SELF, because there is nothing other than the SELF.

Example: The ancient saying that "BRAHMAN and Samsara are ONE" is another way of saying that the Cosmos, though Illusory, is nevertheless ESSENTIALLY the ABSOLUTE SELF. In order to remember that the Cosmos is REALLY nothing but the SELF, we might call Cosmos and All within It, the SELF-as-Illusion.

SELF-as-Not-SELF

By the *SELF-as-Not-Self* is meant that the Not-SELF, though apparently *not* the SELF is, ESSENTIALLY and REALLY the SELF.

Example: While to advance in-Cosmos (according to the Laws and Regulations of the Divine Plan as It reflects Original Intent) it is necessary at times to repudiate the Not-SELF, it is necessary to remember during the act of repudiation, that even that which is repudiated IS ESSENTIALLY the ONE AND ONLY SELF. Every repulsed presentation is REALLY the SELF-as-Not-SELF. This realization allows for proper synthetic reclamation after initial repudiation.

SELF, the extra-Cosmic

By *the extra-Cosmic SELF* is meant the INFINITE SELF considered as either predating the formation of Cosmos, postdating the dissolution of Cosmos, or as abstracted from Cosmos and dwelling fully as ITSELF and 'within' ITSELF in Universal Pralaya, and, as well, *abiding* even 'while' the 'Illusory Happening' called Cosmos is transpiring.

Example: All intra-Cosmic selves are *formally* illusory though *substantially* REAL. The ONE AND ONLY SELF IS *always* the extra-Cosmic SELF, WHICH, though invested fully *in* and, even, *as* all intra-Cosmic forms, IS nevertheless abidingly *transcendental* to ITS Cosmos, and hence *extra-Cosmic*.

SELF, the ONE AND ONLY

By *the ONE AND ONLY SELF* is meant the INFINITE SELF considered as the ONE WITHOUT A SECOND. Although nothing can REALLY be predicated about the SELF, it is helpful to consider IT in various ways in relation to the phenomena of the World of Becoming. In the Islamic Religion there are at least ninety-nine "Names of God". Surely God (or rather, GOD) is ESSENTIALLY *nameless*, but these Names allow the human mind to relate the DEITY more meaningfully to the multitude of relationships in the World of Becoming.

Example: Though, in-Cosmos, apparent *selves* are multitudinous, they are all ESSENTIALLY and in REALITY the ONE AND ONLY SELF.

SELF-as-Point

By the *SELF-as-Point* is indicated the Pre-Cosmic State of the SELF-as-Self 'Seeing' ITSELF (as Itself) from an *infinite* number of Points of View and, then, (after Condensation) from the *singular* n-'potenti-dimensional' Point of View. The sequence would be reversed were the 'Point' being considered Post-Cosmically.

Pre-Cosmically considered, the term 'Point' in the SELF-as-Point signifies, Really, (and at first) an infinite number of Points of View, all of them revealing the Infinite Potential Objectivity of the 'Seen' SELF—i.e., the SELF-as-Mulaprakriti. Then, the 'Point' passes through the Stage of Condensation, and 'Becomes' the Condensed Point, which 'Sees' only the Potential Cosmos from an 'n'-'potenti-dimensional' perspective, in which the value for 'n' may be different for each Cosmos, dependent upon the *infinitessentially* 'EXTRUDED' 'IDEA' which must be *cosmified* (i.e., *reified* into a Cosmos), Again, the process of Cosmic Obscuration, leading to a Post-Cosmic State, reverses the sequence.

Example: The 'FLASHING FORTH' of the 'RAY' of the ABSOLUTE creates the SELF-as-Point (first, the Infinite Subject). The 'FLASHING FORTH' is the Beginning of the Primal Subject/Object Duality. All 'Seeing' is done (at first) through the infinitely diversified Point, (which in this treatise is called the 'Infinified Point').

The Infinified Point/Infinite Subject may well 'See' all the generalized *Potential for Objectification* of Mulaprakriti, but does It also 'See' (simultaneously and from an infinitude of perspectives) the *Infinite Objectification* which Mulaprakriti has *already* 'Become' in all Cosmoses past? Mulaprakriti may not only be the Infinite Object but, perhaps, the *Infinite Memory of All Nature* (or, perhaps, the Objectification of that Memory)? This Vision by the Infinified Point would be a *Vision of Infinitude-in-Extension,* utterly different (because 'partite') from the *Infinitization of All Possibility* which *noumenessentially inheres* 'within' the INFINITESSENCE.

Although this is a possibility, the Vision of Mulaprakriti by the Infinified Point may be utterly 'impartite' and homogeneous. Some other Pre-Cosmic Phrase may be required to reveal the 'specifiable particular potential' of Mulaprakriti—probably the Phase of 'De-Infinitizing' or 'De-Infinifying' characteristic of the Condensing Point or Focusing Point.

SELF, the EGOLESS

By *the EGOLESS SELF* is meant the ALL-SELF, the emphasis being laid through this descriptor upon the idea that the SELF has no specifiable, delimitable, definable IDENTITY. The term *ego* always means definition, specification, or delimitation of identity.

Example: When the term SELF is used to denote the ABSOLUTE, one must always remember that the SELF IS <u>the *EGOLESS* SELF</u>. In the realm of egoistic selfhood, ego is constantly compared to ego in order to establish identity. <u>The EGOLESS SELF</u> IS, however, the INCOMPARABLE ONE.

The sense of EGOLESS SELFHOOD is completely devoid of any *objectification.* Normal ego-hood is built upon objectification—seeing a self as this or that. The sense of ULTIMATE SELFHOOD is devoid of the sense of ego, hence, the term <u>the EGOLESS SELF</u>.

SELF, PER SE

By the *SELF, PER SE,* is meant the SELF in its ABSOLUTENESS, and *not* at all as it appears in-Cosmos. Of course, everything in-Cosmos is necessarily nothing but the SELF, but the factor of Illusion is present. With respect to The SELF, PER SE, nothing *apparently other* than the SELF can ever be considered.

Example: The <u>SELF, PER SE</u>, is the SELF-AS-SELF. About IT none can speak in TRUTH.

SELF-as-Self-in-Universe

By *the SELF-as <u>S</u>elf-in-Universe* is meant the Universal Logos as the Reflection of the SUPER-Cosmic SELF. This term may also be applied to advancing Entities who are increasingly group-conscious.

Example: Every <u>S</u>elf is Really the <u>SELF-as-Self-in-Universe</u>. There comes a time when every Self-Realized spiritual Self must take a still greater step and discover its *truer nature* as the <u>SELF-as-Self-in-Universe</u>. This step upon the evolutionary arc could be called *'the cosmification of consciousness'* (not to be confused with the term 'cosmification' as it is used to denote the Cosmic Incarnation of the INFINITE SELF).

SELF-'REDUCTION'

By *SELF-'REDUCTION'* is meant the First *apparent* 'ACT' of the SELF. To the SELF, any 'ACT' is necessarily a 'REDUCTION'. This first 'REDUCTION' leads to a whole chain of extra-Cosmic and intra-Cosmic Reductions:

- The Super-Cosmic Self (the Infinite Subject) Is a SELF-Reduction.
- Mulaprakriti (for all Its Infinitude as an Object) Is a SELF-Reduction.
- The Universal Logos imbedded in an 'n-dimensional' Sphere of Cosmic Prakriti Is a SELF-Reduction.
- The Three Sub-Universal Logoi are Self-Reductions.
- The Four Sub-Universal Logoi of Attribute—the Highest Correspondences to the Four Rays of Attribute—are Self-Reductions, and so forth.

Example: Notice the difference between SELF-'REDUCTION' and Self-Reduction. The *SELF* 'REDUCES' ITSELF through 'CHANGE' of 'ATTENTION' and, then, (though ITS 'RADIATED' Pre-Cosmic Agents) by a *narrowing of focus.* The *Self* Reduces Itself by means of the Divine Emanatory Stream. All forms of SELF/Self-Reduction are forms of Maya. Maya (REALLY, the SELF-as-Maya) *Is* the *Reducer.* All SELF/Self-Reduction proceeds through the Veiling Process of Maya. Maya *reduces* the IMMEASURABLE to *measurability.*

SELF-'REFLECTION'

By *SELF-'REFLECTION'* is meant first SELF-'CONSCIOUSNESS', and hence 'MAYA'. SELF-'REFLECTION' is the 'SEEING' of the SELF by the SELF. In one respect, SELF-'REFLECTION' *Is* The 'RAY' which 'FLASHES FORTH' from THAT. The 'FLASHING FORTH' *Is* the 'ACT' of SELF-'REFLECTION', immediately 'TRANSPORTED' 'outside' the DOMAIN of INFINITE SELFHOOD.

Example: SELF-'REFLECTION' is the very *first* 'ACT' of the ALL-SELF as IT 'EMERGES' from the ALL-IN-ALLNESS of Universal Pralaya. It becomes clear that the 'STATE' of *infinitely absorbed* ALL-IN-ALLNESS ('INFINIDENTIFICATION') is *not* a 'STATE' of SELF-'REFLECTION'. *Reflection*, in general, requires *duality* and within the SELF-AS-SELF there can be none.

SELF-Substanded-Self

By the *SELF-Substanded-Self* is meant the **I-as-8.** This unusual wording calls attention to the idea that *substanding* the Self (whether that Self be the Universal Logos or the human Ego) is the SELF, the INFINITE SELF. When one encounters the word 'Self', it is possible that it will be forgotten that the 'Self' is REALLY the 'SELF'. Although emanations of all kinds proceed from the intra-Cosmic Self-of-All-selves (i.e., the Universal Logos), it must be remembered that the Universal Logos Is REALLY a 'RADIATION' of the INFINITE SELF, and has no REAL Existence apart from this ONE AND ONLY SELF.

Example: Every encounter with the incredible diversity of Selves/selves through patterns high and low is simply an encounter with a SELF-Substanded-Self. All the various encounters are pervaded by an irreducible *sameness* regardless of their *apparently* astonishing variety.

SELF, SUPER-Universal

By the *Super-Universal SELF* is meant simply the SELF, with emphasis upon the idea that the SELF is *necessarily* SUPER-Universal.

Example: Even *while* there is a Universe, the ONE AND ONLY SELF 'REMAINS' the SUPER-Universal SELF, and abides as the UNPERTURBED HOMOGENEITY in a Transcendental, SUPER-Universal 'STATE'.

Selves, the Emanatory Sequence of Lessening

By *the Emanatory Sequence of Lessening Selves*, is meant that series of Entities, of ever diminishing scope and capacity, which originally Emanated from the Universal Logos or Self, which Logos was Itself 'RADIATED' from the INFINITE SELF. The 'RAY' of the ABSOLUTE *Is* the 'RADIATION' *per se*, and the Universal Logos Is, as it were, a *Condensation* of that 'RADIATION', (a Condensation developing during the Pre-Cosmic Process which *follows* the initial 'RADIATION').

Example: The E/entities found along the Emanatory Sequence of Lessening Selves certainly do not *lessen* ESSENTIALLY, and, as well, T/they retain *qualitatively* (at least to some extent) the Quality of T/their particular Source. They do, however, appear to *lessen* in the power to influence and pervade Cosmic Prakriti, though even this lessening must, fundamentally, be an illusion because of the *indivisibility* of the SELF.

It must be remembered that all E/entities, *apparently* descending into deeper and deeper prakritic immersion along the Divine Emanatory Stream, never Really "leave the Father's Home", but are merely 'extended', as it were, into various denser fields of prakritic expression. The E/entity we *seem* to see is *not* the only apparently distinct E/entity.

If the 'Ray' of the ABSOLUTE (which every 'Ray'/Entity Essentially Is) is 'retracted', as it were, along the Divine Emanatory Stream, a vast and rich participation upon multiple prakritic levels will be discovered to have *always been in place* for each and every E/entity throughout the Cosmic Aeon. In other words, all of us are Really conscious and active upon many levels but 'we' in our presently localized and circumscribed state of consciousness don't know it!

semi-subjective

By the term *semi-subjective* is meant a subtle state or condition which pertains to man's inner world of patterned energies, but not to the world of PURE SUBJECTIVITY (i.e., the ATMAN in man).

Example: Even the world within which the Monad of man is presently focussed must be considered a semi-subjective world. Whenever Cosmic Prakriti (no matter how subtle) is to be considered in the analysis of a particular E/entity, that analysis is necessarily focussed at a semi-subjective level.

The only truly and completely *subjective* level is the level of the SPIRIT. Subjectivity in-Cosmos can go no deeper than SPIRIT-as-Spirit expressing through Cosmic-Prakritic forms. We must remember again that the usual states which the human consciousness considers subjective are extremely *objective* when considered from even the higher intra-Cosmic Perspectives.

sentiency

By *sentiency* is meant that capacity (in any being) to register impact. Where the human being is concerned, the concept of sentiency usually relates to the astral body, but mental sentiency, buddhic sentiency and still more advanced kinds of sentiency certainly exist.

Example: The astral body, because of its great sensitivity and responsiveness is called the "body of sentiency".

sequence

By a *sequence* is meant a flow of distinct and identifiable conditions/configurations coming into registration by consciousness one after the other (apparently), instead of occurring in consciousness simultaneously. Such a flow of conditions/configurations need not be repeated in order to be named a *sequence* but many sequences are repetitive.

Example: As the disciple approaches the second initiation, the sequence of processes he experiences is ever thus: Dedication, Glamor, and Devotion. Dedication leads to Glamor which is dissipated by Devotion.

serious

By the term *serious* is meant an attitude which always is mindful of the deeper issues even while dealing with apparently superficial things. Ultimately the serious thinker forever holds the link between the INFINITE and the Finite.

Example: A thinker can be serious and still have a sense of humor. To be serious requires simply one thing—never to forget the SUBSTANCE of things. From this point of view it would have to be said that most of us are *not* REALLY serious.

Shamballa

By *Shamballa* is meant the head center of the planet Earth. Shamballa is the major Center of Direction upon our planet, and is called the "Center where the Will of God is Known".

Example: The Beings Who create and function within Shamballa, are, as far as Humanity is concerned, *Will*. The Energy of Shamballa promotes Synthesis and the resolution of Subject and Object into Oneness.

Solar Angel

By a *Solar Angel* is meant a member of the Fifth Creative Hierarchy, a high Angelic Being Who (having created and sustained the human Causal Body on the higher systemic mental plane) supervises the evolutionary development of a human being from human incarnation to incarnation.

Example: Every human being 'has' a supervisory Solar Angel. The purpose of occult meditation is to help the meditator contact and identify with the energy and presence of the Solar Angel.

'something'

By a *'something'* is simply meant a *thing.* A 'something' is any object whatsoever. It is a term used to call attention to any registration in consciousness. The term gives extra emphasis to 'thingness'.

Example: Many apperceptions related to the higher worlds have not been properly named. Until this naming is possible, such contacts may be known as 'somethings'. For instance, when we cannot specify what happened, we often say *"something* happened", or *"something* impressed me".

Son, the (Cosmic and Universal)

By *the Son* or *Cosmic Son* is meant That which *appears* when PARABRAHMAN in ITS 'ASPECT' as Cosmic Father 'Sees' ITSELF *(in reduction)* within ITS 'ASPECT' as Cosmic Mother

By *the Cosmic Son* is meant the specific Form which Arises when the Cosmic Father 'Sees' Itself reflected-in-reduction in the Cosmic Mother.

Example: The Son is 'in' the Father before It is 'in' the Mother. There are two phases of the Son:

1. The Son when it is 'invisible' in the Father, at which time the Son and the Father are One.
2. The Son when it Appears within the Cosmic Mother, because the Father has 'Seen' what is 'within' Himself.

Example: The Son is always a reduced-but-faithful-Reflection of the Father. As the 'Mother' the Father 'Sees' Himself as He *Is.* As the 'Son', the Father 'Sees' Himself as a *reduced embodiment* of Himself. The Father's Essence is *not* reduced within the Son. The Father's *scope* and *extent is* reduced in the Son.

Example: The Father *Is* the Mother. The Son is the *specificity* and *finitude,* first *resident* within the Father and then 'shown forth' by the Mother.

Example: The Universal Son is the "only begotten Son". The Universal Father/Universal Identity has but One Son. That Son, however, (as a Father) may have multiple Sons—very likely, three, but perhaps more. If the Universal Son has 'Sons', then the Universal Father has 'Grandsons'.

Source-Entity (in relation to emanation theory)

By a *Source-Entity* is meant an authentic E/entity-in-Cosmos capable of emanating lesser though still authentic E/entities, which themselves may be capable of becoming Source-Entities. This Emanative Sequence naturally stops at the level of some least unit capable of forming a vehicle for a 'Ray' of the ABSOLUTE, one of the many *seeming* divisions of the One Monad.

Whether this smallest unit is an ultimate particle/event or not is open for debate. Probably it is *not,* as ultimate particles are strictly Fohatic, may *not* be able to *evolve* (because they were immediately *in-volved* by and as Fohat) and belong to a different 'Outpouring' what we usually call the Monads. (But, Really, is there any life in Cosmos which is not, ultimately, the One Cosmic Monad?) At any rate, there *is* a smallest entity in the Divine Emanatory Sequence which *cannot* be, in turn, *emanative*, as Cosmic Law (the Willed Intention of the Universal Logos of a given Cosmos) will allow for no lesser entity, than that smallest entity, to receive and express the 'Ray' of the One Monad.

It could become important to examine the ways in which E/entities *reproduce* themselves. A great many higher Entities reproduce themselves through the Process of Emanation, and their 'offspring' are "Mind-Born" (such as the "Mind-Born Sons of Brahma"). Other lower entities, having "fallen" 'lower' into "generation", do not reproduce themselves by the Process of Emanation. However, even in these cases, it could be said that the strictly *spiritual* aspect of such lower entities (for instance, humans, animals, vegetables, even mineral atoms), come forth from a higher Spiritual Source *emanatively*. The Monads of such life forms are, in fact, emanated 'Rays' of the One 'RAY' of the ABSOLUTE.

The detailed structure of the Emanatory Sequence is *extremely difficult to conceive* nor (at this point in evolution) does it have to be conceived in detail. It is enough to realize that when considering the strictly spiritual nature of all authentic E/entities, the dynamics of *emanation* describe their descent from Unity into Multiplicity (materially) and their re-ascent from Multiplicity into Unity (spiritually).

Example: Our Solar Logos can be considered a Source-Entity with respect to many orders of Solar Angels, though it is disputable whether to consider the Solar Logos a Source-Entity with respect to the Planetary Logoi, as Their Relationship is that of "Brothers".

Example: A Source-Entity is always homogeneous and identical with Itself, and yet, It expresses Itself through an (apparent) emanative heterogeneity. The Source-Entity is a 'Ray' of the ABSOLUTE, which *radiates forth* other 'Rays' of the ABSOLUTE, which, though apparently different from Itself, are identical in Essence with Itself though attenuated in material expression.

space

By *a space* is meant an apparent but ESSENTIALLY illusory interval between apparent things.

By *space* is meant the discontinuous medium in which the relationship between apparently 'extended' things occurs. The illusion of space is created through the illusion of extension. The illusion of extension is created by the nature of consciousness; consciousness always places a limitation upon the registration of REALITY. What we usually call Space is discontinuous, because the *reflected subjective particulations of Fohat* in Cosmic Prakriti are discontinuous. Only 'infinitely dense Space' (i.e., Mulaprakriti) is utterly continuous, and, thus, It is 'impartite'.

Example: Although space as an interval *seems* to exist, and although in the World of Becoming this *seeming interval* must be acknowledged and respected (in order to "Play by the Rules" within that Illusory World), nevertheless, the appearance of a *seeming interval* 'between' things is a function of the deceptive nature of consciousness which objectifies REALITY by inducing the 'appearance of extension'. With the appearance of extension the arising of the illusion of space is inevitable. Space is an *appearance;* extension is an *appearance.*

Example: The Field of Space is a "playing field". The Field is un-REAL and what happens upon it is equally un-REAL. It is all a 'Cosmic Game'.

Example: Amazing revelations begin to dawn when it is realized that there is no REAL space between communicants.

Space

By *Space* (in its Cosmic and intra-Cosmic Modes) is meant the *Arena of Finitization* in which the SELF-as-Triple Point-as-Universal Logos-as-Self finds Objectified Expression.

Super-Cosmic Space is infinite (though periodical) in duration (for it has been recurring cyclically forever) and, before the occurrence of the Pre-Cosmic Process which we might call the 'Centralization of the Infinified Point', Space is infinite in *extent*, as well (though, Really, 'extension' is an illusion, and, thus, we could say that Space is 'concentrated' at an *infinified point).*

Intra-Cosmic Space (Cosmic Prakriti), on the other hand, is limited in duration and limited in extent. Mulaprakriti, however, has associated with It a kind of *infiniteness,* but this is not to say that It is *absolutely infinite.* Nothing *absolutely infinite* can *actually* exist in the other-than-SELF. Only the ABSOLUTE INFINITUDE is *absolutely infinite.*

Example: Space ceases to exist periodically. Intra-Cosmic Space is the Focus of the Attention of the Universal Logos, hence, intra-Cosmic Space *is* Cosmic Prakriti. Such Space is Really Matter (considering Matter as the Root of Objectivity). In fact, in general, Space is Matter. Space is the Content of the Consciousness is Matter. Mulaprakriti Is Infinite Space. Space does not exist when there is no Subject to be Conscious of Itself. The 'Relation' between the Infinite Subject and the Infinite Object is, ESSENTIALLY, the SELF in the 'ACT' of SELF-'REFLECTION' (this 'ACT' having been 'TRANSFERRED' from SUPER-Cosmos to Super-Cosmos. When the SELF-as-Infinite Self 'Sees' Itself, It 'Sees' Infinite Space (i.e., Mulaprakriti).

Example: Cosmic Space is both particulated and unparticulated. In the World of Fabrication, Space is particulated; in the World of Being, Space is 'partite' but unparticulated. This is the difference between discontinuous and relatively continuous Space.

SPACE, ABSOLUTE

By *ABSOLUTE SPACE* is meant no *space* at all. ABSOLUTE SPACE is the ABSOLUTE, ITSELF, in which there is no 'TIME', no 'SPACE' and no 'MOTION'. Time, Space and Motion have existed *forever* but not *always.* (i.e., not at *all* 'Times' in Infinite Duration).

Example: To think of ABSOLUTE SPACE may be useful because it suggests a profound *abstraction.* Such an image may deepen one's 'feeling' for what the GREAT ABSTRACTION IS. But ABSOLUTE SPACE is REALLY the EVER-FULL VOID and not *space* at all, as we usually understand the concept of *space.* If space (as interval) is to exist there must always be at least two, and within the VOID there is but ONE.

Space, Infinite

By *Infinite Space* (not ABSOLUTE SPACE) is meant, from one perspective, Mulaprakriti. Infinite Space is the Unbounded Fundamental Abstract Matter and, also, the possibility for Differentiated Matter—'matter' as we know it. Mulaprakriti, as Infinite Space, is the *possibility* of an infinitude of objective 'things'. Infinite Space (Mulaprakriti) is an Objective Continuum.

Cosmic Space (Cosmic Prakriti) becomes progressively discontinuous through the Agency of Fohat in its various Modes assisting the Self-Reflective Imaging Process (which 'Creates' varieties of Cosmic Prakritic). While pure Mulaprakriti is always a Continuum, perhaps Cosmic Prakriti is relatively *discontinuous* from its inception—the Pre-Cosmic 'Moment' of the *bounding* of the Universe-to-Come. The 'Condensation/Centralization of the Infinified Point' producing the Condensed Point, may contribute to a *species* of *discontinuity* in Cosmic Prakriti. At least it represents a *change* from a thoroughly homogeneous condition, and any change signals discontinuity.

The kind of discontinuity in Cosmic Prakriti before the action of Fohat in its Fourth (particulative) Mode, has to do with 'Stratification of Self-Reflected Images'. Cosmic Prakriti (in relation to the World of Being) consists of a series of 'Image Veils', each Image separated from the one 'above' and the one 'below' by difference in scope and depth. Fohat, in its intra-Cosmic Mode, begins the 'Active Generation of Differentiated Matter', which introduces a new and revolutionary factor into Cosmic Prakriti. From that time, the World of Being (with Its kind of Cosmic Prakriti) and the Fohatically Particulate World stand distinct.

Example: Perhaps no pure Mulaprakriti (no purely Infinite Self-Image) is to be found within the Cosmic Ring-Pass-Not. After all, Mulaprakriti is <u>Infinite Space</u>. Can <u>Infinite Space</u> be found within the Bounded Space which Cosmic Prakriti Is? (Another way of asking this is, Can the largest of all Finite Objects 'Contain' the Infinite Object? Perhaps Cosmic Prakriti is Created from a Pre-Cosmic Condition in which the Infinified Point is transformed into the Condensing Point and, finally, the Condensed Point (which 'Sees' Mulaprakriti in such a 'Specific and Bounded Way' that It becomes Cosmic Prakriti). <u>R</u>eally, the Infinite Subject 'Sees' Itself in such a deliberately limited way that the Image of Itself which *was* Mulaprakriti, the Infinite Object, then becomes Cosmic Prakriti which is the Universal Object (albeit, at that first Stage, unarticulated, undifferentiated).

When the infinitude of Points of View of the Infinified Point become *concentrated* into One Point of View, as it were, and a *boundary* for the New Cosmos is 'Self-Seen', then what we call Cosmic Prakriti arises as *Bounded* <u>Infinite Space</u>, which is, simply, Cosmic Space, the Condensed Point (Self-'Seen' as a Field of Space/Consciousness) as a distinct Singularity of a definite (hence *bounded)* scope.

This *concentration* of the Infinified Point into the single Point which is to be the 'Center of the New Cosmos' is what might be called the 'Second Contraction'. The First Contraction' was the 'FLASHING FORTH' of the 'RAY' of the ABSOLUTE which instantaneously 'GENERATED' (after the instantaneously 'EVANESCENT SUPER-COSMIC TRINITY') the Infinite Subject, Infinite Object, and Infinite Consciousness to relate them—the first <u>S</u>uper-Cosmic Trinity.

space-point

By a *space-point* is meant a dimensionless unit-of-location within-Cosmos. It differs from a time-point.

Example: Where is a thing 'located' in-Cosmos if there is no such thing as an *actual* (i.e., manifested) Real point? If all points (because of the illusoriness of *space)* are REALLY the same Point which (in its turn) has no *actuality* at all, how can there be location? REALLY, 'INFINISPECTIVELY', there *is* no location. However, in the World of Illusion (and even though space-points are only *idea structures)* the concept of the space-point (the point-in-space) is required to understand (albeit illusory) relations.

A space-point is simply a *convention in thought* and a rough approximation to Reality. A space-point is the thought-designated 'center of location' for any located thing. Never mind that such a point cannot *actually* exist! We could also call a space-point the ever-infinitesimalizing core of a Real point, which though indeterminate, would be small enough to satisfy estimations of what a 'center' should be.

Example: Space is an 'expanded' *point*, or a *point*-in-Extension. Expansion or contraction of a point is dependent upon the position and speed of a hypothetical Viewer.

space-in-time

By the term *space-in-time* is meant the interval of Cosmo-Subjective time between ultimate moments within a given Cosmos. A more accurate phrase would be, perhaps, 'spaces in the discontinuous flow of Cosmo-Objective Time'. Whereas no Cosmo-Objectively measurable time exists between such ultimate moments, the World of Fabrication is not 'on' at all 'times', so a 'time' (does exist when Cosmo-Objective Time is not— an Cosmo-Subjective Time *is)*.

Example: The duration of space-in-time does not affect the consciousness of any E/entities Fohatically Fabricated Cosmos. E/entities immersed in that lower World (in that part of their awareness that does not dwell in the World of Being) do not *know* when the Cosmo-Objective World is 'off', their experience is only one of constant 'on-ness'. Never for more that the 'splitest'-split of a second are we separated from realization within the World of Being; however, veiled as we are, and vibratorily obtuse, we do not *know* this.

What goes on during a space-in-time? No-thing in Objective Cosmos! Another way of saying this, is that *all limitation* (as limitation is known within the World of Fabrication) *ceases* (let us say) 'x-tillions' of times a second. The number is unimaginably huge and we have no way, at present (so great is our ignorance of Cosmic Structure) to calculate it.

spirit

By *spirit* is meant a *subject*—a 'Ray' of the ABSOLUTE—which is ESSENTIALLY, *the* 'RAY' of the ABSOLUTE.

Example: A Monad is an authentic subjective Self, which is ESSENTIALLY a 'Ray' of the ABSOLUTE SELF. A Monad is, Really, a Spirit at-one with the One Universal Spirit, at-one with SPIRIT.

Example: Spirit is always identical with ESSENCE.

spirit, divisibility of the

By the *divisibility of the spirit* is meant the capacity of the trained identity to bi-locate, tri-locate, multi-locate and, eventually, omni-locate in-Cosmos. Conscious Omni-Location is the capacity of the Universal Logos alone.

Example: Through the divisibility of the spirit, God watches and knows when "the sparrow falls". Due to the factor of the Omnipresence in-Cosmos of God-the-Universal-Logos, it is reasonable to say that 'God *Is* the falling sparrow' as well as the Observer of the fall. The further one 'retracts' the principal focus of consciousness 'upwards' along the Divine Emanatory Stream, the more one participates in all life events through identification, and, simultaneously, the more one is *abstracted* into the Attitude of the Observer.

Example: The phrase, "the divisibility of the spirit" does *not* mean that spirit is REALLY or ESSENTIALLY divisible. While spirit has capacities for Omnipresence, it can never be fragmented or, REALLY, 'split' into multiple identities. All identities (despite appearances to the contrary) remain One Universal Identity-as-IDENTITY.

SPIRIT-as-Spirit-as-spirit, indivisibility of the

By the *indivisibility of the SPIRIT-as-Spirit-as-spirit* is meant the fact that a *subject* or a *spirit* (no matter what the *prakritic mode* through which it may be manifesting) cannot REALLY be divided in any way. The spirit is the Spirit Is the SPIRIT (i.e., the Monad is the Universal Monad Is *the* ULTIMATE MONAD) and the SPIRIT IS the GREAT INDIVISIBILITY.

Example: One realizes the indivisibility of the SPIRIT-as-Spirit-as-spirit when one realizes that ESSENTIAL IDENTITY cannot be divided, or changed in any way. How can an indivisible SPIRIT-as-Spirit-as-spirit *appear* to be divisible? Cosmic Prakriti is divisible (or at least can become a Reflection of multiplicity and divisibility), and so the indivisible SPIRIT-as-Spirit-as-spirit can manifest through a great number of prakritic appearances, but always as the Same PRESENCE/Presence. If is as if one and the same Eye saw through a myriad of orifices simultaneously.

state

By a *state* is not meant exactly a *condition,* though in ordinary usage they are similar. By a *state* is meant the nature and quality of the relatively abiding relationships which characterize the Realities-in-Universe. Conditions refer more to *actualities* rather than to the relatively Real, and are far more unstable than states.

Example: On a human level it would be proper to inquire about the condition of the personality, and the state of the causal body. We could as well inquire about the state of Planetary or Universal Archetypes but less profitably about Their condition, for being the stable Formative Patterns of Cosmos, They do not Really change condition.

'STATELESS STATE'

By the paradoxical term *'STATELESS STATE'* is meant the *beginningless, endless* ONGOING SAMENESS of NATURE of THE NO-THING.

Example: There is *no-thing* like IT. The term STATELESS 'STATE' emphasizes ITS *super-rational uniqueness* about which nothing *ultimately true* can be said. Since ITS 'STATE' can never REALLY vary or be defined, that 'STATE' IS, as it were, 'STATELESS'.

Subject

By a *Subject* is meant One Who 'Sees'.

Example: Every Subject is/becomes an Object to the Object It 'Sees'. If an 'Seen' Object is endowed with Subjectivity (because it cannot *be* other than the Subject that 'Sees' It), then that Object (now a Subject as well) must be able to 'See' the Subject which 'Saw' It. (i.e., 'See' that Subject as an Object).

Example: Every Subject (other than the INFINITE SUBJECTIVITY) is also an Object, for every Subject is, as it were, 'Mayavically 'distanced' by 'Sight' from that SUBJECTIVITY. Yet if there is such a thing as an Object (however illusory that Object may be) can It be other than THAT? What 'ELSE' is there? The INFINITE SUBJECT is 'ESSE-IDENTIFIED' with every possible object, even if IT, ITSELF, cannot 'SEE' them, and must 'SEE', as it were, by *proxy.*

SUBJECT, the

By *the SUBJECT* is meant an illimitable, infinite BEING WHO for all ETERNITY IS the SOLE 'GENERATOR' and SOLE 'RECIPIENT' of all possible registrations, including the 'REGISTRATION' of ITSELF by ITSELF.

Example: The SUBJECT has been the one and only *theme* of all that has ever transpired or will transpire. As students of life, the SUBJECT is our *only subject.*

Example: The ONE SUBJECT is subject only unto ITSELF.

SUBJECT, the 'EVANESCENT' INFINITE

By *the 'EVANESCENT INFINITE SUBJECT'* is meant the 'SUBJECT' which the INFINITE SUBJECTIVITY 'BECOMES' for (perhaps) an *infinitesimal instant* (due to the 'ARISING' of 'MAYA') before *instantly* 'BECOMING' the Infinite Subject.

Example: For the *infinitesimal instant* that the 'EVANESCENT INFINITE SUBJECT' 'ARISES', the 'EVANESCENT' INFINITE OBJECT, and 'EVANESCENT INFINITE MAYA' also 'ARISE'. These Three, in virtually "no Time at all" 'BECOME' Their Pre-Cosmic Correspondences:

1. The Infinite Subject
2. The Infinite Object
3. Maya as the Consciousness of Pre-Cosmic Infinitude

Subject, Infinite

By the *Infinite Subject* is meant the First Subject appearing (with *infinitesimal instantaneity)* in Super-Cosmos after the 'FLASHING FORTH' of the 'RAY' of the ABSOLUTE. The Infinite Subject can also be called the Infinified Point, the Infinite Father, and the Infinite Self.

Example: When the Infinite Subject 'Sees' Itself, the result of the 'Seeing' is the appearance of the Infinite Object, Mulaprakriti, (the 'Self-Image' of the Infinite Subject).

Example: The Infinite Subject is a an *infinitude* of a different order from the INFINITE SUBJECTIVITY.

Example: The Infinite Subject (because is has been deprived of the *pure* INFINITUDE of the INFINITE SUBJECTIVITY) should, at one point in the Pre-Cosmic Process, be called, the *De-Infinitizing Subject* once It begins to 'See' 'something' definite other than Its Wholly Infinite Self. In Post-Cosmos, the parallel Entity would be called the *'Re-Infinitizing Subject'.*

Example: In Post-Cosmos, the reapproach of the Infinite Subject/Infinite Object to the INFINITE SUBJECTIVITY is not an Infinitizing/Infinispectivizing approach/approximation at near *infinite speed* (i.e., a speed which only *infinitesimally* less than *infinite speed),* but, presumably, rather a 'COUNTER-FLASH' of 'REABSORPTION, paralleling the 'FLASH' which, in effect, 'BECAME' them at the Dawn of Pre-Cosmos

Example: If the Infinite Subject 'Sees' anything less than its utter Selfhood (which is infinite) it can no longer remain the Infinite Subject, but must become the 'De-Infinitizing Subject' or the 'Re-Infinitizing Subject'. A truly Infinite Subject cannot 'See' anything other than the Wholeness of Its Infinite Self. The sight of Particularity means the Infinite Subject can no longer be a truly Infinite Subject.

Subject, the Infinitizing (or 'Re-Infinitizing') and 'De-Infinitizing'

By *the De-Infinitizing Subject* is meant the Super-Cosmic Subject (immediately heretofore Infinite, and now 'Re-Focusing') that can 'See' something other, hence, lesser, than Itself. Another name for this Subject is the Condensing Point. A 'De-Infinitizing' Subject is on Its way (at whatever 'speed' appropriate to that Pre-Cosmic Process) to becoming definitely Finite, and capable of 'Seeing' Itself as One Definite Thing (i.e., as Cosmic Prakriti, rather than Mulaprakriti or the unimaginable fecundity of Mulaprakriti).

By *the Infinitizing* or *'Re-Infinitizing Subject'* is meant the a Universal Logos, *after* the fulfillment of Cosmos, who is returning from 'Seeing' Itself as Singularity/Individuality, to 'Seeing' Itself as Infinitude. Another name for this Subject would be the 'Un-Condensing Point'. The Re-Infinitizing Subject is on Its way to becoming the Infinite Subject capable of 'Seeing' only Its Own homogeneous Infinitude as Mulaprakriti (the Infinite Object).

Example: A Real Point can be 'Seen' only by an Infinitizing Subject or a 'De-Infinitizing Subject'. An Infinite Subject can 'See' only the 'Pointless Infinified Object-Point' which It-as-Mulaprakriti, *Is.*

Example: That which we call the 'Infinispectivizing' is the Perspective of either the 'De-Infinitizing' Subject or the 'Infinitizing'/'Re-Infinitizing' Subject.

SUBJECTIVITY, the INFINITE

By *the INFINITE SUBJECTIVITY* is meant BE-NESS ITSELF. The INFINITE SUBJECTIVITY is not the Infinite Subject, because the Infinite Subject is part of the Primal Pre-Cosmic Duality which exists because the Infinite Subject ongoingly 'Sees' the Infinite Object. There is no REAL 'SEEING' 'within' the INFINITE SUBJECTIVITY, although such 'SEEING' (in a 'FLASH') *inaugurates* the Pre-Cosmic Process in which a more prolonged 'Seeing' of Object by Subject (a 'Seeing' subject to Pre-Cosmic Time) can occur. The INFINITE SUBJECTIVITY IS the ALL-SELF 'in' ITS ultimate 'STATELESS-STATE' of *preoccupied* ALL-IN-ALLNESS.

Example: The INFINITE SUBJECTIVITY has no Object Consciousness 'in' IT, but rather is 'INFINIDENTIFIED' or 'ESSEDENTIFIED' (if words can be put to the description of ITS ETERNAL 'SELF-PREOCCUPATION'). In a way, however, all *consciousness* is object *consciousness*, if by 'object' is meant any registration whatsoever.

If we speak of "consciousness without an object" we are Really speaking of the Field of Consciousness as the 'object'. Such is Really a state of Self-awareness in which the Point of Selfhood is 'Seen' as the Field of Consciousness. Further this State can be said to involve *identification*, absorption in the state of *Being*. There may be consciousness without 'delimited', shaped objects possessing normal 'extension', but there is no *consciousness, per se*, "without an *object*". In what usually passes for "consciousness without an object", the mayavically projected, Presence-Image of the Self has become the Field of Consciousness.

Note: The problem, as ever, is to specify the *Mayavic* 'MOMENT' when 'SEEING' or 'CONSCIOUSNESS' 'AROSE' in IT. The Primal 'ACT' is simply a new SELF-'ORIENTATION' differing from *infinitized SELF-'PREOCCUPATION'*. The 'ACT' *is* effectively SELF-'DIVISION'—the Arising 'outside' the ABSOLUTE of the SELF-as-Infinite Subject 'brooding over' the SELF-as-Infinite Object. One could almost say that the 'ACT' 'OCCURS' as the SELF-as-'MAYA' arises within ITSELF, instantly becoming the Super Cosmic Trinity. 'MAYA'-instantly-Maya is the SELF 'LAUNCHED' upon the Path of SELF-'LIMITATION'-as-Limitation in Super-Cosmos and Cosmos.

'Subjectobjectivity' (Infinite)

By *Infinite 'Subjectobjectivity'* is meant the state of Infinite Spirit-Matter, or the *mutually cognizant* 'Embrace' of the Infinite Subject and Infinite Object. In 'Subjectobjectivity', Infinite Vision prevails and no 'things' can be seen in-'particular', or *all* 'things' in-'particular' (as *all* 'points-in-potential' or as *all* 'points-occurred') the whole of which 'Seeing' is equivalent to 'Seeing' *no*—'things'. In 'Subjectobjectivity' no *point* appears—no specificity; only Infinitude in general.

Example: In 'Subjectobjectivity' only the unarticulated Infinitude of Mulaprakriti is 'Seen'. If the 'Seen' 'Becomes' the 'Seer', then, in 'Subjectobjectivity', Mulaprakriti 'Sees' the Infinitude of the Infinite Subject.

Super-Cosmic

By the term *Super-Cosmic* is indicated a context which is greater than the context of a particular Cosmos, but of an Order less than the INFINITUDE.

Example: The considerations of the infinitude of Cosmoses past and to come, as well as of inter-Cosmic intervals past and to come, are Super-Cosmic Considerations.

Super-Cosmic Consciousness

By *Super-Cosmic Consciousness* is meant a 'REDUCTION' in ABSOLUTE 'CONSCIOUSNESS'. ABSOLUTE 'CONSCIOUSNESS' (which cannot be like any kind of consciousness with which we are familiar and should, perhaps be called 'INFINIDENTIFICATION') is the *infinitization* of *consciousness* which, with all other Noumena, is *absolutized* in the ONE NOUMENON called the INFINITESSENCE. Super-Cosmic Consciousness, however, must be considered a more exalted (though less specific) State of Consciousness than that possessed by the Universal Logos. Remember, the Universal Logos is, ESSENTIALLY a 'RADIATION' of the ONE SELF and can be illuminatively thought of as the SELF-as-Condensed Point. The Infinite Subject can be thought of as the SELF-as-Infinified Point.

Example: Super-Cosmic Consciousness Is *'Pointness'*, and 'Pointness' is, *at first*, before *condensation,* Infinitely 'Pointed'—the Infinite Subject (the Reflection of the Infinitude of Infinitessentialized Points 'within' the INFINITE SELF) *Observing* Its own Reflection as the Infinite Object through an infinity of points, i.e., from all possible points of view.

Super-Cosmic Consciousness (which is the Infinite Subject's Consciousness of the Infinite Object) emerges with the appearance of the Infinified Point of Observation. The following is an *equivalence:* the arising of the Infinified Point Is the (apparent) Finitization of the SELF, and Is a Pre-Cosmic Process/Event which is identical to the arising of the Infinite Subject (Who Is the Infinite Observer with Its Super-Cosmic Consciousness).

It must be understood that, for the INFINITE SELF, the Super-Cosmic Consciousness of the Infinite Subject, by means of which Mulaprakriti is Registered, is a great *limitation.* To the SELF, *any* form of *consciousness* is a limitation. Consciousness is always infinitely less that INFINITIZED BE-NESS.

Super-Cosmic Family

By the *Super-Cosmic Family* is meant principally the Universal Logos, the Universal Son, Intra-Cosmic Fohat, Per Se, as the Holy Spirit, and undifferentiated/unarticulated Cosmic Prakriti, as the Cosmic Birther of Forms. It can be questioned whether the INFINITE SELF and ITS immediate 'EXTENSIONS' via the 'RAY' of the ABSOLUTE (namely, the Infinite Subject and the Infinite Object, or Mulaprakriti) should be included.

Example: The Members of the Super Cosmic Family are Really the only Actors or Players in the Cosmic Drama. Every one of the Them is ESSENTIALLY none other than the ONE AND ONLY INFINITE SELF.

SUPER-Cosmic 'IDEATION'

By *SUPER-Cosmic 'IDEATION'* is meant the 'BIRTH' of the 'IDEA' of the Universe-to-Come 'within' the INFINITE SELF. The IDEAS'/'THEMES' 'EXTRUDED' from the INFINITESSENCE—'IDEAS'/'THEMES' that are to be enacted through Cosmic Prakriti (i.e., through Self-Reflection) by the Universal Logos in the Universe-to-Come—are 'Carried' by the 'RAY' of the ABSOLUTE, Pre-Cosmic 'FOHAT', Who Is the 'AGENT' of the ABSOLUTE. The 'RAY', it seems, *splits* the BE-NESS of the ABSOLUTE (first into the

'EVANESCENT INFINITE TRINITY') and, immediately, thence into Infinite Subject/Infinite Object (and the Infinite Relation Between) and, in fact, *IS* that *splitting.*

The 'RAY' is the 'CHANGE' which 'INAUGURATES' Duality. The 'RAY' is the 'ACTION' of 'BECOMING' Two. The 'RAY' (which we are calling SUPER-Cosmic FOHAT) Is as much in the Infinite *Object* as the Infinite Subject! Yet, despite this 'CHANGE' which is the first 'DIVISION', BE-NESS 'REMAINS' *as ever.* Via the 'RAY', the Infinite Subject is 'ENDOWED' with the 'IDEAS'/'THEMES' for the Cosmos-to-Come and the Infinite Object is 'ENDOWED' with the Reflectivity which is the Power of Objectification (the Power to Objectify that with which the Infinite Object is 'ENDOWED')—the *distinguishing feature* of Root Matter. The 'IDEAS'/'THEMES' (now, Ideas/Themes, since they have *gone forth* from the ABSOLUTE) are 'Carried' by the Infinite Subject through all Pre-Cosmic Processes until the Universal Logos is 'Born' from the SELF-as-Condensed Point.

The question may arise, Who or What really develops or unfolds the Ideas and Themes which are to be manifested in a Cosmos? It may be that throughout the Pre-Cosmic Process, the Infinite Subject-as-Infinified Point and later the Infinite Subject-as-Condensed Point is *instinct* with the Ideas and Themes but they are not *developed* and *unfolded* until there is a Cosmos.

So, the 'FLASHING FORTH' of the 'RAY' 'GENERATES' (after the briefest of 'INTRA-SOURCE' 'GENERATION') an Infinite Subject with an infinite, infini-directional point of view of an Infinite Object, which is also simultaneously 'GENERATED'. The Infinite Subject 'Sees' Infinite Space utterly dense—reflective, ESSENTIALLY, of the UTTER HOMOGENEITY of PARABRAHMAN. This infinite Point of View is 'Seen' from what we are calling the SELF-as-Infinified Point, or, more simply, the Infinified Point. Whether, as earlier suggested, the SELF-as-Infinified Point *sees in particular* the entire Memory of Mulaprakriti, as well as ALL that *might* appear Reflected within Mulaprakriti-as-Cosmic Prakriti (if the 'FOHATIC INSTRUCTIONS' to so Reflect were given, both of which are doubtful, or whether such a Vision, if it exists, is the 'Province' of the 'De-Infinifying Point' or 'Re-Infinifying Point') is interesting to consider. However, this is completely speculative, at best, and certainly cannot be solved at this time by human beings.

The Infinified Point 'Contracts' and 'Condenses' to a single focus. As it does so it creates an Objective 'Spherical' Boundary—the Pre-Cosmic Egg, which is Really a Spherical Field of Consciousness. Instead of all possible points of view (including an infinite Vision of all possible experiences past and to come in an *articulated state),* there is now instituted (in the 'Movement' towards the formation of Cosmos) one 'Spherical Point of View'.

The Infinite Subject has 'Now' 'Become' the SELF-as-Condensed Point, and the Infinite Object is now the much-restricted *Bounded* Object. The SELF-as-Condensed Point now has limited Vision, i.e., Cosmic Vision. The contraction of the Infinified Point to the Condensed Point (the Single Spherical Point of View) is the Formation of the Universal Logos. Cosmic Prakriti is still perhaps (or perhaps not) as dense as Mulaprakriti (which theoretically means that the Self-Perception of the Focused Subject is *homogeneous and unarticulated* just as was the Infinite Self-Perception of the Infinite Subject, but there has been a change). It is the Contraction and Sphericalization (with the Consciousness of the Infinite Subject becoming the Consciousness of the Focusing Subject) that have changed

Mulaprakriti into Cosmic Prakriti—from Infinite Homogeneity to 'Bounded' Homogeneity.

In Pre-Cosmic Psycho-Spiritual terms, this means that there has been a de-emphasis upon Mulaprakriti as a Reflection of Infinitude of the Infinite Subject, and an emphasis upon a *Cosmo-Pertinent Prakritic Reflection* of a Focused Pre-Universal Subject, which Reflection can, in all accuracy, no longer be called Mulaprakriti, but must be called *Cosmic* Prakriti. The two 'Images' (for that is what they Really are) of Mulaprakriti or Cosmic Prakriti are different, though there are similarities in terms of unarticulated homogeneous potential. Mulaprakriti, however, ever remains the Root of all lesser reflections which arise in the ever-narrowing, Cosmically-Intent, Pre-Cosmic Consciousness of That which will 'Become' the Universal Logos.

It must be said with regard to both the Super-Cosmic Process and the intra-Cosmic Process that movements towards precipitation and concretion do not negate the existence of the more abstract state out of which such precipitations and concretions emerged. Mulaprakriti must *remain* as It *Is* even though Cosmic Prakriti is formed, just as the INFINITE SELF 'REMAINS' as IT IS, even though the 'RAY' has 'FLASHED FORTH'. Indeed, the Infinite Subject (under the sway of the Reflective Power of Maya) continues to 'See' Itself as the Infinite Object, and so Mulaprakriti persists in purity no matter what other lesser prakritic reflections arise. We are seeing a vision of 'layers within layers'; 'layers' of Objectivity within 'layers' of Objectivity; 'layers' of Subjectivity within 'layers' of Subjectivity. All *reductions* are only *apparent reductions.*

This same dynamic holds in-Cosmos—the Universal Logos, Remaining exactly what It Is, no matter what kind of Emanations may Go Forth from It. The analogy is in the *flame*, which retains is form and vitality no matter how many other flames are ignited from it.

Example: The Ideas and Themes to be developed in Cosmos as a result of Super-Cosmic 'IDEATION' are 'IMPLANTED' within the Infinite Subject-as-Infinified Point with the 'FLASHING FORTH' of the 'RAY' of the ABSOLUTE. The Infinite Subject-as-Infinified Point (although capable of 'Seeing' the fullness of the Potential Infinite Productivity of Mulaprakriti, as this Point proceeds through the Process of 'Condensation' and 'De-Infinification') remains *instinct with* Its 'INSTRUCTIONS' *conveyed* to It by the 'RAY' of the ABSOLUTE.

Super-Cosmic Realm, the

By *the Super-Cosmic Realm* is meant the 'Field' of the Infinite Subject and Infinite Object, or, worded otherwise, the Field of the SELF-as-Infinified Point and of Mulaprakriti. The Super-Cosmic Realm is the Domain of Pre-Cosmic Subject/Object Relations. The SELF-as-Condensing and then Condensed Point as well as incipient Cosmic Prakriti should also be included.

Example: What transpires in the Super-Cosmic Realm is, on the one hand, the preparation for the Creation of the Universe-to-Come, and, on the other (a Cosmic Aeon later), the resolution of the dissolved Universe into the Primal Duality of Infinite Subject/Infinite Object prior to Their reabsorption into the ABSOLUTE—the *epiphany of the returning 'Ray'-as-'RAY'.*

SUPER-Cosmic REALM, the

By *the SUPER-Cosmic REALM,* is meant the REALM of the ABSOLUTE.

Example: SUPER-Cosmic 'IDEATION' Really 'BEGINS' (and 'ENDS') *apparently* in a 'FLASH' within the SUPER-Cosmic REALM, but is specified and elaborated as Super-Cosmic Ideation-Become-Cosmic Ideation within Cosmos, Itself.

Super-Cosmic Self, the

By *the Super-Cosmic Self* is meant what we will call the Infinite Subject (or the SELF-as-Infinified Point). It is an Infinite Being ESSENTIALLY identical with the ONE AND ONLY BEING/NON-BEING, but *actually,* 'midway', as it were, between the INFINITE SELF and the Universal Logos (which may also be called SELF-as-Condensed Point). The Super-Cosmic Self is the SELF *before* IT becomes the *Pervading Agent/Director/Informer* of the New Universe—the SELF-as-Universal Logos.

The Super-Cosmic Self, however, has another Aspect to It. It should not be thought that there is no inhering SELFHOOD in Mulaprakriti. MULAPRAKRITI is the Reflected SELF and, therefore, Mulaprakriti can be considered the Receptive Pole of the Super-Cosmic Self. SELFHOOD is All-Pervading. There is no factor in the UTTER-ALLNESS which is not fully and completely the ONE AND ONLY SELF.

Example: The Super-Cosmic Self Is an Infinite Subject 'Seeing' an Infinite Object, and the Super-Cosmic Self is the Infinite Object being 'Seen' by the Infinite Subject (and, hypothetically, 'Seeing' that Subject as well), but the Super-Cosmic Self is not the INFINITE SELF in WHICH there is no 'OBJECT' to 'SEE'.

SUPER-SYSTEM; SUPREME-SYSTEM

By the *SUPER-SYSTEM* is meant both the ALL-SELF *and* the particular Universe of the Moment. The SUPER-SYSTEM is the combination of the ALL-SELF and the All. A still greater SYSTEM, the SUPREME-SYSTEM would be the combination of the ALL-SELF and the ALL.

Example: An infinitude of particularized details resides in the SUPREME-SYSTEM composed of the ALL-SELF and the ALL (i.e., the contents of the infinitude of Universes which have been manifested). The contents of the SUPREME-SYSTEM are still, ESSENTIALLY, reducible to the ALL-IN-ALLNESS. The SUPREME SYSTEM and the UTTER ALLNESS are equivalent.

Example: The concepts of a SUPER-SYSTEM and a SUPREME-SYSTEM suggest the existence of Duality, but the INFINITE SELF and ALL that 'EMERGES' from IT are naught but the INFINITE SELF. In other words, if Infinity is added to INFINITY the result is simply INFINITY. An infinity of particularities is, ESSENTIALLY, nothing but the INFINITE HOMOGENEITY.

system

By *system* is meant a coherent and dynamic pattern of items-in-Universe held in relationship by the purposeful intent of some Self-conscious being(s). Systems may also be Super-Cosmic, and by some modes of thought, inclusive of the INFINITE SELF.

Example: The Grand Design of the Cosmos is revealed through an understanding of virtually countless, hierarchically organized, interrelated systems. The entire Cosmos as well as the tiniest atom are both systems.

Example: There is a Super-Cosmic System (*not* the SUPER-SYSTEM) as well as an immense number of Intra-Cosmic Systems. The main Participants in the Super-Cosmic System are the 'RAY' of the ABSOLUTE-as-Super-Cosmic 'Fohat', the Infinite Subject-as-Infinified Point, the Infinite Subject-as-Condensing and Condensed Point, and Mulaprakriti.

Of course, every one of these 'Players' or Participants is, ESSENTIALLY, one and the same ENTITY/NON-ENTITY. This is the *great simplicity* lying behind all these complex differentiations of *Actors*—the ONE AND ONLY IS 'BEING'-it-*All* and 'DOING'-it-*All!* If terminology ever gets too burdensome (as well it may!) this great and simple fact must be remembered.

- *T* -

Tetraktys, the Supernal

By *the Supernal Tetraktys* is meant, the Ten Emanations who are in the Company of the Universal Son. These include the Three Persons or Sub-Logoi Who immediately 'surround' the Son, and then, the Seven Subsidiary Logoi. These Ten Beings/Emanations are to be found focused within the World of Being (along with the 'implicate', 'enfolded' Spirit Aspect of every authentic E/entity').

Example: The Emanation of the Supernal Tetraktys could occur is several ways:

1. The Universal Son could Emanate all Ten, discovering Them all as Sons within His Nature.
2. The Universal Son could emanate the Principal Three, Who, Themselves, would emanate the Subsidiary Seven.
3. The Universal Son could emanate the First of the Principal Three, Who, then, would emanate the Second of the Three, which, then, would emanate the Third of the Three, and so forth.

There are other possibilities as well. As, according to analogy, a Father and Mother can have more than One Son (except in the case of the "Only Begotten Son"), option number two seems reasonable and attractive. Again, these models are proposed for the sake of discussion: there is no attempt to state that any model *must* be correct.

that

The term *that* is a word-symbol which denotes a presence.

Example: The use of the word 'that' always reveals that an object has been registered. If this is so, there is great difficulty in calling THAT, 'THAT'!

thing

By the term *thing* is meant any object appearing as a *presentation in consciousness.* As there are no objects without consciousness, there are no *things* without some manner of perceiving/apperceiving consciousness to 'create' them through focus. This means that *things* are created by the focus of some consciousness. Without this kind of focus there is only NOTHING—a 'STATE' of PURE 'THINGLESS' REALITY.

By a *thing* is meant a point (either a virtual point or a Real point depending upon the perspective of the Perceiver).

Example: By a thing is meant a perceived/apperceived item.

Example: There are many 'things' to be noticed in the interior worlds, but they are for many as if non-existent, because such consciousnesses fail to notice them or focus upon them.

Example: One cannot properly speak of things occurring *at* a point in Space but only *in* a point in Space. Points are things, objects. Where there are no things/objects, there are no points! This is why things cannot move through points in space—because (within the Domain of Extension) things cannot move through things! From another perspective, points are 'within' points, and all things are 'within' all things, there being only one Point in Space (even this Point being un-REAL).

thrill

By a *thrill* is meant a process of intensest enlivening caused by identification with LIFE/Life and BEING/Being. The sutratma (life thread) is the organ which imparts the thrill of livingness.

Example: Those who *know* the PRESENCE *are* the PRESENCE and are thrilled to be alive—alive 'within', and identified with and *as* the PRESENCE.

Thrill, Universal

By the *Universal Thrill* is meant the All-Enlivening Presence of the Universal Logos which induces *blissful livingness* in any Self-conscious E/entity who has entered into It.

Example: He who has identified with the Pervader of the Universe *lives* ecstatically within the Universal Thrill.

time

By *time* is meant, the often arbitrary division of Endless Duration into sequential and (probably) equal units, the length of these units usually being determined in relation to various vibratory movements/cycles, great and small, possible and actual, within a particular Finite Universe. The term 'time' (with no capitalization) is used to denote the general concept of time, but not 'Time', as a great Principle.

Example: The length of the unit of time called the 'day' is determined by a cyclic planetary motion—one complete rotation of the Earth upon its axis.

By *time* is meant simply the number of cycles which recur during the span of another cycle chosen as a standard of measurement. The measurement of time, then, is simply based upon the comparison of cycles. If a given cycle (cycle 'a') is chosen as a standard of measurement to be used in all measurements of time, one measures time by counting how many cycles of cycle 'b' and how many cycles of cycle 'c' elapse during (or are 'contained in) *one* cycle of cycle 'a'. There can be no measurement of time unless a standard of measurement is selected.

Example: During that standard of time-measurement called the 'year' (a cycle created by a single revolution of the Earth around the Sun), many other lesser cycles elapse in a predictable way. The natural cycle called the 'day', for instance, occurs approximately 365 times during a single yearly cycle. The natal cycle called the 'month', and marked by successive lunations occurs approximately thirteen times during a single yearly cycle. When we want to know 'how much time' we 'have', or 'how much time' an activity will 'take', we simply compare that activity with certain regularly recurring, well-established cycles which are our standards of measurement—cycles such as the day, the month and the year, or more arbitrary standards such as the week, the hour, the minute, etc.

Example: Every Self-conscious creator in-Cosmos requires time in which to complete its intended cyclic unfoldment.

Example: When some of the permutations of the word, 'Time' are analyzed—words such as, 'Emit', 'Mite', 'Meti-', 'Item', 'It-Me', 'Ti-Me'—a wealth of philosophical understanding can result.

Example: The concept of 'time' is meaningless without reference to the concepts of 'space', 'division', 'comparison'—and 'motion'.

time, 'space' of

By a *space of time* is meant the duration of a unit of time or of a sequence of units of time.

Example: Time, as it were, takes 'space' along the Infinite Time Line. When a given 'space' of time is compared with a different 'space' of time used as a standard of measurement, one can determine the relative duration of the first 'space' of time which was measured against the standard, *in terms of the standard.* If I measure the 'space' of time it takes to travel from one place to another against a standard 'space' of time called the hour, then I can measure the first 'space' of time against the second 'space' of time, and express the first 'space' of time *in terms of the second 'space' of time*. In other words, it takes a specific number of *hours* (a specific number of second 'spaces-of-time') to travel from one place to another (first 'space' of time). In this case, a certain number of standard 'spaces of time' (the hour) can be counted during the 'space' of time to be measured.

Example: The term 'space' of time is useful in relating to the Time/Space Continuum (which, REALLY, is *no true* continuum).

time, ultimate unit of

By an *ultimate unit of time* is meant the minutest quantum of time possible within a given Objective Universe. Presumably, this unit can vary from Universe to Universe, depending upon the 'key measure' ('key frequency') of the Universal System. An ultimate unit of time within an Objective Cosmos is keyed to the most rapid 'unit of change' (probably the change of 'position' of an ultimate particle) possible in that Objective Cosmos. Subjective Cosmos is an entirely 'Ideational Word', in which Time is far more elastic and is probably not quantized as it is in Fohatically Particulated Cosmos.

By an *ultimate unit of time* is meant an 'ultimate moment'. The ultimate unit of time is the 'time it takes' for the most rapidly changing particle in-Cosmos to appear, endure/hold position, and disappear. The particle will then move discontinuously (by quantum leap) from one apparent position-in-Space to another, but the "leap" or change of position *seems* to take no time at all because of the phenomenon of Ontological Oscillation. A particle *begins* each ultimate moment in a *new* position which it seems to 'reach' (due to Ontological Oscillation) instantaneously or in a "totally negligible amount of time", or "in no time at all." Vibration may well be *quantized change of position*. The seeming continuity of vibratory motion or wave motion is, from this perspective, very much an illusion.

Example: The time which elapses during the most rapid single *apparent* movement possible in an Objective Cosmos is an ultimate unit of time for that Objective Cosmos. The most rapid single movement in Cosmos is obviously not the time it takes for an item to go from point 'a' to point 'b' in Space, but, instead, is the duration of the appearance/disappearance of an ultimate particle/event as measured against some more permanent Super-Cosmic Time Standard. It is interesting to realize that Fohat, as 'Speedy' as He Is, must "observe the (cosmic) speed limit" and, in fact, *enforces* it.

Example: With regard to the measurement of ultimate moments the manner of measurement (which may some day available) may be to measure the duration of the particle/event from the onset of Position 1 to the end of Position 1 (which, in Objective Cosmos, will *seem* to be the *beginning* of Position 2—but of course, an inter-moment instant will, according to this theory, intervene between the 'end' and the next 'beginning'). In such a measurement of an ultimate moment in Objective Cosmos, the interval of negation will necessarily not have been considered because it cannot be registered objectively. Time will seem to be 'moving' about twice as fast as it really is, and the duration of the Cosmos will Really take about twice as long as it *seems* to take intra-Cosmically.

So much concerning the exact duration of a Cosmos will depend upon the duration of the 'interval of negation' (and whether it is entirely regular or elastic). If Cosmos is like a Game of Chess, the 'Players' take a long (and variable) time to contemplate their next move. What an interesting Cosmos it would be if this were the case.

What we are saying is that measurements made with respect to Objective Cosmos will only be able to account for 'intervals of affirmation', the onset of which will be noticeable by seemingly instantaneous changes of position of ultimate particles (providing there is any way for them to be detected). There is the curious thought that, due to the fundamental immobility of Objective Cosmos (the Cosmic Chess Board standing still) the kind of measurements we are talking about may have to be assayed from within the Realm of Cosmo-Subjectivity, from which the whole Process can be 'Seen', and something can be 'Done'.

Clearly, we do not have to worry at this time about the either external or 'internal' technology which will make such measurements possible, but a theoretical consideration of the issues involved in certainly promotive of interesting thought.

Time

By *Time* is meant one of the three basic Structural Principles in-Cosmos. Space and Motion are the other two. The capitalization is used when time is regarded as a Cosmic Fundamental.

Example: Time, Space and Motion are three great Categories of Cosmic and Super-Cosmic Perception. Their existence depends upon the existence of Consciousness. Where there is no Consciousness they are not to be found. There is no Consciousness 'within' the ABSOLUTE, therefore Time, Space, and Motion cannot exist 'within' the ABSOLUTE.

Time, Infinite

By the use of the term *Infinite Time* is indicated the fact that Time and Its 'Brothers'—Space, Motion, Number, Object, etc.,—have existed forever (*cyclically*, not *constantly*). Forever have there been Cosmoses in which all these have been *necessary.*

Example: Never was the time when Time was not. Time is Infinite, though periodically non-existent. Yet even while it does not exist, it does (or, conceptually, *seems* to). For instance, during Universal Pralaya, Time is not. Yet Universal Pralayas seem to take a certain amount of *time* to elapse, if compared to the duration of a Universal Manvantara. *Comparison* is the key, and comparison is only sometimes possible. Thus (though Time periodically ceases to exist) never was the time when Time was not.

The Great Breath continues perpetually throughout Infinite Time. The question arises, Is it legitimate to measure the time of a 'period' of utter Timelessness? Who is the Observer/Measurer? Once certainly cannot 'measure' 'during!' the time when there is no Time, but only, afterwards, in *memory*, and simply because the human mind has that rare and inescapable ability to make NOTHING into an apparently measurable 'Something'.

That which is being measured is, indeed, a NOTHINGNESS. So is it legitimate to 'thing' that NOTHINGNESS into 'Somethingness', and then say IT had a 'duration'? The entire existence of the Infinite Time Line (as a Generator of a Super-Cosmic Standard of Measurement) depends upon the answer. Further, if is it not legitimate to 'thing' as 'nothing', we have an endless, infinite material Universe, never quite disappearing. At least, this would be the 'Seeming' to any Mind that thought about the matter from a position *in*-Cosmos or Super-Cosmos. The habit of 'thinging' [sic] NOTHING lies at the root of our problem.

Returning just 'in time' to Time, Time is not the only 'Devourer'; the INFINITE is the 'DEVOURER' of Time. However, Time comes to 'birth' forever, throughout Infinite Duration. So are we, or are we not, led to a paradoxical conclusion that Time exists even when Time is not.

Example: The Standard of Measurement that can be applied to Infinite Time could be considered either the Universal Manvantara or the Universal Pralaya (assuming these to occur at Regular Intervals, and are measurable, which, perhaps, should not be assumed). Perhaps, then, there is no Standard of Measurement that can be applied to Infinite Time.

Example: Time is *motion-dependent.* The Great Breath has been *in motion forever.* Time thus has existed forever, though there are times in the Cycle of the Great Breath when Time is eradicated—not experienced, because there is no experience possible. Yet upon the re-emergence of the Universe *and* Time into perceptibility, it is *seemingly* seen that even 'during' the time that Time ceased to exist, it, necessarily, *did* exist, (at, least as measured against the Infinite Time Line of Infinite Duration) for, by such a 'measure', a certain measurable amount of time elapsed during Universal Pralaya, even though, at the time there was no Time. During Universal Manvantara, comparison (needed to detect time) again becomes possible, and one of the terms in the comparison is the duration (or time elapsed) during the 'STATELESS STATE' during which no comparison was possible.

So, we could come to the conclusion that Time *always* exists (as Infinite Time) although *experientially*, it ceases to exist periodically. Time cannot *always* be detected, and only in a recurring infinitude of 'laters' (*retrospectives* undertaken during Universal Manvantara) is it understood to have existed forever. The *ultimate truth* of the question, divorced from considerations forced upon us by the appearance of the World of Illusion, is that Time, like every-'thing' else that exists, *cannot* REALLY exist at all. Only NOTHING IS. Paradox!

In connection with this consideration concerning Infinite Time another question arises. Can there be any *memory* in any Consciousness of the time when Time did not exist? Does extra-SOURCE Consciousness (which includes such SELF-'MODIFICATIONS' as the 'RAY' which 'FLASHED FORTH', the Infinite Subject-as-Infinified Point, Mulaprakriti, the Infinite Subject-as-Condensing and Condensed Point and the Universal Logos, etc.) suffer from a kind of *amnesia* with respect to the Cosmic Pralayic Interlude of ALL-IN-ALLNESS just elapsed? Is there any *memory* of the 'NON-EXPERIENCE' which IS the INFINITE SELF in ITS 'STATE' of TOTAL SELF-ABSORPTION, 'INFINIDENTIFICATION', INFINITE BE-NESS?

Is there a kind of analogy to the early morning awakening of the human being after a night of profound, dreamless sleep, at which point 'time' is presumed to have elapsed even though the moment of falling asleep and the moment of awakening seem co-joined, as if awakening followed *in the very next moment* the act of falling asleep. If dreams are remembered, that memory gives the sense of time having passed, but if there are no dreams (and it can be presumed that in the NO-THING-NESS of ALL-IN-ALLNESS there would be *no dreams)* there would be no sense at all that time had elapsed during Universal Pralaya, or, for that matter, that there even *was* a Universal Pralaya.

The question is, When *no-thing* happens, is there *any-thing* to *remember?* And yet the Sages of Old speak (through *The Secret Doctrine*, for example) of the Great Breath, of the Universal Manvantara and Universal Pralaya, as if this Great Process is Real. Can it REALLY be that the INFINITE SELF is absolutely 'UNCONSCIOUS' and incapable of 'KNOWING' or 'REMEMBERING'? When the NO-THING-NESS of ALL-IN-ALLNESS is emphasized the human thinker might be inclined to say, Yes, but when the INFINITE SELF as 'INFINITIZED PLENUM' is emphasized, the thought readily arises that it would be impossible for the ONE AND ONLY BEING/NON-BEING (the FOUNT OF INFINITE POSSIBILITY) *not* to be possessed of the POSSIBILITY of 'KNOWING' and 'REMEMBERING'.

One of the great and appalling joys of this study is to continually discover just how *absolute* the ABSOLUTE REALLY *IS*. Perhaps, we will have to leave the resolution of this (and all problems concerning the ABSOLUTE) until the "Day Be With Us" when we will have the privilege of discovering that there REALLY is *no* problem, and never has been.

TIME

By *TIME* is meant infinite, indivisible, imperturbable DURATION. Technically speaking, there is no *TIME*, for TIME means *ABSOLUTE* TIME (i.e., *time* as it would be found 'within' the ABSOLUTE, the SELF, *if* time were REAL). All the necessary concomitants of Time (Space, Number, Item, Motion, etc.) are, however, *not* to be found *independently* within the ABSOLUTE and, thus, neither is Time. Thus there is, REALLY, no TIME. ABSOLUTE TIME would have no moments, for they all would be dissolved in the ETERNAL NOW. Time (as a general principle) is *meaningless* without moments.

Example: <u>TIME</u> is FOREVER.

Time Line, the Infinite

By *the Infinite Time Line* is meant an infinitely regressive or progressive sequence of appearing and disappearing Cosmoses considered as a Super-Cosmic 'Yardstick' along which or against which all units of Time in the UTTER ALLNESS can be measured.

By *the Infinite Time Line* is meant the beginningless/endless Dis-Continuum of Infinite Time (a Dis-Continuum which exists principally because of the appearances and disappearances of Cosmoses). Given the selection of any unit of time for a standard (for instance the year of Earth time), it should theoretically be possible to measure 'backwards' in this Cosmos and into others, in order to find out 'when' they occurred (expressible in the units of time chosen) relative to the present moment.

This type of measurement, impossible now to All except, perhaps, that BEING WHOSE MEMORY is INFINITE (or ITS *memory-capable* Super Cosmic 'EXTENSIONS'/ 'RADIATIONS'/EMANATIONS) cannot even be *begun* by beings in-Cosmos unless there is such thing as a standard Universal Pralayic Interval (and regular Universal Manvantaras may also be required).

Since the duration of the Universal Manvantara may be subject only to the 'WILL' of THAT, why should not the same be so for the Universal Pralaya? Although there is something Cosmically *Elegant* about conceiving an absolute regularity of Universal Manvantaras and Universal Pralayas, they *could* be completely *irregular* (relative to each other) and, still, no 'Schedule of Manifestation' would be upset, for THAT which induces Manifestation has *forever and ever* to do ITS *'JOB'!*—the absolutely endless task of expressing INFINITE POSSIBILITY in Finite Cosmoses.

Example: Theoretically the exact timing (relative to the present moment in this Cosmos) of any event in any of an infinitude of Cosmoses past could be measured along <u>the Infinite Time Line</u>, provided all cycles in our present Cosmos were known, all cycles in a sufficient number of Cosmoses past were known, and the duration of a sufficient number of Universal Pralayas past were known. Who can do this? It must be *possible* simply because there is a FOUNT OF ALL POSSIBILITY. *What* might be learned by the historical researcher of an infinitude of Cosmoses past? Yet, would it matter?

Example: <u>The Infinite Time Line</u> is *generated* by the 'Archetypal Quantum Process'.

time-point

By a *time-point* is meant a specific ultimate moment within the context of a given Cosmos. All events occur exactly 'at' or 'on' a time-point (in fact, in a way, *are* a time point, or event-point). Within a Cosmos, a time-point or point-in-time has duration—the duration of an ultimate moment. 'Within' the CONTINUUM of INFINITE DURATION it is meaningless to speak of a time-point, because the value of such a point would always be zero. In the ETERNAL NOW, all time-points are the *same utterly dimensionless time-point,* the duration of which is zero.

Example: A time-point in Fohatically Particulated Cosmos may be designated, but, conceivably, an infinitude of time-points (conceived from within the World of Being) could occur during that Cosmo-Objective time-point. Again we come to points within points—this time, 'time'-points. If imagination/ideation can occur at any 'speed' within Cosmo-Subjectivity, then ideational/imaginative time is *infinitely divisible.* In such a case, a given time-division of a Cosmo-Subjective Now may be the 'endurance of a thought/image/idea' *'held'* in the World of Being. Perhaps there is no smallest imagistic, ideational, conceptual 'moment' in the World of Being, however definite and defines moments may be in the Cosmo-Objective World.

Example: No movement in Fohatically Particulated Cosmos can occur *during* an ultimate moment. All ultimate particle/events 'hold' their position (or are Fohatically 'Held') in complete stillness during an ultimate moment. A time-point is a *still* point. An ultimate moment is the duration of a 'frozen frame', the duration of one frozen frame in the Cosmic Configuration. Macro-movements in-Cosmos *seem* to occur because ultimate particle/events change their positions *relative to each other.*

The implication is that the entire Cosmo-Objective World alternates between disappearance and *manifest stasis.* Perceived movement results simply in a *quantized change of relationship* between variables. Movement is simply the perception of successive changes of frozen patterns.

A motion picture holds the analogy. The film frames show different 'frozen' positions. The rapid passage of these frames (before the eyes of the viewer) gives the *illusion* of *continuous motion.* But a motion picture is a discontinuity disguised as a continuity. The Cosmo-Objective World, too, is a Great Dis-Continuity disguised as a Great Continuity. A rapid sequence of successive static patterns gives the illusion of movement. In a way, there is no *movement* in Fohatically Particulated Cosmos, only the successive presentation of *static configurations.*

TRUTH

By the *TRUTH* is meant the INFINITE PRINCIPLE ITSELF. This TRUTH is not an idea or mental construct, but IS BE-NESS ITSELF.

Example: The search for Truth must precede the search for TRUTH. While 'Truth' is the destroyer of 'truth', TRUTH is the destroyer of 'Truth'. Reality cannot abide in the face of REALITY. REALITY-as-'SHIVA' destroys even the Real.

- U -

Ultimate Identity

By the *Ultimate Identity* is meant the Ultimate Cosmic Monad, the 'Ray' of the ABSOLUTE emanatively pursuing a path of apparently individualized descent and reascent through all dimensions of Cosmos along the Divine Emanatory Stream.

Example: The Ultimate Identity masquerades as *many* apparently distinct individualities. Yet in all these many Entity/Identity/Individualities, there is but *One* Ultimate Identity.

Example: The Life of an atom is, Essentially, the Ultimate Identity; so is the Life of a human being, the Life of a Planet, the Life of a Solar Logos, etc. Every authentic E/entity in-Cosmos is, in fact, *the* Ultimate Identity (apparently diversified, Really but One Subjective Identity).

unconditional

By *unconditional* is designated a process or quality which is unaffected by those variations in configuration or relationship we call conditions.

Example: Unconditional Love is not affected by personality fluctuations.

UNCONDITIONED, the

By *the UNCONDITIONED* is meant the SELF—THAT in which no conditions of any kind can possibly BE.

Example: The SELF changes not. IT IS forever the UNCONDITIONED—containing no conditions and affected by no conditions. When man, through identification, *rebecomes* the SELF He also is, in his ESSENCE he is affected by *no* conditions—this despite the fact that any number of misfortunes may befall His form nature. He has achieved *conscious immortality.*

unit

By a *unit* is meant an authentic system-in-Cosmos, whether relatively tiny or immense. A unit is a *cosmically sanctioned combination.*

By a *unit* is meant a unity of factors-in-Cosmos—possibly a self-conscious unity but not necessarily so. Units, as here described, have *authenticity* in-Cosmos. They are not secondary or tertiary creations. The patterns constituting units are patterns found within the Design-at-the-Beginning.

Example: Human units and Deva units go to the formation of the chakras of a Heavenly Man.

Example: An evanescent congery of variables is not a unit.

unit-in-Cosmos

By a *unit-in-Cosmos* is simply meant a unit, but with special reference to its intra-Cosmic relations.

Example: The Ashrams of the Masters are units-in-Cosmos which go to the formation of the Great Ashram of Sanat Kumara, Itself a unit in a still greater Combination.

Universal Condition, the

By *the Universal Condition* is meant the immediate configuration of all possible items-in-Universe. The Universal Condition changes, of course, but does not Really change so quickly as from ultimate moment to ultimate moment; the Cosmic Configuration definitely changes from ultimate moment to ultimate moment. The Universal Condition is more abiding than the Cosmic Configuration. It would take many, many changes in the Cosmic Configuration to warrant the statement that the Universal Condition had Really changed. The term *Universal Condition* invites an assessment of Quality, whereas the term *Cosmic Configuration,* being more abstract, does not, and is simply more a 'Map' of the positioning of all factors-in-Cosmos.

Example: The Universal Logos, with Its Universe-encompassing Consciousness is ever precisely aware of the Universal Condition. But man and even the Masters of the Spiritual Hierarchy of the Earth can only be aware of relatively *local* conditions-in-Universe. An accurate assessment of the Universal Condition at a certain time would disclose the degree to which the Universal Evolutionary Process was fulfilling the 'expectations' of the Cosmic Divine Purpose/Plan/Intent for that particular moment/phase of Cosmic time. With respect to our Earth Globe, the Masters may know whether the Global Process is 'on schedule' and performing adequately. What can be said of the Universal Process and Its Quality of Performance? Perhaps only the Universal Logos can be the Judge.

Universal Flux

By the *Universal Flux* is meant the ceaseless (though quantized) *apparent flow* of all movements-in-Fohatically Fabricated Cosmos from one 'instant' or 'ultimate moment' to the next, virtually throughout the entire duration of the Cosmos.

Example: The Universal Flux (as it relates to the Cosmo-Objective World) is not Really a *continuous flow,* but only a *seeming* flow. The Universal Flux results from the virtually instantaneous change (and, perhaps, '*timeless* change') from one Cosmic Configuration to the next. The Universal Flux, therefore, consists of virtually countless quantum ('jerky') repositionings which are 'held' for the duration of an ultimate moment—i.e., until the next change/repositioning in the Cosmic Configuration.

Example: Within the Universal Flux of the World of Approximation, the objective aspect of all E/entities as well as of all artifacts can be classified, when precisely considered, as *motions.* Motions however are only *quantum changes of relationship.*

Example: Were the Universal Flux to include the World of Adjustment, the Flux would be far more complex and virtually continuous.

Universal Identity (as Entity); **Universal Identity** (as the Universal Logos' 'Inperience' of Self)

By the *Universal Identity* is meant *not* the ONE IDENTITY forever PRESENT throughout all DURATION, but a 'special case' of that IDENTITY. The Universal Identity, from one perspective Is an Entity—the Universal Logos. From another perspective the Universal Identity is the Sense of Identity or the Quality of the Self-Realization of the Universal Logos of any Universe/Cosmos. We will focus on this second definition.

The sense of Identity of this Universal Logos (although the Logos *knows* that, ESSENTIALLY, It Is the ONE ABSOLUTE IDENTITY) is, nevertheless, circumscribed by the presence of Universal Prakriti (Universal Logoic Self-'Sight') which provides a kind of Universal Ring-Pass-Not (a *Limiting Object*, or Field of Space, which compels the Attention of Universal Consciousness) and which, in one respect, *finitizes* the sense of Identity of the Universal Logos. (Of course, it is ultimately impossible to *finitize* the ESSENTIAL *identity* of any E/entity.) The ONE IDENTITY in ITS ALL-IN-ALLNESS has no *consciousness of prakriti* (no *consciousness of object)* and REALLY, no 'CONSCIOUSNESS' of any kind, and so IT 'dwells' in ABSOLUTE INFINITIZED IDENTITY.

Example: <u>The Universal Identity</u> is the Universal Logos' Conscious Sense of Self. Universal Identity arises when the Universal Logos focuses Its Consciousness upon All that It *Has* (all that It 'Sees' of Itself—for *having* is *seeing*, and *seeing* is *having)*. All that It *Is*, is, ESSENTIALLY, *infinitely more* than what It *Has*. The Universal Logos, like every other identity/entity in-Cosmos, *Is* ESSENTIALLY, the ONE AND ONLY IDENTITY. But Its *Having* is different. Any self-conscious Entity/Identity, when not identified with its ESSENCE/Essence, 'inperiences' a kind of secondary identity by which 'it *is* through what it *has*'— One *has* an *object;* one *is* a *subject. Seeing* and *having* are inseparable.

Universal Life Unit

By a *Universal Life Unit* is meant a 'Ray' of the ABSOLUTE.

Example: In any particular Cosmos there is a definite, planned and invariable number of <u>Universal Life Units</u>. They emanate from each other, lesser Points from greater Points. There are less in the early 'days' of Cosmos, most at the midway point, and less, again, as Cosmos is preparing for reabsorption. The Monad of the Solar Logos is as equally a <u>Universal Life Unit</u> as is the Monad of the atom or the Monad of the Galactic Logos. They are all the *same* identical Monad, appearing (apparently) as differentiated <u>Universal Life Units</u>.

Universal Logos

By the *Universal Logos* is meant the Principal Self of Cosmos, the Deity of the Universe. In another way, the Universal Logos can be called 'The SELF-as-Condensed Point', as the Universal Logos 'Emerges' from this State of Pre-Cosmic 'Identity-Condensation'.

Example: The <u>Universal Logos</u> is 'Created' by the formation of a Single, *bounded,* 'Spherical' Point of View within the Infinified Point of View of the Infinite Subject. When the Infinified Point is transformed (via the Condensing Point) into the Condensed Point, the <u>Universal Logos</u> is 'Born'.

Universal Prakriti

By *Universal Prakriti* is meant that 'part' of Mulaprakriti (the Infinite Object or the Infinite Potential for Articulated Objectivity) which is 'Pre-Selected' or 'Encompassed' by the SELF-as-Condensed Point to provide the Objectivity/Matter for the Universe being formed.

From another perspective, Universal Prakriti is all 'Matter' within the Universal Ring-Pass-Not.

From a more Cosmo-Psychological Perspective, Universal Prakriti is the Universal Logos' 'Self-Reflected Image of SELF-Selected 'Identideational' Limitation'. This means that Universal Prakriti reflects the Identity of the Universal Logos, that Identity being Essentially a Limited 'IDEA' 'EXTRUDED' from the INFINITESSENCE for 'Cosmo-Objectification'. All Qualified Identity is 'Identideational'; distinct Identity *is* Quality. The Essence of Identity is Spirit.

Example: Universal Prakriti is fashioned into (or, better, Objectively Reflects) innumerable temporary forms inherent in the Fixed Design of the Universal Logos and Created through the Agency of Fohat. Universal Prakriti is an infinitesimal 'portion' of Mulaprakriti, for any 'portion' of an infinite 'Something' is infinitesimal when compared to that 'Something'. Also, Singular Identity is an infinite reduction upon Infinite Identity. Because this is so, Mulaprakriti is infinitely greater than Cosmic Prakriti (which is an Image reflecting a Singular Identity—that of the Universal Logos).

Universal Process, the

By *the Universal Process* is meant a Purposeful Activity in-Cosmos characterized by an intelligent, loving, willed *directing* of all cosmic energies and forces by the Universal Logos of that Cosmos throughout the entire duration of the Cosmos.

Example: The Purpose of the Universal Process is the fulfillment of the Design-at-the-Beginning.

Universal Self, the

By *the Universal Self* is meant the Universal Logos.

Example: The Universal Self Is, as it were, infinitely removed from the ABSOLUTE SELF, and, yet, the Universal Self is *none other* than the ABSOLUTE SELF.

Example: Two *selves* are greater than the Universal Self: the Super-Universal Self and the SUPER-Universal SELF.

UNIVERSAL SOLVENT, the

By *the UNIVERSAL SOLVENT* is meant THAT which holds the Universe *in solution*. IT IS the *ultimate* SOURCE of all form and IS THAT into which all objectivity vanishes.

Example: When the time for the Great Re-Absorption comes, all worlds disappear into the UNIVERSAL SOLVENT.

Universal Subject, the Focusing

By *the Focusing Universal Subject* is meant the Universe-Intent Phase of the Infinite Subject. The Infinite Subject seeing Itself as the Infinite Object (Mulaprakriti) become *intent* upon limiting Its Infinified Point of View so that One, Specific Cosmos may be 'Created' (or, better, 'Become').

Thus the Infinite Subject begins to *focus* upon a 'Limited Vision of Itself' (one which accords with the Cosmos It Must Become) and in so doing, the Infinite Subject 'Becomes' the *Focusing* Universal Subject, and finally the Focused Universal Subject—the Universal Logos. Describing this Process in other terms, the Infinified Point, becomes the Condensing Point and finally the Condensed Point.

Example: The Focusing Universal Subject is treading the Path in Pre-Cosmic Consciousness from Infinitude to Finitude. Is this Path a Gradual Path or a Quantum Path? The theory is that the Path is *gradual.*

Universal Theme, the

By *the Universal Theme* is meant that Collection of Super-Cosmic 'IDEAS'-as-Ideas which are to be enacted through Cosmic Prakriti during the Universe-to-Be.

Example: The Universal Theme is *the* Universal Archetype, the greatest of all Divinely Intended Cosmic Patterns.

Universe

By the term *Universe* is meant the entirety of the *limited* Cosmos, but the term is exclusive of the SELF of which It (i.e., the Universe) is but a 'RADIATION'. The term is *also* inclusive of the Realm of Super-Cosmos, which is an *indefinite* Realm.

Example: While there are a multitude of parallel worlds and dimensions, there cannot be simultaneously existing *parallel* Universes, per se, for all such parallel 'Universes' would necessarily be resolved into the One Universe if the word 'Universe' were to keep Its true meaning—the One and Only, Singular, and Definite Manifestation of THAT occurring at any given 'time' in Infinite Duration.

Universe-to-Be

By the *Universe-to-Be* is meant the Coming Universe as seen from the Superior Perspective of the Pre-Cosmic Phase of the Great Exhalation during which Preparations are being made for the Appearance of the Universe.

Example: The Universe-to-Be may be alternatively named the Universe-to-Come.

Example: The Infinite Subject/Infinified Point as It 'Phases' into the Condensing Point, and, finally, Condensed Point/Universal Logos 'Sees' the emerging Outline of the Universe-to-Be.

un-Real; un-REAL

By *un-Real* is meant that which is not part of the Pattern of Archetypal Energies which structure a given Cosmos.

Example: So many human plans, as influential as they may be within the lower three worlds of human evolution, are totally un-Real, i.e., they have nothing to do with the fulfillment of the Divine Purpose via the Divine Plan.

By *un-REAL* is meant that *other than* the INFINITE SELF. ITSELF alone IS REAL.

Example: The World of Becoming is *formally* un-REAL, though ESSENTIALLY REAL.

unreifiable

By *unreifiable* is meant that which cannot be resolved into a 'thing'.

Example: The adage, "If you can conceive it, you can achieve it" implicitly states that all conceivable energy patterns are reifiable. The FOUNT OF ALL POSSIBILITY in ITS 'ENTIRETY' IS unreifiable in any given Cosmos, but the *phased reification* of the FOUNT OF ALL POSSIBILITY IS the very 'PURPOSE' of the INFINITE SELF. The ESSENCE (in and of ITSELF) is *never* reifiable: ITS SELF-as-Self Projection *is.*

Unrepeatability, Principle of

By the *Principle of Unrepeatability* is meant the following: no form can possibly be exactly duplicated throughout the entire infinitude of Cosmoses. This Principle exists because of the infinite fecundity of the INFINITE SELF.

Example: Even seemingly identical atoms of matter cannot Really be *absolutely* identical, because the Principle of Unrepeatability holds sway.

- V -

value

By the *value* of something is meant the relative importance of that something in relation to the fulfillment of the Original Intent.

Example: Any E/entity or item-in-Cosmos has value to the degree that its presence/quality/activity contribute to the fulfillment of the Original Intent, the Design-at-the-Beginning.

variable

The term *variable* describes the behavior of an item-in-Cosmos when that item can be depended upon to change or remain inconstant relative to a given context. A variable must always be seen in relation to a given context, whether the context be static or dynamic. A variable is a contextual item which can be expected to change either in response to other changes within the context, or simply because of its own changeable nature and, thus, independently of the context.

Example: The variables upon the Path of Discipleship are many—so many, in fact, that it is extremely difficult to manage them all. One policy towards the management of unpredictable variables is voiced in the Biblical adage, "Seek Ye first the Kingdom of Heaven, and all things shall be added unto you."

Example: The Third Ray approach to the Path of Spirituality is often through 'management of variables'. The First Ray approach is through assertion of Principle, ignoring the many contingent but inconsequential variables, the many minor unpredictabilities.

vibration

By *vibration* is meant a regular oscillation of an item-in-Cosmos—an oscillation with a specific frequency per unit of time, and a specific strength or amplitude.

By *vibration* in a Dis-Continuous Universe is meant a series of regular and relatively repetitive changes of quantum positioning, such that the changes contribute to the formation of a (*seemingly* continuous) wave pattern which is measurable in terms of frequency and amplitude.

Example: Vibration can be established by particle/events, the behavior of which, hypothetically, creates wave forms. A vibration can be understood as a wave form consisting of many particle-events. The particle-events however are primary to the wave forms, which are derivative and secondary. In our Cosmos (according to the Theory of Cosmic Discontinuity) there would be no such things as *continuous waves.*

Example: It is open to discussion whether the ontological appearance and disappearance of ultimate particles should be called vibrations.

vibration (in-Cosmos), minimum

By the *minimum vibration in-Cosmos* is meant the smallest possible vibration (and probably the *fastest* possible vibration) in-Cosmos.

Example: The minimum vibration in-Cosmos must necessarily include a number of ultimate particle-events, because it takes many of these to generate the appearance of an oscillating wave form. Therefore, even a minimum vibration (or vibratory cycle) is, relatively, a macro-form when compared to an ultimate particle-event.

The phenomenon of increase/decrease/increase/decrease found in wave forms in Cosmos must be dependent upon the positioning of ultimate particle/events relative to each other. Certain positions must result (on the relatively macro level) in the perception of increase and intensification; certain other positions must result (on the relatively macro level) in the perception of decrease and de-intensification.

void; VOID

By the term *void* is meant empty.

Example: The decision made was utterly void of reason.

By *the VOID* is meant the ABSOLUTE, with emphasis upon an *emptiness* which is *utter fullness.* The VOID is void of all lesser things (particularities) the presence of which would compromise ITS PERFECT HOMOGENEITY.

Example: The VOID IS the PLENUM—such is the NATURE of THAT—the GREAT CONTRADICTION.

- W -

will

By *will* is meant the power which *initiates* and *sustains* the *'being-ness'* of a state or condition, or which *withdraws* and thus *eradicates* the *'being-ness'* of a state or condition. The will may also initiate, sustain, or destroy the 'patternedness' of a state or condition.

By *will* is meant the power which confers or withdraws *life* from any configuration-in-Cosmos (whether that configuration be the form of an authentic E/entity or secondary/tertiary-entity).

Example: The Solar Angel wills to appear upon the higher mental plane and sustain with Its presence the development of the human causal body. The periodic death of the personality occurs when the Angel wills to withdraw Its attention from the 'appearance' It is supervising in the lower three worlds.

Will, Delimiting

By the *Delimiting Will* is meant the Universe-Bounding and Conditioning Will of the Super-Cosmic Self. It is the 'Will-to-See-Limitation', and further, the 'Will-to-Limited-Self-Reflection'.

Example: The Super-Cosmic Self, the Infinite Subject, ('Carrying' the 'Instructions' which 'FLASHED FORTH' with the 'RAY' of the ABSOLUTE) 'Determines', by means of these Instructions the Parameters of the Cosmos-to-Come. Once the Parameters are 'Determined' and 'Enacted' upon Mulaprakriti, the Super-Cosmic Self has 'Become' the Universal Logos, the SELF-as-Condensed Point, taking Its position in the (perhaps *figurative)* 'Center' of the Bounded Sphere of Cosmic Prakriti which forms the Ring-Pass-Not of the Cosmos-to-Be. In forming the 'Center' and 'Ring-Pass-Not' of the Cosmos-to-Be, the Super Cosmic Self 'Exercises' the Delimiting Will. This Will is the 'Will to Specific Singularity'.

Does a Cosmos have a 'Center' or are all 'points' Its Center? Perhaps, *both* are true. In terms of Spirit, all 'points' are Its Center (for in Cosmos, there is but One Point) but in terms of Prakritic Distribution within the Fohatically Particulated World, a Spatial 'Center' is conceivable and even probable.

Will, Free

By *Free Will* is meant the Will of the One Being in-Cosmos (the Universal Logos) untrammeled and unlimited by any external Law or Compulsion *except* by the relatively *infinitessential* amount of 'INFORMATION' 'EXTRUDED' from the INFINITESSENCE concerning the 'IDEA' to be Manifested *as* and *in* the Coming Cosmos. The 'INFORMATION' 'Conveyed' also bears upon the *limiting Parameters* of the Cosmos-to-Be which were 'FIXED' and 'DETERMINED' by the ALL-SELF as FOUNT OF ALL POSSIBILITY.

Example: It is beyond the scope of the Free Will of the Universal Logos of the present Cosmos to substitute an *'octenary'* Principle for the present *septenary* Principle by means of which the structure of the Present Cosmos is organized. Nothing is impossible in relation to the FOUNT OF ALL POSSIBILITY, but such a 'DECISION' regarding Cosmic Structure would have to have been 'MADE' when Cosmic Parameters were being 'DEFINED' (probably 'within' the INFINITE SELF). But if a 'DECISION' was *not* then made, then during the Primal, SUPER-Cosmic and Pre-Cosmic Act(s) of SELF-'LIMITATION' by means of which the ALL-SELF by means of ITS Agents:

1. 'BECAME' the Infinified Point—
2. 'Saw' Mulaprakriti (and, only *perhaps*, its *articulated infinite potential and residual content)* from an infinity of Points of View—
3. 'Became' the Condensed Point (the Universal Logos)—
4. 'Measured Out' (Maya is the *act of measuring)* the Boundaries of the Cosmos-to-Be, forming the Cosmic-Prakritic 'Sphere'.

In our Present Cosmos substituting *eights* for *sevens* would not lead to Cosmically-Sanctioned Configurations, and would, therefore, be *actually* impossible.

WILL, FREE

By *FREE WILL* is meant the 'SELF-DETERMINATION' of the ONE BEING (if one can speak of such a thing), utterly untrammeled by any external *law* or *compulsion*, as there is no *source* external to IT from which such a *law* or *compulsion* could originate.

Example: The 'GENERATION' of each of an infinitude of Cosmoses can be seen as an 'ACT' of utterly FREE WILL arising 'within' the ONE AND ONLY SOURCE of ALL.

Example: A profound metaphysical inquiry would concern itself with the relationship between FREE WILL and LAW 'within' the INFINITE SELF.

'WITNESS' OF ALL, the

By *the 'WITNESS 'OF ALL* is meant the ONE AND ONLY INFINITE SELF, other than WHOM there is no possible 'OBSERVER' in relation to the Infinitude of Cosmoses past and future. Yet, this 'OBSERVATION' may have to be 'PERFORMED' *by proxy*, as 'CONSCIOUSNESS' can *barely* 'ARISE' in the INFINITE SELF, yet alone be *sustained*. The, the 'WITNESSING' of the INFINITE SELF is *accomplished through* a descending Hierarchy of 'Seeing' *Subjects*, each of whom is the INFINITE SELF in ESSENCE. As for the INFINITE SELF, IT 'INFINIDENTIFIES' or 'ESSEDENTIFIES', but it can be strongly questioned whether IT 'SEES'.

Example: The 'WITNESS' OF ALL 'BEHOLDS' the All (our Cosmos and Its Universal Logos) as well as the ALL (all Cosmoses past and future {not excluding our present Cosmos} and their Universal Logoi) as Objects. *Who*, however, is the 'WITNESS'?

Example: The 'WITNESS' OF ALL IS (ESSENTIALLY, but not *actually)* PARABRAHMAN, the ETERNAL INFINITE ATMAN. When the Vedantins called PARABRAHMAN the WITNESS, could they have been meaning PARABRAHMAN as Isvara, the Universal Logos?

Example: Can the 'WITNESS' OF ALL, REALLY, *witness* anything if IT is necessarily (as the INFINITE) out of all relation with any other thing? And yet, if there is to be *witnessing*, as undoubtedly there is, there is *none other than* IT to be the 'WITNESS'.

world

By a *world* is meant a prakritic domain with a characteristic vibratory frequency.

By a *world* is meant a Self-Reflected Image.

Example: The prakriti of the lower three worlds have a vibratory rate far lower than the prakriti of, for instance, the Cosmic Astral Plane. Yes, (according to the Theory of Discontinuity) the Cosmic Astral Plane should probably be considered an Aspect of Fohatically Particulated Cosmos and, thus, would necessarily be both particulately-prakritic and dis-continuous

World of Adjustment

By the *World of Adjustment* is meant a Domain, 'between' the World of Being and the World of Approximation, wherein 'Willed Ideation' is directed towards setting the Coming Pattern for the next ultimate moment to which the World of Approximation must conform. Really, the World of Adjustment is an 'Aspect' of the World of Being. The higher strata of the World of Being simply 'hold' the Ideational Images to which the World of Approximation must conform. Movement in these higher strata is therefore slow and sustained.

There must, however, be a Domain from which the semi-blinded efforts of Fohat and His Host can be assessed, and the 'Means of Conformity' *planned.* This would be the World of Adjustment, which could be considered the simultaneously 'downward/upward' gazing Aspect of the World of Being. Within the World of Adjustment, the Gaze is 'upwards' so that the next intended Pattern can be known; but the Gaze is 'downwards' as well, so that the adjustment to bring about Conformity to A/archetypes can be wisely/intelligently planned.

Example: The World of Adjustment (considered as the lower strata of the World of Being) is a World of Cosmo-Subjectivity, which is particularly active 'during' the intermoment instant called the Cosmo-Subjective Now. The gods within the World of Adjustment have two 'faces' (like Janus, but vertical not horizontal), one gazing 'upwards' to the higher strata of the World of Being, and the other gazing 'downwards' into the World of Cosmo-Objectivity. The World of Adjustment is a *semi-veiled* World, otherwise, there would be a *perfect approximation* of the Cosmic Configuration to the Archetype of the Moment (in the World of Being) from 'at' each ultimate moment, which there is not.

World of Approximation

By the *World of Approximation* is meant the Lower World of Becoming, i.e., the World of Effects created by Fohat and His Host to approximate the Fixed Cosmic Design, the Design at the Beginning *held* by the Supernal Tetraktys (the Cosmic Son and His Company) in the World of Being.

Example: The methods pursued in the World of Approximation are ever *variable* and subject to *free will* though the 'End' is certain. Hypothetically, the Beauty of the Fixed Design of Cosmos will be achieved in exactitude by the time of the "Day Be With Us."

World of becoming

By the *World of becoming* is meant the lowest and most material levels of the World of Becoming This includes those levels of the Cosmic Physical Plane which to the Solar Logos are *not* considered a Principle. The World of becoming comprises, at the very least, our three worlds of human evolution, and perhaps, from a much larger perspective, even the whole of the Cosmic Physical Plane.

Example: The World of *becoming* is formed largely of the unredeemed substance of the First Solar System, a substance now informed by the relatively unsuccessful Creative Hierarchies of that System.

World of Becoming

By the *World of Becoming* is meant the entire Cosmos as the World of Illusion—a Domain of incessant change. Some change in this Domain is extremely rapid (relatively) and other change, relatively slow and sustained. Change is most rapid 'below' and least 'above'.

By the *World of Becoming* is *usually* meant, those levels of Cosmos within which activity is incessant, and in which there ever proceeds an attempt to approximate the configuration of material forms to the Patterns of certain relatively static Archetypes within the World of Being, a World located upon the higher planes of Cosmos.

Example: The World of Becoming is often distinguished from the Archetypal Levels of Cosmos that are called the World of Being, and this distinction is somewhat justified. However, the Cosmos considered as a Whole is very much in a state of 'Becoming', and can, as a Whole, justifiably be called the World of Becoming. Even the World of Being is changing, hence, 'Becoming'.

World of Being

By the *World of Being* is meant the World of Formative Patterns.

Example: The Archetypes that Direct the Universal Process are within the World of Being. Technically, the World of Being is an Aspect of the greater World of Becoming.

Example: It is conceivable that the World of Being, being an *ideational* World is, as well, a *non-vibratory* World. Must there be vibration if Light is to exist? Must there be vibration if Sound is to exist? If Sound and Light exist *ideationally*, vibration as we understand it may not be necessary.

WORLD OF BEING

By the *WORLD OF BEING* is meant the DOMAIN of the INFINITE SELF, the ETERNAL BE-NESS.

Example: If humans want to understand their TRUE IDENTITY they must find a way to identify with the WORLD OF BEING. Though it is a major achievement to understand and identify even with the World of Being, the final goal is not thereby 'reached'.

World of Conditions

By the *World of Conditions* is meant the lower Aspects of the World of Becoming, though, in a sense, naught but *conditions* exists in the entire World of Becoming.

Example: The three worlds of human evolution are, from the human perspective, the World of Conditions. From a larger perspective, the entire Cosmic Physical Plane might be called the World of Conditions, for the Cosmic Physical Plane comes under the rulership of Saturn, the planet most responsible for determining *conditions.*

Example: These are all the *same* World: the World of Conditions, the World of becoming, the World of Effects, the World of Approximation, the World of Fabrication, the World of Particulate Arrangement, the Mosaic World, the Cosmo-Objective World.

World of Effects

The term the *World of Effects* is a relative term, and has a different meaning depending upon the 'altitude' of one's focus upon the many planes and subplanes of Cosmos. In general, the term means a World in which the results of causes set in motion elsewhere are experienced.

Example: To the human being the World of Effects comprises the lower eighteen subplanes of the Cosmic Physical Plane. For the Solar Logos, the World of Effects comprises the entire Cosmic Physical and Cosmic Astral Planes, as well as the lowest four subplanes of the Cosmic Mental Plane. In order to understand the meaning of the World of Effects for any classification of E/entity, the World of Causes for that E/entity must also be understood.

World of Fabrication

By *the World of Fabrication* is meant the Domain of Cosmo-Objectivity which intra-Cosmic Fohat *fabricates* under the Direction of the Universal Son. The World of Fabrication is the "Mosaic World", the World of Particulate Arrangement.

Example: The World of Fabrication stands in contrast with the World of Being. The Goal of Cosmic Evolution is to ensure that the Design within the World of Fabrication reflects as faithfully as possible the Design-at-the-Beginning within the World of Being.

World of Illusion

By *the World of Illusion* is meant the entire World of Becoming, i.e., the Cosmos.

By *the World of Illusion* is *usually* meant those planes upon which erroneous perception and understanding of the Divine Pattern are most frequently found.

Example: Since Cosmos is an Object, in one sense infinitely less REAL than the WORLD OF BEING, it is justifiable to call It the World of Illusion.

World of 'Play'

By *the World of Play* is meant the Cosmos.

Example: In Hindu Mythology one of the great purposes for which Cosmos was brought into existence is to provide a 'playground' for the Gods, or for the ONE LIFE. What happens to the consciousness of man when it considers Cosmos as the World of Play?

World of Relativity

By *the World of Relativity* is meant the Universal Field in which all quantities and qualities found within-Cosmos are related within the Parameters of Cosmically Sanctioned Configurations.

Example: Cosmos is inescapably the World of Relativity, for every item-in-Cosmos is inescapably related to every other item. The ABSOLUTE, however, is void of *relativity*.

- Y -

you; You; YOU

Although, technically considered, there is no *you* in any form, these words are meant to be used in the discourse of relationship in order to stratify the various levels of functioning-in-Cosmos of the 'other'. Sometimes, for the purpose of galvanizing activity, and, other times, simply because of convenience in using pre-established language conventions, it is useful to use a form of direct address involving the words 'you', 'You' and 'YOU'. They should be interpreted in the same way as any number of such triplicities we have used in this treatise—such as for instance: 'me', 'Me', ME'; or 'self', 'Self', 'SELF'; or, finally, 'I', '**8**', '**I**'. (*I* is such an important word that special diacritical attention has been given to it to differentiate one level from another. Still, the symbology is crude.)

- *'YOU'* is a personal pronoun indicating your identicalness with the ONE SELF.
- *'You'* is a personal pronoun indicating your existence as an identifiable spiritual Self in-Cosmos.
- '*you'* is a personal pronoun indicating your existence simply as a personal ego.

Obviously, these words are used so often, that at times is necessary to resort to the use of the conventionalized 'you', simply for the sake of convenience, otherwise the verbiage would become excessively laborious—at such times the reader is advised to use the power of intuition.

Example: 'YOU' *ARE* THE ONE SELF. The older form of this most ancient Formula of Being is, "Thou art That." Although 'You' are the individualized Self and 'you' are simply the person you are for this particular incarnation, both 'you' and 'You' are, ultimately, YOU.

- Z -

zero

By *zero* is meant no quantity—the total absence of enumeration.

Example: A question arises as to whether zero should be considered a number or quantity in the following enumeration: ... 3, 2, 1, 0, -1, -2, -3 ... If it is not a number, what is it?

ZERO, the

By *the ZERO* is meant the ABSOLUTE NO-THING WHICH IS BE-NESS.

Example: All number emerges from the ZERO, but the manner of emergence is mysterious. Number is ESSENTIALLY a sequential, systematic minimization of the fullness of ZERO.

It is wise to become accustomed to there being no rest nor end. But the single realization of Our Brotherhood and Hierarchy already directs the traveler along the shortest path to Infinity.

Verse 652, Agni Yoga, Agni Yoga Society

The Index

8 (as 'I')

A

B

C

D

E

F

G

H

I

M

N

O

P

Q

R

S

T

U

V

W

Y

Z